AF509015

The Molecular and Cellular Basis of Retinal Diseases

The Molecular and Cellular Basis of Retinal Diseases

Editors

Steven J. Pittler
Steven J. Fliesler

MDPI • Basel • Beijing • Wuhan • Barcelona • Belgrade • Manchester • Tokyo • Cluj • Tianjin

Editors
Steven J. Pittler
Department of Optometry and
Vision Science, Vision Science
Research Center, University of
Alabama at Birmingham
USA

Steven J. Fliesler
Departments of Ophthalmology
and Biochemistry and the
Neuroscience Graduate
Program, The State University of
New York (SUNY)—University
at Buffalo
USA

Editorial Office
MDPI
St. Alban-Anlage 66
4052 Basel, Switzerland

This is a reprint of articles from the Special Issue published online in the open access journal *Cells* (ISSN 2073-4409) (available at: https://www.mdpi.com/journal/cells/special_issues/basis_retinal_diseases).

For citation purposes, cite each article independently as indicated on the article page online and as indicated below:

LastName, A.A.; LastName, B.B.; LastName, C.C. Article Title. *Journal Name* **Year**, *Article Number*, Page Range.

ISBN 978-3-03936-654-5 (Hbk)
ISBN 978-3-03936-655-2 (PDF)

Cover image courtesy of Steven Pittler.

Contents

About the Editors

Steven J. Pittler is a professor at the Department of Optometry and Vision Science, a senior scientist and director of the Vision Science Research Center, with secondary appointments in the Departments of Ophthalmology and Biochemistry and Molecular Genetics. He received his undergraduate training in biochemistry from Michigan State University (1983). He completed his doctoral studies in Biochemistry at Michigan State University (1989), after which he was an NRSA Neurobiology Postdoctoral Fellow (Ocular Molecular Genetics) at Cullen Eye Institute, Baylor College of Medicine. In 1991, he was an assistant professor of Biochemistry and Molecular Biology at the University of South Alabama. In 1995, he became Director of the Center for Eye Research, at the College of Medicine, University of South Alabama. Moreover, in 1995, Dr. Pittler was the recipient of the Cogan Award for excellence in vision research from the Association for Research in Vision and Ophthalmology. He came to the University of Alabama in Birmingham in 1999, and was promoted to the rank of professor in 2000. In 2013, he received the National Eye Institute Audacious Idea Award.

Steven J. Fliesler is a SUNY Distinguished Professor, UB Distinguished Professor, the Meyer H. Riwchun Endowed Chair Professor of Ophthalmology, and Vice-Chair/Director of Research in the Department of Ophthalmology, State University of New York (SUNY)—University at Buffalo (UB). He also holds concurrent appointments as a professor in the Department of Biochemistry and in the Neuroscience Graduate Program at UB, as well as being a Department of Veterans Affairs Research Career Scientist at the Buffalo VA Medical Center, VA Western NY Healthcare System. Dr. Fliesler obtained a PhD in biochemistry from Rice University, completed a postdoctoral fellowship at the Cullen Eye Institute/Baylor College of Medicine, and was previously on the faculties of Baylor College of Medicine, Bascom Palmer Eye Institute/University of the Miami School of Medicine, and the Saint Louis University School of Medicine, prior to joining the faculty of SUNY—University at Buffalo in 2008. His research is focused on inborn errors of cholesterol and isoprenoid metabolism and their impact on the development, structure and function of the retina, as well as on blast injury to the eye, using animal models. He has published more than 150 peer-reviewed journal articles, book chapters and review articles, and is the editor of two books. His research program has been funded continuously for more than 35 years by multiple grants from the NEI/NIH and private foundations, as well as, more recently, MERIT Awards from the U.S. Department of Veterans Affairs. Dr. Fliesler recently served on the Board of Trustees of the Association for Research in Vision and Ophthalmology (ARVO), representing the Retinal Cell Biology (RC) Section, as well as being President and immediate past President of ARVO. In 2009, he was inducted as a Silver-tier Fellow of ARVO (FARVO) and, in 2014, became a Gold-tier FARVO. In addition, he is a past councilor for North America, and Treasurer and President of the International Society for Eye Research (ISER). Dr. Fliesler is the Editor-in-Chief of Experimental Eye Research and serves on six other journal editorial boards, including Molecular Vision and the Journal of Lipid Research.

Preface to "The Molecular and Cellular Basis of Retinal Diseases"

Our goal with this series of articles was to bring together many prominent vision scientists to report on various aspects of ocular disease, with a focus on the use of animal models to elucidate the underlying mechanisms of pathobiology involved. Rather than limit the focus to only certain aspects of retinal research, we contacted a broad representation of vision scientists studying the retina, in the hopes of compiling a more comprehensive overview of the field. We believe that you will find, herein, a breadth of studies touching on many of the key areas of current retina research.

Steven J. Pittler, Steven J. Fliesler
Editors

Article

Characterizing the Retinal Phenotype in the High-Fat Diet and Western Diet Mouse Models of Prediabetes

Bright Asare-Bediako [1], Sunil K. Noothi [2], Sergio Li Calzi [2], Baskaran Athmanathan [3], Cristiano P. Vieira [2], Yvonne Adu-Agyeiwaah [1], Mariana Dupont [1], Bryce A. Jones [4], Xiaoxin X. Wang [5], Dibyendu Chakraborty [2], Moshe Levi [5], Prabhakara R. Nagareddy [3] and Maria B. Grant [2,*]

[1] Vision Science Graduate Program, School of Optometry, University of Alabama at Birmingham, Birmingham, AL 35233, USA; basareb@uab.edu (B.A.-B.); yvonnad@uab.edu (Y.A.-A.); mdupont@uab.edu (M.D.)

[2] Department of Ophthalmology and Visual Sciences, School of Medicine, The University of Alabama at Birmingham, Birmingham, AL 35294, USA; sunilnooti@uabmc.edu (S.K.N.); scalzi@uabmc.edu (S.L.C.); cvieira@uabmc.edu (C.P.V.); dchakraborty@uabmc.edu (D.C.)

[3] Division of Cardiac Surgery, Department of Surgery, Ohio State University Wexner Medical Center, Columbus, OH 43210, USA; baskaran.athmanathan@osumc.edu (B.A.); prabhakara.nagareddy@osumc.edu (P.R.N.)

[4] Department of Pharmacology and Physiology, Georgetown University, Washington, DC 20057, USA; baj46@georgetown.edu

[5] Department of Biochemistry and Molecular & Cellular Biology, Georgetown University, Washington, DC 20057, USA; xiaoxin.wang@georgetown.edu (X.X.W.); moshe.levi@georgetown.edu (M.L.)

* Correspondence: mariagrant@uabmc.edu

Received: 9 January 2020; Accepted: 13 February 2020; Published: 18 February 2020

Abstract: We sought to delineate the retinal features associated with the high-fat diet (HFD) mouse, a widely used model of obesity. C57BL/6 mice were fed either a high-fat (60% fat; HFD) or low-fat (10% fat; LFD) diet for up to 12 months. The effect of HFD on body weight and insulin resistance were measured. The retina was assessed by electroretinogram (ERG), fundus photography, permeability studies, and trypsin digests for enumeration of acellular capillaries. The HFD cohort experienced hypercholesterolemia when compared to the LFD cohort, but not hyperglycemia. HFD mice developed a higher body weight (60.33 g vs. 30.17g, $p < 0.0001$) as well as a reduced insulin sensitivity index (9.418 vs. 62.01, $p = 0.0002$) compared to LFD controls. At 6 months, retinal functional testing demonstrated a reduction in a-wave and b-wave amplitudes. At 12 months, mice on HFD showed evidence of increased retinal nerve infarcts and vascular leakage, reduced vascular density, but no increase in number of acellular capillaries compared to LFD mice. In conclusion, the HFD mouse is a useful model for examining the effect of prediabetes and hypercholesterolemia on the retina. The HFD-induced changes appear to occur slower than those observed in type 2 diabetes (T2D) models but are consistent with other retinopathy models, showing neural damage prior to vascular changes.

Keywords: retinal phenotype; neural infarcts; vascular leakage

1. Introduction

Diabetes is now considered a worldwide epidemic [1,2]. Recent reports indicate that over 90% of diabetic individuals have type 2 diabetes (T2D) [3,4]. The most common microvascular complication of diabetes is diabetic retinopathy (DR) [2]. Despite a growing number of different approaches to arrest DR, the incidence and prevalence of DR continues to rise [5]. The understanding of the pathogenesis of DR remains incomplete [4], and this is, in part, due to the lack of readily available models that completely recapitulate the metabolic phenotype [6]. The high-fat diet (HFD) mouse model has been described as a robust model for investigating obesity-associated T2D and its related metabolic complications [7]. Studies have shown that HFD-fed mice develop obesity, impaired glucose tolerance, and reduced insulin sensitivity [8,9] with systemic manifestations involving adipose tissue [10], liver [8], and kidneys [11]. However, the ocular changes associated with the HFD model have not been fully investigated. Moreover, the typical Western diet (WD; 40% fat) has also been given to rodents to recapitulate obesity-driven pathology. However, to mimic the features of T2D, the administration of low-dose Streptozotocin (STZ) is also given to the WD mice [12–14].

The retinal response to high fat exposure would likely involve local changes in the expression of lipid transport proteins, such as the liver X receptors (LXRs). The LXRs are the key transcription factors that regulate lipid and cholesterol metabolism [15]. While liver X receptor alpha (LXRα) is expressed only in some tissues, the expression of liver X receptor beta (LXRβ) is ubiquitous [12]. Previously we showed that whole body LXRα/β deficiency resulted in the generation of increased numbers of acellular capillaries, while LXR agonists improved DR in Streptozotocin (STZ)-induced diabetes [12] and in diabetic Lepr$^{db/db}$ (db/db) mice [16]; however, it is not known if the WD modulates the expression of LXR in the retina.

Retinopathy is typically characterized by macroglia activation and gliosis identified by glial fibrillary acidic protein (GFAP) overexpression, which can be considered as a marker for retinal damage [17,18]. In the healthy mammalian retina, GFAP is expressed only in astrocytes and not in Muller cells. Following inherited or acquired retinal pathology, GFAP is expressed also in Muller cells [19,20]. GFAP expression in Muller cells has been widely used as a cellular marker for retinal pathology [21–25]. Hypoxia-inducible factor 1 alpha (HIF-1α) is known to be a key regulator of a tissue's response to hypoxia [26] and plays a role in obesity-induced metabolic syndrome. It has been shown that HFD leads to gradual increase in HIF-1α and associated pathological changes in the liver [27,28]. However, the role of HIF-1α in the retina of WD-fed mice is not known.

A better understanding of DR in obesity-driven models is needed and may facilitate the optimal choice of disease models for future investigations. Thus, in the present study, we hypothesized that HFD and WD feeding would result in a distinct retinal phenotype and a time course slower than that observed in models of T2D, such as the db/db mouse [29] or the high fructose and high fat fed mouse [30]. For this purpose, we characterized not only systemic endpoints of glucose and lipid metabolism but also the function of the retina and development of retinal pathology, including retinal vascular changes and changes in expression of the critical proteins LXRβ, HIF-1α, and GFAP.

2. Materials and Methods

2.1. Animals

All animal experiments were approved by the University of Alabama at Birmingham (IACUC-20467, approved on 06/16/2016) and Georgetown University (animal project #2017-0059, approved on 10/27/2017), and followed the Association for Research in Vision and Ophthalmology Statement for the Use of Animals. Six to eight-week-old C57BL/6J mice were fed either a low-fat diet (LFD) (10%kcal fat, 70%kcal carbohydrate, 20% protein), a Western diet (40% kcal fat, 43% kcal carbohydrate, 17%kcal protein), or a HFD (60% kcal fat, 20% kcal carbohydrate, 20% kcal protein) for up to 12 months. Diets were purchased from Research Diets, Inc, New Brunswick, NJ, USA. Full details of the composition of each diet is given in Supplementary Table S1.

2.2. Body Composition, Glucose Tolerance, and Insulin Sensitivity Testing

The fat mass, lean mass, and water content of the animals were measured by magnetic resonance imaging using EchoMRI (Echo Medical Systems, LLC, Houston, TX, USA). For glucose and insulin tolerance tests, mice were fasted for 5-6 h, injected intraperitoneally with D-glucose at 1.5 g/kg of lean mass and tail bled for glucose and insulin measurements. Blood glucose and insulin levels were measured 0, 15, 30, 45, 60, and 120 min after glucose administration. The insulin sensitivity index (ISI) was estimated using the Matsuda–Defronzo method [31].

2.3. Electroretinogram (ERG)

ERGs were performed using a LKC Bigshot ERG system. Briefly, mice were dark-adapted overnight. The animals were anesthetized with ketamine (80 mg/kg total body mass) and xylazine (15 mg/kg total body mass), then dilated with atropine/phenylephrine under dim red light. Once dilated, animals were exposed to 5 full-field white light flashes at 0.25 and 2.5 cd.s/m^2 under scotopic conditions. The animals were then light-adapted for 5 min and exposed to 10–15 full-field white light flashes at 10 and 25 cd.s/m^2 under photopic conditions. Responses were averaged and analyzed using the LKC EM software.

2.4. Fundus Photography and Fluorescein Angiography

Fundus photography and fluorescein angiography were performed using the Phoenix Micron IV retinal imaging microscope (Phoenix Technology Group, Pleasanton, CA, USA). Briefly, mice were anesthetized with ketamine and xylazine, then dilated with atropine/phenlylephrine, as described above. Once dilated, the animals were placed on the instrument and fundus photographs were taken. Animals were then given intraperitoneal injection of fluorescein (AK-FLUOR 10%, Sigma Pharmaceuticals, North Liberty, IA, USA) and the retinal vasculature was imaged with blue light illumination after 5–8 min when all the vessels were filled.

2.5. Acellular Capillaries Quantification

Trypsin digestion of the retina was performed according to a previously published protocol [32,33]. Briefly, eyeballs were enucleated and incubated in 4% paraformaldehyde overnight. Retinas were isolated, washed, and digested in elastase solution (40 Units elastase/mL; Sigma-Aldrich, St. Louis, MO, USA) to remove the non-vascular tissue. The vascular beds were mounted on glass slides followed by staining with periodic acid–Schiff's base and hematoxylin. About 5–6 fields from the central to mid-periphery were imaged and the number of acellular capillaries per square millimeter were quantified.

2.6. Immunohistochemistry

Immunohistochemical staining of mouse retinas was performed according to a previously published protocol [34]. Briefly, mice were euthanized and eyes were immediately enucleated and fixed in 4% paraformaldehyde (PFA) solution for 15 min. Cornea and lenses were carefully removed and posterior cups were incubated in 15% sucrose solution in phosphate-buffered saline (PBS) overnight at 4 °C after washing briefly in PBS. Posterior cups were transferred to 30% sucrose in PBS for 3–4 h, then embedded in optimal cutting temperature (O.C.T) medium and immediately frozen on dry ice. The frozen samples were stored at −80 °C until further processing. The sections were thawed at room temperature for 4 h, washed in PBS for 5 min, and permeabilized with 0.25% Triton-X in PBS for 5 min at room temperature. Sections were blocked with 10% horse serum in 1% bovine serum albumin (BSA) for 2 h then incubated with primary antibody diluted in blocking solution (1:100 dilution) overnight at 4 °C. The antibodies used were rabbit anti-GFAP (Abcam, MA, USA), mouse mAB HIF-1α antibody (Novus Biologicals, CO, USA), LXR-β polyclonal antibody (Invitrogen, IL, USA), rabbit anti-Vimentin (Cell Signaling Technology, MA, USA), and isolectin GS-IB4 Alexa Fluor 568 (Life Technologies, OR, USA). Sections were then washed and incubated in fluorescent-labeled secondary antibodies (goat anti-rabbit IgG Alexa Fluor 488, Life Technologies, OR, USA) for 1 h at room temperature, followed by washing and incubation with 4′,6-diamidino-2-phenylindole, dihydrochloride (DAPI) solution (Life Technologies, OR, USA) for 5 min at room temperature. Finally, sections were washed and mounted with anti-fade mounting medium (Vector Laboratories, CA, USA) for imaging. Image analysis was completed in a masked fashion using four images taken at defined positions and quantified using ImageJ software.

The analysis was performed in a masked fashion by three separate observers, then averaged. To achieve unbiased results, positive and negative controls were included alongside experimental test and control groups. Fluorescent microscopy was performed by trained masked operators. To address selection bias in immunofluorescence, the entire areas of retinal cross-sections were imaged.

2.7. Statistics

All experiments were repeated at least 3 times. All data were assessed using one-way ANOVA. When the results were significant, we determined which means differed from each other using Tukey's multiple-comparisons test. Results are expressed as mean ± standard error of the mean (SEM). Statistical analysis was performed using GraphPad Prism, with $p < 0.05$ considered statistically significant. Only significant comparisons are shown in the figures. All the examiners were blinded to the identities of the samples they were analyzing.

3. Results

3.1. HFD Mice Have Normal Glucose Levels but Are Insulin-Resistant

We first sought to validate our model by confirming in our cohort that HFD feeding led to similar degrees of body weight gain as reported in the literature [10,35]. Mice on HFD showed increased body weights by 4 weeks of feeding ($p = 0.0020$). This increase was sustained throughout the 12-month observation period (Figure 1A). At 12 months, HFD mice had moderately higher lean mass (difference of 5.580 ± 1.003g, $p < 0.0001$) and water content (difference of 4.517 ± 0.876, $p = 0.0001$) but a markedly increased fat mass (difference of 21.15 ± 2.362, $p < 0.0001$) compared to LFD controls (Figure 1B). Unexpectedly, chronic high-fat feeding did not cause hyperglycemia. Despite feeding mice with a HFD for 12 months, the HbA1c levels were not different between HFD and LFD mice (Figure 1C). At 12 months, there was no difference in fasting blood glucose levels (Figure 1D, basal) and intraperitoneal glucose tolerance test (IP-GTT) did not show any significant differences between LFD and HFD mice (Figure 1D). However, due to the very high levels of insulin in HFD mice (0.6 ng/mL for LFD and 3.5 ng/mL for HFD; Figure 1E, basal), the insulin sensitivity index demonstrated that HFD mice had much lower insulin sensitivity compared to LFD mice (Figure 1E,F). Also, plasma total cholesterol levels were higher in HFD mice compared to LFD (174.4 vs. 114.9, $p = 0.0008$) (Figure 1G)

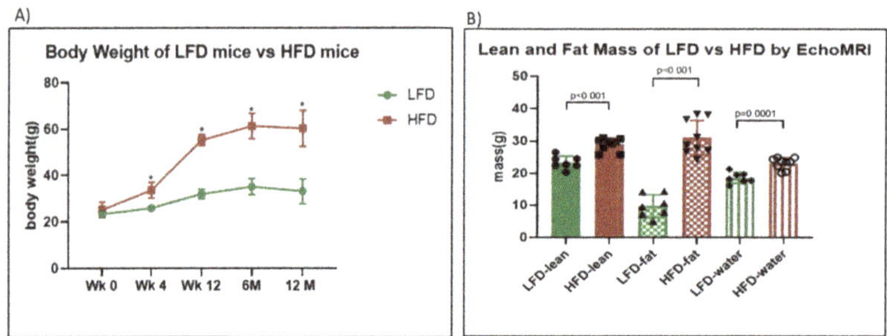

Figure 1. *Cont.*

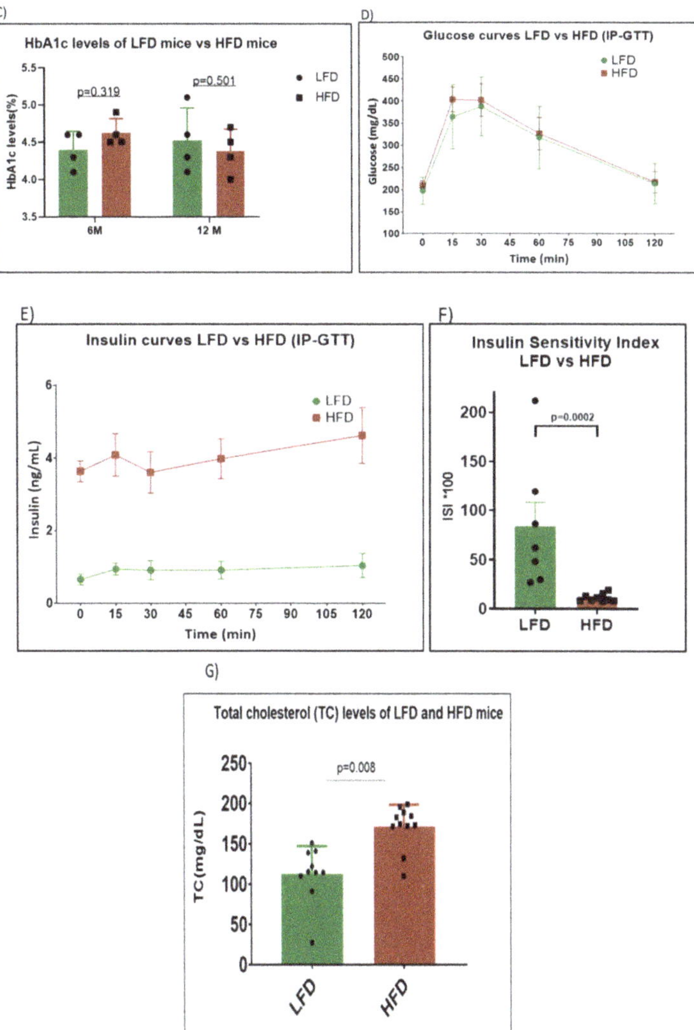

Figure 1. Body weight, glucose levels, and insulin sensitivity of high-fat diet (HFD) mice vs. low-fat diet (LFD) mice. (**A**) Body weights as measured for mice on LFD (green) and HFD (red) for 12 months. (* $p < 0.000001$; $n = 6$). (**B**) Lean mass, fat mass and water content of LFD mice vs. HFD mice. (**C**) Glycated hemoglobin (HbA1c) levels measured for the mice after 6 months and 12 months. (**D–G**) Glucose curves ($p > 0.46$ for all time points), insulin curves ($p < 0.0018$ for all time points), insulin sensitivity index, and total cholesterol levels for LFD mice vs. HFD mice, respectively, following intraperitoneal glucose tolerance test (IP-GTT) after 12 months of feeding.

3.2. HFD Mice Have Functional Deficits in Their Retinas

Full-field ERG under both scotopic and photopic conditions was performed at 6 months and 12 months of HFD feeding (Figure 2A–D). HFD mice at 6 months showed significantly reduced a- and b-wave amplitudes under scotopic conditions ($p = 0.00125$ and $p = 0.000002$ for 0.25 cd.s/m^2 and 2.5 cd.s/m^2 stimulus luminance, respectively) but not photopic conditions when compared to LFD mice. After 12 months of feeding of the respective diets, the difference was not significant ($p = 0.183$ and

$p = 0.154$ for 0.25 cd.s/m^2 and 2.5 cd.s/m^2 stimulus luminance respectively) (Figure 2C,D). Interestingly, when comparing 6 and 12 months of LFD feeding, the mice experienced marked reductions in both the a- and b- waves under both photopic and scotopic conditions at 12 months (Figure 2E,F), but no significant difference was noted in the HFD-fed mice (Figure 2G,H).

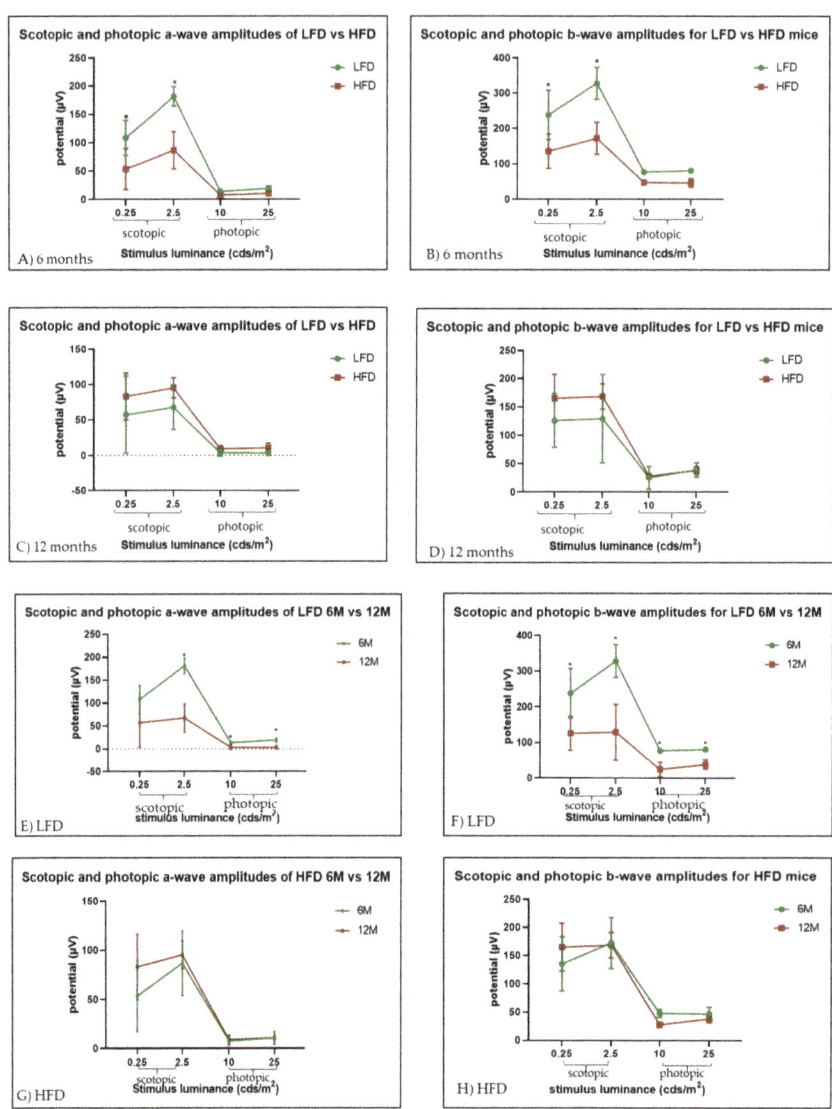

Figure 2. Assessment of retinal function of LFD mice versus HFD mice by electroretinogram (ERG). The amplitudes of a-waves and b-waves were assessed under both scotopic and photopic conditions for LFD mice and HFD mice after 6 months (**A**,**B**) and 12 months (**C**,**D**). LFD mice showed a significant reduction in retinal response between 6 months and 12 months of feeding (**E**,**F**), but HFD mice did not (**G**,**H**); ($n = 4$ for both groups).

3.3. Fundus Photography shows Neural Retinal Lesions in HFD Mice

In humans, DR is associated with retinal lesions such as hemorrhages, microaneurysms, exudates, and "cotton wool spots" [36]. Fundus photography using Micron IV demonstrated retinal pathology in the HFD mice. Though not statistically significant, HFD mice showed a trend of increased numbers of "lipid-laden-like" lesions (Figure 3A) after 6 months ($p = 0.057$). However, with 12 months of feeding, HFD mice showed significantly higher number of lesions in the retina (Figure 3B).

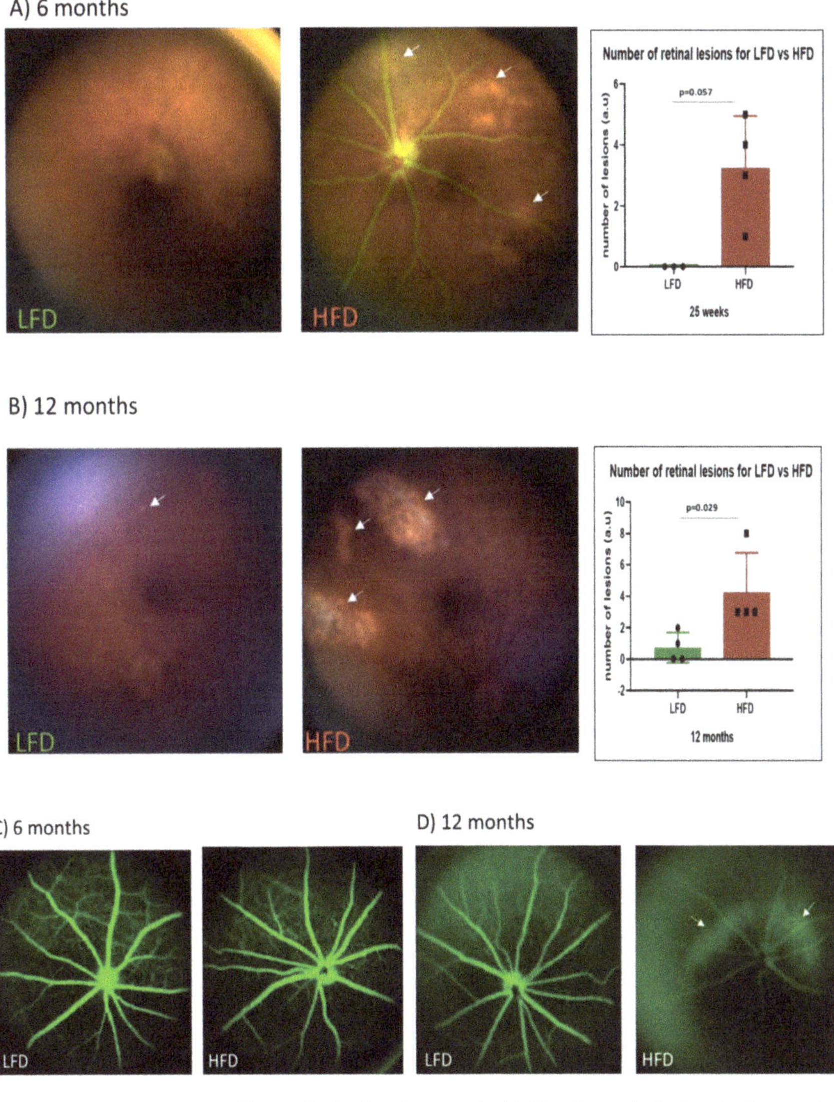

Figure 3. Assessment of retinal lesions by fundus photography (**A**,**B**) and vascular leakage by fluorescein angiography (**C**,**D**). HFD mice developed more neural infarcts ((**A**,**B**), white arrows) than LFD mice. No infarct was observed for LFD after 6 months (**A**). However, vascular leakage was observed in HFD mice after 12 months of feeding ((**D**), white arrows).

3.4. Vascular Permeability Changes in HFD Mice

A hallmark of DR in humans is increased vascular permeability, ultimately leading to diabetic macular edema in humans. To determine if HFD mice developed a breakdown in the blood–retinal barrier, we assessed vascular leakage by fluorescein angiography (FA). At 6 months of HFD feeding, FA did not show any evidence of retinal vascular leakage and were similar to FAs in LFD controls (Figure 3C). However, after 12 months of HFD feeding, increased leakage of fluorescein was observed in the retina compared to LFD control retinas (Figure 3D).

3.5. Acellular Capillary Formation in HFD Mice

A well-established feature of diabetic microvascular dysfunction is an increase in the number of acellular capillaries in the retina, defined as basal membrane tubes lacking endothelial cells and pericyte nuclei. At 12 months of HFD feeding, there was no significant increase in acellular capillary numbers in the HFD mice (Figure 4B,C) compared to the LFD mice (Figure 4A,C). However, the HFD retinas showed lower vascular densities compared to LFD retinas (Figure 4D).

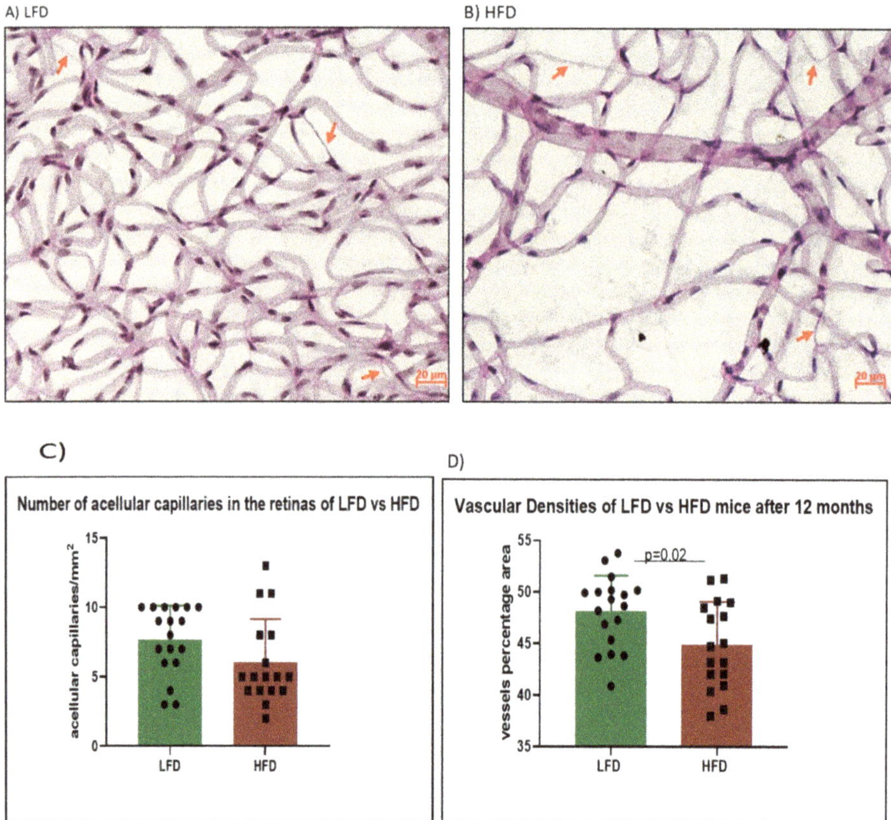

Figure 4. Enumeration of acellular capillaries in LFD and HFD mice after 12 months of feeding. Red arrows indicate acellular capillaries in the retinas of LFD (**A**) and HFD (**B**) mice. There was no significant difference in the number of acellular capillaries between both groups (**C**) ($p = 0.086$). However, HFD retinas showed lesser vascular densities compared to LFD retinas (**D**).

3.6. Retinal Damage, Hypoxia, and Lipid Transport in WD Mice

While the HFD represents a diet with 60% fat content that is used as a model of obesity and T2D, the WD with 40% fat content has garnered popularity as it represents a regimen closer to that actually ingested by humans. Since the WD diet has lower fat content and is not associated with hyperglycemia, we hypothesized that if retinal changes were present they would be subtle compared to those we observed with HFD feeding. To test the validity of our hypothesis, we performed IHC studies and first examined whether there was evidence of glial activation by examining expression of the glial marker GFAP after 6 months of WD feeding. Although there was no statistically significant difference ($p = 0.88$) in the total expression of GFAP between retinas of WD and LFD mice (Figure 5A–C), increased expression of GFAP was observed in selected Vimentin-positive Muller cells in the WD mice (Figure 5G–I) compared to LFD (Figure 5D–F). Increased expression of GFAP in Muller cells is supportive of increased oxidative stress and inflammation in these cells, and suggests that the impact of WD is not experienced uniformly across all Muller cells [37,38].

To assess whether WD feeding induced retinal hypoxia, changes in HIF-1α expression were examined by IHC. After 6 months of WD feeding, a significant increase ($p = 0.025$) in expression of HIF-1α was seen in WD mice (Figure 6D) compared to LFD mice (Figure 6C). This was not observed after 3 months of WD feeding (Figure 6A,B). Quantitation of HIF-1α expression is shown in Figure 6E, demonstrating that WD-fed mice exhibit higher levels than LFD-fed mice. Co-localization with isolectin, a known vascular endothelial cell marker, showed increased expression of HIF- 1α in some endothelial cells in WD mice (I–K) but not in LFD mice (F–H). Higher magnification images from two different WD samples are shown in Figure 6L,M.

Retinal lipid content is regulated in part by liver X receptor beta (LXRβ) expression. We next examined changes in LXRβ expression in the two experimental cohorts. In control mice, LXRβ localized predominantly in the ganglion cell layer, as well as the inner nuclear layer (Figure 7A), which is the location of the bipolar cells, horizontal cells, and amacrine cells. There was a significant reduction in expression of LXRβ in WD only in the ganglion cell layer ($p = 0.0079$) after 3 months of feeding (Figure 7B). However, after 6 months of WD feeding, WD mice (Figure 7E) showed significantly reduced expression of LXRβ in the ganglion cell layer ($p = 0.0374$), inner nuclear layer ($p < 0.0001$), and outer nuclear layer, as well as in the photoreceptors of the outer nuclear layer ($p = 0.0020$). The expression of LXRβ was reduced after 6 months compared to 3 months of feeding in both LFD ($p < 0.0001$) and WD ($p < 0.0001$) in the nuclear and ganglion cell layers, suggesting an age-related loss in LXRβ.

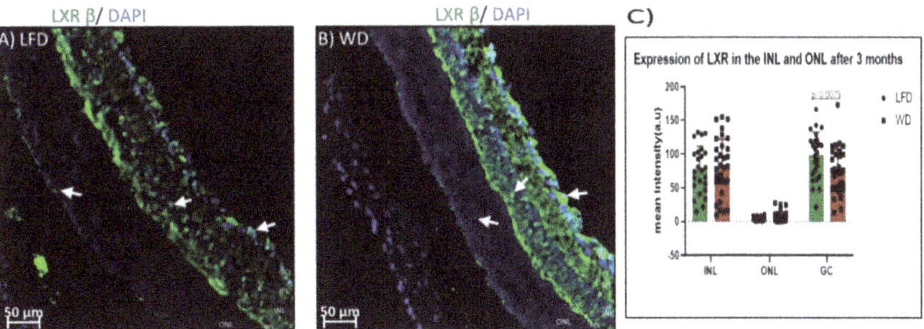

Figure 5. *Cont.*

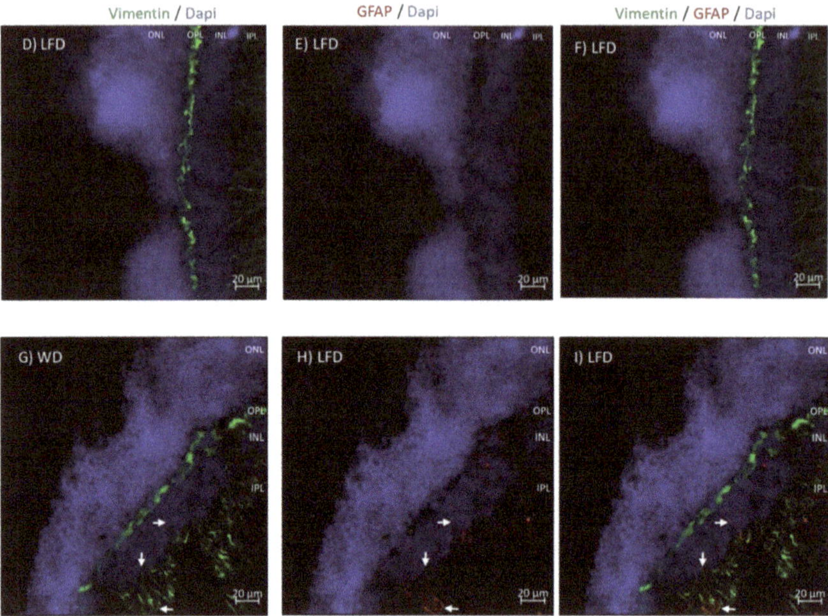

Figure 5. Retinal glial fibrillary acidic protein (GFAP) expression after 6 months of feeding. Some Muller cells in Western diet (WD) retinas express GFAP (**A,C**, white arrows), but not in LFD (**A,B**), indicating that the impact of WD is not uniform across all Muller cells. Co-localization with Vimentin, a known Mueller cell marker, showed increased expression of GFAP in some Mueller cells in WD mice (**G–I**) but not in LFD mice (**D–F**).

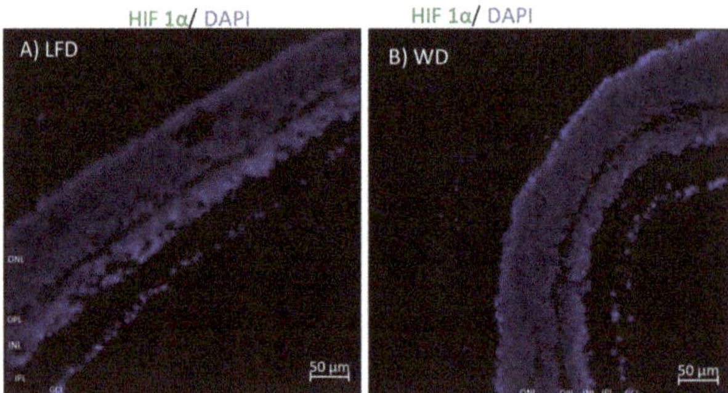

Figure 6. *Cont.*

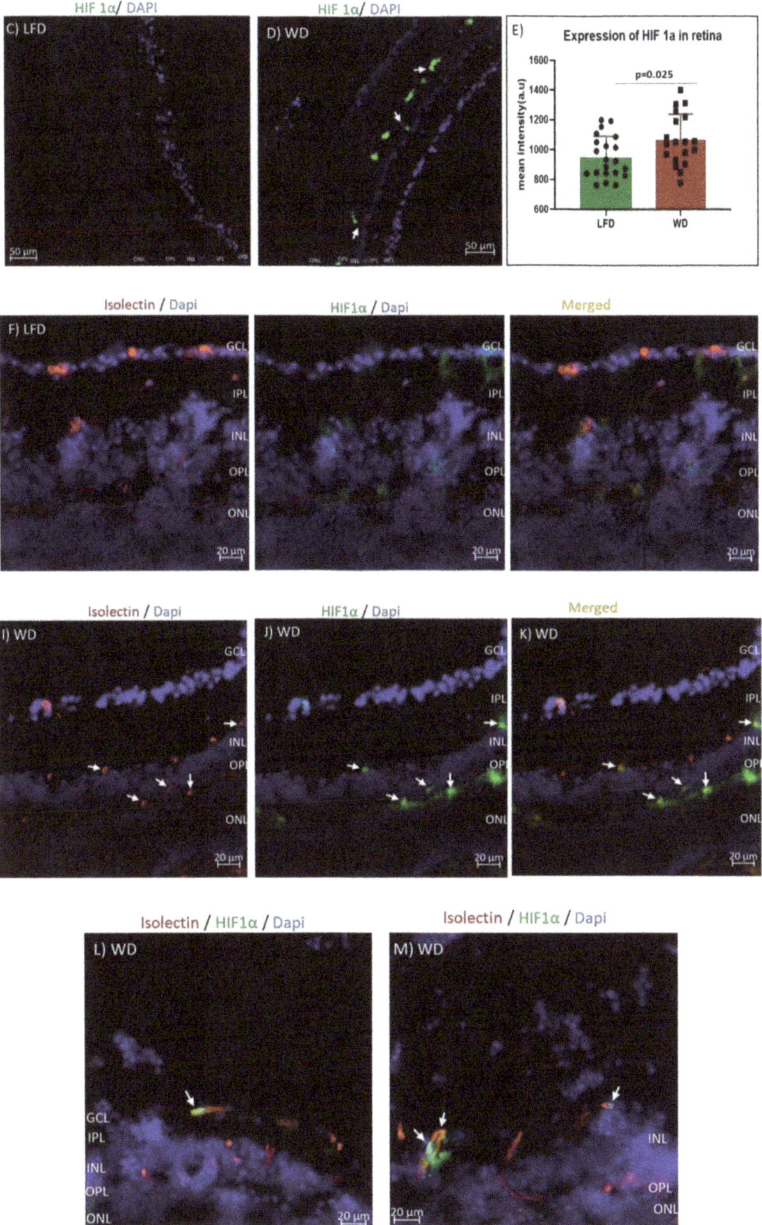

Figure 6. Retinal hypoxia-inducible factor 1 alpha (HIF-1α) expression after 3 and 6 months of WD feeding. There was increased expression of HIF-1α in WD retinas (**D**, white arrows) compared to LFD retinas (**C**), as shown by quantification (**E**). Also, there was no significant difference in expression of HIF-1α after 3 months of feeding (**A,B**). Co-localization with isolectin, a known vascular endothelial cell marker, showed increased expression of HIF-1α in some endothelial cells in WD mice (**I–K**) but not in LFD mice (**F–H**). (**L,M**) Magnified merged images from two different WD samples.

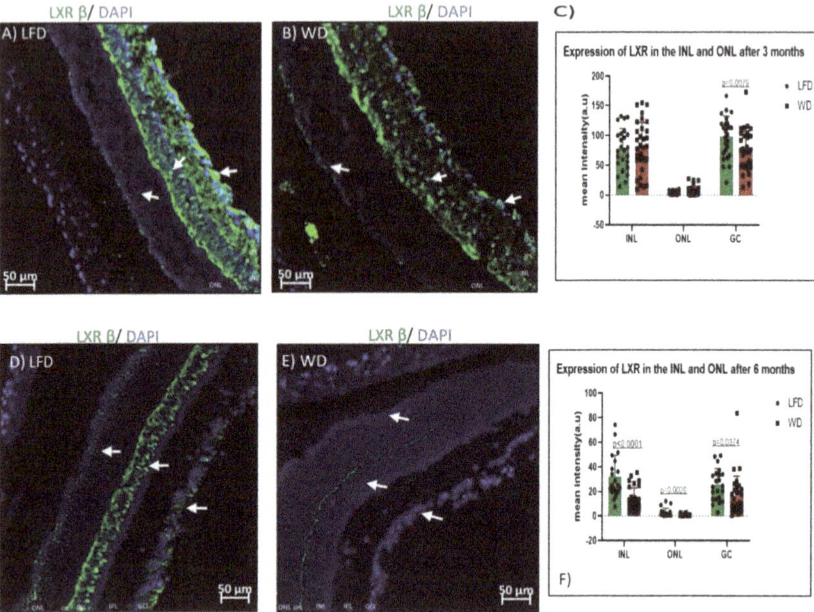

Figure 7. Retinal liver X receptor beta (LXRβ) expression after 3 and 6 months of feeding. After 3 months of either WD or LFD feeding, there was significant reduction in the expression of LXRβ in only the ganglion cell layer of WD mice (**B**) compared to LFD mice (**A**). However, after 6 months of feeding, there was reduced expression of LXRβ in the ganglion cell layer as well as inner and outer nuclear layers of WD mice (**E**, white arrows) compared to LFD mice (**D**). Quantification of LXR in the inner nuclear layer (INL) and outer nuclear layer (ONL) at 3 months shows reductions in the ganglion cell (GC) layer (**C**). At 6 months, reductions are seen in the INL, ONL, and ganglion cell (GC) layer of the WD-fed mice when compared to LFD mice.

4. Discussion

Diabetic retinopathy causes both neural and vascular defects, with neural deficits preceding vascular changes [6,39–42]. Even before the onset of clinically detectable retinopathy, diabetic patients have a reduced ERG implicit time [43] and high-frequency flicker amplitude [44]. Later, they experience decreased vascular density [45]. In this study, we have shown that HFD feeding results in a suitable model of prediabetes, with the HFD cohort exhibiting insulin resistance and hypercholesterolemia without hyperglycemia. The retinopathy that is exhibited occurs over a slower time course than in T2D models, where both hyperglycemia and hyperinsulinemia exist.

The HFD mouse has previously been described as a model for T2D [7,46], as C57BL/6J mice fed HFD develop obesity and insulin resistance [47,48], but as we show in this study, this model has a distinct timeline and different characteristics than those seen in T2D. We show that HFD mice have hypercholesterolemia and insulin resistance but the absence of hyperglycemia, which is typical of T2D models.

In agreement with the literature, our study shows that mice fed a HFD have a sustained increase in body weight [6,49,50]. As confirmed by EchoMRI, the increase in body weight is primarily due to elevated body fat mass. After 12 weeks of feeding, HFD mice showed a two-fold increase in body fat mass over control LFD mice. Despite the marked increase in fat mass, HFD mice did not develop overt hyperglycemia. Glycated hemoglobin levels measured at 6 months and 12 months showed that both groups had normal glycated hemoglobin, thus indicating a key difference between the HFD model and other T2D rodent models, many of which are genetic. However, HFD mice develop hyperinsulinemia (Figure 1E,F), and their insulin production is sufficient to maintain euglycemia, as indicated by their glycated hemoglobin levels. The marked hyperinsulinemia we observed is supported by the literature [6,51–54]. In contrast, T2D in humans is characterized by not only insulin resistance but also the presence of sustained hyperglycemia and elevated HbA1c levels. When only insulin resistance is present, individuals are described as prediabetics [55,56].

Insulin resistance is believed to play a key role in diabetic neuropathy by increasing oxidative stress and mitochondrial dysfunction [57,58], and may also drive the early neural retinal dysfunction that we observed in our HFD mice. Thus, the HFD mice secrete sufficiently elevated insulin to maintain a normal glucose level, and as such the HFD model may be better characterized as a prediabetes model. Importantly, the incidence of prediabetes is often higher than that of diabetes [59]. The prevalence of prediabetes is also increasing; it is estimated that more than 470 million people worldwide will be suffering from prediabetes by 2030 [60]. Most importantly, the three classical microvascular complications, retinopathy, neuropathy, and nephropathy, have all been documented in individuals with prediabetes [61].

While classifications of diabetes remain "glucose-centric", our study draws attention to the importance of earlier events, when glucose levels are still normal. Thus, in our model, hyperinsulinemia with hypercholesterolemia will likely lead to the retinal pathology observed. Not surprisingly, these pathologies take a longer time to develop than those typically seen when hyperglycemia is also present.

Systemic and retinal lipid abnormalities have been shown to promote retinal damage [16,62,63]. Previously, we demonstrated that diabetes-induced disruption of the LXR axis results in abnormal lipid metabolism, inadequate vascular repair, and localized and systemic inflammation [16,64]. The LXRs (LXRα and LXRβ) play important roles in cholesterol homeostasis [65]. They regulate the expression of reverse cholesterol transporters [12]. Activation of LXRs using pharmacological agents repress inflammatory genes such as TNF-α and IL-1β [66], inhibit the expression of pro-apoptotic factors [67], and prevent the development of DR [12]. We showed that use of GW3965, an LXR agonist, resulted in normalization of cholesterol homeostasis and repression of inflammatory genes, such as iNOS, IL-1β, ICAM-1, and CCL2 in the retina [16]. We found that inadequate cholesterol removal due to deficiency in LXR and reduced oxysterol production in the retina due to loss of cytochromes p450 27A1 and 46A1 resulted in widespread retinal pathology [68]. In the current study, we showed that concentrations of 40% fat in the diet were sufficient to reduce expression of LXR in the inner and outer nuclear layers.

Our study showed that HFD mice develop neural retinal deficits after 6 months of feeding, as both a-waves and b- waves were reduced under scotopic conditions. Unexpectedly, the a- and b- wave responses for LFD mice was significantly less after 12 months compared to the response after 6 months of feeding ($p < 0.01$ for both scotopic and photopic conditions), which suggests that the LFD may have detrimental effects on the neural retina. Because the composition of the diets must be isocaloric, when the amount of fat is reduced, some other dietary component needs to be increased to compensate. Inn the LFD, the amount of sucrose increases from 72 g to 354 g and 315 g of corn starch is also added so that the LFD can be isocaloric with the HFD. However, this largely occurs at the expense of making the diet high in carbohydrates. The literature supports that LFD may be detrimental [69–71]. While we were unable to find literature supporting the impact of LFD specifically on ERGs, the systemic consequences of LFD may indirectly affect the retina, for example by reduced availability of fat-soluble vitamins or changing retinal cholesterol metabolism. Moreover, the increased sucrose and cornstarch in the LFD may have direct deleterious effects [72,73]. LFDs promote insulin resistance, and while most of the research has been performed in humans, these findings may have relevance to murine studies. LFD, typically considered a high carbohydrate diet, is known to promote inflammation [74–76]. A recent study compared ERGs in HFD fed rats, Streptozotocin (STZ) rats and type 2 diabetes (T2D) rats at 6 months to controls. Kowluru found differences between the diabetic ERGs and controls, but no differences between the ERGs of the HFD rats compared to controls; however, Kowluru did not look at 12 month tests and the study was performed in rats, not in mice [77]. Thus, it is difficult to compare these findings with our results.

While neural damage was detected at 6 months, the vascular damage was not observed until much later. This is in agreement with Rajagopal et al. [6], who demonstrated that vascular damage was not observed at 6 months of HFD feeding. However, despite the absence of vascular damage after 6 months of HFD, we observed the presence of "lipid-laden like" lesions, and also neural infarcts similar to what is described in humans as "cotton-wool" spots. These lesions, which appeared to increase as the retinopathy progressed in the HFD mice, could become a useful measure of retinal damage and may be sensitive enough to use as a novel endpoint for the preclinical investigation of therapeutic agents.

GFAP is normally expressed in retinal astrocytes in rodents; however, during stress and inflammation, Muller cells [37] respond by increasing GFAP expression. In this study, we show that WD induces GFAP expression in selective Muller cells, supporting the presence of increased stress and inflammation in the retina of these mice. Kim et al. have reported increased inflammation in other tissues such as adipose tissue and intestines [78]. Lee et al. showed increased numbers of activated macrophages in the retina of HFD mice [79]. In both humans and rodents, obesity-induced diabetes is associated with hypoxia in tissues such adipose tissue, and suppression of HIF-1α mitigates tissue-specific pathological changes associated with HFD [80]. The liver, brain, kidney, and heart display tissue-specific regulation of HIF-1α under systemic hypoxia [81]. After 6 months, but not after 3 months, we observed that HIF-1α expression is increased in the WD retinas compared to LFD controls. Similar to our observation in the retina of 3-month-old mice on WD, Prasad et al. showed the absence of pimonidazole staining in the kidneys of 10–11-week old db/db mice [82], also indicating the absence of hypoxia response in the kidneys at this time point.

5. Conclusions

Our study demonstrates that HFD feeding generates a useful prediabetes model. Specifically, the combination of hypercholesterolemia and insulin resistance are sufficient to induce retinal dysfunction with a slower time course of development compared to T2D models such as the db/db mouse. In agreement with reports describing diabetes models, we show that neural functional deficits are the earliest indicator of damage in the retina of this prediabetes model before vascular changes. Key molecular targets such as HIF-1α and the LXRs provide insights into the retinal pathobiology observed in this hypercholesterolemic, hyperinsulinemic model. The appearance and frequency of neural infarcts or "lipid-laden lesions" in the retina of HFD mice could represent a novel endpoint for evaluation of therapeutic interventions.

Supplementary Materials: The following are available online at http://www.mdpi.com/2073-4409/9/2/464/s1, Table S1: Detailed Composition of high-fat diet (HFD), Western diet (WD), and low-fat diet (LFD).

Author Contributions: Conceptualization, M.L., P.R.N., and M.B.G.; data curation, B.A.-B. and S.K.N.; formal analysis, B.A.-B., S.K.N., B.A.J., and M.B.G.; funding acquisition, M.B.G.; investigation, B.A.-B., S.K.N., S.L.C., B.A., C.P.V., Y.A.-A., M.D., B.A.J., X.X.W., D.C., and M.B.G.; methodology, B.A.-B., S.K.N., P.R.N., and M.B.G.; project administration, P.R.N. and M.B.G.; resources, M.L., P.R.N., and M.B.G.; supervision, M.L. and P.R.N.; validation, B.A.-B. and S.L.C.; visualization, B.A.-B.; writing—original draft, B.A.-B., S.K.N., P.R.N., and M.B.G.; writing—review and editing, B.A.-B., S.K.N., S.L.C., B.A., C.P.V., Y.A.-A., B.A.J., M.L., P.R.N., and M.B.G. All authors have read and agreed to the published version of the manuscript.

Funding: M.G is supported by NIH funding (EY012601, EY028037, and EY028858). P.R.N is supported by NIH funding (HL122505, HL137799); M.L is supported by NIH funding (DK116567). BAJ is supported by the National Center for Advancing Translational Sciences of the NIH under award number TL1TR001431.

Conflicts of Interest: The authors declare no conflict of interest.

References

1. Cho, N.; Shaw, J.; Karuranga, S.; Huang, Y.; Fernandes, J.D.R.; Ohlrogge, A.; Malanda, B. IDF Diabetes Atlas: Global estimates of diabetes prevalence for 2017 and projections for 2045. *Diabetes Res. Clin. Pr.* **2018**, *138*, 271–281. [CrossRef] [PubMed]

2. Chatterjee, S.; Khunti, K.; Davies, M.J. Type 2 diabetes. *Lancet* **2017**, *389*, 2239–2251. [CrossRef]

3. Busik, J.V.; Tikhonenko, M.; Bhatwadekar, A.; Opreanu, M.; Yakubova, N.; Caballero, S.; Player, D.; Nakagawa, T.; Afzal, A.; Kielczewski, J.; et al. Diabetic retinopathy is associated with bone marrow neuropathy and a depressed peripheral clock. *J. Exp. Med.* **2009**, *206*, 2897–2906. [CrossRef] [PubMed]

4. Turner, R.C.; Cull, C.A.; Frighi, V.; Holman, R.R.; UK Prospective Diabetes Study (UKPDS) Group. Glycemic Control With Diet Sulfonylurea, Metformin, or Insulin in Patients with Type 2 Diabetes MellitusProgressive Requirement for Multiple Therapies (UKPDS 49). *JAMA* **1999**, *281*, 2005. [CrossRef] [PubMed]

5. Yao, Z.; Gu, Y.; Zhang, Q.; Liu, L.; Meng, G.; Wu, H.; Xia, Y.; Bao, X.; Shi, H.; Sun, S. Estimated daily quercetin intake and association with the prevalence of type 2 diabetes mellitus in chinese adults. *Eur. J. Nutr.* **2019**, *58*, 819–830. [CrossRef]

6. Rajagopal, R.; Bligard, G.W.; Zhang, S.; Yin, L.; Lukasiewicz, P.; Semenkovich, C.F. Functional Deficits Precede Structural Lesions in Mice With High-Fat Diet–Induced Diabetic Retinopathy. *Diabetes* **2016**, *65*, 1072–1084. [CrossRef]

7. Winzell, M.S.; Ahrén, B. The high-fat diet-fed mouse: A model for studying mechanisms and treatment of impaired glucose tolerance and type 2 diabetes. *Diabetes* **2004**, *53*, S215–S219. [CrossRef]

8. Liou, C.-J.; Lee, Y.-K.; Ting, N.-C.; Chen, Y.-L.; Shen, S.-C.; Wu, S.-J.; Huang, W.-C. Protective Effects of Licochalcone A Ameliorates Obesity and Non-Alcoholic Fatty Liver Disease Via Promotion of the Sirt-1/AMPK Pathway in Mice Fed a High-Fat Diet. *Cells* **2019**, *8*, 447. [CrossRef]

9. Collins, S.; Martin, T.L.; Surwit, R.S.; Robidoux, J. Genetic vulnerability to diet-induced obesity in the C57BL/6J mouse: Physiological and molecular characteristics. *Physiol. Behav.* **2004**, *81*, 243–248. [CrossRef]

10. Illesca, P.; Valenzuela, R.; Espinosa, A.; Echeverría, F.; Soto-Alarcon, S.; Ortiz, M.; Videla, L.A. Hydroxytyrosol supplementation ameliorates the metabolic disturbances in white adipose tissue from mice fed a high-fat diet through recovery of transcription factors nrf2, srebp-1c, ppar-γ and nf-κb. *Biomed. Pharmacother.* **2019**, *109*, 2472–2481. [CrossRef]

11. Declèves, A.-E.; Mathew, A.V.; Armando, A.M.; Han, X.; Dennis, E.A.; Quehenberger, O.; Sharma, K.; Declèves, A.-E. AMP-activated protein kinase activation ameliorates eicosanoid dysregulation in high-fat-induced kidney disease in mice. *J. Lipid Res.* **2019**, *60*, 937–952. [CrossRef] [PubMed]

12. Hazra, S.; Rasheed, A.; Bhatwadekar, A.; Wang, X.; Shaw, L.C.; Patel, M.; Caballero, S.; Magomedova, L.; Solis, N.; Yan, Y.; et al. Liver X Receptor Modulates Diabetic Retinopathy Outcome in a Mouse Model of Streptozotocin-Induced Diabetes. *Diabetes* **2012**, *61*, 3270–3279. [CrossRef]

13. Wang, X.X.; Jiang, T.; Shen, Y.; Caldas, Y.; Miyazaki-Anzai, S.; Santamaria, H.; Urbanek, C.; Solis, N.; Scherzer, P.; Lewis, L.; et al. Diabetic Nephropathy Is Accelerated by Farnesoid X Receptor Deficiency and Inhibited by Farnesoid X Receptor Activation in a Type 1 Diabetes Model. *Diabetes* **2010**, *59*, 2916–2927. [CrossRef] [PubMed]

14. Chen, X.; Yuan, H.; Shi, F.; Zhu, Y. Effect of garden cress in reducing blood glucose, improving blood lipids and reducing oxidative stress in a mouse model of diabetes induced by a high fat diet and streptozotocin. *J. Sci. Food Agric.* **2019**. [CrossRef] [PubMed]

15. Zheng, W.; Mast, N.; Saadane, A.; Pikuleva, I.A. Pathways of cholesterol homeostasis in mouse retina responsive to dietary and pharmacologic treatments. *J. Lipid Res.* **2015**, *56*, 81–97. [CrossRef]

16. Hammer, S.S.; Beli, E.; Kady, N.; Wang, Q.; Wood, K.; Lydic, T.A.; Malek, G.; Saban, D.R.; Wang, X.X.; Hazra, S.; et al. The Mechanism of Diabetic Retinopathy Pathogenesis Unifying Key Lipid Regulators, Sirtuin 1 and Liver X Receptor. *EBioMedicine* **2017**, *22*, 181–190. [CrossRef]

17. Li, Q.; Zemel, E.; Miller, B.; Perlman, I. Early Retinal Damage in Experimental Diabetes: Electroretinographical and Morphological Observations. *Exp. Eye Res.* **2002**, *74*, 615–625. [CrossRef]

18. Krady, J.K.; Basu, A.; Allen, C.M.; Xu, Y.; LaNoue, K.F.; Gardner, T.W.; Levison, S.W. Minocycline reduces proinflammatory cytokine expression, microglial activation, and caspase-3 activation in a rodent model of diabetic retinopathy. *Diabetes* **2005**, *54*, 1559–1565. [CrossRef]

19. Dahl, D. The radial glia of Müller in the rat retina and their response to injury. An immunofluorescence study with antibodies to the glial fibrillary acidic (GFA) protein. *Exp. Eye Res.* **1979**, *28*, 63–69. [CrossRef]

20. Osborne, N.N.; Block, F.; Sontag, K.-H. Reduction of ocular blood flow results in glial fibrillary acidic protein (GFAP) expression in rat retinal Müller cells. *Vis. Neurosci.* **1991**, *7*, 637–639. [CrossRef]

21. Penn, J.S.; Thum, A.L.; Rhem, M.N.; Dell, S.J. Effects of oxygen rearing on the electroretinogram and GFA-protein in the rat. *Investig. Ophthalmol. Vis. Sci.* **1988**, *29*, 1623–1630.

22. Tanaka, Y.; Takagi, R.; Ohta, T.; Sasase, T.; Kobayashi, M.; Toyoda, F.; Shimmura, M.; Kinoshita, N.; Takano, H.; Kakehashi, A. Pathological Features of Diabetic Retinopathy in Spontaneously Diabetic Torii Fatty Rats. *J. Diabetes Res.* **2019**, *2019*, 8724818. [CrossRef] [PubMed]

23. Fan, Y.; Lai, J.; Yuan, Y.; Wang, L.; Wang, Q.; Yuan, F. Taurine protects retinal cells and improves synaptic connections in early diabetic rats. *Curr. Eye Res.* **2020**, *45*, 52–63. [CrossRef]

24. Bahr, H.I.; Abdelghany, A.A.; Galhom, R.A.; Barakat, B.M.; Arafa, E.-S.A.; Fawzy, M.S. Duloxetine protects against experimental diabetic retinopathy in mice through retinal GFAP downregulation and modulation of neurotrophic factors. *Exp. Eye Res.* **2019**, *186*, 107742. [CrossRef]

25. Gu, L.; Xu, H.; Zhang, C.; Yang, Q.; Zhang, L.; Zhang, J. Time-dependent changes in hypoxia-and gliosis-related factors in experimental diabetic retinopathy. *Eye* **2019**, *33*, 600. [CrossRef]

26. Chen, J.; Chen, J.; Fu, H.; Li, Y.; Wang, L.; Luo, S.; Lu, H. Hypoxia exacerbates nonalcoholic fatty liver disease via the HIF-2α/PPARα pathway. *Am. J. Physiol. Metab.* **2019**, *317*, E710–E722. [CrossRef]

27. Han, J.; He, Y.; Zhao, H.; Xu, X. Hypoxia inducible factor-1 promotes liver fibrosis in nonalcoholic fatty liver disease by activating PTEN/p65 signaling pathway. *J. Cell. Biochem.* **2019**, *120*, 14735–14744. [CrossRef]

28. Carabelli, J.; Burgueño, A.L.; Rosselli, M.S.; Gianotti, T.F.; Lago, N.R.; Pirola, C.J.; Sookoian, S. High fat diet-induced liver steatosis promotes an increase in liver mitochondrial biogenesis in response to hypoxia. *J. Cell. Mol. Med.* **2011**, *15*, 1329–1338. [CrossRef]

29. Beli, E.; Yan, Y.; Moldovan, L.; Vieira, C.P.; Gao, R.; Duan, Y.; Prasad, R.; Bhatwadekar, A.; White, F.A.; Townsend, S.D.; et al. Restructuring of the Gut Microbiome by Intermittent Fasting Prevents Retinopathy and Prolongs Survival in db/db Mice. *Diabetes* **2018**, *67*, 1867–1879. [CrossRef]

30. Li, M.; Reynolds, C.M.; Gray, C.; Patel, R.; Sloboda, D.M.; Vickers, M.H. Long-term effects of a maternal high-fat: High-fructose diet on offspring growth and metabolism and impact of maternal taurine supplementation. *J. Dev. Orig. Heal. Dis.* **2019**, 1–8. [CrossRef]

31. Matsuda, M.; DeFronzo, R.A. Insulin sensitivity indices obtained from oral glucose tolerance testing: Comparison with the euglycemic insulin clamp. *Diabetes Care* **1999**, *22*, 1462–1470. [CrossRef]

32. Veenstra, A.; Liu, H.; Lee, C.A.; Du, Y.; Tang, J.; Kern, T.S. Diabetic Retinopathy: Retina-Specific Methods for Maintenance of Diabetic Rodents and Evaluation of Vascular Histopathology and Molecular Abnormalities. *Curr. Protoc. Mouse Boil.* **2015**, *5*, 247–270. [CrossRef]

33. Bhatwadekar, A.D.; Duan, Y.; Chakravarthy, H.; Korah, M.; Caballero, S.; Busik, J.V.; Grant, M.B. Ataxia telangiectasia mutated dysregulation results in diabetic retinopathy. *Stem Cells* **2016**, *34*, 405–417. [CrossRef]

34. Léger, H.; Santana, E.; Beltran, A.W.; Luca, F.C. Preparation of Mouse Retinal Cryo-sections for Immunohistochemistry. *J. Vis. Exp.* **2019**, e59683. [CrossRef]

35. Fan, S.; Zhang, Y.; Hu, N.; Sun, Q.; Ding, X.; Li, G.; Zheng, B.; Gu, M.; Huang, F.; Sun, Y.-Q.; et al. Extract of Kuding Tea Prevents High-Fat Diet-Induced Metabolic Disorders in C57BL/6 Mice via Liver X Receptor (LXR) β Antagonism. *PLoS ONE* **2012**, *7*, e51007. [CrossRef]

36. Engerman, R.L. Pathogenesis of diabetic retinopathy. *Diabetes* **1989**, *38*, 1203–1206. [CrossRef]

37. Kumar, B.; Gupta, S.K.; Nag, T.C.; Srivastava, S.; Saxena, R.; Jha, K.A.; Srinivasan, B.P. Retinal neuroprotective effects of quercetin in streptozotocin-induced diabetic rats. *Exp. Eye Res.* **2014**, *125*, 193–202. [CrossRef]

38. Eisenfeld, A.J.; Bunt-Milam, A.H.; Sarthy, P.V. Müller cell expression of glial fibrillary acidic protein after genetic and experimental photoreceptor degeneration in the rat retina. *Investig. Ophthalmol. Vis. Sci.* **1984**, *25*, 1321–1328.

39. Lieth, E.; Gardner, T.W.; Barber, A.J.; Antonetti, D.A. Retinal neurodegeneration: Early pathology in diabetes. *Clin. Exp. Ophthalmol. Viewpoint* **2000**, *28*, 3–8. [CrossRef]

40. Barber, A.J. A new view of diabetic retinopathy: A neurodegenerative disease of the eye. *Prog. Neuro-Psychopharmacol. Biol. Psychiatry* **2003**, *27*, 283–290. [CrossRef]

41. Stratton, I.M.; Kohner, E.M.; Aldington, S.J.; Turner, R.C.; Holman, R.R.; Manley, S.E.; Matthews, D.R.; UKPDS Group UKPDS 50. Risk factors for incidence and progression of retinopathy in Type II diabetes over 6 years from diagnosis. *Diabetologia* **2001**, *44*, 156–163. [CrossRef]

42. Tang, J.; Kern, T.S. Inflammation in diabetic retinopathy. *Prog. Retin. Eye Res.* **2011**, *30*, 343–358. [CrossRef]

43. Fortune, B.; Schneck, E.M.; Adams, A.J. Multifocal electroretinogram delays reveal local retinal dysfunction in early diabetic retinopathy. *Investig. Ophthalmol. Vis. Sci.* **1999**, *40*, 2638–2651.

44. McAnany, J.J.; Park, J.C.; Chau, F.Y.; Leiderman, Y.I.; Lim, J.I.; Blair, N.P. Amplitude loss of the high-frequency flicker electroretinogram in early diabetic retinopathy. *Retin.* **2019**, *39*, 2032–2039. [CrossRef] [PubMed]

45. Zeng, Y.; Cao, D.; Yu, H.; Yang, D.; Zhuang, X.; Hu, Y.; Li, J.; Yang, J.; Wu, Q.; Liu, B.; et al. Early retinal neurovascular impairment in patients with diabetes without clinically detectable retinopathy. *Br. J. Ophthalmol.* **2019**, *103*, 1747–1752. [PubMed]

46. Sone, H.; Kagawa, Y. Pancreatic beta cell senescence contributes to the pathogenesis of type 2 diabetes in high-fat diet-induced diabetic mice. *Diabetologia* **2005**, *48*, 58–67. [CrossRef]
47. Surwit, R.S.; Kuhn, C.M.; Cochrane, C.; McCubbin, A.J.; Feinglos, M.N. Diet-induced type II diabetes in C57BL/6J mice. *Diabetes* **1988**, *37*, 1163–1167. [CrossRef]
48. Surwit, R.S.; Feinglos, M.N.; Rodin, J.; Sutherland, A.; Petro, E.A.; Opara, E.C.; Kuhn, C.M.; Rebuffé-Scrive, M. Differential effects of fat and sucrose on the development of obesity and diabetes in C57BL/6J and A/J mice. *Metabolism* **1995**, *44*, 645–651. [CrossRef]
49. Cani, P.D.; Neyrinck, A.M.; Fava, F.; Knauf, C.; Burcelin, R.G.; Tuohy, K.M.; Gibson, G.R.; Delzenne, N.M. Selective increases of bifidobacteria in gut microflora improve high-fat-diet-induced diabetes in mice through a mechanism associated with endotoxaemia. *Diabetologia* **2007**, *50*, 2374–2383. [CrossRef]
50. Lin, S.; Thomas, T.; Storlien, L.; Huang, X. Development of high fat diet-induced obesity and leptin resistance in C57Bl/6J mice. *Int. J. Obes.* **2000**, *24*, 639–646. [CrossRef]
51. Membrez, M.; Blancher, F.; Jaquet, M.; Bibiloni, R.; Cani, P.D.; Burcelin, R.G.; Corthesy, I.; Chou, C.J.; Macé, K. Gut microbiota modulation with norfloxacin and ampicillin enhances glucose tolerance in mice. *FASEB J.* **2008**, *22*, 2416–2426. [CrossRef] [PubMed]
52. Rabot, S.; Membrez, M.; Bruneau, A.; Gerard, P.; Harach, T.; Moser, M.; Raymond, F.; Mansourian, R.; Chou, C.J. Germ-free C57BL/6J mice are resistant to high-fat-diet-induced insulin resistance and have altered cholesterol metabolism. *FASEB J.* **2010**, *24*, 4948–4959. [CrossRef] [PubMed]
53. Uysal, K.T.; Wiesbrock, S.M.; Marino, M.W.; Hotamisligil, G.S. Protection from obesity-induced insulin resistance in mice lacking TNF-α function. *Nature* **1997**, *389*, 610–614. [CrossRef] [PubMed]
54. Elchebly, M. Increased Insulin Sensitivity and Obesity Resistance in Mice Lacking the Protein Tyrosine Phosphatase-1B Gene. *Sci.* **1999**, *283*, 1544–1548. [CrossRef]
55. Nathan, D.M.; Buse, J.B.; Davidson, M.B.; Ferrannini, E.; Holman, R.R.; Sherwin, R.; Zinman, B. Medical management of hyperglycemia in type 2 diabetes: A consensus algorithm for the initiation and adjustment of therapy: A consensus statement of the American Diabetes Association and the European Association for the Study of Diabetes. *Diabetes Care* **2009**, *32*, 193–203. [CrossRef] [PubMed]
56. Kim, K.; Kim, E.S.; Yu, S.-Y. Longitudinal Relationship Between Retinal Diabetic Neurodegeneration and Progression of Diabetic Retinopathy in Patients With Type 2 Diabetes. *Am. J. Ophthalmol.* **2018**, *196*, 165–172. [CrossRef]
57. Kim, B.; McLean, L.L.; Philip, S.S.; Feldman, E.L. Hyperinsulinemia induces insulin resistance in dorsal root ganglion neurons. *Endocrinology* **2011**, *152*, 3638–3647. [CrossRef]
58. Kim, B.; Feldman, E.L. Insulin resistance in the nervous system. *Trends Endocrinol. Metab.* **2012**, *23*, 133–141. [CrossRef]
59. Das, A.K.; Kalra, S.; Tiwaskar, M.; Bajaj, S.; Seshadri, K.; Chowdhury, S.; Sahay, R.; Indurkar, S.; Unnikrishnan, A.G.; Phadke, U.; et al. Expert Group Consensus Opinion: Role of Anti-inflammatory Agents in the Management of Type-2 Diabetes (T2D). *J. Assoc. Physicians India* **2019**, *67*, 65–74.
60. Tabák, A.G.; Herder, C.; Rathmann, W.; Brunner, E.J.; Kivimäki, M. Prediabetes: A high-risk state for diabetes development. *Lancet* **2012**, *379*, 2279–2290. [CrossRef]
61. Lamparter, J.; Raum, P.; Pfeiffer, N.; Peto, T.; Höhn, R.; Elflein, H.; Wild, P.; Schulz, A.; Schneider, A.; Mirshahi, A. Prevalence and associations of diabetic retinopathy in a large cohort of prediabetic subjects: The Gutenberg Health Study. *J. Diabetes its Complicat.* **2014**, *28*, 482–487. [CrossRef]
62. Tikhonenko, M.; Lydic, T.A.; Opreanu, M.; Calzi, S.L.; Bozack, S.; McSorley, K.M.; Sochacki, A.L.; Faber, M.S.; Hazra, S.; Duclos, S.; et al. N-3 Polyunsaturated Fatty Acids Prevent Diabetic Retinopathy by Inhibition of Retinal Vascular Damage and Enhanced Endothelial Progenitor Cell Reparative Function. *PLoS ONE* **2013**, *8*, e55177. [CrossRef] [PubMed]
63. Busik, J.V.; Esselman, W.J.; Reid, E.G. Examining the role of lipid mediators in diabetic retinopathy. *Clin. Lipidol.* **2012**, *7*, 661–675. [CrossRef] [PubMed]
64. Hammer, S.S.; Busik, J.V. The role of dyslipidemia in diabetic retinopathy. *Vis. Res.* **2017**, *139*, 228–236. [CrossRef] [PubMed]
65. Calkin, A.C.; Tontonoz, P. Liver x receptor signaling pathways and atherosclerosis. *Arter. Thromb. Vasc. Boil.* **2010**, *30*, 1513–1518. [CrossRef] [PubMed]
66. Zelcer, N.; Tontonoz, P. Liver X receptors as integrators of metabolic and inflammatory signaling. *J. Clin. Investig.* **2006**, *116*, 607–614. [CrossRef]

67. Steffensen, K.R.; Jakobsson, T.; Gustafsson, J.-Å. Targeting liver X receptors in inflammation. *Expert Opin. Ther. Targets* **2013**, *17*, 977–990. [CrossRef]

68. Saadane, A.; Mast, N.; Trichonas, G.; Chakraborty, D.; Hammer, S.; Busik, J.V.; Grant, M.B.; Pikuleva, I.A. Retinal Vascular Abnormalities and Microglia Activation in Mice with Deficiency in Cytochrome P450 46A1–Mediated Cholesterol Removal. *Am. J. Pathol.* **2019**, *189*, 405–425. [CrossRef]

69. Sacks, F.M.; Lichtenstein, A.H.; Wu, J.H.; Appel, L.J.; Creager, M.A.; Kris-Etherton, P.M.; Miller, M.; Rimm, E.B.; Rudel, L.L.; Robinson, J.G.; et al. Dietary Fats and Cardiovascular Disease: A Presidential Advisory from the American Heart Association. *Circulation* **2017**, *136*, e1–e23. [CrossRef]

70. Andraski, A.B.; Singh, S.A.; Lee, L.H.; Higashi, H.; Smith, N.; Zhang, B.; Aikawa, M.; Sacks, F.M. Effects of Replacing Dietary Monounsaturated Fat With Carbohydrate on HDL (High-Density Lipoprotein) Protein Metabolism and Proteome Composition in Humans. *Arter. Thromb. Vasc. Boil.* **2019**, *39*, 2411–2430. [CrossRef]

71. Bolla, A.M.; Caretto, A.; Laurenzi, A.; Scavini, M.; Piemonti, L. Low-Carb and Ketogenic Diets in Type 1 and Type 2 Diabetes. *Nutrients* **2019**, *11*, 962. [CrossRef] [PubMed]

72. Gomes, J.A.; Silva, J.F.; Silva, G.C.; Gomes, G.F.; de Oliveira, A.C.; Soares, V.L.; Oliveira, M.C.; Ferreira, A.V.; Aguiar, D.C. High-refined carbohydrate diet consumption induces neuroinflammation and anxiety-like behavior in mice. *J. Nutr. Biochem.* **2019**, *77*, 108317. [CrossRef] [PubMed]

73. Tobias, D.K.; Chen, M.; Manson, J.E.; Ludwig, D.S.; Willett, W.; Hu, F.B. Effect of low-fat diet interventions versus other diet interventions on long-term weight change in adults: A systematic review and meta-analysis. *Lancet Diabetes Endocrinol.* **2015**, *3*, 968–979. [CrossRef]

74. Vega-López, S.; Venn, B.J.; Slavin, J.L. Relevance of the Glycemic Index and Glycemic Load for Body Weight, Diabetes, and Cardiovascular Disease. *Nutrients* **2018**, *10*, 1361. [CrossRef] [PubMed]

75. Livesey, G.; Taylor, R.; Livesey, H.F.; Buyken, A.E.; Jenkins, D.J.A.; Augustin, L.S.A.; Sievenpiper, J.L.; Barclay, A.W.; Liu, S.; Wolever, T.M.S.; et al. Dietary Glycemic Index and Load and the Risk of Type 2 Diabetes: Assessment of Causal Relations. *Nutrients* **2019**, *11*, 1436. [CrossRef]

76. Myette-Côté, É.; Durrer, C.; Neudorf, H.; Bammert, T.D.; Botezelli, J.D.; Johnson, J.D.; DeSouza, C.A.; Little, J.P. The effect of a short-term low-carbohydrate, high-fat diet with or without postmeal walks on glycemic control and inflammation in type 2 diabetes: A randomized trial. *Am. J. Physiol. Integr. Comp. Physiol.* **2018**, *315*, R1210–R1219.

77. Kowluru, R.A. Retinopathy in a Diet-Induced Type 2 Diabetic Rat Model, and Role of Epigenetic Modifications. *Diabetes* **2020**, db191009. [CrossRef]

78. Kim, K.-A.; Gu, W.; Lee, I.-A.; Joh, E.-H.; Kim, N.-H. High Fat Diet-Induced Gut Microbiota Exacerbates Inflammation and Obesity in Mice via the TLR4 Signaling Pathway. *PLoS ONE* **2012**, *7*, e47713. [CrossRef]

79. Lee, J.-J.; Wang, P.-W.; Yang, I.-H.; Huang, H.-M.; Chang, C.-S.; Wu, C.-L.; Chuang, J.-H. High-Fat Diet Induces Toll-Like Receptor 4-Dependent Macrophage/Microglial Cell Activation and Retinal Impairment. *Investig. Opthalmol. Vis. Sci.* **2015**, *56*, 3041. [CrossRef]

80. Krishnan, J.; Danzer, C.; Simka, T.; Ukropec, J.; Walter, K.M.; Kumpf, S.; Mirtschink, P.; Ukropcova, B.; Gasperikova, D.; Pedrazzini, T. Dietary obesity-associated hif1α activation in adipocytes restricts fatty acid oxidation and energy expenditure via suppression of the sirt2-nad+ system. *Genes Dev.* **2012**, *26*, 259–270. [CrossRef]

81. Stroka, D.M.; Burkhardt, T.; Desbaillets, I.; Wenger, R.H.; Neil, D.A.; Bauer, C.; Gassmann, M.; Candinas, D. Hif-1 is expressed in normoxic tissue and displays an organ-specific regulation under systemic hypoxia. *FASEB J.* **2001**, *15*, 2445–2453. [CrossRef] [PubMed]

82. Prasad, P.; Li, L.-P.; Halter, S.; Cabray, J.; Ye, M.; Batlle, D. Evaluation of renal hypoxia in diabetic mice by BOLD MRI. *Investig. Radiol.* **2010**, *45*, 819–822. [CrossRef] [PubMed]

Article

Lack of Overt Retinal Degeneration in a K42E *Dhdds* Knock-In Mouse Model of RP59

Sriganesh Ramachandra Rao [1,2,†], Steven J. Fliesler [1,2,†], Pravallika Kotla [3], Mai N. Nguyen [3] and Steven J. Pittler [3,*]

[1] Research Service, VA Western NY Healthcare System, Buffalo, NY 14215, USA;
    sramacha@buffalo.edu (S.R.R.); fliesler@buffalo.edu (S.J.F.)
[2] Departments of Ophthalmology and Biochemistry and Neuroscience Graduate Program,
    The State University of New York- University at Buffalo, Buffalo, NY 14209, USA
[3] Department of Optometry and Vision Science, Vision Science Research Center, University of Alabama at
    Birmingham, School of Optometry, Birmingham, AL 35294, USA; pkotla@uab.edu (P.K.);
    mnnguyen@uab.edu (M.N.N.)
* Correspondence: pittler@uab.edu; Tel.: +1-205-934-6744
† These authors contributed equally to this work.

Received: 17 February 2020; Accepted: 4 April 2020; Published: 7 April 2020

Abstract: Dehydrodolichyl diphosphate synthase (DHDDS) is required for protein N-glycosylation in eukaryotic cells. A K42E point mutation in the DHDDS gene causes an autosomal recessive form of retinitis pigmentosa (RP59), which has been classified as a congenital disease of glycosylation (CDG). We generated K42E *Dhdds* knock-in mice as a potential model for RP59. Mice heterozygous for the *Dhdds* K42E mutation were generated using CRISPR/Cas9 technology and crossed to generate *Dhdds*[K42E/K42E] homozygous mice. Spectral domain-optical coherence tomography (SD-OCT) was performed to assess retinal structure, relative to age-matched wild type (WT) controls. Immunohistochemistry against glial fibrillary acidic protein (GFAP) and opsin (1D4 epitope) was performed on retinal frozen sections to monitor gliosis and opsin localization, respectively, while lectin cytochemistry, plus and minus PNGase-F treatment, was performed to assess protein glycosylation status. Retinas of *Dhdds*[K42E/K42E] mice exhibited grossly normal histological organization from 1 to 12 months of age. Anti-GFAP immunoreactivity was markedly increased in *Dhdds*[K42E/K42E] mice, relative to controls. However, opsin immunolocalization, ConA labeling and PNGase-F sensitivity were comparable in mutant and control retinas. Hence, retinas of *Dhdds*[K42E/K42E] mice exhibited no overt signs of degeneration, yet were markedly gliotic, but without evidence of compromised protein N-glycosylation. These results challenge the notion of RP59 as a DHDDS loss-of-function CDG and highlight the need to investigate unexplored RP59 disease mechanisms.

Keywords: retinitis pigmentosa; knock-in mouse model; congenital disorder of glycosylation; retina

---

1. Introduction

Retinitis pigmentosa (RP) represents a group of hereditary retinal degenerative disorders of diverse genetic origins that have as their common trait the progressive, irreversible dysfunction, degeneration, and demise of retinal photoreceptor cells, with rods initially undergoing these pathological changes followed eventually by cones [1,2]. Relatively recently, a K42E point mutation in the dehydrodolichyl diphosphate synthase (DHDDS) gene was shown to cause a rare, recessive form of RP (RP59; OMIM #613861) [3–5]. DHDDS catalyzes *cis*-prenyl chain elongation in the synthesis of dolichyl diphosphate (Dol-PP), which is required for protein N-glycosylation [6,7]. DHDDS catalyzes the condensation of multiple units of isopentenyl pyrophosphate (IPP, also called isopentenyl diphosphate) to farnesyl pyrophosphate (FPP, also called farnesyl diphosphate) to produce Dol-PP [8,9]. This is used as the "lipid carrier" onto which oligosaccharide chains are built that are ultimately transferred to specific asparagine (*N*) residues on nascent polypeptide chains in the lumen of the endoplasmic reticulum (ER) to form *N*-linked glycoproteins [10]. The monophosphate (Dol-P) is used as a sugar carrier, transferring sugars from their corresponding sugar-nucleotide adducts (e.g., UDP-glucose, GDP-mannose, etc.) to the growing Dol-PP-linked oligosaccharide chains in the ER. Mutations in rhodopsin that block its glycosylation have been shown to cause retinal degeneration in vertebrate animals [11,12]. In addition, pharmacological inhibition of protein *N*-glycosylation with tunicamycin has been shown to disrupt retinal photoreceptor outer segment (OS) disc membrane morphogenesis in vitro [13], as well as to cause retinal degeneration with progressive shortening and loss of photoreceptor OSs in vivo [14].

In the present study, we created a DHDDS K42E homozygous knock-in mouse model (hereafter called *Dhdds*$^{K42E/K42E}$) of RP59—since K42E is the most prevalent point mutation in the RP59 patient population [3–5]—to study its underlying pathological mechanism, with the working hypothesis that defective protein *N*-glycosylation underlies the retinal dysfunction and degeneration observed in human RP59. Herein, we present a description of the generation and initial characterization of the phenotypic features of the *Dhdds*$^{K42E/K42E}$ mouse model. Surprisingly, although we expected to observe an early onset, progressive, and potentially severe retinal degeneration, this was not the case. The retina appeared histologically intact and normal according to spectral domain optical coherence tomography (SD-OCT) analysis for up to at least one year of age. However, there was evidence of gliotic reactivity (glial fibrillary acidic protein (GFAP) immunostaining), despite the lack of obvious neuronal degeneration or cell death/loss. Also, despite the homozygous mutation in *Dhdds*, we found no evidence of compromised protein *N*-glycosylation in mutant mouse retinas.

2. Materials and Methods

2.1. Animals

Heterozygous (K42E/+) *Dhdds* knock-in (KI) mice were generated on a C57Bl/6J background by Applied StemCell (Milipitas, CA, USA). Briefly, CRISPR guide RNA (5′-TCGCTATGCCAAGAAGTGTC-3′ with PAM site AGG) was generated using in vitro transcription and was used to create a double strand break in the murine *Dhdds* locus to promote introduction of a single-stranded oligodeoxynucleotide (SSO) carrying the K42E mutation and a second silent DNA polymorphism to eliminate the PAM recognition site required for cleavage by CAS9 (5′-ATTATCTGTTCTCTTCTACAGGCTGGCCCAGTACCCAAACATATCGCGTTCATAATGGACGGC AACCGTCGCTATGCCAAGGAGTGTCAAGTGGAGCGCCAGGAGGGCCACACACAGGGCTTCA ATAAGCTTGCTGAGGTGGGTGCGGGTGACAGAGCCTAGA-3′). Mouse zygotes were injected with 100, 100, and 250 ng/µL of Cas9 enzyme, guide RNA, and SSO, respectively, which were then transferred into pseudo pregnant CD-1 females. Three potential founder (F0) pups were identified out of 13 mice tested, and an F0 founder was verified by DNA sequence analysis. Sequence-validated heterozygous (*Dhdds*$^{K42E/+}$) mice were crossed to generate homozygous (*Dhdds*$^{K42E/K42E}$) mice, as confirmed by PCR and DNA sequencing (see below). C57Bl/6J wild type (WT) mice, age- and sex-matched, were used as controls. All procedures conformed to the ARVO Statement for the Use of Animals in Ophthalmic and Vision Research, and were approved by the Institutional Animal Care and Use Committee (IACUC) of the University of Alabama at Birmingham. All animals were maintained on a standard 12/12 h light/dark cycle (20–40 lux ambient room illumination), fed standard rodent chow, provided water ad libitum, and housed in plastic cages with standard rodent bedding.

2.2. PCR Genotyping and DNA Analysis

PCR primers were designed that spanned the targeted region (forward primer, 5′-TCTAGGCTCTGTCACCCGCA-3′ and reverse primer 5′-TCTAGGCTCTGTCACCCGCA-3′) amplifying a 292 bp segment of DNA in both WT and *Dhdds*$^{K42E/K42E}$ mice. For initial verification of the knock-in, PCR products were sequenced in the UAB Heflin Center for Genomic Sciences. The presence of the knock-in sequence was confirmed in subsequent generations by restriction enzyme digestion with StyI, which cleaves the knock-in allele only (data not shown). Knock-in alleles were independently verified by Transnetyx, Inc. (Cordova, TN, USA) using proprietary technology. While the analysis was set up to recognize and differentiate the knock-in mutation and the PAM site polymorphism, only the knock-in mutation was maintained in all subsequent breeding.

2.3. Spectral Domain Optical Coherence Tomography (SD-OCT)

In vivo retinal imaging was performed as previously described in detail by DeRamus et al. [15], using a Bioptigen Model 840 Envisu Class-R high-resolution SD-OCT instrument (Bioptigen/Leica, Inc.; Durham, NC, USA). Data were collected from $Dhdds^{K42E/K42E}$ and WT mice at postnatal day (PN) 1 (KI, n = 5; WT, n = 9), 2 (KI, n = 4; WT, n = 8), 3 (KI, n = 5; WT n = 3), 8 (KI, n = 4; WT n = 5), and 12 months (mos) (KI, n = 3; WT n = 3) to assess retinal structure. Layer thicknesses were determined manually using Bioptigen InVivoVue® and Bioptigen Diver® V. 3.4.4 software and the data were analyzed and graphed using Microsoft Excel software.

2.4. Immunohistochemistry (IHC)

Procedures utilized for fixation, O.C.T. embedment, and sectioning of mouse eyes were as described in detail previously by Ramachandra Rao et al. [16]. In brief, eyes were immersion fixed overnight in phosphate-buffered saline (PBS) containing freshly prepared paraformaldehyde (4% v/v), appropriately cryopreserved, embedded in O.C.T., and cryosectioning was performed on a Leica Model CM3050 S Cryostat (Leica Biosystems, Wetzlar, Germany). Retinal sections were first "blocked" with 0.1% BSA, 0.5% serum (species corresponding to secondary antibody host) in Tris-buffered saline containing 0.1% Tween-20 (TBST), then incubated for 1 h at room temperature with a rabbit polyclonal antibody against glial fibrillary acidic protein (GFAP;, DAKO/Agilent, Santa Clara, CA, USA; 1:500 dilution in TBST) and a mouse monoclonal antibody against the C-terminal epitope of opsin (1D4; Novus Biologicals, Littleton, CO, USA; 1:500 dilution in TBST), followed by incubation with fluor-conjugated secondary antibodies (AlexaFluor®-488 conjugated anti-mouse IgG, AlexaFluor®-568 conjugated anti-rabbit IgG; Thermo Fisher Scientific, Waltham, MA, USA; 1:500 dilution in TBST). Sections were then counterstained with DAPI and cover slipped with anti-fade mounting medium (Vectashield®; Vector Laboratories, Burlingame, CA, USA) and viewed with a Leica TCS SPEII DMI4000 scanning laser confocal microscope (Leica Biosystems). Images were captured using a 40X oil immersion (RI 1.518) objective under normal laser intensity (10% of laser power source), arbitrary gain (850 V) and offset (–0.5) values, to optimize signal-to-noise ratio. Digital images were captured and stored as TIFF files on a PC computer.

2.5. Lectin Cytochemistry

Paraformaldehyde-fixed eyes (as described above) were processed for paraffin embedment. Paraffin sections of mouse eyes were then incubated (45 min at room temperature) with biotinylated Concanavalin-A (ConA, B-1005; Vector Laboratories; 1:200 dilution in PBS), followed by incubation with AlexaFluor®-488 conjugated streptavidin (Thermo Fisher Scientific; 1:500 dilution in PBS) and AlexaFluor®-647-conjugated peanut agglutinin (PNA, L32460; Thermo Fisher Scientific; 1:250 dilution in PBS), with or without pre-treatment (37 °C, overnight) with peptide:N-glycosidase F (PNGase-F, 200 U, P0704S; New England Biolabs, Inc., Ipswich, MA, USA). Sections were DAPI-stained and mounted using Vectashield mounting media, and digital images obtained using scanning laser confocal microscopy as described above [16].

3. Results

3.1. Generation and Validation of K42E DHDDS Knock-In Mutation

K42E knock-in mice were generated commercially using CRISPR-Cas9 technology. The K42E knock-in mutations in both heterozygous and homozygous mice were confirmed by DNA sequence for one of the heterozygous F0 founder mice, which is shown in Figure 1. Both the A-to-G and G-to-A transitions that lead to the K42E mutation and the Q44Q silent polymorphism, respectively, are heterozygous (arrows). Intra-litter mating was done to establish at least fourth generation homozygous mice that were used for all subsequent analyses. Heterozygous mice were initially characterized by SD-OCT and histology and found not to differ from WT (not shown).

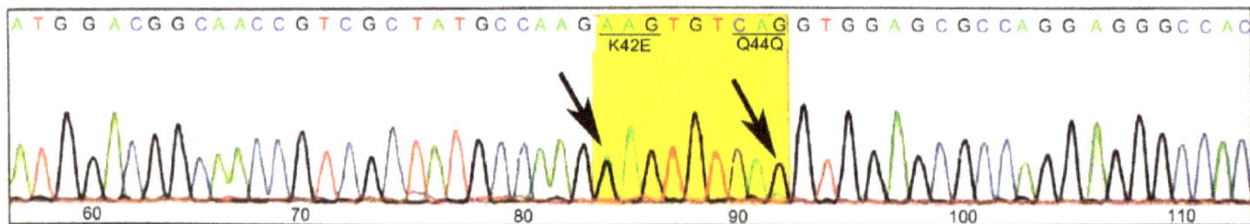

Figure 1. DNA sequence analysis of a tail DNA from a K42E/+ founder mouse. Tail DNA was amplified with primers that cover a 292 bp segment spanning the target region. The sequence analysis confirmed the presence (arrows) of the K42E (A-to-G) mutation and the Q44Q (G-to-A) polymorphism that was included to eliminate the CRISPR-related PAM site.

3.2. SD-OCT Analysis Reveals No Evidence for Retinal Degeneration in Dhdds$^{K42E/K42E}$ Mice

SD-OCT provides a non-invasive means of assessing retinal morphology *in vivo*. Qualitative SD-OCT images obtained from wild type (WT) and *Dhdds*$^{K42E/K42E}$ mice are presented in Figure 2. From these images, it is clear that the gross morphology of the retina in the homozygous knock-in animals, from PN 1 to 12 months of age, are comparable to that observed in fully mature, age-matched WT control mice. All retinal histological layers were intact and of normal appearance. Hence, there was no evidence of retinal degeneration, even up to one year of age.

We used SD-OCT to perform quantitative analysis of retinal morphology to compare ocular tissue layer thicknesses in WT and knock-in mice. Figure 3 compares data obtained at PN 1, 2, 3, 8, and 12 mos for *Dhdds*$^{K42E/K42E}$ mice, compared to age-matched WT control littermates. The data are shown both with respect to outer nuclear layer (ONL) thickness (yellow and gray lines) as well as total neural retina thickness (blue and orange lines) as a function of distance from the optic nerve head (ONH, point 4 in each graph) along the vertical meridian, for both the inferior and superior hemispheres. No differences in these quantitative metrics of retinal morphology were observed with respect to genotype, consistent with the representative OCT images shown in Figure 2.

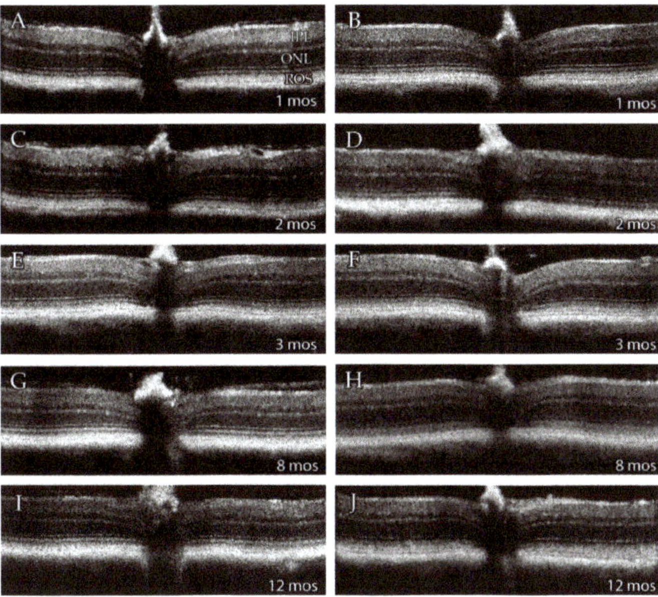

Figure 2. Representative averaged SD-OCT images of retinas from (left panels: **A,C,E,G,I**) 1-, 2-, 3-, 8- and 12-months (mos) old wild type (WT), and (right panels: **B,D,F,H,J**) 1-, 2-, 3-, 8-, and 12-months old *Dhdds*$^{K42E/K42E}$ mice. Abbreviations: IPL, inner plexiform layer; ONL, outer nuclear layer; ROS, rod outer segment layer. No changes were observed at any age in retinas of *Dhdds*$^{K42E/K42E}$ mice compared to WT mice.

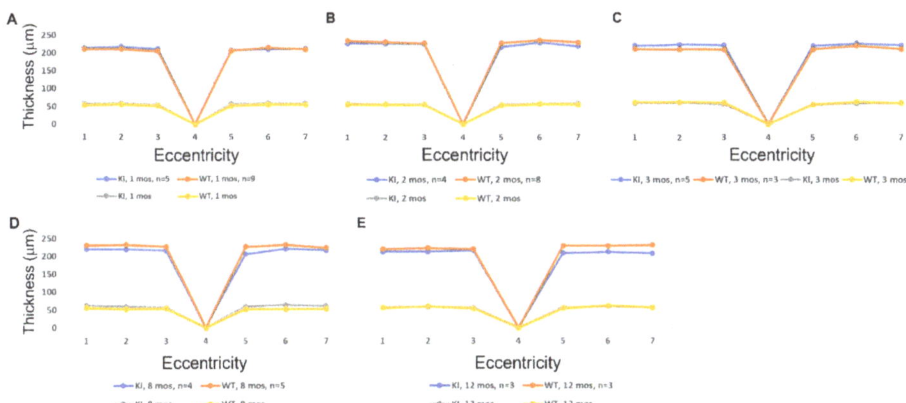

Figure 3. Analysis of the ONL thickness (yellow and gray lines) and total retinal thickness (blue and orange lines) in WT and *Dhdds*$^{K42E/K42E}$ mice ranging in age from PN 1 to 12 months. (**A**) 1 month, (**B**) 2 months, (**C**) 3 months, (**D**) 8 months, (**E**) 12 months. Outer nuclear layer (ONL) thickness and total retina thickness measurements (in microns), as a function of genotype and distance from the optic nerve head (ONH) along the vertical meridian in both the inferior and superior hemispheres. Genotypes: WT and *Dhdds*$^{K42E/K42E}$ mice. No significant differences were observed between the groups.

3.3. Gliotic Reactivity, Despite Lack of Overt Neural Retina Degeneration, in Dhdds$^{K42E/K42E}$ Mice

We performed immunohistochemical analysis on frozen sections of fixed, O.C.T.-embedded WT and *Dhdds*$^{K42E/K42E}$ mouse eyes at PN 2 months of age, probing with antibodies against GFAP (polyclonal) and a C-terminal epitope of rod opsin (1D4 monoclonal). As shown in Figure 4, whereas WT control retinas only exhibited GFAP immunoreactivity (pseudocolored red) along the vitreoretinal interface, corresponding to astrocytes and Müller glia "end feet", retinas from *Dhdds*^K42E/K42E mice exhibited extensive, robust anti-GFAP labeling in a radial pattern. This extended throughout the inner retinal layers to the outer plexiform layer (OPL), in addition to intense labeling along the vitreoretinal interface. The latter results are indicative of massive gliotic activation, which is remarkable considering the lack of overt retinal degeneration or loss of retinal neurons (per the SD-OCT data; see Figures 2 and 3). Gliosis in *Dhdds*^K42E/K42E mouse retinas was also detected at PN one month and persisted even at PN six months of age (data not shown). Anti-opsin immunolabeling (pseudocolored green) was comparable in both WT and *Dhdds*^K42E/K42E retinas. Notably, the label was confined to the OS layer; there was no mislocalization of opsin to the plasma membrane of the cell in the IS or ONL layer—unlike what is often observed in degenerating photoreceptor cells in various animal models—suggesting normal trafficking of opsin to the outer segment, and consistent with a lack of overt photoreceptor degeneration. It is worth noting that the green labeling in a few cells in the inner retina in Figure 4 is due to mouse-on-mouse binding of the monoclonal antibody to endogenous IgG in blood vessels. It does not represent true anti-opsin immunolabeling.

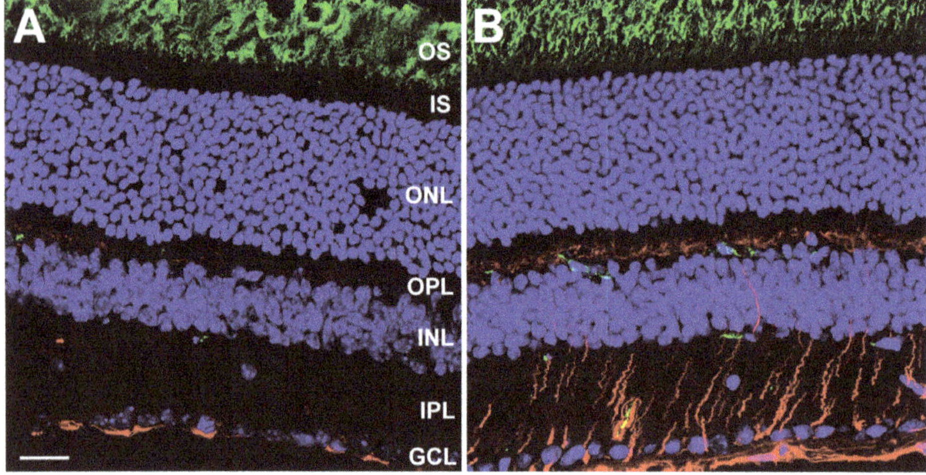

Figure 4. Laser confocal microscopy images of (**A**) WT control and (**B**) *Dhdds*^K42E/K42E mouse retina frozen sections at PN 2 months of age, stained with antibodies to GFAP (pseudocolor: red) and rod opsin (pseudocolor: green), and counterstained with DAPI (blue). Scale bar (both panels) is 20 μm. Abbreviations: OS, outer segment layer; IS, inner segment layer; ONL, outer nuclear layer; OPL, outer plexiform layer; INL, inner nuclear layer; IPL, inner plexiform layer; GCL, ganglion cell layer. Scale bar (both panels) is 20 μm.

3.4. Lack of Defective Protein Glycosylation in Dhdds^K42E/K42E Mouse Retinas

The *N*-linked oligo-saccharides of glycoproteins contain alpha-linked mannose residues as constituents, which are cognate ligands for the lectin concanavalin A (Con A) [17]. Hence, ConA lectin cytochemistry offers a reliable means of detecting the presence (or absence) of N-linked oligo-saccharides in tissue sections of *Dhdds*^K42E/K42E mice, and a way to directly test the current hypothesis that RP59 is driven by lack of glycosylation. This is because the synthesis of oligosaccharide chains in cells and tissues obligatorily depends upon the presence of Dol-PP and Dol-P (which requires upstream DHDDS activity). Furthermore, N-linked oligosaccharide chains are selectively susceptible to hydrolysis by peptide:N-glycosidase F (PNGase-F) [18]; hence, tissue sections treated with PNGase-F should exhibit a marked loss of Con A binding (serving as a true negative control), thereby mimicking the scenario where upstream DHDDS activity may be lacking. We performed ConA lectin cytochemical analysis on retinal sections from WT control and *Dhdds*[K42E/K42E] mice at PN six months of age, with and without pre-treatment with PNGase-F. The results are shown in Figure 5.

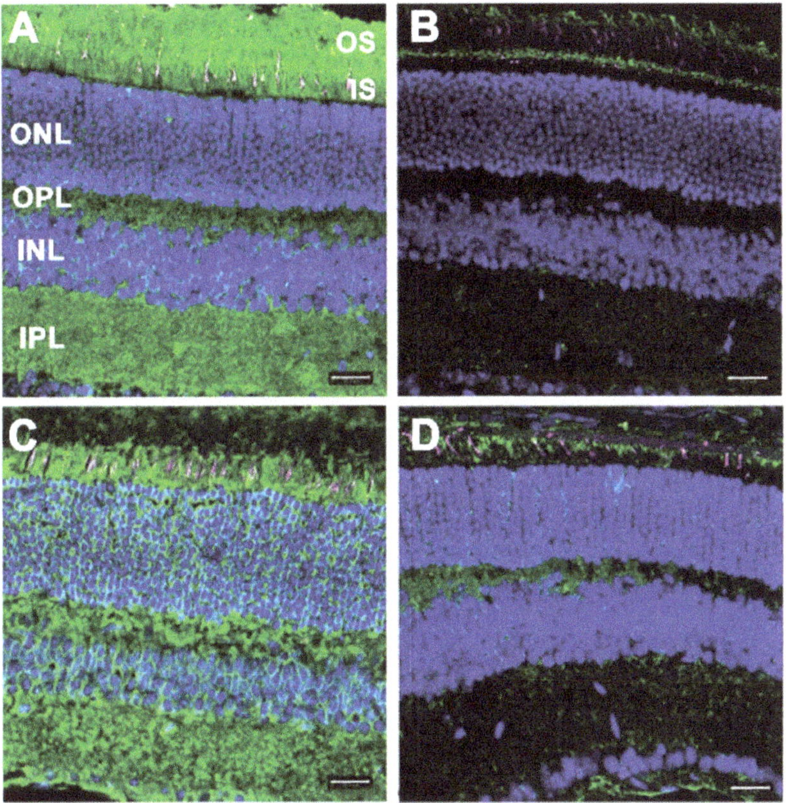

Figure 5. ConA lectin cytochemical analysis of retinas from (**A,B**) WT control and (**C,D**) *Dhdds*[K42E/K42E] mice at PN 6 months of age, with (**B,D**) or without (**A,C**) pretreatment with PNGase-F. ConA binding (green); PNA binding (magenta); DAPI counterstain (blue). Abbreviations are the same as in the Figure 4 legend. Scale bar (all panels): 20 μm.

Normally, *N*-linked glycoproteins are present throughout the retina, being notably enriched in photoreceptor cells and the synaptic endings of neurons (IPL, OPL). Hence, the inner and outer segment layers (IS and OS, respectively), including the glycoconjugate-rich interphotoreceptor matrix (IPM), as well as the inner and outer plexiform layers (IPL and OPL, respectively) were robustly labeled with fluor-tagged ConA in untreated WT retinal sections (Figure 5 A). As expected, treatment of WT retinal tissue sections with PNGase-F dramatically reduced the level of ConA binding throughout the retina (Figure 5B). Notably, retinal sections from *Dhdds*[K42E/K42E] mice also exhibited robust, pan-retinal ConA binding (Figure 5C), comparable to that of WT controls. Upon treatment with PNGase-F, most of the ConA staining was lost (Figure 5D). These results obviate any significant DHDDS loss-of-function in *Dhdds*[K42E/K42E] mice.

Peanut agglutinin (PNA) binds to the disaccharide Gal-β(1-3)-GalNAc in glycoproteins and glycolipids [17]. Oligosaccharides containing this disaccharide are highly enriched in the extracellular matrix surrounding cone photoreceptor outer segments (the "cone matrix sheath") [19,20]. Thus, PNA binding can be used to selectively label cone photoreceptors in retinal tissue sections, since rod photoreceptors and their associated "rod matrix sheath" lack such glycan chains [21]. Furthermore, oligosaccharides containing Gal-β(1-3)-GalNAc are generally *O*-linked (e.g., through Ser or Thr residues), rather than *N*-linked, and their synthesis is not dolichol-dependent in mammalian cells [22]. Furthermore, the PNA-binding disaccharide epitope is not susceptible to PNGase-F hydrolysis [17]. Hence, we expected to observe no appreciable differences in the binding of PNA to retinal tissue sections from *Dhdds*K42E/K42E retinas vs. WT controls, nor effects of PNGase-F treatment on PNA binding, either with regard to labeling intensity or distribution. These expectations were realized, as illustrated in Figure 5 (magenta staining, all four panels). Both *Dhdds*K42E/K42E and control retinas exhibited comparable distribution of PNA-positive cone matrix sheaths, suggesting persistence of viable cone photoreceptors in the K42E mutants.

4. Discussion

Here, we have presented the generation and initial characterization of a novel mouse model of RP59, where we have achieved global homozygous knock-in of the *K42E Dhdds* mutation specifically associated with RP59 [3–5]. Based upon the clinical presentation of RP59 in human patients [3–5], as well as the demonstrable importance of dolichol-dependent protein glycosylation in maintaining the normal structure and function of the vertebrate retina [11–14], we expected to observe retinal degeneration and retinal thinning in *Dhdds*K42E/K42E mice, particularly in mice homozygous for the K42E mutation. This expectation was also predicated on a preliminary report [23], using a similar K42E mouse knock-in model, that claimed nearly 50% loss of OS length and reduction in ONL thickness by about two-thirds at PN 3 months of age, compared to WT mouse retinas. However, we observed no evidence of retinal degeneration in *Dhdds*K42E/K42E mice up to one year of age. Furthermore, despite the confirmed mutation of *Dhdds*, we found no evidence for defective protein *N*-glycosylation in the retinas of these mice. The retinas were labeled robustly with fluor-tagged ConA lectin, irrespective of genotype. These findings are in good agreement with observations made by Sabry et al. [24] who found normal mannose incorporation into *N*-linked oligosaccharides using either siRNA silencing of *DHDDS* in a HepG2 cell line or in RP59 (severe mutation) patient fibroblasts. The ConA binding observed in our study is further consistent with observations made by Wen et al., who observed that rather than any loss in dolichols, there was an alteration in dolichol chain lengths (increased D17:D18 ratio) in RP59 patients compared to normal human subjects, but without obvious hypoglycosylation of serum transferrin [25]. These findings collectively suggest hypoglycosylation-independent retinal degeneration in RP59, the mechanism of which still remains to be elucidated.

Understanding the pathophysiological and biochemical mechanisms underlying RP59 remains limited due to the lack of a validated vertebrate animal model that faithfully mimics the key hallmarks of the disease. Heretofore, only a zebrafish model of RP59 has been documented, using global knock-down of *DHDDS* expression by injection of morpholino oligonucleotides at the one-cell embryo stage [26]. In that case, the fish exhibited defective photoresponses and their cone outer segments (as assessed indirectly by PNA staining) were dramatically shortened, if not nearly absent. It should be noted that zebrafish have a highly cone-rich retina, unlike humans or mice (which have highly rod-dominant retinas). Also, the reduction and loss of PNA binding in the zebrafish knock-down retinas most likely reflects degeneration and death of cone photoreceptors, with concomitant degeneration and loss of their outer segments, due to their requirement for dolichol. Unlike the zebrafish *Dhdds* knock-down model, the murine RP59 model generated in the present study exhibits robust PNA staining in the outer retina, suggesting persistence of viable cone photoreceptors. In a parallel study (Ramachandra Rao et al., unpublished), we have observed *Dhdds* transcript distribution in all retinal nuclear layers by in situ hybridization, consistent with the fact that all cells require dolichol derivatives to support protein *N*-glycosylation. Taken together, these findings suggest that the K42E *Dhdds* mutation does not affect cone photoreceptor viability. Recently (Ramachandra Rao et al., manuscript submitted for publication), we also generated a conditional *Dhdds* knockout mouse model, with targeted ablation of *Dhdds* in retinal rod photoreceptors, using a Cre-lox approach; however, unlike the K42E knock-in model, the rod-specific *Dhdds* knockout model exhibits profound, rapid retinal degeneration, with almost complete loss of photoreceptors by PN 6 weeks. Yet, there was no evidence of compromised protein *N*-glycosylation prior to the onset of photoreceptor degeneration. In addition, as reported in a companion article in this Special Issue of *Cells* [27], targeted ablation of *Dhdds* in retinal pigment epithelium, (RPE) cells in mice also results in a progressive, but somewhat slower, retinal degeneration.

As pointed out by Zelinger et al. [4], the phenotype of RP59 only involves the retina; there is no observable dysfunction or pathology in other tissues and organs in RP59 patients. Hence, those authors speculated that the K42E mutation, "alters, rather than abolishes, enzymatic function, perhaps either by reducing the level of DHDDS protein or by preventing requisite interactions between DHDDS and a photoreceptor-specific protein" [4]. They also suggested, alternatively, that mutation of DHDDS might result in, "a toxic accumulation of isoprenoid compounds," such as occurs in various forms of neuronal ceroid lipofuscinosis (e.g., Batten disease). While such speculations may turn out to be true, there is no direct empirical evidence extant to support this hypothesis. It is also entirely possible, however, that mutations (whether K42E or others) in DHDDS may affect its interactions with its enzymatic partner, Nogo-B receptor (NgBR, encoded by the *Nus1* gene) [8,9], with concomitant alterations in dolichol synthesis and protein *N*-glycosylation [28,29]. At present, nothing is known about the expression of Nogo-B receptor or its interactions with DHDDS, specifically in the retina. Our *Dhdds*$^{K42E/K42E}$ mouse line and retinal cell type-specific conditional DHDDS knockout mice offer potentially valuable model systems in which to pursue further investigations along these lines. In addition, we are currently pursuing studies employing dual, targeted ablation of DHDDS and NgBR in the retina. (See also the article by DeRamus et al., in this Special Issue of *Cells*, regarding an RPE-specific DHDDS knockout mouse model [27].)

Our findings bring into question the current concept that RP59 is a member of a large and diverse class of diseases known as "congenital disorders of glycosylation" (CDGs) [30,31]. While, in principle, it would be reasonable to consider RP59 as a CDG, due to the associated mutation(s) in DHDDS, there is no direct evidence to demonstrate a glycosylation defect in the human retinal disease or in any animal model of RP59 generated to date. The mechanism underlying the DHDDS-dependent retinal degeneration in human arRP patients remains to be elucidated, but is more complex than simply loss-of-function of DHDDS.

Author Contributions: Conceptualization, S.J.P. and S.J.F.; methodology, S.J.P., S.J.F., S.R.R., P.K.; M.N.N.; validation, S.R.R., M.N.N., P.K., S.J.P. and S.J.F.; formal analysis, M.N.N., S.J.P., S.J.F. and S.R.R.; investigation, S.J.P., S.R.R., P.K.; resources, S.J.P. and S.J.F.; data curation, S.J.P. and S.J.F.; writing—original draft preparation, S.J.F.; writing—review and editing, S.J.F., S.J.P., P.K., M.N.N. and S.R.R.; visualization, S.J.F., S.R.R. and S.J.P.; supervision, S.J.P. and S.J.F.; project administration, S.J.P. and S.J.F.; funding acquisition, S.J.P. and S.J.F. All authors have read and agreed to the published version of the manuscript.

Funding: This research was supported by U.S. Department of Health and Human Services (National Institutes of Health (NIH)/National Eye Institute (NEI)) grant R01 EY029341 to S.J.P. and S.J.F., and NIH/NEI core grant P30 003039 to S.J.P.; a Fight for Sight Summer Student Fellowship to S.R.R.; a Career-Starter Research Grant from the Knights Templar Eye Foundation to S.R.R.; as well as support from the UAB Vision Science Research Center (S.J.P., M.N.N., P.K.) and facilities and resources provided by the VA Western NY Healthcare System (S.J.F., S.R.R.).

Acknowledgments: We thank Isaac Cobb for technical assistance with OCT and genotyping. The opinions expressed herein do not reflect those of the Department of Veteran Affairs or the U.S. Government.

Conflicts of Interest: The authors declare no conflict of interest. The funders had no role in the design of the study; in the collection, analyses, or interpretation of data; in the writing of the manuscript, or in the decision to publish the results.

References

1. Hartong, D.T.; Berson, E.L.; Dryja, T.P. Retinitis pigmentosa. *Lancet* **2006**, *368*, 1795–1809. [CrossRef]
2. Zhang, Q. Retinitis Pigmentosa: Progress and Perspective. *Asia Pac. J. Ophthalmol. (Phila)* **2016**, *5*, 265–271. [CrossRef] [PubMed]

3.   Zuchner, S.; Dallman, J.; Wen, R.; Beecham, G.; Naj, A.; Farooq, A.; Kohli, M.A.; Whitehead, P.L.; Hulme, W.; Konidari, I.; et al. Whole-exome sequencing links a variant in DHDDS to retinitis pigmentosa. *Am. J. Hum. Genet.* **2011**, *88*, 201–206. [CrossRef]

4.   Zelinger, L.; Banin, E.; Obolensky, A.; Mizrahi-Meissonnier, L.; Beryozkin, A.; Bandah-Rozenfeld, D.; Frenkel, S.; Ben-Yosef, T.; Merin, S.; Schwartz, S.B.; et al. A missense mutation in DHDDS, encoding dehydrodolichyl diphosphate synthase, is associated with autosomal-recessive retinitis pigmentosa in Ashkenazi Jews. *Am. J. Hum. Genet.* **2011**, *88*, 207–215. [CrossRef] [PubMed]

5.   Lam, B.L.; Zuchner, S.L.; Dallman, J.; Wen, R.; Alfonso, E.C.; Vance, J.M.; Pericak-Vance, M.A. Mutation K42E in dehydrodolichol diphosphate synthase (DHDDS) causes recessive retinitis pigmentosa. *Adv. Exp Med. Biol.* **2014**, *801*, 165–170. [CrossRef] [PubMed]

6.   Parodi, A.J.; Leloir, L.F. The role of lipid intermediates in the glycosylation of proteins in the eucaryotic cell. *Biochim. Biophys. Acta* **1979**, *559*, 1–37. [CrossRef]

7.   Hemming, F.W. Control and manipulation of the phosphodolichol pathway of protein N-glycosylation. *Biosci. Rep.* **1982**, *2*, 203–221. [CrossRef]

8.   Giladi, M.; Edri, I.; Goldenberg, M.; Newman, H.; Strulovich, R.; Khananshvili, D.; Haitin, Y.; Loewenstein, A. Purification and characterization of human dehydrodolychil diphosphate synthase (DHDDS) overexpressed in E. coli. *Protein Expr. Purif.* **2017**, *132*, 138–142. [CrossRef]

9.   Lisnyansky Bar-El, M.; Lee, S.Y.; Ki, A.Y.; Kapelushnik, N.; Loewenstein, A.; Chung, K.Y.; Schneidman-Duhovny, D.; Giladi, M.; Newman, H.; Haitin, Y. Structural Characterization of Full-Length Human Dehydrodolichyl Diphosphate Synthase Using an Integrative Computational and Experimental Approach. *Biomolecules* **2019**, *9*. [CrossRef]

10.   Behrens, N.H.; Leloir, L.F. Dolichol monophosphate glucose: An intermediate in glucose transfer in liver. *Proc. Natl. Acad. Sci. USA* **1970**, *66*, 153–159. [CrossRef]

11.   Tam, B.M.; Moritz, O.L. The role of rhodopsin glycosylation in protein folding, trafficking, and light-sensitive retinal degeneration. *J. Neurosci.* **2009**, *29*, 15145–15154. [CrossRef] [PubMed]

12.   Murray, A.R.; Vuong, L.; Brobst, D.; Fliesler, S.J.; Peachey, N.S.; Gorbatyuk, M.S.; Naash, M.I.; Al-Ubaidi, M.R. Glycosylation of rhodopsin is necessary for its stability and incorporation into photoreceptor outer segment discs. *Hum. Mol. Genet.* **2015**, *24*, 2709–2723. [CrossRef] [PubMed]

13.   Fliesler, S.J.; Rayborn, M.E.; Hollyfield, J.G. Membrane morphogenesis in retinal rod outer segments: Inhibition by tunicamycin. *J. Cell Biol.* **1985**, *100*, 574–587. [CrossRef] [PubMed]

14.   Fliesler, S.J.; Rapp, L.M.; Hollyfield, J.G. Photoreceptor-specific degeneration caused by tunicamycin. *Nature* **1984**, *311*, 575–577. [CrossRef]

15.   DeRamus, M.L.; Stacks, D.A.; Zhang, Y.; Huisingh, C.E.; McGwin, G.; Pittler, S.J. GARP2 accelerates retinal degeneration in rod cGMP-gated cation channel beta-subunit knockout mice. *Sci. Rep.* **2017**, *7*, 42545. [CrossRef]

16.   Ramachandra Rao, S.; Pfeffer, B.A.; Mas Gomez, N.; Skelton, L.A.; Keiko, U.; Sparrow, J.R.; Rowsam, A.M.; Mitchell, C.H.; Fliesler, S.J. Compromised phagosome maturation underlies RPE pathology in cell culture and whole animal models of Smith-Lemli-Opitz Syndrome. *Autophagy* **2018**, 1–22. [CrossRef]

17.   Goldstein, I.J.; Hayes, C.E. The lectins: Carbohydrate-binding proteins of plants and animals. *Adv. Carbohydr. Chem. Biochem.* **1978**, *35*, 127–340. [CrossRef]

18.   Wang, T.; Voglmeir, J. PNGases as valuable tools in glycoprotein analysis. *Protein Pept. Lett.* **2014**, *21*, 976–985. [CrossRef]

19.   Blanks, J.C.; Johnson, L.V. Specific binding of peanut lectin to a class of retinal photoreceptor cells. A species comparison. *Invest. Ophthalmol. Vis. Sci.* **1984**, *25*, 546–557.

20.   Johnson, L.V.; Hageman, G.S.; Blanks, J.C. Interphotoreceptor matrix domains ensheath vertebrate cone photoreceptor cells. *Invest. Ophthalmol. Vis. Sci.* **1986**, *27*, 129–135.

21.   Fariss, R.N.; Anderson, D.H.; Fisher, S.K. Comparison of photoreceptor-specific matrix domains in the cat and monkey retinas. *Exp. Eye Res.* **1990**, *51*, 473–485. [CrossRef]

22.   Gemmill, T.R.; Trimble, R.B. Overview of N- and O-linked oligosaccharide structures found in various yeast species. *Biochim. Biophys. Acta* **1999**, *1426*, 227–237. [CrossRef]

23.   Li, Y.; Lam, B.L.; Guan, Z.; Wang, Z.; Wang, N.; Chen, Y.; Wen, R. Photoreceptor degeneration in the DHDDSK42E/K42E mouse. *Invest. Ophthalmol. Vis. Sci.* **2014**, *55*, 4371.

24. Sabry, S.; Vuillaumier-Barrot, S.; Mintet, E.; Fasseu, M.; Valayannopoulos, V.; Heron, D.; Dorison, N.; Mignot, C.; Seta, N.; Chantret, I.; et al. A case of fatal Type I congenital disorders of glycosylation (CDG I) associated with low dehydrodolichol diphosphate synthase (DHDDS) activity. *Orphanet. J. Rare Dis.* **2016**, *11*, 84. [CrossRef]

25. Wen, R.; Lam, B.L.; Guan, Z. Aberrant dolichol chain lengths as biomarkers for retinitis pigmentosa caused by impaired dolichol biosynthesis. *J. Lipid Res.* **2013**, *54*, 3516–3522. [CrossRef]

26. Wen, R.; Dallman, J.E.; Li, Y.; Zuchner, S.L.; Vance, J.M.; Pericak-Vance, M.A.; Lam, B.L. Knock-down DHDDS expression induces photoreceptor degeneration in zebrafish. *Adv. Exp. Med. Biol.* **2014**, *801*, 543–550. [CrossRef]

27. DeRamus, M.L.; Davis, S.J.; Rao, S.R.; Nyankerh, C.; Stacks, D.; Kraft, T.W.; Fliesler, S.J.; Pittler, S.J. Selective Ablation of Dehydrodolichyl Diphosphate Synthase in Murine Retinal Pigment Epithelium (RPE) Causes RPE Atrophy and Retinal Degeneration. *Cells* **2020**, *9*, 771. [CrossRef]

28. Harrison, K.D.; Park, E.J.; Gao, N.; Kuo, A.; Rush, J.S.; Waechter, C.J.; Lehrman, M.A.; Sessa, W.C. Nogo-B receptor is necessary for cellular dolichol biosynthesis and protein N-glycosylation. *EMBO J.* **2011**, *30*, 2490–2500. [CrossRef]

29. Park, E.J.; Grabinska, K.A.; Guan, Z.; Stranecky, V.; Hartmannova, H.; Hodanova, K.; Baresova, V.; Sovova, J.; Jozsef, L.; Ondruskova, N.; et al. Mutation of Nogo-B receptor, a subunit of cis-prenyltransferase, causes a congenital disorder of glycosylation. *Cell Metab.* **2014**, *20*, 448–457. [CrossRef]

30. Haeuptle, M.A.; Hennet, T. Congenital disorders of glycosylation: An update on defects affecting the biosynthesis of dolichol-linked oligosaccharides. *Hum. Mutat.* **2009**, *30*, 1628–1641. [CrossRef]

31. Ng, B.G.; Freeze, H.H. Perspectives on Glycosylation and Its Congenital Disorders. *Trends Genet.* **2018**, *34*, 466–476. [CrossRef] [PubMed]

 *cells*

Article

Selective Ablation of Dehydrodolichyl Diphosphate Synthase in Murine Retinal Pigment Epithelium (RPE) Causes RPE Atrophy and Retinal Degeneration

Marci L. DeRamus [1], Stephanie J. Davis [1], Sriganesh Ramachandra Rao [2], Cyril Nyankerh [1], Delores Stacks [1], Timothy W. Kraft [1], Steven J. Fliesler [2] and Steven J. Pittler [1,*]

[1] Department of Optometry and Vision Science, Vision Science Research Center, University of Alabama at Birmingham, Birmingham, AL 35294, USA; marcismith1952@gmail.com (M.L.D.); stephanie.davis@bison.howard.edu (S.J.D.); cyrilnya@uab.edu (C.N.); delorez@uab.edu (D.S.); twkraft@uab.edu (T.W.K.)

[2] Departments of Ophthalmology and Biochemistry, State University of New York-University at Buffalo, Buffalo, NY 14209, USA; and Research Service, VA Western NY Healthcare System, Buffalo, NY 14215, USA; sramacha@buffalo.edu (S.R.R.); fliesler@buffalo.edu (S.J.F.)

* Correspondence: pittler@uab.edu; Tel.: +1-205-934-6744

Received: 12 February 2020; Accepted: 17 March 2020; Published: 21 March 2020

Abstract: Patients with certain defects in the dehydrodolichyl diphosphate synthase (DHDDS) gene (RP59; OMIM #613861) exhibit classic symptoms of retinitis pigmentosa, as well as macular changes, suggestive of retinal pigment epithelium (RPE) involvement. The DHDDS enzyme is ubiquitously required for several pathways of protein glycosylation. We wish to understand the basis for selective ocular pathology associated with certain DHDDS mutations and the contribution of specific ocular cell types to the pathology of mutant *Dhdds*-mediated retinal degeneration. To circumvent embryonic lethality associated with *Dhdds* knockout, we generated a Cre-dependent knockout allele of murine *Dhdds* (*Dhdds^{flx/flx}*). We used targeted Cre expression to study the importance of the enzyme in the RPE. Structural alterations of the RPE and retina including reduction in outer retinal thickness, cell layer disruption, and increased RPE hyper-reflectivity were apparent at one postnatal month. At three months, RPE and photoreceptor disruption was observed non-uniformly across the retina as well as RPE transmigration into the photoreceptor layer, external limiting membrane descent towards the RPE, and patchy loss of photoreceptors. Functional loss measured by electroretinography was consistent with structural loss showing scotopic a- and b-wave reductions of 83% and 77%, respectively, at three months. These results indicate that RPE dysfunction contributes to DHDDS mutation-mediated pathology and suggests a more complicated disease mechanism than simply disruption of glycosylation.

Keywords: retinal degeneration; retinitis pigmentosa; retinal pigment epithelium dystrophy; RPE transmigration; Cre-Lox technology; mouse models

---

1. Introduction

Retinitis pigmentosa (RP) and related disorders are characterized by degeneration and loss of photoreceptors, attenuation of retinal blood vessels, pigment deposits, and a waxy pallor of the optic disc that result in impaired night vision and peripheral and central vision loss [1]. Defective protein glycosylation in the retinal pigment epithelium (RPE) has been associated with retinal degeneration elicited by photoreceptor abnormalities and impaired phagocytosis of aging photoreceptor outer segment (OS) membranes [2–5]. One such example, dehydrodolichyl diphosphate synthase (DHDDS; OMIM #608172), is an essential enzyme in the mevalonate pathway where it functions ubiquitously in isoprenoid chain elongation to form dolichols that comprise 17–20 isoprene units. The phosphorylated form of dolichol, dolichol pyrophosphate is necessary for N-glycosylation at specific residues in many membrane proteins [6]. There is no known unique function for DHDDS in the retina, or in any other ocular tissue. There are numerous N-glycosylated proteins in the retina and photoreceptors, such as rhodopsin [7] and the cGMP-gated cation channel in rods. Additionally, in the RPE, there are several glycosylated structural proteins, ion channels and transport proteins that contribute to the transepithelial potential of the polarized RPE monolayer [8–10].

Mutations in the gene encoding DHDDS lead to a recessive form of RP called RP59 (OMIM #613861, DHDDS, K42E), which was first identified in several families of Ashkenazi Jewish origin [11–13]. RP59 is considered to belong to the family of human genetic diseases known as "congenital disorders of glycosylation" (CDGs) [14]. Two other mutations in the DHDDS gene (T206A and R98W), which occur heterozygously with the K42E mutation, were also reported, but have not been studied in detail [15,16]. Patients with the K42E mutation in DHDDS exhibit all the cardinal features of RP as well as macular changes [12], suggesting possible RPE involvement. While the K42E, T206A, and R98W mutations are only associated with known RP symptoms, another mutation in the DHDDS gene has led to infant morbidity at seven months of age [17].

As a first step to assess the role of DHDDS in specific retinal cell types and to understand the molecular mechanism of RP59, we created a Cre-lox dependent line of mice that allows targeted, cell type-specific deletion of *Dhdds* in cells of interest. Using a conditional knockout was necessary to circumvent embryonic lethality associated with global knockout of *Dhdds* [18,19]. Here, we describe the generation and characterization of a *Dhdds*$^{flx/flx}$ CreRPE mouse line (i.e., RPE-specific *Dhdds* knockout) and its validation as a model of RPE atrophy and retinal degeneration. We show that RPE-specific deletion of *Dhdds* induces structural and functional deficits in the RPE and the photoreceptors, which suggests that RPE pathology may be a significant contributor to the retinal degeneration observed in patients with RP59 mutations.

2. Materials and Methods

2.1. Generation of Dhdds$^{flx/flx}$ CreRPE Mice

A construct containing lacZ flanked by FLP-FRT and *Dhdds* exon 3 flanked by loxP sites from the Knockout Mouse Project (KOMP, UC Davis, Davis, CA, USA) was linearized and introduced into mouse ES cells (C57Bl/6J background) at the Roswell Park Cancer Institute (RPCI) Gene Targeting and Transgenic Facility (Buffalo, NY, USA) using standard technology. To confirm the correctly targeted cells, polymerase chain reaction (PCR) was performed with the primers listed below (see *PCR Genotyping*, below). The lacZ cassette was excised with FLP-FRT recombinase, and excision was confirmed by PCR. Mouse lines that carried the *Dhdds* loxP conditional knockout allele were crossed to generate homozygotes and the latter were also crossed to a mouse line (on a C57Bl/6J background) carrying a homozygous transgene expressing Cre recombinase under the control of the RPE-specific VMD2 (vitelliform macular degeneration 2) promoter [20,21]. RPE-specific expression of Cre in the VMD2 promoter-driven mouse line was confirmed by crossing those mice with a ZsGreen reporter mouse line (B6.Cg-Gt(ROSA)26Sor$^{tm6(CAG-ZsGreen1)Hze}$/J, The Jackson Laboratory, Bar Harbor, ME, USA) and examining the retinas by confocal fluorescence microscopy. Genotypes of offspring were confirmed by PCR with *Dhdds*- and Cre RPE-specific primer pairs. All mice used in this study were treated following the ARVO *Statement on the Use of Animals in Ophthalmic and Vision Research* and the policies of the University of Alabama at Birmingham (UAB) Institutional Animal Care and Use Committee (IACUC). This project was approved for animal use on April 2019 by the UAB IACUC and requires updated approvals each year (protocol number IACUC-21270). All animals were maintained on a standard 12/12-h light/dark cycle, fed standard rodent chow, provided water *ad libitum*, and housed in plastic cages with standard rodent bedding.

2.2. PCR Genotyping

Mouse genomic DNA samples obtained from tail snips were verified by PCR using primers *Dhdds*-FWD: 5′-GTGTCATCCCCTGCTGCAGAT-3′ and *Dhdds*-REV: 5′-TGGGTGTAGTG-GCTCAGGTC-3′ for genotype identification of floxed *Dhdds* alleles designed in a region which was conserved in both wild type (WT) and floxed alleles and also in the region flanking the loxP sites. The expected PCR product sizes for the WT and floxed alleles are 393 and 517 bp, respectively, thus differentiating WT, heterozygous floxed, and homozygous floxed alleles. PCR verification of Cre transgene modification was carried out using the following forward and reverse primer sets for Cre RPE 5′-AGGTGTAGAGAAGGCACTTAGC-3′ and 5′-CTAATCGCCATCT-TCCAGCAGG-3′, respectively, yielding a 411 bp product. RPE specific expression and activity of Cre-recombinase in VMD2-RPE Cre was verified by breeding these mouse lines against an ZsGreen reporter mouse strain (B6.Cg-*Gt(ROSA)26Sor*$^{tm6(CAG-ZsGreen1)Hze}$/J, Stock# 007906; The Jackson Laboratory, Bar Harbor, ME, USA) and monitoring ZsGreen expression in the retina.

2.3. Spectral-Domain Optical Coherence Tomography

Spectral-domain optical coherence tomography (SD-OCT) (840 nm; Bioptigen, Inc./Leica, Durham, NC, USA) was used to obtain in vivo images of the retina at one, two, and three months of age. OCT images were collected with Bioptigen InVivoVue® 1.4 software and Bioptigen Diver® 2.0 software was used to analyze outer nuclear layer (ONL) thickness and full retinal thickness (F.R.T.) across all retinal layers at five eccentricities spanning two-thirds of the retina centered at the optic nerve head. A detailed description of this procedure has been reported previously [22]. F.R.T. was calculated from the difference of markers 1 and 10 and ONL thickness was calculated from the difference between 5 and 6.

2.4. Fundus Examination and Fluorescein Angiography

Fundus examination was performed with a Micron IV digital fundus microscope (Phoenix Technology Group, Pleasanton, CA, USA), using the mouse objective and hydroxypropylmethyl cellulose 2.5% on the surface of the cornea. Digital images of the fundus were captured with dedicated StreamPix® 6 digital software (NorPix, Inc., Montreal, Quebec, Canada) and processed using Adobe® Photoshop® 6 (Adobe Systems, Inc., San Jose, CA, USA).

2.5. Visual Function Testing

Full-field scotopic and photopic electroretinograms (ERGs) were obtained as described in detail previously using an OcuScience® HMsERG instrument (OcuScience, Henderson, NV, USA) [22]. Response amplitudes and implicit times of a-waves and b-waves representing the activity of photoreceptors and bipolar cells, respectively, were quantified. In brief, mice were dark-adapted overnight, anesthetized, and then placed on a heating pad to maintain body temperature. Following pupil dilation with 2.5% phenylephrine and 1% tropicamide ophthalmic solutions, a thin silver wire electrode was placed on the cornea (interfaced with methylcellulose and covered with a specially designed contact lens) and referenced to a needle ground electrode in the cheek. The responses to Ganzfeld flash stimuli, spanning a range of five log units, were measured and recorded; following light adaptation to background illumination, cone-driven responses also were recorded.

2.6. Histology/Immunohistochemistry

The methodologies used in this study have been described in detail previously [23]. Briefly, for conventional histology, eyes (n ≥ 4 per condition) were fixed by immersion in freshly prepared 4% paraformaldehyde in 0.125 M Na-phosphate buffer, pH 7.4, at 4 °C overnight, embedded in paraffin, and tissue sections (toluidine blue-stained) were viewed with an Olympus BH2 photomicroscope equipped with a Nikon digital camera. Digitized images were collected and further analyzed with ImagePro

Plus® software, Version 4.1 (Media Cybernetics; Rockville, MD, USA). For immunohistochemistry, frozen sections of retinal tissue (embedded in Tissue-Plus™ Optimal Cutting Temperature (O.C.T.) compound; Thermo Fisher Scientific, Waltham, MA, USA), obtained with a cryostat and collected on glass microscope slides were incubated with suitable primary antibodies, with detection by application of species-specific, fluor-conjugated secondary antibodies, counterstaining nuclei with DAPI, followed by laser confocal immunofluorescence microscopy (Leica TCS SPE scanning confocal microscope; Leica Microsystems Inc., Buffalo Grove, IL, USA), as previously reported [24].

2.7. Fluorescein Angiography

Fluorescein angiography was performed following intraperitoneal injection (i.p., 10 µL/gram body weight) of a 10 mg/mL solution of AK-Fluor 10% (Sigma Pharmaceuticals; Liberty, IA, USA) in PBS. Uptake of fluorescein and fundus imaging was monitored with a Micron IV Retinal Imaging Microscope (Phoenix Technology Group, Inc., Pleasanton, CA, USA).

2.8. Electron Microscopy

Mouse eyes were processed for plastic embedment, ultramicrotomy, and EM analysis essentially as described in detail previously [23]. Immediately after sacrifice, eyes were orientated by marking the superior hemisphere along the vertical meridian at the limbus with a hot needle, before starting the dissection. A cut was made in the superior cornea and the eyes were fixed for 2 h at 4 °C in fresh 0.1 M sodium phosphate buffer (pH 7.4), containing 2.5% (v/v) glutaraldehyde, 2.0% formaldehyde and 0.025% $CaCl_2$. After a 20–30 min primary fixation, the superior cornea and lens were removed, and fixation was continued overnight. The fixed eyes were then rinsed with 0.1 M sodium cacodylate buffer (pH 7.4) containing 0.025% $CaCl_2$, and then post-fixed for 1 h in 1% osmium tetroxide in 0.1 M sodium cacodylate buffer. After post-fixation, the eyes were rinsed twice in 0.1 M sodium cacodylate buffer and once in distilled water, then dehydrated in graded ethanol series followed by propylene oxide and infiltration overnight in Spurr's resin. The eyes were then embedded in resin-filled BEEM® capsules (Polysciences, Warrington, PA, USA) and allowed to polymerize in a 70 °C oven for 48 h. Tissue sections were obtained with a Reichert–Jung Ultracut E® microtome using a diamond knife. Thin (60–80 nm thickness) sections were collected on copper 75/300 mesh grids and stained with 2% (v/v) uranyl acetate and Reynolds' lead citrate. Sections were viewed with a JEOL 100CX electron microscope at an accelerating voltage of 60 keV.

2.9. Serial Block-Face Scanning Electron Microscopy (SBF-SEM)

Samples prepared for TEM as described above were further processed by Thermo Fisher Scientific (Waltham, MA, USA) using an Apreo VolumeScope™ serial block-face scanning electron microscope (SBF-SEM). Excess resin was removed from the tissue using a Leica ultramicrotome. The trimmed blocks were then glued to a SEM stub (Agar Scientific, AGG1092450) using a two-component silver conductive epoxy, H20E EPO-TEK (Ted Pella, Inc.; Redding, CA, USA). To minimize charging of the block by the electron beam, the bottom and sides of the block were sputter-coated with a 30 nm thick gold film layer. The samples were then imaged on the VolumeScope™ operating in low vacuum mode at 50 Pa and using a lens mounted backscattered detector. All of the data sets were imaged with an accelerating voltage of 2.2 kV and a beam current of 100 pA using 1-µs dwell time combined with two-line integration. Two regions of interest (ROIs) were acquired on the knock-out sample (KO-148). For ROI1, 738 sections were collected with the internal microtome set to a 40 nm cutting thickness (z resolution) with an area of 92.9 µm × 89.1 µm at 10 nm/pixel. For ROI2, 745 sections were collected with an area of 97.6 µm × 96.3 µm using the same imaging condition as ROI1. For the wild type sample (WT-146), an area of 92.9 µm × 89.1 µm was imaged at 10 nm/pixel using a cutting thickness (z resolution) of 40 nm. The acquired data sets were finally aligned and visualized using 3D volume rendering to highlight the RPE anomalies using Amira software (Thermo Fisher Scientific).

2.10. Statistical Analysis

For ERG analyses, we evaluated differences between genetically modified vs. WT control mice across flash intensities by performing repeated measures two-way ANOVA with Holm–Sidak post-hoc analysis at each time point. To evaluate functional (visual acuity, a-wave and b-wave amplitudes and implicit times) and structural (SD-OCT retinal layer thickness) parameters across time, we used a two-way repeated measures ANOVA (time X treatment conditions) with Holm–Sidak post-hoc analysis. For biochemical and quantitative immunohistochemical data, binary statistical comparisons between specific genetically modified vs. WT control group data were analyzed using an unpaired Student's *t*-test.

3. Results

3.1. Generation of a Floxed Dhdds Mouse Line

The scheme used for the generation of a *Dhdds* conditional allele, employing a validated *Dhdds* construct (from KOMP) and mouse embryonic stem cells (ESCs), is shown in Figure 1 (see *Materials and Methods*, above). Clones from confirmed flippase recognition target (FRT)-excised alleles were used to generate *Dhdds* heterozygous and homozygous mouse lines on a C57Bl/6J background.

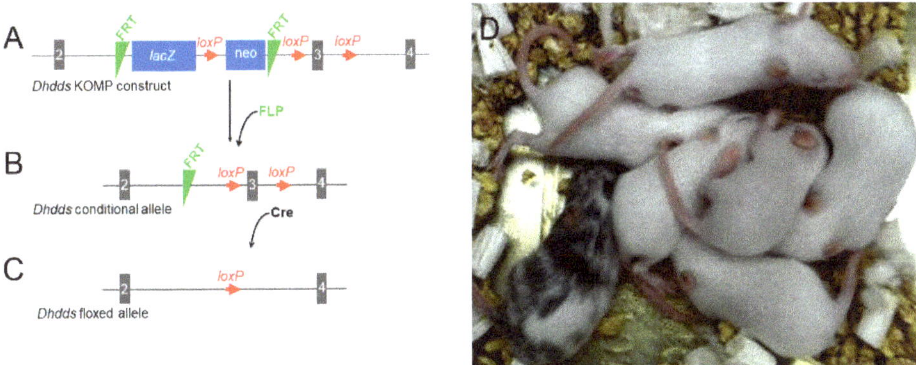

Figure 1. Generation of Cre-dependent *Dhdds* conditional knockout (KO) mice. (**A**) A validated *Dhdds* construct from the Knockout Mouse Project (KOMP) (U.C. Davis) was linearized and introduced into mouse ESCs. (**B**) Transformed cells were treated with FLP-FRT recombinase and PCR was used to verify lacZ cassette excision. (**C**) Clones from confirmed FRT-excised alleles were used to generate *Dhdds$^{flx/flx-}$* mice. (**D**) Pups carrying the *Dhdds* floxed allele were identified by coat appearance.

3.2. Validation of Retinal Cell Type-Specific Cre-Expressing Mouse Lines

To assess the specificity and efficiency of the Cre recombinase in the RPE, we cross-bred transgenic mice expressing Cre recombinase (bred to homozygosity) with a ZsGreen Ai6 reporter mouse line, and then evaluated ZsGreen expression in the retina by confocal fluorescence microscopy. As expected, WT mouse retinas (a negative control) did not exhibit ZsGreen expression (Figure 2A). However, ZsGreen fluorescence was detected in Cre recombinase-positive mice specifically in the RPE layer by 1 postnatal (PN) month (Figure 2B), with >90% of the RPE cells being labeled. We subsequently confirmed that Cre recombinase continued to be expressed robustly and specifically in the RPE for >3 months (data not shown).

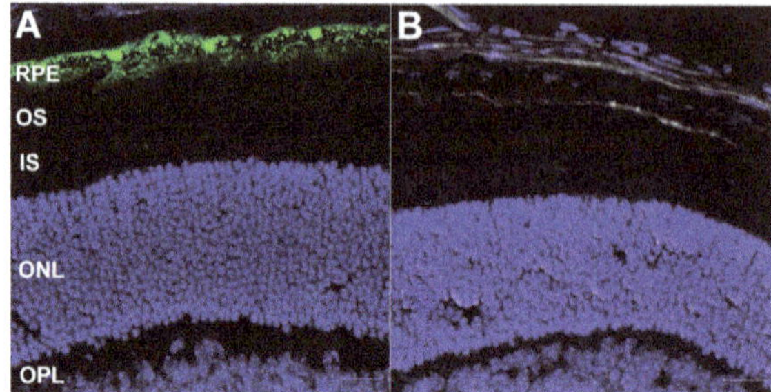

Figure 2. Cre recombinase-dependent ZsGreen expression (green) in mouse retina. (**A**) Retina from a postnatal (PN) one-month old CreRPE × ZsGreen reporter mouse, demonstrating ZsGreen expression specifically in retinal pigment epithelium (RPE) cells. (**B**) Retina from an age-matched, wild type (WT) mouse retina, demonstrating lack of ZsGreen expression. Nuclei counterstained with DAPI (blue). Abbreviations: IS, photoreceptor inner segment layer; OS, photoreceptor outer segment layer; ONL, outer nuclear layer; OPL, outer plexiform layer. Scale bars (both panels): 20 μm.

3.3. RPE-Specific Ablation of Dhdds Causes a Geographic Atrophy-Like Phenotype and Retinal Degeneration, Involving Photoreceptors

In vivo retinal imaging using SD-OCT (Figure 3) showed comparable normal layer stratification in WT, CreRPE and *Dhdds$^{+/flx}$* CreRPE age-matched mice in each group. However, *Dhdds$^{flx/flx}$* CreRPE mice showed altered hyper-reflectivity at all ages (indicated by red arrows, Figure 3), indicative of pathologic changes and a reduction in outer retinal layer thickness.

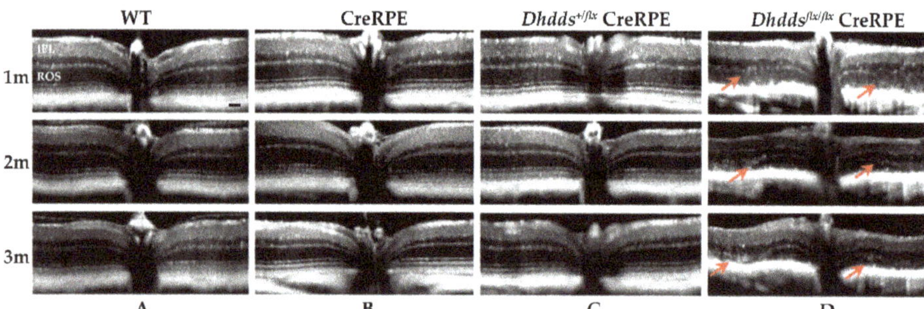

Figure 3. SD-OCT revealed structural changes in *Dhdds$^{flx/flx}$* CreRPE mice at all ages. OCT scans were performed at 1, 2, and 3 months (m) postnatal for each genotype. (**A**) WT, (**B**) CreRPE, (**C**) *Dhdds$^{+/flx}$* CreRPE all showed normal layer stratification. (**D**) *Dhdds$^{flx/flx}$* CreRPE showed alterations of hyper-reflectivity at all ages, indicative of pathologic changes (red arrows) and reduction in layer thickness, particularly in the outer retina. IPL, inner plexiform layer, ROS, rod outer segment. Scale bar (shown in WT 1 m panel): 50 μm, applies to all panels.

"Spidergram" plots of average thickness values (in mm) vs. retinal eccentricity (distance from the optic nerve head (ONH) along with vertical meridian) for the outer nuclear layer (ONL) and FRT are shown in Figure 4. No significant differences were observed when comparing WT, heterozygous, or Cre-only mice for both ONL and FRT measurements. However, *Dhdds$^{flx/flx}$* CreRPE mice showed a significant reduction (vs. WT) in ONL and F.R.T. values at all ages analyzed (n ≥ 4, all $p < 0.001$).

Histologically, at PN 3 months, light micrograph images revealed that all retina layers in mice lacking Cre expression appeared normal (Figure 5A and Supplementary Materials, Figure S1A). In contrast, the age-matched Dhdds$^{flx/flx}$ Cre RPE mice displayed a geographic atrophy-like RPE appearance with the most degeneration observed mid-centrally throughout the retina (Supplementary Materials, Figure S1B). There were regions of well-preserved Dhdds$^{flx/flx}$ Cre RPE retina that showed severe RPE pathology (Figure 5B). Notably, the descent of the external limiting membrane (ELM) [25] towards Bruch's membrane was also observed (Figure 5C,E). There was a near-total loss of photoreceptors and RPE within the most affected regions (Figure 5D). Very well-preserved regions were also observed in the periphery (Figure 5F).

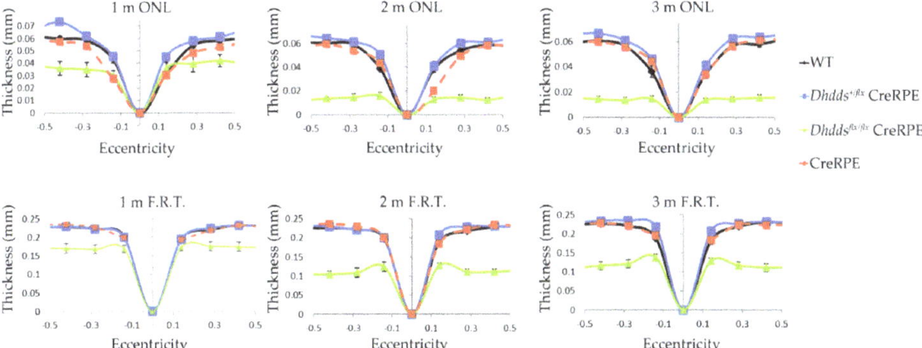

Figure 4. Quantitative morphometric analysis of WT, CreRPE, *Dhdds*^+/flx CreRPE, and *Dhdds*^flx/flx CreRPE mice SD-OCT data. Average OCT measurements at each eccentricity revealed no significant differences when comparing WT, CreRPE and *Dhdds*^+/flx CreRPE mice for both outer nuclear layer (ONL) and F.R.T. thickness measurements (n = 4 for all genotypes, except for WT at one month (1 m; n = 8) and two months (2 m; n = 5). However, significant changes were observed when comparing WT and *Dhdds*^flx/flx CreRPE mice at any age with respect to both ONL and F.R.T. values.

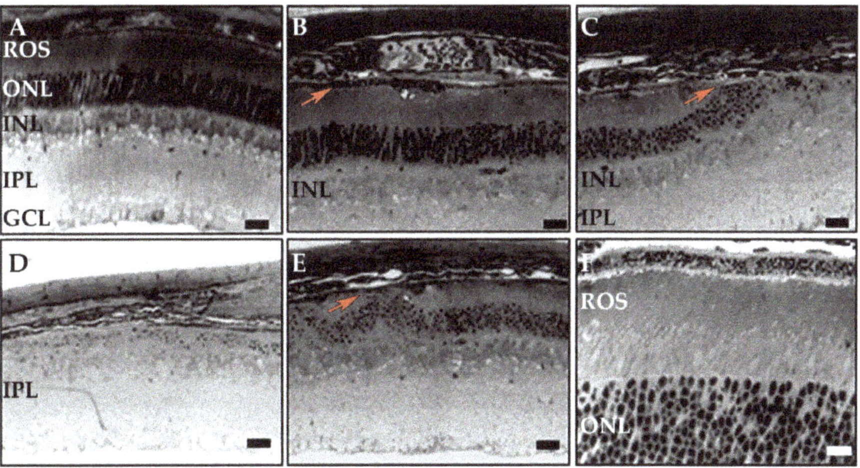

Figure 5. Pathology observed in *Dhdds*^flx/flx Cre RPE mouse retina. (**A**) *Dhdds*^flx/flx retina without Cre expression appears normal. (**B–F**) Five regions of a retina from a *Dhdds*^flx/flx CreRPE mouse expressing Cre in RPE are shown. (**B,F**) Relatively well-preserved peripheral retina with some photoreceptor loss, outer segment (OS) shortening, and differing severity of RPE pathology are shown (arrow in panel **B** points to severely compromised RPE. (**C,E**) Transition zones of severe to mild retinal pathology showing loss of photoreceptors, severe compromise of RPE and external limiting membrane (ELM) descent (arrows). (**D**) A more central region of the retina showing severe cell loss in both the retina and RPE. Scale bars (all panels, except **F**): 20 μm; scale bar, panel **F**: 10 μm. ROS, rod outer segments; ONL, outer nuclear layer, INL, inner nuclear layer; IPL, inner plexiform layer, GCL, ganglion cell layer.

Additional pathology was revealed by higher magnification EM analysis (Figure 6, panels C–O). In contrast, *Dhdds*^{flx/flx} retinas without Cre expression were indistinguishable from WT retinas (Figure 6A,B). The two areas most affected showed severe RPE dystrophy with concomitant degeneration and loss of photoreceptor cells (Figure 6C,D,F,G,J,M–O). ELM descent was apparent in areas where there was a transition from milder to more severe pathology (* in panels D–F,J). RPE cell transmigration was apparent in the outer retina (arrows, panels G,J,O,L). Thus, compared to WT neural retina and RPE, the observed RPE anomalies including migration of nucleus and RPE melanosomes, displacement of the ELM, and shortened misshaped outer segments throughout the retina were seen in the *Dhdds*^{flx/flx} CreRPE mice.

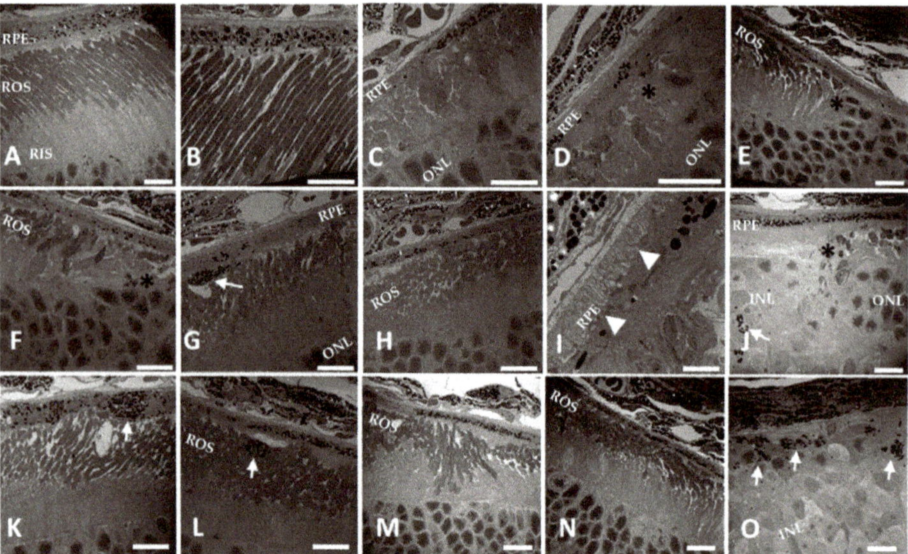

Figure 6. High magnification observation of *Dhdds*^{flx/flx} CreRPE retina. EM images of 3m (**A,B**) *Dhdds*^{flx/flx} and (**C–O**) *Dhdds*^{flx/flx} CreRPE mice. Mice homozygous for the floxed *Dhdds* allele (*Dhdds*^{flx/flx}) are not distinguishable from WT. (**A**) Rod outer segments are properly aligned and all retinal layers are intact (only rod inner and outer segments and the outer nuclear layer are shown). (**B**) The RPE shows normal thickness and melanin distribution. (**C–O**) Examples of mildly and severely compromised RPE and retina in *Dhdds*^{flx/flx} CreRPE mice. Severe RPE and outer retina degeneration is concentrated mid-centrally on both sides of the optic nerve head (see histology, Figure 5, and Supplementary Materials, Figure S1). Severe RPE/PR atrophy is observed in the most affected central regions (**C–F,I,J,O**). External limiting membrane (ELM) descent (*) towards the RPE is apparent in the transition regions from severely affected to more intact regions (**D–F,J**). Transmigration of RPE cellular material into the ROS space (arrows) is also seen (**G,J,L,O**). Extended RPE basal fenestrations (arrowheads, **I**) are apparent in the most affected regions (**D,I**). Outside of the central region of severe degeneration are regions with compromised but still apparent retinal and RPE layers (**K–N**). Scale bars (all panels, except **I**): 10 μm; panel **I**, 2.5 μm.

To further examine the effects of *Dhdds* deletion in the RPE, serial block face-scanning electron microscopy was used to examine a ~100 μ³ region of the *Dhdds*^{flx/flx} retina across an area of transition from milder to more severe pathology (Supplementary Materials, Videos S1–3). In WT mice, the retina appeared normal in all layers across the entire block. Examination of two regions of the *Dhdds*^{flx/flx} Cre RPE retina showed areas of severe compromise as well as areas where retinal histology was more well-preserved. Transmigration of RPE nuclei and melanosomes into the photoreceptor region (subretinal space and photoreceptor outer segment layer, and even deeper, into the ONL) was also apparent.

3.4. Altered Scotopic and Photopic ERG Amplitudes in Dhdds$^{flx/flx}$ CreRPE Mice

Analysis by ERG (Figure 7) showed that both homozygous and heterozygous *Dhdds*$^{flx/flx}$ CreRPE mice exhibited reduced scotopic ERG a- and b-wave responses. At PN 1 month, scotopic a-wave responses (289 ± 42 µV; n = 5) were significantly reduced in *Dhdds*$^{flx/flx}$ CreRPE mice compared to WT mice (395 ± 14 µV; n = 8; $p < 0.05$). Scotopic b-wave responses also were reduced for both *Dhdds*$^{+/flx}$ CreRPE (772 ± 42 µV; n = 20) and *Dhdds*$^{flx/flx}$ CreRPE (479 ± 69 µV; n = 5) mice, compared to WT mice (934 ± 38 µV; n = 8; $p < 0.01$). At PN 2 months of age, *Dhdds*$^{flx/flx}$ CreRPE scotopic ERG a-wave (223 ± 48 µV; n = 10) and b-wave (477 ± 88 µV; n = 10) responses and *Dhdds*$^{+/flx}$ CreRPE a-wave (305 ± 19 µV; n = 10) and b-wave (557 ± 64 µV; n = 10) responses were significantly reduced compared to the WT a-wave (366 ± 10 µV; n = 15) and b-wave (935 ± 27 µV; n = 15) responses ($p < 0.005$ for all comparisons). At PN 3 months, scotopic a-wave amplitudes were reduced only in *Dhdds*$^{flx/flx}$ CreRPE (60 ± 48 µV; n = 6) compared to WT mice (354 ± 16 µV; n = 18; $p < 0.001$); b-wave responses were reduced in both *Dhdds*$^{+/flx}$ CreRPE (662 ± 88 µV; n = 6) and *Dhdds*$^{flx/flx}$ CreRPE (150 ± 105 µV; n = 6) mice compared to WT mice (914 ± 35 µV; n = 18) $p < 0.05$ and $p < 0.01$, respectively).

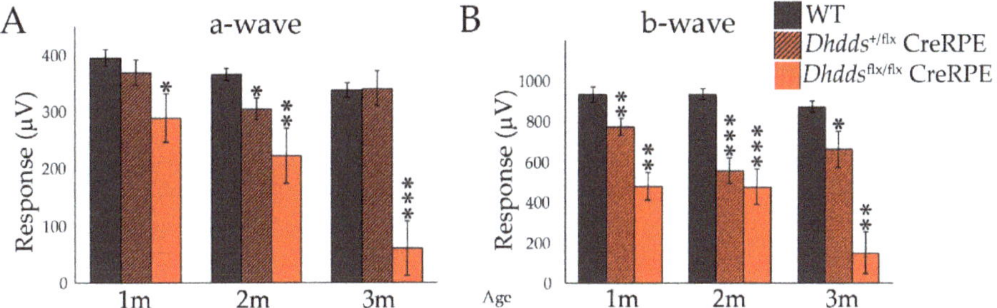

Figure 7. Scotopic ERG analysis of WT, *Dhdds*$^{+/flx}$ CreRPE, *Dhdds*$^{flx/flx}$ CreRPE mice. Maximum responses to saturating light stimuli showed significant decreases in (**A**) a-wave and (**B**) b-wave amplitudes for *Dhdds*$^{+/flx}$ CreRPE and *Dhdds*$^{flx/flx}$ CreRPE when compared to WT mice at 1, 2 and 3 postnatal months. Statistical significance: * $p < 0.05$, ** $p < 0.01$, *** $p < 0.001$.

The photopic a-wave and b-wave responses (Figure 8) at PN 1 month were not different when comparing WT (n = 5) and *Dhdds*$^{+/flx}$ CreRPE (n = 13) mice, however the photopic responses of *Dhdds*$^{flx/flx}$ CreRPE mice (n = 4) were significantly lower. Photopic ERG responses were significantly different only for *Dhdds*$^{+/flx}$ CreRPE a-wave (29 ± 2 µV; n = 4; $p < 0.01$), but not b-wave, responses at PN 2 months of age. In *Dhdds*$^{flx/flx}$ CreRPE mice, b-wave (86 ± 19 µV; n = 6) amplitudes were significantly reduced ($p < 0.01$), compared to WT mice (n = 13), but not the a-wave responses. At PN 3 months, *Dhdds*$^{flx/flx}$ CreRPE mice exhibited significant functional impairment in the photopic a-wave (2 ± 1 µV; n = 4) and b-wave (7 ± 1 µV; n = 4) responses compared to WT a-wave (14 ± 1 µV; n = 9) and b-wave responses (144 ± 8 µV; n = 9; $p < 0.001$ for all comparisons).

Optokinetic reflex (OKR) analysis (Supplementary Materials, Figure S3), a measure of retina-to-brain transmission (i.e., visual capacity), showed reductions in photopic (4–31%) and scotopic (8–29%) contrast sensitivity over the range of 0.031 to 0.272 c/d in PN 3-month old *Dhdds*$^{flx/flx}$ CreRPE mice ($p < 0.05$), compared to WT controls of the same age. However, no differences in spatial frequency (a measure of visual acuity) were observed between the different mouse lines. This is partly evident in the OKR scotopic and photopic plots, which show a similar high-frequency cut-off for all three mouse lines examined.

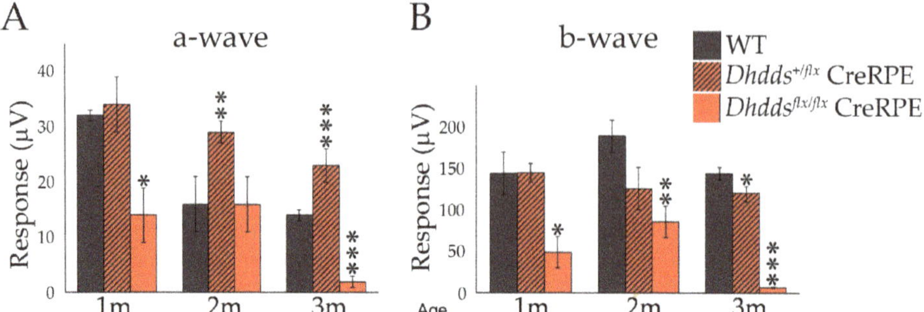

Figure 8. Photopic ERG analysis of WT, CreRPE *Dhdds*[+/flx], CreRPE *Dhdds*[flx/flx] mice. (**A**) a-wave responses were significantly lower in CreRPE *Dhdds*[-/-] at 1 (n = 4) and 3 (n = 4) months postnatal compared to WT (one month (1 m), n = 5; three months (3 m), n = 9) mice. (**B**) Photopic b-wave amplitudes for *Dhdds*[+/flx] CreRPE mice showed no statistically significant differences when compared to WT mice at 1 (n = 13) and 2 (n = 4) postnatal months, but were reduced by 3 months (n = 4). In contrast, the responses of *Dhdds*[flx/flx] CreRPE mice (two months (2 m), n = 6) were significantly lower when compared to WT mice at all postnatal ages examined. Statistical significance: * $p < 0.05$, ** $p < 0.01$, *** $p < 0.001$.

4. Discussion

Studies involving genetic screening of families with autosomal recessive retinitis pigmentosa have implicated a founder missense mutation (K42E) in the gene encoding DHDDS [12,26]. DHDDS is required for N-glycosylation of proteins by adding multiple copies of isopentenyl pyrophosphate (IPP) to farnesyl pyrophosphate (FPP) to produce dehydrodolichyl diphosphate (Dedol-PP), a precursor of dolichol, which is utilized as a sugar carrier in protein glycosylation in the endoplasmic reticulum [6]. Even though the generation of isoprenoid chains is complex, involving multiple enzymes and enzyme complexes [27], only DHDDS and Nogo-B receptor are required for long-chain isoprenoid synthesis (C_{70}-C_{120}) [28]. Previous studies have reported that mutations in the opsin gene that abolish N-linked glycosylation cause retinal degeneration [28–30]. To understand the basis for the ocular pathology associated with DHDDS mutation, we utilized a Cre recombinase conditional knockout (*Dhdds*[flx/flx] CreRPE) driven under the control of the RPE-specific VMD2 promoter to achieve RPE-targeted excision of the loxP-modified *Dhdds* gene, which renders the enzyme non-functional.

The VMD2 construct was originally generated to be conditional on the presence of doxycycline; however, we found that the mice displayed a phenotype in the absence of doxycycline induction. While we cannot rule out the presence of some level of doxycycline in the standard chow mouse diet, it is more likely that Cre expression was due to "leaky" expression, which bypassed doxycycline control, as confirmed in Figure 2. Since the promoters used for cell-specific targeting of Cre expression turn on after development is complete, developmental changes that would otherwise preempt retina/RPE development were not observed.

The expression of Cre recombinase in the Cre lines was confirmed using a ZsGreen reporter strategy. The *Dhdds*[flx/flx] CreRPE mice exhibited about 90% coverage of Cre expression by PN 1 month (Figure 2) which persisted up to 3 months at least. We crossed the conditional *Dhdds*[flx/flx] lines with heterozygous VMD2 Cre lines to generate homozygous *Dhdds*[flx/flx] mice with RPE-specific knockout of DHDDS expression as the model for our study.

In vivo imaging suggested that ONL thickness and F.R.T. were comparable between age-matched WT and *Dhdds*[+/flx] CreRPE mice, but significantly reduced in *Dhdds*[flx/flx] CreRPE mice. While the specific pathology is observed in much greater detail in the histological (light microscopy; Figure 5) and ultrastructural (electron microscopy; Figure 6) images provided, it is clear that the pathology indicated in the SD-OCT tomograms (Figure 3D) is consistent with the histological observations. This altered structural integrity of *Dhdds*[flx/flx] CreRPE retinas also corresponded to a significant decrease in a-wave and b-wave compared to age-matched WT at all flash intensities. Unexpectedly, the scotopic a- and b-waves also were reduced in $Dhdds^{+/flx}$ CreRPE mice, which may suggest a functional change that occurs prior to any obvious retinal structural changes and suggests that 50% DHDDS activity is insufficient in the RPE to maintain the required enzymatic activity level. Thus, carriers of *Dhdds* mutations also may develop visual defects, depending on the nature of the mutation, and other factors, such as genetic background and environment. This agrees with the rod-cone dystrophy reported in patients with autosomal recessive RP [16]. However, we cannot rule out the expression of a truncated protein that leads to a gain of function. This could explain the significant attenuation of the photoresponse in mice heterozygous for the floxed allele (Figures 7 and 8). It could also possibly explain the manifestation of the disease only in ocular tissues, if the gain of function is due to specific targeting of retina-specific protein complexes. Further experimentation will be required to sort this out.

We carried out further experiments to examine the fundus of older $Dhdds^{flx/flx}$ CreRPE mice at 6, 8 and 10 months PN. These fundus images obtained with fluorescein angiography showed the classical signs of RP including abnormal pigmentation in 8-month old $Dhdds^{flx/flx}$ CreRPE mice; vascular changes, including microaneurysms, increased vessel tortuosity and attenuation, were also observed by 6 months of age (Supplementary Materials, Figure S2).

Originally it was proposed that the cause of the pathology in DHDDS-related RP patients (RP59) is defective glycosylation of rod opsin because of reduced DHDDS enzyme activity [12,26]. While this assumption may explain the classic RP symptoms observed, it fails to explain the AMD-like macular involvement. Interestingly, Lam and colleagues [11] found no N-glycosylation deficiency in RP59 patients, based upon isoelectric focusing gel analysis of plasma transferrin (a systemic glycoprotein). In addition, in a related study (Rao et al., manuscript submitted for publication), utilizing a rod photoreceptor-specific knockout of *Dhdds* in mice, we found no evidence for a resulting lack of protein glycosylation in the retina; yet, there was a rapidly progressing photoreceptor degeneration, resulting in complete loss of photoreceptors by PN 6 weeks of age. Also, in a companion article in this Special Issue [31], we generated a mouse model harboring the global K42E homozygous *Dhdds* mutation associated with RP59 patients, but observed no retinal degeneration, even out to 9 postnatal months of age. Clearly, mutations that only partially diminish enzymatic activity would be far less severe than a complete ablation of the gene encoding the enzyme.

5. Conclusions

From these observations, we conclude that targeted ablation of *Dhdds* selectively in RPE cells results in perturbation of DHDDS-dependent processes, resulting in structural and functional deficits in both the RPE and photoreceptors, in a manner resembling geographic atrophy. The degeneration progresses relatively slowly (compared to the rod-specific *Dhdds* ablation model) over the course of a few months, rather than weeks.

Supplementary Materials: The following are available online at http://www.mdpi.com/2073-4409/9/3/771/s1, Figure S1: Stitched images of $Dhdds^{flx/flx}$ and $Dhdds^{flx/flx}$ Cre RPE mouse retina, Figure S2: Fundus imaging and fluorescein angiography (FA), Figure S3: Contrast sensitivity and spatial frequency assessment in WT, $Dhdds^{flx/flx}$, and $Dhdds^{flx/flx}$ Cre RPE mice, Video S1: Serial Block-Face Scanning Electron Microscopy of WT $Dhdds^{flx/flx}$, Video S2: Serial Block-Face Scanning Electron Microscopy of region 1 of $Dhdds^{flx/flx}$ Cre RPE, Video S3: Serial Block-Face Scanning Electron Microscopy of region 2 of $Dhdds^{flx/flx}$ Cre RPE.

Author Contributions: Conceptualization, S.J.P., S.R.R., S.J.F. and M.L.D.; methodology, S.J.D., M.L.D., S.J.F., and S.J.P.; software, S.J.P., M.L.D., and T.W.K; validation, S.J.D., M.L.D., D.S., S.R.R., and S.J.P.; data curation, M.L.D., C.N., S.R.R., S.J.F. and S.J.P.; writing—original draft preparation, M.L.D., C.N., and S.J.P.; writing—review and editing, S.J.F., S.J.P., S.R.R., M.L.D., and D.S., visualization, S.J.P., S.R.R., and S.J.F.; supervision, T.W.K., S.J.F., and S.J.P.; project administration, S.J.F. and S.J.P.; funding acquisition, S.J.P. and S.J.F. All authors have read and agreed to the published version of the manuscript.

Funding: This research was supported by U.S.P.H.S. (National Institutes of Health/National Eye Institute) grant R01 EY029341 to S.J.P. and S.J.F., support from the UAB Vision Science Research Center, and a core grant P30 EY003039 to SJP, as well as facilities and resources provided by the VA Western NY Healthcare System (S.J.F., S.R.R.).

Acknowledgments: We thank Jeffrey Messinger for expert technical assistance with light level histology and TEM, Isaac Cobb for technical assistance with some of the experiments, Amy Stablewski (Roswell Park Cancer Institute (RPCI) Gene Targeting and Transgenic Facility) for the generation of *Dhdds^{flx/flx}* mice, and Yun Le (University of Oklahoma Health Sciences Center) for generously providing the CreRPE mouse line. The opinions expressed herein do not reflect those of the Department of Veteran Affairs or the U.S. Government.

Conflicts of Interest: The authors declare no conflict of interest. The funders of this study had no role in the design; in the collection, analyses, or interpretation of data; in the writing of the manuscript; or in the decision to publish the results.

References

1. Dryja, T.P.; McGee, T.L.; Reichel, E.; Hahn, L.B.; Cowley, G.S.; Yandell, D.W.; Sandberg, M.A.; Berson, E.L. A point mutation of the rhodopsin gene in one form of retinitis pigmentosa. *Nature* **1990**, *343*, 364–366. [CrossRef] [PubMed]

2. Bok, D.; Hall, M.O. The role of the pigment epithelium in the etiology of inherited retinal dystrophy in the rat. *J. Cell Biol.* **1971**, *49*, 664–682. [CrossRef] [PubMed]

3. Clark, V.M.; Hall, M.O. RPE cell surface proteins in normal and dystrophic rats. *Investig. Ophthalmol. Vis. Sci.* **1986**, *27*, 136–144.

4. Clark, V.M.; Zhou, X.Y.; Pfeffer, B.A. Partial characterization of fucosylated cell surface glycoproteins of cultured RPE. *Curr. Eye Res.* **1990**, *9*, 977–986. [CrossRef] [PubMed]

5. Edwards, R.B.; Szamier, R.B. Defective phagocytosis of isolated rod outer segments by RCS rat retinal pigment epithelium in culture. *Science* **1977**, *197*, 1001–1003. [CrossRef]

6. Buczkowska, A.; Swiezewska, E.; Lefeber, D.J. Genetic defects in dolichol metabolism. *J. Inherit. Metab. Dis.* **2015**, *38*, 157–169. [CrossRef]

7. Giladi, M.; Edri, I.; Goldenberg, M.; Newman, H.; Strulovich, R.; Khananshvili, D.; Haitin, Y.; Loewenstein, A. Purification and characterization of human dehydrodolychil diphosphate synthase (DHDDS) overexpressed in E. coli. *Protein Expr. Purif.* **2017**, *132*, 138–142. [CrossRef]

8. Kean, E.L. The dolichol pathway in the retina and its involvement in the glycosylation of rhodopsin. *Biochim. Biophys. Acta* **1999**, *1473*, 272–285. [CrossRef]

9. Nowycky, M.C.; Wu, G.; Ledeen, R.W. Glycobiology of ion transport in the nervous system. *Adv. Neurobiol.* **2014**, *9*, 321–342. [CrossRef]

10. Wimmers, S.; Karl, M.O.; Strauss, O. Ion channels in the RPE. *Prog. Retin. Eye Res.* **2007**, *26*, 263–301. [CrossRef]

11. Lam, B.L.; Zuchner, S.L.; Dallman, J.; Wen, R.; Alfonso, E.C.; Vance, J.M.; Pericak-Vance, M.A. Mutation K42E in dehydrodolichol diphosphate synthase (DHDDS) causes recessive retinitis pigmentosa. *Adv. Exp. Med. Biol.* **2014**, *801*, 165–170. [CrossRef] [PubMed]

12. Zelinger, L.; Banin, E.; Obolensky, A.; Mizrahi-Meissonnier, L.; Beryozkin, A.; Bandah-Rozenfeld, D.; Frenkel, S.; Ben-Yosef, T.; Merin, S.; Schwartz, S.B.; et al. A missense mutation in DHDDS, encoding dehydrodolichyl diphosphate synthase, is associated with autosomal-recessive retinitis pigmentosa in Ashkenazi Jews. *Am. J. Hum. Genet.* **2011**, *88*, 207–215. [CrossRef] [PubMed]

13. Venturini, G.; Koskiniemi-Kuendig, H.; Harper, S.; Berson, E.L.; Rivolta, C. Two specific mutations are prevalent causes of recessive retinitis pigmentosa in North American patients of Jewish ancestry. *Genet. Med.* **2015**, *17*, 285–290. [CrossRef] [PubMed]

14. Ng, B.G.; Freeze, H.H. Perspectives on Glycosylation and Its Congenital Disorders. *Trends Genet.* **2018**, *34*, 466–476. [CrossRef]

15. Biswas, P.; Duncan, J.L.; Maranhao, B.; Kozak, I.; Branham, K.; Gabriel, L.; Lin, J.H.; Barteselli, G.; Navani, M.; Suk, J.J.; et al. Genetic analysis of ten pedigrees with inherited retinal degeneration (IRD) by exome sequencing and phenotype-genotype association. *Physiol. Genom.* **2017**, *49*, 216–229. [CrossRef]

16. Kimchi, A.; Khateb, S.; Wen, R.; Guan, Z.; Obolensky, A.; Beryozkin, A.; Kurtzman, S.; Blumenfeld, A.; Pras, E.; Jacobson, S.G.; et al. Nonsyndromic Retinitis Pigmentosa in the Ashkenazi Jewish Population: Genetic and Clinical Aspects. *Ophthalmology* **2018**, *125*, 725–734. [CrossRef]

17. Sabry, S.; Vuillaumier-Barrot, S.; Mintet, E.; Fasseu, M.; Valayannopoulos, V.; Heron, D.; Dorison, N.; Mignot, C.; Seta, N.; Chantret, I.; et al. A case of fatal Type I congenital disorders of glycosylation (CDG I) associated with low dehydrodolichol diphosphate synthase (DHDDS) activity. *Orphanet J. Rare Dis.* **2016**, *11*, 84. [CrossRef]

18. Ferrara, N.; Carver-Moore, K.; Chen, H.; Dowd, M.; Lu, L.; O'Shea, K.S.; Powell-Braxton, L.; Hillan, K.J.; Moore, M.W. Heterozygous embryonic lethality induced by targeted inactivation of the VEGF gene. *Nature* **1996**, *380*, 439–442. [CrossRef]

19. Rucker, E.B., 3rd; Dierisseau, P.; Wagner, K.U.; Garrett, L.; Wynshaw-Boris, A.; Flaws, J.A.; Hennighausen, L. Bcl-x and Bax regulate mouse primordial germ cell survival and apoptosis during embryogenesis. *Mol. Endocrinol.* **2000**, *14*, 1038–1052. [CrossRef]

20. Le, Y.Z.; Zheng, W.; Rao, P.C.; Zheng, L.; Anderson, R.E.; Esumi, N.; Zack, D.J.; Zhu, M. Inducible expression of cre recombinase in the retinal pigmented epithelium. *Investig. Ophthalmol. Vis. Sci.* **2008**, *49*, 1248–1253. [CrossRef] [PubMed]

21. Le, Y.Z.; Zhu, M.; Anderson, R.E. Cre Recombinase: You Can't Live with It, and You Can't Live Without It. *Adv. Exp. Med. Biol.* **2016**, *854*, 725–730. [CrossRef] [PubMed]

22. DeRamus, M.L.; Stacks, D.A.; Zhang, Y.; Huisingh, C.E.; McGwin, G.; Pittler, S.J. GARP2 accelerates retinal degeneration in rod cGMP-gated cation channel beta-subunit knockout mice. *Sci. Rep.* **2017**, *7*, 42545. [CrossRef] [PubMed]

23. Stricker, H.M.; Ding, X.Q.; Quiambao, A.; Fliesler, S.J.; Naash, M.I. The Cys214–>Ser mutation in peripherin/rds causes a loss-of-function phenotype in transgenic mice. *Biochem. J.* **2005**, *388*, 605–613. [CrossRef] [PubMed]

24. Tu, C.; Li, J.; Jiang, X.; Sheflin, L.G.; Pfeffer, B.A.; Behringer, M.; Fliesler, S.J.; Qu, J. Ion-current-based proteomic profiling of the retina in a rat model of Smith-Lemli-Opitz syndrome. *Mol. Cell. Proteom.* **2013**, *12*, 3583–3598. [CrossRef] [PubMed]

25. Zanzottera, E.C.; Ach, T.; Huisingh, C.; Messinger, J.D.; Spaide, R.F.; Curcio, C.A. Visualizing retinal pigment epithelium phenotypes in the transition to geographic atrophy in age-related macular degeneration. *Retina* **2016**, *36*, S12–S25. [CrossRef]

26. Zuchner, S.; Dallman, J.; Wen, R.; Beecham, G.; Naj, A.; Farooq, A.; Kohli, M.A.; Whitehead, P.L.; Hulme, W.; Konidari, I.; et al. Whole-exome sequencing links a variant in DHDDS to retinitis pigmentosa. *Am. J. Hum. Genet.* **2011**, *88*, 201–206. [CrossRef]

27. Grabińska, K.A.; Park, E.J.; Sessa, W.C. cis-Prenyltransferase: New insights into protein glycosylation, rubber synthesis, and human diseases. *J. Biol. Chem.* **2016**, *291*, 18582–18590. [CrossRef]

28. Park, E.J.; Grabinska, K.A.; Guan, Z.; Stranecky, V.; Hartmannova, H.; Hodanova, K.; Baresova, V.; Sovova, J.; Jozsef, L.; Ondruskova, N.; et al. Mutation of Nogo-B receptor, a subunit of cis-prenyltransferase, causes a congenital disorder of glycosylation. *Cell Metab.* **2014**, *20*, 448–457. [CrossRef]

29. Murray, A.R.; Fliesler, S.J.; Al-Ubaidi, M.R. Rhodopsin: The functional significance of asn-linked glycosylation and other post-translational modifications. *Ophthalmic Genet.* **2009**, *30*, 109–120. [CrossRef]

30. White, D.A.; Fritz, J.J.; Hauswirth, W.W.; Kaushal, S.; Lewin, A.S. Increased sensitivity to light-induced damage in a mouse model of autosomal dominant retinal disease. *Investig. Ophthalmol. Vis. Sci.* **2007**, *48*, 1942–1951. [CrossRef]

31. Ramachanra Rao, S.; Fliesler, S.J.; Nguyen, M.N.; Kotla, P.; Pittler, S.J. Lack of overt retinal degeneration in K42E DHDDS knock-in mouse model of RP59. *Cells* **2020**. submitted.

Article

Diurnal Rhythmicity of Autophagy Is Impaired in the Diabetic Retina

Xiaoping Qi [1], Sayak K. Mitter [1], Yuanqing Yan [2], Julia V. Busik [3], Maria B. Grant [1] and Michael E. Boulton [1,*]

[1] Department of Ophthalmology and Visual Sciences, University of Alabama at Birmingham, AL 35294, USA; xqi@uabmc.edu (X.Q.); sayakmitter@gmail.com (S.K.M.); mariagrant@uabmc.edu (M.B.G.)
[2] Department of Neurosurgery, The University of Texas Health Science Center, Houston, TX 77030, USA; Yuanqing.Yan@uth.tmc.edu
[3] Department of Physiology, Michigan State University, East Lansing, MI 48824, USA; busik@msu.edu
* Correspondence: meboulton@uabmc.edu

Received: 1 March 2020; Accepted: 2 April 2020; Published: 7 April 2020

Abstract: Retinal homeostasis is under both diurnal and circadian regulation. We sought to investigate the diurnal expression of autophagy proteins in normal rodent retina and to determine if this is impaired in diabetic retinopathy. C57BL/6J mice and Bio-Breeding Zucker (BBZ) rats were maintained under a 12h/12h light/dark cycle and eyes, enucleated over a 24 h period. Eyes were also collected from diabetic mice with two or nine-months duration of type 1 diabetes (T1D) and Bio-Breeding Zucker diabetic rat (BBZDR/wor rats with 4-months duration of type 2 diabetes (T2D). Immunohistochemistry was performed for the autophagy proteins Atg7, Atg9, LC3 and Beclin1. These autophagy proteins (Atgs) were abundantly expressed in neural retina and endothelial cells in both mice and rats. A differential staining pattern was observed across the retinas which demonstrated a distinctive diurnal rhythmicity. All Atgs showed localization to retinal blood vessels with Atg7 being the most highly expressed. Analysis of the immunostaining demonstrated distinctive diurnal rhythmicity, of which Atg9 and LC3 shared a biphasic expression cycle with the highest level at 8:15 am and 8:15 pm. In contrast, Beclin1 revealed a 24-h cycle with the highest level observed at midnight. Atg7 was also on a 24-h cycle with peak expression at 8:15 am, coinciding with the first peak expression of Atg9 and LC3. In diabetic animals, there was a dramatic reduction in all four Atgs and the distinctive diurnal rhythmicity of these autophagy proteins was significantly impaired and phase shifted in both T1D and T2D animals. Restoration of diurnal rhythmicity and facilitation of autophagy protein expression may provide new treatment strategies for diabetic retinopathy.

Keywords: diurnal rhythm; autophagy; retina; diabetes; diabetic retinopathy

---

1. Introduction

Macroautophagy (hereafter referred to as autophagy) is an evolutionarily conserved cellular catabolic mechanism that facilitates the degradation of damaged cellular organelles and proteins, by targeting them to the lysosomes and recycling the macromolecules for the rebuilding of cellular machinery [? ? ?]. Autophagy undergoes rhythmic variation in accordance with circadian patterns of rest/activity and feeding in adult mammals [?]. Dysregulated autophagy has been implicated in several neurodegenerative disorders, hepatitis, cancer, aging associated diseases and in the general aging process [? ? ? ? ?]. Recently, a growing body of evidence indicates that dysregulated autophagy is also linked to diabetes [? ? ? ?].

The disruption of circadian rhythm has a profound negative impact on health and is associated with elevated risk for several diseases [?]. The physiological relevance of an altered circadian rhythm in diabetes is evidenced by the observation of a high incidence of myocardial dysfunction, acute coronary syndrome, sudden cardiac death, and ischemic stroke in diabetics during the night, compared to a higher frequency during the day in non-diabetics [? ? ? ?]. In diabetic conditions, Bmal1 and Clock are inactivated, causing deregulated glucose homeostasis and suppressed diurnal variation in glucose and triglycerides, along with reduced gluconeogenesis [?]. Streptozotocin (STZ)-induced type 1 diabetes (T1D) in mice exhibits altered phase of the circadian clock in the heart [?] and a significant reduction of circadian sensitivity to low-intensity light in the retina [?]. Furthermore, STZ-mice develop a deficiency in their ability to re-entrain the circadian rhythm when subjected to a phase advance of the 12L/12D cycle [?]. Bio-Breeding Zucker diabetic rat (BBZDR/Wor) and Goto-Kakizaki rat type 2 diabetes (T2D) models also show impairment of the molecular clock, suggesting that the disruption of the circadian clock is a common phenomenon in both T1D and T2D [? ?].

Several hallmarks of diabetic retinopathy can be recapitulated in rodent retina deficient of clock genes [? ?]. Reduced tube formation and increased senescence of endothelial cells coupled with impaired progenitor-mediated repair is observed in Per2 mutant mice, emphasizing the importance of the circadian clock in retinal homeostasis [?]. Recent studies on autophagy in the retina have shed light on the association of key molecules of the autophagic pathway with phagocytosis of photoreceptor outer segments (POS) by the retinal pigmented epithelium (RPE) [? ?]. Photoreceptor disk shedding has been widely reported to exhibit diurnal rhythmicity in the retina [? ?] and the tight coupling of phagocytic ingestion and autophagic degradation of the POS to this diurnal rhythm is a critical aspect of retinal homeostasis [? ?].

The role of the peripheral clock and diurnal variation on the regulation of autophagy in the normal and diabetic rodent retinas and the fate of autophagy in the diabetic retina remain unexplored. Understanding these control mechanisms may help find effective treatments for diabetic retinopathy. In this study, we demonstrate that the spatial distribution and temporal expression of autophagy proteins show a diurnal rhythm and that this is depressed and phase shifted in the diabetic retina.

2. Materials and Methods

2.1. Experimental Animals

All animal procedures were performed in accordance with a) protocols approved by the Institutional Animal Care and Use Committees at University of Florida, Gainesville, FL, USA (#CR-201106001), Michigan State University, Lansing, MI, USA (#Busik09/14-160-00, and Indiana University, Indianapolis, IN, USA (#10574 MD/R), b, the National Institutes of Health Guide for Care and Use of Laboratory Animals and c) the ARVO Statement for the Use of Animals in Ophthalmic and Vision Research. Male C57BL/6J mice were purchased from Jackson Labs at 6 weeks of age and male BBZDR/wor T2D rats and lean heterozygote nondiabetic control littermates were obtained from biomedical research models (BRM, Inc.) Worcester, Massachusetts at 5 months of age. C57BL/6J mice (n = 240) and BBZ rats (n = 75) were maintained on a standard 12/12 h light/dark cycle (6.00 am lights on/6.00 pm lights off). T1D was induced in eight-week-old C57BL/6 mice (n = 120) by five consecutive intraperitoneal injection of freshly prepared streptozotocin (STZ) solution with the concentration of 40mg/kg body weight in 0.1M/citrate buffer, pH 4.5 as previously described [? ?]. The control group received sodium citrate buffer (vehicle) alone. Diabetes was confirmed by two consecutive blood glucose levels >240 mg/dl, using the AlphaTrak® blood glucose monitor and test strip system (Abbot laboratories, Irvine, CA, USA), according to the manufacturer's instruction. No animal used in this study required insulin injections. Eyes from one group were collected at 2 months following establishment of diabetes and the second cohort 9 months after diabetes induction (the age of the animals at these time points being 4 and 11 months, respectively). We also investigated inbred Bio-Breeding Zucker (BBZDR)/wor T2D rats (n = 37) and age-matched controls (n = 38) [?]. BBZDR/wor rats spontaneously became diabetic at ~10 weeks of age and were maintained for 4 months after the onset of diabetes. A second group of C57BL/6J mice (n = 24) were dark adapted for 48 h before sacrifice and referred to as being on the dark/dark cycle. In all groups, animals were euthanized by deep anesthesia with isoflurane followed by cervical dislocation and eyes were immediately enucleated every 2 or 3 h over a 24 h period and prepared for immunofluorescence staining. For animals in the dark cycle, euthanasia and enucleation were performed in a dark room under a red safe light. For each set of experiments, all animals were started at the same time point to clearly identify any phase shift in peak or trough of ATG expression. The number of biological replicates of the enucleated eyes at each time point is n = 10 for mice and n = 3 for rats. Only one eye from each animal was used in the assessment.

2.2. Immunofluorescence Microscopy

Eyes were processed for standard paraffin embedding and 4 μm sections were prepared. Rodent Decloaker (Biocare Medical LLC, Concord, CA, USA Catalog# RD913L) was used to unmask antigens and non-specific binding was blocked with 10% normal goat sera and 5% BSA for 1 h at room temperature. Sections received either mouse monoclonal anti-Beclin1 (BD Transduction Lab, San Jose, CA, USA, Cat#612112), or rabbit polyclonal antibodies against Atg7, Atg9 and LC3 (provided by Dr. Dunn, Department of Anatomy and Cell Biology, University of Florida, Gainesville, FL, USA), diluted in phosphate buffered saline (PBS) with 1% normal goat sera plus 1% bovine serum albumin (BSA) (Beclin1 1:20; Atg7- 1:300; Atg9 and LC3 - 1:100). After washing, sections were incubated with an appropriate FITC-conjugated secondary antibody for 1 h. In some sections, colocalization of autophagy proteins within the retinal vasculature was confirmed by dual staining with the endothelial cell marker, TRITC-agglutinin (Vector Labs, Burlingame, CA, USA, Cat#RCA120). Sections were covered with Vectashield mounting medium/DAPI (Vector Labs, Inc.). Sections were viewed using a Zeiss Axioplan 2 Upright Fluorescence Microscope with Qimage/QCapture software Version 8 (QImaging Corporate, Surrey, British Columbia, Canada). Omission of the primary antibody was the baseline control and all the fluorescence photographs were obtained under the same scaled conditions.

2.3. Grading of Immunostaining

Three independent masked observers, using a previously described grading system [? ?], graded the intensity of the immunoreactivity for each antibody in transverse retinal sections. The intensity of labeling was graded qualitatively as: 10, strong bright intense immunoreactivity, 9, bright intense; 8, uniformly intense; 7, patchy and intense; 6, uniform and moderate; 5, patchy and moderate; 4, uniform and weak; 3, patchy and weak; 2, uniform and very weak; 1, patchy and very weak; and 0, none. A mean score ±SEM for each group was determined from the scores of all graders for each retina structure.

2.4. Trypsin Digestion and the Detection of Superoxide

The retinal vasculature was prepared as previously described [?]. Briefly, mouse eyes were fixed overnight in 4% paraformaldehyde, freshly made in PBS. The retinas were dissected from the eye cups, kept in water overnight, and digested in 3% trypsin (Invitrogen-Gibco, Grand Island, NY, USA) for 3 h at 37 °Celcius. The tissue was mounted carefully on a glass slide under a dissection microscope. The tissue was stained with PAS-H&E (periodic acid Schiff–hematoxylin and eosin; Gill No.3; Sigma-Aldrich), according to the instruction manual. The images were taken using a Zeiss light microscope equipped with a digital camera (AxioCam MRC5, Axiovert 200; Carl Zeiss Meditec, Inc., Dublin, CA, USA), using 20X objective lenses. Eight to ten representative fields from each quadrant of the retina were imaged and the number of acellular capillaries per square millimeter of retina were quantified.

For detecting the reactive oxygen species (ROS), the superoxide indicator, hydroethidine, was used to detect the production of superoxide radicals, as previously described [?]. Superoxide oxidizes hydroethidine to yield a red fluorescent signal at approximately 600 nm. Mice received two intraperitoneal injections, 15 min apart, of freshly prepared hydroethidine (20 mg/kg; Invitrogen) and were euthanized 18 h after injection. The fluorescence intensity was measured in the neural retina using a fluorescent plate reader (BioTek, Winooski, VT, USA) and a spectrofluorometer (FLUOstar Optima; BMG Labtechnologies, Cary, NC, USA). The relative fluorescence intensity was calculated by normalizing to protein concentration.

2.5. Statistical Analysis

The diurnal data are time series data, and single cosine analysis was employed to fit the data. The data was considered as diurnal oscillation by a zero-amplitude test, with a p-value less than 0.05. All experiments were assessed by comparing two groups mean scores from control animals and diabetic animals using a Student's t-test, plus ANOVA for multiple comparisons by using Prism® statistical software ver. 5 (GraphPad, La Jolla, CA, USA) with differences of p value < 0.05 were considered statistically significant. Results are expressed as mean $\pm$ SEM.

3. Results

3.1. Autophagy Proteins Exhibit Diurnal Expression/Localization in Normal Mouse Neural Retina

To determine if autophagic protein expression in the neural retina had an intrinsic diurnal rhythm, we assessed the expression and localization of four autophagy proteins, Atg7, LC3, Atg9 and Beclin1 in the normal retinas of young mice (age = 3 months) that were maintained under standard 12 h light/12 h dark conditions. When we examined the tissue at two-hour intervals over a 24 h period, these proteins exhibited not only a distinct diurnal rhythm, but remarkably showed distinct staining patterns across the layers of the neural retina in mice (????). Atg9 and LC3 exhibited a biphasic 12/12 h circadian cycle with zenith at 8:15 AM and 8:15 PM and nadir at 2:15 AM and 2:15 PM ($p < 0.05$) (Figure ??A–D). By contrast, Atg7 and Beclin1 expression showed a monophasic rhythm and the overall expression of these proteins was much lower than either Atg9 or LC3 (Figure ??E–H). The expression of Atg7 started to rise at 4:15 AM, peaked at 8:15 AM and gradually decreased until 10:15 AM ($p < 0.05$) (Figure ??E–G). Beclin1 expression was highest at around midnight and reached lowest levels at midday ($p < 0.05$) (Figure ??F–H). Omission of the primary antibody showed no staining in mouse retina (Figure ??D).

An assessment of the staining pattern of the proteins at zenith revealed that Atg9 and LC3 were localized throughout the retina, predominantly within the retinal ganglion cell (RGC) layer, inner (INL) and outer nuclear layer (ONL) (Figure ??A,B). Atg7 was only weakly localized to the neural retina in both the inner and outer nuclear layers and showed stronger localization in the photoreceptor outer segment. Beclin1 staining was detected throughout the retina, with highest intensity in the inner and outer plexiform layers and in the photoreceptors. Closer observation of the retina revealed that ATG7 and 9 showed remarkably strong association within the inner retinal vasculature (Figure ??C).

Using dual staining for endothelial cells and autophagy proteins, we confirmed that Atg9, LC3, Atg7 and Beclin1 were all expressed in both the inner and outer retinal plexus of the retinal vasculature (Figure ??C).

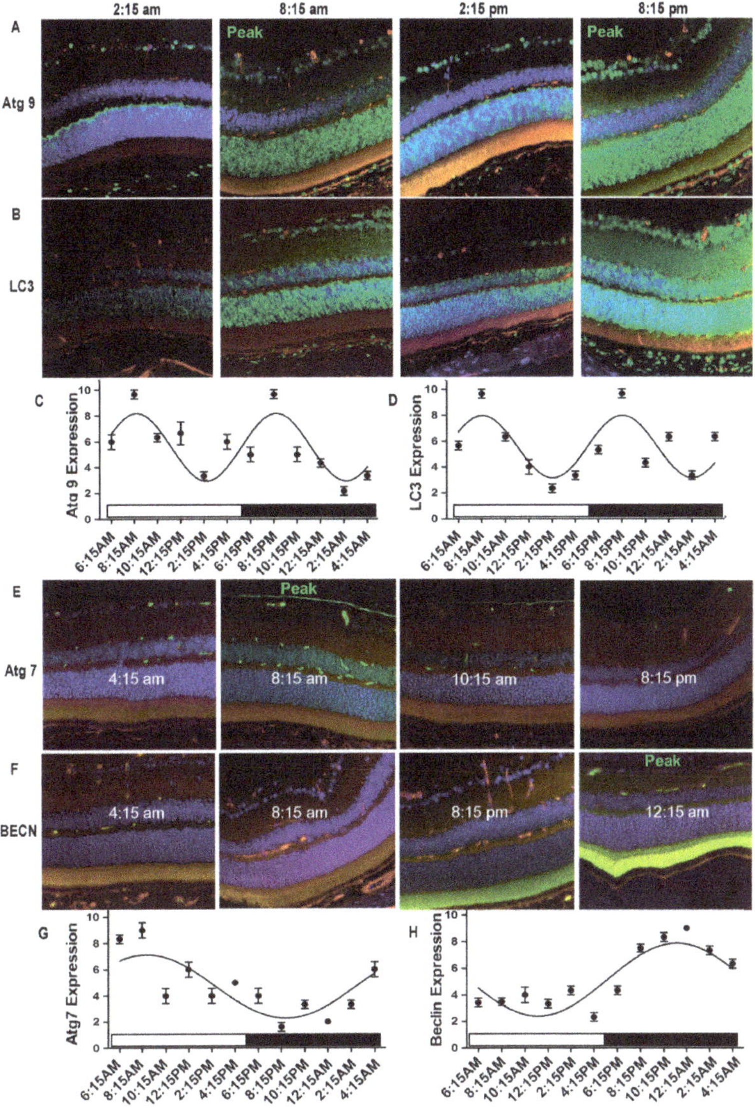

Figure 1. Autophagy proteins exhibited diurnal expression/localization in normal mouse neural retina. Retina were harvested every 2 h over a 24 h time period. Antibodies for autophagy proteins (ATGs) ATG9, Microtubule-associated protein 1A/1B-light chain 3 (LC3), ATG7 and Beclin1 (BECN) were used to detect respective protein expression (green) in the retina, Tetramethylrhodamine (TRITC) agglutinin for vessels (red) and DAPI (blue) shows nuclear staining. Distinct diurnal rhythmic patterns of the autophagic proteins Atg9 and LC3 expressions (**A**,**B**) revealed a biphasic diurnal cycle (12/12 h), with peaks at 8:15 AM and 8:15 PM and lowest levels at 2:15 AM and 2:15 PM (**C**,**D**). Atg7 and Beclin1 (BECN) expressions (**E**,**F**) were on a monophasic 24 h cycle. Atg7 expression peaked at 8:15 AM and lowest levels were at 10:15 AM. The peak of Beclin1 expression was at midnight and lowest levels were observed during the late morning (**H**). All animals were maintained in a standard 12/12 h light/dark phase with lights ON at 6:00 AM and lights OFF at 6:00 PM. The error bars in the diurnal plots represent the mean+SEM and diurnal oscillation had a *p*-value less than 0.05.

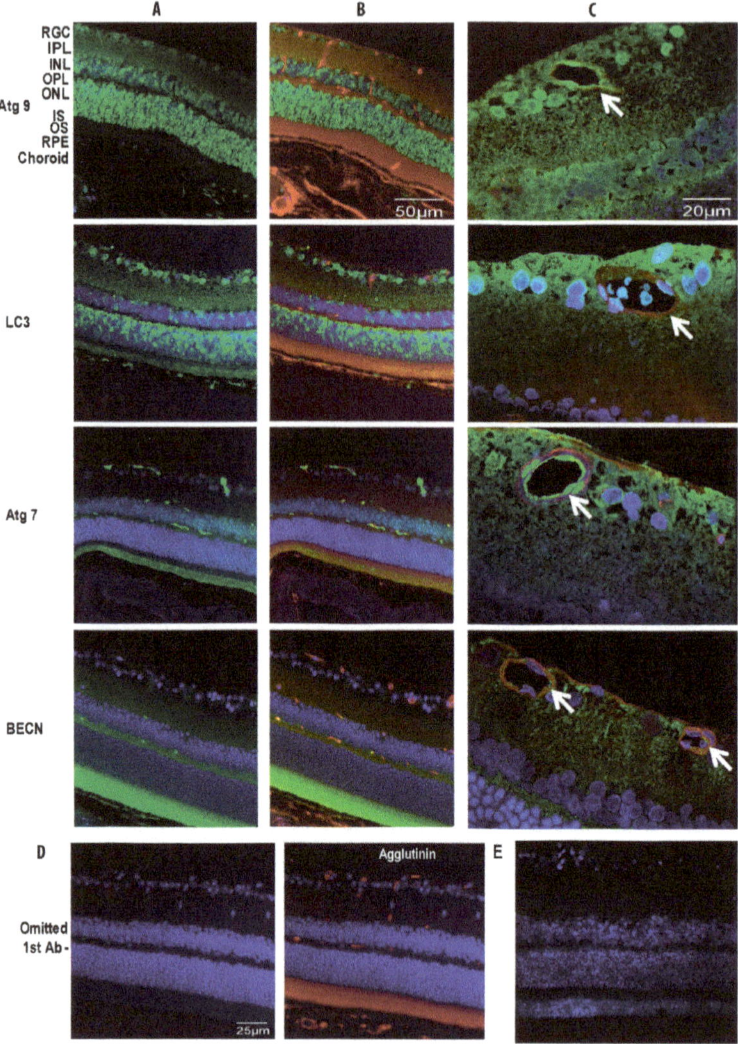

Figure 2. Localization of autophagy proteins in normal mouse retina and vasculature. Animals were kept in tight 12/12-h light/dark cycle before the experiment. Antibodies for ATG9, LC3, ATG7 and Beclin1 (BECN) were used to detect respective protein expression (green) in the retinas with different staining patterns and DAPI was used to stain nuclei (blue) (**A**). Agglutinin, an endothelial cell marker, conjugated with TRITC (red) was used to co-localize with antibodies specific to individual autophagy proteins (FITC) in the retinal vasculature (**B**). High magnification images demonstrated autophagy protein localization to the endothelial cells and pericytes of the retinal vasculature (**C**). No fluorescence was observed with omission of the primary antibody in the mouse retina. Using agglutinin staining, the section displayed retinal vascular patterns with normal architecture (**D**). Omitted autophagy primary antibody in rat retina also was negative (**E**). The arrows indicate autophagy protein localization to retinal vessels. RGC, retina ganglion cell; IPL, inner plexiform layer; INL, inner nuclear layer; OPL, outer plexiform layer; ONL, outer nuclear layer; IS, inner segment; OS, outer segment; RPE, retinal pigment epithelium.

3.2. Autophagic Activity in the Retina Is Tightly Entrained by Light

We repeated immunohistochemistry of ATG9, ATG7, LC3 and Beclin1 in the retina (and the retinal vasculature) of 48 h dark adapted mice and compared the results to those of 12/12 h light/dark adapted mice. The immunohistochemistry data indicated that the periodic oscillations, as well as overall levels of the autophagic proteins ATG9 and LC3, were attenuated in the 48 h dark adapted mice retina when compared to the 12/12 h dark/dark controls, both in the retina and the vasculature (Figure **??**A). Similarly, peaks in Atg7 and Beclin1 were phase shifted compared to the light/dark controls and overall expression levels were reduced (Figure **??**B). We concluded that diurnal oscillations in autophagic activity in the murine retina depend, at least in part, on light-effected entrainment.

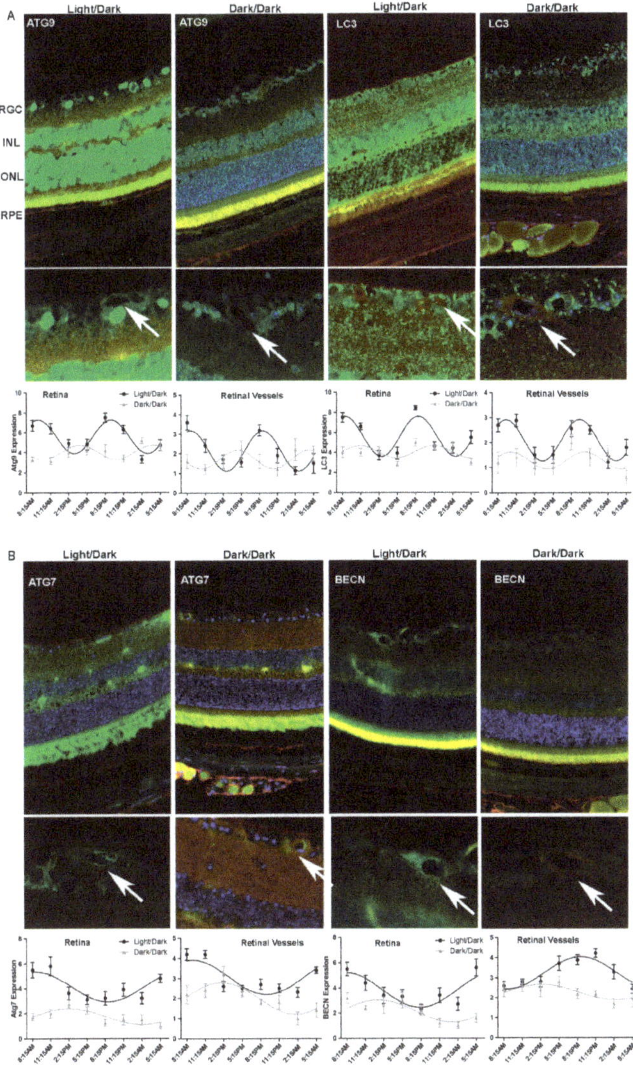

Figure 3. Immunohistochemistry confirmed alterations in diurnal patterns of autophagic gene expression upon light cycle disruption.

Mice were separated into two groups. One group was kept under 12/12-h light/dark cycle and the other group was kept in the dark for 48-h (dark/dark), before being euthanized. Samples were collected every 3 h for 24 h and were analyzed by immunohistochemistry for autophagic markers (**A**) ATG9 (**B**) LC3 (**C**) ATG7 and (**D**) Beclin1 (BECN). Autophagy protein expression is green, TRITC agglutinin for vessels red and DAPI nuclear staining blue. The arrows indicate autophagy protein localization to retinal vessels. RGC, retina ganglion cell; INL, inner nuclear layer; OPL, outer plexiform layer; ONL, outer nuclear layer; RPE, retinal pigment epithelium. The error bars in the circadian plots represent the mean + SEM and diurnal oscillation had a *p*-value less than 0.05.

3.3. Diurnal Rhythmicity of Expression of Autophagic Proteins Is Dampened in T1D

Having established that autophagy in the retina possesses a diurnal rhythm, we next examined the rhythmicity of autophagy proteins in T1D. We chose mice from 2- and 9-months duration of diabetes to assess the expression of autophagic proteins. To confirm that our T1D mice exhibited the expected retinopathy features [? ?], we performed trypsin-digests of the retina from nondiabetic C57BL/6J (Figure ??A, left) and T1D (Figure ??A, right). T1D mice exhibited an increase in the number of acellular capillaries per unit area. Quantitative measurements of acellular capillaries suggested a dramatic 4-fold increase, in contrast to the age-matched nondiabetic mouse of retinas that showed a normal vascular pattern (Figure ??B). These eyes showed significantly higher levels of superoxide anions in the neural retina, typical of retinopathy in this model (Figure ??C).

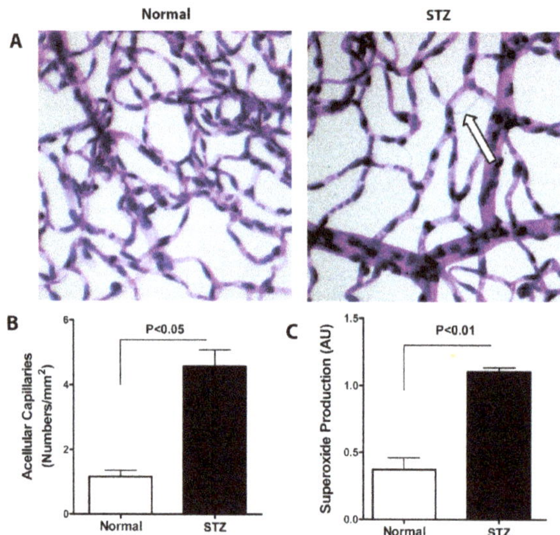

Figure 4. Evaluation of acellular capillary formation and superoxide anion generation in T1D mice. Representative images of trypsin-digested retinal vascular preparations from 4-month old control mice and 4-month old mice with 2-month duration of T1D (**A**) The arrow indicates an acellular capillary. Quantitative measurement of acellular capillaries was significantly increased in diabetic eyes (n = 6). (**B**) STZ-induced diabetic retina showed >1.5 fold increase in superoxide anion levels, *p* < 0.01 (**C**) The errors bars represent SEM.

While immunohistochemistry results from 4-month old normal control mice corresponded with the respective results shown in **????**, we observed a dramatic attenuation of diurnal rhythmicity (amplitudes), as well as overall levels of the autophagic proteins in the T1D mice retina (Figure ??). This loss of amplitude was noted for ATG9, LC3 and Beclin1 expression in the diabetic retina, in either duration of diabetes group when compared to the respective age-matched controls (**????**). ATG7 exhibited a shift in phase in both the retina and the vasculature of mice with 2-month duration of diabetes while in mice with 9-months duration of diabetes the phase-shift was prominent in the retina and there was no significant variation in expression between time-points in the retinal vasculature (Figures ??B and ??B).

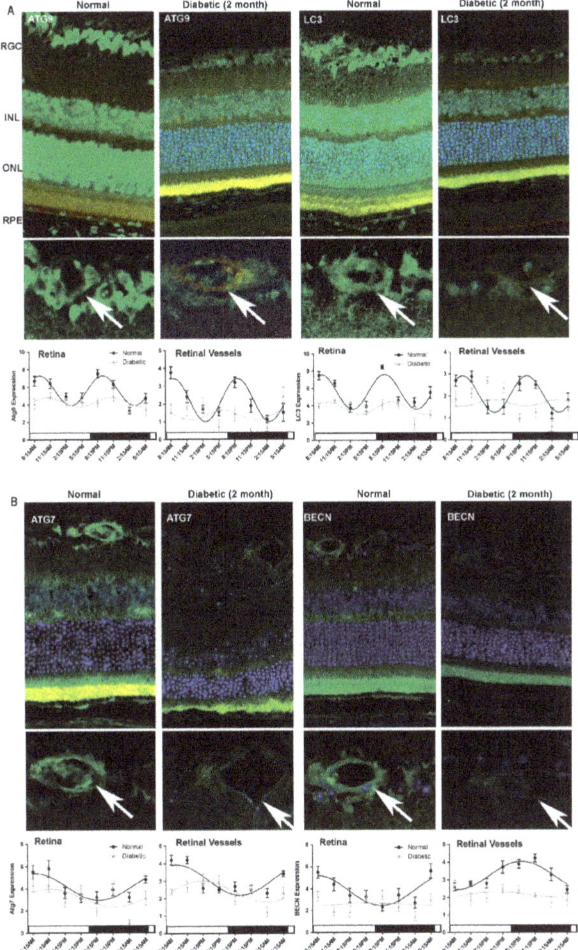

Figure 5. Impairment of diurnal rhythmicity of autophagy in T1D mice with two-months duration of diabetes. Retinas were collected from C57Bl/6J mice with two-months duration of T1D and age-matched control mice. Immunostaining and intensity analyses of retina and retinal vasculature demonstrated a dramatic loss of oscillatory amplitude of autophagic protein expression in the diabetic animals compared to normal mice (**A**,**B**). Autophagy protein expression is green, TRITC agglutinin for vessels red and DAPI nuclear staining blue. The arrows indicate autophagy protein localization to retinal vessels. RGC, retina ganglion cell; INL, inner nuclear layer; OPL, outer plexiform layer; ONL, outer nuclear layer; RPE, retinal pigment epithelium. Loss and phase-shifting of diurnal rhythmicity in diabetic retinopathy is demonstrated in single cosine plots, with ordinary least square fitted ($p < 0.01$, n = 10). All animals were maintained in a standard 12/12-h light/dark phase with lights ON at 6:00 AM and lights OFF at 6:00 PM. The error bars in the circadian plots represent the mean+SEM and diurnal oscillation had a *p*-value less than 0.05.

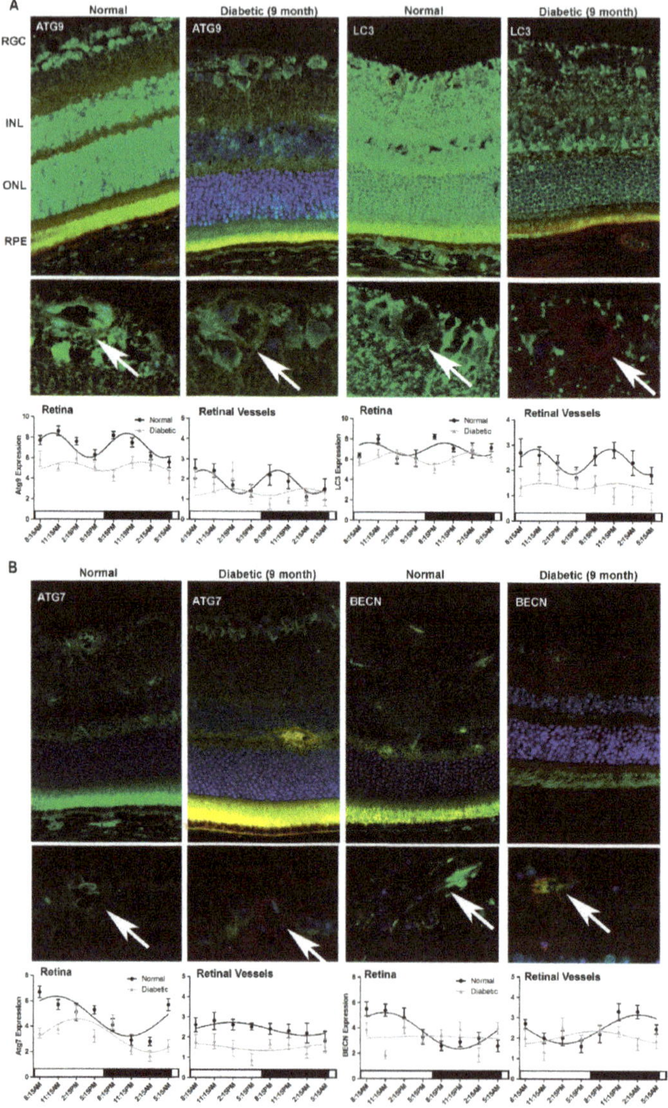

Figure 6. Impairment of diurnal rhythmicity of autophagy in T1D mice with nine-months duration of diabetes. Retinas were collected from C57Bl/6J mice with nine-months duration of T1D and age-matched control mice. Immunostaining and intensity analyses of retina and retinal vasculature demonstrated a dramatic loss of oscillatory amplitude of autophagic protein expression in the diabetic animals compared to normal mice (**A**.**B**). Autophagy protein expression is green, TRITC agglutinin for vessels red and DAPI nuclear staining blue. The arrows indicate autophagy protein localization to retinal vessels. RGC, retina ganglion cell; INL, inner nuclear layer; OPL, outer plexiform layer; ONL, outer nuclear layer; RPE, retinal pigment epithelium. Loss and phase-shifting of diurnal rhythmicity in diabetic retinopathy is demonstrated in single cosine plot with ordinary least square fitted ($p < 0.01$, n = 10). All animals were maintained in a standard 12/12-h light/dark phase with lights ON at 6:00 AM and lights OFF at 6:00 PM. The error bars in the circadian plots represent the mean+SEM and diurnal oscillation had a *p*-value less than 0.05.

3.4. Autophagy Proteins Were Suppressed Severely in T2D Rats

We next examined autophagy in the retina of T2D rats following 4-months duration of diabetes (Figure ??). The normal age-matched rats selected as controls demonstrated a similar staining pattern of immunohistochemistry across the retina as were observed in normal mice (Figure ??A–D). Atg9 (Figure ??A) and LC3 (Figure ??B) were mainly present in the ganglion cell layer, inner and outer nuclear layers in the normal rat retina, similar to what was observed in normal mice (Figure ??) but their expression was dramatically decreased in the retinas of the diabetic animals by 49% to 58%, respectively. For better demonstration of changes in autophagic protein levels in the vasculature of diabetic retinas, we showed higher magnification images. Both Atg9 and LC3 clearly distributed within the retinal endothelia of normal rats but were completely absent from the endothelia of the diabetic retinas (Figure ??E,F). Atg7 was strongly expressed in normal retinal vessels near the ganglion cells layer, meanwhile the small vessels in inner and outer plexus layers were also stained (Figure ??C). Semi-quantitative analysis revealed that the level of Atg7 was deceased by 52%, $p < 0.05$ in diabetic retinas. Diabetic retina also displayed pathologic changes in retina vessels, which appeared to be abnormally protruded toward the intravitreous cavity. Beclin1 staining was detected across the retina, including retinal vessels in the plexus and photoreceptors, however, Beclin1 staining was diminished in diabetic retinas by 53%, $p < 0.01$ (Figure ??D). Quantitative analyses of autophagic proteins in the retina of control and diabetic mice demonstrated, not only suppressed levels, but also revealed impairment in diurnal rhythmicity in T2D rats. Expression of Atg9 and LC3 were severely suppressed with insignificant biphasic oscillatory pattern and ATG7 and Beclin1 were phase-shifted by approximately for 4–6 h. We concluded that disruption in autophagy is a characteristic phenomenon of both T1D and T2D.

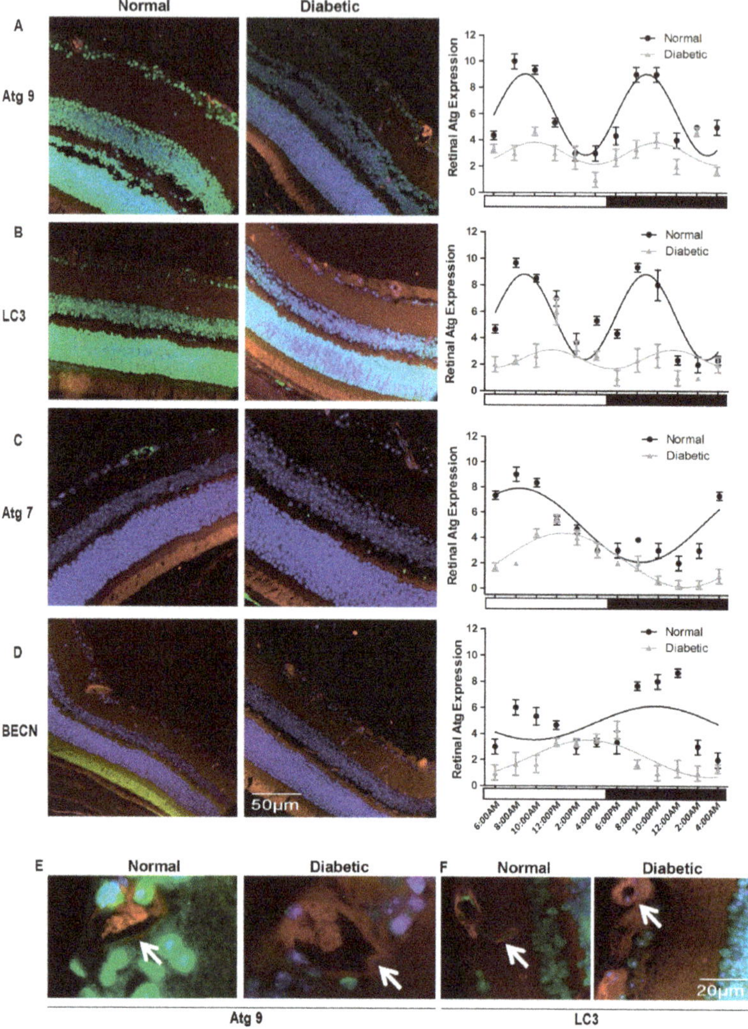

Figure 7. Impairment of diurnal rhythmicity of autophagy in T2D rats. Eyes were collected from 6.5-month old BBZDR/wor type 2 diabetic rats with 4-month duration of diabetes and the normal age-matched rats selected as controls. Immunostaining demonstrated a dramatic decrease in amplitude of diurnal rhythmicity of autophagic protein expression in the retina of the diabetic animals compared to normal rats. Representative micrographs of immunostained sections at 8:15 am are shown for (**A**) ATG9, (**B**) LC3 (**C**) ATG7 (**D**) Beclin1 (BECN). Autophagy protein expression is green, TRITC agglutinin for vessels red and DAPI nuclear staining blue. The arrows indicate autophagy protein localization to retinal vessels. RGC, retina ganglion cell; INL, inner nuclear layer; OPL, outer plexiform layer; ONL, outer nuclear layer; RPE, retinal pigment epithelium. ATG7 and Beclin1 displayed phase-shifting in the diurnal oscillation (**C**) and (**D**) respectively. Loss and phase-shifting of diurnal rhythmicity in diabetic retinopathy is demonstrated in single cosine plot with ordinary least square fitted ($p < 0.01$, n = 10). High magnification images confirmed the loss of autophagy proteins ATG9 and LC3 in the retinal vasculature (**E**). All animals were maintained in a standard 12/12-h light/dark phase with lights ON at 6:00 AM and lights OFF at 6:00 PM. The error bars in the circadian plots represent the mean + SEM and diurnal oscillation had a *p*-value less than 0.05.

4. Discussion

Autophagy is a critical and indispensable housekeeping process in the cells of the retina. Numerous reports support the functional relevance of autophagy in specific cells of the retina and how dysregulated autophagy contributes to retinal malfunction and degeneration [? ? ? ? ? ?]. While acute short-term noxious stimuli (e.g., nutrient deprivation; hypoxia and endoplasmic reticulum (ER) stress; oxidative stress arising from light, lipofuscin, POS phagocytosis by RPE or mitochondrial ROS generated in retinal cells (which have a generally high metabolic rate)) can stimulate autophagy as a generally cytoprotective mechanism, it must be acknowledged that diurnal basal autophagic modulation is a critical factor in abrogating the unavoidable regular cellular damage that occurs during the basic functioning of retinal cells. An efficient regulatory molecular circuit is required in the retina, that would modulate both the intensity and duration of basal autophagic activity in specific cell types and thus meet their regular housekeeping demands [?]. Previous studies have suggested the existence of autophagic rhythm in rodent heart muscles and liver, as well as kidney proximal tubules [? ? ?]. In this study, we establish the existence of diurnal regulation of autophagic activity, in both the neural and vascular cells of the retina. We show evidence that certain autophagic markers like Beclin1, ATG9 and LC3 are highly expressed across the retina, while ATG7 shows preferential staining patterns and is enriched only in certain layers (????). The first report on diurnal rhythmicity in autophagy in the retina recorded increased autophagosome formation following outer disk shedding in rat photoreceptors [?]. Our results regarding the biphasic oscillation of diurnal rhythmicity of LC3 expression agree with the findings in a report by Yao et al. 2014, who also observed two peaks of elevated autophagic activity [?]. Our results in Figure ?? make a strong argument in favor of the influence of external stimuli such as light for the entrainment and maintenance of healthy amplitudes of oscillation of autophagic protein expression in the retina. However, we observed that even in the absence of light entrainment, the retina continued to display rhythmicity in ATG7, ATG9, LC3 and BECN, but were lower in amplitude and out of phase, suggesting that there may be an intrinsic diurnal rhythm of autophagy in the retina not dependent on light entrainment.

Our study encompasses autophagy in aging and diabetic conditions. We successfully demonstrate that perturbed autophagy is a characteristic feature of aging. It has been suggested that macroautophagy suffers a decline with aging and is replaced partly by other forms of autophagy [?]. It remains to be determined if other forms of autophagy i.e., the chaperone mediated autophagy and microautophagy are also under regulation by the circadian system and, if so, whether their rhythmicity is affected in aging and disease. In addition to the above, our study reveals that the level of disruption in T1D and T2D is significant, not only in younger mice, but also in older animals (????). Compared to age-matched control mouse retinas, the dramatic deviations in the oscillatory patterns or the overall levels of one or more Atgs in both two- and nine-month old diabetic mouse retina could be an indication of faulty housekeeping and pathological outcomes, such as the accumulation of damaged mitochondria and ROS production [? ? ? ?].

A limitation of this study is that it relies on immunohistochemistry to determine changes in the expression of autophagy proteins throughout the neural retina. We did not attempt to confirm our findings by assessing gene expression or protein levels using Western blot. Neither did we assess autophagic flux by determining the LC3II:LC3I ratio. The reasons for this were a combination of the large number of animals needed; regional/cell-specific changes would be masked in whole retinal preps and the non-canonical role of LC3 in retinal and immune cell phagocytosis [?] could complicate the analysis of the autophagic flux. Furthermore, based on our data, there is a need to determine the mechanistic link between diurnal changes and autophagy in the neural retina. There is considerable evidence that diurnal/circadian rhythm is associated with the induction of autophagy [? ?], a key regulator of autophagy; the mechanistic target of rapamycin (mTOR) is regulated by the circadian clock [?] and the circadian regulation of metabolism is mediated through reciprocal signaling between the clock and metabolic regulatory networks such as autophagy [?]. Recently, Ryzhikov and colleagues reported that diurnal rhythms spatially and temporarily organize autophagy [?]. They reported that basal autophagy rhythms could be resolved into two antiphase clusters that were distinguished by the subcellular location of targeted proteins. Daytime autophagy was directed towards cytosolic proteins and proteosomal degradation, while nighttime autophagy was directed towards ER and mitochondrial removal. There is now a need to better understand the mechanistic control of these processes.

As mentioned above, autophagy in diabetes and related diseases has become an area of intense research, with focus on how this pathway may be targeted in synergy, with other therapeutic approaches to encourage a better clinical outcome. Our study, along with other recent reports, adds a novel extension to the mammalian diurnal rhythmicity in its relevance in regulation of biological processes. It provides a novel link between dysregulated autophagy and disrupted diurnal rhythm in the aging and diabetic retina and suggests that our analyses of myriad biological processes in the retina should be reconsidered from a diurnal perspective, in order to better comprehend age-related vision loss and disease pathology.

Author Contributions: Conceptualization, M.E.B., M.B.G. and J.V.B; Data curation, X.Q., S.K.M; Formal analysis, X.Q., Y.Y.; Investigation, X.Q., S.K.M., M.E.B. Resources, M.E.B., J.V.B., M.B.G.; Project administration: M.E.B., J.V.B., M.B.G. Writing—original draft preparation, X.Q., M.E.B.; Writing—review and editing, X.Q., S.K.M., Y.Y., J.V.B., M.B.G., M.E.B. All authors have read and agreed to the published version of the manuscript.

Funding: M.E.B. is supported by NIH funding (EY019688, EY021626) and an unrestricted grant from Research to Prevent Blindness. J.V.B. is supported by NIH funding (EY01EY028049, R01EY016077; R01EY025383).

Acknowledgments: The animal experiments were performed over a period of time when MEB, MBG, XQ and SKM were at University of Florida and then Indiana University, before moving to UAB. All data were analyzed at UAB.

Conflicts of Interest: The authors declare no conflict of interest.

References

1. Choi, A.M.; Ryter, S.W.; Levine, B. Autophagy in human health and disease. *N. Engl. J. Med.* **2013**, *368*, 651–662. [CrossRef]
2. Mitter, S.K.; Boulton, M.E. Autophagy in ocular physiology. In *Autophagy in Current Trends in Cellular Physiology and Pathology*; Gorbunov, N.V., Schneider, M., Eds.; InTech: Open Access: London, UK, 2016; pp. 287–327.
3. Sridhar, S.; Botbol, Y.; Macian, F.; Cuervo, A.M. Autophagy and disease: Always two sides to a problem. *J. Pathol.* **2012**, *226*, 255–273. [CrossRef]
4. Sachdeva, U.M.; Thompson, C.B. Diurnal rhythms of autophagy: Implications for cell biology and human disease. *Autophagy* **2008**, *4*, 581–589. [CrossRef]
5. Boya, P.; Esteban-Martinez, L.; Serrano-Puebla, A.; Gomez-Sintes, R.; Villarejo-Zori, B. Autophagy in the eye: Development, degeneration, and aging. *Prog. Retin. Eye Res.* **2016**, *55*, 206–245. [CrossRef]
6. Czaja, M.J.; Ding, W.X.; Donohue, T.M., Jr.; Friedman, S.L.; Kim, J.S.; Komatsu, M.; Lemasters, J.J.; Lemoine, A.; Lin, J.D.; Ou, J.H.; et al. Functions of autophagy in normal and diseased liver. *Autophagy* **2013**, *9*, 1131–1158. [CrossRef]
7. Kiriyama, Y.; Nochi, H. The Function of Autophagy in Neurodegenerative Diseases. *Int. J. Mol. Sci.* **2015**, *16*, 26797–26812. [CrossRef]
8. Mancias, J.D.; Kimmelman, A.C. Mechanisms of Selective Autophagy in Normal Physiology and Cancer. *J. Mol. Biol.* **2016**, *428*, 1659–1680. [CrossRef]
9. Yin, Y.; Sun, G.; Li, E.; Kiselyov, K.; Sun, D. ER stress and impaired autophagy flux in neuronal degeneration and brain injury. *Ageing Res. Rev.* **2016**. [CrossRef]
10. Ebato, C.; Uchida, T.; Arakawa, M.; Komatsu, M.; Ueno, T.; Komiya, K.; Azuma, K.; Hirose, T.; Tanaka, K.; Kominami, E.; et al. Autophagy is important in islet homeostasis and compensatory increase of beta cell mass in response to high-fat diet. *Cell Metab.* **2008**, *8*, 325–332. [CrossRef]
11. Fujitani, Y.; Kawamori, R.; Watada, H. The role of autophagy in pancreatic beta-cell and diabetes. *Autophagy* **2009**, *5*, 280–282. [CrossRef]
12. Goldman, S.; Zhang, Y.; Jin, S. Autophagy and adipogenesis: Implications in obesity and type II diabetes. *Autophagy* **2010**, *6*, 179–181. [CrossRef] [PubMed]

13. Graham, T.E.; Abel, E.D. Autophagy in diabetes and the metabolic syndrome. In *Autophagy in Health and Disease*; Gottlieb, R.A., Ed.; Academic Press: London, UK, 2013; pp. 117–140.
14. Shi, S.Q.; Ansari, T.S.; McGuinness, O.P.; Wasserman, D.H.; Johnson, C.H. Circadian disruption leads to insulin resistance and obesity. *Curr. Biol. Cb* **2013**, *23*, 372–381. [CrossRef] [PubMed]
15. Busik, J.V.; Tikhonenko, M.; Bhatwadekar, A.; Opreanu, M.; Yakubova, N.; Caballero, S.; Player, D.; Nakagawa, T.; Afzal, A.; Kielczewski, J.; et al. Diabetic retinopathy is associated with bone marrow neuropathy and a depressed peripheral clock. *J. Exp. Med.* **2009**, *206*, 2897–2906. [CrossRef] [PubMed]
16. Ikeda, H.; Yong, Q.; Kurose, T.; Todo, T.; Mizunoya, W.; Fushiki, T.; Seino, Y.; Yamada, Y. Clock gene defect disrupts light-dependency of autonomic nerve activity. *Biochem. Biophys. Res. Commun.* **2007**, *364*, 457–463. [CrossRef] [PubMed]
17. Ruiter, M.; Buijs, R.M.; Kalsbeek, A. Hormones and the autonomic nervous system are involved in suprachiasmatic nucleus modulation of glucose homeostasis. *Curr. Diabetes Rev.* **2006**, *2*, 213–226. [CrossRef] [PubMed]
18. Thomas, H.E.; Redgrave, R.; Cunnington, M.S.; Avery, P.; Keavney, B.D.; Arthur, H.M. Circulating endothelial progenitor cells exhibit diurnal variation. *Arterioscler. Thromb. Vasc. Biol.* **2008**, *28*, e21–e22. [CrossRef]
19. Kohsaka, A.; Laposky, A.D.; Ramsey, K.M.; Estrada, C.; Joshu, C.; Kobayashi, Y.; Turek, F.W.; Bass, J. High-fat diet disrupts behavioral and molecular circadian rhythms in mice. *Cell Metab.* **2007**, *6*, 414–421. [CrossRef]
20. Young, M.E.; Razeghi, P.; Taegtmeyer, H. Clock genes in the heart: Characterization and attenuation with hypertrophy. *Circ. Res.* **2001**, *88*, 1142–1150. [CrossRef]
21. Lahouaoui, H.; Coutanson, C.; Cooper, H.M.; Bennis, M.; Dkhissi-Benyahya, O. Clock genes and behavioral responses to light are altered in a mouse model of diabetic retinopathy. *PLoS ONE* **2014**, *9*, e101584. [CrossRef]
22. Ando, H.; Ushijima, K.; Yanagihara, H.; Hayashi, Y.; Takamura, T.; Kaneko, S.; Fujimura, A. Clock gene expression in the liver and adipose tissues of non-obese type 2 diabetic Goto-Kakizaki rats. *Clin. Exp. Hypertens.* **2009**, *31*, 201–207. [CrossRef]
23. Bhatwadekar, A.D.; Yan, Y.; Qi, X.; Thinschmidt, J.S.; Neu, M.B.; Li Calzi, S.; Shaw, L.C.; Dominiguez, J.M.; Busik, J.V.; Lee, C.; et al. Per2 mutation recapitulates the vascular phenotype of diabetes in the retina and bone marrow. *Diabetes* **2013**, *62*, 273–282. [CrossRef]
24. Frost, L.S.; Lopes, V.S.; Bragin, A.; Reyes-Reveles, J.; Brancato, J.; Cohen, A.; Mitchell, C.H.; Williams, D.S.; Boesze-Battaglia, K. The Contribution of Melanoregulin to Microtubule-Associated Protein 1 Light Chain 3 (LC3) Associated Phagocytosis in Retinal Pigment Epithelium. *Mol. Neurobiol.* **2015**, *52*, 1135–1151. [CrossRef] [PubMed]
25. Kim, J.Y.; Zhao, H.; Martinez, J.; Doggett, T.A.; Kolesnikov, A.V.; Tang, P.H.; Ablonczy, Z.; Chan, C.C.; Zhou, Z.; Green, D.R.; et al. Noncanonical autophagy promotes the visual cycle. *Cell* **2013**, *154*, 365–376. [CrossRef] [PubMed]
26. Bosch, E.; Horwitz, J.; Bok, D. Phagocytosis of outer segments by retinal pigment epithelium: Phagosome-lysosome interaction. *J. Histochem. Cytochem. Off. J. Histochem. Soc.* **1993**, *41*, 253–263. [CrossRef] [PubMed]
27. Young, R.W. The daily rhythm of shedding and degradation of cone outer segment membranes in the lizard retina. *J. Ultrastruct. Res.* **1977**, *61*, 172–185. [CrossRef]
28. Rodriguez-Muela, N.; Koga, H.; Garcia-Ledo, L.; de la Villa, P.; de la Rosa, E.J.; Cuervo, A.M.; Boya, P. Balance between autophagic pathways preserves retinal homeostasis. *Aging Cell* **2013**, *12*, 478–488. [CrossRef]
29. Yao, J.; Jia, L.; Shelby, S.J.; Ganios, A.M.; Feathers, K.; Thompson, D.A.; Zacks, D.N. Circadian and noncircadian modulation of autophagy in photoreceptors and retinal pigment epithelium. *Investig. Ophthalmol. Vis. Sci.* **2014**, *55*, 3237–3246. [CrossRef]
30. Feit-Leichman, R.A.; Kinouchi, R.; Takeda, M.; Fan, Z.; Mohr, S.; Kern, T.S.; Chen, D.F. Vascular damage in a mouse model of diabetic retinopathy: Relation to neuronal and glial changes. *Investig. Ophthalmol. Vis. Sci.* **2005**, *46*, 4281–4287. [CrossRef]
31. Hazra, S.; Rasheed, A.; Bhatwadekar, A.; Wang, X.; Shaw, L.C.; Patel, M.; Caballero, S.; Magomedova, L.; Solis, N.; Yan, Y.; et al. Liver X receptor modulates diabetic retinopathy outcome in a mouse model of streptozotocin-induced diabetes. *Diabetes* **2012**, *61*, 3270–3279. [CrossRef]
32. Bhutto, I.A.; McLeod, D.S.; Hasegawa, T.; Kim, S.Y.; Merges, C.; Tong, P.; Lutty, G.A. Pigment epithelium-derived factor (PEDF) and vascular endothelial growth factor (VEGF) in aged human choroid and eyes with age-related macular degeneration. *Exp. Eye Res.* **2006**, *82*, 99–110. [CrossRef]

33. McLeod, D.S.; Lefer, D.J.; Merges, C.; Lutty, G.A. Enhanced expression of intracellular adhesion molecule-1 and P-selectin in the diabetic human retina and choroid. *Am. J. Pathol.* **1995**, *147*, 642–653. [PubMed]

34. Kuwabara, T.; Cogan, D.G. Studies of retinal vascular patterns. I. Normal architecture. *Arch. Ophthalmol.* **1960**, *64*, 904–911. [CrossRef] [PubMed]

35. Komeina, K.; Usui, S.; Shen, J.; Rogers, B.S.; Campochiaro, P.A. Blockade of neuronal nitric oxide synthase reduces cone cell death in a model of retinitis pigmentosa. *Free Radic. Biol. Med.* **2008**, *45*, 905–912. [CrossRef] [PubMed]

36. Hammes, H.P.; Lin, J.; Renner, O.; Shani, M.; Lundqvist, A.; Betsholtz, C.; Brownlee, M.; Deutsch, U. Pericytes and the pathogenesis of diabetic retinopathy. *Diabetes* **2002**, *51*, 3107–3112. [CrossRef] [PubMed]

37. Toda, N.; Nakanishi-Toda, M. Nitric oxide: Ocular blood flow, glaucoma, and diabetic retinopathy. *Prog. Retin. Eye Res.* **2007**, *26*, 205–238. [CrossRef] [PubMed]

38. Krohne, T.U.; Kaemmerer, E.; Holz, F.G.; Kopitz, J. Lipid peroxidation products reduce lysosomal protease activities in human retinal pigment epithelial cells via two different mechanisms of action. *Exp. Eye Res.* **2010**, *90*, 261–266. [CrossRef]

39. Viiri, J.; Amadio, M.; Marchesi, N.; Hyttinen, J.M.; Kivinen, N.; Sironen, R.; Rilla, K.; Akhtar, S.; Provenzani, A.; D'Agostino, V.G.; et al. Autophagy activation clears ELAVL1/HuR-mediated accumulation of SQSTM1/p62 during proteasomal inhibition in human retinal pigment epithelial cells. *PLoS ONE* **2013**, *8*, e69563. [CrossRef]

40. Chu, Y.K.; Lee, S.C.; Byeon, S.H. VEGF rescues cigarette smoking-induced human RPE cell death by increasing autophagic flux: Implications of the role of autophagy in advanced age-related macular degeneration. *Investig. Ophthalmol. Vis. Sci.* **2013**, *54*, 7329–7337. [CrossRef]

41. Mohlin, C.; Taylor, L.; Ghosh, F.; Johansson, K. Autophagy and ER-stress contribute to photoreceptor degeneration in cultured adult porcine retina. *Brain Res.* **2014**, *1585*, 167–183. [CrossRef]

42. Rodriguez-Muela, N.; Hernandez-Pinto, A.M.; Serrano-Puebla, A.; Garcia-Ledo, L.; Latorre, S.H.; de la Rosa, E.J.; Boya, P. Lysosomal membrane permeabilization and autophagy blockade contribute to photoreceptor cell death in a mouse model of retinitis pigmentosa. *Cell Death Differ.* **2015**, *22*, 476–487. [CrossRef]

43. Mitter, S.K.; Song, C.; Qi, X.; Mao, H.; Rao, H.; Akin, D.; Lewin, A.; Grant, M.; Dunn, W., Jr.; Ding, J.; et al. Dysregulated autophagy in the RPE is associated with increased susceptibility to oxidative stress and AMD. *Autophagy* **2014**, *10*, 1989–2005. [CrossRef] [PubMed]

44. Ma, D.; Li, S.; Molusky, M.M.; Lin, J.D. Circadian autophagy rhythm: A link between clock and metabolism? *Trends Endocrinol. Metab. Tem.* **2012**, *23*, 319–325. [CrossRef]

45. Ma, D.; Panda, S.; Lin, J.D. Temporal orchestration of circadian autophagy rhythm by C/EBPbeta. *Embo J.* **2011**, *30*, 4642–4651. [CrossRef] [PubMed]

46. Pfeifer, U.; Scheller, H.; Ormanns, W. Diurnal rhythm of lysosomal organelle decomposition in liver, kidney and pancreas. *Acta Histochemica Suppl.* **1976**, *16*, 205–210.

47. Pfeifer, U.; Strauss, P. Autophagic vacuoles in heart muscle and liver. A comparative morphometric study including circadian variations in meal-fed rats. *J. Mol. Cell. Cardiol.* **1981**, *13*, 37–49.

48. Reme, C.E.; Sulser, M. Diurnal variation of autophagy in rod visual cells in the rat. *Graefes Arch. Clin. Exp. Ophthalmol.* **1977**, *203*, 261–270. [CrossRef]

49. Kanwar, M.; Chan, P.S.; Kern, T.S.; Kowluru, R.A. Oxidative damage in the retinal mitochondria of diabetic mice: Possible protection by superoxide dismutase. *Investig. Ophthalmol. Vis. Sci.* **2007**, *48*, 3805–3811. [CrossRef]

50. Kowluru, R.A.; Chan, P.S. Oxidative stress and diabetic retinopathy. *Exp. Diabetes Res.* **2007**, *2007*, 43603. [CrossRef]

51. Wu, Y.; Tang, L.; Chen, B. Oxidative stress: Implications for the development of diabetic retinopathy and antioxidant therapeutic perspectives. *Oxid. Med. Cell. Longev.* **2014**, *2014*, 752387. [CrossRef]

52. Zhong, Q.; Kowluru, R.A. Diabetic retinopathy and damage to mitochondrial structure and transport machinery. *Investig. Ophthalmol. Vis. Sci.* **2011**, *52*, 8739–8746. [CrossRef]

53. Maiese, K. Moving to the Rhythm with Clock (Circadian) Genes, Autophagy, mTOR, and SIRT1 in Degenerative Disease and Cancer. *Curr. Neurovasc. Res.* **2017**, *14*, 299–304. [CrossRef]

54. Cao, R. mTOR Signaling, Translational Control, and the Circadian Clock. *Front. Genet.* **2018**, *9*, 367. [CrossRef]
55. Ryzhikov, M.; Ehlers, A.; Steinberg, D.; Xie, W.; Oberlander, E.; Brown, S.; Gilmore, P.E.; Townsend, R.R.; Lane, W.S.; Dolinay, T.; et al. Diurnal Rhythms Spatially and Temporally Organize Autophagy. *Cell Rep.* **2019**, *26*, 1880–1892.e6. [CrossRef] [PubMed]

Article

Innate and Autoimmunity in the Pathogenesis of Inherited Retinal Dystrophy

T. J. Hollingsworth [1,2] and Alecia K. Gross [2,3,]

[1]   Hamilton Eye Institute, Department of Ophthalmology, University of Tennessee Health Science Center, Memphis, TN 38163, USA; thollin1@uthsc.edu

[2]   Department of Optometry and Vision Science, University of Alabama at Birmingham, Birmingham, AL 35294, USA

[3]   Department of Neurobiology, University of Alabama at Birmingham, Birmingham, AL 35294, USA

[]   Correspondence: agross@uab.edu; Tel.: +1-205-975-8396

Received: 28 January 2020; Accepted: 3 March 2020; Published: 5 March 2020

Abstract: Inherited retinal dystrophies (RDs) are heterogenous in many aspects including genes involved, age of onset, rate of progression, and treatments. While RDs are caused by a plethora of different mutations, all result in the same outcome of blindness. While treatments, both gene therapy-based and drug-based, have been developed to slow or halt disease progression and prevent further blindness, only a small handful of the forms of RDs have treatments available, which are primarily for recessively inherited forms. Using immunohistochemical methods coupled with electroretinography, optical coherence tomography, and fluorescein angiography, we show that in rhodopsin mutant mice, the involvement of both the innate and the autoimmune systems could be a strong contributing factor in disease progression and pathogenesis. Herein, we show that monocytic phagocytosis and inflammatory cytokine release along with protein citrullination, a major player in forms of autoimmunity, work to enhance the progression of RD associated with a rhodopsin mutation.

Keywords: retinal degeneration; immunity; autoimmunity; rhodopsin; citrullination; retinitis pigmentosa

---

1. Introduction

Inherited retinal dystrophies (RDs) are the result of mutations in genes associated with cells of the outer retina, primarily rod and cone photoreceptors and retinal pigment epithelium (RPE) [1]. RDs are often highly heterogenous in every aspect of the disease, from the age of onset to the rate of progression, to the very mechanisms underlying the pathogenesis [2]. In order to study the molecular mechanisms of pathogenesis underlying retinal disease, numerous animal models, mostly in mouse, have been generated, carrying genetic defects in several genes causing the disease. A major cause of RDs, specifically, autosomal dominant retinitis pigmentosa (RP), is the presence of dominantly inherited mutations in the gene for the rod photoreceptor protein rhodopsin [2]. While most rhodopsin mutations cause protein misfolding and subsequent apoptosis, the more severe mutations result in improper rhodopsin trafficking and can lead to a subsequent loss of outer segment formation [2]. These mutations result in improper localization of both normal and mutant rhodopsin to the cellular membrane of the inner segment, nuclear, and axonal regions of the rod photoreceptor cells [1]. Two rhodopsin mutations resulting in the earliest onset of retinal degeneration, rhodopsin Ter349Glu and Gln344Ter, behave in this manner [2]. In recent years, many studies have shown the involvement of the immune system in the pathogenesis of many retinal diseases including glaucoma, age-related macular degeneration (AMD), RP, and others [3,4]. Also brought to light is the role of autoimmunity in RDs including AMD and RP [3,5–8]. While the eye is immune-privileged, the homeostatic disruptions caused by the progression of RDs can allow for a loss of this immune-privileged status, mainly due to the breakdown of the blood retinal barrier and choroidal neovascularization. In this study, we show the involvement of pro-inflammatory pathways in the progression of rhodopsin-mediated RD using the Ter349Glu rhodopsin knock-in mouse model of the disease, including monocytic phagocytosis and activation of the Janus Kinase/Signal Transducer and Activator of Transcription (JAK/STAT) pathway. We also show the presence of known autoimmunity proteins and protein modifications in the rhodopsin Ter349Glu mutant mouse.

2. Materials and Methods

2.1. Measuring Electrical Function in the Ter349Glu Rhodopsin Knock-in Mouse by Electroretinogram (ERG)

Wild-type (+/+, WT), Ter349Glu rhodopsin heterozygous and homozygous mice at one month of age were dark-adapted overnight (O/N). The following day, the mice were anesthetized using ketamine/xylazine (14.3 mg/mL ketamine/2.8 mg/mL xylazine in PBS, pH7.4), and ERGs were performed. Dark-adapted flashes (505 nm stimulus) of varying intensities were achieved using neutral density (ND) filters of nominal optical densities (OD) 4.8, 1.4, 3.6, 2.4, 1.8, 1.2, 0.6, and 0.0. A dark-adapted green camera flash performed with no attenuation. Data were analyzed using Labview and IgorPro software.

2.2. Monitoring the Ter349Glu Rhodopsin Knock-in Mouse Retina for Vascular and Laminar Abnormalities

WT and Ter349Glu homozygous mice at 4 weeks of age were anesthetized using ketamine/xylazine (14.3 mg/mL ketamine/2.8 mg/mL xylazine in PBS, pH7.4). The mice were subsequently examined by optical coherence tomography (OCT) (Bioptigen 840 nm Spectral Domain-OCT, Durham, NC, USA) to measure retinal thickness and assess for structural anomalies. After OCT, the mice were injected intraperitoneally with 100 μL of 4% fluorescein, and the retinas were imaged approximately 1 to 2 minutes post-fluorescein injection by fluorescein angiography (FA), using a 488 nm lamp co-equipped with a Micron III digital microscope (Phoenix Laboratories, Mukilteo, WA, USA) to assess the retinal vasculature for abnormalities including neovascularization, retinal hemorrhage, and vessel alterations.

2.3. Immunohistochemical Survey for Inflammatory Markers in the C-Terminal Mutant Rhodopsin Knock-in Mouse Retina

Whole eyes from WT, Ter349Glu rhodopsin heterozygous and homozygous mice 4, 8, and/or 12 weeks of age were fixed in 4% paraformaldehyde (PFA) in PBS, pH 7.4, overnight at 4 °C, cryoprotected in 30% sucrose in PBS, pH 7.4, frozen in optimal cutting temperature medium, and cryosectioned into 10 μm-thick sections. After washing away the medium, sections were prepped for immunolabeling using heat-mediated antigen retrieval in 10 mM sodium citrate, pH 6.0, with 0.05% Tween-20 for 1 h. After cooling, the slides were washed in PBS and subsequently blocked in 10% goat serum/5% BSA/0.5% Triton X-100 in PBS for 1 h at RT. Following blocking, labeling for markers of retinal inflammation including activated macrophages (F4/80), phosphorylated STAT3 (pSTAT3), and suppressor of cytokine signaling 3 (SOCS3) was performed using fluorescent immunohistochemistry. For both pSTAT3 (Cell Signaling Technology, Danvers, MA, USA) and SOCS3 (abcam, Cambridge, United Kingdom) labeling, a goat anti-rabbit IgG secondary antibody conjugated to horseradish peroxidase (HRP) was allowed to bind the primary antibodies, and Cy3-Tyramide Signal Amplification (TSA, Perkin Elmer, Waltham, MA, USA) was used to visualize the proteins. Briefly, TSA uses the peroxidase activity of HRP to generate reactive oxygen species from peroxide. These reactive oxygen species then oxidize and covalently bind the Cy3-conjugated tyramide molecule to the nearby proteins including the antigen/antibody complex, thus amplifying the signal. Macrophage labeling was performed using an antibody to F4/80-antigen (Bio-Rad, Hercules, CA, USA) followed by a goat anti-rat IgG secondary antibody conjugated to AlexaFluor488 (Invitrogen, Carlsbad, CA, USA). All sections were co-labeled for rhodopsin using B6-30N or K62-82 (provided by W. Clay Smith, University of Florida, Gainesville, FL, USA) and goat anti-mouse IgG$_1$ or IgG$_3$ conjugated to AlexaFluor488 (Invitrogen). Labeling for Müller cells was performed using an antibody against glial fibrillary acidic protein (GFAP, EMD Millipore,

Burlington, MA, USA) followed by goat anti-mouse secondary IgG_1 conjugated to AlexaFluor647 (Invitrogen). All images were captured as Z-stacks and expressed as maximum intensity projections.

2.4. Analysis of Retinal Citrullination in the Ter349Glu/Ter349Glu Rhodopsin Mouse Retina by Fluorescent Immunohistochemistry

Whole eyes from 10- to 12-week-old mice were fixed in 4% PFA in PBS, pH 7.4, overnight at 4 °C, cryoprotected in 30% sucrose in PBS, pH 7.4, frozen in optimal cutting temperature medium, and cryosectioned into 10 μm-thick sections. After washing away the medium, the sections were treated using heat-mediated antigen retrieval by boiling in 10 mM sodium citrate, pH 6.0, with 0.05% Tween-20 for 1 h. After cooling, the slides were washed in PBS and subsequently blocked in 10% goat serum/5% BSA/0.5% Triton X-100 in PBS for 1 h at RT. Following blocking, the sections were incubated with primary antibodies against peptidyl arginine deiminase 4 (PAD4, ProteinTech, Rosemont, IL, USA) and citrullinated peptides (Clone F95, EMD Millipore, Burlington, MA, USA) O/N at 4 °C. After washing in PBS, the slides were probed using goat anti-rabbit IgG conjugated to AlexaFluor488 (Invitrogen) and goat anti-mouse IgM conjugated to AlexaFluor555 (Invitrogen), and the nuclei were labeled with DAPI (Invitrogen). After washing in PBS, the slides were mounted using Prolong Diamond Anti-Fade Mountant (Invitrogen) and imaged using a Zeiss 710 Laser Scanning Confocal Microscope (Zeiss, Oberkochen, Germany). All images were captured as Z-stacks and expressed as maximum intensity projections.

3. Results

3.1. Loss of Functional ERG in Early-Onset RD

Patients expressing the Ter349Glu mutant rhodopsin experience a loss of photoreceptor function earlier in life, resulting in early and rapid central vision loss, when compared to other mutants of rhodopsin [9]. We tested if the Ter349Glu knock-in mouse displayed a similar early-onset (4 weeks of age) loss of visual capacity by ERG (Figure 1). Compared to +/+ mice ($n = 5$), the *Ter349Glu/Ter349Glu* mice ($n = 5$) showed an increase in the threshold of the dark-adapted b-wave by three orders of magnitude, with a maximum amplitude about 25% that of +/+ mice, while the maximum a-wave amplitude was only about 6% that of +/+ mice. This indicated a drastic loss of rod photoreceptor function. Interestingly, when compared to +/+ and *Ter349Glu/Ter349Glu* mice, *Ter349Glu/+* mice ($n = 4$) appeared to have a gain of function with an increase in sensitivity and a decrease in response latency without significant changes in amplitudes.

3.2. Effects of RD-Associated Photoreceptor Loss on Retinal Vasculature and Laminar Architecture

In cases of RP, retinal degeneration exerts effects on both retinal vasculature and laminar architecture in the forms of attenuated vessels and outer nuclear layer (ONL) thinning, respectively [10]. To examine the *Ter349Glu/Ter349Glu* retina for such abnormalities, FA and OCT were performed on 4-week-old +/+ and *Ter349Glu/Ter349Glu* animals in triplicate (Figure 2). Using FA, when compared to +/+ mice, *Ter349Glu/Ter349Glu* mice exhibited heterogeneous types and degrees of vascular abnormalities. The most common anomalies included attenuated vessels, tortuous vessels indicating hyperoxia, and reduced retinal venous and arterial vessel numbers. Using OCT, *Ter349Glu/Ter349Glu* mice exhibited the expected thinning of the ONL; however, at the interfaces between choroid, RPE, and rod outer segments (ROS)/ rod inner segments (RIS) layers the mice also exhibited a possible edema, likely due to loss of contacts between the RPE and the photoreceptors, associated with the lack of ROS. This edema was observed extending to varying degrees both inferiorly and superiorly to the optic nerve. Representative images were taken from the retina inferior to the optic nerve.

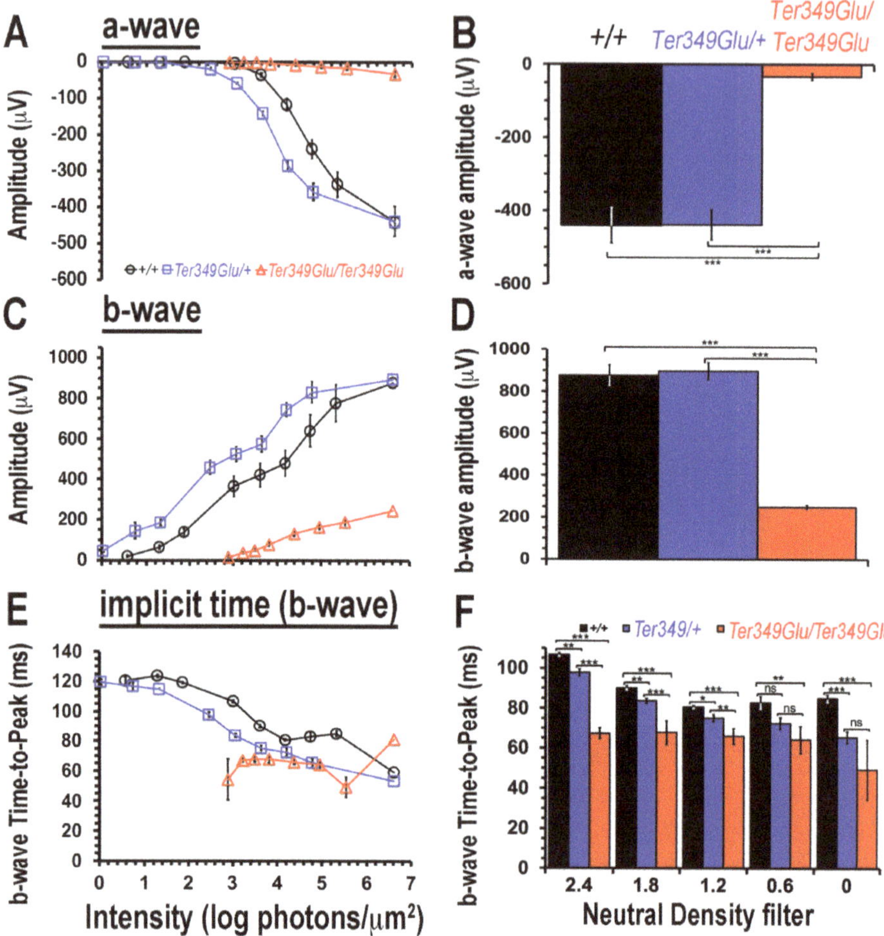

Figure 1. Electroretinogram (ERG) responses decline in mice expressing Ter349Glu rhodopsin. Using ERG to record extracellular potential differences across the retina, the electrophysiological function of +/+, *Ter349Glu/+*, and *Ter349Glu/Ter349Glu* mice was monitored by measuring a-, b-waves, and response latencies (Time-to-Peak, TTP) under increasing stimulus intensities (**A,C,E**). Graphs compare maximum average wave amplitudes and TTP (**B,D,F**) under dark-adapted conditions. Data analyzed using two-tailed T-test and expressed as the mean ± S.E.M. *, $p < 0.05$; **, $p < 0.01$; ***, $p < 0.001$; ns, not significant.

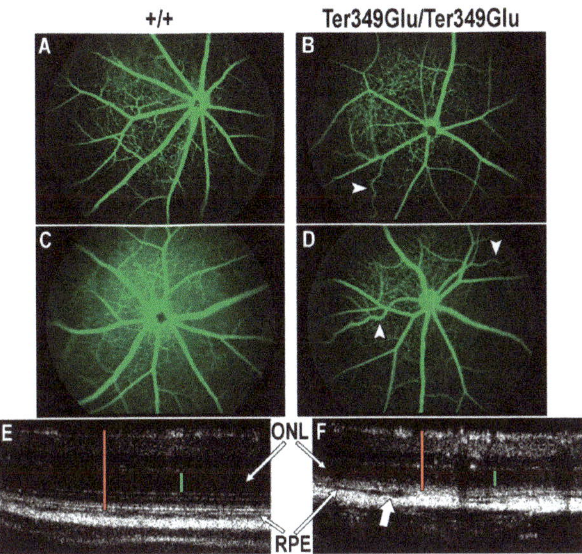

Figure 2. Ter349Glu rhodopsin knock-in mouse retina exhibits both vascular and laminar abnormalities. (**A–D**) Utilizing fluorescein angiography (FA), the state of the retinal vasculature of 4-week-old +/+ (**A,B**) and *Ter349Glu/Ter349Glu* (**C,D**) mice was examined. Abnormal phenotypes varied in severity among mice, with overall attenuated retinal vessels and tortuous retinal vessels (arrowheads) being commonplace amongst all mice examined. (**E,F**) Optical coherence tomography (OCT) was used to examine the retinas of 4-week-old +/+ (**E**) and *Ter349Glu/Ter349Glu* mice (**F**) for architectural abnormalities. *Ter349Glu/Ter349Glu* mice exhibited thinning of the outer nuclear layer (ONL) and patches of varying degrees of separation among the choroid, retinal pigment epithelium (RPE), and photoreceptors (block arrow), indicative of edema. Retinal thickness (red calipers) = 240 µm (**E**) and 180 µm (**F**); ONL (green calipers) = 60 µm (**E**) and 50 µm (**F**).

3.3. Activated Monocytes Are Present in RD Retinas from Rhodopsin Mutant Knock-in Mice

The retina contains resident macrophages similarly to the cortex, known as microglia, and these cells remain in the inner retinal layers under normal physiological conditions. Here, they remain in an inactivated state unless triggered by cytokine signaling or apoptotic signals [11]. Other types of leukocytes are typically not resident in the retina, and evidence of these cells in ocular tissues is indicative of retinal inflammation. Damage to retinal and choroidal vessels can allow leakage of not only blood-borne macrophages into the retina, but also cytokines, antibodies, and a plethora of other inflammatory factors [12]. Unfortunately, no method exists to differentiate between microglia and blood-borne macrophages; however, due to the observation of abnormal vasculature and retinal edemas, we chose to monitor the +/+ and *Ter349Glu/Ter349Glu* retinas for activated macrophages as a whole, using an antibody against F4/80 antigen, a cell surface glycoprotein expressed upon macrophage maturation (Figure 3). F4/80-positive macrophages were found in multiple animals from the *Ter349Glu/Ter349Glu* cohort at 12 weeks of age, with the most labeling observed in sections from *Ter349Glu/Ter349Glu* animals where nearly the whole retina was degenerated. Macrophages remained in the outer retina after almost total rod cell death (12 weeks). These macrophages were not observed in +/+ sections from any animal (*n* = 3 at all ages).

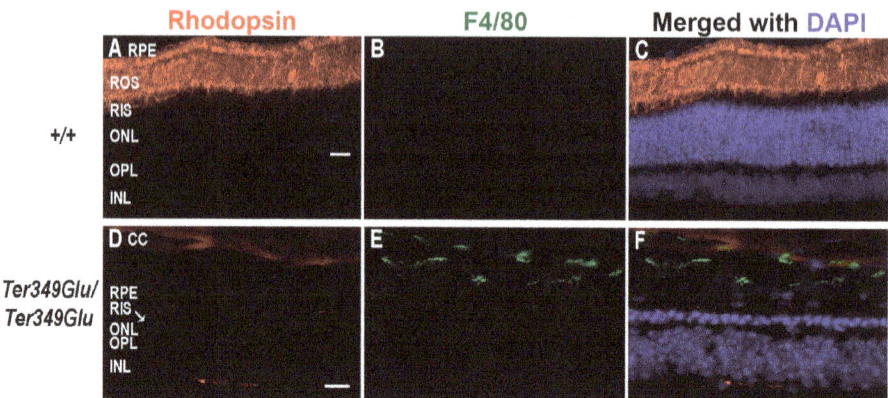

Figure 3. Retinas from retinal dystrophy (RD) mice exhibit monocyte activation. Using fluorescent immunohistochemistry, the presence of activated macrophages was examined in +/+ (**A–C**) and *Ter349Glu/Ter349Glu* (**D–F**) mice at 12 weeks of age. Retinal sections were labeled with anti-F4/80 antigen (green) and K62-82 (rhodopsin, red) antibodies. Nuclei were labeled with DAPI (blue). Autofluorescence in the choroid was observed in the red channel. CC, choriocapillaris; ROS, rod outer segments; RIS, rod inner segments; OPL, outer plexiform layer. Scale bars = 20 μm.

3.4. Activation of the Pro-Inflammatory JAK/STAT Pathway and Its Inhibitor SOCS3 in RD

STAT3 is a downstream signaling partner of JAK. In inflammatory conditions, STAT3 is activated when cytokines such as IL-6, ciliary neurotrophic factor, leukemia inhibitory factor, and others bind to and activate the glycoprotein 130 (gp130) receptor [13]. Gp130 activates JAK, which in turn phosphorylates STAT proteins (pSTAT). Upon phosphorylation, pSTATs form both homo- and heterodimers, allowing nuclear entry and activation of gene transcription. The protein SOCS3 works as a negative regulator of the JAK/STAT pathway by binding the gp130 receptor and JAK together and blocking active sites involved in phosphorylation of STAT3, thus prohibiting further signal transduction [14]. In instances where inflammatory signaling needs to be slowed or stopped, SOCS3 works to perform this task. To examine the Ter349Glu knock-in mouse retina for inflammatory cytokine signaling, retinas from both +/+, *Ter349Glu/+*, and *Ter349Glu/Ter349Glu* mice were labeled with an antibody against pSTAT3 (Figure 4). We found that +/+ retinas showed no STAT3 activation across all ages, while the *Ter349Glu/+* and *Ter349Glu/Ter349Glu* mice exhibited increasing activation with age, beginning at 8 weeks and 4 weeks, respectively ($n = 3$ for all ages).

Early pSTAT3 activation began in the inner nuclear layer (INL) and, by 12 weeks, extended to the RPE. The location of the activated nuclei in the INL suggested the nuclei belonged to Müller cells. Indeed, co-labeling for GFAP, a marker of gliosis in Müller cells, showed Müller cells to be pSTAT3-positive (Figure 5).

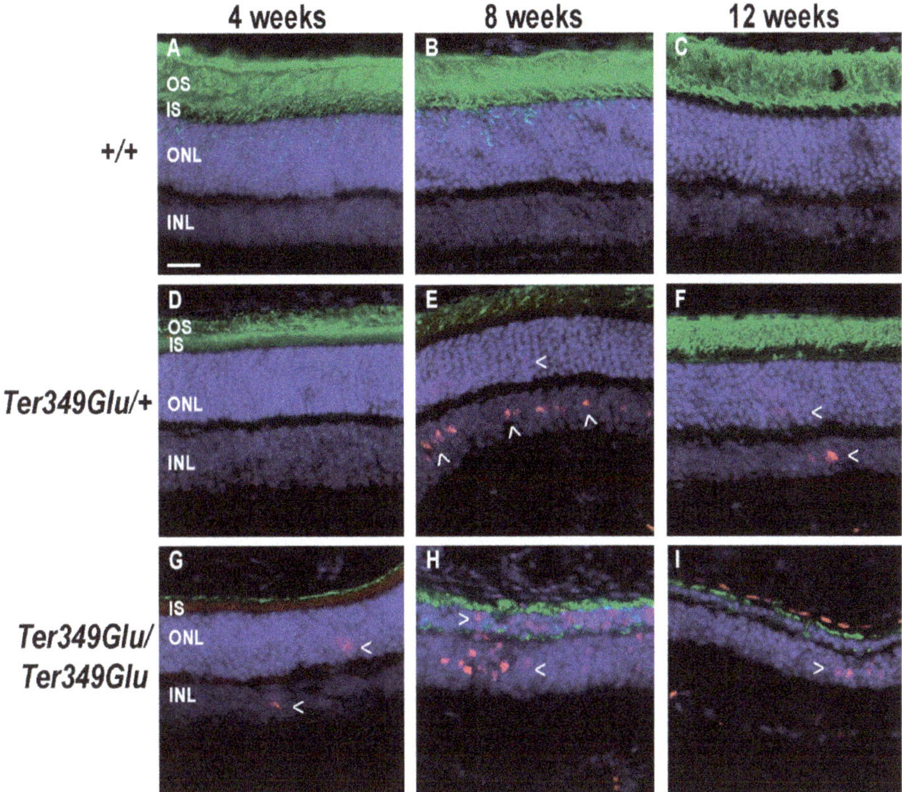

Figure 4. JAK/STAT pathway is activated in Ter349Glu rhodopsin knock-in mouse retina. Activation of the JAK/STAT pathway was examined using fluorescent immunohistochemistry on retinas from WT (+/+, **A–C**), Ter349Glu heterozygous (*Ter349Glu/+*, **D–F**), and Ter349Glu homozygous (*Ter349Glu/Ter349Glu*, **G–H**) mice. Retinal sections were treated for antigen retrieval and labeled for phosphorylated STAT3 (pSTAT3, red) and rhodopsin (green). Nuclei were labeled with DAPI (blue). Arrowheads (>), areas of JAK/STAT activation; OS, outer segments; IS, inner segments; INL, inner nuclear layer. Scale bar = 20 μm.

Since Müller cells maintain retinal homeostasis, this finding suggests they may act to respond to inflammatory cytokines, as STAT3 activation indicates the presence of pro-inflammatory cytokines in the neural retina, likely being released from activated macrophages within the retina. It should be noted that in these images, labeling for rhodopsin can be seen in the ONL even though rhodopsin is not normally localized in large amounts in this region. This is likely an artifact due to a heightened number of anti-rhodopsin epitopes following the antigen retrieval process. Labeling for SOCS3 revealed minimal expression in +/+ retinas while, similarly to pSTAT3, *Ter349Glu/+* and *Ter349Glu/Ter349Glu* mice began expressing SOCS3 in the neural retina and RPE at 4 weeks of age; with time, the expression increased in *Ter349Glu/Ter349Glu* mice and decreased in *Ter349Glu/+* mice (Figure 6, *n* = 3 at all ages).

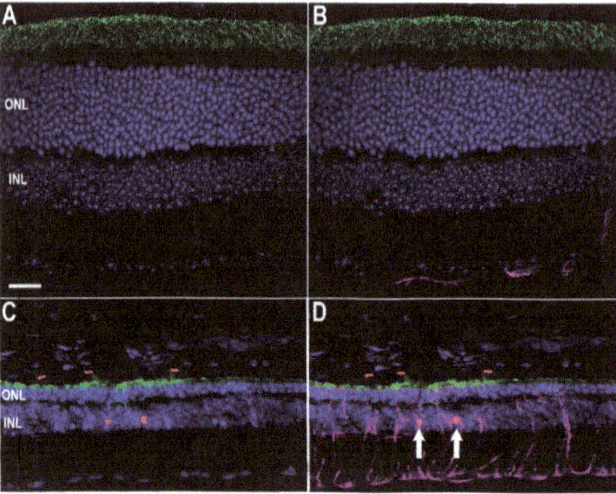

Figure 5. Identification of Müller cell nuclei as INL centers for STAT3 activation. To assess which retinal cells exhibited STAT3 phosphorylation, labeling was performed on +/+ (**A,B**) and *Ter349Glu/Ter349Glu* (**C,D**) animals for glial fibrillary acidic protein (purple), a marker of astrocytes and gliotic Müller cells, rhodopsin (green), and pSTAT3 (red). Nuclei were labeled with DAPI (blue). Arrows show red nuclei surrounded by purple. Scale bar = 20 μm.

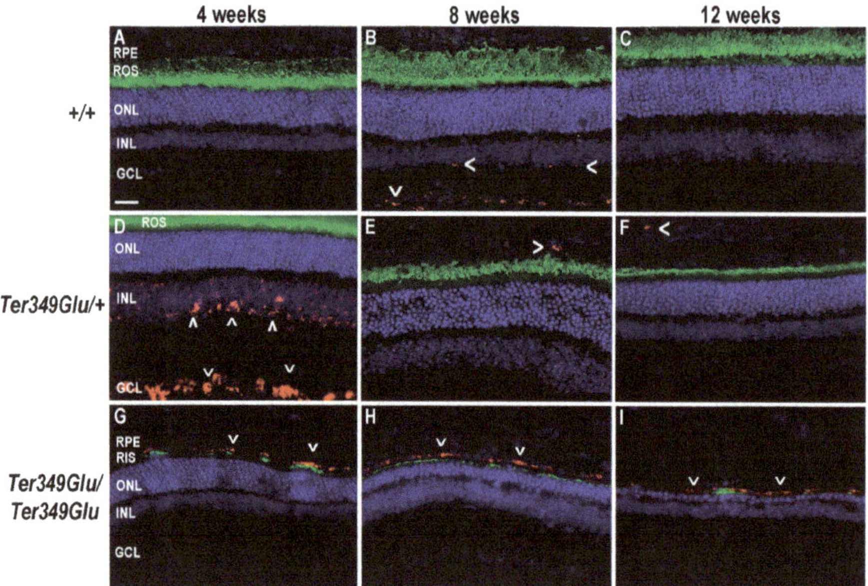

Figure 6. The JAK/STAT antagonist SOCS3 is expressed in Ter349Glu rhodopsin knock-in mouse retina. Using fluorescent immunohistochemistry, the expression of SOCS3, an antagonist to the JAK/STAT pathway, was examined in retinas from wild-type (WT, +/+, **A–C**), Ter349Glu heterozygous (*Ter349Glu/+*, **D–F**), and Ter349Glu homozygous (*Ter349Glu/Ter349Glu*, **G–I**) mice. Retinal sections were treated for antigen retrieval and labeled for SOCS3 (red) and rhodopsin (green). Nuclei were labeled with DAPI (blue). GCL, ganglion cell layer; arrowheads (<), SOCS3 labeling. Scale bar = 20 μm.

3.5. Cell-Specific Expression of PAD4 and Heightened Citrullination in Early-Onset RD

Recent studies have shown an increase in the expression of the deiminating enzyme PAD4 and increased citrullination in the event of ocular insult of the anterior segment [15]. Due to the inherent ability of citrullinated proteins to cause autoimmunity [16–22] and our previous finding that PAD4 is the primary retinal PAD in mouse [23], we tested the Ter349Glu retina for changes in PAD4 expression and citrullination compared with WT retina at 10 to 12 weeks of age (Figure 7). WT retina exhibited expression of PAD4 and exhibited INL nuclear citrullination, as shown previously [23]; however, the Ter349Glu retina exhibited higher levels of citrullination, much of it spanning the entire retina. This was observed in parallel with PAD4 expression increases, especially in the photoreceptors.

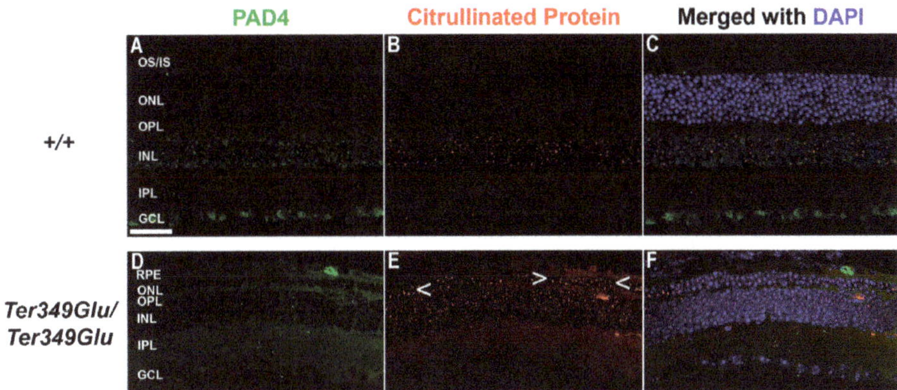

Figure 7. Expression of PAD4 and citrullination of retinal proteins in normal and degenerated states. WT (+/+, **A–C**) and *Ter349Glu/Ter349Glu* (**D–F**) mice at 10 to 12 weeks of age were labeled for PAD4 (green) and citrullinated peptides (red). Nuclei were labeled with DAPI (blue). Arrowheads (<), areas of increased citrullination in ONL; OS/IS, photoreceptor outer and inner segments; IPL, inner plexiform layer; Scale bar = 20 μm.

4. Discussion

Increasing amounts of evidence linking the immune system to ocular disease has emerged in the last decade. Studies focusing on glaucoma, AMD, RP, and other RDs have continued to show increased presence of many immune components including monocytes (blood-borne and resident), pro- and anti-inflammatory cytokines, and autoantibodies in many models of these diseases [24–26]. While all of these diseases present an initial insult of genetic and/or environmental origins, these findings overwhelmingly implicate the immune system in the pathogenesis of RDs. For example, the prevalence of single-nucleotide polymorphisms in the genes coding for complement factor H, complement factor I, as well as many other components of the innate immune system bolster the chances of developing AMD in otherwise normal patients [27,28]. In addition, models of glaucoma have shown the presence of macrophages and T-lymphocytes in the eye along with autoantibodies to proteins of the retina [3,4,29,30]. Models of RP have also shown similar features, including monocytic phagocytosis of both diseased and healthy retinal cells as well as numerous cytokines such as interleukin-1, interleuken-6, vascular endothelial growth factor, and others [4,31,32]. While the molecular mechanisms of the primary genetic assault have been teased out by copious amounts of work using cell culture and animal models for many RDs (i.e., loss of outer segment formation, disrupted visual cycle, etc.) [1,2,33], the full mechanisms of pathogenesis and progression of RDs still need thorough investigation before we fully understand them. In our findings, the *Ter349Glu/Ter349Glu* mice exhibited an almost complete loss of photoreceptor function by ERG; however, the *Ter349Glu/+* mice had heightened a and b waves while also showing a decrease in b wave latency compared to the +/+

animals. This phenomenon might be explained by the expression levels of mutant rhodopsin compared to WT rhodopsin, as Ter349Glu rhodopsin expression levels are less than half of the WT levels [2]. This lower level of expression while still having 50% WT rhodopsin could result in a somewhat thinner outer segment with less rhodopsin protein in the discs, thus allowing for faster rates of diffusion of the phototransduction cascade components, decreasing the response latency, while normal WT rhodopsin activates the cascade, enhancing the signal. We also showed that animals with more advanced RD had significant numbers of macrophages, a feature found in many RDs, as well as activation of a known pro-inflammatory pathway (JAK/STAT). Interestingly, the *Ter349Glu/+* mice showed SOCS3 expression in the inner retinal layers early in life (4 weeks of age), with a dramatic decrease in expression with age, while the +/+ mice had minimal SOCS3 expression, which peaked at 8 weeks of age, and the *Ter349Glu/Ter349Glu* mice exhibited a somewhat constant SOCS3 expression in the outer retina and RPE. While the relatively stable SOCS3 expression in the *Ter349Glu/Ter349Glu* mice is explainable when compared to the levels of STAT3 phosphorylation which was present throughout the first 3 months of the animals' life, the stark difference in SOCS 3 expression the *Ter349Glu/+* mice remains to be elucidated. More experiments examining the activation and deactivation of the proinflammatory JAK/STAT pathway are needed to better understand these expression differences. Further work will aim at pinning down the cytokines responsible for pathway activation; experiments inhibiting the pathway activation using antagonists to the JAK/STAT pathway and/or those inhibiting specific cytokine(s) will better underpin this pathway's role in RD. Due to the nature of STAT3 role in numerous developmental pathways [13], a more targeted approach to delivering inhibitory compounds directly to the eye would be necessary. Future experiments will also work to decrease the number of activated macrophages in RD models to attempt to rescue some photoreceptor degeneration due to excessive phagocytosis and/or macrophage-derived cytokines. Work is already underway testing the inhibition of PAD4 in RD models to rescue retinal cells, retinal function, or both [34]. This work will be achieved using PAD4-deficient mice as well as inhibitors of the enzyme. Due to the excessive citrullination observed in *Ter349Glu/Ter349Glu* mice, it is not far-fetched to think that lowering or preventing this post-translation modification from occurring could lead to a slower rate of disease progression, allowing for not only longer lasting vision in patients affected but also an extended opportunity to correct the genetic insult, thus preventing further retinal degeneration. In all, our work further contributes to the increasing volume of studies indicating a role of the immune system in RDs as well as provides possible targets for the treatment of these debilitating blinding diseases.

Author Contributions: Conceptualization, A.K.G. and T.J.H.; methodology, A.K.G. and T.J.H.; investigation, A.K.G. and T.J.H.; resources, A.K.G. and T.J.H.; data curation, A.K.G. and T.J.H.; writing—original draft preparation, T.J.H.; writing—review and editing, A.K.G. and T.J.H.; visualization, A.K.G. and T.J.H.; project administration, A.K.G.; funding acquisition, A.K.G. All authors have read and agreed to the published version of the manuscript.

Funding: This research was funded by the NIH R01EY019311 and the E. Matilda Ziegler Foundation. Center support for the project was provided by NIH P30EY003039.

Acknowledgments: We thank Brian Simms and Meredith Hubbard for technical assistance and W. Clay Smith for the K62-82 rhodopsin antibody.

Conflicts of Interest: The authors declare no conflicts of interest.

References

1. Hollingsworth, T.J.; Gross, A.K. Defective trafficking of rhodopsin and its role in retinal degenerations. In *International Review of Cell and Molecular Biology*; Academic Press: Cambridge, MA, USA, 2012; Volume 293. [CrossRef]
2. Hollingsworth, T.J.; Gross, A.K. The severe autosomal dominant retinitis pigmentosa rhodopsin mutant Ter349Glu mislocalizes and induces rapid rod cell death. *J. Biol. Chem.* **2013**, *288*, 29047–29055. [CrossRef]
3. Bhattacharya, S.K.; Crabb, J.S.; Bonilha, V.L.; Gu, X.; Takahara, H.; Crabb, J.W. Proteomics implicates peptidyl arginine deiminase 2 and optic nerve citrullination in glaucoma pathogenesis. *Investig. Ophthalmol. Vis. Sci.* **2006**, *47*, 2508–2514. [CrossRef] [PubMed]

4. Wooff, Y.; Man, S.M.; Aggio-Bruce, R.; Natoli, R.; Fernando, N. IL-1 family members mediate cell death, inflammation and angiogenesis in retinal degenerative diseases. *Front. Immunol.* **2019**, *10*, 1618. [CrossRef] [PubMed]

5. Iannaccone, A.; Giorgianni, F.; New, D.D.; Hollingsworth, T.J.; Umfress, A.; Alhatem, A.H.; Neeli, I.; Lenchik, N.I.; Jennings, B.J.; Calzada, J.I.; et al. Circulating autoantibodies in age-related macular degeneration recognize human macular tissue antigens implicated in autophagy, immunomodulation, and protection from oxidative stress and apoptosis. *PLoS ONE* **2015**, *10*, e0145323. [CrossRef] [PubMed]

6. Bhattacharya, S.K. Retinal deimination in aging and disease. *IUBMB Life* **2009**, *61*, 504–509. [CrossRef] [PubMed]

7. Bonilha, V.L.; Shadrach, K.G.; Rayborn, M.E.; Li, Y.; Pauer, G.J.; Hagstrom, S.A.; Bhattacharya, S.K.; Hollyfield, J.G. Retinal deimination and PAD2 levels in retinas from donors with age-related macular degeneration (AMD). *Exp. Eye Res.* **2013**, *111*, 71–78. [CrossRef]

8. Iannaccone, A.; Hollingsworth, T.J.; Koirala, D.; New, D.D.; Lenchik, N.I.; Beranova-Giorgianni, S.; Gerling, I.C.; Radic, M.Z.; Giorgianni, F. Retinal pigment epithelium and microglia express the CD5 antigen-like protein, a novel autoantigen in age-related macular degeneration. *Exp. Eye Res.* **2017**, *155*, 64–74. [CrossRef] [PubMed]

9. Bessant, D.A.; Khaliq, S.; Hameed, A.; Anwar, K.; Payne, A.M.; Mehdi, S.Q.; Bhattacharya, S.S. Severe autosomal dominant retinitis pigmentosa caused by a novel rhodopsin mutation (Ter349Glu). Mutations in brief no. 208. Online. *Hum. Mutat.* **1999**, *13*, 83. [CrossRef]

10. Rezaei, K.A.; Zhang, Q.; Chen, C.L.; Chao, J.; Wang, R.K. Retinal and choroidal vascular features in patients with retinitis pigmentosa imaged by OCT based microangiography. *Graefes Arch. Clin. Exp. Ophthalmol.* **2017**, *255*, 1287–1295. [CrossRef]

11. Fontainhas, A.M.; Wang, M.; Liang, K.J.; Chen, S.; Mettu, P.; Damani, M.; Fariss, R.N.; Li, W.; Wong, W.T. Microglial morphology and dynamic behavior is regulated by ionotropic glutamatergic and GABAergic neurotransmission. *PLoS ONE* **2011**, *6*, e15973. [CrossRef]

12. Zhao, L.; Ma, W.; Fariss, R.N.; Wong, W.T. Minocycline attenuates photoreceptor degeneration in a mouse model of subretinal hemorrhage microglial: inhibition as a potential therapeutic strategy. *Am. J. Pathol.* **2011**, *179*, 1265–1277. [CrossRef] [PubMed]

13. Levy, D.E.; Lee, C.K. What does Stat3 do? *J. Clin. Investig.* **2002**, *109*, 1143–1148. [CrossRef] [PubMed]

14. Kershaw, N.J.; Murphy, J.M.; Liau, N.P.; Varghese, L.N.; Laktyushin, A.; Whitlock, E.L.; Lucet, I.S.; Nicola, N.A.; Babon, J.J. SOCS3 binds specific receptor-JAK complexes to control cytokine signaling by direct kinase inhibition. *Nat. Struct. Mol. Biol.* **2013**, *20*, 469–476. [CrossRef] [PubMed]

15. Wizeman, J.W.; Nicholas, A.P.; Ishigami, A.; Mohan, R. Citrullination of glial intermediate filaments is an early response in retinal injury. *Mol. Vis.* **2016**, *22*, 1137–1155.

16. Acharya, N.K.; Nagele, E.P.; Han, M.; Coretti, N.J.; DeMarshall, C.; Kosciuk, M.C.; Boulos, P.A.; Nagele, R.G. Neuronal PAD4 expression and protein citrullination: possible role in production of autoantibodies associated with neurodegenerative disease. *J. Autoimmun.* **2012**, *38*, 369–380. [CrossRef]

17. Bicker, K.L.; Thompson, P.R. The protein arginine deiminases: Structure, function, inhibition, and disease. *Biopolymers* **2013**, *99*, 155–163. [CrossRef]

18. Dwivedi, N.; Neeli, I.; Schall, N.; Wan, H.; Desiderio, D.M.; Csernok, E.; Thompson, P.R.; Dali, H.; Briand, J.P.; Muller, S.; et al. Deimination of linker histones links neutrophil extracellular trap release with autoantibodies in systemic autoimmunity. *FASEB J.* **2014**, *28*, 2840–2851. [CrossRef]

19. Dwivedi, N.; Radic, M. Citrullination of autoantigens implicates NETosis in the induction of autoimmunity. *Ann. Rheum. Dis.* **2014**, *73*, 483–491. [CrossRef]

20. Lundberg, K.; Nijenhuis, S.; Vossenaar, E.R.; Palmblad, K.; van Venrooij, W.J.; Klareskog, L.; Zendman, A.J.; Harris, H.E. Citrullinated proteins have increased immunogenicity and arthritogenicity and their presence in arthritic joints correlates with disease severity. *Arthritis Res. Ther.* **2005**, *7*, R458–R467. [CrossRef]

21. Nicholas, A.P.; Sambandam, T.; Echols, J.D.; Barnum, S.R. Expression of citrullinated proteins in murine experimental autoimmune encephalomyelitis. *J. Comp. Neurol.* **2005**, *486*, 254–266. [CrossRef]

22. Turunen, S.; Huhtakangas, J.; Nousiainen, T.; Valkealahti, M.; Melkko, J.; Risteli, J.; Lehenkari, P. Rheumatoid arthritis antigens homocitrulline and citrulline are generated by local myeloperoxidase and peptidyl arginine deiminases 2, 3 and 4 in rheumatoid nodule and synovial tissue. *Arthritis Res. Ther.* **2016**, *18*, 239. [CrossRef] [PubMed]

23. Hollingsworth, T.J.; Radic, M.Z.; Beranova-Giorgianni, S.; Giorgianni, F.; Wang, Y.; Iannaccone, A. Murine Retinal Citrullination Declines With Age and is Mainly Dependent on Peptidyl Arginine Deiminase 4 (PAD4). *Investig. Ophthalmol. Vis. Sci.* **2018**, *59*, 3808–3815. [CrossRef] [PubMed]

24. Lu, Z.; Lin, V.; May, A.; Che, B.; Xiao, X.; Shaw, D.H.; Su, F.; Wang, Z.; Du, H.; Shaw, P.X. HTRA1 synergizes with oxidized phospholipids in promoting inflammation and macrophage infiltration essential for ocular VEGF expression. *PLoS ONE* **2019**, *14*, e0216808. [CrossRef]

25. Murakami, Y.; Ishikawa, K.; Nakao, S.; Sonoda, K.H. Innate immune response in retinal homeostasis and inflammatory disorders. *Prog. Retin. Eye Res.* **2019**, *74*, 100778. [CrossRef] [PubMed]

26. Ronning, K.E.; Karlen, S.J.; Miller, E.B.; Burns, M.E. Molecular profiling of resident and infiltrating mononuclear phagocytes during rapid adult retinal degeneration using single-cell RNA sequencing. *Sci. Rep.* **2019**, *9*, 4858. [CrossRef] [PubMed]

27. Mansoor, N.; Wahid, F.; Azam, M.; Shah, K.; den Hollander, A.I.; Qamar, R.; Ayub, H. Molecular mechanisms of complement system proteins and matrix metalloproteinases in the pathogenesis of age-related macular degeneration. *Curr. Mol. Med.* **2019**, *19*, 705–718. [CrossRef]

28. Landowski, M.; Kelly, U.; Klingeborn, M.; Groelle, M.; Ding, J.D.; Grigsby, D.; Bowes Rickman, C. Human complement factor H Y402H polymorphism causes an age-related macular degeneration phenotype and lipoprotein dysregulation in mice. *Proc. Natl. Acad. Sci. USA* **2019**, *116*, 3703–3711. [CrossRef]

29. Reinehr, S.; Kuehn, S.; Casola, C.; Koch, D.; Stute, G.; Grotegut, P.; Dick, H.B.; Joachim, S.C. HSP27 immunization reinforces AII amacrine cell and synapse damage induced by S100 in an autoimmune glaucoma model. *Cell Tissue Res.* **2018**, *371*, 237–249. [CrossRef]

30. Tsai, T.; Grotegut, P.; Reinehr, S.; Joachim, S.C. Role of heat shock proteins in glaucoma. *Int. J. Mol. Sci.* **2019**, *20*, 5160. [CrossRef]

31. Yi, Q.Y.; Wang, Y.Y.; Chen, L.S.; Li, W.D.; Shen, Y.; Jin, Y.; Yang, J.; Wang, Y.; Yuan, J.; Cheng, L. Implication of inflammatory cytokines in the aqueous humour for management of macular diseases. *Acta Ophthalmol.* **2019**. [CrossRef]

32. Zhao, L.; Zabel, M.K.; Wang, X.; Ma, W.; Shah, P.; Fariss, R.N.; Qian, H.; Parkhurst, C.N.; Gan, W.B.; Wong, W.T. Microglial phagocytosis of living photoreceptors contributes to inherited retinal degeneration. *EMBO Mol. Med.* **2015**, *7*, 1179–1197. [CrossRef] [PubMed]

33. den Hollander, A.I.; Roepman, R.; Koenekoop, R.K.; Cremers, F.P. Leber congenital amaurosis: genes, proteins and disease mechanisms. *Prog. Retin. Eye Res.* **2008**, *27*, 391–419. [CrossRef] [PubMed]

34. Iannaccone, A.; Radic, M.Z. Increased protein citrullination as a trigger for resident immune system activation, intraretinal inflammation, and promotion of anti-retinal autoimmunity: intersecting paths in retinal degenerations of potential therapeutic relevance. *Adv. Exp. Med Biol.* **2019**, *1185*, 175–179. [CrossRef] [PubMed]

Article

Single-Cell RNA Sequencing in Human Retinal Degeneration Reveals Distinct Glial Cell Populations

Andrew P. Voigt [1,2], Elaine Binkley [1,2], Miles J. Flamme-Wiese [1,2], Shemin Zeng [1,2], Adam P. DeLuca [1,2], Todd E. Scheetz [1,2], Budd A. Tucker [1,2], Robert F. Mullins [1,2] and Edwin M. Stone [1,2,*]

[1] Department of Ophthalmology and Visual Sciences, The University of Iowa Carver College of Medicine, Iowa City, IA 52242, USA; andrew-voigt@uiowa.edu (A.P.V.); elaine-binkley@uiowa.edu (E.B.); miles-flamme-wiese@uiowa.edu (M.J.F.-W.); shemin-zeng@uiowa.edu (S.Z.); adam-deluca@uiowa.edu (A.P.D.); scheetzt@ivr.uiowa.edu (T.E.S.); budd-tucker@uiowa.edu (B.A.T.); robert-mullins@uiowa.edu (R.F.M.)
[2] Institute for Vision Research, The University of Iowa, Iowa City, IA 52242, USA
* Correspondence: edwin-stone@uiowa.edu

Received: 19 December 2019; Accepted: 10 February 2020; Published: 13 February 2020

Abstract: Degenerative diseases affecting retinal photoreceptor cells have numerous etiologies and clinical presentations. We clinically and molecularly studied the retina of a 70-year-old patient with retinal degeneration attributed to autoimmune retinopathy. The patient was followed for 19 years for progressive peripheral visual field loss and pigmentary changes. Single-cell RNA sequencing was performed on foveal and peripheral retina from this patient and four control patients, and cell-specific gene expression differences were identified between healthy and degenerating retina. Distinct populations of glial cells, including astrocytes and Müller cells, were identified in the tissue from the retinal degeneration patient. The glial cell populations demonstrated an expression profile consistent with reactive gliosis. This report provides evidence that glial cells have a distinct transcriptome in the setting of human retinal degeneration and represents a complementary clinical and molecular investigation of a case of progressive retinal disease.

Keywords: autoimmune retinopathy; retinal degeneration; Müller cell; single-cell

---

1. Introduction

Photoreceptor cells are highly specialized, terminally differentiated neurons that detect photons of light and transmit this information to bipolar cells in the retina. Unfortunately, their exacting structural and metabolic requirements make them very susceptible to a large number of acquired and genetic sources of injury, leading to irreversible vision loss [1]. Degenerative diseases affecting photoreceptor cells have multiple etiologies. For example, genetic variants in over 100 genes have been shown to cause heritable photoreceptor degeneration [2]. However, photoreceptor degeneration can also be immune mediated, as in the case of autoimmune retinopathy (AIR), where circulating retinal autoantibodies lead to inflammation and downstream photoreceptor destruction [3]. Photoreceptor loss can also occur secondary to damage or dysfunction of adjacent cells and extracellular structures; for example, diseases affecting the retinal pigment epithelium (RPE), Bruch's membrane, or choroid can lead to increased oxidative stress and decreased metabolic support to the outer retina [4].

One approach for studying retinal degeneration is to characterize transcriptomic changes within diseased retina using microarrays or, more recently, next-generation sequencing of cDNA libraries (RNA sequencing, or RNA-Seq). Conventional gene expression studies with RNA-Seq have analyzed pools of retinal RNA from numerous cell types [5,6]. However, the high degree of cellular complexity and diversity in the human retina can prevent detection of even large gene expression changes that are restricted to specific classes of cells that are relatively unrepresented in the pool [7]. This concern has been largely obviated by the development of single-cell RNA sequencing, which has recently been employed to characterize the transcriptome of individual retinal cell populations. The neural retina is well suited for dissociation into single-cells, and protocols for recovery of viable, singlet cells are well established [8,9]. Such protocols facilitated the exploration of the murine retina transcriptome in the first report of Drop-Seq single-cell RNA sequencing [10]. Since this initial investigation, several additional studies have described the transcriptome of murine retina [10–12] and more recently, human retina [13–15] at the single-cell level.

In this report, we describe the clinical course of a 70-year-old patient with progressive photoreceptor degeneration attributed to AIR. We perform single-cell RNA sequencing on paired foveal and peripheral retinal samples from this patient and four unaffected control patients to investigate how different populations of retinal cells respond to photoreceptor degeneration. A total of 23,429 cells were recovered in this experiment, including 7189 cells from the AIR patient. This study provides insight into the responses of the retina to a blinding inflammatory condition at the cellular and transcriptional levels.

2. Materials and Methods

Human Donor Eyes: Eyes from the human donors utilized for this study were acquired from the Iowa Lions Eye Bank in accordance with the Declaration of Helsinki and following full consent of the donors' next of kin. The Institutional Review Board at the University of Iowa has judged that experiments performed on the donated eyes of deceased individuals does not fall under human subjects rules. All of the experiments in present paper were on the eyes of deceased individuals donated to science by the donors' next of kin. The work we performed in this paper was not human subjects research. Donor information is presented in Table 1. All tissue was received in the laboratory within 7 h post-mortem and processed immediately. A 2 mm foveal centered punch and an 8 mm peripheral retinal punch from the inferotemporal region centered on the equator were acquired with a disposable trephine from each donor. For the AIR donor, the OS was used for single-cell RNA sequencing and the OD was preserved in freshly generated 4% paraformaldehyde in phosphatidylcholine buffer solution. Frozen sections from the macula and peripheral retina were prepared as described previously [16]. Sections were stained with hematoxylin-eosin stain.

Table 1. Sample information from the donor eyes utilized in this study. Note that donor eyes 1–3 serve as controls for the current study and have been previously published [13].

Donor	Age	Sex	Time Postmortem	Eye	Cause of Death	Ophthalmologic History
Donor 1	89	Male	5:21	OD	Cancer	Early stage glaucoma documented; histologically normal
Donor 2	54	Male	5:29	OD	Cardiac arrest	No records received; histologically normal
Donor 3	82	Female	4:18	OD	Cardiopulmonary arrest	No ophthalmic records; histologically normal
Donor 4	76	Male	5:14	OS	Respiratory failure/cancer	posterior vitreous detachment OU, nuclear sclerosis OU; histologically normal
Donor 5	70	Male	6:36	OS	Chronic obstructive pulmonary disease	AIR, see results 3.1

Dissociation for single-cell analysis: The overlying retinal tissue was peeled off of the retinal pigment epithelium and choroid. Retinal tissue was subsequently dissociated in 20 units/mL of papain with 0.005% DNase I (Worthington Biochemical Corporation, Lakewood NJ) for 1.25 h on a shaker at 37 °C. Dissociated cell suspensions were frozen in DMSO-based Recovery Cell Culture Freezing Media (Life Technologies Corporation, Grand Island NY) in a Cryo-Safe cooler (CryoSafe, Summerville SC) to cool at 1 °C/min at −80 °C for 3–8 h before storage in liquid nitrogen.

Sample Preparation: Cryopreserved retinal samples were rapidly thawed and resuspended in phosphatidylcholine buffer solution with 0.04% non-acetylated bovine serum albumin (New England Biolabs, Ipswich, MA, USA) at a concentration of 1000 cells/μL. Viability analysis was performed with the Annexin V/Dead Cell Apoptosis Kit (Life Technologies Corporation, Eugene, OR, USA), with viability >90% using the Countess II FL Automated Cell Counter (ThermoFisher Scientific, Waltham, MA, USA). Next, single cells were captured and barcoded using the Chromium system v3.0 chemistry kit (10X Genomics, Pleasanton, CA, USA). Barcoded libraries were pooled before sequencing on the HiSeq 4000 platform (Illumina, San Diego, CA, USA), generating 150 base pair paired-end reads.

Immunohistochemistry: Immunohistochemical experiments were performed on frozen tissue sections from donor eyes fixed in 4% paraformaldehyde. Sections were blocked with 1 mg/mL of bovine serum albumin before one-hour incubation with anti-ANXA1 (1:1.7, Developmental Studies Hybridoma Bank, Iowa City, IA) or Blue Cone Opsin (1:200, Millipore, AB5407), Red/Green Cone Opsin (1:200, Millipore, AB5405), and RetP1 (1:1000, Thermo Scientific). Sections were subsequently washed and incubated with Alexa-546-conjugated anti-mouse IgG (1:200, Invitrogen) or Alexa-488-conjugated anti-mouse IgG (1:200, Invitrogen) and Alexa-546-conjugated anti-rabbit IgG (1:200, Invitrogen). Each secondary antibody was supplemented with 100 μg/mL diamidino-phenyl-indole (DAPI, Sigma). Sections were incubated for 30 min before washing and cover slipping. Negative controls were included by omitting each primary antibody. Sections were photographed with an epifluorescent microscope (Olympus BX41) equipped with a digital camera (SPOT-RT; Diagnostic Instruments).

Computational Analysis: In addition to the two new donors sequenced for this study, we recently reported single-cell RNA sequencing on paired foveal (2 mm) and peripheral neural retina isolated from three human donors [13] with identical sample processing. FASTQ files from the previous experiment (n = 3 paired samples, donors 1-3; GSE130636) and the current experiment (n = 2 paired samples, donors 4–5) were utilized for downstream analysis. Briefly, FASTQ files were generated from basecalls with the bcl2fastq software (Illumina, San Diego, CA, USA) by the University of Iowa Institute of Human Genetics. Next, FASTQ files were mapped to the hg19 genome with CellRanger (v3.0.1) [17]. Cells with unique gene counts fewer than 200 were filtered, and cells with greater than 7000 unique genes per cell were removed to eliminate potential doublets. Libraries were aggregated to the same effective sequencing depth, and log-normalization of aggregated reads was performed with Seurat (v2.3.4) using a scale factor of 10,000 [18]. All raw and processed data have been deposited in NCBI's Gene Expression Omnibus (GSE142449).

3. Results

3.1. Patient Description

The patient initially presented to the neuro-ophthalmology service at the age of 51 for evaluation of decreased peripheral visual fields and photopsias in both eyes. He had no family history of inherited retinal degeneration. At presentation, his visual acuity was 20/20 in each eye, but he was found to have peripheral visual field loss. Ophthalmoscopic examination early in his disease course showed granular juxtapapillary pigmentary changes and mild vascular attenuation in both eyes. Electroretinogaphy (ERG) at presentation was consistent with widespread retinal dysfunction affecting both rods and cones. He experienced relatively rapid progression of his visual field loss and was seen by the inherited retinal degeneration service with concern for retinitis pigmentosa versus autoimmune retinopathy (AIR). Cancer-associated retinopathy was also considered, but his workup for malignancy was negative and his ERG was felt to be inconsistent with a paraneoplastic process at that time.

He ultimately developed peripheral bone-spicule-like pigmentary changes in both eyes (Figure 1A,B, 7 years after initial presentation) and progressive visual field constriction (Figure 1C,D, 8 years after initial presentation). Molecular evaluation for an inherited retinal degeneration, including whole exome sequencing, was performed but failed to identify a genetic etiology for his condition (for methods see [2]). He developed colon cancer several years after presentation, but this was thought to be unrelated to his ocular disease. He was given a presumed diagnosis of autoimmune retinopathy and ultimately required treatment for cystoid macular edema with intravitreal steroids.

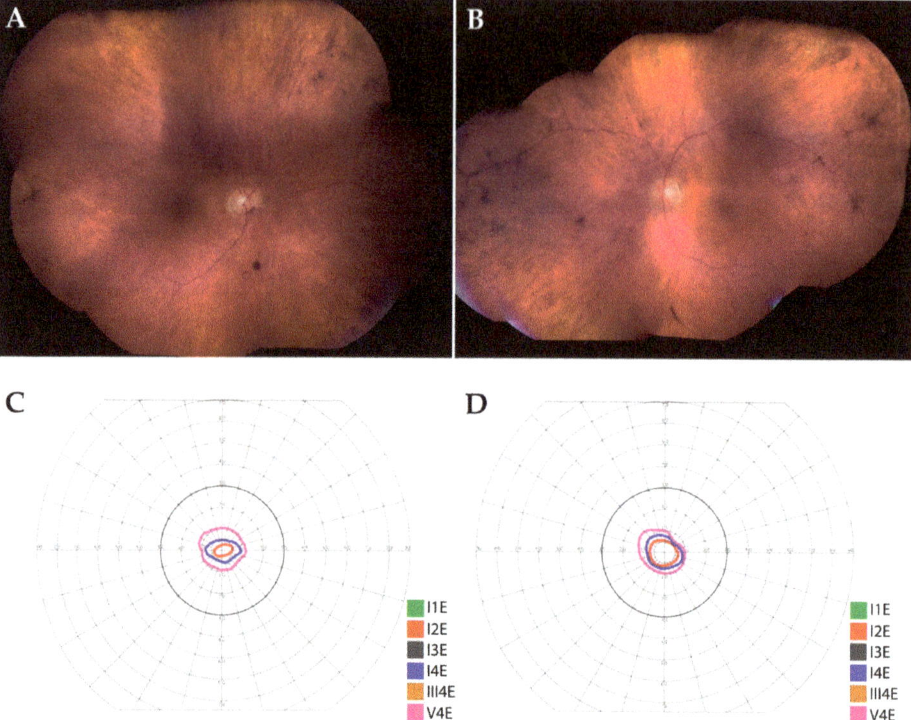

Figure 1. Clinical findings in a patient with retinal degeneration. (**A**,**B**): Montage color fundus photographs of the right (**A**) and left (**B**) eyes. There was granular, retinal pigment epithelial atrophy in the mid-periphery of both eyes, in addition to peripheral bone-spicule-like pigmentary changes and pigment clumps in both eyes. Arteriolar attenuation was notable in both eyes. (**C**,**D**): Goldman visual fields of the right (**C**) and left (**D**) eyes. There was severe constriction of the peripheral visual field in both eyes.

At the time of his last follow up with the retina service at the age of 69, his visual acuity was 20/40 + 1 in the right eye and 20/100 in the left eye with stable peripheral pigmentary changes in both eyes, and no cystoid macular edema. At age 70, the patient expired and donated his eyes for ophthalmic research. Evaluation of the patient's serum with Western blotting revealed the presence of antibodies that reacted with a 23 kilodalton protein in human retina.

3.2. Histological Findings

Sections from an eye with normal ocular history (Figure 2A), from the macula of the AIR donor (in the OD, the eye with better visual acuity) (Figure 2B), and from the periphery of the AIR donor (Figure 2C,D) were acquired. The macula of the AIR donor showed a loss of rod photoreceptors with only a single layer of attenuated cone cells remaining. In spite of the photoreceptor cell loss, the RPE was confluent and the inner retina appeared intact, with discrete inner nuclear layers and ganglion cell layers. In the periphery, the AIR donor showed complete loss of inner and outer segments and of the outer nuclear layer (Figure 2C). Considerable pigment migration into the inner retina was also observed in the periphery of the AIR donor (Figure 2D).

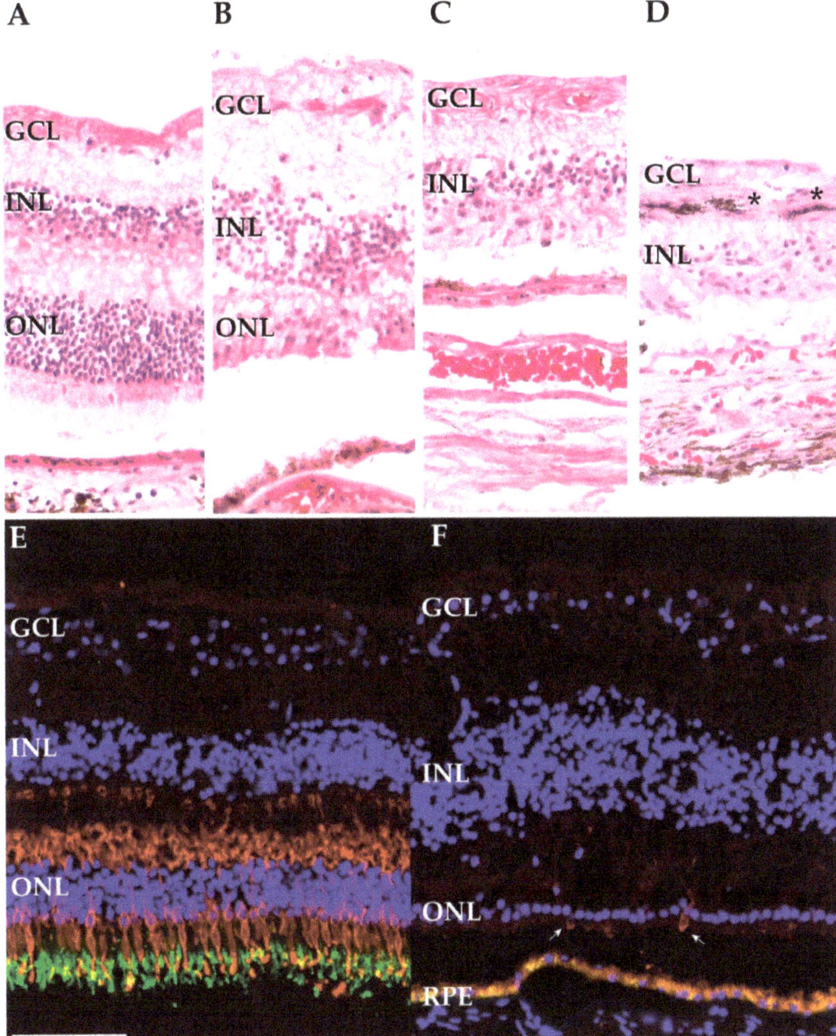

Figure 2. Histological and immunohistochemical investigation of the autoimmune retinopathy (AIR) and control donors. (**A–D**). Hematoxylin and eosin staining of the AIR and control donors. Sections from the periphery of a control donor (donor 2) (**A**), the macula (OD) of the AIR donor (**B**), and the periphery of the AIR donor (**C,D**). The AIR macula demonstrates intact ganglion cell and inner nuclear layers with attenuated cone photoreceptor outer segments. In contrast, in the periphery of the AIR donor complete loss of the outer nuclear layer (**C**) and retinal pigment epithelium (RPE) pigment migration into the inner retina (**D**) is observed (*). (**E,F**): Cone opsins (blue cone opsin and red/green cone opsins) are labeled in red while RetP1 is labeled in green. (**E**): A macula from a donor with normal ocular history demonstrates abundant labeling of cone opsins and rhodopsin. Of note, the RPE below the photoreceptors is out of frame. (**F**) The macula from the AIR donor demonstrates a complete lack of rod photoreceptors with rare, extremely attenuated cone photoreceptors (arrows). Autofluorescent lipofuscin from the RPE appears below the photoreceptor cells. Scalebar (100 microns) for all subpanels is provided in (**E**).

Cone and rod photoreceptor cells were also visualized with fluorescent immunohistochemistry (IHC). Within the macula of a donor with normal ocular history, abundant cone opsin and rhodopsin labeling was observed (Figure 2E). In contrast, the AIR donor demonstrates complete loss of the rod specific opsin rhodopsin as well as extreme attenuation of cone photoreceptors (Figure 2F).

3.3. Single-Cell Gene Profiling of Diseased Cell Populations

Paired foveal and peripheral retinal punches were acquired from each of the five donors. While the four control donors had grossly normal retinas upon examination in the laboratory (Figure 3A), the donor with AIR had abundant peripheral pigmentation with a mostly unaffected macula (Figure 3B). After gentle dissociation, single-cell RNA sequencing was performed on each foveal and peripheral sample, and a total of 23,429 cells were recovered after filtering (Figure 3C). A total of 23 clusters were identified, and expression profiles were used to assign each cluster to its corresponding retinal cell type (Figure 3D). All major populations of retinal neurons, as well as supporting retinal endothelial cells, pericytes, glial cells, and microglia, were identified.

Next, the distribution of recovered cell types was compared between the AIR donor and the four control donors (Table S1). As the cellular composition of the retina varies between the fovea and periphery, comparisons were stratified by region. Within the fovea, cone photoreceptor cells are more abundant than rod photoreceptor cells, and cone photoreceptor cells synapse one-to-one with bipolar cells and upstream retinal ganglion cells (Figure 4A). The fovea centralis comprises the central 0.65-0.70 mm of the retina, consisting exclusively of cone photoreceptor cells and excluding vascular elements [19]. Our use of 2 mm foveal centered punches completely captures the fovea centralis but also includes some central rod photoreceptors and retinal endothelial cells. In the four control donors, all major populations of inner retinal neurons were recovered from the fovea (Figure 4B). No RPE cells were detected, suggesting that the foveal retinal punch was well separated from the underlying RPE and choroid. Unlike the control donors, no rod photoreceptor cells were detected in the foveal punch from the donor with AIR. However, a similar proportion of foveal cone photoreceptor cells were recovered in the AIR donor and the control donors. In addition, the AIR donor demonstrated a moderate increase in the proportion of recovered foveal bipolar and Müller cells.

In the periphery, rod photoreceptor cells were predominant, and peripheral bipolar cells receive input from multiple rod photoreceptor cells (Figure 4D). In the four control donors, peripheral rod photoreceptor cells were much more abundant than cone photoreceptor cells, and relatively few microglia or astrocytes were detected (Figure 4E). In contrast, only a single rod photoreceptor cell was recovered from the periphery of the AIR donor while microglia and astrocyte cells were recovered in much higher frequency. In addition, a small proportion of RPE cells were recovered from the periphery of the AIR donor, consistent with the histological observation of peripheral RPE migration into the retina (Figure 2D).

Next, the transcriptomic consequences of photoreceptor degeneration in the AIR donor were investigated. For each cell type, gene expression was compared between cells originating from the AIR donor and the control donors, and the proportion of significantly differentially expressed genes (adjusted *p*-value < 0.05) that exhibited an absolute log fold-change in expression greater than 0.5 was calculated. As gene expression within a single cell type can vary between the fovea and the periphery [13], this analysis was again stratified by region. Within the fovea, Müller cells and horizontal cells demonstrated modest expression differences, with a total of 1.1% and 1.2% of assayed genes significantly enriched in the AIR Müller cell and horizontal cell populations, respectively (Figure 4C). A greater proportion of differentially expressed genes between the AIR and control donors were identified in the periphery (Figure 4F). Müller and astrocyte glial cells both demonstrated a modest proportion of genes significantly enriched in the periphery of the AIR donor (1.3% and 2.1%, respectively), as did microglia and horizontal cells (2.2% and 2.6%, respectively). Differential expression results for each comparison are shown in detail in Table S2.

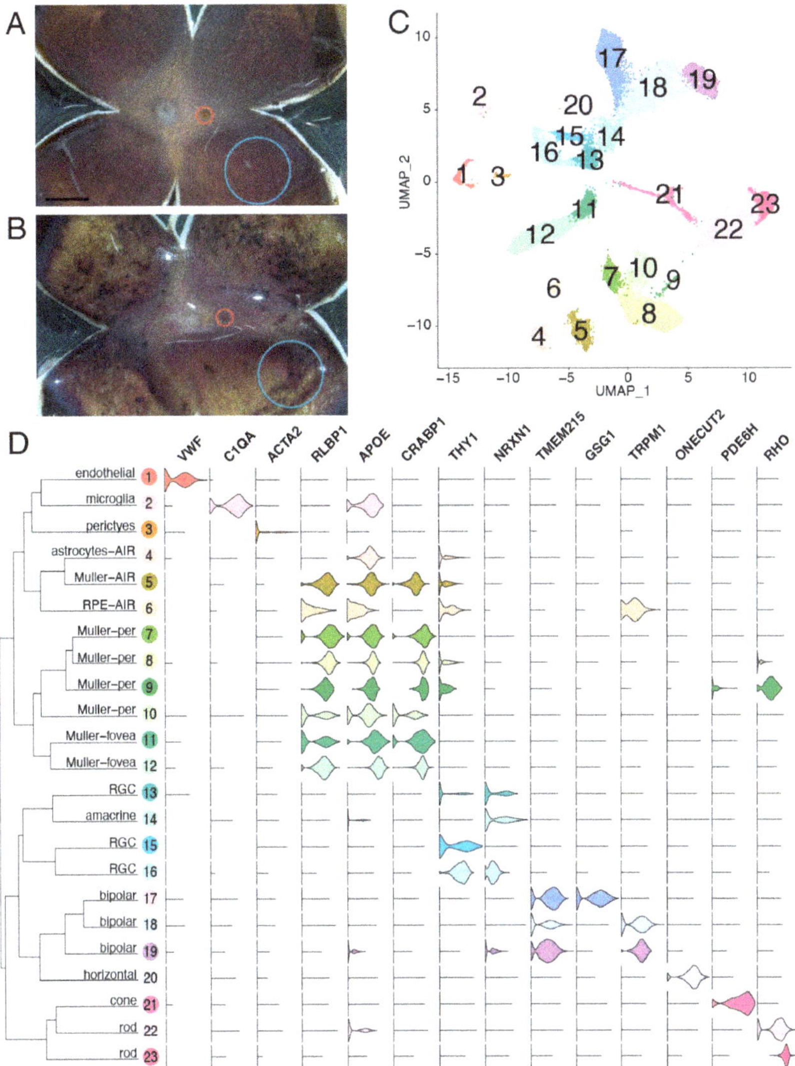

Figure 3. Single-cell RNA sequencing of the AIR donor. (**A**,**B**): Five human donor eyes were used for this study. A gross image of a control eye (donor 4) (**A**) and the AIR eye (donor 5) (**B**) are included. From each eye, a 2 mm foveal centered punch (red) and an 8 mm peripheral punch isolated from the inferotemporal region (blue) were acquired and gently dissociated. Scalebar (**A**) is 5 mm. (**C**): Single-cell RNA sequencing of retinal cells from the AIR donor and four control patients. A total of 23,429 cells were recovered after filtering. Unsupervised clustering of cells resulted in 23 clusters, which are visualized with uniform manifold approximation and projection (UMAP) dimensionality reduction, where each point represents the multidimensional transcriptome of a single-cell and each cluster of cells is depicted in a different color. (**D**): Violin plots depict the expression of cell-type specific genes across the 23 identified clusters. Per = peripheral retina. AIR = autoimmune retinopathy. RPE = retinal pigment epithelium. RGC = retinal ganglion cell.

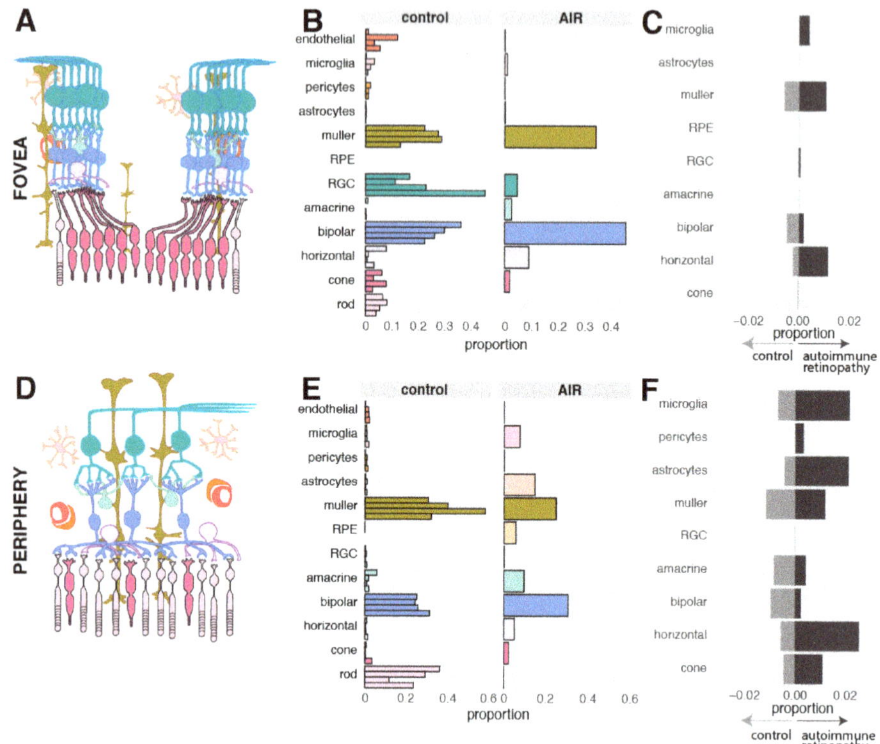

Figure 4. Library composition of recovered cells. (**A**): In the fovea, cone photoreceptor cells synapse with one bipolar cell, which synapse with one retinal ganglion cell. (**B**): The proportion of each cell type recovered from the fovea of the four control donors and the autoimmune retinopathy donor. No foveal rods were recovered from the AIR donor. (**C**): In order to visualize the degree of gene expression differences within each population of cells between the AIR and control donors, differential expression analysis was performed. In each cell type, the number of differentially expressed genes that were enriched in the AIR donor and the control donors were enumerated and divided by the total number of expressed genes (in at least 10% of cells). For example, 1.09% of foveal Müller cell genes were significantly enriched in the AIR donor (dark grey), while 0.57% of foveal Müller cell genes were significantly enriched in the control donors (light grey). (**D**): In the periphery, multiple rod photoreceptor cells synapse with a single bipolar cell. (**E**): The proportion of each cell type recovered from the periphery of the four control donors and the periphery of the AIR donor. (**F**): As in (**C**), the proportion of differentially expressed genes between the AIR and control donors was performed in each cell type. More genes were differentially expressed in the periphery compared to the fovea (**C**). As no RPE cells originated from control donors, differential expression could not be performed in the periphery for this cell type.

While most clusters contained cells from each of the five donors, Clusters 4–6 were comprised predominantly of cells from the periphery of the AIR patient (each cluster possessing >85% of cells from the AIR donor) (Figure 5A). Therefore, gene expression patterns from these clusters were further investigated. Cluster 4 was classified as astrocytes (Figure 5C,D). Cells in this cluster demonstrated high expression of the glial fibrillary acid protein (GFAP), which is widely expressed in astrocytes responding to neuronal injury [20], and the astrocyte-specific inflammatory cytokine IFITM3 [21]. A total of 624 cells were recovered in Cluster 4 and 551 of them (88%) originated from the periphery of the AIR donor. Differential expression analysis was performed to investigate if astrocytes from the AIR donor demonstrated a reactive gene expression profile (Figure S1). Astrocytes from the AIR donor were enriched for SOCS3 [22], SLPI [23], and CH25H [24], genes that have all been previously found to be expressed in astrocytes responding to CNS injury. In addition, reactive glial cells are involved in inflammatory responses, and have been shown to increase the production of pro-inflammatory chemokines [25]. The chemokine CXCL2 was highly enriched (logFC = 1.47) in peripheral astrocytes from the AIR donor.

Cluster 5, with 98% of cells originating from the periphery of the AIR donor, was interpreted as Müller glial cells. Cells in this cluster highly expressed the Müller cell genes RLBP1 and CRALBP1 (Figure 5E). Six additional clusters of Müller glia were identified, which largely separated Müller cells of peripheral (Clusters 6–10) and foveal (Clusters 11–2) origin (Figure 5A). As previously shown in monkey [26] and human [13] retina, foveal and peripheral Müller cells have distinct gene expression profiles (Figure 5G). Interestingly, foveal Müller cells from the AIR donor clustered with foveal Müller cells from the other four donors, and differential expression analysis yielded relatively few expression differences (log fold-change greater than 1.25) (Figure 5H). NFKBAI, which has been previously associated with glial cell degeneration, [27] was the most upregulated gene in the AIR donor's foveal Müller cells.

In contrast, the majority of peripheral Müller cells from the AIR donor formed their own cluster (Cluster 5). Differential expression revealed numerous expression differences between peripheral Müller cells from the AIR donor versus peripheral Müller cells from other donors (Figure 5I). Among the first hallmarks of reactive gliosis is the increased expression of intermediate filament proteins [28]. The intermediate filament gene GFAP (logFC = 2.54) was the most enriched gene in Müller cells from the AIR donor, which was also observed to be more abundant in the AIR donor at the protein level (Figure S2). In addition, peripheral Müller cells from the AIR donor were enriched for ANXA1 and ANXA2, which have been shown to be upregulated in reactive glial populations in the brain [29]. Immunofluorescent IHC also demonstrates increased ANXA1 labeling in cells of the AIR donor (Figure 5B), which co-localizes with GFAP expression (Figure S3).

Cluster 6, with 99% of cells originating from the periphery of the AIR donor, was interpreted as retinal pigment epithelium (RPE) cells (Figure 5F). Cells in this cluster demonstrated high expression of SERPINF1 (the gene encoding PEDF) and RLBP1 (the gene encoding cellular retinaldehyde binding protein). Although retinal samples were dissected away from the underlying RPE and choroid, the recovery of RPE cells suggests that either some RPE cells migrated into the inner retina or remained adhered to the outer retina after dissection, consistent with both the clinical observation of bone spicule like pigmentation in the patient's neurosensory retina (Figure 1A–B) and the morphological finding of pigment migration into the inner retina (Figure 2D).

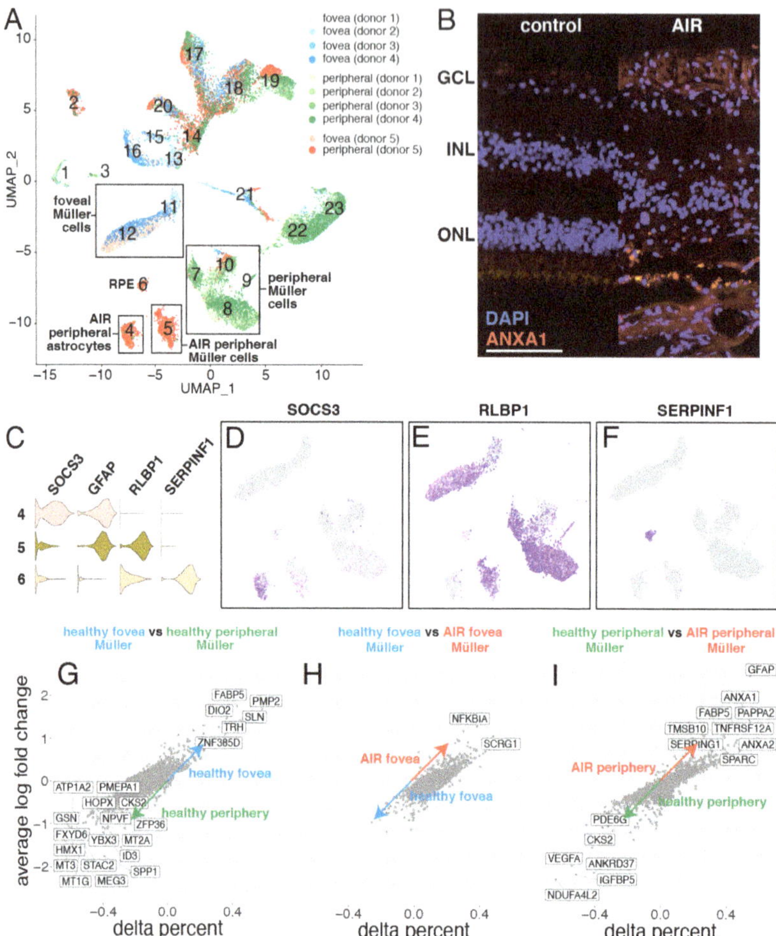

Figure 5. Exploration of autoimmune retinopathy dominant clusters. (**A**): The library composition of each cluster is displayed, with cells originating from the control foveas represented in shades of blue while cells originating from the control peripheries are in shades of green. Cells originating from the AIR donor are colored light red (fovea) or dark red (periphery). Three clusters (Cluster 4–6) consist predominantly of cells from the periphery of the AIR donor. (**B**): Immunofluorescent labeling of ANXA1 in the retina of a control donor (left) and the AIR donor (right). The AIR donor demonstrates increased ANXA1-labeling of the inner retina. Scale bar = 100 microns. (**C**): Violin plots of SOCS3, GFAP, RLBP1, and SERPINF1 expression are used to classify the cell types of clusters 4–6. (**D**): Cluster 4 specifically expressed SOCS3, which is enriched in reactive astrocytes. (**E**): Cluster 5 and Clusters 7-12 express the Müller cell specific gene RLBP1. (**F**): Cluster 6 expresses the RPE-specific gene SERPINF1. (**G**): Healthy foveal and peripheral Müller cells have distinct gene expression profiles. The variable delta percent along the x-axis represents the proportion of foveal Müller cells that express the gene of interest minus the proportion of peripheral Müller cells that express that gene. (**H**): Foveal Müller cells originating from the control donors have similar gene expression profiles to foveal Müller cells originating from the AIR donor. (**I**): In contrast, peripheral Müller cells from control versus the AIR donor demonstrated more transcriptomic differences. Genes with a log fold-change greater than 1.0 and a delta percent greater than 0.35 are labeled in (**G–I**).

4. Discussion

Autoimmune retinopathy (AIR) is a blinding, immune-mediated inflammatory condition in which anti-retinal antibodies result in retinal cell destruction. In most cases, photoreceptor cells are the primarily targeted cell type, although antibodies against bipolar cells have also been reported [30]. The clinical presentation of the patient in this report follows the classical trajectory of sudden, bilateral loss of peripheral vision, consistent with rod photoreceptor cell dysfunction [31]. The etiology of AIR is broadly subdivided into paraneoplastic and non-paraneoplastic disease. Identifying anti-retinal antibodies can support a diagnosis of AIR, however unaffected individuals may also have circulating anti-retinal antibodies, limiting the specificity of this diagnostic assay [32,33].

Consistent with the clinical history of peripheral visual deterioration, histological examination of the AIR patient revealed a profound loss of both cone and rod photoreceptor cells and destruction of the inner and outer photoreceptor segments in the periphery (Figure 2C). However, in the macula, rare attenuated cone photoreceptor cells were still present (Figure 2F). Single-cell RNA sequencing supported these clinical and histological findings. Only a single rod photoreceptor cell was recovered from the periphery of the AIR donor while numerous foveal cones were recovered. In addition, single-cell RNA sequencing identified the presence of RPE cells in the periphery of the AIR donor, as was observed on histological examination (Figure 2D). Gene expression comparisons between the AIR donor and the four control donors were remarkably similar for most cell types. However, peripheral astrocytes and Müller glial cells were more abundant and demonstrated unique expression signatures in the AIR patient. Collectively, these expression data corroborate the clinical and histologic findings and provide evidence that single-cell RNA sequencing can be a complementary tool for investigating the molecular features of a human retinal disease.

The retina contains two major classes of glial cells: Müller cells and astrocytes. Müller cells are elongated cells that extend from the external limiting membrane (apical end) to the internal limiting membrane (basal feet). Müller cells provide metabolic and structural support to retinal neurons, ensheathing neural somas and comprising an important part of the blood retina barrier. Astrocytes also metabolically support the retina, however astrocytes do not originate from the embryonic retinal neuroepithelium but rather enter the retina by migrating along the developing optic nerve [34]. As opposed to Müller cells, astrocytes are star-shaped cells with radiating processes located in the nerve fiber and ganglion cell layers. Both astrocytes and Müller glial cells are capable of responding to retinal injury and exerting neuroprotective effects on the retina in a process known as reactive gliosis [35]. In this wound response process, glial cells proliferate and undergo changes in gene expression for improved neuronal protection and repair [36,37].

In the donor with AIR, the transcriptional response of the glial cells can likely be attributed to their interactions with degenerating retina. Within the fovea, where the retina clinically and histologically was most intact, Müller cells from the AIR donor were transcriptionally similar to foveal Müller cells from the control patients. Yet in the peripheral retina, where the AIR donor experienced progressive visual field loss and a complete loss of the outer nuclear layer, peripheral Müller cells segregated into a distinct cluster and demonstrated a reactive gliotic phenotype (Figure 5I). Likewise, many astrocytes were recovered from the periphery of the AIR donor that expressed genes implicated in reactive gliosis (Figure S1). Reactive astrogliosis is marked by astrocyte proliferation and migration, which may have led to an increased number of peripherally localized astrocytes available for recovery in the AIR donor, consistent with recent single-cell RNA sequencing studies characterizing microglial proliferation in response to retinal damage in mice [38]. Collectively, the gliotic injury response induced by Müller cells and astrocytes has many neuroprotective benefits, yet chronic gliotic activation can further injure retinal neurons and disrupt the blood–retinal barrier, leading to worsening vision [39,40]. In the setting of chronic retinal injury, interventions that modulate gliotic activation may optimize preservation of remaining retinal function [41].

While glial cells from the AIR donor demonstrated reactive transcriptional changes, most inner retinal cell populations from this donor had remarkably similar gene expression profiles to the control donors (Figure 4C,F). Likewise, histological examination revealed preserved inner retinal morphology with discrete inner nuclear and ganglion cell layers (Figure 2B,C). Collectively, these findings suggest that even in the setting of photoreceptor cell degeneration, the inner retinal wiring remains largely undamaged. The presence of morphologically and transcriptomically normal inner retinal cells is promising for prospective photoreceptor degeneration treatments, including autologous retinal cellular replacement strategies [42].

There are several limitations to this study. First, AIR is a rare retinal disease, preventing us from including multiple patients with this condition in this investigation. As a result, gene expression differences between the AIR donor and the four control donors are valuable for hypothesis generation but should be interpreted with caution. Second, while all samples had identical sample processing, certain cell types might have a selective advantage in cellular recovery for single-cell RNA sequencing. Recovered proportions of cells at the single-cell level (Figure 4B,E) should not be interpreted as the true cellularity of the retina.

This study provides a complementary investigation of the clinical and molecular response of the retina in AIR. Clinical, histologic, and transcriptomic evidence identify the loss of cone and rod photoreceptor cells with relative preservation of inner retinal cell types. The gliotic transcriptional profile of astrocyte and Müller glial populations observed in this case provides some new insight into the retina's response to photoreceptor degeneration.

Supplementary Materials: The following are available online at http://www.mdpi.com/2073-4409/9/2/438/s1, Figure S1: AIR astrocytes demonstrate a reactive gene expression profile. Figure S2: Increased GFAP in the AIR donor versus a healthy control donor. Figure S3 ANXA1 and GFAP co-localization in the AIR donor. Table S1: Library Composition, Table S2: Differential expression between the AIR donor versus four control donors.

Author Contributions: Conceptualization, R.F.M. and E.M.S.; methodology, A.P.V., E.B., R.F.M., E.M.S.; software, A.P.V., A.P.D., T.E.S.; validation, E.B., M.J.F.-W., S.Z.; formal analysis, A.P.V., A.P.D., T.E.S.; investigation, A.P.V., E.B., M.J.F.-W., S.Z., A.P.D., T.E.S., B.A.T., R.F.M., E.M.S.; resources, T.E.S., B.A.T., R.F.M., E.M.S. data curation, A.P.V., E.B., T.E.S., R.F.M.; writing—original draft preparation, A.P.V., E.B., B.A.T., R.F.M., E.M.S.; writing—review and editing, M.J.F.-W., S.Z., A.P.D., T.E.S.; visualization, A.P.V., E.B., R.F.M.; supervision, T.E.S., B.A.T., R.F.M., E.M.S.; project administration, E.M.S.; funding acquisition, T.E.S., B.A.T., R.F.M., E.M.S. All authors have read and agreed to the published version of the manuscript.

Funding: This research was funded by NIH grants T32 GM007337, R21 EY027038, and P30 EY025580 with support from Research to Prevent Blindness and the Elmer and Sylvia Sramek Charitable Foundation.

Acknowledgments: We wish to thank the Iowa Lions Eye Bank, the donors, and their families for the generous role in this research. The ANXA1 monoclonal antibody developed by the Clinical Proteomics Technologies for Cancer and was obtained from the Developmental Studies Hybridoma Bank, created by the National Institute of Child Health and Human Development of the NIH and maintained at the Department of Biology, The University of Iowa (Iowa City, IA).

Conflicts of Interest: The authors declare no conflict of interest.

References

1. Wright, A.F.; Chakarova, C.F.; Abd El-Aziz, M.M.; Bhattacharya, S.S. Photoreceptor degeneration: Genetic and mechanistic dissection of a complex trait. *Nat. Rev. Genet.* **2010**, *11*, 273–284. [CrossRef]
2. Stone, E.M.; Andorf, J.L.; Whitmore, S.S.; DeLuca, A.P.; Giacalone, J.C.; Streb, L.M.; Braun, T.A.; Mullins, R.F.; Scheetz, T.E.; Sheffield, V.C.; et al. Clinically Focused Molecular Investigation of 1000 Consecutive Families with Inherited Retinal Disease. *Ophthalmology* **2017**, *124*, 1314–1331. [CrossRef] [PubMed]
3. Adamus, G.; Ren, G.; Weleber, R.G. Autoantibodies against retinal proteins in paraneoplastic and autoimmune retinopathy. *BMC Ophthalmol.* **2004**, *4*, 5. [CrossRef] [PubMed]
4. Kannan, R.; Hinton, D.R. Sodium iodate induced retinal degeneration: New insights from an old model. *Neural Regen. Res.* **2014**, *9*, 2044–2045. [PubMed]
5. Li, M.; Jia, C.; Kazmierkiewicz, K.L.; Bowman, A.S.; Tian, L.; Liu, Y.; Gupta, N.A.; Gudiseva, H.V.; Yee, S.S.; Kim, M.; et al. Comprehensive analysis of gene expression in human retina and supporting tissues. *Hum. Mol. Genet.* **2014**, *23*, 4001–4014. [CrossRef] [PubMed]

6. Whitmore, S.S.; Wagner, A.H.; DeLuca, A.P.; Drack, A.V.; Stone, E.M.; Tucker, B.A.; Zeng, S.; Braun, T.A.; Mullins, R.F.; Scheetz, T.E. Transcriptomic analysis across nasal, temporal, and macular regions of human neural retina and RPE/choroid by RNA-Seq. *Exp. Eye Res.* **2014**, *129*, 93–106. [CrossRef]

7. Hwang, B.; Lee, J.H.; Bang, D. Single-cell RNA sequencing technologies and bioinformatics pipelines. *Exp. Mol. Med.* **2018**, *50*, 96. [CrossRef]

8. Sarthy, P.V.; Lam, D.M. Isolated cells from a mammalian retina. *Brain Res.* **1979**, *176*, 208–212. [CrossRef]

9. Feodorova, Y.; Koch, M.; Bultman, S.; Michalakis, S.; Solovei, I. Quick and reliable method for retina dissociation and separation of rod photoreceptor perikarya from adult mice. *MethodsX* **2015**, *2*, 39–46. [CrossRef]

10. Macosko, E.Z.; Basu, A.; Satija, R.; Nemesh, J.; Shekhar, K.; Goldman, M.; Tirosh, I.; Bialas, A.R.; Kamitaki, N.; Martersteck, E.M.; et al. Highly Parallel Genome-wide Expression Profiling of Individual Cells Using Nanoliter Droplets. *Cell* **2015**, *161*, 1202–1214. [CrossRef]

11. Rheaume, B.A.; Jereen, A.; Bolisetty, M.; Sajid, M.S.; Yang, Y.; Renna, K.; Sun, L.; Robson, P.; Trakhtenberg, E.F. Single cell transcriptome profiling of retinal ganglion cells identifies cellular subtypes. *Nat. Commun.* **2018**, *9*, 2759. [CrossRef] [PubMed]

12. Clark, B.S.; Stein-O'Brien, G.L.; Shiau, F.; Cannon, G.H.; Davis-Marcisak, E.; Sherman, T.; Santiago, C.P.; Hoang, T.V.; Rajaii, F.; James-Esposito, R.E.; et al. Single-Cell RNA-Seq Analysis of Retinal Development Identifies NFI Factors as Regulating Mitotic Exit and Late-Born Cell Specification. *Neuron* **2019**, *102*, 1111.e1115–1126.e1115. [CrossRef] [PubMed]

13. Voigt, A.P.; Whitmore, S.S.; Flamme-Wiese, M.J.; Riker, M.J.; Wiley, L.A.; Tucker, B.A.; Stone, E.M.; Mullins, R.F.; Scheetz, T.E. Molecular characterization of foveal versus peripheral human retina by single-cell RNA sequencing. *Exp. Eye Res.* **2019**, *184*, 234–242. [CrossRef]

14. Hu, Y.; Wang, X.; Hu, B.; Mao, Y.; Chen, Y.; Yan, L.; Yong, J.; Dong, J.; Wei, Y.; Wang, W.; et al. Dissecting the transcriptome landscape of the human fetal neural retina and retinal pigment epithelium by single-cell RNA-seq analysis. *PLoS Biol* **2019**, *17*, e3000365. [CrossRef]

15. Lukowski, S.W.; Lo, C.Y.; Sharov, A.A.; Nguyen, Q.; Fang, L.; Hung, S.S.; Zhu, L.; Zhang, T.; Grunert, U.; Nguyen, T.; et al. A single-cell transcriptome atlas of the adult human retina. *EMBO J.* **2019**, *38*, e100811. [CrossRef] [PubMed]

16. Barthel, L.K.; Raymond, P.A. Improved method for obtaining 3-microns cryosections for immunocytochemistry. *J. Histochem. Cytochem.* **1990**, *38*, 1383–1388. [CrossRef]

17. Zheng, G.X.Y.; Terry, J.M.; Belgrader, P.; Ryvkin, P.; Bent, Z.W.; Wilson, R.; Ziraldo, S.B.; Wheeler, T.D.; McDermott, G.P.; Zhu, J.; et al. Massively parallel digital transcriptional profiling of single cells. *Nat. Commun.* **2017**, *8*, 14049. [CrossRef]

18. Butler, A.; Hoffman, P.; Smibert, P.; Papalexi, E.; Satija, R. Integrating single-cell transcriptomic data across different conditions, technologies, and species. *Nat. Biotechnol.* **2018**, *36*, 411–420. [CrossRef]

19. Yuodelis, C.; Hendrickson, A. A qualitative and quantitative analysis of the human fovea during development. *Vision Res.* **1986**, *26*, 847–855. [CrossRef]

20. Sofroniew, M.V.; Vinters, H.V. Astrocytes: Biology and pathology. *Acta Neuropathol.* **2010**, *119*, 7–35. [CrossRef]

21. Ibi, D.; Nagai, T.; Nakajima, A.; Mizoguchi, H.; Kawase, T.; Tsuboi, D.; Kano, S.; Sato, Y.; Hayakawa, M.; Lange, U.C.; et al. Astroglial IFITM3 mediates neuronal impairments following neonatal immune challenge in mice. *Glia* **2013**, *61*, 679–693. [CrossRef] [PubMed]

22. Okada, S.; Nakamura, M.; Katoh, H.; Miyao, T.; Shimazaki, T.; Ishii, K.; Yamane, J.; Yoshimura, A.; Iwamoto, Y.; Toyama, Y.; et al. Conditional ablation of Stat3 or Socs3 discloses a dual role for reactive astrocytes after spinal cord injury. *Nat. Med.* **2006**, *12*, 829–834. [CrossRef] [PubMed]

23. Ghasemlou, N.; Bouhy, D.; Yang, J.; Lopez-Vales, R.; Haber, M.; Thuraisingam, T.; He, G.; Radzioch, D.; Ding, A.; David, S. Beneficial effects of secretory leukocyte protease inhibitor after spinal cord injury. *Brain* **2010**, *133*, 126–138. [CrossRef] [PubMed]

24. Zhu, Z.; Hu, Y.; Zhou, Y.; Zhang, Y.; Yu, L.; Tao, L.; Guo, A.; Fang, Q. Macrophage Migration Inhibitory Factor Promotes Chemotaxis of Astrocytes through Regulation of Cholesterol 25-Hydroxylase Following Rat Spinal Cord Injury. *Neuroscience* **2019**, *408*, 349–360. [CrossRef]

25. Farina, C.; Aloisi, F.; Meinl, E. Astrocytes are active players in cerebral innate immunity. *Trends Immunol.* **2007**, *28*, 138–145. [CrossRef]

26. Peng, Y.R.; Shekhar, K.; Yan, W.; Herrmann, D.; Sappington, A.; Bryman, G.S.; van Zyl, T.; Do, M.T.H.; Regev, A.; Sanes, J.R. Molecular Classification and Comparative Taxonomics of Foveal and Peripheral Cells in Primate Retina. *Cell* **2019**, *176*, 1222.e1222–1237.e1222. [CrossRef]

27. Haenold, R.; Weih, F.; Herrmann, K.H.; Schmidt, K.F.; Krempler, K.; Engelmann, C.; Nave, K.A.; Reichenbach, J.R.; Lowel, S.; Witte, O.W.; et al. NF-kappaB controls axonal regeneration and degeneration through cell-specific balance of RelA and p50 in the adult CNS. *J. Cell Sci.* **2014**, *127*, 3052–3065. [CrossRef]

28. Lu, Y.B.; Iandiev, I.; Hollborn, M.; Korber, N.; Ulbricht, E.; Hirrlinger, P.G.; Pannicke, T.; Wei, E.Q.; Bringmann, A.; Wolburg, H.; et al. Reactive glial cells: Increased stiffness correlates with increased intermediate filament expression. *FASEB J.* **2011**, *25*, 624–631. [CrossRef]

29. Eberhard, D.A.; Brown, M.D.; VandenBerg, S.R. Alterations of annexin expression in pathological neuronal and glial reactions. Immunohistochemical localization of annexins I, II (p36 and p11 subunits), IV, and VI in the human hippocampus. *Am. J. Pathol.* **1994**, *145*, 640–649.

30. Choi, E.Y.; Kim, M.; Adamus, G.; Koh, H.J.; Lee, S.C. Non-Paraneoplastic Autoimmune Retinopathy: The First Case Report in Korea. *Yonsei Med. J.* **2016**, *57*, 527–531. [CrossRef]

31. Rahimy, E.; Sarraf, D. Paraneoplastic and non-paraneoplastic retinopathy and optic neuropathy: Evaluation and management. *Surv. Ophthalmol.* **2013**, *58*, 430–458. [CrossRef] [PubMed]

32. Braithwaite, T.; Vugler, A.; Tufail, A. Autoimmune retinopathy. *Ophthalmologica* **2012**, *228*, 131–142. [CrossRef] [PubMed]

33. Adamus, G.; Brown, L.; Schiffman, J.; Iannaccone, A. Diversity in autoimmunity against retinal, neuronal, and axonal antigens in acquired neuro-retinopathy. *J. Ophthalmic Inflamm. Infect.* **2011**, *1*, 111–121. [CrossRef] [PubMed]

34. Stone, J.; Dreher, Z. Relationship between astrocytes, ganglion cells and vasculature of the retina. *J. Comp. Neurol.* **1987**, *255*, 35–49. [CrossRef] [PubMed]

35. de Hoz, R.; Rojas, B.; Ramirez, A.I.; Salazar, J.J.; Gallego, B.I.; Trivino, A.; Ramirez, J.M. Retinal Macroglial Responses in Health and Disease. *Biomed. Res. Int.* **2016**, *2016*, 2954721. [CrossRef]

36. Barres, B.A. The mystery and magic of glia: A perspective on their roles in health and disease. *Neuron* **2008**, *60*, 430–440. [CrossRef]

37. Sarthy, V.P.; Sawkar, H.; Dudley, V.J. Endothelin2 Induces Expression of Genes Associated with Reactive Gliosis in Retinal Muller Cells. *Curr. Eye Res.* **2015**, *40*, 1181–1184. [CrossRef]

38. Todd, L.; Palazzo, I.; Suarez, L.; Liu, X.; Volkov, L.; Hoang, T.V.; Campbell, W.A.; Blackshaw, S.; Quan, N.; Fischer, A.J. Reactive microglia and IL1beta/IL-1R1-signaling mediate neuroprotection in excitotoxin-damaged mouse retina. *J. Neuroinflammation* **2019**, *16*, 118. [CrossRef]

39. Coorey, N.J.; Shen, W.; Chung, S.H.; Zhu, L.; Gillies, M.C. The role of glia in retinal vascular disease. *Clin. Exp. Optom.* **2012**, *95*, 266–281. [CrossRef]

40. Baumann, B.; Sterling, J.; Song, Y.; Song, D.; Fruttiger, M.; Gillies, M.; Shen, W.; Dunaief, J.L. Conditional Muller Cell Ablation Leads to Retinal Iron Accumulation. *Invest. Ophthalmol. Vis. Sci.* **2017**, *58*, 4223–4234. [CrossRef]

41. Peng, L.; Parpura, V.; Verkhratsky, A. EDITORIAL Neuroglia as a Central Element of Neurological Diseases: An Underappreciated Target for Therapeutic Intervention. *Curr. Neuropharmacol.* **2014**, *12*, 303–307. [CrossRef] [PubMed]

42. Burnight, E.R.; Giacalone, J.C.; Cooke, J.A.; Thompson, J.R.; Bohrer, L.R.; Chirco, K.R.; Drack, A.V.; Fingert, J.H.; Worthington, K.S.; Wiley, L.A.; et al. CRISPR-Cas9 genome engineering: Treating inherited retinal degeneration. *Prog. Retin. Eye Res.* **2018**, *65*, 28–49. [CrossRef] [PubMed]

Article

Role of FGF and Hyaluronan in Choroidal Neovascularization in Sorsby Fundus Dystrophy

Alyson Wolk [1,2], Dilara Hatipoglu [1], Alecia Cutler [1], Mariya Ali [1], Lestella Bell [1,3], Jian Hua Qi [1], Rupesh Singh [1], Julia Batoki [1], Laura Karle [1], Vera L. Bonilha [1,2,3], Oliver Wessely [2,4], Heidi Stoehr [5], Vincent Hascall [6] and Bela Anand-Apte [1,2,3,*]

1   Cole Eye Institute & Lerner Research Institute, Cleveland Clinic Foundation, Cleveland, OH 44195, USA; wolka@ccf.org (A.W.); Hatipod@ccf.org (D.H.); cutlera@ccf.org (A.C.); alim2@ccf.org (M.A.); belll3@ccf.org (L.B.); qij@ccf.org (J.H.Q.); Singhr4@ccf.org (R.S.); Batokij@ccf.org (J.B.); lauraikarle@gmail.com (L.K.); bonilhav@ccf.org (V.L.B.)
2   Cleveland Clinic Lerner College of Medicine, Department of Molecular Medicine, Case Western Reserve University, Cleveland, OH 44195, USA; Wesselo@ccf.org
3   Cleveland Clinic Lerner College of Medicine at Case Western Reserve University, Department of Ophthalmology, Cleveland, OH 44195, USA
4   Department of Cardiovascular and Metabolic Sciences, Lerner Research Institute, Cleveland Clinic Foundation, Cleveland, OH 44195, USA
5   Institute of Human Genetics, University of Regensburg, 93053 Regensburg, Germany; Heidi.Stoehr@klinik.uni-regensburg.de
6   Department of Biomedical Engineering, Lerner Research Institute, Cleveland Clinic Foundation, Cleveland, OH 44195, USA; Hascalv@ccf.org
*   Correspondence: anandab@ccf.org

Received: 11 January 2020; Accepted: 28 February 2020; Published: 4 March 2020

Abstract: Sorsby's fundus dystrophy (SFD) is an inherited blinding disorder caused by mutations in the tissue inhibitor of metalloproteinase-3 (*TIMP3*) gene. The SFD pathology of macular degeneration with subretinal deposits and choroidal neovascularization (CNV) closely resembles that of the more common age-related macular degeneration (AMD). The objective of this study was to gain further insight into the molecular mechanism(s) by which mutant TIMP3 induces CNV. In this study we demonstrate that hyaluronan (HA), a large glycosaminoglycan, is elevated in the plasma and retinal pigment epithelium (RPE)/choroid of patients with AMD. Mice carrying the S179C-TIMP3 mutation also showed increased plasma levels of HA as well as accumulation of HA around the RPE in the retina. Human RPE cells expressing the *S179C-TIMP3* mutation accumulated HA apically, intracellularly and basally when cultured long-term compared with cells expressing wildtype *TIMP3*. We recently reported that RPE cells carrying the *S179C-TIMP3* mutation have the propensity to induce angiogenesis via basic fibroblast growth factor (FGF-2). We now demonstrate that FGF-2 induces accumulation of HA in RPE cells. These results suggest that the TIMP3-MMP-FGF-2-HA axis may have an important role in the pathogenesis of CNV in SFD and possibly AMD.

Keywords: sorsby's fundus dystrophy; hyaluronan; neovascularization; retina

---

1. Introduction

Sorsby's fundus dystrophy (SFD) is a dominantly inherited, degenerative disease of the macula that is characterized by bilateral loss of central vision as a consequence of choroidal neovascularization (CNV) [1–6]. Specific mutations in the tissue inhibitor of metalloproteinase 3 (*TIMP3*) gene involving exon 5, exon 1 or the intron 4-exon 5 boundary have been shown to be causative [7–14]. In comparative studies using TIMP3 deficient mice, S179C-TIMP3 transgenic mice and in vitro culture experiments we have determined that TIMP3 partially inhibits angiogenesis by blocking the binding of vascular endothelial growth factor (VEGF) to VEGF Receptor 2 (VEGFR2). We have also demonstrated that the S179C-TIMP3 mutant protein induces angiogenesis via VEGF and fibroblast growth factor 2 (FGF-2) [15–21].

TIMP3 is produced constitutively by the retinal pigment epithelium (RPE) and choroidal endothelial cells [2,20]. It is a normal component of Bruch's membrane [22] and binds to sulfated glycosaminoglycans of the extracellular matrix (ECM) [23,24]. Hyaluronan (HA) is a large glycosaminoglycan that is a significant component of peri-cellular and extracellular matrices. HA is essential for numerous physiological functions that are dependent on its chain size and its interactions with various effector proteins and receptors [25]. HA has been implicated in the regulation of neovascularization and endothelial barrier function [26]. While studies have demonstrated that signaling via HA and its cell surface receptor CD44 accentuates CNV in mice using a laser-induced model [27], the exact molecular mechanism by which HA regulates tissue remodeling and neovascularization is unknown.

We have recently reported that RPE cells expressing mutant TIMP3 secrete increased amounts of FGF-2 [28] and that this contributes to increased angiogenesis. FGF-2 has been shown to be important in tumor angiogenesis, but its role in CNV has been less well studied. The most direct evidence for a role of FGF-2 in CNV comes from studies in which Flk1-Cre or Tie2-Cre mediated deletions of FGF receptor 1 (FGFR1) and FGF receptor 2 (FGFR2) in endothelial cells resulted in reduced laser-induced CNV in mice [29]. Extracellular matrix components such as heparan sulfate proteoglycans (HSPGs) bind and regulate the activity of growth factors such as FGF-2 [30] and have a critical role in the regulation of neovascularization [31]. In addition, the observation that activation of the FGFR-STAT3 pathway can induce a hyaluronan-rich microenvironment that can affect tumor growth [32] led us to test the hypothesis that in addition to VEGF, FGF-2 and hyaluronan also have critical roles in the increased neovascularization induced by mutant TIMP3 in Sorsby's fundus dystrophy.

2. Materials and Methods

2.1. Human Samples

Patients with AMD and controls (without AMD or any other retinal disease) were recruited from the eye clinics at Cole Eye Institute under Cleveland Clinic Foundation approved IRB protocols. Plasma samples were prepared and stored at −80 °C. Samples from patients (n = 49, with 26 males and 23 females) given a clinical diagnosis of geographic atrophy or CNV and age-matched controls (n = 59 with 28 males and 31 females) were included in this pilot study to evaluate HA in the plasma. Normal and/or AMD post-mortem eyes were obtained from the Cleveland Eye Bank, the National Disease Research Interchange (Philadelphia, PA, USA) or from the Cole Eye Institute Eye Tissue Repository through the Foundation Fighting Blindness (FFB) Eye Donor Program (Columbia, MD, USA). All post-mortem tissue were obtained in accordance with the policies of the Eye Bank Association of America and the Institutional Review Board of the Cleveland Clinic Foundation (IRB#14-057). Eye bank records accompanying the donor eyes indicated whether the donor had AMD or no known eye diseases. The analyzed tissue included FFB donations #714 (82 y.o.), #781 (80 y.o.), #711 (83 y.o.), #722 (90 y.o.), #716 (80 y.o.) and #739 (90 y.o.), identified as AMD. Postmortem eyes from a 95 (#784), 92 (#979) and a 91 year-old donor without a history of retinal disease were used as controls. Eyes were enucleated 4 to 22 h postmortem and fixed in 4% paraformaldehyde and 0.5% glutaraldehyde in phosphate buffer. The globes were stored in 2% paraformaldehyde in D-PBS.

2.2. Mice

All mice utilized in this study were housed in the Cole Eye Institute vivarium under approved Institutional Animal Care and Use Committee (IACUC) protocols. All procedures on the mice were in accordance with ARVO statement for the Use of Animals in Ophthalmic and Vision Research and conformed to the National Institutes of Health Guide for the Care and Use of Animals in Research and to the ARVO statement for the use of animals in ophthalmic and vision research. Timp3$^{+/S179C}$ mice were generated in the laboratory of Dr. Bernhard Weber using site-directed mutagenesis and homologous recombination in embryonic stem (ES) cells to generate mutant ES cells carrying the *Timp3^{S179C}* allele. Heterozygous breeding of Timp3$^{+/S179C}$ [33] produced homozygous Timp3$^{S179C/S179C}$ mice and age-matched littermate controls in a C57BL6 background. Similarly, heterozygous Timp3$^{+/-}$ mice [34] were bred to generate Timp3$^{-/-}$ knockouts and TIMP3$^{+/+}$ littermate controls. Eyes were enucleated following euthanasia and fresh frozen in tissue-plus optical cutting temperature embedding medium (Scigen, #4583) for sectioning and histology. Blood samples were collected via cardiac puncture and plasma prepared via standard protocols.

2.3. Hyaluronan Enzyme-Linked Immunosorbent Assay (ELISA)

Plasma from human patients with and without AMD and from SFD mouse models were measured for HA contents by solid-phase sandwich ELISA in 96-well plates (Costar, #9018) using the Hyaluronan Duo-Set ELISA kit (R&D Systems, #DY3614-05).

2.4. Immunofluorescence

Retina sections and flat-mounted ARPE-19 cells grown on polyester trans-wells were fixed for 5 min in 4% paraformaldehyde and blocked in 1% bovine serum albumin with 0.1% Triton X-100 in phosphate-buffered saline. Human sections were processed with melanin bleaching kit to remove autofluorescence (Polysciences, Inc., Warrington, PA, USA, #24883A-B). Samples were incubated overnight with biotinylated HA binding protein, (Millipore Sigma, #385911) or primary antibodies (anti-ezrin, clone 3C12, Invitrogen, Carlsbad, CA, USA #MA5-13862) in humidified chambers at 4 °C. Subsequently, secondary antibodies (anti-mouse AlexaFluor 594, streptavidin-AlexaFluor 488, streptavidin-AlexaFluor 647, all from ThermoFisher Scientific, Waltham, MA, USA) were incubated with samples at room temperature for one hour in the dark. Rhodamine-phalloidin (Thermo Fisher Scientific, R415) was incubated together with secondary antibodies. Then, 4′,6-diamidino-2-phenylindole (DAPI) was used to stain nuclei of murine sections and cell culture mounts and SYTOX green (ThermoFisher Scientific, #S7020) was used to stain nuclei in human sections. Imaging by confocal microscopy was performed (Leica TCS-SP8, Exton, PA, USA). The localization of Bruch's membrane was determined by its autofluorescence at 405 nm.

2.5. Hyaluronidase Treatment of Retina Sections

Hyaluronidase from *Streptomyces hyalurolyticus* (Millipore Sigma, Burlington, MA USA, #H1136) was used to treat retina sections as described previously [35]. Streptomyces hyaluronidase was resuspended in 0.1 M sodium acetate buffer, pH 5.0, at 100 U/mL. To prevent any nonspecific digestion, the following protease inhibitors were added to the sodium acetate buffer: 1 mM iodoacetic acid, 1 mM phenylmethyl sulfonylfluoride, 1 mM EDTA, 1 μg/mL pepstatin A, 250 μg/mL ovomucoid. Hyaluronidase solution (100 mU/mL of hyaluronidase in PBS with CaCl$_2$ (0.1 g/L) and MgCl$_2$ (0.1 g/L)) was applied onto the sections for 3 h at 37 °C. Slides were subsequently fixed in 4% paraformaldehyde and examined by fluorescence microscopy.

2.6. Cells and Reagents

ARPE-19 cells stably expressing S179C-TIMP3, wild-type-TIMP3 (WT), or vector alone were reported previously [19]. Cells were expanded in DMEM-F12 with 10% FBS before transfer to polyester inserts coated with mouse laminin (Corning Inc., Corning, NY, USA, #23017). 720,000 cells and 100,000 cells were plated per well in each well of a 12-well plate or 24-well plate, respectively using a previously published protocol [36]. Essentially, ARPE-19 cells were cultured for at least 2 weeks in nicotinamide-supplemented media with 1% FBS. Media were replaced twice per week. Cells were serum-starved for 24 h before treatment with the FGF Receptor inhibitor BGJ-398 (Selleckchem, Houston, TX, USA, #S2183) for 48 h. Similarly, cells were treated with FGF-2 (Gibco from Thermo

Fisher Scientific, #13256-029) with the required cofactor heparin sodium salt (1 µg/mL, Sigma Aldrich, #H3149) for 48 h after serum starving for 24 h.

2.7. Quantitation of Immunofluorescence by Integrated Density Analysis

Fluorescence intensity of HABP staining was quantified using integrated density analysis as previously described [37,38]. For all the RPE cell culture confocal microscopy images, fluorescence was quantitated using a standard measure of integrated density, which is the product of area and mean gray value. A custom written automated image analysis code was developed using Matlab (MATLAB 2019a, The MathWorks, Inc., Natick, MA, USA) for separating the desired color channel from the image, thereby obtaining the total area (in pixels), the mean gray value, and the integrated density.

2.8. In Vivo Imaging and Laser Injury Model

Laser mediated CNV was induced as described previously [28]. Briefly, mice were anesthetized with 65–68 mg/kg sodium pentobarbital delivered intra-peritoneally. Topical 0.5% procaine solution was applied for cornea anesthesia. Following anesthesia, pupils were dilated with 0.5% topical tropicamide/phenylephrine combination drops (Santen Pharmaceuticals, Osaka, Japan).

Four laser spots were placed in the superior, superior-temporal, or superior-nasal quadrants of the fundus using a green solid-state laser (Oculight by Iridex Corp., Mountain View, CA, USA) (532 nm; 2500 mW; 0.50 s pulse duration; 50 µm spot size) using a slit lamp delivery system and a microscope coverslip placed and affixed to the cornea with a drop of Systane Ultra artificial tears (Alcon, Ft Worth, TX, USA). All animals were scanned immediately after laser injury with optical coherence tomography (Envisu R2210 UHR Leica Microsystems Inc., Wetzlar, Germany) to confirm successful RPE-Bruch's membrane rupture, an endpoint in laser-induced CNV models.

2.9. Statistical Analysis

All parameters in the study were distributed normally. Data are expressed as mean ± SEM. Differences were tested by unpaired t-test (Figure 1, Figure 2 and Figure 3) or by using multiple t-tests employing two-stage linear step-up procedure of Benjamini, Krieger and Yekutieli, with a false discovery rate set to 1% (Figure 4 and Figure 5). Each group was analyzed separately without the assumption of consistent standard deviation. $p < 0.05$ values were considered statistically significant. Statistical analysis was performed with GraphPad Prism 7.03 (GraphPad Software, Inc., San Diego, CA, USA).

3. Results

3.1. Hyaluronan is Elevated in Plasma and RPE/Choroid of Patients with AMD

Age-related macular degeneration (AMD) is usually seen as two main types. "Dry" AMD where deposits called drusen develop in the macular region that ultimately progress to a late stage in which there is atrophy of the macula (geographic atrophy). "Wet" AMD describes AMD in which patients develop abnormal growth and leakage of the choroid vessels beneath and into the retina, termed choroidal neovascularization (CNV). HA contents were measured in plasma from patients with late stage AMD (geographic atrophy or choroidal neovascularization) and from age-matched controls without the disease. ELISA analysis (Figure 1A) indicates that HA contents were significantly increased in the plasma of patients with late-stage AMD (mean ± SEM: 111.8 ± 5.78 ng/mL) compared with plasma of controls without AMD (32.91 ± 5.75).

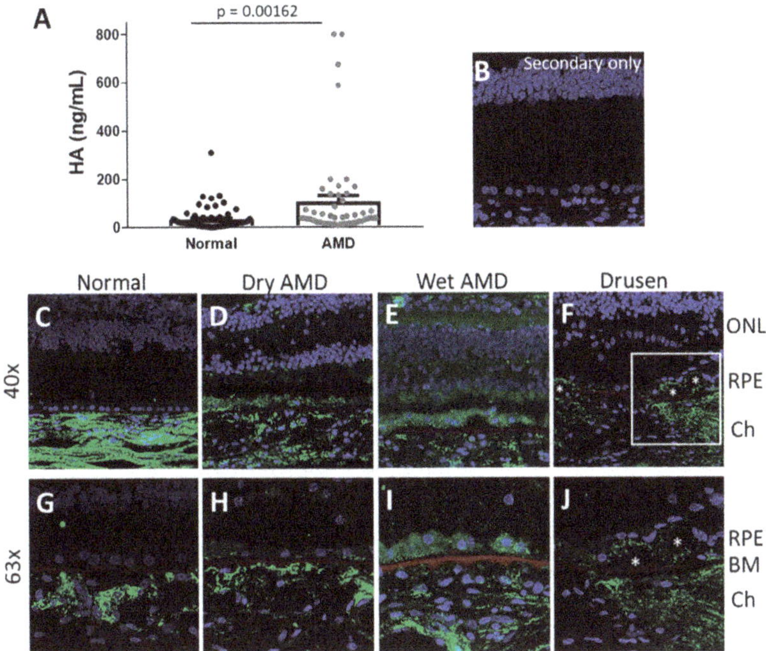

Figure 1. Hyaluronan HA is increased in circulation and in the RPE of age-related macular degeneration (AMD) patients. (**A**) HA was increased in plasma from patients with late-stage AMD (GA or CNV) compared to age-matched, normal controls. Data are presented as mean ± SEM (**B–J**) Representative human retina sections stained with HA binding protein (HABP) (**B**) Human retina section stained with streptavidin-AlexaFluor 647 in the absence of HA binding protein serves as a specificity control. (**C–J**) Human retina sections stained with streptavidin-AlexaFluor 647 in the presence of HA binding protein. HA is increased in the RPE in patients with dry AMD (**D,H**), wet AMD (**E,I**), and around drusen (**F,J**) compared to the RPE from an aged-match normal control (**C,G**). 40× images (**B–F**), 63× images (**G–J**). Green: HA; red: Bruch's membrane determined by its autofluorescence at 405 nm; blue: DAPI. Asterisks indicate drusen (**F,J**). GA—geographic atrophy; CNV—choroidal neovascularization; ONL—outer nuclear layer; RPE—retinal pigment epithelium; Ch—choroid; DAPI—4′,6-diamidino-2-phenylindole.

To evaluate the distribution of HA in the retina under physiological and pathological conditions, sections from post-mortem human donor eyes from 3 controls and 6 AMD (4 dry and 2 wet AMD) patients were stained for HA using biotinylated HA binding protein (HABP). HA was found to be localized predominantly in the choroid of normal eyes (Figure 1C) as described previously [39,40]. Increased deposition of HA was seen around the RPE in AMD eyes (both in the dry (Figure 1D,H) and wet AMD specimens (Figure 1E,I)). HA was particularly enhanced in drusen and in areas of atrophy (Figure 1F,J) in AMD specimens. The sections stained with secondary antibody alone serves as a specificity control and shows minimal staining compared with sections stained with HABP (Figure 1B).

3.2. Increased Plasma HA and Accumulation of HA in the RPE of SFD Mice

We utilized two mouse models to study the potential role of TIMP3 in the regulation of HA in the retina: mice lacking TIMP3 [34] and mice carrying the S179C-TIMP3 SFD mutation [33]. Plasma from S179C-TIMP3 and TIMP3-KO mice at 4–6 weeks of age was collected and HA contents were analyzed by ELISA. HA content of plasma was significantly increased in mice lacking TIMP3 as well as in mice carrying the S179C-TIMP3 mutation (Figure 2A), suggesting that TIMP3 may be important in regulating HA.

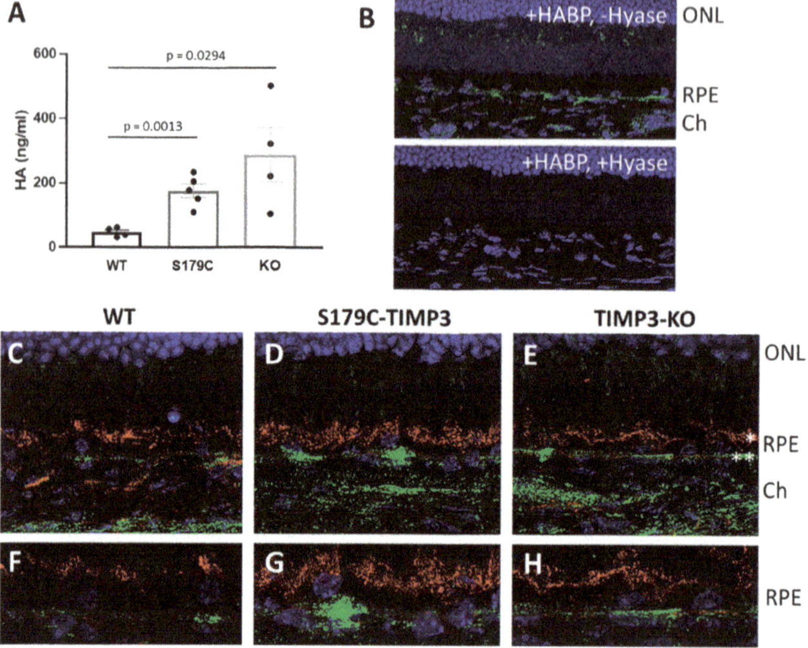

Figure 2. HA is increased in circulation and in the RPE and choroid in mouse models of Sorsby's fundus dystrophy. (**A**) HA was increased in plasma from S179C-TIMP3 knockin mice (Timp3$^{S179C6/S179C}$) and TIMP3-KO (Timp3$^{-/-}$) mice compared to wild-type (WT) littermates. (n ≥ 5). Data are presented as mean ± SEM. (**B**) Mouse retina sections stained with biotinylated HA binding protein (HABP) in the absence (upper panel) or presence (lower panel) of hyaluronidase to detect HA. HABP staining is specific for HA as shown by the absence of staining in sections treated with hyaluronidase (lower panel). Green: HA; blue: DAPI. (**C–H**) Representative images of HA staining of mouse sections from wild-type (WT) mice (**C,F**), S179C-TIMP3 mutant mice (**D,G**) and TIMP3-KO mice (**E,H**). HA is increased in the RPE and choroid of S179C-TIMP3 (**D**) and TIMP3-KO (**E**) mice compared to wild-type (WT) littermate controls (**C**). HA (green) is predominantly localized to the basal surface of the RPE (******) (**F–H**) and not to the apical surface (*****) as shown by co-staining with ezrin, a marker for the apical microvilli of RPE (red). 40× images (**C–E**); 63× images (**F–H**). n ≥ 3 for all immunohistochemistry data.

To determine if there was a similar correlation between the plasma HA levels and the accumulation of HA in the RPE as observed in human sections with AMD, we evaluated accrual of HA in the retinas of mice (8 weeks of age) lacking TIMP3 or carrying the SFD mutation. Cryosections of retina from mice of each specific genotype and wild-type littermates were stained for HA content with HABP. To ascertain that HABP binds HA specifically, sections were treated with hyaluronidase prior to staining with HABP. Indeed, pre-treatment with hyaluronidase resulted in absence of staining with HABP (Figure 2B, lower panel). S179C-TIMP3 mice (Figure 2D) and TIMP3-KO mice (Figure 2E) show increased accumulation of HA beneath the RPE and in the choroid compared to that seen in wildtype littermates (Figure 2C). Staining with antibodies to ezrin served as a marker for RPE apical microvilli (Figure 2C–H), and the higher magnification images (Figure 2F–H) confirmed the RPE localization of HA to the basal surface of the cells.

To identify the potential mechanism by which S179C-TIMP3 regulates HA we utilized stable human RPE lines (ARPE-19) expressing S179C-TIMP3 [20]. ARPE-19 cells (expressing S179C-TIMP3, wildtype TIMP3 (WT-TIMP3) and empty vector (Vector)) were cultured for 2–6 weeks in 1% serum on trans-well inserts and stained for HA. Increased accumulation of HA was observed in RPE cells expressing S179C-TIMP3 (Figure 3C,G) compared with cells transfected with empty vector (Figure 3A,G) or expressing wildtype TIMP3 (Figure 3B,G). The accumulation was predominantly intracellular and apical in the RPE (Figure 3D–F). There appear to be multiple layers of S179C-TIMP3 RPE cells on the transwell compared with a single monolayer for WT-TIMP3 and vector cells, which might suggest epithelial-mesenchymal transition.

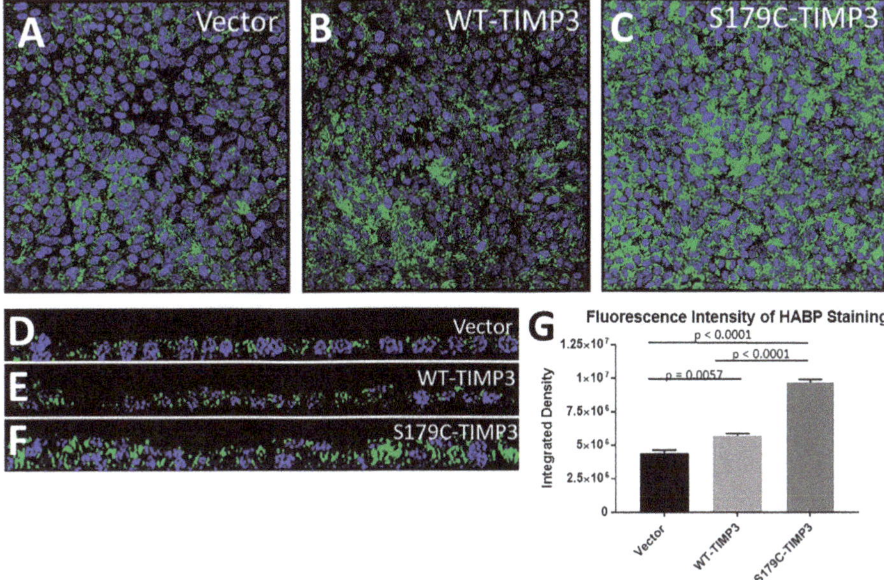

Figure 3. HA is increased in S179C-TIMP3 RPE cells in culture. (**A–C**) ARPE-19 cells expressing S179C-TIMP3 grown in culture for at least 2 weeks on trans-well inserts have increased HA (**C**) compared to WT-TIMP3 expressing cells (**B**) or vector only controls (**A**).(**D–F**) Z-plane images of HA in RPE monolayers grown on trans-well inserts show increased intracellular HA in S179C-TIMP3 cells. Green: HA; blue: DAPI. (**G**) Fluorescence intensity was quantitated by integrated density measurement (n ≥ 4, for each cell line). Data are presented as mean ± SEM.

3.3. FGF-2 Contributes to HA Accumulation in the RPE

We have recently reported that RPE cells expressing S179C-TIMP3 secrete higher amounts of FGF-2 compared with control cells [28]. Previous studies have suggested that FGF signaling has the propensity to increase HA accumulation [32]. To experimentally test this hypothesis in RPE cells, we evaluated the ability of FGF-2 to induce HA accumulation in primary porcine RPE cells. Cells cultured for 3 weeks on trans-well inserts in 1% serum were treated with 0 ng/mL, 10 ng/mL, 25 ng/mL, or 100 ng/mL of FGF-2 in the presence of 1 μg/mL heparin, a cofactor for FGF receptor signaling. FGF-2 induced HA accumulation in a dose-dependent manner (Figure 4A–I) with maximum HA deposits being observed with a dose of 25 and 100 ng/mL (Figure 4D,H). This was confirmed by quantitation of fluorescence by integrated density measurements (Figure 4I). The FGF-2 induced accumulation of HA was seen predominantly on the apical surface of the RPE with increased basal and peri-cellular accumulation at higher doses (Figure 4E–H).

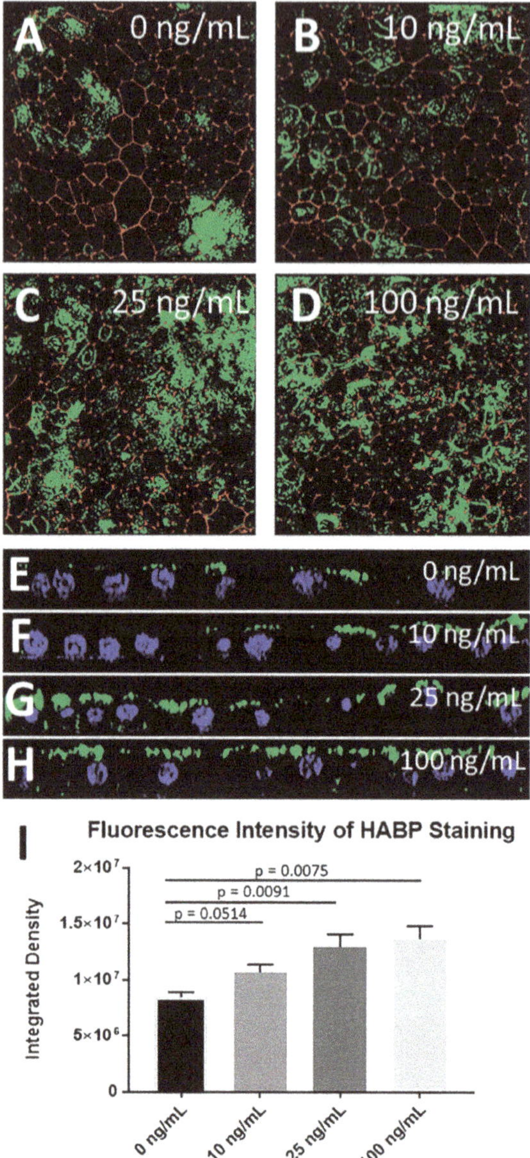

Figure 4. FGF-2 induces HA accumulation in primary RPE cells. (**A–D**) FGF-2 induced HA accumulation in primary porcine RPE cells in a dose-dependent manner (**A**) 0 ng/mL, (**B**) 10 ng/mL, (**C**) 25 ng/mL, (**D**) 100 ng/mL. Fluorescence intensity was quantitated by integrated density measurement (**I**) (n ≥ 4, for each cell line). Data are presented as mean ± SEM (**E–H**) Z-plane images show that increased concentrations of FGF-2 induce increased apical accumulation of HA in addition to some peri-cellular and basal deposits of HA at high doses of FGF-2. Green: HA, red: phalloidin; blue: DAPI.

To determine if the accumulation of HA seen in RPE cells expressing S179C-TIMP3 was a consequence of increased FGF signaling, cells (RPE cells expressing S179C-TIMP3, wildtype TIMP3 or empty vector) were treated with BGJ-398, an FGF receptor inhibitor. 10 µM BGJ-398 decreased HA content in RPE cells expressing S179C-TIMP3 (Figure 5C,F,H,I) and WT-TIMP3 (Figure 5B,E,I) but not vector only transfected cells (Figure 5A,D,I) when compared with their respective untreated cells that served as controls (Figure 5A: vector, 5B: WT-TIMP3 and 5C,G: S179C-TIMP3). Quantitation of fluorescence by integrated density analysis revealed that 10 μM BGJ-398 decreased HA accumulation in S179C-TIMP3 cells 63.1% (SD = 0.241) and only 43.4% (SD = 0.291) in wildtype cells (Figure 5I). Quantitation confirmed that BGJ-398 had no significant effect on vector only cells (Figure 5I), suggesting an FGF-specific mechanism for HA accumulation in S179C-TIMP3 cells.

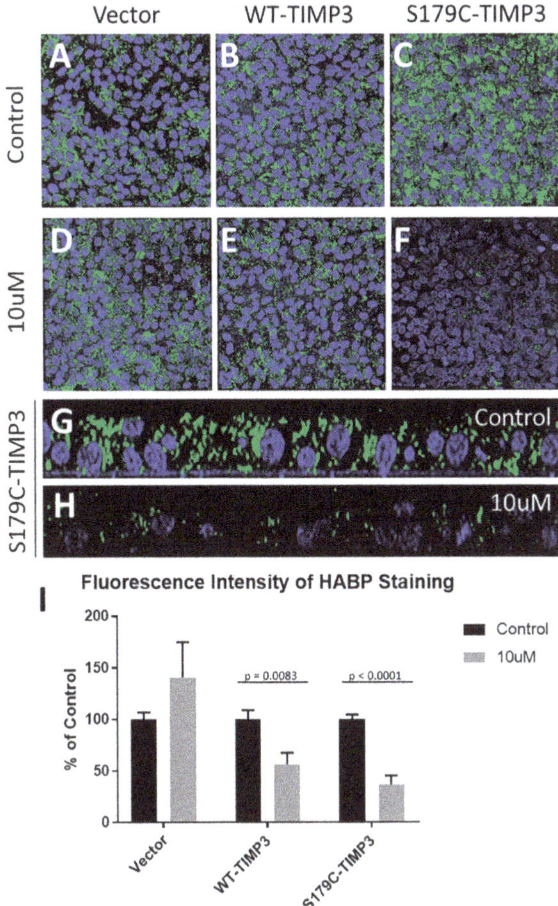

Figure 5. Inhibition of FGF signaling decreases HA accumulation in S179C-TIMP3 RPE cells. Vector only (**A**), WT-TIMP3 (**B**), and S179C-TIMP3 (**C**) expressing ARPE-19 cells were grown on trans-well inserts for 4 weeks before treatment with FGF receptor inhibitor BGJ-398. Treatment with BGJ-398 decreased HA accumulation in S179C-TIMP3 expressing RPE cells (**F,H,I**) and to a lesser extent in WT-TIMP3 expressing cells (**E,I**) but has no effect on control vector RPE cells (**A,D,I**). (**I**) Control (Black Bar) indicates respective untreated cells compared with BGJ-398 treated cells (Grey bar). (**G,H**) Z-plane images of S179C-TIMP3 cells in the absence (**G**) and presence (**H**) of 10 μM BGJ-398 S179C-TIMP3 cells show an overall reduction in HA accumulation, including intracellular HA, after treatment with BGJ-398. Green: HA; blue: DAPI. (**I**) data are presented as mean ± SEM (n ≥ 6).

3.4. Increased HA is Associated with CNV in AMD and SFD

Sections of post-mortem eyes from a patient with CNV showed significant HA deposition around the RPE (Figure 6C,D) when compared with control eyes (Figure 6A,B). While S179C-TIMP3 mice do not demonstrate a florid SFD phenotype as seen in humans, they do have increased susceptibility to experimental laser-induced CNV [28] as do the TIMP3-KO mice [41]. We have previously shown that FGF-2 from S179C-TIMP3 RPE cells can stimulate angiogenesis [28]. Since FGF-2 contributes to HA accumulation in the RPE and to angiogenesis, we evaluated if HA content and distribution was altered in laser-induced CNV lesions in S179C-TIMP3 mice. As described previously lesions in S179C-TIMP3 mice were larger and leakier compared to controls [28]. Increased HA accumulation was observed in CNV in S179C-TIMP3 mice (Figure 6H–J) when compared to lesions in control mice (Figure 6E–G) which appears to be a consequence of altered distribution to the CNV lesions in S179C-TIMP3 mice.

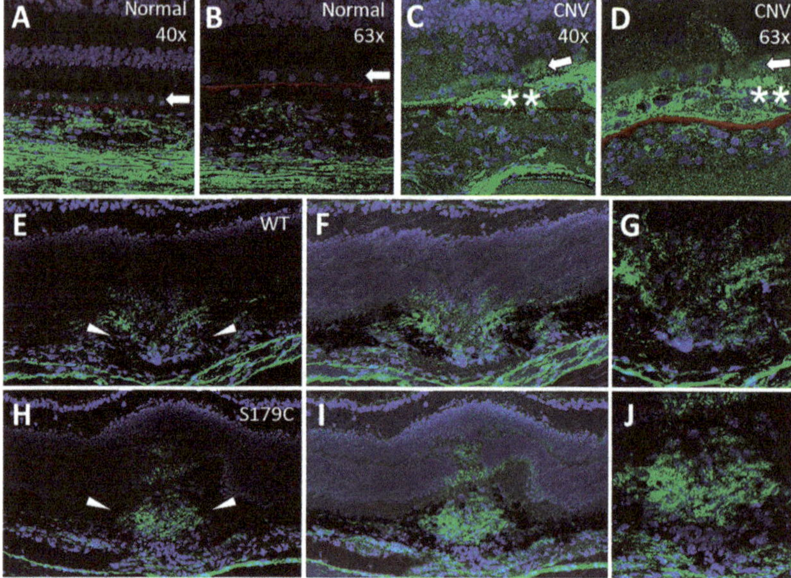

Figure 6. HA is increased in choroidal neovascular lesions in AMD patients and in S179C-TIMP3 mice. (**A–D**) HA is increased in the RPE in a patient with CNV (**C,D**) and in the CNV lesion compared to the RPE of a normal patient (**A,B**). Green: HA; blue: nuclei; red: Bruch's membrane. Arrows indicate RPE, double asterisks indicate CNV lesion. (**E–G**) HA is increased in laser-induced CNV lesions in S179C-TIMP3 mice compared to wild-type (WT) littermates 5 days post injury. (**E,H**) Arrows indicate CNV lesion. (**F,I**) Brightfield image overlaid on fluorescent image shows disruption of RPE and Bruch's membrane. (**G,J**) HA accumulation is diffuse and appears predominantly in the borders of the lesion in WT (**E–G**) compared to dense mass within the lesions of S179C-TIMP3 mice (**H–J**). Green: HA; blue: nuclei.

4. Discussion

Sorsby's fundus dystrophy (SFD) is a rare macular dystrophy characterized by vision loss due to persistent choroidal neovascularization [1–6]. SFD is an autosomal dominant, fully penetrant degenerative disease of the macula and is notable for its similarity in histopathological features to AMD [3–6]. The majority of SFD patients develop CNV, as well as confluent, 20–30 μm thick, amorphous deposits between the basement membrane of the RPE and Bruch's membrane. TIMP3 and/or its downstream substrates have been postulated to have a role in the pathogenesis of both SFD and AMD, because accumulation of TIMP3 has been observed in subretinal deposits in SFD [42] as well as in AMD drusen [43–45]. In this study, we show that hyaluronan accumulates around the RPE in AMD as well as in CNV lesions of mice expressing S179C-TIMP3.

One interesting observation from our studies was the increase in HA in the plasma of patients with AMD as well as in mice lacking TIMP3 or carrying the S179C-TIMP3 mutation. Although we observe significant differences in the plasma levels of HA in patients with advanced AMD (GA and CNV), the number of patients analyzed (n = 49 controls) and n = 59 (AMD) is not sufficient to determine if this is of prognostic value. In addition, we did not have access to samples from patients at different degrees of severity to be able to draw conclusions with this or the rate of progression of the disease. Future studies are warranted to address this question. While patients with SFD generally demonstrate disease localized to the retina, our results may be a consequence of ubiquitous expression of TIMP3 in a variety of tissues in the body [46] which could potentially explain the systemic increase in HA. Whether there is accumulation of HA in other tissues in S179C TIMP3 mice has not been evaluated.

Our studies with RPE cells suggest that the increase in HA in these cells is likely a consequence of increased FGF. A recent study [47] suggests that TIMP proteins can control FGF-2 bioavailability in skeletal tissue and the same might be true of multiple tissues leading to systemic increase in HA in the plasma. The exact mechanism by which TIMP3 regulates FGF bioavailability in the RPE is currently unknown, but it is highly likely that the TIMP-metalloproteinase axis likely has a key role. The extracellular matrix (ECM) serves as a high capacity reservoir for FGF-2 and early studies have demonstrated that matrix metalloproteases (MMPs) have the ability to mobilize FGF-2 to a soluble phase that results in receptor activation [48]. Additional studies identifying the molecular mechanisms by which TIMP3 regulates FGF-2 bioavailability will provide insight into the pathophysiology of the disease.

FGF-2 is sufficient to increase HA accumulation and distribution in the RPE, and blocking FGF signaling in S179C-TIMP3 RPE cells brings HA levels back to normal. The mechanism by which FGF-2 increases HA accumulation is not understood. HA is endogenously synthesized by a family of membrane-integrated glycosyltransferases, called hyaluronan synthases (HAS 1-3) and is exported directly into the ECM [49,50]. Hyaluronidases (HYAL1-2) are a class of enzymes that degrade HA [51]. A balance between HA synthesizing and degrading activity keeps HA at physiological levels. In order to determine the mechanism of accumulation of HA in the RPE in SFD, we performed quantitative PCR analysis of HAS1-3 and HYAL1-2 from RPE isolated from S179C-TIMP3, TIMP3-KO mice and wildtype littermate controls. Interestingly, we observed no changes in gene expression of any of these enzymes in the mutant mice (Supplementary Figure S1). Therefore, at least in the RPE in SFD mice, the differences in HA content are not due to increased expression of the synthases nor decreased expression of canonical degradation enzymes. However, there are other possibilities that need to be explored in the future. HA production could be modulated by decreasing enzyme recycling from endosomes back to the cell surface as seen in keratocytes [52]. Additionally, there is a possibility that other non-canonical hyaluronidases such as KIAA1199 [53] and Tmem2 [54] could be involved. Alternatively, as previously reported degradation might be prevented by increased binding proteins on HA and leading to net accumulation [55].

HA has been shown to exhibit a diverse array of biological functions including a role in the response to tissue damage and inflammation [56]. Our studies demonstrating accumulation of HA in laser-induced CNV lesions corroborates previous studies [27]. This study also reported an increase in *CD44* and *HAS2* mRNA following laser injury [27]. It is possible that the increased accumulation of HA in laser-induced CNV lesions in S179C-TIMP3 mice might result from similar increases in mRNA transcription.

Chronic low-grade inflammation has been suggested to contribute to age-related macular degeneration [57]. In the laser-induced mouse model of CNV, inflammatory processes have been shown to play a role in the development and regression of the lesions. A number of reports link HA remodeling to the modulation of neuroinflammation with low-molecular weight HA being pro-inflammatory and high molecular weight HA being anti-inflammatory [58–60]. While we see increased deposition of HA in and around the RPE, we have not determined its physical properties such as size and molecular weight distribution in the tissue.

The receptor engagement of HA in the retina or it's downstream signaling under physiological or pathological conditions has not yet been identified and will be important as we determine its exact role in the pathology of macular degenerative disease. In our study we demonstrate that primary porcine RPE cells deposited HA predominantly on the apical surface under physiological conditions similar to what had been previously reported for human RPE cells [61]. Our data revealed that FGF-2 induced HA accumulation apically as well as between cells and on the basal surface, suggesting that in addition to increased total HA content, the distribution of HA may be important for disease pathogenesis and warrants further investigation. We have recently reported that the secretion of FGF-2 by RPE cells expressing S179C-TIMP3 led to increased angiogenesis [28]. Whether HA is modified in the endothelial glycocalyx as a consequence of FGF-2 has not been studied and might provide further insight into the pathogenesis of CNV in AMD and SFD leading to the identification of novel therapeutic approaches.

Supplementary Materials: The following are available online at http://www.mdpi.com/2073-4409/9/3/608/s1, Figure S1: RNA was isolated from mouse RPE using the Simultaneous RPE cell Isolation and RNA Stabilization method (SRIRS method) using the RNA Plus Mini Kit (Qiagen). Quantitative PCR was performed following reverse transcription using TaqMan probes for the mouse genes Has1 (A), Has2 (B), Has3 (No signal), Hyal1 (C), Hyal2 (D), and 18S ribosomal RNA (rRNA) (Applied Biosystems). 18S rRNA was used as endogenous control for each gene tested. mRNA expression was calculated using 2-$\Delta\Delta$Ct method and shown relative to expression in wildtype littermate mice.

Author Contributions: Conceptualization, A.W., O.W., V.H. and B.A.-A.; data curation, A.W.; formal analysis, B.A.-A.; funding acquisition, B.A.-A.; methodology, A.W., D.H., A.C., M.A., L.B., J.H.Q., R.S., J.B. and L.K.; project administration, B.A.-A.; resources, V.L.B., O.W. and H.S.; software, R.S.; supervision, B.A.-A.; validation, A.W.; writing—original draft, A.W.; writing—review and editing, A.W., H.S., V.H. and B.A.-A. All authors have read and agreed to the published version of the manuscript.

Funding: This work was supported in part by US National Institute of Health EY027083 (BA-A), EY026181 (BA-A), P30EY025585(BA-A), T32EY024236 (AW), EY022768 (JHQ), EY027750 (VLB), Research to Prevent Blindness (RPB) Challenge Grant and RPB Lew Wasserman award to BA-A, Cleveland Eye Bank Foundation Grant and funds from Cleveland Clinic Foundation.

Acknowledgments: The authors thank the retina specialists at Cole Eye Institute from whose clinics patients with or without AMD were recruited. The authors are grateful for the generous gift of TIMP-3 null mice from Rama Khokha at the Princess Margaret Cancer Centre-Ontario Cancer Institute, Toronto, Canada and to Emma Lessieur who served as the research coordinator at the time the blood samples were collected from patients The authors also acknowledge the support of Foundation Fighting Blindness in setting up the Eye Tissue Repository. We wish to extend a sincere apology to colleagues whose work was not cited due to space limitations.

Conflicts of Interest: The authors declare no conflict of interest. The funders had no role in the design of the study; in the collection, analyses, or interpretation of data; in the writing of the manuscript, or in the decision to publish the results.

References

1. Sorsby, A.; Mason, M.E.; Gardener, N. A fundus dystrophy with unusual features (late onset and dominant inheritance of a central retinal lesion showing oedema, haemorrhage and exudates developing into generalised choroidal atrophy with massive pigment proliferation). *Br. J. Ophthalmol.* **1949**, *33*, 67–97. [CrossRef] [PubMed]

2. Della, N.G.; Campochiaro, P.A.; Zack, D.J. Localization of timp-3 mRNA expression to the retinal pigment epithelium. *Investig. Ophthalmol. Vis. Sci.* **1996**, *37*, 1921–1924.

3. Holz, F.G.; Haimovici, R.; Wagner, D.G.; Bird, A.C. Recurrent choroidal neovascularization after laser photocoagulation in sorsby's fundus dystrophy. *Retina* **1994**, *14*, 329–334. [CrossRef] [PubMed]

4. Jacobson, S.G.; Cideciyan, A.V.; Regunath, G.; Rodriguez, F.J.; Vandenburgh, K.; Sheffield, V.C.; Stone, E.M. Night blindness in sorsby's fundus dystrophy reversed by vitamin a. *Nat. Genet.* **1995**, *11*, 27–32. [CrossRef]

5. Kalmus, H.; Seedburgh, D. Probable common origin of a hereditary fundus dystrophy (sorsby's familial pseudoinflammatory macular dystrophy) in an english and australian family. *J. Med. Genet.* **1976**, *13*, 271–276. [CrossRef]

6. Polkinghorne, P.J.; Capon, M.R.; Berninger, T.; Lyness, A.L.; Sehmi, K.; Bird, A.C. Sorsby's fundus dystrophy. A clinical study. *Ophthalmology* **1989**, *96*, 1763–1768. [CrossRef]

7. Barbazetto, I.A.; Hayashi, M.; Klais, C.M.; Yannuzzi, L.A.; Allikmets, R. A novel timp3 mutation associated with sorsby fundus dystrophy. *Arch. Ophthalmol.* **2005**, *123*, 542–543. [CrossRef]

8. Felbor, U.; Benkwitz, C.; Klein, M.L.; Greenberg, J.; Gregory, C.Y.; Weber, B.H. Sorsby fundus dystrophy: Reevaluation of variable expressivity in patients carrying a timp3 founder mutation. *Arch. Ophthalmol.* **1997**, *115*, 1569–1571. [CrossRef]

9. Felbor, U.; Stohr, H.; Amann, T.; Schonherr, U.; Weber, B.H. A novel ser156cys mutation in the tissue inhibitor of metalloproteinases-3 (timp3) in sorsby's fundus dystrophy with unusual clinical features. *Hum. Mol. Genet.* **1995**, *4*, 2415–2416. [CrossRef]

10. Langton, K.P.; Barker, M.D.; McKie, N. Localization of the functional domains of human tissue inhibitor of metalloproteinases-3 and the effects of a sorsby's fundus dystrophy mutation. *J. Biol. Chem.* **1998**, *273*, 16778–16781. [CrossRef]

11. Langton, K.P.; McKie, N.; Curtis, A.; Goodship, J.A.; Bond, P.M.; Barker, M.D.; Clarke, M. A novel tissue inhibitor of metalloproteinases-3 mutation reveals a common molecular phenotype in sorsby's fundus dystrophy. *J. Biol. Chem.* **2000**, *275*, 27027–27031. [PubMed]

12. Lin, R.J.; Blumenkranz, M.S.; Binkley, J.; Wu, K.; Vollrath, D. A novel his158arg mutation in timp3 causes a late-onset form of sorsby fundus dystrophy. *Am. J. Ophthalmol.* **2006**, *142*, 839–848. [CrossRef] [PubMed]

13. Tabata, Y.; Isashiki, Y.; Kamimura, K.; Nakao, K.; Ohba, N. A novel splice site mutation in the tissue inhibitor of the metalloproteinases-3 gene in sorsby's fundus dystrophy with unusual clinical features. *Hum. Genet.* **1998**, *103*, 179–182. [PubMed]

14. Weber, B.H.; Vogt, G.; Pruett, R.C.; Stohr, H.; Felbor, U. Mutations in the tissue inhibitor of metalloproteinase-3 (timp-3) in patients with sorsby's fundus dystrophy. *Nat. Genet.* **1994**, *8*, 352–356. [CrossRef]

15. Qi, J.H.; Anand-Apte, B. Tissue inhibitor of metalloproteinase-3 (timp3) promotes endothelial apoptosis via a caspase-independent mechanism. *Apoptosis* **2015**, *20*, 523–534. [CrossRef]

16. Qi, J.H.; Dai, G.; Luthert, P.; Chaurasia, S.; Hollyfield, J.; Weber, B.H.; Stohr, H.; Anand-Apte, B. S156c mutation in tissue inhibitor of metalloproteinases-3 induces increased angiogenesis. *J. Biol. Chem.* **2009**, *284*, 19927–19936. [CrossRef]

17. Qi, J.H.; Ebrahem, Q.; Ali, M.; Cutler, A.; Bell, B.; Prayson, N.; Sears, J.; Knauper, V.; Murphy, G.; Anand-Apte, B. Tissue inhibitor of metalloproteinases-3 peptides inhibit angiogenesis and choroidal neovascularization in mice. *PLoS ONE* **2013**, *8*, e55667. [CrossRef]

18. Qi, J.H.; Ebrahem, Q.; Anand-Apte, B. Tissue inhibitor of metalloproteinases-3 and sorsby fundus dystrophy. *Adv. Exp. Med. Biol.* **2003**, *533*, 97–105.

19. Qi, J.H.; Ebrahem, Q.; Moore, N.; Murphy, G.; Claesson-Welsh, L.; Bond, M.; Baker, A.; Anand-Apte, B. A novel function for tissue inhibitor of metalloproteinases-3 (timp3): Inhibition of angiogenesis by blockage of VEGF binding to vegf receptor-2. *Nat. Med.* **2003**, *9*, 407–415. [CrossRef]

20. Qi, J.H.; Ebrahem, Q.; Yeow, K.; Edwards, D.R.; Fox, P.L.; Anand-Apte, B. Expression of sorsby's fundus dystrophy mutations in human retinal pigment epithelial cells reduces matrix metalloproteinase inhibition and may promote angiogenesis. *J. Biol. Chem.* **2002**, *30*, 30. [CrossRef]

21. Stohr, H.; Anand-Apte, B. A review and update on the molecular basis of pathogenesis of sorsby fundus dystrophy. *Adv. Exp. Med. Biol.* **2012**, *723*, 261–267. [PubMed]

22. Fariss, R.N.; Apte, S.S.; Olsen, B.R.; Iwata, K.; Milam, A.H. Tissue inhibitor of metalloproteinases-3 is a component of bruch's membrane of the eye. *Am. J. Pathol.* **1997**, *150*, 323–328. [PubMed]

23. Troeberg, L.; Lazenbatt, C.; Anower, E.K.M.F.; Freeman, C.; Federov, O.; Habuchi, H.; Habuchi, O.; Kimata, K.; Nagase, H. Sulfated glycosaminoglycans control the extracellular trafficking and the activity of the metalloprotease inhibitor timp-3. *Chem. Biol.* **2014**, *21*, 1300–1309. [CrossRef] [PubMed]

24. Yu, W.H.; Yu, S.; Meng, Q.; Brew, K.; Woessner, J.F., Jr. Timp-3 binds to sulfated glycosaminoglycans of the extracellular matrix. *J. Biol. Chem.* **2000**, *275*, 31226–31232. [CrossRef] [PubMed]

25. Itano, N.; Kimata, K. Altered hyaluronan biosynthesis in cancer progression. *Semin. Cancer Biol.* **2008**, *18*, 268–274. [CrossRef] [PubMed]

26. Singleton, P.A. Hyaluronan regulation of endothelial barrier function in cancer. *Adv. Cancer Res.* **2014**, *123*, 191–209.

27. Mochimaru, H.; Takahashi, E.; Tsukamoto, N.; Miyazaki, J.; Yaguchi, T.; Koto, T.; Kurihara, T.; Noda, K.; Ozawa, Y.; Ishimoto, T.; et al. Involvement of hyaluronan and its receptor cd44 with choroidal neovascularization. *Investig. Ophthalmol. Vis. Sci.* **2009**, *50*, 4410–4415. [CrossRef]

28. Qi, J.H.; Bell, B.; Singh, R.; Batoki, J.; Wolk, A.; Cutler, A.; Prayson, N.; Ali, M.; Stoehr, H.; Anand-Apte, B. Sorsby fundus dystrophy mutation in tissue inhibitor of metalloproteinase 3 (timp3) promotes choroidal neovascularization via a fibroblast growth factor-dependent mechanism. *Sci. Rep.* **2019**, *9*, 17429. [CrossRef]

29. Oladipupo, S.S.; Smith, C.; Santeford, A.; Park, C.; Sene, A.; Wiley, L.A.; Osei-Owusu, P.; Hsu, J.; Zapata, N.; Liu, F.; et al. Endothelial cell fgf signaling is required for injury response but not for vascular homeostasis. *Proc. Natl. Acad. Sci. USA* **2014**, *111*, 13379–13384. [CrossRef]

30. Sarrazin, S.; Lamanna, W.C.; Esko, J.D. Heparan sulfate proteoglycans. *Cold Spring Harb. Perspect. Biol.* **2011**, *3*, a004952. [CrossRef]

31. Weis, S.M.; Cheresh, D.A. Tumor angiogenesis: Molecular pathways and therapeutic targets. *Nat. Med.* **2011**, *17*, 1359–1370. [CrossRef] [PubMed]

32. Bohrer, L.R.; Chuntova, P.; Bade, L.K.; Beadnell, T.C.; Leon, R.P.; Brady, N.J.; Ryu, Y.; Goldberg, J.E.; Schmechel, S.C.; Koopmeiners, J.S.; et al. Activation of the fgfr-stat3 pathway in breast cancer cells induces a hyaluronan-rich microenvironment that licenses tumor formation. *Cancer Res.* **2014**, *74*, 374–386. [CrossRef] [PubMed]

33. Weber, B.H.; Lin, B.; White, K.; Kohler, K.; Soboleva, G.; Herterich, S.; Seeliger, M.W.; Jaissle, G.B.; Grimm, C.; Reme, C.; et al. A mouse model for sorsby fundus dystrophy. *Investig. Ophthalmol. Vis. Sci.* **2002**, *43*, 2732–2740.

34. Leco, K.J.; Waterhouse, P.; Sanchez, O.H.; Gowing, K.L.; Poole, A.R.; Wakeham, A.; Mak, T.W.; Khokha, R. Spontaneous air space enlargement in the lungs of mice lacking tissue inhibitor of metalloproteinases-3 (timp-3). *J. Clin. Investig.* **2001**, *108*, 817–829. [CrossRef]

35. Tammi, R.; Ripellino, J.A.; Margolis, R.U.; Maibach, H.I.; Tammi, M. Hyaluronate accumulation in human epidermis treated with retinoic acid in skin organ culture. *J. Investig. Dermatol.* **1989**, *92*, 326–332. [CrossRef]

36. Hazim, R.A.; Volland, S.; Yen, A.; Burgess, B.L.; Williams, D.S. Rapid differentiation of the human rpe cell line, arpe-19, induced by nicotinamide. *Exp. Eye Res.* **2019**, *179*, 18–24. [CrossRef]

37. Venkatareddy, M.; Verma, R.; Kalinowski, A.; Patel, S.R.; Shisheva, A.; Garg, P. Distinct requirements for vacuolar protein sorting 34 downstream effector phosphatidylinositol 3-phosphate 5-kinase in podocytes versus proximal tubular cells. *J. Am. Soc. Nephrol.* **2016**, *27*, 2702–2719. [CrossRef]

38. Scheuring, D.; Lofke, C.; Kruger, F.; Kittelmann, M.; Eisa, A.; Hughes, L.; Smith, R.S.; Hawes, C.; Schumacher, K.; Kleine-Vehn, J. Actin-dependent vacuolar occupancy of the cell determines auxin-induced growth repression. *Proc. Natl. Acad. Sci. USA* **2016**, *113*, 452–457. [CrossRef]

39. Hollyfield, J.G. Hyaluronan and the functional organization of the interphotoreceptor matrix. *Investig. Ophthalmol. Vis. Sci.* **1999**, *40*, 2767–2769.

40. Hollyfield, J.G.; Rayborn, M.E.; Tammi, R. Hyaluronan localization in tissues of the mouse posterior eye wall: Absence in the interphotoreceptor matrix. *Exp. Eye Res.* **1997**, *65*, 603–608. [CrossRef]

41. Ebrahem, Q.; Qi, J.H.; Sugimoto, M.; Ali, M.; Sears, J.E.; Cutler, A.; Khokha, R.; Vasanji, A.; Anand-Apte, B. Increased neovascularization in mice lacking tissue inhibitor of metalloproteinases-3. *Investig. Ophthalmol. Vis. Sci.* **2011**, *52*, 6117–6123. [CrossRef] [PubMed]

42. Fariss, R.N.; Apte, S.S.; Luthert, P.J.; Bird, A.C.; Milam, A.H. Accumulation of tissue inhibitor of metalloproteinases-3 in human eyes with sorsby's fundus dystrophy or retinitis pigmentosa. *Br. J. Ophthalmol.* **1998**, *82*, 1329–1334. [CrossRef] [PubMed]

43. Crabb, J.W.; Miyagi, M.; Gu, X.; Shadrach, K.; West, K.A.; Sakaguchi, H.; Kamei, M.; Hasan, A.; Yan, L.; Rayborn, M.E.; et al. Drusen proteome analysis: An approach to the etiology of age-related macular degeneration. *Proc. Natl. Acad. Sci. USA* **2002**, *99*, 14682–14687. [CrossRef] [PubMed]

44. Kamei, M.; Apte, S.S.; Rayborn, M.E.; Lewis, H.; Hollyfield, J.G. Timp-3 accumulation in drusen and bruch's membrane in eyes from donors with age-related macular degeneration (1997). In *Degenerative Diseases of the Retina*; Luvail, M.M., Anderson, R.E., Hollyfield, J.G., Eds.; Plenum Press: New York, NY, USA, 1997; pp. 11–15.

45. Kamei, M.; Hollyfield, J. Timp-3 in bruch's membrane: Changes during aging and in age-related macular degeneration. *Investig. Ophthalmol. Vis. Sci.* **1999**, *40*, p2367–p2375.

46. Apte, S.S.; Mattei, M.-G.; Olsen, B.R. Cloning of the cdna encoding human tissue inhibitor of metalloproteinase-3 (timp-3) and mapping of the timp-3 gene to chromosome 22. *Genomics* **1994**, *19*, 86–90. [CrossRef]

47. Saw, S.; Aiken, A.; Fang, H.; McKee, T.D.; Bregant, S.; Sanchez, O.; Chen, Y.; Weiss, A.; Dickson, B.C.; Czarny, B.; et al. Metalloprotease inhibitor timp proteins control fgf-2 bioavailability and regulate skeletal growth. *J. Cell Biol.* **2019**, *218*, 3134–3152. [CrossRef]

48. Rifkin, D.B.; Moscatelli, D.; Bizik, J.; Quarto, N.; Blei, F.; Dennis, P.; Flaumenhaft, R.; Mignatti, P. Growth factor control of extracellular proteolysis. *Cell Differ. Dev.* **1990**, *32*, 313–318. [CrossRef]

49. DeAngelis, P.L. Hyaluronan synthases: Fascinating glycosyltransferases from vertebrates, bacterial pathogens, and algal viruses. *Cell Mol. Life Sci.* **1999**, *56*, 670–682. [CrossRef]

50. Tien, J.Y.; Spicer, A.P. Three vertebrate hyaluronan synthases are expressed during mouse development in distinct spatial and temporal patterns. *Dev. Dyn.* **2005**, *233*, 130–141. [CrossRef]

51. Stern, R.; Kogan, G.; Jedrzejas, M.J.; Soltes, L. The many ways to cleave hyaluronan. *Biotechnol. Adv.* **2007**, *25*, 537–557. [CrossRef]

52. Tammi, R.; Rilla, K.; Pienimaki, J.P.; MacCallum, D.K.; Hogg, M.; Luukkonen, M.; Hascall, V.C.; Tammi, M. Hyaluronan enters keratinocytes by a novel endocytic route for catabolism. *J. Biol. Chem.* **2001**, *276*, 35111–35122. [CrossRef] [PubMed]

53. Marella, M.; Jadin, L.; Keller, G.A.; Sugarman, B.J.; Frost, G.I.; Shepard, H.M. Kiaa1199 expression and hyaluronan degradation colocalize in multiple sclerosis lesions. *Glycobiology* **2018**, *28*, 958–967. [CrossRef] [PubMed]

54. De Angelis, J.E.; Lagendijk, A.K.; Chen, H.; Tromp, A.; Bower, N.I.; Tunny, K.A.; Brooks, A.J.; Bakkers, J.; Francois, M.; Yap, A.S.; et al. Tmem2 regulates embryonic VEGF signaling by controlling hyaluronic acid turnover. *Dev. Cell* **2017**, *40*, 123–136. [CrossRef] [PubMed]

55. Nagaoka, A.; Yoshida, H.; Nakamura, S.; Morikawa, T.; Kawabata, K.; Kobayashi, M.; Sakai, S.; Takahashi, Y.; Okada, Y.; Inoue, S. Regulation of hyaluronan (ha) metabolism mediated by hybid (hyaluronan-binding protein involved in ha depolymerization, kiaa1199) and ha synthases in growth factor-stimulated fibroblasts. *J. Biol. Chem.* **2015**, *290*, 30910–30923. [CrossRef]

56. Garantziotis, S.; Savani, R.C. Hyaluronan biology: A complex balancing act of structure, function, location and context. *Matrix Biol.* **2019**, *78–79*, 1–10. [CrossRef]

57. Nita, M.; Grzybowski, A.; Ascaso, F.J.; Huerva, V. Age-related macular degeneration in the aspect of chronic low-grade inflammation (pathophysiological parainflammation). *Mediators Inflamm.* **2014**, *2014*, 930671. [CrossRef]

58. Avenoso, A.; Bruschetta, G.; D'Ascola, A.; Scuruchi, M.; Mandraffino, G.; Saitta, A.; Campo, S.; Campo, G.M. Hyaluronan fragmentation during inflammatory pathologies: A signal that empowers tissue damage. *Mini Rev. Med. Chem.* **2020**, *20*, 54–65. [CrossRef]

59. Mueller, A.M.; Yoon, B.H.; Sadiq, S.A. Inhibition of hyaluronan synthesis protects against central nervous system (cns) autoimmunity and increases cxcl12 expression in the inflamed cns. *J. Biol. Chem.* **2014**, *289*, 22888–22899. [CrossRef]

60. Scuruchi, M.; D'Ascola, A.; Avenoso, A.; Campana, S.; Abusamra, Y.A.; Spina, E.; Calatroni, A.; Campo, G.M.; Campo, S. 6-mer hyaluronan oligosaccharides modulate neuroinflammation and alpha-synuclein expression in neuron-like sh-sy5y cells. *J. Cell Biochem.* **2016**, *117*, 2835–2843. [CrossRef]

61. deS Senanayake, P.; Calabro, A.; Nishiyama, K.; Hu, J.G.; Bok, D.; Hollyfield, J.G. Glycosaminoglycan synthesis and secretion by the retinal pigment epithelium: Polarized delivery of hyaluronan from the apical surface. *J. Cell Sci.* **2001**, *114*, 199–205.

Review

Mouse Models of Inherited Retinal Degeneration with Photoreceptor Cell Loss

Gayle B. Collin [1,†], Navdeep Gogna [1,†], Bo Chang [1], Nattaya Damkham [1,2,3], Jai Pinkney [1], Lillian F. Hyde [1], Lisa Stone [1], Jürgen K. Naggert [1], Patsy M. Nishina [1,*] and Mark P. Krebs [1,*]

[1] The Jackson Laboratory, Bar Harbor, Maine, ME 04609, USA; gayle.collin@jax.org (G.B.C.); navdeep.gogna@jax.org (N.G.); bo.chang@jax.org (B.C.); nattaya.damkham@jax.org (N.D.); jai.pinkney@jax.org (J.P.); Lillian.Hyde@jax.org (L.F.H.); lisa.stone@jax.org (L.S.); juergen.naggert@jax.org (J.K.N.)

[2] Department of Immunology, Faculty of Medicine Siriraj Hospital, Mahidol University, Bangkok 10700, Thailand

[3] Siriraj Center of Excellence for Stem Cell Research, Faculty of Medicine Siriraj Hospital, Mahidol University, Bangkok 10700, Thailand

[*] Correspondence: patsy.nishina@jax.org (P.M.N.); mark.krebs@jax.org (M.P.K.); Tel.: +1-207-2886-383 (P.M.N.); +1-207-2886-000 (M.P.K.)

[†] These authors contributed equally to this work.

Received: 29 February 2020; Accepted: 7 April 2020; Published: 10 April 2020

Abstract: Inherited retinal degeneration (RD) leads to the impairment or loss of vision in millions of individuals worldwide, most frequently due to the loss of photoreceptor (PR) cells. Animal models, particularly the laboratory mouse, have been used to understand the pathogenic mechanisms that underlie PR cell loss and to explore therapies that may prevent, delay, or reverse RD. Here, we reviewed entries in the Mouse Genome Informatics and PubMed databases to compile a comprehensive list of monogenic mouse models in which PR cell loss is demonstrated. The progression of PR cell loss with postnatal age was documented in mutant alleles of genes grouped by biological function. As anticipated, a wide range in the onset and rate of cell loss was observed among the reported models. The analysis underscored relationships between RD genes and ciliary function, transcription-coupled DNA damage repair, and cellular chloride homeostasis. Comparing the mouse gene list to human RD genes identified in the RetNet database revealed that mouse models are available for 40% of the known human diseases, suggesting opportunities for future research. This work may provide insight into the molecular players and pathways through which PR degenerative disease occurs and may be useful for planning translational studies.

Keywords: visual photoreceptor cell loss; mouse genetic models; retinitis pigmentosa; Leber congenital amaurosis; ciliopathies

---

1. Introduction

Inherited forms of retinal degeneration (RD) encompass a genetically and clinically heterogeneous group of disorders estimated to cause vision impairment and loss in more than 5.5 million individuals worldwide [1,2], with 282 mapped and identified retinal degenerative disease genes documented in the RetNet human database [3]. Animal models, such as non-human primates [4], dogs [5], mice [6,7], zebrafish [8], and fruit flies [9], have been used to identify candidates for human retinal disease genes, to elucidate pathological mechanisms, and to serve as a resource for exploring therapeutic approaches. As potential therapies for retinal diseases are investigated, the need for animal models increases. Information about the disease onset and rate of progression, the pathogenic pathways involved, and the genetic background in which the disrupted genes are situated are all factors that must be considered when selecting appropriate models for testing therapeutics. These factors will also play a role in interpreting the outcome of treatment studies.

The purpose of this review is to compile a searchable list of mouse models of inherited retinal diseases caused by single gene mutations that specifically lead to the post-developmental rod and/or cone photoreceptor (PR) cell loss. To identify these models, we reviewed mouse-specific data available in the Mouse Genome Informatics (MGI) and National Center for Biotechnology Information (NCBI) databases at The Jackson Laboratory (JAX) and National Institutes of Health (NIH), respectively. We recorded, when available, PR cell loss data from publications describing mutant alleles of the genes identified. We also included representative fundus photographs and optical coherence tomography (OCT) images of selected mouse models from the Eye Mutant Resource (EMR) and the Translational Vision Research Models (TVRM) programs at JAX as examples of the retinal phenotypes found among mouse models that fit our criteria. We attempted to cluster genes based on the function and then compared the progression of PR cell loss among these clusters to provide potential insights into disease mechanisms. We also compared our list of mouse genes associated with PR cell loss with the RetNet gene list, to highlight mouse models for specific retinal diseases, to reveal opportunities to create novel models, and to identify candidate genes within human loci for which a causative gene is currently unknown. Finally, we coordinated with the MGI team to incorporate our annotations into the MGI database, which will allow future analyses using tools available through that platform. It is hoped that our work will be useful as a resource for investigators to assist in the selection of appropriate mouse models within and across functional clusters in new studies to understand and develop treatments for human retinal degenerative disease.

In the three decades since the genes linked to PR loss phenotype were first identified in the mouse and human [10–12], rapid progress in understanding the genetic basis of inherited RD has been summarized in many excellent reviews. Many of the topics presented in the current article have been discussed previously in reviews of mouse RD models [6,7,13,14] and in summaries of our work at JAX [15–21]. Although we have made every effort to acknowledge the many contributions to this field, we note that there is a large body of relevant literature and apologize in advance to authors whose reviews or articles we may have inadvertently overlooked.

2. Background

2.1. Photoreceptor (PR) Cell Structure

PR cells are sensory neurons within the retina that detect light and signal this event to other cells. Since PR cells are essential for vision, their loss can dramatically and negatively affect the quality of life. PR cells include rod and cone cells (Figure 1a,b) that occupy the outermost layers of the neurosensory retina. Although intrinsically photosensitive retinal ganglion cells have also been described as photoreceptors [22], we did not include them in this review, as their contribution to RD is unknown. Rod and cone photoreceptors possess unique structures that serve to compartmentalize processes that are critical for cell function and maintenance.

- The outer segment (OS), which is cylindrical in rod PR cells and tapered in cones, contains phototransduction proteins that sense light and amplify the ensuing signal, culminating in PR cell hyperpolarization (Figure 1c). Much of the phototransduction apparatus is localized to double-bilayer discs formed by evagination of the plasma membrane at the base of the OS. These discs are largely internalized in rods except at the base of the OS, but remain contiguous with the plasma membrane in cones to yield a highly convoluted OS surface [23].
- The OS is stabilized by a ciliary axoneme, which runs through much of its length (Figure 1d; Ax). At the proximal end of the axoneme, the connecting cilium, analogous to the transition zone in other cilia (Figure 1d; CC-TZ), serves as a conduit through which all membrane and protein components destined for the OS are thought to pass. At the base of the connecting cilium lies the basal body (Figure 1d; BB), a cylindrical organelle derived from the mother centriole. Altogether

these structures represent a modified primary cilium that encompasses an extensive network of protein complexes that transport proteins and lipids and shares characteristics with primary cilia in many other cell types. The ciliary networks also function to prevent the flow of OS components to other parts of the cell and may associate with the intracellular trafficking apparatus to ensure the directed movement of needed components to the OS.

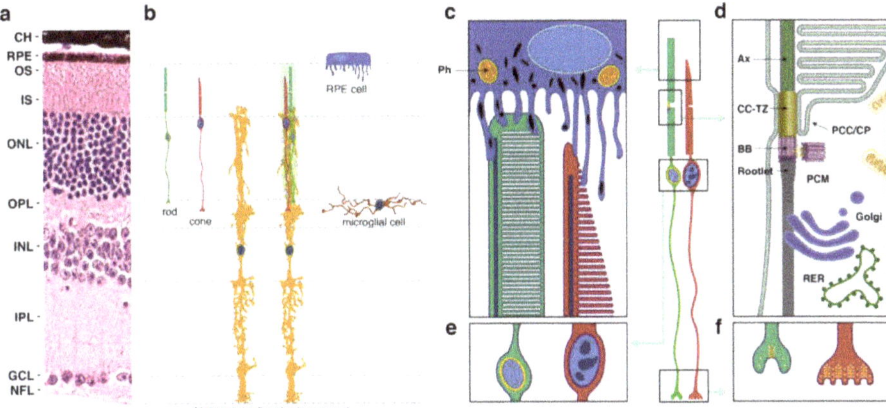

Figure 1. Retinal tissue organization emphasizing cell types and subcellular structures that may be sites of pathological processes in mouse models of photoreceptor (PR) cell loss. (**a**) A radial section of the posterior eye stained with hematoxylin and eosin shows the layered structure of the retina. CH, choroid; RPE, retinal pigment epithelium; IS, inner segment; OS, outer segment; ONL, outer nuclear layer; OPL, outer plexiform layer; INL, inner nuclear layer; IPL, inner plexiform layer; GCL, ganglion cell layer; NFL, nerve fiber layer. (**b**) Two PR cell types (rods and cones) and additional cell types that may be the target of processes implicated in PR cell loss. Dashed lines indicate alignment with retinal layers in (**a**). A columnar unit consisting of one cone and one Müller cell and roughly 20 rod cells is shown. (**c–f**) Details of PR and RPE cells. (**c**) PR cell OSs contain flattened discs (rod, left) or incomplete discs (cone, right) where the light sensing apparatus is located. OS tips engulfed by RPE cell apical processes are digested in phagolysosomes (Ph). (**d**) The base of the OS, the connecting cilium, and the apical portion of a rod cell IS (adapted with permission from [24]). The axoneme (Ax) and rootlet provide physical stability to the cilium. BB, basal bodies; CC-TZ, connecting cilium-transition zone; PCM, pericentriolar matrix; PCC/CP, periciliary complex/ciliary pocket; RER, rough endoplasmic reticulum. (**e**) The PR cell soma is largely occupied by the nucleus. (**f**) Rod and cone synaptic termini include presynaptic ribbons and associated neurotransmitter vesicles.

- The inner segment (IS) contains the biosynthetic machinery and energy sources needed to produce and assemble newly synthesized phototransduction proteins and their associated membranes (Figure 1d). The capacity of this cellular factory is impressive, as up to 10% of the OS is shed daily and removed via phagocytosis by the retinal pigment epithelium (see below) and must be renewed. Most protein and lipid components are synthesized de novo, but the IS also has an extensive recycling machinery that can reassemble components provided from outside the cell.
- The cell body or soma includes the nucleus, which is highly condensed in rod PR cells, but is larger in cones and includes patches of heterochromatin (Figure 1e). To increase the density of rod and cone OSs in the retina, the somas are stacked in columns within the outer nuclear layer (ONL). This arrangement necessitates thin cell extensions reaching from the soma to the IS or to the synapse. PR cell loss is measured by counting ONL nuclei, which are prominently stained in retinal sections (Figure 1a), or in the case of rods, which are more abundant than cones, by measuring ONL thickness from micrographs or by OCT.

- The PR cell terminus contains ribbon synapses close to the presynaptic membrane loaded with vesicles containing the excitatory neurotransmitter glutamate (Figure 1f). In the dark, a steady-state level of glutamate is released at the synapse, which is reduced when the cells are hyperpolarized in the light. Changes in glutamate levels at the synapse signal postsynaptic secondary neurons in the inner nuclear layer, which communicate with ganglion cells on the vitreal surface of the retina that connect through long axons to the visual cortex of the brain.

2.2. Neighboring Cells

Müller glia are radial cells that span much of the neurosensory retina, reaching from the internal limiting membrane at the vitreal surface of the retina to the external limiting membrane on the scleral edge of the ONL [25]. Within the ONL, fine Müller cell extensions appear to ensheath the PR cell soma. As they also interact with vascular layers within the retina, Müller cells may provide essential nutrients to PR cells, which do not directly contact the circulation. They also regulate extracellular volume, ion and water homeostasis, serve to modify neuronal activity through release of neuroactive compounds, and modulate immune and inflammatory responses [26]. At the external limiting membrane, Müller cell endfeet engage rod and cone cell ISs in intercellular adhesion interactions, including tight junctions, which create a diffusion barrier. Notably, the arrangement of an Müller cell, rods, and a cone cell has been proposed to form a columnar unit (Figure 1b), which may result in physiological and functional coordination of these cell types.

RPE cells (Figure 1b,c) constitute an epithelial monolayer that lies between the retina and a capillary bed, the choriocapillaris. The flow of water, ions, small molecules, and metabolites from the blood to the outer retina is thus regulated by RPE cells. Their apical surface features microvilli and microplicae (Figure 1b,c) that contact roughly the outermost third of OSs and play important roles in recycling molecules needed for PR renewal. These apical processes also mediate initial steps in the daily phagocytosis of OS tips. As an epithelium with high-resistance intercellular junctions [27], the RPE performs an important barrier function, disruption of which may cause PR degeneration.

Microglial cells form ramified networks within the same retinal layers as the retinal vasculature, which includes the superficial, intermediate and deep vascular beds. Microglia at the level of the outer plexiform layer in healthy retinas extend dendritic arms into the ONL (Figure 1b), where they contact PR soma as part of a dynamic survey process. During development and in rare events that occur in healthy retinas, these cells engulf and phagocytose PR soma, presumably in response to a defect in PR function.

PRs form synaptic connections in the outer plexiform layer with secondary neurons, including bipolar and horizontal cells. Although these connections are critical for signal transmission, they are not as extensive as the contacts made between PRs and Müller glia, RPE cell apical processes, or the dynamic extensions on microglial cells, and were therefore omitted from the summary diagram in Figure 1. Nevertheless, perturbation of the interactions among these cells may lead to PR degeneration, conceivably due to an alteration of signal transmission.

2.3. Inherited Diseases that Cause PR Cell Loss

Major monogenic inherited RDs in which PR cells are lost include: retinitis pigmentosa [28], Leber congenital amaurosis [29], and syndromic disorders that manifest disease in multiple organs, including the eye, particularly ciliopathies, such as Joubert [30], Bardet–Biedl [31], or Usher [32] syndrome. The remarkable success of gene augmentation therapy for a form of Leber congenital amaurosis has invigorated research efforts to treat these diseases [33].

3. Methods

3.1. Public Database and Literature Searches

The search strategy employed in this review is summarized in Figure 2. Initially, the MGI database [34] was queried to identify mutant protein-coding genes associated with a mammalian disease phenotype indicating a loss of PR cells. MGI is a curated database that includes expert annotation based on full-text searching of 148 selected journals, which is limited compared to literature databases, such as PubMed. Typically, only the first paper describing a new allele is fully curated for phenotype data in the database due to resource constraints. Although our analysis yielded 159 mutant genes that were associated with PR cell loss, a number of mutant genes known to cause this phenotype were absent or not annotated, possibly due to these aforementioned limitations.

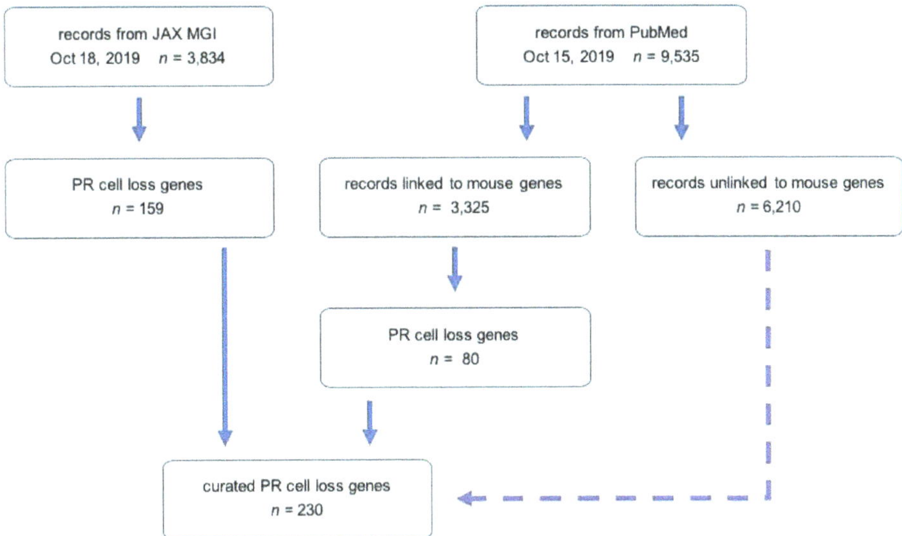

Figure 2. Flow chart depicting the progression of the search strategy utilized in this review. The number of records or genes identified from them is indicated at each stage of the process. The dashed line indicates that some genes were identified from records that remain to be systematically screened.

To expand the search, we used NCBI databases, including PubMed [35] and Gene. We refined a PubMed query by searching with keyword phrases to generate article lists, using the Gene option of Find Related Data to yield mouse genes linked to the articles, and then assessing whether these genes were on our MGI list. The goal was to develop a broad query that included as many genes from the MGI list as possible but also included additional hits. The most successful was: (ONL OR "outer nuclear layer" OR retina* OR PR OR rod OR rods OR cone*) AND (degener* OR loss OR thin* OR thick*) AND (mouse OR mice OR murine), which captured >97% of the MGI list. Restricting this query to entries posted to PubMed on or before October 15, 2019, yielded 9535 articles. To review these articles efficiently we tried two approaches, the first generating a spreadsheet containing hyperlinks in which each linked gene symbol was combined with the Boolean query, and the second using mouse gene identification numbers corresponding to the linked genes and applying an Entrez script that accessed the Gene and PubMed databases to find all articles satisfying the Boolean query for each linked gene.

3.2. Search Strategy

Each MGI database-derived entry was curated manually or automatically to identify candidate models that reported PR degeneration as a phenotype, as described above. In the case of PubMed entries, although the automated approaches were useful for quickly identifying genes that satisfied our criteria, neither was comprehensive, and additional candidate models were identified by review of the title and abstract from some of the remaining articles in the full collection of 9535 articles. Subsequently, an independent coauthor identified an original publication for each candidate gene and determined if PR cell loss was reported. If sufficient evidence for PR cell loss was obtained, the gene and mouse model was assigned to one of 11 categories. Genes within each category were curated further by coauthors who identified alternate alleles and extracted information regarding the disease phenotype induced by the disruption of a gene. Each entry in Table S1 is the result of the examination of an original article (indicated by PubMed ID numbers, PMIDs) and data from MGI to capture information such as mutation type, associated human diseases, and disease onset and progression.

3.3. Comparative Analysis and Updating the MGI Database

Once our final list was completed, we used tools in Excel to compare it to a list constructed from online tables downloaded on 8 December 2019, from RetNet, a public compilation of human genes linked to inherited RD. We also provided our data to the MGI team at JAX, who assigned allele nomenclature, added strain information for newly described mutants, and updated phenotype data for alleles that were present in the MGI database but not yet annotated with respect to PR cell loss. This review has been referenced at MGI so that the alleles documented in the article can be examined using MGI tools or downloaded in tabular format for analysis with other software. The collaborative approach between mouse phenotyping experts and the MGI team may be attractive for ensuring that this useful resource remains current in the face of limited funding, personnel, and time.

3.4. Inclusion/Exclusion Criteria

Monogenic models generated from a variety of sources were included in Table S1. However, in the case of conditional models, only those for which a germline null allele was reported in the MGI database that resulted in embryonic, prenatal, or postnatal lethality were included. We excluded the following models from Table S1: those for which a causative gene had yet to be identified and for which complementation tests were unavailable; those requiring multiple genes for the presentation of the disease phenotype; those based on overexpressing transgenes; and those in which PR degeneration depended on experimental interventions, such as an altered diet, drug treatment, or exposure to bright illumination. Environmental influences on retinal diseases are very important and may affect the progression of PR cell loss, but models that depend on environmental conditions are challenging to compare because of the significant variation among the types of environmental perturbations and the methods used to apply them. We also excluded models that exhibited a reduction in the PR cell number during development but not a progressive loss with age, and those where IS and/or OS dysmorphology or reduction in length was observed without a loss of PR cells, as indicated by a reduction in nuclei number or ONL thickness within the time frame reported in the papers. Although these models were excluded from Table S1, examples are included in the Results.

3.5. Heterogeneity of Data

The type and frequency of data gathered varied greatly among the studies reviewed. In some papers, only one figure with one retinal section was offered as evidence for PR degeneration, while other papers showed extensive quantitation of their data. To document potential sources of variability in the data, we indicate the method by which the degree of degeneration was determined, either by measuring ONL thickness or by count of nuclei in the ONL, typically the number of rows of nuclei spanning the ONL but sometimes a total count of ONL nuclei in a fixed area of a retinal micrograph.

In some instances, when data was quantified in spider plots or bar graphs, mean values obtained from the central retina were used in estimating the PR loss. We normalized the data among studies by recording the percent degeneration as determined by dividing the mutant values by the corresponding values from age-matched controls as reported in each publication.

3.6. Comparison of Progressive PR Cell Loss

To compare progressive PR cell loss among models, we fit normalized data from each article to an exponential decay that includes a delay, or offset [36]. Ranges of either age or photoreceptor numbers, if reported, were averaged. Fitting was performed in Excel Visual Basic using a piecewise equation that modeled the delay with a straight line at 100% and the remaining points with a monoexponential decay to 0%. Two adjustable parameters, the delay and the decay rate constant, were optimized. We calculated the age at which PR cell numbers reached 50% of control values (D_{50}) as a measure of progression. Roughly one third of the datasets contained only a single point within the exponential regime, which was insufficient to calculate D_{50}. In these cases, D_{50} was calculated at the extremes of zero delay and infinite rate, and the mean of these values was used as a D_{50} estimate.

3.7. Generation of Primary Data Using Fundus Imaging and OCT Scans

Fundus photographs of EMR mutants were taken in unanesthetized mice treated with 1% cyclopentolate to dilate or enlarge the pupil with an in vivo bright field retinal imaging microscope equipped with image-guided OCT capabilities (Micron III; Phoenix Laboratories, Inc., Pleasanton, CA, USA) as previously described [20]. This system allows for the visualization of the location of the OCT scan using the real-time Micron III bright-field image. A superimposed line placed directly on the image over the retinal feature being examined delivers precise cross-sectional information, allowing for the assessment of changes in layer thickness and morphological alterations.

Fundus photodocumentation for TVRM mutants and C57BL/6J control mice was performed using a Micron III or IV retinal camera (Phoenix Laboratories, Inc., Pleasanton, CA, USA) as described [37], except that 1% cyclopentolate or 1% atropine was used as a dilating agent, and in some cases, mice were anesthetized with isoflurane. OCT imaging to assess retinal layer thickness in $Nmnat1^{tvrm113}$, $Ctnna1^{Tvrm5}$, and C57BL/6J control mice was performed using a Bioptigen ultrahigh-resolution (UHR) Envisu R2210 spectral domain OCT (SDOCT) imaging system for volume scanning as described [37,38] with ketamine/xylazine (1.6 mL ketamine (100 mg/mL), 1.6 mL xylazine (20 mg/mL), and 6.8 mL sodium chloride (0.9% *w/v*)) as an anesthetic. A representative B-scan through the optic nerve head was derived from the OCT volume dataset. $Rpgrip1^{nmf247}$ and $Alms1^{Gt(XH152)Byg}$ were assessed on the same OCT system by obtaining a linear B-scan with the following parameters: length, 1.9 mm; width, 1.9 mm; angle, 0 degrees; horizontal offset, 0 mm; vertical offset, 0 mm; A-scans/B-scan, 1000 lines; B-scans, 1 line; frames/B-scan, 20 frames; and inactive A-scans/B-scan, 80 lines. Linear scans were registered and averaged in the InVivoVue program to merge the 20 frames into a single image.

4. Results

4.1. Summary of Studies that Report PR Cell Loss

The combined searches of MGI and PubMed databases yielded a total of 230 genes associated with PR cell loss. Ultimately, 3834 reports at MGI and 3325 at PubMed, which most typically characterized one mutant gene but on rare occasions described more than one, were used in the present review. The distribution of retrieved publications sorted by functional categories is summarized in Table S1. The genes identified in these models are summarized in Figure 3. Descriptions of gene and protein symbols used in the text, figures, and Table S1 are provided in Table S2.

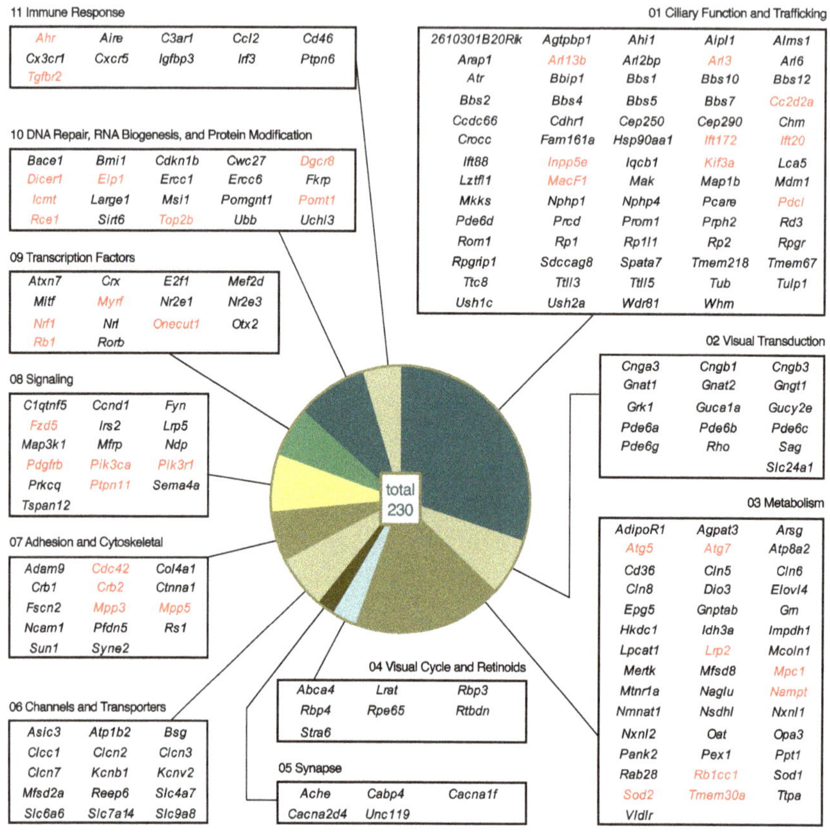

Figure 3. Genes associated with PR cell loss in monogenic mouse models of retinal degeneration (RD). Genes identified by combined review of the Mouse Genome Informatics (MGI) database and articles from a PubMed query were assigned to the indicated functional categories as described in the text. Genes for which mutant alleles are available only in the conditional form are displayed in red. Conditional alleles were included only in instances where germline null alleles resulted in embryonic, prenatal or postnatal lethality. For additional details on inclusion/exclusion criteria, see Section 3.4.

4.1.1. PR Cell Loss Models

The mouse models described in Table S1 were either spontaneous (12%) or chemically induced mutants (11%), or those produced through genetic engineering approaches (77%). This latter group, which was by far the largest, utilized standard homologous recombination, gene-traps, nuclease mediated approaches such as CRISPR/Cas9, and conditionals to mediate genomic changes. Additionally, four models of inadvertent transgene insertion into a unique gene, whose disruption led to PR degeneration, were included within this group. Interesting examples of differences in the disease onset or rate of progression were demonstrated in different models of the same gene (e.g., *Aipl1*) that may be related to allelic differences, null versus missense mutations, or genetic background effects [21,39,40]. Most of the genetically engineered models in Table S1 tended to be in mixed, segregating genetic backgrounds that might impact phenotypic expression (discussed below).

Within the genetically engineered category, a relatively large group of models, 16%, were conditional models, representing 39 genes (Figure 3; red). Generation of conditional mutants is based on the Cre-Lox recombination approach, which requires a floxed gene and a cre-driver to excise the targeted genomic region in a spatial (e.g., cell/tissue specific) or temporal manner (e.g., induction by chemicals such as doxycycline). Since 30% of all null mutations lead to embryonic lethality, as they represent genes that are essential during development, conditionals are often used to examine the adult function of genes [41]. This was the case in 92% of the conditional models described here, as standard organism-wide removal of the genes was reported to be embryonic, perinatal, or postnatal lethal. Thus, conditionals allow us to learn the function of a gene post-developmentally. Conditionals are also sometimes used to determine the cellular contributions to a disease phenotype. If a gene is expressed in multiple retinal cell types, by removing them systematically and examining the consequent phenotype, one can learn how the loss of function of the gene within particular cell types affects the disease phenotype. For example, removal of *Arl3* from rod PRs using a Rho-icre driver shows a later and slower rate of degeneration than that found with Six3-cre, a Cre driver that expresses in early retinal development. This suggests that *Arl3* in rods is necessary for PR survival but that *Arl3* function in other retinal cell types also affects PR survival [42]. The most widely used Cre models include: for targeting retinal progenitor cells, Tg(rx3-icre)1Mjam, Tg(Six3-cre)69Frty, Tg(Chx10-EGFP/cre,-ALPP)2Clc, Tg(Crx-cre)1Tfur, and Tg(Pax6-cre,GFP)2Pgr; for targeting rods, Tg(Rho-icre)1Ck, Tg(RHO-cre)8Eap, and *Pde6g*$^{tm1(cre/ERT2)Eye}$; for targeting M-cone PRs, Tg(OPN1LW-cre)4Yzl (also known as HRGP-cre); for targeting PRs, Tg(Rbp3-cre)528Jxm (also known as IRBP-cre); for targeting RPE, Tg(BEST1-cre)1Jdun, Tg(BEST1-rtTA,tetO-cre)1Yzl and *Foxg1*$^{tm1(cre)Skm}$; and for targeting adult tissues using tamoxifen, Tg(CAG-cre/Esr1*)5Amc.

4.1.2. Mouse Models from Phenotyping Programs

The models listed in Table S1 come from many sources. In addition to individual investigator-initiated efforts, currently the largest contributor to ocular models is the International Mouse Phenotyping consortium, in which 19 phenotyping centers from 11 countries participate to systematically characterize knockout mice generated in a standardized manner [41,43]. All centers do some eye phenotyping, thus providing a window into potential models. Although only a few models from this program are included in Table S1, as most are not yet fully characterized, it is anticipated that this consortium will provide a wealth of models for individual laboratories to study. For example, in the MGI database, 39 IMPC models were identified with "reduced retinal thickness" that with further characterization may reveal PR degeneration.

At The Jackson Laboratory, the Eye Mutant Resource (EMR) and the Translational Vision Research Models (TVRM) programs are dedicated to screen for or generate mouse models with ocular diseases. The EMR has been screening retired breeders by slit lamp biomicroscopy, indirect ophthalmoscopy, and electroretinography since 1988. Retired breeders from the production and genetic resources colonies are screened. Heritable mutants are phenotypically and genetically characterized and the spontaneous mutants are distributed worldwide. The TVRM program arose from the JAX Neuromutagenesis Facility. Mice for this program are generated by chemical mutagenesis or genetic engineering. Carefully characterized mutants are also distributed. Examples of mutants from the EMR and TVRM programs are shown in Figures 4 and 5, respectively.

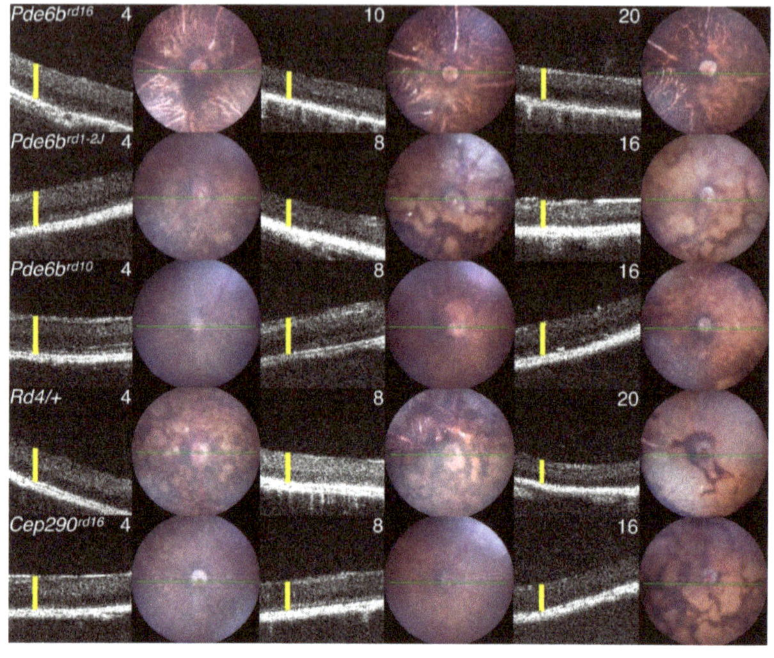

(**a**)

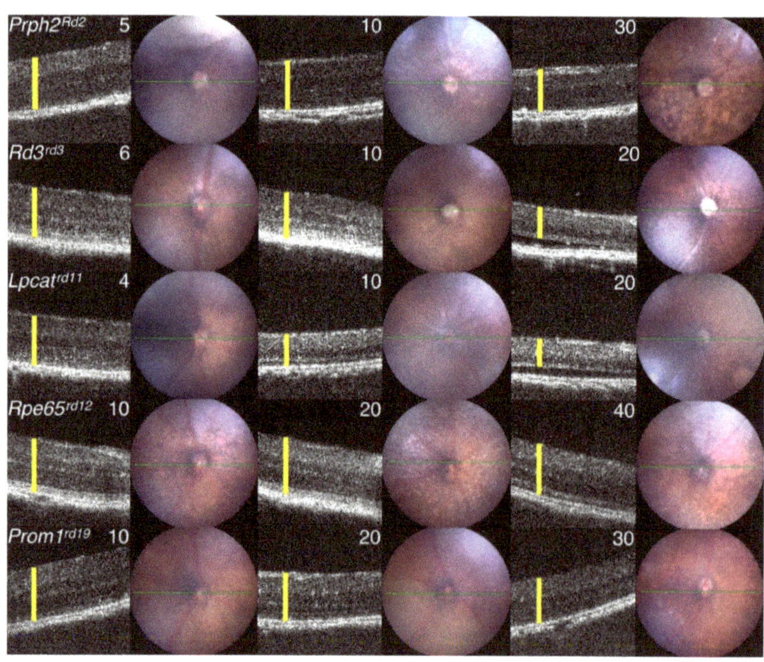

(**b**)

Figure 4. *Cont.*

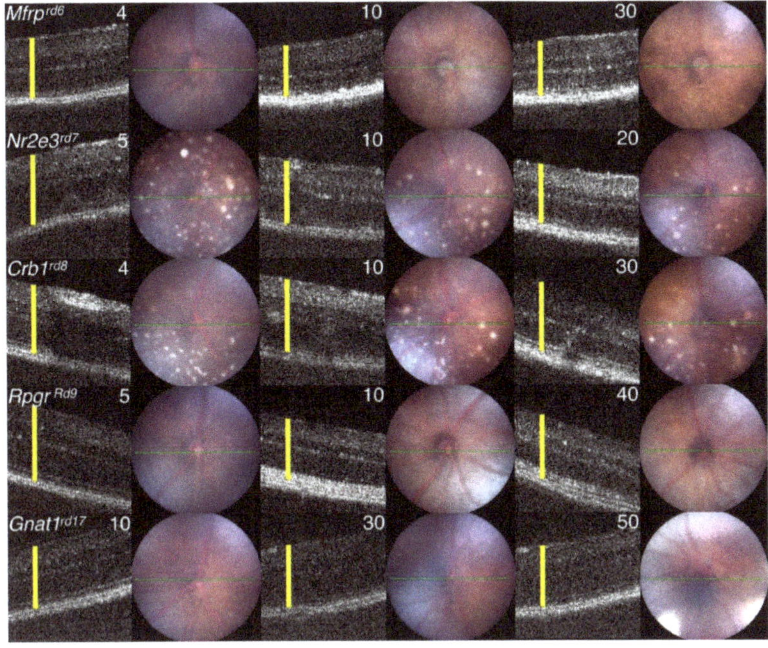

(c)

Figure 4. (**a**) Characterization of mouse models from the Eye Mutant Resource (EMR) program at JAX. Example optical coherence tomography (OCT) and fundus images were taken from rapid RD models: *Pde6b^rd1* (B6.C3-*Pde6b^rd1 Hps4^le*/J, Stock No: 000002), *Pde6b^rd1-2J* (C57BL/6J-*Pde6b^rd1-2J*/J, Stock No: 004766), *Pde6b^rd10* (B6.CXB1-*Pde6b^rd10*/J, Stock No: 004297), *Rd4/+* (STOCK In(4)56Rk/J, Stock No: 001379) and *Cep290^rd16* (B6.Cg-*Cep290^rd16*/Boc, Stock No: 012283) (**b**) Example images from slower RD models: *Prph2^Rd2* (C3A.Cg-*Pde6b^+ Prph2^Rd2*/J, Stock No: 001979), *Rd3^rd3* (B6.Cg-*Rd3^rd3*/Boc, Stock No: 008627), *Lpcat1^rd11* (B6.Cg-*Lpcat1^rd11*/Boc, Stock No: 006947), *Rpe65^rd12* (B6(A)-*Rpe65^rd12*/J, Stock No: 005379) and *Prom1^rd19* (B6.BXD83-*Prom1^rd19*/Boc, Stock No: 026803). (**c**) Examples from slow and very slow RD models: *Mfrp^rd6* (B6.C3Ga-*Mfrp^rd6*/J, Stock No: 003684), *Nr2e3^rd7* (B6.Cg-*Nr2e3^rd7*/J, Stock No: 004643), *Crb1^rd8* (STOCK *Crb1^rd8*/J, Stock No: 003392), *Rpgr^Rd9* (C57BL/6J-*Rpgr^Rd9*/Boc, Stock No: 003391), and *Gnat1^rd17* (B6.Cg-*Gnat1^irdr*/Boc, Stock No: 008811). *Yellow bars* indicate full retinal thickness. Values correspond to the mouse age at the time of imaging (weeks).

5. Analysis

5.1. Progression of PR Cell Loss

While a host of effects can occur as a result of disruptions in genes expressed in PRs and ancillary cell types that functionally impair vision, such as night blindness, or color vision defects, the focus of the models described here are those that bear single gene mutations that lead to actual PR cell loss. From a review of the models in Table S1, significant PR cell loss is reported as early as postnatal day 7 (P7) and can extend throughout the lifetime of animals examined. The progression of PR cells loss is also highly variable and includes models that progress rapidly with complete ablation within several weeks to models with extremely slow progression where only <10% PR cell loss is noted over the span of time in which animals were examined. Generally, while rapid to moderate progression led to almost complete PR ablation, slow and very slow progression, or degeneration in models that primarily affect cone PRs left a substantial number of PR cells intact.

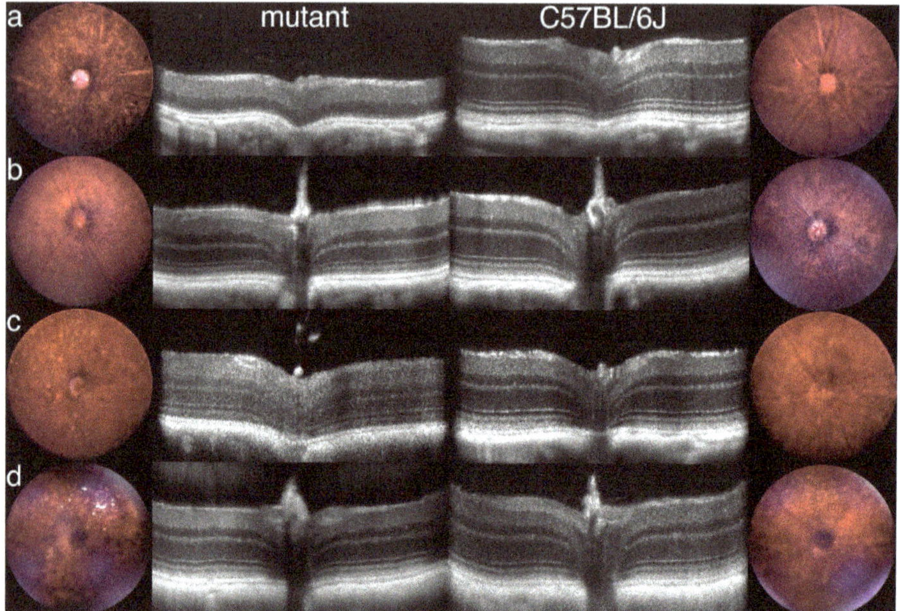

Figure 5. Characterization of mouse models from the Translational Vision Research Models (TVRM) program at JAX. A fundus image (circular panels) and corresponding OCT B-scan are shown for homozygous (**a**) $Rpgrip1^{nmf247}$, at one month of age; (**b**) $Nmnat1^{tvrm113}$, at two months; (**c**) $Alms1^{Gt(XH152)Byg}$, at one year; and (**d**) $Ctnna1^{Tvrm5}$, at two years. Age-matched OCT and fundus images for C57BL6/J control mice are shown to the right of the mutant images. PR cell loss is indicated by a decreased ONL thickness. Fundus images were acquired using a Micron III or IV retinal camera. The vertical dimension of OCT images was doubled to emphasize changes in retinal layer thicknesses.

To compare cell loss among functionally similar genetic models, models in each category of Table S1 were sorted based on the estimated age at which the PR cell population had degenerated by 50% compared to control values, defined as D_{50} (Figure 6). This quantity represents neither the rate of PR cell loss nor the delay before loss commences, although both parameters are used to calculate it. Rather, D_{50} provides a common measure of progression that allows both complete and sparse datasets to be evaluated. With sufficient data, delay, and exponential decay constants were calculated and are reflected by the shaded bars in Figure 6. Many datasets with fewer measurements, some containing a single point, allowed only an estimate of D_{50} (Figure 6; filled circles, range lines indicate estimated limits). Figure 6 may be used to identify models in which overall PR cell loss progresses at an earlier age (lower D_{50}), proceeds at a higher rate once initiated (shorter bar), or is accompanied by a substantial delay (bar starts farther to the right), which may aid in experimental design. Values for D_{50}, the exponential decay constant k, and the delay, when available, are also included in Table S1. We relate qualitative descriptions of progression to D_{50} as follows: rapid, <2 months; moderate, 2 to <6 months; slow, 6 to <12 months; and very slow, ≥12 months.

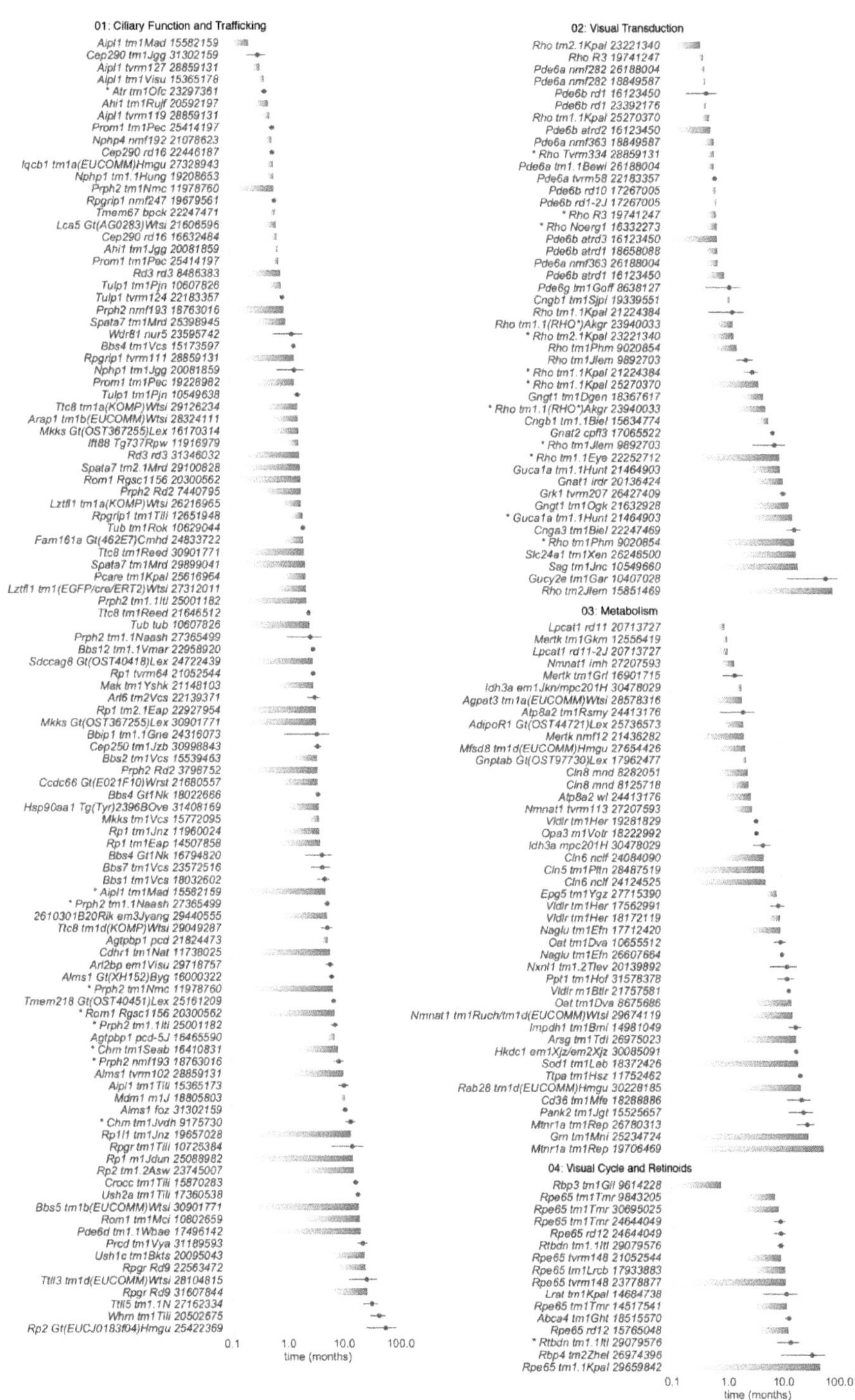

Figure 6. *Cont.*

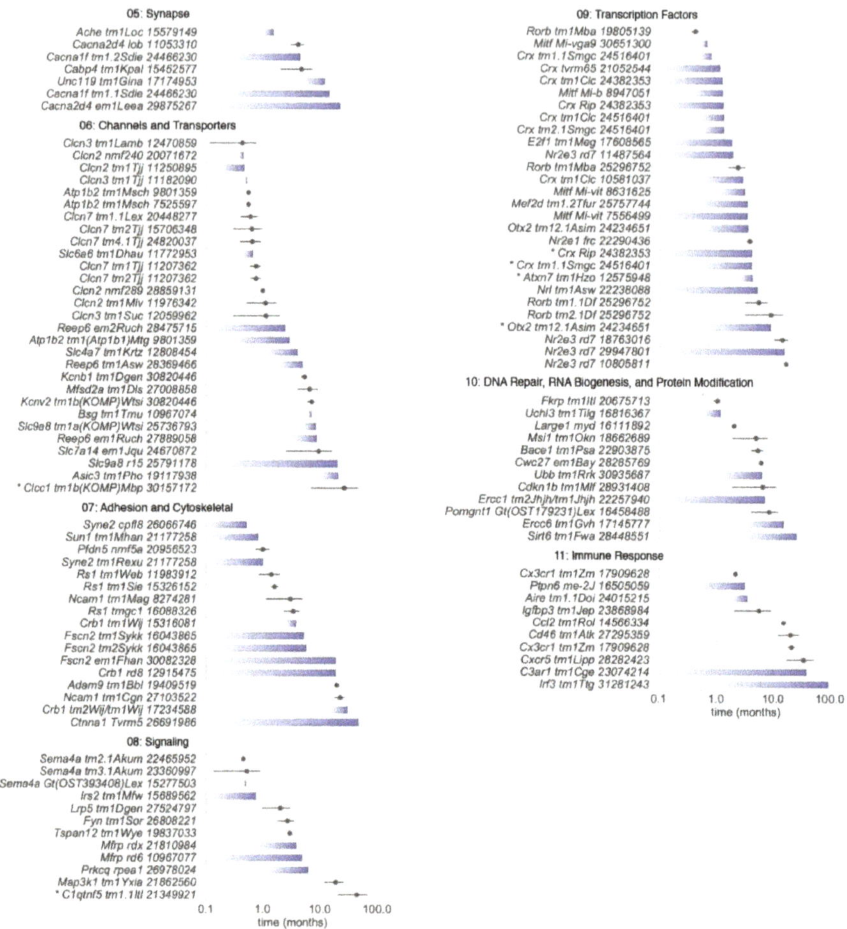

Figure 6. Progression of PR cell loss in mouse models by functional category. Models in each category were sorted by the age at which the estimated number of PR cells reached 50% of wild-type (D_{50}) as determined by fitting reported data to an exponential decay function combined with a delay. Models are identified by gene symbol, MGI allele symbol, and PMID. Models are homozygous except as indicated by * (heterozygous) or by the presence of two allele symbols (compound heterozygous). Left- and right-hand margins of shaded bars represent the delay and D_{50} values, respectively. Bars with a delay value of ≤0.1 month derive from datasets in which no values at 100% of wild type were reported. The midpoint and range of estimated D_{50} values for datasets with only one point in the 5–95% range is indicated by closed circles and range lines, respectively. In cases where the dataset consisted of a single point at 50%, no estimate was needed.

Figure 6 does not include models in which the age of the animal at the time of measurements was not provided, or was reported as "adult", although these models were included in Table S1. We also omitted models where only cone PR cell degeneration was reported, as limited data and a lack of cone nuclei counts made mathematical modeling of cell loss unreliable. Finally, we omitted all conditional alleles from Figure 6, as the efficiency and specificity of inactivation of genes, as well as the temporal expression of different transgenic Cre drivers utilized varies, making comparison of these alleles difficult.

5.2. Biological Processes Affected by Mutations

Categorization of PR cell loss genes (Figure 3) relied on current functional knowledge, similarity to other genes, and localization, if known. Genes are often expressed in multiple cell types and serve different functions within the retina. We placed genes in categories that were most pertinent to their roles in the PR or effect on PR survival. The data in Table S1 can be explored after reassigning models to different categories if desired. In the descriptions below, we provide an overview for the first three categories and then a detailed description of the effect of gene disruptions on biological functions included in these categories. Subsequent categories are described without overview.

5.2.1. Category 01: Ciliary Function and Trafficking

Overview. Disruptions in a vast number of genes associated with PR sensory cilia result in RD [24,44]. The onset and rate of PR cell loss may vary depending on the role of a given gene in maintaining ciliary structural integrity and function during early PR development and maintenance in adulthood. Retinal defects within any one of the key ciliary structures and/or processes such as ciliogenesis, OS morphogenesis, or PR homeostasis may result in the loss of rod and cone PR cells. Understanding the pathophysiological mechanisms involved in PR cell loss requires a detailed examination of the components and functionalities of the PR sensory cilium.

The mouse PR sensory cilium is a specialized structure comprised of a tubulin-rich axoneme with a symmetrical "9+0" arrangement of microtubular doublets [45]. The connecting cilium (CC) is akin to the transition zone (TZ) of primary cilia [46,47] and serves as a passageway for the movement of proteins from the organelle-rich IS to the photosensory OS. The OS is composed of an elongated distal axoneme with adjacent stacked membranous discs decorated with proteins necessary for phototransduction. It has been hypothesized that extension of the PR plasma membrane towards the apical RPE provides a convenient sink in the OS for the storage of a large number of membrane proteins [48]. In the CC-TZ, axonemal microtubule doublets interconnect the distal axoneme and basal body, and also connect to the periciliary membrane via Y-linkers. The CC-TZ harbors membrane-associated and soluble proteins that coordinate as gatekeepers to regulate the entry, retention, and exit of proteins in and out of the OS [47,49,50]. At the axonemal base, the ciliary membrane is anchored by the nine microtubular triplets of the basal body and its associated appendages.

In developing murine PRs, the connecting cilium and OSs assemble through a series of coordinated events. Shortly after cell cycle exit of retinal progenitor cells, the mother centriole docks to a primary ciliary vesicle initiating its expansion and fusion with the plasma membrane. Concurrently, the microtubular axoneme elongates towards the apical end of the neuroblastic layer [51]. In rod PRs, ciliogenesis typically begins shortly after birth with the appearance of a mature centriolar-bound ciliary vesicle at around P4 [52]. Subsequently, OS biogenesis occurs asynchronously between P8 and P14 [52–54] while axoneme and PM extension continue until OS maturation around P19–25 [52,53]. During this process, intraflagellar transport (IFT) provides an efficient mechanism for the movement and delivery of crucial proteins to the developing OS [55,56], as discussed in greater detail below.

Over the past several decades, two disparate mechanisms of rod disc morphogenesis have been debated, a vesicular fusion model [57,58], which postulates that membrane discs originate from rhodopsin bearing vesicles that undergo intracellular membrane fusion, and the classic evagination model [59], which proposes that new discs result from the evagination of the plasma membrane at the base of the OS. Recent ultrastructural studies in mouse rod PRs have provided compelling evidence that support the classic evagination model. By high resolution microscopy, several groups demonstrated the plasma membrane origin of the evaginated rod disc membranes [60], which subsequently flatten, elongate, and become enclosed [52,60,61].

OS membranous discs undergo a rapid and continuous turnover with approximately 75 rod discs being shed daily, corresponding to 10% of the OS [53,62]. At the ciliary tip, aged discs are removed by the adjacent RPE through phagocytosis. Consequently, OS renewal requires a continuous flow of new proteins from IS to the OS through the connecting cilium, a process that requires careful regulation.

For the CC-TZ gate to function properly, efficient mechanisms are needed to prevent the entry of undesired proteins and to remove the non-OS proteins improperly targeted to the OS. At the base of the CC-TZ lie transition fibers, which act as a barrier in conjunction with other CC-TZ components, such as the membrane-associated Meckel syndrome (MKS) complex. The BBSome, an octameric coat complex, is thought to coordinate the delivery and removal of proteins from the OS during ciliary formation and maturation [63,64].

Finally, protein trafficking through the CC-TZ is important for PR development and maintenance [51,65,66]. The movement of protein cargo from the IS to the OS requires a highly regulated passage of vesicles along microtubules that dock and fuse with the periciliary membrane to deliver their cargo at the CC base. Movement of targeted proteins through the cilia to the OS may be facilitated by IFT transport machinery [65] or by a lipidated protein trafficking system [67,68]. During IFT, protein cargo associate with IFT particles that attach to kinesin and dynein motors and move along the axonemal microtubules in both anterograde and retrograde directions, respectively. The BBSome is known to associate with IFT particles and may provide a mechanism for the removal of non-targeted protein accumulation in the OS [69].

Ciliogenesis. As the ciliary axoneme serves as an important conduit for the IFT movement of signaling molecules, it comes as no surprise that the disruption of ciliogenesis genes may result in significant developmental abnormalities causing early lethality and/or rapid PR degeneration. Genes essential for the elongation of the proximal axoneme (A and B tubules) include *Kif3a* [70], which encodes a subunit of kinesin 2, *Iqcb1* [71], *Arl3* [42], and *Arl13b* [72]. While patients with missense mutations in *ARL13B* present with Joubert-associated features [73], a null mutation, *Arl13b^{hnn}*, results in embryonic lethality in mice [74]. Axonemal disturbances and a failure to form OS discs are observed in developing retinas with conditional *Arl13b* disruption [72].

Axonemal and ciliary membrane extension. Disruptions in genes that affect ciliary extension include *Rp1/Rp1l1/Spata7* (distal axoneme), *Mak,* and *Pcare* (*C2orf71*; ciliary membrane). Mice with knockout alleles of *Rp1* (*Rp1^{tm1Eap}* and *Rp1^{tm1Jnz}*) and *Spata7* (*Spata7^{tm1Mrd}*) show a progressive, moderate loss in PR cells through the first year of life. *Rp1^{m1Jdun}* mice homozygous for a Leu66Pro missense mutation experience a much slower degeneration with 30% of PRs left at 26 months of age. Conditional ablation studies of *Spata7* in PRs and in the RPE have shown that the disruption of SPATA7 in rod and cone PRs, but not in the RPE, is the molecular basis of the retinal degenerative phenotype [75].

Ciliary Gate and the CC-TZ. Sensory/primary cilia and their gatekeepers (CC-TZ) are found abundantly in most cell types [76]. Thus, the disease spectrum of ciliary proteins is extensive given their roles in ciliary trafficking, signaling, and development. Disruptions in CC-TZ genes may result in isolated cases of inherited retinal dystrophies such as Leber congenital amaurosis or in multisystemic, ciliopathies such as Joubert, Meckel, or Senior-Løken Syndrome. Such syndromic ciliopathies may include a multitude of disease phenotypes such as brain malformations, renal cysts, nephronophthisis, and retinal dystrophy.

Within the CC-TZ reside MKS and NPHP modules that closely interact and form multiple distinct protein complexes [47,50,77–79]. The MKS complex includes membrane-associated proteins, such as MKS1 and TMEM67, while NPHP complex proteins, such as NPHP1 and NPHP4, associate in closer proximity to the ciliary axoneme. Mice harboring mutations in genes coding for these complex-associated proteins form normal cilia, however, display early abnormalities in OS morphogenesis. After ciliary biogenesis, retinas in these mutant mice quickly degenerate, eliminating most PRs by 3–4 weeks of age. Genes whose disruptions affect the ciliary gate functions of the CC-TZ and cause rapid degeneration include *Nphp1, Nphp4, Ahi1, Iqcb1, Tmem67,* and *Cep290*. In humans, mutations in *CEP290* can lead to primarily single-organ diseases such, as retinitis pigmentosa and nephronophthisis, or pleiotropic diseases, such as the Joubert, Meckel, and Bardet–Biedl syndromes. The most studied allele is *rd16*, which harbors a 297 basepair in-frame deletion in *Cep290*. Compared to *Cep290$^{tm1.1Jgg}$* knockout mice, which show a rapid 78% loss at P14, *Cep290^{rd16}* homozygotes have a longer disease progression with a 60% ONL loss at three weeks of age.

Basal bodies and associated pericentriolar material (PCM). The basal body is a structure derived from the mother centriole and resides at the base of the cilium along with the daughter centriole, neighboring centriolar satellites and other related PCM. Proteins positioned at the ciliary base can also be seen in centrosomes of dividing cells (ALMS1, CEP250, and C8ORF37). Specifically, ALMS1 and CEP250 (CNAP1) localize in close proximity to each other at the proximal ends of centrioles [80]. *ALMS1* encodes a 460kDa protein that when disrupted results in the Alström syndrome (ALMS) [81,82]. Mice with a gene trap, frameshift, and nonsense mutations recapitulate human ALMS disease features such as obesity, diabetes, and neurosensory deficits [21,83,84]. The proper formation of the connecting cilium and the slow progression of PR cell loss in *Alms1*$^{Gt(XH152)Byg}$ [83], *Alms1*^{foz} [84], and *Alms1*^{tvrm102} [21] models suggests that ALMS1 is not essential for ciliary biogenesis but necessary for overall PR homeostasis.

The ciliary base contains supportive structures necessary for the proper docking of cargo to the ciliary membrane. Targeted *Macf1* null mutants fail to develop the ciliary vesicle needed for basal body docking while conditional ablation of *Macf1* in the developing retina disrupts retinal lamination and maturation [85]. Mutations in *Cdcc66*, which encodes a component of centriolar satellites and *Sdccag8*, which encodes a recruiter of PCM, result in an early onset but slow-moderately progressive disease [86,87]. The slower RD makes these alleles attractive models for therapeutic investigations.

Genetic mutations in *CC2D2A*, which encodes a component of the subdistal appendages of mother centrioles and basal bodies [88], have been observed in patients with Meckel Syndrome [89], Joubert Syndrome [90], and non-syndromic rod-cone dystrophy [91]. Mice with null mutations in *Cc2d2a* experience embryonic lethality due to the absence of subdistal appendages and nodal cilia [88]. The retinas of adult mice with tamoxifen-induced deletion of *Cc2d2a* in PRs have a significantly diminished ONL (2–3 layers) 12 weeks post-injection [92], suggesting that CC2D2A is necessary for ciliary homeostasis.

Periciliary membrane complex. At the periciliary membrane complex of PRs lies an Usher protein interactome complex that provides a scaffold for the anchoring of fibers to the periciliary membrane [93]. Mutations in genes encoding members of this complex, *Ush1c*, *Whrn*, and *Ush2a*, result in Usher syndrome, a disease that results in progressive hearing and vision loss. Multiple forms of Usher syndrome exist resulting in different degrees of the onset and severity of disease symptoms. In the mouse, targeted mutations in Usher genes results in a late-onset and very slow progression of PR degeneration. Homozygous *Ush2a*^{tm1Tili} mice have normal retinas at 10 months of age and lose 70% of their PRs by 20 months of age [94]. PR degeneration in *Whrn*^{tm1Tili} retinas is protracted with only 30% loss observed at 28 months of age [95].

Disc morphogenesis. Although the molecular mechanisms involved in OS disc morphogenesis are not completely understood, there has been considerable progress within the past decade with the emergence of refined ultrastructural methods. The OS protein, peripherin-2 (PRPH2), localizes to the rims of rod and cone discs and functions to establish and maintain the membrane rim curvature during disc formation and maintenance [96,97]. Recent investigations using the *Prph2*^{Rd2} (*rds*) mouse model [10] have suggested another role for PRPH2 during disc morphogenesis [52]. Using transmission electron microscopy, Salinas et al. [52] demonstrated that like other forms of cilia [98,99], PR sensory cilia have an innate ability to spew off ectosomes at the OS base. During normal development of the OS, ectosome release is inhibited and the retained membrane at the CC-TZ is transformed into discs upon membrane evagination. In the homozygous *Prph2*^{Rd2} mice, discs fail to form resulting in the accumulation of ectosomes at the OS base. This finding led the authors to propose that PRPH2 may play a role in inhibiting ectosome release during normal rim formation [52].

Knock-in mice carrying heterozygous alleles of *Prph2* (Tyr141Cys [100] and Lys153Δ [101], mimic the dominant RP disease observed in human patients. It is interesting that PR degeneration rates vary among *Prph2* mutant alleles. While homozygous *Prph2*^{Rd2} mice gradually lose their PRs within the first year [102,103], mice harboring a homozygous null mutation, *Prph2*^{tm1Nmc} undergo a faster degeneration with most PRs lost by 4 months of age [104]. Heterozygous *Prph2*^{tm1Nmc} mice also experience PR loss, however the rate of decline is much slower [104]. Comparative studies of *Prph2*^{Rd2} with rhodopsin double-knockout mice have suggested that abnormal accumulation of mislocalized rhodopsin may contribute to PR degeneration in *Prph2*^{Rd2} [105]. Hence, the zygosity differences in degeneration rates may be a result of varying rhodopsin: PRPH2 ratios. In addition, the onset and severity of the disease may be influenced by the location of the mutation, as different PRPH2 domains have been implicated in dual roles during disc morphogenesis, the tetraspanin core in rim membrane curvature, and the *C*-terminal domain in ectosome release suppression [52].

ROM1 is thought to be involved in the regulation of OS disc formation. PRPH2 and ROM1 are closely associated at the disc rims in the OS. In humans, a double heterozygous disruption in both *ROM1* (a PRPH2-interacting protein) and *PRPH2* results in digenic RP [106]. While it is not clear whether defects in ROM1 alone causes RP in humans, mice with a monogenic *Rom1* disruption show signs of dominant RP. At one month of age, *Rom1* knockout OS discs are visible but appear enlarged and slightly disorganized [107]. PRs slowly degenerated, reducing the ONL by 34% at 1 year of age. In contrast, *Rom1*^{Rgsc1156} mice with a heterozygous missense mutation, p.W182R, show a 55% loss of PRs at 35 weeks of age [108]. Furthermore, RD was more pronounced in homozygous mice. The degeneration in *Rom1*^{Rgsc1156} mice may be a consequence of an early reduction in endogenous PRPH2 and ROM1 levels, which may interfere with PRPH2-mediated stabilization of disc outer rims.

PRCD, progressive rod-cone degeneration, is a rhodopsin-binding protein [109] that localizes to the OS disc rims [110]. Patients and canines with *PRCD*^{C2Y} mutations have a slowly progressive form of rod-cone degeneration [111]. The Cys2Tyr mutation results in mislocalization of PRCD from the OS to the ONL where it is actively degraded [109]. In the PRs of mice with homozygous *Prcd*^{tm1Vya} mutations, loss of PRCD results in the formation of bulging discs that do not properly flatten and in the accumulation of extracellular vesicles that originate at the OS base [112]. Interestingly, mutant PRs are able to form membrane discs and the distribution of OS proteins and light response do not appear to be perturbed. While activated microglia infiltrate the interphotoreceptor space to remove extracellular vesicles and debris, removal is insufficient and PRs undergo a very slow degeneration. Homozygous *Prcd*^{tm1Vya} mice show only 36% ONL loss at 17 months [112] while in *Prcd*$^{tm1(KOMP)Mbp}$ homozygotes a similar loss is observed at 30 weeks of age [110]. Both models are knockout alleles that target the 5′ end of *Prcd* but are on different genetic backgrounds. Further investigations are necessary to determine whether gene modifiers affect progressive PR cell loss in these two models.

IFT trafficking. IFT is essential for ciliogenesis in mammals [113] and disruption of this process often leads to abnormalities in embryonic development. In the mouse, null mutations in genes encoding subunits of the IFT-A (*Ift122* [114], *Ift88* [115], and *Ttc21b* [116]) and IFT-B (*Ift172* [117], *Ift80* [118], and *Traf3ip1* [119]) complexes result in embryonic lethalities, many of which are attributable to ciliary-related disturbances in hedgehog signaling [120,121]. These findings further highlight the integral role of cilia and IFT machinery during embryogenesis.

Hypomorphic and conditional alleles have been useful for elucidating the roles of IFT components in retinal disease. The hypomorphic allele, *Ift88*^{Tg737Rpw}, contains a transgenic insertion resulting in a 2.7 kb intronic deletion. Homozygous *Ift88*^{Tg737Rpw} mice exhibit disorganized OSs as early as P10 and a progressive degeneration of PRs that reduces the ONL to one layer at P77 [66]. Rod-specific ablation of *Ift172* [122] leads to mislocalization of rhodopsin, RP1, and TTC21B (IFT139) and rapid degeneration of PRs. Conditional depletion of *Ift20* in M cones and mature rods both results in opsin mislocalization suggesting that proper opsin trafficking hinges on functional IFT components [123]. To gain a clearer understanding of the roles that IFT molecules play in both rod and cone PRs, additional studies using conditional models are warranted to elucidate the contributions of impaired IFT components to PR cell loss.

Lipidated protein trafficking. Lipid modification of proteins, such as prenylation or acylation, helps direct intracellular protein targeting and regulates protein activity [124]. These hydrophobic modifications help tether their protein partners to the surface of specific membranes throughout the cell, such as the ER, Golgi, transport vesicles, or plasma membranes. Improper trafficking of lipidated proteins can result in RD. RP2 is a GTPase activating protein that interacts with ARL3 to regulate assembly and movement of membrane-associated protein complexes [125]. Homozygous mice with mutations in the gene encoding RP2 exhibit a slowly progressive rod-cone degeneration [126,127]. ARL3, a small GTPase, traffics lipidated membrane-associated proteins to the rod OS [128]. Although *Arl3* knockout mice exhibit early postnatal lethality and Joubert-like features [42], mice with hypomorphic mutations survive to post-wean and display OS abnormalities as early as P9 [129]. Conditional ablation of *Arl3* in the developing retina results in the absence of cilia, and therefore PR cells are rapidly lost [42]. In contrast, depletion of *Arl3* in mature rods leads to mislocalization of lipidated OS proteins, shortened OS, and a moderate progressive PR loss. These results are consistent with roles for ARL3 in ciliogenesis during development and cargo displacement during lipidated protein trafficking.

The ciliary TZ-associated protein, RPGR, binds and directs the ciliary targeting of INPP5E [130], a phosphoinositide phosphatase that is important for ciliogenesis [131]. Ciliary localization of RPGR itself requires modification with a prenyl group, which interacts with PDE6D [130], a prenyl-binding protein first discovered as a copurifying component of cGMP phosphodiesterase 6 (PDE6) [132]. Like *Rpgr^{rd9}* [133] and *Rpgr^{tm1Tili}* knockout mice [134], mice with *Pde6d^{tm1.1Wbae}* null mutations [135] undergo a very slow degeneration with at least 50% of PR cells remaining at 20 months of age. Rao et al. have demonstrated reduction of INPP5E in RPGR-deficient axonemal OSs [130]. Altogether, these observations validate RPGRs role in ciliary trafficking and homeostasis and suggest that other players may be involved in ciliary targeting of INPP5E.

Mutations in *Aipl1* result in early and rapid loss of PR cells (Table S1). $D_{50} < 0.55$ months for the four germline alleles shown in Figure 6, *Aipl1^{tm1Mad}*, *Aipl1^{tvrm127}*, *Aipl1^{tm1Visu}*, and *Aipl1^{tvrm119}* [21,39,136]. AIPL1 is a protein chaperone that mediates the folding of phosphodiesterase 6 (PDE6), a key component of the visual transduction pathway that regulates cGMP levels (see Section 5.2.2. below) [137]. AIPL1 binding is promoted by prenylation of PDE6 subunits [137]. In *Aipl1* mutants, PDE6 subunits are greatly diminished [136], providing further evidence for the importance of lipid modification in PR viability and vision.

BBSome assembly and regulation. Disruptions in the octameric BBSome complex or associated chaperonins may cause syndromic ciliopathies such as the Bardet–Biedl syndrome and McKusick–Kaufman syndrome. In mice, most gene disruptions that affect the BBSome [138] (BBS1, BBS2, BBS4, BBS7, BBIP1, TTC8, and ARL6), and its regulators (LZTFL1, MKKS, BBS10, and BBS12) result in a moderate degeneration of PRs. For instance, PRs in homozygous mice harboring gene trap or null mutations of *Bbs4*, *Bbs4^{Gt1Nk}* [139], and *Bbs4^{tm1Vcs}* [140] appear to progressively decline after maturation with >90% loss at 7 months of age. The delay and lack of ciliogenesis defects suggests that there may be some functional redundancy amongst components of the BBSome.

5.2.2. Category 02: Visual Transduction

Overview. Mutant alleles of genes encoding proteins responsible for light detection comprise a second category of models (Category 02: Visual Transduction; Figure 1c, Table S1). The multistep phototransduction process that detects light and amplifies this signal is similar in rod and cone cells, but the specific proteins that catalyze many of the steps are often unique to each cell type [141]. Phototransduction is initiated by the response of opsin-based light-sensitive G protein coupled receptors that are covalently linked to vitamin A retinal as a cofactor. The receptor rhodopsin (RHO) is expressed exclusively in rod cells and is optimized to detect dim green light. Cone pigments that detect short or medium wavelength visible light (OPN1SW and OPN1MW, respectively) are exclusively expressed in cone cells, in some retinal regions coordinately within the same cell. These receptors constitute >90% of OS protein and are localized to the disc membranes.

Light activation of RHO or cone pigments causes the bound retinal to isomerize from an 11-*cis* to an all-*trans* configuration, ultimately leading to its release from the receptor by hydrolysis. Isomerization results in a conformational change in the protein that alters its interaction with a bound heterotrimeric G protein, transducin, activating the exchange of GTP for GDP bound to the α subunit of this protein.

In turn, activated α transducin-GTP binds the inhibitory γ subunits of phosphodiesterase 6, releasing it from the α and β subunits of this complex, which are thereby activated to catalyze the conversion of cGMP to GMP. The ensuing reduction in cGMP levels in the OS closes the cGMP-gated cation channel, slowing the influx of Na^+ and Ca^{2+} ions, which hyperpolarizes the plasma membrane of the OS and, ultimately, the entire cell. Hyperpolarization causes Ca^{2+} channels to close at the cell synapse, which leads to a decrease in the calcium-dependent release of glutamate-containing vesicles into the synapse and activates postsynaptic bipolar neurons.

The process is regulated to ensure the highest sensitivity to illumination. Following its activation, rhodopsin is quenched by the action of arrestin, which binds to bleached opsin molecules that are phosphorylated by rhodopsin kinase. Resetting of the cell following the light flash requires the formation of cGMP from GTP, catalyzed by a membrane-bound guanylate cyclase, the subsequent closing of the cGMP-gated cation channel, and the restoration of electrolyte distribution across the plasma membrane as achieved by ion pumps and transporters. Hydrolyzed retinal is passed from the OS to the RPE as part of the visual cycle (see below), where it is re-isomerized and returned to the PR cell to regenerate bleached opsin. An additional visual cycle involving Müller cells contributes to the regeneration of cone pigments.

Visual pigments. Profound effects on PR viability are observed due to mutations that affect rod cells, which represent 97% of the PR population. Mouse models bearing *Rho* alleles exhibit semidominant and recessive rod cell loss phenotypes that vary greatly in the onset and rate, consistent with the variety of possible disease mechanisms that have been proposed for RHO mutations over decades of study. For example, some missense alleles in Table S1, such as those that encode the Pro23His, Cys110Tyr, Tyr178Cys, and Cys185Arg variants [21,142–146] may support a hypothesis that excessive RHO misfolding in the endoplasmic reticulum induces cellular stress pathways that lead to PR cell loss [147]. Although the pathways linking misfolded RHO to cell death are not fully resolved, recent studies of the Pro23His variant in cultured cells and in rats [148] or mice [145,146] suggest that stress pathways induced by the unfolded protein response are protective, and raise the possibility that increased intracellular calcium due to ER stress may cause cell death [146]. Misfolding may also explain the partial mislocalization of RHO Glu150Lys to the IS [149]. However, in this mutant, much of the protein appears to be correctly exported to the OS, where it leads to irregularly shaped and disorganized discs, possibly due to a defect in higher-order RHO organization [149]. Pro23His RHO also disrupts the orientation of discs during their morphogenesis, possibly through similar effects on higher-order structure [150].

By contrast, the effect of the Gln344Ter variant (Table S1), which is correctly folded but includes sequence extensions at the C-terminus that interfere with export to the OS [151], as well as the graded effect of heterozygous or homozygous knockout alleles *Rho*^{tm1Jlem} and *Rho*^{tm1Phm} [152,153] or the premature truncation mutant Arg107Ter (Table S1), provides evidence that a steady flow of RHO to the OS is essential for PR cell viability. These observations fit an emerging view that a proteostasis network, incorporating not only cellular stress pathways but also protein trafficking and degradation, regulates the cellular protein balance to ensure viability [147,154]. According to this view, a failure to sort vesicles bearing RHO from the Golgi to the periciliary membrane, or a partial or complete loss of the protein, leads to protein imbalance in the IS. This imbalance may induce cellular stress responses and also affect the trafficking of other molecules destined for the OS, such as other phototransduction proteins, lipids, and vitamin A, resulting in cellular toxicity. Finally, the RHO Asp190Asn variant (Table S1) appears to traffic properly to the OSs but may have structural defects that lead to constitutive signaling [155], which has been linked to PR degeneration [156]. The same mechanism may account for the effect of *Rho* mutants that result in rapid degeneration upon bright illumination [157] but were not included in Table S1 due to the dependence of the mutant phenotype on an environmental perturbation (see Discussion). Future studies of these and other models may resolve or converge the many proposed hypotheses to explain RHO-associated RD.

Based on the often profound effect of *Rho* variants on rod cell viability, it might be expected that cone pigment variants would similarly cause cone PR cell loss. However, cones remain viable for more than 1.5 years in homozygous *Opn1sw*^{tm1Pugh} mice, which show a 1000-fold decrease in transcript and produce no detectable OPN1SW by immunoblotting, histochemistry, or single-cell recording of light responses [158]. Likewise, cones are viable for at least 10 months in homozygous *Opn1mw*$^{tm1a(EUCOMM)Wtsi}$ knockout mice, despite an absence of OPN1MW in immunoblotting and immunohistochemical studies [159]. These studies suggest fundamental differences in the cellular sensitivity of rod and cone cells to visual pigment deficiency. They also highlight the concern that reactivity to antibodies against cone opsins or other cone cell markers may be abolished even though the cells remain viable, and therefore may not be as reliable as counting cone nuclei [160] to assess cell loss.

Transducins. Rod transducin subunits α, β, and γ (encoded by *Gnat1*, *Gnb1*, and *Gngt1*, respectively) form the heterotrimeric G protein complex that is essential for propagating the signal from light-activated rhodopsin. *Gnat1* knockout mice have attenuated rod responses and model congenital stationary night blindness (CSNB) [161]. Although slow PR loss was reported for this model, our measurement of ONL thickness at four weeks of age based on reported images yielded a value of 90% of wild type, matching the author's value at 13 weeks [161] and suggesting an early developmental difference rather than progressive cell loss. In support of this finding, others using the same strain reported ONL thickness was 85% of wild type at eight weeks of age with no evidence of significant cell loss up to 52 weeks of age [162]. By contrast, IRD2 mice, which are homozygous for a *Gnat1*^{irdr} allele predicted to yield a prematurely truncated polypeptide, exhibit significant rod PR cell loss (Table S1) accompanied by late cone cell loss and reduced rod-specific ERG responses [163]. Homozygous *Gnat1*^{irdr} mice may recapitulate recessive rod-cone dystrophy, which has recently been linked to human *GNAT1* variants predicted to encode prematurely truncated proteins [164–166]. The *Gnat1*^{irdr} allele was discovered independently in *rd17* mice at JAX, suggesting a founder effect [167,168].

Gnb1 knockout mice have not been studied due to embryonic and perinatal lethality. However knockout alleles of the gene encoding rod γ transducin, *Gngt1*^{tm1Dgen} and *Gngt1*^{tm1Ogk}, result in PR loss that is more rapid than in *Gnat1* mutants [169,170]. In these strains, GNGT1 deficiency is accompanied by a 6- to 50-fold post-translational reduction of GNAT1 and GNB1, indicating a key role of the transducin γ subunit in complex assembly. *Gngt1*^{tm1Dgen}-associated degeneration is rescued by heterozygous *Gnb1*$^{Gt(prvSStrap)4B8Yiw}$ mice [171], which express retinal GNB1 at 50% of wild type levels. This result suggests that the toxicity of GNGT1-deficiency is due to an excess of improperly assembled GNB1, which is targeted for degradation but exceeds the capacity of the proteasome [171]. This observation supports the proteostasis network model of PR degeneration [154].

Among genes encoding cone transducin subunits α, β, and γ (*Gnat2*, *Gnb3*, and *Gngt2*), only *Gnat2* alleles have been reported to cause PR loss. A progressive reduction of cone cell ERG responses and a 27% decrease in PNA-positive cells at 12 months of age in homozygous *Gnat2*^{tm1Erica} mice (Table S1) is consistent with cone PR loss [172]. However, cone nuclei were not counted directly, so it is possible that cone cell loss is less pronounced than reported. The predicted GNAT2 Asp173Gly substitution in this model may alter guanine nucleotide binding [172], although how this change might cause cell loss is unresolved. Interestingly, mislocalized cone opsin OPN1MW in this model suggests endoplasmic reticulum stress, which is often associated with PR degeneration. *Gnat2*^{cpfl3} mice (Table S1) show no cone cell loss for at least 14 weeks but exhibit a slow loss of rod cells [173]. In contrast to these models, a recently developed *Gnat2* knockout strain abolishes GNAT2 function without PR loss or dysmorphology in the oldest mice examined at 9 months of age [174]. Although human *GNAT1*-variants are a rare cause of achromatopsia [175], a stationary congenital colorblindness, the clinical presentation is variable and some cases are associated with a reduction in visual acuity with age [176] that may suggest progressive cone cell loss. The available mouse alleles may help to identify disease mechanisms that contribute to this phenotypic variability.

Phosphodiesterase 6. Rod phosphodiesterase 6 consists of a catalytic $\alpha\beta$ complex encoded by *Pde6a* and *Pde6b* and two inhibitory γ subunits encoded by *Pde6g*. The control of cGMP levels by this enzyme is expected to affect both PR function and viability, as cGMP has a central role in the phototransduction cascade and PR cell metabolism [177], and elevated cGMP levels have been linked to PR cell loss [178]. Indeed, *Pde6a* and *Pde6b* mutants show depressed ERG responses at an early age and rapid PR loss with D_{50} values of 11–30 days (Figure 6, Table S1). A study of *Pde6a* mutations on the same strain background made use of an allelic series that varied in disease severity [179]. The order of disease progression due to the alleles reported in this study, *nmf282* (Val685Met; fastest) > *tm1.1Bewi* (Arg562Trp) > *nmf363* (Asp670Gly; slowest), is the same as assessed by D_{50} (Figure 6). This allelic series led to a correlation of more rapid PR degeneration with an increased number of cGMP-positive PR cells [179]. The same trend in the progression of disease in *Pde6a*^{nmf282} and *Pde6a*^{nmf363} mice was found earlier [180], but an opposite cGMP result was obtained, possibly due to the assessment of total retinal cGMP rather than a count of cGMP-positive PR cells [179] (a 0.1-month difference in the D_{50} of *Pde6a*^{nmf363} mice measured in the two studies may reflect strain differences that might also contribute to the difference in findings). The later study also combined two alleles that matched human *PDE6A* variants to create a compound heterozygote [179], mirroring the more typical situation in human genetic disease. Further, the allelic series highlighted a non-apoptotic cell death mechanism involving calpain rather than the expected caspase-mediated apoptotic process [179]. Both elevated cGMP and calpain activation have been observed in other mouse RD models [181]. Thus, allelic series as used in these studies are informative for assessing disease mechanisms and identifying potential differences in treatment efficacy that may reflect disease severity.

Of the *Pde6b* alleles described, *Pde6b*^{rd1} and *Pde6b*^{rd10} have been used most extensively as PR degeneration models. *Pde6b*^{rd10} disease develops later, providing a longer window of opportunity to test therapeutic efficacy (Figure 6). The *Pde6b*^{atrd1} model has an even slower progression ($D_{50} = 0.71$) than *Pde6b*^{rd10} mice ($D_{50} = 0.65$), which may make it more attractive for assessing the variation in treatment with disease severity (Figure 6, Table S1). Finally, loss of the inhibitory subunit in homozygous *Pde6g*^{tm1Goff} mice did not lead to an expected increase in catalytic activity; instead PDE6G was found to be essential for activation and possibly stable assembly of the holoenzyme [182].

Cone phosphodiesterase 6 includes two catalytic α subunits encoded by *Pde6c* and two inhibitory γ subunits encoded by *Pde6h*. The *Pde6c*^{cpfl1} mutation leads to severely reduced cone ERG response at three weeks and progressive cone PR loss with age [15] as determined by counting cone nuclei (Bo Chang, unpublished data, presented in Table S1). This model mimics achromatopsia in humans, which is sometimes accompanied by cone PR cell loss [183]. Surprisingly, *Pde6h* knockout mice show no detectable functional cone loss or degeneration, likely due to the expression of the *Pde6g* subunit in mouse cones, which may compensate for PDE6H loss [184]. Variants in human *PDE6H* cause achromatopsia [185,186] but cone cell loss has not been reported.

Cyclic nucleotide gated channels and cation exchanger. The decrease in cGMP levels resulting from PDE6 activation leads to the closing of cyclic nucleotide cation channels in the OS plasma membrane of both rods and cones. Channel closing diminishes the inward flux of Na^+ and Ca^{2+} ions that maintain the PR cell in a hyperpolarized state. The rod protein encoded by *Cnga1* and *Cngb1* is an $\alpha_3\beta_1$ heterotetramer, in which the β subunit is a long isoform, CNGB1a [187,188]. *Cnga1* mutations have not yet been described. Rod OSs of homozygous *Cngb1*$^{tm1.1Biel}$ mice yield no detectable CNGB1a or CNGA1, and rapid PR loss is observed [189]. Together with evidence that CNGA1, but not CNGB1a, is capable of self-oligomerizing in heterologous expression systems, this result suggests that CNGB1 plays a critical role in stabilizing CNGA1 for channel assembly during synthesis in the secretory pathway and/or subsequent transport to the OS. Although the mechanisms leading to PR cell loss are unknown, low intracellular Ca^{2+} may overactivate guanylyl cyclase and cause toxicity due to elevated cGMP [189].

The cone channel encoded by *Cnga3* and *Cngb3* functions as an $\alpha_2\beta_2$ tetramer. Due to the absence of downstream synaptic signaling associated with channel defects, mutations in both genes result in a loss of cone ERG responses modeling achromatopsia. In addition, the alleles included in Table S1, *Cnga3^{cpfl5}*, *Cnga3^{tm1Biel}*, *Cngb3^{cpfl10}*, and *Cngb3^{tm1Dgen}* result in cone PR degeneration as assessed by marker analysis, although confirmation of cell loss by a direct nuclear count was lacking in some studies. The mechanism of cell death is unknown in these models, but by analogy may involve elevated cGMP as hypothesized in rods.

A critical component of phototransduction is SLC24A1 (also called NCKX1), which exports sodium and calcium ions in exchange for potassium. This activity is responsible for the decrease in intracellular Ca^{2+} upon closing of the cGMP-gated channels. Homozygous *Slc24a1^{tm1Xen}* mice exhibit slow degeneration, possible due to malformation of OS discs [190].

Guanylyl cyclase and activating proteins. Photoreceptor guanylyl cyclases function as homodimers encoded by two genes in mice, *Gucy2e*, and *Gucy2f*. In the homozygous *Gucy2e^{tm1Gar}* model, D_{50} was >12 months (Figure 6), indicating very slow rod PR cell loss, while cone cell numbers decreased rapidly to 33% of controls in 5 weeks [191]. Cone loss with rod preservation has been observed in Leber congenital amaurosis cases linked to variants of the human *Gucy2e* ortholog, *GUCY2D* [192]. However, *Gucy2e^{tm1Gar}* mice are not considered to model this disease because rod ERG function, though diminished, is still detectable [191]. Although *Gucy2f* knockout did not cause PR cell loss, double knockout of both guanylyl cyclase genes resulted in moderate degeneration [193]. Rod and cone ERG responses were abolished in this model, suggesting that the residual function in *Gucy2e^{tm1Gar}* mice was due to compensatory activity expressed from *Gucy2f*. The mechanism of PR cell loss in these models is unlikely to involve elevated cGMP as the enzymes needed for its production are ablated. The post-translational downregulation of other phototransduction proteins in double-knockout mice [193] may indicate a disruption of the proteostasis network that could explain PR cell loss.

Guanylyl cyclase activator proteins provide a feedback loop to restore cGMP levels. When intracellular Ca^{2+} is high, these proteins inhibit guanylyl cyclase; when Ca^{2+} levels are low, they switch to an activating Mg^{2+}-bound conformation that promotes cGMP synthesis. This Ca^{2+}-sensitive regulation permits PR cells to reestablish cGMP levels following light exposure due to lowered intracellular Ca^{2+}, thereby resetting the cell for another stimulus. Double knockout of *Guca1a* and *Guca1b*, which encode the activator proteins in both rods and cones, had no detectable effect on retinal morphology up to eight months of age [194]. However, homozygous *Guca1a^{tm1.1Hunt}* mice, which have a Glu155Gly missense substitution identical to one found associated with a severe dominant cone dystrophy [195], result in rapid loss of cones and subsequently rods (Figure 6, Table S1). This mutation, like others associated with the human disease, may constitutively activate guanylyl cyclase due to a defect in calcium sensing [196], leading to cytotoxic accumulation of cGMP.

Recovery from light stimuli. Mechanisms to terminate the phototransduction cascade and recover the PR cell for additional stimuli include the phosphorylation of activated RHO by a *Grk1*-encoded kinase and the binding of *Sag*-encoded arrestin to the phosphorylated RHO. The binding of SAG limits transducin access to RHO and thereby prevents further activation of transducin and downstream processes. Significantly, defects in either gene induce photoreceptor cell loss, likely due to the accumulation of excess cGMP arising from unregulated active RHO. Early studies aimed at elaborating the role of the SAG or GRK1 proteins used mice raised in the dark [197,198], as typical vivarium cyclic light–dark rearing conditions were described as leading to rapid degeneration. Subsequent studies of homozygous *Sag^{tm1Jnc}* [199] or homozygous *Grk1^{tvrm207}* mice [200] reveal slow PR cell loss with D_{50} > 10 months under normal rearing conditions.

5.2.3. Category 03: Metabolism

Overview. Inborn errors of metabolism constitute a heterogeneous group of disorders that affect metabolic pathways due to underlying genetic defects [201] and result in abnormalities in the synthesis or catabolism of biomolecules [201,202]. Many such inborn errors of metabolism are known to be associated with PR cell loss, manifested either as a primary ocular defect or as part of a systemic disease [201]. PR cells, with their high metabolic activity, are particularly vulnerable to defects in metabolism of biomolecules such as lipids, carbohydrates, nucleotides, and proteins, which provide energy and serve many other functions described below. Additionally, since organelles such as mitochondria and lysosomes are the major sites for cellular energy production and homeostasis, defects in organellar metabolism and function are also known to cause PR degeneration. The PR cell loss associated with different metabolic diseases varies in the age of onset, severity, and rate of progression (Figure 6, Table S1) and the underlying genetic defects can be categorized based on the type of biomolecular metabolism or the subcellular location of the pathways affected.

Biomolecular metabolism: lipids. PRs are extremely rich in lipids, which make up to 15% of their cellular wet weight as compared to 1% in most other cell types [203,204]. Phospholipids and cholesterol represent 90–95% and 4–6% (*w/w*) of total lipids, respectively [205]. The major phospholipids in rod outer segments include phosphatidylethanolamine, phosphatidylcholine, large amounts of phosphatidylserine, along with small amounts of sphingomyelin, phosphatidylinositol, and phosphatidic acid [205]. It has been suggested that the phospholipids in OS membranes are metabolically active and involved in generation of physiological mediators, and changes in metabolism of glycerolipids have been associated with transduction of visual stimuli [205]. Cholesterol has been reported to modulate the function of rhodopsin, a major protein of the OS membranes, by influencing membrane lipid properties [206]. Low-density lipoproteins (LDLs) are reported to be significant suppliers of PR lipids, especially cholesteryl esters [207,208]. The OSs of PRs are particularly rich in very-long-chain polyunsaturated fatty acids (PUFA), such as docasohexaenoic acid (DHA), which is considered to be essential for visual function [209], and phospholipid-containing DHA is suggested to help in isomerization of 11-*cis*-retinal to the all-*trans* form, which is further reduced for its entry into the visual cycle [210]. Recently, DHA has also been implicated in the maintenance of OS homeostasis [211] and mediating PR cell survival [212,213].

Thus, it is not surprising that disorders of lipid metabolism cause inherited PR degeneration. For example, mouse models for mutations in the elongation of very-long-chain fatty acids-like 4 (*Elovl4*) gene are reported to show features resembling Stargardt-like macular dystrophy in humans with cone degeneration preceding that of rods [214,215]. Mutations in genes involved in phospholipid metabolism such as *Lpcat1* cause rapid PR degeneration (90% and 75% degeneration in *Lpcat1^{rd11}* and *Lpcat1^{rd11-2}* alleles, respectively by 47 days) [216]. Similarly, mutations in genes involved in cholesterol biosynthesis such as *Nsdhl* [217] or in the biosynthesis and regulation of DHA-containing phospholipids, such as *Agpat3* and *Adipor1,* respectively [210], also cause PR degeneration, confirming the importance of lipids in preserving PR integrity. Since membrane phospholipid asymmetry is critical to performing various biological functions, mutations in genes important for its generation and maintenance, also lead to PR degeneration. For example, mutations in *Atp8a2,* a type of P4-ATPase that translocates and maintains phospholipid asymmetry show a 30–40% PR degeneration by two months of age [218]. Similarly, conditional inactivation of *Tmem30a,* known to be required for folding and transport of several P4-ATPases to their plasma membrane destination [219,220], also results in severe PR degeneration [221]. *Tmem30a* knockout mice exhibit a more severe phenotype compared to *Atp8a2* knockout mice, possibly because *Tmem30a* binds multiple P4-ATPases [221].

Biomolecular metabolism: carbohydrate and nucleotide energy metabolism. The retina, and in particular PRs, have a high metabolic rate [222,223] to support functions that are energetically demanding, such as phototransduction during constant illumination, maintenance of ion gradients in darkness, and performing anabolic metabolism to replace the approximately 10% of OSs that are lost every day to phagocytosis by RPE cells [223]. RPE cells also perform many energy demanding functions, such as maintenance of appropriate ionic and fluid composition in the subretinal space, uptake and conversion of all-*trans*-retinol to 11-*cis*-retinal and its transport back to photoreceptor cells, and OS phagocytosis. This high energy requirement makes the retina and RPE particularly vulnerable to functional deficits induced by deficits in energy metabolism [222]. The retina relies on blood-derived glucose and oxygen for its energy requirements. Additionally, PR cells use excess lactate obtained from Müller glial cells and convert it to pyruvate to provide energy via oxidative phosphorylation [222]. In addition to carbohydrates, the retina uses fatty acids [224] and nucleotides for its energy requirements [223].

Thus, neuronal activity and energy metabolism are tightly coupled and any mutations at the level of glucose, fatty acid or nucleotide biosynthesis can lead to PR degeneration. For example, mice lacking *Hkdc1*, which encodes a kinase found in the IS that phosphorylates glucose to glucose-6-phosphate, show 40% PR degeneration by 17 months [225]. Mice mutant for *Vldlr*, which encodes the receptor facilitating the uptake of triglyceride-derived fatty acids, show reduced cellular uptake and availability of fatty acids for energy production [224]. For some alleles of *Vldlr* (*Vldlr^{m1Btlr}* and *Vldlr^{tm1Her}*), more than 50% of PRs are lost by 12–14 months [226,227], with cones being affected more significantly than rods [228]. The decrease in net available energy may lead to greater cone loss, as cones have been reported to require three times more energy than rods [222]. Similarly, while in some cases, mutations in genes involved in nucleotide metabolism such as *Nampt*, show embryonic lethality [229], others such as mutation in *Nmnat1*, show severe PR degeneration by 4–6 months [230].

Biomolecular metabolism: hormones. The physiology of eye is also dependent on the action of several hormones [231]. Mouse models mutant for thyroid hormone metabolizing genes, such as *Dio3*, which is important for local amplification of triiodothyronine (T3), show selectively detrimental effects on cone cells [232]. This confirms the proposed role of thyroid hormone signaling in regulating cone viability and cone opsin expression [232,233]. Melatonin, a hormone that plays a role in sleep patterns, is known to have protective role against oxidative stress and apoptosis, and regulates retinal circadian rhythms [234]. A mouse model, mutant for the melatonin hormone receptor *Mtnr1a*, shows very slow PR degeneration (25% in 18 months) [235].

Biomolecular metabolism: oxidative stress. The eye is constantly subjected to oxidative stress due to daily exposure to light, atmospheric oxygen, and high metabolic activities [236]. Reactive oxygen species (ROS) are derived from diatomic oxygen and processes such as mitochondrial respiration that form superoxide anion radicals, toxic bis-retinoids that undergo photo-oxidation, and lipids, such as PUFAs, that undergo peroxidation [237]. Having unpaired electrons confers a great degree of ROS reactivity that can damage biomolecules such as DNA, lipids and proteins, and organelles including mitochondria and lysosomes [238,239], thereby impairing their biological functions [203,236]. Compared to other cells, non-proliferative postmitotic cells such as PRs and RPE cells are particularly sensitive to oxidative damage due to the apparent absence of a DNA damage detection system [240–242].

Under physiological conditions, cellular redox homeostasis is maintained by a balance between ROS generation and antioxidant systems [236]. Antioxidant enzymes such as *Sod1, Sod2,* and *Gpx4* are known to play a major role in ROS scavenging and changes in their expression or activity or both are reported to cause increased oxidative stress and are associated with diseases such as age-related macular degeneration (AMD) [243]. For example, mutations in the *Sod1* gene, encoding a cytosolic Cu-Zn superoxide dismutase that catalyzes the conversion of superoxide to hydrogen peroxide, are known to cause PR degeneration [244]. *Sod2*, which encodes a mitochondrial Mn superoxide dismutase, is required for survival and mutations in this gene lead to embryonic lethality [244,245]. Genes such as *Nxnl1* and *Nxnl2*, known as rod-derived cone viability factors are also suggested to have antioxidant function and show cone degeneration when mutated [246,247], with *Nxnl1* also showing a progressive rod cell loss [246]. Similarly, a mouse model for loss of *Ttpa*, coding for a protein that transports vitamin E, which is known to have antioxidant function, also shows 40% PR degeneration by 20 months [248].

Organellar metabolism: lysosomes. The lysosome, a subcellular organelle is critical for performing several vital functions such as degradation of extracellular and intracellular material, nutrient sensing, energy metabolism, and maintaining cellular homeostasis [249]. Lysosomes contain a wide variety of hydrolytic enzymes that enzymatically degrade biomolecules such as polysaccharides, lipids, etc. [250]. Defects in lysosomal function results in lysosomal storage disorders, a group of inherited metabolic disorders sharing a common biochemical feature of accumulating incompletely degraded metabolites within the lysosomes. Lysosomal storage disorders are generally classified by the composition of the material accumulated within them and often differ depending on the lysosomal proteins affected, which reflect different cell biological processes that are affected but terminating in a similar pathology of reduced clearance of metabolic aggregates.

RD is an early consequence of lysosomal storage diseases, especially in neuronal ceroid lipofuscinoses (NCL) [251], also called Batten disease, an early-onset neurodegenerative disease with other systemic features such as dementia and epilepsy [252]. NCL may be caused by disruption of genes encoding lysosomal enzymes (*Ppt1* and *Cln5*) and membrane proteins (*Mfsd8*) as well as ER membrane (*Cln6* and *Cln8*) and secretory pathway (*Grn*) proteins, and is characterized by a common lysosomal accumulation of ceroid. Similar to the early retinal phenotype reported for most human NCLs, most mouse models for NCL disease show an early onset of PR degeneration, beginning at 1 month of age and showing greater than 60% degeneration by 6–9 months [253–256]. Additionally, similar to the adult-onset reported for mutations in human GRN, the mouse model for loss of *Grn* also shows a late onset PR degeneration by 12 months [257].

Mouse models for other lysosomal disorders, namely, mucopolysaccharidosis and mucolipidosis due to mutations in lysosomal proteins required for the breakdown of glycosaminoglycans and enzymes required for phosphorylation of glycoproteins, respectively, also develop PR degeneration. For example, mouse models for mucopolysaccharidosis with a mutation in *Naglu* present with a slowly progressive rod-cone degeneration [258], and for mucolipidosis with a mutation in *Gnptab* develop a severe PR degeneration with complete PR loss by 10 months [259].

The lysosome receives materials for degradation via two major pathways, autophagy and phagocytosis. Phagocytosis has an important function in maintaining retinal health since 10% of the OSs are phagocytosed daily by the RPE cells to dispose of waste such as photo-oxidative products while retaining and recycling useful contents back to the PR cells [260]. Phagocytosis by RPE requires its own machinery for processes such as recognition (e.g., *Cd36*), engulfment (e.g., *Mertk*), and degradation (lysosomal enzymes) of the extracellular material. Disruption of the phagocytic machinery due to absence/mutations in proteins involved in the phagocytic pathway, therefore, have severe consequences for PRs and can lead to PR cell death. Mouse models for mutations in genes involved in phagocytosis such as *Mertk*, *Cd36*, and *Rab28* show PR degeneration with the loss of *Mertk* showing a more severe phenotype (>80% degeneration by 60 days for *Mertk^{tm1Grl}* and *Mertk^{tm1Gkm}*) [261,262] than loss of *Cd36* (17% degeneration at 12 month) [263], and the model for *Rab28* loss showing a more cone-specific response [264].

Autophagy is another lysosome-mediated degradation process essential for maintaining cellular homeostasis [265]. Autophagic flux, the complete dynamic process of autophagy, includes multiple steps involving the formation of phagosomes and autophagosomes, autophagosome fusion with lysosomes, the degradation of the intra-autophagosomal contents, and recycling [266]. Thus, both lysosomal function and autophagy are interconnected wherein disruption of the hydrolytic functions of lysosomes impairs autophagic flux and, conversely, lysosomal function requires normal flux through autophagy [267,268]. In the retina, autophagy plays a dual role: promoting cell survival against harmful stress, and cell death. High basal autophagic levels are maintained in RPE and PR cells. RPE cells being post-mitotic phagocytes are not self-renewing; the autophagy of intracellular components is therefore essential for a normal cellular function of the RPE [265]. In PR cells, autophagy occurs during various cellular activities such as OS degeneration [269], rhodopsin protein expression [270], visual cycle function, and PR apoptosis [271]. Mouse models of conditional inactivation of autophagy genes such as *Atg5*, *Atg7*, and *Rb1cc1* in RPE cells show that these genes are indeed important for survival of the animal and show PR degeneration.

Organellar metabolism: mitochondria. Mitochondria, often referred to as "the powerhouse of the cell", are the major site for cellular energy production in the form of ATP via oxidative phosphorylation. They also perform other important functions such as ROS generation and scavenging, calcium regulation, steroid, and nucleotide metabolism, regulation of intermediary metabolism, and initiation of apoptosis [272]. Oxidative phosphorylation is carried out by the mitochondrial respiratory chain, which consists of five complexes located along the inner mitochondrial membrane. These complexes, in an intricately organized series of biochemical events, synthesize ATP from ADP in response to cellular energy demands. A large number of mitochondria are present in the rod and cone IS and in RPE cells. The total surface area of the inner mitochondrial membrane in cones is 3-fold greater than in rods, presumably accommodating more respiratory chain enzymes to generate more ATP. Cones require more ATP than rods as they do not saturate in bright light and use more ATP/sec for light transduction and phosphorylation [222].

Defective cellular energy production due to abnormal oxidative phosphorylation in mitochondria can therefore lead to PR degeneration. A mouse model for the Leu122Pro mutation of OPA3, a protein hypothesized to be important for maintaining the inner mitochondrial membrane, is reported to cause a multisystemic disease characterized by severely reduced vision, loss of ganglion cells and PR degeneration (by 50%) at 3–4 months of age, a much more severe progression than observed in humans [273]. Similarly, a mouse model for a mutation in the gene for NAD-specific mitochondrial enzyme isocitrate dehydrogenase 3 (*Idh3a*), catalyzing the rate limiting step of TCA cycle, also causes an early and severe PR degeneration (more than 90%) by 90 days [274].

Extra-mitochondrial components of the tricarboxylic acid cycle and oxidative phosphorylation machinery have been localized to the rod OS [275]. It has been hypothesized that perturbation of this machinery results in excess ROS production, leading to PR cell death due to oxidative stress [275–277]. Mutations in a subset of mouse RD models in Table S1 alter genes (*Mpc1, Opa3, Idh3a, Impdh1,* and *Oat*) that encode mouse homologs of mitochondria-associated proteins identified in bovine rod OS [275]. Of these, only IDH3A is directly involved in cellular energy production [274]; the others may influence oxidative phosphorylation or the TCA cycle indirectly, possibly altering the generation of ROS. It may be of interest to determine whether PR cell loss in these mouse models correlates with an altered distribution of extra-mitochondrial oxidative phosphorylation proteins in the rod OS [278], or an increased ROS production, which can be measured in retinal explants [279].

Organellar metabolism: peroxisomes. Peroxisomes are subcellular organelles with various catabolic and anabolic functions such as catabolism of long chain fatty acids and biosynthesis of DHA and bile acids [280]. Several childhood multisystem disorders with prominent ophthalmological manifestations have been ascribed to the malfunction of the peroxisomes, either at the level of peroxisomal biogenesis (PBD) or single enzyme deficiencies [281]. While little is known about the metabolic role of these organelles in retina, studies have shown the presence of peroxisomes in nearly all layers of retina and RPE, albeit with differential expression of lipid metabolizing enzymes, suggesting different functions in different cell types [282]. For example, Zellweger spectrum disorder (ZSD) is a disease continuum known to result from inherited defects in *Pex* genes essential for normal peroxisome assembly. Mice homozygous for the G844D point mutation in *Pex1* show a decreased ERG response and loss of cone PRs (up to 80%) by 22 weeks, recapitulating the abnormal retinal function phenotype in ZSD patients with mild disease [283]. The retinal pathology in such disorders suggests the importance of peroxisomes in maintaining retinal homeostasis and function.

5.2.4. Category 04: Visual Cycle and Retinoids

The visual cycle reisomerizes vitamin A retinal that has been released from visual pigments in PR cells, allowing regeneration of the bleached pigments and the subsequent detection of additional light stimuli. The process is catalyzed by enzymes located in PR and RPE cells, so the retinoid intermediates in the process must be transported between them. Mutation of genes involved in the visual cycle pathway cause PR degeneration, in most instances with a moderate to slow progression depending on the allele and the genetic background. Most *Rpe65* mutant alleles show moderately slow PR cell loss (D_{50} = 7–11 months) [284–288]. Allelic effects are observed in models bearing missense mutations, *Rpe65^{tm1Lrcb}* [289] or *Rpe65$^{tm1.1Kpal}$* [290], which cause slower progression than observed in *Rpe65^{tm1Tmr}* knockout mice [285–288]. *Abca4^{tm1Ght}* on the BALB/c strain, which also carries a homozygous *Rpe65* Leu450Met mutation, show a late-onset PR degeneration with 40% loss by 11 months of age [291].

By contrast, the same *Abca4^{tm1Ght}* mutation on a 129S4/SvJae background results in abnormal thickening of Bruch's membrane but normal ONL nuclei count and thickness [292]. Several visual cycle mutant alleles have other retinal abnormalities but normal ONL nuclei/thickness. For example, *Abca4$^{tm1.1Rsmy}$* causes only autofluorescence and A2E accumulation [293] and *Abca4$^{tm2.1Kpal}$* on C57BL/6*129Sv leads to a RPE defect but normal ONL nuclei count and thickness [294]. In addition, PR degeneration in *Abca4* mutants can be induced by light exposure [295] or through interaction with other genes such as *Rdh8* [296–298]. The *Lrat^{tm1Kpal}* mutation on a 129S6/SvEvTac*C57BL/6J background results in mild PR degeneration, with <10% loss at 4–5 months [299]. However, a 35% decrease in rod OS length was also reported in this model, indicating the importance of the visual cycle for OS maintenance. Another allele, *Lrat$^{tm1.1Bok}$*, showed a similar loss of rod OS length and 18% PR degeneration at 6 months of age [300]. The *Rbp3^{tmGil}* mutation results in the most rapid PR cell loss in this category ($D_{50} = 0.79$ months), possibly attributable to an early developmental role of the protein [301]. The *Rbp4^{tm2Zhel}* congenic mutation on C57BL/6J showed 20% PR cell loss in some peripheral areas and 10% in the central retina an age of 40 weeks [302]. Mutations in two genes that play a role in retinoid uptake in the eye also result in PR cell loss. The *Rtbdn$^{tm1.1Itl}$* allele causes a slow degeneration with a 20% and 37% loss of PR nuclei at 240 days of age in heterozygotes and homozygotes, respectively. *Stra6^{tm1Nbg}* mice exhibit a normal number of rod PR nuclei but significant cone PR cell loss as detected by the cone-specific marker peanut agglutinin [303]. PR cell loss in *Stra6$^{tm1.1Jvil}$* mice was more pronounced with vitamin A restriction [304].

5.2.5. Category 05: Synapse

PRs absorb light that passes through the anterior portion of the eye and convert the light to electrochemical signals that are transmitted through the neuroretina via synaptic connections to the optic nerve and visual cortex [305]. Thus, synapses, necessary for proper cell-to-cell communication, are critical for vision. Discussion of the complexity of PR synaptic development and function is reviewed in [306–308], and is beyond the scope of this review. Suffice to say that mutations in many of the components of synapses, such as presynaptic exocytotic proteins, endocytic proteins, calcium channels, postsynaptic receptors, and associated elements, must be properly organized to mediate transmission of signals, or can lead to visual problems [307]. It is interesting to note that disruption of some synaptic components of the secondary neurons (e.g., GRM6, GPR179, TRPM1, NYX, GNAO1, GNB5, and GNB3), while affecting function as assessed by ERG response, does not normally lead to PR degeneration [309]. This is also true of some presynaptic proteins, such as dystrophin [310] or dystroglycan [311]. However, disruption of some synaptic genes such as *Ache, Cabp4, Cacna1f, Cacna2d4,* and *Unc119* does lead to PR degeneration. For example, a null allele of *Ache* [312], causes a 50% loss of PR nuclei between 1.5 and 2 months and >80% by 6–8 months. Although it was initially determined that the ACHE protein played an important role in hydrolyzing acetylcholine at synapses, its isoforms are now recognized to have far reaching structural functions [313]. Additionally, it has been shown that the loss of secondary neurons in the null allele model is likely to cause secondary PR cell loss [312]. Null or spontaneous alleles of synaptic genes that encode subunits of calcium channels that regulate the release of neurotransmitters, and the development and maturation of exocytic function of PR ribbon synapses, *Cacna1f* [314] and *Cacna2d4* [315,316], respectively, show a slower rate of degeneration. By two months, there are approximately 10–25% of PR nuclei that have degenerated. CABP4, a protein that regulates calcium levels and neurotransmitter release at PR synapses, and modulates CACNA1F and other calcium channel activity shows a similar rate of PR degeneration of 10–25% loss at 2 months [317]. UNC119, which localizes to PR synapses (and IS) and is hypothesized to play a role in neurotransmitter release, also leads to a relatively late onset, slower rate of PR degeneration [318]. Interestingly, Haeseleer has described an interaction between synaptic genes *Cabp4* and *Unc119* [319]. It is likely that other synaptic proteins will also lead to PR degeneration through either a primary or secondary effect and that the interactions among the synaptic proteins will play a significant role in determining the relative rate of the degenerative process.

5.2.6. Category 06: Channels and Transporters

Ions such as sodium, potassium, and chloride, play important roles in the visual circuitry [320]. Their intracellular concentrations and movements within the cell, and between cells and the environment are exquisitely regulated by channels and transporters. Due to the importance of maintaining appropriate levels of these ions for proper function and maintenance of PRs, it is not surprising that disruption of these genes can lead to PR degeneration. Members of the ClC family of chloride channels, such as *Clcn2, Clcn3,* and *Clcn7,* show particularly early and significant PR degeneration. Compared to other channels in this section, they appear to have an enriched expression in the RPE. A 50% PR cell loss can be seen as early as 14–16 days in certain models with disruptions in these genes [231,321–323]. Indeed, rapid progression of PR cell loss is observed in mice carrying any of the following alleles: *Clcn2^{nmf240}, Clcn2^{tm1Tjj}, Clcn3^{tm1Lamb}, Clcn3^{tm1Tjj}, Clcn7^{tm1Tjj}, Clcn7^{tm2Tjj},* and *Clcn7^{tm4.1Tjj}* [231,321,322,324–327]. A similar rapid and complete loss of PRs is seen when *Atp1b2,* a Na$^+$/K$^+$-ATPase thought to play a role in cell adhesion, is inactivated [328]. A targeted mutation in *Slc6a6,* which encodes a taurine/beta-alanine transporter, also leads to rapid, complete degeneration [329]. In contrast, inactivation of the bicarbonate, amino acid transporters, and Na$^+$/H$^+$ exchangers, *Slc4a7, Slc7a14,* and *Slc9a8* results in a later onset, but still severe PR degeneration [330–332]. Mouse models bearing mutations affecting BSG, a protein that has a role in targeting monocarboxylate transporters such as SLC16A1 to the plasma membrane, or REEP6, which mediates trafficking of clathrin-coated vesicles from the ER to the plasma membrane at outer plexiform layer sites enriched for synaptic ribbon protein STX3, also fall in this latter late onset/severe category [333,334]. *Asic3,* an acid sensing Na$^+$ channel, *Clcc1,* an intracellular chloride channel, and *Slc7a14,* an intracellular arginine transporter, all cause moderately slow degeneration when mutated [331,335,336]. Slowly progressive PR cell loss is also caused by mutation of *Mfsd2a,* which encodes a sodium-dependent lipid transporter responsible for maintaining a high DHA concentration in the retina is important for OS homeostasis, as discussed in Category 03 [337]. More data are needed to see if sensing and intracellular channels/transporters generally have milder phenotypes.

5.2.7. Category 07: Adhesion and Cytoskeletal

Proper structure of the retina is developed through protein interactions between cells and within cells. The spatial and laminar organization of the retina is maintained through junctional interactions between cells that impart mechanical support to maintain retinal architecture, a means for bidirectional communication (e.g., extracellular changes to the cell and from the cell to its environment), and together can form diffusion barriers. Within cells, cytoskeletal architecture is maintained through interactions of proteins with actin, intermediate filaments, and microtubules that serve to maintain cell morphology and polarity, and as discussed elsewhere, intracellular trafficking, contractility, motility, and cell division. Examples of disrupted proteins that lead to gaps between cell layers, presumably through aberrant adhesion, are mutations in *Adam9* and *Rs1.* The null allele of *Adam9,* a single pass transmembrane protein with disintegrin and metalloprotease domains that has been shown to interact with a number of integrins [338], leads to aberrant adhesion between the apical processes of the RPE and OS and to late onset PR degeneration [339]. Likewise, mutations in RS1, a protein with a discoid domain, which has been implicated in cell adhesion and cell–cell interactions [340] lead to a splitting of the inner retinal layer and progressive PR loss [341–343]. RS1 binding to phospholipids on the membrane surface, together with other proteins [340] may provide a stabilizing scaffold that is important in cell–matrix, cell–cell, and cytoskeletal organization.

CRB1 and its interacting partners, such MPP3, MPP5, and PARD6A, have been shown to be important in establishing proper retinal lamination presumably through their essential roles in establishing cellular apical basal polarity [344]. A primary defect of a disruption in CRB1 is the fragmentation of the outer limiting membrane [345,346]. As reviewed previously [344,347], the outer limiting membrane consists of adherens/tight junctions formed in part by the CRB1 complexes between Müller glia and the rod or cone IS that form a diffusion barrier. Loss of components of the CRB1 complexes (CRB1-MPP5-PATJ, CRB1-MPP5-MPDZ, and CRB1-PARD6A-MPP5-MPP3/MPP4) leads to lamination defects with formation of rosettes and a progressive loss of PRs. Although yet to be reported, it is likely that mutations in PATJ, PARD3, MPDZ, and MPP4 will lead to similar disease phenotypes, as a reduction in ERG response in MPDZ mutants [348] and a reduced ERG with abnormal retinal morphology have been indicated for PARD3 mutants [41,43].

Equally important to the function and maintenance of the retina are the intracellular components that make up the cytoskeletal cell structure. Disruption of proteins that interact with actin intracellularly have been shown to lead to PR degeneration. For example, CDC42, a small GTPase that is a key regulator of actin dynamics [349] leads to an early onset, progressive PR degeneration when disrupted. Models caused by mutations in FSCN2, an actin crosslinking protein [350,351], and by a hypomorphic variant of CTNNA1, a protein that coordinates cell surface cadherins with the intracellular actin filament network [352], show slow-paced PR cell loss. The proper localization of organelles within the cell is also mediated by the cytoskeletal architecture and can have an untoward effect when disrupted. For example, SYNE2, a nuclear outer membrane protein that binds to F-actin, tethers the nucleus to the cytoskeleton and is necessary for the structural integrity of the nucleus [353,354]. Without it, early onset, moderately paced PR cell loss occurs.

5.2.8. Category 08: Signaling

Molecules such as growth factors/cytokines, hormones, neurotransmitters, and extracellular matrix proteins, or alternatively, mechanical stimuli, are examples of signals used to communicate environmental changes to the cell. Surface or intracellular (e.g., nuclear) receptors recognize the signals and effect changes within the cell, often setting in motion amplifying transduction cascades that mediate responses such as activation or inhibition of protein activity or migration to different cellular localizations. Further, signals can also be transmitted from the cell to other cells, for example, through neurotransmitters. Since intra- and intercellular communication is crucial for the proper development or function of cells, it is not surprising that a large number of mutations in cellular signaling lead to defects in retinal development, which in turn affects PR survival. For example, vascular development is affected in mutants bearing disruptions in *Fzd5, Lrp5, Ndp,* and *Tspan12*—all components of the Wnt signaling pathway. Integral membrane frizzled receptors, of which they are 10, together with coreceptors, LRP5 and LRP6, mediate canonical Wnt signaling [355]. Thus, conditional *Fzd5* null mutants develop microphthalmia, coloboma and persistent fetal vasculature, and late-onset progressive RD [356] and *Lrp5* mutants exhibit similar vascular and retinal phenotypes [357]. Mice that are null for NDP, a ligand for FZD4, exhibit delayed retinal vasculature development, retrolental masses, disorganization of the ganglion cell layer, and occasionally focal areas of ONL absence at later stages of the disease [358]. TSPAN12, mediates NDP-FZD4-LRP5 signaling in the retinal vasculature, where it localizes, and a mutation leads to vascular defects that phenocopies disruptions in *Ndp, Fzd4,* and *Lrp5,* and at 3 months exhibits a 50% loss of PRs [359]. In all of these models, it is likely that the loss of PRs is caused by the aberrant retinal vasculature having secondary effects on PRs. A review by Hackam suggests that Wnt signaling may affect the apoptotic pathway and neurotrophin release, dysregulation of which may affect PR survival [360]. MFRP, which bears a CRD domain shared by all frizzled proteins, also leads to PR degeneration when disrupted [361,362], as does the human knock-in allele, p.S163R, of its bicistronic partner, CTRP5 [363]. The exact role or function of either protein has yet to be fully elucidated.

Like the frizzled-associated proteins whose pathological effects on PRs are likely to be mediated through an aberrant retinal vasculature, other signaling molecules, *Ptpn11* and *Fyn*, appear to mediate their effects on PRs through another cell type as well, in this case, Müller glia cells, and PRKQ through the RPE. A *Six3-cre* mediated conditional knockout of *Ptnp11* [364] leads to altered ERK and MAPK signaling in Müller glia and alteration in their adhesive capabilities. FYN, a Src-kinase membrane associated tyrosine kinase, localizes to Müller glia cells, and FYN deficiency leads to altered adhesion properties of Müller cells and retinal dysmorphology [365]. PRKCQ, a serine threonine protein kinase, which localizes to the lateral surface of the RPE cells, causes a reduction in adhesion between the apical processes of the RPE and OSs when it is disrupted. The reduction in adhesion may be responsible for the retinal detachment and subsequent PR loss observed in this model [366].

The family of PI3Ks or phosphoinositide 3-kinases, made up of catalytic and regulatory subunits, function to phosphorylate the inositol ring of phosphatidylinositol and thereby regulate growth, proliferation, differentiation, motility, survival, and intracellular trafficking. For example, it mediates insulin-stimulated increase in glucose uptake and glycogen synthesis and responds to signals such as FGFRs and PDGFRs. Conditional knockouts of *Pik3cb*, encoding a catalytic subunit [367] and *Pik3r1*, encoding a regulatory subunit [368], using the cone-specific CRE, Tg(OPN1LW-cre)4Yzl, lead to progressive cone PR loss. IRS2, necessary for the integration of signals from insulin and IGF1 receptors, causes an early-onset, moderately paced PR loss [369]. Additionally, a targeted conditional allele of PDGFRB developed diabetic retinopathy like features with angiogenesis, proliferative DR-like lesions, pericyte drop out, and eventual PR loss [370]. Disruption of MAP3K1, a serine/threonine kinase, which participates in the ERK, JNK, and NF-κB signaling pathways, leads to retinal laminar and vascular defects, aberrant RPE, and PR cell death [371]. SEMA4A, a transmembrane protein, also causes PR loss, most probably through its effects on endosomal sorting [372].

5.3. Category 09: Transcription Factors

In mice, cone and rod PRs are born and develop between approximately E12 and P0, and approximately E13.5 and P7, respectively, from the same multipotent retinal progenitor cell (RPC) pool [373]. PR development, orchestrated by a network of transcription factors, is divided into five phases: proliferation of multipotent RPCs, restriction of RPC competence, cell fate specification, expression of genes important for PR function, and finally, PR structural maturation [374,375]. RB1 and E2F1 function by controlling the G1 to S phase transition in the cell cycle; RB1 plays an inhibitory role until activated by phosphorylation, balancing cell proliferation and cell fate specification [376]. OTX2 is critical for fate determination, while CRX is necessary for terminal PR differentiation and acts at different steps in PR development. Transcriptional factors important for rod PR subtype specification include RORβ, NRL, and NR2E3, and for generation of the cone subtypes, TRβ2 and RXRγ [374,375]. Further, transcription factors regulate the expression of other transcription factors in the network (e.g., CRX interacts with *Nrl*, *Rorb*, and *Mef2d*, to name a few, to mediate rod differentiation, cone differentiation, and proteins necessary for the maturation of the PR, respectively).

The importance of transcription factors in retinal development has been explored in many studies resulting in a number of mouse models with different disease phenotypes (MGI JAX). In many cases, disruption of transcription factors, especially those affecting earlier phases of PR development lead to a reduction in the total number of retinal cells generated. We have only included within this category those disrupted transcription factors that eventually lead to PR degeneration. Interestingly, the onset of degeneration of the PR transcription factor models is highly variable—14 days to 2 months of age—and appears to be dependent upon the method used to generate the model and possibly background strain, as variation of severity and onset differs among different models of the same gene. Interestingly, the *Crx* models provide a series that recapitulate the clinical diagnoses of autosomal dominant cone-rod dystrophy, Leber congenital amaurosis, and late-onset dominant retinitis pigmentosa. Crx^{Rip} heterozygotes showed 34% degeneration at five weeks, compared to mice homozygous for the mutation, which reached 55% degeneration at the same age [377]. $Crx^{tm1.1Smgc}$ was also noted to have a heterozygous disease presentation more similar to a cone-rod dystrophy, while the homozygous mutant presented with a disease phenotype similar to Leber congenital amaurosis with 70% loss of PRs at one month of age [378].

Other transcriptional factors necessary for the proper maturation of the PRs, such as, MEF2D, shown to be important in regulating transcription of OS and synaptic proteins [379], or NRF1 [380], important in mitochondrial biogenesis also develop PR degeneration when disrupted. Finally, there are transcriptional factors that are important in the development or function of supporting cells such as

ONECUT1 for horizontal cells [381] and MITF for RPE and/or choroidal melanocytes [382], which affect PR survival when disrupted.

5.4. Category 10: DNA Repair, RNA Biogenesis, and Protein Modification

Among the many disrupted genes that lead to PR degeneration, several instances have been documented in genes necessary for producing fully functional proteins, from transcription through post-translational modification. Defects in these genes are likely to impact the function of many other genes that they act upon, and hence, have a greater effect. Since they play a central and basic role, when disrupted they often lead to prenatal lethality in mice, and the adult phenotype is unknown unless a conditional knockout or hypomorphic allele is generated. For example, disruption of DNA repair genes such as *Ercc1*, RNA splicing genes such as *Prpf3, Prpf6, Prpf8, Prpf31*, and *Bnc2*, and miRNA processing genes, *Dicer1* and *Dgcr8* are prenatal lethal in a homozygous state [34,383]. In contrast, homozygous null alleles of *Bmi1* [384] and *Msi1* [385], both involved in repression of regulatory genes in embryonic development, are viable, suggesting potential compensatory mechanisms for the functional loss of these genes. Thus, germline, conditional or hypomorphic models were considered in this category.

Review of genes in this category suggested that a DNA damage response network to ensure transcription in the face of DNA lesions might be required for PR cell maintenance. DNA lesions, such as pyrimidine dimers, interstrand crosslinks, or double-strand breaks (DSBs), are induced by many mechanisms that include UV radiation or free radicals. Repair of such damage is essential for DNA replication and, of particular importance for long-lived post-mitotic neuronal cells, transcription [386–388]. Proteins encoded by *Bmi1, Dgcr8, Dicer1, Elp1, Ercc1, Ercc6, Msi1, Sirt6, Top2b, Ubb*, and *Uchl3* are known to participate in the DNA damage response [387,389–398], some in transcription-coupled DNA repair. For example, BMI1 represses transcription at sites of UV-induced DNA damage to allow repair [389]; ELP1 is a required component of the Elongator complex [399], which couples RNA polymerase II to an alkyladenine glycosylase that initiates base excision repair [392]; ERCC6 promotes DSB repair in actively transcribed regions by displacing RNA polymerase from the lesion site [387], and DGCR8 interacts with both RNA polymerase II and ERCC6 to mediate transcription-coupled nucleotide excision repair of UV-induced DNA lesions [390]. Intriguingly, topoisomerase TOP2B, which creates DSBs during transcriptional activation [396], has been identified as a key regulator of transcription during the last stages of postnatal PR development [400]. Thus, DSBs in PR cells may arise in part from transcriptional activation of genes that encode components destined for the OS. Additionally supporting the importance of DNA repair to PR maintenance, Category 01 gene *Atr* encodes a master regulator of the DNA damage response that has surprisingly been linked to retinal degenerative disease and localized to the cilium [401]. Further, Category 03 gene *Nmnat1* encodes an enzyme that synthesizes nicotinamide adenine dinucleotide in the nucleus, which may regulate the large-scale polyADP-ribosylation of protein targets at sites of DNA damage [402]. Mutations in the genes encoding these proteins all result in PR cell loss [230,384,385,400,401,403–411]. Mutations in five of these genes as included in Figure 6 (*Cwc27, Ercc1, Ercc6, Sirt6*, and *Ubb*) caused moderate to slow progression of PR cell loss ($D_{50} \geq 2$ months), consistent with a steady accumulation of unresolved DNA damage with age. The rapid PR cell loss observed in *Atr*^{tm1Ofc} mice ($D_{50} = 13$ days) may reflect its direct involvement in OS development [401] in addition to the DNA damage response.

Due to the high percentage of alternatively spliced genes in the human retina [412,413], it is not surprising that mutations in mRNA splicing genes: *PRPF3, PRPF4, PRPF6, PRPF8, PRPF31, PDAP1*, and *BNC2* have been shown to lead to PR degeneration in humans [3]. In fact, in human retinal disease, 14% of disease genes are categorized as playing a role in RNA metabolism [383]. Interestingly, heterozygous humanized alleles of *PRPF3* and *PRPF8* and the null allele of *Prpf31* in mice do not recapitulate PR degeneration observed in humans but rather exhibit late-onset RPE degeneration [414]. In contrast, a hypomorphic allele of the mRNA splicing gene, *Cwc27*, with reduced viability, does lead to moderate onset PR degeneration [415]. The differences observed among species require additional studies to unravel the complexities that govern genetic interactions.

Post-translational modification, which occurs by adding modifying molecules to amino acids or removing or altering these modified amino acids, is important for proper folding, transport/trafficking, localization, function, regulation, and/or degradation of proteins. Examples of post-translational modifications include phosphorylation, glycosylation, acetylation, ubiquitination, sumoylation, methylation, and lipidation [416]. Kinases that affect activity by mediating phosphorylation states are described elsewhere, however, post-translational modification genes affecting glycosylation and lipidation/prenylation are prominent among those that lead to PR degeneration. For example, the encoded proteins of *Fkrp*, *Large1*, *Pomt1*, and *PomgnT1*, necessary for the glycosylation of alpha-dystroglycan, essential for formation of the dystroglycan complex and for proper retinal lamination, lead to moderate rates of PR degeneration when disrupted. Prenylation is critical for proper trafficking and localization of retinal proteins. Of the three genes important in the prenylation and postprenylation processes, conditional loss of *Rce1* leads to an absence of phosphodiesterase subunits PDE6A, PDE6B, and PDE6C from the rod OS, probably due to a failure to prenylate one or more of these proteins [417]. By contrast, ablation of *Icmt* does not appear to affect phosphodiesterase transport but rather results in lowered levels of prenylated proteins GNAT1, PDE6G, and GRK1 [418], which are essential PR proteins. The null mutation of farnesyl-diphosphate farnesyltransferase 1, which adds a farnesyl group to the cysteine of the CAAX amino acid motif is prenatal lethal, but as a conditional tissue specific knockout may result in the same PR effects.

Two additional types of post-translational modification involve glycylation and glutamylation of proteins essential for normal connecting cilia function. Disruption of *Ttll3*, encoding a protein-glycine ligase necessary for glycylation of tubulin, results in an absence of glycylation in PR cells, shortening of the connecting cilia, and slow PR cell loss [419]. Interestingly, PR tubulin glutamylation increased in *Ttll3* mutant mice. TTLL5, tubulin tyrosine ligase like 5, adds glutamate residues on proteins. Sun et al. [420] reported that *Ttll5* disruption leads to late onset, slowly progressive PR cell loss that phenocopied retinal disease observed in *Rpgr* mutants. Perhaps this is not surprising as these investigators determined that TTLL5 glutamylates RPGR, a modification that is necessary for normal RPGR function in the PR cilium. *Agtpbp1* encodes a metallocarboxypeptidase that deglutamylates target proteins. Its disruption in *pcd* mutants leads to abnormal tubulin glutamylation [419] and an accumulation of vesicles in the interphotoreceptor space [421], indicating the importance of proper post-translational modification for PR survival.

5.5. Category 11: Immune Response

As resident immune cells, microglia survey the retina constantly, presumably with the goal of removing unwanted debris and responding to damage arising from environmental and/or genetic stressors. They respond to damage by eliciting various responses that can range from regenerative to inflammatory depending on the type of injury. Thus, although microglia are unlikely to be instigators in RD, it may well be the case that microglia influence the severity of responses to ocular damage depending on mutations. Mutations in several genes central to the immune system lead to PR degeneration in mouse models. *Aire*$^{tm1.1Doi}$ show early onset PR degeneration with 20% of ONL thickness loss at 10 weeks with rapid progression to 60% ONL thickness loss by 18 weeks [422]. *C3ar1*^{tm1Cge} mutants show very slow PR degeneration with about 20% loss at 14 months [423]. *Cd46*^{tm1Atk} show different rates of PR nuclei loss in male and female mice with 23% and 31% at 12 months of age, respectively [424]. Mutations in *Cx3cr1*, normally expressed in immune cells including microglia, were associated with PR cell loss. Homozygous *Cx3cr1*^{tm1Litt} [425] and *Cx3cr1*^{tm1Zm} [426] mice on the same C57BL/6J background showed similar rates of PR degeneration with 30% and 40% loss, respectively, at 16–18 months of age. However, *Cx3cr1*^{tm1Zm} mice on the BALB/cJ background show complete nuclei loss at 4 months of age [426]. *Cxcr5*^{tm1Lipp} causes late onset PR degeneration with 20% loss of ONL thickness at 17 months of age and RPE disorganization [427], whereas ablation of *Irf3* and *Igfbp3* showed mild PR degeneration at 2–4 months of age, about 10–14% [428,429]. *Ccl2* and *Ccr2* mutations also led to PR degeneration and fundus lesions, ONL loss in some areas and development of neovascular lesions, resembling phenotypes of AMD [430]. *Cfh*^{tm1Mbo} was shown to have an impairment of rod and cone function by ERG and 29% decreased thickness of Bruch's membrane; however, rod opsin was distributed normally and no significant reduction in the number of PR cells was observed [431]. *Cfh*$^{tm1.1Song}$ demonstrated retinal whitening and cotton wool spots by fundus imaging [432]. Other genes involved in immune function that also showed PR degeneration as conditional knockouts encode transforming growth factor beta receptor II (*Tgfbr2*) [433] and aryl hydrocarbon receptor (*Ahr*) [434].

5.6. Omitted Models with PR Abnormalities that May be of Interest

Based on the exclusion criteria described in the Methods section, a number of models with PR abnormalities caused by single gene mutations were not included in our final Table S1. Since we narrowly defined PR degeneration models as post-developmental loss of PR nuclei, some models, which were described with only OS alterations or ERG differences, were not included. For example, mice bearing a spontaneous point mutation in the *Ttc26*^{hop} [435] that leads to the generation of a stop codon, Tyr430Ter, were reported to show OS shortening at one year of age with no PR loss. Likewise, ectopic expression of cone opsins in rod OSs led to scotopic ERG abnormalities but not PR degeneration in *Samd7*^{tm1TFur} mice at 12 months of age [436]. The many allelic variants that cause ERG abnormalities without PR cell loss are listed in the MGI database and can be accessed through a phenotype query.

5.7. Factors Leading to Phenotypic Variability

5.7.1. Effects of Allelic Heterogeneity

Allelic heterogeneity is frequently a cause of phenotypic variability. For mouse models, this is often encountered when comparing a knockout model with spontaneous or induced mutations that still allow a protein to be produced. The latter would primarily be hypomorphic alleles due to amino acid substitutions, some splicing mutations that leave alternate splice forms intact and some C-terminal truncating mutations, which may retain some protein function. Often the knockout allele will be the more severe, presumably because in addition to the loss of protein function, the loss of the protein itself may cause secondary defects such as the failure to form a molecular complex that normally needs the native protein to form.

Mutations in the voltage gated calcium channel, *Cacna1f*, cause congenital stationary night blindness in humans due to abnormal neurotransmitter release in PR synapses. A null mutation in the *Cacna1f* gene (ΔEx14–17) leads to an absent b-wave, abnormal PR synapses, lack of Ca^{2+} response in PR terminals and PR degeneration to 8 rows in the ONL at 8 months [314]. In contrast, an Ile756Thr amino acid substitution found in human patients and introduced into mouse, led to a different phenotype with reduced b-wave, some intact ribbon synapses, a strong abnormal Ca^{2+} response, and a more severe degeneration (3–4 rows at 8 months of age [314]). Here the human allele represents a gain-of-function mutation that in addition to the loss of the original enzyme activity results in a new activity, or causes cell stress, which then induces additional phenotypes and makes the disease presentation more severe.

Within an allelic series of amino acid substitutions there are also frequently gradations of phenotypic severity. If a protein has several functional domains, mutations in different domains may lead to distinct phenotypes. In addition, some mutations can lead to an abnormal tertiary structure of the protein. Such structural changes can lead to a failure to interact with binding partners or substrates/ligands or change the nature of such interactions [437]. Structural changes can also affect export of the protein from the endoplasmic reticulum (ER) and result in ER stress and eventually apoptosis of the cell [438].

One of the larger allelic series available is for human PRPH2 with more than 150 disease causing mutations reported [439]. Although only the secondary structure of the protein is available, some clustering of disease phenotypes is apparent. For example, the area around amino acids 190–220 on the intradiscal loop 2 is enriched for mutations causing autosomal dominant retinitis pigmentosa. This area is thought to interact with ROM1. Mutations leading to macular degeneration are more frequently present between amino acids 142 and 172. However, some macular degeneration and autosomal dominant retinitis pigmentosa mutations are also found elsewhere in the protein [439,440]. Once a 3D structure is available, we may find that the disease specific mutations may well be in spatial proximity and a clearer picture of the genotype-phenotype relation may be revealed.

Allelic heterogeneity can also arise from the intron/exon structure of the gene itself. Many genes produce several distinct transcripts through alternative splicing of their exons [441]. These differing transcripts can each produce proteins, which possess unique functions. For example, the *Rpgrip1* gene produces two splice variants that code for proteins that differ at their C-terminus, a full-length transcript and a shorter transcript encompassing exons 1–13 plus three additional C-terminal amino acids. An insertion between exons 14 and 15 of the full-length transcript leads to PRs with vertically stacked OS discs [442], whereas, a chemically induced mutation in the splice acceptor site in intron 6 that leads to a loss of both splice variant forms results in a failure to develop OSs altogether [443].

Despite the promise of genotype–phenotype correlation analyses to aid in the functional annotation of retinal proteins as well as in the diagnosis and prognosis of retinal degenerative diseases, few allelic series are yet available. In humans the analysis is complicated by the fact that environment and genetic background effects can confound the allelic effect. In animal models, large allelic series are not yet available.

Until recently allelic heterogeneity posed a problem for the generation of mouse models for human retinal diseases because only transgenesis and the generation of knockout models by homologous recombination were available. The removal of the gene products using knockouts can only model recessive or haploinsufficiency diseases, and often the complete lack of the protein will lead to embryonic lethality.

Transgenic models are associated with their own set of problems. Depending on the transgene integration site, the expression of the transgene can be reduced or cellularly restricted. Integration into an unrelated gene can disrupt expression of that gene and cause a phenotype that is not related to the transgene. The use of directed transgene insertion into safe sites, such as the Rosa26 locus (*Gt(ROSA)26Sor*) provides a workaround for some of these problems, although the choice of a promoter that faithfully mimics the native expression is still a difficult process. For these reasons, transgenic mouse models were not included in this review.

With the advent of CRISPR/Cas9 technology to produce precise cuts in genomic DNA, and the ability to perform gene editing through homology directed repair, it is now feasible to recreate human mutations in the mouse and directly probe for the phenotypic effects of allelic heterogeneity [444]. Comitato et al. present an interesting phenotype comparison of transgenic and knock-in rhodopsin P23H models [445].

5.7.2. Effects of Genetic Interactions

Gene interaction, or epistasis, is frequently observed during genetic analysis when two or more alleles at different loci combine to alter the onset, type, or severity of disease phenotypes. Such phenotype altering interactions arise from the organization of proteins and RNAs into macromolecular complexes and/or biochemical and regulatory pathways and networks. For example, consider hypomorphic mutations in two proteins that are components of a linear enzymatic pathway. Individually the reduced activity may not greatly impact the flux through the pathway, but combined in the same cell, the pathway flux may be reduced and become severe enough to induce a disease phenotype due to a lack of sufficient pathway product. Alternatively, a mutation may impair Pathway A, so that a disease phenotype arises. A second mutation may arise in a Pathway B that allows it to compensate for the malfunction in Pathway A and thus reduce the severity of the original disease phenotype. Mutations of this latter type of interacting mutations are called suppressor mutations and are extremely useful because they directly identify potential drug targets whose manipulation may be used to treat disease.

In general, identification of genetic interactors can be useful for placing the primary mutated gene in a biological context and help to define its cellular and organismal function. Often, the known function of a gene and its biology can suggest candidate interacting genes. Similar to the first hypothetical interaction case above, mutations in two proteins involved in iron homeostasis, ceruloplasmin (CP), a ferroxidase associated with transferrin transport across the plasma membrane, and hephaestin (HEPH), implicated in iron transport across cells, individually do not show obvious PR degeneration. Combined in a double mutant mouse model, however, they lead to iron overload in the retina and subsequent RPE abnormalities and PR degeneration [446]. Another example involves two proteins necessary for retinoid recycling, ABCA4 and RDH8. Mutations in each alone do not show any phenotype; combined they cause all-*trans*-retinoid accumulation and PR degeneration [296]. Since previous studies had suggested that activation of TLR3 may lead to inflammation and mediating apoptosis [447], the authors explored the role of *Tlr3* in their *Abca4/Rdh8* double mutant model. Importantly, adding a targeted mutation of *Tlr3* to make a triple mutant mouse resulted in rescue of PR cells [448]. Here then the *Tlr3* mutation acts as a suppressor of the degenerative phenotype of the *Abca4/Rdh8* double mutant.

Additional interacting gene pairs have been found that affect PR degeneration, among them *Mertk*^{tm1Grl}; *Tyro3*^{tm1Grl} [449], *Cep290*^{rd16}; *Bbs4*^{tm1Vcs} [450], *Cep290*^{rd16}; *Mkks*^{tm1Vcs} [451], *Rpgr*^{tm1Tili}; *Cep290*^{rd16} [452], *Cngb1*$^{tm1.1Biel}$; *Cnga3*^{tm1Biel}; *Hcn1*^{tm2Kndl} [453], *Crb1*$^{tm1.1Wij}$; *Crb2*$^{tm1.1Wij}$ [454], *Dio3*^{tm1Stg}; *Dio2*^{tm1Vag} [455], and *Ercc6*^{tm1Gvh}; and *Xpa*^{tm1Hvs} [456].

In addition to testing candidate interacting genes, methods have been developed to identify such interactors in an unbiased fashion that is illustrated below.

Effects of genetic background. For the calcium channel gene *Cacna1f* mentioned above, there is a third allele available. Chang et al. [457] reported the phenotype of the *nob2* mutation, an out-of-frame insertion of a transposable element into the *Cacna1f* gene, which is predicted to cause a truncation after 32 amino acids. The authors demonstrated by western blot that this is a null mutation and no protein is detected. Compared to the ΔEx14–17 null mutation, however, the phenotype of *nob2* is much milder with no apparent PR degradation [457]. The most likely explanation for this discrepancy can be deduced from the fact that the *nob2* mutation arose on the AxB6 recombinant inbred strain, a strain whose DNA is composed of alternate segments derived from C57BL/6J and A/J. It is likely that the A/J strain carries one or more modifier loci that suppress the PR degeneration induced by a *Cacna1f* null mutation.

Upon outcrossing an inbred strain carrying a mutation that leads to a particular phenotype with a different inbred strain, it is frequently observed that the phenotype of the offspring differs from that of the parents. This was often encountered in the past when knockout alleles were created in embryonic stem cells derived from strain 129/Sv and the founder animals were then made congenic on the C57BL/6 background. An early example is a study of a homozygous *Rho* knockout that was shown to lose PR nuclei significantly faster on the 129Sv background than on the C57BL/6 background [458]. Corresponding differences were also found in the number of apoptotic nuclei and in ERG responses. It was concluded that the B6 strain carries protective alleles of modifier genes that lead to a slower rate of PR degeneration [458]. Alternatively, it is also possible that 129Sv carries modifier alleles that accelerate degeneration.

Other inbred strains have also been reported to modify retinal phenotypes. For example, a targeted mutation of *Rp1* (*Rp1*^{tm1Eap}) only showed moderate PR degeneration as an incipient congenic (N6) on the A/J strain background, but not on C57BL/6J or DBA/1J backgrounds [459]. ONL dysplasia and excess blue cone formation caused by loss of *Nr2e3* in C57BL/6J are suppressed by the genetic backgrounds of CAST/EiJ, AKR/J, and NOD.NON-*H2*^{nb1} strains [460].

In principle, all inbred strains will carry modifier alleles. However, which strain modifies a particular mutation will depend on the primary mutation. It should be emphasized that an inbred strain represents a single genotype. In order to model the phenotypic spectrum of a human disease-causing mutation, many inbred strain backgrounds would have to be examined. Recently, advanced genetically diverse mouse populations have become available, such as the collaborative cross (CC) or the diversity outcross (DO) populations, that allow for more efficient modeling of human populations compared to the classical inbred strains [461,462].

Modifier screens. Modifier screens are a tool to identify genes that modify phenotypic traits caused by a particular mutation. The disease modifying properties of inbred strains have been used for many decades to identify the underlying modifier genes by using genetic crosses, marker assisted genetic mapping of modifying loci, and positional cloning or more recently high throughput whole exome or whole genome sequencing approaches. For example, when B6.Cg-*Nr2e3^{rd7}* homozygotes are outcrossed to CAST/EiJ, AKR/J, or NOD.NON-*H2^{nb1}* and then the F1 mice intercrossed, homozygous *Nr2e3^{rd7}* mice of the F2 generation are found that unlike the parental B6.Cg-*Nr2e3^{rd7}* homozygotes have fewer spots on fundus examination and no PR layer dysplasia in histological sections [460]. This phenotypic variability is caused by the genetic interaction between the *Nr2e3^{rd7}* disease allele and variants of so-called modifier genes that are specific to the outcross partner strain. Several quantitative trait loci (QTL) on chromosomes 7, 8, 11, and 19 were mapped [460]. Generation of a congenic line carrying the Chr11 modifier, along with further fine mapping, reduced the critical genomic interval to 3.3 cM. Several candidate genes were sequenced and a single nucleotide polymorphism was found in a nuclear receptor gene, *Nr1d1*, that is predicted to lead to an Arg409Gln amino acid change. Causality was confirmed by phenotypic rescue of the *rd7*-associated phenotypes by in vivo electroporation of a wild-type *Nr1d1* expression construct [463].

Several other modifiers have been mapped and identified based on inbred strain differences. For example, mapping crosses have been carried out for *rd3* (BALB/cJ and C57BL/6J, [464]), *rd1* (C3H/HeOu and FVB/N, [465]), *Crb1* (C57BL/6N and C57BL/6JOlaHsd, Chr15, [466]), *Mfrp* (B6.C3Ga and CAST/EiJ, Chr 1, 6, and 11 [467]), and *Tub* and *Tulp1* (C57BL/6J and AKR/J, *Mtap1a*, [468]).

Although not yet widely used as a means to explore retinal biology, a very efficient way to identify modifier genes is the use of a sensitized mutagenesis screen in which a male mouse carrying a mutation of interest is given a chemical mutagen and its offspring are examined for any change in the original phenotype. Offspring carrying a potential mutation is backcrossed to the unmutagenized parental inbred strain to test for heritability and to reduce the mutational load. Mutations are identified using whole exome sequencing of the pheno-deviant mouse. This approach avoids the limited genetic diversity of inbred strains since in principle all genes can be mutated. An example of the utility of mutagenesis to search for modifier genes is the identification of a suppressor mutation in *Frmd4b* that prevents the PR dysplasia and external limiting membrane fragmentation observed in *Nr2e3^{rd7}* mutant mice [469].

5.7.3. Effects of Environment on PR Degeneration

PR cell loss has been shown to be induced by a number of environmental factors such as light, diet, and smoking in combination with particular genotypes. Perhaps not surprisingly, light exposure in some models bearing mutations in genes that function directly or in an ancillary fashion in the visual transduction pathway trend toward hastening PR degeneration [470,471]. For example, transgenic mice bearing the rhodopsin VPP mutation, widely used in visual transduction studies, is susceptible to light-exacerbated PR degeneration [472]. Likewise, mice carrying a homozygous *Prom1* null mutation are particularly susceptible to light-induced degeneration. At eye opening, with exposure to light, degeneration initiates at P14, and all PRs are gone by P20, whereas dark rearing from P8 to P30 leads to significant preservation of PRs [471]. Dark-rearing has also been demonstrated to delay PR degeneration in *Slc6a6^{tm1Dhau}* (10% loss vs. 90% loss in normal vivarium lighting at three weeks of age) [473] or have no effect in C57BL/6-*Mitf^{mi-vit}*/J homozygotes [474]. In some situations, light may actually trigger the disease phenotype, as is the case in *Sag* knockout mice [198,199], with three Class B1 Rhodopsin missense mutations, *Tvrm1* and *Tvrm4* [157] or *Tvrm144* [18], and in null mutation models of *Rdh12* [475], *Asic2* [476], *Myo7a* [477], *Whrn* [478], or *Akt2* [479]. *Sag* mutants must be reared in the dark to observe any PR cells. Under normal vivarium lighting conditions, the other light-sensitive mouse models do not show PR degeneration or only a slight shortening of OS at one year of age, as in the those carrying *Rho* alleles *Tvrm1*, *Tvrm4*, or *Tvrm144*, and in retinol dehydrogenase (*Rdh12*) mutant mice. However, exposure to bright light or rearing under cyclic moderate-lighting, even subjecting mice to fundus examination, leads to PR degeneration. A comprehensive list of animal models and the effects of dark-rearing or light exposure can be found in reference [470].

Like light exposure, smoking and high fat intake have been proposed to have a negative impact on retinal function by increasing oxidative stress and inflammation in PR and RPE cells [480]. Smoking has been implicated as a major risk factor in the development of age-related macular degeneration in humans [481,482], and the results have been replicated in mouse models as well. Smoking leads to increased oxidative stress and inflammation in B6 mice [483] and in the presence of *Nfe2l2* deficiency [484]. Likewise, combinations of smoking and high fat intake in the presence of an *ApoB* mutation that promotes production of the APOB100 isoform [485] leads to significant loss of PRs [484]. Further, high-fat diet intake for certain genotypes, such as mutations of *Ldlr* [486] or certain alleles of *Apoe* [487], has been shown to compromise PR integrity in mice.

The majority of pharmacological or dietary interventions that have been reported in the relationship to PR degeneration in mouse models are associated with the goal of increasing vitamin A derivative availability [488–490] or reducing oxidative stress [491,492] in the retina. Heritable mutations in enzymes, such as LRAT or RPE65, required for processing of vitamin A within the retina are known to cause early onset RD due a deficiency of the 11-*cis*-retinal chromophore. Efficacy of treatment with 9-*cis*-retinal derivatives of mice with null mutations in *Lrat* and *Rpe65* mice is thoroughly discussed in a review by Perusek and Maeda [488,489]. Administration of antioxidants has in some cases improved PR survival. *Rs1^{tm1Web}* homozygous females or hemizygous males fed a diet high in DHA [493] or *Pde6b^{rd10}* mice fed lutein and zeaxanthin [494] showed a significant PR preservation. Further, injections of a mixture of antioxidants—alpha tocopheral, ascorbic acid, alpha-lipoic acid, and/or Mn(III)tetrakis porphyrin—were able to slow the loss of cone/rod PRs in *Pde6b^{rd1}* [495], and *Pde6b^{rd10}* mice and in mice with a rhodopsin Q344ter mutation [492]. Environmental enhancement of *Pde6b^{rd10}* mice was able to significantly reduce PR loss presumably by reducing retinal oxidative stress [496].

5.8. Relationship to Human Disease Genes

Of the 273 retinal degenerative disease genes in RetNet [3] for which mouse homologs exist, mouse models are available for 110 or 40% of them, including both germline and conditional mutants (Figure 7). Through our survey, we found 120 additional genes, in which mutations lead to PR degeneration. These genes could serve as candidates for yet to be identified human retinal diseases. The available mouse models, for the most part, recapitulate the human disease phenotype well and permit mechanistic and therapeutic studies. However, apparent failures of mouse models do occur. When mutations in *MFRP* were first identified in humans [497], mice were thought to be a poor model because unlike humans [498], mice were previously reported to develop PR degeneration [499], and the microphthalmia and hyperopia found in human patients had not been reported in homozygous *Mfrp^{rd6}* mice. In subsequent years, numerous human patients have been identified that do show a degenerative phenotype [500] and hyperopia was detected both in a mouse model carrying a human *MFRP* c.498_499insC allele [501] and the original *Mfrp^{rd6}* mouse (our unpublished observations). An important family of deaf–blindness diseases, Usher syndrome, was also thought to be poorly recapitulated in mice, because early models like the shaker-1 mouse had only the characteristic hearing loss, but no retinal degeneration [502]. Later, however, it was found that moderate light exposure does result in photoreceptor degeneration in shaker-1 mice [477]. In addition, a knock-in of the Acadian *USH1C* c.216G>A mutation into the mouse *Ush1c* gene recapitulates both deafness and retinal degeneration phenotypes [503]. In many cases, discordance between the human and mouse phenotypes can be attributed to insufficient information about variation in the human disease, or to allelic effects (knockout vs. hypomorph or gain of function, expression of alternatively splice isoforms), or strain background (modifier genes) in the mouse models. Such shortcomings in mouse models can often be addressed by testing multiple models, including human disease alleles, and by using multiple genetic backgrounds.

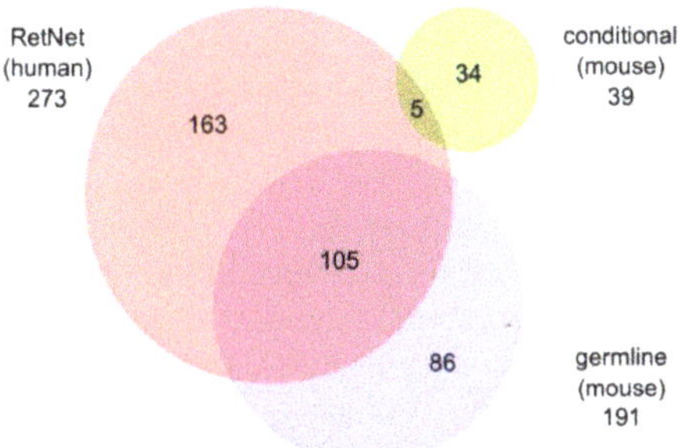

Figure 7. Comparison of the number of RD genes identified in human (RetNet) and mouse as listed in Table S1 and summarized in Figure 3. The total number of genes in the RetNet database that cause monogenic disease and have mouse homologs is indicated, as is the total for which conditional or germline mutations have been associated with PR cell loss in mice, as described in this review. Numbers within the overlapping areas of the diagram represent genes present in both RetNet and Table S1; the remaining numbers represent genes that are unique to the indicated category.

Although humans and mice share about 98% of their genes, species differences do exist and need to be considered when selecting a model. Examples of vision-related genes that mice lack are *EYS*, *ARMS2*, and *CETP*. Species differences are the result of different evolutionary histories; humans and mice have encountered different pathogens, resulting in adaptations of our respective immune systems. Mice have different nutritional requirements, resulting in differences in lipid metabolism. Additionally, mouse eyes are adapted to a nocturnal life, resulting in a rod dominated retina with no macula. Nevertheless, mice possess all of the same retinal cell types necessary for vision and the vast majority of the same genes, and even when missing genes are introduced into mice they result in relevant phenotypes. For example, in a transgenic mouse model for Stargardt-like macular degeneration 3 due to a mutation in *Elovl4*, PR cell loss occurs in the central retina in a pattern that resembles the human disease [504]. For the many retinal diseases still in need of models, including complex diseases such as AMD or diabetic retinopathy, it remains the case that valuable new insights into disease mechanism and basic eye biology can still be obtained from mouse studies.

6. Discussion

6.1. Variability in Measuring PR Cell Loss

We noted an extremely large variability in PR cell loss data presented in the various reports, which were based on several different types of measures, such as ONL thickness obtained from toluidine-stained plastic sections, hematoxylin and eosin-stained paraffin sections, or DAPI-stained cryosections; counts of rows of ONL nuclei in the same preparations; counts of the total number of ONL nuclei in a fixed retinal area; spider plots assessing ONL thickness over the full perimeter; ONL thickness from OCT B-scans. Cone numbers were assessed in sections or whole mounts stained with cone opsins, peanut agglutinin lectin, or cone arrestin. Methods to count cone cell nuclei efficiently might benefit from studies that examine the effect of mutations in genes that specifically affect cones. Perhaps more challenging is that many studies provided insufficient sample sizes or number of time points to assess the progression of PR cell loss. Some of this variability reflects the evolution of methods and quantitative approaches over several decades, and some may be attributed to the different choices each laboratory makes depending on what works best given resources and interests. However, to aid future efforts to compare PR cell loss between studies, we recommend that at least three ages be assessed, one prior to the onset of PR cell loss, and at least two within the age range where PR cells are declining exponentially (that is, between 95% and 5% of wild-type values), and that the data be quantified relative to gender-matched controls at these same ages.

Although a decrease in PR cell numbers as estimated above is widely accepted as evidence for RD, an alternative explanation may apply in some instances. *Mcoln1*^{tm1Sasl} homozygotes exhibit an apparent decrease in PR cells to 54% of wild-type at one month of age, a value that remained unchanged at two and six months [505]. TUNEL analysis revealed no increase in apoptosis at any age compared to controls, and both histology and OCT indicated a decrease in total retinal thickness. It is possible that the rapid initial decrease to a stable value may result from retinal thinning as the eye enlarges due to myopia during postnatal development, an occasionally observed feature of *MCOLN1*-associated disease in humans [506]. Methods are available to measure axial length in mice [507], which might be used to determine whether the observed change in nuclear layer thickness is due to ocular enlargement. Models in Table S1 with a similar phenotype include *Guc1a*$^{tm1.1Hunt}$ and *Ctnna1*^{Tvrm5}.

6.2. Correlation of PR Cell Loss with Gene Function

As a fortuitous consequence of our inquiry, in some cases, the progress of PR cell loss could be correlated with gene function within categories. For example, in the ciliary function and trafficking section we see trends in the onset and progression of RD depending on the role and location of gene products within the PR. Mice with mutations in genes involved in ciliogenesis or in transition zone protein complexes, typically result in a PR degeneration that begins at 2 weeks during OS biogenesis and progresses rapidly through OS maturation. Null mutations in genes that encode components of the IFT machinery tend to result in premature death or embryonic lethality, and conditional ablation of these genes in the retina typically leads to early and rapid PR loss. Models with disruptions in protein complexes with roles in BBSome assembly or regulation, protein/lipid trafficking, axoneme extension, or disc morphogenesis tend to have a moderate to slow degeneration. Lastly, mice with mutations that disturb basal body and pericentriolar anchoring and integrity such as the Usher periciliary membrane complex undergo a very slow form of RD, which results in partial PR function throughout the life of the mouse. It remains possible that the slower progression of PR cell loss, especially when associated with members of protein complexes, may be the result of genetic redundancy where multiple genes encode proteins that have similar biochemical functions.

It was also interesting to note that disruption of chloride channels appeared to be particularly deleterious to PR survival. Many of the models we identified with early and rapid progression of PR cell loss, including those affecting a subset of the ciliary genes discussed above or the OS components *Prom1*, *Prph2*, and *Rho*, appear to be required for OS assembly. Chloride channel defects resulted in similar progression, raising the possibility that chloride homeostasis is important for OS development. This idea is supported by evidence that chloride transport by the chloride channel ANO1 is required for ciliogenesis [508] and that control of intracellular chloride ion levels by this channel regulates the membrane organization of phosphatidylinositol 4,5-bisphosphate [509], a prominent lipid that regulates ciliary development [131,510]. Characterization of early OS development in mouse models defective in chloride channels CLCN2, CLCN3, or CLCN7 may provide mechanistic clues on the role of intracellular chloride in ciliogenesis.

Finally, our analysis revealed links between PR cell loss and a network of 13 genes known to participate in the cellular response to DNA damage, four of which have been directly associated with transcription-coupled DNA repair. Based on canonical indicators of DNA repair, such as the colocalization of phosphorylated histone H2AX and TRP53BP1 (also known as 53BP1) at DSBs, it has been reported that rod PR cells in mice lack a robust canonical DNA damage response, [241,511]. An attenuated response may reflect an adaptation to improve rod cell survival [241,511]. Nevertheless, mechanisms are present in rod cells to repair DNA damage, as evidenced by the robust levels of DNA repair factors [241] and by our results indicating that a network of DNA damage response genes is required for maintaining PR cell viability. Together, these results support the idea that a non-canonical

DNA damage response pathway exists in rod PR cells [512]. Further study of the DNA damage response genes linked to PR cell loss in mice may be useful for elucidating this pathway.

7. Conclusions

This review highlights mouse models of monogenic retinal degenerative diseases that cause rod or cone PR cell loss. The models include germline mutations and conditional alleles, in which characterization of retinal phenotypes in germline mutations was not possible due to embryonic or perinatal lethality. By providing an extensive list of these models as well as a means of comparing their progression, we hope to benefit researchers who seek to optimize their experimental approaches.

Supplementary Materials: The following are available online at http://www.mdpi.com/2073-4409/9/4/931/s1, Figure S1: OCT and fundus images of C57BL/6J control mice at various ages, Table S1: Monogenic mouse RD models with PR cell loss, Table S2: Description of gene/protein symbols used in the text and figures.

Author Contributions: Conceptualization, P.M.N., M.P.K.; methodology, M.P.K., B.C., P.M.N.; software, M.P.K.; formal analysis, M.P.K.; investigation, G.B.C., N.G., N.D., J.P., L.F.H., L.S., J.K.N., M.P.K., P.M.N.; resources, B.C., J.K.N., P.M.N.; data curation, G.B.C., N.G., N.D., J.P., L.F.H., L.S., J.K.N., M.P.K., P.M.N.; writing—original draft preparation, G.B.C., N.G., N.D., J.P., L.H., J.K.N., M.P.K., P.M.N.; writing—review and editing, G.B.C., N.G., J.K.N., M.P.K., P.M.N.; visualization, B.C., M.P.K.; supervision, J.K.N., P.M.N.; project administration, P.M.N.; funding acquisition, M.P.K., B.C., J.K.N., P.M.N. All authors have read and agree to the published version of the manuscript.

Funding: This work was funded in part by National Institutes of Health grants 5R01EY011996, 5R01EY027860, R01EY027305 (to P.M.N), 5R01EY028561 (to J.K.N.), 5R01EY019943 (to B.C.), and 5R21EY027894 (to M.P.K). N.D. was supported by the grant from the Royal Golden Jubilee (RGJ) Scholarship (PHD/0102/2559), Thailand Research Fund.

Acknowledgments: The authors thank James Kadin and Grace Stafford for help with scripts to filter PubMed data, Cynthia Smith for her contributions to Table S1 and for coordinating our efforts with those at MGI, Melissa Berry for assistance with nomenclature, Bernard Fitzmaurice and Wanda Hicks for fundus and/or OCT imaging, and Jane Cha for drawing Figure 1.

Conflicts of Interest: The authors declare no conflict of interest. The funders had no role in the design of the study; in the collection, analyses, or interpretation of data; in the writing of the manuscript, or in the decision to publish the results.

References

1. Cremers, F.P.M.; Boon, C.J.F.; Bujakowska, K.; Zeitz, C. Special Issue Introduction: Inherited Retinal Disease: Novel Candidate Genes, Genotype-Phenotype Correlations, and Inheritance Models. *Genes (Basel)* **2018**, *9*, 215. [CrossRef] [PubMed]

2. Hanany, M.; Rivolta, C.; Sharon, D. Worldwide carrier frequency and genetic prevalence of autosomal recessive inherited retinal diseases. *Proc. Natl. Acad. Sci. USA* **2020**, *117*, 2710–2716. [CrossRef] [PubMed]

3. RetNet—Retinal Information Network. Available online: https://sph.uth.edu/retnet/home.htm (accessed on 8 December 2019).

4. Picaud, S.; Dalkara, D.; Marazova, K.; Goureau, O.; Roska, B.; Sahel, J.A. The primate model for understanding and restoring vision. *Proc. Natl. Acad. Sci. USA* **2019**, *116*, 26280–26287. [CrossRef] [PubMed]

5. Bunel, M.; Chaudieu, G.; Hamel, C.; Lagoutte, L.; Manes, G.; Botherel, N.; Brabet, P.; Pilorge, P.; Andre, C.; Quignon, P. Natural models for retinitis pigmentosa: Progressive retinal atrophy in dog breeds. *Hum. Genet.* **2019**, *138*, 441–453. [CrossRef] [PubMed]

6. Fletcher, E.L.; Jobling, A.I.; Vessey, K.A.; Luu, C.; Guymer, R.H.; Baird, P.N. Animal models of retinal disease. *Prog. Mol. Biol. Transl. Sci.* **2011**, *100*, 211–286. [CrossRef]

7. Veleri, S.; Lazar, C.H.; Chang, B.; Sieving, P.A.; Banin, E.; Swaroop, A. Biology and therapy of inherited retinal degenerative disease: Insights from mouse models. *Dis. Model. Mech.* **2015**, *8*, 109–129. [CrossRef]

8. Angueyra, J.M.; Kindt, K.S. Leveraging Zebrafish to Study Retinal Degenerations. *Front. Cell Dev. Biol.* **2018**, *6*, 110. [CrossRef]

9. Lehmann, M.; Knust, E.; Hebbar, S. Drosophila melanogaster: A Valuable Genetic Model Organism to Elucidate the Biology of Retinitis Pigmentosa. *Methods Mol. Biol.* **2019**, *1834*, 221–249. [CrossRef]

10. Travis, G.H.; Brennan, M.B.; Danielson, P.E.; Kozak, C.A.; Sutcliffe, J.G. Identification of a photoreceptor-specific mRNA encoded by the gene responsible for retinal degeneration slow (rds). *Nature* **1989**, *338*, 70–73. [CrossRef]
11. Dryja, T.P.; McGee, T.L.; Reichel, E.; Hahn, L.B.; Cowley, G.S.; Yandell, D.W.; Sandberg, M.A.; Berson, E.L. A point mutation of the rhodopsin gene in one form of retinitis pigmentosa. *Nature* **1990**, *343*, 364–366. [CrossRef]
12. Bowes, C.; Li, T.; Danciger, M.; Baxter, L.C.; Applebury, M.L.; Farber, D.B. Retinal degeneration in the rd mouse is caused by a defect in the beta subunit of rod cGMP-phosphodiesterase. *Nature* **1990**, *347*, 677–680. [CrossRef] [PubMed]
13. Wheway, G.; Parry, D.A.; Johnson, C.A. The role of primary cilia in the development and disease of the retina. *Organogenesis* **2014**, *10*, 69–85. [CrossRef] [PubMed]
14. Baehr, W.; Hanke-Gogokhia, C.; Sharif, A.; Reed, M.; Dahl, T.; Frederick, J.M.; Ying, G. Insights into photoreceptor ciliogenesis revealed by animal models. *Prog. Retin. Eye Res.* **2019**, *71*, 26–56. [CrossRef] [PubMed]
15. Chang, B.; Hawes, N.L.; Hurd, R.E.; Davisson, M.T.; Nusinowitz, S.; Heckenlively, J.R. Retinal degeneration mutants in the mouse. *Vis. Res.* **2002**, *42*, 517–525. [CrossRef]
16. Chang, B.; Hawes, N.L.; Hurd, R.E.; Wang, J.; Howell, D.; Davisson, M.T.; Roderick, T.H.; Nusinowitz, S.; Heckenlively, J.R. Mouse models of ocular diseases. *Vis. NeuroSci.* **2005**, *22*, 587–593. [CrossRef] [PubMed]
17. Won, J.; Shi, L.Y.; Hicks, W.; Wang, J.; Hurd, R.; Naggert, J.K.; Chang, B.; Nishina, P.M. Mouse model resources for vision research. *J. Ophthalmol.* **2011**, *2011*, 391384. [CrossRef]
18. Won, J.; Shi, L.Y.; Hicks, W.; Wang, J.; Naggert, J.K.; Nishina, P.M. Translational vision research models program. *Adv. Exp. Med. Biol.* **2012**, *723*, 391–397. [CrossRef]
19. Chang, B. Mouse models for studies of retinal degeneration and diseases. *Methods Mol. Biol.* **2013**, *935*, 27–39. [CrossRef]
20. Chang, B. Mouse Models as Tools to Identify Genetic Pathways for Retinal Degeneration, as Exemplified by Leber's Congenital Amaurosis. *Methods Mol. Biol.* **2016**, *1438*, 417–430. [CrossRef]
21. Krebs, M.P.; Collin, G.B.; Hicks, W.L.; Yu, M.; Charette, J.R.; Shi, L.Y.; Wang, J.; Naggert, J.K.; Peachey, N.S.; Nishina, P.M. Mouse models of human ocular disease for translational research. *PLoS ONE* **2017**, *12*, e0183837. [CrossRef]
22. Do, M.T.; Yau, K.W. Intrinsically photosensitive retinal ganglion cells. *Physiol. Rev.* **2010**, *90*, 1547–1581. [CrossRef] [PubMed]
23. Goldberg, A.F.; Moritz, O.L.; Williams, D.S. Molecular basis for photoreceptor outer segment architecture. *Prog. Retin. Eye Res.* **2016**, *55*, 52–81. [CrossRef] [PubMed]
24. Rachel, R.A.; Li, T.; Swaroop, A. Photoreceptor sensory cilia and ciliopathies: Focus on CEP290, RPGR and their interacting proteins. *Cilia* **2012**, *1*, 22. [CrossRef] [PubMed]
25. Wang, J.; O'Sullivan, M.L.; Mukherjee, D.; Punal, V.M.; Farsiu, S.; Kay, J.N. Anatomy and spatial organization of Muller glia in mouse retina. *J. Comp. Neurol.* **2017**, *525*, 1759–1777. [CrossRef]
26. Reichenbach, A.; Bringmann, A. Glia of the human retina. *Glia* **2019**, *68*, 768–796. [CrossRef]
27. Strauss, O. The retinal pigment epithelium in visual function. *Physiol. Rev.* **2005**, *85*, 845–881. [CrossRef]
28. Verbakel, S.K.; van Huet, R.A.C.; Boon, C.J.F.; den Hollander, A.I.; Collin, R.W.J.; Klaver, C.C.W.; Hoyng, C.B.; Roepman, R.; Klevering, B.J. Non-syndromic retinitis pigmentosa. *Prog. Retin. Eye Res.* **2018**, *66*, 157–186. [CrossRef]
29. Kumaran, N.; Moore, A.T.; Weleber, R.G.; Michaelides, M. Leber congenital amaurosis/early-onset severe retinal dystrophy: Clinical features, molecular genetics and therapeutic interventions. *Br. J. Ophthalmol.* **2017**, *101*, 1147–1154. [CrossRef]
30. Parisi, M.A. The molecular genetics of Joubert syndrome and related ciliopathies: The challenges of genetic and phenotypic heterogeneity. *Transl. Sci. Rare Dis.* **2019**, *4*, 25–49. [CrossRef]
31. Suspitsin, E.N.; Imyanitov, E.N. Bardet-Biedl Syndrome. *Mol. Syndromol.* **2016**, *7*, 62–71. [CrossRef]
32. Mathur, P.; Yang, J. Usher syndrome: Hearing loss, retinal degeneration and associated abnormalities. *Biochim. Biophys. Acta* **2015**, *1852*, 406–420. [CrossRef] [PubMed]
33. Kumaran, N.; Michaelides, M.; Smith, A.J.; Ali, R.R.; Bainbridge, J.W.B. Retinal gene therapy. *Br. Med. Bull.* **2018**, *126*, 13–25. [CrossRef] [PubMed]

34. Mouse Genome Database (MGD) at the Mouse Genome Informatics website, The Jackson Laboratory, Bar Harbor, Maine. Available online: http://www.informatics.jax.org (accessed on 18 October 2019).

35. PubMed [Internet]. Bethesda (MD): National Library of Medicine (US), National Center for Biotechnology Information. Available online: https://www.ncbi.nlm.nih.gov/pubmed/ (accessed on 15 October 2019).

36. Clarke, G.; Collins, R.A.; Leavitt, B.R.; Andrews, D.F.; Hayden, M.R.; Lumsden, C.J.; McInnes, R.R. A one-hit model of cell death in inherited neuronal degenerations. *Nature* **2000**, *406*, 195–199. [CrossRef] [PubMed]

37. Krebs, M.P.; Xiao, M.; Sheppard, K.; Hicks, W.; Nishina, P.M. Bright-Field Imaging and Optical Coherence Tomography of the Mouse Posterior Eye. *Methods Mol. Biol.* **2016**, *1438*, 395–415. [CrossRef] [PubMed]

38. Krebs, M.P. Using Vascular Landmarks to Orient 3D Optical Coherence Tomography Images of the Mouse Eye. *Curr. Protoc. Mouse Biol.* **2017**, *7*, 176–190. [CrossRef]

39. Dyer, M.A.; Donovan, S.L.; Zhang, J.; Gray, J.; Ortiz, A.; Tenney, R.; Kong, J.; Allikmets, R.; Sohocki, M.M. Retinal degeneration in Aipl1-deficient mice: A new genetic model of Leber congenital amaurosis. *Brain Res. Mol. Brain Res.* **2004**, *132*, 208–220. [CrossRef]

40. Liu, X.; Bulgakov, O.V.; Wen, X.H.; Woodruff, M.L.; Pawlyk, B.; Yang, J.; Fain, G.L.; Sandberg, M.A.; Makino, C.L.; Li, T. AIPL1, the protein that is defective in Leber congenital amaurosis, is essential for the biosynthesis of retinal rod cGMP phosphodiesterase. *Proc. Natl. Acad. Sci. USA* **2004**, *101*, 13903–13908. [CrossRef]

41. Dickinson, M.E.; Flenniken, A.M.; Ji, X.; Teboul, L.; Wong, M.D.; White, J.K.; Meehan, T.F.; Weninger, W.J.; Westerberg, H.; Adissu, H.; et al. High-throughput discovery of novel developmental phenotypes. *Nature* **2016**, *537*, 508–514. [CrossRef]

42. Hanke-Gogokhia, C.; Wu, Z.; Gerstner, C.D.; Frederick, J.M.; Zhang, H.; Baehr, W. Arf-like Protein 3 (ARL3) Regulates Protein Trafficking and Ciliogenesis in Mouse Photoreceptors. *J. Biol. Chem.* **2016**, *291*, 7142–7155. [CrossRef]

43. International Mouse Phenotyping Consortium: Home—IMPC. Available online: https://www.mousephenotype.org/ (accessed on 22 December 2019).

44. Bujakowska, K.M.; Liu, Q.; Pierce, E.A. Photoreceptor Cilia and Retinal Ciliopathies. *Cold Spring Harb. Perspect. Biol.* **2017**, *9*, a028274. [CrossRef]

45. Gilliam, J.C.; Chang, J.T.; Sandoval, I.M.; Zhang, Y.; Li, T.; Pittler, S.J.; Chiu, W.; Wensel, T.G. Three-dimensional architecture of the rod sensory cilium and its disruption in retinal neurodegeneration. *Cell* **2012**, *151*, 1029–1041. [CrossRef] [PubMed]

46. Rohlich, P. The sensory cilium of retinal rods is analogous to the transitional zone of motile cilia. *Cell Tiss. Res.* **1975**, *161*, 421–430. [CrossRef] [PubMed]

47. Goncalves, J.; Pelletier, L. The Ciliary Transition Zone: Finding the Pieces and Assembling the Gate. *Mol. Cells* **2017**, *40*, 243–253. [CrossRef] [PubMed]

48. Seo, S.; Datta, P. Photoreceptor outer segment as a sink for membrane proteins: Hypothesis and implications in retinal ciliopathies. *Hum. Mol. Genet.* **2017**, *26*, R75–R82. [CrossRef]

49. Khanna, H. Photoreceptor Sensory Cilium: Traversing the Ciliary Gate. *Cells* **2015**, *4*, 674–686. [CrossRef]

50. Garcia-Gonzalo, F.R.; Corbit, K.C.; Sirerol-Piquer, M.S.; Ramaswami, G.; Otto, E.A.; Noriega, T.R.; Seol, A.D.; Robinson, J.F.; Bennett, C.L.; Josifova, D.J.; et al. A transition zone complex regulates mammalian ciliogenesis and ciliary membrane composition. *Nat. Genet.* **2011**, *43*, 776–784. [CrossRef]

51. Sedmak, T.; Wolfrum, U. Intraflagellar transport proteins in ciliogenesis of photoreceptor cells. *Biol. Cell* **2011**, *103*, 449–466. [CrossRef]

52. Salinas, R.Y.; Pearring, J.N.; Ding, J.D.; Spencer, W.J.; Hao, Y.; Arshavsky, V.Y. Photoreceptor discs form through peripherin-dependent suppression of ciliary ectosome release. *J. Cell Biol.* **2017**, *216*, 1489–1499. [CrossRef]

53. LaVail, M.M. Kinetics of rod outer segment renewal in the developing mouse retina. *J. Cell Biol.* **1973**, *58*, 650–661. [CrossRef]

54. De Robertis, E. Morphogenesis of the retinal rods; an electron microscope study. *J. Biophys. Biochem. Cytol.* **1956**, *2*, 209–218. [CrossRef]

55. Insinna, C.; Besharse, J.C. Intraflagellar transport and the sensory outer segment of vertebrate photoreceptors. *Dev. Dyn.* **2008**, *237*, 1982–1992. [CrossRef] [PubMed]

56. Rosenbaum, J.L.; Cole, D.G.; Diener, D.R. Intraflagellar transport: The eyes have it. *J. Cell Biol.* **1999**, *144*, 385–388. [CrossRef]

57. Chuang, J.Z.; Zhao, Y.; Sung, C.H. SARA-regulated vesicular targeting underlies formation of the light-sensing organelle in mammalian rods. *Cell* **2007**, *130*, 535–547. [CrossRef] [PubMed]

58. Chuang, J.Z.; Hsu, Y.C.; Sung, C.H. Ultrastructural visualization of trans-ciliary rhodopsin cargoes in mammalian rods. *Cilia* **2015**, *4*, 4. [CrossRef] [PubMed]

59. Steinberg, R.H.; Fisher, S.K.; Anderson, D.H. Disc morphogenesis in vertebrate photoreceptors. *J. Comp. Neurol.* **1980**, *190*, 501–508. [CrossRef]

60. Ding, J.D.; Salinas, R.Y.; Arshavsky, V.Y. Discs of mammalian rod photoreceptors form through the membrane evagination mechanism. *J. Cell Biol.* **2015**, *211*, 495–502. [CrossRef]

61. Burgoyne, T.; Meschede, I.P.; Burden, J.J.; Bailly, M.; Seabra, M.C.; Futter, C.E. Rod disc renewal occurs by evagination of the ciliary plasma membrane that makes cadherin-based contacts with the inner segment. *Proc. Natl. Acad. Sci. USA* **2015**, *112*, 15922–15927. [CrossRef]

62. Young, R.W. The renewal of photoreceptor cell outer segments. *J. Cell Biol.* **1967**, *33*, 61–72. [CrossRef]

63. Jin, H.; White, S.R.; Shida, T.; Schulz, S.; Aguiar, M.; Gygi, S.P.; Bazan, J.F.; Nachury, M.V. The conserved Bardet-Biedl syndrome proteins assemble a coat that traffics membrane proteins to cilia. *Cell* **2010**, *141*, 1208–1219. [CrossRef]

64. Liew, G.M.; Ye, F.; Nager, A.R.; Murphy, J.P.; Lee, J.S.; Aguiar, M.; Breslow, D.K.; Gygi, S.P.; Nachury, M.V. The intraflagellar transport protein IFT27 promotes BBSome exit from cilia through the GTPase ARL6/BBS3. *Dev. Cell* **2014**, *31*, 265–278. [CrossRef]

65. Taub, D.G.; Liu, Q. The Role of Intraflagellar Transport in the Photoreceptor Sensory Cilium. *Adv. Exp. Med. Biol.* **2016**, *854*, 627–633. [CrossRef] [PubMed]

66. Pazour, G.J.; Baker, S.A.; Deane, J.A.; Cole, D.G.; Dickert, B.L.; Rosenbaum, J.L.; Witman, G.B.; Besharse, J.C. The intraflagellar transport protein, IFT88, is essential for vertebrate photoreceptor assembly and maintenance. *J. Cell Biol.* **2002**, *157*, 103–113. [CrossRef] [PubMed]

67. Jensen, V.L.; Leroux, M.R. Gates for soluble and membrane proteins, and two trafficking systems (IFT and LIFT), establish a dynamic ciliary signaling compartment. *Curr. Opin. Cell Biol.* **2017**, *47*, 83–91. [CrossRef] [PubMed]

68. Baehr, W. Membrane protein transport in photoreceptors: The function of PDEdelta: The Proctor lecture. *Investig. Ophthalmol. Vis. Sci.* **2014**, *55*, 8653–8666. [CrossRef]

69. Hsu, Y.; Garrison, J.E.; Kim, G.; Schmitz, A.R.; Searby, C.C.; Zhang, Q.; Datta, P.; Nishimura, D.Y.; Seo, S.; Sheffield, V.C. BBSome function is required for both the morphogenesis and maintenance of the photoreceptor outer segment. *PLoS Genet.* **2017**, *13*, e1007057. [CrossRef]

70. Jiang, L.; Wei, Y.; Ronquillo, C.C.; Marc, R.E.; Yoder, B.K.; Frederick, J.M.; Baehr, W. Heterotrimeric kinesin-2 (KIF3) mediates transition zone and axoneme formation of mouse photoreceptors. *J. Biol. Chem.* **2015**, *290*, 12765–12778. [CrossRef]

71. Ronquillo, C.C.; Hanke-Gogokhia, C.; Revelo, M.P.; Frederick, J.M.; Jiang, L.; Baehr, W. Ciliopathy-associated IQCB1/NPHP5 protein is required for mouse photoreceptor outer segment formation. *FASEB J.* **2016**, *30*, 3400–3412. [CrossRef]

72. Hanke-Gogokhia, C.; Wu, Z.; Sharif, A.; Yazigi, H.; Frederick, J.M.; Baehr, W. The guanine nucleotide exchange factor Arf-like protein 13b is essential for assembly of the mouse photoreceptor transition zone and outer segment. *J. Biol. Chem.* **2017**, *292*, 21442–21456. [CrossRef]

73. Cantagrel, V.; Silhavy, J.L.; Bielas, S.L.; Swistun, D.; Marsh, S.E.; Bertrand, J.Y.; Audollent, S.; Attie-Bitach, T.; Holden, K.R.; Dobyns, W.B.; et al. Mutations in the cilia gene ARL13B lead to the classical form of Joubert syndrome. *Am. J. Hum. Genet.* **2008**, *83*, 170–179. [CrossRef]

74. Caspary, T.; Larkins, C.E.; Anderson, K.V. The graded response to Sonic Hedgehog depends on cilia architecture. *Dev. Cell* **2007**, *12*, 767–778. [CrossRef]

75. Eblimit, A.; Agrawal, S.A.; Thomas, K.; Anastassov, I.A.; Abulikemu, T.; Moayedi, Y.; Mardon, G.; Chen, R. Conditional loss of Spata7 in photoreceptors causes progressive retinal degeneration in mice. *Exp. Eye Res.* **2018**, *166*, 120–130. [CrossRef] [PubMed]

76. Singla, V.; Reiter, J.F. The primary cilium as the cell's antenna: Signaling at a sensory organelle. *Science* **2006**, *313*, 629–633. [CrossRef]

77. Garcia-Gonzalo, F.R.; Reiter, J.F. Open Sesame: How Transition Fibers and the Transition Zone Control Ciliary Composition. *Cold Spring Harb. Perspect. Biol.* **2017**, *9*, a028134. [CrossRef]

78. Shi, X.; Garcia, G., 3rd; Van De Weghe, J.C.; McGorty, R.; Pazour, G.J.; Doherty, D.; Huang, B.; Reiter, J.F. Super-resolution microscopy reveals that disruption of ciliary transition-zone architecture causes Joubert syndrome. *Nat. Cell Biol.* **2017**, *19*, 1178–1188. [CrossRef] [PubMed]

79. Williams, C.L.; Li, C.; Kida, K.; Inglis, P.N.; Mohan, S.; Semenec, L.; Bialas, N.J.; Stupay, R.M.; Chen, N.; Blacque, O.E.; et al. MKS and NPHP modules cooperate to establish basal body/transition zone membrane associations and ciliary gate function during ciliogenesis. *J. Cell Biol.* **2011**, *192*, 1023–1041. [CrossRef] [PubMed]

80. Knorz, V.J.; Spalluto, C.; Lessard, M.; Purvis, T.L.; Adigun, F.F.; Collin, G.B.; Hanley, N.A.; Wilson, D.I.; Hearn, T. Centriolar association of ALMS1 and likely centrosomal functions of the ALMS motif-containing proteins C10orf90 and KIAA1731. *Mol. Biol. Cell* **2010**, *21*, 3617–3629. [CrossRef] [PubMed]

81. Hearn, T.; Renforth, G.L.; Spalluto, C.; Hanley, N.A.; Piper, K.; Brickwood, S.; White, C.; Connolly, V.; Taylor, J.F.; Russell-Eggitt, I.; et al. Mutation of ALMS1, a large gene with a tandem repeat encoding 47 amino acids, causes Alstrom syndrome. *Nat. Genet.* **2002**, *31*, 79–83. [CrossRef]

82. Collin, G.B.; Marshall, J.D.; Ikeda, A.; So, W.V.; Russell-Eggitt, I.; Maffei, P.; Beck, S.; Boerkoel, C.F.; Sicolo, N.; Martin, M.; et al. Mutations in ALMS1 cause obesity, type 2 diabetes and neurosensory degeneration in Alstrom syndrome. *Nat. Genet.* **2002**, *31*, 74–78. [CrossRef]

83. Collin, G.B.; Cyr, E.; Bronson, R.; Marshall, J.D.; Gifford, E.J.; Hicks, W.; Murray, S.A.; Zheng, Q.Y.; Smith, R.S.; Nishina, P.M.; et al. Alms1-disrupted mice recapitulate human Alstrom syndrome. *Hum. Mol. Genet.* **2005**, *14*, 2323–2333. [CrossRef]

84. Brun, A.; Yu, X.; Obringer, C.; Ajoy, D.; Haser, E.; Stoetzel, C.; Roux, M.J.; Messaddeq, N.; Dollfus, H.; Marion, V. In vivo phenotypic and molecular characterization of retinal degeneration in mouse models of three ciliopathies. *Exp. Eye Res.* **2019**, *186*, 107721. [CrossRef]

85. May-Simera, H.L.; Gumerson, J.D.; Gao, C.; Campos, M.; Cologna, S.M.; Beyer, T.; Boldt, K.; Kaya, K.D.; Patel, N.; Kretschmer, F.; et al. Loss of MACF1 Abolishes Ciliogenesis and Disrupts Apicobasal Polarity Establishment in the Retina. *Cell Rep.* **2016**, *17*, 1399–1413. [CrossRef] [PubMed]

86. Gerding, W.M.; Schreiber, S.; Schulte-Middelmann, T.; de Castro Marques, A.; Atorf, J.; Akkad, D.A.; Dekomien, G.; Kremers, J.; Dermietzel, R.; Gal, A.; et al. Ccdc66 null mutation causes retinal degeneration and dysfunction. *Hum. Mol. Genet.* **2011**, *20*, 3620–3631. [CrossRef] [PubMed]

87. Insolera, R.; Shao, W.; Airik, R.; Hildebrandt, F.; Shi, S.H. SDCCAG8 regulates pericentriolar material recruitment and neuronal migration in the developing cortex. *Neuron* **2014**, *83*, 805–822. [CrossRef] [PubMed]

88. Veleri, S.; Manjunath, S.H.; Fariss, R.N.; May-Simera, H.; Brooks, M.; Foskett, T.A.; Gao, C.; Longo, T.A.; Liu, P.; Nagashima, K.; et al. Ciliopathy-associated gene Cc2d2a promotes assembly of subdistal appendages on the mother centriole during cilia biogenesis. *Nat. Commun.* **2014**, *5*, 4207. [CrossRef]

89. Tallila, J.; Jakkula, E.; Peltonen, L.; Salonen, R.; Kestila, M. Identification of CC2D2A as a Meckel syndrome gene adds an important piece to the ciliopathy puzzle. *Am. J. Hum. Genet.* **2008**, *82*, 1361–1367. [CrossRef]

90. Gorden, N.T.; Arts, H.H.; Parisi, M.A.; Coene, K.L.; Letteboer, S.J.; van Beersum, S.E.; Mans, D.A.; Hikida, A.; Eckert, M.; Knutzen, D.; et al. CC2D2A is mutated in Joubert syndrome and interacts with the ciliopathy-associated basal body protein CEP290. *Am. J. Hum. Genet.* **2008**, *83*, 559–571. [CrossRef]

91. Mejecase, C.; Hummel, A.; Mohand-Said, S.; Andrieu, C.; El Shamieh, S.; Antonio, A.; Condroyer, C.; Boyard, F.; Foussard, M.; Blanchard, S.; et al. Whole exome sequencing resolves complex phenotype and identifies CC2D2A mutations underlying non-syndromic rod-cone dystrophy. *Clin. Genet.* **2019**, *95*, 329–333. [CrossRef]

92. Lewis, W.R.; Bales, K.L.; Revell, D.Z.; Croyle, M.J.; Engle, S.E.; Song, C.J.; Malarkey, E.B.; Uytingco, C.R.; Shan, D.; Antonellis, P.J.; et al. Mks6 mutations reveal tissue- and cell type-specific roles for the cilia transition zone. *FASEB J.* **2019**, *33*, 1440–1455. [CrossRef]

93. Sorusch, N.; Bauss, K.; Plutniok, J.; Samanta, A.; Knapp, B.; Nagel-Wolfrum, K.; Wolfrum, U. Characterization of the ternary Usher syndrome SANS/ush2a/whirlin protein complex. *Hum. Mol. Genet.* **2017**, *26*, 1157–1172. [CrossRef]

94. Liu, X.; Bulgakov, O.V.; Darrow, K.N.; Pawlyk, B.; Adamian, M.; Liberman, M.C.; Li, T. Usherin is required for maintenance of retinal photoreceptors and normal development of cochlear hair cells. *Proc. Natl. Acad. Sci. USA* **2007**, *104*, 4413–4418. [CrossRef]

95. Yang, J.; Liu, X.; Zhao, Y.; Adamian, M.; Pawlyk, B.; Sun, X.; McMillan, D.R.; Liberman, M.C.; Li, T. Ablation of whirlin long isoform disrupts the USH2 protein complex and causes vision and hearing loss. *PLoS Genet.* **2010**, *6*, e1000955. [CrossRef] [PubMed]

96. Khattree, N.; Ritter, L.M.; Goldberg, A.F. Membrane curvature generation by a C-terminal amphipathic helix in peripherin-2/rds, a tetraspanin required for photoreceptor sensory cilium morphogenesis. *J. Cell Sci.* **2013**, *126*, 4659–4670. [CrossRef] [PubMed]

97. Molday, R.S.; Hicks, D.; Molday, L. Peripherin. A rim-specific membrane protein of rod outer segment discs. *Investig. Ophthalmol. Vis. Sci.* **1987**, *28*, 50–61. [PubMed]

98. Wood, C.R.; Huang, K.; Diener, D.R.; Rosenbaum, J.L. The cilium secretes bioactive ectosomes. *Curr. Biol.* **2013**, *23*, 906–911. [CrossRef]

99. Wood, C.R.; Rosenbaum, J.L. Ciliary ectosomes: Transmissions from the cell's antenna. *Trends Cell Biol.* **2015**, *25*, 276–285. [CrossRef] [PubMed]

100. Stuck, M.W.; Conley, S.M.; Naash, M.I. The Y141C knockin mutation in RDS leads to complex phenotypes in the mouse. *Hum. Mol. Genet.* **2014**, *23*, 6260–6274. [CrossRef]

101. Chakraborty, D.; Conley, S.M.; Zulliger, R.; Naash, M.I. The K153Del PRPH2 mutation differentially impacts photoreceptor structure and function. *Hum. Mol. Genet.* **2016**, *25*, 3500–3514. [CrossRef]

102. Sanyal, S.; De Ruiter, A.; Hawkins, R.K. Development and degeneration of retina in rds mutant mice: Light microscopy. *J. Comp. Neurol.* **1980**, *194*, 193–207. [CrossRef]

103. Sanyal, S.; Hawkins, R.K. Development and degeneration of retina in rds mutant mice: Effects of light on the rate of degeneration in albino and pigmented homozygous and heterozygous mutant and normal mice. *Vis. Res.* **1986**, *26*, 1177–1185. [CrossRef]

104. McNally, N.; Kenna, P.F.; Rancourt, D.; Ahmed, T.; Stitt, A.; Colledge, W.H.; Lloyd, D.G.; Palfi, A.; O'Neill, B.; Humphries, M.M.; et al. Murine model of autosomal dominant retinitis pigmentosa generated by targeted deletion at codon 307 of the rds-peripherin gene. *Hum. Mol. Genet.* **2002**, *11*, 1005–1016. [CrossRef]

105. Chakraborty, D.; Conley, S.M.; Al-Ubaidi, M.R.; Naash, M.I. Initiation of rod outer segment disc formation requires RDS. *PLoS ONE* **2014**, *9*, e98939. [CrossRef] [PubMed]

106. Kajiwara, K.; Berson, E.L.; Dryja, T.P. Digenic retinitis pigmentosa due to mutations at the unlinked peripherin/RDS and ROM1 loci. *Science* **1994**, *264*, 1604–1608. [CrossRef] [PubMed]

107. Clarke, G.; Goldberg, A.F.; Vidgen, D.; Collins, L.; Ploder, L.; Schwarz, L.; Molday, L.L.; Rossant, J.; Szel, A.; Molday, R.S.; et al. Rom-1 is required for rod photoreceptor viability and the regulation of disk morphogenesis. *Nat. Genet.* **2000**, *25*, 67–73. [CrossRef] [PubMed]

108. Sato, H.; Suzuki, T.; Ikeda, K.; Masuya, H.; Sezutsu, H.; Kaneda, H.; Kobayashi, K.; Miura, I.; Kurihara, Y.; Yokokura, S.; et al. A monogenic dominant mutation in Rom1 generated by N-ethyl-N-nitrosourea mutagenesis causes retinal degeneration in mice. *Mol. Vis.* **2010**, *16*, 378–391.

109. Spencer, W.J.; Pearring, J.N.; Salinas, R.Y.; Loiselle, D.R.; Skiba, N.P.; Arshavsky, V.Y. Progressive Rod-Cone Degeneration (PRCD) Protein Requires N-Terminal S-Acylation and Rhodopsin Binding for Photoreceptor Outer Segment Localization and Maintaining Intracellular Stability. *Biochemistry* **2016**, *55*, 5028–5037. [CrossRef]

110. Allon, G.; Mann, I.; Remez, L.; Sehn, E.; Rizel, L.; Nevet, M.J.; Perlman, I.; Wolfrum, U.; Ben-Yosef, T. PRCD is Concentrated at the Base of Photoreceptor Outer Segments and is Involved in Outer Segment Disc Formation. *Hum. Mol. Genet.* **2019**, *28*, 4078–4088. [CrossRef]

111. Zangerl, B.; Goldstein, O.; Philp, A.R.; Lindauer, S.J.; Pearce-Kelling, S.E.; Mullins, R.F.; Graphodatsky, A.S.; Ripoll, D.; Felix, J.S.; Stone, E.M.; et al. Identical mutation in a novel retinal gene causes progressive rod-cone degeneration in dogs and retinitis pigmentosa in humans. *Genomics* **2006**, *88*, 551–563. [CrossRef]

112. Spencer, W.J.; Ding, J.D.; Lewis, T.R.; Yu, C.; Phan, S.; Pearring, J.N.; Kim, K.Y.; Thor, A.; Mathew, R.; Kalnitsky, J.; et al. PRCD is essential for high-fidelity photoreceptor disc formation. *Proc. Natl. Acad. Sci. USA* **2019**, *116*, 13087–13096. [CrossRef]

113. Pedersen, L.B.; Rosenbaum, J.L. Intraflagellar transport (IFT) role in ciliary assembly, resorption and signalling. *Curr. Top. Dev. Biol.* **2008**, *85*, 23–61. [CrossRef]

114. Cortellino, S.; Wang, C.; Wang, B.; Bassi, M.R.; Caretti, E.; Champeval, D.; Calmont, A.; Jarnik, M.; Burch, J.; Zaret, K.S.; et al. Defective ciliogenesis, embryonic lethality and severe impairment of the Sonic Hedgehog pathway caused by inactivation of the mouse complex A intraflagellar transport gene Ift122/Wdr10, partially overlapping with the DNA repair gene Med1/Mbd4. *Dev. Biol.* **2009**, *325*, 225–237. [CrossRef]

115. Murcia, N.S.; Richards, W.G.; Yoder, B.K.; Mucenski, M.L.; Dunlap, J.R.; Woychik, R.P. The Oak Ridge Polycystic Kidney (orpk) disease gene is required for left-right axis determination. *Development* **2000**, *127*, 2347–2355. [PubMed]

116. Stottmann, R.W.; Tran, P.V.; Turbe-Doan, A.; Beier, D.R. Ttc21b is required to restrict sonic hedgehog activity in the developing mouse forebrain. *Dev. Biol.* **2009**, *335*, 166–178. [CrossRef] [PubMed]

117. Gorivodsky, M.; Mukhopadhyay, M.; Wilsch-Braeuninger, M.; Phillips, M.; Teufel, A.; Kim, C.; Malik, N.; Huttner, W.; Westphal, H. Intraflagellar transport protein 172 is essential for primary cilia formation and plays a vital role in patterning the mammalian brain. *Dev. Biol.* **2009**, *325*, 24–32. [CrossRef] [PubMed]

118. Rix, S.; Calmont, A.; Scambler, P.J.; Beales, P.L. An Ift80 mouse model of short rib polydactyly syndromes shows defects in hedgehog signalling without loss or malformation of cilia. *Hum. Mol. Genet.* **2011**, *20*, 1306–1314. [CrossRef]

119. Berbari, N.F.; Kin, N.W.; Sharma, N.; Michaud, E.J.; Kesterson, R.A.; Yoder, B.K. Mutations in Traf3ip1 reveal defects in ciliogenesis, embryonic development, and altered cell size regulation. *Dev. Biol.* **2011**, *360*, 66–76. [CrossRef]

120. Bangs, F.; Anderson, K.V. Primary Cilia and Mammalian Hedgehog Signaling. *Cold Spring Harb. Perspect. Biol.* **2017**, *9*, a028175. [CrossRef]

121. Ko, H.W.; Liu, A.; Eggenschwiler, J.T. Analysis of hedgehog signaling in mouse intraflagellar transport mutants. *Methods Cell Biol.* **2009**, *93*, 347–369. [CrossRef]

122. Gupta, P.R.; Pendse, N.; Greenwald, S.H.; Leon, M.; Liu, Q.; Pierce, E.A.; Bujakowska, K.M. Ift172 conditional knock-out mice exhibit rapid retinal degeneration and protein trafficking defects. *Hum. Mol. Genet.* **2018**, *27*, 2012–2024. [CrossRef]

123. Keady, B.T.; Le, Y.Z.; Pazour, G.J. IFT20 is required for opsin trafficking and photoreceptor outer segment development. *Mol. Biol. Cell* **2011**, *22*, 921–930. [CrossRef]

124. Resh, M.D. Trafficking and signaling by fatty-acylated and prenylated proteins. *Nat. Chem. Biol.* **2006**, *2*, 584–590. [CrossRef]

125. Schwarz, N.; Hardcastle, A.J.; Cheetham, M.E. Arl3 and RP2 mediated assembly and traffic of membrane associated cilia proteins. *Vis. Res.* **2012**, *75*, 2–4. [CrossRef] [PubMed]

126. Li, L.; Khan, N.; Hurd, T.; Ghosh, A.K.; Cheng, C.; Molday, R.; Heckenlively, J.R.; Swaroop, A.; Khanna, H. Ablation of the X-linked retinitis pigmentosa 2 (Rp2) gene in mice results in opsin mislocalization and photoreceptor degeneration. *Investig. Ophthalmol. Vis. Sci.* **2013**, *54*, 4503–4511. [CrossRef] [PubMed]

127. Zhang, H.; Hanke-Gogokhia, C.; Jiang, L.; Li, X.; Wang, P.; Gerstner, C.D.; Frederick, J.M.; Yang, Z.; Baehr, W. Mistrafficking of prenylated proteins causes retinitis pigmentosa 2. *FASEB J.* **2015**, *29*, 932–942. [CrossRef] [PubMed]

128. Wright, Z.C.; Singh, R.K.; Alpino, R.; Goldberg, A.F.; Sokolov, M.; Ramamurthy, V. ARL3 regulates trafficking of prenylated phototransduction proteins to the rod outer segment. *Hum. Mol. Genet.* **2016**, *25*, 2031–2044. [CrossRef]

129. Schrick, J.J.; Vogel, P.; Abuin, A.; Hampton, B.; Rice, D.S. ADP-ribosylation factor-like 3 is involved in kidney and photoreceptor development. *Am. J. Pathol.* **2006**, *168*, 1288–1298. [CrossRef]

130. Rao, K.N.; Zhang, W.; Li, L.; Anand, M.; Khanna, H. Prenylated retinal ciliopathy protein RPGR interacts with PDE6delta and regulates ciliary localization of Joubert syndrome-associated protein INPP5E. *Hum. Mol. Genet.* **2016**, *25*, 4533–4545. [CrossRef]

131. Xu, W.; Jin, M.; Hu, R.; Wang, H.; Zhang, F.; Yuan, S.; Cao, Y. The Joubert Syndrome Protein Inpp5e Controls Ciliogenesis by Regulating Phosphoinositides at the Apical Membrane. *J. Am. Soc. Nephrol.* **2017**, *28*, 118–129. [CrossRef]

132. Gillespie, P.G.; Prusti, R.K.; Apel, E.D.; Beavo, J.A. A soluble form of bovine rod photoreceptor phosphodiesterase has a novel 15-kDa subunit. *J. Biol. Chem.* **1989**, *264*, 12187–12193.

133. Thompson, D.A.; Khan, N.W.; Othman, M.I.; Chang, B.; Jia, L.; Grahek, G.; Wu, Z.; Hiriyanna, S.; Nellissery, J.; Li, T.; et al. Rd9 is a naturally occurring mouse model of a common form of retinitis pigmentosa caused by mutations in RPGR-ORF15. *PLoS ONE* **2012**, *7*, e35865. [CrossRef]

134. Hong, D.H.; Pawlyk, B.S.; Shang, J.; Sandberg, M.A.; Berson, E.L.; Li, T. A retinitis pigmentosa GTPase regulator (RPGR)-deficient mouse model for X-linked retinitis pigmentosa (RP3). *Proc. Natl. Acad. Sci. USA* **2000**, *97*, 3649–3654. [CrossRef]

135. Zhang, H.; Li, S.; Doan, T.; Rieke, F.; Detwiler, P.B.; Frederick, J.M.; Baehr, W. Deletion of PrBP/delta impedes transport of GRK1 and PDE6 catalytic subunits to photoreceptor outer segments. *Proc. Natl. Acad. Sci. USA* **2007**, *104*, 8857–8862. [CrossRef] [PubMed]

136. Ramamurthy, V.; Niemi, G.A.; Reh, T.A.; Hurley, J.B. Leber congenital amaurosis linked to AIPL1: A mouse model reveals destabilization of cGMP phosphodiesterase. *Proc. Natl. Acad. Sci. USA* **2004**, *101*, 13897–13902. [CrossRef] [PubMed]

137. Yadav, R.P.; Artemyev, N.O. AIPL1: A specialized chaperone for the phototransduction effector. *Cell Signal.* **2017**, *40*, 183–189. [CrossRef] [PubMed]

138. Scheidecker, S.; Etard, C.; Pierce, N.W.; Geoffroy, V.; Schaefer, E.; Muller, J.; Chennen, K.; Flori, E.; Pelletier, V.; Poch, O.; et al. Exome sequencing of Bardet-Biedl syndrome patient identifies a null mutation in the BBSome subunit BBIP1 (BBS18). *J. Med. Genet.* **2014**, *51*, 132–136. [CrossRef] [PubMed]

139. Eichers, E.R.; Abd-El-Barr, M.M.; Paylor, R.; Lewis, R.A.; Bi, W.; Lin, X.; Meehan, T.P.; Stockton, D.W.; Wu, S.M.; Lindsay, E.; et al. Phenotypic characterization of Bbs4 null mice reveals age-dependent penetrance and variable expressivity. *Hum. Genet.* **2006**, *120*, 211–226. [CrossRef] [PubMed]

140. Mykytyn, K.; Mullins, R.F.; Andrews, M.; Chiang, A.P.; Swiderski, R.E.; Yang, B.; Braun, T.; Casavant, T.; Stone, E.M.; Sheffield, V.C. Bardet-Biedl syndrome type 4 (BBS4)-null mice implicate Bbs4 in flagella formation but not global cilia assembly. *Proc. Natl. Acad. Sci. USA* **2004**, *101*, 8664–8669. [CrossRef] [PubMed]

141. Koch, K.W.; Dell'Orco, D. Protein and Signaling Networks in Vertebrate Photoreceptor Cells. *Front. Mol. Neurosci.* **2015**, *8*, 67. [CrossRef]

142. Pinto, L.H.; Vitaterna, M.H.; Shimomura, K.; Siepka, S.M.; McDearmon, E.L.; Fenner, D.; Lumayag, S.L.; Omura, C.; Andrews, A.W.; Baker, M.; et al. Generation, characterization, and molecular cloning of the Noerg-1 mutation of rhodopsin in the mouse. *Vis. Neurosci.* **2005**, *22*, 619–629. [CrossRef]

143. Liu, H.; Wang, M.; Xia, C.H.; Du, X.; Flannery, J.G.; Ridge, K.D.; Beutler, B.; Gong, X. Severe retinal degeneration caused by a novel rhodopsin mutation. *Investig. Ophthalmol. Vis. Sci.* **2010**, *51*, 1059–1065. [CrossRef]

144. Sakami, S.; Maeda, T.; Bereta, G.; Okano, K.; Golczak, M.; Sumaroka, A.; Roman, A.J.; Cideciyan, A.V.; Jacobson, S.G.; Palczewski, K. Probing mechanisms of photoreceptor degeneration in a new mouse model of the common form of autosomal dominant retinitis pigmentosa due to P23H opsin mutations. *J. Biol. Chem.* **2011**, *286*, 10551–10567. [CrossRef]

145. Chiang, W.C.; Kroeger, H.; Sakami, S.; Messah, C.; Yasumura, D.; Matthes, M.T.; Coppinger, J.A.; Palczewski, K.; LaVail, M.M.; Lin, J.H. Robust Endoplasmic Reticulum-Associated Degradation of Rhodopsin Precedes Retinal Degeneration. *Mol. Neurobiol.* **2015**, *52*, 679–695. [CrossRef] [PubMed]

146. Comitato, A.; Schiroli, D.; Montanari, M.; Marigo, V. Calpain Activation Is the Major Cause of Cell Death in Photoreceptors Expressing a Rhodopsin Misfolding Mutation. *Mol. Neurobiol.* **2020**, *57*, 589–599. [CrossRef] [PubMed]

147. Athanasiou, D.; Aguila, M.; Bevilacqua, D.; Novoselov, S.S.; Parfitt, D.A.; Cheetham, M.E. The cell stress machinery and retinal degeneration. *FEBS Lett.* **2013**, *587*, 2008–2017. [CrossRef] [PubMed]

148. Athanasiou, D.; Aguila, M.; Bellingham, J.; Kanuga, N.; Adamson, P.; Cheetham, M.E. The role of the ER stress-response protein PERK in rhodopsin retinitis pigmentosa. *Hum. Mol. Genet.* **2017**, *26*, 4896–4905. [CrossRef]

149. Zhang, N.; Kolesnikov, A.V.; Jastrzebska, B.; Mustafi, D.; Sawada, O.; Maeda, T.; Genoud, C.; Engel, A.; Kefalov, V.J.; Palczewski, K. Autosomal recessive retinitis pigmentosa E150K opsin mice exhibit photoreceptor disorganization. *J. Clin. Investig.* **2013**, *123*, 121–137. [CrossRef]

150. Sakami, S.; Kolesnikov, A.V.; Kefalov, V.J.; Palczewski, K. P23H opsin knock-in mice reveal a novel step in retinal rod disc morphogenesis. *Hum. Mol. Genet.* **2014**, *23*, 1723–1741. [CrossRef]

151. Hollingsworth, T.J.; Gross, A.K. The severe autosomal dominant retinitis pigmentosa rhodopsin mutant Ter349Glu mislocalizes and induces rapid rod cell death. *J. Biol. Chem.* **2013**, *288*, 29047–29055. [CrossRef]

152. Humphries, M.M.; Rancourt, D.; Farrar, G.J.; Kenna, P.; Hazel, M.; Bush, R.A.; Sieving, P.A.; Sheils, D.M.; McNally, N.; Creighton, P.; et al. Retinopathy induced in mice by targeted disruption of the rhodopsin gene. *Nat. Genet.* **1997**, *15*, 216–219. [CrossRef]

153. Lem, J.; Krasnoperova, N.V.; Calvert, P.D.; Kosaras, B.; Cameron, D.A.; Nicolo, M.; Makino, C.L.; Sidman, R.L. Morphological, physiological, and biochemical changes in rhodopsin knockout mice. *Proc. Natl. Acad. Sci. USA* **1999**, *96*, 736–741. [CrossRef]

154. Faber, S.; Roepman, R. Balancing the Photoreceptor Proteome: Proteostasis Network Therapeutics for Inherited Retinal Disease. *Genes* **2019**, *10*, 557. [CrossRef]

155. Sancho-Pelluz, J.; Cui, X.; Lee, W.; Tsai, Y.T.; Wu, W.H.; Justus, S.; Washington, I.; Hsu, C.W.; Park, K.S.; Koch, S.; et al. Mechanisms of neurodegeneration in a preclinical autosomal dominant retinitis pigmentosa knock-in model with a Rho(D190N) mutation. *Cell Mol. Life Sci.* **2019**, *76*, 3657–3665. [CrossRef] [PubMed]

156. Park, P.S. Constitutively active rhodopsin and retinal disease. *Adv. Pharmacol.* **2014**, *70*, 1–36. [CrossRef] [PubMed]

157. Budzynski, E.; Gross, A.K.; McAlear, S.D.; Peachey, N.S.; Shukla, M.; He, F.; Edwards, M.; Won, J.; Hicks, W.L.; Wensel, T.G.; et al. Mutations of the opsin gene (Y102H and I307N) lead to light-induced degeneration of photoreceptors and constitutive activation of phototransduction in mice. *J. Biol. Chem.* **2010**, *285*, 14521–14533. [CrossRef] [PubMed]

158. Daniele, L.L.; Insinna, C.; Chance, R.; Wang, J.; Nikonov, S.S.; Pugh, E.N., Jr. A mouse M-opsin monochromat: Retinal cone photoreceptors have increased M-opsin expression when S-opsin is knocked out. *Vis. Res.* **2011**, *51*, 447–458. [CrossRef] [PubMed]

159. Zhang, Y.; Deng, W.T.; Du, W.; Zhu, P.; Li, J.; Xu, F.; Sun, J.; Gerstner, C.D.; Baehr, W.; Boye, S.L.; et al. Gene-based Therapy in a Mouse Model of Blue Cone Monochromacy. *Sci. Rep.* **2017**, *7*, 6690. [CrossRef] [PubMed]

160. Carter-Dawson, L.D.; LaVail, M.M.; Sidman, R.L. Differential effect of the rd mutation on rods and cones in the mouse retina. *Investig. Ophthalmol. Vis. Sci.* **1978**, *17*, 489–498. [PubMed]

161. Calvert, P.D.; Krasnoperova, N.V.; Lyubarsky, A.L.; Isayama, T.; Nicolo, M.; Kosaras, B.; Wong, G.; Gannon, K.S.; Margolskee, R.F.; Sidman, R.L.; et al. Phototransduction in transgenic mice after targeted deletion of the rod transducin alpha -subunit. *Proc. Natl. Acad. Sci. USA* **2000**, *97*, 13913–13918. [CrossRef]

162. Barber, A.C.; Hippert, C.; Duran, Y.; West, E.L.; Bainbridge, J.W.; Warre-Cornish, K.; Luhmann, U.F.; Lakowski, J.; Sowden, J.C.; Ali, R.R.; et al. Repair of the degenerate retina by photoreceptor transplantation. *Proc. Natl. Acad. Sci. USA* **2013**, *110*, 354–359. [CrossRef]

163. Miyamoto, M.; Aoki, M.; Sugimoto, S.; Kawasaki, K.; Imai, R. IRD1 and IRD2 mice, naturally occurring models of hereditary retinal dysfunction, show late-onset and progressive retinal degeneration. *Curr. Eye Res.* **2010**, *35*, 137–145. [CrossRef]

164. Mejecase, C.; Laurent-Coriat, C.; Mayer, C.; Poch, O.; Mohand-Said, S.; Prevot, C.; Antonio, A.; Boyard, F.; Condroyer, C.; Michiels, C.; et al. Identification of a Novel Homozygous Nonsense Mutation Confirms the Implication of GNAT1 in Rod-Cone Dystrophy. *PLoS ONE* **2016**, *11*, e0168271. [CrossRef]

165. Carrigan, M.; Duignan, E.; Humphries, P.; Palfi, A.; Kenna, P.F.; Farrar, G.J. A novel homozygous truncating GNAT1 mutation implicated in retinal degeneration. *Br. J. Ophthalmol.* **2016**, *100*, 495–500. [CrossRef] [PubMed]

166. Zenteno, J.C.; Garcia-Montano, L.A.; Cruz-Aguilar, M.; Ronquillo, J.; Rodas-Serrano, A.; Aguilar-Castul, L.; Matsui, R.; Vencedor-Meraz, C.I.; Arce-Gonzalez, R.; Graue-Wiechers, F.; et al. Extensive genic and allelic heterogeneity underlying inherited retinal dystrophies in Mexican patients molecularly analyzed by next-generation sequencing. *Mol. Genet. Genomic Med.* **2020**, *8*. [CrossRef] [PubMed]

167. Deng, W.T.; Sakurai, K.; Liu, J.; Dinculescu, A.; Li, J.; Pang, J.; Min, S.H.; Chiodo, V.A.; Boye, S.L.; Chang, B.; et al. Functional interchangeability of rod and cone transducin alpha-subunits. *Proc. Natl. Acad. Sci. USA* **2009**, *106*, 17681–17686. [CrossRef] [PubMed]

168. Chang, B.; Hawes, N.L.; Hurd, R.E.; Wang, J.; Davisson, M.T.; Nusinowitz, S.; Heckenlively, J.R. A New Mouse Model of Retinal Degeneration (rd17). Proceedings of ARVO Annual Meeting Abstract, Fort Lauderdale, FL, USA, 6–10 May 2007.

169. Lobanova, E.S.; Finkelstein, S.; Herrmann, R.; Chen, Y.M.; Kessler, C.; Michaud, N.A.; Trieu, L.H.; Strissel, K.J.; Burns, M.E.; Arshavsky, V.Y. Transducin gamma-subunit sets expression levels of alpha- and beta-subunits and is crucial for rod viability. *J. Neurosci.* **2008**, *28*, 3510–3520. [CrossRef] [PubMed]

170. Kolesnikov, A.V.; Rikimaru, L.; Hennig, A.K.; Lukasiewicz, P.D.; Fliesler, S.J.; Govardovskii, V.I.; Kefalov, V.J.; Kisselev, O.G. G-protein betagamma-complex is crucial for efficient signal amplification in vision. *J. Neurosci.* **2011**, *31*, 8067–8077. [CrossRef]

171. Lobanova, E.S.; Finkelstein, S.; Skiba, N.P.; Arshavsky, V.Y. Proteasome overload is a common stress factor in multiple forms of inherited retinal degeneration. *Proc. Natl. Acad. Sci. USA* **2013**, *110*, 9986–9991. [CrossRef]

172. Jobling, A.I.; Vessey, K.A.; Waugh, M.; Mills, S.A.; Fletcher, E.L. A naturally occurring mouse model of achromatopsia: Characterization of the mutation in cone transducin and subsequent retinal phenotype. *Investig. Ophthalmol. Vis. Sci.* **2013**, *54*, 3350–3359. [CrossRef]

173. Chang, B.; Dacey, M.S.; Hawes, N.L.; Hitchcock, P.F.; Milam, A.H.; Atmaca-Sonmez, P.; Nusinowitz, S.; Heckenlively, J.R. Cone photoreceptor function loss-3, a novel mouse model of achromatopsia due to a mutation in Gnat2. *Investig. Ophthalmol. Vis. Sci.* **2006**, *47*, 5017–5021. [CrossRef]

174. Ronning, K.E.; Allina, G.P.; Miller, E.B.; Zawadzki, R.J.; Pugh, E.N., Jr.; Herrmann, R.; Burns, M.E. Loss of cone function without degeneration in a novel Gnat2 knock-out mouse. *Exp. Eye Res.* **2018**, *171*, 111–118. [CrossRef]

175. Hirji, N.; Aboshiha, J.; Georgiou, M.; Bainbridge, J.; Michaelides, M. Achromatopsia: Clinical features, molecular genetics, animal models and therapeutic options. *Ophthalmic Genet.* **2018**, *39*, 149–157. [CrossRef]

176. Michaelides, M.; Aligianis, I.A.; Holder, G.E.; Simunovic, M.; Mollon, J.D.; Maher, E.R.; Hunt, D.M.; Moore, A.T. Cone dystrophy phenotype associated with a frameshift mutation (M280fsX291) in the alpha-subunit of cone specific transducin (GNAT2). *Br. J. Ophthalmol.* **2003**, *87*, 1317–1320. [CrossRef]

177. Du, J.; An, J.; Linton, J.D.; Wang, Y.; Hurley, J.B. How Excessive cGMP Impacts Metabolic Proteins in Retinas at the Onset of Degeneration. *Adv. Exp. Med. Biol.* **2018**, *1074*, 289–295. [CrossRef]

178. Tolone, A.; Belhadj, S.; Rentsch, A.; Schwede, F.; Paquet-Durand, F. The cGMP Pathway and Inherited Photoreceptor Degeneration: Targets, Compounds, and Biomarkers. *Genes* **2019**, *10*, 453. [CrossRef] [PubMed]

179. Sothilingam, V.; Garcia Garrido, M.; Jiao, K.; Buena-Atienza, E.; Sahaboglu, A.; Trifunovic, D.; Balendran, S.; Koepfli, T.; Muhlfriedel, R.; Schon, C.; et al. Retinitis pigmentosa: Impact of different Pde6a point mutations on the disease phenotype. *Hum. Mol. Genet.* **2015**, *24*, 5486–5499. [CrossRef]

180. Sakamoto, K.; McCluskey, M.; Wensel, T.G.; Naggert, J.K.; Nishina, P.M. New mouse models for recessive retinitis pigmentosa caused by mutations in the Pde6a gene. *Hum. Mol. Genet.* **2009**, *18*, 178–192. [CrossRef]

181. Power, M.; Das, S.; Schutze, K.; Marigo, V.; Ekstrom, P.; Paquet-Durand, F. Cellular mechanisms of hereditary photoreceptor degeneration—Focus on cGMP. *Prog. Retin. Eye Res.* **2020**, *74*, 100772. [CrossRef] [PubMed]

182. Tsang, S.H.; Gouras, P.; Yamashita, C.K.; Kjeldbye, H.; Fisher, J.; Farber, D.B.; Goff, S.P. Retinal degeneration in mice lacking the gamma subunit of the rod cGMP phosphodiesterase. *Science* **1996**, *272*, 1026–1029. [CrossRef] [PubMed]

183. Thiadens, A.A.; Somervuo, V.; van den Born, L.I.; Roosing, S.; van Schooneveld, M.J.; Kuijpers, R.W.; van Moll-Ramirez, N.; Cremers, F.P.; Hoyng, C.B.; Klaver, C.C. Progressive loss of cones in achromatopsia: An imaging study using spectral-domain optical coherence tomography. *Investig. Ophthalmol. Vis. Sci.* **2010**, *51*, 5952–5957. [CrossRef] [PubMed]

184. Brennenstuhl, C.; Tanimoto, N.; Burkard, M.; Wagner, R.; Bolz, S.; Trifunovic, D.; Kabagema-Bilan, C.; Paquet-Durand, F.; Beck, S.C.; Huber, G.; et al. Targeted ablation of the Pde6h gene in mice reveals cross-species differences in cone and rod phototransduction protein isoform inventory. *J. Biol. Chem.* **2015**, *290*, 10242–10255. [CrossRef]

185. Kohl, S.; Coppieters, F.; Meire, F.; Schaich, S.; Roosing, S.; Brennenstuhl, C.; Bolz, S.; van Genderen, M.M.; Riemslag, F.C.; European Retinal Disease, C.; et al. A nonsense mutation in PDE6H causes autosomal-recessive incomplete achromatopsia. *Am. J. Hum. Genet.* **2012**, *91*, 527–532. [CrossRef]

186. Pedurupillay, C.R.; Landsend, E.C.; Vigeland, M.D.; Ansar, M.; Frengen, E.; Misceo, D.; Stromme, P. Segregation of Incomplete Achromatopsia and Alopecia Due to PDE6H and LPAR6 Variants in a Consanguineous Family from Pakistan. *Genes* **2016**, *7*, 41. [CrossRef]

187. Weitz, D.; Ficek, N.; Kremmer, E.; Bauer, P.J.; Kaupp, U.B. Subunit stoichiometry of the CNG channel of rod photoreceptors. *Neuron* **2002**, *36*, 881–889. [CrossRef]

188. Zheng, J.; Trudeau, M.C.; Zagotta, W.N. Rod cyclic nucleotide-gated channels have a stoichiometry of three CNGA1 subunits and one CNGB1 subunit. *Neuron* **2002**, *36*, 891–896. [CrossRef]

189. Huttl, S.; Michalakis, S.; Seeliger, M.; Luo, D.G.; Acar, N.; Geiger, H.; Hudl, K.; Mader, R.; Haverkamp, S.; Moser, M.; et al. Impaired channel targeting and retinal degeneration in mice lacking the cyclic nucleotide-gated channel subunit CNGB1. *J. Neurosci.* **2005**, *25*, 130–138. [CrossRef] [PubMed]

190. Vinberg, F.; Wang, T.; Molday, R.S.; Chen, J.; Kefalov, V.J. A new mouse model for stationary night blindness with mutant Slc24a1 explains the pathophysiology of the associated human disease. *Hum. Mol. Genet.* **2015**, *24*, 5915–5929. [CrossRef] [PubMed]

191. Yang, R.B.; Robinson, S.W.; Xiong, W.H.; Yau, K.W.; Birch, D.G.; Garbers, D.L. Disruption of a retinal guanylyl cyclase gene leads to cone-specific dystrophy and paradoxical rod behavior. *J. Neurosci.* **1999**, *19*, 5889–5897. [CrossRef] [PubMed]

192. Bouzia, Z.; Georgiou, M.; Hull, S.; Robson, A.G.; Fujinami, K.; Rotsos, T.; Pontikos, N.; Arno, G.; Webster, A.R.; Hardcastle, A.J.; et al. GUCY2D-Associated Leber Congenital Amaurosis: A Retrospective Natural History Study in Preparation for Trials of Novel Therapies. *Am. J. Ophthalmol.* **2020**, *210*, 59–70. [CrossRef]

193. Baehr, W.; Karan, S.; Maeda, T.; Luo, D.G.; Li, S.; Bronson, J.D.; Watt, C.B.; Yau, K.W.; Frederick, J.M.; Palczewski, K. The function of guanylate cyclase 1 and guanylate cyclase 2 in rod and cone photoreceptors. *J. Biol. Chem.* **2007**, *282*, 8837–8847. [CrossRef]

194. Mendez, A.; Burns, M.E.; Sokal, I.; Dizhoor, A.M.; Baehr, W.; Palczewski, K.; Baylor, D.A.; Chen, J. Role of guanylate cyclase-activating proteins (GCAPs) in setting the flash sensitivity of rod photoreceptors. *Proc. Natl. Acad. Sci. USA* **2001**, *98*, 9948–9953. [CrossRef]

195. Buch, P.K.; Mihelec, M.; Cottrill, P.; Wilkie, S.E.; Pearson, R.A.; Duran, Y.; West, E.L.; Michaelides, M.; Ali, R.R.; Hunt, D.M. Dominant cone-rod dystrophy: A mouse model generated by gene targeting of the GCAP1/Guca1a gene. *PLoS ONE* **2011**, *6*, e18089. [CrossRef]

196. Marino, V.; Dal Cortivo, G.; Oppici, E.; Maltese, P.E.; D'Esposito, F.; Manara, E.; Ziccardi, L.; Falsini, B.; Magli, A.; Bertelli, M.; et al. A novel p.(Glu111Val) missense mutation in GUCA1A associated with cone-rod dystrophy leads to impaired calcium sensing and perturbed second messenger homeostasis in photoreceptors. *Hum. Mol. Genet.* **2018**, *27*, 4204–4217. [CrossRef] [PubMed]

197. Chen, C.K.; Burns, M.E.; Spencer, M.; Niemi, G.A.; Chen, J.; Hurley, J.B.; Baylor, D.A.; Simon, M.I. Abnormal photoresponses and light-induced apoptosis in rods lacking rhodopsin kinase. *Proc. Natl. Acad. Sci. USA* **1999**, *96*, 3718–3722. [CrossRef] [PubMed]

198. Xu, J.; Dodd, R.L.; Makino, C.L.; Simon, M.I.; Baylor, D.A.; Chen, J. Prolonged photoresponses in transgenic mouse rods lacking arrestin. *Nature* **1997**, *389*, 505–509. [CrossRef] [PubMed]

199. Chen, J.; Simon, M.I.; Matthes, M.T.; Yasumura, D.; LaVail, M.M. Increased susceptibility to light damage in an arrestin knockout mouse model of Oguchi disease (stationary night blindness). *Investig. Ophthalmol. Vis. Sci.* **1999**, *40*, 2978–2982. [PubMed]

200. Charette, J.R.; Samuels, I.S.; Yu, M.; Stone, L.; Hicks, W.; Shi, L.Y.; Krebs, M.P.; Naggert, J.K.; Nishina, P.M.; Peachey, N.S. A Chemical Mutagenesis Screen Identifies Mouse Models with ERG Defects. *Adv. Exp. Med. Biol.* **2016**, *854*, 177–183. [CrossRef] [PubMed]

201. Rajappa, M.; Goyal, A.; Kaur, J. Inherited metabolic disorders involving the eye: A clinico-biochemical perspective. *Eye* **2010**, *24*, 507–518. [CrossRef] [PubMed]

202. Poll-The, B.T.; Maillette de Buy Wenniger-Prick, C.J. The eye in metabolic diseases: Clues to diagnosis. *Eur. J. Paediatr. Neurol.* **2011**, *15*, 197–204. [CrossRef]

203. Wright, A.F.; Chakarova, C.F.; Abd El-Aziz, M.M.; Bhattacharya, S.S. Photoreceptor degeneration: Genetic and mechanistic dissection of a complex trait. *Nat. Rev. Genet.* **2010**, *11*, 273–284. [CrossRef]

204. Fliesler, S.J.; Anderson, R.E. Chemistry and metabolism of lipids in the vertebrate retina. *Prog. Lipid Res.* **1983**, *22*, 79–131. [CrossRef]

205. Giusto, N.M.; Pasquare, S.J.; Salvador, G.A.; Ilincheta de Boschero, M.G. Lipid second messengers and related enzymes in vertebrate rod outer segments. *J. Lipid Res.* **2010**, *51*, 685–700. [CrossRef]

206. Niu, S.L.; Mitchell, D.C.; Litman, B.J. Manipulation of cholesterol levels in rod disk membranes by methyl-beta-cyclodextrin: Effects on receptor activation. *J. Biol. Chem.* **2002**, *277*, 20139–20145. [CrossRef] [PubMed]

207. Bretillon, L.; Thuret, G.; Gregoire, S.; Acar, N.; Joffre, C.; Bron, A.M.; Gain, P.; Creuzot-Garcher, C.P. Lipid and fatty acid profile of the retina, retinal pigment epithelium/choroid, and the lacrimal gland, and associations with adipose tissue fatty acids in human subjects. *Exp. Eye Res.* **2008**, *87*, 521–528. [CrossRef]

208. Fliesler, S.J.; Bretillon, L. The ins and outs of cholesterol in the vertebrate retina. *J. Lipid Res.* **2010**, *51*, 3399–3413. [CrossRef]

209. German, O.L.; Agnolazza, D.L.; Politi, L.E.; Rotstein, N.P. Light, lipids and photoreceptor survival: Live or let die? *Photochem. Photobiol. Sci.* **2015**, *14*, 1737–1753. [CrossRef] [PubMed]

210. Shindou, H.; Koso, H.; Sasaki, J.; Nakanishi, H.; Sagara, H.; Nakagawa, K.M.; Takahashi, Y.; Hishikawa, D.; Iizuka-Hishikawa, Y.; Tokumasu, F.; et al. Docosahexaenoic acid preserves visual function by maintaining correct disc morphology in retinal photoreceptor cells. *J. Biol. Chem.* **2017**, *292*, 12054–12064. [CrossRef] [PubMed]

211. Lobanova, E.S.; Schuhmann, K.; Finkelstein, S.; Lewis, T.R.; Cady, M.A.; Hao, Y.; Keuthan, C.; Ash, J.D.; Burns, M.E.; Shevchenko, A.; et al. Disrupted Blood-Retina Lysophosphatidylcholine Transport Impairs Photoreceptor Health But Not Visual Signal Transduction. *J. Neurosci.* **2019**, *39*, 9689–9701. [CrossRef] [PubMed]

212. Pham, T.L.; He, J.; Kakazu, A.H.; Jun, B.; Bazan, N.G.; Bazan, H.E.P. Defining a mechanistic link between pigment epithelium-derived factor, docosahexaenoic acid, and corneal nerve regeneration. *J. Biol. Chem.* **2017**, *292*, 18486–18499. [CrossRef]

213. Comitato, A.; Subramanian, P.; Turchiano, G.; Montanari, M.; Becerra, S.P.; Marigo, V. Pigment epithelium-derived factor hinders photoreceptor cell death by reducing intracellular calcium in the degenerating retina. *Cell Death Dis.* **2018**, *9*, 560. [CrossRef]

214. Bernstein, P.S.; Tammur, J.; Singh, N.; Hutchinson, A.; Dixon, M.; Pappas, C.M.; Zabriskie, N.A.; Zhang, K.; Petrukhin, K.; Leppert, M.; et al. Diverse macular dystrophy phenotype caused by a novel complex mutation in the ELOVL4 gene. *Investig. Ophthalmol. Vis. Sci.* **2001**, *42*, 3331–3336.

215. Vasireddy, V.; Jablonski, M.M.; Mandal, M.N.; Raz-Prag, D.; Wang, X.F.; Nizol, L.; Iannaccone, A.; Musch, D.C.; Bush, R.A.; Salem, N., Jr.; et al. Elovl4 5-bp-deletion knock-in mice develop progressive photoreceptor degeneration. *Investig. Ophthalmol. Vis. Sci.* **2006**, *47*, 4558–4568. [CrossRef]

216. Friedman, J.S.; Chang, B.; Krauth, D.S.; Lopez, I.; Waseem, N.H.; Hurd, R.E.; Feathers, K.L.; Branham, K.E.; Shaw, M.; Thomas, G.E.; et al. Loss of lysophosphatidylcholine acyltransferase 1 leads to photoreceptor degeneration in rd11 mice. *Proc. Natl. Acad. Sci. USA* **2010**, *107*, 15523–15528. [CrossRef] [PubMed]

217. Perkovic, T.; Duh, D.; Peterlin, B.; Gregoric, J. The Str mouse as a model for incontinentia pigmenti. *Pflugers Arch.* **2000**, *440*, R53–R54. [CrossRef] [PubMed]

218. Coleman, J.A.; Zhu, X.; Djajadi, H.R.; Molday, L.L.; Smith, R.S.; Libby, R.T.; John, S.W.; Molday, R.S. Phospholipid flippase ATP8A2 is required for normal visual and auditory function and photoreceptor and spiral ganglion cell survival. *J. Cell Sci.* **2014**, *127*, 1138–1149. [CrossRef] [PubMed]

219. Bryde, S.; Hennrich, H.; Verhulst, P.M.; Devaux, P.F.; Lenoir, G.; Holthuis, J.C. CDC50 proteins are critical components of the human class-1 P4-ATPase transport machinery. *J. Biol. Chem.* **2010**, *285*, 40562–40572. [CrossRef]

220. van der Velden, L.M.; Wichers, C.G.; van Breevoort, A.E.; Coleman, J.A.; Molday, R.S.; Berger, R.; Klomp, L.W.; van de Graaf, S.F. Heteromeric interactions required for abundance and subcellular localization of human CDC50 proteins and class 1 P4-ATPases. *J. Biol. Chem.* **2010**, *285*, 40088–40096. [CrossRef]

221. Zhang, L.; Yang, Y.; Li, S.; Zhang, S.; Zhu, X.; Tai, Z.; Yang, M.; Liu, Y.; Guo, X.; Chen, B.; et al. Loss of Tmem30a leads to photoreceptor degeneration. *Sci. Rep.* **2017**, *7*, 9296. [CrossRef]

222. Wong-Riley, M.T. Energy metabolism of the visual system. *Eye Brain* **2010**, *2*, 99–116. [CrossRef] [PubMed]

223. Du, J.; Rountree, A.; Cleghorn, W.M.; Contreras, L.; Lindsay, K.J.; Sadilek, M.; Gu, H.; Djukovic, D.; Raftery, D.; Satrustegui, J.; et al. Phototransduction Influences Metabolic Flux and Nucleotide Metabolism in Mouse Retina. *J. Biol. Chem.* **2016**, *291*, 4698–4710. [CrossRef]

224. Joyal, J.S.; Sun, Y.; Gantner, M.L.; Shao, Z.; Evans, L.P.; Saba, N.; Fredrick, T.; Burnim, S.; Kim, J.S.; Patel, G.; et al. Retinal lipid and glucose metabolism dictates angiogenesis through the lipid sensor Ffar1. *Nat. Med.* **2016**, *22*, 439–445. [CrossRef]

225. Zhang, L.; Sun, Z.; Zhao, P.; Huang, L.; Xu, M.; Yang, Y.; Chen, X.; Lu, F.; Zhang, X.; Wang, H.; et al. Whole-exome sequencing revealed HKDC1 as a candidate gene associated with autosomal-recessive retinitis pigmentosa. *Hum. Mol. Genet.* **2018**, *27*, 4157–4168. [CrossRef]

226. Xia, C.H.; Lu, E.; Liu, H.; Du, X.; Beutler, B.; Gong, X. The role of Vldlr in intraretinal angiogenesis in mice. *Investig. Ophthalmol. Vis. Sci.* **2011**, *52*, 6572–6579. [CrossRef] [PubMed]

227. Hu, W.; Jiang, A.; Liang, J.; Meng, H.; Chang, B.; Gao, H.; Qiao, X. Expression of VLDLR in the retina and evolution of subretinal neovascularization in the knockout mouse model's retinal angiomatous proliferation. *Investig. Ophthalmol. Vis. Sci.* **2008**, *49*, 407–415. [CrossRef] [PubMed]

228. Chen, Y.; Hu, Y.; Moiseyev, G.; Zhou, K.K.; Chen, D.; Ma, J.X. Photoreceptor degeneration and retinal inflammation induced by very low-density lipoprotein receptor deficiency. *Microvasc. Res.* **2009**, *78*, 119–127. [CrossRef] [PubMed]

229. Lin, J.B.; Kubota, S.; Ban, N.; Yoshida, M.; Santeford, A.; Sene, A.; Nakamura, R.; Zapata, N.; Kubota, M.; Tsubota, K.; et al. NAMPT-Mediated NAD(+) Biosynthesis Is Essential for Vision In Mice. *Cell Rep.* **2016**, *17*, 69–85. [CrossRef]

230. Greenwald, S.H.; Charette, J.R.; Staniszewska, M.; Shi, L.Y.; Brown, S.D.M.; Stone, L.; Liu, Q.; Hicks, W.L.; Collin, G.B.; Bowl, M.R.; et al. Mouse Models of NMNAT1-Leber Congenital Amaurosis (LCA9) Recapitulate Key Features of the Human Disease. *Am. J. Pathol.* **2016**, *186*, 1925–1938. [CrossRef]

231. Bosl, M.R.; Stein, V.; Hubner, C.; Zdebik, A.A.; Jordt, S.E.; Mukhopadhyay, A.K.; Davidoff, M.S.; Holstein, A.F.; Jentsch, T.J. Male germ cells and photoreceptors, both dependent on close cell-cell interactions, degenerate upon ClC-2 Cl(-) channel disruption. *EMBO J.* **2001**, *20*, 1289–1299. [CrossRef]

232. Ng, L.; Lyubarsky, A.; Nikonov, S.S.; Ma, M.; Srinivas, M.; Kefas, B.; St Germain, D.L.; Hernandez, A.; Pugh, E.N., Jr.; Forrest, D. Type 3 deiodinase, a thyroid-hormone-inactivating enzyme, controls survival and maturation of cone photoreceptors. *J. Neurosci.* **2010**, *30*, 3347–3357. [CrossRef]

233. Ng, L.; Hurley, J.B.; Dierks, B.; Srinivas, M.; Salto, C.; Vennstrom, B.; Reh, T.A.; Forrest, D. A thyroid hormone receptor that is required for the development of green cone photoreceptors. *Nat. Genet.* **2001**, *27*, 94–98. [CrossRef]

234. Gianesini, C.; Hiragaki, S.; Laurent, V.; Hicks, D.; Tosini, G. Cone Viability Is Affected by Disruption of Melatonin Receptors Signaling. *Investig. Ophthalmol. Vis. Sci.* **2016**, *57*, 94–104. [CrossRef]

235. Baba, K.; Pozdeyev, N.; Mazzoni, F.; Contreras-Alcantara, S.; Liu, C.; Kasamatsu, M.; Martinez-Merlos, T.; Strettoi, E.; Iuvone, P.M.; Tosini, G. Melatonin modulates visual function and cell viability in the mouse retina via the MT1 melatonin receptor. *Proc. Natl. Acad. Sci. USA* **2009**, *106*, 15043–15048. [CrossRef]

236. Chen, Y.; Mehta, G.; Vasiliou, V. Antioxidant defenses in the ocular surface. *Ocul. Surf.* **2009**, *7*, 176–185. [CrossRef]

237. Nita, M.; Grzybowski, A. The Role of the Reactive Oxygen Species and Oxidative Stress in the Pathomechanism of the Age-Related Ocular Diseases and Other Pathologies of the Anterior and Posterior Eye Segments in Adults. *Oxid. Med. Cell Longev.* **2016**, *2016*, 3164734. [CrossRef] [PubMed]

238. Chen, H.; Lukas, T.J.; Du, N.; Suyeoka, G.; Neufeld, A.H. Dysfunction of the retinal pigment epithelium with age: Increased iron decreases phagocytosis and lysosomal activity. *Investig. Ophthalmol. Vis. Sci.* **2009**, *50*, 1895–1902. [CrossRef] [PubMed]

239. Blasiak, J.; Glowacki, S.; Kauppinen, A.; Kaarniranta, K. Mitochondrial and nuclear DNA damage and repair in age-related macular degeneration. *Int. J. Mol. Sci.* **2013**, *14*, 2996–3010. [CrossRef]

240. Tan, B.L.; Norhaizan, M.E.; Liew, W.P.; Sulaiman Rahman, H. Antioxidant and Oxidative Stress: A Mutual Interplay in Age-Related Diseases. *Front. Pharmacol.* **2018**, *9*, 1162. [CrossRef]

241. Frohns, A.; Frohns, F.; Naumann, S.C.; Layer, P.G.; Lobrich, M. Inefficient double-strand break repair in murine rod photoreceptors with inverted heterochromatin organization. *Curr. Biol.* **2014**, *24*, 1080–1090. [CrossRef]

242. Blasiak, J.; Petrovski, G.; Vereb, Z.; Facsko, A.; Kaarniranta, K. Oxidative stress, hypoxia, and autophagy in the neovascular processes of age-related macular degeneration. *Biomed. Res. Int.* **2014**, *2014*, 768026. [CrossRef]

243. Tokarz, P.; Kaarniranta, K.; Blasiak, J. Role of antioxidant enzymes and small molecular weight antioxidants in the pathogenesis of age-related macular degeneration (AMD). *Biogerontology* **2013**, *14*, 461–482. [CrossRef]

244. Hashizume, K.; Hirasawa, M.; Imamura, Y.; Noda, S.; Shimizu, T.; Shinoda, K.; Kurihara, T.; Noda, K.; Ozawa, Y.; Ishida, S.; et al. Retinal dysfunction and progressive retinal cell death in SOD1-deficient mice. *Am. J. Pathol.* **2008**, *172*, 1325–1331. [CrossRef]

245. Biswal, M.R.; Ildefonso, C.J.; Mao, H.; Seo, S.J.; Wang, Z.; Li, H.; Le, Y.Z.; Lewin, A.S. Conditional Induction of Oxidative Stress in RPE: A Mouse Model of Progressive Retinal Degeneration. *Adv. Exp. Med. Biol.* **2016**, *854*, 31–37. [CrossRef]

246. Cronin, T.; Raffelsberger, W.; Lee-Rivera, I.; Jaillard, C.; Niepon, M.L.; Kinzel, B.; Clerin, E.; Petrosian, A.; Picaud, S.; Poch, O.; et al. The disruption of the rod-derived cone viability gene leads to photoreceptor dysfunction and susceptibility to oxidative stress. *Cell Death Differ.* **2010**, *17*, 1199–1210. [CrossRef] [PubMed]

247. Jaillard, C.; Mouret, A.; Niepon, M.L.; Clerin, E.; Yang, Y.; Lee-Rivera, I.; Ait-Ali, N.; Millet-Puel, G.; Cronin, T.; Sedmak, T.; et al. Nxnl2 splicing results in dual functions in neuronal cell survival and maintenance of cell integrity. *Hum. Mol. Genet.* **2012**, *21*, 2298–2311. [CrossRef] [PubMed]

248. Yokota, T.; Igarashi, K.; Uchihara, T.; Jishage, K.; Tomita, H.; Inaba, A.; Li, Y.; Arita, M.; Suzuki, H.; Mizusawa, H.; et al. Delayed-onset ataxia in mice lacking alpha -tocopherol transfer protein: Model for neuronal degeneration caused by chronic oxidative stress. *Proc. Natl. Acad. Sci. USA* **2001**, *98*, 15185–15190. [CrossRef] [PubMed]

249. Mukherjee, A.B.; Appu, A.P.; Sadhukhan, T.; Casey, S.; Mondal, A.; Zhang, Z.; Bagh, M.B. Emerging new roles of the lysosome and neuronal ceroid lipofuscinoses. *Mol. Neurodegener.* **2019**, *14*, 4. [CrossRef] [PubMed]

250. Schulze, H.; Kolter, T.; Sandhoff, K. Principles of lysosomal membrane degradation: Cellular topology and biochemistry of lysosomal lipid degradation. *Biochim. Biophys. Acta* **2009**, *1793*, 674–683. [CrossRef] [PubMed]

251. Birch, D.G. Retinal degeneration in retinitis pigmentosa and neuronal ceroid lipofuscinosis: An overview. *Mol. Genet. Metab.* **1999**, *66*, 356–366. [CrossRef]

252. Ostergaard, J.R. Juvenile neuronal ceroid lipofuscinosis (Batten disease): Current insights. *Degener. Neurol. Neuromuscul. Dis.* **2016**, *6*, 73–83. [CrossRef]

253. Leinonen, H.; Keksa-Goldsteine, V.; Ragauskas, S.; Kohlmann, P.; Singh, Y.; Savchenko, E.; Puranen, J.; Malm, T.; Kalesnykas, G.; Koistinaho, J.; et al. Retinal Degeneration In A Mouse Model Of CLN5 Disease Is Associated With Compromised Autophagy. *Sci. Rep.* **2017**, *7*, 1597. [CrossRef]

254. Bartsch, U.; Galliciotti, G.; Jofre, G.F.; Jankowiak, W.; Hagel, C.; Braulke, T. Apoptotic photoreceptor loss and altered expression of lysosomal proteins in the nclf mouse model of neuronal ceroid lipofuscinosis. *Investig. Ophthalmol. Vis. Sci.* **2013**, *54*, 6952–6959. [CrossRef]

255. Jankowiak, W.; Brandenstein, L.; Dulz, S.; Hagel, C.; Storch, S.; Bartsch, U. Retinal Degeneration in Mice Deficient in the Lysosomal Membrane Protein CLN7. *Investig. Ophthalmol. Vis. Sci.* **2016**, *57*, 4989–4998. [CrossRef]

256. Chang, B.; Bronson, R.T.; Hawes, N.L.; Roderick, T.H.; Peng, C.; Hageman, G.S.; Heckenlively, J.R. Retinal degeneration in motor neuron degeneration: A mouse model of ceroid lipofuscinosis. *Investig. Ophthalmol. Vis. Sci.* **1994**, *35*, 1071–1076. [PubMed]

257. Hafler, B.P.; Klein, Z.A.; Jimmy Zhou, Z.; Strittmatter, S.M. Progressive retinal degeneration and accumulation of autofluorescent lipopigments in Progranulin deficient mice. *Brain Res.* **2014**, *1588*, 168–174. [CrossRef] [PubMed]

258. Heldermon, C.D.; Hennig, A.K.; Ohlemiller, K.K.; Ogilvie, J.M.; Herzog, E.D.; Breidenbach, A.; Vogler, C.; Wozniak, D.F.; Sands, M.S. Development of sensory, motor and behavioral deficits in the murine model of Sanfilippo syndrome type B. *PLoS ONE* **2007**, *2*, e772. [CrossRef] [PubMed]

259. Gelfman, C.M.; Vogel, P.; Issa, T.M.; Turner, C.A.; Lee, W.S.; Kornfeld, S.; Rice, D.S. Mice lacking alpha/beta subunits of GlcNAc-1-phosphotransferase exhibit growth retardation, retinal degeneration, and secretory cell lesions. *Investig. Ophthalmol. Vis. Sci.* **2007**, *48*, 5221–5228. [CrossRef] [PubMed]

260. Kevany, B.M.; Palczewski, K. Phagocytosis of retinal rod and cone photoreceptors. *Physiology* **2010**, *25*, 8–15. [CrossRef] [PubMed]

261. Prasad, D.; Rothlin, C.V.; Burrola, P.; Burstyn-Cohen, T.; Lu, Q.; Garcia de Frutos, P.; Lemke, G. TAM receptor function in the retinal pigment epithelium. *Mol. Cell Neurosci.* **2006**, *33*, 96–108. [CrossRef] [PubMed]

262. Duncan, J.L.; LaVail, M.M.; Yasumura, D.; Matthes, M.T.; Yang, H.; Trautmann, N.; Chappelow, A.V.; Feng, W.; Earp, H.S.; Matsushima, G.K.; et al. An RCS-like retinal dystrophy phenotype in mer knockout mice. *Investig. Ophthalmol. Vis. Sci.* **2003**, *44*, 826–838. [CrossRef]

263. Houssier, M.; Raoul, W.; Lavalette, S.; Keller, N.; Guillonneau, X.; Baragatti, B.; Jonet, L.; Jeanny, J.C.; Behar-Cohen, F.; Coceani, F.; et al. CD36 deficiency leads to choroidal involution via COX2 down-regulation in rodents. *PLoS Med.* **2008**, *5*, e39. [CrossRef]

264. Ying, G.; Boldt, K.; Ueffing, M.; Gerstner, C.D.; Frederick, J.M.; Baehr, W. The small GTPase RAB28 is required for phagocytosis of cone outer segments by the murine retinal pigmented epithelium. *J. Biol. Chem.* **2018**, *293*, 17546–17558. [CrossRef]

265. Lin, W.; Xu, G. Autophagy: A Role in the Apoptosis, Survival, Inflammation, and Development of the Retina. *Ophthalmic Res.* **2019**, *61*, 65–72. [CrossRef]

266. Seranova, E.; Connolly, K.J.; Zatyka, M.; Rosenstock, T.R.; Barrett, T.; Tuxworth, R.I.; Sarkar, S. Dysregulation of autophagy as a common mechanism in lysosomal storage diseases. *Essays Biochem.* **2017**, *61*, 733–749. [CrossRef] [PubMed]

267. Byrne, S.; Jansen, L.; JM, U.K.-I.; Siddiqui, A.; Lidov, H.G.; Bodi, I.; Smith, L.; Mein, R.; Cullup, T.; Dionisi-Vici, C.; et al. EPG5-related Vici syndrome: A paradigm of neurodevelopmental disorders with defective autophagy. *Brain* **2016**, *139*, 765–781. [CrossRef] [PubMed]

268. Smucker, W.D.; Kontak, J.R. Adverse drug reactions causing hospital admission in an elderly population: Experience with a decision algorithm. *J. Am. Board Fam. Pract.* **1990**, *3*, 105–109. [PubMed]

269. Kim, J.Y.; Zhao, H.; Martinez, J.; Doggett, T.A.; Kolesnikov, A.V.; Tang, P.H.; Ablonczy, Z.; Chan, C.C.; Zhou, Z.; Green, D.R.; et al. Noncanonical autophagy promotes the visual cycle. *Cell* **2013**, *154*, 365–376. [CrossRef] [PubMed]

270. Mohlin, C.; Taylor, L.; Ghosh, F.; Johansson, K. Autophagy and ER-stress contribute to photoreceptor degeneration in cultured adult porcine retina. *Brain Res.* **2014**, *1585*, 167–183. [CrossRef]

271. Chen, Y.; Sawada, O.; Kohno, H.; Le, Y.Z.; Subauste, C.; Maeda, T.; Maeda, A. Autophagy protects the retina from light-induced degeneration. *J. Biol. Chem.* **2013**, *288*, 7506–7518. [CrossRef]

272. Falk, M.J. Neurodevelopmental manifestations of mitochondrial disease. *J. Dev. Behav. Pediatr.* **2010**, *31*, 610–621. [CrossRef]

273. Davies, V.J.; Powell, K.A.; White, K.E.; Yip, W.; Hogan, V.; Hollins, A.J.; Davies, J.R.; Piechota, M.; Brownstein, D.G.; Moat, S.J.; et al. A missense mutation in the murine Opa3 gene models human Costeff syndrome. *Brain* **2008**, *131*, 368–380. [CrossRef]

274. Findlay, A.S.; Carter, R.N.; Starbuck, B.; McKie, L.; Novakova, K.; Budd, P.S.; Keighren, M.A.; Marsh, J.A.; Cross, S.H.; Simon, M.M.; et al. Mouse Idh3a mutations cause retinal degeneration and reduced mitochondrial function. *Dis. Model. Mech.* **2018**, *11*, 036426. [CrossRef]

275. Bruschi, M.; Petretto, A.; Caicci, F.; Bartolucci, M.; Calzia, D.; Santucci, L.; Manni, L.; Ramenghi, L.A.; Ghiggeri, G.; Traverso, C.E.; et al. Proteome of Bovine Mitochondria and Rod Outer Segment Disks: Commonalities and Differences. *J. Proteome Res.* **2018**, *17*, 918–925. [CrossRef]

276. Calzia, D.; Barabino, S.; Bianchini, P.; Garbarino, G.; Oneto, M.; Caicci, F.; Diaspro, A.; Tacchetti, C.; Manni, L.; Candiani, S.; et al. New findings in ATP supply in rod outer segments: Insights for retinopathies. *Biol. Cell* **2013**, *105*, 345–358. [CrossRef] [PubMed]

277. Funk, R.H.; Schumann, U.; Engelmann, K.; Becker, K.A.; Roehlecke, C. Blue light induced retinal oxidative stress: Implications for macular degeneration. *World J. Ophthalmol.* **2014**, *4*, 29–34. [CrossRef]

278. Calzia, D.; Garbarino, G.; Caicci, F.; Manni, L.; Candiani, S.; Ravera, S.; Morelli, A.; Traverso, C.E.; Panfoli, I. Functional expression of electron transport chain complexes in mouse rod outer segments. *Biochimie* **2014**, *102*, 78–82. [CrossRef] [PubMed]

279. Roehlecke, C.; Schumann, U.; Ader, M.; Brunssen, C.; Bramke, S.; Morawietz, H.; Funk, R.H. Stress reaction in outer segments of photoreceptors after blue light irradiation. *PLoS ONE* **2013**, *8*, e71570. [CrossRef] [PubMed]

280. Wanders, R.J.; Waterham, H.R.; Ferdinandusse, S. Metabolic Interplay between Peroxisomes and Other Subcellular Organelles Including Mitochondria and the Endoplasmic Reticulum. *Front. Cell Dev. Biol.* **2015**, *3*, 83. [CrossRef]

281. Folz, S.J.; Trobe, J.D. The peroxisome and the eye. *Surv. Ophthalmol.* **1991**, *35*, 353–368. [CrossRef]

282. Das, Y.; Roose, N.; De Groef, L.; Fransen, M.; Moons, L.; Van Veldhoven, P.P.; Baes, M. Differential distribution of peroxisomal proteins points to specific roles of peroxisomes in the murine retina. *Mol. Cell Biochem.* **2019**, *456*, 53–62. [CrossRef]

283. Hiebler, S.; Masuda, T.; Hacia, J.G.; Moser, A.B.; Faust, P.L.; Liu, A.; Chowdhury, N.; Huang, N.; Lauer, A.; Bennett, J.; et al. The Pex1-G844D mouse: A model for mild human Zellweger spectrum disorder. *Mol. Genet. Metab.* **2014**, *111*, 522–532. [CrossRef]

284. Pang, J.J.; Chang, B.; Hawes, N.L.; Hurd, R.E.; Davisson, M.T.; Li, J.; Noorwez, S.M.; Malhotra, R.; McDowell, J.H.; Kaushal, S.; et al. Retinal degeneration 12 (rd12): A new, spontaneously arising mouse model for human Leber congenital amaurosis (LCA). *Mol. Vis.* **2005**, *11*, 152–162.

285. Wright, C.B.; Chrenek, M.A.; Feng, W.; Getz, S.E.; Duncan, T.; Pardue, M.T.; Feng, Y.; Redmond, T.M.; Boatright, J.H.; Nickerson, J.M. The Rpe65 rd12 allele exerts a semidominant negative effect on vision in mice. *Investig. Ophthalmol. Vis. Sci.* **2014**, *55*, 2500–2515. [CrossRef]

286. Redmond, T.M.; Yu, S.; Lee, E.; Bok, D.; Hamasaki, D.; Chen, N.; Goletz, P.; Ma, J.X.; Crouch, R.K.; Pfeifer, K. Rpe65 is necessary for production of 11-cis-vitamin A in the retinal visual cycle. *Nat. Genet.* **1998**, *20*, 344–351. [CrossRef] [PubMed]

287. Tanabu, R.; Sato, K.; Monai, N.; Yamauchi, K.; Gonome, T.; Xie, Y.; Takahashi, S.; Ishiguro, S.I.; Nakazawa, M. The findings of optical coherence tomography of retinal degeneration in relation to the morphological and electroretinographic features in RPE65-/- mice. *PLoS ONE* **2019**, *14*, e0210439. [CrossRef] [PubMed]

288. Woodruff, M.L.; Wang, Z.; Chung, H.Y.; Redmond, T.M.; Fain, G.L.; Lem, J. Spontaneous activity of opsin apoprotein is a cause of Leber congenital amaurosis. *Nat. Genet.* **2003**, *35*, 158–164. [CrossRef] [PubMed]

289. Samardzija, M.; von Lintig, J.; Tanimoto, N.; Oberhauser, V.; Thiersch, M.; Reme, C.E.; Seeliger, M.; Grimm, C.; Wenzel, A. R91W mutation in Rpe65 leads to milder early-onset retinal dystrophy due to the generation of low levels of 11-cis-retinal. *Hum. Mol. Genet.* **2008**, *17*, 281–292. [CrossRef] [PubMed]

290. Choi, E.H.; Suh, S.; Sander, C.L.; Hernandez, C.J.O.; Bulman, E.R.; Khadka, N.; Dong, Z.; Shi, W.; Palczewski, K.; Kiser, P.D. Insights into the pathogenesis of dominant retinitis pigmentosa associated with a D477G mutation in RPE65. *Hum. Mol. Genet.* **2018**, *27*, 2225–2243. [CrossRef]

291. Radu, R.A.; Yuan, Q.; Hu, J.; Peng, J.H.; Lloyd, M.; Nusinowitz, S.; Bok, D.; Travis, G.H. Accelerated accumulation of lipofuscin pigments in the RPE of a mouse model for ABCA4-mediated retinal dystrophies following Vitamin A supplementation. *Investig. Ophthalmol. Vis. Sci.* **2008**, *49*, 3821–3829. [CrossRef]

292. Weng, J.; Mata, N.L.; Azarian, S.M.; Tzekov, R.T.; Birch, D.G.; Travis, G.H. Insights into the function of Rim protein in photoreceptors and etiology of Stargardt's disease from the phenotype in abcr knockout mice. *Cell* **1999**, *98*, 13–23. [CrossRef]

293. Molday, L.L.; Wahl, D.; Sarunic, M.V.; Molday, R.S. Localization and functional characterization of the p.Asn965Ser (N965S) ABCA4 variant in mice reveal pathogenic mechanisms underlying Stargardt macular degeneration. *Hum. Mol. Genet.* **2018**, *27*, 295–306. [CrossRef]

294. Zhang, N.; Tsybovsky, Y.; Kolesnikov, A.V.; Rozanowska, M.; Swider, M.; Schwartz, S.B.; Stone, E.M.; Palczewska, G.; Maeda, A.; Kefalov, V.J.; et al. Protein misfolding and the pathogenesis of ABCA4-associated retinal degenerations. *Hum. Mol. Genet.* **2015**, *24*, 3220–3237. [CrossRef]

295. Wu, L.; Ueda, K.; Nagasaki, T.; Sparrow, J.R. Light damage in Abca4 and Rpe65rd12 mice. *Investig. Ophthalmol. Vis. Sci.* **2014**, *55*, 1910–1918. [CrossRef]

296. Maeda, A.; Maeda, T.; Golczak, M.; Palczewski, K. Retinopathy in mice induced by disrupted all-trans-retinal clearance. *J. Biol. Chem.* **2008**, *283*, 26684–26693. [CrossRef]

297. Chen, Y.; Okano, K.; Maeda, T.; Chauhan, V.; Golczak, M.; Maeda, A.; Palczewski, K. Mechanism of all-trans-retinal toxicity with implications for stargardt disease and age-related macular degeneration. *J. Biol. Chem.* **2012**, *287*, 5059–5069. [CrossRef] [PubMed]

298. Okano, K.; Maeda, A.; Chen, Y.; Chauhan, V.; Tang, J.; Palczewska, G.; Sakai, T.; Tsuneoka, H.; Palczewski, K.; Maeda, T. Retinal cone and rod photoreceptor cells exhibit differential susceptibility to light-induced damage. *J. Neurochem.* **2012**, *121*, 146–156. [CrossRef] [PubMed]

299. Batten, M.L.; Imanishi, Y.; Maeda, T.; Tu, D.C.; Moise, A.R.; Bronson, D.; Possin, D.; Van Gelder, R.N.; Baehr, W.; Palczewski, K. Lecithin-retinol acyltransferase is essential for accumulation of all-trans-retinyl esters in the eye and in the liver. *J. Biol. Chem.* **2004**, *279*, 10422–10432. [CrossRef] [PubMed]

300. Ruiz, A.; Ghyselinck, N.B.; Mata, N.; Nusinowitz, S.; Lloyd, M.; Dennefeld, C.; Chambon, P.; Bok, D. Somatic ablation of the Lrat gene in the mouse retinal pigment epithelium drastically reduces its retinoid storage. *Investig. Ophthalmol. Vis. Sci.* **2007**, *48*, 5377–5387. [CrossRef]

301. Liou, G.I.; Fei, Y.; Peachey, N.S.; Matragoon, S.; Wei, S.; Blaner, W.S.; Wang, Y.; Liu, C.; Gottesman, M.E.; Ripps, H. Early onset photoreceptor abnormalities induced by targeted disruption of the interphotoreceptor retinoid-binding protein gene. *J. Neurosci.* **1998**, *18*, 4511–4520. [CrossRef]

302. Shen, J.; Shi, D.; Suzuki, T.; Xia, Z.; Zhang, H.; Araki, K.; Wakana, S.; Takeda, N.; Yamamura, K.; Jin, S.; et al. Severe ocular phenotypes in Rbp4-deficient mice in the C57BL/6 genetic background. *Lab. Investig.* **2016**, *96*, 680–691. [CrossRef]

303. Ruiz, A.; Mark, M.; Jacobs, H.; Klopfenstein, M.; Hu, J.; Lloyd, M.; Habib, S.; Tosha, C.; Radu, R.A.; Ghyselinck, N.B.; et al. Retinoid content, visual responses, and ocular morphology are compromised in the retinas of mice lacking the retinol-binding protein receptor, STRA6. *Investig. Ophthalmol. Vis. Sci.* **2012**, *53*, 3027–3039. [CrossRef]

304. Amengual, J.; Zhang, N.; Kemerer, M.; Maeda, T.; Palczewski, K.; Von Lintig, J. STRA6 is critical for cellular vitamin A uptake and homeostasis. *Hum. Mol. Genet.* **2014**, *23*, 5402–5417. [CrossRef]

305. Wu, S.M. Synaptic transmission in the outer retina. *Annu. Rev. Physiol.* **1994**, *56*, 141–168. [CrossRef]

306. Wu, S.M. Synaptic organization of the vertebrate retina: General principles and species-specific variations: The Friedenwald lecture. *Investig. Ophthalmol. Vis. Sci.* **2010**, *51*, 1263–1274. [CrossRef] [PubMed]

307. Mercer, A.J.; Thoreson, W.B. The dynamic architecture of photoreceptor ribbon synapses: Cytoskeletal, extracellular matrix, and intramembrane proteins. *Vis. Neurosci.* **2011**, *28*, 453–471. [CrossRef] [PubMed]

308. Furukawa, T.; Ueno, A.; Omori, Y. Molecular mechanisms underlying selective synapse formation of vertebrate retinal photoreceptor cells. *Cell Mol. Life Sci.* **2019**, *77*, 1251–1266. [CrossRef] [PubMed]

309. Pardue, M.T.; Peachey, N.S. Mouse b-wave mutants. *Doc. Ophthalmol.* **2014**, *128*, 77–89. [CrossRef] [PubMed]

310. Pillers, D.A.; Weleber, R.G.; Woodward, W.R.; Green, D.G.; Chapman, V.M.; Ray, P.N. mdxCv3 mouse is a model for electroretinography of Duchenne/Becker muscular dystrophy. *Investig. Ophthalmol. Vis. Sci.* **1995**, *36*, 462–466.

311. Satz, J.S.; Philp, A.R.; Nguyen, H.; Kusano, H.; Lee, J.; Turk, R.; Riker, M.J.; Hernandez, J.; Weiss, R.M.; Anderson, M.G.; et al. Visual impairment in the absence of dystroglycan. *J. Neurosci.* **2009**, *29*, 13136–13146. [CrossRef]

312. Bytyqi, A.H.; Lockridge, O.; Duysen, E.; Wang, Y.; Wolfrum, U.; Layer, P.G. Impaired formation of the inner retina in an AChE knockout mouse results in degeneration of all photoreceptors. *Eur. J. Neurosci.* **2004**, *20*, 2953–2962. [CrossRef]

313. Grisaru, D.; Sternfeld, M.; Eldor, A.; Glick, D.; Soreq, H. Structural roles of acetylcholinesterase variants in biology and pathology. *Eur. J. Biochem.* **1999**, *264*, 672–686. [CrossRef]

314. Regus-Leidig, H.; Atorf, J.; Feigenspan, A.; Kremers, J.; Maw, M.A.; Brandstatter, J.H. Photoreceptor degeneration in two mouse models for congenital stationary night blindness type 2. *PLoS ONE* **2014**, *9*, e86769. [CrossRef]

315. Kerov, V.; Laird, J.G.; Joiner, M.L.; Knecht, S.; Soh, D.; Hagen, J.; Gardner, S.H.; Gutierrez, W.; Yoshimatsu, T.; Bhattarai, S.; et al. alpha2delta-4 Is Required for the Molecular and Structural Organization of Rod and Cone Photoreceptor Synapses. *J. Neurosci.* **2018**, *38*, 6145–6160. [CrossRef]

316. Ruether, K.; Grosse, J.; Matthiessen, E.; Hoffmann, K.; Hartmann, C. Abnormalities of the photoreceptor-bipolar cell synapse in a substrain of C57BL/10 mice. *Investig. Ophthalmol. Vis. Sci.* **2000**, *41*, 4039–4047. [PubMed]

317. Haeseleer, F.; Imanishi, Y.; Maeda, T.; Possin, D.E.; Maeda, A.; Lee, A.; Rieke, F.; Palczewski, K. Essential role of Ca2+-binding protein 4, a Cav1.4 channel regulator, in photoreceptor synaptic function. *Nat. Neurosci.* **2004**, *7*, 1079–1087. [CrossRef] [PubMed]

318. Ishiba, Y.; Higashide, T.; Mori, N.; Kobayashi, A.; Kubota, S.; McLaren, M.J.; Satoh, H.; Wong, F.; Inana, G. Targeted inactivation of synaptic HRG4 (UNC119) causes dysfunction in the distal photoreceptor and slow retinal degeneration, revealing a new function. *Exp. Eye Res.* **2007**, *84*, 473–485. [CrossRef] [PubMed]

319. Haeseleer, F. Interaction and colocalization of CaBP4 and Unc119 (MRG4) in photoreceptors. *Investig. Ophthalmol. Vis. Sci.* **2008**, *49*, 2366–2375. [CrossRef]

320. Giblin, J.P.; Comes, N.; Strauss, O.; Gasull, X. Ion Channels in the Eye: Involvement in Ocular Pathologies. *Adv. Protein Chem. Struct. Biol.* **2016**, *104*, 157–231. [CrossRef]

321. Edwards, M.M.; Marin de Evsikova, C.; Collin, G.B.; Gifford, E.; Wu, J.; Hicks, W.L.; Whiting, C.; Varvel, N.H.; Maphis, N.; Lamb, B.T.; et al. Photoreceptor degeneration, azoospermia, leukoencephalopathy, and abnormal RPE cell function in mice expressing an early stop mutation in CLCN2. *Investig. Ophthalmol. Vis. Sci.* **2010**, *51*, 3264–3272. [CrossRef]

322. Stobrawa, S.M.; Breiderhoff, T.; Takamori, S.; Engel, D.; Schweizer, M.; Zdebik, A.A.; Bosl, M.R.; Ruether, K.; Jahn, H.; Draguhn, A.; et al. Disruption of ClC-3, a chloride channel expressed on synaptic vesicles, leads to a loss of the hippocampus. *Neuron* **2001**, *29*, 185–196. [CrossRef]

323. Rajan, I.; Read, R.; Small, D.L.; Perrard, J.; Vogel, P. An alternative splicing variant in Clcn7-/- mice prevents osteopetrosis but not neural and retinal degeneration. *Vet. Pathol.* **2011**, *48*, 663–675. [CrossRef]

324. Dickerson, L.W.; Bonthius, D.J.; Schutte, B.C.; Yang, B.; Barna, T.J.; Bailey, M.C.; Nehrke, K.; Williamson, R.A.; Lamb, F.S. Altered GABAergic function accompanies hippocampal degeneration in mice lacking ClC-3 voltage-gated chloride channels. *Brain Res.* **2002**, *958*, 227–250. [CrossRef]

325. Kornak, U.; Kasper, D.; Bosl, M.R.; Kaiser, E.; Schweizer, M.; Schulz, A.; Friedrich, W.; Delling, G.; Jentsch, T.J. Loss of the ClC-7 chloride channel leads to osteopetrosis in mice and man. *Cell* **2001**, *104*, 205–215. [CrossRef]

326. Kasper, D.; Planells-Cases, R.; Fuhrmann, J.C.; Scheel, O.; Zeitz, O.; Ruether, K.; Schmitt, A.; Poet, M.; Steinfeld, R.; Schweizer, M.; et al. Loss of the chloride channel ClC-7 leads to lysosomal storage disease and neurodegeneration. *EMBO J.* **2005**, *24*, 1079–1091. [CrossRef] [PubMed]

327. Weinert, S.; Jabs, S.; Hohensee, S.; Chan, W.L.; Kornak, U.; Jentsch, T.J. Transport activity and presence of ClC-7/Ostm1 complex account for different cellular functions. *EMBO Rep.* **2014**, *15*, 784–791. [CrossRef] [PubMed]

328. Weber, P.; Bartsch, U.; Schachner, M.; Montag, D. Na,K-ATPase subunit beta1 knock-in prevents lethality of beta2 deficiency in mice. *J. Neurosci.* **1998**, *18*, 9192–9203. [CrossRef] [PubMed]

329. Heller-Stilb, B.; van Roeyen, C.; Rascher, K.; Hartwig, H.G.; Huth, A.; Seeliger, M.W.; Warskulat, U.; Haussinger, D. Disruption of the taurine transporter gene (taut) leads to retinal degeneration in mice. *FASEB J.* **2002**, *16*, 231–233. [CrossRef] [PubMed]

330. Bok, D.; Galbraith, G.; Lopez, I.; Woodruff, M.; Nusinowitz, S.; BeltrandelRio, H.; Huang, W.; Zhao, S.; Geske, R.; Montgomery, C.; et al. Blindness and auditory impairment caused by loss of the sodium bicarbonate cotransporter NBC3. *Nat. Genet.* **2003**, *34*, 313–319. [CrossRef]

331. Jin, Z.B.; Huang, X.F.; Lv, J.N.; Xiang, L.; Li, D.Q.; Chen, J.; Huang, C.; Wu, J.; Lu, F.; Qu, J. SLC7A14 linked to autosomal recessive retinitis pigmentosa. *Nat. Commun.* **2014**, *5*, 3517. [CrossRef]

332. Jadeja, S.; Barnard, A.R.; McKie, L.; Cross, S.H.; White, J.K.; Sanger Mouse Genetics, P.; Robertson, M.; Budd, P.S.; MacLaren, R.E.; Jackson, I.J. Mouse slc9a8 mutants exhibit retinal defects due to retinal pigmented epithelium dysfunction. *Investig. Ophthalmol. Vis. Sci.* **2015**, *56*, 3015–3026. [CrossRef]

333. Hori, K.; Katayama, N.; Kachi, S.; Kondo, M.; Kadomatsu, K.; Usukura, J.; Muramatsu, T.; Mori, S.; Miyake, Y. Retinal dysfunction in basigin deficiency. *Investig. Ophthalmol. Vis. Sci.* **2000**, *41*, 3128–3133.

334. Veleri, S.; Nellissery, J.; Mishra, B.; Manjunath, S.H.; Brooks, M.J.; Dong, L.; Nagashima, K.; Qian, H.; Gao, C.; Sergeev, Y.V.; et al. REEP6 mediates trafficking of a subset of Clathrin-coated vesicles and is critical for rod photoreceptor function and survival. *Hum. Mol. Genet.* **2017**, *26*, 2218–2230. [CrossRef]

335. Ettaiche, M.; Deval, E.; Pagnotta, S.; Lazdunski, M.; Lingueglia, E. Acid-sensing ion channel 3 in retinal function and survival. *Investig. Ophthalmol. Vis. Sci.* **2009**, *50*, 2417–2426. [CrossRef]

336. Li, L.; Jiao, X.; D'Atri, I.; Ono, F.; Nelson, R.; Chan, C.C.; Nakaya, N.; Ma, Z.; Ma, Y.; Cai, X.; et al. Mutation in the intracellular chloride channel CLCC1 associated with autosomal recessive retinitis pigmentosa. *PLoS Genet.* **2018**, *14*, e1007504. [CrossRef] [PubMed]

337. Wong, B.H.; Chan, J.P.; Cazenave-Gassiot, A.; Poh, R.W.; Foo, J.C.; Galam, D.L.; Ghosh, S.; Nguyen, L.N.; Barathi, V.A.; Yeo, S.W.; et al. Mfsd2a Is a Transporter for the Essential omega-3 Fatty Acid Docosahexaenoic Acid (DHA) in Eye and Is Important for Photoreceptor Cell Development. *J. Biol. Chem.* **2016**, *291*, 10501–10514. [CrossRef] [PubMed]

338. Mahimkar, R.M.; Visaya, O.; Pollock, A.S.; Lovett, D.H. The disintegrin domain of ADAM9: A ligand for multiple beta1 renal integrins. *Biochem. J.* **2005**, *385*, 461–468. [CrossRef] [PubMed]

339. Parry, D.A.; Toomes, C.; Bida, L.; Danciger, M.; Towns, K.V.; McKibbin, M.; Jacobson, S.G.; Logan, C.V.; Ali, M.; Bond, J.; et al. Loss of the metalloprotease ADAM9 leads to cone-rod dystrophy in humans and retinal degeneration in mice. *Am. J. Hum. Genet.* **2009**, *84*, 683–691. [CrossRef] [PubMed]

340. Vijayasarathy, C.; Ziccardi, L.; Sieving, P.A. Biology of retinoschisin. *Adv. Exp. Med. Biol.* **2012**, *723*, 513–518. [CrossRef]

341. Jablonski, M.M.; Dalke, C.; Wang, X.; Lu, L.; Manly, K.F.; Pretsch, W.; Favor, J.; Pardue, M.T.; Rinchik, E.M.; Williams, R.W.; et al. An ENU-induced mutation in Rs1h causes disruption of retinal structure and function. *Mol. Vis.* **2005**, *11*, 569–581.

342. Han, J.; Farmer, S.R.; Kirkland, J.L.; Corkey, B.E.; Yoon, R.; Pirtskhalava, T.; Ido, Y.; Guo, W. Octanoate attenuates adipogenesis in 3T3-L1 preadipocytes. *J. Nutr.* **2002**, *132*, 904–910. [CrossRef]

343. Zeng, Y.; Takada, Y.; Kjellstrom, S.; Hiriyanna, K.; Tanikawa, A.; Wawrousek, E.; Smaoui, N.; Caruso, R.; Bush, R.A.; Sieving, P.A. RS-1 Gene Delivery to an Adult Rs1h Knockout Mouse Model Restores ERG b-Wave with Reversal of the Electronegative Waveform of X-Linked Retinoschisis. *Investig. Ophthalmol. Vis. Sci.* **2004**, *45*, 3279–3285. [CrossRef]

344. Quinn, P.M.; Pellissier, L.P.; Wijnholds, J. The CRB1 Complex: Following the Trail of Crumbs to a Feasible Gene Therapy Strategy. *Front. Neurosci.* **2017**, *11*, 175. [CrossRef]

345. Mehalow, A.K.; Kameya, S.; Smith, R.S.; Hawes, N.L.; Denegre, J.M.; Young, J.A.; Bechtold, L.; Haider, N.B.; Tepass, U.; Heckenlively, J.R.; et al. CRB1 is essential for external limiting membrane integrity and photoreceptor morphogenesis in the mammalian retina. *Hum. Mol. Genet.* **2003**, *12*, 2179–2189. [CrossRef]

346. van de Pavert, S.A.; Kantardzhieva, A.; Malysheva, A.; Meuleman, J.; Versteeg, I.; Levelt, C.; Klooster, J.; Geiger, S.; Seeliger, M.W.; Rashbass, P.; et al. Crumbs homologue 1 is required for maintenance of photoreceptor cell polarization and adhesion during light exposure. *J. Cell Sci.* **2004**, *117*, 4169–4177. [CrossRef] [PubMed]

347. Quinn, P.M.J.; Wijnholds, J. Retinogenesis of the Human Fetal Retina: An Apical Polarity Perspective. *Genes* **2019**, *10*, 987. [CrossRef]

348. Moore, B.A.; Leonard, B.C.; Sebbag, L.; Edwards, S.G.; Cooper, A.; Imai, D.M.; Straiton, E.; Santos, L.; Reilly, C.; Griffey, S.M.; et al. Identification of genes required for eye development by high-throughput screening of mouse knockouts. *Commun. Biol.* **2018**, *1*, 236. [CrossRef] [PubMed]

349. Watson, J.R.; Owen, D.; Mott, H.R. Cdc42 in actin dynamics: An ordered pathway governed by complex equilibria and directional effector handover. *Small GTPases* **2017**, *8*, 237–244. [CrossRef]

350. Yokokura, S.; Wada, Y.; Nakai, S.; Sato, H.; Yao, R.; Yamanaka, H.; Ito, S.; Sagara, Y.; Takahashi, M.; Nakamura, Y.; et al. Targeted disruption of FSCN2 gene induces retinopathy in mice. *Investig. Ophthalmol. Vis. Sci.* **2005**, *46*, 2905–2915. [CrossRef] [PubMed]

351. Liu, X.; Zhao, M.; Xie, Y.; Li, P.; Wang, O.; Zhou, B.; Yang, L.; Nie, Y.; Cheng, L.; Song, X.; et al. Null Mutation of the Fascin2 Gene by TALEN Leading to Progressive Hearing Loss and Retinal Degeneration in C57BL/6J Mice. *G3* **2018**, *8*, 3221–3230. [CrossRef]

352. Saksens, N.T.; Krebs, M.P.; Schoenmaker-Koller, F.E.; Hicks, W.; Yu, M.; Shi, L.; Rowe, L.; Collin, G.B.; Charette, J.R.; Letteboer, S.J.; et al. Mutations in CTNNA1 cause butterfly-shaped pigment dystrophy and perturbed retinal pigment epithelium integrity. *Nat. Genet.* **2016**, *48*, 144–151. [CrossRef]

353. Maddox, D.M.; Collin, G.B.; Ikeda, A.; Pratt, C.H.; Ikeda, S.; Johnson, B.A.; Hurd, R.E.; Shopland, L.S.; Naggert, J.K.; Chang, B.; et al. A Mutation in Syne2 Causes Early Retinal Defects in Photoreceptors, Secondary Neurons, and Muller Glia. *Investig. Ophthalmol. Vis. Sci.* **2015**, *56*, 3776–3787. [CrossRef]

354. Yu, J.; Lei, K.; Zhou, M.; Craft, C.M.; Xu, G.; Xu, T.; Zhuang, Y.; Xu, R.; Han, M. KASH protein Syne-2/Nesprin-2 and SUN proteins SUN1/2 mediate nuclear migration during mammalian retinal development. *Hum. Mol. Genet.* **2011**, *20*, 1061–1073. [CrossRef]

355. Gordon, M.D.; Nusse, R. Wnt signaling: Multiple pathways, multiple receptors, and multiple transcription factors. *J. Biol. Chem.* **2006**, *281*, 22429–22433. [CrossRef]

356. Liu, C.; Nathans, J. An essential role for frizzled 5 in mammalian ocular development. *Development* **2008**, *135*, 3567–3576. [CrossRef] [PubMed]

357. Wang, Z.; Liu, C.H.; Sun, Y.; Gong, Y.; Favazza, T.L.; Morss, P.C.; Saba, N.J.; Fredrick, T.W.; He, X.; Akula, J.D.; et al. Pharmacologic Activation of Wnt Signaling by Lithium Normalizes Retinal Vasculature in a Murine Model of Familial Exudative Vitreoretinopathy. *Am. J. Pathol.* **2016**, *186*, 2588–2600. [CrossRef] [PubMed]

358. Berger, W.; van de Pol, D.; Bachner, D.; Oerlemans, F.; Winkens, H.; Hameister, H.; Wieringa, B.; Hendriks, W.; Ropers, H.H. An animal model for Norrie disease (ND): Gene targeting of the mouse ND gene. *Hum. Mol. Genet.* **1996**, *5*, 51–59. [CrossRef] [PubMed]

359. Junge, H.J.; Yang, S.; Burton, J.B.; Paes, K.; Shu, X.; French, D.M.; Costa, M.; Rice, D.S.; Ye, W. TSPAN12 regulates retinal vascular development by promoting Norrin- but not Wnt-induced FZD4/beta-catenin signaling. *Cell* **2009**, *139*, 299–311. [CrossRef] [PubMed]

360. Hackam, A.S. The Wnt signaling pathway in retinal degenerations. *IUBMB Life* **2005**, *57*, 381–388. [CrossRef]

361. Hawes, N.L.; Chang, B.; Hageman, G.S.; Nusinowitz, S.; Nishina, P.M.; Schneider, B.S.; Smith, R.S.; Roderick, T.H.; Davisson, M.T.; Heckenlively, J.R. Retinal degeneration 6 (rd6): A new mouse model for human retinitis punctata albescens. *Investig. Ophthalmol. Vis. Sci.* **2000**, *41*, 3149–3157.

362. Hawkes, W.G. Bibliography: Nurses and smoking. *J. N. Y. State Nurses Assoc.* **1990**, *21*, 14.

363. Chavali, V.R.; Khan, N.W.; Cukras, C.A.; Bartsch, D.U.; Jablonski, M.M.; Ayyagari, R. A CTRP5 gene S163R mutation knock-in mouse model for late-onset retinal degeneration. *Hum. Mol. Genet.* **2011**, *20*, 2000–2014. [CrossRef]

364. Cai, Z.; Simons, D.L.; Fu, X.Y.; Feng, G.S.; Wu, S.M.; Zhang, X. Loss of Shp2-mediated mitogen-activated protein kinase signaling in Muller glial cells results in retinal degeneration. *Mol. Cell Biol.* **2011**, *31*, 2973–2983. [CrossRef]

365. Chavez-Solano, M.; Ibarra-Sanchez, A.; Trevino, M.; Gonzalez-Espinosa, C.; Lamas, M. Fyn kinase genetic ablation causes structural abnormalities in mature retina and defective Muller cell function. *Mol. Cell Neurosci.* **2016**, *72*, 91–100. [CrossRef]

366. Ji, X.; Liu, Y.; Hurd, R.; Wang, J.; Fitzmaurice, B.; Nishina, P.M.; Chang, B. Retinal Pigment Epithelium Atrophy 1 (rpea1): A New Mouse Model With Retinal Detachment Caused by a Disruption of Protein Kinase C, theta. *Investig. Ophthalmol. Vis. Sci.* **2016**, *57*, 877–888. [CrossRef]

367. Ivanovic, I.; Anderson, R.E.; Le, Y.Z.; Fliesler, S.J.; Sherry, D.M.; Rajala, R.V. Deletion of the p85alpha regulatory subunit of phosphoinositide 3-kinase in cone photoreceptor cells results in cone photoreceptor degeneration. *Investig. Ophthalmol. Vis. Sci.* **2011**, *52*, 3775–3783. [CrossRef] [PubMed]

368. Rajala, R.V.; Ranjo-Bishop, M.; Wang, Y.; Rajala, A.; Anderson, R.E. The p110alpha isoform of phosphoinositide 3-kinase is essential for cone photoreceptor survival. *Biochimie* **2015**, *112*, 35–40. [CrossRef] [PubMed]

369. Yi, X.; Schubert, M.; Peachey, N.S.; Suzuma, K.; Burks, D.J.; Kushner, J.A.; Suzuma, I.; Cahill, C.; Flint, C.L.; Dow, M.A.; et al. Insulin receptor substrate 2 is essential for maturation and survival of photoreceptor cells. *J. Neurosci.* **2005**, *25*, 1240–1248. [CrossRef] [PubMed]

370. Kitahara, H.; Kajikawa, S.; Ishii, Y.; Yamamoto, S.; Hamashima, T.; Azuma, E.; Sato, H.; Matsushima, T.; Shibuya, M.; Shimada, Y.; et al. The Novel Pathogenesis of Retinopathy Mediated by Multiple RTK Signals is Uncovered in Newly Developed Mouse Model. *EBioMedicine* **2018**, *31*, 190–201. [CrossRef]

371. Mongan, M.; Wang, J.; Liu, H.; Fan, Y.; Jin, C.; Kao, W.Y.; Xia, Y. Loss of MAP3K1 enhances proliferation and apoptosis during retinal development. *Development* **2011**, *138*, 4001–4012. [CrossRef]

372. Toyofuku, T.; Nojima, S.; Ishikawa, T.; Takamatsu, H.; Tsujimura, T.; Uemura, A.; Matsuda, J.; Seki, T.; Kumanogoh, A. Endosomal sorting by Semaphorin 4A in retinal pigment epithelium supports photoreceptor survival. *Genes Dev.* **2012**, *26*, 816–829. [CrossRef]

373. Carter-Dawson, L.D.; LaVail, M.M. Rods and cones in the mouse retina. II. Autoradiographic analysis of cell generation using tritiated thymidine. *J. Comp. Neurol.* **1979**, *188*, 263–272. [CrossRef]

374. Swaroop, A.; Kim, D.; Forrest, D. Transcriptional regulation of photoreceptor development and homeostasis in the mammalian retina. *Nat. Rev. Neurosci.* **2010**, *11*, 563–576. [CrossRef]

375. Brzezinski, J.A.; Reh, T.A. Photoreceptor cell fate specification in vertebrates. *Development* **2015**, *142*, 3263–3273. [CrossRef]

376. Khidr, L.; Chen, P.L. RB, the conductor that orchestrates life, death and differentiation. *Oncogene* **2006**, *25*, 5210–5219. [CrossRef] [PubMed]

377. Roger, J.E.; Hiriyanna, A.; Gotoh, N.; Hao, H.; Cheng, D.F.; Ratnapriya, R.; Kautzmann, M.A.; Chang, B.; Swaroop, A. OTX2 loss causes rod differentiation defect in CRX-associated congenital blindness. *J. Clin. Investig.* **2014**, *124*, 631–643. [CrossRef] [PubMed]

378. Tran, N.M.; Zhang, A.; Zhang, X.; Huecker, J.B.; Hennig, A.K.; Chen, S. Mechanistically distinct mouse models for CRX-associated retinopathy. *PLoS Genet.* **2014**, *10*, e1004111. [CrossRef] [PubMed]

379. Omori, Y.; Kitamura, T.; Yoshida, S.; Kuwahara, R.; Chaya, T.; Irie, S.; Furukawa, T. Mef2d is essential for the maturation and integrity of retinal photoreceptor and bipolar cells. *Genes Cells* **2015**, *20*, 408–426. [CrossRef]

380. Kiyama, T.; Chen, C.K.; Wang, S.W.; Pan, P.; Ju, Z.; Wang, J.; Takada, S.; Klein, W.H.; Mao, C.A. Essential roles of mitochondrial biogenesis regulator Nrf1 in retinal development and homeostasis. *Mol. Neurodegener.* **2018**, *13*, 56. [CrossRef] [PubMed]

381. Wu, F.; Li, R.; Umino, Y.; Kaczynski, T.J.; Sapkota, D.; Li, S.; Xiang, M.; Fliesler, S.J.; Sherry, D.M.; Gannon, M.; et al. Onecut1 is essential for horizontal cell genesis and retinal integrity. *J. Neurosci.* **2013**, *33*, 13053–13065. [CrossRef]

382. Goding, C.R.; Arnheiter, H. MITF-the first 25 years. *Genes Dev.* **2019**, *33*, 983–1007. [CrossRef]

383. Zelinger, L.; Swaroop, A. RNA Biology in Retinal Development and Disease. *Trends Genet.* **2018**, *34*, 341–351. [CrossRef]

384. Barabino, A.; Plamondon, V.; Abdouh, M.; Chatoo, W.; Flamier, A.; Hanna, R.; Zhou, S.; Motoyama, N.; Hebert, M.; Lavoie, J.; et al. Loss of Bmi1 causes anomalies in retinal development and degeneration of cone photoreceptors. *Development* **2016**, *143*, 1571–1584. [CrossRef]

385. Susaki, K.; Kaneko, J.; Yamano, Y.; Nakamura, K.; Inami, W.; Yoshikawa, T.; Ozawa, Y.; Shibata, S.; Matsuzaki, O.; Okano, H.; et al. Musashi-1, an RNA-binding protein, is indispensable for survival of photoreceptors. *Exp. Eye Res.* **2009**, *88*, 347–355. [CrossRef]

386. McKinnon, P.J. Maintaining genome stability in the nervous system. *Nat. Neurosci.* **2013**, *16*, 1523–1529. [CrossRef] [PubMed]

387. Gregersen, L.H.; Svejstrup, J.Q. The Cellular Response to Transcription-Blocking DNA Damage. *Trends Biochem. Sci.* **2018**, *43*, 327–341. [CrossRef] [PubMed]

388. Lans, H.; Hoeijmakers, J.H.J.; Vermeulen, W.; Marteijn, J.A. The DNA damage response to transcription stress. *Nat. Rev. Mol. Cell Biol.* **2019**, *20*, 766–784. [CrossRef] [PubMed]

389. Sanchez, A.; De Vivo, A.; Uprety, N.; Kim, J.; Stevens, S.M., Jr.; Kee, Y. BMI1-UBR5 axis regulates transcriptional repression at damaged chromatin. *Proc. Natl. Acad. Sci. USA* **2016**, *113*, 11243–11248. [CrossRef] [PubMed]

390. Calses, P.C.; Dhillon, K.K.; Tucker, N.; Chi, Y.; Huang, J.W.; Kawasumi, M.; Nghiem, P.; Wang, Y.; Clurman, B.E.; Jacquemont, C.; et al. DGCR8 Mediates Repair of UV-Induced DNA Damage Independently of RNA Processing. *Cell Rep.* **2017**, *19*, 162–174. [CrossRef] [PubMed]

391. Burger, K.; Schlackow, M.; Potts, M.; Hester, S.; Mohammed, S.; Gullerova, M. Nuclear phosphorylated Dicer processes double-stranded RNA in response to DNA damage. *J. Cell Biol.* **2017**, *216*, 2373–2389. [CrossRef]

392. Montaldo, N.P.; Bordin, D.L.; Brambilla, A.; Rosinger, M.; Fordyce Martin, S.L.; Bjoras, K.O.; Bradamante, S.; Aas, P.A.; Furrer, A.; Olsen, L.C.; et al. Alkyladenine DNA glycosylase associates with transcription elongation to coordinate DNA repair with gene expression. *Nat. Commun.* **2019**, *10*, 5460. [CrossRef]

393. Faridounnia, M.; Folkers, G.E.; Boelens, R. Function and Interactions of ERCC1-XPF in DNA Damage Response. *Molecules* **2018**, *23*, 3205. [CrossRef]

394. de Araujo, P.R.; Gorthi, A.; da Silva, A.E.; Tonapi, S.S.; Vo, D.T.; Burns, S.C.; Qiao, M.; Uren, P.J.; Yuan, Z.M.; Bishop, A.J.; et al. Musashi1 Impacts Radio-Resistance in Glioblastoma by Controlling DNA-Protein Kinase Catalytic Subunit. *Am. J. Pathol.* **2016**, *186*, 2271–2278. [CrossRef]

395. Onn, L.; Portillo, M.; Ilic, S.; Cleitman, G.; Stein, D.; Kaluski, S.; Shirat, I.; Slobodnik, Z.; Einav, M.; Erdel, F.; et al. SIRT6 is a DNA double-strand break sensor. *Elife* **2020**, *9*. [CrossRef]

396. Calderwood, S.K. A critical role for topoisomerase IIb and DNA double strand breaks in transcription. *Transcription* **2016**, *7*, 75–83. [CrossRef] [PubMed]

397. Aleksandrov, R.; Dotchev, A.; Poser, I.; Krastev, D.; Georgiev, G.; Panova, G.; Babukov, Y.; Danovski, G.; Dyankova, T.; Hubatsch, L.; et al. Protein Dynamics in Complex DNA Lesions. *Mol. Cell* **2018**, *69*, 1046.e1045–1061.e1045. [CrossRef]

398. Nishi, R.; Wijnhoven, P.W.G.; Kimura, Y.; Matsui, M.; Konietzny, R.; Wu, Q.; Nakamura, K.; Blundell, T.L.; Kessler, B.M. The deubiquitylating enzyme UCHL3 regulates Ku80 retention at sites of DNA damage. *Sci. Rep.* **2018**, *8*, 17891. [CrossRef] [PubMed]

399. Xu, H.; Lin, Z.; Li, F.; Diao, W.; Dong, C.; Zhou, H.; Xie, X.; Wang, Z.; Shen, Y.; Long, J. Dimerization of elongator protein 1 is essential for Elongator complex assembly. *Proc. Natl. Acad. Sci. USA* **2015**, *112*, 10697–10702. [CrossRef] [PubMed]

400. Li, Y.; Hao, H.; Swerdel, M.R.; Cho, H.Y.; Lee, K.B.; Hart, R.P.; Lyu, Y.L.; Cai, L. Top2b is involved in the formation of outer segment and synapse during late-stage photoreceptor differentiation by controlling key genes of photoreceptor transcriptional regulatory network. *J. Neurosci. Res.* **2017**, *95*, 1951–1964. [CrossRef] [PubMed]

401. Valdes-Sanchez, L.; De la Cerda, B.; Diaz-Corrales, F.J.; Massalini, S.; Chakarova, C.F.; Wright, A.F.; Bhattacharya, S.S. ATR localizes to the photoreceptor connecting cilium and deficiency leads to severe photoreceptor degeneration in mice. *Hum. Mol. Genet.* **2013**, *22*, 1507–1515. [CrossRef]

402. Bian, C.; Zhang, C.; Luo, T.; Vyas, A.; Chen, S.H.; Liu, C.; Kassab, M.A.; Yang, Y.; Kong, M.; Yu, X. NADP(+) is an endogenous PARP inhibitor in DNA damage response and tumor suppression. *Nat. Commun.* **2019**, *10*, 693. [CrossRef]

403. Sundermeier, T.R.; Sakami, S.; Sahu, B.; Howell, S.J.; Gao, S.; Dong, Z.; Golczak, M.; Maeda, A.; Palczewski, K. MicroRNA-processing Enzymes Are Essential for Survival and Function of Mature Retinal Pigmented Epithelial Cells in Mice. *J. Biol. Chem.* **2017**, *292*, 3366–3378. [CrossRef]

404. Damiani, D.; Alexander, J.J.; O'Rourke, J.R.; McManus, M.; Jadhav, A.P.; Cepko, C.L.; Hauswirth, W.W.; Harfe, B.D.; Strettoi, E. Dicer inactivation leads to progressive functional and structural degeneration of the mouse retina. *J. Neurosci.* **2008**, *28*, 4878–4887. [CrossRef] [PubMed]

405. Ueki, Y.; Ramirez, G.; Salcedo, E.; Stabio, M.E.; Lefcort, F. Loss of Ikbkap Causes Slow, Progressive Retinal Degeneration in a Mouse Model of Familial Dysautonomia. *eNeuro* **2016**, *3*. [CrossRef] [PubMed]

406. Spoor, M.; Nagtegaal, A.P.; Ridwan, Y.; Borgesius, N.Z.; van Alphen, B.; van der Pluijm, I.; Hoeijmakers, J.H.; Frens, M.A.; Borst, J.G. Accelerated loss of hearing and vision in the DNA-repair deficient Ercc1(delta/-) mouse. *Mech. Ageing Dev.* **2012**, *133*, 59–67. [CrossRef] [PubMed]

407. Gorgels, T.G.; van der Pluijm, I.; Brandt, R.M.; Garinis, G.A.; van Steeg, H.; van den Aardweg, G.; Jansen, G.H.; Ruijter, J.M.; Bergen, A.A.; van Norren, D.; et al. Retinal degeneration and ionizing radiation hypersensitivity in a mouse model for Cockayne syndrome. *Mol. Cell Biol.* **2007**, *27*, 1433–1441. [CrossRef] [PubMed]

408. Eblimit, A.; Zaneveld, S.A.; Liu, W.; Thomas, K.; Wang, K.; Li, Y.; Mardon, G.; Chen, R. NMNAT1 E257K variant, associated with Leber Congenital Amaurosis (LCA9), causes a mild retinal degeneration phenotype. *Exp. Eye Res.* **2018**, *173*, 32–43. [CrossRef]

409. Peshti, V.; Obolensky, A.; Nahum, L.; Kanfi, Y.; Rathaus, M.; Avraham, M.; Tinman, S.; Alt, F.W.; Banin, E.; Cohen, H.Y. Characterization of physiological defects in adult SIRT6-/- mice. *PLoS ONE* **2017**, *12*, e0176371. [CrossRef] [PubMed]

410. Lim, D.; Park, C.W.; Ryu, K.Y.; Chung, H. Disruption of the polyubiquitin gene Ubb causes retinal degeneration in mice. *Biochem. Biophys. Res. Commun.* **2019**, *513*, 35–40. [CrossRef]

411. Semenova, E.; Wang, X.; Jablonski, M.M.; Levorse, J.; Tilghman, S.M. An engineered 800 kilobase deletion of Uchl3 and Lmo7 on mouse chromosome 14 causes defects in viability, postnatal growth and degeneration of muscle and retina. *Hum. Mol. Genet.* **2003**, *12*, 1301–1312. [CrossRef]

412. Pinelli, M.; Carissimo, A.; Cutillo, L.; Lai, C.H.; Mutarelli, M.; Moretti, M.N.; Singh, M.V.; Karali, M.; Carrella, D.; Pizzo, M.; et al. An atlas of gene expression and gene co-regulation in the human retina. *Nucleic Acids Res.* **2016**, *44*, 5773–5784. [CrossRef]

413. Hoshino, A.; Ratnapriya, R.; Brooks, M.J.; Chaitankar, V.; Wilken, M.S.; Zhang, C.; Starostik, M.R.; Gieser, L.; La Torre, A.; Nishio, M.; et al. Molecular Anatomy of the Developing Human Retina. *Dev. Cell* **2017**, *43*, 763.e764–779.e764. [CrossRef]

414. Graziotto, J.J.; Farkas, M.H.; Bujakowska, K.; Deramaudt, B.M.; Zhang, Q.; Nandrot, E.F.; Inglehearn, C.F.; Bhattacharya, S.S.; Pierce, E.A. Three gene-targeted mouse models of RNA splicing factor RP show late-onset RPE and retinal degeneration. *Investig. Ophthalmol. Vis. Sci.* **2011**, *52*, 190–198. [CrossRef]

415. Xu, M.; Xie, Y.A.; Abouzeid, H.; Gordon, C.T.; Fiorentino, A.; Sun, Z.; Lehman, A.; Osman, I.S.; Dharmat, R.; Riveiro-Alvarez, R.; et al. Mutations in the Spliceosome Component CWC27 Cause Retinal Degeneration with or without Additional Developmental Anomalies. *Am. J. Hum. Genet.* **2017**, *100*, 592–604. [CrossRef]

416. Chen, B.J.; Lam, T.C.; Liu, L.Q.; To, C.H. Post-translational modifications and their applications in eye research (Review). *Mol. Med. Rep.* **2017**, *15*, 3923–3935. [CrossRef] [PubMed]

417. Christiansen, J.R.; Kolandaivelu, S.; Bergo, M.O.; Ramamurthy, V. RAS-converting enzyme 1-mediated endoproteolysis is required for trafficking of rod phosphodiesterase 6 to photoreceptor outer segments. *Proc. Natl. Acad. Sci. USA* **2011**, *108*, 8862–8866. [CrossRef] [PubMed]

418. Christiansen, J.R.; Pendse, N.D.; Kolandaivelu, S.; Bergo, M.O.; Young, S.G.; Ramamurthy, V. Deficiency of Isoprenylcysteine Carboxyl Methyltransferase (ICMT) Leads to Progressive Loss of Photoreceptor Function. *J. Neurosci.* **2016**, *36*, 5107–5114. [CrossRef]

419. Bosch Grau, M.; Masson, C.; Gadadhar, S.; Rocha, C.; Tort, O.; Marques Sousa, P.; Vacher, S.; Bieche, I.; Janke, C. Alterations in the balance of tubulin glycylation and glutamylation in photoreceptors leads to retinal degeneration. *J. Cell Sci.* **2017**, *130*, 938–949. [CrossRef] [PubMed]

420. Sun, X.; Park, J.H.; Gumerson, J.; Wu, Z.; Swaroop, A.; Qian, H.; Roll-Mecak, A.; Li, T. Loss of RPGR glutamylation underlies the pathogenic mechanism of retinal dystrophy caused by TTLL5 mutations. *Proc. Natl. Acad. Sci. USA* **2016**, *113*, E2925–E2934. [CrossRef] [PubMed]

421. Blanks, J.C.; Spee, C. Retinal degeneration in the pcd/pcd mutant mouse: Accumulation of spherules in the interphotoreceptor space. *Exp. Eye Res.* **1992**, *54*, 637–644. [CrossRef]

422. Chen, J.; Qian, H.; Horai, R.; Chan, C.C.; Falick, Y.; Caspi, R.R. Comparative analysis of induced vs. spontaneous models of autoimmune uveitis targeting the interphotoreceptor retinoid binding protein. *PLoS ONE* **2013**, *8*, e72161. [CrossRef]

423. Yu, M.; Zou, W.; Peachey, N.S.; McIntyre, T.M.; Liu, J. A novel role of complement in retinal degeneration. *Investig. Ophthalmol. Vis. Sci.* **2012**, *53*, 7684–7692. [CrossRef]

424. Lyzogubov, V.V.; Bora, P.S.; Wu, X.; Horn, L.E.; de Roque, R.; Rudolf, X.V.; Atkinson, J.P.; Bora, N.S. The Complement Regulatory Protein CD46 Deficient Mouse Spontaneously Develops Dry-Type Age-Related Macular Degeneration-Like Phenotype. *Am. J. Pathol.* **2016**, *186*, 2088–2104. [CrossRef]

425. Jobling, A.I.; Waugh, M.; Vessey, K.A.; Phipps, J.A.; Trogrlic, L.; Greferath, U.; Mills, S.A.; Tan, Z.L.; Ward, M.M.; Fletcher, E.L. The Role of the Microglial Cx3cr1 Pathway in the Postnatal Maturation of Retinal Photoreceptors. *J. Neurosci.* **2018**, *38*, 4708–4723. [CrossRef]

426. Combadiere, C.; Feumi, C.; Raoul, W.; Keller, N.; Rodero, M.; Pezard, A.; Lavalette, S.; Houssier, M.; Jonet, L.; Picard, E.; et al. CX3CR1-dependent subretinal microglia cell accumulation is associated with cardinal features of age-related macular degeneration. *J. Clin. Investig.* **2007**, *117*, 2920–2928. [CrossRef]

427. Huang, H.; Liu, Y.; Wang, L.; Li, W. Age-related macular degeneration phenotypes are associated with increased tumor necrosis-alpha and subretinal immune cells in aged Cxcr5 knockout mice. *PLoS ONE* **2017**, *12*, e0173716. [CrossRef] [PubMed]

428. Zhang, X.; Zhu, J.; Chen, X.; Jie-Qiong, Z.; Li, X.; Luo, L.; Huang, H.; Liu, W.; Zhou, X.; Yan, J.; et al. Interferon Regulatory Factor 3 Deficiency Induces Age-Related Alterations of the Retina in Young and Old Mice. *Front. Cell Neurosci.* **2019**, *13*, 272. [CrossRef] [PubMed]

429. Zhang, Q.; Jiang, Y.; Miller, M.J.; Peng, B.; Liu, L.; Soderland, C.; Tang, J.; Kern, T.S.; Pintar, J.; Steinle, J.J. IGFBP-3 and TNF-alpha regulate retinal endothelial cell apoptosis. *Investig. Ophthalmol. Vis. Sci.* **2013**, *54*, 5376–5384. [CrossRef] [PubMed]

430. Ambati, J.; Anand, A.; Fernandez, S.; Sakurai, E.; Lynn, B.C.; Kuziel, W.A.; Rollins, B.J.; Ambati, B.K. An animal model of age-related macular degeneration in senescent Ccl-2- or Ccr-2-deficient mice. *Nat. Med.* **2003**, *9*, 1390–1397. [CrossRef] [PubMed]

431. Coffey, P.J.; Gias, C.; McDermott, C.J.; Lundh, P.; Pickering, M.C.; Sethi, C.; Bird, A.; Fitzke, F.W.; Maass, A.; Chen, L.L.; et al. Complement factor H deficiency in aged mice causes retinal abnormalities and visual dysfunction. *Proc. Natl. Acad. Sci. USA* **2007**, *104*, 16651–16656. [CrossRef]

432. Ueda, Y.; Mohammed, I.; Song, D.; Gullipalli, D.; Zhou, L.; Sato, S.; Wang, Y.; Gupta, S.; Cheng, Z.; Wang, H.; et al. Murine systemic thrombophilia and hemolytic uremic syndrome from a factor H point mutation. *Blood* **2017**, *129*, 1184–1196. [CrossRef]

433. Ma, W.; Silverman, S.M.; Zhao, L.; Villasmil, R.; Campos, M.M.; Amaral, J.; Wong, W.T. Absence of TGFbeta signaling in retinal microglia induces retinal degeneration and exacerbates choroidal neovascularization. *Elife* **2019**, *8*. [CrossRef]

434. Zhou, Y.; Li, S.; Huang, L.; Yang, Y.; Zhang, L.; Yang, M.; Liu, W.; Ramasamy, K.; Jiang, Z.; Sundaresan, P.; et al. A splicing mutation in aryl hydrocarbon receptor associated with retinitis pigmentosa. *Hum. Mol. Genet.* **2018**, *27*, 2563–2572. [CrossRef]

435. Swiderski, R.E.; Nakano, Y.; Mullins, R.F.; Seo, S.; Banfi, B. A mutation in the mouse ttc26 gene leads to impaired hedgehog signaling. *PLoS Genet.* **2014**, *10*, e1004689. [CrossRef]

436. Omori, Y.; Kubo, S.; Kon, T.; Furuhashi, M.; Narita, H.; Kominami, T.; Ueno, A.; Tsutsumi, R.; Chaya, T.; Yamamoto, H.; et al. Samd7 is a cell type-specific PRC1 component essential for establishing retinal rod photoreceptor identity. *Proc. Natl. Acad. Sci. USA* **2017**, *114*, E8264–E8273. [CrossRef] [PubMed]

437. Behnen, P.; Felline, A.; Comitato, A.; Di Salvo, M.T.; Raimondi, F.; Gulati, S.; Kahremany, S.; Palczewski, K.; Marigo, V.; Fanelli, F. A Small Chaperone Improves Folding and Routing of Rhodopsin Mutants Linked to Inherited Blindness. *iScience* **2018**, *4*, 1–19. [CrossRef] [PubMed]

438. Kroeger, H.; Chiang, W.C.; Felden, J.; Nguyen, A.; Lin, J.H. ER stress and unfolded protein response in ocular health and disease. *FEBS J.* **2019**, *286*, 399–412. [CrossRef] [PubMed]

439. Stuck, M.W.; Conley, S.M.; Naash, M.I. PRPH2/RDS and ROM-1: Historical context, current views and future considerations. *Prog. Retin. Eye Res.* **2016**, *52*, 47–63. [CrossRef]

440. Boon, C.J.; den Hollander, A.I.; Hoyng, C.B.; Cremers, F.P.; Klevering, B.J.; Keunen, J.E. The spectrum of retinal dystrophies caused by mutations in the peripherin/RDS gene. *Prog. Retin. Eye Res.* **2008**, *27*, 213–235. [CrossRef]

441. Liu, M.M.; Zack, D.J. Alternative splicing and retinal degeneration. *Clin. Genet.* **2013**, *84*, 142–149. [CrossRef]

442. Zhao, Y.; Hong, D.H.; Pawlyk, B.; Yue, G.; Adamian, M.; Grynberg, M.; Godzik, A.; Li, T. The retinitis pigmentosa GTPase regulator (RPGR)- interacting protein: Subserving RPGR function and participating in disk morphogenesis. *Proc. Natl. Acad. Sci. USA* **2003**, *100*, 3965–3970. [CrossRef]

443. Won, J.; Gifford, E.; Smith, R.S.; Yi, H.; Ferreira, P.A.; Hicks, W.L.; Li, T.; Naggert, J.K.; Nishina, P.M. RPGRIP1 is essential for normal rod photoreceptor outer segment elaboration and morphogenesis. *Hum. Mol. Genet.* **2009**, *18*, 4329–4339. [CrossRef]

444. Zhang, C.; Quan, R.; Wang, J. Development and application of CRISPR/Cas9 technologies in genomic editing. *Hum. Mol. Genet.* **2018**, *27*, R79–R88. [CrossRef]

445. Comitato, A.; Schiroli, D.; La Marca, C.; Marigo, V. Differential Contribution of Calcium-Activated Proteases and ER-Stress in Three Mouse Models of Retinitis Pigmentosa Expressing P23H Mutant RHO. *Adv. Exp. Med. Biol.* **2019**, *1185*, 311–316. [CrossRef]

446. Hahn, P.; Qian, Y.; Dentchev, T.; Chen, L.; Beard, J.; Harris, Z.L.; Dunaief, J.L. Disruption of ceruloplasmin and hephaestin in mice causes retinal iron overload and retinal degeneration with features of age-related macular degeneration. *Proc. Natl. Acad. Sci. USA* **2004**, *101*, 13850–13855. [CrossRef]

447. Kumar, M.V.; Nagineni, C.N.; Chin, M.S.; Hooks, J.J.; Detrick, B. Innate immunity in the retina: Toll-like receptor (TLR) signaling in human retinal pigment epithelial cells. *J. Neuroimmunol.* **2004**, *153*, 7–15. [CrossRef] [PubMed]

448. Shiose, S.; Chen, Y.; Okano, K.; Roy, S.; Kohno, H.; Tang, J.; Pearlman, E.; Maeda, T.; Palczewski, K.; Maeda, A. Toll-like receptor 3 is required for development of retinopathy caused by impaired all-trans-retinal clearance in mice. *J. Biol. Chem.* **2011**, *286*, 15543–15555. [CrossRef] [PubMed]

449. Vollrath, D.; Yasumura, D.; Benchorin, G.; Matthes, M.T.; Feng, W.; Nguyen, N.M.; Sedano, C.D.; Calton, M.A.; LaVail, M.M. Tyro3 Modulates Mertk-Associated Retinal Degeneration. *PLoS Genet.* **2015**, *11*, e1005723. [CrossRef] [PubMed]

450. Zhang, Y.; Seo, S.; Bhattarai, S.; Bugge, K.; Searby, C.C.; Zhang, Q.; Drack, A.V.; Stone, E.M.; Sheffield, V.C. BBS mutations modify phenotypic expression of CEP290-related ciliopathies. *Hum. Mol. Genet.* **2014**, *23*, 40–51. [CrossRef] [PubMed]

451. Rachel, R.A.; May-Simera, H.L.; Veleri, S.; Gotoh, N.; Choi, B.Y.; Murga-Zamalloa, C.; McIntyre, J.C.; Marek, J.; Lopez, I.; Hackett, A.N.; et al. Combining Cep290 and Mkks ciliopathy alleles in mice rescues sensory defects and restores ciliogenesis. *J. Clin. Investig.* **2012**, *122*, 1233–1245. [CrossRef] [PubMed]

452. Rao, K.N.; Zhang, W.; Li, L.; Ronquillo, C.; Baehr, W.; Khanna, H. Ciliopathy-associated protein CEP290 modifies the severity of retinal degeneration due to loss of RPGR. *Hum. Mol. Genet.* **2016**, *25*, 2005–2012. [CrossRef]

453. Schon, C.; Asteriti, S.; Koch, S.; Sothilingam, V.; Garcia Garrido, M.; Tanimoto, N.; Herms, J.; Seeliger, M.W.; Cangiano, L.; Biel, M.; et al. Loss of HCN1 enhances disease progression in mouse models of CNG channel-linked retinitis pigmentosa and achromatopsia. *Hum. Mol. Genet.* **2016**, *25*, 1165–1175. [CrossRef]

454. Quinn, P.M.; Mulder, A.A.; Henrique Alves, C.; Desrosiers, M.; de Vries, S.I.; Klooster, J.; Dalkara, D.; Koster, A.J.; Jost, C.R.; Wijnholds, J. Loss of CRB2 in Muller glial cells modifies a CRB1-associated retinitis pigmentosa phenotype into a Leber congenital amaurosis phenotype. *Hum. Mol. Genet.* **2019**, *28*, 105–123. [CrossRef]

455. Ng, L.; Liu, H.; St Germain, D.L.; Hernandez, A.; Forrest, D. Deletion of the Thyroid Hormone-Activating Type 2 Deiodinase Rescues Cone Photoreceptor Degeneration but Not Deafness in Mice Lacking Type 3 Deiodinase. *Endocrinology* **2017**, *158*, 1999–2010. [CrossRef]

456. van der Pluijm, I.; Garinis, G.A.; Brandt, R.M.; Gorgels, T.G.; Wijnhoven, S.W.; Diderich, K.E.; de Wit, J.; Mitchell, J.R.; van Oostrom, C.; Beems, R.; et al. Impaired genome maintenance suppresses the growth hormone–insulin-like growth factor 1 axis in mice with Cockayne syndrome. *PLoS Biol.* **2007**, *5*, e2. [CrossRef]

457. Chang, B.; Heckenlively, J.R.; Bayley, P.R.; Brecha, N.C.; Davisson, M.T.; Hawes, N.L.; Hirano, A.A.; Hurd, R.E.; Ikeda, A.; Johnson, B.A.; et al. The nob2 mouse, a null mutation in Cacna1f: Anatomical and functional abnormalities in the outer retina and their consequences on ganglion cell visual responses. *Vis. Neurosci.* **2006**, *23*, 11–24. [CrossRef] [PubMed]

458. Humphries, M.M.; Kiang, S.; McNally, N.; Donovan, M.A.; Sieving, P.A.; Bush, R.A.; Machida, S.; Cotter, T.; Hobson, A.; Farrar, J.; et al. Comparative structural and functional analysis of photoreceptor neurons of Rho-/- mice reveal increased survival on C57BL/6J in comparison to 129Sv genetic background. *Vis. Neurosci.* **2001**, *18*, 437–443. [CrossRef]

459. Liu, Q.; Saveliev, A.; Pierce, E.A. The severity of retinal degeneration in Rp1h gene-targeted mice is dependent on genetic background. *Investig. Ophthalmol. Vis. Sci.* **2009**, *50*, 1566–1574. [CrossRef] [PubMed]

460. Haider, N.B.; Zhang, W.; Hurd, R.; Ikeda, A.; Nystuen, A.M.; Naggert, J.K.; Nishina, P.M. Mapping of genetic modifiers of Nr2e3 rd7/rd7 that suppress retinal degeneration and restore blue cone cells to normal quantity. *Mamm. Genome* **2008**, *19*, 145–154. [CrossRef]

461. Threadgill, D.W.; Miller, D.R.; Churchill, G.A.; de Villena, F.P. The collaborative cross: A recombinant inbred mouse population for the systems genetic era. *ILAR J.* **2011**, *52*, 24–31. [CrossRef]

462. Churchill, G.A.; Gatti, D.M.; Munger, S.C.; Svenson, K.L. The Diversity Outbred mouse population. *Mamm. Genome* **2012**, *23*, 713–718. [CrossRef]

463. Cruz, N.M.; Yuan, Y.; Leehy, B.D.; Baid, R.; Kompella, U.; DeAngelis, M.M.; Escher, P.; Haider, N.B. Modifier genes as therapeutics: The nuclear hormone receptor Rev Erb alpha (Nr1d1) rescues Nr2e3 associated retinal disease. *PLoS ONE* **2014**, *9*, e87942. [CrossRef]

464. Danciger, M.; Ogando, D.; Yang, H.; Matthes, M.T.; Yu, N.; Ahern, K.; Yasumura, D.; Williams, R.W.; Lavail, M.M. Genetic modifiers of retinal degeneration in the rd3 mouse. *Investig. Ophthalmol. Vis. Sci.* **2008**, *49*, 2863–2869. [CrossRef]

465. van Wyk, M.; Schneider, S.; Kleinlogel, S. Variable phenotypic expressivity in inbred retinal degeneration mouse lines: A comparative study of C3H/HeOu and FVB/N rd1 mice. *Mol. Vis.* **2015**, *21*, 811–827.

466. Luhmann, U.F.; Carvalho, L.S.; Holthaus, S.M.; Cowing, J.A.; Greenaway, S.; Chu, C.J.; Herrmann, P.; Smith, A.J.; Munro, P.M.; Potter, P.; et al. The severity of retinal pathology in homozygous Crb1rd8/rd8 mice is dependent on additional genetic factors. *Hum. Mol. Genet.* **2015**, *24*, 128–141. [CrossRef] [PubMed]

467. Won, J.; Charette, J.R.; Philip, V.M.; Stearns, T.M.; Zhang, W.; Naggert, J.K.; Krebs, M.P.; Nishina, P.M. Genetic modifier loci of mouse Mfrp(rd6) identified by quantitative trait locus analysis. *Exp. Eye Res.* **2014**, *118*, 30–35. [CrossRef] [PubMed]

468. Maddox, D.M.; Ikeda, S.; Ikeda, A.; Zhang, W.; Krebs, M.P.; Nishina, P.M.; Naggert, J.K. An allele of microtubule-associated protein 1A (Mtap1a) reduces photoreceptor degeneration in Tulp1 and Tub Mutant Mice. *Investig. Ophthalmol. Vis. Sci.* **2012**, *53*, 1663–1669. [CrossRef]

469. Kong, Y.; Zhao, L.; Charette, J.R.; Hicks, W.L.; Stone, L.; Nishina, P.M.; Naggert, J.K. An FRMD4B variant suppresses dysplastic photoreceptor lesions in models of enhanced S-cone syndrome and of Nrl deficiency. *Hum. Mol. Genet.* **2018**, *27*, 3340–3352. [CrossRef] [PubMed]

470. Paskowitz, D.M.; LaVail, M.M.; Duncan, J.L. Light and inherited retinal degeneration. *Br. J. Ophthalmol.* **2006**, *90*, 1060–1066. [CrossRef] [PubMed]

471. Dellett, M.; Sasai, N.; Nishide, K.; Becker, S.; Papadaki, V.; Limb, G.A.; Moore, A.T.; Kondo, T.; Ohnuma, S. Genetic background and light-dependent progression of photoreceptor cell degeneration in Prominin-1 knockout mice. *Investig. Ophthalmol. Vis. Sci.* **2014**, *56*, 164–176. [CrossRef]

472. Naash, M.L.; Peachey, N.S.; Li, Z.Y.; Gryczan, C.C.; Goto, Y.; Blanks, J.; Milam, A.H.; Ripps, H. Light-induced acceleration of photoreceptor degeneration in transgenic mice expressing mutant rhodopsin. *Investig. Ophthalmol. Vis. Sci.* **1996**, *37*, 775–782.

473. Rascher, K.; Servos, G.; Berthold, G.; Hartwig, H.G.; Warskulat, U.; Heller-Stilb, B.; Haussinger, D. Light deprivation slows but does not prevent the loss of photoreceptors in taurine transporter knockout mice. *Vis. Res.* **2004**, *44*, 2091–2100. [CrossRef]

474. Smith, S.B.; Cope, B.K.; McCoy, J.R. Effects of dark-rearing on the retinal degeneration of the C57BL/6-mivit/mivit mouse. *Exp. Eye Res.* **1994**, *58*, 77–84. [CrossRef]

475. Maeda, A.; Maeda, T.; Imanishi, Y.; Sun, W.; Jastrzebska, B.; Hatala, D.A.; Winkens, H.J.; Hofmann, K.P.; Janssen, J.J.; Baehr, W.; et al. Retinol dehydrogenase (RDH12) protects photoreceptors from light-induced degeneration in mice. *J. Biol. Chem.* **2006**, *281*, 37697–37704. [CrossRef]

476. Ettaiche, M.; Guy, N.; Hofman, P.; Lazdunski, M.; Waldmann, R. Acid-sensing ion channel 2 is important for retinal function and protects against light-induced retinal degeneration. *J. Neurosci.* **2004**, *24*, 1005–1012. [CrossRef] [PubMed]

477. Peng, Y.W.; Zallocchi, M.; Wang, W.M.; Delimont, D.; Cosgrove, D. Moderate light-induced degeneration of rod photoreceptors with delayed transducin translocation in shaker1 mice. *Investig. Ophthalmol. Vis. Sci.* **2011**, *52*, 6421–6427. [CrossRef] [PubMed]

478. Tian, M.; Wang, W.; Delimont, D.; Cheung, L.; Zallocchi, M.; Cosgrove, D.; Peng, Y.W. Photoreceptors in whirler mice show defective transducin translocation and are susceptible to short-term light/dark changes-induced degeneration. *Exp. Eye Res.* **2014**, *118*, 145–153. [CrossRef] [PubMed]

479. Li, G.; Anderson, R.E.; Tomita, H.; Adler, R.; Liu, X.; Zack, D.J.; Rajala, R.V. Nonredundant role of Akt2 for neuroprotection of rod photoreceptor cells from light-induced cell death. *J. Neurosci.* **2007**, *27*, 203–211. [CrossRef]

480. Datta, S.; Cano, M.; Ebrahimi, K.; Wang, L.; Handa, J.T. The impact of oxidative stress and inflammation on RPE degeneration in non-neovascular AMD. *Prog. Retin. Eye Res.* **2017**, *60*, 201–218. [CrossRef]

481. Khan, J.C.; Thurlby, D.A.; Shahid, H.; Clayton, D.G.; Yates, J.R.; Bradley, M.; Moore, A.T.; Bird, A.C.; Genetic Factors in AMD Study. Smoking and age related macular degeneration: The number of pack years of cigarette smoking is a major determinant of risk for both geographic atrophy and choroidal neovascularisation. *Br. J. Ophthalmol.* **2006**, *90*, 75–80. [CrossRef]

482. Cano, M.; Thimmalappula, R.; Fujihara, M.; Nagai, N.; Sporn, M.; Wang, A.L.; Neufeld, A.H.; Biswal, S.; Handa, J.T. Cigarette smoking, oxidative stress, the anti-oxidant response through Nrf2 signaling, and Age-related Macular Degeneration. *Vis. Res.* **2010**, *50*, 652–664. [CrossRef]

483. Fujihara, M.; Nagai, N.; Sussan, T.E.; Biswal, S.; Handa, J.T. Chronic cigarette smoke causes oxidative damage and apoptosis to retinal pigmented epithelial cells in mice. *PLoS ONE* **2008**, *3*, e3119. [CrossRef]

484. Ebrahimi, K.B.; Cano, M.; Rhee, J.; Datta, S.; Wang, L.; Handa, J.T. Oxidative Stress Induces an Interactive Decline in Wnt and Nrf2 Signaling in Degenerating Retinal Pigment Epithelium. *Antioxid. Redox Signal.* **2018**, *29*, 389–407. [CrossRef]

485. Farese, R.V., Jr.; Veniant, M.M.; Cham, C.M.; Flynn, L.M.; Pierotti, V.; Loring, J.F.; Traber, M.; Ruland, S.; Stokowski, R.S.; Huszar, D.; et al. Phenotypic analysis of mice expressing exclusively apolipoprotein B48 or apolipoprotein B100. *Proc. Natl. Acad. Sci. USA* **1996**, *93*, 6393–6398. [CrossRef]

486. Schmidt-Erfurth, U.; Rudolf, M.; Funk, M.; Hofmann-Rummelt, C.; Franz-Haas, N.S.; Aherrahrou, Z.; Schlotzer-Schrehardt, U. Ultrastructural changes in a murine model of graded Bruch membrane lipoidal degeneration and corresponding VEGF164 detection. *Investig. Ophthalmol. Vis. Sci.* **2008**, *49*, 390–398. [CrossRef] [PubMed]

487. Malek, G.; Johnson, L.V.; Mace, B.E.; Saloupis, P.; Schmechel, D.E.; Rickman, D.W.; Toth, C.A.; Sullivan, P.M.; Bowes Rickman, C. Apolipoprotein E allele-dependent pathogenesis: A model for age-related retinal degeneration. *Proc. Natl. Acad. Sci. USA* **2005**, *102*, 11900–11905. [CrossRef] [PubMed]

488. Perusek, L.; Maeda, T. Vitamin A derivatives as treatment options for retinal degenerative diseases. *Nutrients* **2013**, *5*, 2646–2666. [CrossRef] [PubMed]

489. Perusek, L.; Maeda, A.; Maeda, T. Supplementation with vitamin a derivatives to rescue vision in animal models of degenerative retinal diseases. *Methods Mol. Biol.* **2015**, *1271*, 345–362. [CrossRef]

490. Guadagni, V.; Novelli, E.; Piano, I.; Gargini, C.; Strettoi, E. Pharmacological approaches to retinitis pigmentosa: A laboratory perspective. *Prog. Retin. Eye Res.* **2015**, *48*, 62–81. [CrossRef] [PubMed]

491. Cai, X.; McGinnis, J.F. Oxidative stress: The achilles' heel of neurodegenerative diseases of the retina. *Front. Biosci. Landmark. Ed.* **2012**, *17*, 1976–1995. [CrossRef]

492. Komeima, K.; Rogers, B.S.; Campochiaro, P.A. Antioxidants slow photoreceptor cell death in mouse models of retinitis pigmentosa. *J. Cell Physiol.* **2007**, *213*, 809–815. [CrossRef]

493. Ebert, S.; Weigelt, K.; Walczak, Y.; Drobnik, W.; Mauerer, R.; Hume, D.A.; Weber, B.H.; Langmann, T. Docosahexaenoic acid attenuates microglial activation and delays early retinal degeneration. *J. Neurochem.* **2009**, *110*, 1863–1875. [CrossRef]

494. Yu, M.; Yan, W.; Beight, C. Lutein and Zeaxanthin Isomers Reduce Photoreceptor Degeneration in the Pde6b (rd10) Mouse Model of Retinitis Pigmentosa. *Biomed. Res. Int.* **2018**, *2018*, 4374087. [CrossRef]

495. Komeima, K.; Rogers, B.S.; Lu, L.; Campochiaro, P.A. Antioxidants reduce cone cell death in a model of retinitis pigmentosa. *Proc. Natl. Acad. Sci. USA* **2006**, *103*, 11300–11305. [CrossRef]

496. Barone, I.; Novelli, E.; Piano, I.; Gargini, C.; Strettoi, E. Environmental enrichment extends photoreceptor survival and visual function in a mouse model of retinitis pigmentosa. *PLoS ONE* **2012**, *7*, e50726. [CrossRef]

497. Sundin, O.H.; Leppert, G.S.; Silva, E.D.; Yang, J.M.; Dharmaraj, S.; Maumenee, I.H.; Santos, L.C.; Parsa, C.F.; Traboulsi, E.I.; Broman, K.W.; et al. Extreme hyperopia is the result of null mutations in MFRP, which encodes a Frizzled-related protein. *Proc. Natl. Acad. Sci. USA* **2005**, *102*, 9553–9558. [CrossRef] [PubMed]

498. Sundin, O.H. The mouse's eye and Mfrp: Not quite human. *Ophthalmic Genet.* **2005**, *26*, 153–155. [CrossRef] [PubMed]

499. Kameya, S.; Hawes, N.L.; Chang, B.; Heckenlively, J.R.; Naggert, J.K.; Nishina, P.M. Mfrp, a gene encoding a frizzled related protein, is mutated in the mouse retinal degeneration 6. *Hum. Mol. Genet.* **2002**, *11*, 1879–1886. [CrossRef] [PubMed]

500. Almoallem, B.; Arno, G.; De Zaeytijd, J.; Verdin, H.; Balikova, I.; Casteels, I.; de Ravel, T.; Hull, S.; Suzani, M.; Destree, A.; et al. The majority of autosomal recessive nanophthalmos and posterior microphthalmia can be attributed to biallelic sequence and structural variants in MFRP and PRSS56. *Sci. Rep.* **2020**, *10*, 1289. [CrossRef] [PubMed]

501. Chekuri, A.; Sahu, B.; Chavali, V.R.M.; Voronchikhina, M.; Soto-Hermida, A.; Suk, J.J.; Alapati, A.N.; Bartsch, D.U.; Ayala-Ramirez, R.; Zenteno, J.C.; et al. Long-Term Effects of Gene Therapy in a Novel Mouse Model of Human MFRP-Associated Retinopathy. *Hum. Gene Ther.* **2019**, *30*, 632–650. [CrossRef] [PubMed]

502. Gibson, F.; Walsh, J.; Mburu, P.; Varela, A.; Brown, K.A.; Antonio, M.; Beisel, K.W.; Steel, K.P.; Brown, S.D. A type VII myosin encoded by the mouse deafness gene shaker-1. *Nature* **1995**, *374*, 62–64. [CrossRef]

503. Lentz, J.J.; Gordon, W.C.; Farris, H.E.; MacDonald, G.H.; Cunningham, D.E.; Robbins, C.A.; Tempel, B.L.; Bazan, N.G.; Rubel, E.W.; Oesterle, E.C.; et al. Deafness and retinal degeneration in a novel USH1C knock-in mouse model. *Dev. Neurobiol.* **2010**, *70*, 253–267. [CrossRef]

504. Karan, G.; Lillo, C.; Yang, Z.; Cameron, D.J.; Locke, K.G.; Zhao, Y.; Thirumalaichary, S.; Li, C.; Birch, D.G.; Vollmer-Snarr, H.R.; et al. Lipofuscin accumulation, abnormal electrophysiology, and photoreceptor degeneration in mutant ELOVL4 transgenic mice: A model for macular degeneration. *Proc. Natl. Acad. Sci. USA* **2005**, *102*, 4164–4169. [CrossRef]

505. Grishchuk, Y.; Stember, K.G.; Matsunaga, A.; Olivares, A.M.; Cruz, N.M.; King, V.E.; Humphrey, D.M.; Wang, S.L.; Muzikansky, A.; Betensky, R.A.; et al. Retinal Dystrophy and Optic Nerve Pathology in the Mouse Model of Mucolipidosis IV. *Am. J. Pathol.* **2016**, *186*, 199–209. [CrossRef]

506. Amir, N.; Zlotogora, J.; Bach, G. Mucolipidosis type IV: Clinical spectrum and natural history. *Pediatrics* **1987**, *79*, 953–959. [PubMed]

507. Park, H.; Qazi, Y.; Tan, C.; Jabbar, S.B.; Cao, Y.; Schmid, G.; Pardue, M.T. Assessment of axial length measurements in mouse eyes. *Optom. Vis. Sci.* **2012**, *89*, 296–303. [CrossRef]

508. Ruppersburg, C.C.; Hartzell, H.C. The Ca2+-activated Cl- channel ANO1/TMEM16A regulates primary ciliogenesis. *Mol. Biol. Cell* **2014**, *25*, 1793–1807. [CrossRef] [PubMed]

509. He, M.; Ye, W.; Wang, W.J.; Sison, E.S.; Jan, Y.N.; Jan, L.Y. Cytoplasmic Cl(-) couples membrane remodeling to epithelial morphogenesis. *Proc. Natl. Acad. Sci. USA* **2017**, *114*, E11161–E11169. [CrossRef] [PubMed]

510. Gupta, A.; Fabian, L.; Brill, J.A. Phosphatidylinositol 4,5-bisphosphate regulates cilium transition zone maturation in Drosophila melanogaster. *J. Cell Sci.* **2018**, *131*. [CrossRef]

511. Bhatia, V.; Valdes-Sanchez, L.; Rodriguez-Martinez, D.; Bhattacharya, S.S. Formation of 53BP1 foci and ATM activation under oxidative stress is facilitated by RNA:DNA hybrids and loss of ATM-53BP1 expression promotes photoreceptor cell survival in mice. *F1000Res* **2018**, *7*, 1233. [CrossRef]

512. Muller, B.; Ellinwood, N.M.; Lorenz, B.; Stieger, K. Detection of DNA Double Strand Breaks by gammaH2AX Does Not Result in 53bp1 Recruitment in Mouse Retinal Tissues. *Front. Neurosci.* **2018**, *12*, 286. [CrossRef]

Review

Large Animal Models of Inherited Retinal Degenerations: A Review

Paige A. Winkler, Laurence M. Occelli and Simon M. Petersen-Jones *

Department of Small Animal Clinical Sciences, Veterinary Medical Center, Michigan State University, East Lansing, MI 48824, USA; winkle38@msu.edu (P.A.W.); occelli@msu.edu (L.M.O.)
* Correspondence: peter315@msu.edu; Tel.: +1-517-353-3278

Received: 1 March 2020; Accepted: 31 March 2020; Published: 3 April 2020

Abstract: Studies utilizing large animal models of inherited retinal degeneration (IRD) have proven important in not only the development of translational therapeutic approaches, but also in improving our understanding of disease mechanisms. The dog is the predominant species utilized because spontaneous IRD is common in the canine pet population. Cats are also a source of spontaneous IRDs. Other large animal models with spontaneous IRDs include sheep, horses and non-human primates (NHP). The pig has also proven valuable due to the ease in which transgenic animals can be generated and work is ongoing to produce engineered models of other large animal species including NHP. These large animal models offer important advantages over the widely used laboratory rodent models. The globe size and dimensions more closely parallel those of humans and, most importantly, they have a retinal region of high cone density and denser photoreceptor packing for high acuity vision. Laboratory rodents lack such a retinal region and, as macular disease is a critical cause for vision loss in humans, having a comparable retinal region in model species is particularly important. This review will discuss several large animal models which have been used to study disease mechanisms relevant for the equivalent human IRD.

Keywords: large animal model; inherited retinal disease; progressive retinal atrophy; retinitis pigmentosa; Leber congenital amaurosis; achromatopsia; congenital stationary night blindness

---

1. Introduction

Large animal models for inherited retinal degenerations (IRDs) have been identified within populations of dogs, cats, sheep, horses and, more recently, non-human primates (NHP). Many different IRDs have been identified in pedigree dogs, most of which mimic retinitis pigmentosa (RP) or Leber congenital amaurosis (LCA). The term "progressive retinal atrophy" (PRA) is used to describe this group of photoreceptor degenerations. In some instances, more detailed descriptive terms such as rod–cone dysplasia or progressive rod cone degeneration have been used. Common dog breeding practices have tended to bring out recessive conditions and have made the pedigree dog a rich source of models for inherited disease, including IRDs. Engineered large animal IRD models such as transgenic pigs have also been produced [1]. With advances in gene editing technologies, further models are likely to be produced in species such as pigs and NHPs and possibly even cats and dogs. The advantages of large animal models over laboratory rodent models of IRDs include the similarity in the size of the eye to that of man [2]. This is of particular importance for the development of translational therapies because it allows identical surgical delivery approaches to be used in the animal model to those that will be eventually used in human patients. NHPs are obviously close relatives to humans, making them attractive models. However, only a few spontaneous IRDs in primates have been identified [3,4], although steps are being taken to identify more animals with disease-causing mutations and to use genome editing to generate additional models with mutations in genes of importance, either for germline transmission (see [5] for a review) or somatic gene knockout [6]. Another major advantage of large animal models is the presence of a retinal region equivalent to the macula. Laboratory mice and rats are nocturnal rodents that do not have a macula equivalent. The macula is of major importance for high acuity vision and some conditions specifically or differentially affect that retinal region compared to the peripheral retina. Macular dystrophies have been associated with mutations in a number of different genes (see [7,8] for reviews), some of which have relevant large animal models which are discussed below, including Stargardt Disease (*ABCA4* mutations), Best Disease (*BEST1* mutations) and in some patients with *RDH5* mutations. Age-related macular degeneration (AMD) is a major cause of vision loss and has genetic and environmental contributors [9,10]. Screening of primate colonies for animals with lesions comparable to AMD have been performed [11] (for a review, including primate models, see Pennesi et al. [12]). The large animal model species considered here have an area centralis with high photoreceptor density, including, importantly, cones that are equivalent to the human macula [13]. While NHP also have a fovea, most of the other model species do not, although the dog has been reported to have a small concentration of cones in the center of the area centralis, referred to as a "bouquet" of cones [14].

There are several important instances where laboratory rodent engineered models fail to recapitulate the human disease; important examples including *ABCA4*-Stargardt disease, *RDH5*-retinopathy, or where the gene involved is not present in the mouse or rat genome e.g., *EYS* [15].

The disadvantages of the large animal model species tend to be cost, generation of sufficient animals due to the slower reproduction, and because of the longer lifespan (compared to laboratory rodents), the disease course may be longer.

Large animal models have also been important in therapy development. The first proof-of-concept gene augmentation therapy that eventually led to the FDA approval of the first gene therapy product, Luxturna, for the treatment of LCA due to *RPE65* mutations, was performed in a dog model [16].

Table 1 shows a list of IRDs in large animal models and their identified mutations. Retinal layers and the genes discussed in detail within the text are shown in Figure 1. This review will cover the models in which the studies have contributed to the understanding of the mechanism of disease and/or protein structure and function in greater detail.

Table 1. List of mutations in large animal models of inherited retinal disease. Detailed information for some genes (indicated with*) are displayed in the text, highlighting the role of large animal models in advancing the understanding of the disease mechanism and/or providing insights into normal protein structure and function. References are available in Supplementary Material Table S1.

Mechanism	Gene	Species	Mutation
Phototransduction	*RHO**	dog pig pig pig	c.11C>G, p.Thr4Arg p.Pro23His c.1040C>T, p.Pro347Leu p.Pro347Ser
	*PDE6A**	dog	c.1939delA, p.Asn616ThrfsTer39
	*PDE6B**	dog	c.2420G>A, p.Trp807Ter; c.2449_2450insTGAAGTCC; p.Lys816Terfs817 c.2404_2406delAAC, p.Asn802del
	*PDE6C**	NHP	c.1694G>A, p.Arg565Gln
	SAG	dog	c.1216T>C; p.Ter406extArg*25
Visual Cycle	*ABCA4**	dog	c.4176insC, p.Phe1393LeufsTer3
	*RPE65**	dog	c.487_490delAAGA; p.Lys154LeufsTer53
	*RDH5**	cat	unpublished
Channelopathies/ channel related	*CNGA1**	dog	c.1752_1755delAACT, p.Thr584SerfsTer9
	*CNGB1**	dog	c.2387delA;2389_2390insAGCTAC, p.Ser791ArgfsTer2
	*CNGA3**	dog dog sheep sheep	c.1270C>T; p.Arg424Trp c.1931_1933delTGG, p.Val644del p.Arg236Ter p.Gly540Ser
	*CNGB3**	dog	c.784G>A; p.Asp262Asn CFA29:g.35,699,378-36,104,197del, c.0

Table 1. *Cont.*

Mechanism	Gene	Species	Mutation
	BEST1*	dog	c.73C>T, p.Arg25Ter; c.482G>A, p.Gly161Asp; c.C1388del and c.1466G>T, p.Pro463fs and p.Gly489Val
	BBS4*	dog	c.58A>T, p.Lys20Ter
	BBS7*	NHP	c.160delG, p.Ala54GlnfsTer18
	c2orf71	dog	c.3149_3150insC, p.Lys1051ValfsTer91
	CCDC66	dog	c.521_522insA, p.Asn174LysfsTer2
	CEP290*	cat	c.6966+9T>G, p.Ile2323AlafsTer3
	FAM161A	dog	c.1758-15_1758-16ins238, p.Ser588MetfsTer14
Ciliopathies	NPHP4*	dog	c.462_526del, p.Leu155LysfsTer2
	NPHP5(IQCB1)*	dog / cat	c.952-953insC, p.Ser319IlefsTer13 c.1282delCT, p.Leu428Ter
	RPGR*	dog	c.1084-1085delGA, c.1028-1032delGAGAA CFAX:g. 33106747+190-33102324del
	RGRIP1*	dog	CFA15:g.8228_8229insA29GGAAGCAACAGGATG
	TTC8	dog	c.669delA, p.Lys223ArgfsTer15
Photoreceptor development	CRX*	cat	c.546delC, p.Pro185LysfsTer2
	STK38L*	dog	c.299_300ins [218;285_299]; p.Lys63_Glu103del

Table 1. *Cont.*

Mechanism	Gene	Species	Mutation
Photoreceptor to Bipolar Cell	LRIT3*	dog	c.762_763delG, p.Lys246AsnfsTer5
	TRPM1*	horse	ECA1g.108,297,929_108,297,930ins1378
	Whippet*	dog	unpublished
Structural/Other	ADAM9*	dog	c.1592_1881del p.Lys531AsnfsTer3
	AIPL1*	cat	c.577C>T, p.Arg193Ter
	MERTK*	dog	CFA17:g.36338057_36338058ins[(6401);36338043-36338057]
	PRCD*	dog	c.5G>A, p.Cys2tyr
	RD3*	dog	c.418_419ins[22]
	NECAP1	dog	c.544G>A, p.Gly182Arg
	PPT1	dog	CFA15:g.[2,866,454_2,877,574dup; 2,874,661_2,875,048con2,877,563-2,877,607inv]
	SLC4A3	dog	c.2601_2602insC, p.Glu868ArgfsTer104

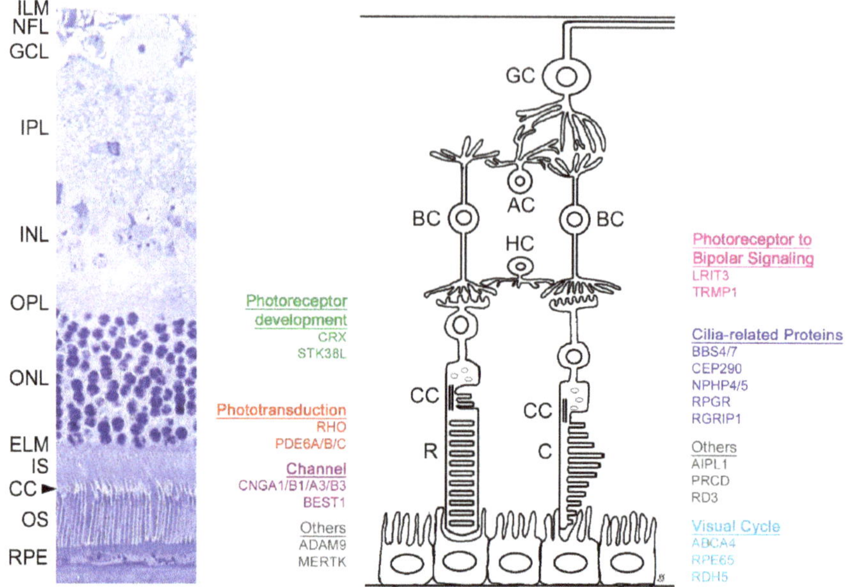

Figure 1. Schematic of retinal layers and associated genes discussed within this review. The left image shows a histologic section of a feline retina (with comparable anatomy to the human retina). The right panel depicts a schematic showing the genes detailed in this review and their site of expression, grouped per biological process. Inner limiting membrane (ILM), nerve fiber layer (NFL), ganglion cell layer (GCL), inner plexiform layer (IPL), inner nuclear layer (INL), outer plexifom layer (OPL), outer nuclear layer (ONL), external limiting membrane (ELM), photoreceptor inner segment (IS), connecting cilium (CC), photoreceptor outer segment (OS), retinal pigmentary epithelium (RPE). Ganglion cell (GC), amacrine cell (AC), bipolar cell (BC), horizontal cell (HC), rod (R), cone (C).

2. Mutations in Phototransduction Genes

There are spontaneously occurring large animal models with mutations in the genes of the phototransduction cascade, and rhodopsin (*RHO*) transgenic pig models have been generated. Studies of these models have contributed to the understanding of how the mutation of these genes leads to photoreceptor death.

2.1. RHO

Mutations in *RHO* can result in a range of phenotypes, most commonly autosomal dominant RP (adRP), but also autosomal recessive RP (arRP) and congenital stationary night blindness (CSNB). Animal models and in vitro studies have allowed *RHO* mutations to be divided into seven classes (see [17] for a recent review).

2.1.1. Dog Model

A spontaneous dog model of *RHO* adRP has been identified [18]. The c.11C>G, p.Thr4Arg mutant dog (Rho^{T4R}) develops a retinal degeneration that is greatly exacerbated by light exposure [19]. The phenotype of this dog model closely resembles that of the human p.Thr4Lys *RHO* mutation which results in dominant RP [20]. Studies of the Rho^{T4R} dog have advanced the understanding of the disease mechanism underlying class four *RHO* mutations (those with altered post-translational modification and stability [17]).

RHO has seven transmembrane loops with intradiscal and cytoplasmic loops and in the dark-adapted state is combined with the chromophore 11-cis-retinal. The N-terminal of the protein, which is affected by the p.Thr4Lys mutation, is positioned within the lumen of the outer segment discs and creates a "cap" over one of the extracellular loops. This cap contributes towards thermal stability and receptor activation of the protein; it also protects the chromophore to opsin protein covalent bond from hydrolysis. Important for the cap role of the N-terminal is glycosylation at N2 and N15. The p.Thr4Arg mutation interferes with glycosylation at the N2 site, altering the cap role. The monoglycosylated rhodopsin is expressed and is trafficked to the outer segment; however, it loses the chromophore faster than the normal meta-rhodopsin II and interacts poorly with the G-protein [21]. The Rho^{T4R} dog has light dependent degeneration similar to the sector RP seen with some rhodopsin mutations. In sector RP, the inferior retina is more severely affected as this region gets more light exposure [22]. There is increasing awareness of the need to reduce light exposure to patients with certain *RHO* mutations [23]. Studies suggested that the unliganded form of the mutant opsin has a detrimental effect because of the loss of its structural integrity. Further evidence to support this was provided by cross breeding Rho^{T4R} dogs with the $Rpe65^{-/-}$ dog to produce $Rho^{T4R/+}$ $Rpe65^{-/-}$ dogs which lack 11-cis-retinal chromophore (due to the lack of Rpe65 function) and thus have only unliganded mutant rod opsin (i.e., have a lack of rod opsin combined with 11-cis-retinal) and show a greatly accelerated rate of retinal degeneration compared to $Rho^{T4R/+}Rpe65^{+/+}$ dogs (see details on the $Rpe65^{-/-}$ dog below) [21].

2.1.2. Pig Models

A number of *Rho* transgenic pigs have been generated, representing different human mutations: p.Pro23His [24], p.Pro347Leu [25] and p.Pro347Ser [26]. Studies using the p.Pro347Leu pigs showed the development of ectopic cone to rod bipolar cell synapses [27] and also interference with the cone to OFF-bipolar cell connection maturation [28]. The potential contribution of oxidative stress to cone death was demonstrated in the model [29]. The light responses of single rod photoreceptors of p.Pro347Leu and p.Pro347Ser transgenic pigs have been studied by suction pipette recording. The recording revealed protracted recovery of the photoresponse and a progressive reduction in the time to peak of the response with reduced sensitivity. This work suggests that the mutant rhodopsin reaches the outer segment and that the substitution at Pro347 interferes with inactivation of the activated form of Rho. The resulting hypothesis was that the carboxyl end of Rho may be involved in the binding of rhodopsin kinase. Mutations at Pro347 may reduce the stability of the carboxyl end attachment to rhodopsin kinase, potentially slowing phosphorylation and the subsequent binding of arrestin [30].

The p.Pro23His pig model has been used to study factors associated with cones developing dormancy and whether they can be reactivated. There are stages of retinal degeneration where the degenerating cones lose inner and outer segments. The remaining cell bodies are described as dormant cones. One hypothesis is that a lack of glucose supply to the cones as a result of loss of surrounding rods leads to cone dormancy. Experiments to either introduce rod precursors or to supply glucose to the subretinal space resulted in reactivation of the dormant cones suggesting mechanisms for the treatment of later-stage IRDs [31].

2.2. Phosphodiesterase 6 Genes

Mutations in genes encoding for the subunits forming the rod specific cyclic guanosine monophosphate (cGMP) phosphodiesterase (PDE) holoenzyme cause about 36,000 cases of autosomal recessive RP worldwide in humans, leading to the early onset of night blindness and retinal degeneration [32].

The rod PDE heteromeric holoenzyme has a catalytic core made of PDE6A and PDE6B subunits combined with two inhibitory gamma subunits [33]. Mutation in the gene encoding the alpha subunit of cGMP-PDE (Pde6a) causes PRA in the Cardigan Welsh corgi dog [34,35] and is a model of RP43 in humans, which accounts for 3% to 4% of recessive RP in North America [36,37]. Mutations in the gene encoding the beta subunit of cGMP-PDE (Pde6b) have been identified in a few dog breeds (see below), including the Irish setter dog [38], which is a model for RP40 in humans and represents about 3% to 5% of the recessive forms of RP [39,40]. Disease mechanisms for these canine models are detailed below.

2.2.1. PDE6A

A dog model with a null mutation in the gene encoding for Pde6a has been identified. It has a relatively severe phenotype and is a model for RP43 in humans [34,35]. This form of PRA was given the name rod–cone dysplasia type 3 (*rcd3*). The frameshift mutation, c.1939delA. p.Asn616ThrfsTer39, results in the absence of Pde6a in the affected dog retina (*Pde6a$^{-/-}$*) [35]. Western blot analysis shows the absence of all Pde6 subunits, showing the requirement of Pde6a for stability and normal trafficking of Pde6b [35]. In the absence of the Pde6 holoenzyme, the cGMP hydrolyzing activity is absent and cGMP accumulates in the rod photoreceptors [41]. Increased cGMP is a well-established cause of photoreceptor cell death, likely due to the increased influx of calcium ions into the outer segment [42], triggering apoptosis [43]. The rod outer segments fail to mature in *Pde6a$^{-/-}$* dogs and the genetically unaffected cones have stunted outer segments, which is reflected in a reduction in cone electroretinogram (ERG) a-waves early in the disease process [35]. Following the death of rod photoreceptors, there is a progressive loss of cones, which is reflected in the declining cone ERG amplitudes which eventually become undetectable at about one year of age (SMPJ unpublished data). Adeno-associated gene therapy was able to rescue rod function and promote cone function, as well as preserve retinal morphology [41,44].

2.2.2. PDE6B

The catalytic beta subunit (PDE6B) of the cGMP-PDE heteromeric holoenzyme is another essential component of the rod photoreceptor phototransduction cascade located in the outer segments [33]. Mutations in the gene encoding for Pde6b causes PRA (rod–cone dysplasia type 1, *rcd1*) in the Irish Setter dog [38,45,46] which is a model for RP40 in humans. RP40 is one of the most common autosomal recessive RPs leading to blindness in midlife in humans. As with the *Pde6a$^{-/-}$* dog, rod photoreceptors are affected first, then cones later in the disease [47,48]. The *Pde6b$^{-/-}$* dog phenotype has an autosomal recessive mode of inheritance and is caused by a nonsense mutation (c.2420G>A, p.Trp807Ter) [38,45]. This truncated protein would lack the C-terminal domain that is required for posttranslational processing and membrane association. The failure of phosphodiesterase activity due to a lack of function in Pde6b leads to elevated cGMP levels from an early age [47,49]. The elevation of cGMP in rods as they develop outer segments results in the halting of outer segment elongation followed by rod degeneration, starting in the central retina first, then spreading to the entire retina following the same pattern of rod maturation. The cone outer segment development is also halted and, with the loss of the rods, the genetically unaffected cone photoreceptors also progressively degenerate [46,47]. Interestingly, in contrast to the *Pde6a$^{-/-}$* dog retina where the lack of Pde6a leads to the absence of both alpha and beta Pde6 subunits, the *Pde6b$^{-/-}$* dog retina has a detectable Pde6a subunit in Western blot, prior to rod degeneration [38]. Therefore, it appears that the alpha subunit is necessary for the beta subunit to be maintained, while the beta subunit is not essential for the maintenance of the alpha subunit. Adeno-associated gene therapy has been shown to halt retinal degeneration in the *rcd1* dog [50].

Mutations in *Pde6b* have been identified in at least two other breeds of dogs (Sloughi and American Staffordshire terriers; see Table 1 for mutation information) [51,52].

2.2.3. PDE6C

Recently, a NHP spontaneous achromatopsia model was identified with a missense mutation in a cone phosphodiesterase subunit gene (*Pde6c*; c.1694G>A, p.Arg565Gln) [4]. Affected animals had behavioral changes, reflecting the photophobia seen in human subjects. They also had macula changes including foveal thinning and a subtle bullseye maculopathy. In vitro studies suggested that the mutant protein was expressed and colocalized with its chaperones, Aipl1 and P. However, the mutation alters the catalytic domain, meaning that the mutant protein fails to hydrolyze cGMP.

3. Visual Cycle

3.1. ABCA4

Stargardt disease is an inherited macular dystrophy which affects one in 8,000–10,000 people. It is the most common inherited macular dystrophy and there is currently no cure. It results from mutations in the gene *ABCA4*. Mutations in *ABCA4* also result in cone–rod dystrophies and RP. ABCA4 is an ATP-dependent flippase expressed in the photoreceptor disc membrane and is necessary in the visual cycle for its transport of N-retinylidene-phosphatidylethanolamine and phosphatidylethanolamine out of the lumen into the cytoplasm [53].

A 1 bp insertion in *Abca4* was identified in a family group of Labrador retriever dogs resulting in a frameshift and premature stop codon [54]. This mutation causes a decrease in the mRNA transcript and the loss of the full-length protein. As seen in humans, there is an accumulation of lipofuscin in the RPE cells. Cone and rod photoreceptors both had abnormal function and were decreased in number in older affected dogs (10+ years). The development of a colony of these dogs as a model for therapy will have a substantial impact on the treatment of humans with Stargardt disease because mouse *Abca4*$^{-/-}$ models lack a phenotype [54,55]. The phenotype in the dog appears to be milder than that seen in human subjects, with *ABCA4* mutations reflecting species differences [56]. Early biomarkers of retinal changes in the affected dogs will facilitate the use of the *Abca4* mutant dog as a model for human Stargardt disease.

3.2. RPE65

Large animal models with visual cycle gene mutations include the *Rpe65*$^{-/-}$ dog. This is a model for LCA2. This model was crucial in the development of translational gene augmentation therapy which led to the first FDA-approved gene therapy product. The first animal injected with a therapeutic vector for LCA2 was an *Rpe65*$^{-/-}$ dog. Therapies were developed by three independent groups and consisted in each instance of recombinant adeno-associated virus vectors packaged with *RPE65* cDNA. The precise details of promoters and other features such as polyadenylation signals and enhancers differed between the groups. Four groups with colonies of *Rpe65*$^{-/-}$ dogs reported successful restoration of rod and cone function [16,57–59]. Loss of rod photoreceptors in the *Rpe65*$^{-/-}$ dog was slow and gene therapy showed ERG rescue even in middle-age [60]. Studies showed that S-cones were sensitive to the lack of normal 11-cis-retinal supply and s-cone opsin immunoreactivity was lost at an early age [61]. There were some phenotypic differences between the *Rpe65*$^{-/-}$ dog colonies, with one showing early photoreceptor degeneration in the area centralis (canine equivalent of the human macula) [62]. Trials in human subjects have not resulted in the same restoration of function shown by the dramatic improvement in ERG responses seen in dog and mouse models. A possible explanation for this species difference in therapy efficacy was provided by a comparison of the Rpe65 function of primates and dogs. This suggested that primates require a higher level of Rpe65 than dogs for the function of the visual cycle and that the current therapy might not result in adequate levels of enzymatic function in humans [63].

3.3. RDH5

Recently, a cat model with a mutation in another visual cycle gene, *Rdh5*, has been identified by our group [64]. Rdh5 functions to convert 11-cis-retinol to 11-cis-retinal for transport to photoreceptors for reforming the visual pigments. In humans, *RDH5* mutations cause a variety of phenotypes. Fundus albipuncatatus is the predominant phenotype [65] but a subset of patients have macular atrophy [66–69] or cone dystrophy [70]. The *Rdh5*-mutant cat promises to be a valuable model because the *Rdh5*$^{-/-}$ mouse lacks a phenotype and does not recapitulate *RDH5*-retinopathy in human patients [71]. In contrast, the cat model, similar to affected humans, shows a very delayed recovery of photoreceptor function following light exposure and recapitulates the *RDH5*-macular atrophy phenotype.

4. Channelopathies/Channel-Related Mutations

Mutations that affect channel protein structure and function resulting in disease are termed channelopathies. Spontaneously occurring retinal channelopathies have been identified in dogs resulting from mutations in the cyclic nucleotide-gated ion channels (CNG) of the rod and cone photoreceptors and the anion channel bestrophin 1 (BEST1).

CNG channels in the rod and cone photoreceptors are directly involved in phototransduction. The CNG channels expressed in the rod photoreceptor consist of four subunits: three CNGA1 and one CNGB1 [72–74]. The CNG channel in cone photoreceptors consists of CNGA3 and CNGB3 proteins, in a 3:1 or 2:2 ratio (the stoichiometry of the CNGA3/CNGB3 channel is under debate) [72]. Damaging mutations in the rod CNG channels result in RP, while mutations in cone CNG channels result in achromatopsia.

BEST1 is a homopentameric channel expressed in the retinal pigment epithelium (RPE) and involved in anion transport and intracellular calcium homeostasis [75]. Mutations in *BEST1* result in a collection of retinal diseases. Mutations in *BEST1* often result in Best Vitelliform Macular Dystrophy, but the age of onset, mode of inheritance, disease characteristics and prognosis can vary [75].

4.1. CNGA1

In humans, damaging mutations in *CNGA1* result in RP49 representing 1% or less of arRP cases [39]. A 4 bp deletion in *Cnga1* was identified in Shetland Sheepdogs with PRA. The mutation causes a frameshift and a premature stop codon in a highly conserved region of the protein [76].

4.2. CNGB1

There are three splice variants of *CNGB1* expressed in the retina, glutamic acid rich proteins (GARPs) GARP1, GARP2 and CNGB1a. The full-length protein, CNGB1a, is part of the heterotetrameric CNG channel of the rod photoreceptor [72]. A complex mutation in *Cngb1* was identified in Papillon dogs with PRA. The mutation identified in these dogs is a 6 bp insertion accompanied by a 1 bp deletion, predicted to result in a frameshift and a premature stop codon, six amino acids downstream [77]. Upon further analysis of the *Cngb1a* transcript in the *Cngb1*$^{-/-}$ dogs, it was found that the mutation led to the skipping of exon 26, resulting in a premature stop codon early in exon 27. A truncated Cngb1a protein is produced, but does not form channels with Cnga1 and remains in the inner segments of the rod photoreceptors [78]. Mutations in *CNGB1* cause RP45 in humans, which represents less than 4% of arRP cases [39].

Human RP45 patients with mutations downstream of the GARP splice variants, the mouse knockout model (*Cngb1-X26*) and the *Cngb1*$^{-/-}$ dog have similar phenotypes [78,79]. The canine model shows a slow loss of rod photoreceptors and the relative preservation of cones, particularly in the area centralis and visual streak. Cone function decreases as the disease progresses, as assessed by ERG, but cone-led vision remains normal, at least up to four years [77]. Preliminary gene therapy trials have shown that treatment at 3.5–6.5 months of age in affected dogs rescues vision and slows the progression of the disease in the treated areas [78].

Both the slow disease progression and the large treatment window (in dogs and anticipated in humans [80]) make the *Cngb1*$^{-/-}$ dog an ideal model for studying therapies and outcome measures.

4.3. CNGA3

About 25% of human achromatopsia cases are caused by damaging mutations in the alpha subunit of the cone CNG channel (*CNGA3*) [81]. Two canine models of achromatopsia were identified with mutations in *CNGA3*. The mutations, p.Arg424Trp and p.Val644del (R424W and V644del, respectively), provided intriguing mutation sites for in vitro testing. Modeling of these mutations lead to further insights into CNG channel gating and subunit interactions [82]. The residue R424 is located in the S6 transmembrane helix and forms a salt bridge with the residue E306 in the S4–S5 linker. Protein modeling of the canine R424W mutation found the disruption of this salt bridge plays an important role in protein folding, subunit assembly and channel gating. In vitro expression studies of the R424W mutation showed increased mislocalization of the mutant CNGA3 protein and the mutant channels did not produce cyclic nucleotide-activated currents [82].

The canine V644del mutation is in the C-terminal leucine zipper (CLZ) domain, which is involved in channel assembly and stability. The one amino acid deletion shifts all subsequent residue interactions within the domain. In vitro studies in a heterologous expression system showed evidence of mislocalization and a decrease in the total number of active channels, but some mutant channels (~60%) reached the cell membrane and had normal cyclic nucleotide-activated currents [82].

Two spontaneous sheep models of *Cnga3* achromatopsia have been identified [83]. The first identified has a nonsense mutation (p.Arg236Ter) and the second a missense mutation (p.Gly540Ser) [84]. Adeno-associated virus gene therapy was able to restore cone function in both models [84,85].

4.4. CNGB3

Two unique mutations were identified in the canine *Cngb3* gene—a genomic deletion that removes the entire *Cngb3* gene (*Cngb3$^{-/-}$*) and a missense mutation (p. Asp262Asn; D262N). The missense mutation was identified in German Shorthaired Pointers, while the genomic deletion was initially found in Alaskan Malamutes and later found in other breeds [86,87]. Mutations in *CNGB3* are responsible for at least 50% of achromatopsia cases in humans, making it an ideal target for therapies [87,88].

The D262N mutation in *Cngb3* inspired investigation into the Tri-Asp motif that is conserved in CNG channels [89]. When this mutation was introduced to heterologously expressed CNGA3, it abolished homomeric channel function and was responsible for mislocalization of the protein. When the mutation was introduced in CNGB3 in coexpression studies with CNGA3, the response to cyclic nucleotides was reduced but still present and consistent with CNGA3 homomeric channel function. These experiments show the Tri-Asp motif is necessary for proper cone CNG channel formation, but raises the question why mutations in *CNGB3* result in the loss of cone function and achromatopsia when CNGA3 can form a functioning homomeric channel in the absence of CNGB3. Further work in retinal tissue and/or cone photoreceptors will be needed to elucidate this [89].

Delivery of a human *CNGB3* gene via rAAV5 vector to both *Cngb3$^{-/-}$* and the D262N mutant dogs showed that the vector could be targeted to the medium and long wavelength (M/L) cones via a human red cone opsin promoter and the rescue was dependent on the age of the dog and the promotor, not on the mutation type [90].

4.5. BEST1

Best vitelliform macular dystrophy (Best disease, BVMD) has autosomal dominant inheritance and is the most common disease associated with mutations in the gene *BEST1*. Four other disease phenotypes have been described in association with *BEST1* mutations: adult onset vitelliform macular dystrophy, autosomal recessive bestrophinopathy, autosomal dominant vitreoretinochoroidopathy and retinitis pigmentosa. BVMD is characterized by at least one vitelliform lesion in the macula but can present with multiple lesions. The disease slowly progresses to degenerate the RPE and retina in the affected regions, resulting in vision impairment [75]. In dogs, a similar disease has been described, termed canine multifocal retinopathy (CMR) and is caused by mutations in the *Best1* gene. In contrast to the BVMD in humans, CMR due to *Best1* mutations is an autosomal recessive disease and has a consistent and predictable disease phenotype. This consistency and the detailed natural history of the disease lends well to measuring the outcomes of translatable therapies.

Initially, two *Best1* mutations were identified in dogs: p.Arg25Ter (Great Pyrenees and mastiff-related breeds, *cmr1*) and p.Gly161Asp (Coton de Tulear, *cmr2*), with the former resulting in a premature stop codon and, presumably, a lack of Best1 protein [91]. Analysis of additional breeds with CMR identified another breed (Lapponian herder, *cmr3*) with two deleterious mutations in exon 10, a 1 bp deletion leading to a frameshift and premature stop codon (p.Pro463fs) and a missense mutation (p.Gly489Val) [92]. Interestingly, the phenotype resulting from the three mutations in dogs is indistinguishable. The dogs present with multifocal regions of retinal separation with a pink or tan-colored subretinal fluid, which eventually leads to retinal degeneration. Using optical coherence tomography to image eyes from *cmr1/cmr1*, *cmr3/cmr3* and *cmr1/cmr3* dogs in vivo, the earliest detectable sign of the disease is at ~11 weeks of age in which there is a retinal elevation in the fovea-like region of the area centralis. As the disease progresses, this retinal elevation becomes a macrodetachment, surrounded by microdetachments. Rates of progression, detachment location and number varied, but the disease was typically localized to the more cone-rich regions of the retina [93].

Immunohistochemistry of CMR canine tissues showed a lack of RPE apical microvilli at the cone photoreceptor/RPE interface along with accumulated lipofuscin within the RPE. The loss of the RPE apical processes results in a loss of all direct contact of the cones to the RPE, severely impacting the physiological role the RPE has on retinal maintenance. It is hypothesized that the lack of RPE apical processes and subsequent weakened interphotoreceptor matrix is instrumental in the characteristic detachments and lesions observed in diseases caused by *BEST1* mutations [93,94]. Interestingly, microdetachments were identified in response to light exposure in pre-clinical CMR dogs. The detachments occurred between the photoreceptor inner/outer segments and the RPE/tapetum interface. These light-induced detachments occurred within minutes and increased in response to time of light exposure and would resolve within 24 h.

Using a rAAV2 vector delivered subretinally, CMR dogs were treated by gene augmentation with wildtype canine or human *BEST1*. Macro and microdetachments were resolved and RPE microvilli ensheathment of cone photoreceptors returned within the treatment area. This positive outcome was retained in the dogs for as long as 207 weeks post injection. The same outcome was present regardless of age of treatment (within 27–69 weeks of age), the stage of detachments or mutation (*cmr1/cmr1*, *cmr3/cmr3* or *cmr1/cmr3*). These results show promise for the treatment of human patients with *BEST1* mutations [93].

5. Cilia-Related Proteins (Ciliopathies)

Photoreceptors are non-motile sensory ciliated cells. The connecting cilium between inner and outer segment is critical for the transport of proteins to and from the outer segment. Many proteins are expressed in this region of photoreceptors and act to control the trafficking of proteins. Ciliopathies is a general term used to group conditions resulting from mutations in cilia proteins. As cilia are present in many cell types, mutations in cilia proteins can cause syndromic disease where retinal degeneration is present, accompanied by disorders in other organ systems. Mutations that have a milder effect on function may result in a photoreceptor-only phenotype such as LCA or RP. A number of large animal models with mutations in cilia related genes have been identified (Table 1).

5.1. BBS4

A mutation in the *Bbs4*, (Bardet–Biedl syndrome 4) gene, was identified in Hungarian Puli dogs with PRA. This mutation (c.58A>T, p.Lys20Ter) is predicted to result in nonsense mediated decay of the *Bbs4* transcript. *BBS4* is one of the BBS genes involved in cilia function. These dogs showed a typical PRA phenotype along with noted obesity and abnormal sperm, although more affected dogs may need to be examined before confirming that this is a syndromic BBS model [95].

5.2. BBS7

An NHP model of Bardet–Biedl syndrome has been recently described. A mutation in exon 3 of the *Bbs7* gene c.160delG (p.Ala54GlnfsTer18) that is predicted to produce a truncated non-functional protein was identified. As with BBS4, BBS7 is involved in cilia function. Typically in humans, mutations in BBS genes cause a syndromic condition. The affected NHPs had several features of BBS, including retinal atrophy, which was most severe centrally; the affected animals had smaller brains, renal disease, and the males had small testes. This is the first described model of BBS in NHP and shares many characteristics with BBS patients with truncating mutations [3].

5.3. CEP290

This cat model, *rdAc* (retinal degeneration, Abyssinian cat) has been studied in detail for many years. Compared to most *CEP290* (centrosomal protein 290) mutations in other species, the cat model has a mild phenotype. The onset and rate of photoreceptor degeneration in the cat with the sparing of the area centralis makes it a model for RP [96,97]. In humans, *CEP290* mutations can result in a spectrum of phenotypes including lethality, severe syndromic disease (e.g., Joubert syndrome) and most commonly LCA; see [98] for a review. Milder phenotypes such as RP are less common. In the cat, an intronic mutation (c.6966+9T>G) leads to the introduction of a stronger splice acceptor site (the wildtype splice acceptor is a GC rather than the much more commonly used GT). The mutation strengthens an adjacent GT, which, when used, adds 4 bps to the exon 50 sequence, the resulting frameshift introduces a premature stop codon [99]. The truncated protein escapes nonsense mediated decay and is expressed. In addition, the wildtype acceptor site is still used for a percentage of the transcripts (unpublished data). The combination of a truncated protein with some remaining function combined with a low-level production of full-length transcripts explains the mild phenotype. *CEP290* is too large to be packaged in an adeno-associated virus vector and one therapeutic approach being investigated is the use of a truncated transcript so that a miniprotein may have partial function and convert a severe phenotype into a milder phenotype similar to that of the cat.

5.4. NPHP4

Wire-haired dachshunds have a cone–rod dystrophy resulting from a truncating mutation in *Nphp4*. The identified *Nphp4* mutation was a deletion involving exon 5/intron 5 that led to skipping of exon 5 and a premature truncation in exon 6 of 30 (c.462_526del, p.Leu155LysfsTer2) [100]. A colony was established from a single founder male [101]. Puppies had miotic pupils and cone-mediated ERGs were reduced prior to retinal maturation. Furthermore, they did not show the normal increase in amplitudes with retinal maturation and further declined in amplitude rapidly. The amplitudes of the rod-driven responses were less severely affected, but were lost with age [102,103]. Interestingly, the condition is non-syndromic in dogs, just presenting as a cone–rod dystrophy. Human patients with *NPHP4* mutations do not always develop a retinal phenotype, but typically have nephronophthisis, with some patients having Senior–Løken syndrome, which combines the renal phenotype with a retinal dystrophy (see for [104] a review). This phenotype difference may represent a species difference, with *Nphp4*$^{-/-}$ mice also only having an ocular phenotype with no renal abnormalities, but also showing male infertility [105].

5.5. NPHP5 (IQCB1)

A mutation was identified in the *NPHP5* gene (aka *IQCB1*) resulting in a cone–rod dystrophy in American pit bull terrier dogs (*crd2*) modeling non-syndromic LCA. This mutation (c.952-953insC, p.Ser319IlefsTer13) results in a frameshift and a premature stop codon [52]. At 6 weeks of age, the *crd2* dogs had no cone function and abnormal rod function. Morphologically, the cones completely lacked outer segments, while the rods developed disorganized outer segments. Despite the lack of cone function, the cones are retained within the central retina up to 33 weeks of age [106]. Adeno-associated gene therapy restores and improves photoreceptor function and preserves the outer retina layer [107].

5.6. RPGR

Retinitis pigmentosa GTPase regulator (RPGR) localizes to the connecting cilium and mutations within the *RPGR* gene account for >80% of X-linked RP (XLRP). There are two major retinal isoforms, one encoded by exons 1–19 and a second isoform that consists of exons 1–15 and a retained portion of intron

15 (ORF15; see [108] for a summary). ORF15 is a mutation hotspot [109]. Three different *Rpgr* mutations cause X-linked retinal degeneration in dogs. One is caused by a genomic deletion [110] and two are due to microdeletions in ORF15 and provide two distinct mechanistic models for RPGR-XLRP [111]. The first, known in the dog as XLPRA1, has a 5 bp deletion and a premature stop codon [111]. This is a relatively late onset and slowly progressive degeneration [112]. The second form, XLPRA2, has a 2 bp deletion with a frameshift resulting in the addition of 34 basic residues. In vitro studies showed that the mutant protein aggregated in the endoplasmic reticulum and is hypothesized to have a toxic effect [111]. The phenotype presents as an early onset degeneration with early changes being outer segment disruption and opsin (rod and cone) mislocalization. Photoreceptor cell death was shown to occur in a biphasic manner with two distinct phases of cell death with evidence of remodeling occurring [113]. Studies of the heterozygous females revealed that there are patches of diseased retina, presumably resulting from regions where random X-inactivation has resulted in expression of the mutant allele and patches of unaffected retina (where the wildtype allele is expressed). In the earlier onset XLPRA2 migration of adjacent photoreceptors into regions where rod photoreceptors died occurs, showing retinal plasticity in the younger animals. This remodeling was not described in the XLPRA1, where patches of degeneration occur at a later age [114].

5.7. RPGRIP1

Retinitis pigmentosa GTPase regulator-interacting protein 1 (RPGRIP1) localizes to the connecting cilium, where it interacts with RPGR. Mutations in *RPGRIP1* are associated with autosomal recessive LCA. A cone–rod dystrophy form of PRA in a colony of miniature longhaired Dachshunds was reported to be due to an insertion in *Rpgrip1* [115] and the rescue of the phenotype by gene therapy was achieved [116]. When miniature longhaired Dachshunds in pet homes were investigated, the *Rpgrip1* insertion did not appear to segregate with disease status [117]. Further studies have shown that two other loci influence the development of the phenotype [118–120]. This is an example of the potential effect of modifying loci on phenotype.

6. Photoreceptor Development

6.1. CRX

The *CRX* gene encodes an OTX-like homeodomain transcription factor which is essential for photoreceptor development, maturation and survival [121–124]. Transcription factors are essential for control of the maturation of progenitor cells [123–126]. Only one spontaneously occurring large animal model of a retinal transcription factor homeobox mutation has been described, the Crx^{Rdy} cat [127,128]. The heterozygous animal has a severe phenotype of cone-led retinal dystrophy and is a model for one form of severe early childhood onset blindness (LCA7). While most mutations in *CRX* result in LCA7, other phenotypes including cone–rod dystrophy, RP and macular degeneration have been described [129–131]. *CRX* mutations have been reported to account for between 0.6% and 2.35% of LCA [131–134].

As with other transcription factors, CRX has a characteristic structure with a DNA binding domain (homeodomain) near its N-terminal and a transactivation domain at the C-terminal. The $Crx^{Rdy/+}$ mutant cat models Class III *CRX* mutations, which are antimorphic frameshift/nonsense mutations with intact DNA binding, but a lack of target gene transactivation [135]. The $Crx^{Rdy/+}$ cat mutation is caused by a 1 bp deletion (c.546delC; p.Pro185LysfsTer2) in the final exon of the *Crx* gene [127]. This mutant mRNA avoids nonsense mediated decay and produces a truncated Crx protein with an intact DNA-binding domain but disrupted transactivation domain [128]. Both the mutant transcript and protein accumulate at higher levels than the wildtype versions, possibly due to increased stability of the mutant mRNA compared to the wildtype. In vitro studies showed that the mutant mRNA fails to activate its own promoter [128]. This suggests that the mutant protein has a dominant negative effect by binding promoter recognition sites, but fails in its transactivation function. The result is misregulated gene expression; for example, *Rho* and cone arrestin (*Arr3*) mRNA levels are severely decreased. Truncated photoreceptor outer segments are produced, but the photoreceptors fail to fully mature [128]. A decreased rod ERG is detectable, which shows evidence of partial maturation before this is halted and the ERG amplitudes decline, paralleling rod degeneration. Cone function is more severely affected, with no cone ERG responses being detectable [128].

6.2. STK38L

An early retinal degeneration was described in Norwegian Elkhounds in 1987 [136] in which the dogs present with vision impairment in low light as early as 6 weeks of age. The causal mutation was found to be a short interspersed element (SINE) insertion in exon 4 of *Stk38l*, a gene that had not been associated with retinal degenerations before its discovery in the Norwegian Elkhounds [137]. The mRNA transcript is produced in the affected dog but does not contain exon 4. Interestingly, at ~6 weeks of age, the expression of the mutant *Stk38l* transcript is similar to control dogs, but expression increases at ~8 and ~12 weeks [138].

Detailed histological experiments were performed on affected canine retina to understand the role of Stk38l in retinal development and how the lack of the functional protein results in retinal degeneration. The retina in ~4 week old *Stk38l*-affected dogs is normal, but by ~8 weeks of age the rod photoreceptors begin to show mislocalization of Rho, irregularities in the outer segments and increased TUNEL labeling. However, this increase in apoptotic cells is not accompanied by a thinning of the outer nuclear layer (ONL), as would be expected from loss of rod photoreceptors. The maintenance of ONL thickness is a result of proliferation of rod-like photoreceptors which express both Rho and cone S-opsin. After ~14 weeks of age, this proliferation ceases and the ONL begins to thin as the photoreceptors die, resulting in the loss of at least 50% of the ONL rows by 48 weeks [138].

The *Stk38l*-affected dogs provide an interesting and unique model of retinal degeneration in which differentiated photoreceptors either apoptose or proliferate for a short amount of time before retinal degeneration progresses.

7. Photoreceptor to Bipolar Cell Signaling

7.1. LRIT3

Recently, a dog model of recessively inherited CSNB has been identified due to a mutation in leucine-rich-repeat, immunoglobulin-like transmembrane-domain 3 gene (*LRIT3*). Mutations in *LRIT3* in humans causes a form of CSNB [139]. The ERG of the affected dogs shows a lack of ON-bipolar cell function with preservation of cone OFF-bipolar contributions [140–142]. There is currently debate about the positioning of the LRIT3 protein which had been described as being in the synaptic tips of the bipolar cells; however, a recent publication showed that it was presynaptic, being present in photoreceptors, but bridged the synapse to influence the positioning of post-synaptic glutamate signaling proteins [143]. The new availability of a large animal model may facilitate further investigations.

7.2. TRPM1

Studies of Appaloosa horses with CSNB contributed to the identification of the role of transient receptor potential cation channel subfamily M, member 1 (*Trpm1*) in bipolar cell signaling [144,145]. The night-blind Appaloosa horse was first recognized as an animal model for the Schubert–Bornschein type of CSNB in the 1970s [146]. The study published at that time reported a lack of an ERG b-wave and night blindness but no retinal degeneration or obvious morphological abnormality of the photoreceptor synapses with second order neurons [146]. The correlation of CSNB in the Appaloosa with the Leopard complex spotting coat color was recognized [147]. This coat color is governed at a single gene locus with animals homozygous for the Leopard complex spotting associated allele (LP) also having CSNB [147]. When the LP locus was mapped, *Trpm1*, which is also expressed in melanocytes, was identified as a positional candidate gene and showed markedly reduced expression in retina and skin from affected animals [148]. Investigation of *TRPM1* in humans with complete CSNB identified mutations in *TRPM1* [149–151] and its role in ON-bipolar cell signaling was identified.

The LP mutation was identified as a retroviral insertion in intron 1 of equine *Trpm1* that disrupts gene transcription by causing premature polyadenylation [152].

7.3. Whippet Dog Model of Incomplete CSNB with Retinal Degeneration

A dog model of cone–rod synaptic dysfunction, which has been described by some authors as a form of incomplete CSNB, has been identified [153]. The affected dogs lack an ERG b-wave and also lack cone OFF-bipolar cell attributable ERG components [154]. Interestingly, the dog model develops a progressive retinal degeneration, which is not reported as a feature of the condition in humans [155].

8. Structural/Other

Retinal degenerations can occur due to mutations in genes involved in the structure or maintenance of the retina.

8.1. ADAM9

ADAM9 (a disintegrin and metalloprotease domain, family member 9) mutations are associated with cone–rod dystrophies in humans and dogs. A genomic deletion in *Adam9* was identified in Irish Glen of Imaal Terriers, resulting in a premature stop codon and the loss of the full-length protein. Similar to the mouse model, histological sections show that the RPE cells do not invest in the outer segments of the photoreceptors. This mutation effects both the rod and cone photoreceptors, but the cones are more severely affected throughout the course of the disease [156,157]. Further work is needed to understand the role of ADAM9 in the retinal structure and RPE maintenance of photoreceptors.

8.2. AIPL1

Mutations in the gene encoding aryl hydrocarbon receptor-interacting protein-like 1 (*AIPL1*) causes LCA4 accounting for about 3% to 7% of autosomal recessive LCA [132,134,158–160]. The *AIPL1* gene encodes for a protein expressed in the photoreceptor and pineal gland [161]. In photoreceptors, AIPL1 is concentrated in the tip of the inner segment in proximity to the connecting cilia, acting as a co-chaperone involved in the folding, assembly and transport of the retinal cGMP phosphodiesterase (PDE6) heteromeric holoenzyme within photoreceptor outer segments [162–166].

One spontaneous occurring feline large animal model of LCA4 currently exists, the Persian cat [167]. This feline model, similar to human patients with LCA4, presents as an autosomal recessive severe retinal dystrophy. The disease is characterized by a very early photoreceptor loss, progressing to severe retinal degeneration before adulthood. This leads to very early blindness [167]. The feline phenotype is caused by a nonsense mutation at position c.577C>T producing a stop codon (p.Arg193Ter), leading to the production of a truncated non-functional protein [168]. The precise mechanisms of the severe phenotype in the feline model are yet to be investigated. The feline model represents a valuable large animal model for mechanistic studies underlying *AIPL1*-LCA in humans.

8.3. MERTK

Mutations in *MERTK* (MER proto-concogene, tyrosine kinase) account for about 1% of arRP. *Mertk* was found to be overexpressed (six-fold) in Swedish Vallhund dogs with a retinal dystrophy [169]. After whole genome sequencing of an affected dog, an intact LINE-1 insertion was identified in intron one of *Mertk*. A LINE-1 retrotransposon is a long interspersed repetitive element that is found in all mammalian genomes [170]. It is unclear how this insertion results in overexpression of the *Mertk* transcript and subsequent development of retinal disease. More in-depth investigation of this mechanism will aid a better understanding of how retrotransposon insertions can modify gene regulation/expression, ultimately leading to disease, in addition to understanding the specific role of overexpression of the *MERTK* transcript in the context of retinal disease [171].

8.4. PRCD

Progressive rod–cone degeneration (*Prcd*) was a formerly unknown gene that was mapped and sequenced first in dogs with progressive rod–cone degeneration (PRCD) PRA [172]. Dogs with this form of PRA had been studied in detail over many years and many different breeds of dog are affected. The mutation is p.Cys2Tyr, which is in a highly conserved region of the gene. An identical mutation was found in a human RP patient. Additional mutations also located in exon 1 of *PRCD* have been identified in RP patients [172,173]. Since the discovery of *Prcd*, a mouse knockout model has been generated and experiments have shown Prcd's involvement in photoreceptor disc formation and maintenance [174–176]. Studies in the PRCD dog showed an early ultrastructural change that may have resulted from this purported function. This was the development of vesicular profiles adjacent to outer segments being consistent with abnormality in disc formation, furthermore studies also showed affected dogs had slower photoreceptor disc renewal than normal controls [177].

8.5. RD3

Mutations in *RD3* result in LCA12 in humans, a severe retinal degeneration in mice and rod–cone dysplasia type 2 (*rcd2*) in Collie dogs [178–180]. RD3 is thought to act as competitor for guanylyl cyclase-activating proteins, preventing premature cyclase activity in inner segments [181].

Collie dogs have a 22 bp insertion in the *Rd3* gene resulting in altered amino acids and an extended open reading frame [182]. Three splice variants of the gene were identified in the canine retina. In the *Rd3*-mutant Collie, the rod and cone photoreceptors outer segments do not fully develop and the retina degenerates rapidly. By 6 months, there are no photoreceptor inner or outer segments and the outer nuclear layer contains only one layer of nuclei [183].

9. Conclusions

IRDs that lead to visual impairment and blindness show considerable heterogeneity, where even similar disease phenotypes can result from mutations in a wide range of different genes. Closer examination of the broad phenotypes such as RP and LCA show the genetic heterogeneity with altered function of a range of genes leading to photoreceptor loss. Large numbers of different mutations within a single gene exist and can result in a range of phenotypic differences.

The mechanisms by which the gene mutation leads to cell death differs between genes and between mutations within individual genes. Laboratory rodents have been the workhorses for studying disease mechanisms because of the ease with which their genome can be engineered, the short reproductive time and relatively low costs of maintenance. They have allowed identification of disease pathways and allowed the study of how and why degeneration occurs. However, laboratory rodents have very significant shortcomings. Rodents are nocturnal animals with a different pattern of distribution of photoreceptors to humans. The large animal models discussed in this review have retinal photoreceptor distribution that is much closer to that of humans. This is particularly important when considering conditions that have a major effect on the macula. Large animal models have also been of value for the development and testing of translational therapies. They are particularly valuable for this because the larger eye size and proportions of lens and vitreous is much closer to that in humans, allowing for identical surgical approaches for trials of translational therapies. The several, mostly spontaneously occurring, large animal models mentioned here have provided valuable information about disease phenotypes and disease mechanisms. They offer great potential for even more detailed studies to understand how photoreceptor function and survivability is affected by the gene mutations and to study the extensive inner retinal remodeling and glial cell activation that occurs in IRDs. Modeling of IRDs using retinal organoids developed from induced pluripotent stem cells from IRD patients is an exciting field of research. These potentially allow the study of the effect of the exact mutation that is present in the patient and in a human cultured tissue. However, they do not yet fully recapitulate the specific environment within the retina of a living animal, so, while showing great promise, they cannot yet replace whole animal studies. Improving our understanding of how and why photoreceptors die may suggest novel therapies to preserve function and slow down vision loss. There remain untapped populations of companion animals (dogs and cats) with spontaneous IRDs, with new potential models being identified with increasing frequency. The advances in gene editing also expand the opportunity to develop additional large animal models with specific mutations that even more accurately model human IRDs.

Supplementary Materials: The following are available online at http://www.mdpi.com/2073-4409/9/4/882/s1, Table S1: List of mutations in large animal models of inherited retinal disease and references.

Author Contributions: All authors contributed to the conception, design and writing of the manuscript. All authors have read and agreed to the published version of the manuscript.

Funding: The NIH (EY027285) and the Donald R. Myers and William E. Dunlap Endowment for Canine Health, to Simon Petersen-Jones, provided support for the work done in this review.

Conflicts of Interest: The authors declare no conflict of interest.

References

1. Fan, N.; Lai, L. Genetically Modified Pig Models for Human Diseases. *J. Genet. Genom.* **2013**, *40*, 67–73. [CrossRef] [PubMed]

2. Petersen-Jones, S.M. Drug and gene therapy of hereditary retinal disease in dog and cat models. *Drug Discov. Today Dis. Model.* **2013**, *10*, e215–e223. [CrossRef]

3. Peterson, S.M.; McGill, T.J.; Puthussery, T.; Stoddard, J.; Renner, L.; Lewis, A.D.; Colgin, L.M.; Gayet, J.; Wang, X.; Prongay, K.; et al. Bardet-Biedl Syndrome in rhesus macaques: A nonhuman primate model of retinitis pigmentosa. *Exp. Eye Res.* **2019**, *189*, 107825. [CrossRef]

4. Moshiri, A.; Chen, R.; Kim, S.; Harris, R.A.; Li, Y.; Raveendran, M.; Davis, S.; Liang, Q.; Pomerantz, O.; Wang, J.; et al. A nonhuman primate model of inherited retinal disease. *J. Clin. Investig.* **2019**, *129*, 863–874. [CrossRef]

5. Luo, X.; Li, M.; Su, B. Application of the genome editing tool CRISPR/Cas9 in non-human primates. *Zool. Res.* **2016**, *37*, 214–219.

6. Kang, Y.; Chu, C.; Wang, F.; Niu, Y. CRISPR/Cas9-mediated genome editing in nonhuman primates. *Dis. Model. Mech.* **2019**, *12*, dmm039982. [CrossRef] [PubMed]

7. Rahman, N.; Georgiou, M.; Khan, K.N.; Michaelides, M. Macular dystrophies: Clinical and imaging features, molecular genetics and therapeutic options. *Br. J. Ophthalmol.* **2019**, *104*, 451–460. [CrossRef]

8. Michaelides, M.; Hunt, D.M.; Moore, A.T. The genetics of inherited macular dystrophies. *J. Med. Genet.* **2003**, *40*, 641–650. [CrossRef]

9. Fritsche, L.G.; Igl, W.; Bailey, J.N.C.; Grassmann, F.; Sengupta, S.; Bragg-Gresham, J.L.; Burdon, K.P.; Hebbring, S.J.; Wen, C.; Gorski, M.; et al. A large genome-wide association study of age-related macular degeneration highlights contributions of rare and common variants. *Nat. Genet.* **2015**, *48*, 134–143. [CrossRef]

10. Heesterbeek, T.J.; Lorés-Motta, L.; Hoyng, C.B.; Lechanteur, Y.T.E.; Hollander, A.I.D. Risk factors for progression of age-related macular degeneration. *Ophthalmic Physiol. Opt.* **2020**, *40*, 140–170. [CrossRef]

11. Nishiguchi, K.; Yokoyama, Y.; Fujii, Y.; Furukawa, T.; Ono, F.; Shimozawa, N.; Togo, M.; Suzuki, M.; Nakazawa, T. Association between drusen and blood test results in a colony of 1174 monkeys. *Acta Ophthalmol.* **2015**, *93*, 93. [CrossRef]

12. Pennesi, M.E.; Neuringer, M.; Courtney, R.J. Animal models of age related macular degeneration. *Mol. Asp. Med.* **2012**, *33*, 487–509. [CrossRef] [PubMed]

13. Mowat, F.M.; Petersen-Jones, S.M.; Williamson, H.; Williams, D.L.; Luthert, P.J.; Ali, R.R.; Bainbridge, J.W. Topographical characterization of cone photoreceptors and the area centralis of the canine retina. *Mol. Vis.* **2008**, *14*, 2518–2527. [PubMed]

14. Beltran, W.A.; Cideciyan, A.V.; Guziewicz, K.E.; Iwabe, S.; Swider, M.; Scott, E.M.; Savina, S.V.; Ruthel, G.; Stefano, F.; Zhang, L.; et al. Canine Retina Has a Primate Fovea-Like Bouquet of Cone Photoreceptors Which Is Affected by Inherited Macular Degenerations. *PLoS ONE* **2014**, *9*, e90390. [CrossRef]

15. El-Aziz, M.M.A.; Barragan, I.; O'Driscoll, C.A.; Goodstadt, L.; Prigmore, E.; Borrego, S.; Mena, M.; Pieras, J.I.; El-Ashry, M.F.; Abu Safieh, L.; et al. EYS, encoding an ortholog of Drosophila spacemaker, is mutated in autosomal recessive retinitis pigmentosa. *Nat. Genet.* **2008**, *40*, 1285–1287. [CrossRef]

16. Acland, G.M.; Aguirre, G.D.; Ray, J.; Zhang, Q.; Aleman, T.S.; Cideciyan, A.V.; Pearce-Kelling, S.E.; Anand, V.; Zeng, Y.; Maguire, A.M.; et al. Gene therapy restores vision in a canine model of childhood blindness. *Nat. Genet.* **2001**, *28*, 92–95. [CrossRef]

17. Athanasiou, D.; Aguila, M.; Bellingham, J.; Li, W.; McCulley, C.; Reeves, P.J.; Cheetham, M.E. The molecular and cellular basis of rhodopsin retinitis pigmentosa reveals potential strategies for therapy. *Prog. Retin. Eye Res.* **2017**, *62*, 1–23. [CrossRef]

18. Kijas, J.W.; Cideciyan, A.V.; Aleman, T.S.; Pianta, M.; Pearce-Kelling, S.E.; Miller, B.J.; Jacobson, S.G.; Aguirre, G.D.; Acland, G.M. Naturally occurring rhodopsin mutation in the dog causes retinal dysfunction and degeneration mimicking human dominant retinitis pigmentosa. *Proc. Natl. Acad. Sci. USA* **2002**, *99*, 6328–6333. [CrossRef]

19. Iwabe, S.; Ying, G.-S.; Aguirre, G.D.; Beltran, W.A. Assessment of visual function and retinal structure following acute light exposure in the light sensitive T4R rhodopsin mutant dog. *Exp. Eye Res.* **2016**, *146*, 341–353. [CrossRef]

20. Born, L.I.V.D.; Van Schooneveld, M.J.; De Jong, L.A.M.S.; Riemslag, F.C.C.; DeJong, P.T.V.M.; Gal, A.; Bleeker-Wagemakers, E.M. Thr4Lys rhodopsin mutation is associated with autosomal dominant retinitis pigmentosa of the cone-rod type in a small Dutch family. *Ophthalmic Genet.* **1994**, *15*, 51–60. [CrossRef]

21. Zhu, L.; Jang, G.-F.; Jastrzebska, B.; Filipek, S.; Pearce-Kelling, S.E.; Aguirre, G.D.; Stenkamp, R.E.; Acland, G.M.; Palczewski, K. A naturally occurring mutation of the opsin gene (T4R) in dogs affects glycosylation and stability of the G protein-coupled receptor. *J. Boil. Chem.* **2004**, *279*, 53828–53839. [CrossRef] [PubMed]

22. Heckenlively, J.R.; Rodriguez, J.A.; Daiger, S.P. Autosomal Dominant Sectoral Retinitis Pigmentosa. *Arch. Ophthalmol.* **1991**, *109*, 84. [CrossRef] [PubMed]

23. Orlans, H.O.; MacLaren, R.E. Comment on: 'Sector retinitis pigmentosa caused by mutations of the RHO gene'. *Eye* **2019**, 1–2. [CrossRef] [PubMed]

24. Ross, J.W.; De Castro, J.P.F.; Zhao, J.; Samuel, M.; Walters, E.; Rios, C.; Bray-Ward, P.; Jones, B.W.; Marc, R.E.; Wang, W.; et al. Generation of an Inbred Miniature Pig Model of Retinitis Pigmentosa. *Investig. Opthalmol. Vis. Sci.* **2012**, *53*, 501–507. [CrossRef] [PubMed]

25. Petters, R.M.; Alexander, C.A.; Wells, K.D.; Collins, E.B.; Sommer, J.; Blanton, M.R.; Rojas, G.; Hao, Y.; Flowers, W.L.; Banin, E.; et al. Genetically engineered large animal model for studying cone photoreceptor survival and degeneration in retinitis pigmentosa. *Nat. Biotechnol.* **1997**, *15*, 965–970. [CrossRef]

26. Kraft, T.; Allen, D.; Petters, R.M.; Hao, Y.; Peng, Y.-W.; Wong, F. Altered light responses of single rod photoreceptors in transgenic pigs expressing P347L or P347S rhodopsin. *Mol. Vis.* **2005**, *11*, 1246–1256.

27. Peng, Y.-W.; Hao, Y.; Petters, R.M.; Wong, F. Ectopic synaptogenesis in the mammalian retina caused by rod photoreceptor-specific mutations. *Nat. Neurosci.* **2000**, *3*, 1121–1127. [CrossRef]

28. Banin, E.; Cideciyan, A.V.; Aleman, T.S.; Petters, R.M.; Wong, F.; Milam, A.H.; Jacobson, S.G. Retinal Rod Photoreceptor–Specific Gene Mutation Perturbs Cone Pathway Development. *Neuron* **1999**, *23*, 549–557. [CrossRef]

29. Shen, J.; Yang, X.; Dong, A.; Petters, R.M.; Peng, Y.-W.; Wong, F.; Campochiaro, P.A. Oxidative damage is a potential cause of cone cell death in retinitis pigmentosa. *J. Cell. Physiol.* **2005**, *203*, 457–464. [CrossRef]

30. Sommer, J.R.; Wong, F.; Petters, R.M. Phenotypic stability of Pro347Leu rhodopsin transgenic pigs as indicated by photoreceptor cell degeneration. *Transgenic Res.* **2011**, *20*, 1391–1395. [CrossRef]

31. Wang, W.; Lee, S.J.; Scott, P.A.; Lu, X.; Emery, D.; Liu, Y.; Ezashi, T.; Roberts, M.R.; Ross, J.W.; Kaplan, H.J.; et al. Two-Step Reactivation of Dormant Cones in Retinitis Pigmentosa. *Cell Rep.* **2016**, *15*, 372–385. [CrossRef] [PubMed]

32. Daiger, S.P.; Bowne, S.J.; Sullivan, L.S. Perspective on Genes and Mutations Causing Retinitis Pigmentosa. *Arch. Ophthalmol.* **2007**, *125*, 151–158. [CrossRef] [PubMed]

33. Baehr, W.; Devlin, M.J.; Applebury, M.L. Isolation and characterization of cGMP phosphodiesterase from bovine rod outer segments. *J. Boil. Chem.* **1979**, *254*, 11669–11677.

34. Petersen-Jones, S.M.; Entz, D.D.; Sargan, D.R. cGMP phosphodiesterase-α mutation causes progressive retinal atrophy in the Cardigan Welsh corgi dog. *Investig. Ophthalmol. Vis. Sci.* **1999**, *40*, 1637–1644.

35. Tuntivanich, N.; Pittler, S.J.; Fischer, A.J.; Omar, G.; Kiupel, M.; Weber, A.; Yao, S.; Steibel, J.P.; Khan, N.W.; Petersen-Jones, S.M. Characterization of a canine model of autosomal recessive retinitis pigmentosa due to a PDE6A mutation. *Investig. Opthalmol. Vis. Sci.* **2008**, *50*, 801–813. [CrossRef] [PubMed]

36. Dryja, T.P.; Finn, J.T.; Peng, Y.W.; McGee, T.L.; Berson, E.L.; Yau, K.W. Mutations in the gene encoding the alpha subunit of the rod cGMP-gated channel in autosomal recessive retinitis pigmentosa. *Proc. Natl. Acad. Sci. USA* **1995**, *92*, 10177–10181. [CrossRef]

37. Huang, S.H.; Pittler, S.J.; Huang, X.; Oliveira, L.; Berson, E.L.; Dryja, T.P. Autosomal recessive retinitis pigmentosa caused by mutations in the α subunit of rod cGMP phosphodiesterase. *Nat. Genet.* **1995**, *11*, 468–471. [CrossRef]

38. Suber, M.L.; Pittler, S.J.; Qin, N.; Wright, G.C.; Holcombe, V.; Lee, R.H.; Craft, C.M.; Lolley, R.N.; Baehr, W.; Hurwitz, R.L. Irish setter dogs affected with rod/cone dysplasia contain a nonsense mutation in the rod cGMP phosphodiesterase beta-subunit gene. *Proc. Natl. Acad. Sci. USA* **1993**, *90*, 3968–3972. [CrossRef]

39. Hartong, D.T.; Berson, E.L.; Dryja, T.P. Retinitis pigmentosa. *Lancet* **2006**, *368*, 1795–1809. [CrossRef]

40. McLaughlin, M.E.; Ehrhart, T.L.; Berson, E.L.; Dryja, T.P. Mutation spectrum of the gene encoding the beta subunit of rod phosphodiesterase among patients with autosomal recessive retinitis pigmentosa. *Proc. Natl. Acad. Sci. USA* **1995**, *92*, 3249–3253. [CrossRef]

41. Occelli, L.M.; Schön, C.; Seeliger, M.W.; Biel, M.; Michalakis, S.; Petersen-Jones, S.M.; The RD-Cure Consortium. Gene Supplementation Rescues Rod Function and Preserves Photoreceptor and Retinal Morphology in Dogs, Leading the Way Toward Treating Human PDE6A-Retinitis Pigmentosa. *Hum. Gene Ther.* **2017**, *28*, 1189–1201. [CrossRef] [PubMed]

42. Arango-Gonzalez, B.; Trifunović, D.; Sahaboglu, A.; Kranz, K.; Michalakis, S.; Farinelli, P.; Koch, S.; Koch, F.; Cottet, S.; Janssen-Bienhold, U.; et al. Identification of a Common Non-Apoptotic Cell Death Mechanism in Hereditary Retinal Degeneration. *PLoS ONE* **2014**, *9*, e112142. [CrossRef] [PubMed]

43. Wensel, T.G.; Zhang, Z.; Anastassov, I.; Gilliam, J.C.; He, F.; Schmid, M.F.; Robichaux, M.A. Structural and molecular bases of rod photoreceptor morphogenesis and disease. *Prog. Retin. Eye Res.* **2016**, *55*, 32–51. [CrossRef] [PubMed]

44. Mowat, F.M.; Occelli, L.M.; Bartoe, J.T.; Gervais, K.J.; Bruewer, A.R.; Querubin, J.; Dinculescu, A.; Boye, S.L.; Hauswirth, W.; Petersen-Jones, S.M. Gene Therapy in a Large Animal Model of PDE6A-Retinitis Pigmentosa. *Front. Mol. Neurosci.* **2017**, *11*, 342. [CrossRef]

45. Clements, P.J.M.; Gregory, C.Y.; Petersen-Jones, S.M.; Sargan, D.R.; Bhattacharya, S.S. Confirmation of the rod cGMP phophodiesterase á-subunit (PDEá) nonsense mutation in affected rcd-1 Irish setters in the UK and development of a diagnostic test. *Curr. Eye Res.* **1993**, *12*, 861–866. [CrossRef]

46. Farber, D.B.; Danciger, J.S.; Aguirre, G. The beta subunit of cyclic GMP phosphodiesterase mRNA is deficient in canine rod-cone dysplasia 1. *Neuron* **1992**, *9*, 349–356. [CrossRef]

47. Aguirre, G.; Farber, D.; Lolley, R.; O'Brien, P.; Alligood, J.; Fletcher, R.T.; Chader, G. Retinal degenerations in the dog III abnormal cyclic nucleotide metabolism in rod-cone dysplasia. *Exp. Eye Res.* **1982**, *35*, 625–642. [CrossRef]

48. Aguirre, G.D.; Rubin, L.F. Rod-cone dysplasia (progressive retinal atrophy) in Irish setters. *J. Am. Veter Med. Assoc.* **1975**, *166*, 157–164.

49. Aquirre, G.; Farber, D.; Lolley, R.; Fletcher, R.; Chader, G. Rod-cone dysplasia in Irish setters: A defect in cyclic GMP metabolism in visual cells. *Science* **1978**, *201*, 1133–1134. [CrossRef]

50. Pichard, V.; Provost, N.; Mendes-Madeira, A.; Libeau, L.; Hulin, P.; Tshilenge, K.-T.; Biget, M.; Ameline, B.; Deschamps, J.-Y.; Weber, M.; et al. AAV-mediated Gene Therapy Halts Retinal Degeneration in PDE6β-deficient Dogs. *Mol. Ther.* **2016**, *24*, 867–876. [CrossRef]

51. Dekomien, G.; Runte, M.; Gödde, R.; Epplen, J.T. Generalized progressive retinal atrophy of Sloughi dogs is due to an 8-bp insertion in exon 21 of the PDE6B gene. *Cytogenet. Cell Genet.* **2000**, *90*, 261–267. [CrossRef] [PubMed]

52. Goldstein, O.; Mezey, J.G.; Schweitzer, P.A.; Boyko, A.R.; Gao, C.; Bustamante, C.D.; Jordan, J.A.; Aguirre, G.D.; Acland, G.M. IQCB1 and PDE6B Mutations Cause Similar Early Onset Retinal Degenerations in Two Closely Related Terrier Dog Breeds. *Investig. Opthalmol. Vis. Sci.* **2013**, *54*, 7005–7019. [CrossRef] [PubMed]

53. Quazi, F.; Lenevich, S.; Molday, R.S. ABCA4 is an N-retinylidene-phosphatidylethanolamine and phosphatidylethanolamine importer. *Nat. Commun.* **2012**, *3*, 925. [CrossRef] [PubMed]

54. Mäkeläinen, S.; Gòdia, M.; Hellsand, M.; Viluma, A.; Hahn, D.; Makdoumi, K.; Zeiss, C.J.; Mellersh, C.; Ricketts, S.L.; Narfström, K.; et al. An ABCA4 loss-of-function mutation causes a canine form of Stargardt disease. *PLoS Genet.* **2019**, *15*, e1007873. [CrossRef] [PubMed]

55. Maeda, A.; Maeda, T.; Golczak, M.; Palczewski, K. Retinopathy in mice induced by disrupted all-trans-retinal clearance. *J. Boil. Chem.* **2008**, *283*, 26684–26693. [CrossRef]

56. Tsang, S.H.; Sharma, T. Stargardt Disease. In *Atlas of Inherited Retinal Diseases*; Tsang, S.H., Sharma, T., Eds.; Springer International Publishing: Cham, Switzerland, 2018; pp. 139–151. [CrossRef]

57. Le Meur, G.; Stieger, K.; Smith, A.J.; Weber, M.; Deschamps, J.Y.; Nivard, D.; Mendes-Madeira, A.; Provost, N.; Péréon, Y.; Cherel, Y.; et al. Restoration of vision in RPE65-deficient Briard dogs using an AAV serotype 4 vector that specifically targets the retinal pigmented epithelium. *Gene Ther.* **2006**, *14*, 292–303. [CrossRef]

58. Narfström, K.; Katz, M.L.; Bragadottir, R.; Seeliger, M.; Boulanger, A.; Redmond, T.M.; Caro, L.; Lai, C.-M.; Rakoczy, P.E. Functional and structural recovery of the retina after gene therapy in the RPE65 null mutation dog. *Investig. Opthalmol. Vis. Sci.* **2003**, *44*, 1663–1672. [CrossRef]

59. Annear, M.J.; Bartoe, J.T.; Barker, S.E.; Smith, A.J.; Curran, P.G.; Bainbridge, J.W.; Ali, R.R.; Petersen-Jones, S.M. Gene therapy in the second eye of RPE65-deficient dogs improves retinal function. *Gene Ther.* **2010**, *18*, 53–61. [CrossRef]

60. Annear, M.J.; Mowat, F.M.; Bartoe, J.T.; Querubin, J.; Azam, S.A.; Basche, M.; Curran, P.G.; Smith, A.; Bainbridge, J.W.; Ali, R.R.; et al. Successful Gene Therapy in Older Rpe65-Deficient Dogs Following Subretinal Injection of an Adeno-Associated Vector Expressing RPE65. *Hum. Gene Ther.* **2013**, *24*, 883–893. [CrossRef]

61. Mowat, F.M.; Breuwer, A.R.; Bartoe, J.T.; Annear, M.J.; Zhang, Z.; Smith, A.J.; Bainbridge, J.W.; Petersen-Jones, S.M.; Ali, R.R. RPE65 gene therapy slows cone loss in Rpe65-deficient dogs. *Gene Ther.* **2012**, *20*, 545–555. [CrossRef]

62. Mowat, F.M.; Gervais, K.J.; Occelli, L.M.; Annear, M.J.; Querubin, J.; Bainbridge, J.W.; Smith, A.J.; Ali, R.R.; Petersen-Jones, S.M. Early-Onset Progressive Degeneration of the Area Centralis in RPE65-Deficient Dogs. *Investig. Opthalmol. Vis. Sci.* **2017**, *58*, 3268. [CrossRef] [PubMed]

63. Bainbridge, J.W.; Mehat, M.S.; Sundaram, V.; Robbie, S.J.; Barker, S.E.; Ripamonti, C.; Georgiadis, A.; Mowat, F.M.; Beattie, S.G.; Gardner, P.; et al. Long-Term Effect of Gene Therapy on Leber's Congenital Amaurosis. *N. Engl. J. Med.* **2015**, *372*, 1887–1897. [CrossRef] [PubMed]

64. Petersen-Jones, S.M.; Occelli, L.; Winkler, P.; Minella, A.; Sun, K.; Lyons, L.; Daruwalla, A.; Kiser, P.; Palczewski, K. New large animal model for RDH5-associated retinopathies. *Investig. Ophthalmol. Vis. Sci.* **2019**, *60*, 458.

65. Gonzalez-Fernandez, F.; Kurz, D.; Bao, Y.; Newman, S.; Conway, B.P.; Young, J.E.; Han, D.P.; Khani, S.C. 11-cis retinol dehydrogenase mutations as a major cause of the congenital night-blindness disorder known as fundus albipunctatus. *Mol. Vis.* **1999**, *5*, 41. [PubMed]

66. Hotta, K.; Nakamura, M.; Kondo, M.; Ito, S.; Terasaki, H.; Miyake, Y.; Hida, T. Macular dystrophy in a Japanese family with fundus albipunctatus. *Am. J. Ophthalmol.* **2003**, *135*, 917–919. [CrossRef]

67. Nakamura, M.; Miyake, Y. Macular dystrophy in a 9-year-old boy with fundus albipunctatus. *Am. J. Ophthalmol.* **2002**, *133*, 278–280. [CrossRef]

68. Nakamura, M.; Skalet, J.; Miyake, Y. RDH5 gene mutations and electroretinogram in fundus albipunctatus with or without macular dystrophy: RDH5 mutations and ERG in fundus albipunctatus. *Doc. Ophthalmol.* **2003**, *107*, 3–11. [CrossRef]

69. Yamamoto, H.; Yakushijin, K.; Kusuhara, S.; Escaño, M.F.T.; Nagai, A.; Negi, A. A novel RDH5 gene mutation in a patient with fundus albipunctatus presenting with macular atrophy and fading white dots. *Am. J. Ophthalmol.* **2003**, *136*, 572–574. [CrossRef]

70. Kuehlewein, L.; Nasser, F.; Gloeckle, N.; Kohl, S.; Zrenner, E. Fundus albipunctatus associated with cone dysfunction. *Retin. Cases Brief Rep.* **2017**, *11*, S73–S76. [CrossRef]

71. Kim, T.S.; Maeda, A.; Maeda, T.; Heinlein, C.; Kedishvili, N.; Palczewski, K.; Nelson, P.S. Delayed dark adaptation in 11-cis-retinol dehydrogenase-deficient mice: A role of RDH11 in visual processes in vivo. *J. Boil. Chem.* **2005**, *280*, 8694–8704. [CrossRef]

72. Kaupp, U.B.; Seifert, R. Cyclic Nucleotide-Gated Ion Channels. *Physiol. Rev.* **2002**, *82*, 769–824. [CrossRef] [PubMed]

73. Zheng, J.; Trudeau, M.C.; Zagotta, W.N. Rod cyclic nucleotide-gated channels have a stoichiometry of three CNGA1 subunits and one CNGB1 subunit. *Neuron* **2002**, *36*, 891–896. [CrossRef]

74. Zhong, H.; Molday, L.L.; Molday, R.S.; Yau, K.-W. The heteromeric cyclic nucleotide-gated channel adopts a 3A:1B stoichiometry. *Nature* **2002**, *420*, 193–198. [CrossRef] [PubMed]

75. Johnson, A.A.; Guziewicz, K.E.; Lee, C.J.; Kalathur, R.C.; Pulido, J.S.; Marmorstein, L.Y.; Marmorstein, A.D. Bestrophin 1 and retinal disease. *Prog. Retin. Eye Res.* **2017**, *58*, 45–69. [CrossRef] [PubMed]

76. Wiik, A.C.; Ropstad, E.O.; Ekesten, B.; Karlstam, L.; Wade, C.; Lingaas, F. Progressive retinal atrophy in Shetland sheepdog is associated with a mutation in theCNGA1gene. *Anim. Genet.* **2015**, *46*, 515–521. [CrossRef] [PubMed]

77. Winkler, P.A.; Ekenstedt, K.J.; Occelli, L.M.; Frattaroli, A.V.; Bartoe, J.T.; Venta, P.J.; Petersen-Jones, S.M. A Large Animal Model for CNGB1 Autosomal Recessive Retinitis Pigmentosa. *PLoS ONE* **2013**, *8*, 72229. [CrossRef]

78. Petersen-Jones, S.M.; Occelli, L.M.; Winkler, P.A.; Lee, W.; Sparrow, J.R.; Tsukikawa, M.; Boye, S.L.; Chiodo, V.; Capasso, J.E.; Becirovic, E.; et al. Patients and animal models of CNGβ1-deficient retinitis pigmentosa support gene augmentation approach. *J. Clin. Investig.* **2017**, *128*, 190–206. [CrossRef]

79. Hüttl, S.; Michalakis, S.; Seeliger, M.; Luo, N.-G.; Acar, N.; Geiger, H.; Hudl, K.; Mader, R.; Haverkamp, S.; Moser, M.; et al. Impaired channel targeting and retinal degeneration in mice lacking the cyclic nucleotide-gated channel subunit CNGB1. *J. Neurosci.* **2005**, *25*, 130–138. [CrossRef]

80. Hull, S.; Attanasio, M.; Arno, G.; Carss, K.; Robson, A.; Thompson, D.; Plagnol, V.; Michaelides, M.; Holder, G.E.; Henderson, R.H.; et al. Clinical Characterization of CNGB1-Related Autosomal Recessive Retinitis Pigmentosa. *JAMA Ophthalmol.* **2017**, *135*, 137–144. [CrossRef]

81. Wissinger, B.; Gamer, D.; Jägle, H.; Giorda, R.; Marx, T.; Mayer, S.; Tippmann, S.; Broghammer, M.; Jurklies, B.; Rosenberg, T.; et al. CNGA3 Mutations in Hereditary Cone Photoreceptor Disorders. *Am. J. Hum. Genet.* **2001**, *69*, 722–737. [CrossRef]

82. Tanaka, N.; Dutrow, E.; Miyadera, K.; Delemotte, L.; MacDermaid, C.; Reinstein, S.L.; Crumley, W.R.; Dixon, C.J.; Casal, M.L.; Klein, M.L.; et al. Canine CNGA3 Gene Mutations Provide Novel Insights into Human Achromatopsia-Associated Channelopathies and Treatment. *PLoS ONE* **2015**, *10*, e0138943. [CrossRef] [PubMed]

83. Reicher, S.; Seroussi, E.; Gootwine, E. A mutation in gene CNGA3 is associated with day blindness in sheep. *Genomics* **2010**, *95*, 101–104. [CrossRef] [PubMed]

84. Gootwine, E.; Abu-Siam, M.; Obolensky, A.; Rosov, A.; Honig, H.; Nitzan, T.; Shirak, A.; Ezra-Elia, R.; Yamin, E.; Banin, E.; et al. Gene Augmentation Therapy for a Missense Substitution in the cGMP-Binding Domain of Ovine CNGA3 Gene Restores Vision in Day-Blind Sheep. *Investig. Opthalmol. Vis. Sci.* **2017**, *58*, 1577–1584. [CrossRef] [PubMed]

85. Gootwine, E.; Ofri, R.; Banin, E.; Obolensky, A.; Averbukh, E.; Ezra-Elia, R.; Ross, M.; Honig, H.; Rosov, A.; Yamin, E.; et al. Safety and Efficacy Evaluation of rAAV2tYF-PR1.7-hCNGA3 Vector Delivered by Subretinal Injection in CNGA3 Mutant Achromatopsia Sheep. *Hum. Gene Ther. Clin. Dev.* **2017**, *28*, 96–107. [CrossRef] [PubMed]

86. Sidjanin, D.J.; Lowe, J.K.; McElwee, J.; Milne, B.S.; Phippen, T.M.; Sargan, D.R.; Aguirre, G.D.; Acland, G.M.; Ostrander, E.A. Canine CNGB3 mutations establish cone degeneration as orthologous to the human achromatopsia locus ACHM3. *Hum. Mol. Genet.* **2002**, *11*, 1823–1833. [CrossRef]

87. Yeh, C.Y.; Goldstein, O.; Kukekova, A.; Holley, D.; Knollinger, A.M.; Huson, H.J.; Pearce-Kelling, S.E.; Acland, G.M.; Komáromy, A.M. Genomic deletion of CNGB3 is identical by descent in multiple canine breeds and causes achromatopsia. *BMC Genet.* **2013**, *14*, 27. [CrossRef]

88. Kohl, S.; Varsányi, B.; Antunes, G.A.; Baumann, B.; Hoyng, C.B.; Jägle, H.; Rosenberg, T.; Kellner, U.; Lorenz, B.; Salati, R.; et al. CNGB3 mutations account for 50% of all cases with autosomal recessive achromatopsia. *Eur. J. Hum. Genet.* **2004**, *13*, 302–308. [CrossRef]

89. Tanaka, N.; Delemotte, L.; Klein, M.L.; Komáromy, A.M.; Tanaka, J.C. A Cyclic Nucleotide-Gated Channel Mutation Associated with Canine Daylight Blindness Provides Insight into a Role for the S2 Segment Tri-Asp motif in Channel Biogenesis. *PLoS ONE* **2014**, *9*, e88768. [CrossRef]

90. Ye, G.-J.; Komáromy, A.M.; Zeiss, C.; Calcedo, R.; Harman, C.D.; Koehl, K.L.; Stewart, G.A.; Iwabe, S.; Chiodo, V.A.; Hauswirth, W.; et al. Safety and Efficacy of AAV5 Vectors Expressing Human or Canine CNGB3 in CNGB3-Mutant Dogs. *Hum. Gene Ther. Clin. Dev.* **2017**, *28*, 197–207. [CrossRef]

91. Guziewicz, K.E.; Zangerl, B.; Lindauer, S.J.; Mullins, R.F.; Sandmeyer, L.S.; Grahn, B.H.; Stone, E.M.; Acland, G.M.; Aguirre, G.D. Bestrophin gene mutations cause canine multifocal retinopathy: A novel animal model for best disease. *Investig. Opthalmol. Vis. Sci.* **2007**, *48*, 1959–1967. [CrossRef]

92. Zangerl, B.; Wickström, K.; Slavik, J.; Lindauer, S.J.; Ahonen, S.; Schelling, C.; Lohi, H.; Guziewicz, K.E.; Aguirre, G.D. Assessment of canine BEST1 variations identifies new mutations and establishes an independent bestrophinopathy model (cmr3). *Mol. Vis.* **2010**, *16*, 2791–2804. [PubMed]

93. Guziewicz, K.E.; Cideciyan, A.V.; Beltran, W.A.; Komaromy, A.M.; Dufour, V.L.; Swider, M.; Iwabe, S.; Sumaroka, A.; Kendrick, B.T.; Ruthel, G.; et al. BEST1 gene therapy corrects a diffuse retina-wide microdetachment modulated by light exposure. *Proc. Natl. Acad. Sci. USA* **2018**, *115*, E2839–E2848. [CrossRef] [PubMed]

94. Guziewicz, K.E.; McTish, E.; Dufour, V.L.; Zorych, K.; Dhingra, A.; Boesze-Battaglia, K.; Aguirre, G.D. Underdeveloped RPE Apical Domain Underlies Lesion Formation in Canine Bestrophinopathies. *Single Mol. Single Cell Seq.* **2018**, *1074*, 309–315.

95. Chew, T.; Haase, B.; Bathgate, R.; Willet, C.; Kaukonen, M.K.; Mascord, L.J.; Lohi, H.; Wade, C. A Coding Variant in the Gene Bardet-Biedl Syndrome 4 (BBS4) Is Associated with a Novel Form of Canine Progressive Retinal Atrophy. *G3 Genes Genomes Genet.* **2017**, *7*, 2327–2335. [CrossRef] [PubMed]

96. Narfström, K. Progressive retinal atrophy in the Abyssinian cat. Clinical characteristics. *Investig. Ophthalmol. Vis. Sci.* **1985**, *26*, 193–200.

97. Minella, A.L.; Occelli, L.M.; Narfström, K.; Petersen-Jones, S.M. Central retinal preservation in rdAc cats. *Veter Ophthalmol.* **2017**, *21*, 224–232. [CrossRef]

98. Coppieters, F.; Lefever, S.; Leroy, B.P.; De Baere, E. CEP290, a gene with many faces: Mutation overview and presentation of CEP290base. *Hum. Mutat.* **2010**, *31*, 1097–1108. [CrossRef]

99. Menotti-Raymond, M.; David, V.A.; Schäffer, A.A.; Stephens, R.; Wells, D.; Kumar-Singh, R.; O'Brien, S.; Narfström, K. Mutation in CEP290 Discovered for Cat Model of Human Retinal Degeneration. *J. Hered.* **2007**, *98*, 211–220. [CrossRef]

100. Wiik, A.C.; Wade, C.; Biagi, T.; Ropstad, E.-O.; Bjerkås, E.; Lindblad-Toh, K.; Lingaas, F. A deletion in nephronophthisis 4 (NPHP4) is associated with recessive cone-rod dystrophy in standard wire-haired dachshund. *Genome Res.* **2008**, *18*, 1415–1421. [CrossRef]

101. Ropstad, E.O.; Narfström, K.; Lingaas, F.; Wiik, C.; Bruun, A.; Bjerkas, E. Functional and Structural Changes in the Retina of Wire-Haired Dachshunds with Early-Onset Cone-Rod Dystrophy. *Investig. Opthalmol. Vis. Sci.* **2008**, *49*, 1106–1115. [CrossRef]

102. Ropstad, E.O.; Bjerkås, E.; Narfström, K. Electroretinographic findings in the Standard Wire Haired Dachshund with inherited early onset cone–rod dystrophy. *Doc. Ophthalmol.* **2006**, *114*, 27–36. [CrossRef] [PubMed]

103. Ropstad, E.O.; Bjerkås, E.; Narfström, K. Clinical findings in early onset cone-rod dystrophy in the Standard Wire-haired Dachshund. *Veter Ophthalmol.* **2007**, *10*, 69–75. [CrossRef] [PubMed]

104. Ronquillo, C.; Bernstein, P.S.; Baehr, W. Senior-Løken syndrome: A syndromic form of retinal dystrophy associated with nephronophthisis. *Vis. Res.* **2012**, *75*, 88–97. [CrossRef] [PubMed]

105. Won, J.; De Evsikova, C.M.; Smith, R.S.; Hicks, W.L.; Edwards, M.M.; Longo-Guess, C.; Li, T.; Naggert, J.K.; Nishina, P.M. NPHP4 is necessary for normal photoreceptor ribbon synapse maintenance and outer segment formation, and for sperm development. *Hum. Mol. Genet.* **2010**, *20*, 482–496. [CrossRef]

106. Downs, L.M.; Scott, E.M.; Cideciyan, A.V.; Iwabe, S.; Dufour, V.L.; Gardiner, K.L.; Genini, S.; Marinho, L.F.; Sumaroka, A.; Kosyk, M.S.; et al. Overlap of abnormal photoreceptor development and progressive degeneration in Leber congenital amaurosis caused by NPHP5 mutation. *Hum. Mol. Genet.* **2016**, *25*, 4211–4226. [CrossRef]

107. Aguirre, G.D.; Cideciyan, A.V.; Boye, S.L.; Iwabe, S.; Dufour, V.; Swider, M.; Roszak, K.; Hauswirth, W.W.; Jacobson, S.G.; Beltran, W.A. Long-term preservation of photoreceptor function and structure following early-stage treatment by AAV-mediated gene augmentation in canine model of NPHP5 Leber congenital amaurosis. *Investig. Ophthalmol. Vis. Sci.* **2018**, *59*, 6006.

108. Khanna, H. More Than Meets the Eye: Current Understanding of RPGR Function. In *Advances in Experimental Medicine and Biology*; Springer: Cham, Switzerland, 2018; pp. 521–538.

109. Vervoort, R.; Lennon, A.; Bird, A.C.; Tulloch, B.; Axton, R.; Miano, M.G.; Meindl, A.; Meitinger, T.; Ciccodicola, A.; Wright, A.F. Mutational hot spot within a new RPGR exon in X-linked retinitis pigmentosa. *Nat. Genet.* **2000**, *25*, 462–466. [CrossRef]

110. Kropatsch, R.; Akkad, D.A.; Frank, M.; Rosenhagen, C.; Altmüller, J.; Nürnberg, P.; Epplen, J.T.; Dekomien, G. A large deletion in RPGR causes XLPRA in Weimaraner dogs. *Canine Genet. Epidemiol.* **2016**, *3*, 7. [CrossRef]

111. Zhang, Q.; Acland, G.M.; Wu, W.X.; Johnson, J.; Pearce-Kelling, S.; Tulloch, B.; Vervoort, R.; Wright, A.F.; Aguirre, G.D. Different RPGR exon ORF15 mutations in Canids provide insights into photoreceptor cell degeneration. *Hum. Mol. Genet.* **2002**, *11*, 993–1003. [CrossRef]

112. Zeiss, C.J.; Acland, G.M.; Aguirre, G.D. Retinal pathology of canine X-linked progressive retinal atrophy, the locus homologue of RP3. *Investig. Ophthalmol. Vis. Sci.* **1999**, *40*, 3292–3304.

113. Beltran, W.A.; Hammond, P.; Acland, G.M.; Aguirre, G.D. A Frameshift Mutation in RPGR Exon ORF15 Causes Photoreceptor Degeneration and Inner Retina Remodeling in a Model of X-Linked Retinitis Pigmentosa. *Investig. Opthalmol. Vis. Sci.* **2006**, *47*, 1669–1681. [CrossRef] [PubMed]

114. Beltran, W.A.; Acland, G.M.; Aguirre, G.D. Age-dependent disease expression determines remodeling of the retinal mosaic in carriers of RPGR exon ORF15 mutations. *Investig. Opthalmol. Vis. Sci.* **2009**, *50*, 3985–3995. [CrossRef] [PubMed]

115. Mellersh, C.; Boursnell, M.; Pettitt, L.; Ryder, E.; Holmes, N.; Grafham, D.; Forman, O.; Sampson, J.; Barnett, K.; Blanton, S.; et al. Canine RPGRIP1 mutation establishes cone–rod dystrophy in miniature longhaired dachshunds as a homologue of human Leber congenital amaurosis. *Genomics* **2006**, *88*, 293–301. [CrossRef] [PubMed]

116. Lhériteau, E.; Petit, L.; Weber, M.; Le Meur, G.; Deschamps, J.-Y.; Libeau, L.; Mendes-Madeira, A.; Guihal, C.; François, A.; Guyon, R.; et al. Successful Gene Therapy in the RPGRIP1-deficient Dog: A Large Model of Cone–Rod Dystrophy. *Mol. Ther.* **2013**, *22*, 265–277. [CrossRef] [PubMed]

117. Kuznetsova, T.; Iwabe, S.; Boesze-Battaglia, K.; Pearce-Kelling, S.; Chang-Min, Y.; McDaid, K.; Miyadera, K.; Komaromy, A.; Aguirre, G.D. Exclusion of RPGRIP1 ins44 from Primary Causal Association with Early-Onset Cone–Rod Dystrophy in Dogs. *Investig. Opthalmol. Vis. Sci.* **2012**, *53*, 5486–5501. [CrossRef]

118. Forman, O.P.; Hitti, R.J.; Boursnell, M.; Miyadera, K.; Sargan, D.R.; Mellersh, C. Canine genome assembly correction facilitates identification of a MAP9 deletion as a potential age of onset modifier for RPGRIP1-associated canine retinal degeneration. *Mamm. Genome* **2016**, *27*, 237–245. [CrossRef]

119. Miyadera, K.; Kato, K.; Boursnell, M.; Mellersh, C.; Sargan, D.R. Genome-wide association study in RPGRIP1$^{-/-}$ dogs identifies a modifier locus that determines the onset of retinal degeneration. *Mamm. Genome* **2011**, *23*, 212–223. [CrossRef]

120. Miyadera, K.; Murgiano, L.; Spector, C.; Marinho, F.P.; Dufour, V.; Das, R.G.; Brooks, M.; Swaroop, A.; Aguirre, G.D. Isolated population helps tease out a third locus underlying a multigenic form of canine RPGRIP1 cone-rod dystrophy. *Investig. Ophthalmol. Vis. Sci.* **2018**, *59*, 1438.

121. Chau, K.-Y.; Chen, S.; Zack, D.J.; Ono, S.J. Functional Domains of the Cone-Rod Homeobox (CRX) Transcription Factor. *J. Boil. Chem.* **2000**, *275*, 37264–37270. [CrossRef]

122. Morrow, E.M.; Furukawa, T.; Raviola, E.; Cepko, C.L. Synaptogenesis and outer segment formation are perturbed in the neural retina of Crx mutant mice. *BMC Neurosci.* **2005**, *6*, 5. [CrossRef]

123. Chen, S.; Wang, Q.-L.; Nie, Z.; Sun, H.; Lennon, G.; Copeland, N.; Gilbert, D.J.; Jenkins, N.A.; Zack, D.J. Crx, a novel Otx-like paired-homeodomain protein, binds to and transactivates photoreceptor cell-specific genes. *Neuron* **1997**, *19*, 1017–1030. [CrossRef]

124. Furukawa, T.; Morrow, E.M.; Cepko, C.L. Crx, a novel otx-like homeobox gene, shows photoreceptor-specific expression and regulates photoreceptor differentiation. *Cell* **1997**, *91*, 531–541. [CrossRef]

125. Hennig, A.K.; Peng, G.-H.; Chen, S. Regulation of photoreceptor gene expression by Crx-associated transcription factor network. *Brain Res.* **2007**, *1192*, 114–133. [CrossRef] [PubMed]

126. Peng, G.-H.; Chen, S. Crx activates opsin transcription by recruiting HAT-containing co-activators and promoting histone acetylation. *Hum. Mol. Genet.* **2007**, *16*, 2433–2452. [CrossRef] [PubMed]

127. Menotti-Raymond, M.; Deckman, K.H.; David, V.; Myrkalo, J.; O'Brien, S.J.; Narfström, K. Mutation discovered in a feline model of human congenital retinal blinding disease. *Investig. Opthalmol. Vis. Sci.* **2010**, *51*, 2852–2859. [CrossRef] [PubMed]

128. Occelli, L.M.; Tran, N.M.; Narfström, K.; Chen, S.; Petersen-Jones, S.M. CrxRdy Cat: A Large Animal Model for CRX-Associated Leber Congenital Amaurosis. *Investig. Opthalmol. Vis. Sci.* **2016**, *57*, 3780–3792. [CrossRef]

129. Sohocki, M.M.; Sullivan, L.S.; Mintz-Hittner, H.A.; Birch, D.G.; Heckenlively, J.R.; Freund, C.L.; McInnes, R.R.; Daiger, S.P. A range of clinical phenotypes associated with mutations in CRX, a photoreceptor transcription-factor gene. *Am. J. Hum. Genet.* **1998**, *63*, 1307–1315. [CrossRef]

130. Hull, S.; Arno, G.; Plagnol, V.; Chamney, S.; Russell-Eggitt, I.; Thompson, D.; Ramsden, S.C.; Black, G.C.; Robson, A.; Holder, G.E.; et al. The Phenotypic Variability of Retinal Dystrophies Associated With Mutations in CRX, With Report of a Novel Macular Dystrophy Phenotype. *Investig. Opthalmol. Vis. Sci.* **2014**, *55*, 6934–6944. [CrossRef]

131. Huang, L.; Xiao, X.; Li, S.; Jia, X.; Wang, P.; Guo, X.; Zhang, Q. CRX variants in cone–rod dystrophy and mutation overview. *Biochem. Biophys. Res. Commun.* **2012**, *426*, 498–503. [CrossRef]

132. Hanein, S.; Perrault, I.; Gerber, S.; Tanguy, G.; Barbet, F.; Ducroq, D.; Calvas, P.; Dollfus, H.; Hamel, C.; Löppönen, T.; et al. Leber congenital amaurosis: Comprehensive survey of the genetic heterogeneity, refinement of the clinical definition, and genotype-phenotype correlations as a strategy for molecular diagnosis. *Hum. Mutat.* **2004**, *23*, 306–317. [CrossRef]

133. Kumaran, N.; Pennesi, M.E.; Yang, P.; Trzupek, K.M.; Schlechter, C.; Moore, A.T.; Weleber, R.G.; Michaelides, M. Leber Congenital Amaurosis/Early-Onset Severe Retinal Dystrophy Overview. In *GeneReviews((R))*; Adam, M.P., Ardinger, H.H., Pagon, R.A., Wallace, S.E., Bean, L.J.H., Stephens, K., Amemiya, A., Eds.; University of Washington: Seattle, WA, USA, 1993.

134. Hollander, A.I.D.; Roepman, R.; Koenekoop, R.K.; Cremers, F.P. Leber congenital amaurosis: Genes, proteins and disease mechanisms. *Prog. Retin. Eye Res.* **2008**, *27*, 391–419. [CrossRef] [PubMed]

135. Tran, N.M.; Chen, S. Mechanisms of blindness: Animal models provide insight into distinct CRX-associated retinopathies. *Dev. Dyn.* **2014**, *243*, 1153–1166. [CrossRef]

136. Acland, G.M.; Aguirre, G.D. Retinal degenerations in the dog: IV. Early retinal degeneration (erd) in the Norwegian elkhound. *Exp. Eye Res.* **1987**, *44*, 491–521. [CrossRef]

137. Goldstein, O.; Kukekova, A.V.; Aguirre, G.D.; Acland, G.M. Exonic SINE insertion in STK38L causes canine early retinal degeneration (erd). *Genomics* **2010**, *96*, 362–368. [CrossRef] [PubMed]

138. Berta, Á.I.; Boesze-Battaglia, K.; Genini, S.; Goldstein, O.; O'Brien, P.J.; Szél, Á.; Acland, G.M.; Beltran, W.A.; Aguirre, G.D. Photoreceptor Cell Death, Proliferation and Formation of Hybrid Rod/S-Cone Photoreceptors in the Degenerating STK38L Mutant Retina. *PLoS ONE* **2011**, *6*, e24074. [CrossRef] [PubMed]

139. Zeitz, C.; Jacobson, S.G.; Hamel, C.P.; Bujakowska, K.M.; Neuillé, M.; Orhan, E.; Zanlonghi, X.; Lancelot, M.-E.; Michiels, C.; Schwartz, S.B.; et al. Whole-Exome Sequencing Identifies LRIT3 Mutations as a Cause of Autosomal-Recessive Complete Congenital Stationary Night Blindness. *Am. J. Hum. Genet.* **2013**, *92*, 67–75. [CrossRef]

140. Kondo, M.; Das, R.; Imai, R.; Santana, E.; Nakashita, T.; Imawaka, M.; Ueda, K.; Ohtsuka, H.; Sakai, K.; Aihara, T.; et al. A Naturally Occurring Canine Model of Autosomal Recessive Congenital Stationary Night Blindness. *PLoS ONE* **2015**, *10*, e0137072. [CrossRef]

141. Das, R.; Becker, D.; Jagannathan, V.; Goldstein, O.; Santana, E.; Carlin, K.; Sudharsan, R.; Leeb, T.; Nishizawa, Y.; Kondo, M.; et al. Genome-wide association study and whole-genome sequencing identify a deletion in LRIT3 associated with canine congenital stationary night blindness. *Sci. Rep.* **2019**, *9*, 14166. [CrossRef]

142. Oh, A.; Loew, E.R.; Foster, M.L.; Davidson, M.G.; English, R.V.; Gervais, K.J.; Herring, I.; Mowat, F.M. Phenotypic characterization of complete CSNB in the inbred research beagle: How common is CSNB in research and companion dogs? *Doc. Ophthalmol.* **2018**, *137*, 87–101. [CrossRef]

143. Hasan, N.; Pangeni, G.; Cobb, C.A.; Ray, T.A.; Nettesheim, E.R.; Ertel, K.J.; Lipinski, D.M.; McCall, M.A.; Gregg, R.G. Presynaptic Expression of LRIT3 Transsynaptically Organizes the Postsynaptic Glutamate Signaling Complex Containing TRPM1. *Cell Rep.* **2019**, *27*, 3107–3116. [CrossRef]

144. Morgans, C.; Zhang, J.; Jeffrey, B.G.; Nelson, S.M.; Burke, N.S.; Duvoisin, R.M.; Brown, R.L. TRPM1 is required for the depolarizing light response in retinal ON-bipolar cells. *Proc. Natl. Acad. Sci. USA* **2009**, *106*, 19174–19178. [CrossRef] [PubMed]

145. Morgans, C.; Brown, R.L.; Duvoisin, R.M. TRPM1: The endpoint of the mGluR6 signal transduction cascade in retinal ON-bipolar cells. *BioEssays* **2010**, *32*, 609–614. [CrossRef] [PubMed]

146. Witzel, D.A.; Smith, E.L.; Wilson, R.D.; Aguirre, G.D. Congenital stationary night blindness: An animal model. *Investig. Ophthalmol. Vis. Sci.* **1978**, *17*, 788–795.

147. Sandmeyer, L.S.; Breaux, C.B.; Archer, S.; Grahn, B.H. Clinical and electroretinographic characteristics of congenital stationary night blindness in the Appaloosa and the association with the leopard complex. *Veter Ophthalmol.* **2007**, *10*, 368–375. [CrossRef] [PubMed]

148. Bellone, R.; Brooks, S.A.; Sandmeyer, L.; Murphy, B.A.; Forsyth, G.; Archer, S.; Bailey, E.; Grahn, B. Differential Gene Expression of TRPM1, the Potential Cause of Congenital Stationary Night Blindness and Coat Spotting Patterns (LP) in the Appaloosa Horse (Equus caballus). *Genetics* **2008**, *179*, 1861–1870. [CrossRef] [PubMed]

149. Audo, I.; Kohl, S.; Leroy, B.P.; Munier, F.L.; Guillonneau, X.; Mohand-Saïd, S.; Bujakowska, K.M.; Nandrot, E.F.; Lorenz, B.; Preising, M.; et al. TRPM1 Is Mutated in Patients with Autosomal-Recessive Complete Congenital Stationary Night Blindness. *Am. J. Hum. Genet.* **2009**, *85*, 720–729. [CrossRef]

150. Li, Z.; Sergouniotis, P.I.; Michaelides, M.; Mackay, D.; Wright, G.A.; Devery, S.; Moore, A.T.; Holder, G.E.; Robson, A.; Webster, A.R. Recessive Mutations of the Gene TRPM1 Abrogate ON Bipolar Cell Function and Cause Complete Congenital Stationary Night Blindness in Humans. *Am. J. Hum. Genet.* **2009**, *85*, 711–719. [CrossRef]

151. Van Genderen, M.M.; Bijveld, M.M.; Claassen, Y.B.; Florijn, R.J.; Pearring, J.N.; Meire, F.M.; McCall, M.A.; Riemslag, F.C.; Gregg, R.G.; Bergen, A.A.; et al. Mutations in TRPM1 Are a Common Cause of Complete Congenital Stationary Night Blindness. *Amer. J. Human Genet.* **2009**, *85*, 730–736. [CrossRef]

152. Bellone, R.; Holl, H.; Setaluri, V.; Devi, S.; Maddodi, N.; Archer, S.; Sandmeyer, L.; Ludwig, A.; Förster, D.; Pruvost, M.; et al. Evidence for a Retroviral Insertion in TRPM1 as the Cause of Congenital Stationary Night Blindness and Leopard Complex Spotting in the Horse. *PLoS ONE* **2013**, *8*, e78280. [CrossRef]

153. Littink, K.W.; Van Genderen, M.M.; Collin, R.W.; Roosing, S.; De Brouwer, A.P.M.; Riemslag, F.C.C.; Venselaar, H.; Thiadens, A.A.H.J.; Hoyng, C.; Rohrschneider, K.; et al. A Novel Homozygous Nonsense Mutation inCABP4Causes Congenital Cone–Rod Synaptic Disorder. *Investig. Opthalmol. Vis. Sci.* **2009**, *50*, 2344–2350. [CrossRef]

154. Marinho, L.L.P.; Occelli, L.M.; Pasmanter, N.; Somma, A.T.; Montiani-Ferreira, F.; Petersen-Jones, S.M. Autosomal recessive night blindness with progressive photoreceptor degeneration in a dog model. *Investig. Ophthalmol. Vis. Sci.* **2019**, *60*, 465.

155. Somma, A.T.; Moreno, J.C.D.; Sato, M.T.; Rodrigues, B.D.; Occelli, L.M.; Bacellar-Galdino, M.; Petersen-Jones, S.M.; Montiani-Ferreira, F. Characterization of a novel form of progressive retinal atrophy in Whippet dogs: A clinical, electroretinographic, and breeding study. *Veter Ophthalmol.* **2016**, *20*, 450–459. [CrossRef] [PubMed]

156. Kropatsch, R.; Petrasch-Parwez, E.; Seelow, D.; Schlichting, A.; Gerding, W.M.; Akkad, D.A.; Epplen, J.T.; Dekomien, G. Generalized progressive retinal atrophy in the Irish Glen of Imaal Terrier is associated with a deletion in the ADAM9 gene. *Mol. Cell. Probes* **2010**, *24*, 357–363. [CrossRef] [PubMed]

157. Goldstein, O.; Mezey, J.G.; Boyko, A.R.; Gao, C.; Wang, W.; Bustamante, C.D.; Anguish, L.J.; Jordan, J.A.; Pearce-Kelling, S.E.; Aguirre, G.D.; et al. An ADAM9 mutation in canine cone-rod dystrophy 3 establishes homology with human cone-rod dystrophy 9. *Mol. Vis.* **2010**, *16*, 1549–1569. [PubMed]

158. Sohocki, M.M.; Perrault, I.; Leroy, B.P.; Payne, A.; Dharmaraj, S.; Bhattacharya, S.S.; Kaplan, J.; Maumenee, I.H.; Koenekoop, R.; Meire, F.M.; et al. Prevalence of AIPL1 Mutations in Inherited Retinal Degenerative Disease. *Mol. Genet. Metab.* **2000**, *70*, 142–150. [CrossRef]

159. Van Der Spuy, J.; Kim, J.H.; Yu, Y.S.; Szel, A.; Luthert, P.J.; Clark, B.J.; Cheetham, M.E. The expression of the Leber congenital amaurosis protein AIPL1 coincides with rod and cone photoreceptor development. *Investig. Opthalmol. Vis. Sci.* **2003**, *44*, 5396–5403. [CrossRef]

160. Kumaran, N.; Moore, A.T.; Weleber, R.G.; Michaelides, M. Leber congenital amaurosis/early-onset severe retinal dystrophy: Clinical features, molecular genetics and therapeutic interventions. *Br. J. Ophthalmol.* **2017**, *101*, 1147–1154. [CrossRef]

161. Sohocki, M.M.; Bowne, S.J.; Sullivan, L.S.; Blackshaw, S.; Cepko, C.L.; Payne, A.M.; Bhattacharya, S.S.; Khaliq, S.; Mehdi, S.Q.; Birch, D.G.; et al. Mutations in a new photoreceptor-pineal gene on 17p cause Leber congenital amaurosis. *Nat. Genet.* **2000**, *24*, 79–83. [CrossRef]

162. Gopalakrishna, K.N.; Boyd, K.; Yadav, R.P.; Artemyev, N.O. Aryl Hydrocarbon Receptor-interacting Protein-like 1 Is an Obligate Chaperone of Phosphodiesterase 6 and Is Assisted by the γ-Subunit of Its Client. *J. Boil. Chem.* **2016**, *291*, 16282–16291. [CrossRef]

163. Hidalgo-De-Quintana, J.; Evans, R.J.; Cheetham, M.E.; Van Der Spuy, J. The Leber congenital amaurosis protein AIPL1 functions as part of a chaperone heterocomplex. *Investig. Opthalmol. Vis. Sci.* **2008**, *49*, 2878–2887. [CrossRef]

164. Ramamurthy, V.; Niemi, G.A.; Reh, T.A.; Hurley, J.B. Leber congenital amaurosis linked to AIPL1: A mouse model reveals destabilization of cGMP phosphodiesterase. *Proc. Natl. Acad. Sci. USA* **2004**, *101*, 13897–13902. [CrossRef]

165. Yadav, R.P.; Artemyev, N.O. AIPL1: A specialized chaperone for the phototransduction effector. *Cell. Signal.* **2017**, *40*, 183–189. [CrossRef]

166. Kolandaivelu, S.; Singh, R.K.; Ramamurthy, V. AIPL1, A protein linked to blindness, is essential for the stability of enzymes mediating cGMP metabolism in cone photoreceptor cells. *Hum. Mol. Genet.* **2013**, *23*, 1002–1012. [CrossRef]

167. Rah, H.; Maggs, D.J.; Blankenship, T.N.; Narfström, K.; Lyons, L.A. Early-Onset, Autosomal Recessive, Progressive Retinal Atrophy in Persian Cats. *Investig. Opthalmol. Vis. Sci.* **2005**, *46*, 1742–1747. [CrossRef] [PubMed]

168. Lyons, L.A.; Creighton, E.K.; Alhaddad, H.; Beale, H.; Grahn, R.; Rah, H.; Maggs, D.J.; Helps, C.; Gandolfi, B. Whole genome sequencing in cats, identifies new models for blindness in AIPL1 and somite segmentation in HES7. *BMC Genom.* **2016**, *17*, 265. [CrossRef] [PubMed]

169. Ahonen, S.; Arumilli, M.; Seppälä, E.; Hakosalo, O.; Kaukonen, M.K.; Komaromy, A.M.; Lohi, H. Increased Expression of MERTK is Associated with a Unique Form of Canine Retinopathy. *PLoS ONE* **2014**, *9*, e114552. [CrossRef] [PubMed]

170. Fanning, T.; Singer, M. LINE-1: A mammalian transposable element. *Biochim. Biophys. Acta (BBA) Gene Struct. Expr.* **1987**, *910*, 203–212. [CrossRef]

171. Everson, R.; Pettitt, L.; Forman, O.P.; Dower-Tylee, O.; McLaughlin, B.; Ahonen, S.; Kaukonen, M.; Komaromy, A.M.; Lohi, H.; Mellersh, C.S.; et al. An intronic LINE-1 insertion in MERTK is strongly associated with retinopathy in Swedish Vallhund dogs. *PLoS ONE* **2017**, *12*, e0183021. [CrossRef] [PubMed]

172. Zangerl, B.; Goldstein, O.; Philp, A.R.; Lindauer, S.J.; Pearce-Kelling, S.E.; Mullins, R.F.; Graphodatsky, A.; Ripoll, D.; Felix, J.S.; Stone, E.M.; et al. Identical mutation in a novel retinal gene causes progressive rod-cone degeneration in dogs and retinitis pigmentosa in humans. *Genomics* **2006**, *88*, 551–563. [CrossRef]

173. Nevet, M.J.; Shalev, S.A.; Zlotogora, J.; Mazzawi, N.; Ben-Yosef, T. Identification of a prevalent founder mutation in an Israeli Muslim Arab village confirms the role of PRCD in the aetiology of retinitis pigmentosa in humans. *J. Med. Genet.* **2010**, *47*, 533–537. [CrossRef]

174. Allon, G.; Mann, I.; Remez, L.; Sehn, E.; Rizel, L.; Nevet, M.J.; Perlman, I.; Wolfrum, U.; Ben-Yosef, T. PRCD is Concentrated at the Base of Photoreceptor Outer Segments and is Involved in Outer Segment Disc Formation. *Hum. Mol. Genet.* **2019**, *28*, 4078–4088. [CrossRef] [PubMed]

175. Spencer, W.J.; Pearring, J.N.; Salinas, R.Y.; Loiselle, D.R.; Skiba, N.P.; Arshavsky, V.Y. Progressive Rod–Cone Degeneration (PRCD) Protein Requires N-Terminal S-Acylation and Rhodopsin Binding for Photoreceptor Outer Segment Localization and Maintaining Intracellular Stability. *Biochemistry* **2016**, *55*, 5028–5037. [CrossRef]

176. Spencer, W.J.; Ding, J.-D.; Lewis, T.R.; Yu, C.; Phan, S.; Pearring, J.N.; Kim, K.-Y.; Thor, A.; Mathew, R.; Kalnitsky, J.; et al. PRCD is essential for high-fidelity photoreceptor disc formation. *Proc. Natl. Acad. Sci. USA* **2019**, *116*, 13087–13096. [CrossRef]

177. Aguirre, G.D.; Alligood, J.; O'Brien, P.; Buyukmihci, N. Pathogenesis of progressive rod-cone degneration in miniature poodles. *Investig. Ophthalmol. Vis. Sci.* **1982**, *23*, 610–630.

178. Friedman, J.S.; Chang, B.; Kannabiran, C.; Chakarova, C.; Singh, H.; Jalali, S.; Hawes, N.L.; Branham, K.; Othman, M.; Filippova, E.; et al. Premature Truncation of a Novel Protein, RD3, Exhibiting Subnuclear Localization Is Associated with Retinal Degeneration. *Am. J. Hum. Genet.* **2006**, *79*, 1059–1070. [CrossRef] [PubMed]

179. Wolf, E.D.; Vainisi, S.J.; Santos-Anderson, R. Rod-cone dysplasia in the collie. *J. Am. Veter Med Assoc.* **1978**, *173*, 1331–1333.

180. Woodford, B.; Liu, Y.; Fletcher, R.; Chader, G.; Farber, D.; Santos-Anderson, R.; Tso, M.O. Cyclic nucleotide metabolism in inherited retinopathy in collies: A biochemical and histochemical study. *Exp. Eye Res.* **1982**, *34*, 703–714. [CrossRef]

181. Dizhoor, A.M.; Olshevskaya, E.V.; Peshenko, I.V. Retinal guanylyl cyclase activation by calcium sensor proteins mediates photoreceptor degeneration in an rd3 mouse model of congenital human blindness. *J. Boil. Chem.* **2019**, *294*, 13729–13739. [CrossRef]

182. Kukekova, A.V.; Goldstein, O.; Johnson, J.; Richardson, M.A.; Pearce-Kelling, S.E.; Swaroop, A.; Friedman, J.S.; Aguirre, G.D.; Acland, G.M. Canine RD3 mutation establishes rod-cone dysplasia type 2 (rcd2) as ortholog of human and murine rd3. *Mamm. Genome* **2009**, *20*, 109–123. [CrossRef]

183. Santos-Anderson, R.M.; Tso, M.O.; Wolf, E.D. An inherited retinopathy in collies. A light and electron microscopic study. *Investig. Ophthalmol. Vis. Sci.* **1980**, *19*, 1282–1294.

Review

The Interplay between Peripherin 2 Complex Formation and Degenerative Retinal Diseases

Lars Tebbe, Mashal Kakakhel, Mustafa S. Makia, Muayyad R. Al-Ubaidi * and Muna I. Naash *

Department of Biomedical Engineering, University of Houston, Houston, TX 77204, USA;
ltebbe@Central.UH.EDU (L.T.); mkakakhe@central.uh.edu (M.K.); msmakia@Central.UH.EDU (M.S.M.)
* Correspondence: malubaid@central.uh.edu (M.R.A.-U.); mnaash@central.uh.edu (M.I.N.);
  Tel.: +1-713-743-1651 (M.I.N.)

Received: 4 February 2020; Accepted: 20 March 2020; Published: 24 March 2020

Abstract: Peripherin 2 (Prph2) is a photoreceptor-specific tetraspanin protein present in the outer segment (OS) rims of rod and cone photoreceptors. It shares many common features with other tetraspanins, including a large intradiscal loop which contains several cysteines. This loop enables Prph2 to associate with itself to form homo-oligomers or with its homologue, rod outer segment membrane protein 1 (Rom1) to form hetero-tetramers and hetero-octamers. Mutations in *PRPH2* cause a multitude of retinal diseases including autosomal dominant retinitis pigmentosa (RP) or cone dominant macular dystrophies. The importance of Prph2 for photoreceptor development, maintenance and function is underscored by the fact that its absence results in a failure to initialize OS formation in rods and formation of severely disorganized OS membranous structures in cones. Although the exact role of Rom1 has not been well studied, it has been concluded that it is not necessary for disc morphogenesis but is required for fine tuning OS disc size and structure. Pathogenic mutations in *PRPH2* often result in complex and multifactorial phenotypes, involving not just photoreceptors, as has historically been reasoned, but also secondary effects on the retinal pigment epithelium (RPE) and retinal/choroidal vasculature. The ability of Prph2 to form complexes was identified as a key requirement for the development and maintenance of OS structure and function. Studies using mouse models of pathogenic *Prph2* mutations established a connection between changes in complex formation and disease phenotypes. Although progress has been made in the development of therapeutic approaches for retinal diseases in general, the highly complex interplay of functions mediated by Prph2 and the precise regulation of these complexes made it difficult, thus far, to develop a suitable Prph2-specific therapy. Here we describe the latest results obtained in Prph2-associated research and how mouse models provided new insights into the pathogenesis of its related diseases. Furthermore, we give an overview on the current status of the development of therapeutic solutions.

Keywords: peripherin 2; retinal degeneration; retina; tetraspanin; photoreceptor

---

1. Introduction

Tetraspanins represent a family of highly conserved membrane proteins involved in a variety of functions. These functions include membrane organization and compartmentalization, cell signaling, adhesion and migration [1–4]. Structurally, tetraspanins consist of a short cytoplasmic N-terminus, four transmembrane domains (TM), one small and one large extracellular loop (EC1 and EC2, respectively), as well as a short cytoplasmic C-terminus [3,5,6]. The EC2 loop can be further divided in a conserved and a variable region [7,8]. The highly conserved part of the EC2 loop was shown to mediate the dimerization of tetraspanin proteins, while the variable region is required for specific interactions of the tetraspanins with their variable interaction partners [7–10]. Tetraspanin interactions can be divided into three levels of interactions [1]. The primary level consists of direct associations between two tetraspanin proteins as well as between tetraspanin and a non-tetraspanin interacting partners, interactions that resist the treatment with strong detergents [1,4]. Binding of these primary complexes to each other represents the secondary level which is indirect and in some cases supported by palmitoylation [1,4,11]. In the third level, tetraspanins can indirectly associate with their interacting partners to form large insoluble complexes, which only resist mild detergents like CHAPS [1,4,12,13]. The ability of tetraspanins to interact with each other and with other partners allows them to form tetraspanin-enriched microdomains known as the tetraspanin web [4,14].

Peripherin 2 (Prph2, formally known as retinal degeneration slow, RDS) represents a photoreceptor-specific tetraspanin. It is a structural protein that is critical for the proper development of rod and cone outer segments (ROS and COS, respectively) and thus for vision [15,16]. The main function of Prph2, in the morphogenesis of ROS and COS, is to promote membrane curvature, flattening and fusion; processes that are required for the rim formation of the OS discs and lamella [15,17–20]. Like the other members of the tetraspanin family, Prph2 contains two loops, referred to as intradiscal loops (D1 and D2, respectively). The D2 loop is equivalent to the EC2 loop of other tetraspanins and mediates the bulk of interactions of Prph2 [21–23]. Prph2 interacts with another photoreceptor-specific tetraspanin, called rod outer segment membrane protein 1 (Rom1) [24]. Prph2 forms non-covalently associated homo- and hetero-tetramers with Rom1. These tetramers associate and form covalently linked intermediate complexes. In addition to that, Prph2 is able to form higher order complexes consisting of several Prph2 homo-tetramers. Interestingly, hetero-tetramers including Rom-1 are excluded from these higher order complexes [21,25,26]. Formation of these complexes is required for Prph2 to promote the development of rim domains essential for OS formation [18,27,28]. In line with its key role in OS formation, mutations in *PRPH2* are connected with a multitude of retinal diseases ranging from autosomal dominant retinitis pigmentosa (ADRP) to macular degeneration (MD) [29]. The associated phenotypes show a high degree of variability of onset and phenotype even between patients carrying the same mutation. In addition to the degeneration of the retina, some Prph2 related diseases were found to cause secondary effects on the retinal pigment epithelium (RPE) or the choroid [29–31].

This review describes the importance of Prph2 in the development of both ROS and COS. Recent data provided new insight into the steps necessary for OS morphogenesis and the role of Prph2 in orchestrating these events. Furthermore, we will summarize how the precise formation and regulation of Prph2/Rom1 complex affects the function of Prph2 in rods versus cones. Finally, we will discuss the animal models expressing pathogenic Prph2 mutations and provide a short summary about therapeutic approaches so far aimed to treat Prph2 related diseases.

2. The Role of Prph2 in Photoreceptor Outer Segment Morphogenesis

Since mice lacking Prph2 (*Prph2*$^{-/-}$) fail to form ROS, it became clear that the major function of Prph2 is in the initialization and elaboration of ROS [32]. Recent data confirmed this function and identified the C-terminus of Prph2 as the motif responsible for the initialization of ROS formation [33,34]. The photoreceptor outer segment (OS) represents a highly specialized form of a modified primary cilium [35]. The initial steps in the development of the OS are similar to the development of other primary cilia, starting with the attachment of the ciliary vesicle to the basal body [36–38]. After this initial step, the elongation of the ciliary axoneme and the expansion of the plasma membrane happen in a synchronous fashion. A recent study showed that these initial steps in ciliogenesis take place in both wild type (WT) and *Prph2*$^{-/-}$ photoreceptors, an observation which was in line with previous data showing the presence of an intact connecting cilium (CC) in the *Prph2*$^{-/-}$ photoreceptors [32,33,39]. Additionally, the same study revealed that the Prph2 C-terminus mediates suppression of the formation of ectosomes at the distal part of the CC in rod photoreceptors [33]. Ectosomes are thought to play an important role in the processes of ciliary resorption, regulation of the ciliary protein content and regulation of signal transduction [40–44]. The ability of Prph2 C-terminus to suppress the release of ectosomes offers an explanation to why photoreceptor cilia are able to form the highly complexed intra-ciliary membranous structures like that of the ROS [33]. Further evidence for a key role of the Prph2 C-terminus in the initialization of ROS formation was provided by a recent study that generated a chimeric protein composed of the C-terminus of Prph2 fused to the body of Rom1, the non-glycosylated homologue of Prph2 [34]. Expression of the chimeric protein was sufficient to initialize the formation of ROS structures but failed to elaborate them proving that the Prph2 C-terminus is responsible for the formation of the initial membrane outgrowth while the other parts of Prph2 are required for the development of the highly organized structure of the OS.

Prph2$^{+/-}$ mice show defects in both function and structure of photoreceptor OS, which proves that the role of Prph2 is not restricted to the initialization of OS formation but also important for its structural development and maintenance [16]. In mature ROS discs, Prph2 is restricted to the rim [15]. In addition, previous studies demonstrated that Prph2 displayed membrane fusion activity, which led to the theory that Prph2 is involved in the shaping of the closed rim of fully developed discs in ROS [45,46]. Indeed, a function of Prph2 in inducing the curvature of the plasma membrane was observed in several studies [25,33,47,48]. While all these studies agree that Prph2 is involved in the generation of membrane curvature, they are contradictory in defining the exact region of Prph2 responsible for this activity. Two studies pinpointed the membrane curvature inducing activity to an α-helix motif located in the C-terminus of Prph2 [47,48], while a third study argues that this motif is rather preventing membrane curvature than inducing it [25]. A recent study found evidence for a function of the tetraspanin core of Prph2 in generating membrane curvature while the C-terminus was unable to perform this function [33]. The localization of Prph2 in newly formed ROS discs provided further evidence for the importance of Prph2 in membrane curvature and disc closure [17]. Newly formed discs in the ROS have a close rim at the site adjacent to the axoneme of the photoreceptor cilium while the opposing side of the disc remains open [49]. In these newly formed discs, Prph2 is restricted to the axonemal closed side of the disc [17]. A recent study confirmed the localization of Prph2 on the axonemal side of newly formed discs in the ROS [50]. An additional finding in this study was that Prph2 is more abundant at the closed rim of newly formed discs than at the rim of mature discs, indicating that Prph2 is not only involved in the closure of this rim but also required for the initialization of the formation of new discs.

The membrane curvature of the open end of newly formed discs in ROS is opposite to the curvature on the closed rim, and thus unlikely to allow the binding of Prph2 [26]. This led to the question of whether a different transmembrane protein may promote the curvature of the open ends. Prominin-1 was identified as a potential candidate involved in the shaping of newly synthetized discs in ROS. Prominin-1 is a membrane protein with five transmembrane domains that localizes at the open end of newly synthesized discs in ROS, thus opposite to the localization of Prph2 [51–55]. The vital role of Prominin-1 in the structural integrity of the photoreceptor OS is supported by the finding that mutations in Prominin-1 are connected to retinal diseases causing degeneration of the photoreceptors [54,56–59]. Earlier studies on Prominin-1 demonstrated that it shows a binding preference for curved membranes, but does not induce the curvature itself [60,61]. Thus, Prominin-1 is a promising candidate for the maintenance of the curvature on the open end, while the protein responsible for inducing the curvature at these ends remains elusive. A second question, which remains open thus far, is what mechanism restricts Prominin-1 to the newly synthesized discs? When ROS discs mature and reach their final diameter, the open ends close; presumably through the activity of Prph2 and become separated from the plasma membrane of ROS. While Prph2 is located on the entire rim of these closed discs, Prominin-1 is no longer found after the closure of the now matured disc [55].

While rod OS discs are closed and separated from the plasma membrane, cone OSs on the other hand contain open discs also referred to as lamellae and are contiguous with the plasma membrane. Although amphibian COS consists exclusively of open discs, the number of open discs in mammalian COS varies depending on the species [38,62–65]. The localization of Prph2 in the open discs of COS is comparable with its localization in the newly synthesized ROS discs, with Prph2 being restricted to the side adjacent to the axoneme of the photoreceptor cilium [55]. The discs in COS display a closed rim formation on the axonemal sides. Prominin-1 is localized on the open side, in line with its localization in the newly formed ROS. Studying the role of Prph2 in the development of cones in mice has proven to be difficult due to the low percentage of cone photoreceptors in the murine retina, whereby only ~3–5% of photoreceptors are cones. We have analyzed the effect of loss of Prph2 in the background of an Nrl knockout (*Nrl*$^{-/-}$), in which all rods are converted to S-cone-like cells [66–68]. Here the cones in the double knockout mice (*Prph2*$^{-/-}$/*Nrl*$^{-/-}$) displayed disorganized COSs lacking the flattened lamellae characteristic of WT COS [69]. Furthermore, this study provided evidence for the trafficking of many OS proteins to this altered OS structure, indicating that the transport of those proteins towards the OS does not rely on Prph2 [68]. The ability of *Prph2*$^{-/-}$ retinas to form disorganized COSs argues for a different role of Prph2 in cones than in rods. As mentioned above, ROSs fail to develop in the *Prph2*$^{-/-}$ mice, thus demonstrating a vital role for Prph2 in the initialization of the formation of this complex intra-ciliary membrane structure in ROS. The discovery of a role for the C-terminal region of Prph2 in inhibiting the release of ectosomes from the rod photoreceptor cilia [33] added new interesting aspect to the process of the rod OS formation. Future studies are needed to determine if this function of Prph2 is preserved in cones.

The identification of Rom1 as a Prph2 interactor initiated a multitude of studies investigating the complex formation between Prph2 and Rom1 [24,26,70]. Like Prph2, Rom1 is a photoreceptor-specific membrane protein and a member of the tetraspanin superfamily [71]. While only sharing a sequence identity of 35% with Prph2, both proteins form a highly conserved and similar secondary and tertiary structures with four transmembrane domains and two intradiscal loops [24]. Prph2 and Rom1 form heteromeric and homomeric tetrameric complexes, held together by non-covalent bonding mediated by the second intradiscal (D2) loop [21–23]. These tetramers can assemble into intermediate complexes consisting of both hetero- and homotetramers to establish a mix of covalent and non-covalent linked complexes [72]. These intermediate complexes were found to include at least two tetramers. In addition to the intermediate complexes, higher order complexes containing only Prph2 were also identified [72]. The core of both intermediate and higher order complexes consists of tetramers [21,25]. Interestingly Rom1 is excluded from these higher order complexes [21,25]. How this exclusion is mediated is not fully understood. A possible explanation is that Prph2 can utilize both the conventional secretory pathway through the Golgi apparatus as well as the unconventional pathway which bypasses the Golgi apparatus [26,73]. The presence of Rom1 results in more Prph2 being transported through the conventional pathway [34]. It is likely that this level of sorting during the initial steps of protein trafficking provides a possible mechanism mediating the exclusion of Rom1 from higher order complexes. Studies verifying this mechanism are a possible direction for future work. The higher order complexes are localized at the closed rim of discs indicating an important role for them in the membrane folding necessary to close the disc rim [25,74]. Biochemical evidence supported the idea that during the formation of Prph2/Rom1 complexes, intermolecular disulfide bonding between Prph2 and Rom1 are formed [72,75,76].

The ability of Prph2 and Rom1 to form complexes was proven to be vital for their function in OS formation [18,27,74]. The cysteine at position 150 (C150) in the D2 loop of Prph2 was found to be critical for the formation of higher order Prph2/Rom1 complexes [21,28,77]. C150 of Prph2 forms an intermolecular disulfide bond with C150 of another Prph2 to make homodimers or with C153 of Rom1, thus allowing the formation of Prph2/Rom1 heterodimers and heteromeric tetramers. A knock-in mouse expressing a point mutation at C150 of Prph2 (C150S) provided further evidence for the importance of this cysteine in the formation of higher order complexes and thus for the role of Prph2 and Rom1 in the development of the OS [78]. The formation of intermediate as well as higher order complexes was fully abolished by the C150S mutation and both ROS and COS were severely disorganized. Despite being disorganized, the OS structure of the C150S mice were somewhat better when compared to *Prph2*$^{+/-}$ while no functional improvement was observed, indicating that the ability of Prph2 to form intermediate complexes with Rom-1 as well as its ability to form higher order complexes consisting of Prph2 homo-tetramers is not only relevant for maintaining the OS structure but also for its function [78].

The development of higher order complexes seems to be expendable for the initiation of OS formation, since generation of both ROS and COS is initiated in the C150S mutant retinas that are devoid of these complexes. Comparable results were obtained for a mutation deleting lysine at position 153 in the D2 loop of Prph2 (K153Δ) [79]. K153Δ represents a mutation found in patients resulting in variable phenotypes ranging from ADRP to more cone specific MD [80]. This mutation also obliterates the ability of Prph2 to form higher order complexes [79]. The initiation of ROS and COS structures was unaffected while both displayed structural and functional decline. While the C150S and K153Δ mutations prevent the formation of higher order complexes, some pathogenic mutations of *Prph2* lead to the formation of an abnormal high molecular weight aggregates that include both Prph2 and Rom1 [81,82]. A patient mutation causing the exchange of the tyrosine at position 141 for a cysteine (Y141C), effectively adding another cysteine in the D2 loop, causes mostly defects in the macula but some cases of RP were also reported [30,82,83]. The Prph2 and Rom1 complexes in Y141C knockin retinas were significantly altered, showing the formation of abnormal high molecular weight aggregates containing both Prph2 and Rom1. This is a remarkable finding considering that Rom1 is normally excluded from higher order complexes [82]. These mice display structural and functional defects in ROS and COS. The formation of abnormal high molecular weight aggregates could also be found in another patient mutation in which tryptophan in Prph2 is substituted by arginine at position 172 (R172W) [81]. Transgenic mice carrying this mutation displayed the formation of high molecular weight aggregates of Rom1 while Prph2 complexes were not affected. Additionally, these high molecular weight Rom1 aggregates were more abundant in cones, a result that is in line with the decline of cone function in this mouse model [81].

Taken together, these results demonstrate that the exact regulation of the Prph2/Rom1 complex formation is critical for proper OS development, maintenance and disc size in both rods and cones as well as for their function. For OS initiation, proper complex formation seems to be irrelevant. It is worth noting that these studies point to a striking difference between rods and cones with regard to how they are affected by the different *Prph2* mutations. Mutations, which primarily affect the rods, seem to be resulting from haploinsufficiency or loss-of-function effects [78,84,85], while those effecting cones seem to be more susceptible to changes in the complex formation, likely due to gain-of-function defects [78,79,81,82,86–88]. Another differentiation between rods and cones is how they process Rom1. While ablation of Prph2 results in the complete loss of ROS and severely disorganized COS, elimination of Rom1 ($Rom1^{-/-}$) is less severe, indicating that Rom1's role is more in fine tuning of disc sizing [27]. It is worth noting that most studies considering these functions were performed in WT murine rods. A recent study performed on a $Nrl^{-/-}$ background using the K153Δ mutation of Prph2 demonstrated a loss of Prph2/Rom1 interaction specifically in cones [79]. In order to unravel the precise differences in the roles of Rom1 in rods versus cones, further studies are needed.

The D2 loop of Prph2 contains a motif for N-linked glycosylation, a motif which is absent in the D2 loop of Rom1 [24]. In rods, the N-glycosylation on Prph2 was found to be expendable for ROS [89]. In transgenic mice expressing the unglycosylated form of Prph2 specifically in rods, the interaction with Rom1 or the formation of complexes was not affected. Furthermore, expressing unglycosylated Prph2 on a $Prph2^{-/-}$ background achieved a full rescue of the phenotype, further proving that the N-glycosylation of Prph2 is unnecessary for its function in rods [89]. A knockin mouse model expressing unglycosylated Prph2 [75] recapitulated the rod results observed by Kedzierski et al. [89]. However, while rods were unaffected, cone function was found to decrease significantly in these mice. Furthermore, a significant decrease in Prph2 and Rom1 levels in the cones of the knockin mice was also observed. These results further highlight the differential functional roles of Prph2 in rods versus cones.

Apart from its interaction with Rom1, Prph2 was found to interact with additional OS specific proteins. One of those interactors is with cyclic nucleotide gated channel B1a (CNGB1a) [90,91]. CNGB1a was shown to be relevant for the correct sizing and alignment of the rod discs, supporting the hypothesis that the interaction between Prph2 and CNGB1a plays a role in the proper shaping of the discs [92,93]. Additionally, Prph2 was also found to directly interact with rhodopsin (Rho), thus forming a complex of Prph2, CNGB1a and Rho to anchor the rim of the discs to the OS plasma membrane and the rims of adjacent discs. It has been hypothesized that these interactions are responsible for correct alignment and stacking of the discs [94]. This was supported by findings that co-depletion of these three proteins exacerbated both functional and structural defects in rod photoreceptors [95].

3. Insight from Mouse Models into the Pathophysiology of *Prph2* Mutations

3.1. Transgenic Mouse Models of Prph2 Mutations

Genetically engineered mouse models expressing pathogenic mutations of Prph2 have proven to be valuable tools in studying the pathophysiology of Prph2 related blinding diseases. These models have been generated as transgenic expressing *Prph2* under the control of heterologous promoters that in some cases are specific to rods, cones or both.

3.1.1. Prph2^{R172W}

One of the most common disease-causing mutations of *PRPH2* is the substitution of tryptophan to arginine at position 172 (R172W) [96–99]. Patients carrying this mutation display a decrease in cone function, while rods' function remains unaffected [96]. The R172W mutation is located in the D2 loop, which is the region that promotes complex formation and where the vast majority of pathogenic *PRPH2* mutations are located [22,87]. Interestingly, the R172W mutation is located outside the part of the D2 loop involved in the formation of complexes [22]. Expressing the R172W mutation in the presence of the full complement of WT Prph2 led initially to normal rod function and structure [87]. In contrast, the retina of these mice exhibited a significant loss in the number of blue and green cones and associated with a decrease in their photopic responses. When expressed on *Prph2*$^{+/-}$ background, the R172W mutation led to a late-onset reduction in the number of rod cells and in scotopic responses, besides the cone phenotype. This varies from the cone exclusive phenotype observed in patients carrying this mutation and is most likely due to haploinsufficiency of Prph2. Expression of the R172W mutation in the *Nrl*$^{-/-}$ retina reproduced the functional decline in the photopic responses and provided evidence as to how the R172W mutation disrupted COS structures by causing the formation of abnormal high molecular weight Prph2/Rom1 aggregates [81]. This result demonstrates how the functional decline of the cones occurs in patients carrying this mutation.

3.1.2. Prph2^{C214S}

The substitution of a cysteine at position 214 in Prph2 with a serine (C214S) was shown to cause ADRP, a degeneration of rods followed by a late-onset degeneration of cones [100]. Expression of the C214S transgene in the presence of the full complement of WT Prph2 mice failed to produce any phenotype, proving that the phenotype observed in patients is through a mechanism other than dominant gain-of-function [85]. However, when expressed in *Prph2*$^{+/-}$ or *Prph2*$^{-/-}$ retinas, the C214S protein failed to rescue the functional and structural phenotypes in both rods and cones, indicating that the C214S mutation is resulting in a loss of function. An interesting finding was that the C214S mutant Prph2 was unable to interact with Rom1, indicating an alteration in the formation of complexes caused by this mutation. Although an ample amount of Prph2 C214S transgene message was detected, only a trace amount of the mutant protein was identified, which led to the conclusion of a loss-of-function phenotype associated with the C214S mutation. It is unclear whether the low levels of mutant protein are due to a low rate of synthesis or instability of the mutant protein.

3.1.3. Prph2^{P216L}

A substitution of a proline at position 216 in Prph2 with a leucine (P216L) was found to cause RP in patients [101]. Expression of P216L in mice in presence of WT Prph2 (*Prph2*$^{+/+}$) levels led to an age-dependent significant decrease in scotopic responses and shortened rod OSs with normal disc alignments, indicative of a dominant-gain-of-function effect [84]. Moreover, in the presence of one

WT allele of *Prph2* (*Prph2*$^{+/-}$), expression of the P216L caused a significant decrease in rod function compared to *Prph2*$^{+/-}$ mice. In these mice, a more severe OS shortening and disorganization of the discs could be observed in presence of the P216L mutant protein. Interestingly, expression of the P216L protein in a *Prph2*$^{-/-}$ retina caused the formation of small, highly ROS, which represents a limited rescue when compared to the complete absence of ROS in *Prph2*$^{-/-}$. Taken together, these results prove that the P216L mutation in Prph2 results in a dominant-gain-of effect, in agreement with the rod dominant phenotype observed in patients.

3.1.4. Prph2^{L185P}

The substitution of leucine at position 185 to proline (L185P) in Prph2 causes a rare digenic form of RP. Patients heterozygous for both the L185P mutation and a null mutation in Rom1 exhibit symptoms of RP, while carriers of any of the two mutations do not show a phenotype [102,103]. Transgenic mice expressing the L185P mutation in a *Prph2*$^{+/-}$ or a digenic *Prph2*$^{+/-}$/*Rom1*$^{+/-}$ background displayed a late onset thinning of the outer nuclear layer (ONL) concomitant with reduced scotopic electroretinograms (ERG) [104].

The transgenic mouse models described above provided valuable insights into the pathophysiology of Prph2 related diseases. Since the promoters used to express the transgene are heterologous, often levels of expression of the transgene differ considerably from that of the endogenous leading to a varied phenotype.

3.2. Prph2 Knockin Mouse Models

The variation in expression levels of the transgene observed in transgenic mouse models is particularly problematic in cases of reduced levels of the expressed protein since reductions in the expression level of WT Prph2 were shown to result in haploinsufficiency leading to severe retinal defects [16,105]. Haploinsufficiency makes it hard to distinguish between the effects seen in transgenic mice that are due to the mutation or those resulting from the reduced levels of expressed protein. In order to overcome this, recent studies relied on Prph2 knockin mouse models for a set of mutations found in patients of Prph2 related diseases. Below are the models currently presented in the literature and comparison of their retinal phenotypes to patient's phenotype carrying the same mutation.

3.2.1. Prph2$^{307/+}$ and Prph2$^{307/307}$

A deletion of a single base pair at codon 307 in human *PRPH2* results in a slow progressing form of ADRP [106]. A mouse model in which a targeted single base deletion at codon 307 in *Prph2* was introduced and showed a severe decline in photoreceptor survival and function [107]. ERG revealed a decrease in photoreceptor function in heterozygous animals starting at two months of age, while no ERG responses were detected in the homozygous animals even at one month of age. Retinal phenotypes in the heterozygous or homozygous mice were more severe than in age-matched *Prph2*$^{+/-}$ and *Prph2*$^{-/-}$ mice, respectively. Thus, the deletion mutation resulted in a strong dominant gain-of-function effect. This knockin model represents a drastic case of a retinal phenotype that differs greatly from that observed in the patients whereby the structural and functional decline is slow, starting in the fifth or sixth decades of life. The mouse model, on the other hand, displays a rapid degeneration with an early onset. The deletion at codon 307 in the human *PRPH2* gene causes a frameshift and creates a stop codon after the addition of 16 amino acids. If the resulting mutant protein is stable, it is expected to be 26 amino acid shorter than the wildtype. It is likely the case since the phenotype seen in patients is mild due to gain-of-function defect, but if the mutant protein is unstable, then the phenotype would be loss-of-function defect similar to that of haploinsufficiency. However, in the mouse genome, such one base deletion at codon 307 is predicted to result in the alteration of the last 40 amino acids of the C terminus of the protein and the addition of 11 extra amino acids. This results in the translation of 51 amino acids after codon 307 in the mouse [107]. The different effects on the translated protein caused by the deletion in codon 307 observed between human and mouse explains the differences in the severity of the retinal phenotypes. It is important to note that retinal phenotype in the homozygous mouse is reported to be worse than the complete-loss-of-function phenotype seen in the Prph2 null mice. This observation suggests that the mouse phenotype is likely a combination of loss- and gain-of-function defects and that the toxicity beyond that seen in the Prph2 knockout is probably the outcome of the latter. Since the mouse model was not assessed for the presence of the mutant protein, at this point, it is unclear whether the toxicity arises from the absence of the protein, the 51 altered amino acids at the C-terminus or the lack of endogenous 40 amino acids at the C-terminus. Obviously, the addition of 16 altered amino acids in the human PRPH2 and shortening the protein by 26 amino acids at the C-terminus have less drastic effect on the retina than those modifications occurred in the mouse protein.

3.2.2. Prph2$^{C213Y/+}$ and Prph2$^{C213Y/C213Y}$

The C213Y mutation is located in the D2 loop of Prph2, in the motif that mediates intramolecular and intermolecular disulfide bonds, thus responsible for the formation of Prph2/Prph2 and Prph2/Rom1 tetramers as well as for the formation of intermediate and higher order complexes [77]. The heterozygous mice (*Prph2$^{C213Y/+}$*) displayed a shortened and disorganized ROS. When compared to *Prph2$^{+/-}$* mice, the ROS of *Prph2$^{C213Y/+}$* mice showed a slight improvement in the stacking and alignment of the discs (Figure 1A,B) [76].

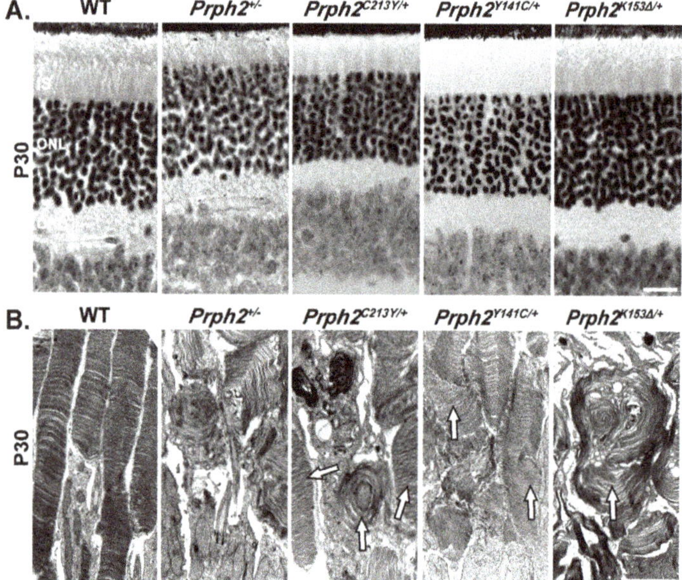

Figure 1. Mutations in the mouse *Prph2* gene lead to varying degrees of photoreceptor degeneration. (**A**) Representative light microscopic images from hematoxylin and eosin stained retinal sections at P30 aligned at the upper edge of the retinal pigment epithelium (RPE). (**B**) Transmission electron microscopic (TEM) images of the interface between the IS and OS of photoreceptors of the indicated genotypes. OS, outer segments; IS, inner segment; ONL, outer nuclear layer; OPL, outer plexiform layer. Scale bars: 20 μm for A and 2 μm for B. Eyes used in this study were dissected, fixed and embedded as previously described [79]. Images were captured at 40× and converted to black and white using ZEN Image Analysis software. The plastic-embedding and TEM methods were as described previously [85]. Images were adjusted and cropped using Adobe Photoshop CS5.

This is evident from the well stacked OS discs seen in some photoreceptors while others looked like whorls similar to those seen in the $Prph2^{+/-}$ (arrows in Figure 1). Scotopic ERG responses of the $Prph2^{C213Y/+}$ mice were significantly reduced starting at P30 (Figure 2) and persist all the way to P365. The scotopic response was nearly completely abolished in $Prph2^{C213Y/C213Y}$ animals and the OS was almost non-existent. The photopic responses were significantly decreased in both heterozygous and homozygous animals at P30. At later time points, the photopic response decreased further in the $Prph2^{C213Y/+}$ mice and was completely absent in the $Prph2^{C213Y/C213Y}$ animals. A key finding in this study was that the $Prph2^{C213Y/+}$ mice, that represent the genotype present in patients, showed better retinal structure despite the fact that both of rod and cone ERG responses were reduced when compared to $Prph2^{+/-}$ animals.

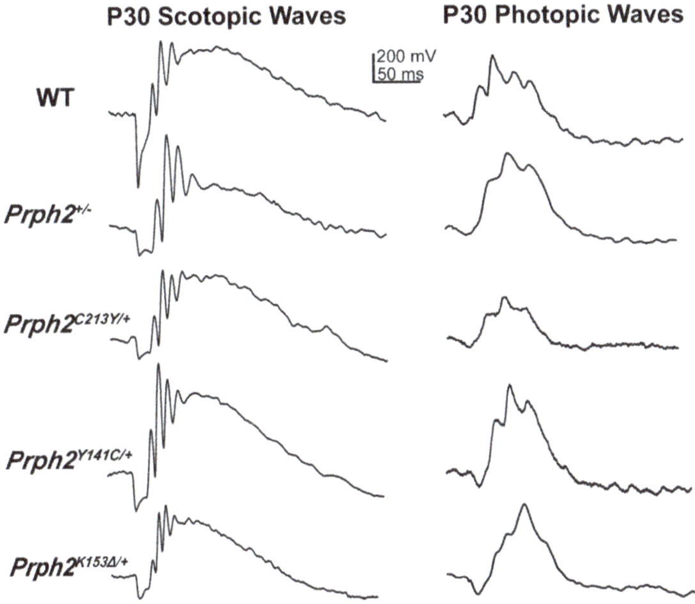

Figure 2. Mutations in *Prph2* hinder OS function assessed via scotopic and photopic electroretinograms (ERGs) at P30. Full-field ERGs were recorded under scotopic and photopic conditions. Shown are representative ERG waveforms from the indicated genotypes at P30. Full-field ERG tests were performed as previously described [85]. After overnight dark adaptation, mice were anesthetized and their pupils dilated. ERGs were recorded with a UTAS system (LKC, Gaithersburg, MD, USA). Waveforms were exported into GraphPad software to obtain wave traces and then exported into Photoshop using a uniform scale.

A key requirement for the function of Prph2 is its ability to interact with Rom1 and to form oligomeric complexes [18,27,74]. Co-immunoprecipitation (co-IP) experiments using retinal lysate from $Prph2^{C213Y/C213Y}$ mice revealed the inability of $Prph2^{C213Y}$ to interact with Rom1. The homomeric interaction of Prph2 was also reduced in these mice as evident from reduced ability of $Prph2^{C213Y}$ to form intermediate and higher order complexes as determined by sucrose gradient velocity sedimentation [76]. The reduction in oligomeric complexes offers an explanation for the functional decline observed in both $Prph2^{C213Y/+}$ and $Prph2^{C213Y/C213Y}$ mice and no abnormal high molecular weight aggregates were observed under non-reducing conditions. $Prph2^{C213Y/+}$ retina retains some Prph2 in the IS, shown by immunofluorescence (IF) staining of Prph2 (green and arrows in Figure 3) and IS maker syntaxin 3B (STX3B) (red, Figure 3). However, $Prph2^{C213Y/C213Y}$ lacks the ability to form complexes which leads to complete retention of Prph2 in the inner segments (IS), and perinuclear region while a smaller amount of Rom1 was retained in the IS [76].

Patients carrying the C213Y mutation in *PRPH2* display a butterfly-shaped pattern/macular dystrophy, while a rod dominant RP phenotype is absent [29,108–110]. The knockin mouse model for the C213Y displays functional defects in both rods and cones. While the defects in the cones are in line with the defects observed in the cone rich macula of the patients, the defects in the rods displayed by the model are not. The difference between phenotypes in patients and the mouse model may be due to the fact that the mouse retina consists almost exclusively of rod photoreceptors. Since the murine retina lacks a macula, it was not possible to reproduce the butterfly shaped macular dystrophy in the mouse model. One interesting finding in the Prph2^{C213Y} mouse model was the observation of a yellow flecking in the fundus of both *Prph2*$^{C213Y/+}$ and *Prph2*$^{C213Y/C213Y}$ mice at P180. This phenotype mimics funduscopic anomalies found in patients carrying the C213Y mutation [110].

Gene supplementation was performed by crossing a WT Prph2 overexpressing mouse line (NMP) onto hetero- (*Prph2*$^{C213Y/+}$) or homozygous (*Prph2*$^{C213Y/C213Y}$) mice. The defects observed in the protein trafficking as well as in OS structure were rescued in *Prph2*^{C213Y} mice. A rescue on the functional level however could not be observed [76]. These results show that the presence of the mutant protein has a detrimental effect on photoreceptor function.

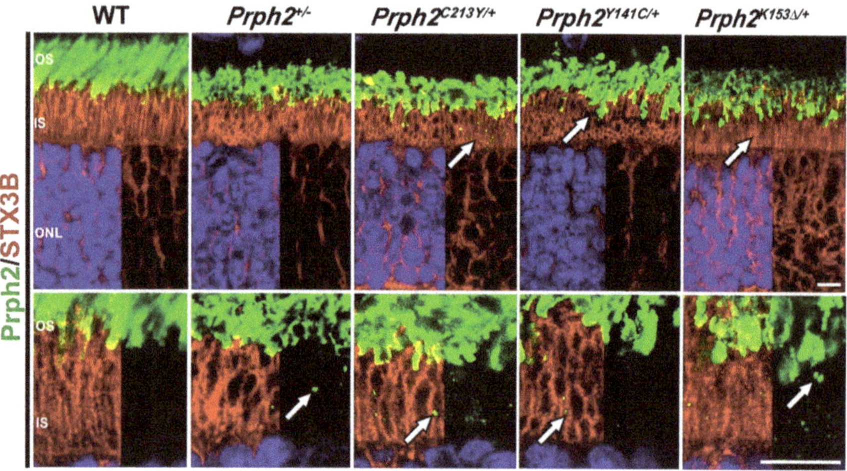

Figure 3. Mutated Prph2 protein traffics to the OS while a small pool is retained in the inner segment. Retinal sections at P30 from the indicated genotypes were probed with antibodies against Prph2 (green) and syntaxin 3B (STX3B) (red). Arrows indicate regions of mislocalization of Prph2. OS, outer segments; IS, inner segment; ONL, outer nuclear layer. Scale bar: 20 µm. Primary antibodies used for immunostaining were polyclonal antibody against Prph2 C-terminus (Prph2-CT) [22] and monoclonal antibody against STX3B [111] (inner segment marker) diluted at (1:1000). AlexaFluor conjugated secondary antibodies (Alexa 488 Rabbit and Alexa 555 Mouse, Life Technologies/ThermoFisher) were used at a dilution of 1:1000 for 2 hours at room temperature. Images were captured on a ZEISS Confocal LSM 900 microscope equipped with a Zeiss Axiocam (Zeiss, Jena, Germany) using a 63× (oil, 1.4 NA) objective. Images were then processed using ZEN Image Analysis software (Zeiss, Jena, Germany). All images shown are orthogonally projected from an eight slice confocal z-stack.

3.2.3. Prph2$^{Y141C/+}$ and Prph2$^{Y141C/Y141C}$

Like the C213Y mutation, the Y141C mutation of Prph2 is also located in the D2 loop. The OSs in the Y141C knockin mouse model (*Prph2*$^{Y141C/+}$ and *Prph2*$^{Y141C/Y141C}$ for heterozygous and homozygous animals, respectively) displayed structural anomalies. In *Prph2*$^{Y141C/+}$ mice, the OS was shortened with some structural alterations of the discs, including lengthening and vesicular structures [82]. When compared to the $Prph2^{+/-}$ mice, the structure was better conserved in the $Prph2^{Y141C/+}$ retina (Figure 1B). In the $Prph2^{Y141C/Y141C}$ mice, OS formation is initialized but neither mature discs nor lamellae could be observed. Instead, the OS contained flattened whorl shaped membranous structures with vesicular arrangements lining up adjacent to them. Scotopic ERG revealed a significant decrease in rod function displayed by the $Prph2^{Y141C/+}$ mice at P30 (Figure 2). Interestingly, the scotopic ERG did not deteriorate further at P180 [82].

Non-reducing SDS-PAGE/immunoblots of $Prph2^{Y141C/+}$ retinal lysates showed abnormal high molecular weight aggregates at 250kDA. This was further evaluated by sucrose gradient velocity sedimentation experiments which revealed the formation of the expected intermediate and higher order complexes with an aberrant high molecular weight band indicating that the Y141C mutation results in the formation of abnormal large aggregates [82]. In the $Prph2^{Y141C/Y141C}$ retinas, the observed phenotype was even more pronounced with an increase in the aberrant high molecular weight aggregates at the expense of the formation of the normal intermediate and higher order complexes. While in WT mice Rom1 is normally excluded from higher order complexes, in the mutant retina, a portion of Rom1 was incorporated in the aberrant high molecular weight aggregates. Surprisingly though, mutant Prph2 was transported correctly to the OS which is shown by IF staining of Prph2 (green) and IS marker STX3B (red) (Figure 3). However, the formation of these high molecular weight aggregates likely interfered with the normal function of Prph2 and likely Rom1 in the OS. This offers an explanation for the structural and functional phenotypes observed in the knockin mice.

Patients carrying the Y141C mutation in *PRPH2* display primarily defects in the macula with some reported cases of more rod-associated phenotypes such as night blindness and RP [30,82,83]. The $Prph2^{Y141C/+}$ mice, which represent patients' genotype, exhibit a decline in rods' function, mimicking the observed rod phenotype in some patients. Again, the difference between the knockin model and the patient's phenotypes can be seen in the cone function. The most prominent phenotype found in patients with the Y141C mutation is the functional and structural decline in cones. The $Prph2^{Y141C/+}$ mice on the other hand displayed only a slight functional decline in the photopic ERG which was not found to be significant [82], indicating that the cone function in these mice is not severely affected. While the heterozygous mice do not show significant decline in cone function, they do display a flecking in funduscopic analyses [82], which mimics findings in the fundus of patients. Recently, a study was undertaken to identify a potential explanation for the huge variation in phenotypes observed in patients by determining the role Rom1 plays in the observed phenotype [88]. Mice heterozygous or homozygous for the Y141C mutation were crossed with Rom1 knockout mice to produce mice that express mutant Prph2 in absence of Rom1 ($Prph2^{Y141C/+}/Rom1^{-/-}$ and $Prph2^{Y141C/Y141C}/Rom1^{-/-}$). Absence of Rom1 abolished the formation of the abnormal high molecular weight aggregates and accumulation of mutant Prph2 in the IS and ONL in $Prph2^{Y141C/Y141C}/Rom1^{-/-}$ retinas [88]. The depletion of Rom1 also changed the symptoms seen in the Y141C model. While the photopic ERG amplitudes in $Prph2^{Y141C/+}/Rom1^{-/-}$ mice were comparable to that of the WT, a significant reduction in the scotopic ERG responses were observed [88].

When compared to the cone-rod functional defects noted for the $Prph2^{Y141C/+}$ mice, $Prph2^{Y141C/+}/Rom1^{-/-}$ mice mainly displayed a rod-dominant functional defect. In addition, the funduscopic anomalies were almost completely abolished in the $Prph2^{Y141C/+}/Rom1^{-/-}$ mice. The results obtained in this study prove that alteration in the level of Rom1 can change the phenotype caused by a pathogenic *Prph2* mutation, and thus potentially provide an explanation to the variable phenotypes seen in patients carrying the same *PRPH2* mutation. Further studies combining *Prph2* mutations and mutations in *Rom1* should provide further insights into the complexity of phenotypes seen in the various *PRPH2* related diseases.

3.2.4. Prph2$^{K153\Delta/+}$ and Prph2$^{K153\Delta/K153\Delta}$

Another mutation in *PRPH2* that leads to variable phenotypes among patients is the deletion of codon 153 (K153Δ) that results in the elimination of the lysine at position 153 in the D2 loop of Prph2. This mutation is found to associate with RP, pattern dystrophy and fundus flavimaculatus [80]. K153Δ-Prph2 knockin mouse model was generated and provided evidence that the mutant protein cannot form the complexes required for OS formation [79]. The heterozygous knockin mice (*Prph2$^{K153\Delta/+}$*) displayed a shortened OS with minor structural defects at P30 [79] (Figure 1). At P180, these animals also exhibited a reduction in ONL thickness without further structural deterioration of the OS [79]. The overall Prph2 protein level in these mice was around 80% compared to the level in WT mice, thus demonstrating that the observed structural defects are most likely caused by the dominant effect of the mutant protein rather than due to haploinsufficiency.

The heterozygous animals displayed a significant progressive reduction in scotopic responses that started as early as P30 (Figure 2) and worsened with age. The photopic response in the *Prph2$^{K153\Delta/+}$* was also significantly reduced at P30, albeit this reduction did not worsen as the animals aged (Figure 2). In the homozygous animals, the scotopic and photopic responses were minimal at P30. Biochemical analysis showed that the formation of covalently linked Prph2 dimers in *Prph2$^{K153\Delta/K153\Delta}$* mice was abolished while Rom1 homodimers were present [79].

Due to the observed effects on cone function, *Prph2$^{K153\Delta/+}$* animals were crossed into the *Nrl$^{-/-}$* background in order to assess the effects of the K153Δ mutation on cones. Photoreceptor cells in the resulting *Prph2$^{K153\Delta/+}$/Nrl$^{-/-}$* retina displayed a highly disrupted structure with many photoreceptors having no lamellae what so ever. Interestingly, the mutant Prph2 was able to interact with Rom1 in the rod-dominant *Prph2$^{K153\Delta/K153\Delta}$* retinas, but this interaction was abolished in *Prph2$^{K153\Delta/K153\Delta}$/Nrl$^{-/-}$* retina, indicating a defect in the Prph2/Rom1 interaction specific to cones [79]. Sucrose gradient velocity sedimentation showed no significant alteration in complex formation in *Prph2$^{K153\Delta/+}$* retina when compared to WT while in the *Prph2$^{K153\Delta/K153\Delta}$* retina, the formation of intermediate and higher order complexes is abolished [79]. Here, both Prph2 and Rom1 are restricted to the tetramer fractions. The same was observed in the *Prph2$^{K153\Delta/K153\Delta}$/Nrl$^{-/-}$* retinas. Sedimentation profile showed that the amount of higher order complexes in the *Prph2$^{K153\Delta/+}$/Nrl$^{-/-}$* retina was reduced, while unaffected in *Prph2$^{K153\Delta/+}$* retina [79]. This provides further evidence for a differential role of the lysine at position 153 in rods and cones, and hence emphasizes the notion for potential varied roles for Prph2 in rods versus cones. Localization studies performed in *Prph2$^{K153\Delta/+}$* mice demonstrated that most of Prph2 and Rom1 were successfully transported to the OS with some amount of Prph2 mislocalized to the IS (arrows in Figure 3). Small amount of rhodopsin (Rho) and M-opsin were also found to be mislocalized to the ONL and outer plexiform layer (OPL) in this model [79].

In the *Prph2$^{K153\Delta/K153\Delta}$* retinas, the majority of Rho and Prph2, but not Rom1, were found to be mislocalized to the ONL. As stated above, patients with the K153Δ mutation exhibit a highly variable phenotype, ranging from rod dominant RP to more cone related defects in the macula [80]. The K153Δ knockin mouse model displayed functional and structural defects in both rods and cones, and thus mimics the phenotype seen in patients carrying this mutation. While the lack of a macula in the murine retina made it impossible to observe the macular pattern dystrophy often found in patients [80], the knockin mouse still showed funduscopic anomalies [79] which are characteristic of the pattern dystrophy in patients. *Prph2$^{K153\Delta/+}$* mice show a flecking in the fundus at P180 which is more severe in the *Prph2$^{K153\Delta/K153\Delta}$* mice. This phenotype does not deteriorate further in P365 heterozygous animals while the flecking was replaced by large splotches in the homozygous animals [79].

Gene supplementation using the NMP mouse that over-express Prph2 (*NMP/Prph2$^{K153\Delta/+}$* and *NMP/Prph2$^{K153\Delta/K153\Delta}$*) rescued the structural defects but failed to rescue the functional decline seen in scotopic and photopic ERGs [79]. This indicates that the presence of the mutant protein alone is sufficient to deteriorate the photoreceptor function and suggest that gene silencing along with gene augmentation is the best strategy for this model.

The knockin models for *PRPH2* related patient mutations proved to be very useful in highlighting dominant-effects of the *Prph2* mutations. In general, the models were more successful in mimicking patient phenotypes related to a decline in rod function. Patient phenotypes related to functional defects in the cones or pattern dystrophies in the macula were more difficult to reproduce in mice due to the lack of a macula and a lower overall percentage of cones in the murine retina. Crossing the knockin mouse models with $Nrl^{-/-}$ mice, as done in the studies with the K153Δ model [79], has proven to be a successful approach in studying the effects of the mutation on cones in detail. While the knockin models could not reproduce the macular pattern dystrophy, they successfully reproduced the funduscopic aberrations, which connect with the pattern dystrophy.

3.2.5. Prph2$^{N229S/+}$ and Prph2$^{N229S/N229}$

The knockin mouse that alters the N-linked glycosylation at asparagine 229 (N229S) in the D2 loop was used to study if Prph2 glycosylation plays a role in its interaction with Rom1. Heterozygous mice ($Prph2^{N229S/+}$) did not have any significant changes in structure or function of the OS [75]. However, homozygous mice ($Prph2^{N229S/N229S}$) displayed a late onset thinning of the outer nuclear layer (ONL) and occasional abnormal disc staking in cones and a slightly reduced photopic ERG at P180. Since Prph2 could not be glycosylated, higher order complexes were decreased and there was an increase of Prph2 and Rom-1 in the intermediate complexes [75]. Therefore, it was concluded that the glycosylation plays a major part in regulating the interaction between Prph2 and Rom1, which is critical for cone health.

Tables 1 and 2 summarize the phenotypes associated with the Prph2 knockin models, highlighting rod and cone structure, Scotopic and Photopic ERG, complex formation and protein localization. It also summarizes fundus observations, and patient's phenotypes whether mainly rod- or cone-specific defects or the combination of the two. Figures 1 and 2 show structural and functional differences between Prph2 heterozygous knockin mutations, while Figure 3 is a representative IF showing retinal localization of Prph2 among the models.

Table 1. *Prph2* mutations and correlating phenotypes.

Genotype	*Prph2*$^{+/-}$	*Prph2*$^{-/-}$	*Prph2*$^{K153Δ/+}$	*Prph2*$^{K153Δ/K153Δ}$	*Prph2*$^{Y141C/+}$	*Prph2*$^{Y141C/Y141C}$
Mouse Rod Structure	Short OSs with whorl structures and ONL thinning at P180.	No OS structures.	Short OSs with whorl structures and ONL thinning at P180.	Almost no Oss with rare whorl structures and ONL thinning at P30, more severe at P180.	Short OS with longer discs and accumulation of vesicular structures, no ONL thinning at P30, ONL thinning at P180.	Small OS with flattened whorls and vesicular structures at P30, no OS present at P180, ONL thinning at P30, more severe at P180.
Mouse Cone Structure	Oss with whorl structures	Open OS with no lamella	Occasional whorl shaped OS, mostly open OS with no lamella	Occasional COS seen but mostly absent.	Abnormal and short COS structures.	Occasional COS seen but mostly absent.
Scotopic ERG	57% and 33% reduction in a- and b-wave at P30. 74% and 48% reduction in a- and b-waves at P180, respectively.	96% and 93% reduction in a- and b-wave at P30, respectively.	63% and 49% reduction in a- and b-wave at P30. 83% and at 60% reduction in a- and b-wave at P180, respectively.	90% and 89% reduction in a- and b-wave at P30, respectively.	54% and 27% reduction in a- and b-wave at P30 50% and 25% reduction in a- and b-wave at P180, respectively.	90% and 78% reduction in a- and b-wave at P30. 95% and 94% reduction in a- and b-wave at P180, respectively.

Table 1. *Cont.*

Genotype	$Prph2^{+/-}$	$Prph2^{-/-}$	$Prph2^{K153\Delta/+}$	$Prph2^{K153\Delta/K153\Delta}$	$Prph2^{Y141C/+}$	$Prph2^{Y141C/Y141C}$
Photopic ERG	Photopic b-wave comparable to WT at P30. 35% reduction in photopic b-wave at P180.	91% reduction in b-wave at P30.	24% reduction in b-wave P30 and 50% at P180.	64% reduction in b-wave at P30.	10% reduction in b-wave at P30 and P180.	64% reduction in b-wave at P30 and 90% at P180.
Complex formation	Prph2 complexes and distribution unchanged, 50% less Prph2 and Rom1. On NR$^{L-/-}$ background, higher order complexes decreased.	No Prph2 present, Rom1 still present but at lesser amount.	Prph2 complexes and distribution unchanged, while Rom1 shifted towards tetramers. On NRL$^{-/-}$ background, higher order complexes decreased, shift of Prph2 and Rom1 towards intermediate complexes.	No Prph2 dimers were formed, while Rom1 dimers were still formed. Prph2 interacted with Rom1. Prph2 and Rom1 restricted to tetramers. On NRL$^{-/-}$ background, no Prph2 dimers were formed, while Rom1 dimers were still formed. Prph2 did not interact with Rom1. Prph2 and Rom1 restricted to tetramers.	Prph2 occasionally found in abnormal high molecular weight aggregates. Abnormal aggregates were held together by intermolecular disulfide bonds. Rom1 also present in abnormal aggregates. Intermediate and higher order complexes formed but abnormal high molecular weight aggregates also present.	Prph2 almost exclusively found in abnormal high molecular weight aggregates. Abnormal aggregates were held together by intermolecular disulfide bonds. Rom1 also present in abnormal aggregates. Intermediate and higher order complexes reduced in favor of the abnormal high molecular weight aggregates.
Protein localization	Some rhodopsin detected in the IS and ONL.	Rhodopsin mislocalized to IS and ONL.	Small amount of rhodopsin and M-opsin mislocalized in the IS and ONL.	Huge amount of rhodopsin and Prph2 mislocalized in the IS and ONL.	NA	NA
Fundus	No abnormality	Flecking and splotches at P360 and older.	Flecking at P180 and no change at P365.	Severe flecking at P180 and big splotches at P365.	Flecking at P180.	Flecking at P180.
Rod defect in patients	NA	NA	RP	NA	Night blindness and RP reported in some patients.	NA
Cone defect in patients	NA	NA	Pattern dystrophy and fundus flavimaculs.	NA	Pattern dystrophy changed fundus in macula.	NA
Reference	[26,76,79]	[26,76,79]	[79]	[79]	[82]	[82]

Table 2. *Prph2* mutations and correlating phenotypes (*continued*).

Genotype	Prph2$^{C213Y/+}$	Prph2$^{C213Y/C213Y}$	Prph2$^{N229S/+}$	Prph2$^{N229S/N229S}$	Prph2$^{C150S/+}$	Prph2$^{C150S/C150S}$	Prph2$^{307/+}$	Prph2$^{307/307}$
Rod structure mouse	Shortened OS, irregular structure, some disc structure better organized than in *Prph2$^{+/-}$*. ONL thinning at P30.	Short OS formed with highly disorganized discs. Severe ONL thinning at P30.	Structure unaffected.	Modest ONL thinning at P180.	Modest elongation of discs, occasional formation of whorls. No ONL thinning.	Shortened OS, elongated discs curving into whorls. No ONL thinning.	ONL thinning starting at P60. Rod OS shortened, formation of whorls.	ONL thinning starting at P30 and rod OSs and absent.
Cone structure mouse	Well organized lamella at P30 but slightly shorter.	Abnormally stacked lamella with whorl shaped structures.	Normal COS structures.	Occasional abnormal disc stacking at P180	NA	Shortened OS, elongated discs curving into whorls.	NA	Cone OS are absent.
Scotopic ERG	60% and 47% reduction in a- and b-wave P30. 89% and 67% reduction in a- and b-wave at P180, respectively.	96% and 94% reduction in a- and b-wave P30, respectively.	ERG a- and b-wave comparable to WT at P30 and P180.	ERG a- and b-wave comparable to WT at P30 and P180.	50% and 30% reduction in a- and b-wave at P30, respectively.	75% and 56% reduction in a- and b-wave P30, respectively.	60% and 62% reduction in a- and b-wave at P180. 80% and 75% reduction in a- and b-wave at P300, respectively.	Scotopic ERG absent
Photopic ERG	33% and 46% reduction in b-wave at P30 and P180, respectively.	79% reduction in b-wave at P30 ERG UV b-wave.	ERG b-wave comparable to WT at P30 and P180	Normal at P30 but 22% reduction in b-wave at P180.	29% reduction in b-wave at P30.	64% reduction in b-wave at P30.	60% and 80% reduction in b-wave at P180 and P300, respectively.	Photopic ERG absent
Complex formation rods	Mutant Prph2 unable to interact with Rom1 in rods. Intermediate and higher order complex formation slightly impaired. Rom1 occasionally found in higher order complexes.	Mutant Prph2 unable to interact with Rom1 in rods. Protein levels of Rom1 decreased No intermediate or higher order complexes were formed Prph2 and Rom1 restricted to tetramers	Prph2 could not be glycosylated and formed normal Prph2/Rom1 complexes.	Prph2 could not be glycosylated. Amount of higher order complexes decreased, while the amount of intermediate complexes was increased.	Reduction in Prph2 protein levels, less pronounced than in *Prph2$^{+/-}$*. Rom1 protein levels not affected.	Strong decrease in Prph2 and Rom1 protein levels. No intermediate or higher order complexes were formed. Prph2 and Rom1 restricted to tetramers.		
Protein localization	Prph2 partially mislocalized to IS and ONL. Small amount of Rom1 mislocalized to IS.	Prph2 almost completely mislocalized to IS and ONL. Small amount of Rom1 mislocalized to IS. M- and S-Opsin as well as rhodopsin mislocalized to the IS and ONL.	None	None	None	None	NA	NA
Fundus	Flecking at P180, persisted till P365.	Flecking at P180, more pronounced than in C213Y/+, replaced by splotches at P365.	None	None	NA	NA	NA	NA
Rod defect patient	NA	NA	No patient model	NA	No patient model	NA	NA	NA
Cone defect patient	Pattern dystrophy	NA	No patient model	NA	No patient model	NA	NA	NA
Reference	[76]	[76]	[75]	[75]	[78]	[78]	[107]	[107]

4. Gene Therapy of *PRPH2* Mutations

There is a vast variety of *PRPH2* mutations associated with autosomal dominant retinal degenerative diseases, including RP, several forms of macular dystrophy and cone rod dystrophies [29]. Gene therapy seems to be a promising approach in the treatment of those *PRPH2* associated diseases. The small size of the *Prph2* cDNA, which is approximately 1.1 kb, is advantageous since many vectors used for gene therapy can only carry DNA of a limited size. Despite the variety of feasible approaches available for the gene therapy of *Prph2*, thus far no treatment ready for clinical trials has been developed.

The expression of the WT Prph2 in the background of pathogenic *Prph2* mutations represents one of the most promising approaches. Pioneer studies using transgenic mice expressing WT Prph2 under different promoters in a $Prph2^{+/-}$ or $Prph2^{-/-}$ background revealed a significant increase in both structural and functional properties of rods and cones [105,112]. Replacing Prph2 in $Prph2^{-/-}$ mice utilizing adeno-associated virus (AAV) carrying *Prph2* regulated under the rhodopsin promoter provided promising results [113]. Here the subretinal injection of the AAV resulted in a partial rescue of ROS structure as well as scotopic ERG response. A follow up study was able to show that repeated injections with the AAV resulted in an even more pronounced rescue of the phenotype with an increase in the scotopic b-wave response observed in the injected mice [114]. In this study, several time points after the injections were analyzed in order to validate whether the observed rescue was long lasting. This analyses revealed that the rescue achieved by the AAV injection was lost after 15 weeks post injection [114]. In addition to the loss of the rescue with time, the amplitude of the scotopic b-wave was significantly lower than in WT mice and the scotopic a-wave was not improved by the injection of the AAV. A reason for this is that the injection of the AAV in mice only resulted in a low transduction rate of roughly 30% [114]. An improvement in the transduction rate will thus be necessary in order to develop a viable AAV based gene therapy.

Nanoparticles (NP) represent a second promising approach in transferring WT Prph2. These particles were found to be well tolerated by the retina, even after multiple injections, and have a high DNA capacity up to 14 kb, as tested in the eye [115–120]. NP carrying full-length murine *Prph2* cDNA under the control of either rod or cone specific IRBP promoters or ubiquitous chicken beta actin promoter was injected in the retina of $Prph2^{+/-}$ mice [121]. The injection resulted in a partial rescue of OS structure and also prevented the thinning of the ONL. These rescue effects lasted 15 months post injection and were most pronounced near the site of injection [121]. While the NP treatment could overcome the decline in rescue overtime, the effect on the photoreceptor function remained limited. NP injection resulted in a small yet insignificant improvement in the scotopic a-wave but a significant improvement in the photopic b-wave. The results obtained with the different promoters used were comparable. The small benefit of the NP injection observed in the functional tests might be due to an incomplete distribution of the NPs in the retina. In line with this, the structural improvements observed after NP injection were best close to the site of injection. Improving on the distribution and uptake of the used vector in the retina, regardless if AAVs or NPs are used, seems to be a necessary next step in order to achieve a more pronounced functional rescue.

The studies above described the replacement of Prph2 in either $Prph2^{+/-}$ or $Prph2^{-/-}$ mice, thus in a scenario whereby Prph2 is either absent or haploinsufficient. However, most patients suffer from a dominant mutation in *PRPH2*. In order to analyze the efficiency of treatment in a scenario closer to actual *PRPH2* related diseases, the knockin mouse models carrying a disease related *Prph2* mutation were analyzed. Both $Prph2^{K153\Delta/+}$ and $Prph2^{C213Y/+}$ mice were crossed with a normal-Prph2-overexpressing mouse line (NMP) [76,79]. The resulting $Prph2^{K153\Delta/+}/NMP^{+/-}$ and $Prph2^{C213Y/+}/NMP^{+/-}$ mice displayed a rescue in OS structure, protein expression levels and trafficking, but no rescue in both rod and cone functions [76]. In both cases, the presence of the mutant protein continues to affect the photoreceptor function. This is due to dominant-gain-of-function effects caused by these mutations. These results show that, in addition to gene augmentation therapy, silencing the mutant *Prph2* allele is essential for ultimate rescue. A combined approach of short hairpin (sh)-RNA mediated knockdown and expression of a shRNA resistant protein has been performed in models for ADRP carrying pathogenic mutations of rhodopsin [122,123]. Here, both WT and mutant Rhodopsin were knocked down by sh-RNA carried in AAV together with a sh-RNA resistant rhodopsin. The expression of the sh-RNA resistant rhodopsin following the knockdown rescued the phenotype. Studies combining the knockdown of mutant Prph2 with the expression of WT Prph2 provided first promising results thus far showing partial functional and structural rescue in mice expressing pathogenic mutations of Prph2 [124,125].

The examples above show that while progress was made in the gene therapy of *PRPH2* related diseases, there are many factors, which need to be considered for the development of a successful therapy. Further complications might arise from the fact that the functional role of Prph2 seems to vary in rods and cones. In addition to that, secondary effects of *PRPH2* mutations on the RPE and the choroid could be observed [29–31]. A high variability in the clinical phenotypes displayed by patients, even when carrying the same mutation, represents another challenge, which has to be overcome in order to treat *PRPH2* related diseases.

5. Conclusions and Perspectives

Prph2 plays a key role in the maintenance as well as the development of photoreceptor OS. While Prph2 is found to be vital for both rods and cones, there seem to be distinct differences in its function in these two types of photoreceptor cells. A complete knockout of Prph2 results in the complete absence of ROS while COS formation is still initialized even though the resulting COS are severely disorganized and lack lamellae and discs. This begs the question, which protein is mediating the formation of the COS and if it works in an interplay with Prph2. The OS of both rods and cones represent a highly modified primary cilium. The Prph2 mediated suppression of ectosomes shedding from the photoreceptor cilium was found to be a prerequisite for the formation of the ROS. Thus far there is no study demonstrating a similar mechanism in the formation of COS. The investigation of the formation of ectosomes and membrane dynamics in the development of COS in WT and *Prph2*$^{-/-}$ mice might be a potential approach for further studies aiming to unravel the different function of Prph2 in ROS versus COS. While the function of Prph2 in the initialization of ROS and COS varies, it seems to be indispensable for the correct shaping, sizing and stacking of discs and lamellae in both. Furthermore, not only the presence of Prph2 is needed for developing and maintaining these structures but also the precise regulation in the formation of the different Prph2/Rom1 and Prph2/Prph2 complexes. Changes in the ratio of the tetramers, intermediate complexes and higher order complexes were proven detrimental to the function as well as the structure of the photoreceptors. The transgenic and knockin mouse models discussed here proved the importance of the precise regulation of the complex formation. Each of the knockin mouse models carrying a pathogenic *Prph2* mutation displayed an altered ratio of the different Prph2/Rom1 and Prph2 complexes resulting in disorganized OS structure and decreased response found in ERG measurements. Interestingly rods and cones were differently affected by the alterations in complex formation, providing further evidence for the differential use of Prph2 by the two photoreceptor cell types. In addition to that, the mouse models used helped identify whether a mutation results in a loss-of-function or a gain-of-function effect.

While there is agreement in the numerous studies concerning the ratio between tetramers, intermediate and higher order complexes and its vital importance for OS structure and function, the precise roles of the different complexes are still not fully understood. In the C150S knockin mouse model, both Prph2 and Rom1 are found exclusively as tetramers, while intermediate and higher order complexes are absent. Still the formation of both ROS and COS is initiated, proving that the tetramers alone are sufficient to initiate OS formation. Homozygous animals fail to form disc and lamellae, indicating that the intermediate and higher order complexes are more likely to be involved in the formation of the rim, membrane closure and disc stacking. The knockout of Rom1 showed minor effects on disc alignment, sizing and stacking while the disc rim formation is unaffected. Rom1 is found in non-covalent hetero-tetramers and in intermediate complexes but it is excluded from higher order Prph2 complexes. It seems plausible that the Prph2 higher order complexes, which are unaffected in the Rom1 knockout retinas, mediate the membrane curvature and rim closure, while the intermediate hetero complexes support disc spacing, sizing and alignment. Additional studies targeting the formation of the different complexes specifically and analyzing their function in rods and cones are needed to further pinpoint the roles of the different Prph2 and Rom1 complexes.

Progress in the development of a feasible gene therapy of Prph2 related diseases has been made in recent years. Transducing the photoreceptors of $Prph2^{-/-}$ and $Prph2^{+/-}$ mice either via AAVs or NPs showed a partial rescue of the knockout phenotype, even though the magnitude of the rescue and (in case of the AAVs) the persistence of the rescue effect is still low. The small magnitude of the rescue is most likely due to a low transduction rate of photoreceptors by the vectors used. Improving the transduction rate achieved by the vector as well as its distribution in the eye is one obstacle which needs to be overcome in future attempts. A second obstacle is that pathogenic *PRPH2* mutations do not result in a knockout scenario, but instead in a scenario where the mutant protein is still expressed. The knockin mouse models summarized in this review have proven to be a valid approach in understanding how the presence of the mutant protein affects both rod and cone photoreceptors, and also how stable the mutant protein is. In addition, these models helped to identify mutations, which result in dominant effects. Using a sh-RNA-mediated knockdown of both WT and mutant proteins followed by the expression of a sh-RNA resistant form of the WT protein provided a promising approach in the therapy of dominant gain-of-function mutations of Rho and Prph2. The fact that rods and cones utilize Prph2 differently could lead to further complications; this issue needs to be taken into consideration when developing effective therapeutic strategies for *PRPH2* related diseases. Testing these strategies in the murine models adds another layer of complication due to the low percentage of cones, which makes it difficult to address cone-dominant phenotypes in mouse models. Analyzing *Prph2* mutations in the $Nrl^{-/-}$ background provides an approach for studying functional and structural effects of mutation specifically on cones. The lack of a macula in the murine retina continues to be the rate limiting factor which prevents conclusive studies aimed at developing therapies for mutations associated with macular defects and pattern dystrophy using mouse models. In addition to this, several *PRPH2* mutations were found to cause secondary effects in the choroid and RPE. The reasons for these secondary defects as well as their impact on the pathogenesis of *PRPH2* associated diseases are not well understood. Characterizing these secondary defects has to be achieved in future studies in order to devise a successful therapy.

Author Contributions: L.T. wrote original draft of the manuscript. M.S.M. performed electroretinography and histology. M.K. performed immunohistochemistry, developed the tables and participate in the writing. M.I.N., M.R.A.-U. designed the experiments, reviewed and edited the manuscript. All authors have read and agreed on the final version of the manuscript.

Funding: This research is supported by a grant from the National Institutes of Health (R01 EY10609-MIN, MRA).

Conflicts of Interest: The authors declare no conflicts of interest.

References

1. Hemler, M.E. Tetraspanin functions and associated microdomains. *Nat. Rev. Mol. Cell Biol.* **2005**, *6*, 801–811. [CrossRef]

2. Levy, S.; Shoham, T. The tetraspanin web modulates immune-signalling complexes. *Nat. Rev. Immunol.* **2005**, *5*, 136–148. [CrossRef] [PubMed]

3. Charrin, S.; Jouannet, S.; Boucheix, C.; Rubinstein, E. Tetraspanins at a glance. *J. Cell Sci.* **2014**, *127*, 3641–3648. [CrossRef]

4. Van Deventer, S.J.; Dunlock, V.E.; van Spriel, A.B. Molecular interactions shaping the tetraspanin web. *Biochem. Soc. Trans.* **2017**, *45*, 741–750. [CrossRef] [PubMed]

5. Boucheix, C.; Rubinstein, E. Tetraspanins. *Cell. Mol. Life Sci. Cmls* **2001**, *58*, 1189–1205. [CrossRef] [PubMed]

6. Hemler, M.E. Tetraspanin proteins mediate cellular penetration, invasion, and fusion events and define a novel type of membrane microdomain. *Annu. Rev. Cell Dev. Biol.* **2003**, *19*, 397–422. [CrossRef]

7. Kitadokoro, K.; Bordo, D.; Galli, G.; Petracca, R.; Falugi, F.; Abrignani, S.; Grandi, G.; Bolognesi, M. CD81 extracellular domain 3D structure: Insight into the tetraspanin superfamily structural motifs. *Embo. J.* **2001**, *20*, 12–18. [CrossRef]

8. Murru, L.; Moretto, E.; Martano, G.; Passafaro, M. Tetraspanins shape the synapse. *Mol. Cell. Neurosci.* **2018**, *91*, 76–81. [CrossRef]

9. Stipp, C.S.; Kolesnikova, T.V.; Hemler, M.E. Functional domains in tetraspanin proteins. *Trends Biochem. Sci.* **2003**, *28*, 106–112. [CrossRef]

10. DeSalle, R.; Mares, R.; Garcia-España, A. Evolution of cysteine patterns in the large extracellular loop of tetraspanins from animals, fungi, plants and single-celled eukaryotes. *Mol. Phylogenetics Evol.* **2010**, *56*, 486–491. [CrossRef]

11. Yang, X.; Kovalenko, O.V.; Tang, W.; Claas, C.; Stipp, C.S.; Hemler, M.E. Palmitoylation supports assembly and function of integrin-tetraspanin complexes. *J. Cell Biol.* **2004**, *167*, 1231–1240. [CrossRef]

12. Skubitz, K.M.; Campbell, K.D.; Skubitz, A.P. CD63 associates with CD11/CD18 in large detergent-resistant complexes after translocation to the cell surface in human neutrophils. *Febs Lett.* **2000**, *469*, 52–56. [CrossRef]

13. Claas, C.; Stipp, C.S.; Hemler, M.E. Evaluation of prototype transmembrane 4 superfamily protein complexes and their relation to lipid rafts. *J. Biol. Chem.* **2001**, *276*, 7974–7984. [CrossRef] [PubMed]

14. Hemler, M.E. Specific tetraspanin functions. *J. Cell Biol.* **2001**, *155*, 1103–1107. [CrossRef] [PubMed]

15. Molday, R.S.; Hicks, D.; Molday, L. Peripherin. A rim-specific membrane protein of rod outer segment discs. *Invest. Ophthalmol. Vis. Sci.* **1987**, *28*, 50–61. [PubMed]

16. Cheng, T.; Peachey, N.S.; Li, S.; Goto, Y.; Cao, Y.; Naash, M.I. The effect of peripherin/rds haploinsufficiency on rod and cone photoreceptors. *J. Neurosci. Off. J. Soc. Neurosci.* **1997**, *17*, 8118–8128. [CrossRef]

17. Arikawa, K.; Molday, L.L.; Molday, R.S.; Williams, D.S. Localization of peripherin/rds in the disk membranes of cone and rod photoreceptors: Relationship to disk membrane morphogenesis and retinal degeneration. *J. Cell Biol.* **1992**, *116*, 659–667. [CrossRef] [PubMed]

18. Wrigley, J.D.; Ahmed, T.; Nevett, C.L.; Findlay, J.B. Peripherin/rds influences membrane vesicle morphology. Implications for retinopathies. *J. Biol. Chem.* **2000**, *275*, 13191–13194. [CrossRef]

19. Boesze-Battaglia, K. Fusion between retinal rod outer segment membranes and model membranes: Functional assays and role for peripherin/rds. *Methods Enzymol.* **2000**, *316*, 65–86.

20. Conley, S.M.; Cai, X.; Naash, M.I. Nonviral ocular gene therapy: Assessment and future directions. *Curr. Opin. Mol. Ther.* **2008**, *10*, 456–463.

21. Goldberg, A.F.; Molday, R.S. Subunit composition of the peripherin/rds-rom-1 disk rim complex from rod photoreceptors: Hydrodynamic evidence for a tetrameric quaternary structure. *Biochemistry* **1996**, *35*, 6144–6149. [CrossRef] [PubMed]

22. Ding, X.Q.; Stricker, H.M.; Naash, M.I. Role of the second intradiscal loop of peripherin/rds in homo and hetero associations. *Biochemistry* **2005**, *44*, 4897–4904. [CrossRef] [PubMed]

23. Chakraborty, D.; Conley, S.M.; Stuck, M.W.; Naash, M.I. Differences in RDS trafficking, assembly and function in cones versus rods: Insights from studies of C150S-RDS. *Hum. Mol. Genet.* **2010**, *19*, 4799–4812. [CrossRef] [PubMed]

24. Bascom, R.A.; Manara, S.; Collins, L.; Molday, R.S.; Kalnins, V.I.; McInnes, R.R. Cloning of the cDNA for a novel photoreceptor membrane protein (rom-1) identifies a disk rim protein family implicated in human retinopathies. *Neuron* **1992**, *8*, 1171–1184. [CrossRef]

25. Kevany, B.M.; Tsybovsky, Y.; Campuzano, I.D.; Schnier, P.D.; Engel, A.; Palczewski, K. Structural and functional analysis of the native peripherin-ROM1 complex isolated from photoreceptor cells. *J. Biol. Chem.* **2013**, *288*, 36272–36284. [CrossRef]

26. Stuck, M.W.; Conley, S.M.; Naash, M.I. PRPH2/RDS and ROM-1: Historical context, current views and future considerations. *Prog Retin Eye Res.* **2016**, *52*, 47–63. [CrossRef]

27. Clarke, G.; Goldberg, A.F.; Vidgen, D.; Collins, L.; Ploder, L.; Schwarz, L.; Molday, L.L.; Rossant, J.; Szel, A.; Molday, R.S.; et al. Rom-1 is required for rod photoreceptor viability and the regulation of disk morphogenesis. *Nat. Genet.* **2000**, *25*, 67–73. [CrossRef]

28. Chakraborty, D.; Ding, X.Q.; Conley, S.M.; Fliesler, S.J.; Naash, M.I. Differential requirements for retinal degeneration slow intermolecular disulfide-linked oligomerization in rods versus cones. *Hum. Mol. Genet.* **2009**, *18*, 797–808. [CrossRef]

29. Boon, C.J.; den Hollander, A.I.; Hoyng, C.B.; Cremers, F.P.; Klevering, B.J.; Keunen, J.E. The spectrum of retinal dystrophies caused by mutations in the peripherin/RDS gene. *Prog Retin Eye Res.* **2008**, *27*, 213–235. [CrossRef]

30. Francis, P.J.; Schultz, D.W.; Gregory, A.M.; Schain, M.B.; Barra, R.; Majewski, J.; Ott, J.; Acott, T.; Weleber, R.G.; Klein, M.L. Genetic and phenotypic heterogeneity in pattern dystrophy. *Br. J. Ophthalmol.* **2005**, *89*, 1115–1119. [CrossRef]

31. Gocho, K.; Akeo, K.; Itoh, N.; Kameya, S.; Hayashi, T.; Katagiri, S.; Gekka, T.; Ohkuma, Y.; Tsuneoka, H.; Takahashi, H. High-Resolution Adaptive Optics Retinal Image Analysis at Early Stage Central Areolar Choroidal Dystrophy With PRPH2 Mutation. *Ophthalmic Surg. Lasers Imaging Retin.* **2016**, *47*, 1115–1126. [CrossRef] [PubMed]

32. Jansen, H.G.; Sanyal, S. Development and degeneration of retina in rds mutant mice: Electron microscopy. *J. Comp. Neurol.* **1984**, *224*, 71–84. [CrossRef] [PubMed]

33. Salinas, R.Y.; Pearring, J.N.; Ding, J.D.; Spencer, W.J.; Hao, Y.; Arshavsky, V.Y. Photoreceptor discs form through peripherin-dependent suppression of ciliary ectosome release. *J. Cell Biol.* **2017**, *216*, 1489–1499. [CrossRef] [PubMed]

34. Conley, S.M.; Stuck, M.W.; Watson, J.N.; Zulliger, R.; Burnett, J.L.; Naash, M.I. Prph2 initiates outer segment morphogenesis but maturation requires Prph2/Rom1 oligomerization. *Hum. Mol. Genet.* **2019**, *28*, 459–475. [CrossRef]

35. May-Simera, H.; Nagel-Wolfrum, K.; Wolfrum, U. Cilia - The sensory antennae in the eye. *Prog. Retin. Eye Res.* **2017**, *60*, 144–180. [CrossRef]

36. Sorokin, S. Centrioles and the formation of rudimentary cilia by fibroblasts and smooth muscle cells. *J. Cell Biol.* **1962**, *15*, 363–377. [CrossRef]

37. Sedmak, T.; Wolfrum, U. Intraflagellar transport proteins in ciliogenesis of photoreceptor cells. *Biol. Cell* **2011**, *103*, 449–466. [CrossRef]

38. Pearring, J.N.; Salinas, R.Y.; Baker, S.A.; Arshavsky, V.Y. Protein sorting, targeting and trafficking in photoreceptor cells. *Prog. Retin. Eye Res.* **2013**, *36*, 24–51. [CrossRef]

39. Chakraborty, D.; Conley, S.M.; Al-Ubaidi, M.R.; Naash, M.I. Initiation of rod outer segment disc formation requires RDS. *PLoS ONE* **2014**, *9*, e98939. [CrossRef]

40. Wang, J.; Silva, M.; Haas, L.A.; Morsci, N.S.; Nguyen, K.C.; Hall, D.H.; Barr, M.M. C. elegans ciliated sensory neurons release extracellular vesicles that function in animal communication. *Curr. Biol.* **2014**, *24*, 519–525. [CrossRef]

41. Cao, M.; Ning, J.; Hernandez-Lara, C.I.; Belzile, O.; Wang, Q.; Dutcher, S.K.; Liu, Y.; Snell, W.J. Uni-directional ciliary membrane protein trafficking by a cytoplasmic retrograde IFT motor and ciliary ectosome shedding. *eLife* **2015**, *4*, e05242. [CrossRef] [PubMed]

42. Wang, J.; Barr, M.M. Ciliary Extracellular Vesicles: Txt Msg Organelles. *Cell. Mol. Neurobiol.* **2016**, *36*, 449–457. [CrossRef] [PubMed]

43. Nager, A.R.; Goldstein, J.S.; Herranz-Perez, V.; Portran, D.; Ye, F.; Garcia-Verdugo, J.M.; Nachury, M.V. An Actin Network Dispatches Ciliary GPCRs into Extracellular Vesicles to Modulate Signaling. *Cell* **2017**, *168*, 252–263.e214. [CrossRef] [PubMed]

44. Phua, S.C.; Chiba, S.; Suzuki, M.; Su, E.; Roberson, E.C.; Pusapati, G.V.; Setou, M.; Rohatgi, R.; Reiter, J.F.; Ikegami, K.; et al. Dynamic Remodeling of Membrane Composition Drives Cell Cycle through Primary Cilia Excision. *Cell* **2017**, *168*, 264–279.e215. [CrossRef] [PubMed]

45. Boesze-Battaglia, K.; Lamba, O.P.; Napoli, A.A., Jr.; Sinha, S.; Guo, Y.; Boesze-Battagliaa, K.; Stefano, F.P.; Milstein, M.L.; Kimler, V.A.; Ghatak, C.; et al. Fusion between retinal rod outer segment membranes and model membranes: A role for photoreceptor peripherin/rds Peripherin/rds fusogenic function correlates with subunit assembly An inducible amphipathic helix within the intrinsically disordered C terminus can participate in membrane curvature generation by peripherin-2/rds. *Biochemistry* **1998**, *37*, 9477–9487. [PubMed]

46. Boesze-Battagliaa, K.; Stefano, F.P. Peripherin/rds fusogenic function correlates with subunit assembly. *Exp. Eye Res.* **2002**, *75*, 227–231. [CrossRef]

47. Khattree, N.; Ritter, L.M.; Goldberg, A.F. Membrane curvature generation by a C-terminal amphipathic helix in peripherin-2/rds, a tetraspanin required for photoreceptor sensory cilium morphogenesis. *J. Cell Sci.* **2013**, *126*, 4659–4670. [CrossRef]

48. Milstein, M.L.; Kimler, V.A.; Ghatak, C.; Ladokhin, A.S.; Goldberg, A.F.X. An inducible amphipathic helix within the intrinsically disordered C terminus can participate in membrane curvature generation by peripherin-2/rds. *J. Biol. Chem.* **2017**, *292*, 7850–7865. [CrossRef]

49. Steinberg, R.H.; Fisher, S.K.; Anderson, D.H. Disc morphogenesis in vertebrate photoreceptors. *J. Comp. Neurol.* **1980**, *190*, 501–508. [CrossRef]

50. Ding, J.D.; Salinas, R.Y.; Arshavsky, V.Y. Discs of mammalian rod photoreceptors form through the membrane evagination mechanism. *J. Cell Biol.* **2015**, *211*, 495–502. [CrossRef]

51. Miraglia, S.; Godfrey, W.; Yin, A.H.; Atkins, K.; Warnke, R.; Holden, J.T.; Bray, R.A.; Waller, E.K.; Buck, D.W. A novel five-transmembrane hematopoietic stem cell antigen: Isolation, characterization, and molecular cloning. *Blood* **1997**, *90*, 5013–5021. [CrossRef] [PubMed]

52. Weigmann, A.; Corbeil, D.; Hellwig, A.; Huttner, W.B. Prominin, a novel microvilli-specific polytopic membrane protein of the apical surface of epithelial cells, is targeted to plasmalemmal protrusions of non-epithelial cells. *Proc. Natl. Acad. Sci. USA* **1997**, *94*, 12425–12430. [CrossRef]

53. Yin, A.H.; Miraglia, S.; Zanjani, E.D.; Almeida-Porada, G.; Ogawa, M.; Leary, A.G.; Olweus, J.; Kearney, J.; Buck, D.W. AC133, a novel marker for human hematopoietic stem and progenitor cells. *Blood* **1997**, *90*, 5002–5012. [CrossRef]

54. Maw, M.A.; Corbeil, D.; Koch, J.; Hellwig, A.; Wilson-Wheeler, J.C.; Bridges, R.J.; Kumaramanickavel, G.; John, S.; Nancarrow, D.; Roper, K.; et al. A frameshift mutation in prominin (mouse)-like 1 causes human retinal degeneration. *Hum. Mol. Genet.* **2000**, *9*, 27–34. [CrossRef] [PubMed]

55. Han, Z.; Anderson, D.W.; Papermaster, D.S. Prominin-1 localizes to the open rims of outer segment lamellae in Xenopus laevis rod and cone photoreceptors. *Investig. Ophthalmol. Vis. Sci.* **2012**, *53*, 361–373. [CrossRef] [PubMed]

56. Yang, Z.; Chen, Y.; Lillo, C.; Chien, J.; Yu, Z.; Michaelides, M.; Klein, M.; Howes, K.A.; Li, Y.; Kaminoh, Y.; et al. Mutant prominin 1 found in patients with macular degeneration disrupts photoreceptor disk morphogenesis in mice. *J. Clin. Investig.* **2008**, *118*, 2908–2916. [CrossRef]

57. Zacchigna, S.; Oh, H.; Wilsch-Brauninger, M.; Missol-Kolka, E.; Jaszai, J.; Jansen, S.; Tanimoto, N.; Tonagel, F.; Seeliger, M.; Huttner, W.B.; et al. Loss of the cholesterol-binding protein prominin-1/CD133 causes disk dysmorphogenesis and photoreceptor degeneration. *J. Neurosci. Off. J. Soc. Neurosci.* **2009**, *29*, 2297–2308. [CrossRef]

58. Collison, F.T.; Fishman, G.A.; Nagasaki, T.; Zernant, J.; McAnany, J.J.; Park, J.C.; Allikmets, R. Characteristic Ocular Features in Cases of Autosomal Recessive PROM1 Cone-Rod Dystrophy. *Investig. Ophthalmol. Vis. Sci.* **2019**, *60*, 2347–2356. [CrossRef]

59. Liang, J.; She, X.; Chen, J.; Zhai, Y.; Liu, Y.; Zheng, K.; Gong, Y.; Zhu, H.; Luo, X.; Sun, X. Identification of novel PROM1 mutations responsible for autosomal recessive maculopathy with rod-cone dystrophy. *Graefe's Arch. Clin. Exp. Ophthalmol.* **2019**, *257*, 619–628. [CrossRef]

60. Huttner, W.B.; Zimmerberg, J. Implications of lipid microdomains for membrane curvature, budding and fission. *Curr. Opin. Cell Biol.* **2001**, *13*, 478–484. [CrossRef]

61. Iglic, A.; Hagerstrand, H.; Veranic, P.; Plemenitas, A.; Kralj-Iglic, V. Curvature-induced accumulation of anisotropic membrane components and raft formation in cylindrical membrane protrusions. *J. Theor. Biol.* **2006**, *240*, 368–373. [CrossRef] [PubMed]

62. Cohen, A.I. Further studies on the question of the patency of saccules in outer segments of vertebrate photoreceptors. *Vis. Res.* **1970**, *10*, 445–453. [CrossRef]

63. Anderson, D.H.; Fisher, S.K.; Steinberg, R.H. Mammalian cones: Disc shedding, phagocytosis, and renewal. *Investig. Ophthalmol. Vis. Sci.* **1978**, *17*, 117–133. [PubMed]

64. Bunt, A.H. Fine structure and radioautography of rabbit photoreceptor cells. *Investig. Ophthalmol. Vis. Sci.* **1978**, *17*, 90–104.

65. Carter-Dawson, L.D.; LaVail, M.M. Rods and cones in the mouse retina. I. Structural analysis using light and electron microscopy. *J. Comp. Neurol.* **1979**, *188*, 245–262. [CrossRef]

66. Mears, A.J.; Kondo, M.; Swain, P.K.; Takada, Y.; Bush, R.A.; Saunders, T.L.; Sieving, P.A.; Swaroop, A. Nrl is required for rod photoreceptor development. *Nat. Genet.* **2001**, *29*, 447–452. [CrossRef]

67. Daniele, L.L.; Lillo, C.; Lyubarsky, A.L.; Nikonov, S.S.; Philp, N.; Mears, A.J.; Swaroop, A.; Williams, D.S.; Pugh, E.N., Jr. Cone-like morphological, molecular, and electrophysiological features of the photoreceptors of the Nrl knockout mouse. *Investig. Ophthalmol. Vis. Sci.* **2005**, *46*, 2156–2167. [CrossRef]

68. Conley, S.M.; Al-Ubaidi, M.R.; Han, Z.; Naash, M.I. Rim formation is not a prerequisite for distribution of cone photoreceptor outer segment proteins. *FASEB J.* **2014**, *28*, 3468–3479. [CrossRef]

69. Farjo, R.; Skaggs, J.S.; Nagel, B.A.; Quiambao, A.B.; Nash, Z.A.; Fliesler, S.J.; Naash, M.I. Retention of function without normal disc morphogenesis occurs in cone but not rod photoreceptors. *J. Cell Biol.* **2006**, *173*, 59–68. [CrossRef]

70. Moritz, O.L.; Molday, R.S. Molecular cloning, membrane topology, and localization of bovine rom-1 in rod and cone photoreceptor cells. *Investig. Ophthalmol. Vis. Sci* **1996**, *37*, 352–362.

71. Conley, S.M.; Stuck, M.W.; Naash, M.I. Structural and functional relationships between photoreceptor tetraspanins and other superfamily members. *Cell. Mol. Life Sci.* **2012**, *69*, 1035–1047. [CrossRef] [PubMed]

72. Loewen, C.J.; Molday, R.S. Disulfide-mediated oligomerization of Peripherin/Rds and Rom-1 in photoreceptor disk membranes. Implications for photoreceptor outer segment morphogenesis and degeneration. *J. Biol. Chem.* **2000**, *275*, 5370–5378. [CrossRef] [PubMed]

73. Tian, G.; Ropelewski, P.; Nemet, I.; Lee, R.; Lodowski, K.H.; Imanishi, Y. An unconventional secretory pathway mediates the cilia targeting of peripherin/rds. *J. Neurosci. Off. J. Soc. Neurosci.* **2014**, *34*, 992–1006. [CrossRef] [PubMed]

74. Chakraborty, D.; Ding, X.Q.; Fliesler, S.J.; Naash, M.I. Outer segment oligomerization of Rds: Evidence from mouse models and subcellular fractionation. *Biochemistry* **2008**, *47*, 1144–1156. [CrossRef] [PubMed]

75. Stuck, M.W.; Conley, S.M.; Naash, M.I. Retinal Degeneration Slow (RDS) Glycosylation Plays a Role in Cone Function and in the Regulation of RDS.ROM-1 Protein Complex Formation. *J. Biol. Chem.* **2015**, *290*, 27901–27913. [CrossRef] [PubMed]

76. Chakraborty, D.; Strayve, D.G.; Makia, M.S.; Conley, S.M.; Kakahel, M.; Al-Ubaidi, M.R.; Naash, M.I. Novel molecular mechanisms for Prph2-associated pattern dystrophy. *FASEB J.* **2020**, *34*, 1211–1230. [CrossRef]

77. Goldberg, A.F.; Loewen, C.J.; Molday, R.S. Cysteine residues of photoreceptor peripherin/rds: Role in subunit assembly and autosomal dominant retinitis pigmentosa. *Biochemistry* **1998**, *37*, 680–685. [CrossRef]

78. Zulliger, R.; Conley, S.M.; Mwoyosvi, M.L.; Al-Ubaidi, M.R.; Naash, M.I. Oligomerization of Prph2 and Rom1 is essential for photoreceptor outer segment formation. *Hum. Mol. Genet.* **2018**, *27*, 3507–3518. [CrossRef]

79. Chakraborty, D.; Conley, S.M.; Zulliger, R.; Naash, M.I. The K153Del PRPH2 mutation differentially impacts photoreceptor structure and function. *Hum. Mol. Genet.* **2016**, *25*, 3500–3514. [CrossRef]

80. Weleber, R.G.; Carr, R.E.; Murphey, W.H.; Sheffield, V.C.; Stone, E.M. Phenotypic variation including retinitis pigmentosa, pattern dystrophy, and fundus flavimaculatus in a single family with a deletion of codon 153 or 154 of the peripherin/RDS gene. *Arch. Ophthalmol.* **1993**, *111*, 1531–1542. [CrossRef]

81. Conley, S.M.; Stuck, M.W.; Burnett, J.L.; Chakraborty, D.; Azadi, S.; Fliesler, S.J.; Naash, M.I. Insights into the mechanisms of macular degeneration associated with the R172W mutation in RDS. *Hum. Mol. Genet.* **2014**, *23*, 3102–3114. [CrossRef] [PubMed]

82. Stuck, M.W.; Conley, S.M.; Naash, M.I. The Y141C knockin mutation in RDS leads to complex phenotypes in the mouse. *Hum. Mol. Genet.* **2014**, *23*, 6260–6274. [CrossRef] [PubMed]

83. Khani, S.C.; Karoukis, A.J.; Young, J.E.; Ambasudhan, R.; Burch, T.; Stockton, R.; Lewis, R.A.; Sullivan, L.S.; Daiger, S.P.; Reichel, E.; et al. Late-onset autosomal dominant macular dystrophy with choroidal neovascularization and nonexudative maculopathy associated with mutation in the RDS gene. *Investig. Ophthalmol. Vis. Sci.* **2003**, *44*, 3570–3577. [CrossRef] [PubMed]

84. Kedzierski, W.; Lloyd, M.; Birch, D.G.; Bok, D.; Travis, G.H. Generation and analysis of transgenic mice expressing P216L-substituted rds/peripherin in rod photoreceptors. *Investig. Ophthalmol. Vis. Sci.* **1997**, *38*, 498–509.

85. Stricker, H.M.; Ding, X.Q.; Quiambao, A.; Fliesler, S.J.; Naash, M.I. The $Cys^{214}{\rightarrow}Ser$ mutation in peripherin/rds causes a loss-of-function phenotype in transgenic mice. *Biochem. J.* **2005**, *388*, 605–613. [CrossRef]

86. Ding, X.Q.; Nour, M.; Ritter, L.M.; Goldberg, A.F.; Fliesler, S.J.; Naash, M.I. The R172W mutation in peripherin/rds causes a cone-rod dystrophy in transgenic mice. *Hum. Mol. Genet.* **2004**, *13*, 2075–2087. [CrossRef]

87. Conley, S.; Nour, M.; Fliesler, S.J.; Naash, M.I. Late-onset cone photoreceptor degeneration induced by R172W mutation in Rds and partial rescue by gene supplementation. *Investig. Ophthalmol. Vis. Sci.* **2007**, *48*, 5397–5407. [CrossRef]

88. Conley, S.M.; Stuck, M.W.; Watson, J.N.; Naash, M.I. Rom1 converts Y141C-Prph2-associated pattern dystrophy to retinitis pigmentosa. *Hum. Mol. Genet.* **2017**, *26*, 509–518. [CrossRef]

89. Kedzierski, W.; Bok, D.; Travis, G.H. Transgenic analysis of rds/peripherin N-glycosylation: Effect on dimerization, interaction with rom1, and rescue of the rds null phenotype. *J. Neurochem.* **1999**, *72*, 430–438. [CrossRef]

90. Poetsch, A.; Molday, L.L.; Molday, R.S. The cGMP-gated channel and related glutamic acid-rich proteins interact with peripherin-2 at the rim region of rod photoreceptor disc membranes. *J. Biol. Chem.* **2001**, *276*, 48009–48016. [CrossRef]

91. Ritter, L.M.; Khattree, N.; Tam, B.; Moritz, O.L.; Schmitz, F.; Goldberg, A.F. In situ visualization of protein interactions in sensory neurons: Glutamic acid-rich proteins (GARPs) play differential roles for photoreceptor outer segment scaffolding. *J. Neurosci. Off. J. Soc. Neurosci.* **2011**, *31*, 11231–11243. [CrossRef] [PubMed]

92. Zhang, Y.; Molday, L.L.; Molday, R.S.; Sarfare, S.S.; Woodruff, M.L.; Fain, G.L.; Kraft, T.W.; Pittler, S.J. Knockout of GARPs and the beta-subunit of the rod cGMP-gated channel disrupts disk morphogenesis and rod outer segment structural integrity. *J. Cell Sci.* **2009**, *122*, 1192–1200. [CrossRef] [PubMed]

93. Gilliam, J.C.; Chang, J.T.; Sandoval, I.M.; Zhang, Y.; Li, T.; Pittler, S.J.; Chiu, W.; Wensel, T.G. Three-dimensional architecture of the rod sensory cilium and its disruption in retinal neurodegeneration. *Cell* **2012**, *151*, 1029–1041. [CrossRef] [PubMed]

94. Becirovic, E.; Nguyen, O.N.; Paparizos, C.; Butz, E.S.; Stern-Schneider, G.; Wolfrum, U.; Hauck, S.M.; Ueffing, M.; Wahl-Schott, C.; Michalakis, S.; et al. Peripherin-2 couples rhodopsin to the CNG channel in outer segments of rod photoreceptors. *Hum. Mol. Genet.* **2014**, *23*, 5989–5997. [CrossRef] [PubMed]

95. Chakraborty, D.; Conley, S.M.; Pittler, S.J.; Naash, M.I. Role of RDS and Rhodopsin in Cngb1-Related Retinal Degeneration. *Investig. Ophthalmol. Vis. Sci.* **2016**, *57*, 787–797. [CrossRef] [PubMed]

96. Wroblewski, J.J.; Wells, J.A., 3rd; Eckstein, A.; Fitzke, F.; Jubb, C.; Keen, T.J.; Inglehearn, C.; Bhattacharya, S.; Arden, G.B.; Jay, M.; et al. Macular dystrophy associated with mutations at codon 172 in the human retinal degeneration slow gene. *Ophthalmology* **1994**, *101*, 12–22. [CrossRef]

97. Piguet, B.; Heon, E.; Munier, F.L.; Grounauer, P.A.; Niemeyer, G.; Butler, N.; Schorderet, D.F.; Sheffield, V.C.; Stone, E.M. Full characterization of the maculopathy associated with an Arg-172-Trp mutation in the RDS/peripherin gene. *Ophthalmic Genet.* **1996**, *17*, 175–186. [CrossRef]

98. Payne, A.M.; Downes, S.M.; Bessant, D.A.; Bird, A.C.; Bhattacharya, S.S. Founder effect, seen in the British population, of the 172 peripherin/RDS mutation-and further refinement of genetic positioning of the peripherin/RDS gene. *Am. J. Hum. Genet.* **1998**, *62*, 192–195. [CrossRef]

99. Downes, S.M.; Fitzke, F.W.; Holder, G.E.; Payne, A.M.; Bessant, D.A.; Bhattacharya, S.S.; Bird, A.C. Clinical features of codon 172 RDS macular dystrophy: Similar phenotype in 12 families. *Arch. Ophthalmol.* **1999**, *117*, 1373–1383. [CrossRef]

100. Saga, M.; Mashima, Y.; Akeo, K.; Oguchi, Y.; Kudoh, J.; Shimizu, N. A novel Cys-214-Ser mutation in the peripherin/RDS gene in a Japanese family with autosomal dominant retinitis pigmentosa. *Hum. Genet.* **1993**, *92*, 519–521. [CrossRef]

101. Kajiwara, K.; Hahn, L.B.; Mukai, S.; Travis, G.H.; Berson, E.L.; Dryja, T.P. Mutations in the human retinal degeneration slow gene in autosomal dominant retinitis pigmentosa. *Nature* **1991**, *354*, 480–483. [CrossRef]

102. Kajiwara, K.; Berson, E.L.; Dryja, T.P. Digenic retinitis pigmentosa due to mutations at the unlinked peripherin/RDS and ROM1 loci. *Science* **1994**, *264*, 1604–1608. [CrossRef] [PubMed]

103. Dryja, T.P.; Hahn, L.B.; Kajiwara, K.; Berson, E.L. Dominant and digenic mutations in the peripherin/RDS and ROM1 genes in retinitis pigmentosa. *Investig. Ophthalmol. Vis. Sci.* **1997**, *38*, 1972–1982. [PubMed]

104. Kedzierski, W.; Nusinowitz, S.; Birch, D.; Clarke, G.; McInnes, R.R.; Bok, D.; Travis, G.H. Deficiency of rds/peripherin causes photoreceptor death in mouse models of digenic and dominant retinitis pigmentosa. *Proc. Natl. Acad. Sci. USA* **2001**, *98*, 7718–7723. [CrossRef] [PubMed]

105. Nour, M.; Ding, X.Q.; Stricker, H.; Fliesler, S.J.; Naash, M.I. Modulating expression of peripherin/rds in transgenic mice: Critical levels and the effect of overexpression. *Investig. Ophthalmol. Vis. Sci.* **2004**, *45*, 2514–2521. [CrossRef] [PubMed]

106. Apfelstedt-Sylla, E.; Theischen, M.; Ruther, K.; Wedemann, H.; Gal, A.; Zrenner, E. Extensive intrafamilial and interfamilial phenotypic variation among patients with autosomal dominant retinal dystrophy and mutations in the human RDS/peripherin gene. *Br. J. Ophthalmol.* **1995**, *79*, 28–34. [CrossRef]

107. McNally, N.; Kenna, P.F.; Rancourt, D.; Ahmed, T.; Stitt, A.; Colledge, W.H.; Lloyd, D.G.; Palfi, A.; O'Neill, B.; Humphries, M.M.; et al. Murine model of autosomal dominant retinitis pigmentosa generated by targeted deletion at codon 307 of the rds-peripherin gene. *Hum. Mol. Genet.* **2002**, *11*, 1005–1016. [CrossRef]
108. Nichols, B.E.; Sheffield, V.C.; Vandenburgh, K.; Drack, A.V.; Kimura, A.E.; Stone, E.M. Butterfly-shaped pigment dystrophy of the fovea caused by a point mutation in codon 167 of the RDS gene. *Nat. Genet.* **1993**, *3*, 202–207. [CrossRef]
109. Fossarello, M.; Bertini, C.; Galantuomo, M.S.; Cao, A.; Serra, A.; Pirastu, M. Deletion in the peripherin/RDS gene in two unrelated Sardinian families with autosomal dominant butterfly-shaped macular dystrophy. *Arch. Ophthalmol.* **1996**, *114*, 448–456. [CrossRef]
110. Zhang, K.; Garibaldi, D.C.; Li, Y.; Green, W.R.; Zack, D.J. Butterfly-shaped pattern dystrophy: A genetic, clinical, and histopathological report. *Arch. Ophthalmol.* **2002**, *120*, 485–490. [CrossRef]
111. Zulliger, R.; Conley, S.M.; Mwoyosvi, M.L.; Stuck, M.W.; Azadi, S.; Naash, M.I. SNAREs Interact with Retinal Degeneration Slow and Rod Outer Segment Membrane Protein-1 during Conventional and Unconventional Outer Segment Targeting. *PLoS ONE* **2015**, *10*, e0138508. [CrossRef] [PubMed]
112. Travis, G.H.; Groshan, K.R.; Lloyd, M.; Bok, D. Complete rescue of photoreceptor dysplasia and degeneration in transgenic retinal degeneration slow (rds) mice. *Neuron* **1992**, *9*, 113–119. [CrossRef]
113. Ali, R.R.; Sarra, G.M.; Stephens, C.; Alwis, M.D.; Bainbridge, J.W.; Munro, P.M.; Fauser, S.; Reichel, M.B.; Kinnon, C.; Hunt, D.M.; et al. Restoration of photoreceptor ultrastructure and function in retinal degeneration slow mice by gene therapy. *Nat. Genet.* **2000**, *25*, 306–310. [CrossRef] [PubMed]
114. Schlichtenbrede, F.C.; da Cruz, L.; Stephens, C.; Smith, A.J.; Georgiadis, A.; Thrasher, A.J.; Bainbridge, J.W.; Seeliger, M.W.; Ali, R.R. Long-term evaluation of retinal function in Prph2Rd2/Rd2 mice following AAV-mediated gene replacement therapy. *J. Gene Med.* **2003**, *5*, 757–764. [CrossRef] [PubMed]
115. Ding, X.Q.; Quiambao, A.B.; Fitzgerald, J.B.; Cooper, M.J.; Conley, S.M.; Naash, M.I. Ocular delivery of compacted DNA-nanoparticles does not elicit toxicity in the mouse retina. *PLoS ONE* **2009**, *4*, e7410. [CrossRef]
116. Cai, X.; Conley, S.M.; Nash, Z.; Fliesler, S.J.; Cooper, M.J.; Naash, M.I. Gene delivery to mitotic and postmitotic photoreceptors via compacted DNA nanoparticles results in improved phenotype in a mouse model of retinitis pigmentosa. *FASEB J.* **2010**, *24*, 1178–1191. [CrossRef]
117. Han, Z.; Conley, S.M.; Makkia, R.S.; Cooper, M.J.; Naash, M.I. DNA nanoparticle-mediated ABCA4 delivery rescues Stargardt dystrophy in mice. *J. Clin. Investig.* **2012**, *122*, 3221–3226. [CrossRef]
118. Han, Z.; Conley, S.M.; Makkia, R.; Guo, J.; Cooper, M.J.; Naash, M.I. Comparative analysis of DNA nanoparticles and AAVs for ocular gene delivery. *PLoS ONE* **2012**, *7*, e52189. [CrossRef]
119. Han, Z.; Koirala, A.; Makkia, R.; Cooper, M.J.; Naash, M.I. Direct gene transfer with compacted DNA nanoparticles in retinal pigment epithelial cells: Expression, repeat delivery and lack of toxicity. *Nanomedicine* **2012**, *7*, 521–539. [CrossRef]
120. Conley, S.M.; Naash, M.I. Gene therapy for PRPH2-associated ocular disease: Challenges and prospects. *Cold Spring Harb. Perspect. Med.* **2014**, *4*, a017376. [CrossRef]
121. Cai, X.; Nash, Z.; Conley, S.M.; Fliesler, S.J.; Cooper, M.J.; Naash, M.I. A partial structural and functional rescue of a retinitis pigmentosa model with compacted DNA nanoparticles. *PLoS ONE* **2009**, *4*, e5290. [CrossRef] [PubMed]
122. Mao, H.; Gorbatyuk, M.S.; Rossmiller, B.; Hauswirth, W.W.; Lewin, A.S. Long-term rescue of retinal structure and function by rhodopsin RNA replacement with a single adeno-associated viral vector in P23H RHO transgenic mice. *Hum. Gene Ther.* **2012**, *23*, 356–366. [CrossRef] [PubMed]
123. Cideciyan, A.V.; Sudharsan, R.; Dufour, V.L.; Massengill, M.T.; Iwabe, S.; Swider, M.; Lisi, B.; Sumaroka, A.; Marinho, L.F.; Appelbaum, T.; et al. Mutation-independent rhodopsin gene therapy by knockdown and replacement with a single AAV vector. *Proc. Natl. Acad. Sci. USA* **2018**, *115*, E8547–E8556. [CrossRef] [PubMed]

124. Palfi, A.; Ader, M.; Kiang, A.S.; Millington-Ward, S.; Clark, G.; O'Reilly, M.; McMahon, H.P.; Kenna, P.F.; Humphries, P.; Farrar, G.J. RNAi-based suppression and replacement of rds-peripherin in retinal organotypic culture. *Hum. Mutat.* **2006**, *27*, 260–268. [CrossRef]
125. Petrs-Silva, H.; Yasumura, D.; Matthes, M.T.; LaVail, M.M.; Lewin, A.S.; Hauswirth, W.W. Suppression of rds expression by siRNA and gene replacement strategies for gene therapy using rAAV vector. *Adv. Exp. Med. Biol.* **2012**, *723*, 215–223.

Review

Phosphoinositides in Retinal Function and Disease

Theodore G. Wensel

Verna and Marrs McLean Department of Biochemistry and Molecular Biology, Baylor College of Medicine, Houston, TX 77030, USA; twensel@bcm.edu

Received: 1 March 2020; Accepted: 30 March 2020; Published: 2 April 2020

Abstract: Phosphatidylinositol and its phosphorylated derivatives, the phosphoinositides, play many important roles in all eukaryotic cells. These include modulation of physical properties of membranes, activation or inhibition of membrane-associated proteins, recruitment of peripheral membrane proteins that act as effectors, and control of membrane trafficking. They also serve as precursors for important second messengers, inositol (1,4,5) trisphosphate and diacylglycerol. Animal models and human diseases involving defects in phosphoinositide regulatory pathways have revealed their importance for function in the mammalian retina and retinal pigmented epithelium. New technologies for localizing, measuring and genetically manipulating them are revealing new information about their importance for the function and health of the vertebrate retina.

Keywords: phosphoinositides; retinal lipids; membrane trafficking

---

1. Introduction

Phosphoinositides are membrane phospholipids with the six-member cyclic polyol *myo*-inositol, $(CHOH)_6$, or O-phosphorylated forms of inositol, as their headgroup (Figure 1). The phosphorylated forms are of low abundance in eukaryotic cells of all types, generally comprising 1% or less of total phospholipid. Nevertheless, they play critical roles in cellular regulation, and defects in their synthesis and regulation lead to devastating diseases [1–6]. The retina is clearly no exception to this generality, but surprisingly few details have been worked out in what is arguably one of the most extensively studied tissues in our bodies, about the regulation and regulatory roles of retinal phosphoinositides. Recent technological advances make it possible to make substantial advances in this field in the next few years.

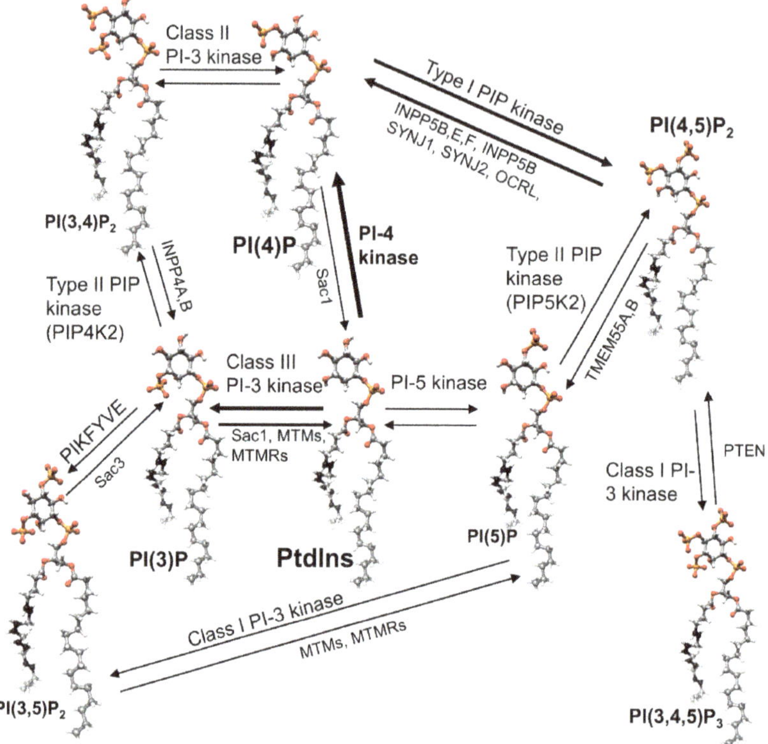

Figure 1. Structures of the cellular phosphoinositides and enzymes responsible for their synthesis and interconversion. Phosphoinositides (PI) species are shown with the acyl chains most commonly found on phosphatidylinositol, which is the starting point for all the others, arachidonic acid (20:4) and stearic acid (C18). Relative font sizes correlate with relative abundance.

2. Chemical Structures of Phosphoinositides

Phosphoinositides (abbreviated here as "PI") include phosphatidylinositol (PtdIns), in which the 1′-position of the inositol ring is attached via a phosphodiester bond to the *sn*-3 position of (1, 2) diacyl glycerol, and derivatives of PtdIns with one, two, or three phosphates attached in various combinations to the 4′, 5′ or 3′ hydroxyls of PtdIns. Including PtdIns, a total of eight different PI head groups are commonly found in eukaryotic cells (Figure 1) [4,5].

3. Phosphatidylinositol Content in the Retina and RPE

PtdIns is a relatively substantial component of the membranes of most cells in metazoans, with mole fractions ranging from ~4%–20% of total phospholipid [4]. Early reports stated that retina phospholipids contained 4.4%–6.4% PtdIns across six different mammalian species [7–9]. Interestingly, rod outer segment (ROS) membranes were found to have much lower levels, 1.5% to 2.5% in bovine and 2.1% in frog [10–13]. Retinal pigment epithelial cells (RPE), in contrast, have higher levels of PtdIns, at 6.5% of total phospholipid [14], while ER membranes isolated from bovine retinas contain 9.6% PtdIns [10]. The general picture that emerges is that total PtdIns content in total retinal membranes outside of outer segments and that of RPE are more or less "normal" as compared to other tissues and cell types, consistent with the notion that widespread roles of PI are likely to be conserved in many retinal cell types. In contrast, ROS have an unusually low PtdIns content, suggesting different roles for this lipid class in that organelle. This idea is consistent with findings discussed below, indicating different lipid compositions in plasma membranes as compared to ciliary membranes.

4. Content of Minor Phosphoinositides in the Retina and RPE

As a negatively charged lipid, which typically contains one polyunsaturated side chain fatty acid, such as arachidonic acid [8], PtdIns has a major impact on the physical properties of the membrane domains containing it. In contrast, the phosphorylated forms are much less abundant, and exert their influence on cell physiology largely through interactions with proteins with high-affinity and high-specificity PI-binding domains [15–17]. There have been few measurements of phosphorylated PI in retina or cells isolated from retina, although there have been several papers addressing the turnover of PtdIns and other PI, or activity of enzymes involved in PI metabolism [18–21]. The likely reason for the absence of information on PI levels is their low abundance and the lack of sensitivity provided by conventional methods for lipid analysis. More sensitive techniques have been developed recently, including one based on recombinant phosphoinositide-binding domains fused to an epitope tag, allowing sensitive detection by enzyme-linked immunosorbent assays (ELISAs) and measurement of chemiluminescence [22]. This technique was used to quantify PI(3)P and PI(3,4,5)P$_3$ in preparations of rod cells that contain both outer segments and fragments of the inner segments and demonstrated levels of PI(3)P at 0.0035 mol% of total phospholipid under illumination conditions that yielded the highest levels of that lipid, and at least 10-fold lower (i.e., undetectable) levels of PI(3,4,5)P$_3$. PI(4)P and PI(4,5)P$_2$ are, in general, found at much higher levels than the 3-phosphorylated PI, but even those appear to be present in rods at very low levels, which are only about 10-fold higher than the levels of PI(3)P, i.e., on the order of 0.04 mol% (He and Wensel, unpublished observations).

Comparison to Other Tissues and Cell Types

These numbers are comparable to those found in other eukaryotic cells, reported as PtdIns3P, 0.002% of phospholipid mass; PtdIns4P, 0.05%; PtdIns5P, 0.002%; PtdIns(4,5)P, 0.05%; PtdIns(3,4)P, 0.0001%; PtdIns(3,5)P, 0.0001% [17], see also references in [23]. A more recent mass spectrometry study reported PI(3,4,5)3 levels as 50-fold or more lower than those of the more common PIs in mammalian U87MG cells [24]. An interesting observation derived from measurements of PI levels in cultured cells is that, generally, they do not change greatly upon activation with extracellular stimuli; for the more common forms, PtdIns, PI(4)P, PI(4,5)P$_2$ and PI(3)P, the change is generally less than 30% [17]. The implication, for PI(4,5)P$_2$, which is rapidly degraded to form InsP$_3$ and diacylglycerol upon activation of phospholipase C isozymes by G-protein-coupled receptors or growth factor receptors [25,26], is that homeostatic mechanisms are in place to regenerate rapidly the pools of both PI(4,5)P$_2$ and its precursor PI(4)P upon PLC activation. In contrast, PI(3,4,5)P$_3$ level changes due to receptor stimulation are generally too low to have a major impact on levels of its precursor, PI(4,5)P$_2$.

5. Importance of PI Generally

Despite their low abundance, phosphoinositides play major roles in regulation of cell signaling and membrane dynamics, and are essential for control of a wide range of processes including development, proliferation and differentiation, membrane excitability, exocytosis, phagocytosis, cell motility, and detection of extracellular signals [4]. These functional roles are primarily mediated by a plethora of enzymes, scaffold proteins and complex-nucleating proteins containing phosphoinositide-binding domains of high affinity and high specificity. Genetic defects in the enzymes responsible for their regulation are, in most cases, lethal at the embryo stage, except for some cases of redundancy (i.e., more than one gene encoding enzymes or PI-binding proteins with similar activities). Cell-type specific deletion of these enzymes leads to more specific pathologies, e.g., neurodegeneration in the case of the type III PI-3 kinase, Vps34 (also known as PIK3C3) [27,28]. Some of these enzymes, such as type I PI-3 kinases, can act as oncogenes when mutations disrupt their regulation, and are considered prime targets for cancer chemotherapy, whereas others, such as the phosphoinositide phosphatase, PTEN, are considered tumor suppressor genes [29]; these therapeutic targets are present in the retina, where their roles in retinal function, and any effects of drugs targeting them are unknown.

Importance of Phosphoinositides for Dynamics and Functions of Membranes in Retina and RPE

Among the most important roles of phosphoinositides and their protein effectors are those mediating intracellular membrane traffic, by directing membrane proteins and lipids from one compartment to another in response to cellular needs and changing environments. For example, PI(3)P is found in early and recycling endosomes, and is important for recruiting key proteins that regulate trafficking to these compartments [30]. It is also important for autophagy, a survival-promoting pathway leading to the lysosomal degradation of organelles [31]. PI(4)P is enriched in the Golgi apparatus, and is thought to be important for membrane trafficking through the Golgi compartments and from the Golgi to the plasma membrane and other subcellular compartments [32], likely including disk membranes. PI(4,5)P$_2$ plays a key role in clathrin-mediated endocytosis, and serves to direct many effector proteins to the plasma membrane where it is primarily found [33].

PI(4,5)P$_2$ and other phosphoinositides directly regulate the activity of ion channels, transporters and enzymes in membranes [34,35]. In addition, PI(4,5)P$_2$ serves as the substrate for phosphoinositide-specific phospholipase C, leading to production of the important second messengers, inositol (1,4,5) trisphosphate (InsP$_3$) and diacylglycerol [25,26,36,37]. Phosphoinositides are also critically important for regulating interactions between membranes and cytoskeletal elements [38,39]. Changes in these interactions are critical for cell growth and mobility, and for remodeling of intracellular structures. They may well play a role in cytoskeleton-dependent disk morphogenesis in rods.

6. Membrane Trafficking in Retina and RPE

Every aspect of retinal biology depends heavily on the correct organization and composition of highly specialized membranes, from the unique disc membranes of the photoreceptor sensory cilia, to the ribbon synapses of rods, cones, and bipolar cells, to the apical processes of RPE (retinal pigmented epithelium) cells, uniquely tuned to the detection and engulfment of shed outer segment fragments. The formation, maintenance, and functions of these membranes rely heavily on the phosphorylated phosphoinositides [2–4,40]. Despite the intense interest in phosphoinositide research in recent decades, surprisingly little is known about their regulation and functional roles in the retina, although it is known that disruption of phosphoinositide regulation can lead to blindness in human patients and animal models [41–48]. The membranes and pathways they regulate are known to be essential for the function and health of the retina as well as for disease processes and cellular responses to disease states.

One of the reasons for the dearth of knowledge has been a lack of tools for studying these very low-abundance lipids within the multiple cell types of the retina and the adjacent retinal pigment epithelium. Recently, tools developed by the broader PI field have begun to be applied to the unique challenges and opportunities posed by the retina [22,49]. Doing so will have enormous impact on our understanding of the cell biology of the retina and its disruption in disease. This can help to inform the design and optimization of therapies aimed at treating and preventing retinal dysfunction and degeneration.

7. Features of the Retina that Make It Ideal for Studies of Phosphoinositide Regulation In Vivo

The field of phosphoinositide regulation has long been dominated by studies in cultured immortalized cell lines, giving rise to a critical need for elucidation of their physiological regulation in terminally differentiated neurons. For this reason, in vivo studies have broad significance for this field. There are several features of the retina that make it particularly amenable to studies of phosphoinositide regulation in vivo. These include the ability to assay both structure and function non-invasively, the ease of making cell-type-specific knockouts, the ability to isolate rod cells for either biochemical analysis or ultrastructure determination, an extensive understanding of biochemistry and cell biology, especially of rods, which exceeds that of any of other vertebrate neurons, and a wealth of knowledge of RPE cell biology.

8. Importance of Phosphoinositides for Membrane Trafficking in Retina

Despite years of study, no convincing evidence has accumulated for an important role for phosphoinositides in the phototransduction cascade. In contrast, a steady stream of evidence supports a central role for these lipids in membrane trafficking and sorting in all mammalian cell types, just as a critical role in photoreceptors for membrane sorting and trafficking has long been established [3,50,51]. A review article in 2011 covered advances in understanding the roles of phosphoinositides in photoreceptors [6]. It is highly likely that events such as endocytosis and exocytosis, endosomal sorting, membrane budding, post-Golgi vesicle trafficking, and disk morphogenesis all depend on phosphoinositide dynamics, and published reports support a role for PI(3)P and PI(4,5)P_2 in rhodopsin trafficking [52,53]. For example, PI(4,5)P_2-binding proteins, ezrin and moesin, were reported to colocalize with Rac1 and Rab8 on rhodopsin transport carrier vesicles at the site of their fusion with the plasma membrane. A recent report suggesting the involvement of actin-nucleating proteins Arp2/Arp3 in basal disc extension [54] potentially implicates local pools or PI(4,5)P_2, which are known to be critical for their function [55,56].

In cone photoreceptors, ablation of a type I PI-3 kinase leads to enhanced sensitivity to light damage [57]. Mutations in the phosphoinositide phosphatase, synaptojanin 1 lead to defects in synaptic vesicle trafficking in cone cells [44].

9. Retinal Cilia and Phosphoinositides

Phosphoinositides, and especially PI(4,5)P_2 and PI(4)P, have been proposed to play important roles in assembly, disassembly, and regulation of primary cilia [52,58–62]. The lipid content of cilia is different from that of the plasma membrane, and a membrane diffusion barrier surrounding the cilium has been demonstrated (see [63] for a review). Two PI-5-phosphatases, INPP5E and OCRL, essential for cilium function have been reported to be localized to the cilium [41,64–66]

Phosphoinositides and the BBSome

The BBSome, a heterooctameric protein complex involved in ciliary trafficking, whose defects lead to the blinding ciliopathy, Bardet–Biedl syndrome, binds to membranes, and shows a preference for acidic lipids, including phosphoinositides [67,68], and isolated BBS5 which contains pleckstrin homology domains, binds to phosphoinositides, especially PI(3)P [68]. More recently, it was shown that a core BBSome complex containing BBS 1, 4, 5, 8, 9 and 18 and a smaller sub-complex lacking BBS1 and BBS5 bind phosphoinositides with similar specificities. A caveat for these studies is that the commercial "PIP strips" used have local surface densities of phosphoinositides that far exceed anything found under physiological conditions, so further investigation of phosphoinositide-binding of the BBSome and its sub-complexes is warranted.

10. Autophagy and other Stress Responses Involving Redirection of Membrane Traffic and Phosphoinositides

It has been reported that light exposure induces elements of the autophagy pathway in rods and that autophagy plays an important role in photoreceptor homeostasis [69–73]. This pathway has been suggested to be a neuroprotective one that forestalls apoptosis under conditions of stress [69,74]. This process may be part of a more general neuroprotective response involving re-direction of membrane traffic and phosphoinositides. As discussed below, Synaptojanin-1 has been implicated in autophagy in zebrafish cones [75].

11. Evidence for Effects of Light on Phosphoinositide Metabolism

A number of early reports in the 1980s suggested that light had measurable impacts on PI metabolism in photoreceptors or the retina generally. Based on measurement of ^{32}Pi incorporated into PtdIns by metabolic labeling, it was reported that exposure of isolated frog retina to light decreased levels of PIP$_2$ by 14% after 5 s and 37% after 15 s, while levels of PI(4)P, PtdIns and other acidic lipids remained essentially constant [76]. Subsequent publications from the same group reported increased levels of ^{3}H inositol and ^{32}P into phosphoinositides upon illumination [77]. They also reported PI(4,5)P$_2$-specific phospholipase C (PLC) activity in frog photoreceptors, and PLC immunoreactivity in bovine rod outer segments [78,79], as well as PI-kinase and PIP-kinase activities in frog ROS [19]. Another group found that exposure of rat retinas to light led to decreased staining of rod outer segments with anti-PI(4,5)P$_2$ antibodies [80,81]. The caveats of those experiments are that it is known that physical properties of outer segment membranes and their protein composition are altered by bright light exposure, and that it is difficult to establish the specificity of such antibody staining. In addition to light, reports on regulation of photoreceptor PLC by Ca^{2+} [82] or by subunits of the phototransduction G protein, transducin [83], were published, also suggesting a possible influence of light exposure, which is known to control levels of Ca^{2+} and active and inactive forms of transducin subunits.

In contrast, another group [84] reported that bovine rod outer segments have very little PIP kinase activity as compared to the rest of the retina (consistent with the previously reported low level of PI(4,5)P$_2$ in outer segment membranes), and that light adaptation had no measurable effect on phosphoinositide metabolism as compared to in vivo dark adaptation. Yet another group, using metabolic labeling with [^{3}H]inositol, reported that light led to decreases in PIP$_2$ levels without generation of InsP$_3$, suggesting light-dependent activity of a phosphatase rather than of phospholipase C [85]. In vitro studies demonstrated that the presence of PI(4,5)P$_2$ in membranes could affect the activity of components of the phototransduction cascade [86,87], including the cGMP-gated cation channel and the cGMP phosphodiesterase-transducin complex. The physiological relevance of effects of PI(4,5)P$_2$ on phototransduction or of effects of light on PI(4,5)P$_2$ metabolism remains untested, and it seems to be possible to explain the entire time-course of rod light responses without invoking any participation by phosphoinositide metabolism [88]. Antibodies specific for isoforms of phospholipase C or G$_{\alpha q/11}$ PLC-coupled subunits revealed the presence of PLCβ4 and G$_{\alpha q/11}$ in rod outer segments, and G$_{\alpha q}$ and other PKC isoforms elsewhere in the retina [89]. The functional roles of these proteins in outer segments are not known. There have been suggestions that slower effects of prolonged light exposure, such as arrestin translocation from the inner to outer segments of rod cells, may be mediated by a phospholipase C cascade [90].

Light Regulation of PI-3 Kinase

Interest has turned toward Type I PI-3 kinase and PI(3,4,5)P$_3$, with reports of effects of light on the activity of this enzyme in bovine rod outer segments [91,92], which were reported to be mediated by light-stimulated activation and tyrosine phosphorylation of the insulin receptor [93–95]. Deletion of the p85α regulatory subunit of Type I PI-3 kinase in cone cells resulted in progressive degeneration of cones, without observable effects on rod survival [57]. Likewise, cone-specific inactivation of the gene encoding the p110α catalytic subunit also resulted in defects in cone survival [96].

In contrast, Type I PI-3 kinase and its product, PI(3,4,5)P$_3$, seem to be much less important for rod function. Rod-specific ablation of the p85α gene using two different rod-specific Cre transgenes yielded no obvious defects in retinal morphology or rod cell survival [22,97], although modest effects on kinetics of light response recovery and arrestin translocation were reported in one case [97]. Quantitative analysis of PI(3,4,5)P$_3$ levels in rods isolated from dark-adapted or light-adapted retinas revealed levels of this phosphoinositide of more than one order magnitude lower than those found for PI(3)P in the light, or at least two orders of magnitude lower than light-stimulated levels of PI(4,5)P$_2$ [22].

12. PI(4,5)P$_2$ and Phospholipase C in Intrinsically Photosensitive Ganglion Cells

In addition to image-forming light detection mediated by rods and cones, the vertebrate retina also contains intrinsically photosensitive retinal ganglion cells. These contain phototransduction cascades reminiscent of that found in invertebrate rhabdomeric photoreceptors [98–104]. Light activates melanopsin, encoded by the *Opn4* gene, a visual pigment which is more closely related to invertebrate opsins than to vertebrate opsin [105]. In M1-type ganglion cells, melanopsin photoisomerization leads to activation of a G$_{\alpha q/11/14}$ class G-protein, which activates the phosphoinositide-specific phospholipase C isoform, PLCβ4; the phospholipase presumably acts on PI(4,5)P$_2$ as in other cell types, including *Drosophila* photoreceptors, which contain a homologous phospholipase, to produce diacylglycerol and InsP$_3$. PI(4,5)P$_2$ hydrolysis, in turn, leads to the activation of the cation channels TRPC6 and TRPC7 [104]. In M4 ganglion cells, a different phototransduction cascade involving cyclic nucleotides and cyclic nucleotide-regulated HCN channels, whereas in M2 ganglion cells, both of these cascades operate [102].

13. Studies of PI Metabolism in the RPE

A variety of extracellular stimuli acting on tyrosine kinase-associated receptors or G protein-coupled-receptors have been reported to stimulate release of inositol phosphates in cultured RPE cells, presumably derived from PLC action of PI(4,5)P$_2$, on a timescale of tens of minutes; effective stimuli included fetal bovine serum, agonists for muscarinic, histamine, and serotonin, peptides, including bradykinin, arginine vasopressin, bombesin and oxytocin, [106–109]. The physiological relevance of these observations was not explored, but an in vivo study using frogs demonstrated dramatic acceleration of inositol phosphate release, especially of InsP$_3$, following stimulation by light [21]. Acutely isolated rat RPE cells were reported to release InsP$_3$ in response to induction of phagocytosis by addition of isolated rod outer segments [110]; this InsP$_3$ release was not observed in cells from Royal College of Surgeon (RCS) rats, which have a defect in OS phagocytosis due to a deficiency in the receptor tyrosine kinase, MERTK [111]. As in many other cell types, insulin has been reported to stimulate activity of Type I PI-3 kinase to produce PI(3,4,5)P$_3$ [95,112,113]. Responses to hypoxia [112,114,115] and elevated glucose [116,117] are also reported to involve this pathway in RPE.

A number of important processes in RPE are known to rely on phosphoinositides, but how they are regulated in these cells is not well understood. Phagocytosis, autophagy, endocytosis and endosome processing, establishment of epithelial cell polarity and extension of microvilli membranes are all known to critically depend on phosphoinositides. For example, both autophagy and phagocytosis involve the recruitment of the ubiquitin-like protein, LC3 [118–120], whose recruitment to membranes depends on PI(3)P. Phagocytosis is also thought to require PI(4,5)P$_2$ and lysosomal fusion may involve other phosphoinositides such as PI(3,5)P$_2$ and PI(5)P [22,121–125].

14. Phosphoinositide Kinases and Phosphatase

In mammals, there are 47 genes encoding 19 PI-kinase and PIP-kinases and 28 PIP phosphatases [126].

14.1. Kinases

The kinase isoforms are divided into three major families: PI 4-kinases (PI4Ks), the PI 3-kinases (PI3Ks), and PIP (PIP) kinases (PIPKs) [127]. The nomenclature for these is a bit confusing, as some enzymes termed "PI-kinases" actually act primarily as PIP kinases. For example, Type I PI-3 kinases primarily use PI(4,5)P$_2$ as their substrate to produce PI(3,4,5)P$_3$, and there is little evidence for a substantial portion of cellular PI(3)P being formed by these enzymes. When the Type III PI-3 kinase, Vps34, was knocked out in mouse rod cells and phosphoinositide levels measured, the results suggested complete ablation of PI(3)P despite the presence of the Type I enzyme [22]. Several of these have been reported to be expressed in retina or RPE at the level of protein, mRNA or enzyme activity and most, if not all, are likely present at some level; however, as far as their functions, only the Type I and Type III PI-3 kinase have been studied using gene knockouts in the retina and RPE [22,49,57,128]. Global knockouts have been produced for the α, β, and γ isoforms of Type I PIP-kinases, which are the major source of PI(4,5)P$_2$ in most cells [16]. Of these, the γ isoform seems to have the highest expression in the retina [129] and in other neurons [130,131] and leads to severe neuronal phenotypes and early postnatal mortality when knocked out [132]. Type II PIP-kinases have been observed in retina and are reported to be regulated by tyrosine phosphorylation [133]. The proteins encoded by the mouse genes, Pip4k2a, Pip4k2b, Pip4k2c, Pip5k1a, and Pip5k1c were all observed in a proteomic study of mouse retina [134].

14.2. Phosphatases

The phosphoinositide phosphatases encoded by Inpp1, Inpp4a, Inpp4b, Inpp5e, Mtmr2 and Synj1 (synaptojanin-1) have also been detected in retina by proteomics [134]. As noted below, defects in synaptojanin-1 and INPP5E are associated with retinal defects, as is the phosphoinositide 5-phosphatase, OCRL (oculocerebrorenal syndrome of Lowe) [135,136]. Among the other enzymes detected, almost none were enriched in the rod outer segment fraction as compared to the rest of the retina, consistent with the relatively low PI content of that organelle. One exception was Pip4k2c, the PI(5)P-4-kinase Type II γ isoform. Inactivation of the mouse *Pip4kc2* gene was found to lead to hyperactivation of the immune system, but the retinal phenotype was not examined [137].

15. Retinal Phenotypes of Genetic Defects in Genes Related to Phosphoinositide Metabolism and Signaling

Rods and cones have a very high rate of metabolism and biosynthesis of membrane components, due to the energetic demands of the phototransduction cascade and the daily shedding of ~10% of the disk membranes (based on observation of a rate of 9%–13% in the rhesus monkey [138], which have to be engulfed and recycled by the RPE. As highly polarized cells, their function and health depend critically on efficient and accurate transport of the correct proteins and lipids to the correct compartments. A host of human blinding diseases have been linked to defects in membrane transport and sorting [139,140]. In RPE cells, massive amounts of membrane traffic are associated with their role as professional phagophores. Defects in RPE phagocytosis, as in MERTK deficiency [111,141] and Bestrophin deficiency [142–145], cause retinal degeneration in humans and animal models and have been proposed to play a role in age-related macular degeneration [146].

15.1. Phosphatases and Inherited Retinopathies

Inherited defects in the phosphoinositide phosphatase, INPP5E, are associated with the multi-syndromic ciliopathies, Joubert syndrome and Bardet–Biedl syndrome [147–150], and with retinal degeneration. The substrates for this phosphatase are PI(4,5)P$_2$ and PI(3,4,5)P$_3$, and its critical role in the cilia suggests that regulating levels of PI(4,5)P$_2$ may be important in ciliogenesis and cilium stability.

Synaptojanins are phosphoinositide phosphatases associated with synaptic function [151–153], including vesicle uncoating and endocytosis. Synaptojanin-1 deficiency causes severe cone defects in zebrafish and has been implicated in the regulation of autophagy in cones [45,46,75].

Defects in another phosphoinositide phosphatase, OCRL, are associated with a severe ciliopathy, oculocerebrorenal syndrome of Lowe [136]. Symptoms of this disease include glaucoma, a disease of the retina.

15.2. PI3NM3

The phosphatidyl inositol transfer protein PIT3NM3 (aka CCL118) is the mammalian homologue of the *Drosophila* retinal degeneration gene, *rdgB*, and defects in it are associated with autosomal dominant cone-rod dystrophy CORD5 [154–159]. The protein has also been implicated in cancer metastasis [160]. In zebrafish, the Class I phosphatidylinositol transfer protein β isoform, Pitpnb, is essential for biogenesis and maintenance of double cones [161].

15.3. PI Binding Proteins

Proteins with phosphoinositide-binding domains have also been associated with retinal disease. These include the product of the *tub/TUB* gene [162,163] and a related protein, TULP1 [47,164–166]. Tub was discovered as the protein encoded by a gene whose mutation in *tubby* mice causes obesity, deafness and blindness. The widely expressed members of the tubby family, including TULP1-TULP4, have a characteristic carboxyl-terminal tubby domain consisting of an alpha helix surrounded by a beta barrel which has been shown to bind to specific phosphoinositides [167]. Both Tub and Tulp1 are expressed in the retina, and TULP1 mutations cause retinitis pigmentosa [47,168].

16. Conclusions

While the overall field of phosphoinositide regulation and signaling has made dramatic advances in recent years, our understanding of specific regulatory pathways in specific cell types of the retina remains limited. The emergence of new technologies for cell-type-specific gene manipulation, along with imaging techniques for subcellular localization of phosphoinositides, their regulatory enzymes and their effector proteins, and the development of more sensitive methods for measuring PI levels, bodes well for rapid progress in understanding the roles of retinal phosphoinositides in the coming years.

Funding: The author's research in this area has been supported by NIH grants R01-EY007981, R01-EY025218 and the Robert A. Welch Foundation, Q0035.

Conflicts of Interest: The author declares no conflicts of interest relevant to this paper.

References

1. Phan, T.K.; Williams, S.A.; Bindra, G.K.; Lay, F.T.; Poon, I.K.H.; Hulett, M.D. Phosphoinositides: Multipurpose cellular lipids with emerging roles in cell death. *Cell Death Differ.* **2019**, *26*, 781–793. [CrossRef] [PubMed]
2. Dickson, E.J.; Hille, B. Understanding phosphoinositides: Rare, dynamic, and essential membrane phospholipids. *Biochem. J.* **2019**, *476*, 1–23. [CrossRef] [PubMed]
3. Schink, K.O.; Tan, K.W.; Stenmark, H. Phosphoinositides in Control of Membrane Dynamics. *Annu. Rev. Cell Dev. Biol.* **2016**, *32*, 143–171. [CrossRef] [PubMed]
4. Balla, T. Phosphoinositides: Tiny lipids with giant impact on cell regulation. *Physiol. Rev.* **2013**, *93*, 1019–1137. [CrossRef]
5. Maffucci, T. An introduction to phosphoinositides. *Curr. Top. Microbiol. Immunol.* **2012**, *362*, 1–42. [CrossRef]
6. Brockerhoff, S.E. Phosphoinositides and photoreceptors. *Mol. Neurobiol.* **2011**, *44*, 420–425. [CrossRef]
7. Broekhuyse, R.M. Phospholipids in tissues of the eye. I. Isolation, characterization and quantitative analysis by two-dimensional thin-layer chromatography of diacyl and vinyl-ether phospholipids. *Biochim. Biophys. Acta* **1968**, *152*, 307–315. [CrossRef]
8. Anderson, R.E.; Feldman, L.S.; Feldman, G.L. Lipids of ocular tissues. II. The phospholipids of mature bovine and rabbit whole retina. *Biochim. Biophys. Acta* **1970**, *202*, 367–373. [CrossRef]
9. Anderson, R.E. Lipids of ocular tissues. IV. A comparison of the phospholipids from the retina of six mammalian species. *Exp. Eye Res.* **1970**, *10*, 339–344. [CrossRef]
10. Anderson, R.E.; Maude, M.B.; Zimmerman, W. Lipids of ocular tissues–X. Lipid composition of subcellular fractions of bovine retina. *Vision Res.* **1975**, *15*, 1087–1090. [CrossRef]
11. Anderson, R.E.; Maude, M.B. Phospholipids of bovine outer segments. *Biochemistry* **1970**, *9*, 3624–3628. [CrossRef] [PubMed]
12. Nielsen, N.C.; Fleischer, S.; McConnell, D.G. Lipid composition of bovine retinal outer segment fragments. *Biochim. Biophys. Acta* **1970**, *211*, 10–19. [CrossRef]
13. Anderson, R.E.; Risk, M. Lipids of ocular tissues. IX. The phospholipids of frog photoreceptor membranes. *Vision Res.* **1974**, *14*, 129–131. [CrossRef]

14. Anderson, R.E.; Lissandrello, P.M.; Maude, M.B.; Matthes, M.T. Lipids of bovine retinal pigment epithelium. *Exp. Eye Res.* **1976**, *23*, 149–157. [CrossRef]

15. Balla, T. Inositol-lipid binding motifs: Signal integrators through protein-lipid and protein-protein interactions. *J. Cell Sci.* **2005**, *118*, 2093–2104. [CrossRef] [PubMed]

16. Choi, S.; Thapa, N.; Tan, X.; Hedman, A.C.; Anderson, R.A. PIP kinases define PI4,5P(2)signaling specificity by association with effectors. *Biochim. Biophys. Acta* **2015**, *1851*, 711–723. [CrossRef]

17. Lemmon, M.A. Membrane recognition by phospholipid-binding domains. *Nat. Rev. Mol. Cell Biol.* **2008**, *9*, 99–111. [CrossRef]

18. Anderson, R.E.; Maude, M.B.; Kelleher, P.A. Metabolism of phosphatidylinositol in the frog retina. *Biochim. Biophys. Acta* **1980**, *620*, 236–246. [CrossRef]

19. Choe, H.G.; Ghalayini, A.J.; Anderson, R.E. Phosphoinositide metabolism in frog rod outer segments. *Exp. Eye Res.* **1990**, *51*, 167–176. [CrossRef]

20. Gehm, B.D.; Mc Connell, D.G. Phosphoinositide synthesis in bovine rod outer segments. *Biochemistry* **1990**, *29*, 5442–5446. [CrossRef]

21. Rodriguez de Turco, E.B.; Gordon, W.C.; Bazan, N.G. Light stimulates in vivo inositol lipid turnover in frog retinal pigment epithelial cells at the onset of shedding and phagocytosis of photoreceptor membranes. *Exp. Eye Res.* **1992**, *55*, 719–725. [CrossRef]

22. He, F.; Agosto, M.A.; Anastassov, I.A.; Tse, D.Y.; Wu, S.M.; Wensel, T.G. Phosphatidylinositol-3-phosphate is light-regulated and essential for survival in retinal rods. *Sci. Rep.* **2016**, *6*, 26978. [CrossRef]

23. Stephens, L.R.; Jackson, T.R.; Hawkins, P.T. Agonist-stimulated synthesis of phosphatidylinositol(3,4,5)-trisphosphate: A new intracellular signalling system? *Biochim. Biophys. Acta* **1993**, *1179*, 27–75. [CrossRef]

24. Bui, H.H.; Sanders, P.E.; Bodenmiller, D.; Kuo, M.S.; Donoho, G.P.; Fischl, A.S. Direct analysis of PI(3,4,5)P3 using liquid chromatography electrospray ionization tandem mass spectrometry. *Anal. Biochem.* **2018**, *547*, 66–76. [CrossRef]

25. Kadamur, G.; Ross, E.M. Mammalian phospholipase C. *Annu. Rev. Physiol.* **2013**, *75*, 127–154. [CrossRef]

26. Trebak, M.; Lemonnier, L.; Smyth, J.T.; Vazquez, G.; Putney, J.W., Jr. Phospholipase C-coupled receptors and activation of TRPC channels. *Handb Exp. Pharmacol.* **2007**, 593–614.

27. Wang, L.; Budolfson, K.; Wang, F. Pik3c3 deletion in pyramidal neurons results in loss of synapses, extensive gliosis and progressive neurodegeneration. *Neuroscience* **2011**, *172*, 427–442. [CrossRef]

28. Zhou, X.; Wang, L.; Hasegawa, H.; Amin, P.; Han, B.X.; Kaneko, S.; He, Y.; Wang, F. Deletion of PIK3C3/Vps34 in sensory neurons causes rapid neurodegeneration by disrupting the endosomal but not the autophagic pathway. *Proc. Natl. Acad. Sci. USA* **2010**, *107*, 9424–9429. [CrossRef]

29. Carnero, A.; Blanco-Aparicio, C.; Renner, O.; Link, W.; Leal, J.F. The PTEN/PI3K/AKT signalling pathway in cancer, therapeutic implications. *Curr. Cancer Drug Targets* **2008**, *8*, 187–198. [CrossRef]

30. Carpentier, S.; N'Kuli, F.; Grieco, G.; Van Der Smissen, P.; Janssens, V.; Emonard, H.; Bilanges, B.; Vanhaesebroeck, B.; Gaide Chevronnay, H.P.; Pierreux, C.E.; et al. Class III phosphoinositide 3-kinase/VPS34 and dynamin are critical for apical endocytic recycling. *Traffic* **2013**, *14*, 933–948. [CrossRef]

31. Dall'Armi, C.; Devereaux, K.A.; Di Paolo, G. The role of lipids in the control of autophagy. *Curr. Biol.* **2013**, *23*, R33–R45. [CrossRef]

32. Lenoir, M.; Overduin, M. PtdIns(4)P signalling and recognition systems. *Adv. Exp. Med. Biol.* **2013**, *991*, 59–83. [CrossRef]

33. McLaughlin, S.; Wang, J.; Gambhir, A.; Murray, D. PIP(2) and proteins: Interactions, organization, and information flow. *Annu. Rev. Biophys. Biomol. Struct.* **2002**, *31*, 151–175. [CrossRef]

34. Falkenburger, B.H.; Jensen, J.B.; Dickson, E.J.; Suh, B.C.; Hille, B. Phosphoinositides: Lipid regulators of membrane proteins. *J. Physiol.* **2010**, *588*, 3179–3185. [CrossRef]

35. Hilgemann, D.W.; Feng, S.; Nasuhoglu, C. The complex and intriguing lives of PIP2 with ion channels and transporters. *Sci. STKE* **2001**, *2001*, re19. [CrossRef]

36. Fukami, K.; Inanobe, S.; Kanemaru, K.; Nakamura, Y. Phospholipase C is a key enzyme regulating intracellular calcium and modulating the phosphoinositide balance. *Prog. Lipid Res.* **2010**, *49*, 429–437. [CrossRef]

37. Vines, C.M. Phospholipase C. *Adv. Exp. Med. Biol.* **2012**, *740*, 235–254. [CrossRef]

38. Shewan, A.; Eastburn, D.J.; Mostov, K. Phosphoinositides in cell architecture. *Cold Spring Harb. Perspect. Biol.* **2011**, *3*, a004796. [CrossRef]

39. Yin, H.L.; Janmey, P.A. Phosphoinositide regulation of the actin cytoskeleton. *Annu. Rev. Physiol.* **2003**, *65*, 761–789. [CrossRef]

40. Hille, B.; Dickson, E.J.; Kruse, M.; Vivas, O.; Suh, B.C. Phosphoinositides regulate ion channels. *Biochim. Biophys. Acta* **2015**, *1851*, 844–856. [CrossRef]

41. Bielas, S.L.; Silhavy, J.L.; Brancati, F.; Kisseleva, M.V.; Al-Gazali, L.; Sztriha, L.; Bayoumi, R.A.; Zaki, M.S.; Abdel-Aleem, A.; Rosti, R.O.; et al. Mutations in INPP5E, encoding inositol polyphosphate-5-phosphatase E, link phosphatidyl inositol signaling to the ciliopathies. *Nat. Genet.* **2009**, *41*, 1032–1036. [CrossRef]

42. Jacoby, M.; Cox, J.J.; Gayral, S.; Hampshire, D.J.; Ayub, M.; Blockmans, M.; Pernot, E.; Kisseleva, M.V.; Compere, P.; Schiffmann, S.N.; et al. INPP5E mutations cause primary cilium signaling defects, ciliary instability and ciliopathies in human and mouse. *Nat. Genet.* **2009**, *41*, 1027–1031. [CrossRef]

43. Saar, K.; Al-Gazali, L.; Sztriha, L.; Rueschendorf, F.; Nur, E.K.M.; Reis, A.; Bayoumi, R. Homozygosity mapping in families with Joubert syndrome identifies a locus on chromosome 9q34.3 and evidence for genetic heterogeneity. *Am. J. Hum. Genet.* **1999**, *65*, 1666–1671. [CrossRef] [PubMed]

44. George, A.A.; Hayden, S.; Holzhausen, L.C.; Ma, E.Y.; Suzuki, S.C.; Brockerhoff, S.E. Synaptojanin 1 is required for endolysosomal trafficking of synaptic proteins in cone photoreceptor inner segments. *PLoS ONE* **2014**, *9*, e84394. [CrossRef] [PubMed]

45. Holzhausen, L.C.; Lewis, A.A.; Cheong, K.K.; Brockerhoff, S.E. Differential role for synaptojanin 1 in rod and cone photoreceptors. *J. Comp. Neurol.* **2009**, *517*, 633–644. [CrossRef]

46. Van Epps, H.A.; Hayashi, M.; Lucast, L.; Stearns, G.W.; Hurley, J.B.; De Camilli, P.; Brockerhoff, S.E. The zebrafish nrc mutant reveals a role for the polyphosphoinositide phosphatase synaptojanin 1 in cone photoreceptor ribbon anchoring. *J. Neurosci.* **2004**, *24*, 8641–8650. [CrossRef]

47. Hagstrom, S.A.; North, M.A.; Nishina, P.L.; Berson, E.L.; Dryja, T.P. Recessive mutations in the gene encoding the tubby-like protein TULP1 in patients with retinitis pigmentosa. *Nat. Genet.* **1998**, *18*, 174–176. [CrossRef]

48. Jacobson, S.G.; Cideciyan, A.V.; Huang, W.C.; Sumaroka, A.; Roman, A.J.; Schwartz, S.B.; Luo, X.; Sheplock, R.; Dauber, J.M.; Swider, M.; et al. TULP1 mutations causing early-onset retinal degeneration: Preserved but insensitive macular cones. *Invest. Ophthalmol. Vis. Sci.* **2014**, *55*, 5354–5364. [CrossRef]

49. He, F.; Nichols, R.M.; Kailasam, L.; Wensel, T.G.; Agosto, M.A. Critical Role for Phosphatidylinositol-3 Kinase Vps34/PIK3C3 in ON-Bipolar Cells. *Invest. Ophthalmol. Vis. Sci.* **2019**, *60*, 2861–2874. [CrossRef]

50. Hammond, G.R.; Hong, Y. Phosphoinositides and Membrane Targeting in Cell Polarity. *Cold Spring Harb. Perspect. Biol.* **2018**, *10*. [CrossRef]

51. Mayinger, P. Phosphoinositides and vesicular membrane traffic. *Biochim. Biophys. Acta* **2012**, *1821*, 1104–1113. [CrossRef] [PubMed]

52. Chuang, J.Z.; Zhao, Y.; Sung, C.H. SARA-regulated vesicular targeting underlies formation of the light-sensing organelle in mammalian rods. *Cell* **2007**, *130*, 535–547. [CrossRef] [PubMed]

53. Deretic, D.; Traverso, V.; Parkins, N.; Jackson, F.; Rodriguez de Turco, E.B.; Ransom, N. Phosphoinositides, ezrin/moesin, and rac1 regulate fusion of rhodopsin transport carriers in retinal photoreceptors. *Mol. Biol. Cell* **2004**, *15*, 359–370. [CrossRef] [PubMed]

54. Spencer, W.J.; Lewis, T.R.; Phan, S.; Cady, M.A.; Serebrovskaya, E.O.; Schneider, N.F.; Kim, K.Y.; Cameron, L.A.; Skiba, N.P.; Ellisman, M.H.; et al. Photoreceptor disc membranes are formed through an Arp2/3-dependent lamellipodium-like mechanism. *Proc. Natl. Acad. Sci. USA* **2019**. [CrossRef]

55. Bucki, R.; Wang, Y.H.; Yang, C.; Kandy, S.K.; Fatunmbi, O.; Bradley, R.; Pogoda, K.; Svitkina, T.; Radhakrishnan, R.; Janmey, P.A. Lateral distribution of phosphatidylinositol 4,5-bisphosphate in membranes regulates formin- and ARP2/3-mediated actin nucleation. *J. Biol. Chem.* **2019**, *294*, 4704–4722. [CrossRef]

56. Daste, F.; Walrant, A.; Holst, M.R.; Gadsby, J.R.; Mason, J.; Lee, J.E.; Brook, D.; Mettlen, M.; Larsson, E.; Lee, S.F.; et al. Control of actin polymerization via the coincidence of phosphoinositides and high membrane curvature. *J. Cell Biol.* **2017**, *216*, 3745–3765. [CrossRef]

57. Ivanovic, I.; Anderson, R.E.; Le, Y.Z.; Fliesler, S.J.; Sherry, D.M.; Rajala, R.V. Deletion of the p85alpha regulatory subunit of phosphoinositide 3-kinase in cone photoreceptor cells results in cone photoreceptor degeneration. *Invest. Ophthalmol. Vis. Sci.* **2011**, *52*, 3775–3783. [CrossRef]

58. Xu, W.; Jin, M.; Huang, W.; Wang, H.; Hu, R.; Li, J.; Cao, Y. Apical PtdIns(4,5)P2 is required for ciliogenesis and suppression of polycystic kidney disease. *FASEB J.* **2019**, *33*, 2848–2857. [CrossRef]

59. Phua, S.C.; Nihongaki, Y.; Inoue, T. Autonomy declared by primary cilia through compartmentalization of membrane phosphoinositides. *Curr. Opin. Cell Biol.* **2018**, *50*, 72–78. [CrossRef]

60. Park, J.; Lee, N.; Kavoussi, A.; Seo, J.T.; Kim, C.H.; Moon, S.J. Ciliary Phosphoinositide Regulates Ciliary Protein Trafficking in Drosophila. *Cell Rep.* **2015**, *13*, 2808–2816. [CrossRef]

61. Garcia-Gonzalo, F.R.; Phua, S.C.; Roberson, E.C.; Garcia, G., 3rd; Abedin, M.; Schurmans, S.; Inoue, T.; Reiter, J.F. Phosphoinositides Regulate Ciliary Protein Trafficking to Modulate Hedgehog Signaling. *Dev. Cell* **2015**, *34*, 400–409. [CrossRef] [PubMed]

62. Vieira, O.V.; Gaus, K.; Verkade, P.; Fullekrug, J.; Vaz, W.L.; Simons, K. FAPP2, cilium formation, and compartmentalization of the apical membrane in polarized Madin-Darby canine kidney (MDCK) cells. *Proc. Natl. Acad. Sci. USA* **2006**, *103*, 18556–18561. [CrossRef] [PubMed]

63. Nachury, M.V.; Seeley, E.S.; Jin, H. Trafficking to the ciliary membrane: How to get across the periciliary diffusion barrier? *Annu. Rev. Cell Dev. Biol.* **2010**, *26*, 59–87. [CrossRef] [PubMed]

64. Luo, N.; West, C.C.; Murga-Zamalloa, C.A.; Sun, L.; Anderson, R.M.; Wells, C.D.; Weinreb, R.N.; Travers, J.B.; Khanna, H.; Sun, Y. OCRL localizes to the primary cilium: A new role for cilia in Lowe syndrome. *Hum. Mol. Genet.* **2012**, *21*, 3333–3344. [CrossRef] [PubMed]

65. Coon, B.G.; Hernandez, V.; Madhivanan, K.; Mukherjee, D.; Hanna, C.B.; Barinaga-Rementeria Ramirez, I.; Lowe, M.; Beales, P.L.; Aguilar, R.C. The Lowe syndrome protein OCRL1 is involved in primary cilia assembly. *Hum. Mol. Genet.* **2012**, *21*, 1835–1847. [CrossRef] [PubMed]

66. Humbert, M.C.; Weihbrecht, K.; Searby, C.C.; Li, Y.; Pope, R.M.; Sheffield, V.C.; Seo, S. ARL13B, PDE6D, and CEP164 form a functional network for INPP5E ciliary targeting. *Proc. Natl. Acad. Sci. USA* **2012**, *109*, 19691–19696. [CrossRef] [PubMed]

67. Jin, H.; White, S.R.; Shida, T.; Schulz, S.; Aguiar, M.; Gygi, S.P.; Bazan, J.F.; Nachury, M.V. The conserved Bardet-Biedl syndrome proteins assemble a coat that traffics membrane proteins to cilia. *Cell* **2010**, *141*, 1208–1219. [CrossRef]

68. Nachury, M.V.; Loktev, A.V.; Zhang, Q.; Westlake, C.J.; Peranen, J.; Merdes, A.; Slusarski, D.C.; Scheller, R.H.; Bazan, J.F.; Sheffield, V.C.; et al. A core complex of BBS proteins cooperates with the GTPase Rab8 to promote ciliary membrane biogenesis. *Cell* **2007**, *129*, 1201–1213. [CrossRef]

69. Chinskey, N.D.; Zheng, Q.D.; Zacks, D.N. Control of Photoreceptor Autophagy after Retinal Detachment: The Switch from Survival to Death. *Invest. Ophthalmol. Vis. Sci.* **2014**, *55*, 688–695. [CrossRef]

70. Reme, C.; Drinker, C.K.; Aeberhard, B. Modification of autophagic degradation by medium- and illumination conditions in frog visual cells in vitro. *Doc. Ophthalmol.* **1984**, *56*, 377–383. [CrossRef]

71. Reme, C. Autophagy in rods and cones of the vertebrate retina. *Dev. Ophthalmol.* **1981**, *4*, 101–148.

72. Reme, C.E.; Sulser, M. Diurnal variation of autophagy in rod visual cells in the rat. *Albrecht Von Graefes Arch. Klin. Exp. Ophthalmol.* **1977**, *203*, 261–270. [CrossRef]

73. Chen, Y.; Sawada, O.; Kohno, H.; Le, Y.Z.; Subauste, C.; Maeda, T.; Maeda, A. Autophagy protects the retina from light-induced degeneration. *J. Biol. Chem.* **2013**, *288*, 7506–7518. [CrossRef] [PubMed]

74. Besirli, C.G.; Chinskey, N.D.; Zheng, Q.D.; Zacks, D.N. Autophagy activation in the injured photoreceptor inhibits fas-mediated apoptosis. *Invest. Ophthalmol. Vis. Sci.* **2011**, *52*, 4193–4199. [CrossRef] [PubMed]

75. George, A.A.; Hayden, S.; Stanton, G.R.; Brockerhoff, S.E. Arf6 and the 5′phosphatase of synaptojanin 1 regulate autophagy in cone photoreceptors. *Bioessays* **2016**, *38* (Suppl. 1), S119–S135. [CrossRef]

76. Ghalayini, A.; Anderson, R.E. Phosphatidylinositol 4,5-bisphosphate: Light-mediated breakdown in the vertebrate retina. *Biochem. Biophys. Res. Commun.* **1984**, *124*, 503–506. [CrossRef]

77. Anderson, R.E.; Maude, M.B.; Pu, G.A.; Hollyfield, J.G. Effect of light on the metabolism of lipids in the rat retina. *J. Neurochem.* **1985**, *44*, 773–778. [CrossRef] [PubMed]

78. Tarver, A.P.; Anderson, R.E. Phospholipase C activity and substrate specificity in frog photoreceptors. *Exp. Eye Res.* **1988**, *46*, 29–35. [CrossRef]

79. Ghalayini, A.J.; Tarver, A.P.; Mackin, W.M.; Koutz, C.A.; Anderson, R.E. Identification and immunolocalization of phospholipase C in bovine rod outer segments. *J. Neurochem.* **1991**, *57*, 1405–1412. [CrossRef]

80. Das, N.D.; Yoshioka, T.; Samuelson, D.; Shichi, H. Immunocytochemical localization of phosphatidylinositol-4,5-bisphosphate in dark- and light-adapted rat retinas. *Cell Struct. Funct.* **1986**, *11*, 53–63. [CrossRef]

81. Das, N.D.; Yoshioka, T.; Samuelson, D.; Cohen, R.J.; Shichi, H. Immunochemical evidence for the light-regulated modulation of phosphatidylinositol-4,5-bisphosphate in rat photoreceptor cells. *Cell Struct. Funct.* **1987**, *12*, 471–481. [CrossRef] [PubMed]

82. Panfoli, I.; Morelli, A.; Pepe, I. Calcium ion-regulated phospholipase C activity in bovine rod outer segments. *Biochem. Biophys. Res. Commun.* **1990**, *173*, 283–288. [CrossRef]

83. Grigorjev, I.V.; Grits, A.I.; Artamonov, I.D.; Baranova, L.A.; Volotovski, I.D. betagamma-Transducin stimulates hydrolysis and synthesis of phosphatidylinositol 4,5-bisphosphate in bovine rod outer segment membranes. *Biochim. Biophys. Acta* **1996**, *1310*, 131–136. [CrossRef]

84. Van Rooijen, L.A.; Bazan, N.G. The inositide cycle in bovine photoreceptor membranes. *Life Sci.* **1986**, *38*, 1685–1693. [CrossRef]

85. Millar, F.A.; Fisher, S.C.; Muir, C.A.; Edwards, E.; Hawthorne, J.N. Polyphosphoinositide hydrolysis in response to light stimulation of rat and chick retina and retinal rod outer segments. *Biochim. Biophys. Acta* **1988**, *970*, 205–211. [CrossRef]

86. He, F.; Mao, M.; Wensel, T.G. Enhancement of phototransduction g protein-effector interactions by phosphoinositides. *J. Biol. Chem.* **2004**, *279*, 8986–8990. [CrossRef]

87. Womack, K.B.; Gordon, S.E.; He, F.; Wensel, T.G.; Lu, C.C.; Hilgemann, D.W. Do phosphatidylinositides modulate vertebrate phototransduction? *J. Neurosci.* **2000**, *20*, 2792–2799. [CrossRef]

88. Gross, O.P.; Pugh, E.N., Jr.; Burns, M.E. Spatiotemporal cGMP dynamics in living mouse rods. *Biophys J.* **2012**, *102*, 1775–1784. [CrossRef]

89. Peng, Y.W.; Rhee, S.G.; Yu, W.P.; Ho, Y.K.; Schoen, T.; Chader, G.J.; Yau, K.W. Identification of components of a phosphoinositide signaling pathway in retinal rod outer segments. *Proc. Natl. Acad. Sci. USA* **1997**, *94*, 1995–2000. [CrossRef]

90. Orisme, W.; Li, J.; Goldmann, T.; Bolch, S.; Wolfrum, U.; Smith, W.C. Light-dependent translocation of arrestin in rod photoreceptors is signaled through a phospholipase C cascade and requires ATP. *Cell Signal.* **2010**, *22*, 447–456. [CrossRef]

91. Guo, X.; Ghalayini, A.J.; Chen, H.; Anderson, R.E. Phosphatidylinositol 3-kinase in bovine photoreceptor rod outer segments. *Invest. Ophthalmol. Vis. Sci.* **1997**, *38*, 1873–1882. [PubMed]

92. Guo, X.X.; Huang, Z.; Bell, M.W.; Chen, H.; Anderson, R.E. Tyrosine phosphorylation is involved in phosphatidylinositol 3-kinase activation in bovine rod outer segments. *Mol. Vis.* **2000**, *6*, 216–221. [PubMed]

93. Rajala, R.V.; McClellan, M.E.; Chan, M.D.; Tsiokas, L.; Anderson, R.E. Interaction of the retinal insulin receptor beta-subunit with the p85 subunit of phosphoinositide 3-kinase. *Biochemistry* **2004**, *43*, 5637–5650. [CrossRef] [PubMed]

94. Rajala, R.V.; Anderson, R.E. Light regulation of the insulin receptor in the retina. *Mol. Neurobiol.* **2003**, *28*, 123–138. [CrossRef]

95. Rajala, R.V.; McClellan, M.E.; Ash, J.D.; Anderson, R.E. In vivo regulation of phosphoinositide 3-kinase in retina through light-induced tyrosine phosphorylation of the insulin receptor beta-subunit. *J. Biol. Chem.* **2002**, *277*, 43319–43326. [CrossRef] [PubMed]

96. Rajala, R.V.; Ranjo-Bishop, M.; Wang, Y.; Rajala, A.; Anderson, R.E. The p110alpha isoform of phosphoinositide 3-kinase is essential for cone photoreceptor survival. *Biochimie* **2015**, *112*, 35–40. [CrossRef] [PubMed]

97. Ivanovic, I.; Allen, D.T.; Dighe, R.; Le, Y.Z.; Anderson, R.E.; Rajala, R.V. Phosphoinositide 3-kinase signaling in retinal rod photoreceptors. *Invest. Ophthalmol. Vis. Sci.* **2011**, *52*, 6355–6362. [CrossRef]

98. Detwiler, P.B. Phototransduction in Retinal Ganglion Cells. *Yale J. Biol. Med.* **2018**, *91*, 49–52.

99. Graham, D.M.; Wong, K.Y.; Shapiro, P.; Frederick, C.; Pattabiraman, K.; Berson, D.M. Melanopsin ganglion cells use a membrane-associated rhabdomeric phototransduction cascade. *J. Neurophysiol.* **2008**, *99*, 2522–2532. [CrossRef]

100. Hardie, R.C. Photosensitive TRPs. *Handb. Exp. Pharmacol.* **2014**, *223*, 795–826. [CrossRef]

101. Isoldi, M.C.; Rollag, M.D.; Castrucci, A.M.; Provencio, I. Rhabdomeric phototransduction initiated by the vertebrate photopigment melanopsin. *Proc. Natl. Acad. Sci. USA* **2005**, *102*, 1217–1221. [CrossRef] [PubMed]

102. Jiang, Z.; Yue, W.W.S.; Chen, L.; Sheng, Y.; Yau, K.W. Cyclic-Nucleotide- and HCN-Channel-Mediated Phototransduction in Intrinsically Photosensitive Retinal Ganglion Cells. *Cell* **2018**, *175*, 652–664 e612. [CrossRef]

103. Montell, C. Drosophila visual transduction. *Trends Neurosci.* **2012**, *35*, 356–363. [CrossRef] [PubMed]

104. Xue, T.; Do, M.T.; Riccio, A.; Jiang, Z.; Hsieh, J.; Wang, H.C.; Merbs, S.L.; Welsbie, D.S.; Yoshioka, T.; Weissgerber, P.; et al. Melanopsin signalling in mammalian iris and retina. *Nature* **2011**, *479*, 67–73. [CrossRef] [PubMed]

105. Arendt, D. Evolution of eyes and photoreceptor cell types. *Int. J. Dev. Biol.* **2003**, *47*, 563–571. [PubMed]

106. Kuriyama, S.; Ohuchi, T.; Yoshimura, N.; Honda, Y. Growth factor-induced cytosolic calcium ion transients in cultured human retinal pigment epithelial cells. *Invest. Ophthalmol. Vis. Sci.* **1991**, *32*, 2882–2890.
107. Osborne, N.N.; FitzGibbon, F.; Schwartz, G. Muscarinic acetylcholine receptor-mediated phosphoinositide turnover in cultured human retinal pigment epithelium cells. *Vision Res.* **1991**, *31*, 1119–1127. [CrossRef]
108. Feldman, E.L.; Randolph, A.E.; Johnston, G.C.; DelMonte, M.A.; Greene, D.A. Receptor-coupled phosphoinositide hydrolysis in human retinal pigment epithelium. *J. Neurochem.* **1991**, *56*, 2094–2100. [CrossRef]
109. York, N.; Halbach, P.; Chiu, M.A.; Bird, I.M.; Pillers, D.M.; Pattnaik, B.R. Oxytocin (OXT)-stimulated inhibition of Kir7.1 activity is through PIP2-dependent Ca(2+) response of the oxytocin receptor in the retinal pigment epithelium in vitro. *Cell Signal.* **2017**, *37*, 93–102. [CrossRef]
110. Heth, C.A.; Marescalchi, P.A. Inositol triphosphate generation in cultured rat retinal pigment epithelium. *Invest. Ophthalmol. Vis. Sci.* **1994**, *35*, 409–416.
111. D'Cruz, P.M.; Yasumura, D.; Weir, J.; Matthes, M.T.; Abderrahim, H.; LaVail, M.M.; Vollrath, D. Mutation of the receptor tyrosine kinase gene Mertk in the retinal dystrophic RCS rat. *Hum. Mol. Genet.* **2000**, *9*, 645–651. [CrossRef] [PubMed]
112. Treins, C.; Giorgetti-Peraldi, S.; Murdaca, J.; Semenza, G.L.; Van Obberghen, E. Insulin stimulates hypoxia-inducible factor 1 through a phosphatidylinositol 3-kinase/target of rapamycin-dependent signaling pathway. *J. Biol. Chem.* **2002**, *277*, 27975–27981. [CrossRef] [PubMed]
113. Geraldes, P.; Yagi, K.; Ohshiro, Y.; He, Z.; Maeno, Y.; Yamamoto-Hiraoka, J.; Rask-Madsen, C.; Chung, S.W.; Perrella, M.A.; King, G.L. Selective regulation of heme oxygenase-1 expression and function by insulin through IRS1/phosphoinositide 3-kinase/Akt-2 pathway. *J. Biol. Chem.* **2008**, *283*, 34327–34336. [CrossRef]
114. Mwaikambo, B.R.; Yang, C.; Chemtob, S.; Hardy, P. Hypoxia up-regulates CD36 expression and function via hypoxia-inducible factor-1- and phosphatidylinositol 3-kinase-dependent mechanisms. *J. Biol. Chem.* **2009**, *284*, 26695–26707. [CrossRef]
115. Yang, X.M.; Wang, Y.S.; Zhang, J.; Li, Y.; Xu, J.F.; Zhu, J.; Zhao, W.; Chu, D.K.; Wiedemann, P. Role of PI3K/Akt and MEK/ERK in mediating hypoxia-induced expression of HIF-1alpha and VEGF in laser-induced rat choroidal neovascularization. *Invest. Ophthalmol. Vis. Sci.* **2009**, *50*, 1873–1879. [CrossRef]
116. Kim, D.I.; Lim, S.K.; Park, M.J.; Han, H.J.; Kim, G.Y.; Park, S.H. The involvement of phosphatidylinositol 3-kinase/Akt signaling in high glucose-induced downregulation of GLUT-1 expression in ARPE cells. *Life Sci.* **2007**, *80*, 626–632. [CrossRef]
117. Qin, D.; Zhang, G.M.; Xu, X.; Wang, L.Y. The PI3K/Akt signaling pathway mediates the high glucose-induced expression of extracellular matrix molecules in human retinal pigment epithelial cells. *J. Diabetes Res.* **2015**, *2015*, 920280. [CrossRef]
118. Ferguson, T.A.; Green, D.R. Autophagy and phagocytosis converge for better vision. *Autophagy* **2014**, *10*, 165–167. [CrossRef]
119. Kim, J.Y.; Zhao, H.; Martinez, J.; Doggett, T.A.; Kolesnikov, A.V.; Tang, P.H.; Ablonczy, Z.; Chan, C.C.; Zhou, Z.; Green, D.R.; et al. Noncanonical autophagy promotes the visual cycle. *Cell* **2013**, *154*, 365–376. [CrossRef]
120. Muniz-Feliciano, L.; Doggett, T.A.; Zhou, Z.; Ferguson, T.A. RUBCN/rubicon and EGFR regulate lysosomal degradative processes in the retinal pigment epithelium (RPE) of the eye. *Autophagy* **2017**, *13*, 2072–2085. [CrossRef]
121. Shaw, J.D.; Hama, H.; Sohrabi, F.; DeWald, D.B.; Wendland, B. PtdIns(3,5)P2 is required for delivery of endocytic cargo into the multivesicular body. *Traffic* **2003**, *4*, 479–490. [CrossRef]
122. Ketel, K.; Krauss, M.; Nicot, A.S.; Puchkov, D.; Wieffer, M.; Muller, R.; Subramanian, D.; Schultz, C.; Laporte, J.; Haucke, V. A phosphoinositide conversion mechanism for exit from endosomes. *Nature* **2016**, *529*, 408–412. [CrossRef]
123. Isobe, Y.; Nigorikawa, K.; Tsurumi, G.; Takemasu, S.; Takasuga, S.; Kofuji, S.; Hazeki, K. PIKfyve accelerates phagosome acidification through activation of TRPML1 while arrests aberrant vacuolation independent of the Ca2+ channel. *J. Biochem.* **2019**, *165*, 75–84. [CrossRef]
124. Ikonomov, O.C.; Sbrissa, D.; Shisheva, A. Localized PtdIns 3,5-P2 synthesis to regulate early endosome dynamics and fusion. *Am. J. Physiol. Cell Physiol.* **2006**, *291*, C393–C404. [CrossRef]

125. Dong, X.P.; Shen, D.; Wang, X.; Dawson, T.; Li, X.; Zhang, Q.; Cheng, X.; Zhang, Y.; Weisman, L.S.; Delling, M.; et al. PI(3,5)P(2) controls membrane trafficking by direct activation of mucolipin Ca(2+) release channels in the endolysosome. *Nat. Commun.* **2010**, *1*, 38. [CrossRef]

126. Sasaki, T.; Takasuga, S.; Sasaki, J.; Kofuji, S.; Eguchi, S.; Yamazaki, M.; Suzuki, A. Mammalian phosphoinositide kinases and phosphatases. *Prog. Lipid Res.* **2009**, *48*, 307–343. [CrossRef]

127. Brown, J.R.; Auger, K.R. Phylogenomics of phosphoinositide lipid kinases: Perspectives on the evolution of second messenger signaling and drug discovery. *BMC Evol. Biol.* **2011**, *11*, 4. [CrossRef]

128. Azadi, S.; Brush, R.S.; Anderson, R.E.; Rajala, R.V. Class I Phosphoinositide 3-Kinase Exerts a Differential Role on Cell Survival and Cell Trafficking in Retina. *Adv. Exp. Med. Biol.* **2016**, *854*, 363–369. [CrossRef]

129. Sakagami, H.; Katsumata, O.; Hara, Y.; Tamaki, H.; Fukaya, M. Preferential localization of type I phosphatidylinositol 4-phosphate 5-kinase gamma at the periactive zone of mouse photoreceptor ribbon synapses. *Brain Res.* **2014**, *1586*, 23–33. [CrossRef]

130. Wenk, M.R.; Pellegrini, L.; Klenchin, V.A.; Di Paolo, G.; Chang, S.; Daniell, L.; Arioka, M.; Martin, T.F.; De Camilli, P. PIP kinase Igamma is the major PI(4,5)P(2) synthesizing enzyme at the synapse. *Neuron* **2001**, *32*, 79–88. [CrossRef]

131. Wright, B.D.; Loo, L.; Street, S.E.; Ma, A.; Taylor-Blake, B.; Stashko, M.A.; Jin, J.; Janzen, W.P.; Frye, S.V.; Zylka, M.J. The lipid kinase PIP5K1C regulates pain signaling and sensitization. *Neuron* **2014**, *82*, 836–847. [CrossRef]

132. Di Paolo, G.; Moskowitz, H.S.; Gipson, K.; Wenk, M.R.; Voronov, S.; Obayashi, M.; Flavell, R.; Fitzsimonds, R.M.; Ryan, T.A.; De Camilli, P. Impaired PtdIns(4,5)P2 synthesis in nerve terminals produces defects in synaptic vesicle trafficking. *Nature* **2004**, *431*, 415–422. [CrossRef]

133. Huang, Z.; Guo, X.X.; Chen, S.X.; Alvarez, K.M.; Bell, M.W.; Anderson, R.E. Regulation of type II phosphatidylinositol phosphate kinase by tyrosine phosphorylation in bovine rod outer segments. *Biochemistry* **2001**, *40*, 4550–4559. [CrossRef]

134. Zhao, L.; Chen, Y.; Bajaj, A.O.; Eblimit, A.; Xu, M.; Soens, Z.T.; Wang, F.; Ge, Z.; Jung, S.Y.; He, F.; et al. Integrative subcellular proteomic analysis allows accurate prediction of human disease-causing genes. *Genome Res.* **2016**, *26*, 660–669. [CrossRef]

135. Prosseda, P.P.; Luo, N.; Wang, B.; Alvarado, J.A.; Hu, Y.; Sun, Y. Loss of OCRL increases ciliary PI(4,5)P2 in Lowe oculocerebrorenal syndrome. *J. Cell Sci.* **2017**, *130*, 3447–3454. [CrossRef]

136. Attree, O.; Olivos, I.M.; Okabe, I.; Bailey, L.C.; Nelson, D.L.; Lewis, R.A.; McInnes, R.R.; Nussbaum, R.L. The Lowe's oculocerebrorenal syndrome gene encodes a protein highly homologous to inositol polyphosphate-5-phosphatase. *Nature* **1992**, *358*, 239–242. [CrossRef]

137. Shim, H.; Wu, C.; Ramsamooj, S.; Bosch, K.N.; Chen, Z.; Emerling, B.M.; Yun, J.; Liu, H.; Choo-Wing, R.; Yang, Z.; et al. Deletion of the gene Pip4k2c, a novel phosphatidylinositol kinase, results in hyperactivation of the immune system. *Proc. Natl. Acad. Sci. USA* **2016**, *113*, 7596–7601. [CrossRef]

138. Young, R.W. The renewal of rod and cone outer segments in the rhesus monkey. *J. Cell Biol.* **1971**, *49*, 303–318. [CrossRef]

139. Adams, N.A.; Awadein, A.; Toma, H.S. The retinal ciliopathies. *Ophthalmic Genet.* **2007**, *28*, 113–125. [CrossRef]

140. Wheway, G.; Parry, D.A.; Johnson, C.A. The role of primary cilia in the development and disease of the retina. *Organogenesis* **2013**, *10*, 69–85. [CrossRef]

141. Gal, A.; Li, Y.; Thompson, D.A.; Weir, J.; Orth, U.; Jacobson, S.G.; Apfelstedt-Sylla, E.; Vollrath, D. Mutations in MERTK, the human orthologue of the RCS rat retinal dystrophy gene, cause retinitis pigmentosa. *Nat. Genet.* **2000**, *26*, 270–271. [CrossRef]

142. Marmorstein, A.D.; Johnson, A.A.; Bachman, L.A.; Andrews-Pfannkoch, C.; Knudsen, T.; Gilles, B.J.; Hill, M.; Gandhi, J.K.; Marmorstein, L.Y.; Pulido, J.S. Mutant Best1 Expression and Impaired Phagocytosis in an iPSC Model of Autosomal Recessive Bestrophinopathy. *Sci Rep.* **2018**, *8*, 4487. [CrossRef]

143. Strauss, O.; Reichhart, N.; Gomez, N.M.; Muller, C. Contribution of Ion Channels in Calcium Signaling Regulating Phagocytosis: MaxiK, Cav1.3 and Bestrophin-1. *Adv. Exp. Med. Biol.* **2016**, *854*, 739–744. [CrossRef]

144. Muller, C.; Mas Gomez, N.; Ruth, P.; Strauss, O. CaV1.3 L-type channels, maxiK Ca(2+)-dependent K(+) channels and bestrophin-1 regulate rhythmic photoreceptor outer segment phagocytosis by retinal pigment epithelial cells. *Cell Signal.* **2014**, *26*, 968–978. [CrossRef]

145. Xiao, Q.; Hartzell, H.C.; Yu, K. Bestrophins and retinopathies. *Pflugers Arch.* **2010**, *460*, 559–569. [CrossRef]

146. Strauss, O. The retinal pigment epithelium in visual function. *Physiol. Rev.* **2005**, *85*, 845–881. [CrossRef]

147. Luo, N.; Lu, J.; Sun, Y. Evidence of a role of inositol polyphosphate 5-phosphatase INPP5E in cilia formation in zebrafish. *Vision Res.* **2012**, *75*, 98–107. [CrossRef]

148. Travaglini, L.; Brancati, F.; Silhavy, J.; Iannicelli, M.; Nickerson, E.; Elkhartoufi, N.; Scott, E.; Spencer, E.; Gabriel, S.; Thomas, S.; et al. Phenotypic spectrum and prevalence of INPP5E mutations in Joubert syndrome and related disorders. *Eur. J. Hum. Genet.* **2013**, *21*, 1074–1078. [CrossRef]

149. Plotnikova, O.V.; Seo, S.; Cottle, D.L.; Conduit, S.; Hakim, S.; Dyson, J.M.; Mitchell, C.A.; Smyth, I.M. INPP5E interacts with AURKA, linking phosphoinositide signaling to primary cilium stability. *J. Cell Sci.* **2015**, *128*, 364–372. [CrossRef]

150. Xu, W.; Jin, M.; Hu, R.; Wang, H.; Zhang, F.; Yuan, S.; Cao, Y. The Joubert Syndrome Protein Inpp5e Controls Ciliogenesis by Regulating Phosphoinositides at the Apical Membrane. *J. Am. Soc. Nephrol.* **2017**, *28*, 118–129. [CrossRef]

151. Verstreken, P.; Koh, T.W.; Schulze, K.L.; Zhai, R.G.; Hiesinger, P.R.; Zhou, Y.; Mehta, S.Q.; Cao, Y.; Roos, J.; Bellen, H.J. Synaptojanin is recruited by endophilin to promote synaptic vesicle uncoating. *Neuron* **2003**, *40*, 733–748. [CrossRef]

152. Chi, Y.; Zhou, B.; Wang, W.Q.; Chung, S.K.; Kwon, Y.U.; Ahn, Y.H.; Chang, Y.T.; Tsujishita, Y.; Hurley, J.H.; Zhang, Z.Y. Comparative mechanistic and substrate specificity study of inositol polyphosphate 5-phosphatase Schizosaccharomyces pombe Synaptojanin and SHIP2. *J. Biol. Chem.* **2004**, *279*, 44987–44995. [CrossRef]

153. Chang-Ileto, B.; Frere, S.G.; Chan, R.B.; Voronov, S.V.; Roux, A.; Di Paolo, G. Synaptojanin 1-mediated PI(4,5)P2 hydrolysis is modulated by membrane curvature and facilitates membrane fission. *Dev. Cell* **2011**, *20*, 206–218. [CrossRef] [PubMed]

154. Lev, S.; Hernandez, J.; Martinez, R.; Chen, A.; Plowman, G.; Schlessinger, J. Identification of a novel family of targets of PYK2 related to Drosophila retinal degeneration B (rdgB) protein. *Mol. Cell Biol.* **1999**, *19*, 2278–2288. [CrossRef]

155. Tian, D.; Lev, S. Cellular and developmental distribution of human homologues of the Drosophilia rdgB protein in the rat retina. *Invest. Ophthalmol. Vis. Sci.* **2002**, *43*, 1946–1953.

156. Kohn, L.; Kadzhaev, K.; Burstedt, M.S.I.; Haraldsson, S.; Hallberg, B.; Sandgren, O.; Golovleva, I. Mutation in the PYK2-binding domain of PITPNM3 causes autosomal dominant cone dystrophy (CORD5) in two Swedish families. *Eur. J. Hum. Genet.* **2007**, *15*, 664–671. [CrossRef]

157. Kohn, L.; Kadzhaev, K.; Burstedt, M.S.I.; Haraldsson, S.; Sandgren, O.; Golovleva, I. Mutation in the PYK2-binding domain of PITPNM3 causes autosomal dominant cone dystrophy (CORD5) in two Swedish families. *Rencent Adv. Exp. Med. Biol.* **2008**, *613*, 229–234. [CrossRef]

158. Kohn, L.; Kohl, S.; Bowne, S.J.; Sullivan, L.S.; Kellner, U.; Daiger, S.P.; Sandgren, O.; Golovleva, I. PITPNM3 is an uncommon cause of cone and cone-rod dystrophies. *Ophthalmic Genet.* **2010**, *31*, 139–140. [CrossRef]

159. Bakhoum, M.F.; Sengillo, J.D.; Cui, X.; Tsang, S.H. Autoimmune retinopathy in a patient with a missense mutation in PITPNM3. *Retin Cases Brief. Rep.* **2018**, *12* (Suppl. 1), S72–S75. [CrossRef]

160. Lin, Z.; Li, W.; Zhang, H.; Wu, W.; Peng, Y.; Zeng, Y.; Wan, Y.; Wang, J.; Ouyang, N. CCL18/PITPNM3 enhances migration, invasion, and EMT through the NF-kappaB signaling pathway in hepatocellular carcinoma. *Tumour Biol.* **2016**, *37*, 3461–3468. [CrossRef]

161. Ile, K.E.; Kassen, S.; Cao, C.; Vihtehlic, T.; Shah, S.D.; Mousley, C.J.; Alb, J.G., Jr.; Huijbregts, R.P.; Stearns, G.W.; Brockerhoff, S.E.; et al. Zebrafish class 1 phosphatidylinositol transfer proteins: PITPbeta and double cone cell outer segment integrity in retina. *Traffic* **2010**, *11*, 1151–1167. [CrossRef] [PubMed]

162. Chang, B.; Hawes, N.L.; Hurd, R.E.; Davisson, M.T.; Nusinowitz, S.; Heckenlively, J.R. Retinal degeneration mutants in the mouse. *Vision Res.* **2002**, *42*, 517–525. [CrossRef]

163. Borman, A.D.; Pearce, L.R.; Mackay, D.S.; Nagel-Wolfrum, K.; Davidson, A.E.; Henderson, R.; Garg, S.; Waseem, N.H.; Webster, A.R.; Plagnol, V.; et al. A homozygous mutation in the TUB gene associated with retinal dystrophy and obesity. *Hum. Mutat.* **2014**, *35*, 289–293. [CrossRef] [PubMed]

164. Hagstrom, S.A.; Watson, R.F.; Pauer, G.J.; Grossman, G.H. Tulp1 is involved in specific photoreceptor protein transport pathways. *Adv. Exp. Med. Biol.* **2012**, *723*, 783–789. [CrossRef] [PubMed]

165. Xi, Q.; Pauer, G.J.; Traboulsi, E.I.; Hagstrom, S.A. Mutation screen of the TUB gene in patients with retinitis pigmentosa and Leber congenital amaurosis. *Exp. Eye Res.* **2006**, *83*, 569–573. [CrossRef] [PubMed]

166. Xi, Q.; Pauer, G.J.; West, K.A.; Crabb, J.W.; Hagstrom, S.A. Retinal degeneration caused by mutations in TULP1. *Adv. Exp. Med. Biol.* **2003**, *533*, 303–308. [CrossRef]
167. Mukhopadhyay, S.; Jackson, P.K. The tubby family proteins. *Genome Biol.* **2011**, *12*, 225. [CrossRef]
168. Hagstrom, S.A.; Duyao, M.; North, M.A.; Li, T. Retinal degeneration in tulp1-/- mice: Vesicular accumulation in the interphotoreceptor matrix. *Invest. Ophthalmol. Vis. Sci.* **1999**, *40*, 2795–2802.

Review

From Rust to Quantum Biology: The Role of Iron in Retina Physiopathology

Emilie Picard [1,*], Alejandra Daruich [1,2], Jenny Youale [1], Yves Courtois [1] and Francine Behar-Cohen [1,3]

[1] Centre de Recherche des Cordeliers, INSERM, Sorbonne Université, USPC, Université Paris Descartes, Team 17, F-75006 Paris, France; adaruich.matet@gmail.com (A.D.); youale.j@hotmail.fr (J.Y.); courtois.yves@numericable.com (Y.C.); francine.behar@gmail.com (F.B.-C.)

[2] Ophthalmology Department, Necker-Enfants Malades University Hospital, APHP, 75015 Paris, France

[3] Ophtalmopole, Cochin Hospital, AP-HP, Assistance Publique Hôpitaux de Paris, 24 rue du Faubourg Saint-Jacques, 75014 Paris, France

[*] Correspondence: picardemilie@gmail.com; Tel.: +331-44-27-81-82

Received: 17 February 2020; Accepted: 9 March 2020; Published: 13 March 2020

Abstract: Iron is essential for cell survival and function. It is a transition metal, that could change its oxidation state from Fe^{2+} to Fe^{3+} involving an electron transfer, the key of vital functions but also organ dysfunctions. The goal of this review is to illustrate the primordial role of iron and local iron homeostasis in retinal physiology and vision, as well as the pathological consequences of iron excess in animal models of retinal degeneration and in human retinal diseases. We summarize evidence of the potential therapeutic effect of iron chelation in retinal diseases and especially the interest of transferrin, a ubiquitous endogenous iron-binding protein, having the ability to treat or delay degenerative retinal diseases.

Keywords: iron; retina; transferrin

---

1. Introduction

Iron is a major element in biology. Besides its well-known role in prebiotic conditions after the rise of oxygen in the atmosphere, its insolubility led to the development of many mechanisms to allow the primitive cells and organisms to use it. They are driven by the transition of ferrous iron (Fe^{2+}) to ferric iron (Fe^{3+}) involving an electron which is particularly available and is the basis of vital functions and dysfunctions in the organs.

In this review, we analyze several of the well-known or recently discovered functions of iron in the eye, mainly in the retina, and the most promising approaches to regulate it and improve a large number of its negative side effects which can lead to vision impairment. We will focus on the main functions of transferrin (TF) as a partner in the systemic and cellular mechanisms that underlie the regulation of iron homeostasis and its disorders.

2. The Retina Structure and Retinal Oxygen Supply

The eye is a complex and confined organ formed by different compartments and structures essential for the transmission and focus of photons from the cornea to the photoreceptors (PRs), which convert them into an electrical signal transmitted to the brain. The neural retina comprises the PRs, cones, rods, the interneurons, the ganglion cells, the glial cells such as retinal Müller's glial cells (MGC), astrocytes, and microglia (Figure 1). The retina is vascularized by two separate vascular systems, the retinal vessels, branches of the central retinal artery that vascularize the inner retinal layers, and the choroidal vessels, branches of the ciliary arteries that supply the avascular PR layer through the retinal pigment epithelium (RPE) cells. In primates and human, visual acuity, photopic vision, and color vision are ensured by the macula, a highly specialized retinal area that comprises less than 5% of the total retinal surface, located at the center of the visual axis. The center of the macula, the fovea, is devoid of retinal vessels and composed exclusively of cones and MGC cells.

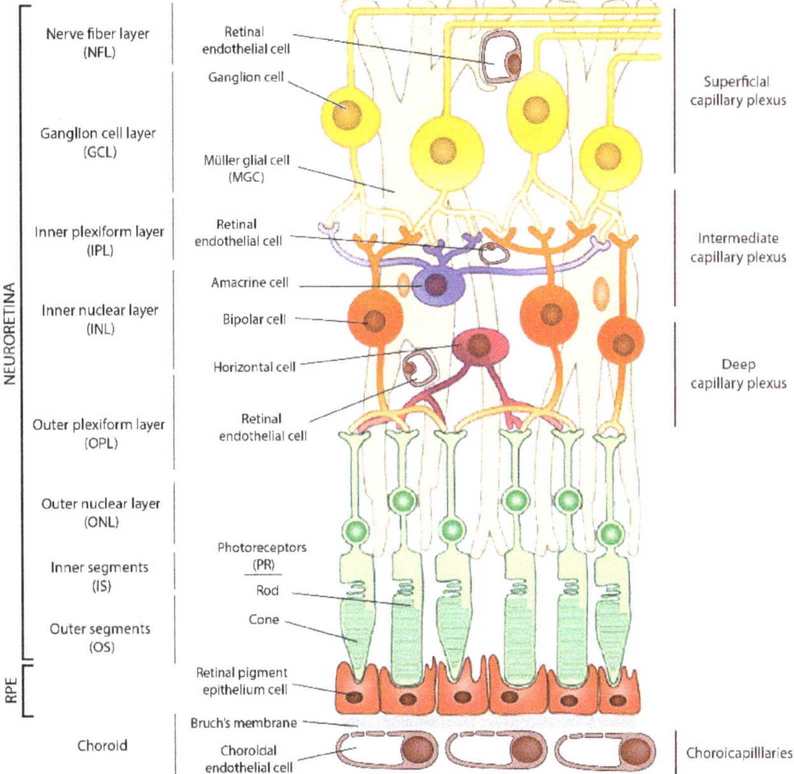

Figure 1. Schematic drawing of the cellular components of the retina. **Legend Figure 1:** There are three retinal vascular plexuses tightly coordinated with retinal neurons and a choroid plexus underlying RPE. GCL: Ganglion cell layer; NFL: Nerve fiber layer; INL: Inner nuclear layer; IPL: Inner plexiform layer; MGC: Müller glial cell; ONL: Outer nuclear layer; OPL: Outer plexiform layer; OS: Outer segments; IS: Inner segments; PR: Photoreceptors; RPE, retinal pigment epithelium.

The retina is separated from the circulation by two blood-retinal barriers (BRB). The inner BRB consists of a neuro-glio-vascular complex formed by tight junctions between endothelial cells of the retinal capillaries, pericytes, astrocytes, MGC, and microglia [1]. The outer BRB is formed by the tight-junction monolayer of RPE cells that are in close contact with the choriocapillaries, which control exchanges through diaphragmed fenestrations [1]. Oxygen is the most supply-limited metabolite in the retina [2]. Its supply to the retina is ensured by the choroid, which provides oxygen to the outer retina, whilst the retinal circulation provides the oxygen requirements of the inner retina. In normal condition, the level of oxygen tension (Po_2) in the outer retina is ten times lower than in the inner retina [3]. Oxygen and glucose consumption are metabolized to lactate, while aerobic glycolysis dominates energy production in the outer retina. Several factors modify Po_2 level and utilization at the cellular level: the retinal depth, the light, and hyperoxia [4,5]. PRs have almost all mitochondria in their inner segments far from blood vessels. Light decreases oxygen utilization on the outer retina as much as by a factor of two and increase Po_2. Hyperoxia dramatically increases Po_2 in the retina with the increase higher in outer retina compared to inner retina. The development and maintenance of retinal vasculature is regulated by α subunits of hypoxia-inducible factor (HIF), which induce genes required for retinal homeostasis, such as vascular endothelial growth factor (VEGF) under hypoxic conditions [6]. HIF proteins, which act as regulators of oxygen homeostasis, also depend on iron for their activity, and they regulate genes involved in iron metabolism [7]. Hyperoxia is deleterious to the outer retina as the oxygen leads to the formation of reactive oxygen species (ROS) according to the Fenton and Haber–Weiss reaction catalyzed by iron to generate $RO°$ radical (review in [3]). Iron and oxygen are thus closely linked in retinal metabolism in health and disease conditions.

3. Iron Homeostasis in the Retina

3.1. Distribution of Iron in the Retina

Iron is widely and unevenly distributed throughout the adult rat retina. The highest concentrations of iron were observed by proton-induced X-ray emission in the choroid, the RPE, and the inner segments of the PR. PR outer segments also contain iron, as inclusions inside the discs [8]. Iron and iron-related parameter (total iron binding capacity, TF and TF saturation percentage) distribution in the eye are different between diurnal and nocturnal animals. In cow and pig retina, iron concentration is higher than in rat retina, suggesting that the nocturnal habit of living could influence iron-related parameters in the retina [9]. The iron level also varies during retinal development and aging. Moos et al. have shown that in rats, iron entry is very high during retinal development and maturation, then decreases in adulthood, and increases again with aging [10]. In rodents, there are gender and strain-specific influences on iron regulation in the neural retina [11]. Human sex-associated differences in iron levels have also been reported, women having more retinal iron than men at all ages [12]. With aging, iron deposits are found in the RPE/choroid complex in rats and in the stroma of the choroid in non-human primate regardless of serum iron concentration [13]. Increased iron levels in the retina have also been reported in human eyes with age [12]. In rodent eyes, both neural retina and RPE/choroid present an increase of iron concentration which is associated with modifications of iron-related proteins mRNA and protein levels [14,15].

3.2. Proteins Involved in Retinal Iron Homeostasis

3.2.1. General Iron Homeostasis

Under physiological conditions, almost all non-heme iron (Fe^{3+}) in the circulation is transported bound to TF (transferrin-bound iron: TBI) with high affinity (for review on cellular iron metabolism [16]). At cellular level, the TF with two Fe^{3+} (holo-TF) is bound by its receptor (TFR1), and the complex is internalized. The Fe^{3+} is released from TF in the endosome under the effect of acidification through the action of an ATP-dependent proton pump. Iron is then reduced into the ferrous form, Fe^{2+}, by endosomal ferrireductase six transmembrane epithelial antigen of the prostate 3 (STEAP3) and exported by divalent metal transporter 1 (DMT1) in the cytosol where it contributes to the labile iron pool (LIP). In case of iron overload, TF is saturated with iron, and the non-transferrin-bound ferrous iron (NTBI) could be up-taken by iron importers such as the ZRT/IRT-like proteins (ZIPs) or DMT1 and joined intracellular LIP. The LIP consists of a transitory pool of iron species associated with a variety of ligands with low affinity (citrate, phosphate, inorganics irons) and easily oxidized, in transit to be distributed to the organelles (in particularly in mitochondria or nucleus) for cell metabolism in iron requiring proteins, stored or released.

In non-erythroid cells, the majority of iron is stored in ferritin (FT). Composed of 24 subunits of both heavy (HFT) and light (LFT) chains, FT forms a tissue-specific heterocomplex which can store up to 4500 Fe^{3+}. Fe^{2+} from LIP is transported to FT by Poly(rC)-binding proteins (PCBP1 and PCBP2) and is oxidized by ferroxidase activity of HFT in an oxygen-dependent manner. Then Fe^{3+} is stored in the cavity formed by LFT. Iron is released from FT in a controlled manner by autophagy involving a cargo receptor, the nuclear receptor coactivator 4 (NCOA4).

The only known mammalian iron exporter is ferroportin (FPN), a transmembrane protein which requires a multicopper ferroxidase to convert Fe^{2+} to Fe^{3+}, allowing binding to TF. Hephaestin (HEPC), ceruloplasmin (CP), amyloid-beta precursor protein (APP), and the newly identified zyloklopen (ZP) are co-localized with FPN at the surface membrane or secreted.

3.2.2. Iron Flux in Retina

A local retinal homeostasis of iron, independent from the systemic regulation, is suspected by the fact that main proteins involved in iron homeostasis, previously confined to systemic expression, are locally synthesized in the retina (Table 1). In addition, the outer and inner BRB prevents the entry of large quantities of iron into the eye in case of systemic iron overload. The study of mouse models invalidated for iron homeostasis proteins (Table 1) led to a hypothetical model in which iron entry and homeostasis in the retina is divided into two compartments, carried by the RPE and MGC, delimited by the external limiting membrane, and having limited exchanges in physiological condition.

Table 1. Proteins involved in iron homeostasis of the retina.

	Proteins	Expression	Functions	Knock-Out Rodent Models	Human Pathologies
Iron uptake/export	Transferrin (TF)	RPE, PR, MGC [8]	Extracellular transporter binding two ferric iron ions (Fe^{3+}) [holoTF]. Kd = $10^{22}M^{-1}$	$Hpx^{-/-}$: Decrease of the electroretinogram. Decrease TF, CP, TFR1 [17]	Congenital atransferrinemia
	Transferrin receptor 1 (TFR1)	RPE, IS, OPL, INL, GCL, endothelial cells [8]	Transmembrane receptor of holoTF	ND	ND
	Lactoferrin (LF)	RPE [18]	Extracellular transporter binding two Fe^{3+}	$Lf^{-/-}$: Higher susceptibility for laser induced choroidal neovascularization [19]	ND
	Lipocalin 2 (LCN2) or (neutrophil gelatinase-associated lipocalin (NGAL) or 24p3	RPE, MGC, neural retina, microglia [20]	Extracellular transporter which binds Fe^{3+} by sequestering bacterial and mammalian siderophores (2,5-dihydroxybenzoic acid).	$Lcn2^{-/-}$: Expression of LFT, TF et TFR1 unchanged (personal data)	ND
	24p3R or (the solute carrier family 22 member 17 (SLC22A17)	RPE [21]	Transmembrane receptor of LCN2 under holo- and apo-forms		
	Megalin or (low density lipoprotein receptor-related protein 2 (LRP2)	RPE [22]	Transmembrane multiligands receptor (such lipocalin 2, lactoferrin, transferrin), co-receptor Cubulin	$Lrp2^{F/F}$ (FoxG1^{Cre+}): myopia, hypertrophic RPE and retinal degeneration [22]. Increase FT and decrease TFR1 (personal data)	Donnai-Barrow syndrome: high myopia, retinal detachment
	Ferroportin (FPN) or (SLC40A1)	RPE, IS, OPL, IPL, MGC, REC [23]	Transmembrane transporter which exports ferrous iron (Fe^{2+}) outside the cell in cooperation with ferroxidases	Fpn^{C326S}: HEPC-resistant FPN mice. Increase FT, iron deposits in RPE and choroid [24]	Hemochromatosis type 4
	Ceruloplasmin (CP)	RPE, MGC [25]	Extracellular ferroxidase that oxidizes Fe^{2+} in Fe^{3+}. Exists a glycosylphosphatidylinositol-anchored form	$Cp^{-/-}Heph^{sla/sla}$: Neovessels, deposits under RPE, loss of RPE and PR. Iron accumulation in RPE and PR. Increase FT and decrease TFR1. $Cp^{-/-}Heph^{F/F}$ (Bestrophin1^{Cre+}): no iron deposits and no retinal dystrophy [26]	Aceruloplasminemia: iron deposits in drusen and RPE. Dry-AMD like phenotype [27]
	Hephaestin (HEPH)	RPE, PR, MGC [25]	Extracellular ferroxidase. 50% of homology with CP		ND
	Amyloid-beta precursor protein (APP)	RPE, IS and OS, MGC, GCL [26]	Membrane ferroxidase	$App^{-/-}$: Disturbances of synaptic development of secondary neurons [28].	Associated with Alzheimer disease and cerebral amyloid angiopathy
	Zyklopen (ZP)	RPE, GCL [26,29]	Membrane ferroxidase	ND	ND
	DMT1 (*Divalent Metal Transporter 1*) or Natural resistance-associated macrophage protein (NRAM2) or SLC11A2	IS, horizontal and rod bipolar cells [30]	Transmembrane import of Fe^2 Iron exit from endosomal vesicle or cytosol under acidic pH (5.5).	ND	Microcytic hypochromic anemia with iron overload
	ZIP14 (*Zinc transporter 14*) or SLC39A14	CEC, RPE, PR, MGC, GCL, REC [31]	Transmembrane zinc transporter which uptakes unbound Fe^{2+} in cytosol. Optimal at physiological pH (7.4).	ND	Hypermanganesemia with dystonia 2; Hyperostosis cranialis interna.
	ZIP8 (*Zinc transporter 8*) or SLC39A8				Congenital disorder of glycosylation 2N

Table 1. *Cont.*

	Proteins	Expression	Functions	Knock-Out Rodent Models	Human Pathologies
Storage	Ferritin (FT)	Ubiquitous and highly express in RPE, IS, bipolar cells [8]	Cytosolic complex of 24 subunits of heavy (H) and light (L) chains, which can store 4, 500 Fe^{3+}. The H subunits have ferroxidase activity. FT has also a nuclear localization	$Hft^{+/-}$: Higher sensibility for stress [32]	HFT: Hemochromatosis type 5LFT: Hyperferritinemia with or without cataract; Neuroferritinopathy; L-ferritin deficiency.
	Mitochondrial ferritin (FtMt)	All retina layers, with higher expression in RPE and ellipsoids of IS [30]	Mitochondria iron transporter. Share 79% of homology with HFT and has ferroxidase activity	ND	ND
	Transferrin receptor 2 (TFR2)	RPE, IS, OPL, IPL [33]	Transmembrane receptor of holo-TF which regulates transcription of HEPC in cooperation with HFE under TF iron-saturation	ND	Hemochromatosis type 3
	Hereditary hemochromatosis protein (HFE)	RPE [33]	Membrane protein which bind β2M to TFR1 or TFR2 in function of TF iron-saturation	$Hfe^{-/-}$: Hypertrophy/Hyperplasia of RPE, PR degeneration. Increase FT [34]	Hemochromatosis type 1: Dysmorphism of RPE, drusen and alteration of vision [30]. Variegate porphyria. Microvascular complications of diabetes 7
	β-2-Microglobulin (β2M)	RPE, OS, IS, OPL, INL, IPL [33]	Membrane protein involved in HFE-TFR1/2 interaction	ND	Immunodeficiency 43; Amyloidosis 8.
Regulation	Bone Morphogenetic protein 6 (BMP6)	RPE, IS, OPL, IPL, GCL	Extracellular protein which regulates HEPC transcription. Bind Activin A receptor (Acvr1A) and BMP receptor type II (BMPR2) and HJV as coreceptor, all expressed in retina	$Bmp6^{-/-}$: Iron accumulation in RPE and retina. RPE hypertrophy and PR degeneration. Decrease TFR1 and increase LFT [35]	ND
	Hemojuvelin (HJV)	RPE, PR, MGC, GCL [36]	Regulation of HEPC transcription	$Hjv^{-/-}$: Neovessels in retina, gliosis, inner BRB leakage, PR degeneration. Increase LFT [37]	Hemochromatosis type 2A
	Transmembrane serine protease 6. (TMPRSS6) or Serine protease matriptase-2	RPE, MGC GCL [38]	Membrane protein with serine protease activity which cleave HVJ	$Tmprss6^{msk/msk}$: No visual alteration [38]. Increase HEPC, HJV, TFR1	Iron-refractory iron deficiency anemia
	Hepcidin (HEPC)	RPE, IS, MGC, OPL [36]	Peptide hormone which transcription is activated by TF saturation or inflammation. Induces the degradation of FPN reducing iron export	$Hamp^{-/-}$: Iron accumulation in RPE/choroid and in retina. Decrease TFR1 and increase FPN [39]	Hemochromatosis type 2B and juvenile
	Iron regulatory protein (IRP1) or cytoplasmic aconitase hydratase (ACO1)	Ubiquitous	Iron sensor protein with cluster iron-sulfur. Bind Iron responsive element (IRE) in target mRNA when intracellular iron levels are low. Under high iron condition, IRP1 is converted into an aconitase whereas IRP2 is degraded in proteasome	$Ireb1^{+/-}Ireb2^{-/-}$: No retinal alteration. Increase FPN and LFT [30]	ND
	IRP2 or Iron-responsive element-binding protein 2 (IREB2)				ND
	Hypoxia Inducible Factor (HIF)	RPE, PR ONL, INL, GCL [40,41]	Transcriptional regulator. Oxygen sensor sensitive to iron level. Bind Hypoxia responsive element (HRE) in target mRNA under hypoxia or when intracellular iron levels are low	ND	Familial erythrocytosis (HIF2α)

Legend Table 1: Proteins localization were obtained from immunostaining on sections of mouse/rat retinas. The mouse models presented are limited to those with retinal changes in iron homeostasis and retinal abnormalities, if any. The corresponding human diseases were obtained by searching the UniProt site. ND: Not determined. Legends: β2M: β-2-Microglobulin; CP: ceruloplasmin; CEC: choroidal endothelial cell; FT: ferritin; GCL: ganglion cells layer; HFE: hereditary hemochromatosis protein, HFT: ferritin heavy chain; IPL: inner plexiform layer; IRE: iron responsive element; IS: inner segments; LFT: ferritin light chain; MGC: Müller's glial cell; OPL: outer plexiform layer; OS: outer segments; PR: photoreceptor; REC: retinal endothelial cell; RPE: retinal pigment epithelium; TFR1: transferrin receptor 1.

Transferrin-Bound Iron Transport in the Retina

The RPE imports iron bound to TF from the choriocapillaries through the transcytosis of TFR1 present at the basal membrane of the RPE (Figure 2). The transcytosis of the TF/TFR1 complex along microtubules via galectin 4 and Rab11a [42] has been described in vitro. The presence of TFR1 on the apical side of RPE is ambiguous and suggests that TF/TFR1 transcytosis or a potential iron-TF uptake by RPE could egress iron from the outer retina to choriocapillaries. Six hours after an intravitreal injection of holoTF tagged with a fluorochrome, TF is localized in RPE and choroid, which favors the later hypothesis [43]. Another iron entry in RPE is the phagocytosis of PR outer segments which contains high quantities of iron [8]. Once in RPE cytosol, iron is stored in FT and in melanosomes [44]. The release of iron from cell is possible through FPN present at the basal membrane of RPE and a multicopper ferroxidase. HEPH, CP, and APP but not ZP are expressed in RPE [26].

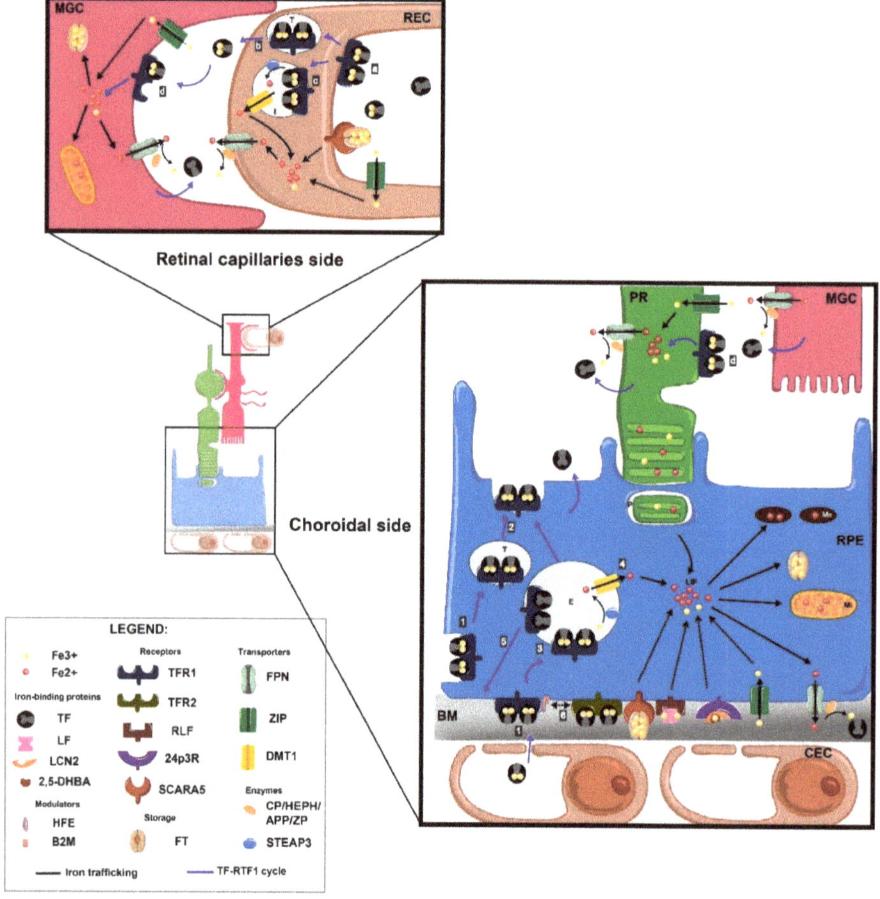

Figure 2. Iron uptake from capillaries and transport in the retina. **Legend Figure 2:** Under physiological condition, non-heme iron (Fe^{3+}) in the circulation is transported bound to transferrin (TF). A. At the

choroidal side, Fe^{3+} linked to TF is captured by its receptor 1 (TFR1) (1) at the basolateral level of the retinal pigment epithelium (RPE) (blue arrows). The internalized TF/TFR complex is transported to the apical pole by transcytosis (T) (2) or to the endosome (E) (3). In this case, Fe^{3+} is released from TF and reduced by the metaloreductase six transmembrane epithelial antigen of the prostate 3 (STEAP3) to ferrous iron (Fe^{2+}) and then exported to the cytosol by the transporter divalent metal transporter 1 (DMT1) where it constitutes the free iron pool (LIP) (4). TF and TRF1 are recycled to membrane (5). Iron is then transported from LIP to the organelles as needed, either stored in ferritin (FT) and melanosomes (Me), or exported by ferroportin (FPN) coupled to ferroxidases such as ceruloplasmin (CP) or hephaestin (HEPH), amyloid-beta precursor protein (APP) or zyloklopen (ZP) (black arrows). Hemochromatosis protein (HFE) and beta-2 microglobulin (B2M) associated to TFR1, shift to TFR2 in case of iron overload (saturation of TF) and activate hepcidin (HEPC) transcription (6). B. At retinal capillaries side, Fe^{3+} bound to TF is up-taken by TFR1 at the luminal side of retinal endothelial cells (REC) (a), and TF/TFR1 pass directly through transcytosis into the retina (b) or endocytosed then exported by FPN (c). TF synthetized by RPE, Müller glial cells (MGC) or photoreceptors (PR) up-taken retinal iron (d) and distributed it throughout the retina, especially to PR. Phototransduction performed on the outer segments of PR is a highly iron-dependent process. PR uptake Fe^{3+} bound to TF by TFR1 presents in inner segments and export it by FPN or by phagocytosis (P) of the outer segments of PR by RPE. TF-independent iron delivery to the retina can occur, especially in case of systemic iron dysregulation (black dotted lines). Serum FT has a specific receptor, the scavenger receptor class A, member 5 (SCARA5) localizes at the basal membrane of RPE, luminal side of REC, PR and MGC. Lactoferrin (LF), a member of TF superfamily and its receptors (LFR) are present in RPE. Fe^{3+} captured by a siderophore (2,5-dihydroxybenzoic acid (2,5-DHBA)) is bound by lipocalin 2 (LCN2) and its receptors (24p3R) in RPE. The non-TF-bound iron (NTBI) is up-taken by MGC, REC, PR and RPE by DMT1 or ZRT/IRT-like proteins (ZIP) importers. BM: Bruch's membrane; CEC: choroidal endothelial cell; E: endosome; Me: melanosome; MGC: Müller's glial cell; MI: mitochondria; P: phagosome; PR: photoreceptor; REC: retinal endothelial cell; RPE: retinal pigment epithelium; T: transcytosis.

In the inner retinal layers, iron is imported through retinal endothelial cells (REC) which express TFR1 at their luminal side [45]. Two mechanisms of iron transfer across the abluminal membrane of REC into the retina are evoked: TF/TFR1 transcytosis or/and TF/TFR1 endocytosis following by iron released from endosome and iron export by FPN. The abluminal membrane of REC expresses FPN colocalized with HEPC, CP, and APP [24,31]. Iron is bound by TF and distributed to the retina or the vitreous. With its unique position extending from the vitreous to PRs and its capacity to synthetize TF [46] and to express FPN [24], MGC plays a crucial role in the distribution of iron from the inner retina to the inner segment of the PR.

Iron presents in the non-vascularized subretinal space, between the apical side of RPE and the PR, is mainly bound to TF, secreted by PRs and RPE, and up-taken by PRs for their highly metabolism activities, by TFR1 express at inner segments. The PR inner segment is the iron storage compartment for PR segments, where both FT chains and mitochondrial FT are highly concentrated [30]. Iron export from the PR is ensured mostly by FPN also present in inner segments, CP and HEPH being poorly involved in favor of APP [26].

Non-Transferrin-Bound Iron Transport in the Retina

Whereas iron-bound TF is the main transport system to cross the BRB, transferrin-independent iron delivery to the retina can occur (Figure 1). Serum FT, exclusively composed of LFT has specific receptor, the scavenger receptor class A, member 5 (SCARA5) expressed in cytoplasm and nucleus in retinal endothelial cells, ganglion cells, astrocytes, the inner nuclear layer, MGC, microglia, outer nuclear layer, cones segments and RPE. After intravenous injection, serum FT remained confined in retinal endothelial cells in the inner retina [47]. Another protein possibly involved in iron transport is lactoferrin (LF), a multifunctional protein which shares 65% of homology with TF. It is synthesized in various human ocular tissues mainly in the RPE but not in the neural retina [18]. LF receptors have not been studied in the eye but are present in the brain [48]. Lipocalin 2 (LCN2) does not bind to iron directly, but through interaction with siderophores (catecholate and carboxylate) as cofactors [49] could also be implicated in iron transport in the retina [20]. β-hydroxybutyrate dehydrogenase-2 (Bdh2), an enzyme that is critical for the synthesis of 2,5-dihydroxybenzoic acid (2,5-DHBA), the mammalian siderophore, is found throughout the retina in all cell layers, including ganglion cells, MGC, and RPE cells [50]. Two major membrane-bound receptors for LCN2, megalin and 24p3R, have been identified in RPE [21,22]. Although LCN2 is being recognized as an important factor in retinal diseases [21], its exact contribution in iron retinal transport in health and diseases remain to be determined.

In case of iron overload, the NTBI could be up taken by iron importer ZIP or DMT1. This could explain why in retinal iron overload models, iron continues to accumulate despite the reduced expression of TFR1 in retina and RPE [31]. The specific localization of DMT1 in PRs and bipolar and horizontal cells suggests that it could be involved in providing iron to these cells, for phototransduction or neurotransmitter synthesis [30] ZIP8 and ZIP14 are expressed in RPE, choroid, REC, choroidal endothelial cells (CEC), ganglion cells, PRs, and MGC. At a high degree of TF saturation in the retina, there is a decreased ZIP14 expression whereas ZIP8 expression remains stable [31].

3.2.3. Iron Regulation in Retina

Cellular iron uptake and release and the intracellular LIP size are tightly controlled. Transcriptional, post-transcriptional, and post-translational processes regulate iron homeostatic proteins (for a review, see [51]). The main mechanisms of intra- and extra-cellular regulation of iron levels are limited to two extremely controlled systems.

The first system includes iron regulatory proteins (IRP) 1 and 2—intracellular iron regulatory proteins which, depending on the amount of iron, bind iron responsive element (IRE) sequences present on the mRNAs of iron homeostasis proteins such as FPN, TFR1, FT, and DMT1. Depending on the position of the IRE site, IRP controls their translation or degradation. Under conditions of increased cellular iron, IRP1 loses its IRE-binding activity by acquiring an iron in the 4Fe–4S cluster, whereas IRP2, is degraded by proteasome. In this condition, *tfr1* and *dmt1* mRNA are degraded, whereas *ft*, *fpn*, and *hif-2α* mRNA are translated. The localization of IRP1 and IRP2 has not yet been identified in the retina but their expressions are ubiquitous in mammalian cells. Mice with *Irp1*$^{+/-}$ *Irp2*$^{-/-}$ genotype show more severe neurodegenerative disease than *Irp2*$^{-/-}$ animals [30]. These IRP deficient retinas have increased FPN and FT in the inner segments, MGC endfeet, and inner retina compared to age and strain matched wild type retinas, suggesting that FPN and FT levels are regulated by IRPs in the retina [23]. In a model of light induced retinal degeneration, 2 h after light exposure, *Irp2* but not *Irp1* mRNA increased in the retina [32].

The second system focuses on hepcidin (HEPC), a peptide hormone principally synthetized by the liver. However, HEPC is also synthesized by PR, RPE, and MGC [39]. It is activated by two cellular signaling pathways induced by excess of iron, the transferrin receptor 2 (TFR2)/Human homeostatic iron regulator protein (HFE) pathway and the Bone Morphogenetic protein (BMP6)/Mothers against decapentaplegic homolog 1 (SMAD) pathway. When the TF saturation is high at the basolateral level of the RPE, the HFE is released from TFR1 and binds to TFR2, which activates the transcription of HEPC. BMP6 secreted by the retina and the RPE, binds to its receptors coupled to hemojuvelin (HJV) protein at the apical level of RPE in order to activate the synthesis of HEPC [35]. HEPC binds to the extracellular domain of FPN on the cell surface, leading to its internalization and degradation, effectively preventing cellular iron export and limiting the amount of iron that gets into the extracellular fluid. The specific deletion of HEPC in the retina does not lead to age-associated retinal iron accumulation, whereas liver-specific HEPC silencing leads to early serum, RPE, and retina iron accumulation followed by retinal degeneration [52].

Finally, the hypoxia inducible factor (HIF) acts as a transcription factor for certain iron homeostasis genes such as the *Tf*, *tfr1*, *Dmt1*, *Fpn*, and *Cp* genes by binding to a specific hypoxia-responsive element (HRE) site present on their mRNAs. Expression and degradation of HIF are also dependent on iron. In fact, Fe^{2+} is the cofactor of prolyl hydroxylase involved in the degradation of HIF-1α, and at the same time HIF-2α has an IRE sequence in the 5'UTR of its mRNA, which in the condition of iron deficiency, inhibits its translation. Nuclear staining of HIF-1α was observed in the GCL, the inner nuclear layer and the outer nuclear layer in human and rat [41]. Under retinal hypoxia, both HIF-1α and HIF-2α are activated but have cell specific expression within the inner retina. Specifically HIF-2α activation seems to play a key role in regulating the response of MGC to hypoxia [53].

4. Physiopathological Role of Iron in the Retina

4.1. Iron in Cellular Metabolism/Functions

4.1.1. Iron as a Fe-S Structural Motif Involved in Various Cellular Machinery Proteins

Iron sulfur (Fe-S) proteins are characterized by the presence of Fe-S clusters localized in different cell compartments (for review [54]). IRP1 is a Fe-S cluster that participates in sensing and regulating iron homeostasis in the retina. Frataxin is a nuclear-encoded mitochondrial protein involved in Fe-S cluster assembly, heme synthesis, and intracellular iron homeostasis. Frataxin is an allosteric activator which binds to this assembly complex [55]. It is present in the retina [56] and in the RPE [57] and could be responsible for retinal neurodegeneration induced by defective mitochondrial function [58]. In addition, Fe-S clusters may act as biological sensors by their binding properties to molecular oxygen and nitric oxide [59] both critical for the retinal physiology and pathology.

4.1.2. Iron in Nucleic Acids Machinery, Cell Proliferation, and DNA Repair

A recent review has reported the multiple implications of iron in DNA synthesis and repair, as well as in RNA metabolism [60]. Cytosolic and nuclear Fe-S proteins intervene in the genome stability [61]. Iron has been implicated in DNA synthesis and repair as a cofactor of sirtuin 2, an histone deacetylase, involved in iron homeostasis [62]. Sirtuin 2 maintains cellular iron levels by binding the nuclear factor erythroid-2-related factor 2 (NRF2) leading to a reduction in total and nuclear NRF2 levels. NRF2 is a transcription factor that plays key roles in retinal antioxidant and detoxification responses and has been linked with the development of age-related macular degeneration (AMD) [63].

Mitochondria are a major source of ROS and mitochondrial DNA is very susceptible to oxidative damage [64]. In RPE cells, mitochondrial DNA is damaged by hydrogen peroxide [65]. Deletions in mitochondrial DNA occurred in function of age in human neural retina [66], and the accumulation of age-related mitochondrial mutations in the eye has been correlated with a decrease in ATP production and increase ROS output, leading to oxidative stress, inflammation, and degradation [67].

4.1.3. Iron in Oxygen Transport and Regulation

Hemoglobin is synthetized in the retina [68]. It is one of the main protein synthesized in primary cultures of human RPE and secreted in vivo through the basolateral membrane [69].

Under physiological condition, free hemoglobin is bound by haptoglobin, but in case of massive hemolysis, hemoglobin releases free heme which binds hemopexin. Both hemopexin and haptoglobin have been described in the human retina [70,71]. The mRNAs for both haptoglobin and hemopexin were detected in the neural retina and PR as well as ganglion cells but not in RPE cells.

Neuroglobin is a highly conserved oxygen-binding protein reviewed in [72] and highly expressed in the retina. Its role is to facilitate oxygen metabolism, being localized in mitochondria. Hemin, the ferric chloride salt of heme enhances neuroglobin expression and protects animal model of N-methyl-N-nitrosourea-induced retinal degeneration [73]. In this model, hemin protects also cones from apoptosis. Neuroglobin has also been associated with retinal damage induced by light [74] which may reflect the changes in iron metabolism first described with light on retina [32]. It has also been associated with VEGF expression and thus could participate in retinal angiogenesis [75].

Heme, Fe^{2+} protoporphyrin IX, the prosthetic group of hemoproteins including hemoglobin, neuroglobin, oxidases/peroxidases, or cytochromes can be released after auto-oxidation. Heme transporter proteins also intervene in iron metabolism in the retina, and their dysregulation could potentially cause oxidative cell damage. All three heme transporters feline leukemia virus subgroup C receptor (FLVCR), breast cancer resistance protein (BCRP), and proton-coupled folate transporter (PCFT/HCP-1) are expressed in the retina and RPE. In the RPE, the expression of FLVCR is restricted to the apical membrane and the expression of BCRP and PCFT to the basolateral membrane. In cases of iron overload, the expression of FLVCR and PCFT is upregulated and BCRP is downregulated, suggesting an important role of heme transporter proteins in retinal iron regulation [76].

4.1.4. Iron and Visual Function

The involvement of iron in the vision cycle was discovered with the characterization of the enzyme RPE65, as an iron-dependent isomerohydrolase [77]. RPE65, abundant in the RPE [78], ensures the isomerization and hydrolysis of all-*trans* retinyl ester to 11-*cis* retinol. RPE65 is essential for vision, and mutations in *rpe65* genes induce Leber congenital amaurosis, a form of retinitis pigmentosa that leads to blindness [79]. Recently, RPE65 was also shown to intervene in the production of meso-zeaxantin, an ocular specific carotenoid which protects the fovea from oxidative stress [80].

An alternate pathway for 11-*cis* retinol recycling has been described in MGC by isomerases 1 or 2 that also appear to be iron dependent [81]. Few studies have analyzed the iron flux in the retina with the diurnal cycle conversely to what has been performed in brain in mice [82,83]. Among the sensory guanylate cyclase proteins and signaling network, guanylyl cyclase activating protein 5 is the only protein that binds strongly Fe^{2+} in zebrafish [84]. It is proposed as redox sensor in visual transduction.

Phototransduction depends on the phagocytosis of outer segments from PR by the RPE. The constant release of the outer segments from PR and their digestion during phagocytosis by RPE implies membrane biogenesis, a process which needs iron as a cofactor of fatty acid desaturase [85]. Royal College Surgeon (RCS) rats invalided for the phagocytosis protein Myeloid-epithelial-reproductive tyrosine kinase (MERTK) have increased iron in retina and particularly in RPE phagosomes and also increased retinal FT and TF expression [86].

Iron is also involved in neurotransmitters secretion as it regulates glutamate secretion by RPE cells via the cytosolic aconitase pathway [87]. Dopamine biosynthesis in specialized amacrine cells results from the conversion of the amino acid L-tyrosine in L-3,4-dihydroxyphenylalanine (L-DOPA) using oxygen and Fe^{2+} [88]. Synaptosomal nerve-associated protein 25 (SNAP-25) is a Fe-S protein involved in synapse vesicle fusion with plasma membranes highly present in retina [89].

A significant number of ATP binding cassette (ABC) transporters, involved in lipid trafficking in retinal cells, have been linked to severe genetic ocular diseases [90]. ABCA4 is present in the PR and transports 11-*cis* and all-*trans* isomers of *N*-retinylidene-phosphatidylethanolamine across disc membranes, preventing the accumulation of toxic bisretinoid lipofuscin compounds in PR and RPE cells. In *Abca4* null mutant mouse which presents accumulation of N-retinylidineN-ethanolamine (A2E) bisretinoids and lipofuscin in the RPE, intracellular iron accumulation is also observed which contributes to enhancing oxidative cell death [91]. The intracellular accumulation of iron in cells of the RPE in culture decreases the expression of the transporters of cholesterol ABCA1/ABCG1, increasing the level of pro-inflammatory cholesterol in retina [50].

4.2. The Dark Side of Iron

4.2.1. The Crucial Role of Iron in Oxidative Stress-Mediated Damages in the Retina

The ability of iron to change easily its valence and switch between the Fe^{2+} and Fe^{3+} forms, providing or accepting electrons, respectively, ensures a privileged position in living matter as mediator of key biochemical reactions. However, the presence of free labile iron in cell or NTBI in circulation is prone to generate highly ROS in the Fenton/Haber–Weiss reaction.

$$Fe^{2+} + H_2O_2 \rightarrow Fe^{3+} + OH^- + HO^{\cdot} : \text{Fenton reaction}$$

$$O_2^{\cdot -} + Fe^{3+} \rightarrow O_2 + Fe^{2+} : \text{Haber–Weiss reaction}$$

$$O_2^{\cdot -} + H_2O_2 \xrightarrow{Fe^{2+};\, Fe^{3+}} O_2 + OH^- + HO^{\cdot} : \text{Fenton/Haber–Weiss reaction}$$

The toxicity of free iron has been extensively studied on neuronal and retinal cells, and they are not sensitive to the same doses of iron [46,92,93], the cones being the most sensitive to iron [94]. In RPE cells, the interaction of iron with bisretinoids and lipofuscin induces cell damage and retinal degeneration [91]. Conversely, melanin can bind large amounts of iron to preserve the RPE and the choroid from a pro-oxidant environment, intensified by light exposure. However, with age, the accumulation of iron in melanosomes associated with a reduction in the amount of melanin in RPE promotes the formation of free radicals [95]. Exposure of RPE cells to high non-lethal doses of iron leads to a decrease in phagocytic and lysosomal activity [15], favoring the accumulation of breakdown products of Vitamin A (lipofuscin) leading to the formation of glycation end products (AGE) present in drusen, RPE, and Bruch's membrane of AMD patients [91]. In addition, phagocytosis of PR discs, peroxidized by ferrous ions, damage the membranes of phagosomes and lysosomes in RPE cells in culture [15,96].

In hypoxic conditions, an efflux of iron from RPE to the basolateral direction [97] could explain, at least in part, that PRs tolerate better hypoxia than hyperoxia [98]. Fe^{2+} contributes also to light-induced PR cell death through the production of hydroxyl radicals [99]. The ascorbate-Fe^{2+} complex induce lipid peroxidation in rod outer segment membranes and subsequently damage proteins such as rhodopsin by carbonylation or loss of thiol groups [100]. Finally, free heme can be also a source of redox-active iron and therefore highly toxic for the retina and for RPE cells [101].

In optic neuropathy, such as glaucoma, several mechanisms involved in ganglion cell death seem to be enhanced by iron-dependent oxidative stress [102,103].

Iron is thus a key component of oxidative-induced damages in the retina and in the RPE and involved in major cell death mechanisms.

4.2.2. Retinal Cells Death Mechanisms in Iron Overload

Iron overload, induced experimentally by the implantation of iron particles in rat vitreous cavity caused apoptosis (TUNEL-positive nuclei) in the outer nuclear layer after only 2 days [104]. Rat retinal explant exposed to iron showed an early increase of necrotic markers, such as lactate deshydrogenase, receptor-interacting serine/threonine-protein (RIP) kinase, and incorporation of propidium iodide, even before intraretinal iron accumulation was detected. Using retinal organo-culture, it was observed that iron deposits in retinal explants induced a shift from necrosis to apoptosis with activation of caspase 3 and TUNEL-positive nuclei [105]. Increased intraocular iron levels following intravitreal $FeSO_4$ injection caused oxidative damage of PR, as shown by the increase of superoxide radicals; hydroxynonenal, a marker of lipid peroxidation; and increased expression of heme oxygenase 1 [94]. Retinal iron overload also activates the NOD-like receptor family, pyrin domain containing 3 (NLRP3) inflammasome signaling pathway. In fact, the expression levels of NLRP3, activated caspase-1, a downstream target of NLRP3, and interleukin (IL) 1ß were higher in the retinas of HFE KO mice, a model of genetic iron overload [106]. Ferroptosis, a newly characterized form of necrosis, is induced by the accumulation of iron in degenerative diseases and has been described in RPE cells in culture subjected to oxidative stress [107]. Glutathione depletion also induced ferroptosis, autophagy, and premature senescence in RPE cells [108].

4.2.3. Inflammation

The general implication of iron in inflammation has been recently reviewed, and it will not be detailed here [109]. In RPE cells, the intracellular accumulation of iron activated the NRLP3 inflammasome pathway via the repression of the degradation of aluRNA by double-stranded RNA-specific endoribonuclease (DICER1). This mechanism involved the sequestration of the cofactor

PCBP2 [110] and has been advocated in AMD. It has been also reported that iron induces the synthesis of complement C3 by activation of the Extracellular signal-regulated kinases (ERK)/SMAD3/CCAAT Enhancer Binding Protein Delta (CEBPD) 48 pathway [111]. The complement factor C5 carries between 13 and 15 iron atoms necessary for its conversion into an active form C5b by C5 convertase, a complex formed from the cleavage products of C3 [112], which shows the importance of iron in complement pathways activation, a recognized risk factor for AMD [113].

The prion protein (PrP^C), the principal protein implicated in the pathogenesis of human and animal prion disorders, is also implied in retinal degeneration due to iron metabolism dysfunction. This neuronal protein is expressed in many tissues of the eye, such as the retina and the cornea trabeculum. PrP^C is also expressed on the basolateral membrane of RPE, where it facilitates uptake of iron from choriocapillaries to neuroretina by functioning as a ferrireductase partner for divalent metal transporters. PrP-scrapie (PrP(Sc)), a misfolded isoform of this PrP^C accumulates in the neuroretina resulting in iron accumulation [114].

In the brain, IL6 produced by microglia in response to lipopolysaccharide, induced the production of HEPC by astrocytes [115]. HEPC prevented the iron overload-activated neuronal apoptosis [115]. LPS induced also an HFE-independent expression of HEPC in MGC and in the RPE, both in vitro and in vivo. The increase in HEPC levels in retinal cells, occurring with a decrease in FPN levels, led to oxidative stress and apoptosis within the retina in vivo [116]. On the other hand, in both in vitro and in vivo models of amyloid β-induced pathology, HEPC downregulates the inflammatory and pro-oxidant processes in astrocytes and microglia and protected neurons from cell death [117]. Microglia and MGC activation associated with reactive gliosis has been observed in HJV knockout mice (*Hjv$^{-/-}$* mice) with aging and subsequent retinal iron accumulation [37].

The role of LCN2 has been suspected in AMD where its expression is increased in aqueous humor and in the infiltrating cells present in the retina and choroid [118]. An age-related increase in LCN2 was described in RPE cells of Beta-crystallin A3 (Cryba1) conditional knockout mouse, a model of AMD associated with chronic inflammation response [49], but the exact implication of LCN2 in iron metabolism in these models remains to be studied.

4.2.4. Angiogenesis

Increased iron levels in the retina could also have a role in the development of new vessels by inhibiting the anti-angiogenic effect of cleaved high molecular weight kininogen (Hka) [119], promoting the expression of succinate receptor 1 (SUCNR1 or GPR91) [120] which stimulates production of pro-angiogenic factors VEGF and angiopoietin [121]. *In Hjv$^{-/-}$* mice, that leads to abnormal retinal iron overload. Proliferation of new leaky blood vessels in the vitreous was associated with reactive gliosis involving MGC and microglia [37]. In addition to proliferation by migratory cells, intravitreal hemoglobin also stimulates a transient proliferation in cells of the RPE and possibly in some supportive cells of the neural retina, such as MGC and astrocytes [122]. Iron also plays a role in HIF transcriptional regulation of pro-angiogenic genes [51]

5. Role of Iron in Retinal Diseases

5.1. Siderosis and Retinal Hemorrhages

Eye siderosis is probably the first known manifestation of iron toxicity for the eye. The presence of a foreign body containing iron inside the eyeball leads to various clinical complications including heterochromia of the iris, mydriasis, cataract, and retinal and RPE atrophy. Electroretinography analysis shows a decrease in a and b wave amplitudes, due to the progressive degeneration of the cones and rods (for a review, see [123]). The increase in iron can be observed histologically as a granular structure with FT or hemosiderin into cells. The level of vitreous iron also increases [124].

Retinal hemorrhages are present in several retinal pathologies, such as exudative AMD, diabetic retinopathy, or myopic degeneration, and they are particularly deleterious for vision when located in the subretinal space. Vision loss is dependent on the size of the hemorrhage and the ability of the tissue to shed blood [125,126]. During sub-macular hemorrhage early PR damage has been reported within 24 h [127]. In rabbits, the injection of their own blood into the subretinal space leads to a progressive degeneration of the PRs from one day after injection until a total destruction at 7 days, with an accumulation of iron in the outer segments of PR and in RPE [128]. The increased release of iron from hemoglobin induces peroxidation of unsaturated phospholipids, which are extensively present in the retina and affects particularly retinal neurons compared to the retinal glial cells [129].

5.2. Retinal Manifestations of Inherited Iron Disorders

Iron can accumulate in the retina of patients with inherited diseases involving mutations in genes encoding proteins of iron homeostasis, which can cause an imbalance in the metabolism of retinal iron. The most common hereditary hemochromatosis is related to a mutation in the *Hfe* gene resulting in excessive absorption of iron by the intestine and its accumulation in the organs. Mutations in the *Tfr2*, *Fpn*, *Hjv*, and *Hepc* genes are also involved in the development of hemochromatosis. The clinical findings associated to the retinal iron accumulation in these patients, as well as the impact on visual function, are quite rarely reported due to the variability of penetrance and the existence of a treatment reducing systemic iron overload. However, iron deposits and other changes in the RPE as well as visual acuity loss have been already reported [130].

Aceruloplasminemia is an autosomal recessive disorder caused by mutations in the Cp gene, resulting in a defect in the export of iron from cells. The retina, brain, and pancreas are overloaded with iron, leading to the clinical consequences as retinal degeneration, dementia, and diabetes. Several cases associating yellow discoloration of the fundus, atrophy of the RPE, and drusen-like deposits in the macula have been described. In post-mortem sections, an accumulation of iron associated with an enlarged RPE and loss of pigment of the RPE was observed (for a review, see [27]).

In animal models invalidated for the genes coding for iron-related proteins, an accumulation of iron in the RPE and PR is systematically observed as well as abnormalities in the RPE and PR degeneration [25,35,39,52,131].

Studies carried out in aging rodents have shown that the increase in iron intakes in food or by intravenous injection leads to local iron deposits in the choroid, the RPE, and the segments of PR, as well as deposits of complement C3 in the Bruch's membrane, hypertrophy, and vacuolation of RPE and changes in the choriocapillaries [132,133].

5.3. Age-Related Macular Degeneration

AMD is a leading cause of worldwide blindness in the elderly population, affecting 200 million individuals by 2020 and nearly 300 million by 2040 [134]. The pathological aging of the macula can cause dry or non-neovascular and wet or neovascular AMD. At the early stage, accumulation of extracellular material forms drusen between the basal lamina of the RPE and the inner layer of Bruch's membrane in the eye. At the late stages, degeneration of the PR overlying the drusen can cause severe central vision loss in the dry form, whilst formation of new abnormal blood vessels from the choroid growing into the retina can cause subretinal fluid accumulation and bleeding. The wet form progresses rapidly and is responsible for 90% of severe vision loss associated with AMD. The pathogenesis of AMD is multifactorial, with genetic and environmental factors such as smoking. It is associated with dysregulations in the angiogenic, oxidative stress, lipid, inflammatory, and complement pathways [135]. Patients with early AMD have more iron in the macula than healthy patients. Iron deposits are found in the melanosomes of the choroid and the RPE, in the central layer of the calcified Bruch's membrane, in the drusen, and at the level of the PR [136]. Part of this iron, found in the pathological retina of AMD patients, is in the toxic free form [137]. Patients with dry AMD have more than twice the concentration of iron in their aqueous humor than in patients with cataract surgery [138]. The macular region of AMD patients with geographic atrophy showed an increase in the expression of proteins involved in iron homeostasis such as TF, FT, and FPN in the PR layer and feet MGC [139]. TF and CP mRNAs are increased in the two advanced forms of AMD [140]. In the serum of patients with the different forms of AMD, a significant increase in TF and TFR1 and a significant decrease in the concentration of soluble FT were observed while iron levels were unchanged [141]. Several polymorphisms of the iron homeostasis genes have been associated with risk factors for AMD: *Tfr1, Tfr2* (obesity, tobacco) [142], *Dmt1* [143], *Irp1* and *Irp2* [143], and *heme oxygenases 1* and *2* (HO1/2) [144]. A recent study has shown that the expression of several miRNA, small non-coding RNA molecules binding in 3′UTR genes, was modified in the serum of AMD patients, especially those controlling the translation of the TFR1 and DMT1 proteins [145].

5.4. Diabetic Retinopathy

Diabetic retinopathy is a vision-threatening complication of diabetes affecting approximately 93 million in the middle-aged and elderly populations [146]. Chronic hyperglycemia causes progressive damage to retinal cells and to the retinal capillaries, leading to ischemia, VEGF-mediated retinal vascular abnormalization, and neovascular vessels that leak and bleed into the retina. Macular edema is also a major cause of vision loss in diabetic retinopathy [1]. Clinical reports have shown the link between iron levels in the vitreous and proliferative diabetic retinopathy [124,147]. A strong iron label was observed in the RPE and outer plexiform layer of patients with diabetic retinopathy [148]. In a mouse model of diabetic retinopathy, higher iron concentrations in the retina led to an increased expression of renin by a mechanism dependent on the GPR91 receptor [106].

5.5. Glaucoma Neuropathy

Glaucoma is increasingly a cause of irreversible blindness in the world. Its global prevalence is expected to be 76 million by 2020 and 112 million by 2040. Progressive damage to the optic nerve, leading to severe vision loss results from increased ocular pressure and other multiple favoring factors [149,150]. Although the link between iron and glaucoma is not yet fully understood, there is a change in iron homeostasis in glaucomatous eyes. TF concentration is increased in the aqueous humor [151], and mRNA of TF are increased in retina [152]. Whilst no differences were found in iron levels in aqueous humor of patients with primary open-angle glaucoma [153], serum levels of iron and FT were significantly increased [154,155], and serum CP level was lower [156]. A glaucomatous mice model had lower retinal iron concentrations than pre-glaucomatous DBA/2J and age-matched C57Bl/6J mice [157]. The expression of FT, CP, and TF was increased in monkey and rat glaucoma models [152,158]. In addition, the role of glutamate excitotoxicity in the pathogenesis of glaucoma is well documented; yet, there seems to be a link between the toxicity of glutamate and the increase in the entry of iron into neurons [159], and iron chelation seems to protect neurons against excitotoxicity and intraocular pressure-induced toxicity [160,161]. A mutation in the autophagy receptor optineurin is associated with the pathogenesis of glaucoma. It induces the degradation of the TFR1 and the Rab12-dependent autophagy mechanism leading to retinal ganglion cell death. The addition of iron in this model reduces cell death [162]. It seems that iron metabolism is dysregulated in glaucoma, but the exact role of iron is optic nerve damage and remains to be studied in the pathogenesis of glaucoma.

5.6. Inherited Retinal Dystrophies and Associated Diseases

Retinitis pigmentosa affects approximately 1.8 to 2.4 million people around the world. The disease is characterized by degeneration of the PR and progressive complete blindness [163]. Although iron has been shown to accumulate in several models of retinal degeneration, as in rd10 mouse or RCS rat [86,164], the direct link between iron and retinitis pigmentosa has not been established in human disease.

Macular telangiectasia type 2 (MacTel 2) is a complex macular disease, characterized by abnormal perifoveal vessels (telangiectasia), loss of retinal organization, and ultimately loss of macula function. MacTel2 is the only human disease recognized as primarily associated with MGC cells loss. It has been shown that iron accumulates in the retina of patients with MacTel 2. In a murine model of MGC ablation that mimics part of MacTel 2 phenotype, there is also an accumulation of iron in retina and in the RPE [148]. Knowing the importance of MGC cells in the regulation of iron levels in the retina, it could be hypothesized that iron accumulates in MacTel 2 as a consequence of MGC loss in the fovea [165].

6. Iron Neutralization as a Therapeutic Strategy for Retinal Diseases

6.1. Chemical Chelators

Whether iron dysmetabolism in the retina is a cause or a consequence of various retinal diseases, iron accumulation is pathogenic, and its neutralization was shown to protect the retina from oxidative damage and retinal cell death in various models using different neutralizing strategies [133] (Table 2). As early as in the 1970s, an iron chelator, Deferroxamine, was used in humans to reduce the amount of "rust" deposited on the eye with satisfactory results. Used in many other models of retinal degeneration (retinitis pigmentosa [166] or light-induced retinal damage models [167]), this chelator reduces the iron load and preserves the retina. Other chelators, such as Deferriprone, have shown significant protection of the retina in mice with impaired mechanisms of iron homeostasis [168–171]. These chemical chelators are mainly used clinically to treat hemosiderosis induced by frequent transfusions. Administered orally, subcutaneously or intramuscularly, they could led to several eye side effects, including vision loss [172,173]. These side effects could be explained because chemical iron chelators also bind the iron necessary for RPE and PR function [133,174].

As highlighted in a recent review [175], the clinical use of chemical chelators is complex because they should (1) target only the organ or tissue which is affected by the iron excess; (2) have a sufficient half-life; (3) cross the different barriers that surround the tissue; and (4) have a rapid elimination route.

Table 2. Comparison between chemical iron chelators and transferrin in clinical use.

	Deferoxamine	Deferiprone	Deferasirox	Transferrin
Iron Binding	1:1	3:1	2:1	2:1
Route of administration	Sub-cutaneous (every 8–12h) Intravenous (IV) (5 days/week)	Oral (t.i.d)	Oral (q.d)	Intravenous [176]
Half-Life(after IV administration)	20–30 min	3–4 h	8–16 h	4–8 d [176]
Excretion	Urinary/fecal	Urinary	Fecal	Unknown
Usual Doses (mg/Kg/d)	25–60 [177]	75–100 [177]	20–40 [177]	100 [176]
Clinical Use	Acute iron intoxicationChronic iron overload	Chronic iron overload	Chronic iron overload	Atransferrinemia [178] Haematological stem cell transplant [176]
Ocular Side effects	Pigmentary retinopathy [173], visual loss [179], impaired night vision [180], optic neuritis [173] and cataract [172].	Diplopia [181], cataract [182] and possible retinal toxicity [183].	Lens opacities [184] and retinal disorders [185]	No adverse effects observed

Legend Table 2: Tid: 3 times a day; q.d: once a day.

6.2. Natural Chelators

Other natural molecules generally coming from plants, such as curcumin, polyphenols, and flavonoids, are iron chelators and have shown effectiveness in mouse models of retinal degeneration (for a review, see [186,187]).

6.3. Transferrin

TF is part of the TF superfamily, which also includes lactoferrin, melanotransferrin, and ovotransferrin, which are found in many species of both mammals and invertebrates. It consists in two lobes, each binding a Fe^{3+} atom with a very high affinity ($10^{22}M^{-1}$). Its primary role is to maintain an environment devoid of free iron. TF synthesized by RPE, PR, and neuronal cells is found in the aqueous and vitreous humors [8,105]. By single-cell RNA sequencing of human neural retina, mRNA for TF was enriched in peripheral retina compared to fovea [188]. Its expression is amplified during inflammation or immunity to increase the buffering capacity of iron. In light-induced retinal degeneration, TF and TFR1 mRNA increased in retina immediately after light exposure and then decreased at basal level. One day after light exposure, TF was increased, whereas TFR1 was reduced compared to not illuminated mice [32]. TF has long been of therapeutic interest due to its antimicrobial capacity and the ubiquitous presence of TFR1 allowing penetration of the blood-brain barrier [189]. TF has also been used successfully in humans in iron metabolism pathologies and for its cytoprotective capacity [190].

Our laboratory is interested in the potential of TF for the treatment of retinal pathologies (Table 3). Our work has shown that administration of the iron-free form (apoTF) by intraperitoneal injections in rd10 mice, a model of retinitis pigmentosa, preserves PRs better compared to the use of other chelators or antioxidants [46,164]. Injected into the vitreous, TF is present throughout the neural retina (MGC) and is eliminated via its receptors by RPE and the choroid without any immunogenic or toxic effect on the retina [46,105]. Thus, TF administered in a model of light-induced degeneration, allows the restoration of iron homeostasis, decreases iron accumulation, reduces inflammation and apoptosis, and preserves PRs and visual function [43]. In an ex vivo model of retinal detachment, TF inhibits the degenerative processes activated by the iron excess by reducing necrosis, apoptosis, gliosis, and oxidative stress. In vivo, human TF constitutively expressed in transgenic mice (TG) reduces loss of cones, cleavage of caspase 3, an apoptosis effector, DNA breaks, and necrosis (Figure 3). In rats, TF injected at the time of the detachment, reduces retinal edema, cell death and preserves PRs. In addition to its ability to reduce the accumulation of iron in the retina following detachment, TF also acts on other cellular pathways, no doubt through its interaction with molecular partners which remain to be discovered [105].

Table 3. Transferrin as a therapeutic drug in retinal diseases models.

Model Experiment	Physiopathology	Administration Mode	Therapeutic Action of Transferrin	References
Primary culture of Müller glial cells.	Iron exposure	Cell isolation from transgenic mice carrying the human transferrin gene (TghTF)	Cell number preservation. Lower necrosis revealed by lactate dehydrogenase release. Inhibition of mRNA TF diminution.	[46]
Primary culture of Müller glial cells	Iron exposure	Addition of apo- or holo-human TF	Dose-dependent cell number preservation by apo- but not holo-human TF	[46]
rd10 mice	Model of retinitis pigmentosa presenting iron accumulation in photoreceptors (PR)	Crossing rd10 mice with TghTF mice	Preservation of retinal histology (outer and inner nuclear layers thickness). Less apoptotic-positive retinal cells.Conservation of rods and cones morphology	[164]
rd10 mice	Model of retinitis pigmentosa presenting iron accumulation in PR	Daily intraperitoneal injections of apo-human TF	Dose-dependent preservation of retinal histology (outer and inner nuclear layers thickness). Less apoptotic-positive retinal cells. Conservation of rods and cones morphology	[164]
Light-induced degeneration	Model of acute degenerative retina	Intravitreal injection of apo-human TF before and after light-induced degeneration	Preservation of retinal histology and functions. Preservation of ONL thickness and PR morphology. Lower ONL apoptotic- positive cells.Regulation of iron homeostasis balance. Lower retinal iron accumulation and oxidative stress. Regulation of retina inflammation and diminution of microglial cells activation in outer retina.	[43]
Light-induced degeneration	Model of acute degenerative retina	Electrotransfer of cDNA of human TF for *in oculo* production	Preservation of retinal histology and ONL layer thickness.	[43]
P23H rats	Model of retinitis pigmentosa	Electrotransfer of cDNA of human TF for *in oculo* production	Preservation of retinal histology and ONL layer thickness.	[43]
Bone morphogenetic protein 6 mice	Model of hemochromatosis with retinal iron accumulation	Intraperitoneal and intravitreal injections of apo-human TF	Diminution of iron accumulation in retina pigment epithelium	[43]

Table 3. *Cont.*

Model Experiment	Physiopathology	Administration Mode	Therapeutic Action of Transferrin	References
Retinal explant of mice	Retinal detachment with iron exposure	Retinas from TghTF	Preservation of cones number and rod outer segments length. Lower necrosis Prevention of iron retinal accumulation	[105]
Retinal explant of rats	Retinal detachment with iron exposure	Addition of apo-human TF after iron exposure	Preservation of rhodopsin expression level and cones number Lower necrosis and apoptosis Prevention of retinal iron accumulation	[105]
Subretinal injection of hyaluronic acid in mice	Retinal detachment presenting iron accumulation in subretinal space	TghTF mice	Preservation of retinal histology, rods outer segments length and number of cones Diminution of retinal oedema and Müller glial cells activation Lower apoptosis and necrosis Regulation of pathways involved in biological functions	[105]
Subretinal injection of hyaluronic acid in rats	Retinal detachment presenting iron accumulation in subretinal space	Intravitreal injection of apo-hTF	Preservation of retinal histology, rods outer segments length Diminution of retinal oedema	[105]

Legend Table 3: ApoTF: transferrin without iron; HoloTF: transferrin binding iron; INL: inner nuclear layer; ONL: outer nuclear layer; PR: photoreceptors; TF: transferrin; TghTf: transgenic mice carrying the complete human transferrin gene.

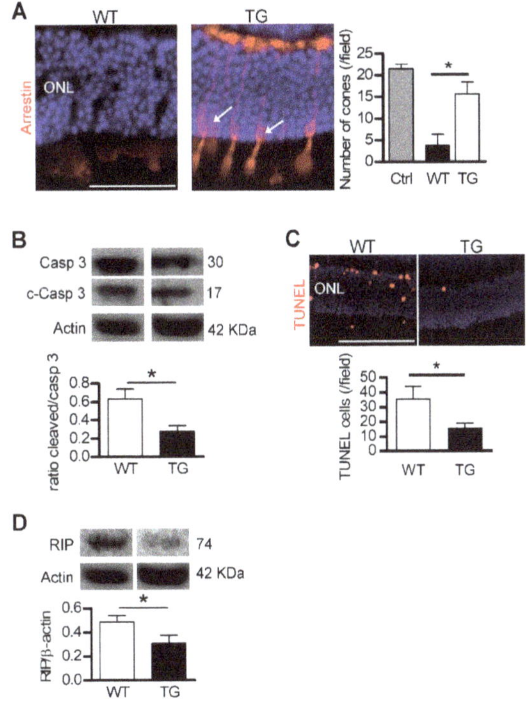

Figure 3. Transferrin expression preserves the detached retina. **Legend Figure 3:** After retinal detachment (RD), photoreceptors died by apoptosis and necrosis. Transgenic mice (TG) expressing human transferrin (TF) were used to demonstrate the protective effects of TF. (**A**) Arrestin staining revealed cones in retinal sections of TG mice (arrows) after RD. Cone number was higher in TG compared with WT mice. (**B**) The ratio of cleaved/pro–caspase 3 protein level was lower in TG mice compared to WT mice after RD. (**C**) The number of nuclei positive apoptotic-DNA breaks, stained by TUNEL, was reduced in TG mice compared to WT mice. (**D**) Necrotic RIP kinase protein level was reduced in TG mice compared with WT mice. All values are represented as the mean ± SEM. Mann–Whitney test (n = 3–6), * $p \leq 0.025$. ONL: Outer nuclear layer. Scale bar, 100 μm. From [105]. Reprinted with permission from AAAS.

7. Conclusions

Iron is one of the most common elements on Earth. Two-hundred years ago, it was discovered that, after desiccation, the residual ashes of an aged human retina could be mobilized by a magnet (as quoted in [191]). Nowadays the chemical study of iron structure and its outer electrons has been revealed by the discovery of quantum effects of iron electron in biology. The best illustrations have been described by Cedric Weber who demonstrated that very specific quantum effects are involved to explain the energy in the binding of iron to oxygen and CO to hemoglobin [192]. This transition metal plays a main role in retinal physiology, but overload leads to retinal degeneration and loss of function. Iron chelation is a potential therapeutic target to prevent retinal degeneration. TF, as an endogenous iron binding protein, avoids toxic effects of iron depletion and activates additional neuroprotective pathways.

Author Contributions: Conceptualization: E.P., Y.C., F.B.-C.; literature search: E.P., A.D.; J.Y., Y.C., F.B.-C.; figure and tables: E.P., J.Y.; writing—original draft preparation E.P., A.D.; Y.C., F.B.-C.; writing—review and editing: E.P., A.D.; Y.C., F.B.-C. All authors have read and agreed to the published version of the manuscript.

Funding: This research received no external funding.

Acknowledgments: We sincerely thank Jean-Claude Jeanny and Marina Yefimova for their significant contribution in the initial description of iron and iron-related proteins in retina.

Conflicts of Interest: Y.C., E.P. and F.B.C. are cited as inventor on a patent for the use of transferrin for the treatment of eye diseases.

Abbreviations

ABC	ATP binding cassette
AMD	Age-Related Macular Degeneration
APP	amyloid-beta precursor protein
BMP	Bone Morphogenetic Protein
BRB	Blood Retinal Barrier
CEC	Choroidal Endothelial Cell
CP	Ceruloplasmin
DFO	Deferroxamine
DMT1	divalent metal transporter 1
Fe-S	Cluster Iron-Sulfur
FPN	Ferroportin
FT	Ferritin
HEPC	Hepcidin
HEPH	Hephaestin
HFE	Hemochromatosis protein
HFT	Heavy Ferritin chin
HIF	Hypoxia Inducible Factor
HJV	Hemojuveline
HRE	Hypoxia Responsive Element
IL	Interleukin
IRE	Iron Responsive Element
IRP	Iron Regulatory Protein
LCN2	Lipocalin 2
LF	Lactoferrin
LFT	Light Ferritin chain
LIP	Labile Iron Pool
MacTel 2	Macular telangiectasia type 2
MGC	Muller Glial cell
NTBI	Non-Transferrin Bound Iron

NLRP3	NOD-like receptor family, pyrin domain containing 3
PCBP	Poly(rC)-binding proteins
PrP^C	Prion protein
PR	photoreceptor
REC	Retinal Endothelial Cell
ROS	Reactive Oxygen Species
RPE	Retinal Pigment Epithelium
SCARA5	Scavenger receptor class A, member 5
SMAD	Mothers against decapentaplegic homolog 1
TBI	Transferrin Bound Iron
TF	Transferrin
TFR	Transferrin Receptor
VEGF	Vascular Endothelial Growth Factor
ZP	Zyloklopen
ZIP	ZRT/IRT-like proteins

References

1. Daruich, A.; Matet, A.; Moulin, A.; Kowalczuk, L.; Nicolas, M.; Sellam, A.; Rothschild, P.-R.; Omri, S.; Gélizé, E.; Jonet, L.; et al. Mechanisms of macular edema: Beyond the surface. *Prog. Retin. Eye Res.* **2018**, *63*, 20–68. [CrossRef] [PubMed]

2. Anderson, B.; Saltzman, H.A. RETINAL OXYGEN UTILIZATION MEASURED BY HYPERBARIC BLACKOUT. *Arch. Ophthalmol.* **1964**, *72*, 792–795. [CrossRef] [PubMed]

3. Linsenmeier, R.A.; Zhang, H.F. Retinal oxygen: From animals to humans. *Prog. Retin. Eye Res.* **2017**, *58*, 115–151. [CrossRef] [PubMed]

4. Wang, L.; Törnquist, P.; Bill, A. Glucose metabolism in pig outer retina in light and darkness. *Acta Physiol. Scand.* **1997**, *160*, 75–81. [CrossRef] [PubMed]

5. Hurley, J.B.; Lindsay, K.J.; Du, J. Glucose, lactate, and shuttling of metabolites in vertebrate retinas. *J. Neurosci. Res.* **2015**, *93*, 1079–1092. [CrossRef] [PubMed]

6. Kurihara, T. Development and pathological changes of neurovascular unit regulated by hypoxia response in the retina. *Prog. Brain Res.* **2016**, *225*, 201–211.

7. Yang, L.; Wang, D.; Wang, X.-T.; Lu, Y.-P.; Zhu, L. The roles of hypoxia-inducible Factor-1 and iron regulatory protein 1 in iron uptake induced by acute hypoxia. *Biochem. Biophys. Res. Commun.* **2018**, *507*, 128–135. [CrossRef]

8. Yefimova, M.G.; Jeanny, J.C.; Guillonneau, X.; Keller, N.; Nguyen-Legros, J.; Sergeant, C.; Guillou, F.; Courtois, Y. Iron, ferritin, transferrin, and transferrin receptor in the adult rat retina. *Investig. Ophthalmol. Vis. Sci.* **2000**, *41*, 2343–2351.

9. Garcia-Castineiras, S. Iron, the retina and the lens: A focused review. *Exp. Eye Res.* **2010**, *90*, 664–678. [CrossRef]

10. Moos, T.; Bernth, N.; Courtois, Y.; Morgan, E.H. Developmental iron uptake and axonal transport in the retina of the rat. *Mol. Cell. Neurosci.* **2011**, *46*, 607–613. [CrossRef]

11. Hahn, P.; Song, Y.; Ying, G.S.; He, X.; Beard, J.; Dunaief, J.L. Age-dependent and gender-specific changes in mouse tissue iron by strain. *Exp. Gerontol.* **2009**, *44*, 594–600. [CrossRef] [PubMed]

12. Hahn, P.; Ying, G.S.; Beard, J.; Dunaief, J.L. Iron levels in human retina: Sex difference and increase with age. *Neuroreport* **2006**, *17*, 1803–1806. [CrossRef] [PubMed]

13. Ugarte, M.; Osborne, N.N.; Brown, L.A.; Bishop, P.N. Iron, zinc, and copper in retinal physiology and disease. *Surv. Ophthalmol.* **2013**, *58*, 585–609. [CrossRef] [PubMed]

14. Chen, H.; Liu, B.; Lukas, T.J.; Suyeoka, G.; Wu, G.; Neufeld, A.H. Changes in iron-regulatory proteins in the aged rodent neural retina. *Neurobiol. Aging* **2009**, *30*, 1865–1876. [CrossRef] [PubMed]

15. Chen, H.; Lukas, T.J.; Du, N.; Suyeoka, G.; Neufeld, A.H. Dysfunction of the retinal pigment epithelium with age: Increased iron decreases phagocytosis and lysosomal activity. *Investig. Ophthalmol. Vis. Sci.* **2009**, *50*, 1895–1902. [CrossRef] [PubMed]

16. Lane, D.J.R.; Merlot, A.M.; Huang, M.L.-H.; Bae, D.-H.; Jansson, P.J.; Sahni, S.; Kalinowski, D.S.; Richardson, D.R. Cellular iron uptake, trafficking and metabolism: Key molecules and mechanisms and their roles in disease. *Biochim. Biophys. Acta (BBA) - Mol. Cell Res.* **2015**, *1853*, 1130–1144. [CrossRef]

17. Lederman, M.; Obolensky, A.; Grunin, M.; Banin, E.; Chowers, I. Retinal Function and Structure in the Hypotransferrinemic Mouse. *Investig. Opthalmol. Vis. Sci.* **2012**, *53*, 605. [CrossRef]

18. Rageh, A.A.; Ferrington, D.A.; Roehrich, H.; Yuan, C.; Terluk, M.R.; Nelson, E.F.; Montezuma, S.R. Lactoferrin Expression in Human and Murine Ocular Tissue. *Curr. Eye Res.* **2016**, *41*, 883–889. [CrossRef]

19. Montezuma, S.R.; Dolezal, L.D.; Rageh, A.A.; Mar, K.; Jordan, M.; Ferrington, D.A. Lactoferrin Reduces Chorioretinal Damage in the Murine Laser Model of Choroidal Neovascularization. *Curr. Eye Res.* **2015**, *40*, 946–953. [CrossRef]

20. Parmar, T.; Parmar, V.M.; Arai, E.; Sahu, B.; Perusek, L.; Maeda, A. Acute Stress Responses Are Early Molecular Events of Retinal Degeneration in Abca4−/−Rdh8−/− Mice After Light Exposure. *Investig. Ophthalmol. Vis. Sci.* **2016**, *57*, 3257–3267. [CrossRef]

21. Parmar, T.; Parmar, V.M.; Perusek, L.; Georges, A.; Takahashi, M.; Crabb, J.W.; Maeda, A. Lipocalin 2 Plays an Important Role in Regulating Inflammation in Retinal Degeneration. *J. Immunol.* **2018**, *200*, 3128–3141. [CrossRef] [PubMed]

22. Cases, O.; Obry, A.; Ben-Yacoub, S.; Augustin, S.; Joseph, A.; Toutirais, G.; Simonutti, M.; Christ, A.; Cosette, P.; Kozyraki, R. Impaired vitreous composition and retinal pigment epithelium function in the FoxG1::LRP2 myopic mice. *Biochim. Biophys. Acta* **2017**, *1863*, 1242–1254. [CrossRef] [PubMed]

23. Hahn, P.; Dentchev, T.; Qian, Y.; Rouault, T.; Harris, Z.L.; Dunaief, J.L. Immunolocalization and regulation of iron handling proteins ferritin and ferroportin in the retina. *Mol. Vis.* **2004**, *10*, 598–607. [PubMed]

24. Theurl, M.; Song, D.; Clark, E.; Sterling, J.; Grieco, S.; Altamura, S.; Galy, B.; Hentze, M.; Muckenthaler, M.U.; Dunaief, J.L. Mice with hepcidin-resistant ferroportin accumulate iron in the retina. *FASEB J.* **2016**, *30*, 813–823. [CrossRef] [PubMed]

25. Hahn, P.; Qian, Y.; Dentchev, T.; Chen, L.; Beard, J.; Harris, Z.L.; Dunaief, J.L. Disruption of ceruloplasmin and hephaestin in mice causes retinal iron overload and retinal degeneration with features of age-related macular degeneration. *Proc. Natl. Acad. Sci. USA* **2004**, *101*, 13850–13855. [CrossRef]

26. Wolkow, N.; Song, D.; Song, Y.; Chu, S.; Hadziahmetovic, M.; Lee, J.C.; Iacovelli, J.; Grieco, S.; Dunaief, J.L. Ferroxidase hephaestin's cell-autonomous role in the retinal pigment epithelium. *Am. J. Pathol.* **2012**, *180*, 1614–1624. [CrossRef]

27. Wolkow, N.; Song, Y.; Wu, T.-D.; Qian, J.; Guerquin-Kern, J.-L.; Dunaief, J.L. Aceruloplasminemia: Retinal histopathologic manifestations and iron-mediated melanosome degradation. *Arch. Ophthalmol.* **2011**, *129*, 1466–1474. [CrossRef]

28. Dinet, V.; An, N.; Ciccotosto, G.D.; Bruban, J.; Maoui, A.; Bellingham, S.A.; Hill, A.F.; Andersen, O.M.; Nykjaer, A.; Jonet, L.; et al. APP involvement in retinogenesis of mice. *Acta Neuropathol.* **2011**, *121*, 351–363. [CrossRef]

29. Chen, H.; Attieh, Z.K.; Syed, B.A.; Kuo, Y.; Stevens, V.; Fuqua, B.K.; Andersen, H.S.; Naylor, C.E.; Evans, R.W.; Gambling, L.; et al. Identification of Zyklopen, a New Member of the Vertebrate Multicopper Ferroxidase Family, and Characterization in Rodents and Human Cells123. *J. Nutr.* **2010**, *140*, 1728–1735. [CrossRef]

30. He, X.; Hahn, P.; Iacovelli, J.; Wong, R.; King, C.; Bhisitkul, R.; Massaro-Giordano, M.; Dunaief, J.L. Iron homeostasis and toxicity in retinal degeneration. *Prog. Retin. Eye Res.* **2007**, *26*, 649–673. [CrossRef]

31. Sterling, J.; Guttha, S.; Song, Y.; Song, D.; Hadziahmetovic, M.; Dunaief, J.L. Iron importers Zip8 and Zip14 are expressed in retina and regulated by retinal iron levels. *Exp. Eye Res.* **2017**, *155*, 15–23. [CrossRef] [PubMed]

32. Picard, E.; Ranchon-Cole, I.; Jonet, L.; Beaumont, C.; Behar-Cohen, F.; Courtois, Y.; Jeanny, J.-C. Light-induced retinal degeneration correlates with changes in iron metabolism gene expression, ferritin level, and aging. *Investig. Ophthalmol. Vis. Sci.* **2011**, *52*, 1261–1274. [CrossRef] [PubMed]

33. Martin, P.M.; Gnana-Prakasam, J.P.; Roon, P.; Smith, R.G.; Smith, S.B.; Ganapathy, V. Expression and polarized localization of the hemochromatosis gene product HFE in retinal pigment epithelium. *Investig. Ophthalmol. Vis. Sci.* **2006**, *47*, 4238–4244. [CrossRef] [PubMed]

34. Gnana-Prakasam, J.P.; Thangaraju, M.; Liu, K.; Ha, Y.; Martin, P.M.; Smith, S.B.; Ganapathy, V. Absence of iron-regulatory protein Hfe results in hyperproliferation of retinal pigment epithelium: Role of cystine/glutamate exchanger. *Biochem. J.* **2009**, *424*, 243–252. [CrossRef] [PubMed]

35. Hadziahmetovic, M.; Song, Y.; Wolkow, N.; Iacovelli, J.; Kautz, L.; Roth, M.P.; Dunaief, J.L. Bmp6 regulates retinal iron homeostasis and has altered expression in age-related macular degeneration. *Am. J. Pathol.* **2011**, *179*, 335–348. [CrossRef] [PubMed]

36. Gnana-Prakasam, J.P.; Zhang, M.; Martin, P.M.; Atherton, S.S.; Smith, S.B.; Ganapathy, V. Expression of the iron-regulatory protein haemojuvelin in retina and its regulation during cytomegalovirus infection. *Biochem. J.* **2009**, *419*, 533–543. [CrossRef] [PubMed]

37. Tawfik, A.; Gnana-Prakasam, J.P.; Smith, S.B.; Ganapathy, V. Deletion of hemojuvelin, an iron-regulatory protein, in mice results in abnormal angiogenesis and vasculogenesis in retina along with reactive gliosis. *Investig. Ophthalmol. Vis. Sci.* **2014**, *55*, 3616–3625. [CrossRef] [PubMed]

38. Gnana-Prakasam, J.P.; Baldowski, R.B.; Ananth, S.; Martin, P.M.; Smith, S.B.; Ganapathy, V. Retinal expression of the serine protease matriptase-2 (Tmprss6) and its role in retinal iron homeostasis. *Mol. Vis.* **2014**, *20*, 561–574. [PubMed]

39. Hadziahmetovic, M.; Song, Y.; Ponnuru, P.; Iacovelli, J.; Hunter, A.; Haddad, N.; Beard, J.; Connor, J.R.; Vaulont, S.; Dunaief, J.L. Age-Dependent Retinal Iron Accumulation and Degeneration in Hepcidin Knockout Mice. *Investig. Ophthalmol. Vis. Sci.* **2011**, *52*, 109–118. [CrossRef] [PubMed]

40. Kast, B.; Schori, C.; Grimm, C. Hypoxic preconditioning protects photoreceptors against light damage independently of hypoxia inducible transcription factors in rods. *Exp. Eye Res.* **2016**, *146*, 60–71. [CrossRef]

41. Hughes, J.M.; Groot, A.J.; van der Groep, P.; Sersansie, R.; Vooijs, M.; van Diest, P.J.; Van Noorden, C.J.F.; Schlingemann, R.O.; Klaassen, I. Active HIF-1 in the normal human retina. *J. Histochem. Cytochem.* **2010**, *58*, 247–254. [CrossRef]

42. Perez Bay, A.E.; Schreiner, R.; Benedicto, I.; Rodriguez-Boulan, E.J. Galectin-4-mediated transcytosis of transferrin receptor. *J. Cell. Sci.* **2014**, *127*, 4457–4469. [CrossRef] [PubMed]

43. Picard, E.; Le Rouzic, Q.; Oudar, A.; Berdugo, M.; El Sanharawi, M.; Andrieu-Soler, C.; Naud, M.C.; Jonet, L.; Latour, C.; Klein, C.; et al. Targeting iron-mediated retinal degeneration by local delivery of transferrin. *Free Radic. Biol. Med.* **2015**, *89*, 1105–1121. [CrossRef] [PubMed]

44. Kaczara, P.; Zaręba, M.; Herrnreiter, A.; Skumatz, C.M.B.; Żądło, A.; Sarna, T.; Burke, J.M. Melanosome-iron interactions within retinal pigment epithelium-derived cells. *Pigment Cell Melanoma Res.* **2012**, *25*, 804–814. [CrossRef] [PubMed]

45. Baumann, B.H.; Shu, W.; Song, Y.; Simpson, E.M.; Lakhal-Littleton, S.; Dunaief, J.L. Ferroportin-mediated iron export from vascular endothelial cells in retina and brain. *Exp. Eye Res.* **2019**, *187*, 107728. [CrossRef] [PubMed]

46. Picard, E.; Fontaine, I.; Jonet, L.; Guillou, F.; Behar-Cohen, F.; Courtois, Y.; Jeanny, J.C. The protective role of transferrin in Muller glial cells after iron-induced toxicity. *Mol. Vis.* **2008**, *14*, 928–941. [PubMed]

47. Mendes-Jorge, L.; Ramos, D.; Valença, A.; López-Luppo, M.; Pires, V.M.R.; Catita, J.; Nacher, V.; Navarro, M.; Carretero, A.; Rodriguez-Baeza, A.; et al. Correction: L-Ferritin Binding to Scara5: A New Iron Traffic Pathway Potentially Implicated in Retinopathy. *PLoS ONE* **2017**, *12*, e0180288. [CrossRef]

48. Rousseau, E.; Michel, P.P.; Hirsch, E.C. The Iron-Binding Protein Lactoferrin Protects Vulnerable Dopamine Neurons from Degeneration by Preserving Mitochondrial Calcium Homeostasis. *Mol. Pharm.* **2013**, *84*, 888–898. [CrossRef]

49. Valapala, M.; Edwards, M.; Hose, S.; Grebe, R.; Bhutto, I.A.; Cano, M.; Berger, T.; Mak, T.W.; Wawrousek, E.; Handa, J.T.; et al. Increased Lipocalin-2 in the retinal pigment epithelium of *Cryba1* cKO mice is associated with a chronic inflammatory response. *Aging Cell* **2014**, *13*, 1091–1094. [CrossRef]

50. Ananth, S.; Gnana-Prakasam, J.P.; Bhutia, Y.D.; Veeranan-Karmegam, R.; Martin, P.M.; Smith, S.B.; Ganapathy, V. Regulation of the cholesterol efflux transporters ABCA1 and ABCG1 in retina in hemochromatosis and by the endogenous siderophore 2,5-dihydroxybenzoic acid. *Biochim. Biophys. Acta (BBA) - Mol. Basis Dis.* **2014**, *1842*, 603–612. [CrossRef]

51. Anderson, C.P.; Shen, M.; Eisenstein, R.S.; Leibold, E.A. Mammalian iron metabolism and its control by iron regulatory proteins. *Biochim. Biophys. Acta* **2012**, *1823*, 1468–1483. [CrossRef] [PubMed]

52. Baumann, B.H.; Shu, W.; Song, Y.; Sterling, J.; Kozmik, Z.; Lakhal-Littleton, S.; Dunaief, J.L. Liver-Specific, but Not Retina-Specific, Hepcidin Knockout Causes Retinal Iron Accumulation and Degeneration. *Am. J. Pathol.* **2019**, *189*, 1814–1830. [CrossRef] [PubMed]

53. Mowat, F.M.; Luhmann, U.F.O.; Smith, A.J.; Lange, C.; Duran, Y.; Harten, S.; Shukla, D.; Maxwell, P.H.; Ali, R.R.; Bainbridge, J.W.B. HIF-1alpha and HIF-2alpha Are Differentially Activated in Distinct Cell Populations in Retinal Ischaemia. *PLoS ONE* **2010**, *5*, e11103. [CrossRef] [PubMed]

54. Maio, N.; Rouault, T.A. Iron-sulfur cluster biogenesis in mammalian cells: New insights into the molecular mechanisms of cluster delivery. *Biochim. Biophys. Acta* **2015**, *1853*, 1493–1512. [CrossRef]

55. Das, D.; Patra, S.; Bridwell-Rabb, J.; Barondeau, D.P. Mechanism of frataxin "bypass" in human iron–sulfur cluster biosynthesis with implications for Friedreich's ataxia. *J. Biol. Chem.* **2019**, *294*, 9276–9284. [CrossRef] [PubMed]

56. Efimova, M.G.; Trottier, Y. Distribution of frataxin in eye retina of normal mice and of transgenic R7E mice with retinal degeneration. *J. Evol. Biochem. Phys.* **2010**, *46*, 414–417. [CrossRef]

57. Crombie, D.E.; Van Bergen, N.; Davidson, K.C.; Anjomani Virmouni, S.; Mckelvie, P.A.; Chrysostomou, V.; Conquest, A.; Corben, L.A.; Pook, M.A.; Kulkarni, T.; et al. Characterization of the retinal pigment epithelium in Friedreich ataxia. *Biochem. Biophys. Rep.* **2015**, *4*, 141–147. [CrossRef]

58. Rouault, T.A.; Maio, N. Biogenesis and functions of mammalian iron-sulfur proteins in the regulation of iron homeostasis and pivotal metabolic pathways. *J. Biol. Chem.* **2017**, *292*, 12744–12753. [CrossRef]

59. Crack, J.C.; Green, J.; Thomson, A.J.; Brun, N.E.L. Iron–Sulfur Clusters as Biological Sensors: The Chemistry of Reactions with Molecular Oxygen and Nitric Oxide. *Acc. Chem. Res.* **2014**, *47*, 3196–3205. [CrossRef]

60. Puig, S.; Ramos-Alonso, L.; Romero, A.M.; Martínez-Pastor, M.T. The elemental role of iron in DNA synthesis and repair. *Metallomics* **2017**, *9*, 1483–1500. [CrossRef]

61. Paul, V.D.; Lill, R. Biogenesis of cytosolic and nuclear iron-sulfur proteins and their role in genome stability. *Biochim. Biophys. Acta* **2015**, *1853*, 1528–1539. [CrossRef] [PubMed]

62. Luo, H.; Zhou, M.; Ji, K.; Zhuang, J.; Dang, W.; Fu, S.; Sun, T.; Zhang, X. Expression of Sirtuins in the Retinal Neurons of Mice, Rats, and Humans. *Front. Aging Neurosci.* **2017**, *9*, 366. [CrossRef] [PubMed]

63. Zhao, Z.; Chen, Y.; Wang, J.; Sternberg, P.; Freeman, M.L.; Grossniklaus, H.E.; Cai, J. Age-Related Retinopathy in NRF2-Deficient Mice. *PLoS ONE* **2011**, *6*, e19456. [CrossRef]

64. Alexeyev, M.; Shokolenko, I.; Wilson, G.; LeDoux, S. The Maintenance of Mitochondrial DNA Integrity–Critical Analysis and Update. *Cold Spring Harb. Perspect. Biol.* **2013**, *5*, a012641. [CrossRef] [PubMed]

65. Ballinger, S.W.; Van Houten, B.; Jin, G.F.; Conklin, C.A.; Godley, B.F. Hydrogen peroxide causes significant mitochondrial DNA damage in human RPE cells. *Exp. Eye Res.* **1999**, *68*, 765–772. [CrossRef] [PubMed]

66. Barreau, E.; Brossas, J.-Y.; Courtois, Y.; Treton, J.A. Accumulation of Mitochondrial DNA Deletions in Human Retina During Aging. *Investig. Ophthalmol. Vis. Sci.* **1996**, *37*, 384–391.

67. Gkotsi, D.; Begum, R.; Salt, T.; Lascaratos, G.; Hogg, C.; Chau, K.-Y.; Schapira, A.H.V.; Jeffery, G. Recharging mitochondrial batteries in old eyes. Near infra-red increases ATP. *Exp. Eye Res.* **2014**, *122*, 50–53. [CrossRef]

68. Tezel, T.H.; Geng, L.; Lato, E.B.; Schaal, S.; Liu, Y.; Dean, D.; Klein, J.B.; Kaplan, H.J. Synthesis and Secretion of Hemoglobin by Retinal Pigment Epithelium. *Investig. Opthalmol. Vis. Sci.* **2009**, *50*, 1911. [CrossRef]

69. Promsote, W.; Makala, L.; Li, B.; Smith, S.B.; Singh, N.; Ganapathy, V.; Pace, B.S.; Martin, P.M. Monomethylfumarate Induces γ-Globin Expression and Fetal Hemoglobin Production in Cultured Human Retinal Pigment Epithelial (RPE) and Erythroid Cells, and in Intact Retina. *Investig. Opthalmol. Vis. Sci.* **2014**, *55*, 5382. [CrossRef]

70. Hunt, R.C.; Hunt, D.M.; Gaur, N.; Smith, A. Hemopexin in the human retina: Protection of the retina against heme-mediated toxicity. *J. Cell. Physiol.* **1996**, *168*, 71–80. [CrossRef]

71. Chen, W.; Lu, H.; Dutt, K.; Smith, A.; Hunt, D.M.; Hunt, R.C. Expression of the protective proteins hemopexin and haptoglobin by cells of the neural retina. *Exp. Eye Res.* **1998**, *67*, 83–93. [CrossRef] [PubMed]

72. Ascenzi, P.; di Masi, A.; Leboffe, L.; Fiocchetti, M.; Nuzzo, M.T.; Brunori, M.; Marino, M. Neuroglobin: From structure to function in health and disease. *Mol. Asp. Med.* **2016**, *52*, 1–48. [CrossRef] [PubMed]

73. Tao, Y.; Ma, Z.; Liu, B.; Fang, W.; Qin, L.; Huang, Y.F.; Wang, L.; Gao, Y. Hemin supports the survival of photoreceptors injured by N-Methyl-N-nitrosourea: The contributory role of neuroglobin in photoreceptor degeneration. *Brain Res.* **2018**, *1678*, 47–55. [CrossRef] [PubMed]

74. Yu, Z.-L.; Qiu, S.; Chen, X.-C.; Dai, Z.-H.; Huang, Y.-C.; Li, Y.-N.; Cai, R.-H.; Lei, H.-T.; Gu, H.-Y. Neuroglobin – A potential biological marker of retinal damage induced by LED light. *Neuroscience* **2014**, *270*, 158–167. [CrossRef]

75. Jin, K.; Mao, X.; Xie, L.; Greenberg, D.A. Interactions between Vascular Endothelial Growth Factor and Neuroglobin. *Neurosci. Lett.* **2012**, *519*, 47–50. [CrossRef]

76. Gnana-Prakasam, J.P.; Reddy, S.K.; Veeranan-Karmegam, R.; Smith, S.B.; Martin, P.M.; Ganapathy, V. Polarized distribution of heme transporters in retinal pigment epithelium and their regulation in the iron-overload disease hemochromatosis. *Investig. Ophthalmol. Vis. Sci.* **2011**, *52*, 9279–9286. [CrossRef]

77. Moiseyev, G.; Takahashi, Y.; Chen, Y.; Gentleman, S.; Redmond, T.M.; Crouch, R.K.; Ma, J.-X. RPE65 is an iron(II)-dependent isomerohydrolase in the retinoid visual cycle. *J. Biol. Chem.* **2006**, *281*, 2835–2840. [CrossRef]

78. Hamel, C.P.; Tsilou, E.; Pfeffer, B.A.; Hooks, J.J.; Detrick, B.; Redmond, T.M. Molecular cloning and expression of RPE65, a novel retinal pigment epithelium-specific microsomal protein that is post-transcriptionally regulated in vitro. *J. Biol. Chem.* **1993**, *268*, 15751–15757.

79. Marlhens, F.; Bareil, C.; Griffoin, J.M.; Zrenner, E.; Amalric, P.; Eliaou, C.; Liu, S.Y.; Harris, E.; Redmond, T.M.; Arnaud, B.; et al. Mutations in RPE65 cause Leber's congenital amaurosis. *Nat. Genet.* **1997**, *17*, 139–141. [CrossRef]

80. Shyam, R.; Gorusupudi, A.; Nelson, K.; Horvath, M.P.; Bernstein, P.S. RPE65 has an additional function as the lutein to *meso* -zeaxanthin isomerase in the vertebrate eye. *Proc. Natl. Acad. Sci. USA* **2017**, *114*, 10882–10887. [CrossRef]

81. Betts-Obregon, B.S.; Gonzalez-Fernandez, F.; Tsin, A.T. Interphotoreceptor retinoid-binding protein (IRBP) promotes retinol uptake and release by rat Müller cells (rMC-1) in vitro: Implications for the cone visual cycle. *Investig. Ophthalmol. Vis. Sci.* **2014**, *55*, 6265–6271. [CrossRef]

82. Unger, E.L.; Earley, C.J.; Beard, J.L. Diurnal cycle influences peripheral and brain iron levels in mice. *J. Appl. Physiol.* **2009**, *106*, 187–193. [CrossRef] [PubMed]

83. Unger, E.L.; Jones, B.C.; Bianco, L.E.; Allen, R.P.; Earley, C.J. Diurnal variations in brain iron concentrations in BXD RI mice. *Neuroscience* **2014**, *263*, 54–59. [CrossRef] [PubMed]

84. Lim, S.; Scholten, A.; Manchala, G.; Cudia, D.; Zlomke-Sell, S.-K.; Koch, K.-W.; Ames, J.B. Structural Characterization of Ferrous Ion Binding to Retinal Guanylate Cyclase Activator Protein 5 from Zebrafish Photoreceptors. *Biochemistry* **2017**, *56*, 6652–6661. [CrossRef]

85. Shichi, H. Microsomal electron transfer system of bovine retinal pigment epithelium. *Exp. Eye Res.* **1969**, *8*, 60–68. [CrossRef]

86. Yefimova, M.G.; Jeanny, J.-C.; Keller, N.; Sergeant, C.; Guillonneau, X.; Beaumont, C.; Courtois, Y. Impaired retinal iron homeostasis associated with defective phagocytosis in Royal College of Surgeons rats. *Investig. Ophthalmol. Vis. Sci.* **2002**, *43*, 537–545.

87. McGahan, M.C.; Harned, J.; Mukunnemkeril, M.; Goralska, M.; Fleisher, L.; Ferrell, J.B. Iron alters glutamate secretion by regulating cytosolic aconitase activity. *Am. J. Physiol. Cell Physiol.* **2005**, *288*, C1117–C1124. [CrossRef]

88. Kaushik, P.; Gorin, F.; Vali, S. Dynamics of tyrosine hydroxylase mediated regulation of dopamine synthesis. *J. Comput. Neurosci.* **2007**, *22*, 147–160. [CrossRef]

89. Huang, Q.; Hong, X.; Hao, Q. SNAP-25 is also an iron-sulfur protein. *FEBS Lett.* **2008**, *582*, 1431–1436. [CrossRef]

90. Molday, R.S. Insights into the Molecular Properties of ABCA4 and Its Role in the Visual Cycle and Stargardt Disease. In *Progress in Molecular Biology and Translational Science*; Elsevier: Amsterdam, The Netherlands, 2015; Volume 134, pp. 415–431. ISBN 978-0-12-801059-4.

91. Ueda, K.; Kim, H.J.; Zhao, J.; Song, Y.; Dunaief, J.L.; Sparrow, J.R. Iron promotes oxidative cell death caused by bisretinoids of retina. *Proc. Natl. Acad. Sci. USA* **2018**, *115*, 4963–4968. [CrossRef]

92. Lucius, R.; Sievers, J. Postnatal retinal ganglion cells in vitro: Protection against reactive oxygen species (ROS)-induced axonal degeneration by cocultured astrocytes. *Brain Res.* **1996**, *743*, 56–62. [CrossRef]

93. Kurz, T.; Karlsson, M.; Brunk, U.T.; Nilsson, S.E.; Frennesson, C. ARPE-19 retinal pigment epithelial cells are highly resistant to oxidative stress and exercise strict control over their lysosomal redox-active iron. *Autophagy* **2009**, *5*, 494–501. [CrossRef] [PubMed]

94. Rogers, B.S.; Symons, R.C.A.; Komeima, K.; Shen, J.; Xiao, W.; Swaim, M.E.; Gong, Y.Y.; Kachi, S.; Campochiaro, P.A. Differential sensitivity of cones to iron-mediated oxidative damage. *Investig. Ophthalmol. Vis. Sci.* **2007**, *48*, 438–445. [CrossRef] [PubMed]

95. Różanowski, B.; Burke, J.M.; Boulton, M.E.; Sarna, T.; Różanowska, M. Human RPE Melanosomes Protect from Photosensitized and Iron-Mediated Oxidation but Become Pro-oxidant in the Presence of Iron upon Photodegradation. *Investig. Ophthalmol. Vis. Sci.* **2008**, *49*, 2838–2847. [CrossRef]

96. Akeo, K.; Hiramitsu, T.; Yorifuji, H.; Okisaka, S. Membranes of retinal pigment epithelial cells in vitro are damaged in the phagocytotic process of the photoreceptor outer segment discs peroxidized by ferrous ions. *Pigment Cell Res.* **2002**, *15*, 341–347. [CrossRef]

97. Harned, J.; Nagar, S.; McGahan, M.C. Hypoxia controls iron metabolism and glutamate secretion in retinal pigmented epithelial cells. *Biochim. Biophys. Acta* **2014**, *1840*, 3138–3144. [CrossRef]

98. Reiner, A.; Fitzgerald, M.E.C.; Del Mar, N.; Li, C. Neural control of choroidal blood flow. *Prog. Retin. Eye Res.* **2018**, *64*, 96–130. [CrossRef]

99. Imamura, T.; Hirayama, T.; Tsuruma, K.; Shimazawa, M.; Nagasawa, H.; Hara, H. Hydroxyl radicals cause fluctuation in intracellular ferrous ion levels upon light exposure during photoreceptor cell death. *Exp. Eye Res.* **2014**, *129*, 24–30. [CrossRef]

100. Guajardo, M.H.; Terrasa, A.M.; Catalá, A. Lipid-protein modifications during ascorbate-Fe2+ peroxidation of photoreceptor membranes: Protective effect of melatonin. *J. Pineal Res.* **2006**, *41*, 201–210. [CrossRef]

101. Hunt, R.C.; Handy, I.; Smith, A. Heme-mediated reactive oxygen species toxicity to retinal pigment epithelial cells is reduced by hemopexin. *J. Cell. Physiol.* **1996**, *168*, 81–86. [CrossRef]

102. Tian, Y.; He, Y.; Song, W.; Zhang, E.; Xia, X. Neuroprotective effect of deferoxamine on N-methyl-d-aspartate-induced excitotoxicity in RGC-5 cells. *Acta Biochim. Biophys. Sin. (Shanghai)* **2017**, *49*, 827–834. [CrossRef] [PubMed]

103. Thaler, S.; Fiedorowicz, M.; Rejdak, R.; Choragiewicz, T.J.; Sulejczak, D.; Stopa, P.; Zarnowski, T.; Zrenner, E.; Grieb, P.; Schuettauf, F. Neuroprotective effects of tempol on retinal ganglion cells in a partial optic nerve crush rat model with and without iron load. *Exp. Eye Res.* **2010**, *90*, 254–260. [CrossRef] [PubMed]

104. Wang, Z.J.; Lam, K.W.; Lam, T.T.; Tso, M.O. Iron-induced apoptosis in the photoreceptor cells of rats. *Investig. Ophthalmol. Vis. Sci.* **1998**, *39*, 631–633. [PubMed]

105. Daruich, A.; Le Rouzic, Q.; Jonet, L.; Naud, M.-C.; Kowalczuk, L.; Pournaras, J.-A.; Boatright, J.H.; Thomas, A.; Turck, N.; Moulin, A.; et al. Iron is neurotoxic in retinal detachment and transferrin confers neuroprotection. *Sci. Adv.* **2019**, *5*, eaau9940. [CrossRef]

106. Chaudhary, K.; Promsote, W.; Ananth, S.; Veeranan-Karmegam, R.; Tawfik, A.; Arjunan, P.; Martin, P.; Smith, S.B.; Thangaraju, M.; Kisselev, O.; et al. Iron Overload Accelerates the Progression of Diabetic Retinopathy in Association with Increased Retinal Renin Expression. *Sci. Rep.* **2018**, *8*, 3025. [CrossRef]

107. Totsuka, K.; Ueta, T.; Uchida, T.; Roggia, M.F.; Nakagawa, S.; Vavvas, D.G.; Honjo, M.; Aihara, M. Oxidative stress induces ferroptotic cell death in retinal pigment epithelial cells. *Exp. Eye Res.* **2019**, *181*, 316–324. [CrossRef]

108. Sun, Y.; Zheng, Y.; Wang, C.; Liu, Y. Glutathione depletion induces ferroptosis, autophagy, and premature cell senescence in retinal pigment epithelial cells. *Cell Death Dis.* **2018**, *9*, 753. [CrossRef]

109. Muckenthaler, M.U.; Rivella, S.; Hentze, M.W.; Galy, B. A Red Carpet for Iron Metabolism. *Cell* **2017**, *168*, 344–361. [CrossRef]

110. Gelfand, B.D.; Wright, C.B.; Kim, Y.; Yasuma, T.; Yasuma, R.; Li, S.; Fowler, B.J.; Bastos-Carvalho, A.; Kerur, N.; Uittenbogaard, A.; et al. Iron Toxicity in the Retina Requires Alu RNA and the NLRP3 Inflammasome. *Cell Rep.* **2015**, *11*, 1686–1693. [CrossRef]

111. Li, Y.; Song, D.; Song, Y.; Zhao, L.; Wolkow, N.; Tobias, J.W.; Song, W.; Dunaief, J.L. Iron-induced Local Complement Component 3 (C3) Up-regulation via Non-canonical Transforming Growth Factor (TGF)-beta Signaling in the Retinal Pigment Epithelium. *J. Biol. Chem* **2015**, *290*, 11918–11934. [CrossRef]

112. Vogi, W.; Nolte, R.; Brunahl, D. Binding of iron to the 5th component of human complement directs oxygen radical-mediated conversion to specific sites and causes nonenzymic activation. *Complement Inflamm* **1991**, *8*, 313–319.

113. Toomey, C.B.; Johnson, L.V.; Bowes Rickman, C. Complement factor H in AMD: Bridging genetic associations and pathobiology. *Prog. Retin. Eye Res.* **2018**, *62*, 38–57. [CrossRef] [PubMed]

114. Asthana, A.; Baksi, S.; Ashok, A.; Karmakar, S.; Mammadova, N.; Kokemuller, R.; Greenlee, M.H.; Kong, Q.; Singh, N. Prion protein facilitates retinal iron uptake and is cleaved at the β-site: Implications for retinal iron homeostasis in prion disorders. *Sci. Rep.* **2017**, *7*, 9600. [CrossRef] [PubMed]

115. You, L.-H.; Yan, C.-Z.; Zheng, B.-J.; Ci, Y.-Z.; Chang, S.-Y.; Yu, P.; Gao, G.-F.; Li, H.-Y.; Dong, T.-Y.; Chang, Y.-Z. Astrocyte hepcidin is a key factor in LPS-induced neuronal apoptosis. *Cell Death Dis.* **2017**, *8*, e2676. [CrossRef] [PubMed]

116. Gnana-Prakasam, J.P.; Martin, P.M.; Mysona, B.A.; Roon, P.; Smith, S.B.; Ganapathy, V. Hepcidin expression in mouse retina and its regulation via lipopolysaccharide/Toll-like receptor-4 pathway independent of Hfe. *Biochem. J.* **2008**, *411*, 79–88. [CrossRef] [PubMed]

117. Urrutia, P.J.; Hirsch, E.C.; González-Billault, C.; Núñez, M.T. Hepcidin attenuates amyloid beta-induced inflammatory and pro-oxidant responses in astrocytes and microglia. *J. Neurochem.* **2017**, *142*, 140–152. [CrossRef]

118. Ghosh, S.; Shang, P.; Yazdankhah, M.; Bhutto, I.; Hose, S.; Montezuma, S.R.; Luo, T.; Chattopadhyay, S.; Qian, J.; Lutty, G.A.; et al. Activating the AKT2-nuclear factor-κB-lipocalin-2 axis elicits an inflammatory response in age-related macular degeneration: Lipocalin-2 as an indicator of early AMD. *J. Pathol.* **2017**, *241*, 583–588. [CrossRef]

119. Coffman, L.G.; Brown, J.C.; Johnson, D.A.; Parthasarathy, N.; D'Agostino, R.B.; Lively, M.O.; Hua, X.; Tilley, S.L.; Muller-Esterl, W.; Willingham, M.C.; et al. Cleavage of high-molecular-weight kininogen by elastase and tryptase is inhibited by ferritin. *Am. J. Physiol. Lung Cell Mol. Physiol.* **2008**, *294*, L505–L515. [CrossRef]

120. Gnana-Prakasam, J.P.; Ananth, S.; Prasad, P.D.; Zhang, M.; Atherton, S.S.; Martin, P.M.; Smith, S.B.; Ganapathy, V. Expression and iron-dependent regulation of succinate receptor GPR91 in retinal pigment epithelium. *Investig. Ophthalmol. Vis. Sci.* **2011**, *52*, 3751–3758. [CrossRef]

121. Arjunan, P.; Gnanaprakasam, J.P.; Ananth, S.; Romej, M.A.; Rajalakshmi, V.-K.; Prasad, P.D.; Martin, P.M.; Gurusamy, M.; Thangaraju, M.; Bhutia, Y.D.; et al. Increased Retinal Expression of the Pro-Angiogenic Receptor GPR91 via BMP6 in a Mouse Model of Juvenile Hemochromatosis. *Investig. Ophthalmol. Vis. Sci.* **2016**, *57*, 1612–1619. [CrossRef]

122. Burke, J.M.; Smith, J.M. Retinal proliferation in response to vitreous hemoglobin or iron. *Investig. Ophthalmol. Vis. Sci.* **1981**, *20*, 582–592. [PubMed]

123. Loporchio, D.; Mukkamala, L.; Gorukanti, K.; Zarbin, M.; Langer, P.; Bhagat, N. Intraocular foreign bodies: A review. *Surv. Ophthalmol.* **2016**, *61*, 582–596. [CrossRef] [PubMed]

124. Konerirajapuram, N.S.; Coral, K.; Punitham, R.; Sharma, T.; Kasinathan, N.; Sivaramakrishnan, R. Trace elements iron, copper and zinc in vitreous of patients with various vitreoretinal diseases. *Indian J. Ophthalmol.* **2004**, *52*, 145–148. [PubMed]

125. Conart, J.-B.; Berrod, J.-P. [Non-traumatic vitreous hemorrhage]. *J. Fr. Ophtalmol.* **2016**, *39*, 219–225. [CrossRef] [PubMed]

126. Levin, A.V. Retinal hemorrhage in abusive head trauma. *Pediatrics* **2010**, *126*, 961–970. [CrossRef]

127. Casini, G.; Loiudice, P.; Menchini, M.; Sartini, F.; De Cillà, S.; Figus, M.; Nardi, M. Traumatic submacular hemorrhage: Available treatment options and synthesis of the literature. *Int. J. Retin. Vitr.* **2019**, *5*, 48. [CrossRef]

128. Bhisitkul, R.B.; Winn, B.J.; Lee, O.-T.; Wong, J.; de Souza Pereira, D.; Porco, T.C.; He, X.; Hahn, P.; Dunaief, J.L. Neuroprotective effect of intravitreal triamcinolone acetonide against photoreceptor apoptosis in a rabbit model of subretinal hemorrhage. *Investig. Ophthalmol. Vis. Sci.* **2008**, *49*, 4071–4077. [CrossRef]

129. Chen-Roetling, J.; Regan, K.A.; Regan, R.F. Protective effect of vitreous against hemoglobin neurotoxicity. *Biochem. Biophys. Res. Commun.* **2018**, *503*, 152–156. [CrossRef]

130. Zerbib, J.; Pierre-Kahn, V.; Sikorav, A.; Oubraham, H.; Sayag, D.; Lobstein, F.; Massonnet-Castel, S.; Haymann-Gawrilow, P.; Souied, E.H. Unusual retinopathy associated with hemochromatosis. *Retin Cases Brief Rep.* **2015**, *9*, 190–194. [CrossRef]

131. Gnana-Prakasam, J.P.; Tawfik, A.; Romej, M.; Ananth, S.; Martin, P.M.; Smith, S.B.; Ganapathy, V. Iron-mediated retinal degeneration in haemojuvelin-knockout mice. *Biochem. J.* **2012**, *441*, 599–608. [CrossRef]

132. Kumar, P.; Nag, T.C.; Jha, K.A.; Dey, S.K.; Kathpalia, P.; Maurya, M.; Gupta, C.L.; Bhatia, J.; Roy, T.S.; Wadhwa, S. Experimental oral iron administration: Histological investigations and expressions of iron handling proteins in rat retina with aging. *Toxicology* **2017**, *392*, 22–31. [CrossRef]

133. Shu, W.; Dunaief, J.L. Potential Treatment of Retinal Diseases with Iron Chelators. *Pharmaceuticals (Basel)* **2018**, *11*, 112. [CrossRef] [PubMed]

134. Wong, W.L.; Su, X.; Li, X.; Cheung, C.M.G.; Klein, R.; Cheng, C.-Y.; Wong, T.Y. Global prevalence of age-related macular degeneration and disease burden projection for 2020 and 2040: A systematic review and meta-analysis. *Lancet Glob. Health* **2014**, *2*, e106–e116. [CrossRef]

135. Handa, J.T.; Bowes Rickman, C.; Dick, A.D.; Gorin, M.B.; Miller, J.W.; Toth, C.A.; Ueffing, M.; Zarbin, M.; Farrer, L.A. A systems biology approach towards understanding and treating non-neovascular age-related macular degeneration. *Nat. Commun.* **2019**, *10*, 3347. [CrossRef] [PubMed]

136. Biesemeier, A.; Yoeruek, E.; Eibl, O.; Schraermeyer, U. Iron accumulation in Bruch's membrane and melanosomes of donor eyes with age-related macular degeneration. *Exp. Eye Res.* **2015**, *137*, 39–49. [CrossRef]

137. Hahn, P.; Milam, A.H.; Dunaief, J.L. Maculas Affected by Age-Related Macular Degeneration Contain Increased Chelatable Iron in the Retinal Pigment Epithelium and Bruch's Membrane. *Arch. Ophthalmol.* **2003**, *121*, 1099–1105. [CrossRef]

138. Junemann, A.G.; Stopa, P.; Michalke, B.; Chaudhri, A.; Reulbach, U.; Huchzermeyer, C.; Schlotzer-Schrehardt, U.; Kruse, F.E.; Zrenner, E.; Rejdak, R. Levels of aqueous humor trace elements in patients with non-exsudative age-related macular degeneration: A case-control study. *PLoS ONE* **2013**, *8*, e56734. [CrossRef]

139. Dentchev, T.; Hahn, P.; Dunaief, J.L. Strong labeling for iron and the iron-handling proteins ferritin and ferroportin in the photoreceptor layer in age-related macular degeneration. *Arch Ophthalmol.* **2005**, *123*, 1745–1746. [CrossRef]

140. Chowers, I.; Wong, R.; Dentchev, T.; Farkas, R.H.; Iacovelli, J.; Gunatilaka, T.L.; Medeiros, N.E.; Presley, J.B.; Campochiaro, P.A.; Curcio, C.A.; et al. The iron carrier transferrin is upregulated in retinas from patients with age-related macular degeneration. *Investig. Ophthalmol. Vis. Sci.* **2006**, *47*, 2135–2140. [CrossRef]

141. Čolak, E.; Žorić, L.; Radosavljević, A.; Ignjatović, S. The Association of Serum Iron-Binding Proteins and the Antioxidant Parameter Levels in Age-Related Macular Degeneration. *Curr. Eye Res.* **2018**, *43*, 659–665. [CrossRef]

142. Wysokinski, D.; Danisz, K.; Pawlowska, E.; Dorecka, M.; Romaniuk, D.; Robaszkiewicz, J.; Szaflik, M.; Szaflik, J.; Blasiak, J.; Szaflik, J.P. Transferrin receptor levels and polymorphism of its gene in age-related macular degeneration. *Acta Biochim. Pol.* **2015**, *62*, 177–184. [CrossRef] [PubMed]

143. Synowiec, E.; Pogorzelska, M.; Blasiak, J.; Szaflik, J.; Szaflik, J.P. Genetic polymorphism of the iron-regulatory protein-1 and -2 genes in age-related macular degeneration. *Mol. Biol. Rep.* **2012**, *39*, 7077–7087. [CrossRef] [PubMed]

144. Synowiec, E.; Szaflik, J.; Chmielewska, M.; Wozniak, K.; Sklodowska, A.; Waszczyk, M.; Dorecka, M.; Blasiak, J.; Szaflik, J.P. An association between polymorphism of the heme oxygenase-1 and -2 genes and age-related macular degeneration. *Mol. Biol. Rep.* **2012**, *39*, 2081–2087. [CrossRef] [PubMed]

145. Szemraj, M.; Oszajca, K.; Szemraj, J.; Jurowski, P. MicroRNA Expression Analysis in Serum of Patients with Congenital Hemochromatosis and Age-Related Macular Degeneration (AMD). *Med. Sci. Monit.* **2017**, *23*, 4050–4060. [CrossRef] [PubMed]

146. Ding, J.; Wong, T.Y. Current epidemiology of diabetic retinopathy and diabetic macular edema. *Curr. Diab. Rep.* **2012**, *12*, 346–354. [CrossRef] [PubMed]

147. Ciudin, A.; Hernández, C.; Simó, R. Iron overload in diabetic retinopathy: A cause or a consequence of impaired mechanisms? *Exp. Diabetes Res.* **2010**, *2010*, 714108. [CrossRef]

148. Baumann, B.; Sterling, J.; Song, Y.; Song, D.; Fruttiger, M.; Gillies, M.; Shen, W.; Dunaief, J.L. Conditional Müller Cell Ablation Leads to Retinal Iron Accumulation. *Investig. Ophthalmol. Vis. Sci.* **2017**, *58*, 4223–4234. [CrossRef]

149. Weinreb, R.N.; Aung, T.; Medeiros, F.A. The pathophysiology and treatment of glaucoma: A review. *JAMA* **2014**, *311*, 1901–1911. [CrossRef]

150. Wang, H.-W.; Sun, P.; Chen, Y.; Jiang, L.-P.; Wu, H.-P.; Zhang, W.; Gao, F. Research progress on human genes involved in the pathogenesis of glaucoma (Review). *Mol. Med. Rep.* **2018**, *18*, 656–674. [CrossRef]

151. Tripathi, R.C.; Borisuth, N.S.; Tripathi, B.J.; Gotsis, S.S. Quantitative and qualitative analyses of transferrin in aqueous humor from patients with primary and secondary glaucomas. *Investig. Ophthalmol. Vis. Sci.* **1992**, *33*, 2866–2873.

152. Farkas, R.H.; Chowers, I.; Hackam, A.S.; Kageyama, M.; Nickells, R.W.; Otteson, D.C.; Duh, E.J.; Wang, C.; Valenta, D.F.; Gunatilaka, T.L.; et al. Increased expression of iron-regulating genes in monkey and human glaucoma. *Investig. Ophthalmol. Vis. Sci.* **2004**, *45*, 1410–1417. [CrossRef] [PubMed]

153. Hohberger, B.; Chaudhri, M.A.; Michalke, B.; Lucio, M.; Nowomiejska, K.; Schlötzer-Schrehardt, U.; Grieb, P.; Rejdak, R.; Jünemann, A.G.M. Levels of aqueous humor trace elements in patients with open-angle glaucoma. *J. Trace Elem. Med. Biol.* **2018**, *45*, 150–155. [CrossRef] [PubMed]

154. Fick, A.; Jünemann, A.; Michalke, B.; Lucio, M.; Hohberger, B. Levels of serum trace elements in patients with primary open-angle glaucoma. *J. Trace Elem. Med. Biol.* **2019**, *53*, 129–134. [CrossRef] [PubMed]

155. Lin, S.-C.; Wang, S.Y.; Yoo, C.; Singh, K.; Lin, S.C. Association between serum ferritin and glaucoma in the South Korean population. *JAMA Ophthalmol.* **2014**, *132*, 1414–1420. [CrossRef] [PubMed]

156. Sarnat-Kucharczyk, M.; Rokicki, W.; Zalejska-Fiolka, J.; Pojda-Wilczek, D.; Mrukwa-Kominek, E. Determination of Serum Ceruloplasmin Concentration in Patients with Primary Open Angle Glaucoma with Cataract and Patients with Cataract Only: A Pilot Study. *Med. Sci. Monit.* **2016**, *22*, 1384–1388. [CrossRef] [PubMed]

157. DeToma, A.S.; Dengler-Crish, C.M.; Deb, A.; Braymer, J.J.; Penner-Hahn, J.E.; van der Schyf, C.J.; Lim, M.H.; Crish, S.D. Abnormal metal levels in the primary visual pathway of the DBA/2J mouse model of glaucoma. *Biometals* **2014**, *27*, 1291–1301. [CrossRef]

158. Anders, F.; Teister, J.; Funke, S.; Pfeiffer, N.; Grus, F.; Solon, T.; Prokosch, V. Proteomic profiling reveals crucial retinal protein alterations in the early phase of an experimental glaucoma model. *Graefes Arch. Clin. Exp. Ophthalmol.* **2017**, *255*, 1395–1407. [CrossRef]

159. Cheah, J.H.; Kim, S.F.; Hester, L.D.; Clancy, K.W.; Patterson, S.E.; Papadopoulos, V.; Snyder, S.H. NMDA receptor-nitric oxide transmission mediates neuronal iron homeostasis via the GTPase Dexras1. *Neuron* **2006**, *51*, 431–440. [CrossRef]

160. Liu, P.; Zhang, M.; Shoeb, M.; Hogan, D.; Tang, L.; Syed, M.F.; Wang, C.Z.; Campbell, G.A.; Ansari, N.H. Metal chelator combined with permeability enhancer ameliorates oxidative stress-associated neurodegeneration in rat eyes with elevated intraocular pressure. *Free Radic. Biol. Med.* **2014**, *69*, 289–299. [CrossRef]

161. Sakamoto, K.; Suzuki, T.; Takahashi, K.; Koguchi, T.; Hirayama, T.; Mori, A.; Nakahara, T.; Nagasawa, H.; Ishii, K. Iron-chelating agents attenuate NMDA-Induced neuronal injury via reduction of oxidative stress in the rat retina. *Exp. Eye Res.* **2018**, *171*, 30–36. [CrossRef]

162. Sirohi, K.; Chalasani, M.L.S.; Sudhakar, C.; Kumari, A.; Radha, V.; Swarup, G. M98K-OPTN induces transferrin receptor degradation and RAB12-mediated autophagic death in retinal ganglion cells. *Autophagy* **2013**, *9*, 510–527. [CrossRef] [PubMed]

163. Hamel, C. Retinitis pigmentosa. *Orphanet. J. Rare Dis.* **2006**, *1*, 40. [CrossRef] [PubMed]

164. Picard, E.; Jonet, L.; Sergeant, C.; Vesvres, M.H.; Behar-Cohen, F.; Courtois, Y.; Jeanny, J.C. Overexpressed or intraperitoneally injected human transferrin prevents photoreceptor degeneration in rd10 mice. *Mol. Vis.* **2010**, *16*, 2612–2625. [PubMed]

165. Scerri, T.S.; Quaglieri, A.; Cai, C.; Zernant, J.; Matsunami, N.; Baird, L.; Scheppke, L.; Bonelli, R.; Yannuzzi, L.A.; Friedlander, M.; et al. Genome-wide analyses identify common variants associated with macular telangiectasia type 2. *Nat. Genet.* **2017**, *49*, 559–567. [CrossRef] [PubMed]

166. Obolensky, A.; Berenshtein, E.; Lederman, M.; Bulvik, B.; Alper-Pinus, R.; Yaul, R.; Deleon, E.; Chowers, I.; Chevion, M.; Banin, E. Zinc-desferrioxamine attenuates retinal degeneration in the rd10 mouse model of retinitis pigmentosa. *Free Radic. Biol. Med.* **2011**, *51*, 1482–1491. [CrossRef]

167. Li, Z.L.; Lam, S.; Tso, M.O. Desferrioxamine ameliorates retinal photic injury in albino rats. *Curr. Eye Res.* **1991**, *10*, 133–144. [CrossRef]

168. Hadziahmetovic, M.; Song, Y.; Wolkow, N.; Iacovelli, J.; Grieco, S.; Lee, J.; Lyubarsky, A.; Pratico, D.; Connelly, J.; Spino, M.; et al. The Oral Iron Chelator Deferiprone Protects against Iron Overload–Induced Retinal Degeneration. *Investig. Ophthalmol. Vis. Sci.* **2011**, *52*, 959–968. [CrossRef]

169. Song, D.; Zhao, L.; Li, Y.; Hadziahmetovic, M.; Song, Y.; Dunaief, J.L. The oral iron chelator deferiprone protects against iron overload-induced retinal degeneration in Hepcidin knockout mice. *Investig. Ophthalmol. Vis. Sci.* **2014**, *55*, 4525–4532. [CrossRef]

170. Song, D.; Song, Y.; Hadziahmetovic, M.; Zhong, Y.; Dunaief, J.L. Systemic administration of the iron chelator deferiprone protects against light-induced photoreceptor degeneration in the mouse retina. *Free Radic. Biol. Med.* **2012**, *53*, 64–71. [CrossRef]

171. Hadziahmetovic, M.; Pajic, M.; Grieco, S.; Song, Y.; Song, D.; Li, Y.; Cwanger, A.; Iacovelli, J.; Chu, S.; Ying, G.-S.; et al. The Oral Iron Chelator Deferiprone Protects Against Retinal Degeneration Induced through Diverse Mechanisms. *Transl. Vis. Sci. Technol.* **2012**, *1*, 7. [CrossRef]

172. Arora, A.; Wren, S.; Gregory Evans, K. Desferrioxamine related maculopathy: A case report. *Am. J. Hematol.* **2004**, *76*, 386–388. [CrossRef] [PubMed]
173. Lakhanpal, V.; Schocket, S.S.; Jiji, R. Deferoxamine (Desferal)-induced toxic retinal pigmentary degeneration and presumed optic neuropathy. *Ophthalmology* **1984**, *91*, 443–451. [CrossRef]
174. Jauregui, R.; Park, K.S.; Bassuk, A.G.; Mahajan, V.B.; Tsang, S.H. Deferoxamine-induced electronegative ERG responses. *Doc. Ophthalmol.* **2018**, *137*, 15–23. [CrossRef] [PubMed]
175. Mobarra, N.; Shanaki, M.; Ehteram, H.; Nasiri, H.; Sahmani, M.; Saeidi, M.; Goudarzi, M.; Pourkarim, H.; Azad, M. A Review on Iron Chelators in Treatment of Iron Overload Syndromes. *Int. J. Hematol. Oncol. Stem Cell Res.* **2016**, *10*, 239–247. [PubMed]
176. Sahlstedt, L.; von Bonsdorff, L.; Ebeling, F.; Ruutu, T.; Parkkinen, J. Effective binding of free iron by a single intravenous dose of human apotransferrin in haematological stem cell transplant patients. *Br. J. Haematol.* **2002**, *119*, 547–553. [CrossRef]
177. Brittenham, G.M. Iron-chelating therapy for transfusional iron overload. *N. Engl. J. Med.* **2011**, *364*, 146–156. [CrossRef]
178. Goya, N.; Miyazaki, S.; Kodate, S.; Ushio, B. A family of congenital atransferrinemia. *Blood* **1972**, *40*, 239–245. [CrossRef]
179. Simon, S.; Athanasiov, P.A.; Jain, R.; Raymond, G.; Gilhotra, J.S. Desferrioxamine-related ocular toxicity: A case report. *Indian J. Ophthalmol.* **2012**, *60*, 315–317. [CrossRef]
180. Di Nicola, M.; Barteselli, G.; Dell'Arti, L.; Ratiglia, R.; Viola, F. Functional and Structural Abnormalities in Deferoxamine Retinopathy: A Review of the Literature. *Biomed. Res. Int.* **2015**, *2015*, 249617. [CrossRef]
181. Beau-Salinas, F.; Guitteny, M.A.; Donadieu, J.; Jonville-Bera, A.P.; Autret-Leca, E. High doses of deferiprone may be associated with cerebellar syndrome. *BMJ* **2009**, *338*, a2319. [CrossRef]
182. Mehdizadeh, M.; Nowroozzadeh, M.H. Posterior subcapsular opacity in two patients with thalassaemia major following deferiprone consumption. *Clin. Exp. Optom.* **2009**, *92*, 392–394. [CrossRef]
183. Taneja, R.; Malik, P.; Sharma, M.; Agarwal, M.C. Multiple transfused thalassemia major: Ocular manifestations in a hospital-based population. *Indian J. Ophthalmol.* **2010**, *58*, 125–130. [PubMed]
184. Masera, N.; Rescaldani, C.; Azzolini, M.; Vimercati, C.; Tavecchia, L.; Masera, G.; De Molfetta, V.; Arpa, P. Development of lens opacities with peculiar characteristics in patients affected by thalassemia major on chelating treatment with deferasirox (ICL670) at the Pediatric Clinic in Monza, Italy. *Haematologica* **2008**, *93*, e9–e10. [CrossRef] [PubMed]
185. Pan, Y.; Keane, P.A.; Sadun, A.A.; Fawzi, A.A. Optical coherence tomography findings in deferasirox-related maculopathy. *Retin Cases Brief Rep.* **2010**, *4*, 229–232. [CrossRef]
186. Farajipour, H.; Rahimian, S.; Taghizadeh, M. Curcumin: A new candidate for retinal disease therapy? *J. Cell. Biochem.* **2018**, *120*, 6886–6893. [CrossRef]
187. Majumdar, S.; Srirangam, R. Potential of the Bioflavonoids in the Prevention/Treatment of Ocular Disorders. *J. Pharm. Pharmacol.* **2010**, *62*, 951–965. [CrossRef] [PubMed]
188. Voigt, A.P.; Whitmore, S.S.; Flamme-Wiese, M.J.; Riker, M.J.; Wiley, L.A.; Tucker, B.A.; Stone, E.M.; Mullins, R.F.; Scheetz, T.E. Molecular characterization of foveal versus peripheral human retina by single-cell RNA sequencing. *Exp. Eye Res.* **2019**, *184*, 234–242. [CrossRef]
189. Qian, Z.M.; Li, H.; Sun, H.; Ho, K. Targeted drug delivery via the transferrin receptor-mediated endocytosis pathway. *Pharm. Rev.* **2002**, *54*, 561–587. [CrossRef]
190. Gomme, P.T.; McCann, K.B.; Bertolini, J. Transferrin: Structure, function and potential therapeutic actions. *Drug Discov. Today* **2005**, *10*, 267–273. [CrossRef]
191. de Jong, P.T.V.M. A Historical Analysis of the Quest for the Origins of Aging Macula Disorder, the Tissues Involved, and Its Terminology. *Ophthalmol. Eye. Dis.* **2016**, *8*, 5–14. [CrossRef]
192. Weber, C.; Cole, D.J.; O'Regan, D.D.; Payne, M.C. Renormalization of myoglobin–ligand binding energetics by quantum many-body effects. *PNAS* **2014**, *111*, 5790–5795. [CrossRef] [PubMed]

Review

Noble Metals and Soft Bio-Inspired Nanoparticles in Retinal Diseases Treatment: A Perspective

Valeria De Matteis [1,*] and Loris Rizzello [2,3,4]

[1] Department of Mathematics and Physics "Ennio De Giorgi", University of Salento, Via Arnesano, 73100 Lecce, Italy

[2] Department of Chemistry, University College London, 20 Gordon Street, London WC1H 0AJ, UK; lrizzello@ibecbarcelona.eu

[3] Institute for Bioengineering of Catalonia (IBEC), The Barcelona Institute of Science and Technology, Baldiri Reixac 10-12, 08028 Barcelona, Spain

[4] Department of Pharmaceutical Sciences, University of Milan, via Mangiagalli 25, 20133 Milano, Italy

* Correspondence: valeria.dematteis@unisalento.it

Received: 31 January 2020; Accepted: 8 March 2020; Published: 10 March 2020

Abstract: We are witnessing an exponential increase in the use of different nanomaterials in a plethora of biomedical fields. We are all aware of how nanoparticles (NPs) have influenced and revolutionized the way we supply drugs or how to use them as therapeutic agents thanks to their tunable physico-chemical properties. However, there is still a niche of applications where NP have not yet been widely explored. This is the field of ocular delivery and NP-based therapy, which characterizes the topic of the current review. In particular, many efforts are being made to develop nanosystems capable of reaching deeper sections of the eye such as the retina. Particular attention will be given here to noble metal (gold and silver), and to polymeric nanoparticles, systems consisting of lipid bilayers such as liposomes or vesicles based on nonionic surfactant. We will report here the most relevant literature on the use of different types of NPs for an efficient delivery of drugs and bio-macromolecules to the eyes or as active therapeutic tools.

Keywords: retinal diseases; noble metals NPs; bio-inspired NPs; drug delivery

---

1. Introduction

The human eye is a unique and fragile structure prone to different types of pathologies. It is characterized by an anterior and the posterior segment [1]. Pupil, cornea, iris, ciliary body, aqueous humor and lens make up the anterior segment, whereas vitreous humor, macula, retina, choroid, and optic nerve are parts of the posterior segment [1]. The macula is a small area of the retina that contains photoreceptor, special light-sensitive cells, connected to a neuronal network [2]. Retina contains five types of neurons: photoreceptor, bipolar, ganglion, horizontal, and amacrine cells (Figure 1) [3,4]. The retina is the nerve layer that receives visual information, creating impulses that are transmitted through the optic nerve to the visual cortex in the brain [5]. Two types of photoreceptor cells exist: the rods and the cones. It is estimated that there are approximately 120 million rods and about six million cones in the human eye. The unique properties of photoreceptor cells expose them to a variety of genetic and environmental threats. Any dysfunction or deterioration of vital ocular tissue has a dramatic impact on a patient's quality of life. Therefore, the retina is susceptible to several diseases that result in visual loss and blindness. As countries become wealthier and the per capita income increases, prevalence of blindness decreases, and causes of blindness change. In a poor African country, the major of blinding conditions are likely to be cataract and corneal scar. In the middle-income countries such as Latin America, the leading causes of blindness are glaucoma and diabetic retinopathy. Because cataract surgery is more readily available, fewer people become blind from cataract [6]. In wealthy countries, glaucoma and cataract continue to be very common and important conditions, but most of the blindness is due to retinal degenerative diseases (RDDs) [7], an heterogeneous group of phatologies characterized by progressive death of photoreceptors. RDDs can be caused by defects in proteins involved in photo transduction, synaptic transmission, retinal pigment epithelial (RPE) integrity function, intracellular trafficking and cilia function. It can be also related to toxic accumulation of retinoids. Each of these RDDs can lead to visual loss or complete blindness [8].

The most common RDDs include age-related macular degeneration (AMD), diabetic retinopathy (DR), and retinitis pigmentosa (RP). Cytomegalovirus (CMV) retinitis, proliferative vitreoretinopathy, Stargardt disease, and retinoblastoma can also occur and are usually associated with toxic retinoid biogenesis and accumulation [9]. Retinal diseases are already the most common cause of childhood blindness [10] with an incidence of 1.5 million people suffering worldwide. RP is characterized by a progressive loss of photoreceptors and its specific mechanisms are not fully understood. Following the loss of photoreceptor functionality, a retinal remodeling phenomenon is triggered as the final step, which then culminates with cell death and topological restructuring of the retina. The progression of retinal remodeling is similar to negative plasticity that occurs in some pathologies like trauma and epilepsy constituting substantial impediments to rescue strategies of all types [11]. Another process that involve the retina segment is neovascularization, a pathological phenomenon that provokes vision loss causing blindness due to its correlation with diabetic retinopathy and AMD [12]. In the neovascularization phenomenon, endothelial growth factor (VEGF) pathway has a key role [13,14].

Ocular diseases, and retinal diseases in particular, are still approached using rather invasive treatments, which are also not completely effective. The development of efficacious therapeutics is thus a primary research goal. So far, many delivery strategies have been explored, and the progress in the field of molecular genetics has led to the identification of genes involved in retinal disruption. For this reason, gene delivery systems represent one of the most promising approaches [15,16]. Delivery approaches can be broadly classified in two groups, whether they are based on a viral or non-viral method of delivery. Each system comes with advantages and disadvantages. For instance, although Adeno-Associated Virus (AAV) based vectors have high gene transduction efficiency, they face a major obstacle in terms of biosafety for clinical applications. The development of effective non-viral approaches for ocular disease is thus crucial. However, how to adequately deliver therapeutic agents to the back of eye (i.e., retina) remains a challenge. With the development of nanotechnology and bioengineering science, a wide range of innovations in this area have been achieved. This holds the promise for developing more effective ocular drug delivery systems.

In this review, we carefully report the last studies in which noble metal and soft (bioinspired) NPs were applied as new delivery agents for treatment of ocular disease in vitro and in vivo. We also provide perspective on opportunities and challenges for future nanotechnology in retinal drug delivery and therapies.

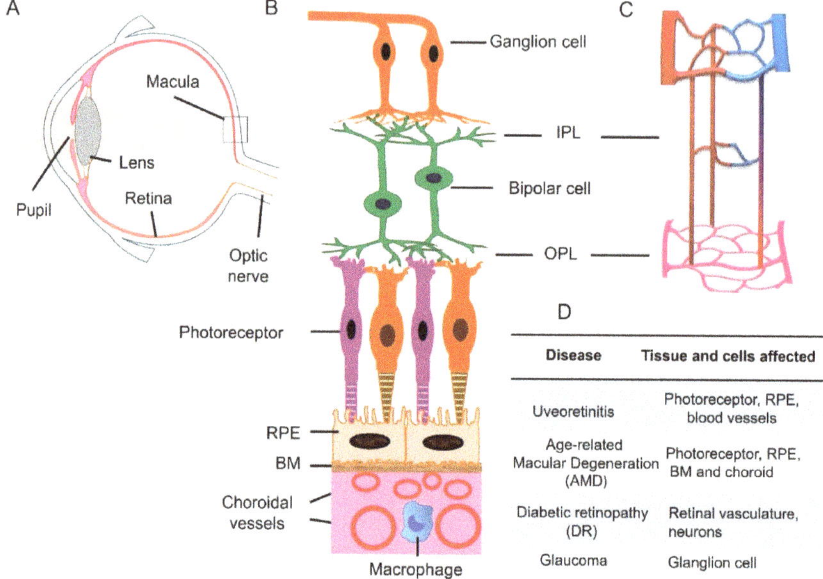

Figure 1. Retinal neuronal and vascular structure and retinal disease. (**A**) Diagram of a human eye. Light passes through the pupil and is focused by lens onto macula of the retinal layer at the back of the eye. (**B**) Retina consists of three layers of neurons, photoreceptor, bipolar, and ganglion cells. The retinal pigment epithelial (RPE) monolayer together with Bruch's membrane (BM) form outer blood retinal barrier that separates the neuroretina from the choroid. Choroidal circulation provides oxygen and nutrients to outer retina. (**C**) The retina has an interconnected network of three vascular layers located in the ganglion cell/nerve fiber layer, inner plexiform layer (IPL), and outer plexiform layer (OPL). (**D**) retinal tissue and cells that are affected under different disease conditions. Adapted from reference [17].

2. Drug Ocular Administrations

The delivery of drugs in the eyes presents many difficulties. As in the case of the brain, it presents multiple physical boundaries: corneal and conjunctival epithelium, blood-aqueous barriers (BAB), and blood-retinal barriers (BRB). These barriers finely control the transit of molecules and fluids across eye and, in most cases, the drug is unable to enrich the deeper layers. The BRB consists of two different barriers: (i) the outer BRB characterized by endothelial cells lining the choroidal vasculature (fenestrated blood vessels) and by tight-junction-coupled RPE cells, and (ii) inner BRB. The latter is made by endothelial cells in conjunction with pericytes, astrocytes, and Müller glial cells. The inner BRB shows a lot of similarities with blood–brain barrier (BBB) with the exception of Müller glia [18]. In general, the most common drug administration is through eye drops and suspensions' that target only the anterior segment of eye, mainly due to the patient compliance. With such a method, less than 5% of topically applied dose is delivered to deeper ocular tissues, such as in vitreous cavity [19]. The chemical nature of bioactive molecules also influences the uptake. For example, BRB is selectively permeable to lipophilic molecules in the presence of specific eye drugs transporters. In this way, the formulation of new therapies should take in consideration this limitation [20]. Ocular transporters are divided in two groups: the solute carrier (SLC) family and the ATP-binding cassette (ABC) family. The electrochemical gradient was used by SLC transporters to induce the uptake of molecules through cell membrane, whereas ABC transporters employ ATP [21]. However, physiological eye phenomena, such as tear turnover, reflex blinking, ocular static, or lacrimal drainage reduce the action of the drug and consequently the eye drop administration fails to treat retina disease [22]. As a matter of fact, in unstimulated conditions, total aqueous tear volume is 7 μL and the normal tear turn-over is about 1.2 μL/min [23]. Thus, the precorneal drug half-life is about 3 minutes [24]. Therefore, treatment of posterior segment injuries still remains a challenge. For all the reasons explained above, medical research seeks alternative routes such as intravitreal injections. This method consists of direct administration into vitreus via *pars plana* using a needle. In this way, a high concentration of drug is guaranteed at the retina level, but the molecule half-life depends on its molecular weight. In fact, proteins and peptides characterized by high molecular weight (ranging from 40 kDa to 70 kDa) and steric hindrance showed longer retention [25,26]. However, a lot of complications (hemorrhages, retinal detachment, cataracts) can manifest after injection [27]. An alternative method is periocular route (Figure 2) that was demonstrated to be an effective route to direct drugs to the posterior eye segment consisting of subconjunctival, sub-tenon, retrobulbar, peribulbar, and posterior juxtascleral [28]. Administration of drug by sub-tenon injection was found to be more suitable as demonstrated by Ghate et al. [29], where sodium fluorescein was used in rabbits' eyes by subocular administration. The study concluded that injection of drug via sub-tenon resulted in highest vitreous concentration of sodium fluorescein (two sodium fluorescein (NaF) concentrations, 2.5 mg in 0.1 mL (c1) and 2.5 mg in 0.5 mL (c2)) compared to the other routes. However, also in this case, a lot of adverse effects have been observed, such as strabismus, hyphema, and intraocular pressure [30].

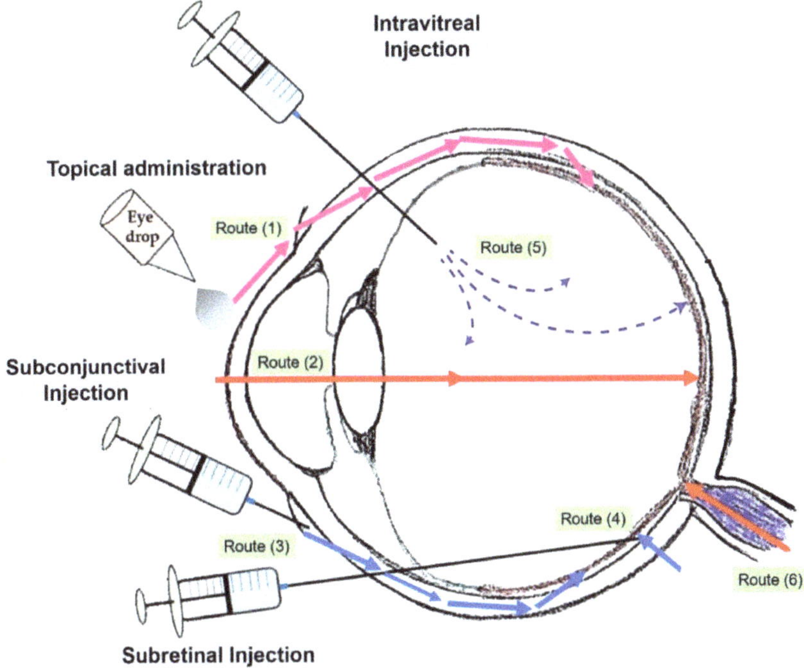

Figure 2. Common drug administration routes through the eye. Topical administration (**1,2**), subconjunctival injection (periocular route) (**3**), subretinal injection (**4**), and intravitreal injection (**5**). Adapted from reference [31].

3. Physico-Chemical Properties of Nanomaterials

A "Nanomaterial" is defined as a "material with any external dimension in the nanoscale or having internal structure or surface structure in the nanoscale" (ISO, 2010) [32]. Similarly, a definition of "nanoparticle" is a "nano-object with all three external dimensions in the nanoscale" where nanoscale was defined as size ranging from 1 nm to 100 nm (ISO, 2008) [33]. NPs can be obtained by many synthetic routes, starting from chemical elements such as carbon, metals, metal oxide, biological molecules, and polymers [34]. The biological effects are strongly affected by physico-chemical properties of NPs such as surface charge, size, shape, and solubility. In addition, they exhibit greater surface area per unit mass compared to bulk materials [35]. Because of the above mentioned properties and of the material they are made, NPs are ideal tools to treat retinal disease as active components of the therapy without the help of drugs [36,37]. In addition, NPs with a smaller size (<20 nm) are demonstrated to have the ability to cross eye barriers including cornea, conjunctiva, and BRB [38,39]. NPs for ocular therapy include inorganic NPs (metal oxide and noble metal NPs), as well as soft-biopolymer-based NPs. Metal NPs are more suitable as active therapy tools (for their intrinsic properties), and soft-NPs are more efficient in encapsulating drugs and macromolecules due to their ability to form aqueous-suspended vesicles having a hollow lumen. All these nanomaterials are different in charge, shape, and size, but they are able to be internalized by cells to treat retinal disease [40]. Finally, recent applications of soft nanorobot (with size range in the molecular scale) have been studied. These structures have the capability to deliver active biomolecules in ocular sections due to their ability to make changes in a controlled and predictable manner to the environment following external stimuli [41].

3.1. Noble Metal NPs: The Case of Gold (Au) and Silver (Ag)

Noble metals, especially AuNPs and AgNPs, are characterized by unique optical properties. For this reason they are used for a variety of applications owing to collective oscillations of conduction electrons coupled with incident light [42–45]. This phenomenon, known as Localized Surface Plasmon Resonance (LSPR) is strongly influenced by NPs' shape and the metal they are formed. The surface plasmon resonance bands of metal NPs can be tuned from visible to Near Infrared Region (NIR), which is a typical wavelength to penetrate and analyze biological tissues [46]. Even if AuNPs are the most studied in nanomedicine field due to their good chemical stability and well-controlled size/surface functionalization and biocompatibility, AgNPs have the advantage of possessing an antibacterial and antiangiogenesis effect [47,48].

3.2. Au and AgNPs: Safety Studies in In Vitro and In Vivo Retinal Models

3.2.1. AuNPs

The use of novel nanomaterials necessitates a better understanding of potential adverse effects on biological entities [49]. These effects are strictly dependent on physico-chemical properties of NPs and evaluation of cytotoxicity is critical, especially from a medical field application perspective [50]. A lot of studies were conducted to assess biocompatibility in order to ensure safe drug release with low toxicity. In general, high degree of biocompatibility is verified when NPs interact with the living organisms without triggering unacceptable toxic, thrombogenic, carcinogenic, and immunogenic responses [51].

The unique physico-chemical properties of AuNPs make them a powerful tool in the biomedical field. They are widely used in photothermal treatment of cancer and in drug and gene delivery. For their synthesis, auric salts such as $HAuCl_4$ are used as starting materials to obtain stable AuNPs with a size ranging from 1–100 nm (Brust or Turkevish [52,53]). In recent years, green syntheses of AuNPs has also been developed to reduce solvents and reduce/cap agents' toxicity so that AuNPs can be synthetized from plant extracts or biomass [54,55]. AuNPs are easy to functionalize with a wide range of ligands ranging to polymer, DNA, peptides, RNA, and fluorescent molecules particularly suitable for bio-imaging applications [56]. Furthermore, various cytotoxicity studies demonstrated their low toxicity due to the inert nature of Au [57,58].

Regarding the biodistribution, a recent work demonstrated a rapid elimination of colloidal AuNPs from blood circulation in a rats model, and liver and spleen accumulation without inducing adverse effects or inflammation [59]. Au comprised in passion fruit-like nanoarchitectures (NAs) were studied in murine models demonstrating the excretion by renal and biliary pathways without nephrotoxicity up to 150 mg kg^{-1} [60]. Some studies reported pro-inflammatory cytokines expressions induced by

AuNPs in different organs of rats [61] and mice [62]. However, the cytotoxic effects were temporary and normalized in a short time. The biocompatibility of Au is strictly correlated to their size, shape, concentration, as well as surface area as demonstrated by Biswas's group [63]. They investigated the effects of Au spheres (5, 10, 20, 30, 50, and 100 nm), cubes (50 nm), and rods (10 × 90 nm) in cells in adult retinal pigment epithelial cell line 19 (ARPE-19). The cubes with an average diameter of 50 nm and spheres (50–100 nm) resulted in being biocompatible up to 750 μg of concentration. NPs with a smaller size (<30 nm) and nanorods (10 nm) showed higher toxicity, but they are more able to be uptaked by cells. Kim et al. [64] used C57BL/6 mice to demonstrate the ability of AuNPs (20 nm) to cross BRB and enriched retinal layers after intravenous administration. NPs greatly accumulated in neurons (75 ± 5%), followed by endothelial cells (17 ± 6%) and peri-endothelial glial cells (8 ± 3%) without affecting both viability of endothelial cells, astrocytes, retinoblastoma cells, and expression of functional biomolecules. After five weeks of post administration, the thickness of retinal layers was undisturbed without any damage. Bakri et al. [65] used Dutch-belted rabbits to test biocompatibility of AuNPs. They used two groups of rabbits receiving a 67 μmol/0.1 mL and 670 μmol/0.1 mL of AuNPs into one eye respectively by intravitreal injection. After one month, evident inflammation or adverse effects were not detected also at a high dosage. Soderstjerna et al. [66] used AuNPs and AgNPs with 19 and 89 nm citrate capped to test biocompatibility in in vitro cultured retinas using 800 NPs/cell. The retina tissue model offered a sustained long-term retinal organization and controlled serum-free conditions. This study demonstrated that low concentrations of Au and AgNPs have adverse effects on retina and can induce apoptosis and oxidative stress. A strong connection between NPs hydrodynamic diameter, surface functionalization/concentration and the biological effects is evident. In particular, the authors showed that NPs size should be large enough to avoid their fast flow into circulation, but small enough to be administrated and easy cross the BRB [67].

Starting from this assumption, the application of AuNPs is influenced by these parameters that should be taken in consideration to customize nanotools for a specific biomedical application.

3.2.2. AgNPs

Numerous in vitro and in vivo studies demonstrated the toxicity of AgNPs that perturbs several cellular pathways [4]. The most accepted mechanism through which they induce side effects, despite still under investigation, seems to be related to Ag ions released from NPs surface [68–70]. In terms of production process, AgNPs can be synthetized by different chemical routes such as chemical-based reduction [71], sol-gel, or green synthesis [72,73].

Jun et al. [74] used zebrafish embryos as model organism to test their delivery at an ocular level. They investigated the toxicity of AgNPs at concentration of 0.4 mg/L and observed down- expressions of different lens crystalline genes by unknown mechanisms that either induced apoptosis or blocked nuclear DNA or RNA export.

Sriram et al. [75] demonstrated the cytotoxicity and Reactive Oxygen Species (ROS)-induced apoptosis after exposure of AgNPs with two different size (22.4 and 42.5 nm) in bovine retinal endothelial cells (BRECs) finding size-dependent toxicity. As a matter of fact, smaller NPs induced adverse effects on cells. Kim and colleagues [76] exposed New Zealand white rabbits (according to OECD test guidelines) to AgNPs in order to understand the possible eye irritation effect. Oedema and conjunctival redness have been observed after 1 h of AgNPs incubation, but the cornea, iris, or conjunctiva resulted unaffected after 24, 48, and 72 h of NPs exposure. In similar experiments, the instillation of AgNPs (5000 ppm) were tested on another in vivo model, the guinea pigs, observing a transient conjunctivae irritation during 24 h of early observation [77].

4. Noble-Metal NPs as Therapy and Diagnostic for Retinal Disease

AuNPs scatter or absorb light at specific wavelengths as a function of their physico-chemical properties. This characteristic is particularly suitable for bioimaging and to treat some diseases like cancer by NIR-triggered photothermal therapy (PTT), in which light is converted to heat in the tumor site where NPs accumulate [78]. In the NIR (650–900 nm), the absorption coefficients of haemoglobin and water are very low, and for this reason, the penetration of NIR wavelength in tissues is very high, allowing NPs stimulation without causing damage [79]. Song et al. [80] reported application of gold nanodisks (GNDs) on human retinal microvascular endothelial cells (HRMEC) and mouse oxygen-induced retinopathy model. The GNDs were optically suitable for nanomedicine applications due to their ability to produce detectable signals regardless of polarization or direction of light source and to scatter in NIR region (830 nm). In addition, GNDs were internalized less efficiently in cells with respect to AuNPs or Au rods (AuNRs), thus minimizing the ROS production and other several adverse effects. The authors used differently charged anisotropic NPs (nanorods or nanodisks) and compared them with AuNPs. After simulating the role of different sizes to select the best diameter (160 nm), they showed that GNDs were able to bind vascular endothelial growth factor (VEGF) inhibiting in vitro angiogenesis with high performance and abolished the VEGF-induced migration of endothelial cells. When these nanostructures were applied to mice eyes, a strong signal in optical coherence tomography (OCT) was detected with a consequent reduction of retinal degeneration (Figure 3). The OCT system involves the use of light energy that, once absorbed by tissues, causes a thermos-elastic expansion generating ultrasonic waves detected by a transducer.

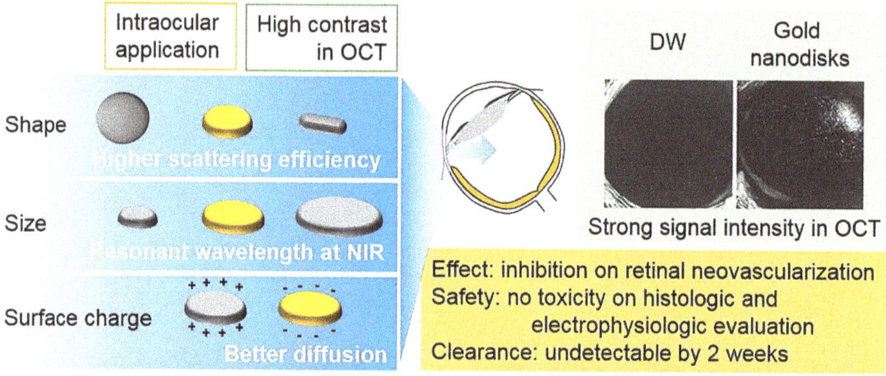

Figure 3. Schematic representation of gold nanodisks (GNDs) effects in retinal degeneration and Optical Coherence Tomography (OCT) imaging. The GNDs were particularly suitable for their optical properties connected to the shape: GNDs absorbed in the Near Infrared Region (NIR). The size and surface charge also influenced the eye diffusion. In OCT images, GNDs exhibited a strong signal compared to the dual window (DW) processing method, which is used to detect modulation of OCT signals due to scattering or absorption. Reprinted from [80] Copyright (2017), with permission from Elsevier.

Recent therapies to treat some diseases strictly connected to retinal degeneration consider the transplantation of photoreceptor precursors (PRPs) as efficient support improving visual acuity [81]. The analysis of cell survivor in the host retina is a key factor in this kind of treatment. Chemla et al. [82] used AuNPs in order to improve the tracking of transplanted PRPs. Cells were labelled with photothermic AuNPs and transplanted in the vitreous and sub-retinal space of Long-Evans pigmented rats (4–8 weeks old) by gauge needle.

By using coherence tomography, computed tomography and fluorescence fundus imaging, NPs have been tracked for 1 month without recording adverse effects. Real-time tracking is essential for evaluating the efficiency of retinal cell replacement therapies. These studies confirmed the possibility to use AuNPs to improve retinal cell therapy in humans. AuNPs-based photothermal therapy has been also used to treat eye tumors in vitro and in vivo. Some eye cancers provoke retinal detachment leading to a decreased visual eye activity.

In this framework, Kim et al. [83] tested anticancer properties of Doxorubicin (Dox)-conjugated Fucoidan (Fu)-encapsulated AuNPs (Dox-Fu@AuNPs) for the treatment of VX2 squamous carcinoma cells and xenograft tumors in New Zealand White rabbits. After intra-tumoral injection of Dox-Fu@AuNPs in the eye, a complete suppression of the tumor has been observed by dual-chemotherapy and PTT treatment. No damage to retina has been also detected. These composite nanostructures enabled the improvement of the OCT imaging technique (thanks to the selective light absorption of AuNPs) [84].

The quality of photoacoustic-based imaging was also improved thanks to selective light absorption of AuNPs. In this way, a double effect in terms of therapy and diagnostic was achieved.

As reported above, OCT is largely used for imaging in ocular clinical research due to the possibility to take structural information based on the reflectivity of tissue. In this respect, Gordon and colleagues [85] confirmed that Au nanorods (with a peak absorption of 750 nm and diameter of 10–35 nm) injected in the retina of Wild-type C57BL/6 mice are able to enhance the contrast, thus significantly improving the quality of a scanned lesion. Additionally, intravitreal injection of Au nanorods functionalized by anti-VEGF antibody triggered a reduction in the extent of anatomic laser damage and lesion-associated photothermal signal density in mice treated in laser induced choroidal neovascularization (LCNV) model and injected with ICAM2-targeted GNR. AuNPs (3–5 nm) were also used as inhibitor of VEGF on choroid-retina endothelial cells (RF/6A) [86]. A similar effect was showed using AgNPs, which have been found to reduce the VEGF activation and induced cell proliferation and migration in bovine retinal endothelial cells [87]. In addition, AgNPs were shown to inhibit the angiogenesis, typical of metastasising cells and of retinal neovascularization development. Ag is widely used because it has the great advantage of being a cost-effective material. For this reason, it is a good candidate for the replacement of the current therapies used for the treatment of various retinal angiogenic disorders. In this respect, Sheikpranbabu and co-workers [88] investigated the effect of green AgNPs, obtained from *Bacillus licheniformis*, on VEGF-and Interleukin 1 β (IL-1β) expression—that induced retinal endothelial cell permeability. AgNPs were able to stop VEGF-and IL-1β-induced permeability in retinal endothelial cells from porcine retina.

5. Bio-Inspired NPs for Ocular Drug Delivery

The use of bio-inspired materials mimicking natural components for applications in living organisms has widely spread recently [89]. In fact, inorganic NPs have the serious drawback of not being degradable and are also difficult to get excreted from the body [90]. Bio-inspired nanomaterials, including liposomes, niosomes, and chitosan/alginate NPs have been then explored as valid alternatives to metal NPs for treatment of several diseases [91]. In addition, these 'soft' NPs have been applied for drug delivery in the posterior segment of eye, including retinal photoreceptors.

5.1. Liposomes as Drug Delivery Carrier

Liposomes consist of a lipid bilayer, resembling a cellular membrane, that surrounds an aqueous core. These structures are widely used in pre-clinical studies to improve drug delivery, for example through Blood Brain Barrier (BBB) [92]. The BBB and BRB have the same functional properties [93,94]: so liposomes improve drug delivery to the retina and its photoreceptors due to their high half-life, permitting the long term drug absorption [95]. Liposomes also offer the ability to significantly improve pharmacokinetics and pharmacodynamics of any loaded drug compared to its free-circulating counterpart. They are also considered as highly biocompatible and biodegradable as well up to some extent both in vitro and in vivo. For this reason, several anticancer drugs such as doxorubicin, daunorubicin, and epirubicin were encapsulated in liposomes core with results demonstrating a significant decrease of many severe adverse effects (nausea, weakness, vomiting) with respect to the same concentration of free (non-encapsulated) drug. In this context, Zhang et al. [96] subcutaneously injected tacrolimus-encapsulated liposomes in Lewis rats with autoimmune uveoretinitis (EAU) by subcutaneous injection. Confocal analysis over rhodamine-conjugated liposomes (containing also the tacrolimus) revealed that such vesicles remained in ocular fluids even after 14 days. They found a reduced intraocular inflammation and decrease of EAU development without observing significant adverse effects.

Liposomes encapsulated with the anti-inflammatory and anti-angiogenic drugs berberine hydrochloride and chrysophanol were applied in myoblasts and in ischemic animal models. These drugs have the ability to contrast oxidative stress by enhancing the activities of antioxidant enzymes such as superoxide dismutase (SOD) and glutathione peroxidase (GSH-Px) [97,98]. Berberine hydrochloride and chrysophanol have been used in ocular liposome-mediated delivery to overcome their low stability. In this respect, a recent study [99] showed a liposome system that can entrap the two drugs at same times using the third polyamidoamine dendrimer (PAMAM G3.0) as a delivery carrier for age-related macular degeneration (AMD) therapy in the ocular posterior chamber.

PAMAM G3.0-coated compound liposomes were used to evaluated HCECs (Human Corneal Epithelial Cells) internalization, trans-corneal permeability in New Zealand white rabbits (weighing 2.0–2.5 kg) and male Sprague–Dawley rats (weighing 210–250 g). Pharmacokinetics were evaluated together with in vitro assessments of anti-reactive oxygen species (ROS) efficacy. The PAMAM G3.0-coated compound liposomes enhanced bio-adhesion on rabbit corneal epithelium after topical administration and exhibited great permeability in HCECs cells. The possibility to deliver a protective agent against photo-oxidative retinal damage in a light-damaged animal model was also investigated through the use of dendrimers (Figure 4A). Results showed that dendrimers encapsulated with chrysophanol–berberine hydrochloride induced the greatest protection of retinal function in light-exposed rats compared to unloaded dendrimers and free drug. In addition, no inflammation and damage of cornea, iris, and conjunctiva was observed after drug-loaded dendrimers instillation for 14 days. Upon exploring the fundus of the eye (interior surface of eye opposite the lens including the retina, optic disc, macula, fovea, and posterior pole) [100], modifications in the vessel structure and distribution have not been observed after different treatments. (Figure 4B).

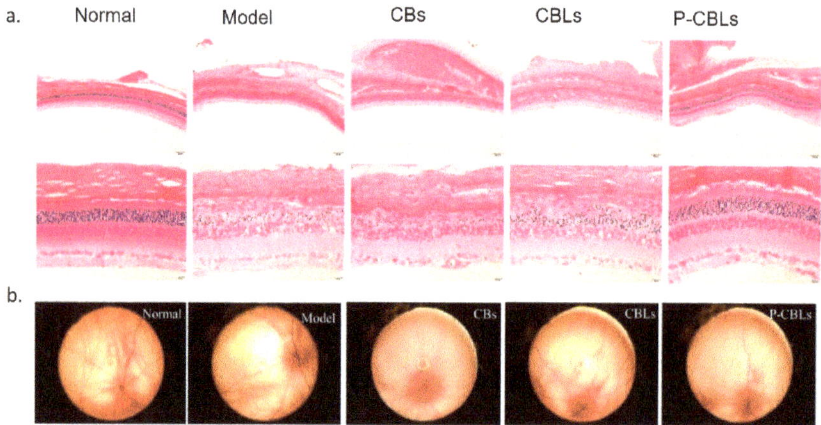

Figure 4. (**A**) Protection evaluation against photo-oxidative retinal damage in a light-damaged animal model by the use of different formulation: chrysophanol–berberine hydrochloride suspension (CBs), compound liposomes (CBLs), and PAMAM coated compound liposomes (P-CBLs). Retina was stained with hematoxilyn/eosin (scale bare = 20 µm). (**B**) Fundus retinography after incubation with the different formulations. Adapted from [99] under the terms of the Creative Commons Attribution 4.0 International License.

Liposomes offer also the great ability of an easy surface functionalization with a plethora of active macromolecules and targeting moieties that bind in a selective manner some specific receptors constitutive of different cell lines [101,102].

The surface modification of liposomes has also been widely studied to improve their specificity of targeting and to deliver them in different eye compartments. It has been reported [103] that the cationic PEGylated liposomes having an average zeta potentials below +20 mV were able to efficiently penetrate the murine retina. In contrast, liposome having surface zeta potential values above +20 mV were retained on vitreous section. This did not change whether PEGylated or non-PEGylated liposomes were used. Another critical factor for a liposome-based therapy is the phospholipids composition characterizing them. For example, edaravone-loaded egg L-phosphatidylcholine liposomes prevent retinal ganglion cells (RGCs) from undesired apoptotic processes [104]. This drug is a free radical scavenger used for acute brain infarction and amyotrophic lateral sclerosis treatment [105]. Surprisingly, minor RGC protection was observed with edaravone-loaded distearoyl phosphatidylcholine (DSPC) liposomes. This evidence suggest a link between the viscoelastic properties of the lipid bilayer and the final biological effect. In this specific case, the higher fluidity was a key factor to improve the ocular therapy. Another study [106] also tested the use of liposomes for the treatment of different kinds of hereditary degeneration of the retina. The degeneration of retina affects rod photoreceptors important for capturing light under low illumination conditions. Because of the pathology development, the patients affected by retina degeneration become aware of defects only when lesions are in an advanced stage. A key factor for this disease seems to be rods-specific Pde6b gene mutation in the transduction pathway that triggers an excessive accumulation of cGMP (that have two effectors: cyclic nucleotide gated ion channels (CNGCs) and cGMP-dependent protein kinase (PKG) [107]. Consequently, rod cell death was induced and in some cases the secondary loss of cones was observed [108]. The authors identified a liposomal cGMP analogue that was able to overcome the blood–retinal barrier (BRB) while strongly preserving retinal function and reducing rod photoreceptor loss in three different in vivo models for retina degeneration [106,109]. The vesicle structure of liposomes was suitable to confine proteins and peptides into aqueous core overcoming potential agglomeration and protein misfolding events. However, the surface functionalization with peptides is the most common method used for retina therapy. For example, liposomes functionalized with tanticoagulant annexin A5 have been demonstrated to have enhanced uptake levels through corneal epithelial barriers [110].

This vector has been injected topically to deliver 127 ng/g of bevacizumab to posterior segment of both rats and rabbit eye. Annexin A5-functionalized liposomes loaded with bevacizumab were also rapidly delivered to retina of the two animal models. Annexin A5 unilamellar liposomes made by different polymers (2-dipalmitoyl-sn-glycero-3-phosphocholine,1,2-dioleoyl-sn-glycero-3-phospho-L-serine and cholesterol) have been also used to target transforming growth factor (TGF)-ß1 into the vitreous of rabbits [111]. It has been found that a higher concentration of molecule with respect to the experimental group treated with the unloaded drug only. The choice for human recombinant TGF-β1 relied on the drug's ability to prevent retinal damage elicited by Aβ oligomers.

5.2. Niosomes for Drug and Gene Delivery

Gene therapies is an approach where diseases are cured by modifying the expression of specific genes regulating pathology. This happens through the administration of nucleic acids aiming to replace abnormal gene sequence [112]. Liposomes and lipoplexes have been taken into consideration as suitable non-viral vectors platform for gene editing. Non-viral gene delivery approaches were widely applied in clinical trials since 2004 to attempt replacing the use of viral vectors that possess high risk of inducing oncogenesis, immunogenicity, mutagenicity phenomena [113].

Among non-viral vectors, niosomes have been recently tested as a potential candidate for gene delivery. Niosomes are vesicular systems comprising of bilayer made up of nonionic surfactants and they can internalize several types of bio-macromolecules such as nucleic acids and drugs. The most commonly used surfactant are Sorbitan fatty acid esters, Spans (Span 20, 40, 60, 65, 80, and 85), Tweens (20, 40, 60, and 80) [114] and brij (30, 35, 52, 58, 72, and 76) [115].

In a recent study [116], niosomes made by cationic lipid 1,2-di-octadecenyl-3-trimethylammonium propane, Squalene, and Tween 20 have been combined with GFP-encoding genetic materials consisting on minicircle (MC-GFP, 2.3 kb), its parental plasmid (pGFP, 3.5 kb), and a larger plasmid (pEGFP, 5.5 kb). Their efficacy in retinal disease treatment has been investigated. These noisome-DNA complexes, named as nioplexes, demonstrated good transfection ability in human ARPE19 retinal pigment epithelium cells (in vitro) and in embrionary rat retinal primary cells (in vivo). Fluorescence showed GFP positive signal emission after transfection with the nioplexes carrying plasmidic DNA. In addition, such transfected cells showed physiological glial cell morphology confirming good cell viability in all tested cases. (Figure 5). The successive injections of nioplexes in rat retinas confirmed the higher capacity of cationic niosomes vectoring minicircle DNA to deliver genetic material into retina cells. Therefore, nioplexes based on cationic niosomes vectoring minicircle DNA are demonstrated as powerful agents in retinal diseases therapies. Cationic formulations made by a combination of Tween-60:DOTMA:lycopene have been used to transfect ARPE-19 cells with pCMS-EGFP plasmid with a transfection efficiency close to 35% [117]. Also in case, in vivo experiments in rat eye showed the efficiency of nioplexes to transfect outer segments of the retina, Niosomes constituted by Span 40 or Span 60 and cholesterol were produced to encapsulate and topically deliver in vivo acetazolamide. The intraocular pressure (IOP) after drug delivery system application was reduced in rabbits. In particular, multilamellar acetazolamide niosomes developed with Span 60 and cholesterol (7:4 molar ratio) were more efficient in reducing IOP, and at same time, no irritation and retinal damage were observed [118]. GuglLeva et al. [119] prepared niosomes by different surfactants to deliver doxycycline hyclate in rats eye. Doxycycline hyclate is an antibiotic used for the treatment of eye infections. It was demonstrated to inhibit effect on matrix metalloprotease 9 (MMP-9) activity and mitogen activated protein kinase in experimental murine dry eye [120]. Niosomes were synthesized using several surfactants (Span 20, Span 60, Span 80, Tween 60) and cholesterol in different molar ratios and tested on male albino rabbits. The experiments clearly showed that the prepared niosomal formulations were high tolerated and did not induce any irritations.

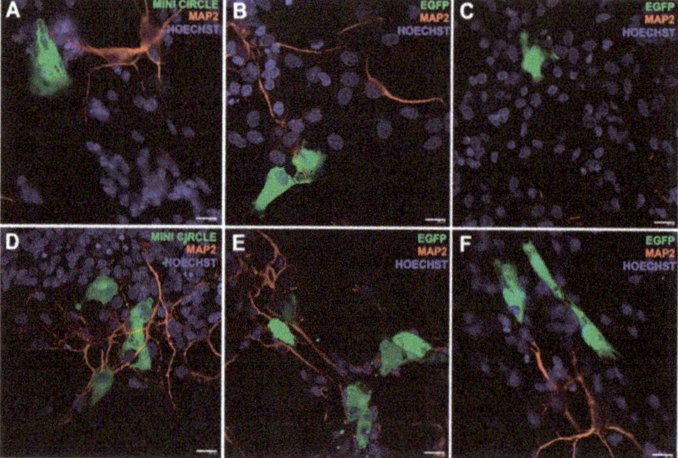

Figure 5. Fluorescence immonostaining by GFP expression in embrionary rat retinal primary cells. The green signal was referred to the transfection event of DST20 nioplexes with MC-GFP, (**B**) pGFP 3.5 kb, or (**C**) pEGFP 5.5 kb. (**D–F**) Positive controls incubated with Lipofectamine™ 2000. Nuclei were stained with Hoechst 33,342 (blue) and neuronal dendrites with MAP2 (red). Scale bars: 20 μm. Reprinted from [116] Copyright (2019), with permission from Elsevier.

5.3. Hydrogels-Based Nanocarriers

Hydrogels are three-dimensional, cross-linked networks of natural or synthetic water-soluble polymers that are suitable for applications like tissue engineering and controlled/sustained drugs delivery [121,122]. The aqueous environment is particularly advantageous to confine soluble active molecules, oligonucleotides, and proteins. For this reason, they are powerful tools in retinal treatment disease [123]. In addition, smart or environment-responsive hydrogels were synthetized to improve drug delivery efficacy [124].

5.3.1. Natural Hydrogels: Chitosan NPs (CHNPs)

Chitosan is a natural hydrogel cationic polysaccharide of co-polymers glucosamine and N-acetyl glucosamine [125] obtained by alkaline deacetylation of chitin that is constitutive of crustacean shells of shrimps, lobster, and crab. It presents a positive charge that can be exploited to form polyelectrolyte vesicular complexes able to act as delivery systems. The positive charged chitosan is also ideal to bind a negatively charged corneal surface, to improve precorneal retention time and to decreases clearance [126]. In addition to chitosan, also the biocompatible alginate is extensively employed in biomedical applications. Alginate can also be a useful biopolymers for designing responsive hydrogels [127]. Because of its characteristics, alginate is also used for ophthalmic applications often in combination with chitosan [128]. For example, chitosan coated sodium alginate NPs [129] loaded with Fluorouracil (5-FU) have been successfully tested for eye delivery demonstrating a high 5-FU release in aqueous humor compared to free drug in rabbit eye. In another approach, chitosan NPs were encapsulated with levofloxacin for the treatment of ocular infection [130]. The formulation of NPs were optimized to enhance corneal retention without inducing eye irritation. These nanovectors demonstrated strong antibacterial agents against *Pseudomonas aeruginosa* and *Stafiloccocus aureus*.

Andrei and co-workers [131] designed biocompatible chitosan and alginate-based micro/NPs with a specific morphology aiming to deliver cefuroxim, a second-generation cephalosporin antibiotic. The authors performed an in vivo assessment to evaluate the biodistribution following an intravitreal administration of fluorescein-chitosan NPs on wistar rats. They observed an efficient NPs distribution in lens and retina. Similarly, chitosan-dextran NPs have been employed to deliver the antibiotic moxifloxacin that has been retained for a long time in vivo thanks to the mucoadhesive properties of NPs [132].

In recent work, humanin, a neuronal peptide that mediated inflammatory responses, was used in the treatment of age-related macular degeneration (AMD) [133]. This peptide was a suppressor of the IL-6 cytokine receptors and protects neuronal cells from necrosis [134]. The authors studied the effect on the inflammatory response reduction using chitosan NPs as vector vesicle (with a size of 346 ± 52 nm) in which humanin (a micropeptide encoded in the mitochondrial genome) derivative AGA-(CR8) HNG17 (AGA-HNG) were encapsulated. The effects of NPs were assessed in retinal pigment epithelial cells. The percentage release profile of free protein after five days was 60%, after which it enriched a plateau whereas the nanostructured exhibited a slow and substantial release with 80% of protein released after fifteen days. Cheng et al. [135] developed a thermosensitive hydrogel containing latanoprost and curcumin-loaded NPs (CUR-NPs). Curcumin was selected because of its antioxidative and antiinflammatory properties that made it suitable also for treatment of eye pathologies such as chronic anterior uveitis and light-induced retinal degeneration [136]. Latanoprost was chosen because it acted on IOP restore due to uveoscleral outflow increase [137]. In glaucoma disease, IOP is generated by loss of aqueous humor drainage through the uveoscleral or trabecular outflow pathway that can be damaged by oxidative stress at trabecular meshwork level [138]. In this condition, Latanoprost and CUR-NPs were demonstrated to decrease the oxidative stress in trabecular meshwork (TM) cells that resulted in decreased levels of apoptosis, inflammation-related gene expression, and mitochondrial damage [135]. In another work [139], neuroprotective and neuro regenerative agent Erythropoietin (EPO) was encapsulated in chitosan/hyaluronic acid (HA) nanoparticulate (size ≤300 nm). Different formulations have been tested and delivered to porcine corneas, scleras, and conjunctivas by ex vivo permeation, finding out a good combination between targeting efficiency and biocompatibility.

Pandit et al. [140] encapsulated high levels of bevacizumab (an anti-angiogenesis drug) within chitosan-coated PLGA NPs (size 222.28 ± 7.45) and demonstrated its efficient deliver to the retina followed by efficient release. As matter of fact, CS-coated NPs showed better permeability of the bevacizumab across sclera as compared to the free drug solution. Another category of carrier is made by glycol chitosan (GCS) NPs [141]. These biocompatible NPs, demonstrated to load high amount of plasmid DNA, have been successfully delivered to retinal pigment epithelium of adult wild-type albino mice. More importantly, the encapsulated pDNA was intact and maintained a proper conformation.

5.3.2. Synthetic Hydrogel Based Nanocarriers

The synthetic hydrogels are easy to customize and indeed are a close model of biological matrices. Hydrogels exhibit high hydrophilicity and high hydration [142]. The most commonly used are poly(hydroxyethyl methacrylate) (PHEMA), poly(ethylene oxide) (PEO, also called poly(ethyleneglycol) (PEG)), polyvinylpyrrolidone (PVP), poly(acrylic acid)(PAA), polyacrylamide (PAM), and poly(vinyl alcohol) (PVA) [143]. These polymers are all used in several ophthalmic applications with good results [144–147].

Anti-VEGF antibodies (anibizumab or aflibercept) were loaded within poly(lactic-co-glycolic acid) microspheres, and successively suspended within poly(N-isopropylacrylamide)-based hydrogel and injected into a laser-induced rat model of choroidal neovascularization (CNV). Upon ocular administration, the rats showed smaller CNV lesions (less than 60%) with respect to non-treated animals [148].

Agrahari et al. [149] obtained novel pentablock (PB) copolymer (PB-1: PCL–PLA–PEG–PLA–PCL) based nanoformulations suspended in a thermosensitive gelling copolymer (PB-2: mPEG–PCL–PLA–PCL–PEGm) with a size of 150 nm. This composite nanoformulation was used to deliver macromolecules to slowly release drugs without toxic effects on the posterior segment of the eye. In vitro tests performed on ocular and mouse macrophage (RAW 264.7) demonstrated their biocompatibility and potential to be clinically translated.

Yang et al. [150] reported the production of non-toxic polyamidoamine (PAMAM) dendrimer hydrogel/poly(lactic-co-glycolic acid) (PLGA) NP platform (HDNP) to treat glaucoma. The dendrimers were loaded with brimonidine and timolol maleat, which underwent an enhanced uptake together with a slow release in vitro up to 28–35 days. The topical administration of the same drug-loaded denndrimers in normotensive adult Dutch-belted male rabbits strongly reduced the introcular pressure (IOP) of about 18% maintaining a glaucoma reduction for four days. Conversely, Individual Dendrimer Hydrogel Formulations (DH) and NP Formulations (NP) decreased the disease symptoms only for 48 h (Figure 6).

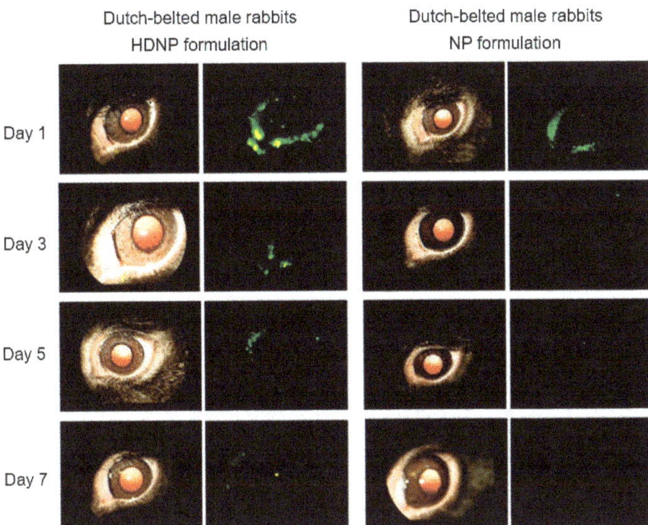

Figure 6. Fundus camera acquisitions of rabbits' eyes after hybrid polyamidoamine (PAMAM) dendrimer hydrogel/poly(lacticco-glycolic acid) (PLGA) nanoparticle platform (HDNP) and NP Formulations (NP) topical administration FluoSpheres were confined in nanosistems to follow the distribution and retention in the eye. The analysis was performed at the end of 1, 3, 5, and 7 days of formulations instillation. Left: regular fundus camera images of the eyes; right: fluorescent fundus camera images. Adapted with permission from [150], copyright (2012), American Chemical Society.

5.4. Active and Self-Propelling Nanoparticles: New Generation of Nanobots in Retinal Disease

The customization of NPs formulation at a molecular level has interesting outcomes for reproduction of fundamental phenomena such as macro- and microorganisms' survival mechanisms [151]. For example, bacteria, uni- or multicellular organisms, and sperm cells are able to sense and actively respond to external stimuli by the generation of movement. In particular, they convert an external energy sources (temperature, magnetic fields, adhesion forces, chemical gradients) into mechanical work that induce migration. This is typical in processes such as thermotaxis, magnetotaxis, haptotaxis, chemotaxis [152]. By taking inspiration from nature, we have also now a panel of NPs able to self-propel towards a gradient of nutrients and to mimic biological taxis [153]. However, the development of such NPs still has many challenges because at nanoscale level, the physical principles of propulsion are different due to the presence of the Brownian effect that can interfere with nanomptors by collisions with solvent molecules. One of these challenges is the generation of particles having asymmetrical geometries at the nanoscale, as this provides guidance for the movement directionality. Bacteria like *Escherichia coli*, in fact, achieve propulsion by non-time-reversible motion of long flagella, and this asymmetric shape of the body is critical for motion generation in a specific direction [154].

It is clear that self-propelling NPs represent a new powerful tool in drug delivery, which is usually dominated by NPs passive diffusion. It should not come as a surprise that such strategies have been employed also for delivery in ocular medicine. For example, Wu et al. [155] designed micro-vehicles that can actively propel through the vitreous humor to enrich retina. The movement can be activated by helical magnetic micropropellers (~120 nm in diameter and ~400 nm in length) with liquid layer coating to reduce the adhesion of biopolymeric network. These nanovectors propelled in nanoporous hyaluronan solution, a fluid mimicking the vitreous. The inspiration for nanotool development is due to the liquid layer that characterizes carnivorous *Nepenthes* pitcher plant, which presents a slippery surface on the peristome to catch insects [156,157]. Porcine eyes were used as a model of human eyes:

applying the external magnetic field, micropropellers exhibited controllable propulsion from eyeball to retina within 30 minutes. Instead, passive silica microparticles did not penetrate the vitreus (Figure 7). Nanorobots were also used to perform microsurgery on the retina and surrounding membranes thanks to the possibility of injecting them elsewhere in the body and bringing the drug to the target eye region [158]. Some intravitreal implants have recently developed to perform more clinical-inspired experiments. These nanotools, constituted by polymeric containers (silicon, poly(vinyl alcohol), or ethylene vinyl acetate), were placed in the posterior segment of the eye and they are permeable to lipophilic drugs. They did not display any toxicity and had a long retention time. These nanosystems have been extremely useful to treat proliferative diabetic retinopathy, retinal vascular occlusion, and uveitis [159].

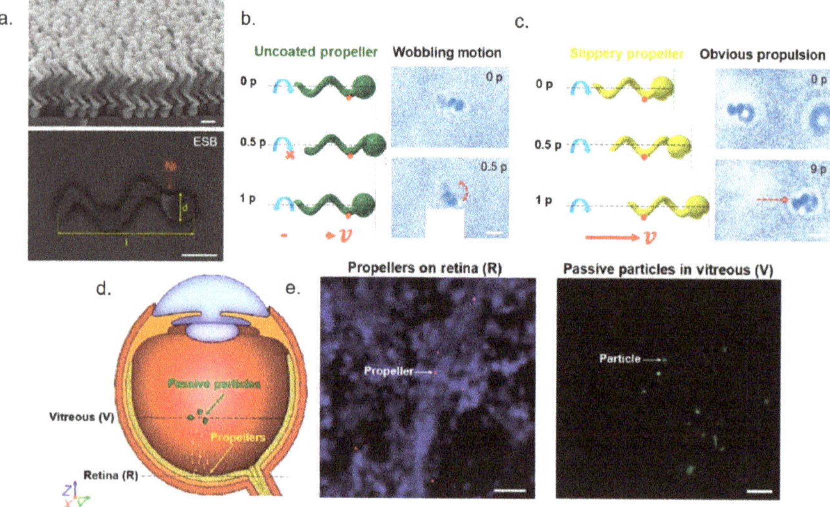

Figure 7. (**a**) SEM and ESB-SEM images of the micropropellers. (**b**,**c**) incomplete rotation of an uncoated micropropeller for one period (p) in the vitreous was showed in schematic and time-lapse microscopy, whereas the coated propeller exhibited propulsion under magnetic stimuli in vitreous for 1.5 s (nine periods). Red arrows showed the propulsion direction of propeller. The original position of nanotools and after the rotating magnetic field application were showed with green and red dashed lines. Scale bars, 1 μm. (**d**) Schematic illustration of micropropeller movement in the vitreous. (**e**) Left: fluorescent acquisitions showed the micropropellers labelled red on retina. Right: passive fluorescent particles were accumulated near center of vitreous. Nuclei of cells were stained with DAPI. Scale bar, 20 μm. Adapted from reference [155] under the terms of the Creative Commons Attribution-Non Commercial license.

6. Conclusions and Perspectives

It is established that eye damage is a leading cause of irreversible vision loss resulting from ocular diseases. The current standard treatments helped in both preventing and curing several pathological conditions, especially when related to the deep sections of the eye. However, many of these methods are rather invasive in terms of procedures, with a consequent possibility of promoting infection, inflammation, or even retinal detachment with potential risk of vision loss. These limitations impose considerable costs on patient quality of life and have an enormous economic impact on the health care system. The progress in the nano-engineering field can help boosting the discovery of a new approach for ocular diagnosis and therapy. Noble metals NPs and bio-inspired NPs represent a powerful tool to overcome all the limitations coming with standard treatments due to their physico-chemical properties. NPs can in fact penetrate deeper inside the ocular segments and also locally deliver high dosages of drugs, enriching local drug concentration when applied topically. In addition, the plasmonic properties of Au and Ag can have an added value to contrast eye diseases. However, numerous challenges still remain to be addressed as there are still limited numbers of in vivo studies. In addition, a deeper toxicology assessment is required for the development of new systems for ocular treatments.

Author Contributions: Conceptualization, V.D.M.; writing—original draft preparation, V.D.M.; review and editing, L.R. All authors have read and agreed to the published version of the manuscript.

Funding: This research received no external funding.

Acknowledgments: VDM kindly acknowledges Programma Operativo Nazionale (PON) Ricerca e Innovazione 2014-2020 Asse I "Capitale Umano", Azione I.2, Avviso "A.I.M: Attraction and International Mobility" CUP F88D18000070001.

Conflicts of Interest: The authors declare no conflict of interest.

References

1. Davson, H. *The Physiology of The Eye*; Elsevier: Amsterdam, The Netherlands, 2012.
2. Stephen, R.; Wilkinson, C.; Schachat, A.; Hinton, D.; Wilkinson, C. *Retina*; Elsevier: Amsterdam, The Netherlands, 2012.
3. Günhan, E.; van der List, D.; Chalupa, L.M. Ectopic Photoreceptors and Cone Bipolar Cells in the Developing and Mature Retina. *J. Neurosci.* **2003**, *15*, 1383–1389. [CrossRef]
4. Reichenbach, A.; Schnitzer, J.; Friedrich, A.; Brückner, G.; Schober, W. Development of the rabbit retina. *Anat. Embryol.* **1991**, *183*, 287–297. [CrossRef] [PubMed]
5. Dilnawaz, F.S.S.K. Nanotechnology-Based Ophthalmic Drug Delivery System. In *Focal Controlled Drug Delivery*; Domb, A., Khan, W., Eds.; Advances in Delivery Science and Technology; Springer: Boston, MA, USA, 2014.
6. Rakhi Dandonaa, L.D. Socioeconomic status and blindness. *Br. J. Ophthalmol.* **2001**, *85*, 1484–1488. [CrossRef] [PubMed]
7. Yorston, D. The Retina Retinal Diseases and VISION 2020. *Comm Eye Health* **2003**, *16*, 19–20.
8. Veleri, S.; Lazar, C.H.; Chang, B.; Sieving, P.A.; Banin, E.; Swaroop, A. Biology and therapy of inherited retinal degenerative disease: insights from mouse models. *Dis. Models Mech.* **2015**, *8*, 109–129. [CrossRef] [PubMed]
9. Jiang, S.; Franco, Y.L.; Zhou, Y.; Chen, J. Nanotechnology in retinal drug delivery. *Int. J. Ophthalmol.* **2018**, *11*. [CrossRef]
10. Gilbert, C.; Foster, A. Childhood blindness in the context of VISION 2020—The Right to Sight. *Bull. World Health Organ.* **2001**, *79*, 227–232.
11. Jones, B.W.; Marc, R.E.; Pfeiffer, R.L. Retinal Degeneration, Remodeling and Plasticity. In *Webvision: The Organization of the Retina and Visual System*; Salt Lake City (UT): University of Utah Health Sciences Center, UT, USA, 2016.
12. Rajappa, M.; Saxena, P.K.J. Ocular angiogenesis: Mechanisms and recent advances in therapy. *Adv. Clin. Chem.* **2010**, *50*, 103–121.
13. Aiello, L.P.; Avery, R.L.; Arrigg, P.G.; Keyt, B.A.; Jampel, H.D.; Shah, S.T.; Pasquale, L.R.; Thieme, H.; Iwamoto, M.A.; Park, J.E.; et al. Vascular endothelial growth factor in ocular fluid of patients with diabetic retinopa-thy and other retinal disorders. *N. Engl. J. Med.* **1994**, *331*, 480–1487. [CrossRef]
14. Weis, S.M.; Cheresh, D.A. Pathophysiological consequences of VEGF-induced vascular permeability. *Nature* **2005**, *437*, 497–504. [CrossRef]
15. Kim, S.J. Novel Approaches for Retinal Drug and Gene Delivery. *Transl. Vis. Sci. Technol.* **2014**, *3*, 7. [CrossRef] [PubMed]
16. Bennett, J.; Chung, D.C.; Maguire, A. Gene delivery to the retina: From mouse to man. *Methods Enzym.* **2012**, *507*, 255–257.
17. Xu, H.; Chen, M. Targeting the complement system for the management of retinal in fl ammatory and degenerative diseases. *Eur. J. Pharmacol.* **2016**, *787*, 94–104. [CrossRef] [PubMed]
18. Díaz-Coránguez, M.; Ramos, C.; Antonetti, D.A. The inner Blood-Retinal Barrier: Cellular Basis and Development. *Vis. Res.* **2017**, *139*, 123–137. [CrossRef]

19. Gaudana, R.; Jwala, J.; Boddu, S.H.S.; Mitra, A.K. Recent Perspectives in Ocular Drug Delivery. *Pharm. Res.* **2009**, *26*, 1197–1216. [CrossRef]

20. Kati-Sisko, V.; Hellinen, L. Expression, activity and pharmacokinetic impact of ocular transporters. *Adv. Drug Deliv. Rev.* **2018**, *126*, 3–22.

21. Souto, E.B.; Dias-ferreira, J.; Ana, L.; Ettcheto, M.; Elena, S. Advanced Formulation Approaches for Ocular Drug Delivery: State-Of-The-Art and Recent Patents. *Pharmaceutics* **2019**, *11*, 460. [CrossRef]

22. Le Bourlais, C.; Acar, L.; Zia, H.; Sado, P.A.; Needham, T.; Leverge, R. Ophthalmic drug delivery systems—Recent advances. *Prog. Retin. Eye Res.* **1998**, *17*, 33–58. [CrossRef]

23. Sugrue, M.F. The pharmacology of antiglaucoma drugs. *Pharmacol. Ther.* **1989**, *43*, 91–1989. [CrossRef]

24. Himawan, E.; Ekström, P.; Buzgo, M.; Gaillard, P.; Stefánsson, E.; Marigo, V.; Loftsson, T.; Paquet-Durand, F. Drug delivery to retinal photoreceptors. *Drug Discov. Today* **2019**, *24*, 1637–1643. [CrossRef]

25. Marmor, M.F.; Akira Negi, D.M.M. Kinetics of macromolecules injected into the subretinal space. *Exp. Eye Res.* **1985**, *40*, 687–696. [CrossRef]

26. Meyer, C.H.; Krohne, T.U.; Issa, P.C.; Liu, Z. Routes for Drug Delivery to the Eye and Retina: Intravitreal Injections. *Retin. Pharmacother. Dev. Ophthalmol.* **2016**, *55*, 63–70.

27. Falavarjani, K.G.; Nguyen, Q.D. Adverse events and complications associated with intravitreal injection of anti-VEGF agents: A review of literature. *Eye* **2013**, *27*, 787–794. [CrossRef] [PubMed]

28. Raghava, S.; Hammond, M.K.U. Periocular routes for retinal drug delivery. *Expert Opin. Drug Deliv.* **2004**, *1*, 99–144. [CrossRef]

29. Ghate, D.; Brooks, W.; McCarey, B.E.; Edelhauser, H.F. Pharmacokinetics of Intraocular Drug Delivery by Periocular Injections Using Ocular Fluorophotometry. *Invest. Ophthalmol Vis. Sci.* **2007**, *48*, 2230–2237. [CrossRef] [PubMed]

30. Castellarinand, A.; Pieramici, D.J. Anterior segment complications following periocular and intraocular injections. *Ophthalmol. Clin. N. Am.* **2004**, *17*, 583–590. [CrossRef]

31. Tsai, H.-H.; Wang, P.-Y.; Lin, I.-C.; Huang, H.; Liu, G.; Tseng, C.-L. Ocular Drug Delivery: Role of Degradable Polymeric Nanocarriers for Ophthalmic Application. *Int. J. Mol. Sci.* **2018**, *19*, 2830. [CrossRef]

32. ISO, Int. Organ Stand Nanotechnologies Vocab Part 1 Core Terms ISO/TS 80004-12010. 2010. Available online: https://www.iso.org/standard/51240.html (accessed on 1 March 2020).

33. ISO International Organization for Standardization. *Technical Specification: Nanotechnologies Terminology and Definitions for Nanoobjects Nanoparticle, Nanofibre and Nanoplate.* ISO/TS 80004-2:2008. 2008. Available online: https://www.iso.org/standard/44278.html (accessed on 1 March 2020).

34. De Matteis, V.; Rinaldi, R. Toxicity Assessment in the Nanoparticle Era. In *Cellular and Molecular Toxicology of Nanoparticles*; Springer: Cham, Switzerland, 2018; pp. 1–19.

35. Mu, Q.; Jiang, G.; Chen, L.; Zhou, H.; Fourches, D.; Tropsha, A.; Yan, B. Chemical Basis of Interactions Between Engineered Nanoparticles and Biological Systems. *Chem. Rev.* **2016**, 7740–7781. [CrossRef]

36. Bhatia, S. Nanoparticles Types, Classification, Characterization, Fabrication Methods and Drug Delivery Applications. In *Natural Polymer Drug Delivery Systems*; Springer: Cham, Switzerland, 2016.

37. Ruth Duncan, R.G. Nanomedicine(s) under the Microscope. *Mol. Pharm.* **2011**, *8*, 2101–2214. [CrossRef]

38. Diebold, Y.; Calonge, M. Applications of nanoparticles in ophthalmology. *Prog. Retin. Eye Res.* **2010**, *29*, 596–609. [CrossRef]

39. Farjo, K.M.; Ma, J. The potential of nanomedicine therapies to treat neovascular disease in the retina. *J. Angiogenes Res.* **2010**, *2*, 21. [CrossRef] [PubMed]

40. JeffreyAdijanto, M.I.N. Nanoparticle-based technologies for retinal gene therapy. *Eur. J. Pharm. Biopharm.* **2015**, *95*, 353–367.

41. Soto, F.; Chrostowski, R.; Soto, F. Frontiers of Medical Micro/Nanorobotics: In vivo Applications and Commercialization Perspectives Toward Clinical Uses. *Front. Bioeng. Biotechnol.* **2018**, *6*, 1–12. [CrossRef]

42. Khlebtsovab, N.G.; Dykman, L.A. Optical properties and biomedical applications of plasmonic nanoparticles. *J. Quant. Spectrosc. Radiat. Transf.* **2010**, *111*, 1–35. [CrossRef]

43. Anker, J.N.; Hall, W.P.; Lyandres, O.; Shah, N.C.; Zhao, J.V. Biosensing with plasmonic nanosensors. *Nat. Mater.* **2008**, *7*, 442–453. [CrossRef] [PubMed]

44. Daniel, M.; Astruc, D. Gold Nanoparticles: Assembly, Supramolecular Chemistry, Quantum-Size-Related Properties, and Applications toward Biology, Catalysis, and Nanotechnology. *Chem. Rev.* **2004**, *104*, 293–346. [CrossRef]

45. Asahi, T.; Uwada, T.; Masuhara, H. Single particle spectroscopic study on surface plasmon resonance probing local environmental conditions. *Handai Nanophotonics* **2006**, *2*, 219–228.

46. Boken, J.; Khurana, P.; Thatai, S. Plasmonic nanoparticles and their analytical applications: A review. *Appl. Spectrosc. Rev.* **2017**, *52*, 774–820. [CrossRef]

47. Loiseau, A.; Asila, V.; Boitel-aullen, G.; Lam, M. Silver-Based Plasmonic Nanoparticles for and Their Use in Biosensing. *Biosens* **2019**, *2019*, 9. [CrossRef]

48. Rizzello, L.; Pompa, P.P. Nanosilver-based antibacterial drugs and devices: Mechanisms, methodological drawbacks, and guidelines. *Chem. Soc. Rev.* **2014**, *43*, 1501–1518. [CrossRef]

49. Draslera, B.; Sayreb, P.; Günter, K.; Steinhäuserc, A.-R. In vitro approaches to assess the hazard of nanomaterials. *NanoImpact* **2018**, *9*, 51. [CrossRef]

50. De Matteis, V. Exposure to Inorganic Nanoparticles: Routes of Entry, Immune Response, Biodistribution and In Vitro/In Vivo Toxicity Evaluation. *Toxics* **2017**, *5*, 29. [CrossRef] [PubMed]

51. Naahidi, S.; Jafari, M.; Edalat, F.; Raymond, K.; Khademhosseini, A.C.P. Biocompatibility of engineered nanoparticles for drug delivery. *J. Control. Release* **2013**, *10*, 182–194. [CrossRef] [PubMed]

52. Turkevich, J.; Stevenson, P.C.; Hillier, J. A study of the nucleation and growth processes in the synthesis of colloidal gold. *Faraday Soc.* **1951**, *11*, 55–75. [CrossRef]

53. Brust, M.; Walker, M.; Bethell, D.; Schiffrin, D.J. Synthesis of thiol-derivatised gold nanoparticles in a two-phaseliquid–liquid system. *J. Chem. Soc. Chem. Commun.* **1994**, *7*, 801–802. [CrossRef]

54. Kumar, P.V.; Kala, S.M.J.; Prakash, K.S. Green synthesis of gold nanoparticles using Croton Caudatus Geisel leaf extract and their biological studies. *Mater. Lett.* **2019**, *236*, 19–22. [CrossRef]

55. Molnár, Z.; Bódai, V.; Szakacs, G.; Erdélyi, B.; Fogarassy, Z. Green synthesis of gold nanoparticles by thermophilic filamentous fungi. *Sci. Rep.* **2018**, *8*, 1–12. [CrossRef]

56. Giljohann, D.A.; Seferos, D.S.; Daniel, W.L.; Massich, M.D.; Patel, P.C. Gold Nanoparticles for Biology and Medicine. *Angew. Chem. Int. Ed. Engl.* **2010**, *26*, 3280–3294. [CrossRef]

57. Orlando, A.; Colombo, M.; Prosperi, D.; Corsi, F.; Panariti, A.; Rivolta, I.; Masserini, M.; Cazzinga, E. Evaluation of gold nanoparticles biocompatibility: A multiparametric study on cultured endothelial cells and macrophages. *J. Nanopart. Res.* **2018**, *18*, 58. [CrossRef]

58. Aghaie, T.; Jazayeri, M.H.; Manian, M.; Khani, l.; Erfani, M.; Rezayi, M.; Ferns, G.A.; Avan, A. Gold nanoparticle and polyethylene glycol in neural regeneration in the treatment of neurodegenerative diseases. *J. Cell. Biochem.* **2019**, *120*, 2749–2755. [CrossRef]

59. Bailly, A.-L.; Correard, F.; Popov, A.; Gleb Tselikov, F.; Appay, R.; Al-Kattan, A.; Andrei, V.; Kabashin, D.B.; Esteve, M.-A. In vivo evaluation of safety, biodistribution and pharmacokinetics of lasersynthesized gold nanoparticles. *Sci. Rep.* **2019**, *9*, 12890. [CrossRef] [PubMed]

60. Cassano, D.; Summa, M.; Pocoví-Martínez, S.; Mapanao, A.-K.; Catelani, T.; Bertorelli, R.; Voliani, V. Biodegradable Ultrasmall-in-Nano Gold Architectures: Mid-Period In Vivo Distribution and Excretion Assessment. *Part. Part. Syst. Charact.* **2019**, *36*, 1800464. [CrossRef]

61. Khan, H.A.; Abdelhalim, M.A.K.; Alhomida, A.S.; Al-Ayed, M.S. Effects of Naked Gold Nanoparticles on Proinflammatory Cytokines mRNA Expression in Rat Liver and Kidney. *Biomed. Res. Int.* **2013**, *2013*, 590730. [CrossRef] [PubMed]

62. Khan, H.A.; Alamery, S.; Ibrahim, K.E.; El-Nagar, D.M.; Al-Harbi, N.; Rusop, M.; Alrokayan, S.H. Size and time-dependent induction of proinflammatory cytokines expression in brains of mice treated with gold nanoparticle. *Saudi J. Biol. Sci.* **2019**, *26*. [CrossRef]

63. Karakoçak, B.B.; Raliya, R.; Davis, J.T.; Chavalmane, S.; Wang, W.N.; Ravi, N. Biocompatibility of gold nanoparticles in retinal pigment epithelial cell line. *Toxicol. Vitr.* **2016**, *37*, 61–69. [CrossRef]

64. Kim, J.H.; Kim, J.H.; Kim, K.-W.; Kim, M.H. Intravenously administered gold nanoparticles pass through the blood–retinal barrier depending on the particle size, and induce no retinal toxicity. *Nanotechnology* **2009**, *20*, 505101. [CrossRef]

65. Bakri, S.J.; Pulido, J.S.; Mukherjee, P.; Marler, R.J. Absence of histologic retinal toxicity of intravitreal nanogold in a rabbit model. *Retin* **2008**, *2008*, 147–149. [CrossRef]

66. Söderstjerna, E.; Bauer, P.; Cedervall, T.; Abdshill, H.; Johansson, F.; Johansson, U.E. Silver and Gold Nanoparticles Exposure to In Vitro Cultured Retina—Studies on Nanoparticle Internalization, Apoptosis, Oxidative Stress, Glial- and Microglial Activity. *PLoS ONE* **2014**, *9*, e0105359. [CrossRef]

67. Nakhlband, A. Impacts of nanomedicines in ocular pharmacotherapy. *Bioimpacts* **2011**, *1*, 7–22.

68. Kittler, S.; Greulich Diendorf, M.; Köller, M.E. Toxicity of Silver Nanoparticles Increases during Storage Because of Slow Dissolution under Release of Silver Ions. *Chem. Mater.* **2010**, *22*, 4548–4554. [CrossRef]

69. Yang, X.; Gondikas, A.P.; Marinakos, S.M.; Auffan, M.; Li, J.; Hsu-Kim, H. Mechanism of Silver Nanoparticle Toxicity Is Dependent on Dissolved Silver and Surface Coating in Caenorhabditis elegans. *Environ. Sci. Technol.* **2012**, *46*, 1119–1127. [CrossRef] [PubMed]

70. De Matteis, V.; Malvindi, M.A.; Galeone, A.; Brunetti, V.; De Luca, E.; Kote, S.; Kshirsagar, P.; Sabella, S.; Bardi, G.; Pompa, P.P. Negligible particle-specific toxicity mechanism of silver nanoparticles: The role of Ag + ion release in the cytosol. Nanomedicine Nanotechnology. *Biol. Med.* **2015**, *11*, 731–739. [CrossRef] [PubMed]

71. Abbasi, E.; Milani, M.; Fekri Aval, S.; Kouhi, M.; Akbarzadeh, A.; Tayefi Nasrabadi, H.; Nikasa, P.; Joo, S.W.; Hanifehpour, Y.; Nejati-Koshki, K.; et al. Silver nanoparticles: Synthesis methods, bio- applications and properties. *Crit. Rev. Microbiol.* **2016**, *42*, 173–180. [CrossRef]

72. Zhang, T.; Song, Y.J.; Zhang, X.Y.; Wu, J.Y. Synthesis of silver nanostructures by multistep methods. *Sensors (Switzerland)* **2014**, *14*, 5860–5889. [CrossRef] [PubMed]

73. Singh, C.; Kumar, J.; Kumar, P.; Singh Chauhan, B.; Tiwari, K.N.; Mishra, S.K.; Srikrishna, S.; Saini, R.; Nath, G.; Singth, J. Green synthesis of silver nanoparticles using aqueous leaf extract of *Premna integrifolia* (L.) rich in polyphenols and evaluation of their antioxidant, antibacterial and cytotoxic activity. *Biotechnol. Biotechnol. Equip.* **2019**, 359–371. [CrossRef]

74. Jun, Y.; Zi, Z.; Wanga, Y.; Zhaoa, G.; Liu, J.-X. Silver nanoparticles affect lens rather than retina development in zebrafish embryos. *Ecotoxicol. Environ. Saf.* **2018**, *163*, 279–288.

75. Sriram, M.I.; Kalishwaralal, K.; Barathmanikanth, S. Size-based cytotoxicity of silver nanoparticles in bovine retinal endothelial cells. *Nanosci. Methods* **2012**, *1*, 56–77. [CrossRef]

76. Kim, J.S.; Song, K.S.; Sung, J.H.; Ryu, H.R.; Choi, B.G.; Cho, H.S.; Lee, J.K. Genotoxicity, acute oral and dermal toxicity, eye and dermal irritation and corrosion and skin sensitisation evaluation of silver nanoparticles. *Nanotoxicology* **2013**, *7*, 953. [CrossRef]

77. Maneewattanapinyo, P.; Banlunara, W.; Thammacharoen, C.; Ekgasit, S.; Kaewamatawong, T. An evaluation of acute toxicity of colloidal silver nanoparticles. *J. Vet. Med. Sci.* **2011**, *73*, 1417–1423. [CrossRef]

78. De Matteis, V.; Cascione, M.; Cristina, C.; Rinaldi, R. Engineered Gold Nanoshells Killing Tumor Cells: New Perspectives. *Curr. Pharm. Des.* **2019**, *25*, 1477–1489. [CrossRef]

79. Weissleder, R. A clearer vision for in vivo imaging. *Nat. Biotechnol.* **2001**, *19*, 316–317. [CrossRef] [PubMed]

80. Song, H.B.; Wi, J.S.; Jo, D.H.; Kim, J.H.; Lee, S.W. Intraocular application of gold nanodisks optically tuned for optical coherence tomography: inhibitory effect on retinal neovascularization without unbearable toxicity. *Nanomedicine* **2017**, *13*, 1901–1911. [CrossRef] [PubMed]

81. Gasparini, S.J.; Llonch, S.; Borsch, O. Transplantation of photoreceptors into the degenerative retina: Current state and future perspectives. *Prog. Retin. Eye Res.* **2019**, *69*, 1–37. [CrossRef]

82. Chemla, Y.; Betzer, O.; Markus, A.; Farah, N.; Motiei, M. Gold nanoparticles for multimodal high-resolution imaging of transplanted cells for retinal replacement therapy. *Nanomedicine* **2019**, *14*, 1857–1871. [CrossRef]

83. Kim, H.; Nguyen, V.; Manivasagan, P.; Kim, S.W.; Oh, J. Doxorubicin-fucoidan-gold nanoparticles composite for dualchemo-photothermal treatment on eye tumors. *Oncotarget* **2017**, *8*, 113719–113733. [CrossRef] [PubMed]

84. Valluru, K.S.; Chinni, B.K.; Rao, N.A.; Bhatt, S.; Dogra, V.S. Basics and Clinical Applications of Photoacoustic Imaging. *Ultrasound Clin.* **2009**, *4*, 403–429. [CrossRef]

85. Gordon, A.Y.; Lapierre-Landry, M.; Skala, M.C.; Penn, J.S. Photothermal Optical Coherence Tomography of AntiAngiogenic Treatment in the Mouse Retina Using Gold Nanorods as Contrast Agents. *Transl. Vis. Sci. Technol.* **2019**, *8*, 18. [CrossRef] [PubMed]

86. Chan, C.-M.; Hsiao, C.-Y.; Li, H.-J.; Fang, J.-Y.; Chang, D.-C.; Hung, C.-F. The Inhibitory Effects of Gold Nanoparticles on VEGF-A-Induced Cell Migration in Choroid-Retina Endothelial Cells. *Int. J. Mol. Sci.* **2020**, *21*, 109. [CrossRef]

87. Gurunathanab, S.; Leeb, K.; Kalishwaralala, K.; Sheikpranbabua, S.; Vaidyanathana, R.; Eom, S.H. Antiangiogenic properties of silver nanoparticles. *Biomaterials* **2009**, *30*, 6341–6350. [CrossRef]

88. Sheikpranbabu, S.; Kalishwaralal, K.; Venkataraman, D.; Eom, S.H.; Park, J.; Gurunathan, S. Silver nanoparticles inhibit VEGF-and IL-1β-induced vascular permeability via Src dependent pathway in porcine retinal endothelial cells. *Nanobiotechnol.* **2009**, *7*, 8. [CrossRef]

89. Ngandeu Neubi, G.M.; Opoku-Damoah, Y.; Gu, X.; Han, Y.; Zhou, J.; Ding, Y. Bio-inspired drug delivery systems: an emerging platform for targeted cancer therapy. *Biomater. Sci.* **2018**, *6*, 958–973. [CrossRef] [PubMed]

90. Yang, L.; Kuang, H.; Zhang, W.; Aguilar, Z.P.; Wei, H.; Xu, H. Comparisons of the biodistribution and toxicological examinations after repeated intravenous administration of silver and gold nanoparticles in mice. *Sci. Rep.* **2017**, *7*, 3303. [CrossRef] [PubMed]

91. Madamsetty, V.S.; Mukherjee, A.; Mukherjee, S. Recent Trends of the Bio-Inspired Nanoparticles in Cancer Theranostics. *Front. Pharmacol.* **2019**, *10*, 1264. [CrossRef] [PubMed]

92. Birngruber, T.; Raml, R.; Gladdines, W.; Gatschelhofer, C.; Gander, E.; Ghosh, A.; Kroath, T.; Gaillard, P.J.; Pieber, T.R. Enhanced Doxorubicin Delivery to the Brain Administered Through Glutathione PEGylated Liposomal Doxorubicin (2B3-101) as Compared with Generic Caelyx,®/Doxil®—A Cerebral Open Flow Microperfusion Pilot Study. *J. Pharm. Sci.* **2014**, *203*, 1945–1948. [CrossRef]

93. Trost, A.; Lange, S.; Schroedl, F.; Bruckner, D.; Motloch, K.A.; Bogner, B.; Kaser-Eichberger, A.; Strohmaier, C.; Runge, C.; Aigner, L.; et al. Brain and Retinal Pericytes: Origin, Function and Role. *Front. Cell Neurosci.* **2016**, *4*, 20. [CrossRef] [PubMed]

94. Haque, S.; Shadab, M.D.; Alam, I.; Sahni, J.K.; Ali, J. Nanostructure-based drug delivery systems for brain targeting. *Drug Dev. Ind. Pharm.* **2012**, *38*, 387–411. [CrossRef] [PubMed]

95. Diebold, Y.; Jarrín, M.; Sáez, V.; Carvalho, E.L.; Orea, M.; Calonge, M.; Seijo, B. Ocular drug delivery by liposome-chitosan nanoparticle complexes (LCS-NP). *Biomaterials* **2007**, *28*, 1553–1564. [CrossRef] [PubMed]

96. Zhang, R.; He, R.; Qian, J.; Guo, J.; Xue, K.; Yuan, Y. Immunology and Microbiology Treatment of Experimental Autoimmune Uveoretinitis with Intravitreal Injection of Tacrolimus (FK506) Encapsulated in Liposomes. *Investig. Ophthalmol. Vis. Sci.* **2010**, *51*, 3575–3582. [CrossRef] [PubMed]

97. Yan, J.; Zheng, M. Chrysophanol liposome preconditioning protects against cerebral ischemia–reperfusion injury by inhibiting oxidative stress and apoptosis in mice. *Int. J. Pharmacol.* **2014**, *10*, 55–68. [CrossRef]

98. Choi, Y.H. Berberine hydrochloride protects C_2C_{12} myoblast cells against oxidative stress-induced damage via induction of Nrf-2-mediated HO-1 expression. *Drug Dev. Res.* **2016**, *77*, 8. [CrossRef]

99. Lai, S.; Wei, Y.; Wu, Q.; Zhou, K.; Liu, T.; Zhang, Y.; Jiang, N.; Xiao, W.; Chen, J.; Liu, Q.; et al. Liposomes for efFective drug delivery to the ocular posterior chamber. *J. Nanobiotechnol.* **2019**, *17*, 64. [CrossRef] [PubMed]

100. Wolin, L.R.; Massopust, L.C. Characteristics of the ocular fundus in primates. *J. Anat.* **1967**, *101*, 693–699. [PubMed]

101. Wang, J.-L.; Liu, Y.-L.; Li, Y.; Dai, W.-B.; Guo, Z.-M.; Wang, Z.-H.; Zhang, Q. EphA2 targeted doxorubicin stealth liposomes as a therapy system for choroidal neovascularization in rats. *Ophthalmol. Vis. Sci.* **2012**, *53*, 7348. [CrossRef] [PubMed]

102. Ravar, F.; Saadat, E.; Gholami, M.; Dehghankelishadi, P.; Mahdavi, M.; Azami, S. Hyaluronic acid-coated liposomes for targeted delivery of paclitaxel, in-vitro characterization and in-vivo evaluation. *J. Control. Release* **2016**, *10*, 10–22. [CrossRef]

103. Lee, J.; Goh, U.; Lee, H.J.; Kim, J.; Jeong, M. Effective retinal penetration of lipophilic and lipid-conjugated hydrophilic agents delivered by engineered liposomes. *Mol. Pharm.* **2017**, *14*, 423–430. [CrossRef]

104. Hironaka, K.; Inokuchi, Y.; Fujisawa, T.; Shimazaki, H.; Akane, M.; Tozuka, Y.; Tsuruma, K.; Shimazawa, M.; Hara, H. Edaravone-loaded liposomes for retinal protection against oxidative stress-induced retinal damage. *Eur. J. Pharm. Biopharm.* **2011**, *79*, 119–125. [CrossRef]

105. Yamamoto, Y.; Kuwahara, T.; Watanabe, K. Antioxidant activity of 3-methyl-1-phenyl-2-pyrazolin-5-one. *Redox Rep.* **1996**, *2*, 333–338. [CrossRef]

106. Vighi, E.; Trifunović, D.; Veiga-Crespo, P.; Rentsch, A.; Hoffmann, D.; Sahaboglu, A.; Strasser, T.; Kulkarni, M.; Bertolotti, E.; van den Heuvel, A.; et al. Combination of cGMP analogue and drug delivery system provides functional protection in hereditary retinal degeneration. *Proc. Natl. Acad. Sci. USA* **2018**, *27*, E2997–E3006. [CrossRef]

107. Lucas, K.A.; Pitari, G.M.; Kazerounian, S.; Ruiz-Stewart, I.; Park, J.; Schulz, S.; Chepenik, K.P.; Waldman, S.A. Guanylyl cyclases and signaling by cyclic GMP. *Pharmacol. Rev.* **2000**, *52*, 375–414.

108. Farber, D.B.; Lolley, R.N. Cyclic guanosine monophosphate: Elevation in degenerating photoreceptor cells of the C3H mouse retina. *Science* **1974**, *186*, 449–451. [CrossRef]

109. Campbell, M.; Humphries, M.M.; Kiang, A.-S.; Nguyen, A.T.H.; Gobbo, O.L.; Tam, L.C.S.; Suzuki, M.; Hanrahan, F.; Ozaki, E.; Farrar, G.-J.; et al. Systemic low-molecular weight drug delivery to pre-selected neuronal regions. *EMBO Mol. Med.* **2011**, *3*, 235–245. [CrossRef]

110. Davis, B.M.; Normando, E.M.; Guo, L.; Turner, L.A.; Nizari, S. Topical Delivery of Avastin to the Posterior Segment of the Eye In Vivo Using Annexin A5-associated Liposomes. *Small* **2014**, *24*, 1575–1584. [CrossRef] [PubMed]

111. Platania, C.B.M.; Fisichella, V.; Fidilio, A.; Geraci, F.; Lazzara, F.; Leggio, G.M.; Salomone, S.; Drago, F.; Pignatello, R.; Caraci, F.; et al. Topical Ocular Delivery of TGF- β 1 to the Back of the Eye: Implications in Age-Related Neurodegenerative Diseases. *Int. J. Mol. Sci.* **2017**, *18*, 2076. [CrossRef] [PubMed]

112. Boulyjenkov, V.; Berg, K. *Gene Therapy: Promises, Problems and Prospects*; Genes Resist. to Dis.; Springer: Berlin/Heidelberg, Germany, 2000.

113. Ramamoorth, M.; Narvekar, A. Non viral vectors in gene therapy—An overview. *J. Clin. Diagn. Res.* **2015**, *9*, GE01. [CrossRef]

114. Das, K.G.; Ram, A.R. Niosome ad a novel drug delivery system: A review. *Appl. Pharm.* **2013**, *5*, 1–7.

115. Seleci, D.A.; Seleci, M.; Walter, J.G. Niosomes as Nanoparticular Drug Carriers: Fundamentals and Recent Applications. *J. Nanomat.* **2016**, *2016*, 13. [CrossRef]

116. Gallego, I.; Villate-Beitia, I.; Martínez-Navarrete, G.; Menéndez, M.; López-Méndez, T.; Soto-Sánchez, C.; Zárate, J.; Puras, G.; Fernández, E.; Pedraz, J.L. Non-viral vectors based on cationic niosomes and minicircle DNA technology enhance gene delivery efficiency for biomedical applications in retinal disorders. *Nanomedicine.* **2019**, *17*, 308–318. [CrossRef] [PubMed]

117. Mashal, M.; Attia, N.; Puras, G.; Martinez-Navarrete, G.; Fernandez, E.P. Retinal gene delivery enhancement by lycopene incorporation into cationic niosomes based on DOTMA and polysorbate 60. *J. Control. Release* **2017**, *254*, 55–64. [CrossRef]

118. Guinedi, A.S.; Mortada, N.D.; Mansour, S. Preparation and evaluation of reverse-phase evaporation and multilamellar niosomes as ophthalmic carriers of acetazolamide. *Int. J. Pharm.* **2005**, *306*, 71–82. [CrossRef]

119. Guglevaa, V.; Titevaa, S.; Rangelovb, S.; Momekova, D. Design and in vitro evaluation of doxycycline hyclate niosomes as a potential ocular delivery system. *Int. J. Pharm.* **2019**, *5*, 567. [CrossRef]

120. Cintia, S.; De Paivaa Rosa, M.; Corralesa Arturo, L.; Villarreala William, J.; De-Quan, F.; Lia Michael, E.; Sternb Stephen, C. Corticosteroid and doxycycline suppress MMP-9 and inflammatory cytokine expression, MAPK activation in the corneal epithelium in experimental dry eye. *Exp. Eye Res.* **2006**, *83*, 526–535.

121. Sapino, S.; Chirio, D.; Peira, E.; Abellán Rubio, E.; Brunella, V.; Jadhav, S.A.; Chindamo, G.; Gallarate, M. Ocular Drug Delivery: A Special Focus on the Thermosensitive Approach. *Nanomaterials (Basel)* **2019**, *9*, 884. [CrossRef] [PubMed]

122. Hoare, T.R.; Kohane, D.S. Hydrogels in drug delivery: Progress and challenges. *Polymer (Guildf)* **2008**, *49*, 1993–2007. [CrossRef]

123. Wang, K.; Han, Z. Injectable hydrogels for ophthalmic applications. *J. Control. Release* **2017**, *268*, 212–224. [CrossRef]

124. Ferreira, N.N.; Ferreira, L.M.B.; Cardoso, V.M.O.; Boni, F.I.; Souza, A.L.R.; Gremião, M.P.D. Recent advances in smart hydrogels for biomedical applications: From self-assembly to functional approaches. *Eur. Polym. J.* **2018**, *99*, 117–133. [CrossRef]

125. Elieh-Ali-Komi, D.; Hamblin, M.R. Chitin and Chitosan: Production and Application of Versatile Biomedical Nanomaterials. *Int. J. Adv. Res. (Indore)* **2016**, *4*, 411–427.

126. ValérieDodane, V.D. Pharmaceutical applications of chitosan. *Pharm. Sci. Technol. Today* **1998**, *1*, 246–253.

127. Lee, K.Y.; Mooney, D.J. Alginate: Properties and biomedical applications. *Prog. Polym. Sci.* **2012**, *37*, 106–126. [CrossRef]

128. Paliwal, R.; Paliwal, S.R.; Sulakhiya, K.; Kurmi, B.D.; Kenwat, R.; Mamgain, A. *Chitosan-Based Nanocarriers for Ophthalmic Applications*; Elsevier: Amsterdam, The Netherlands, 2019.

129. Nagarwal, R.C.; Kumar, R. Chitosan coated sodium alginate-chitosan nanoparticles loaded with 5-FU for ocular delivery: in vitro characterization and in vivo study in rabbit eye. *Eur. J. Pharm. Sci.* **2012**, *20*, 678–685. [CrossRef]

130. Ameeduzzafar Imam, S.S.; Abbas Bukhari, S.N.; Ahmad, J. Formulation and optimization of levofloxacin loaded chitosan nanoparticle for ocular delivery: In-vitro characterization, ocular tolerance and antibacterial activity. *Int. J. Biol. Macromol.* **2018**, *108*, 650–659. [CrossRef]

131. Andrei, G.; Peptu, C.A.; Popa, M.; Desbrieres, J.; Peptu, C.; Gardikiotis, F.; Costuleanu, M.; Costin, D.; Dupin, J.C.; Uhart, A. Formulation and evaluation of cefuroxim loaded submicron particles for ophthalmic delivery. *Int. J. Pharm.* **2015**, *30*, 16–29. [CrossRef]

132. Kaskoos, R. Investigation of moxifloxacin loaded chitosan-dextran nanoparticles for topical instillation into eye: In-vitro and ex-vivo evaluation. *Int. J. Pharm. Investig.* **2014**, *4*, 164–173. [CrossRef] [PubMed]

133. Solanki, A.; Smalling, R.; Parola, A.H.; Nathan, I.; Kasher, R.; Pathak, Y.; Sutariya, V. Humanin Nanoparticles for Reducing Pathological Factors Characteristic of Age-related Macular Degeneration. *Curr. Drug Deliv.* **2019**, *16*, 226–232. [CrossRef]

134. Cohen, A.; Lerner-Yardeni, J.; Meridor, D.; Kasher, R.; Nathan, I.; Parola, A. Humanin derivatives inhibit necrotic cell death in neurons. *Mol. Med.* **2015**, *21*, 505. [CrossRef] [PubMed]

135. Cheng, Y.; Ko, Y.; Chang, Y.; Huang, S. Thermosensitive chitosan-gelatin-based hydrogel containing curcumin-loaded nanoparticles and latanoprost as a dual-drug delivery system for glaucoma treatment. *Exp. Eye Res.* **2019**, *179*, 179–187. [CrossRef] [PubMed]

136. Radomska-leśniewska, D.M.; Osiecka-iwan, A.; Hyc, A.; Góźdź, A.; Dąbrowska, A.M.; Skopiński, P. Therapeutic potential of curcumin in eye diseases. *Cent. Eur. J. Immunol.* **2019**, *44*, 2–4. [CrossRef] [PubMed]

137. Nathan Gooch, S.A.M.; Condie, R.; Michael, B.R.; Bonnie Archer, B.K.A.; Ocular, B. Drug Delivery for Glaucoma Management. *Pharmaceutics* **2012**, *4*, 197–211. [CrossRef]

138. Michael, P.; Fautsch, D.H.J. Aqueous Humor Outflow: What Do We Know? Where Will It Lead Us? *Investig. Ophthalmol. Vis. Sci.* **2006**, *47*, 4181–4187.

139. Silva, B.; Marto, J.; Braz, B.S.; Delgado, E.; Almeida, A.J.; Gonçalves, L. New nanoparticles for topical ocular delivery of erythropoietin. *Int. J. Pharm.* **2020**, 119020. [CrossRef]

140. Pandit, J.; Sultana, Y.; Aqil, M. Chitosan-coated PLGA nanoparticles of bevacizumab as novel drug delivery to target retina: optimization, characterization, and in vitro toxicity evaluation. *Artif. Cells Nanomed. Biotechnol.* **2016**, *45*, 1397–1407. [CrossRef]

141. Mitra, R.N.; Han, Z.; Merwin, M.; Taai, M.; Al Conley, S.M. Synthesis and Characterization of Glycol Chitosan DNA Nanoparticles for Retinal Gene Delivery. *ChemMedChem* **2014**, *9*, 189–196. [CrossRef]

142. Lee, J.-H.; Kim, H.-W. Emerging properties of hydrogels in tissue engineering. *J. Tissue Eng.* **2018**, *9*, 2041731418768285. [CrossRef] [PubMed]

143. Chang, D.; Kinam Park, A. Hydrogels for sustained delivery ofbiologics to the back of the eye. *Drug Discov. Today* **2019**, *24*, 1470–1482. [CrossRef] [PubMed]

144. Van Hovea, A.H.; Beltejar, M.G.; Benoit, D.S.W. Development and in vitro assessment of enzymatically-responsive poly(ethylene glycol) hydrogels for the delivery of therapeutic peptides. *Biomaterials* **2014**, *35*, 9719–9730. [CrossRef] [PubMed]

145. Drapala, P.W.; Jiang, B.; Chiu, Y.-C.; Mieler, W.F.; Brey, E.M. The Effect of Glutathione as Chain Transfer Agent in PNIPAAm-Based Thermo-responsive Hydrogels for Controlled Release of Proteins. *Pharm. Res.* **2014**, *31*, 742–753. [CrossRef] [PubMed]

146. Schultz, C.; Breaux, J.; Pharmd, J.S.; Douglas Morck, D.V.M. Drug delivery to the posterior segment of the eye through hydrogel contact lenses. *Clin. Exp. Optom.* **2011**, *94*, 212–218. [CrossRef]

147. Kabiri, M.; Kamal, S.H.; Pawar, S.V.; Roy, P.R.; Derakhshandeh, M.; Kumar, U.; Hatzikiriakos, S.G.; Hossain, S.; Yadav, V.G. A stimulus-responsive, in situ-forming, nanoparticle-laden hydrogel for ocular drug delivery. *Drug Deliv. Transl. Res.* **2018**, *8*, 484–495. [CrossRef]

148. Osswald, C.R.; Kang-Mieler, J.J. Controlled and Extended In Vitro Release of Bioactive Anti-Vascular Endothelial Growth Factors from a Microsphere-Hydrogel Drug Delivery System. *Curr. Eye Res.* **2016**, *41*, 1216–1222. [CrossRef]

149. Agrahari, V.; Agrahari, V.; Hung, W.-T.; Christenson, L.K. Composite Nanoformulation Therapeutics for Long-Term Ocular Delivery of Macromolecules. *Mol. Pharm.* **2016**, *13*, 2912–2922. [CrossRef]

150. Yang, H.; Tyagi, P.; Kadam, R.S.; Holden, C.A. Hybrid dendrimer hydrogel/PLGA nanoparticle platform sustains drug delivery for one week and antiglaucoma effects for four days following one-time topical administration. *ACS Nano* **2012**, *25*, 7595–7606. [CrossRef]

151. Wadhams, G.H.; Armitage, J.P. Making sense of it all: Bacterial chemotaxis. *Nat. Rev. Mol. Cell. Biol.* **2004**, *5*, 1024–1037. [CrossRef] [PubMed]

152. Micali, G.; Endres, R.G. Bacterial chemotaxis: information processing, thermodynamics, and behavior. *Curr. Opin. Microbiol.* **2016**, *30*, 8–15. [CrossRef] [PubMed]

153. Mei, Y.; Alexander, A. Solovev SS and OGS. Rolled-up nanotech on polymers: from basic perception to self-propelled catalytic microengines. *Chem. Soc. Rev.* **2011**, *40*, 2109–2119. [CrossRef] [PubMed]

154. Kumar, M.S.; Philominathan, P. The physics of flagellar motion of E. coli during chemotaxis. *Biophys. Rev.* **2010**, *2*, 13–20. [CrossRef] [PubMed]

155. Wu, Z.; Troll, J.; Jeong, H.-H.; Wei, Q.; Stang, M.; Ziemssen, F.; Wang, Z.; Dong, M.; Schnichels, S.; Qui, T.; et al. A swarm of slippery micropropellers penetrates the vitreous body of the eye. *Sci. Adv.* **2018**, *4*, eaat4388. [CrossRef] [PubMed]

156. Bohn, H.F.; Federle, W. Insect aquaplaning: Nepenthes pitcher plants capture prey with the peristome, a fully wettable water-lubricated anisotropic surface. *Proc. Natl. Acad. Sci. USA* **2004**, *101*, 14138–14143. [CrossRef]

157. Wong, T.-S.; Kang, S.H.; Tang, S.K.Y.; Smythe, E.J.; Hatton, B.D.; Grinthal, A. Bioinspired self-repairing slippery surfaces with pressure-stable omniphobicity. *Nature* **2011**, *477*, 443–447. [CrossRef]

158. Nistor, M.T.; Rusu, A.G. *Nanorobots with Applications in Medicine*; Elsevier: Amsterdam, The Netherlands, 2019.

159. Alhalafi, A.M. Applications of polymers in intraocular drug delivery systems. *Oman J. Ophtalmol.* **2017**, *10*, 3–8. [CrossRef]

Review

The Intersection of Serine Metabolism and Cellular Dysfunction in Retinal Degeneration

Tirthankar Sinha [†], Larissa Ikelle [†], Muna I. Naash * and Muayyad R. Al-Ubaidi *

Department of Biomedical Engineering, University of Houston, Houston, TX 77204, USA;
tsinha2@Central.UH.EDU (T.S.); likelle@Central.UH.EDU (L.I.)
* Correspondence: mnaash@central.uh.edu (M.I.N.); malubaid@central.uh.edu (M.R.A.-U.);
  Tel.: +1-713-743-1651 (M.I.N.); Fax: +1-713-743-0226 (M.I.N.)
† These authors contributed equally to this work.

Received: 4 February 2020; Accepted: 6 March 2020; Published: 10 March 2020

Abstract: In the past, the importance of serine to pathologic or physiologic anomalies was inadequately addressed. Omics research has significantly advanced in the last two decades, and metabolomic data of various tissues has finally brought serine metabolism to the forefront of metabolic research, primarily for its varied role throughout the central nervous system. The retina is one of the most complex neuronal tissues with a multitude of functions. Although recent studies have highlighted the importance of free serine and its derivatives to retinal homeostasis, currently few reviews exist that comprehensively analyze the topic. Here, we address this gap by emphasizing how and why the de novo production and demand for serine is exceptionally elevated in the retina. Many basic physiological functions of the retina require serine. Serine-derived sphingolipids and phosphatidylserine for phagocytosis by the retinal pigment epithelium (RPE) and neuronal crosstalk of the inner retina via D-serine require proper serine metabolism. Moreover, serine is involved in sphingolipid–ceramide balance for both the outer retina and the RPE and the reductive currency generation for the RPE via serine biosynthesis. Finally and perhaps the most vital part of serine metabolism is free radical scavenging in the entire retina via serine-derived scavengers like glycine and GSH. It is hard to imagine that a single tissue could have such a broad and extensive dependency on serine homeostasis. Any dysregulation in serine mechanisms can result in a wide spectrum of retinopathies. Therefore, most critically, this review provides a strong argument for the exploration of serine-based clinical interventions for retinal pathologies.

Keywords: serine; retinal degeneration; diabetic retinopathy; macular degeneration; macular telangiectasia; oxidative stress; sphingolipids; retina; RPE; Müller cells

---

1. Why Is Serine Important to the Entire Retina?

Serine is a non-essential amino acid directly involved in cellular homeostasis, proliferation, and differentiation [1,2]. The cells of the neural retina are no exception, and, in fact, exhibit a great dependence on serine and its exhaustive variety of metabolic intermediates [3]. Serine uptake occurs either by delivery from the bloodstream or it can be synthesized by the anabolism of the glycolytic intermediate, 3-phosphoglycerate (3-PG) [4] in the retina (neural retina-RPE, Figure 1). After uptake or synthesis, serine becomes available and serves as a central node in many cellular processes [5].

Serine Biosynthesis Pathway

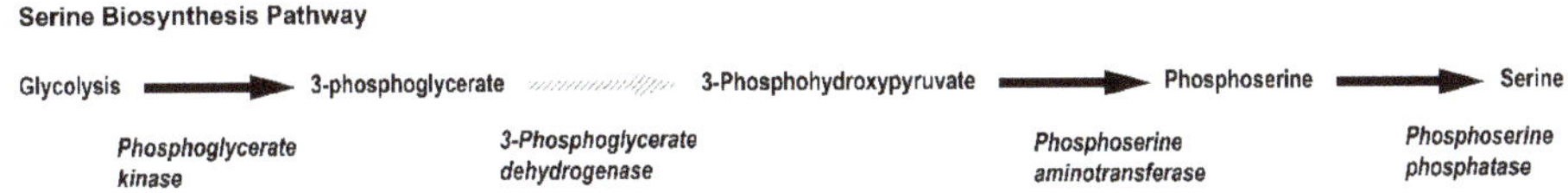

Figure 1. Pathway depicting serine biosynthesis from glycolysis. Metabolic intermediates involved in the enzymatic synthesis of L-serine from glycolysis is shown here with the rate limiting step marked with a dashed grey arrow. The enzymes involved in respective steps are shown in bold italics below the arrow for the individual reaction.

Other than being an integral amino acid in multiple essential proteins, free serine is essential for generating cysteine, glycine, methionine, and sphingolipids [5]. Glycine and cysteine are essential intermediates. Glycine is a precursor molecule for porphyrins and purine nucleotides [6]. Cysteine on the other hand is important in the protection of neuronal cells and for the production of taurine [7]. Together they form glutathione (GSH), a critical anti-oxidant in the retina [8]. Sphingolipids are elemental components of the phospholipid bilayer and are indispensable to cellular viability, homeostasis and function [2,3,5]. Furthermore, serine derived-metabolites have proven essential to methylation [2], apoptosis, and synaptic receptor activation [6].

The integrality of serine to cellular function has been appreciated by dysregulatory events appearing in many retinopathies. Reduced serine levels have been implicated in the etiology of macular related diseases such as macular telangiectasia type 2 (Mac Tel), age related macular degeneration, and diabetic retinopathy (DR) [9,10]. The retina is a complex stratified tissue consisting of the retinal pigment epithelium (RPE), a critical layer of cells for retinal homeostasis, and the neural retina containing the two types of photoreceptor cells. The neural retina also harbors the second order neurons and the retinal ganglion cells (RGC) that form the optic nerve. Serine proves to be a vital intermediate in many of these processes. Consequently, our goal is to provide a focused review of the role of serine homeostasis in maintaining optimum retinal function and proper oxidative balance.

2. Why Does the Retina Need to Synthesize Serine?

In most tissues, serine uptake from either blood or proteolysis sufficiently meets cellular metabolic requirements. However, there are tissues that mandate an elevated level of serine and, accordingly, upregulate enzymes for serine biosynthesis. As previously highlighted, this biosynthetic pathway (Figure 1) helps in maintaining the redox potential of the cell [11]. The primary source of serine biosynthesis is glucose [12], which in most cells is utilized via glycolysis. Serine biosynthesis branches from one of the glycolytic intermediates, 3-phosphoglycerate. The rate limiting step for serine biosynthesis is the reaction involving phosphoglycerate dehydrogenase (PHGDH), which converts 3-phosphoglycerate into phosphoserine. This is followed by the removal of the phosphate to generate l-serine.

It has been previously shown that the retina expresses high levels of all enzymes involved in serine biosynthesis. However, the RPE appears to express even higher levels than the neural retina [13] as has been shown by flux studies, whereby the RPE readily converts glucose into serine [10,14]. Further experiments verified that the neural retina possess an efficient system for serine uptake. [15].

De Novo Serine Synthesis in Müller Cells

Transport of serine, a neutral amino acid, across the tight blood–retinal barrier into the RPE or across the endothelial cells and to the neural retina is supposedly inadequate [16]. So the retina increases the levels of intracellular serine through de novo synthesis [12]. Further supply of serine to photoreceptors and the inner retina is provided by the RPE and retinal Müller cells [15]. The latter was demonstrated by co-immunofluorescence with anti-PHGDH and anti-cellular retinaldehyde–binding protein (CRALBP) antibodies showing that the serine de novo synthesis pathway is indeed present in the Müller cells [17]. Since serine metabolism is central to maintaining redox and oxidative balance, ion flux, glutamate levels, and many other support functions [18], Müller cells through their de novo synthesis of serine play major roles in these functions [17].

Many retinopathies are associated with loss of Müller glia [17]. Mac Tel, is a pathology of the retina recently characterized by significant reductions in serine synthesis and loss of central vision [16]. Although the macula (anatomic) is a small cone-dense region of the retina occupying only 1.4% of the total area of the retina, it harbors approximately 8% of the total cone population and 60% of all RGCs [19]. The macula is incredibly metabolically active and relies heavily on Müller glia [18]. Not only has localization of serine synthesis in the neural retina been demonstrated in Müller cells, but relative to peripheral Müller cells, those of the macula seem to show increased expression of PHGDH [10]. Moreover, the macular Müller cells show increased GSH and glycine production and are more susceptible to induced stress [10].

3. Serine Homeostasis Plays an Important Role in the Maintenance of the Retina

3.1. Serine and Sphingolipids

One of the many fates of biosynthesized L-serine is to combine with palmitoyl-Co-A to form sphingolipids (Figure 2) catalyzed by the enzyme serine palmitoyl-Co-A transferase (SPT) [20]. Even though there are multiple interconnected pathways, which can control the formation of various sphingolipids, the most common pathway is serine incorporation [21]. Neural retinal sphingolipids have been well characterized addressing their beneficial and toxic capacities [3,22–25]. It is very well known that the most vital role of sphingolipids is aiding in sphingomyelin formation, which enables efficient synaptic transmission [3]. Perhaps, that is one reason why the sphingolipid content in the inner retina is quite high [3]. Moreover, sphingolipids in the form of sphingosine-1-phosphate are thought to have anti-apoptotic role, further rationalizing the abundance of sphingolipids in the outer retina [3]. It is important to note that the role of sphingolipid levels and their derivatives have not yet been assessed in the RPE. Given that the RPE may be a vital source of serine for the neural retina, it is imperative to determine how its transport might be facilitated.

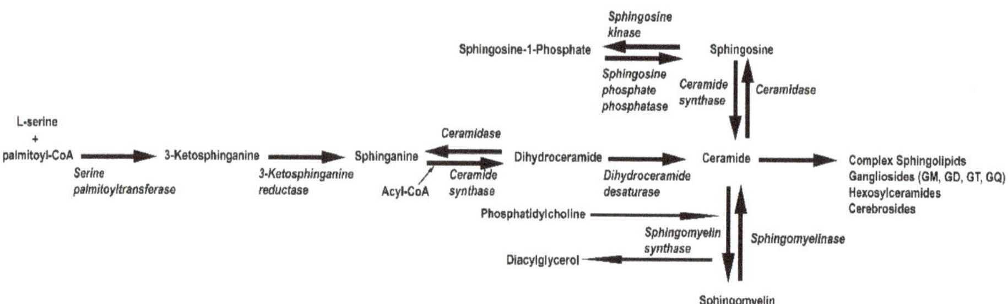

Figure 2. Pathway depicting sphingolipid biosynthesis from L-serine. Metabolic intermediates involved in the enzymatic synthesis of sphingolipids from L-serine are shown here. The enzymes involved in respective steps are shown in bold italics below the arrow for the individual reaction.

3.2. Serine and RPE Phagocytosis

The efficient phagocytosis of photoreceptor outer segments by the RPE is elemental to retinal health. From extensive examination of the process, it has been determined that any delay or inefficiency in the phagocytosis can lead to gross abnormalities for both the neural retina and the RPE [26]. The metabolism of the RPE is largely dependent on recycling the outer segment phospholipids, as they provide an important source of fuel [27–29]. The neural retina eliminates older disks in order to maintain optimal function. So how is serine vital for maintaining such an important step? In order for the RPE to recognize and phagocytose the correct portion of the outer segments, phosphatidylserine will localize to the extracellular surfaces of those digestible regions [30]. Since almost 10% of the entire outer segment is daily phagocytosed from each photoreceptor, it helps us appreciate the enormous amount of serine that needs to be available. Furthermore, it has been suggested that sphingolipids may also play a regulatory role in order to ensure efficient phagocytosis, since disruption of sphingolipid metabolism via SPT and ceramide synthase inhibition impaired phagocytosis [31,32]. In addition, glycosphingolipids like lactosylceramides and gangliosides are major lipid raft components and assist in cell adhesion, membrane polarity and phagocytosis initiation [33–36]. This raises the possibility that the sphingolipid pool in RPE may assist in facilitating outer segment phagocytosis as well as maintaining the tight junction barrier and cellular polarity. Since serine is an integral component of sphingolipids, it goes without saying that this presents another facet of serine availability contributing to efficient neural retina:RPE interdependence.

3.3. The Role of D-Serine

D-serine, an enantiomer of L-serine, is primarily released by glial cells, Müller cells and astrocytes [13]. Initially, it was discovered in the brain but has more recently been studied in the neural retina [13]. D-serine, formed by the racemization of L-serine by serine racemase (SRR) [13], functions as a neurotrophic factor and as a co-agonist for *N*-methyl-D-aspartate receptors (NMDARs) [37]. In the brain, NMDARs normally bind glutamate and glycine, but data suggests preferential or higher affinity binding to D-serine in the retina [13,38]. NMDARs locate at the synaptic terminals of RGCs and sparingly to photoreceptor terminals [39]. The role of NMDARs in the synaptic terminals has not been entirely substantiated, however, studies have indicated that they function in an excitatory role [13]. In the presence of D-serine, NMDA-mediated currents and light-evoked response showed increased amplitude over the D-serine absent control [13]. Irrespective of NMDAR functions, the importance of D-serine is evidenced by pathologies that are directly linked to D-serine irregularities. D-serine insufficiency has been associated with psychiatric and neurodevelopmental disorders [37]. Supplementation with D-serine has proven to mitigate some of the symptoms of psychosis [37]. Patients with DR, by contrast, suffer from increased SRR activity and over production of D-serine, which elicits an excitotoxic effect on RGCs and ultimately contributes to cell death [40]. This will be addressed in more detail in the following sections.

3.4. Serine and Epigenetic Regulation

One carbon metabolism is a fundamental process for purine, thymidine, and amino acid biosynthesis that involves the transfer of a one carbon unit to generate critical metabolites [8]. Serine is essential in this group of metabolic reactions [41] and, in this context, is a precursor for the synthesis of methionine. Briefly, serine donates a methyl group which reacts with homocysteine, originally derived from aspartate [42,43]. This reaction is catalyzed by serine transhydroxymethylase to ultimately form methionine [42,43].

An adenosylation reaction of adenosine triphosphate (ATP) and methionine will yield *S*-adenosyl-L-methionine, which is the critical methyl donor for global methylation of DNA [41,44,45]. DNA methylation, an absolutely essential process to maintain cellular homeostasis [46], occurs on CpG islands of DNA and suppresses gene expression [47]. Therefore, irregularities in serine levels cause downstream hypo- or hyper-methylation of DNA [48], effecting stress responses, proliferation, metabolism, and even responses to extrinsic stimuli [46].

The link between epigenetic modulations and retinal disease is currently at a nascent phase [49], so correlations between aberrant serine metabolism, DNA methylation, and retinal disease remain unclear. However, recent studies performed on three pairs of monozygotic twins with different presentations of AMD indicated significant changes to DNA methylation patterns of genes that may be implicated in disease pathogenesis [50]. More critically, the diet of the studied twins also highlighted nutritional significance in epigenetic regulation. Subjects with reduced dietary methionine, vitamin D, and betaine had worse disease prognosis [50], implicating the importance of nutrient bioavailability and epigenetics. As serine is involved in the synthesis of methionine, further studies should explore serine levels in retinal disease in relation with epigenetic changes that may contribute to disease onset and/or progression.

4. Serine Is an Anti-Oxidant and Mediator of Inflammation

The role that serine plays as a central junction for critical intermediates also extends to anti-oxidative and inflammatory mechanisms. Experimental evidence has indicated that the retina shares some metabolic patterns with neoplastic cells [16]. In these systems, the Warburg effect predominates [51], and energy is generated primarily through glycolysis to ensure accelerated ATP production [16]. As explained earlier, the retina is one of the highest energy consuming tissues and thus a source of excessive reactive oxygen species (ROS). Emerging data has shed light on extra-mitochondrial sources of ROS specifically in the outer retina and even in the outer segment of photoreceptors [52,53]. It was previously shown that oxygen is mainly absorbed at the level of the photoreceptors [54] and thus this site is most prone to oxidative stress, both in physiology and pathology [55]. To build on that, it seems that blue light exposure on outer segment discs as well as ectopic oxidative phosphorylation in the outer segment both pose a greater need for effective free radical scavengers in the outer segment–RPE interface [56–58]. Glycine and GSH and thus serine supply are essential for this ROS mitigation system of the rods and cones [59].

Fundamentally, intracellular redox homeostasis represents the equilibrium between oxidative and anti-oxidative species, creating a balance in which the environment is not cytotoxic and is sensitive enough to redox changes that may mandate an intracellular or extracellular response. Clearly, oxidative stress is caused by aberrations to this balance. Serine metabolism contributes to redox homeostasis through synthesis of glycine and its essential downstream products GSH and nicotinamide adenine dinucleotide phosphate (NADPH), as well as nicotinamide adenine dinucleotide (NADH) generation during serine biosynthesis [60]. In addition, GSH and NADPH deactivate ROS and other oxidative molecules [60]. GSH is the direct result of combining cysteine and glycine, both of which, as previously explored, are synthesized from serine [61]. NADPH is generated in many pathways, but recent studies suggest that metabolism of serine is a significant contributor of NADPH to the mitochondria, especially during hypoxic conditions [11]. During serine synthesis from 3-phosphoglycerate, serine donates a single carbon to folate forming glycine as well as tetrahydrofolate (THF). THF reductase forms $NADP^+$. Then, GSH mediated-reduction of electrophilic molecules helps to maintain the balance of NADPH to $NADP^+$, which is principle to redox homeostasis.

GSH anti-oxidative activity is vital for proper retinal function [62]. So in malignancies where serine is deficient, there is an obvious reduction in GSH level and activity, and, as a consequence, the supply of the precursor molecules have been significantly affected. However, it is not as simple as reduced supply. 5′ adenosine monophosphate-activated protein kinase (AMP-K) can also influence the availability of GSH. AMP-K, like nuclear factor erythroid-derived 2-like 2 (NRF-2), is a "cellular sensor", and responds to metabolic and redox irregularities [63]. AMP-K supports cell survival by upregulating anti-oxidant molecules such as GSH. Elevating serum serine levels in mice fed high fat diet showed increased levels of phosphorylated AMP-K, which resulted in reduced oxidative side-effects and increased GSH expression [63]. AMP-K and NRF-2 are, by no means, the only transcription factors involved in moderating the anti-oxidative response, but they represent the profound integration of serine metabolism and retinal oxidative homeostasis.

NRF-2 is an important redox-activated transcription factor under conditions of stress and imbalance of damaging oxidative species [64]. NRF-2 targets critical anti-oxidant genes such as superoxide dismutase, catalase, and GSH by upregulating transcription. In culture of non-small cell lung cancer cells (NSCLCs), gene enrichment analysis has demonstrated a correlation between PHGDH, the enzyme necessary for shunting 3-phosphoglycerate to serine synthesis, and genes that target the activation of NRF-2 [64]. The correlation suggests that biosynthesis of serine is involved in the expression and regulation of essential anti-oxidant proteins [64]. In addition to the cohort of anti-oxidant proteins, NRF-2 also has some regulatory involvement in the bioavailability of nitric oxide (NO) [65], which has the ability to mitigate the effects of H_2O_2 and superoxide [66]. Treatment with serine increases NO levels in culture which directly links NO production to serine [65], whether endogenously or exogenously presented.

Oxidative stress can also elicit tremendous damage to membranes, DNA, and mitochondria. However, in many pathologies, oxidative stress and inflammation are intimately interlinked [67]. In particular, AMD and DR have been characterized by the slow infiltration of pro-inflammatory constituents of the innate immune system [68]. Serine is currently being explored as a possible therapy for addressing inflammation in these pathologies. However, the involvement of serine metabolism in inflammation and innate immunity remains unclear [69].

Interesting observations have demonstrated a contradictory picture. Exogenously administered serine has been shown to reduce levels of inflammatory cytokines such as interleukin-1β and interleukin-6 [63]. Serine has also been shown to weaken the pro-inflammatory response necessary for macrophage recruitment after bacterial infections in mice [70]. The conflict emanates from the upregulation of inflammatory elements by way of increased glycolytic flux. In DR and AMD, dietary mismanagement and genetic mutations lead to metabolic dysregulation [71–74], which may increase glycolysis in cells. This may lead to increased synthesis of serine since it is produced from the PHGDH shunt. The increase in serine glycolytic synthesis is linked to the amplified activation of toll-like receptor 4 (TLR-4) after activation by lipopolysaccharide (LPS) [69] or hydrogen peroxide [75]. TLR-4 then induces a cytokine response, namely interleukin-1β [69]. Furthermore, GSH, whose activity and expression is upregulated by stress and higher levels of serine, further contributes to cytokine production and maintenance of redox balance in the cellular environment [69].

Serine deprivation was shown to be effective in reducing the activation of TLR-4, and reducing cytokine levels [69]. Activation of TLR-4 has also been implicated in the etiology of AMD [76]. Contrarily, serine supplementation is being explored as a potential therapy. Exogenous supplementation has proven to be effective in many cases reducing oxidative stress and reducing cytokine levels, but endogenous synthesis (as discussed above) augments the inflammatory response [77]. Therefore, balancing the anti-oxidant effects of serine with the pro-inflammatory nature of its endogenous synthesis is something to consider when developing a serine-based therapy for retinopathies in which inflammation contributes to the pathogenesis.

5. Consequences of Aberrations in Serine Metabolism

Our exploration has thus far presented the extensive involvement of serine metabolism in retinal homeostasis. Therefore, aberrations to this important keg in the metabolic machinery can negatively influence retinal pathologies. Glycolytic serine synthesis provides the largest contribution to intracellular serine stores, so deficiencies generally result from abnormalities in that synthetic process. For instance, deleterious mutations in PHGDH have been associated with microcephaly, reduced cognition, and psychomotor abnormalities [78]. Neu-Laxova is a fatal congenital disease marked by serious systemic abnormalities and is attributed to homozygous mutations in enzymes involved in serine biosynthesis [1]. This is recapitulated in the PHGDH$^{-/-}$ mouse model, where mice suffer from embryonic lethality [1]. These irregularities can affect every part of the cell, from mitochondrial biosynthesis to oxidative imbalance.

An extensive comparative metabolomics study was performed by Gao et al., analyzing the metabolic and mitochondrial changes of colon cancer cell lines that occur as a result of serine deprivation [79]. Primarily, glycine, serine, threonine, pyrimidine, and sphingolipid pathways were the most significantly affected. Cells exhibited reduced fatty acid synthesis, and reduced TCA intermediates, consequently cells had a 45% reduction in ATP. Serine deprivation also caused major changes in mitochondrial membrane potential and increased fragmentation. However, phospholipid and phosphatidylserine levels remained comparably similar to standard culture conditions; sphingosine and ceramide levels on the other hand were significantly lowered [79]. It was concluded that mitochondrial fragmentation was exacerbated by reduced production of ceramides that serve as major constituents of the mitochondrial membrane. Ultimately, Gao et al. was able to isolate effected pathways that attenuated proper mitochondrial functions critical for cellular viability [79]. Considering the metabolic demands in the retina, such deficits would be extremely detrimental and could play a significant role in the etiology of many retinal diseases associated with metabolic dysregulation.

Cytotoxic aggregation of deoxysphingolipids is another important feature of serine deficiency [78]. As plasma serine levels decline, alanine flux increases and promotes the production of deoxysphingolipids [78]. As previously indicated, serine sphingolipids are synthesized through the condensation of serine and palmitoyl-CoA mediated by SPT. In cases of serine deficiency, SPT may incorporate alanine or glycine into the formation of the lipid, forming a cytotoxic analogue [78]. These lipids have been shown to aggregate and induce apoptosis in in vitro and in vivo models [78]. Retinal organoids treated with deoxysphingolipids exhibited dose-dependent apoptosis [9].

6. Retinal Degeneration and Dysregulated Serine Metabolism

6.1. Inherited Retinal Degeneration

Inherited retinal degeneration (IRD) can arise from mutations in various genes linked to photoreceptor development and structure or genes involved in phototransduction. To add to the complexity of IRDs, patients with similar mutations may present with very different phenotypes [80]. This is indicative of other underlying factors that can contribute to disease manifestation. It has been shown that metabolic dysregulation is a common element of various IRD models [81]. One example that has recently emerged is the dysregulation of sphingolipid metabolism [3]. Many investigators have shown that ceramide toxicity is elevated in various models of IRD [22]. In contrast, levels of the protective sphingolipid, sphingosine-1-phosphate, are significantly reduced [22]. What is yet to be determined is why and how the toxic ceramides are increased while sphingosine-1-phosphate levels diminish. One mechanism that has been postulated suggests that serine deficiency forces the sphingolipid metabolism to switch to the incorporation of alanine for sphingolipid synthesis [22]. However, in doing so, the reaction is skewed towards the formation of ceramides. Further investigation is required to determine why there is reduced serine availability. Since the glycolytic precursor, phosphoglycerate, is the primary source for serine biosynthesis, it is possible that dysregulation in glucose metabolism is the cause for the reduced serine availability.

6.2. Diabetic Retinopathy

Studies have demonstrated a correlation between serine deficiency and systemic diabetes [77]. Inflammation potentially increases the expression and activity of SRR, and consequently increases the availability of D-serine for receptor binding [40]. The elevated levels of D-serine contribute to glutamate toxicity and induce RGC apoptosis [40]. More precisely, as more D-serine binds to NMDARs, an excitotoxic response is induced [40]. As part of glutamate signal propagation in the retina, NMDAR excitation may be involved in decreasing neurotransmitter sensitivity [40]. Nevertheless, when D-serine levels are substantially elevated, the NMDAR-mediated excitation has a toxic effect in the retinal environment and induces cell death in RGCs [40]. In SRR null models with streptozocin (STZ) induced DR, degeneration of RGCs appeared attenuated, there was a reduction in neovascularization and the retina appeared significantly healthier than the control [40]. Most interestingly, multiple studies have found that if diabetic patients are treated with a serine supplement, blood glucose levels were reduced [77,82]. Currently, no mechanism is known to explain how this happens. But it is known that upon onset of diabetes, similar to serine deficiency, toxic sphingolipid accumulation starts occurring [83]. In fact, deoxysphingolipids have been shown to be potent biomarkers for diabetes mellitus [84]. As in IRD, DR results in similar toxic build-up of these aberrant sphingolipids, but this occurs far before any retinopathic complications start to develop [25]. These ceramides can actually be used as an early indicator of disease. It appears as if serine deprivation and toxic ceramide accumulation occur concomitantly. Perhaps, as the retina relies most heavily on serine, any deficiencies may cause the tissue to exhibit the first signs of disease in DR.

7. Novel Insights: Macular Telangiectasia

It has been elucidated recently that a dysregulated lipid metabolism can lead to increased and leaky vasculature [16]. Given the fact that serine homeostasis closely regulates lipid metabolism, it becomes a prime candidate for such monitoring. Interestingly, it was found in patients suffering from Mac Tel that their PHGDH activity is also compromised [85]. This led to severe serine deficiency, which contributed to vascular abnormalities [9]. As explained earlier, serine deficiency further leads to toxic deoxysphingolipid accumulation, which indeed plays a role in the etiology of Mac Tel type 2 [86]. However, administering exogenous serine to patients as a promising therapeutic approach has been marginally successful [87,88]. Importantly, how serine deficiency may affect the RPE and whether that has a role in Mac Tel is not yet known. Nevertheless, it is to be determined whether the restoration of physiological levels of serine in the retina can prevent further neovascularization in Mac Tel patients.

8. Concluding Remarks

Serine metabolism has vast interconnectivity with many of the homeostatic mechanisms that work in concert to maintain retinal health and function, as depicted in Figure 3. As reviewed above, considerable evidence has indicated that both the RPE and Müller cells of the inner retina have the requisite enzymes for serine biosynthesis. Besides the significant contribution of these two tissues to the serine pool of the entire retina, we also established the role de novo serine synthesis plays in redox currency generation and in mitigating free radical stress for both the neural retina and RPE. Given the retina is a metabolically high functioning tissue with high-energy demands and recent advancements indicating extra-mitochondrial contribution to the elevated presence of free radicals compared to other tissues, it calls for such extensive measures. Thus we show that while the pentose phosphate pathway might be sufficient to maintain the redox balance for other tissues, the retina depends upon additional tools, which it obtains primarily from serine metabolism: like NADPH from de novo biosynthesis, glycine, GSH and NADH. In addition, a review of the literature on serine-based lipid derivatives like phosphatidylserine, sphingolipids, and their toxic form, i.e., ceramides, helped us conclude the essential role these play in both RPE phagocytosis and membrane integrity in retinal homeostasis, while their imbalance is a critical factor in inherited retinal dystrophies. Coupled with the above observations and the recent advance on using D-serine as therapeutic candidate for diabetic retinopathy, we further postulate that there are a multitude of potential therapies targeting serine metabolism that hold tremendous promise against retinal diseases. We highlighted the prospect of using PHGDH replacement gene therapy or serine supplement therapy for Mac Tel patients and serine racemase and ceramide synthase inhibition for DR patients. Careful evaluation of recent literature also helped us align with the growing consensus that metabolic vulnerabilities add an extra layer of susceptibility for IRD patients. Thus, boosting serine biosynthesis and serine metabolism by pharmacological activators or complimentary gene therapy in these patients may reduce this risk factor and help postpone their onset of degeneration. Consequently, and in this review, our goal was to delineate some of the important roles of serine and demonstrate how they are imperative to the health of the retina. By providing a comprehensive view of the relationship between serine and retinal health, we hope to bring more awareness to the importance of serine to the retina so it can be further assessed for treatment options, and proteins that mediate its metabolic processes can be considered as viable targets for gene therapy.

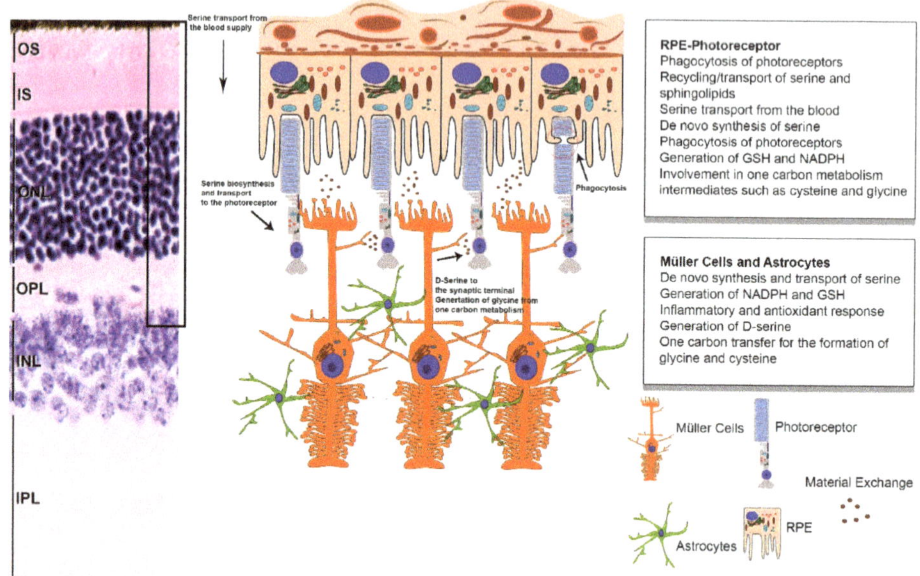

Figure 3. Graphical summary of serine metabolism in the retina. Serine homeostasis is primarily maintained by the RPE and retinal glia. The RPE transports and generates serine, which is ultimately transported to the photoreceptors. Additionally, important serine metabolic products such as glycine and cysteine are transported or catabolized to be used as fuel, as the energetic requirements of the RPE are very high. Photoreceptors also receive serine from the Müller glia and astrocytes. Glial cells are vital to the macula and generate serine from glycolysis, which is crucial in maintaining the redox balance in the photoreceptors, controlling neurotransmission, and mediating inflammation response elements. (IS, inner segment; OS, outer segment; ONL, outer nuclear layer; OPL, outer plexiform layer; INL, inner nuclear layer; IPL, inner plexiform layer).

Funding: This study was supported by a grant from the National Eye Institute (EY026499) to MIN and MRA.

Conflicts of Interest: The authors declare no conflict of interest.

References

1. El-Hattab, A.W. Serine biosynthesis and transport defects. *Mol. Genet. Metab.* **2016**, *118*, 153–159. [CrossRef] [PubMed]
2. Kalhan, S.C.; Hanson, R.W. Resurgence of serine: An often neglected but indispensable amino Acid. *J. Biol. Chem.* **2012**, *287*, 19786–19791. [CrossRef] [PubMed]
3. Simon, M.V.; Spalm, F.H.P.; Vera, M.S.; Rotstein, N.P. Sphingolipids as Emerging Mediators in Retina Degeneration. *Front. Cell Neurosci.* **2019**, *13*. [CrossRef] [PubMed]
4. Metcalf, J.S.; Dunlop, R.A.; Powell, J.T.; Banack, S.A.; Cox, P.A. L-Serine: A Naturally-Occurring Amino Acid with Therapeutic Potential. *Neurotox. Res.* **2018**, *33*, 213–221. [CrossRef]
5. Mattaini, K.R.; Sullivan, M.R.; Vander Heiden, M.G. The importance of serine metabolism in cancer. *J. Cell Biol.* **2016**, *214*, 249–257. [CrossRef]
6. Amelio, I.; Cutruzzolá, F.; Antonov, A.; Agostini, M.; Melino, G. Serine and glycine metabolism in cancer. *Trends Biochem. Sci.* **2014**, *39*, 191–198. [CrossRef]
7. Ripps, H.; Shen, W. Review: Taurine: A "very essential" amino acid. *Mol. Vis.* **2012**, *18*, 2673–2686.
8. Locasale, J.W. Serine, glycine and one-carbon units: Cancer metabolism in full circle. *Nat. Rev. Cancer* **2013**, *13*, 572–583. [CrossRef]

9. Gantner, M.L.; Eade, K.; Wallace, M.; Handzlik, M.K.; Fallon, R.; Trombley, J.; Bonelli, R.; Giles, S.; Harkins-Perry, S.; Heeren, T.F.C.; et al. Serine and Lipid Metabolism in Macular Disease and Peripheral Neuropathy. *N. Engl. J. Med.* **2019**, *381*, 1422–1433. [CrossRef]

10. Zhang, T.; Zhu, L.; Madigan, M.C.; Liu, W.; Shen, W.; Cherepanoff, S.; Zhou, F.; Zeng, S.; Du, J.; Gillies, M.C. Human macular Müller cells rely more on serine biosynthesis to combat oxidative stress than those from the periphery. *eLife* **2019**, *8*, e43598. [CrossRef]

11. Ye, J.; Fan, J.; Venneti, S.; Wan, Y.-W.; Pawel, B.R.; Zhang, J.; Finley, L.W.S.; Lu, C.; Lindsten, T.; Cross, J.R.; et al. Serine Catabolism Regulates Mitochondrial Redox Control during Hypoxia. *Cancer Discovery* **2014**, *4*, 1406–1417. [CrossRef]

12. de Koning, T.J.; Snell, K.; Duran, M.; Berger, R.; Poll-The, B.-T.; Surtees, R. L-serine in disease and development. *Biochem. J.* **2003**, *371*, 653–661. [CrossRef]

13. Stevens, E.R.; Esguerra, M.; Kim, P.M.; Newman, E.A.; Snyder, S.H.; Zahs, K.R.; Miller, R.F. D-serine and serine racemase are present in the vertebrate retina and contribute to the physiological activation of NMDA receptors. *Proc. Natl. Acad. Sci. USA* **2003**, *100*, 6789–6794. [CrossRef]

14. Chao, J.R.; Knight, K.; Engel, A.L.; Jankowski, C.; Wang, Y.; Manson, M.A.; Gu, H.; Djukovic, D.; Raftery, D.; Hurley, J.B.; et al. Human retinal pigment epithelial cells prefer proline as a nutrient and transport metabolic intermediates to the retinal side. *J. Biol. Chem.* **2017**, *292*, 12895–12905. [CrossRef]

15. Yam, M.; Engel, A.L.; Wang, Y.; Zhu, S.; Hauer, A.; Zhang, R.; Lohner, D.; Huang, J.; Dinterman, M.; Zhao, C.; et al. Proline mediates metabolic communication between retinal pigment epithelial cells and the retina. *J. Biol. Chem.* **2019**, *294*, 10278–10289. [CrossRef]

16. Joyal, J.-S.; Gantner, M.L.; Smith, L.E.H. Retinal energy demands control vascular supply of the retina in development and disease: The role of neuronal lipid and glucose metabolism. *Prog. Retin Eye Res.* **2018**, *64*, 131–156. [CrossRef]

17. Zhang, T.; Gillies, M.C.; Madigan, M.C.; Shen, W.; Du, J.; Grünert, U.; Zhou, F.; Yam, M.; Zhu, L. Disruption of De Novo Serine Synthesis in Müller Cells Induced Mitochondrial Dysfunction and Aggravated Oxidative Damage. *Mol. Neurobiol.* **2018**, *55*, 7025–7037. [CrossRef]

18. Poitry, S.; Poitry-Yamate, C.; Ueberfeld, J.; MacLeish, P.R.; Tsacopoulos, M. Mechanisms of Glutamate Metabolic Signaling in Retinal Glial (Müller) Cells. *J. Neurosci.* **2000**, *20*, 1809–1821. [CrossRef]

19. Curcio, C.A.; Allen, K.A. Topography of ganglion cells in human retina. *J. Comp. Neurol.* **1990**, *300*, 5–25. [CrossRef]

20. Gault, C.R.; Obeid, L.M.; Hannun, Y.A. An overview of sphingolipid metabolism: From synthesis to breakdown. *Adv. Exp. Med. Biol.* **2010**, *688*, 1–23. [CrossRef]

21. Merrill, A.H., Jr.; Schmelz, E.M.; Dillehay, D.L.; Spiegel, S.; Shayman, J.A.; Schroeder, J.J.; Riley, R.T.; Voss, K.A.; Wang, E. Sphingolipids–the enigmatic lipid class: Biochemistry, physiology, and pathophysiology. *Toxicol. Appl. Pharmacol.* **1997**, *142*, 208–225. [CrossRef]

22. Rotstein, N.P.; Miranda, G.E.; Abrahan, C.E.; German, O.L. Regulating survival and development in the retina: Key roles for simple sphingolipids. *J. Lipid Res.* **2010**, *51*, 1247–1262. [CrossRef]

23. Brush, R.S.; Tran, J.T.A.; Henry, K.R.; McClellan, M.E.; Elliott, M.H.; Mandal, M.N.A. Retinal Sphingolipids and Their Very-Long-Chain Fatty Acid-Containing Species. *Invest. Ophth. Vis. Sci.* **2010**, *51*, 4422–4431. [CrossRef]

24. Stiles, M.; Qi, H.; Sun, E.; Tan, J.; Porter, H.; Allegood, J.; Chalfant, C.E.; Yasumura, D.; Matthes, M.T.; LaVail, M.M.; et al. Sphingolipid profile alters in retinal dystrophic P23H-1 rats and systemic FTY720 can delay retinal degeneration. *J. Lipid Res.* **2016**, *57*, 818–831. [CrossRef]

25. Fox, T.E.; Han, X.; Kelly, S.; Merrill, A.H., 2nd; Martin, R.E.; Anderson, R.E.; Gardner, T.W.; Kester, M. Diabetes alters sphingolipid metabolism in the retina: A potential mechanism of cell death in diabetic retinopathy. *Diabetes* **2006**, *55*, 3573–3580. [CrossRef]

26. Kevany, B.M.; Palczewski, K. Phagocytosis of retinal rod and cone photoreceptors. *Physiology* **2010**, *25*, 8–15. [CrossRef]

27. Deigner, P.S.; Law, W.C.; Canada, F.J.; Rando, R.R. Membranes as the energy source in the endergonic transformation of vitamin A to 11-cis-retinol. *Science* **1989**, *244*, 968–971. [CrossRef]

28. Rando, R.R. Membrane phospholipids as an energy source in the operation of the visual cycle. *Biochemistry* **1991**, *30*, 595–602. [CrossRef]

29. Rando, R.R. Membrane phospholipids and the dark side of vision. *J. Bioenerg. Biomembr.* **1991**, *23*, 133–146. [CrossRef]

30. Sparrow, J.R.; Wu, Y.; Kim, C.Y.; Zhou, J. Phospholipid meets all-trans-retinal: The making of RPE bisretinoids. *J. Lipid Res.* **2010**, *51*, 247–261. [CrossRef]

31. Fadok, V.A.; Bratton, D.L.; Frasch, S.C.; Warner, M.L.; Henson, P.M. The role of phosphatidylserine in recognition of apoptotic cells by phagocytes. *Cell Death Differ.* **1998**, *5*, 551–562. [CrossRef]

32. Tafesse, F.G.; Rashidfarrokhi, A.; Schmidt, F.I.; Freinkman, E.; Dougan, S.; Dougan, M.; Esteban, A.; Maruyama, T.; Strijbis, K.; Ploegh, H.L. Disruption of Sphingolipid Biosynthesis Blocks Phagocytosis of Candida albicans. *PLoS Pathog.* **2015**, *11*, e1005188. [CrossRef]

33. Bryan, A.M.; Del Poeta, M.; Luberto, C. Sphingolipids as Regulators of the Phagocytic Response to Fungal Infections. *Mediat. Inflamm.* **2015**. [CrossRef]

34. Sarantis, H.; Grinstein, S. Monitoring phospholipid dynamics during phagocytosis: Application of genetically-encoded fluorescent probes. *Methods Cell Biol.* **2012**, *108*, 429–444. [CrossRef]

35. Regina Todeschini, A.; Hakomori, S.I. Functional role of glycosphingolipids and gangliosides in control of cell adhesion, motility, and growth, through glycosynaptic microdomains. *Biochim. Biophys. Acta* **2008**, *1780*, 421–433. [CrossRef]

36. Hoekstra, D.; Maier, O.; van der Wouden, J.M.; Slimane, T.A.; van IJzendoorn, S.C. Membrane dynamics and cell polarity: The role of sphingolipids. *J. Lipid Res.* **2003**, *44*, 869–877. [CrossRef]

37. MacKay, M.-A.B.; Kravtsenyuk, M.; Thomas, R.; Mitchell, N.D.; Dursun, S.M.; Baker, G.B. D-Serine: Potential Therapeutic Agent and/or Biomarker in Schizophrenia and Depression? *Front. Psychiatry* **2019**, *10*. [CrossRef]

38. Kleckner, N.W.; Dingledine, R. Requirement for glycine in activation of NMDA-receptors expressed in Xenopus oocytes. *Science* **1988**, *241*, 835–837. [CrossRef]

39. Fletcher, E.L.; Hack, I.; Brandstätter, J.H.; Wässle, H. Synaptic localization of NMDA receptor subunits in the rat retina. *J. Comp. Neurol.* **2000**, *420*, 98–112. [CrossRef]

40. Ozaki, H.; Inoue, R.; Matsushima, T.; Sasahara, M.; Hayashi, A.; Mori, H. Serine racemase deletion attenuates neurodegeneration and microvascular damage in diabetic retinopathy. *PLoS ONE* **2018**, *13*, e0190864. [CrossRef]

41. Maddocks, O.D.K.; Labuschagne, C.F.; Adams, P.D.; Vousden, K.H. Serine Metabolism Supports the Methionine Cycle and DNA/RNA Methylation through De Novo ATP Synthesis in Cancer Cells. *Mol. Cell* **2016**, *61*, 210–221. [CrossRef]

42. Botsford, J.; Parks, L. Serine transhydroxymethylase in methionine biosynthesis in Saccharomyces cerevisiae. *J. Bacterial.* **1969**, *97*, 1176–1183. [CrossRef]

43. Stauffer, G.V.; Brenchley, J.E. Influence of methionine biosynthesis on serine transhydroxymethylase regulation in Salmonella typhimurium LT2. *J. Bacterial.* **1977**, *129*, 740–749. [CrossRef]

44. Lombardini, J.B.; Talalay, P. Formation, functions and regulatory importance of S-adenosyl-l-methionine. *Adv. Enzyme Regul.* **1971**, *9*, 349–384. [CrossRef]

45. Grundy, F.J.; Henkin, T.M. Synthesis of Serine, Glycine, Cysteine, and Methionine. In *Bacillus subtilis and Its Closest Relatives*; Sonenshein, A., Losick, R., Hoch, J., Eds.; ASM Press: Washington, DC, USA, 2002; pp. 245–254. [CrossRef]

46. Jobe, E.M.; Zhao, X. DNA Methylation and Adult Neurogenesis. *Brain Plast.* **2017**, *3*, 5–26. [CrossRef] [PubMed]

47. Kowluru, R.A.; Shan, Y.; Mishra, M. Dynamic DNA methylation of matrix metalloproteinase-9 in the development of diabetic retinopathy. *Lab. Invest.* **2016**, *96*, 1040–1049. [CrossRef]

48. Zeng, J.-D.; Wu, W.K.K.; Wang, H.-Y.; Li, X.-X. Serine and one-carbon metabolism, a bridge that links mTOR signaling and DNA methylation in cancer. *Pharmacol. Res.* **2019**, *149*, 104352. [CrossRef]

49. Li, W.; Liu, J.; Galvin, J.A. Epigenetics and Common Ophthalmic Diseases. *Yale J. Biol. Med.* **2016**, *89*, 597–600.

50. Liu, M.M.; Chan, C.C.; Tuo, J. Genetic mechanisms and age-related macular degeneration: Common variants, rare variants, copy number variations, epigenetics, and mitochondrial genetics. *Hum. Genomics* **2012**, *6*, 13. [CrossRef]

51. Ng, S.K.; Wood, J.P.M.; Chidlow, G.; Han, G.; Kittipassorn, T.; Peet, D.J.; Casson, R.J. Cancer-like metabolism of the mammalian retina. *Clin. Exp. Ophthalmol.* **2015**, *43*, 367–376. [CrossRef]

52. Panfoli, I.; Calzia, D.; Ravera, S.; Bruschi, M.; Tacchetti, C.; Candiani, S.; Morelli, A.; Candiano, G. Extramitochondrial tricarboxylic acid cycle in retinal rod outer segments. *Biochimie* **2011**, *93*, 1565–1575. [CrossRef] [PubMed]

53. Calzia, D.; Panfoli, I.; Heinig, N.; Schumann, U.; Ader, M.; Traverso, C.E.; Funk, R.H.W.; Roehlecke, C. Impairment of extramitochondrial oxidative phosphorylation in mouse rod outer segments by blue light irradiation. *Biochimie* **2016**, *125*, 171–178. [CrossRef] [PubMed]

54. Alder, V.A.; Ben-Nun, J.; Cringle, S.J. PO2 profiles and oxygen consumption in cat retina with an occluded retinal circulation. *Invest. Ophthalmol. Vis. Sci.* **1990**, *31*, 1029–1034. [PubMed]

55. McHugh, K.J.; Li, D.A.; Wang, J.C.; Kwark, L.; Loo, J.; Macha, V.; Farsiu, S.; Kim, L.A.; Saint-Geniez, M. Computational modeling of retinal hypoxia and photoreceptor degeneration in patients with age-related macular degeneration. *PloS ONE* **2019**, *14*, e0216215. [CrossRef]

56. Roehlecke, C.; Schumann, U.; Ader, M.; Brunssen, C.; Bramke, S.; Morawietz, H.; Funk, R.H.W. Stress Reaction in Outer Segments of Photoreceptors after Blue Light Irradiation. *PloS ONE* **2013**, *8*, e71570. [CrossRef]

57. Roehlecke, C.; Schumann, U.; Ader, M.; Knels, L.; Funk, R.H.W. Influence of blue light on photoreceptors in a live retinal explant system. *Mol. Vis.* **2011**, *17*, 876–884.

58. Calzia, D.; Oneto, M.; Caicci, F.; Bianchini, P.; Ravera, S.; Bartolucci, M.; Diaspro, A.; Degan, P.; Manni, L.; Traverso, C.E.; et al. Effect of polyphenolic phytochemicals on ectopic oxidative phosphorylation in rod outer segments of bovine retina. *Brit. J. Pharmacol.* **2015**, *172*, 3890–3903. [CrossRef]

59. Panieri, E.; Santoro, M.M. ROS homeostasis and metabolism: A dangerous liason in cancer cells. *Cell Death Dis.* **2016**, *7*, e2253. [CrossRef]

60. Vučetić, M.; Cormerais, Y.; Parks, S.K.; Pouysségur, J. The Central Role of Amino Acids in Cancer Redox Homeostasis: Vulnerability Points of the Cancer Redox Code. *Front. Oncol.* **2017**, *7*, 319. [CrossRef]

61. Dringen, R.; Pfeiffer, B.; Hamprecht, B. Synthesis of the antioxidant glutathione in neurons: Supply by astrocytes of CysGly as precursor for neuronal glutathione. *J. Neurosci.* **1999**, *19*, 562–569. [CrossRef]

62. Winkler, B.S.; Giblin, F.J. Glutathione oxidation in retina: Effects on biochemical and electrical activities. *Exp. Eye Res.* **1983**, *36*, 287–297. [CrossRef]

63. Zhou, X.; He, L.; Zuo, S.; Zhang, Y.; Wan, D.; Long, C.; Huang, P.; Wu, X.; Wu, C.; Liu, G.; et al. Serine prevented high-fat diet-induced oxidative stress by activating AMPK and epigenetically modulating the expression of glutathione synthesis-related genes. *Biochim. Biophys. Acta, Mol. Basis Dis.* **2018**, *1864*, 488–498. [CrossRef] [PubMed]

64. DeNicola, G.M.; Chen, P.-H.; Mullarky, E.; Sudderth, J.A.; Hu, Z.; Wu, D.; Tang, H.; Xie, Y.; Asara, J.M.; Huffman, K.E.; et al. NRF2 regulates serine biosynthesis in non–small cell lung cancer. *Nature Genet.* **2015**, *47*, 1475–1481. [CrossRef] [PubMed]

65. Maralani, M.N.; Movahedian, A.; Javanmard, S.H. Antioxidant and cytoprotective effects of L-Serine on human endothelial cells. *Res. Pharm. Sci.* **2012**, *7*, 209–215.

66. Wink, D.A.; Miranda, K.M.; Espey, M.G. Cytotoxicity related to oxidative and nitrosative stress by nitric oxide. *Exp. Biol. Med.* **2001**, *226*, 621–623. [CrossRef]

67. Khansari, N.; Shakiba, Y.; Mahmoudi, M. Chronic Inflammation and Oxidative Stress as a Major Cause of Age- Related Diseases and Cancer. *Recent Pat. Inflamm. Allergy Drug Discov.* **2009**, *3*, 73–80. [CrossRef]

68. Perez, V.L.; Caspi, R.R. Immune mechanisms in inflammatory and degenerative eye disease. *Trends Immunol.* **2015**, *36*, 354–363. [CrossRef]

69. Rodriguez, A.E.; Ducker, G.S.; Billingham, L.K.; Martinez, C.A.; Mainolfi, N.; Suri, V.; Friedman, A.; Manfredi, M.G.; Weinberg, S.E.; Rabinowitz, J.D.; et al. Serine Metabolism Supports Macrophage IL-1β Production. *Cell Metab.* **2019**, *29*, 1003–1011. [CrossRef]

70. He, F.; Yin, Z.; Wu, C.; Xia, Y.; Wu, M.; Li, P.; Zhang, H.; Yin, Y.; Li, N.; Zhu, G.; et al. l-Serine Lowers the Inflammatory Responses during Pasteurella multocida Infection. *Infect. Immun.* **2019**, *87*. [CrossRef]

71. Dornan, T.L.; Ting, A.; McPherson, C.K.; Peckar, C.O.; Mann, J.I.; Turner, R.C.; Morris, P.J. Genetic susceptibility to the development of retinopathy in insulin-dependent diabetics. *Diabetes* **1982**, *31*, 226–231. [CrossRef]

72. Zarbin, M.A. Current concepts in the pathogenesis of age-related macular degeneration. *Arch. Ophthalmol.* **2004**, *122*, 598–614. [CrossRef] [PubMed]

73. Lambert, N.G.; ElShelmani, H.; Singh, M.K.; Mansergh, F.C.; Wride, M.A.; Padilla, M.; Keegan, D.; Hogg, R.E.; Ambati, B.K. Risk factors and biomarkers of age-related macular degeneration. *Prog. Retin Eye Res.* **2016**, *54*, 64–102. [CrossRef] [PubMed]

74. Wong, M.Y.Z.; Man, R.E.K.; Fenwick, E.K.; Gupta, P.; Li, L.J.; van Dam, R.M.; Chong, M.F.; Lamoureux, E.L. Dietary intake and diabetic retinopathy: A systematic review. *PLoS ONE* **2018**, *13*, e0186582. [CrossRef] [PubMed]

75. Yin, Q.; Jiang, D.; Li, L.; Yang, Y.; Wu, P.; Luo, Y.; Yang, R.; Li, D. LPS Promotes Vascular Smooth Muscle Cells Proliferation Through the TLR4/Rac1/Akt Signalling Pathway. *Cell Physiol. Biochem.* **2017**, *44*, 2189–2200. [CrossRef]

76. Yi, H.; Patel, A.K.; Sodhi, C.P.; Hackam, D.J.; Hackam, A.S. Novel role for the innate immune receptor Toll-like receptor 4 (TLR4) in the regulation of the Wnt signaling pathway and photoreceptor apoptosis. *PLoS ONE* **2012**, *7*, e36560. [CrossRef]

77. Holm, L.J.; Buschard, K. L-serine: A neglected amino acid with a potential therapeutic role in diabetes. *APMIS* **2019**, *127*, 655–659. [CrossRef]

78. Esaki, K.; Sayano, T.; Sonoda, C.; Akagi, T.; Suzuki, T.; Ogawa, T.; Okamoto, M.; Yoshikawa, T.; Hirabayashi, Y.; Furuya, S. L-Serine Deficiency Elicits Intracellular Accumulation of Cytotoxic Deoxysphingolipids and Lipid Body Formation. *J. Biol. Chem.* **2015**, *290*, 14595–14609. [CrossRef]

79. Gao, X.; Lee, K.; Reid, M.A.; Sanderson, S.M.; Qiu, C.; Li, S.; Liu, J.; Locasale, J.W. Serine Availability Influences Mitochondrial Dynamics and Function through Lipid Metabolism. *Cell Rep.* **2018**, *22*, 3507–3520. [CrossRef]

80. Duncan, J.L.; Pierce, E.A.; Laster, A.M.; Daiger, S.P.; Birch, D.G.; Ash, J.D.; Iannaccone, A.; Flannery, J.G.; Sahel, J.A.; Zack, D.J.; et al. Inherited Retinal Degenerations: Current Landscape and Knowledge Gaps. *Transl. Vis. Sci. Techn.* **2018**, *7*. [CrossRef]

81. Punzo, C.; Xiong, W.; Cepko, C.L. Loss of daylight vision in retinal degeneration: Are oxidative stress and metabolic dysregulation to blame? *J. Biol. Chem.* **2012**, *287*, 1642–1648. [CrossRef]

82. Holm, L.J.; Haupt-Jorgensen, M.; Larsen, J.; Giacobini, J.D.; Bilgin, M.; Buschard, K. L-serine supplementation lowers diabetes incidence and improves blood glucose homeostasis in NOD mice. *PLoS ONE* **2018**, *13*, e0194414. [CrossRef]

83. Ross, J.S.; Russo, S.B.; Chavis, G.C.; Cowart, L.A. Sphingolipid regulators of cellular dysfunction in Type 2 diabetes mellitus: A systems overview. *Clin. Lipidol.* **2014**, *9*, 553–569. [CrossRef] [PubMed]

84. Bertea, M.; Rutti, M.F.; Othman, A.; Marti-Jaun, J.; Hersberger, M.; von Eckardstein, A.; Hornemann, T. Deoxysphingoid bases as plasma markers in diabetes mellitus. *Lipids Health Dis.* **2010**, *9*, 84. [CrossRef] [PubMed]

85. Scerri, T.S.; Quaglieri, A.; Cai, C.; Zernant, J.; Matsunami, N.; Baird, L.; Scheppke, L.; Bonelli, R.; Yannuzzi, L.A.; Friedlander, M.; et al. Genome-wide analyses identify common variants associated with macular telangiectasia type 2. *Nat. Genet.* **2017**, *49*, 559–567. [CrossRef] [PubMed]

86. Wang, L.; Miao, H.; Li, X. Tamoxifen retinopathy: A case report. *Springerplus* **2015**, *4*, 501. [CrossRef] [PubMed]

87. Othman, A.; Benghozi, R.; Alecu, I.; Wei, Y.; Niesor, E.; von Eckardstein, A.; Hornemann, T. Fenofibrate lowers atypical sphingolipids in plasma of dyslipidemic patients: A novel approach for treating diabetic neuropathy? *J. Clin. Lipidol.* **2015**, *9*, 568–575. [CrossRef]

88. Garofalo, K.; Penno, A.; Schmidt, B.P.; Lee, H.J.; Frosch, M.P.; von Eckardstein, A.; Brown, R.H.; Hornemann, T.; Eichler, F.S. Oral L-serine supplementation reduces production of neurotoxic deoxysphingolipids in mice and humans with hereditary sensory autonomic neuropathy type 1. *J. Clin. Investig.* **2011**, *121*, 4735–4745. [CrossRef]

Review

Effects of Mitochondrial-Derived Peptides (MDPs) on Mitochondrial and Cellular Health in AMD

Sonali Nashine [1] and M. Cristina Kenney [1,2,*]

[1] Department of Ophthalmology, Gavin Herbert Eye Institute, University of California Irvine, Irvine, CA 92697, USA; snashine@uci.edu
[2] Department of Pathology and Laboratory Medicine, University of California Irvine, Irvine, CA 92697, USA
* Correspondence: mkenney@uci.edu

Received: 2 February 2020; Accepted: 23 April 2020; Published: 29 April 2020

Abstract: Substantive evidence demonstrates the contribution of mitochondrial dysfunction in the etiology and pathogenesis of Age-related Macular Degeneration (AMD). Recently, extensive characterization of Mitochondrial-Derived Peptides (MDPs) has revealed their cytoprotective role in several diseases, including AMD. Here we summarize the varied effects of MDPs on cellular and mitochondrial health, which establish the merit of MDPs as therapeutic targets for AMD. We argue that further research to delve into the mechanisms of action and delivery of MDPs may advance the field of AMD therapy.

Keywords: MDPs; mitochondrial-derived peptides; Humanin; HNG; SHLPs; MOTS-c; Age-related Macular Degeneration (AMD); neuroprotection; RPE; mitochondria

1. Introduction

In the United States, geographic atrophy in dry AMD (Age-related Macular Degeneration) is a primary cause of vision loss in the elderly [1], and it has limited treatment options compared to those available for wet AMD [2,3]. Among the wide variety of factors that are instrumental in the etiology and pathogenesis of AMD, mitochondrial damage in the Retinal Pigment Epithelium (RPE), leading to RPE dysfunction, contributes significantly. AMD mitochondria are fragmented, have a higher number of lesions, altered ATP synthase activity, as well as compromised protein expression and nuclear-encoded protein import [4–6]. Furthermore, as confirmed by ATAC-sequencing, chromatin accessibility is decreased significantly in the RPE in AMD retinas [7].

The human mitochondrial genome is ~16.5 kilobases, circular, double-stranded, and consists of 37 genes that code for 13 proteins of the respiratory chain complexes. These 13 proteins are a part of the electron transport chain and aid in oxidative metabolism and ATP production [8,9]. Some of the critical functions of mitochondria include, but are not limited to, ATP production via respiration, promoting thermogenesis via proton leak, regulation of cellular metabolism and calcium signaling, ROS generation, regulation of apoptosis, ion homeostasis, and heme synthesis [10]. The retina is a part of the central nervous system and is one of the highest energy-demanding tissues in the human body. Glycolysis in the cytosol and mitochondrial oxidative phosphorylation are the two primary sources of energy, i.e., ATP (adenosine triphosphate) generation. Retinal neurons derive their energy mostly from oxidative phosphorylation, which has a substantially higher ATP yield than glycolysis [11]. Unmet retinal energy demand puts the retinal neurons at a high risk of cell death that in turn causes impairment or loss of vision [12]. Mitochondrial function declines with age as a result of accumulated mtDNA damage/mutations [13]. The majority of proteins that support mitochondrial health and function are encoded by the nuclear genome. Therefore, coordinated regulation of mitochondrial and nuclear gene expression is essential to maintain cellular homeostasis [14]. Mammalian mitochondrial retrograde signaling, i.e., transmission of information from the mitochondria to the nucleus, is mediated by G-Protein Pathway Suppressor 2 (GPS2), which also functions as a transcriptional activator of nuclear-encoded mitochondrial genes. GPS2-regulated direct translocation from the mitochondria to the nucleus is essential for the transcriptional activation of a nuclear stress response to mitochondrial depolarization and for supporting basal mitochondrial biogenesis [15].

The mitochondrial genome encodes 22 tRNAs and two ribosomal RNAs, i.e., 12S rRNA and 16S rRNA, both of which are essential in synthesis of mitochondrial proteins. The 12S rRNA gene is 954 base pairs (bp) in length and spans 648–1601 bp of the mtDNA; it occupies 1/16 of the entire mtDNA and has 297 nucleotide substitutions, which account for 31% of the 12S rRNA gene. Furthermore, the 16S rRNA gene is 1559 bp long, spanning 1671–3229 bp of the mtDNA; it occupies 1/10 of the entire mtDNA and has 413 nucleotide substitutions, which account for 25% of the 16S rRNA gene [16]. Both 16S rRNA and 12S rRNA genes carry ORFs (Open Reading Frames) that encode mitochondrial-derived peptides.

2. Mitochondrial-Derived Peptides (MDPs)

Novel mitochondrial-derived peptides (MDPs), which are encoded within the mtDNA, serve as signals for organism cytoprotection and energy regulation. The MDPs encoded from the 16S rRNA region of the mtDNA include Humanin and SHLPs, which regulate cell survival and growth via distinct pathways. MOTS-c, encoded from the 12S rRNA region of the mtDNA, plays a key role in regulation of muscle and fat metabolism and prevents hepatic steatosis. Numerous MDPs are well-characterized and are in preclinical development for aging-related diseases [17–20].

3. Humanin

Humanin was the first MDP discovered within the mammalian mitochondrial genome and is encoded from the 16S rRNA coding region of the mtDNA. Humanin cDNA was first discovered in 2001 by functional expression screening of a cDNA library preparation from the occipital cortex of an Alzheimer's disease patient brain [21]. The study identified clones that protected against neuronal cell death induced by neurotoxic amyloid-β peptides and by mutants of FAD (Familial Alzheimer's Disease) genes, namely *APP* (Amyloid Precursor Protein), *PS1* (Presenilin 1), and *PS2* (Presenilin 2). A 75 bp open reading frame that codes for a 24-residue peptide was identified and its sequence was found to be similar to that of a human cDNA and 99% identical to the mtDNA sequence; it was therefore named Humanin. Humanin antagonized the cytotoxicity caused by *APP* (Amyloid Precursor Protein (APP), *PS1* (Presenilin 1), and *PS2* (Presenilin 2) mutants, by suppressing amyloid-$\beta_{1\text{-}42}$-triggered neurotoxicity and attenuating neuronal cell death.

3.1. Structure

Humanin is a secretory 24-amino acid peptide with the amino acid sequence *H-Met-Ala-Pro-Arg-Gly-Phe-Ser-Cys-Leu-Leu-Leu-Leu-Thr-Ser-Glu-Ile-Asp-Leu-Pro-Val-Lys-Arg-Arg-Ala-OH (H-MAPRGFSCL LLLTSEIDLPVKRRA-OH)* and a molecular weight of 2687.26 Da (Figure 1). Following its discovery, the structure and biological functions of Humanin have been very well characterized. Using a combination of oligonucleotide synthesis, dimerization experiments, and bioinformatics, Yamagishi et al. conducted a comprehensive investigation that led to the identification of the amino acids essential for Humanin secretion and for its neuroprotective function [22]. This study revealed that Humanin is a signal peptide and the core domain of Humanin is formed by Leu9-Leu10-Leu11 residues with Leu10 being the key player. Leu9-Leu10-Leu11 and Pro19-Val20 are absolutely essential for the extracellular secretion of full-length Humanin. The neuroprotective action of Humanin requires Pro19, Ser14, Thr13, Leu12, Leu9, Cys8, Ser7, and Pro3. Ser7 and Leu9 residues are required for the self-dimerization of Humanin, which is necessary for its neuroprotective function. Providing critical insights into the amino acid requirements of Humanin would be useful in synthesis of peptides tailored to perform a specific biological function and for targeted improvement of particular cellular functions.

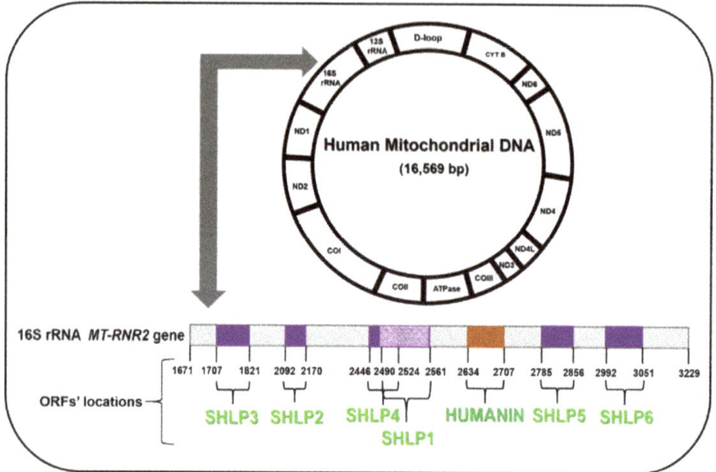

Figure 1. Humanin and Small Human-Like Peptides (SHLPs) Open Reading Frames (ORFs) in the human mitochondrial DNA.

3.2. Tissue Distribution

Measurable levels of Humanin are detectable in plasma, seminal fluid, and cerebrospinal fluid. A circulating Humanin pool has been demonstrated in various tissues of the human body, including hypothalamus, liver, heart, kidney, colon, testes, vasculature, and skeletal muscle [23–28]. Circulating plasma levels of Humanin decline with age in both humans and mice [29]. Demonstrating a proportional relationship between Humanin levels and human aging is of particular importance, as this could promote research that aims to boost Humanin levels and thereby delay aging in humans.

3.3. Humanin Analogs

Amino acid substitutions in Humanin have led to the development of synthetic Humanin analogs, which are more potent than the endogenous Humanin. Humanin G (HNG) is formed by a Ser to Gly amino acid substitution at position 14. HNGF6A is formed by the Phe to Ala amino acid substitution at position 6 [30].

3.4. Humanin Receptors and Regulation

Humanin exerts its cytoprotective effects by binding to receptors intracellularly/extracellularly, regulating the intrinsic mitochondrial pathway, and mediating downstream signaling. Humanin binds to the Formyl-Peptide Receptor-Like Receptor-1 (FPRL-1) and the IL-6 (Interleukin-6) receptor family trimeric receptor complex, comprised of CNTFRα (Ciliary Neurotrophic Factor Receptor α), WSX-1, and GP130 (glycoprotein 130kDa), which are key contributors to the neuroprotective action of Humanin [31–33]. Substantial evidence suggests that mitochondrial retrograde signaling is involved in the endocrine regulation of aging and age-related pathologies. IGF-1 (Insulin Growth Factor-1) is a key player in the conserved endocrine pathway, which contributes to lifespan and healthspan. Humanin, a potent mediator of mitochondrial retrograde signaling, is directly regulated by IGF-1, and the levels of Humanin and IGF-1 simultaneously decline with age [34]. Humanin and IGF-1 are known to have distinct yet overlapping functions. By interacting with the C-terminal domain of IGFBP-3 (Insulin-like Growth Factor Binding Protein-3), Humanin interferes with the binding of Importin-β to IGFBP-3, thereby suppressing IGFBP-3-mediated apoptosis [35]. Humanin regulates peripheral insulin action. In this study, peripherally administered HNGF6A in rats conferred insulin sensitivity via hypothalamic STAT3 (Signal transducer and activator of transcription 3) activation [29].

4. Humanin Functions

4.1. Prevents Apoptosis

Humanin binds to the pro-apoptotic protein BAX and prevents its translocation to the nucleus, thereby antagonizing the apoptotic activity of BAX and inhibiting the release of cytochrome c [36]. This finding reported by Guo et al. in the journal *Nature* was the first thorough study that delineated and reported the interactions of Humanin with the pro-apoptotic BAX protein. Since Humanin is encoded by the mitochondria, this report also suggested a possible retrograde signaling mechanism between the mitochondria and the nucleus that might be contributing to regulation of BAX-mediated apoptosis.

Humanin also binds and inactivates BAX-like proteins, such as Bid and BidEL [37]. Exogenously added HNG exerts substantive protective effects in transmitochondrial AMD RPE cells in vitro by (a) rescue of mitochondrial structure and function, (b) inhibiting the action of pro-apoptotic genes/proteins, and (c) intracellular and extracellular humanin receptor modulation (Figure 2) [38].

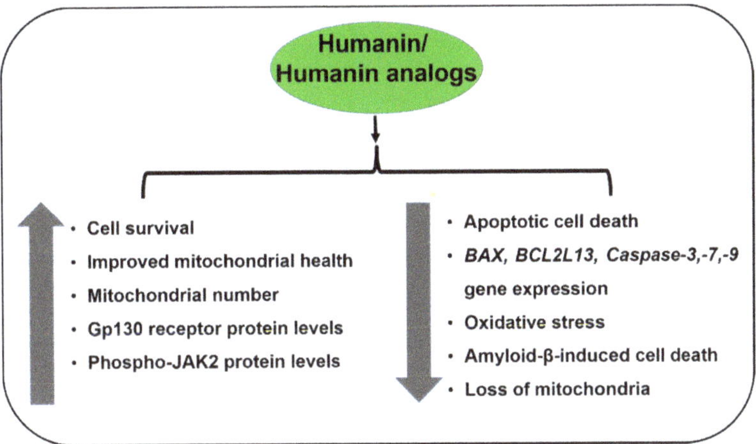

Figure 2. Effects of Humanin/Humanin analogs in RPE/AMD.

Humanin is cytoprotective against amyloid-β-mediated toxicity in neuronal cells, both in vitro and in vivo [17,18]; against cerebral ischemia and cardiac damage in mouse models [19,20]; and in numerous neurodegenerative disease models for Alzheimer's disease, Parkinson's disease, Huntington's disease, and Prion diseases [18,19,39,40]. Moreover, Hinton et al. in a comprehensive study published in the highly reputed *IOVS* eye journal demonstrated that Humanin rescues primary RPE cells from oxidative damage and subsequent death in vitro [41]. This showed that Humanin conferred RPE cell protection via two mechanisms: by enhancing mitochondrial biogenesis and function and by activation of STAT3. Further, Humanin mediated suppression of oxidative stress-induced cell senescence and maintained transepithelial resistance in human RPE monolayers. In summary, this study by Hinton et al. suggested the potential of Humanin as a therapeutic candidate for the treatment of geographic atrophy in dry AMD.

Each of the 24 amino acids in the Humanin peptide have a specific function. Serine at position 14 confers neuroprotection [22], but its substitution with glycine generates a variant called Humanin G/HNG that is 1000-fold more potent than its parent analog Humanin [16]. HNG is known to protect against cell death by preventing mitochondrial dysfunction [26].

Humanin potentially inhibits silver nanoparticles-induced cell death in human neuroblastoma cells by (a) protecting against redox imbalance and oxidative stress-induced DNA damage, (b) increasing mitochondrial membrane potential and enhancing the activity of mitochondrial succinate dehydrogenase, and (c) deactivation of the ER stress pathways that were upregulated by the silver nanoparticles [42].

4.2. Prevents Amyloid-β-Induced Toxicity

Amyloid-β is a key component of drusen deposits that are formed in the AMD retinas. Administration of Humanin G reduces amyloid-β loads and inhibits amyloid-β-induced cell apoptosis by (a) restoring amyloid-β-mediated decline in calcium homeostasis, (b) suppressing amyloid-β-induced membrane fluidity changes, (c) restoring mitochondrial membrane potential, and (d) decreasing intracellular reactive oxygen species [43]. This finding is crucial as Humanin G's ability to mitigate amyloid-β-induced cytotoxicity might be used as a therapeutic approach to delay the progression of dry AMD.

4.3. Stress Resistance Against ER Stress-Induced Apoptosis

Humanin exerts its therapeutic benefits by antagonizing the action of a plethora of cellular insults, thereby protecting the RPE cells against cytotoxicity. Treatment with Humanin downregulates the expression of an ER stress marker CHOP (C/EBP Homologous Protein), inhibits ROS (Reactive Oxygen Species) production, and regulates intracellular calcium influx, thereby preventing RPE cell apoptosis [44,45]. Treatment of primary human RPE cells with three ER stress sensors, i.e., Tunicamycin, Brefeldin A, and Thapsigargin, induced mitochondrial damage and dose-dependent loss of RPE cells. However, pretreatment with Humanin provided significant cytoprotection against ER stress-induced cell death by restoring the depleted mitochondrial glutathione (GSH) levels to normal, reducing mitochondrial superoxide generation, and downregulating ER stress-induced apoptotic Caspase 4 and Caspase 3 [46].

4.4. Activation of the ERK, AKT, and STAT3 Signaling Pathways

As a secretory peptide, Humanin regulates both intracellular and extracellular signaling pathways. Exogenously administered Humanin both in vitro and in vivo rapidly increases AKT-1 phosphorylation and activates the PI3K (Phosphoinositide 3-Kinase)/AKT pathway; AKT-1 is the AKT Serine/Threonine Kinase-1 protein that plays a role in cell motility, metabolism, and proliferation [47]. Moreover, intraperitoneal injection of Humanin in vivo elevates endothelial NOS (Nitric Oxide Synthase) phosphorylation and also increases the phosphorylation of AMPK (5′ Adenosine Monophosphate-Activated Protein Kinase), a protein which contributes to cellular energy homeostasis [25]. The signaling pathway mediated by Humanin involves its interaction with various molecular entities, including protein kinases, integral membrane receptors, and transcription regulators. In neuronal cell lines, Humanin interacts with and activates the GP130 receptor and mediates its effects via its canonical AKT, ERK1/2 (Extracellular Signal-Regulated Kinase 1/2), and STAT3 signaling cascades. Humanin acts as a primary agonist for the GP130 receptor in neuronal cell lines in vitro, and Humanin treatment transiently activates GP130, thereby mediating cellular protective effects, such as anti-apoptosis, enhanced insulin sensitivity, and protection from hypoxic and ischemic stressors [48].

4.5. Preserves Endothelial Function in Atherosclerosis

Oh et al. demonstrated that sustained administration of Humanin G (a more potent analog of Humanin) to ApoE-knockout mice inhibited the progression of atherosclerotic plaques and preserved endothelial function [49].

4.6. Prevents Vascular Remodeling and Inflammation

Exogenously added Humanin attenuated angiogenesis, inflammation, apoptosis, and microvascular remodeling in an ApoE-deficient mouse model of atherosclerosis [23].

4.7. Cytoprotective Against LDL-Induced Oxidative Stress

Bachar et al. demonstrated that Humanin is expressed in the endothelial cell layer of human blood vessels. In human endothelial cells in vitro, exogenous supplementation of Humanin attenuated oxidized (Ox)-LDL-induced ROS generation and apoptotic cell death by 50% and reduced the levels of cellular ceramide, which is a lipid second messenger that initiates apoptotic signaling cascades. Therefore, Humanin renders protection against Ox-LDL-mediated oxidative stress and is cytoprotective in human vasculature [50].

4.8. Protects Germ Cells/Leukocytes—Reduces Cancer Metastases

A recent study by Jia et al. provided evidence of the role of Humanin in maintaining germ cell homeostasis. Substantive Humanin-mediated protection against chemotherapy-induced male germ cell apoptosis both in vivo and ex vivo was observed. The study also demonstrated that this anti-apoptotic effect is primarily mediated via STAT3 and BAX signaling [51].

4.9. Germ Cell Apoptosis by Chemo Drugs

Another similar study by Lue et al. demonstrated that Humanin G prevents cyclophosphamide-induced toxicity and death of male germ cells and leukocytes. In this study, HNG protected normal cells against stress and suppressed cancer metastases [52].

4.10. Cytoprotection in Carotid Atherosclerotic Plaques

In addition, it has been reported that Humanin is present in carotid atherosclerotic plaques and higher expression of Humanin in symptomatic patients compared to asymptomatic patients could be a part of a stress-response defense mechanism to delay the progression of the disease [53].

5. SHLPs (Small Humanin-Like Peptides)

The 16S rRNA region of the mitochondrial DNA also codes for another category of MDPs called Small Humanin-Like Peptides (SHLPs), which include SHLP1, SHLP2, SHLP3, SHLP4, SHLP5, and SHLP6. SHLPs are 24–38 amino acids long and each SHLP may differentially regulate mitochondrial and cellular health and functions. Extensive characterization of SHLP2, SHLP3, and SHLP6 has been detailed in recent literature but the specific biological functions of the SHLPs are still being studied.

5.1. SHLP2

SHLP2 has a molecular weight of 3017.54 Da and is 26 amino acids long with the sequence *H-MGVKFFTLSTRFFPSVQRAVPLWTNS-OH* (Figure 1). The plasma SHLP2 levels decline with age, suggesting its critical role in aging. SHLP2 stabilizes the AMD mitochondria by preserving the mitochondrial oxidative phosphorylation protein complex subunits (I-V) in AMD RPE transmitochondrial cells, and also promotes mitochondrial metabolism. This study also highlighted the potential of an exogenously added SHLP2 peptide to enhance mitochondria-specific mtGFP fluorescence staining, increase mtDNA copy numbers, and upregulate the *PGC-1α* gene, which is a master regulator of mitochondrial biogenesis [54]. This in vitro study by Nashine et al. is the first study that demonstrates an association between SHLP2 and AMD transmitochondrial RPE cells, and therefore provides a basis for further explorative studies in the field of AMD. Pretreatment with SHLP2 inhibited apoptotic cell death as evidenced by higher live cell numbers and substantive downregulation of effector caspases, namely *Caspase-3* and *Caspase-7*, in AMD RPE cells. Moreover, SHLP2 rescued and protected AMD RPE cybrid cells against amyloid-β-induced toxicity in vitro (Figure 3). Therefore, SHLP2 could be considered a potential therapeutic candidate for macular degeneration. SHLP2 also mediates chaperone-like activity, prevents the misfolding of amyloid polypeptides, and prevents neuronal cell death [55].

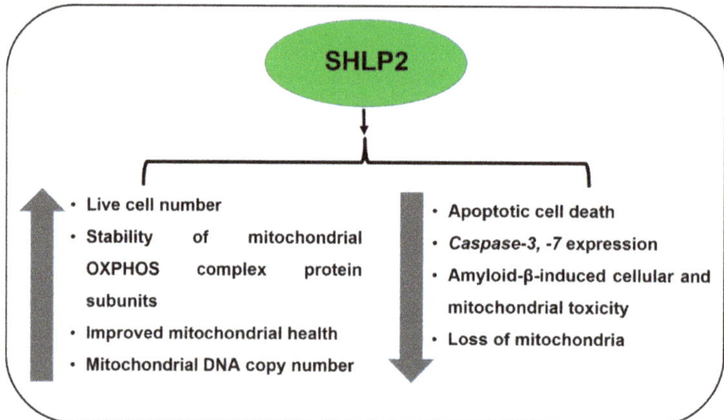

Figure 3. Effects of SHLP2 in AMD.

5.2. SHLP3

SHLP3 is a 38 amino acids-long peptide with the sequence *H-MLGYNFSSFPCGTISIAPGFN FYRLYFIWVNGLAKVVW-OH* and has a molecular weight of 4380.15 Da. SHLP3 increases cellular ATP levels and mitochondrial oxygen consumption rate (OCR) in vitro. SHLP3 suppresses ROS generation, mediates ERK signaling, promotes adipocyte cellular differentiation, and blocks staurosporine-induced apoptosis, thereby promoting mitochondrial and cellular survival and functions. SHLP2 and SHLP3 have insulin-sensitizing effects both in vivo and in vitro, and are known to increase leptin levels. SHLP2 functions as an insulin sensitizer both peripherally and centrally as it enhances peripheral glucose uptake and inhibits glucose production in the liver. SHLP3 also increases IL-6 and MCP-1 levels.

SHLP4 increases cell proliferation in the NIT-1 and 22RV-1 cell lines in vitro. SHLP6 drastically enhances apoptotic cell death in the NIT-1 and 22RV-1 cell lines, and exerts an effect opposite of SHLP2 and SHLP3, which are cytoprotective molecules [56]. Among all the SHLPs, SHLP2 is the only peptide whose cytoprotective function has been established in AMD (Figure 3).

6. MOTS-c

MOTS-c (Mitochondrial ORF (Open Reading Frame) within the Twelve S rRNA c) is encoded in the 12S rRNA region of the mtDNA, and is a 16-amino acid peptide with the sequence *H-MRWQEMG YIFYPRKLR-OH* and a molecular weight 2174.7 Da (Figure 4). It is associated with insulin resistance and is found in plasma, brain, liver, and muscle tissues. MOTS-c enhances insulin sensitivity and regulates plasma metabolites in three metabolic pathways, namely sphingolipid metabolism, monoacylglycerol metabolism, and dicarboxylate metabolism. MOTS-c indirectly decreases cellular oxidative stress by reducing plasma oxidized glucose levels. MOTS-c also activates AKT phosphorylation and AMPK pathways. Moreover, MOTS-c reduces skeletal muscle fatigue and improves performance in vivo [57]. This is the first report by Cohen et al. that delineates the specific roles of MOTS-c and opens new avenues for further research with this MDP. MOTS-c enhances mitochondrial respiration in senescent cells by regulating fatty acid oxidation and increasing the senescence-related effectors [58]. Since MOTS-c levels also decline with age, it has been implicated in the regulation of lifespan and healthspan in organisms. This lifespan/healthspan prolonging ability of MOTS-c may be attributed to (1) MOTS-c-mediated increase in intracellular NAD$^+$, a key redox metabolic coenzyme that activates sirtuins, which in turn regulate aging; and (2) MOTS-c-mediated reduction in methionine metabolism via inhibition of the folate/methionine cycle [59]. Although the function of MOTS-c as a crucial player in cell longevity, mitochondrial function, and metabolic homeostasis has been well-established, its specific protective function in the eye is yet to be reported. Since the AMD etiology involves a metabolic component as well as mitochondrial dysregulation, it would be interesting to investigate the cytoprotective effects of MOTS-c in AMD pathology (Figure 5).

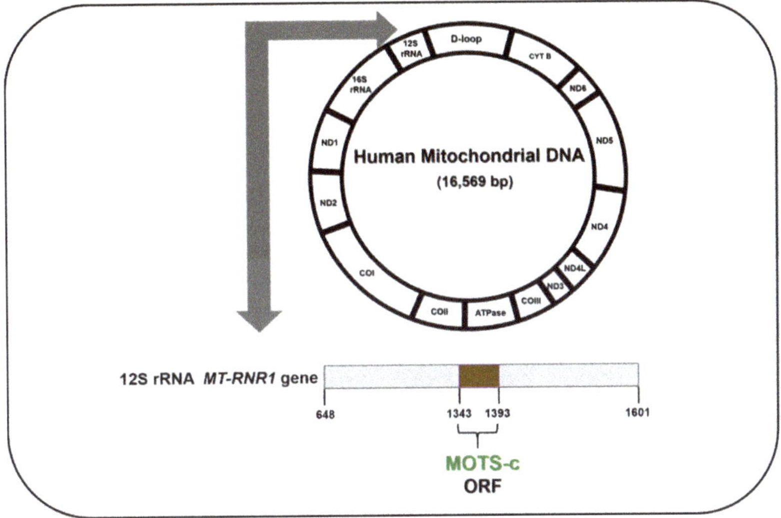

Figure 4. MOTS-c ORF in the human mitochondrial DNA.

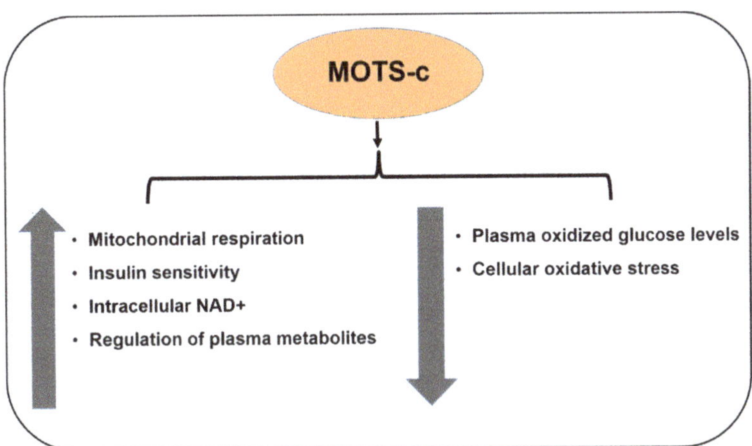

Figure 5. Effects of MOTS-c.

CohBar, a clinical-stage biotechnology company focused on the research and development of mitochondria-based therapeutics, has successfully completed a Phase 1a clinical study and initiated the Phase 1b stage of a double-blind, placebo-controlled clinical trial of CB4211 as a potential treatment for nonalcoholic steatohepatitis (NASH) and obesity. CB4211 is the first therapeutic candidate based on a mitochondrial-derived peptide to enter clinical testing in humans. The completed Phase 1a stage of the CB4211 clinical study evaluated safety and tolerability, and the drug was safe and well tolerated after seven days of dosing. The Phase 1b stage of the study will be an assessment of safety, tolerability, and activity in obese subjects with nonalcoholic fatty liver diseases (NAFLD). Assessments will include changes in liver fat assessed by MRI-PDFF, body weight, and biomarkers relevant to NASH and obesity [60].

7. Conclusions and Future Directions

The dry and wet forms of AMD have some common denominators in terms of AMD disease pathology, since the characteristic features of dry AMD, i.e., RPE cell apoptosis and accumulation of drusen deposits (in which amyloid-β is a key component), may subsequently lead to choroidal neovascularization observed in wet AMD pathology. To our knowledge, differential effects of MDPs in dry AMD versus wet AMD have not been delineated and reported yet.

Since this manuscript presents the literature that establishes the role of MDPs in mitigating RPE cell apoptosis and reducing amyloid-β-induced toxicity, it suggests that MDPs are potential therapeutic candidates for treatment of dry AMD, and may delay its progression to the late form, i.e., wet AMD. However, the specific therapeutic effects of MDPs in reducing or preventing choroidal neovascularization in wet AMD need to be characterized. Experimentally, the effects of MDPs on the angiogenesis-promoting factors should be tested to evaluate if MDPs are able to downregulate the pro-angiogenic factors and related pathways, and thereby prevent or suppress neovascularization in wet AMD.

In conclusion, MDPs, especially Humanin (and its analogs) and SHLPs, provide cytoprotection in ocular diseases, including AMD, and may be considered as potential therapeutic targets for AMD. However, the use of MDPs as therapeutic agents for AMD will require development of appropriate delivery techniques and formulations. Currently, nanoparticles encapsulating Humanin are being tested in some labs to subsequently facilitate its efficient delivery and sustained action. In a recent study by Li et al., it was demonstrated that the Humanin peptide mediates elastin-like polypeptide nano-assembly and protects human RPE cells against oxidative stress-induced apoptotic cell death. This technology may facilitate cellular delivery of biodegradable nanoparticles with potential protection against AMD [61]. Furthermore, with the development of precision medicine, the use of MDPs can be customized for AMD therapy to match patient needs.

Author Contributions: S.N.: Wrote and edited the manuscript; M.C.K.: Corresponding author. All authors have read and agreed to the published version of the manuscript.

Funding: This work is supported by Arnold and Mabel Beckman foundation, NEI R01 EY0127363, Discovery Eye Foundation, Polly and Michael Smith, Edith and Roy Carver, Iris and B. Gerald Cantor Foundation, Unrestricted Departmental Grant from Research to Prevent Blindness, UCI School of Medicine, and support of the Institute for Clinical and Translational Science (ICTS) at University of California Irvine. S.N. is a recipient of the 2017 Genentech/ ARVO AMD Translational Research Fellowship and the 2016 RPB pilot research grant.

Conflicts of Interest: The authors declare no conflict of interest.

References

1. Klein, R.; Chou, C.F.; Klein, B.E.; Zhang, X.; Meuer, S.M.; Saaddine, J.B. Prevalence of age-related macular degeneration in the US population. *Arch Ophthalmol.* **2011**, *129*, 75–80. [CrossRef] [PubMed]
2. McCusker, M.M.; Durrani, K.; Payette, M.J.; Suchecki, J. An eye on nutrition: The role of vitamins, essential fatty acids, and antioxidants in age-related macular degeneration, dry eye syndrome, and cataract. *Clin. Dermatol.* **2016**, *34*, 276–285. [CrossRef] [PubMed]
3. Au, A.; Parikh, V.S.; Singh, R.P.; Ehlers, J.P.; Yuan, A.; Rachitskaya, A.V.; Sears, J.E.; Srivastava, S.K.; Kaiser, P.K.; Schachat, A.P.; et al. Comparison of anti-VEGF therapies on fibrovascular pigment epithelial detachments in age-related macular degeneration. *Br. J. Ophthalmol.* **2016**, *101*, 970–975. [CrossRef] [PubMed]
4. Karunadharma, P.P.; Nordgaard, C.L.; Olsen, T.W.; Ferrington, D.A. Mitochondrial DNA. Damage as a potential mechanism for age-related macular degeneration. *Invest Ophthalmol Vis. Sci.* **2010**, *51*, 5470–5479. [CrossRef]
5. Nordgaard, C.L.; Karunadharma, P.P.; Feng, X.; Olsen, T.W.; Ferrington, D.A. Mitochondrial proteomics of the retinal pigment epithelium at progressive stages of age-related macular degeneration. *Invest Ophthalmol. Vis. Sci.* **2008**, *49*, 2848–2855. [CrossRef]
6. Nordgaard, C.L.; Berg, K.M.; Kapphahn, R.J.; Reilly, C.; Feng, X.; Olsen, T.W.; Ferrington, D.A. Proteomics of the retinal pigment epithelium reveals altered protein expression at progressive stages of age-related macular degeneration. *Invest Ophthalmol. Vis. Sci.* **2006**, *47*, 815–822. [CrossRef]

7.   Wang, J.; Zibetti, C.; Shang, P.; Sripathi, S.R.; Zhang, P.; Cano, M.; Hoang, T.; Xia, S.; Ji, H.; Merbs, S.L.; et al. ATAC-Seq analysis reveals a widespread decrease of chromatin accessibility in age-related macular degeneration. *Nat. Commun.* **2018**, *9*, 1364. [CrossRef]

8.   Clayton, D.A. Structure and function of the mitochondrial genome. *J. Inherit. Metab. Dis.* **1992**, *15*, 439–447. [CrossRef]

9.   Holt, I.J.; Reyes, A. Human mitochondrial DNA replication. *Cold Spring Harb. Perspect. Biol.* **2012**, *4*, a012971. [CrossRef]

10.   Brand, M.D.; Orr, A.L.; Perevoshchikova, I.V.; Quinlan, C.L. The role of mitochondrial function and cellular bioenergetics in ageing and disease. *Br. J. Dermatol.* **2013**, *169*, 1–8. [CrossRef]

11.   Wong-Riley, M.T. Energy metabolism of the visual system. *Eye Brain* **2010**, *2*, 99–116. [CrossRef]

12.   Eells, J.T. Mitochondrial Dysfunction in the Aging Retina. *Biology* **2019**, *8*, 31. [CrossRef]

13.   Chistiakov, D.A.; Sobenin, I.A.; Revin, V.V.; Orekhov, A.N.; Bobryshev, Y.V. Mitochondrial aging and age-related dysfunction of mitochondria. *Biomed Res. Int.* **2014**, *2014*, 238463. [CrossRef]

14.   Taanman, J.W. The mitochondrial genome: Structure, transcription, translation and replication. *Biochim. Biophys. Acta* **1999**, *1410*, 103–123. [CrossRef]

15.   Cardamone, M.D.; Tanasa, B.; Cederquist, C.T.; Huang, J.; Mahdaviani, K.; Li, W.; Rosenfeld, M.G.; Liesa, M.; Perissi, V. Mitochondrial Retrograde Signaling in Mammals Is Mediated by the Transcriptional Cofactor GPS2 via Direct Mitochondria-to-Nucleus Translocation. *Mol. Cell* **2018**, *69*, 757–772.e7. [CrossRef]

16.   Yang, L.; Tan, Z.; Wang, D.; Xue, L.; Guan, M.X.; Huang, T.; Li, R. Species identification through mitochondrial rRNA genetic analysis. *Sci. Rep.* **2014**, *4*, 4089. [CrossRef]

17.   Yen, K.; Lee, C.; Mehta, H.; Cohen, P. The emerging role of the mitochondrial-derived peptide humanin in stress resistance. *J. Mol. Endocrinol.* **2013**, *50*, R11–R19. [CrossRef]

18.   Fuku, N.; Pareja-Galeano, H.; Zempo, H.; Alis, R.; Arai, Y.; Lucia, A.; Hirose, N. The mitochondrial-derived peptide MOTS-c: A player in exceptional longevity? *Aging Cell* **2015**, *14*, 921–923. [CrossRef]

19.   Kim, K.H.; Son, J.M.; Benayoun, B.A.; Lee, C. The Mitochondrial-Encoded Peptide MOTS-c Translocates to the Nucleus to Regulate Nuclear Gene Expression in Response to Metabolic Stress. *Cell Metab.* **2018**, *28*, 516–524.e7. [CrossRef]

20.   Mehta, H.H.; Xiao, J.; Ramirez, R.; Miller, B.; Kim, S.J.; Cohen, P.; Yen, K. Metabolomic profile of diet-induced obesity mice in response to humanin and small humanin-like peptide 2 treatment. *Metabolomics* **2019**, *15*, 88. [CrossRef]

21.   Hashimoto, Y.; Niikura, T.; Tajima, H.; Yasukawa, T.; Sudo, H.; Ito, Y.; Kita, Y.; Kawasumi, M.; Kouyama, K.; Doyu, M.; et al. A rescue factor abolishing neuronal cell death by a wide spectrum of familial Alzheimer's disease genes and Abeta. *Proc. Natl. Acad. Sci. USA* **2001**, *98*, 6336–6341. [CrossRef]

22.   Yamagishi, Y.; Hashimoto, Y.; Niikura, T.; Nishimoto, I. Identification of essential amino acids in Humanin, a neuroprotective factor against Alzheimer's disease-relevant insults. *Peptides* **2003**, *24*, 585–595. [CrossRef]

23.   Zhang, X.; Urbieta-Caceres, V.H.; Eirin, A.; Bell, C.C.; Crane, J.A.; Tang, H.; Jordan, K.L.; Oh, Y.K.; Zhu, X.Y.; Korsmo, M.J.; et al. Humanin prevents intra-renal microvascular remodeling and inflammation in hypercholesterolemic ApoE deficient mice. *Life Sci.* **2012**, *91*, 199–206. [CrossRef]

24.   Widmer, R.J.; Flammer, A.J.; Herrmann, J.; Rodriguez-Porcel, M.; Wan, J.; Cohen, P.; Lerman, L.O.; Lerman, A. Circulating humanin levels are associated with preserved coronary endothelial function. *Am. J. Physiol. Heart Circ. Physiol.* **2013**, *304*, H393–H397. [CrossRef]

25.   Muzumdar, R.H.; Huffman, D.M.; Calvert, J.W.; Jha, S.; Weinberg, Y.; Cui, L.; Nemkal, A.; Atzmon, G.; Klein, L.; Gundewar, S.; et al. Acute humanin therapy attenuates myocardial ischemia and reperfusion injury in mice. *Arterioscler. Thromb. Vasc. Biol.* **2010**, *30*, 1940–1948. [CrossRef]

26.   Moretti, E.; Giannerini, V.; Rossini, L.; Matsuoka, M.; Trabalzini, L.; Collodel, G. Immunolocalization of humanin in human sperm and testis. *Fertil. Steril.* **2010**, *94*, 2888–2890. [CrossRef]

27.   Colón, E.; Strand, M.L.; Carlsson-Skwirut, C.; Wahlgren, A.; Svechnikov, K.V.; Cohen, P.; Söder, O. Anti-apoptotic factor humanin is expressed in the testis and prevents cell-death in leydig cells during the first wave of spermatogenesis. *J. Cell Physiol.* **2006**, *208*, 373–385. [CrossRef]

28.   Tajima, H.; Niikura, T.; Hashimoto, Y.; Ito, Y.; Kita, Y.; Terashita, K.; Yamazaki, K.; Koto, A.; Aiso, S.; Nishimoto, I. Evidence for in vivo production of Humanin peptide, a neuroprotective factor against Alzheimer's disease-related insults. *Neurosci. Lett.* **2002**, *324*, 227–231. [CrossRef]

29. Muzumdar, R.H.; Huffman, D.M.; Atzmon, G.; Buettner, C.; Cobb, L.J.; Fishman, S.; Budagov, T.; Cui, L.; Einstein, F.H.; Poduval, A.; et al. Humanin: A novel central regulator of peripheral insulin action. *PLoS ONE* **2009**, *4*, e6334. [CrossRef]

30. Chin, Y.P.; Keni, J.; Wan, J.; Mehta, H.; Anene, F.; Jia, Y.; Lue, Y.H.; Swerdloff, R.; Cobb, L.J.; Wang, C.; et al. Pharmacokinetics and tissue distribution of humanin and its analogues in male rodents. *Endocrinology* **2013**, *154*, 3739–3744. [CrossRef]

31. Hashimoto, Y.; Kurita, M.; Aiso, S.; Nishimoto, I.; Matsuoka, M. Humanin inhibits neuronal cell death by interacting with a cytokine receptor complex or complexes involving CNTF receptor alpha/WSX-1/gp130. *Mol. Biol. Cell.* **2009**, *20*, 2864–2873. [CrossRef] [PubMed]

32. Hashimoto, Y.; Suzuki, H.; Aiso, S.; Niikura, T.; Nishimoto, I.; Matsuoka, M. Involvement of tyrosine kinases and STAT3 in Humanin-mediated neuroprotection. *Life Sci.* **2005**, *77*, 3092–3104. [CrossRef]

33. Taga, T.; Kishimoto, T. Gp130 and the interleukin-6 family of cytokines. *Annu. Rev. Immunol.* **1997**, *15*, 797–819. [CrossRef]

34. Lee, C.; Wan, J.; Miyazaki, B.; Fang, Y.; Guevara-Aguirre, J.; Yen, K.; Longo, V.; Bartke, A.; Cohen, P. IGF-I regulates the age-dependent signaling peptide humanin. *Aging Cell.* **2014**, *13*, 958–961. [CrossRef]

35. Njomen, E.; Evans, H.G.; Gedara, S.H.; Heyl, D.L. Humanin Peptide Binds to Insulin-Like Growth Factor-Binding Protein 3 (IGFBP3) and Regulates Its Interaction with Importin-β. *Protein Pept. Lett.* **2015**, *22*, 869–876. [CrossRef]

36. Guo, B.; Zhai, D.; Cabezas, E.; Welsh, K.; Nouraini, S.; Satterthwait, A.C.; Reed, J.C. Humanin peptide suppresses apoptosis by interfering with Bax activation. *Nature* **2003**, *423*, 456–461. [CrossRef]

37. Luciano, F.; Zhai, D.; Zhu, X.; Bailly-Maitre, B.; Ricci, J.E.; Satterthwait, A.C.; Reed, J.C. Cytoprotective peptide humanin binds and inhibits proapoptotic Bcl-2/Bax family protein BimEL. *J. Biol. Chem.* **2005**, *280*, 15825–15835. [CrossRef]

38. Nashine, S.; Cohen, P.; Chwa, M.; Lu, S.; Nesburn, A.B.; Kuppermann, B.D.; Kenney, M.C. Humanin G (HNG) protects age-related macular degeneration (AMD) transmitochondrial ARPE-19 cybrids from mitochondrial and cellular damage. *Cell Death Dis.* **2017**, *8*, e2951. [CrossRef]

39. Ikonen, M.; Liu, B.; Hashimoto, Y.; Ma, L.; Lee, K.W.; Niikura, T.; Nishimoto, I.; Cohen, P. Interaction between the Alzheimer's survival peptide humanin and insulin-like growth factor-binding protein 3 regulates cell survival and apoptosis. *Proc. Natl. Acad. Sci. USA* **2003**, *100*, 13042–13047. [CrossRef]

40. Gong, Z.; Tasset, I. Humanin enhances the cellular response to stress by activation of chaperone-mediated autophagy. *Oncotarget* **2018**, *9*, 10832–10833. [CrossRef]

41. Sreekumar, P.G.; Ishikawa, K.; Spee, C.; Mehta, H.H.; Wan, J.; Yen, K.; Cohen, P.; Kannan, R.; Hinton, D.R. The Mitochondrial-Derived Peptide Humanin Protects RPE Cells from Oxidative Stress, Senescence, and Mitochondrial Dysfunction. *Invest Ophthalmol. Vis. Sci.* **2016**, *57*, 1238–1253. [CrossRef]

42. Gurunathan, S.; Jeyaraj, M.; Kang, M.H.; Kim, J.H. Mitochondrial Peptide Humanin Protects Silver Nanoparticles-Induced Neurotoxicity in Human Neuroblastoma Cancer Cells (SH-SY5Y). *Int. J. Mol. Sci.* **2019**, *20*, 4439. [CrossRef]

43. Li, X.; Zhao, W.; Yang, H.; Zhang, J.; Ma, J. S14G-humanin restored cellular homeostasis disturbed by amyloid-beta protein. *Neural Regen. Res.* **2013**, *8*, 2573–2580. [PubMed]

44. Minasyan, L.; Sreekumar, P.G.; Hinton, D.R.; Kannan, R. Protective Mechanisms of the Mitochondrial-Derived Peptide Humanin in Oxidative and Endoplasmic Reticulum Stress in RPE Cells. *Oxid. Med. Cell Longev.* **2017**, *2017*, 1675230. [CrossRef] [PubMed]

45. Knapp, A.; Czech, U.; Polus, A.; Chojnacka, M.; Śliwa, A.; Awsiuk, M.; Zapała, B.; Malińska, D.; Szewczyk, A.; Dembińska-Kieć, A. Humanin Peptides Regulate Calcium Flux in the Mammalian Neuronal, Glial and Endothelial Cells under Stress Conditions. *J. Cell Sci. Ther.* **2012**, *3*, 128. [CrossRef]

46. Matsunaga, D.; Sreekumar, P.G.; Ishikawa, K.; Terasaki, H.; Barron, E.; Cohen, P.; Kannan, R.; Hinton, D.R. Humanin Protects RPE Cells from Endoplasmic Reticulum Stress-Induced Apoptosis by Upregulation of Mitochondrial Glutathione. *PLoS ONE* **2016**, *11*, e0165150. [CrossRef]

47. Xu, X.; Chua, C.C.; Gao, J.; Chua, K.W.; Wang, H.; Hamdy, R.C.; Chua, B.H. Neuroprotective effect of humanin on cerebral ischemia/reperfusion injury is mediated by a PI3K/Akt pathway. *Brain Res.* **2008**, *1227*, 12–18. [CrossRef]

48. Kim, S.J.; Guerrero, N.; Wassef, G.; Xiao, J.; Mehta, H.H.; Cohen, P.; Yen, K. The mitochondrial-derived peptide humanin activates the ERK1/2, AKT, and STAT3 signaling pathways and has age-dependent signaling differences in the hippocampus. *Oncotarget* **2016**, *7*, 46899–46912. [CrossRef]

49. Oh, Y.K.; Bachar, A.R.; Zacharias, D.G.; Kim, S.G.; Wan, J.; Cobb, L.J.; Lerman, L.O.; Cohen, P.; Lerman, A. Humanin preserves endothelial function and prevents atherosclerotic plaque progression in hypercholesterolemic ApoE deficient mice. *Atherosclerosis* **2011**, *219*, 65–73. [CrossRef]

50. Bachar, A.R.; Scheffer, L.; Schroeder, A.S.; Nakamura, H.K.; Cobb, L.J.; Oh, Y.K.; Lerman, L.O.; Pagano, R.E.; Cohen, P.; Lerman, A. Humanin is expressed in human vascular walls and has a cytoprotective effect against oxidized LDL-induced oxidative stress. *Cardiovasc. Res.* **2010**, *88*, 360–366. [CrossRef]

51. Jia, Y.; Ohanyan, A.; Lue, Y.H.; Swerdloff, R.S.; Liu, P.Y.; Cohen, P.; Wang, C. The effects of humanin and its analogues on male germ cell apoptosis induced by chemotherapeutic drugs. *Apoptosis* **2015**, *20*, 551–561. [CrossRef] [PubMed]

52. Lue, Y.; Swerdloff, R.; Wan, J.; Xiao, J.; French, S.; Atienza, V.; Canela, V.; Bruhn, K.W.; Stone, B.; Jia, Y.; et al. The Potent Humanin Analogue (HNG) Protects Germ Cells and Leucocytes While Enhancing Chemotherapy-Induced Suppression of Cancer Metastases in Male Mice. *Endocrinology* **2015**, *156*, 4511–4521. [CrossRef] [PubMed]

53. Zacharias, D.G.; Kim, S.G.; Massat, A.E.; Bachar, A.R.; Oh, Y.K.; Herrmann, J.; Rodriguez-Porcel, M.; Cohen, P.; Lerman, L.O.; Lerman, A. Humanin, a cytoprotective peptide, is expressed in carotid atherosclerotic [corrected] plaques in humans. *PLoS ONE.* **2012**, *7*, e31065. [CrossRef]

54. Nashine, S.; Cohen, P.; Nesburn, A.B.; Kuppermann, B.D.; Kenney, M.C. Characterizing the protective effects of SHLP2, a mitochondrial-derived peptide, in macular degeneration. *Sci. Rep.* **2018**, *8*, 15175. [CrossRef]

55. Okada, A.K.; Teranishi, K.; Lobo, F.; Isas, J.M.; Xiao, J.; Yen, K.; Cohen, P.; Langen, R. The Mitochondrial-Derived Peptides, HumaninS14G and Small Humanin-like Peptide 2, Exhibit Chaperone-like Activity. *Sci. Rep.* **2017**, *7*, 7802. [CrossRef]

56. Cobb, L.J.; Lee, C.; Xiao, J.; Yen, K.; Wong, R.G.; Nakamura, H.K.; Mehta, H.H.; Gao, Q.; Ashur, C.; Huffman, D.M.; et al. Naturally occurring mitochondrial-derived peptides are age-dependent regulators of apoptosis, insulin sensitivity, and inflammatory markers. *Aging (Albany NY)* **2016**, *8*, 796–809. [CrossRef]

57. Kim, S.J.; Miller, B.; Mehta, H.H.; Xiao, J.; Wan, J.; Arpawong, T.E.; Yen, K.; Cohen, P. The mitochondrial-derived peptide MOTS-c is a regulator of plasma metabolites and enhances insulin sensitivity. *Physiol. Rep.* **2019**, *7*, e14171. [CrossRef]

58. Kim, S.J.; Mehta, H.H.; Wan, J.; Kuehnemann, C.; Chen, J.; Hu, J.F.; Hoffman, A.R.; Cohen, P. Mitochondrial peptides modulate mitochondrial function during cellular senescence. *Aging (Albany NY)* **2018**, *10*, 1239–1256. [CrossRef]

59. Lee, C.; Kim, K.H.; Cohen, P. MOTS-c: A novel mitochondrial-derived peptide regulating muscle and fat metabolism. *Free Radic. Biol. Med.* **2016**, *100*, 182–187. [CrossRef]

60. CohBar to Resume its Phase 1a/1b Clinical Trial. Available online: https://www.cohbar.com/news-media/press-releases/detail/81/cohbar-to-resume-its-phase-1a1b-clinical-trial (accessed on 3 May 2019).

61. Li, Z.; Sreekumar, P.G.; Peddi, S.; Hinton, D.R.; Kannan, R.; MacKay, J.A. The humanin peptide mediates ELP nanoassembly and protects human retinal pigment epithelial cells from oxidative stress. *Nanomedicine* **2019**, *24*, 102111. [CrossRef]

MDPI
St. Alban-Anlage 66
4052 Basel
Switzerland
Tel. +41 61 683 77 34
Fax +41 61 302 89 18
www.mdpi.com

Cells Editorial Office
E-mail: cells@mdpi.com
www.mdpi.com/journal/cells

Natural-Hazards Risk Assessment for Disaster Mitigation

Editors

Andrea Chiozzi
Elena Benvenuti
Željana Nikolić

MDPI • Basel • Beijing • Wuhan • Barcelona • Belgrade • Manchester • Tokyo • Cluj • Tianjin

Editors
Andrea Chiozzi
Department of Environmental
and Prevention Sciences
University of Ferrara
Ferrara
Italy

Elena Benvenuti
Engineering Department
University of Ferrara
Ferrara
Italy

Željana Nikolić
Faculty of Civil Engineering,
Architecture and Geodesy
University of Split
Split
Croatia

Editorial Office
MDPI
St. Alban-Anlage 66
4052 Basel, Switzerland

This is a reprint of articles from the Special Issue published online in the open access journal *Applied Sciences* (ISSN 2076-3417) (available at: www.mdpi.com/journal/applsci/special_issues/ nhra_disaster_mitigation).

For citation purposes, cite each article independently as indicated on the article page online and as indicated below:

LastName, A.A.; LastName, B.B.; LastName, C.C. Article Title. *Journal Name* **Year**, *Volume Number*, Page Range.

ISBN 978-3-0365-8017-3 (Hbk)
ISBN 978-3-0365-8016-6 (PDF)

Contents

About the Editors

Andrea Chiozzi

Andrea Chiozzi is an Assistant Professor of Solid and Structural Mechanics at the Department of Environmental and Prevention Sciences of the University of Ferrara, Italy. Since earning his Ph.D. in 2014, he has been actively contributing to several research topics in computational mechanics, mainly addressing analysis of the mechanical behavior of curved three-dimensional masonry structures and the development of innovative numerical methods for elliptic problems in solid mechanics.

Elena Benvenuti

Elena Benvenuti is an Associate Professor of Solid and Structural Mechanics at the Engineering Department of the University of Ferrara with consolidated expertise in nonlinear computational and theoretical mechanics and civil engineering risk assessment. She has authored and co-authored more than 150 scientific contributions published in international journals and conference proceedings. Prof. Benvenuti has also directed PMO-GATE, the Italy–Croatia Interreg project focusing on risk assessment and management that has inspired many of the papers published in the present Special Issue.

Željana Nikolić

Željana Nikolić is a Professor of Structural Mechanics, Computational Mechanics, and Earthquake Engineering and head of the Laboratory for Numerical Modelling at the Faculty of Civil Engineering, Architecture and Geodesy, University of Split, Croatia. Her research focuses on the development of methodologies, advanced numerical techniques, and new non-linear numerical models for the simulation of problems of structural mechanics and earthquake engineering. She has authored and co-authored more than 230 scientific contributions and participated in numerous research projects.

Preface to "Natural-Hazards Risk Assessment for Disaster Mitigation"

Knowledge and awareness of the risks generated by natural hazards are essential requirements for the enhancement of communities' resilience to disasters. United Nations directives have recently pointed out the necessity of undertaking actions aimed at anticipating, managing, and mitigating disaster risks to reduce their economic and social impact and protect the health, socioeconomic assets, cultural heritage, and ecosystems of communities and countries. While the increasing occurrence of disasters caused by meteorological events, such as floods, storms, and droughts, can be directly ascribed to the consequence of climate change, disasters induced by earthquakes and tsunamis are increasing even if their frequency of occurrence is historically unchanged. Therefore, other anthropogenic causes intervene to determine an increment of risk exposure and community vulnerability, such as land misuse in densely populated areas and coastal zones.

This Special Issue addresses concepts, methods, and predictive methodologies for assessing natural hazard risks. It presents fifteen articles focusing on the single-risk assessment of a broad range of natural hazards, such as earthquakes, river/sea floods, meteotsunamis, tornados, hydrological and meteorological drought, liquefaction, as well as on multirisk assessment in the presence of multiple hazards. The adopted methodologies rely on (a) quantitative, semi-quantitative, and qualitative methods for the assessment of the risks related to natural hazards; (b) risk analysis at different scales; (c) multi-hazard risk assessment techniques; (d) real-time hazard monitoring and warning systems; (e) disaster mitigation strategies; and (f) risk management and emergency planning on multiple scales.

Andrea Chiozzi, Elena Benvenuti, and Željana Nikolić
Editors

*applied
sciences*

Editorial

Special Issue on Natural Hazards Risk Assessment for Disaster Mitigation

Željana Nikolić [1,*], Elena Benvenuti [2] and Andrea Chiozzi [2]

[1] Faculty of Civil Engineering, Architecture and Geodesy, University of Split, 21000 Split, Croatia
[2] Engineering Department, University of Ferrara, 44121 Ferrara, Italy
* Correspondence: zeljana.nikolic@gradst.hr

Citation: Nikolić, Ž.; Benvenuti, E.; Chiozzi, A. Special Issue on Natural Hazards Risk Assessment for Disaster Mitigation. *Appl. Sci.* **2023**, *13*, 1940. https://doi.org/10.3390/app13031940

Received: 17 January 2023
Accepted: 28 January 2023
Published: 2 February 2023

Knowledge and awareness of the risks generated by natural hazards are essential requirements for the enhancement of communities' resilience to disasters. The Sendai Framework for Disaster Risk Reduction 2015–2030 has recently pointed out the necessity of undertaking actions aimed to anticipate, manage, and mitigate disaster risks, to reduce their economic and social impact and protect health, socioeconomic assets, cultural heritage, and ecosystems of communities and countries. While the increasing occurrence of disasters caused by meteorological events, such as floods, storms, and droughts, can be directly ascribed to the consequence of climate change, disasters induced by earthquakes and tsunamis are increasing even if the frequency of occurrence is historically unchanged. Therefore, other anthropogenic causes intervene to determine an increment of a risk exposure and community vulnerability, such as land misuse in densely populated areas and coastal zones.

This Special Issue addresses concepts, methods, and predictive methodologies for assessing natural hazards risks. This Special Issue presents fifteen articles focusing on the single-risk assessment of a broad range of natural hazards, such as earthquakes, river, and see floods, meteotsunamis, tornados, hydrological and meteorological drought, liquefaction, as well as on multirisk assessment in the presence of multiple hazards. The adopted methodologies rely on: (a) quantitative, semi-quantitative, and qualitative methods for the assessment of the risks related to natural hazards; (b) risk analysis at different scales; (c) multi-hazard risk assessment techniques; (d) real-time hazard monitoring and warning systems; (e) disaster mitigation strategies; and (f) risk management and emergency planning at multiple scales.

Ahmad et al. [1] propose a Gaussian process regression (GPR) model for analyzing liquefaction-induced lateral displacement based on an impressive amount (247) of case studies of post-liquefaction events. The performance of the GPR model is assessed using statistical parameters, including the coefficient of determination, coefficient of correlation, Nash–Sutcliffe efficiency coefficient, root mean square error (RMSE), and ratio of the RMSE to the standard deviation of the measured data. It was shown that the GPR model can accurately capture the complicated nonlinear relationships between lateral displacements and their influencing factors.

Da Col et al. [2] present two case studies of the seismic surveys to estimate the elastic properties of the soil and rock in the shallow subsurface: (1) a town on the Croatian coast, near the city of Split, built on hard rock and (2) a site located in the Italian town of Ferrara, in an alluvial plain. A Multichannel Analysis of Surface Waves and a first-break tomography were carried out to obtain P-, SH-, and SV-velocity profiles. This acquisition allowed for computing the equivalent shear-wave velocity of the first 30 m of the subsurface (VS30) from the SH profiles, as well as it made it possible to deduce other useful parameters such as the VP/VS, and to estimate the soil's stratigraphy through the analysis of the VSV/VSH profiles.

Gacu et al. [3] studied the spatial distribution of the flood risk of the Municipality of Odiongan using the analytical hierarchy process (AHP) and the geographic information system (GIS), considering the disaster risk factors based on the data collected from various government agencies. The weights of the hazard, exposure, and vulnerability parameters were drawn from the experts' judgment. These weights were subsequently integrated into a flood risk assessment computation which resulted in a flood risk map. The study will guide local government units in developing flood management plans able to reduce flood risk and vulnerability.

Işık et al. [4] carried out a comparative study of the effects of earthquakes in different countries based on target displacement in mid-rise regular RC structures. Five different earthquakes from six countries with a high seismic risk were selected. The measured PGA for each earthquake was compared with the suggested PGA for the respective region. Target displacements specified in the Eurocode-8 were obtained for both the suggested and measured PGA values. It was concluded that both the seismic risk and target displacements were adequately represented for some earthquakes, while not adequately represented for others.

Lin et al. [5] established a method of hazard assessment for the river terraces along the Chenyulan River in Nantou County, Taiwanand. Using GIS, the authors extracted nine parameters and identified the weightings by AHP analysis. Hazard assessment for the river terraces then proceeded via totaling the potential trends of the considered factors and the protected objects, as well as through comparing the historical disaster conditions and satellite images. The results showed different distributions of the relevant risks. Thus, the assessment can be used in reducing the disaster's impact induced by the risks inherent in the riverine terrace settlements.

Maramai et al. [6] developed the database of Adriatic tsunamis and meteotsunamis along the Adriatic coasts, providing an overview of the events and a detailed description of the effects observed at each affected location, and defining a picture of the geographical distribution of the effects for each tsunami and meteotsunami. The database contains 57 observations of tsunami effects related to 27 tsunamis along the Italian, Croatian, Montenegrin, and Albanian coasts and 102 observations of meteotsunami effects related to 33 meteotsunamis. The database can be accessed through a GIS WebApp, which allows the user to visualize the georeferenced information on a map.

Mladineo et al. [7] proposes a methodology for the multi-hazard risk assessment of the urban area of Kaštel Kambelovac, located on the Croatian coast of the Adriatic Sea. The procedure, based on spatial multi-criteria decision making and the PROMETHEE method, was used to assess the multi-hazard risks caused by seismic, flood due to sea level rises, and extreme sea waves impact. The multi-hazard risk is assessed for different scenarios and different levels was based on an exposure and vulnerability for each of the natural hazards and the influence of additional criteria to the overall risk in homogenous zones.

Nikolić et al. [8] developed a methodology for the seismic risk assessment of urban areas based on a hybrid empirical-analytical procedure that combines seismic vulnerability indices with critical peak ground accelerations computed through a non-linear pushover analysis. The procedure's outcomes are the computation of a relationship linking vulnerability indices to the peak ground acceleration for a series of limit states. The methodology was used to estimate the damage index and the index of seismic risk for the selected return periods for masonry buildings in the Croatian settlement Kaštel Kambelovac.

Nikolić et al. [9] present a unique procedure for the real-time assessment of the sea water elevation at the Kaštela Bay in Croatia to ensure a priori warning in the case of expected coastal flooding along the site area caused by barometric pressure, wind-generated waves, and tidal-induced oscillations. The procedure relies on relevant datasets which are site-specific and locally observed. The given information is visualized in a form of mobile application that implements the algorithm and allows end users to set the notifications based on the given ruleset.

Tornadoes are associated with damages, injuries, and even fatalities in Europe. Pîrloagă et al. [10] analyzed a problem of a population bias on tornado reporting in Europe. To account for this bias, a Bayesian modeling approach was used based on tornado observations and the population density for relatively small regions of Europe. The results indicated that the number of tornadoes could be 53% higher than are currently reported. The largest adjustments produced by the model pertain to Northern Europe and some Mediterranean regions.

Rocchi et al. [11] developed a machine learning framework for the assessment of a combined seismic and hydraulic risk at the regional scale. The machine learning techniques were used to aggregate large datasets made of many variables different in nature. The framework is applied to the case study of the Emilia Romagna region, for which the different municipalities are grouped into four homogeneous clusters ranked in terms of the relative levels of combined risk. The proposed approach proved to be robust and delivered a very useful tool for multi-hazard modeling at the regional scale.

Sarwar et al. [12] analyzed meteorological and hydrological drought risk at a regional scale in the Soan basin in Pakistan. The spatiotemporal analysis, statistical approaches, including regression analysis, trend analysis using Mann–Kendall, and moving average, were used to find a linkage between these drought types, the significance of the variations, and the lag time identification, respectively. The overall analysis indicated an increase in the frequency of both hydrological and meteorological droughts during the last three decades.

Shin et al. [13] systematically analyzed the National Flood Insurance Program (NFIP) claims hazard data in Florida. The claims with a presumably incorrect cause of loss fields were identified and revised by adding a variety of other available information. These datasets included tropical cyclone events, rainfall maxima, and distances to the nearest coast. The revised NFIP claims data will be intensively used to validate the outcomes from flood hazard (surge, wave, and inland flooding) models and to develop a flood vulnerability model in the forthcoming Florida Public Flood Loss Model (FPFLM).

Soldati et al. [14] proposes a qualitative multi-hazard risk analysis methodology in the case of combined seismic and flood risk, using PROMETHEE, a multiple-criteria decision analysis technique. The present case study is a multi-hazard risk assessment of the Ferrara province (Italy). The proposed approach provides an original and flexible methodology to qualitatively prioritize the urban centers affected by multi-hazard risks at the regional scale. It delivers a useful tool to stakeholders involved in the processes of hazard management and disaster mitigation.

Vlachogiannis et al. [15] analyzed a climatic multi-hazard risk for Greece, as the first-ever attempt to enhance scientific knowledge for the identification and definition of hazards, a critical element of risk-informed decision making. Many hazards (heatwaves, cold spells, torrential rainfall, snowstorms, and windstorms) were considered to correctly capture the country's susceptibility to climate extremes. The findings highlighted the areas that are exposed to multiple climate hazards in the country, considering the influence of the highly complex topography.

All the contributions gathered in the present special issue contribute to the crucial societal challenge of reducing human and material losses induced by natural hazards caused by extreme climate changes and earthquakes. For this purpose, the continuous upgrade of comprehensive databases of individual risks and the development of new procedures and methodologies for hazard and risk assessment are key to the implementation of effective actions. The present studies show the usefulness of modern technologies, such as multi-criteria decision-making methods coupled with GIS, machine learning, and artificial intelligence. These technologies allowed the authors of the collected contributions to provide robust solutions to several necessities, such as the implementation of numerical models for single-hazard and multi-hazard modeling with a low computational cost, fast training, validation, testing, evaluation, visualization of the results, and fast notification to end users. Furthermore, the presented approaches provide valuable operational tools which can be readily exploited by end users, whether modelers or decision makers, to urgently allocate resources and increase the coping capacity of communities confronting catastrophic events. Therefore, the Editors believe that the Special Issue may significantly contribute to enhance the insight into technological and analytical procedures aimed to "Natural Hazards Risk Assessment for Disaster Mitigation".

Author Contributions: Writing—original draft preparation, Ž.N., E.B. and A.C.; writing—review and editing, E.B and Ž.N. All authors have read and agreed to the published version of the manuscript.

Funding: This research was funded by the EUROPEAN UNION, Programme Interreg Italy-Croatia, Project "Preventing, managing and overcoming natural hazards risks to mitigate economic and social impact"—PMO-GATE ID 10046122. The research is also partially supported through project KK.01.1.1.02.0027, co-financed by the CROATIAN GOVERNMENT and the EUROPEAN UNION through the European Regional Development Fund—the Competitiveness and Cohesion Operational Programme.

Acknowledgments: The Editors are deeply grateful to all authors and peer reviewers for their valuable contributions to the Special Issue 'Natural Hazards Risk Assessment for Disaster Mitigation'. Karen Men is also acknowledged for the helpful assistance in the publication process.

Conflicts of Interest: The authors declare no conflict of interest.

References

1. Ahmad, M.; Amjad, M.; Al-Mansob, R.A.; Kamiński, P.; Olczak, P.; Khan, B.J.; Alguno, A.C. Prediction of Liquefaction-Induced Lateral Displacements Using Gaussian Process Regression. *Appl. Sci.* **2022**, *12*, 1977. [CrossRef]
2. Da Col, F.; Accaino, F.; Böhm, G.; Meneghini, F. Analysis of the Seismic Properties for Engineering Purposes of the Shallow Subsurface: Two Case Studies from Italy and Croatia. *Appl. Sci.* **2022**, *12*, 4535. [CrossRef]
3. Gacu, J.G.; Monjardin, C.E.F.; Senoro, D.B.; Tan, F.J. Flood Risk Assessment Using GIS-Based Analytical Hierarchy Process in the Municipality of Odiongan, Romblon, Philippines. *Appl. Sci.* **2022**, *12*, 9456. [CrossRef]
4. Işık, E.; Hadzima-Nyarko, M.; Bilgin, H.; Ademović, N.; Büyüksaraç, A.; Harirchian, E.; Bulajić, B.; Özmen, H.B.; Aghakouchaki Hosseini, S.E. A Comparative Study of the Effects of Earthquakes in Different Countries on Target Displacement in Mid-Rise Regular RC Structures. *Appl. Sci.* **2022**, *12*, 12495. [CrossRef]
5. Lin, J.-Y.; Chao, J.-C.; Hsu, Y.-M. Risk Assessment of Riverine Terraces: The Case of the Chenyulan River Watershed in Nantou County, Taiwan. *Appl. Sci.* **2022**, *12*, 1375. [CrossRef]
6. Maramai, A.; Brizuela, B.; Graziani, L. A Database for Tsunamis and Meteotsunamis in the Adriatic Sea. *Appl. Sci.* **2022**, *12*, 5577. [CrossRef]
7. Mladineo, N.; Mladineo, M.; Benvenuti, E.; Kekez, T.; Nikolić, Ž. Methodology for the Assessment of Multi-Hazard Risk in Urban Homogenous Zones. *Appl. Sci.* **2022**, *12*, 12843. [CrossRef]
8. Nikolić, Ž.; Benvenuti, E.; Runjić, L. Seismic Risk Assessment of Urban Areas by a Hybrid Empirical-Analytical Procedure Based on Peak Ground Acceleration. *Appl. Sci.* **2022**, *12*, 3585. [CrossRef]
9. Nikolić, Ž.; Srzić, V.; Lovrinović, I.; Perković, T.; Šolić, P.; Kekez, T. Coastal Flooding Assessment Induced by Barometric Pressure, Wind-Generated Waves and Tidal-Induced Oscillations: Kaštela Bay Real-Time Early Warning System Mobile Application. *Appl. Sci.* **2022**, *12*, 12776. [CrossRef]
10. Pîrloagă, R.; Ene, D.; Antonescu, B. Population Bias on Tornado Reports in Europe. *Appl. Sci.* **2021**, *11*, 11485. [CrossRef]
11. Rocchi, A.; Chiozzi, A.; Nale, M.; Nikolic, Z.; Riguzzi, F.; Mantovan, L.; Gilli, A.; Benvenuti, E. A Machine Learning Framework for Multi-Hazard Risk Assessment at the Regional Scale in Earthquake and Flood-Prone Areas. *Appl. Sci.* **2022**, *12*, 583. [CrossRef]
12. Sarwar, A.N.; Waseem, M.; Azam, M.; Abbas, A.; Ahmad, I.; Lee, J.E.; Haq, F.u. Shifting of Meteorological to Hydrological Drought Risk at Regional Scale. *Appl. Sci.* **2022**, *12*, 5560. [CrossRef]
13. Shin, D.W.; Cocke, S.; Kim, B.-M. A Systematic Revision of the NFIP Claims Hazard Data in Florida for Flood Risk Assessment. *Appl. Sci.* **2022**, *12*, 3537. [CrossRef]
14. Soldati, A.; Chiozzi, A.; Nikolić, Ž.; Vaccaro, C.; Benvenuti, E. A PROMETHEE Multiple-Criteria Approach to Combined Seismic and Flood Risk Assessment at the Regional Scale. *Appl. Sci.* **2022**, *12*, 1527. [CrossRef]
15. Vlachogiannis, D.; Sfetsos, A.; Markantonis, I.; Politi, N.; Karozis, S.; Gounaris, N. Quantifying the Occurrence of Multi-Hazards Due to Climate Change. *Appl. Sci.* **2022**, *12*, 1218. [CrossRef]

Article

Methodology for the Assessment of Multi-Hazard Risk in Urban Homogenous Zones

Nenad Mladineo [1], Marko Mladineo [2], Elena Benvenuti [3], Toni Kekez [1] and Željana Nikolić [1,*]

[1]   Faculty of Civil Engineering, Architecture and Geodesy, University of Split, 21000 Split, Croatia
[2]   Faculty of Electrical Engineering, Mechanical Engineering and Naval Architecture, University of Split, 21000 Split, Croatia
[3]   Engineering Department, University of Ferrara, 44121 Ferrara, Italy
*   Correspondence: zeljana.nikolic@gradst.hr

Abstract: The multi-hazard risk assessment of urban areas represents a comprehensive approach that can be used to reduce, manage and overcome the risks arising from the combination of different natural hazards. This paper presents a methodology for multi-hazard risk assessment based on Spatial Multi-Criteria Decision Making. The PROMETHEE method was used to assess multi-hazard risks caused by seismic, flood and extreme sea waves impact. The methodology is applied for multi-hazard risk evaluation of the urban area of Kaštel Kambelovac, located on the Croatian coast of the Adriatic Sea. The settlement is placed in a zone of high seismic risk with a large number of old stone historical buildings which are vulnerable to the earthquakes. Being located along the low-lying coast, this area is also threatened by floods due to climate change-induced sea level rises. Furthermore, the settlement is exposed to flooding caused by extreme sea waves generated by severe wind. In the present contribution, the multi-hazard risk is assessed for different scenarios and different levels, based on exposure and vulnerability for each of the natural hazards and the influence of additional criteria to the overall risk in homogenous zones. Single-risk analysis has shown that the seismic risk is dominant for the whole pilot area. The results of multi-hazard assessment have shown that in all combinations the highest risk is present in the historical part of Kaštel Kambelovac. This is because the historical part is most exposed to sea floods and extreme waves, as well as due to the fact that a significant number of historical buildings is located in this area.

Keywords: risk assessment; multi hazard; multi-hazard risk assessment; multi-criteria decision-making; GIS; PROMETHEE method

Citation: Mladineo, N.; Mladineo, M.; Benvenuti, E.; Kekez, T.; Nikolić, Ž. Methodology for the Assessment of Multi-Hazard Risk in Urban Homogenous Zones. *Appl. Sci.* **2022**, *12*, 12843. https://doi.org/10.3390/app122412843

Academic Editor: Jianbo Gao

Received: 27 October 2022
Accepted: 9 December 2022
Published: 14 December 2022

Publisher's Note: MDPI stays neutral with regard to jurisdictional claims in published maps and institutional affiliations.

1. Introduction

Natural hazards are threatening the population throughout the world more than ever. Efficient planning and preparation are vital, since the question "will the disaster happen?", has changed into "when will it happen?". Enhancing the safety and resilience for disasters requires knowledge about individual territorial hazards, vulnerabilities and risks. Appropriate multi-risk methodology based on existing data and knowledge should produce an interactive and easily understanding map that will enable the visualization of individual and combined risks. Integration of this methodology into the Geographic Information System gives important information to local and regional authorities for preventing, managing and overcoming multi-hazard natural disasters [1], such as river and sea floods, meteotsunamis (or extreme sea waves) and earthquakes. To reduce the possible loss of life and damage to property caused by hazards [2], it is crucial to conduct risk assessments and make decisions pertaining to natural hazards before the hazards occur [3].

A common practice in the hazard risk assessment is to focus on the hazard frequency and intensity in combination with area vulnerability or severity of damage caused by the hazard [4]. Furthermore, the severity is not just the result of hazard intensity and area vulnerability, but it is also influenced by the coping capacity of the emergency units in the area [5]. Hazard occurs in some periods with particular intensity and causes damage in relation to area vulnerability and coping capacity. A brief literature survey was made among scientific and professional papers to investigate what are the common factors used to calculate the risk of natural hazards (Table 1).

Table 1. Brief literature survey on factors used to calculate the risk of natural hazards.

Approach	Factors				Output	Source
	Frequency	Intensity	Vulnerability	Other		
Single hazard	Yes	No	No	Damage	Risk	Di Mauro et al. [6]
Multi-hazard	Yes	Yes	Yes	Coping capacity	Risk	Fleischhauer et al. [7]
Multi-hazard	No	Yes	Yes	Coping capacity	Integrated risk	Greiving et al. [5]
Single hazard	Yes	Yes	No	No	Risk	Kunz et al. [8]
Multi-hazard	Yes	Yes	Yes	Consequence (loss)	Risk	Liu et al. [9]
Multi-hazard	Yes	No	No	Aggregated losses	Risk	Mignan et al. [10]
Single hazard	No	No	Yes	Hazard exposure, Exposed value	Risk index	Munich Re Group [11]
Single hazard	Yes	Yes	No	Area impact	Hazard score	Odeh Engineers, Inc. [12]
Multi-hazard	Yes	Yes	Yes	Elements at risk, Temporal/Spatial probability	Risk	Van Westen [13]

Table 1 shows that hazard frequency (or probability of occurrence) is the most common factor used in risk assessment. Some approaches are focused on hazard intensity and some of them on vulnerability in combination with damage or loss. However, half of the papers dealing with multi-hazard approach are taking into account all three emphasized factors—frequency, intensity and vulnerability—including other factors. In the context of the multi-hazard risk assessment, many factors are used. Since each factor can represent one criterion, evaluation of these factors can be used as an input matrix for the Multi-Criteria Analysis (MCA). The reason for using so many criteria (factors) for multi-hazard risk assessments is due to its complexity. According to the standard for risk assessment IEC 31010:2009 [14], the risk assessment is the overall following process: risk identification, risk analysis and risk evaluation, and it is recommended to use multi-criteria analysis or multi-criteria decision-analysis method for the risk assessment.

The problem of the multi-hazard risk assessment is not just in designing a proper calculation to aggregate all hazard risks in one area [15,16], but also to take into account hazards' mutual correlation [9], since one hazard can trigger another one. For instance, fire is usually spread after earthquakes, and earthquakes can produce tsunamis, thus flooding the area.

Although the world is dealing with hazards that have an increasing frequency, like flood disasters that are caused by extreme climate and urbanization processes [17], another problem of the single-hazard or multi-hazard risk assessment lies in a specific type of hazards. There are specific hazards, earthquakes, for instance, that can be represented as low-probability/high-consequence events [10]. This issue represents a large problem, as earthquakes can have high intensity while their frequency is usually very low, so the risk calculation of multiplying intensity and frequency will result in a low risk level. Therefore, additional factors need to be taken into calculation to emphasize the risk from earthquakes and similar hazards that are low-probability/high-consequence events.

Vulnerability is one of the most important factors in the risk assessment [18]. Namely, high vulnerability of some areas can result in severe losses during a low-intensity hazard, and low vulnerability can result in minor losses during a high-intensity hazard. Many different criteria are used for vulnerability calculation, because the criteria set is also defined by the type of hazard [4]. However, assessing the vulnerability to natural hazards such as earthquakes can be characterized as an ill-structured problem or a problem without unique, identifiable and objectively optimal solution. A review of the literature indicates a number of contrasting definitions of what vulnerability means, as well as numerous conflicting perspectives on what should or should not be included within the broad assessment of vulnerability in cities [19]. For instance, some authors also include coping capacity (emergency units) of the area in the vulnerability analysis [5]. But this is not necessarily a good approach, since coping capacity is something that can dynamically change each year, and other common vulnerability criteria (building age, building structure, building height, etc.) are less or more static. Furthermore, it is important to mention that different vulnerability analyses are used at different scales [20]. The different criteria sets (factors) are used and sometimes different methods must be used, as well.

The aim of this study is the assessment of multi-hazard risk for Kaštel Kambelovac, a small city placed along the Adriatic Sea near the city of Split, Croatia (Figure 1).

(a)

(b)

Figure 1. The City of Kaštela: (**a**) Historical center of Kaštel Kambelovac [21]; (**b**) Coastal flooding events in the City of Kaštela [22,23].

Due to its location in a zone of high seismic risk and considering the large number of stone-made and several-centuries-old historical buildings, Kaštela is a settlement with pronounced seismic vulnerability and risk [18]. The historical center, built right next to the low-lying coast, is also threatened by floods due to rising sea levels caused by climate change. Furthermore, the settlement is exposed to flooding caused by extreme sea waves generated by severe wind. Therefore, the present multi-hazard risk assessment is aimed at determining the combined risk of the settlement caused by earthquakes, sea floods and extreme sea waves. The multi-hazard risk investigation of the Kaštel Kambelovac is a part of the project "Preventing, managing, and overcoming natural-hazards risks to mitigate economic and social impact" (PMO-GATE) [24].

The main challenge in the multi-hazard risk assessment is to evaluate, use and mutually compare different mathematical variables that describe the hazard's frequency, intensity and vulnerability. In this research, this issue is addressed by using the multi-criteria analysis that is commonly used to evaluate and compare quantitative and qualitative criteria in completely different units and the order of magnitude. However, a proper multi-criteria analysis must be selected. An additional challenge in the multi-hazard risk assessment is the data collection and evaluation, which can become complex on the settlement or regional level. Therefore, a proper spatial analysis must be used to organize and aggregate spatial thematic layers. Accordingly, the Geographic Information System is used in combination with multi-criteria analysis in order to establish a Spatial Multi-Criteria Decision Making system.

2. Methodology

The methodology used in this research relies on the combination of the Multi-Criteria Analysis/Multi-Criteria Decision-Making (MCA/MCDM) and Geographic Information System (GIS) to evaluate and visualize this risk assessment in this particular area. The advantage of GIS is in its ability to visualize spatial data (Figure 2) and enhance the spatial decision-making in the risk assessment [25–27].

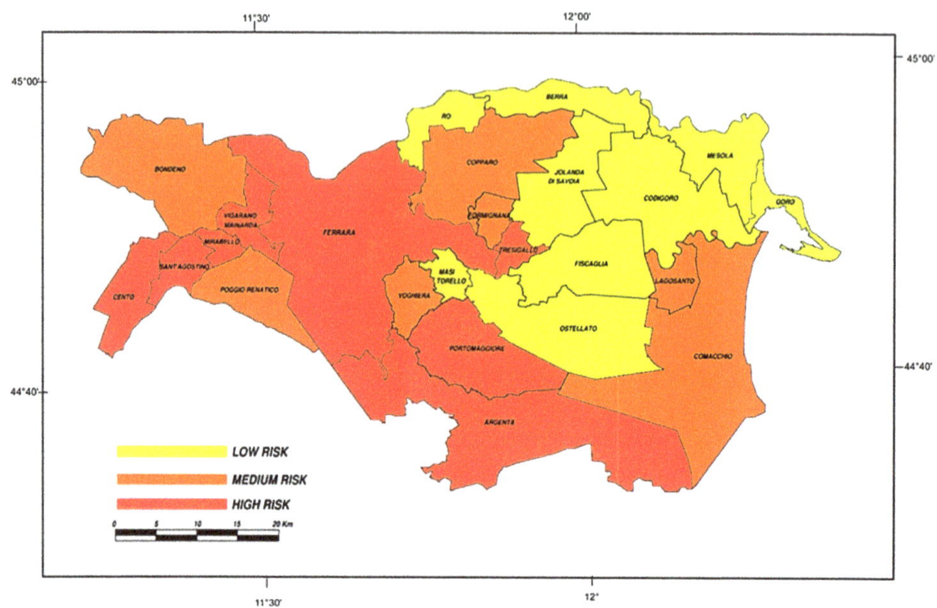

Figure 2. An example of multiple-risk map for the Ferrara province [15].

Furthermore, GIS supports the usage of the multi-scale approach and different criteria sets: the vulnerability analysis can be made for each building, the whole settlement, the settlement's municipality, and for the whole region. The multi-scale approach was discussed in the already-mentioned paper by Vicente et al. [20], but greater contribution was given by Aubrecht et al. [28] that presented multi-level geospatial modeling of vulnerability indicators from building level to country level. Another example of a multi-level and multi-criteria approach is presented in Figure 3.

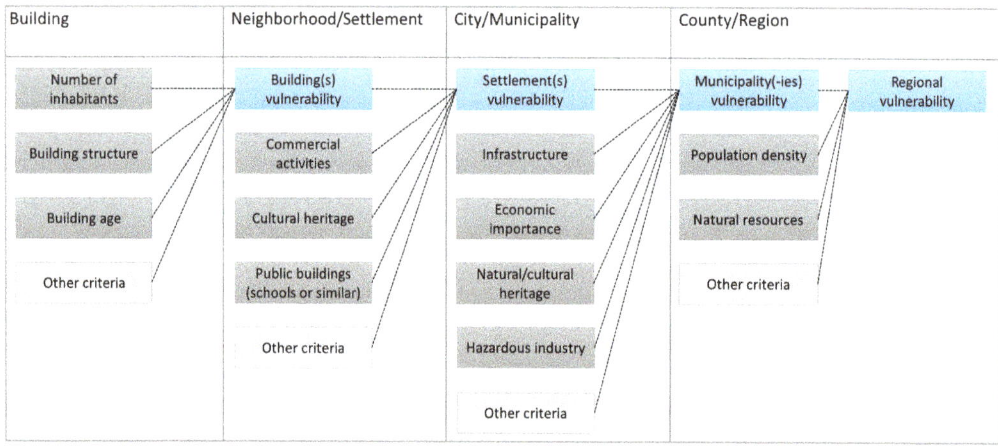

Figure 3. Vulnerability analysis at different scales with multi-criteria approach.

Regarding the risk assessment, it is usually made for a particular assessment area. These areas need to be defined by mutual spatial characteristics or by some already defined urban entity.

The first and simplest approach is to use some administrative areas as assessment areas: settlements, municipalities, provinces, counties, etc. This approach was used in the combined seismic and flood risk assessment for municipalities in the province of Ferrara [15].

The second approach is the definition of assessment areas as "working units" by using a grid of blocks. The working unit is the geographical entity in which the calculations will be computed, hereby controlling the geographical resolution of the study. The definition of the working unit depends strongly on two factors: the geographical unit in which the original data is expressed and the scale of the study. For instance, in the urban-scale seismic risk study for city Almería, a 200 m squared grid was considered appropriate to cover the entire city of Almería, totaling an amount of 400 equal cells or working units [27].

The third approach is to define assessment areas as "homogeneous zones", which are generated by intersecting relevant thematic layers in the assessment area. The intersection of the defined number of layers becomes an assessment area (zone). This approach has been used in this research for Croatian settlement Kaštel Kambelovac, a part of Kaštela City.

2.1. Multi-Criteria Analysis and Decision-Making Approach to Risk Assessment

In the analysis of natural hazards, impacts are often expressed in terms of hazard, vulnerability and exposure. A hazard (H) presents the probability that a harmful event will appear in a particular area and in a certain time interval. Vulnerability (V) is defined as the characteristics and circumstances of a community, system or asset that make it susceptible to the damaging effects of a hazard. Exposure (E) is the totality of people, property, systems or other elements present in hazard zones that are thereby subject to potential losses.

In this research, the hazard (H) is presented by seismic hazard maps with a Peak Ground Acceleration—PGA for the earthquake, while, in the case of sea floods and extreme sea waves, it is based on a flood depth and a wave height in inundation areas, respectively. The exposure (E) that corresponds to the measure of hazard will be presented by intensity and Area Impact (e.g., area exposed to earthquake, flood or extreme waves). Furthermore, vulnerability is evaluated by vulnerability index on multiple levels (for the particular building, for the settlement, etc.) with the use of additional criteria set which is different for each level. Therefore, in this research, function f represents mathematical PROMETHEE (Preference Ranking Organisation METHod for Enrichment Evaluations) method [29] that will connect all criteria and assess the risk for the observed area.

Obviously, there are many other mathematical methods for multi-criteria analysis and decision-making, but some of them are better accepted and more widely used. Three of them have recently become the most popular: AHP [25], TOPSIS [26] and PROMETHEE [30,31]. There is also a need to decide which of the available methods is the most adequate for a particular problem, but very often, the outranking methods like PROMETHEE are the most suitable choice [32]. This is especially because PROMETHEE method can be simplified to be used by non-expert users [33,34].

Furthermore, using the concept of vulnerability makes it more explicit that the impacts of a hazard are also a function of the preventive and preparatory measures that are employed to reduce the risk. Depending on the particular risk analyzed, the measurement of risk can be carried out with a greater number of different variables and factors, depending inter alia on the complexity of the chain of impacts, the number of impact factors considered and the requisite level of precision. The scheme of assessment of single-hazard and multi-hazard exposure for the investigated coastal urban area is shown in Figure 4.

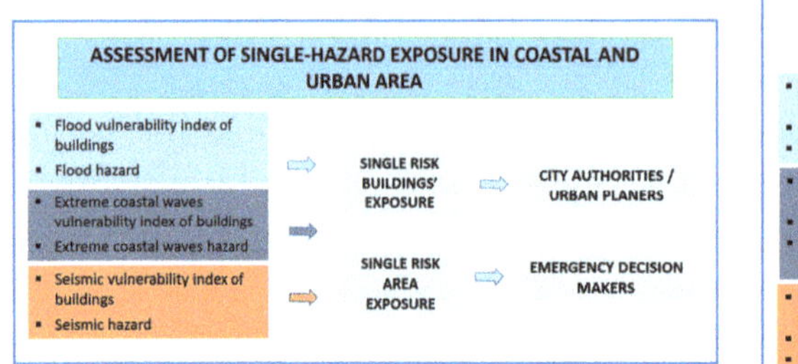

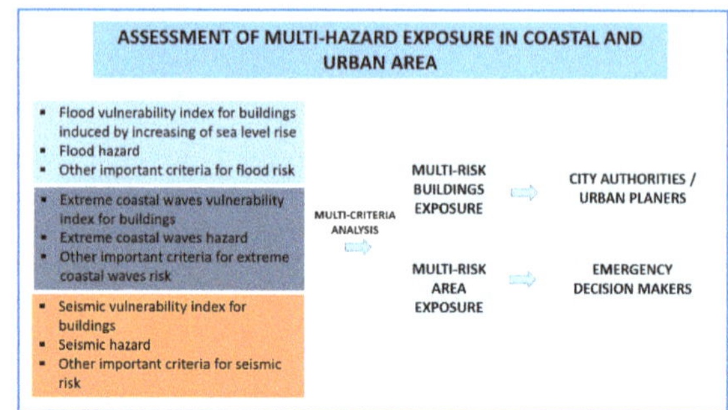

(a)   (b)

Figure 4. The scheme of the risk assessment for investigated area: (**a**) single-hazard exposure; (**b**) multi-hazard exposure.

2.2. Risk Assessment of Buildings

In this research, the risk assessment is made for different levels, starting from the lowest level (micro level), i.e., an individual building. At this micro level, the risk assessment of a single-hazard exposure is based on the calculation of vulnerability indexes of buildings for individual natural disasters:

- Flood vulnerability index for buildings,
- Extreme coastal waves vulnerability index for buildings,
- Seismic vulnerability index for buildings,

as well as on assessment of the single hazards:

- Flood hazard,
- Extreme coastal waves hazard,
- Seismic hazard.

Seismic vulnerability indexes of the buildings for investigated area are calculated according to the seismic vulnerability method [35]. The method is based on the evaluation of 11 geometrical, structural and non-structural vulnerability parameters of the building. They consider the influence of the type and quality of the structural system, the shear resistance in two horizontal directions, the position and the foundations, the properties of floors, the configuration in plan and elevation, the maximum wall spacing, the roof's typology and weight, the existence of non-structural elements, and the state of preservation. Four possibilities for each parameter were decided: from "A", indicating an optimal state, to "D", indicating a poor state. The relative importance of each parameter in the overall vulnerability is computed by using weight coefficients relating to each parameter. Finally, the vulnerability index I_v is calculated in a form $I_v = \sum_i s_{vi} w_i$, where s_{vi} is the numerical score for each class, and w_i is the weight of each parameter. The vulnerability index is normalized in a 0–100% range; a low index indicates high seismic resistance and low vulnerability, while a high vulnerability index is characteristic of the buildings with low seismic resistance and high vulnerability. Vulnerability indexes of the buildings located in the pilot area are presented in Figure 5 [35].

Figure 5. Seismic vulnerability index of buildings divided into 10% intervals.

The flood vulnerability index and the extreme coastal waves vulnerability index are calculated according to the methodology developed in the PMO-GATE project [25]. This methodology is based on the approach of Miranda and Ferreira [36], which takes into consideration different parameters such as building material, overall object condition, number of storeys, building age, importance of exposed objects and level of exposure. The approach has been modified for application in the multicriteria analysis, with each vulnerability index calculated as the weighted sum of set of parameters and evaluated through vulnerability classes (Figure 6).

(a)

Figure 6. *Cont.*

(**b**)

Figure 6. Vulnerability index of buildings for: (**a**) sea floods; (**b**) extreme sea waves.

The seismic hazard for Croatia is presented with two maps for return periods of 475 and 95 years, expressed in terms of the peak ground acceleration during an earthquake for a soil class A [37]. According to HRN EN 1998-1:2011 [36], the soil types A, B, C, D and E, may be used to account for the influence of local ground conditions on the seismic action. The site can be classified according to the value of the average shear wave velocity $v_{s,30}$. An investigation of the deep geology and characteristics of the terrains, performed at pilot area [38], has shown that shear wave velocity $v_{s,30}$ is higher than 800 m/s at the wholearea, which define soil class A. Therefore, local ground conditions do not influence to seismic hazard in the investigated area, i.e., the seismic hazard for all buildings at the pilot area has been assumed to be constant [35].

The flood hazard caused by climate-induced sea level rise is estimated according to IPCC Fifth Assessment Report (AR5) [39] and Strategy for climate change adaptation for The Republic of Croatia [40], considering changes in the mean sea level. However, the EU Flood Directive [41] requires an analysis of a high, moderate and low probability scenario in the flood hazard assessment. The sea level in the Adriatic Sea is dominantly caused by sea tides and the effect of the barometric pressure [42]. The tidal component can be represented by the set of periodic functions [43], and the residual sea level can be represented with a probability distribution due to its randomness and stochastic behavior. In order to fulfil the Flood Directive requirements, probability scenarios are estimated from the particular probability distribution, corresponding to the return periods of 25, 100 and 250 years, respectively [44]. Finally, sea level is estimated as a superposition of mean sea level, maximum estimated tide and each probabilistic scenario. The distribution of the critical zones most prone to flood due to the impacts of climate change on sea level rise for the most critical scenario for year 2100 is shown in Figure 7 [45].

The extreme coastal waves hazard is determined based on the evaluation of wave heights and their propagation toward the coast (Figure 8). The methodology for computing the wave heights by using values of wind speeds in critical wind directions for the investigated area has been developed [46], where probability distribution function is used to fit wind speed histograms and to evaluate return period values.

Figure 7. Distribution of critical zones most prone to flood due to impact of climate changes on sea level rise: scenario for year 2100.

Figure 8. Distribution of critical zones most prone to extreme sea waves exposure.

Vulnerability indexes of buildings for individual natural disasters coupled by the corresponding hazards are the basis for the multi-hazard risk assessment of the area [47].

2.3. Risk Assessment of Homogenous Zones

At the higher level of analysis, which is the intermediate level, the previously-analyzed objects are grouped into spatial units (assessment area) that are called "homogeneous zones" [47]. The process of creation of homogenous zones for the pilot site is presented in Figure 9. In this case, three different layers are intersected: a layer of specific urban characteristics, a layer of areas surrounded by the main roads and a layer of terrain height.

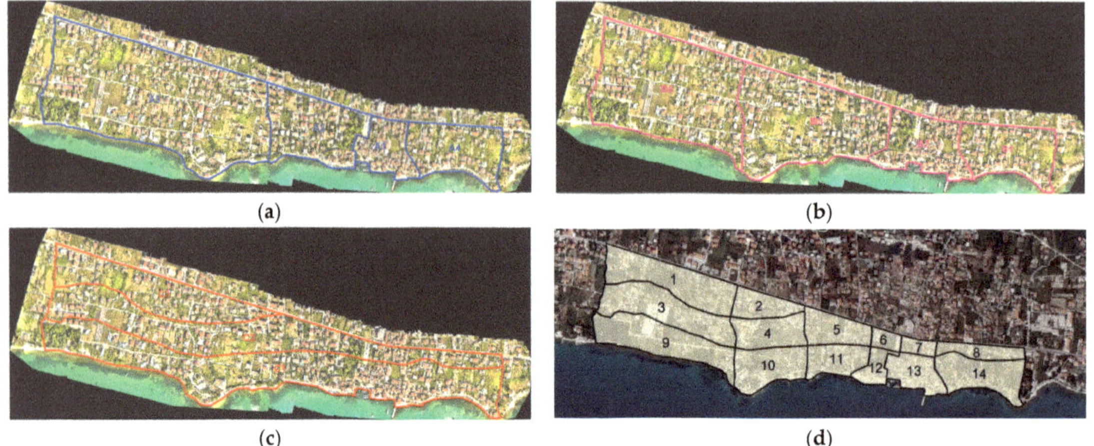

Figure 9. The creation of homogenous zones: (**a**) four areas defined by specific urban characteristics; (**b**) four areas defined by main roads; (**c**) three areas defined by terrain (contours 5 and 10 m); (**d**) intersection of layers resulted with 15 homogenous zones.

The seismic vulnerability indexes of individual buildings are used to calculate the seismic vulnerability index of a homogeneous zone. In the same way, the other vulnerabilities indexes are evaluated. However, it is important to highlight that the additional criteria are very important at this level of analysis. Since homogeneous zones have complex characteristics, additional criteria are used for both single-hazard and a multi-hazard approach, beside vulnerability and hazard.

The literature review shows that the level of risk to the community depends on a number of other parameters whose activation in a particular hazard reduces the resilience to extraordinary events. Each area due to difference in size requires a special approach in identifying the relevant parameters. For example, the most common parameters (criteria) for seismic hazard are grouped into area characteristics (geology, soil, slope, historical earthquake events, fault line, etc.), the characteristics of human intervention in space (land use, built communal infrastructure and roads, etc.), and social characteristics (housing density, social purpose of buildings, social structure, etc.).

For the pilot site at the intermediate level, additional parameters that can be quantified are detected, different for each homogeneous zone. They represent additional criteria for the risk assessment of homogenous zones: communal infrastructure, road network, construction density (distance between buildings), inhabitation density, importance factor (public building, school, etc.), and historical buildings.

3. Results

3.1. Single-Hazard Risk Asssessment of Homogenous Zones

Single-hazard risk assessment is made for 14 homogeneous zones of the pilot site. Since the flood and extreme waves are affecting only a small coastal area, the seismic risk assessment will be presented here. The input data are calculated, or expert estimates are given for the following criteria:

- Seismic hazard—PGA,
- Buildings' seismic vulnerability,
- Geology,
- Communal infrastructure—electricity supply,
- Communal infrastructure—water supply and drainage,
- Road network,
- Construction density (distance between buildings),

- Inhabitation density,
- Importance factor (public, school, etc.),
- Historical buildings.

Since seismic hazard (PGA) and geology have the same values in each of the 14 homogeneous zones, they do not need be included in the numerical processing. The average seismic vulnerability has been calculated for each homogenous zone (Table 2).

Table 2. The basic data of homogenous zones and average seismic vulnerability index.

Homogenous Zone (HZ)	Area (m^2)	Number of Buildings	Seismic Vulnerability Index of Homogenous Zone
HZ1	58.627	56	0.133
HZ2	21.865	29	0.174
HZ3	57.189	54	0.116
HZ4	30.925	25	0.120
HZ5	26.972	38	0.132
HZ6	7.763	4	0.194
HZ7	7.767	20	0.435
HZ8	16.168	19	0.162
HZ9	60.068	38	0.136
HZ10	38.133	14	0.171
HZ11	24.972	35	0.156
HZ12	12.696	17	0.448
HZ13	24.903	71	0.493
HZ14	40.782	48	0.187

The criterion inhabitation density is generated from a digitized population census. For all other additional criteria a profound GIS analysis has been made, and criteria evaluations for each homogenous zone have been calculated or estimated (Figure 10).

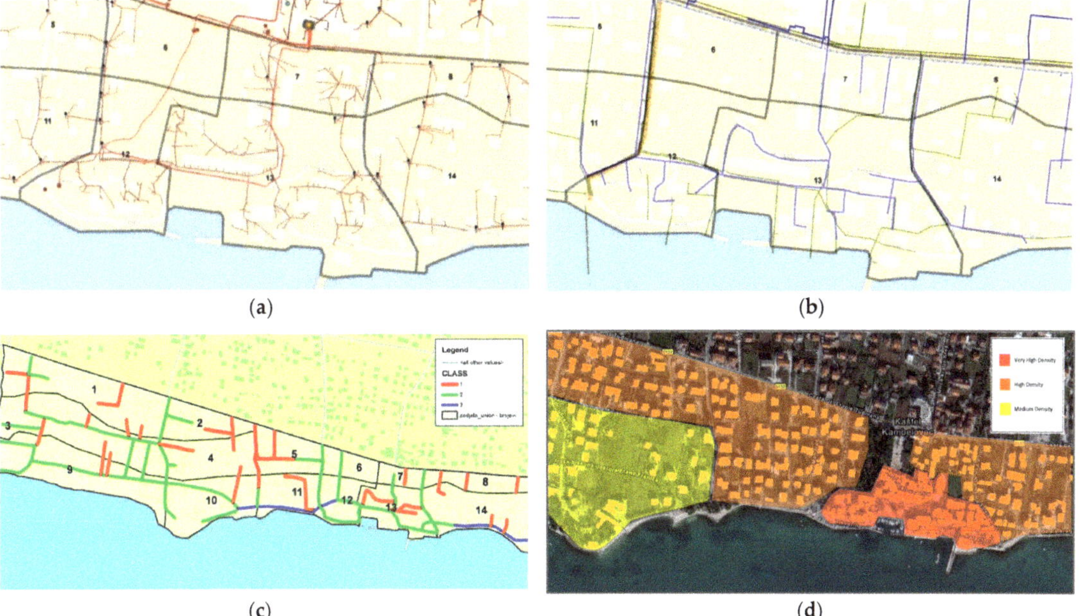

(a)

(b)

(c)

(d)

Figure 10. *Cont.*

(e)   (f)

Figure 10. GIS analysis of additional criteria: (**a**) electricity supply; (**b**) water supply and drainage; (**c**) road network; (**d**) construction density; (**e**) importance factor (public, school, etc.); (**f**) historical buildings.

All the above-mentioned data have been collected into the decision matrix to be used by the PROMETHEE method with the help of Visual PROMETHEE software (Figure 11).

Scenario1	criterion1	criterion2	criterion3	criterion4	criterion5	criterion6	criterion7	criterion8
Unit	unit	unit	unit	unit	unit	unit	unit	unit
Cluster/Group								
Preferences								
Min/Max	max	max	max	min	max	max	max	max
Weight	40,00	10,00	8,00	8,00	14,00	10,00	4,00	6,00
Preference Fn.	Linear	Linear	Linear	Linear	Linear	Linear	Linear	Linear
Thresholds	absolute	absolute	absolute	absolute	absolute	absolute	absolute	absolute
- Q: Indifference	1,00	1,00	1,00	10,00	1,00	1,00	1,00	1,00
- P: Preference	100,00	20,00	20,00	1000,00	10,00	500,00	10,00	20,00
- S: Gaussian	n/a	n/a	n/a	n/a	n/a	n/a	n/a	n/a
Evaluations								
HZ1	13,30	11,00	8,00	-216,00	0,00	252,00	0,00	0,00
HZ2	17,40	13,00	7,00	80,00	0,00	130,00	0,00	0,00
HZ3	11,60	11,00	8,00	492,00	0,00	243,00	0,00	0,00
HZ4	12,00	9,00	11,00	-80,00	0,00	112,00	0,00	0,00
HZ5	13,20	8,00	9,00	-121,00	0,00	171,00	0,00	0,00
HZ6	19,40	2,00	2,00	60,00	0,00	18,00	0,00	0,00
HZ7	43,50	5,00	4,00	39,00	0,00	90,00	0,00	0,00
HZ8	16,20	7,00	9,00	-137,00	0,00	86,00	0,00	0,00
HZ9	13,60	9,00	7,00	633,00	0,00	171,00	0,00	0,00
HZ10	17,10	7,00	12,00	463,00	0,00	63,00	5,00	0,00
HZ11	15,60	10,00	13,00	283,00	0,00	158,00	2,00	0,00
HZ12	44,80	9,00	15,00	325,00	2,00	76,00	0,00	4,00
HZ13	49,30	15,00	18,00	211,00	9,00	319,00	7,00	15,00
HZ14	18,70	9,00	12,00	317,00	0,00	216,00	3,00	0,00

Figure 11. Homogenous zones input data for PROMETHEE method (decision matrix).

The preliminary results of the PROMETHEE method are given in Figure 12, in which better rank represents higher risk. It means that the best-ranked homogenous zone HZ13 has the highest seismic risk in this case.

Rank	action		Phi	Phi+	Phi-
1	HZ13		0,3993	0,4104	0,0111
2	HZ12		0,0975	0,1512	0,0537
3	HZ7		0,0302	0,1110	0,0808
4	HZ1		0,0025	0,0684	0,0659
5	HZ14		-0,0142	0,0518	0,0659
6	HZ2		-0,0269	0,0451	0,0720
7	HZ11		-0,0317	0,0395	0,0713
8	HZ5		-0,0318	0,0418	0,0735
9	HZ4		-0,0401	0,0387	0,0787
10	HZ8		-0,0420	0,0375	0,0795
11	HZ10		-0,0644	0,0349	0,0993
12	HZ3		-0,0654	0,0347	0,1000
13	HZ9		-0,0975	0,0166	0,1141
14	HZ6		-0,1156	0,0250	0,1407

Figure 12. Results of the PROMETHEE method for 14 homogeneous zones (better rank represents higher risk).

Visual representation of the results has been made in GIS, where green represents low risk and red represents high risk (Figure 13).

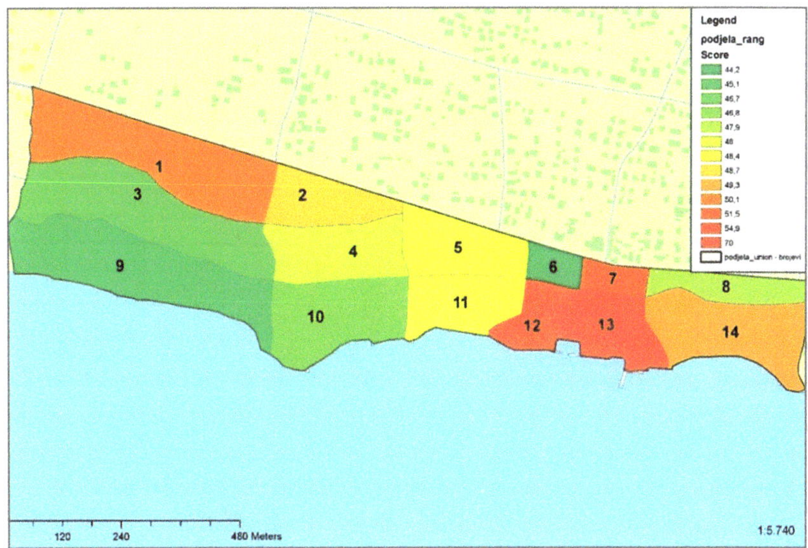

Figure 13. Seismic risk assessment for homogenous zones (green represents low risk, red represents high risk).

3.2. Multi-Hazard Risk Asssessment of Homogenous Zones

Three natural-hazards—seismic, flood and extreme waves—are combined and evaluated together to assess the multi-hazard risk, and the analysis is made on the level of homogenous zones.

The two combined risk analysis are made: combination of two risks for seismic and flood hazard; and combination of three risks for seismic, flood and extreme waves hazard. Each analysis is made on three levels. The first level of analysis is based on hazard and vulnerability data aggregated for each homogenous zone. A second and third level of analysis are using additional criteria for each homogenous zone (Table 3).

Table 3. Combined risks analysis and criteria for each level of analysis.

Level	Combined Seismic-Flood Risk: Scenario S-F	Combined Seismic-Flood-Extreme Waves Risk: Scenario S-F-EW
Level 1 criteria	Seismic hazard (1.1) Seismic vulnerability (1.2) Flood hazard (1.3) Flood vulnerability (1.4)	Seismic hazard (1.1) Seismic vulnerability (1.2) Flood hazard (1.3) Flood vulnerability (1.4) Extreme waves hazard (1.5) Extreme waves vulnerability (1.6)
Level 2 criteria	Seismic hazard (1.1) Seismic vulnerability (1.2) Flood hazard (1.3) Flood vulnerability (1.4) Construction density (2.1) Inhabitation density (2.2) Importance factor (2.3) Historical buildings (2.4)	Seismic hazard (1.1) Seismic vulnerability (1.2) Flood hazard (1.3) Flood vulnerability (1.4) Extreme waves hazard (1.5) Extreme waves vulnerability (1.6) Construction density (2.1) Inhabitation density (2.2) Importance factor (2.3) Historical buildings (2.4)
Level 3 criteria	Seismic hazard (1.1) Seismic vulnerability (1.2) Flood hazard (1.3) Flood vulnerability (1.4) Construction density (2.1) Inhabitation density (2.2) Importance factor (2.3) Historical buildings (2.4) Electrical infrastructure (3.1) Water supply infrastructure (3.2) Road network (3.3)	Seismic hazard (1.1) Seismic vulnerability (1.2) Flood hazard (1.3) Flood vulnerability (1.4) Extreme waves hazard (1.5) Extreme waves vulnerability (1.6) Construction density (2.1) Inhabitation density (2.2) Importance factor (2.3) Historical buildings (2.4) Electrical infrastructure (3.1) Water supply infrastructure (3.2) Road network (3.3)

Therefore, six multicriteria analyses are made on three different levels for each of the two scenarios. These multi-criteria analyses are classifying homogenous zones in accordance with multi-hazard risk.

The first analysis is the combined seismic-flood risk (Scenario S-F) on three different levels. Each level represents different criteria sets. Criteria are grouped in two major groups: main criteria, which are related to hazard and vulnerability and additional criteria, which are related to some important spatial data. Each criteria group has its own weight. In this case, an equal weight is given to each group—50%. An example of distribution of the criteria weights within the group are presented in Table 4. Criteria weights in this particular application are estimated comparing the estimated Expected Annual Damage values for each observed natural hazard. The Expected Annual Damage concept is based on the combination of occurrence probability and corresponding damage caused by each natural hazard [48], and it has proved to be an effective method since it enables a practical comparison of significantly different natural phenomena.

Table 4. An example of criteria weights for combined seismic-flood risk for Scenario S-F Level 2.

Criteria Group	Group Weight	Criteria	Criteria Weight
Main criteria	50%	Seismic hazard (1.1)	21.7%
		Seismic vulnerability (1.2)	21.7%
		Flood hazard (1.3)	3.3%
		Flood vulnerability (1.4)	3.3%
Additional criteria (*n—number of additional criteria*)	50%	Construction density (2.1)	$50/n = 12.5\%$
		Inhabitation density (2.2)	$50/n = 12.5\%$
		Importance factor (2.3)	$50/n = 12.5\%$
		Historical buildings (2.4)	$50/n = 12.5\%$

The input data for analysis is presented as a matrix with alternatives, in this case 14 homogenous zones (HZ) and up to 11 criteria depending on the level of analysis (Figure 14).

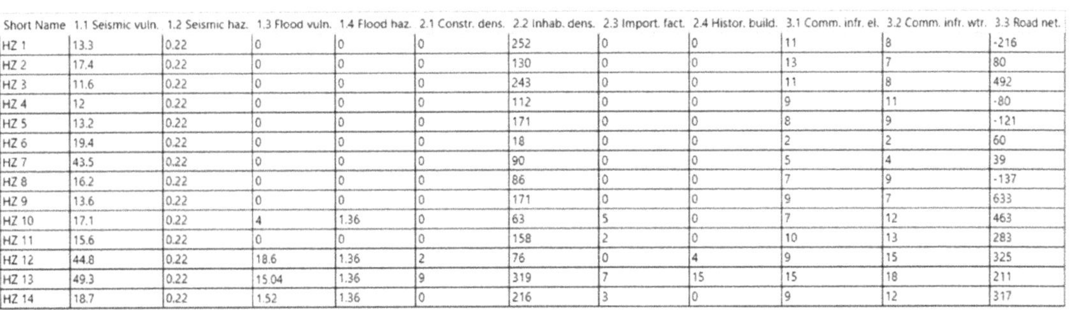

Short Name	1.1 Seismic vuln.	1.2 Seismic haz.	1.3 Flood vuln.	1.4 Flood haz.	2.1 Constr. dens.	2.2 Inhab. dens.	2.3 Import. fact.	2.4 Histor. build.	3.1 Comm. infr. el.	3.2 Comm. infr. wtr.	3.3 Road net.
HZ 1	13.3	0.22	0	0	0	252	0	0	11	8	-216
HZ 2	17.4	0.22	0	0	0	130	0	0	13	7	80
HZ 3	11.6	0.22	0	0	0	243	0	0	11	8	492
HZ 4	12	0.22	0	0	0	112	0	0	9	11	-80
HZ 5	13.2	0.22	0	0	0	171	0	0	8	9	-121
HZ 6	19.4	0.22	0	0	0	18	0	0	2	2	60
HZ 7	43.5	0.22	0	0	0	90	0	0	5	4	39
HZ 8	16.2	0.22	0	0	0	86	0	0	7	9	-137
HZ 9	13.6	0.22	0	0	0	171	0	0	9	7	633
HZ 10	17.1	0.22	4	1.36	0	63	5	0	7	12	463
HZ 11	15.6	0.22	0	0	0	158	2	0	10	13	283
HZ 12	44.8	0.22	18.6	1.36	2	76	0	4	9	15	325
HZ 13	49.3	0.22	15.04	1.36	9	319	7	15	15	18	211
HZ 14	18.7	0.22	1.52	1.36	0	216	3	0	9	12	317

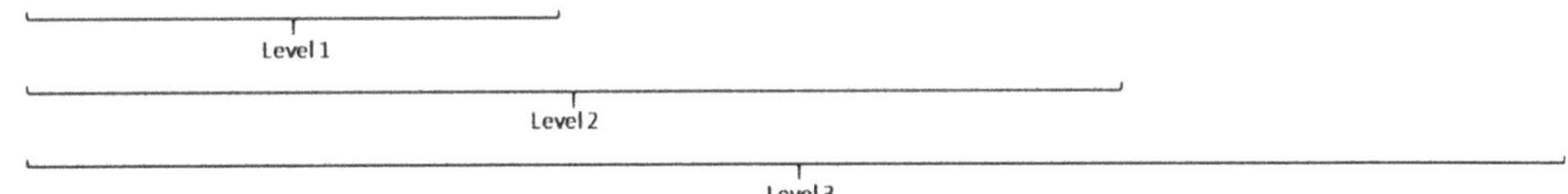

Figure 14. Input matrix for combined seismic-flood risk (Scenario S-F) multi-criteria analysis with criteria evaluation for all three levels.

The input data (Figure 14) and criteria weights (Table 4) are imported in a multi-criteria analysis application based on the PROMETHEE method and results have been calculated for all three levels of analysis. The first analysis was made for the criteria set defined as Level 1, a second analysis for Level 2 and a third for Level 3. The criteria for each analysis were submitted to PROMETHEE method and the results for all three levels are presented in Figures 15–17, respectively. There are no significant variations in results except zone HZ 13, which becomes more exposed when additional criteria are used (Level 1 and 2). At the end, the results are exported into GIS for better visualization and a further analysis of results (Figure 18).

The second analysis is a combined seismic–flood–extreme waves risk (Scenario S-F-EW) on three different levels. Again, each level represents different criteria sets, and criteria are grouped into two groups: main criteria, which are related to hazard and additional criteria, which are related to some important spatial data. Each criteria group has its own weight. In this case, an equal weight is given to each group: 50%. Other criteria weights are presented in Table 5. The input data for analysis are presented as a matrix with alternatives, in this case 14 homogenous zones (HZ) and up to 13 criteria depending on the level of analysis (Figure 19).

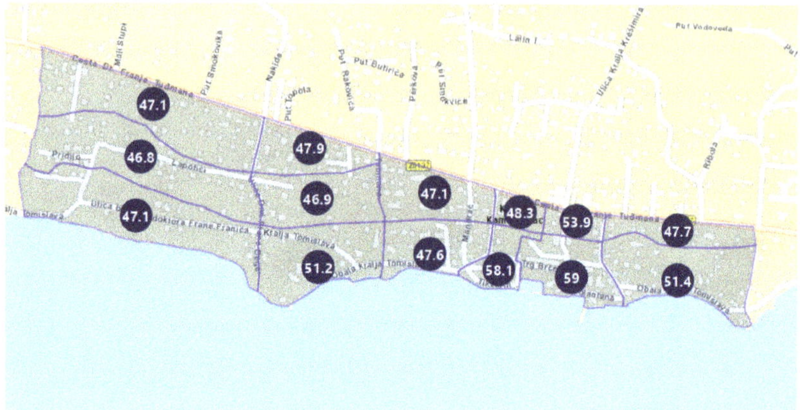

Figure 15. Results of risk analysis for seismic and flood hazard for Level 1 (Scenario S-F L1).

Figure 16. Results of risk analysis for seismic and flood hazard for Level 2 (Scenario S-F L2).

Figure 17. Results of risk analysis for seismic and flood hazard for Level 3 (Scenario S-F L3).

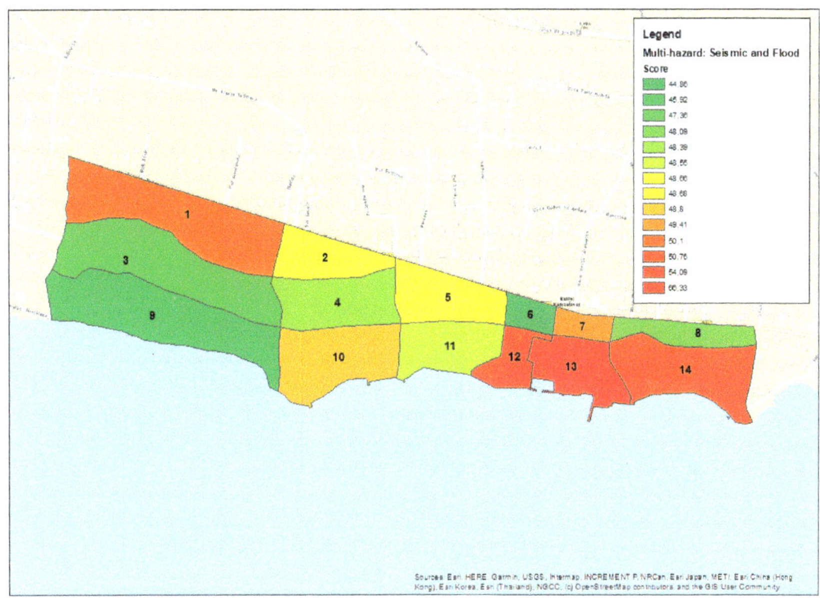

Figure 18. GIS visualization of risk for seismic and flood hazard for Level 3 (Scenario S-F L3).

Table 5. An example of criteria weights for combined seismic–flood–extreme waves risk for Scenario S-F-EW Level 2.

Criteria Group	Group Weight	Criteria	Criteria Weight
Main criteria	50%	Seismic hazard (1.1)	19.5%
		Seismic vulnerability (1.2)	19.5%
		Flood hazard (1.3)	2.5%
		Flood vulnerability (1.4)	2.5%
		Extreme waves hazard (1.5)	3.0%
		Extreme waves vulnerability (1.6)	3.0%
Additional criteria *(n—number of additional criteria)*	50%	Construction density (2.1)	$50/n = 12.5\%$
		Inhabitation density (2.2)	$50/n = 12.5\%$
		Importance factor (2.3)	$50/n = 12.5\%$
		Historical buildings (2.4)	$50/n = 12.5\%$

The input data (Figure 19) and criteria weights (Table 5) are imported in a multi-criteria analysis application based on PROMETHEE method and results have been calculated for all three levels of analysis. The first analysis was made for a criteria set defined as Level 1, a second analysis for Level 2 and a third for Level 3. The criteria for each analysis were submitted to PROMETHEE method and the results for all 3 levels are presented in Figures 20–22, respectively. Again, there are no significant variations in results except zone HZ 13, which becomes more exposed when additional criteria are used (Level 1 and 2). At the end, the results are exported into GIS for better visualization and a further analysis of results (Figure 23).

Short Name	1.1 Seismic vuln.	1.2 Seismic haz.	1.3 Flood vuln.	1.4 Flood haz.	1.5 Extr. wav. vuln.	1.6 Extr. wav. haz.	2.1 Constr. dens.	2.2 Inhab. dens.	2.3 Import. fact.	2.4 Histor. build.	3.1 Comm. infr. el.	3.2 Comm. infr. wtr.	3.3 Road net.
HZ 1	13.3	0.22	0	0	0	0	0	252	0	0	11	8	-216
HZ 2	17.4	0.22	0	0	0	0	0	130	0	0	13	7	80
HZ 3	11.6	0.22	0	0	0	0	0	243	0	0	11	8	492
HZ 4	12	0.22	0	0	0	0	0	112	0	0	9	11	-80
HZ 5	13.2	0.22	0	0	0	0	0	171	0	0	8	9	-121
HZ 6	19.4	0.22	0	0	0	0	0	18	0	0	2	2	60
HZ 7	43.5	0.22	0	0	0	0	0	90	0	0	5	4	39
HZ 8	16.2	0.22	0	0	0	0	0	86	0	0	7	9	-137
HZ 9	13.6	0.22	0	0	0	0	0	171	0	0	9	7	633
HZ 10	17.1	0.22	4	1.36	0	0	0	63	5	0	7	12	463
HZ 11	15.6	0.22	0	0	0	0	0	158	2	0	10	13	283
HZ 12	44.8	0.22	18.6	1.36	5.14	1.639	2	76	0	4	9	15	325
HZ 13	49.3	0.22	15.04	1.36	14.82	1.539	9	319	7	15	15	18	211
HZ 14	18.7	0.22	1.52	1.36	1.52	1.359	0	216	3	0	9	12	317

Level 1

Level 2

Level 3

Figure 19. Input matrix for combined seismic–flood–extreme waves risk (Scenario S-F-EW) multi-criteria analysis with criteria evaluation for all three levels.

Figure 20. Results of the risk analysis for seismic, flood and extreme waves hazards for Level 1 (Scenario S-F-EW L1).

Figure 21. Results of the risk analysis for seismic, flood and extreme waves hazards for Level 2 (Scenario S-F-EW L2).

Figure 22. Results of the combined risk analysis for seismic, flood and extreme waves hazards for Level 3 (Scenario S-F-EW L3).

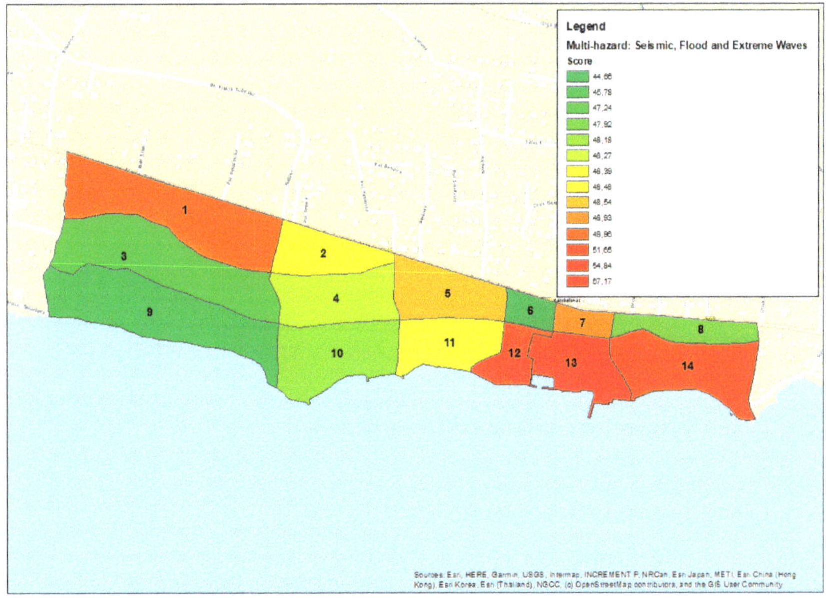

Figure 23. GIS visualization of risk for seismic, flood and extreme waves hazard for Level 3 (Scenario S-F-EW L3).

4. Conclusions

The multi-hazard risk assessment of urban areas is an important step in the risk management process. It can be used to reduce, manage and overcome the risks arising from the combination of different multiple hazards. This paper uses Spatial Multi-Criteria Decision-Making based on PROMETHEE method, coupled with the Geographic Information System, to assess the single-hazard and multi-hazard risk caused by seismic, sea floods and extreme sea waves. A case study for the application of the method is Kaštel Kambelovac, an urban settlement placed at the Croatian part of the Adriatic coast. The observed area has been divided into the homogenous zones that have been identified as areas of the test site with some mutual spatial characteristics. The homogeneity was identified by intersecting spatial layers in the GIS. The multi-hazard risk assessment method is based on the validation of the main components of the risk caused by natural hazard phenomena, such us vulnerability expressed in terms of the vulnerability index, hazard and influence of additional criteria, to overall risk in homogenous zones. The main specificity of this research is multi-hazard risk evaluation based on a previous detailed calculation of vulnerability indexes of each building in the test area for all observed threats.

The methodology ranks the various homogenous zones in terms of the relative proneness to coupled seismic, sea floods and extreme sea waves hazards. It enables a risk analysis for different scenarios for both single and multiple hazards at the level of buildings and homogenous zones. In the case study presented here, the analysis has shown that the seismic risk is a dominant threat in all scenarios. The results of a multi-hazard analysis for the combination of seismic and sea floods hazard have shown that the area with the highest risk is related to the historical part of Kaštel Kambelovac. This is due to the fact that this area, along with high seismic risk, has the highest level of exposure to flooding. Furthermore, the vulnerability of exposed objects in the historical part is highest for both hazards. Likewise, the results of the multi-hazard analysis for the combination of seismic, sea floods and extreme waves hazard are analogous to previous ones, showing that the highest risk is again in the historical part of Kaštel Kambelovac. The vulnerability of exposed objects to extreme waves is the highest in the historical part, and this particular area is most exposed to extreme waves due its low-lying coast. Performed analyses provide useful information for decision makers and public authorities to define priorities in future interventions through the process of risk management planning.

Author Contributions: Conceptualization, Ž.N. and N.M.; data curation, M.M., N.M. and T.K.; funding acquisition, Ž.N. and E.B.; investigation, Ž.N., N.M., M.M., E.B. and T.K.; methodology, N.M. and M.M. and Ž.N.; software, N.M. and M.M.; supervision, Ž.N. and E.B.; validation, Ž.N., N.M., M.M., E.B. and T.K.; visualization, M.M. and T.K.; writing—original draft, Ž.N., N.M., M.M., E.B. and T.K. All authors have read and agreed to the published version of the manuscript.

Funding: This research was funded by the European Union, Programme Interreg Italy–Croatia, Project "Preventing, managing and overcoming natural-hazards risks to mitigate economic and social impact"—PMO-GATE ID 10046122. The research is also partially supported through project KK.01.1.1.02.0027, co-financed by the Croatian Government and the European Union through the European Regional Development Fund—the Competitiveness and Cohesion Operational Programme.

Institutional Review Board Statement: Not applicable.

Informed Consent Statement: Not applicable.

Data Availability Statement: Some of the datasets that were analyzed in this research are publicly available on the website of PMO-GATE project: https://www.italy-croatia.eu/web/pmo-gate (accessed on 25 October 2022).

Acknowledgments: This research was part of the Work Packages 4 and 5 of the European Union, Programme Interreg Italy–Croatia, Project "Preventing, managing and overcoming natural-hazards risks to mitigate economic and social impact"—PMO-GATE ID 10046122. The results of this research were published in the Deliverables: "4.1.1. Methodology for provision assessment indexes based on Spatial Multi-Criteria Decision-Making—Croatia" and "5.1.6. Map of spatial distribution of the critical zones most prone to flood, extreme waves and seismic risks for HR test site".

Conflicts of Interest: The authors declare no conflict of interest.

References

1.   Zerger, A. Examining GIS decision utility for natural hazard risk modelling. *Environ. Model. Softw.* **2002**, *17*, 287–294. [CrossRef]
2.   Yum, S.-G.; Son, K.; Son, S.; Kim, J.-M. Identifying Risk Indicators for Natural Hazard-Related Power Outages as a Component of Risk Assessment: An Analysis Using Power Outage Data from Hurricane Irma. *Sustainability* **2020**, *12*, 7702. [CrossRef]
3.   Chen, P. On the Diversity-Based Weighting Method for Risk Assessment and Decision-Making about Natural Hazards. *Entropy* **2019**, *21*, 269. [CrossRef] [PubMed]
4.   Kappes, M.S.; Keiler, M.; Von Elverfeldt, K.; Glade, T. Challenges of analyzing multi-hazard risk: A review. *Nat. Hazards* **2012**, *64*, 1925–1958. [CrossRef]

5. Greiving, S.; Fleischhauer, M.; Luckenkotter, J. A methodology for an integrated risk assessment of spatially relevant hazards. *J. Environ. Plan. Manag.* **2006**, *49*, 1–19. [CrossRef]

6. Di Mauro, C.; Bouchon, S.M.; Carpignano, A.; Golia, E.; Peressin, S. Definition of Multi-Risk Maps at Regional Level as Management Tool: Experience Gained by Civil Protection Authorities of Piemonte Region. In *Atti del 5° Convegno sulla Valutazione e Gestione del Rischio negli Insediamenti Civili ed Industriali*; University of Pisa: Pisa, Italy, 2006.

7. Fleischhauer, M.; Greiving, S.; Schlusemann, B.; Schmidt-Thomé, P.; Kallio, H.; Tarvainen, T.; Jarva, J. Multi-risk assessment of spatially relevant hazards in Europe. In Proceedings of the ESPON, ESMG Symposium, Nürnberg, Germany, 11–13 October 2005.

8. Kunz, M.; Hurni, L. Hazard maps in Switzerland: State-of-the-art and potential improvements. In Proceedings of the 6th ICA Mountain Cartography Workshop, Lenk, Switzerland, 11–15 February 2008; ICA: Lenk, Switzerland, 2008.

9. Liu, B.; Siu, Y.L.; Mitchell, G. A quantitative model for estimating risk from multiple interacting natural hazards: An application to northeast Zhejiang, China. *Stoch. Environ. Res. Risk Assess.* **2017**, *31*, 1319–1340. [CrossRef]

10. Mignan, A.; Wiemer, S.; Giardini, D. The quantification of low-probability–high-consequences events: Part I. A generic multi-risk approach. *Naural. Hazards* **2014**, *73*, 1999–2022. [CrossRef]

11. Munich Re Group. *Topics—Annual Review: Natural Catastrophes 2002*; Munich Re Group: Munich, Germany, 2003.

12. Odeh Engineers, Inc. Statewide Hazard Risk and Vulnerability Assessment for the State of Rhode Island. Tech. Rep., NOAA Coastal Services Center. Available online: http://www.csc.noaa.gov/rihazard/pdfs/rhdisl_hazard_report.pdf (accessed on 23 October 2019).

13. Van Westen, C.J. Multi-hazard risk assessment and decision making. *Environ. Hazards Methodol. Risk Assess. Manag.* **2017**, 31–94. [CrossRef]

14. *IEC 31010:2009*; Risk Management—Risk Assessment Techniques. International Organization for Standardization ISO: Geneva, Switzerland, 2009.

15. Soldati, A.; Chiozzi, A.; Nikolić, Ž.; Vaccaro, C.; Benvenuti, E. A PROMETHEE Multiple-Criteria Approach to Combined Seismic and Flood Risk Assessment at the Regional Scale. *Appl. Sci.* **2022**, *12*, 1527. [CrossRef]

16. Rocchi, A.; Chiozzi, A.; Nale, M.; Nikolic, Z.; Riguzzi, F.; Mantovan, L.; Gilli, A.; Benvenuti, E. A Machine Learning Framework for Multi-Hazard Risk Assessment at the Regional Scale in Earthquake and Flood-Prone Areas. *Appl. Sci.* **2022**, *12*, 583. [CrossRef]

17. Li, Z.; Song, K.; Peng, L. Flood Risk Assessment under Land Use and Climate Change in Wuhan City of the Yangtze River Basin, China. *Land* **2021**, *10*, 878. [CrossRef]

18. Nikolić, Ž.; Runjić, L.; Ostojić Škomrlj, N.; Benvenuti, E. Seismic Vulnerability Assessment of Historical Masonry Buildings in Croatian Coastal Area. *Appl. Sci.* **2021**, *11*, 5997. [CrossRef]

19. Rashed, T.; Weeks, J. Assessing vulnerability to earthquake hazards through spatial multicriteria analysis of urban areas. *Int. J. Geogr. Inf. Sci.* **2003**, *17*, 547–576. [CrossRef]

20. Vicente, R.; Parodi, S.; Lagomarsino, S.; Varum, H.; Silva, J.A.R.M. Seismic vulnerability and risk assessment: Case study of the historic city centre of Coimbra, Portugal. *Bull. Earthq. Eng.* **2011**, *9*, 1067–1096. [CrossRef]

21. Marinas.com. Kastel Kambelovac Harbour. Available online: https://marinas.com/view/marina/eyc39lv_Kastel_Kambelovac_Harbour_Kastel_Gomilica_Croatia (accessed on 26 October 2022).

22. Portal Grada Kaštela. Novosti. Available online: https://www.kastela.org/novosti/aktualnosti/43639-zbog-podizanja-mora-u-vitturiju-pojedinci-odustali-od-glasanja-policija-zatvorila-promet-rivom-u-sucurcu (accessed on 26 October 2022).

23. Jutarnji List. Vijesti. Available online: https://www.jutarnji.hr/vijesti/hrvatska/orkansko-jugo-poplavilo-kastel-stafilic-9607921 (accessed on 26 October 2022).

24. EU Interreg Italy-Croatia, PMO-GATE Project. Preventing, Managing and Overcoming Natural-Hazards Risks to Mitigate Economic and Social Impact (PMO-GATE). Available online: https://www.italy-croatia.eu/web/pmo-gate/site (accessed on 3 October 2022).

25. Palchaudhuri, M.; Biswas, S. Application of AHP with GIS in drought risk assessment for Puruliya district, India. *Nat. Hazards* **2016**, *84*, 1905–1920. [CrossRef]

26. Nyimbili, P.H.; Erden, T.; Karaman, H. Integration of GIS, AHP and TOPSIS for earthquake hazard analysis. *Nat. Hazards* **2018**, *92*, 1523–1546. [CrossRef]

27. Rivas-Medina, A.; Gaspar-Escribano, J.M.; Benito, B.; Bernabé, M.A. The role of GIS in urban seismic risk studies: Application to the city of Almería (southern Spain). *Nat. Hazards Earth Syst. Sci.* **2013**, *13*, 2717–2725. [CrossRef]

28. Aubrecht, C.; Ozceylan, D.; Steinnocher, K.; Freire, S. Multi-level geospatial modeling of human exposure patterns and vulnerability indicators. *Nat. Hazards* **2013**, *68*, 147–163. [CrossRef]

29. Brans, J.P.; Vincke, P.; Mareschal, B. How to select and how to rank projects: The Promethee method. *Eur. J. Oper. Res.* **1986**, *24*, 228–238. [CrossRef]

30. Mladineo, N.; Margeta, J.; Brans, J.P.; Mareschal, B. Multicriteria ranking of alternative locations for small scale hydro plants. *Eur. J. Oper. Res.* **1987**, *31*, 215–222. [CrossRef]

31. Mladineo, N.; Lozic, I.; Stosic, S.; Mlinaric, D.; Radica, T. An evaluation of multicriteria analysis for DSS in public policy decision. *Eur. J. Oper. Res.* **1992**, *61*, 219–229. [CrossRef]

32. Nemery, P. On the Use of Multicriteria Ranking Methods in Sorting Problems. Ph.D. Thesis, Université Libre de Bruxelles, Brussels, Belgium, 2008.

33. Mladineo, N.; Mladineo, M.; Knezic, S. Web MCA-based decision support system for incident situations in maritime traffic: Case study of Adriatic Sea. *J. Navig.* **2017**, *70*, 1312. [CrossRef]
34. Mladineo, M.; Mladineo, N.; Jajac, N. Project Management in mine actions using Multi- Criteria-Analysis-based decision support system. *Croat. Oper. Res. Rev.* **2014**, *5*, 415–425. [CrossRef]
35. Nikolić, Ž.; Benvenuti, E.; Runjić, L. Seismic Risk Assessment of Urban Areas by a Hybrid Empirical-Analytical Procedure Based on Peak Ground Acceleration. *Appl. Sci.* **2022**, *12*, 3585. [CrossRef]
36. *HRN EN 1998-1:2011*; Eurocode 8: Design of Structures for Earthquake Resistance. Part 1: General Rules, Seismic Actions and Rules for Buildings. Croatian Standards Institute: Zagreb, Croatia, 2011.
37. Miranda, F.N.; Ferreira, T.M. A simplified approach for flood vulnerability assessment of historic sites. *Nat. Hazards* **2019**, *96*, 713–730. [CrossRef]
38. Da Col, F.; Accaino, F.; Bohm, G.; Meneghini, F. Characterization of shallow sediments by processing of P, SH and SV wave-fields in Kaštela (HR). *Eng. Geol.* **2021**, *293*, 106336. [CrossRef]
39. Intergovernmental Panel on Climate Change (IPCC) Fifth Assessment Report (AR5). Available online: https://www.ipcc.ch/assessment-report/ar5/ (accessed on 26 October 2022).
40. Strategy for Climate Change Adaptation for Republic of Croatia (Official Gazette NN 46/2020; in Croatian). Available online: https://narodne-novine.nn.hr/clanci/sluzbeni/2020_04_46_921.html (accessed on 26 October 2022).
41. Directive 2007/60/EC of the European Parliament and of the Council of 23 October 2007 on the Assessment and Management of Flood Risks. Available online: https://eur-lex.europa.eu/legal-content/EN/TXT/PDF/?uri=CELEX:32007L0060&from=EN (accessed on 26 October 2022).
42. Srzić, V.; Lovrinović, I.; Racetin, I.; Pletikosić, F. Hydrogeological Characterization of Coastal Aquifer on the Basis of Observed Sea Level and Groundwater Level Fluctuations: Neretva Valley Aquifer, Croatia. *Water* **2020**, *12*, 348. [CrossRef]
43. Janeković, I.; Kuzmić, M. Numerical Simulation of the Adriatic Sea Principal Tidal Constituents. *Ann. Geophys.* **2005**, *23*, 3207–3218. [CrossRef]
44. EU Interreg Italy-Croatia, PMO-GATE Project. Deliverable 3.1.2. Definition of Flood Exposure Indexes for the HR Test Site. Available online: https://www.italy-croatia.eu/web/pmo-gate/docs-and-tools-details?id=1877461&nAcc=4&file=1 (accessed on 20 October 2022).
45. EU Interreg Italy-Croatia, PMO-GATE Project. Deliverable 5.1.6. Map of Spatial Distribution of the Critical Zones Most Prone to Flood, Extreme Waves and Seismic Risks for HR Test Site. Available online: https://www.italy-croatia.eu/web/pmo-gate/docs-and-tools-details?id=1877461&nAcc=6&file=8 (accessed on 20 October 2022).
46. EU Interreg Italy-Croatia, PMO-GATE Project. Deliverable 3.2.1. Definition of Extreme Waves Exposure Indexes for the HR Test Site. Available online: https://www.italy-croatia.eu/web/pmo-gate/docs-and-tools-details?id=1877461&nAcc=4&file=9 (accessed on 20 October 2022).
47. EU Interreg Italy-Croatia, PMO-GATE Project. Deliverable 4.1.1. Methodology for Provision Assessment Indexes based on Spatial Multi-Criteria Decision Making. Available online: https://www.italy-croatia.eu/web/pmo-gate/docs-and-tools-details?id=1877461&nAcc=5&file=1 (accessed on 20 October 2022).
48. Wang, Q.; Liu, K.; Wang, M.; Koks, E.E. A River Flood and Earthquake Risk Assessment of Railway Assets along the Belt and Road. *Int. J. Disaster Risk Sci.* **2021**, *12*, 553–567. [CrossRef]

Article

Coastal Flooding Assessment Induced by Barometric Pressure, Wind-Generated Waves and Tidal-Induced Oscillations: Kaštela Bay Real-Time Early Warning System Mobile Application

Željana Nikolić [1,*], Veljko Srzić [1], Ivan Lovrinović [1], Toni Perković [2], Petar Šolić [2] and Toni Kekez [1]

1   Faculty of Civil Engineering, Architecture and Geodesy, University of Split, 21000 Split, Croatia
2   Faculty of Electrical Engineering, Mechanical Engineering and Naval Architecture, University of Split, 21000 Split, Croatia
*   Correspondence: zeljana.nikolic@gradst.hr

Abstract: Our work presents a reliable procedure to obtain real-time assessment of the sea water elevation at the Kaštela Bay site to ensure the a priori warning in the case of expected coastal flooding along the site area. In its origin, the presented procedure relies on relevant data sets which are site-specific and locally observed. Observed data sets are used within the procedure to assess sea water surface elevation when induced by barometric pressure changes and wind-generated waves. Tidal-induced changes are introduced into the assessment procedure by a pre-learned algorithm which relies on long-term sea level oscillations from the relevant tidal gauge. Wind-generated wave heights are determined in the near shore area, following the features of the depth and reflection of the shoreline subsections. By coupling three mechanisms, this paper offers a unique real-time procedure to determine the sea water elevation and assess the possibility for coastline structure to be flooded by the sea. Given information is visualized in a form of mobile application that implements the algorithm and allows end users to set the notifications based on the given ruleset.

Keywords: coastal flooding; real-time warning system; tides; wind-generated waves; barometric pressure; mobile application

Citation: Nikolić, Ž.; Srzić, V.; Lovrinović, I.; Perković, T.; Šolić, P.; Kekez, T. Coastal Flooding Assessment Induced by Barometric Pressure, Wind-Generated Waves and Tidal-Induced Oscillations: Kaštela Bay Real-Time Early Warning System Mobile Application. *Appl. Sci.* **2022**, *12*, 12776. https://doi.org/10.3390/app122412776

Academic Editor: Jürgen Reichardt

Received: 31 October 2022
Accepted: 8 December 2022
Published: 13 December 2022

Publisher's Note: MDPI stays neutral with regard to jurisdictional claims in published maps and institutional affiliations.

1. Introduction

Coastal flooding events, including their occurrence and triggers for the appearance of waves, have been investigated in the past to contribute to a better understanding of flood events in coastal areas as well as efficient risk management. Recent investigations have studied the occurrence of storm and sea wave events [1,2], the numerical modelling of wave propagation in coastal areas [3], flood water movement over land areas [4], the impact of storms on coasts [5] and flood vulnerability [6].

Special attention has been given to the development of an early warning system (EWS) for the timely warning of residents in exposed areas as well. Early warning systems are also an integral part of coastal flood management plans and contribute to the development of long-term management strategies [7]. Existing sea-state monitoring technology, historical databases, numerical forecasting models and computer science have been parts of the operational coastal flood early warning system [8]. A EWS modelling framework based on a Bayesian network has been used to link coastal hazards to their socio-economic and environmental factors [9] and connect available field measurements, data obtained from numerical wave simulations and an empirical wave run-up approach [10]. An EWS that determines the total sea level height by combining predictions of tides and sea level anomalies with wave runup estimates has been presented in [11].

Technological progress in computation and communication sciences in the last decade allows for work with large databases that can store registered meteorological data that affect the occurrence of sea floods. This type of data represents a basis for developing different forecasting platforms, which use input data and boundary conditions from global or regional scales and open sea and weather forecast databases, providing wave propagation in the ports [12]. These platforms use a number of effective hydrodynamic numerical models for the simulation of storm surge and the prediction of sea water level [13]. It is important to notice that the sea water level prediction procedure in real time is a time-consuming process which needs significant computational resources, as well as the development and monitoring of a software architecture [14] to model coupling and integration. The models based on machine learning [15] and artificial intelligence provide easy implementation of numerical models with low computational cost, as well as fast training, validation, testing and evaluation [16].

In the present study, coastal flooding caused by the simultaneous combination of several mechanisms, such as tidal-induced sea level oscillations resulting from barometric pressure changes and wind-generated waves, have been investigated in the Kaštela Bay. A special focus has been given to the area of the city of Kaštela, placed at the north part of the bay (Figure 1), with many cultural and historical buildings and/or areas located near the coastline, which are potentially endangered by coastal flooding and subject to significant consequences and damage (Figure 2).

The design of the early warning system for sea flooding risk in the city of Kaštela includes the following components:

- *A monitoring system* consisting of an installation of hardware at the site location and appropriate software for real-time data collection. This system observes the wind speed and the barometric pressure.
- *Application of mathematical and numerical models for sea level prediction on the coastline:* Numerical model is used to analyze the wave transform mechanism. A specific procedure has been established to incorporate simultaneous effects of wind-generated waves, tides and barometric pressure-induced changes in the sea level to determine sea level at the coastline based on the systematic observations of the abovementioned parameters via a real-time monitoring system (wind speed and direction and barometric pressure). Tidal-induced sea level changes have been obtained based on the past observations of the long-term sea level tidal oscillations from the relevant tidal gauge.
- *Estimation of the risk of flooding for humans:* Using a digital terrain model where each pixel is georeferenced (X, Y coordinates) and assigned altitude Z, the calculated sea heights will be compared with the altitudes, and if the sea elevation is greater than the altitude Z of a pixel, that pixel will be marked as flooded. According to the depth of the sea on land, the risk of flooding for humans will be defined based on an analytical function so that it can be easily integrated into the rest of the system. This information is necessary for the input into the early warning system. For a simple and understandable presentation of the flooding risk, the analyzed coastal area is divided into zones in order to define the total sea level. This division is made according to the criteria of the coast type and the coast height, which have a direct impact on the height of the waves.
- *Dissemination and communication of risk information by mobile application:* Flood warnings will be given to people who have the mobile application, which was developed for the purpose of the dissemination of information about flood risk in the observed area. Those that have installed the application on their mobile phone and enabled that app to send them push notifications will receive the information.

Recently, Internet of Things (IoT) systems that monitor in different scenarios to ensure a green, sustainable future have been widely used to create smart environments tailored to particular human needs [17,18]. Affordable equipment, miniaturized in its deployment, is ensured a long lifetime through solar/battery power; the communication ranges are also extremely increased and can be measured in kilometers in urban areas, which is especially pertinent in meteorological scenarios [19,20]. In this study, the established sea level monitoring system uses an IoT-based, solar-powered anemometer (Barani design—Meteo Wind) and solar-powered meteo-station (Barani design Meteo-helix) for monitoring air pressure.

The system is deployed to acquire current meteorological information used as an input for the developed model to estimate the sea level. The system is based on the LoRaWAN IoT radio, which delivers the information to The Things Network (TTN) cloud system, which is then used by the cross-platform web/mobile application called Waves, which is used as an early warning monitoring system. The multilanguage system itself implements: (1) the algorithm that defines sea level based on the tidal, wind speed/direction and air pressure information; (2) information on the shore height at the dedicated measurement and early warning zones; (3) push notification logics that can be separately activated based on user's zone of the interest.

(a)

(b)

Figure 1. (**a**) Location of the city of Kaštela in Kaštela Bay; (**b**) A view of an old historical core, Kaštel Kambelovac.

Figure 2. Coastal flooding events in the city of Kaštela.

This paper presents a methodological approach for the development of an early warning system for coastal flooding. It is based on frequently measured meteorological data and sea level predictions integrated on a single platform to provide real-time information about potential risk for the citizens of the city of Kaštela.

Novelties of the manuscript are summarized as follows: (i) to our best knowledge, the site of the application has been used for the very first time to demonstrate the application of the coastal flooding early warning system; (ii) the relevance of the selected site relies on the fact that the area is faced with coastal flooding more than 15 days per year on average, thus increasing the need for early warning system development, (iii) the manuscript couples three mechanisms contributing to the coastal flooding vulnerability assessment, those being barometric pressure-induced sea level changes, wind-generated waves and tidal-induced fluctuations, (iv) barometric- and tidal-induced sea level changes are obtained in real time, fully relying on local conditions arising from the observations, (v) wind-generated wave heights are site specific, taking into consideration bathymetric features and coastal structure type and (vi) the whole procedure has been implemented in the form of a real-time mobile application, thus resulting in a reliable tool for the end users.

Compared to relevant publications [16,21], this paper couples three of the mechanisms leading to the sea water level rise and refers to a site-specific area. Although it does not offer general findings, the procedure shows the potential to be applied to other sites all over the Mediterranean basin after prior modifications. The latter refers to bathymetric features, the reflection of the coastline, the determination of incident deep water wave parameters and tidal observations.

The paper is structured as follows: Section 2 is the Methodology and Study area section, which couples the methodology used in the study with representative site related data. Section 3 offers Results from the study, while Section 4 consists of relevant Discussion topics and Section 5 summarizes main conclusion points.

2. Methodology and Site Description

2.1. Sea Water Elevation Prediction

Sea water level is a random process resulting from the simultaneous combination of several mechanisms, three of which are found to be dominant, those being tidal-induced oscillations and oscillations due to barometric pressure changes and wind-generated waves. Although characterized by different time scales, those three mechanisms simultaneously contribute to the absolute sea water elevation definition.

Tidal-induced sea water oscillations are driven by tidal forcing, which is induced by simultaneous inter gravity forces between the Sun, Moon and Earth. Tidal-induced changes are characterized as a mixed semidiurnal type at the location of interest, with main periodic intervals corresponding to both semi diurnal and diurnal ones. Compared to tidal oscillations, sea level changes induced by barometric pressure changes are aperiodic with time scales corresponding to two main factors: (i) daily barometric pressure changes corresponding to daily scale air temperature change and (ii) time scales corresponding to the time necessary for the air mass transfer from different geographic locations to the location of interest to occur. The latter corresponds to time scales usually equal to several hours. Wind-generated waves are characterized by very small time scales, up to 8 s in the area of interest, and these are generated as a result of the air mass kinematic energy transfer to the sea surface.

Previous research [22] has shown that tides characterizing the Adriatic Sea basin consist mainly of seven dominant constituents, of which three are diurnal (O1, P1 and K1) and four are semidiurnal (N2, M2, S2 and K2). In its origin, each constituent represents a sinusoidal function with an associated amplitude, period, and phase, contributing to the full tidal signal. To determine the unknown values of the amplitudes, periods and phases, the original signal is initially transferred from the time to frequency domain by applying the Discrete Fourier Transform (DFT) [22]. DFT results are often plotted as Amplitude Spectral Density (ASD). By normalizing magnitudes with number of samples, the amplitude of each frequency can be easily obtained. Due to the fact that DFT calculations are time demanding [23], Fast Fourier Transform (FFT) [24] incorporated in Python in SciPy library [23] has been used for faster transfer of signals from the time to frequency domain. Based on the determined tidal constituent parameters, the tidal-induced sea level can be simulated using a linear superposition of sine functions corresponding to the number of relevant constituents:

$$h_t = \sum_{i=1}^{7} A_i \times \sin\left(\frac{2\pi t}{t_{pi}} + \frac{2\pi \varphi_i}{360}\right), \ t = 0,\ 1,\ \dots,\ M \tag{1}$$

where h_t is the simulated sea level [m], A_i is the amplitude [m], t_{pi} is the period [h] and φ_i phase [°] of i-th constituent, M is the sample size and t is relative time [h].

Simulated sea level h_{sea} can be calculated as a superposition of tidal harmonics from Equation (2) by adding the mean sea level value calculated from the observed sea level signal:

$$h_{sea} = h_t + \overline{h} \tag{2}$$

where h_t represents tidally induced sea level oscillations and $\overline{h}$ represents the mean sea level value.

Increasing the barometric pressure by 1 [cm] of the saltwater column leads to a decrease in the sea level by approximately 1 [cm] and inversely, decreasing the barometric pressure by 1 [cm] raises the sea level by 1 [cm] [25]. This effect is called the Invert Barometric effect (IB). The sea level rise caused by the change of the barometric pressure can be calculated as follows:

$$h_{at} = \frac{-p_a}{\rho g} \tag{3}$$

where h_{at} is the sea level change resulting from a change in barometric pressure, p_a is the barometric pressure [Pa] change, ρ is the density of sea water [kg/m^3] and g is the gravitational acceleration [m/s^2]. Therefore, the simulated sea level incorporating both tidal oscillations and barometric pressure-induced changes is updated from Equation (2) for the value of h_{at}, as shown in Equation (3):

$$h_{sea} = h_t + h_{at} + \overline{h} \tag{4}$$

where the third right hand term ($\overline{h}$) represents mean sea level values as calculated from the observed signal, h_t represents tidally induced sea level oscillations and h_{at} stands for barometric pressure-induced sea level change.

To assess wind-generated wave height in front of the coastline, we start with a determination of deep water wave parameters. For the location of interest and relevant incident wave directions, fetch length has been assessed by applying the Saville method [26]. In order to determine deep water wave parameter values, data sets from the Section 2.3 climatological station have been used. Fully developed sea conditions have been checked by applying the Wilson criteria [27] prior to the determination of wave parameters. After defining relevant fetch length for a given wind duration and wind velocity, the Groen−Dorrestein nomogram is used to determine deep water wave parameters [28].

Wave transform analysis has been performed by applying SMS CGWAVE software [29]. The wave phenomena that can be simulated by CGWAVE are: bathymetric refraction, diffraction by structures (e.g., breakwaters) and bathymetry, reflection (from structures, natural boundaries (seawalls, coastlines, etc.) and bed slopes), friction, wave breaking and floating (fixed) docks influence the wave field. The model is based on the use of a triangular finite element formulation to solve the two-dimensional elliptic mild slope equation, with grid sizes varying throughout the domain based on the local wavelength [30]. The grid can be efficiently generated using the SMS graphical interface when a bathymetry file is provided. The model allows one to specify the desired reflection properties along the coastline and other internal boundaries. While the basic equation is intended for monochromatic waves, irregular (i.e., spectral) wave conditions are simulated in CGWAVE through a linear superposition of monochromatic simulations [31,32].

The procedure or the algorithm for the absolute sea water elevation assessment is based on three steps which are presented above and summarized as: (i) sea level change caused by tidal forcing prediction, (ii) sea level change caused by a barotrophic pressure forcing assessment and (iii) assessment of the wind-generated wave height in front of the coastline. Sea water elevation assessment as a consequence of simultaneous effects for those three mechanisms is assessed as follows:

- From observed tidally induced oscillations, the harmonic parameters (amplitude, period and phase) are initially determined based on the Least Square Method application;
- After all harmonic parameters have been determined, tidal-induced sea level oscillation is determined by using Equation (1);
- From observed barometric pressure data, the change in sea level induced by the drop/rise of the barometric pressure is determined from Equation (3);
- Deep water wave parameters are determined depending on incident direction and wind velocity and duration parameters;
- For relevant incident directions, numerical simulation of the wave transform has been performed by incorporating shoreline reflection coefficients determined on site;

- The study area has been divided into ten zones fundamentally different with regard to reflection features;
- For each zone wave, parameters have been determined within the zone close to the shoreline (4 m away from the shoreline);
- Final determination of the sea water elevation by incorporating three abovementioned mechanisms is done by:

$$h_{sea} = h_t + h_{at} + h_{S(d=4\text{m})} \tag{5}$$

which offers an easy-to-implement way to assess absolute sea level, where h_t represents the tidally induced component, h_{at} represents the barometric pressure-induced component and $h_{S(d=4\text{m})}$ stands for the significant wind-generated height as found 4 m offshore. The value of h_{sea} is expressed relative to HVRS71 datum [33].

2.2. Site Description

The location of the study area with definition of the tidal gauge and meteorological stations are shown in Figure 3. The meteorological station Split with geographical coordinates $\varphi = 43°31'$ N, $\lambda = 16°26'$ E is located northwest of the city Split, on the Marjan hill, at the altitude of 122 m a.s.l. The terrain is a slope towards the sea on the SW-W-NW-N side. To the north is the bay of Kaštela and to the east is the city of Split. At a distance of 7 km toward the north is the mountain Kozjak with the highest peak at 779 m a.s.l. The measuring system Fuess [34] is installed at a 5 m high terrace column on the building roof (12 m above ground). The only obstacles found near the meteorological station building are found to the west and northwest side (trees and terrain elevation overgrown with trees more than 10 m tall). The roughness class of the terrain is 2.5 ($z_0 = 0.2$) [34]. The wind speed and direction data obtained by the classic Feuss measurement consist of an average hourly wind speed with corresponding wind.

The nearest tidal gauge near Kaštela Bay is located within the Institute of Oceanography and Fisheries (IOR) in Split, at the western border of peninsula Marjan. The time series of measured sea surface elevations over a total duration of five years have been obtained from the Marjan tide gauge for the period from 1 January 2010 to 31 December 2014. The sampling frequency of the tide gauge was set to 1 h for the entire time series. The measured water level elevation is referred to as the HVRS71 vertical datum [33]. Due to the maintenance and malfunction of the tide gauge IOR Marjan, the observed time series are characterized by periods without recorded values of the sea level. For the purpose of this work, a continuous time series is required so the longest continuous time series has been found from 25 January 2010 at 12:00 h to 27 June 2011 at 23:00 h with a total of 12,444 h of data.

Barometric pressure data are obtained from two meteorological stations: Split-Airport and Split-Marjan. The meteorological station Split-Airport is located at 21 m a.s.l. and the barometric pressure values are recorded three times a day at 7 am, 2 pm and 9 pm. The meteorological station Split-Marjan is located at 122 m a.s.l. and has an hourly measurement frequency including barometric pressure values. Logs of barometric pressure were obtained from both meteorological stations for the period from 1 January 2010 to 31 December 2014. Due to the location of Kaštela Bay, the meteorological station Split-Airport was considered as relevant, but due to its low measurement frequency (3 times per day), the data from Split-Marjan was used for further analysis. First, the data from Split-Marjan was compared with the data from Split-Airport and a mean difference of 12.67 hPa with a standard deviation of 0.67 hPa was found. All data from Split-Marjan were corrected for the values of 12.67 hPa. The available data from the Split-Marjan station has a continuous record with no missing data for the entire 5-year period. However, since the sea level signal does not have the same continuity as the barometric pressure, the same period of barometric pressure as sea level is included in the further analysis.

Insight into bathymetric features emphasize depth values up to 37 m with a pretty uniform decrease towards the shoreline (Figure 4). A bathymetric survey for the purpose of this paper has been performed in April 2021 with a single beam setup. Shoreline determination of the reflection coefficient has been done at the site by the inspection of both shoreline type and depth in front (Figure 5).

Figure 3. Location of the study area with the definition of the tidal gauge and meteorological stations (IOR—Institute of Oceanography and Fisheries).

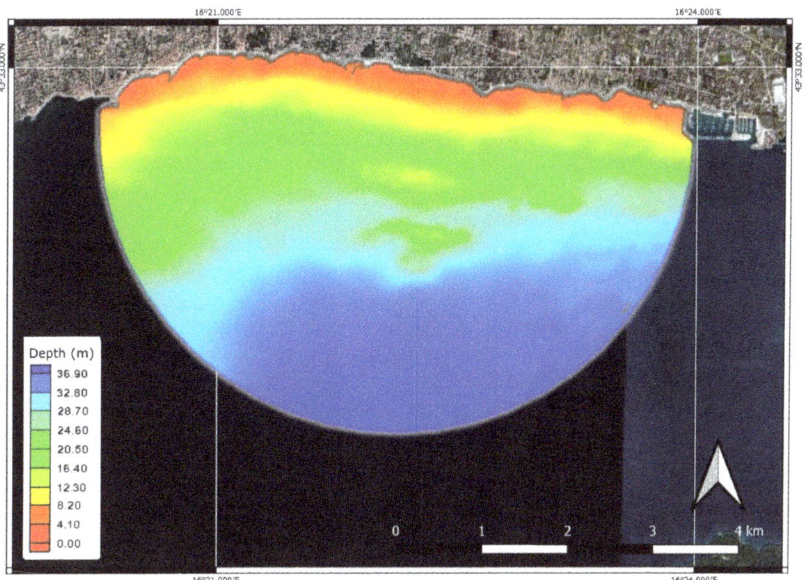

Figure 4. Bathymetric features at the location of interest.

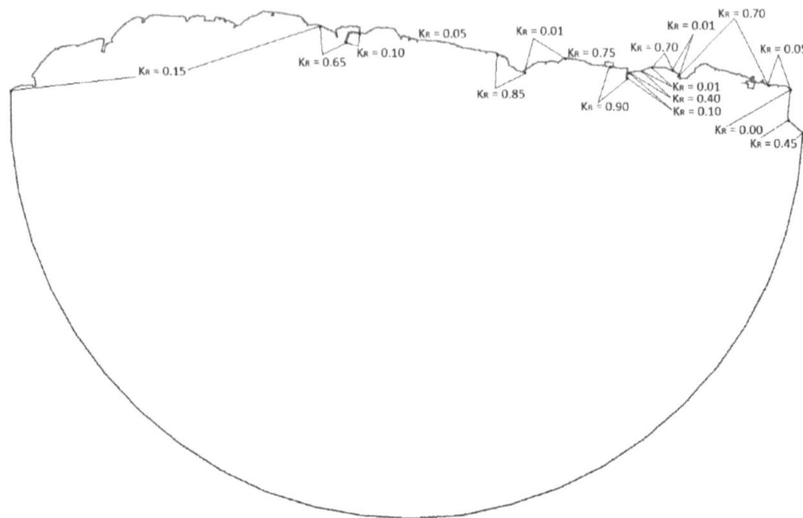

Figure 5. Shoreline reflection coefficient along the coastline.

2.3. Data Collection from LoRaWAN-Based Sensor Device

LoRaWAN is one of the most widely used Low Power Wide Area technologies that aims to collect and communicate data from an end sensor device at a large distance, making it perfect for a scenario in which sensor devices collect wind and barometric data and convey them to a centralized system. LoRaWAN employs a typical star-of-star topology where end devices communicate data in a single hop to one or more gateway devices. These messages are further forwarded to the network and application server for further processing, allowing authorized data to be forwarded to other external services (e.g., using MQTT message forwarding). LoRaWAN allows battery-operated devices to periodically transmit sensor data over large distances, while minimizing consumption during inactive periods. LoRaWAN technology finds its application in smart city/smart agriculture environments where there is no need for real-time (every second) transmission from end devices. During the inactive period, end devices simply cut off the consumption, allowing a battery lifetime up to a couple of years without any external power source.

Figure 6 depicts the architecture of the MeteoHelix weather station of Barani design that utilizes LoRaWAN as a radio technology. MeteoHelix [35] is an automatic all-in-one microweather station which is solar powered and can be active for up to 6 months without sun. It measures air temperature to WMO accuracy, air humidity to WMO accuracy with dew and frost point output, barometric pressure and solar irradiation (pyranometer). Another sensor of Barani design was also installed that utilizes LoRaWAN communication—MeteoWind IoT PRO. MeteoWind [36] is used for wind monitoring and employs two sensors: a separate wind vane and anemometer. MeteoWind allows 4+ months of battery life without sun and a maintenance-free service life with long-term measurement stability due to its elliptical cup and metal construction. As depicted in Figure 7, both devices are placed at a 10 m height without any object around within 150 m so that both the wind speed and wind direction are not distorted, while both are located in location of Kaštela Bay. Since both MeteoHelix and MeteoWind IoT PRO employ LoRaWAN communications to convey data over the air to the centralized system, The Things Network as a service provider was used to collect data for further processing. As a LoRaWAN gateway, an indoor Sentrius RG1xx LoRaWAN gateway device placed around 150 m from the sensor devices was employed that forward messages to The Things Network (TTN) cloud infrastructure. Once the message arrives at the gateway, it is forwarded to the TTN Network and Application server. Furthermore, TTN allows message forwarding from TTN infrastructure to our dedicated Waves Seafront Monitoring app using MQTT protocol, which is described more in detail in the following section.

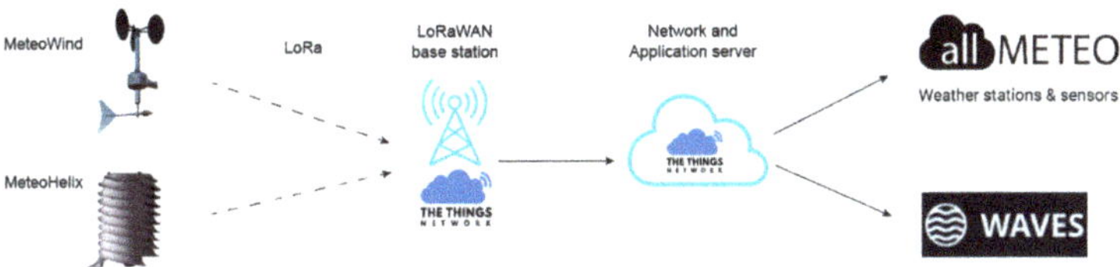

Figure 6. Architecture of LoRaWAN-enabled MeteoHelix and MeteoWind devices from Barani Design.

Figure 7. Installation of MeteoHelix and MeteoWind IoT Pro devices.

2.4. Waves Seafront Monitoring Architecture—Overview and Functionalities

Waves Seafront Monitoring is a cross-platform mobile application that allows users to access and interactively view the current sea level, discretized by coastal segments with respect to coastal height. As shown in Figure 8a, it comprises a cloud and client side. At the cloud side, the Google Firebase server component executes an evaluation algorithm based on data from the sensor it communicates with, serves as a server for these results and performs user authentication. Sensor data comprising air pressure, wind speed and direction are sent to The Things Network (TTN) cloud via the LoRaWAN communication channel, which is forwarded via MQTT protocol to the TTN microservice. Once the packet with sensor data arrives, the TTN microservice captures and stores the LoRaWAN uplink data into the database and forwards the data through the PMO algorithm to estimate sea level according to the data arriving from LoRaWAN sensors (wind speed and direction, barometric pressure and sea tide level). As can be seen, an alarm notification can be sent to the application if the sea level exceeds a predefined level.

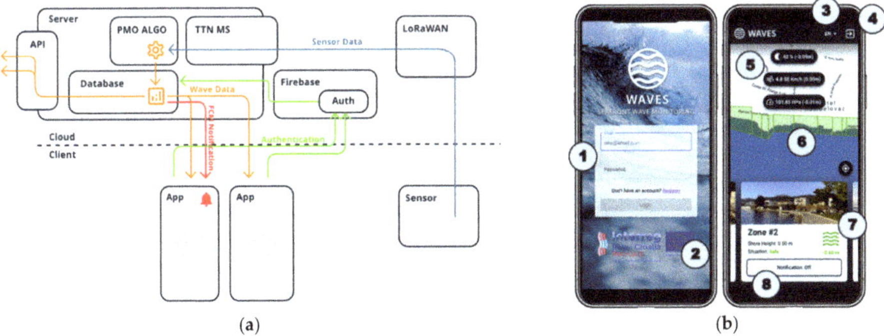

Figure 8. (**a**) High-level overview of the implementation of the solution, (**b**) Front-end interface of the mobile application.

The front-end of the application is comprised of the following elements, as depicted in Figure 8b:

A. Login interface (Figure 8b)—initial interface for users when opening the application:

1. Form—the user enters his login details (e-mail and password) and has the option of registration if he does not have an account by clicking on the appropriate link (Figure 8b, number 1),
2. Prominent logo—the project was realized within the Interreg PMO GATE collaboration (Figure 8b, number 2).

B. Registration Interface—the registration interface is analogous to the login interface, only it allows the user to create an account to access the application by entering login information. To simplify logging into the system, there is no special validation of user accounts, such as email confirmation.

C. Map interface (Figure 8b) —the central part of the application, which contains the following basic functionalities:

1. At the top is a drop-down menu where the user has a choice between three languages: Croatian, English and Italian. By clicking on an option, the interface text adjusts the interface language (Figure 8b, number 3).
2. Log out button from the application (Figure 8b, number 4).
3. Values read from the sensor are updated over time depending on how often the data from the TTN arrives. From top to bottom, sea level is the sum of altitudes caused by sea tide, wind speed and direction and barometric pressure (Figure 8b, number 5). The total amount is calculated based on the algorithm submitted by the Client.
4. Google maps with plotted polygons that correspond to discretized segments of the coast according to their heights (Figure 8b, number 6). The color of the zone corresponds to the early warning status: green—safe, i.e., the sea level is below the coast level; yellow—warning, i.e., the sea can exceed 20 cm above the height of the shore; orange—dangerous, i.e., the sea can rise up to 50 cm above the height of the shore; red—flooded, i.e., the sea exceeds 50 cm above the height of the coast. By clicking on an individual zone, the cards will position themselves next to the corresponding zone and the map will be centered on the selected zone. Based on the coastal height data, the monitoring area is divided into 10 zones.
5. Zone maps showing the names and photos of coastal zones. Here, the user can see exactly the height of this segment of the coast, the estimated sea level in relation to the zone and, consequently, the situation in that zone, which is coded in colors analogous to the zones on the map (Figure 8b, number 7). Each zone information contains the sea level with respect to the coastal height, where the number with the minus sign shows how much the sea level is below the coastal height. Once the number becomes positive, the zones change color since this result corresponds with estimated flood. By moving the tabs left or right, the user can focus on a specific zone, and the map will center on that zone.
6. Notification button (Figure 8b, number 8). By clicking on this button individually for each zone, the user can indicate whether he wants to receive notifications when the situation in a particular zone changes.

3. Results

3.1. Sea Level Determination Based on the Tidal Fluctuations

To enable the assessment of tidal-induced sea level oscillations, a total of 12,444 h of observed time series data was used to perform DFT and obtain an amplitude spectrum (Figure 9). Due to the nature of the observed signal, inspection of the amplitude spectrum offers the presence of the trend, visible within the bins corresponding to the lowest frequencies. In total, seven tidal harmonics has been identified as dominant, thus ensuring the tidal-induced sea level oscillation characterization.

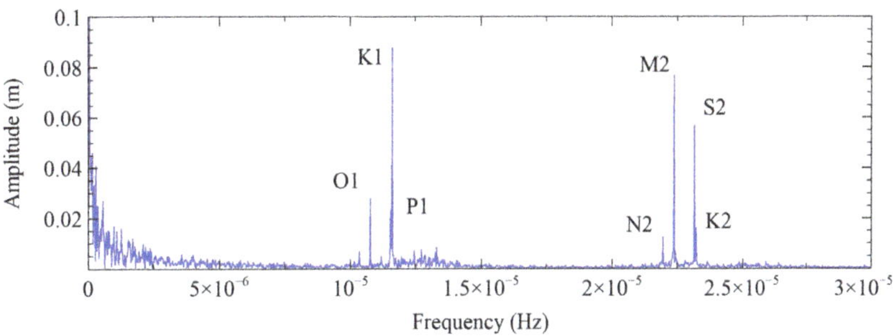

Figure 9. Amplitude Spectral Density for observed sea level reduced for the mean value.

While tidal amplitudes and periods were determined directly from the spectrum shown in Figure 9, phases were determined from the complex component of each of the seven tidal harmonics. All characteristic values for the relevant tidal constituents are shown in Table 1. The quality of the determined harmonic values were checked by the application of the correlation coefficient when applied to both the simulated and observed time series.

Table 1. Tidal constituents with corresponding values of amplitude, period and phase.

Constituent	Amplitude (m)	Period (h)	Phase (°)
O1	0.02821644	25.81743	−160.6065
K1	0.02339150	24.06963	105.3624
P1	0.08773618	23.93077	153.1416
N2	0.01279412	12.65921	65.3833
M2	0.07688095	12.41916	−31.55806
S2	0.05696792	12.00000	−126.8397
K2	0.01636003	11.96538	116.0807

Both the observed and simulated sea level signals, one for each 1000 h of the available time series, are compared by using Pearson correlation coefficient and root mean square error. When simulating the sea level by using Equation (2), the simulated signal discovers the absence of the trend within, which is incorporated in the next step as a response to long-term barometric pressure changes. The residual was determined as a difference between the observed and simulated signals. Both signals are shown relative to the HVRS71 vertical datum, as explained in the Methodology section. The same procedure was repeated for each 1000 h data of the total 12,444 h data representing the total sample. Both Pearson correlation coefficient and root mean square error values are shown in Table 2.

Table 2. Pearson coefficient and root mean square values obtained between the measured and simulated sea water elevation by Equation (2).

Data Sets (h)	0–1000	1000–2000	2000–3000	3000–4000	4000–5000	5000–6000	6000–7000	7000–8000	8000–9000	10,000–11,000	11,000–12,444	0–12,444
RMSE	0.198	0.141	0.083	0.082	0.075	0.071	0.147	0.241	0.091	0.159	0.095	0.136
Pear. Corr. Coef.	0.551	0.597	0.749	0.782	0.923	0.802	0.664	0.639	0.752	0.834	0.791	0.578

For the entire observed and simulated set of 12,444 h data, the root mean square value equals to 0.136 [m], while Pearson's correlation coefficient equals 0.578 [-]. The high value of RMSE and low value of the correlation coefficient indicate the sea level fluctuations are not only subjected to tidal variations, but also to other factors.

3.2. Sea Level Determination Based on the Tidal Fluctuations and Barometric Pressure Changes

Figure 10 shows the effect of barometric pressure on the simulated signal with seven frequencies. The simulated sea level signal has a similar trend to the observed sea level signal with a residual remaining between the simulated and observed signal. When the sea level signal is simulated using the tidal components and the inverted barometric effect, the RMSE decreases to 0.1057 [m] and the Pearson correlation coefficient increases to 0.777 [-], further highlighting the effect of barometric pressure and its contribution to the observed sea level definition at the location of interest (Table 3). The decrease in the residual values when both the barometric pressure and tidal effects are involved in the procedure for the sea level determination implies the relevance of those two mechanisms in the vertical movement of sea surface elevation.

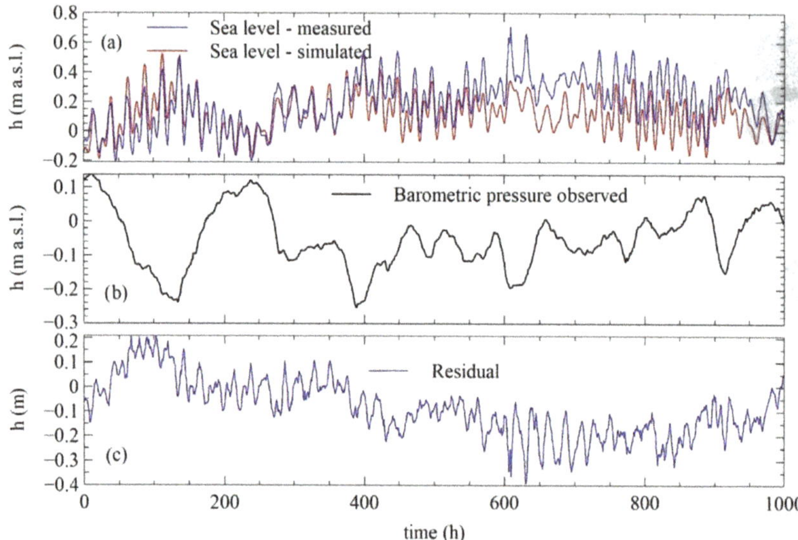

Figure 10. (**a**) Observed and simulated sea level by Equation (4) for the first 1000 h of observed sea level signals, (**b**) Observed barometric pressure, (**c**) Residual between simulated and observed sea level signals.

Table 3. Pearson coefficient and root mean square values obtained between the measured and simulated sea water elevation by Equation (4).

Data Sets (h)	0–1000	1000–2000	2000–3000	3000–4000	4000–5000	5000–6000	6000–7000	7000–8000	8000–9000	10,000–11,000	11,000–12,444	0–12,444
RMSE	0.144	0.103	0.061	0.062	0.083	0.061	0.113	0.189	0.091	0.126	0.076	0.106
Pear. Corr. Coef.	0.735	0.776	0.886	0.886	0.936	0.865	0.886	0.794	0.812	0.891	0.872	0.777

3.3. Wind-Generated Waves

Following the knowledge from the location of interest, two relevant wave incident directions have been identified, SE and SSW. The reason for the selection of those two incident directions relies on three facts: (i) longest possible fetch lengths compared to other incident directions, (ii) longest wind duration and wind velocities and (iii) relative position of the shoreline almost perpendicular to the incident directions. Table 4 offers the highest significant wave height equal to 2.78 m with a corresponding period of 4.76 s. For the SSW incident direction, Table 5 shows the highest wave parameters equal to 2.35 m and 4.15 s.

Table 4. Deepwater wave parameters—incident direction SE.

V (ms^{-1})	t (h)	F_{min} (km)	t_{min} (h)	F_{EFF} (km)	F_{MJ} (km)	H_S (m)	T_S (s)	L_0 (m)
1.5	13	43.36	8.65	24.8	24.8			0
3.3	13	71.25	6.02	24.8	24.8			0
5.4	13	97.17	4.8	24.8	24.8	0.5	2.6	10.55
7.9	13	123.48	4.03	24.8	24.8	0.85	3.15	15.49
10.7	17	215.89	3.5	24.8	24.8	1.21	3.65	20.8
13.8	15	213.49	3.12	24.8	24.8	1.7	4	24.98
17.1	11	159.77	2.82	24.8	24.8	2.19	4.35	29.54
20.7	4	45.07	2.59	24.8	24.8	2.78	4.76	35.38

Table 5. Deepwater wave parameters—incident direction SSW.

V (ms^{-1})	t (h)	F_{min} (km)	t_{min} (h)	F_{EFF} (km)	F_{MJ} (km)	H_S (m)	T_S (s)	L_0 (m)
1.5	5	11.71	6.68	17.4	11.71			0
3.3	5	19.24	4.65	17.4	17.4			0
5.4	5	26.24	3.7	17.4	17.4	0.5	2.5	9.76
7.9	5	33.35	3.11	17.4	17.4	0.77	2.95	13.59
10.7	5	40.37	2.7	17.4	17.4	1.05	3.15	15.49
13.8	5	34.91	2.41	17.4	17.4	1.41	3.6	20.23
17.1	5	26.94	2.18	17.4	17.4	1.92	3.98	24.73
20.7	5	6.75	2	17.4	6.95	2.35	4.15	26.89

The procedure for wave height determination at the distance of 4 m offshore starts with known wind velocity and duration. For the selected incident wave direction and fetch length, deep wave parameters have been selected from Tables 4 and 5. For the purpose of wave height determination in front of the shoreline, the JONSWAP spectrum is used to incorporate a real wave into the wave transform model. Spectrum parameters have been set up as shown in Table 6, where the obtained deep water wave values correspond to a 100-year return period.

Table 6. Deepwater wave parameters—JONSWAP spectrum parameters.

Incident Direction	Hs (m)	T (s)	γ	nn
SE (165°)	3.05	6.40	3.30	4.00
SSW (202.5°)	2.60	5.90	3.30	4.00

Wave height spatial distribution and direction for the SE incident wave direction is presented in Figure 11. Insight into the inner zone of interest discovers the presence of refraction as a main wave transform mechanism. The insight into significant wave height changes at 4 m offshore is shown in Figures 12 and 13. The same has been shown for the SSW incident direction in Figures 12 and 13. By comparing modeled wave heights, it is obvious the SE incident direction results in more significant wave height values.

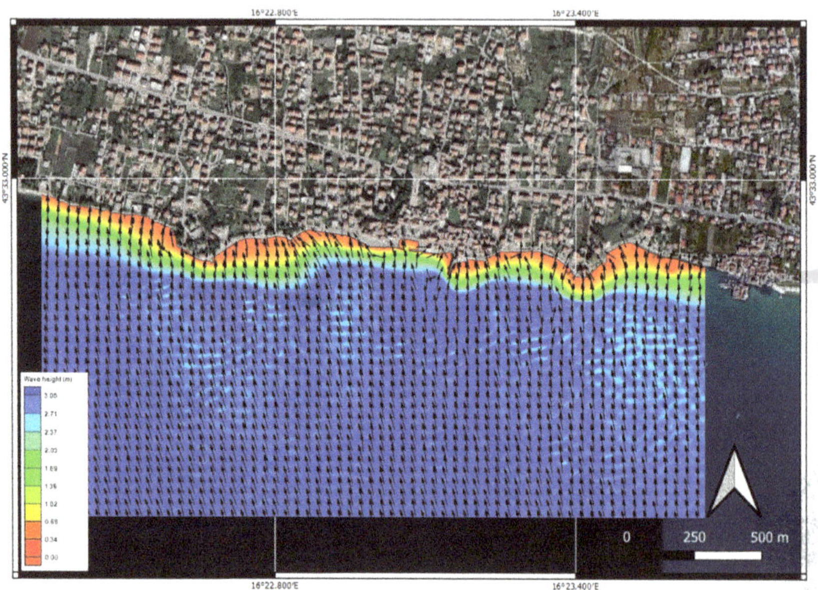

Figure 11. Wave heights and directions for the SE incident direction—local study area.

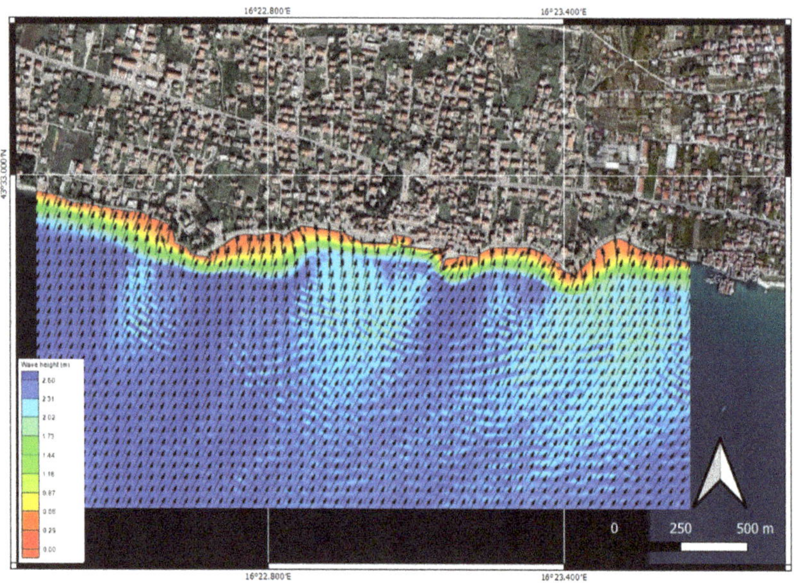

Figure 12. Wave heights and directions for the SSW incident direction—local study area.

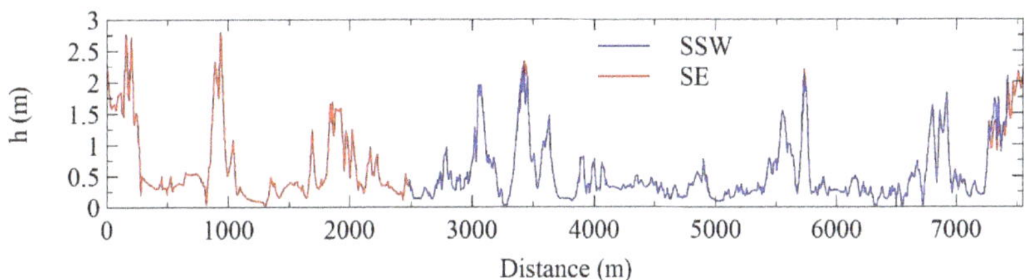

Figure 13. Modeled wave heights along the control line—incident directions: SE and SSW.

3.4. Real-Time Prediction of Sea Water Elevation and Warning

The monitoring system has been installed and put into operation at the Kaštela Bay test site to provide real-time measurement of wind speed, wind direction and barometric pressure. The observed data is used for the real-time prediction of sea level based on the developed algorithm for the estimation of sea level heights and information obtained from the monitoring system, which has been integrated into the algorithm. The real-time prediction of sea level values is obtained based on tidal fluctuations, barometric pressure values and wind characteristics.

Resulting sea level values are compared with coastline elevation in order to identify the potential danger of flooding and issue on-time warnings. Figure 14 shows wind characteristics and barometric pressure as observed during a period of 7 days during February 2022. The data containing barometric pressure along with wind direction and speed were collected into a database for the application. The data were used as an input for an algorithm to estimate sea level height.

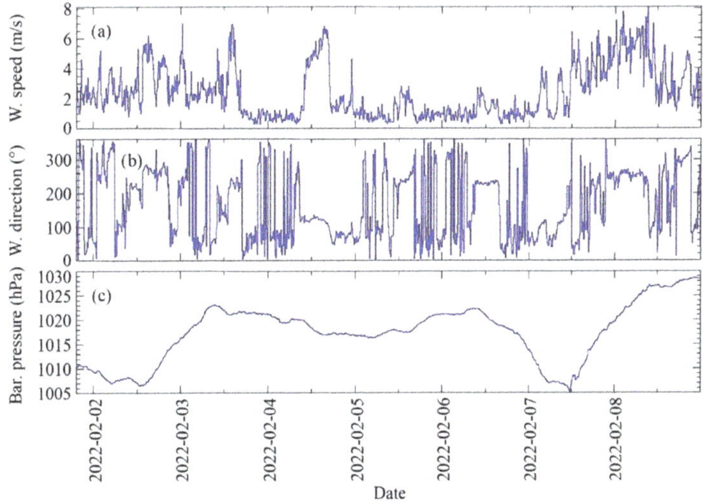

Figure 14. LoRaWAN-obtained observed time series: (**a**) wind speed, (**b**) wind direction and (**c**) barometric pressure.

Figure 15 shows sea water elevation as a consequence of each of the components (tidal effect, barometric pressure and wind-generated waves) as well as absolute sea water elevation as a superposition of the abovementioned components for the period from 15 December 2021–7 February 2022. The results are given for one randomly selected zone along the study area to demonstrate the capacity of the proposed procedure to assess the sea water elevation. The estimated water surface elevation is referred to the HVRS71 vertical reference datum determined on the basis of mean sea level, as stated in the Methodology section.

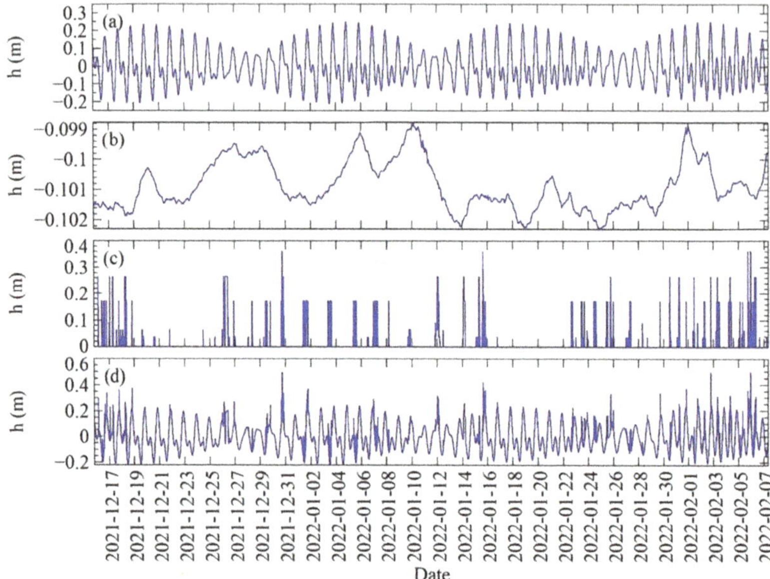

Figure 15. Calculated sea level height during the period from 15 December 2021–7 February 2022 caused by: (**a**) tidal effect, (**b**) barometric pressure-induced sea level changes, (**c**) wind-generated wave height and (**d**) absolute sea level height.

If the resulting sea level value estimated from the algorithm exceeds the coastline elevation, the warning is immediately issued by the mobile application to alert the endangered population. Flood hazard warnings in the mobile application are primarily focused on citizens' safety and their classification is based on the particular threat level. Threat levels are based on tests of the instability of a human body in floodwater and different stability thresholds [37]. Following the results of the tests, the first threat level, issuing a yellow alert, is defined for flood water depth on the coastline up to 20 cm, which is mostly threatening for children. The second level of threat corresponds to an orange warning, which is now becoming dangerous for adults and is issued when the water depth is between 20 cm and 50 cm. Finally, an exceedance of flood water depth over 50 cm on the coastline corresponds to the third threat level. In this case, severe injuries can be caused by flood water and the red warning for the population is issued.

4. Discussion

In this paper, a procedure to assess sea water elevation caused by coastal flooding at Kaštela Bay is presented, together with the monitoring system infrastructure scheme and sea water elevation prediction module. In its origin, the procedure relies on the real-time monitoring IoT-based system for sea level-relevant parameter monitoring to capture wind-generated wave heights and barometric pressure-induced sea level changes.

The latter offers the possibility to determine wind-generated wave heights in front of the coast, with an incorporation of shoreline reflection features as assessed by the sea water depth and type of structure found at the site, and long- and short-term changes in the mean sea water level. Tidal-induced sea level changes are forecasted based on the data series observation from the nearest relevant tidal gauge.

After wind speed/direction and barometric pressure sensors are installed at the site to enable real-time monitoring, a LoRaWAN-based IoT monitoring system was employed, along with the dedicated cross-platform mobile application, which was used as an early warning notification system. Therein, registered end users can set early warning notifications based on their zone of interest.

As stated, the origin of the presented approach relies on the capability of the monitoring system to observe the wind speed/direction and barometric pressure at the site. The selection of the location of the monitoring infrastructure is performed at the location of interest to neglect spatial variations in barometric pressure variations and offer reliable input for barometric pressure-induced sea level changes.

Relevant mechanisms acting to assess the sea water elevation are selected based on the phenomena characterizing the study area and fit suitable for the pre-assumed conditions mostly found on site. The effect of barometric pressure changes and their influence on mean sea water elevation refers to the sea in a calm state and static conditions when no significant changes in the barometric pressure field are observed. In case this requirement is not met, such as when the storm is moving, a difference in predicted sea level changes compared to real ones can be obtained.

Our results demonstrate the capability of the proposed methodology to enable a realistic and reliable forecast of the sea water elevation by taking into consideration local conditions: (i) wind information, (ii) barometric pressure, (iii) tidal forcing, (iv) water depth and (v) coastal structure or type information.

When coupled with the shoreline's absolute height, one is enabled to assess flooding risk or the vulnerability of the coast to sea level-induced flooding. The developed monitoring system, coupled with data collection from LoRaWAN-based sensor device and assessment procedure, presents a unique and robust approach to be used along different coastal areas of the Adriatic Sea.

Apart from this, different coastal structures found along the specific site can easily be taken into consideration by assessing the reflectance features from the structure type and bathymetric features in front of the coastline.

5. Conclusions

This paper presents a procedure to assess the coastal flooding induced by the sea level in the Kaštela study area. In its origin, it consisted of three modes: (i) the LoRaWAN-based sensor device monitoring infrastructure located at the site to capture relevant parameters in real time, (ii) a step-by-step procedure to assess the sea water elevation and (iii) Waves Seafront Monitoring cross-platform mobile application to enable the real-time insight into coastal flooding occurrence. The main conclusions obtained from the study can be summarized as:

- The presented procedure to assess coastal flooding induced by seawater has been shown to be efficient when applied to the Kaštela study area.
- The origin of the assessment for the seawater elevation relies on three basic mechanisms: tidal oscillations, barometric pressure-induced oscillations and wind-generated waves. Despite the potential limitation, no significant deflection from the sea water elevation occurred at the site during the period from 15 December 2021 to 7 February 2022 was observed.
- The presented procedure shows robustness in the potential to incorporate other relevant mechanisms influencing sea water elevation and site-specific features. It can be extended to the relevant mechanisms acting toward the sea level definition and adjusted to the local conditions reflecting the site-specific features. In case of need, additional effects can be added: (i) submerged structures overtopping, (ii) wave run-up, (iii) seiche, (iv) beach or shoreline friction effects, (v) dynamic inverse barometric effects, (vi) coastal flooding induced by the precipitation, especially during the cyclone and its superposition with the sea-induced flooding. The latter not only contributes to enhanced capacity of the coastal flooding awareness at the Kaštela site, but also enables the procedure to be applied at other sites.

Author Contributions: Conceptualization, Ž.N., V.S. and P.Š.; Data curation, V.S., I.L., T.P., P.Š. and T.K.; Funding acquisition, Ž.N.; Investigation, Ž.N., V.S., I.L., T.P., P.Š. and T.K.; Methodology, Ž.N., V.S., P.Š. and T.K.; Software, V.S., I.L. and T.P.; Supervision, Ž.N., V.S. and P.Š.; Validation, Ž.N., V.S., I.L., T.P. and P.Š.; Visualization, I.L. and T.K.; Writing—original draft, Ž.N., V.S., I.L., T.P., P.Š. and T.K. All authors have read and agreed to the published version of the manuscript.

Funding: This research was funded by the EUROPEAN UNION, Programme Interreg Italy-Croatia, the project "Preventing, managing and overcoming natural-hazards risks to mitigate economic and social impact"—PMO-GATE ID 10046122. The research is also partially supported through project KK. 01.1.1.02.0027, co-financed by the CROATIAN GOVERNMENT and the EUROPEAN UNION through the European Regional Development Fund—the Competitiveness and Cohesion Operational Programme and the project "Internet of Things: Research and Applications", UIP-2017-05-4206, financed by the Croatian Science Foundation.

Institutional Review Board Statement: Not applicable.

Informed Consent Statement: Not applicable.

Conflicts of Interest: The authors declare no conflict of interest.

References

1. IPCC. Changes in climate extremes and their impacts on the natural physical environment. In *Managing the Risks of Extreme Events and Disasters to Advance Climate Change Adaptation*; Field, C.B., Barros, V., Stocker, T.F., Dahe, Q., Dokken, D.J., Ebi, K.L., Mastrandrea, M., Mach, K.J., Plattner, G.K., Allen, S.K., et al., Eds.; Cambridge University Press: Cambridge, UK; New York, NY, USA, 2012; pp. 109–230.
2. Petroliagkis, T.I.; Voukouvalas, E.; Disperati, J.; Bidlot, J. Joint Probabilities of Storm Surge, Significant Wave Height and River Discharge Components of Coastal Flooding Events. In *Utilising Statistical Dependence Methodologies and Techniques, JCR Technical Reports*; Publications Office of the European Union: Luxembourg, 2016.
3. Tsoukala, V.K.; Chondros, M.; Kapelonis, Z.G.; Martzikos, N.; Lykou, A.; Belibassakis, K.; Makropoulos, C. An integrated wave modelling framework for extreme and rare events for climate change in coastal areas—The case of Rethymno, Crete. *Oceanologia* **2016**, *58*, 71–89. [CrossRef]
4. Gallien, T.; Sanders, B.; Flick, R. Urban coastal flood prediction: Integrating wave overtopping, flood defenses and drainage. *Coast. Eng.* **2014**, *91*, 18–28. [CrossRef]
5. Barnard, P.L.; Van Ormondt, M.; Erikson, L.H.; Eshleman, J.; Hapke, C.; Ruggiero, P.; Adams, P.; Foxgrover, A.C. Development of the Coastal Storm Modeling System (CoSMoS) for predicting the impact of storms on high-energy, active-margin coasts. *Nat. Hazards* **2014**, *74*, 1095–1125. [CrossRef]
6. Mendoza, E.T.; Jimenez, J.A. Regional vulnerability analysis of Catalan beaches to storms. In Proceedings of the Institution of Civil Engineers—Maritime Engineering; Thomas Telford Ltd.: London, UK, 2009; Volume 162, pp. 127–135.
7. De Kleermaeker, S.; Verlaan, M.; Kroos, J.; Zijl, F. A new coastal flood forecasting system for the Netherlands. In Proceedings of the Hydro12 Conference Proceedings: Taking Care of the Sea, Rotterdam, The Netherlands, 13–15 November 2012; Dorst, L., van Dijk, T., van Ree, R., Boers, J., Brink, W., Kinneging, N., van Lancker, V., de Wulf, A., Nolte, H., Eds.; Hydrographic Society Benelux: Heeg, The Netherlands, 2012. [CrossRef]
8. Doong, D.-J.; Chuang, L.Z.H.; Wu, L.-C.; Fan, Y.-M.; Kao, C.C.; Wang, J.-H. Development of an operational coastal flooding early warning system. *Nat. Hazards Earth Syst. Sci.* **2012**, *12*, 379–390. [CrossRef]
9. Bogaard, T.; De Kleermaeker, S.; Jaeger, W.S.; Van Dongeren, A. Development of Generic Tools for Coastal Early Warning and Decision Support. In Proceedings of the E3S Web of Conferences—3rd European Conference on Flood Risk Management (FLOODrisk 2016), Lyon, France, 17–21 October 2016; EDP Sciences: Paris, France, 2016; Volume 7, p. 18017. [CrossRef]
10. Dreier, N.; Fröhle, P. Operational wave forecast in the German Bight as part of a sensor and risk based early warning system. *J. Coast. Res.* **2018**, *85*, 1161–1165. [CrossRef]
11. Merrifield, M.A.; Johnson, M.; Guza, R.T.; Fiedler, J.W.; Young, A.P.; Henderson, C.S.; Lange, A.M.Z.; O'Reilly, W.C.; Ludka, B.C.; Okihiro, M.; et al. An early warning system for wave-driven coastal flooding at Imperial Beach, CA. *Nat. Hazards* **2021**, *108*, 2591–2612. [CrossRef]
12. Memos, C.; Makris, C.; Metallinos, A.; Karambas, T.; Zissis, D.; Chondros, M.; Spiliopoulos, G.; Emmanouilidou, M.; Papadimitriou, A.; Baltikas, V.; et al. Accu-Waves: A decision support tool for navigation safety in ports. In Proceedings of the 1st International Conference Design and Management of Port, Coastal and Offshore Works (DMPCO), Athens, Greece, 8–11 May 2019.
13. Makris, C.; Androulidakis, Y.; Karambas, T.; Papadimitriou, A.; Metallinos, A.; Kontos, Y.; Baltikas, V.; Chondros, M.; Krestenitis, Y.; Tsoukala, V.; et al. Integrated modelling of sea-state forecasts for safe navigation and operational management in ports: Application in the Mediterranean Sea. *Appl. Math. Model.* **2020**, *89*, 1206–1234. [CrossRef]

14. Spiliopoulos, G.; Bereta, K.; Zissis, D.; Memos, C.; Makris, C.; Metallinos, A.; Ka-rambas, T.; Chondros, M.; Emmanouilidou, M.; Papadimitriou, A.; et al. A Big Data framework for Modelling and Simulating high-resolution hydrodynamic models in sea harbours. In Proceedings of the Global Oceans 2020: Singapore–USA Gulf Coast, Biloxi, MS, USA, 5–30 October 2020; IEEE: New York, NY, USA, 2020; pp. 1–5.
15. Mosavi, A.; Ozturk, P.; Chau, K.-W. Flood Prediction Using Machine Learning Models: Literature Review. *Water* **2018**, *10*, 1536. [CrossRef]
16. Chondros, M.; Metallinos, A.; Papadimitriou, A.; Memos, C.; Vasiliki, T. A Coastal Flood Early-Warning System Based on Offshore Sea State Forecasts and Artificial Neural Networks. *J. Mar. Sci. Eng.* **2021**, *9*, 1272. [CrossRef]
17. Atzori, L.; Iera, A.; Morabito, G. The Internet of Things: A survey. *Comput. Netw.* **2010**, *54*, 2787–2805. [CrossRef]
18. Nižetić, S.; Šolić, P.; López-de-Ipiña González-de-Artaza, D.; Patrono, L. Internet of Things (IoT): Opportunities, issues and challenges towards a smart and sustainable future. *J. Clean. Prod.* **2020**, *274*, 122877. [CrossRef] [PubMed]
19. Reda, H.T.; Daely, P.T.; Kharel, J.; Shin, S.Y. On the application of IoT: Meteorological information display system based on LoRa wireless communication. *IETE Tech. Rev.* **2018**, *35*, 256–265. [CrossRef]
20. LÓpez-Vargas, A.; Fuentes, M.; Vivar, M. On the application of IoT for real-time monitoring of small stand-alone PV systems: Results from a new smart datalogger. In Proceedings of the 2018 IEEE 7th World Conference on Photovoltaic Energy Conversion (WCPEC) (A Joint Conference of 45th IEEE PVSC, 28th PVSEC & 34th EU PVSEC), Waikoloa, HI, USA, 10–15 June 2018; pp. 605–607. [CrossRef]
21. Idier, D.; Aurouet, A.; Bachoc, F.; Baills, A.; Betancourt, J.; Gamboa, F.; Klein, T.; López-Lopera, A.F.; Pedreros, R.; Rohmer, J.; et al. A User-Oriented Local Coastal Flooding Early Warning System Using Metamodelling Techniques. *J. Mar. Sci. Eng.* **2021**, *9*, 1191. [CrossRef]
22. Janeković, I.; Kuzmić, M. Numerical simulation of the Adriatic Sea principal tidal constituents. *Ann. Geophys.* **2005**, *23*, 3207–3218. [CrossRef]
23. Press, W.; Teukolsky, S.; Vetterline, W.T.; Flannery, B.P. *Numerical Recipes: The Art of Scientific Computing*; Cambridge University Press: Cambridge, UK, 2007.
24. Cooley, J.W.; Tukey, J.W. An algorithm for the machine calculation of complex Fourier series. *Math. Comput.* **1965**, *19*, 297–301. [CrossRef]
25. Ponte, R.M. Variability in a homogeneous global ocean forced by barometric pressure. *Dyn. Atmos. Ocean.* **1993**, *18*, 209–234. [CrossRef]
26. The Effect of Fetch Width on Wave Generation, US Corps of Engineers, Technical Memorandum No. 70. 1954. Available online: https://usace.contentdm.oclc.org/digital/api/collection/p266001coll1/id/7549/download (accessed on 27 July 2021).
27. Goda, J. Revisiting Wilson's Formulas for Simplified Wind-Wave Prediction. *J. Waterw. Port Coast.Ocean Eng.* **2003**, *129*, 2003. [CrossRef]
28. World Meteorological Organization, Guide to Wave Analysis and Forecasting, Geneve, Switzerland. 2018. Available online: https://library.wmo.int/doc_num.php?explnum_id=10979 (accessed on 31 July 2021).
29. SMS 13.2 Tutorial CGWAVE Analysis, Aquaveo, v13.2. USA. 2012. Available online: https://www.aquaveo.com/software/sms-learning-tutorials (accessed on 17 August 2021).
30. Demirbilek, Z.; Panchang, V. *CGWAVE: A Coastal Surface Water Wave Model of the Mild Slope Equation*; Technical Report CHL-98-26; US Army Corps of Engineering: Washington, DC, USA, 1998.
31. Panchang, G.V.; Wei, G.; Pearce, R.B.; Briggs, J.M. Numerical Simulation of Irregular Wave Propagation over Shoal. *J. Waterw. Port Coast. Ocean. Eng.* **1990**, *116*, 324–340. [CrossRef]
32. Zhao, L.; Panchang, V.; Chen, W.; Demirbilek, Z.; Chhabbra, N. Simulation of wave breaking effects in two-dimensional elliptic harbor wave models. *Coast. Eng.* **2001**, *42*, 359–373. [CrossRef]
33. Brockmann, E.; Harsson, B.-G.; Ihde, J. Geodetic Reference System of the Republic of Croatia—Consultants Final Report on Horizontal and Vertical Datum Definition, Map Projection and Basic Networks for the state geodetic administration of the Republic of Croatia. 2001; 1–35.
34. Croatian Meteorological and Hydrological Service. Report of Measured Wind Properties on Gauging Station Marjan. 2011. (In Croatian)
35. BARANI DESIGN Technologies, "MeteoHelix®IoT Pro". Available online: https://static1.squarespace.com/static/597dc44391 4e6bed5fd30dcc/t/5f5f9ce8e52f900a7f1982af/1600101615389/MeteoHelix+IoT+Pro+DataSheet.pdf (accessed on 30 March 2022).
36. BARANI DESIGN Technologies, "MeteoWind®IoT Pro". Available online: https://static1.squarespace.com/static/597dc44391 4e6bed5fd30dcc/t/5f1813e0fd8f9a4a19feb781/1595413481804/MeteoWind+IoT+Pro+DataSheet.pdf (accessed on 30 March 2022).
37. Xia, J.; Falconer, R.A.; Xiao, X.; Wang, Y. New criterion for stability of a human body in floodwaters. *J. Hydraul. Res.* **2014**, *52*, 93–104. [CrossRef]

 *applied sciences*

Article

A Comparative Study of the Effects of Earthquakes in Different Countries on Target Displacement in Mid-Rise Regular RC Structures

Ercan Işık [1,*], Marijana Hadzima-Nyarko [2,*], Hüseyin Bilgin [3], Naida Ademović [4], Aydın Büyüksaraç [5], Ehsan Harirchian [6], Borko Bulajić [7], Hayri Baytan Özmen [8] and Seyed Ehsan Aghakouchaki Hosseini [9]

1 Department of Civil Engineering, Bitlis Eren University, Bitlis 13100, Turkey
2 Faculty of Civil Engineering and Architecture Osijek, Josip Juraj Strossmayer University of Osijek, Vladimira Preloga 3, 31000 Osijek, Croatia
3 Department of Civil Engineering, Epoka University, 1001 Tirana, Albania
4 Faculty of Civil Engineering in Sarajevo, University of Sarajevo, 71 000 Sarajevo, Bosnia and Herzegovina
5 Çan Vocational School, Çanakkale 18 Mart University, Çanakkale 17400, Turkey
6 Institute of Structural Mechanics (ISM), Bauhaus-Universität Weimar, 99423 Weimar, Germany
7 Faculty of Technical Sciences, University of Novi Sad, Trg Dositeja Obradovića 6, 21000 Novi Sad, Serbia
8 Department of Civil Engineering, Uşak University, Uşak 64300, Turkey
9 Department of Future Environments, Built Environment Engineering, Faculty of Design and Creative Technologies, Auckland University of Technology (AUT), Auckland 1010, New Zealand
* Correspondence: eisik@beu.edu.tr (E.I.); mhadzima@gfos.hr (M.H.-N.)

Citation: Işık, E.; Hadzima-Nyarko, M.; Bilgin, H.; Ademović, N.; Büyüksaraç, A.; Harirchian, E.; Bulajić, B.; Özmen, H.B.; Aghakouchaki Hosseini, S.E. A Comparative Study of the Effects of Earthquakes in Different Countries on Target Displacement in Mid-Rise Regular RC Structures. *Appl. Sci.* **2022**, *12*, 12495. https://doi.org/10.3390/app122312495

Academic Editors: Andrea Chiozzi and Dario De Domenico

Received: 31 October 2022
Accepted: 30 November 2022
Published: 6 December 2022

Publisher's Note: MDPI stays neutral with regard to jurisdictional claims in published maps and institutional affiliations.

Abstract: Data from past earthquakes is an important tool to reveal the impact of future earthquakes on engineering structures, especially in earthquake-prone regions. These data are important indicators for revealing the seismic loading effects that structures will be exposed to in future earthquakes. Five different earthquakes from six countries with high seismic risk were selected and were within the scope of this study. The measured peak ground acceleration (PGA) for each earthquake was compared with the suggested PGA for the respective region. Structural analyzes were performed for a reinforced-concrete (RC) building model with four different variables, including the number of storeys, local soil types, building importance class and concrete class. Target displacements specified in the Eurocode-8 were obtained for both the suggested and measured PGA values for each earthquake. The main goal of this study is to reveal whether the proposed and measured PGA values are adequately represented in different countries. We tried to reveal whether the seismic risk was taken into account at a sufficient level. In addition, target displacements have been obtained separately in order to demonstrate whether the measured and suggested PGA values for these countries are adequately represented in structural analysis and evaluations. It was concluded that both seismic risk and target displacements were adequately represented for some earthquakes, while not adequately represented for others. Comments were made about the existing building stock of the countries considering the obtained results.

Keywords: target displacement; earthquake; peak ground acceleration; reinforced-concrete; pushover

1. Introduction

Significant loss of life and property after earthquakes increases the consequence of efforts to reduce the effects of earthquakes. The studies on structural and seismic risk analyzes are carried out on both pre-earthquake and post-earthquake in order to prevent and minimize earthquake damages [1–9]. Such studies have special importance in regions with high seismic risk [10]. Ground motion parameters are needed to determine and evaluate the effects of earthquakes in a particular region [11–13]. These parameters are important in terms of both revealing earthquake characteristics and analyzing the behavior of structures under the influence of earthquakes [14–16]. Fault geometry, seismic waves, and earthquake characteristics should be known while the determination of the ground motion parameters by considering local ground conditions. The amplitude parameter is one of the engineering aspects of ground motion parameters. The ground velocity, acceleration and displacement values are known as amplitude parameters [17,18]. Knowing that the earthquake ground motions measurements as a function of time or frequency constitutes an important database for engineering applications and scientific studies for earthquake-resistant structure design [19–21]. In this context, many different programs are used to predict earthquake threats. Openquake Engine [22], Earthquake Loss Estimation Routine (ELER) [23], HAZUS [24], Ez-Frisk [25], PSHRisk-Tool [26], FRISK [27], CRISIS2007 [28], SEISRISK III [29] and OpenSHA [30] are some of the software that are commonly used programs for predicting earthquake threat.

The obtained ground acceleration records from strong ground motion measurements can be used to both determine seismic risk and to monitor the performance of structures during earthquakes. Acceleration records can also be used for the design of earthquake-resistant structures and for the development of attenuation relationships. In addition, the expected damage estimation and intensity distribution in the settlements at different distances from the station can be determined by using attenuation relationships. Earthquake ground motions can be quite complex from this perspective. It is possible to define earthquake motion with three components of linear motion [31,32]. The Peak Ground Acceleration (PGA) is the most common measure used to determine the amplitude of strong ground motion. Any accelerometer used for acceleration records has two horizontal (EW and NS) components and one vertical component. The maximum horizontal ground acceleration is either the geometric mean of the maximum values of the component in both directions or the largest one of them regardless of direction [33–35]. Therefore, obtained PGA values from any earthquake are used to determine seismic and structural risks. Different types of analyzes can be used to decide the performance levels of structures in performance-based design [36–38].

Pushover analysis is a widely used nonlinear analysis technique to estimate the dynamic demands imposed on a structure under earthquake impact. The maximum roof displacements, known as target displacement, are one of the results obtained from this analysis [39–42]. The earthquake performances and damage estimation of the structures can be predicted using the target displacements [43–46]. It is then required to decide the structural performance by comparing the demand values to the deformation capacity for the expected performance levels [47]. Adequate demand displacement values will better reflect real values for the damage estimation of structures and building earthquake performance [48].

In this study, seismic risk and target displacements were compared, taking into account the measured and suggested PGA values for different earthquakes in different countries. Six countries with different seismicity were selected, including as Bosnia and Herzegovina, Albania, Croatia, Iran, Türkiye, and Serbia, and these were within the scope of this study. Two different country groups were selected in this study. In the first group, neighboring Bosnia and Herzegovina, Serbia, Croatia, and Albania were taken into account, while in the second group, neighboring Türkiye and Iran were taken into account. Bulgaria, Macedonia, and Greece are located between these two groups of countries. Earthquakes that occur in both groups of countries also affect other countries within the group. Therefore, seismic and structural parameters were obtained for two different country groups. For this purpose, five different earthquakes were selected for each country. The earthquakes whose data can be accessed were taken into account in the selection of these earthquakes. First, the measured and suggested PGA values were compared for selected earthquakes. Information is provided about the seismicity and the selected earthquakes for each country, respectively. Structural analyzes were made for a sample reinforced concrete (RC) structure to reveal the effect of PGA values. In order to make the structural results more understandable, the RC building has been taken into account with three different numbers of stories, including four, six, and eight-storeys. In order to reveal the effect of different structural conditions, four different variables, namely: the number of storeys, local soil class, building importance class and concrete class were selected. Within the scope of this study, regular mid-rise RC building models were taken into account. In addition, the natural fundamental periods obtained with the empirical formulas used in the earthquake regulations for each country were compared with the period values obtained from the structural analysis. The target displacement values used to determine the performance level and damage estimation of the structures were obtained separately for each number of storeys and each earthquake. In addition, information is given about the building stocks of these countries at the point of the earthquake-structure relationship. The main purpose of this study is to reveal if the suggested PGA values for the building design in seismic design codes and earthquake hazard maps meet the measured PGA values. The novelty of the study is the detailed comparison of both seismic parameters and structural analysis results for six different countries. This study will contribute to the development of seismic hazard maps and seismic design codes for the selected countries. This study will make important contributions to this and similar studies in many different countries and earthquakes.

2. Seismicity of the Selected Countries

Within the scope of this study, six different countries with different seismic characteristics, including Albania, Bosnia and Herzegovina, Croatia, Serbia, Türkiye, and Iran were selected. Comparisons were made by considering the suggested and measured peak acceleration values for the five different earthquakes in each country. In addition to the information about the selected earthquakes, brief information about the seismicity of these countries is given in this section. The locations of selected countries in the active tectonic map were shown in Figure 1.

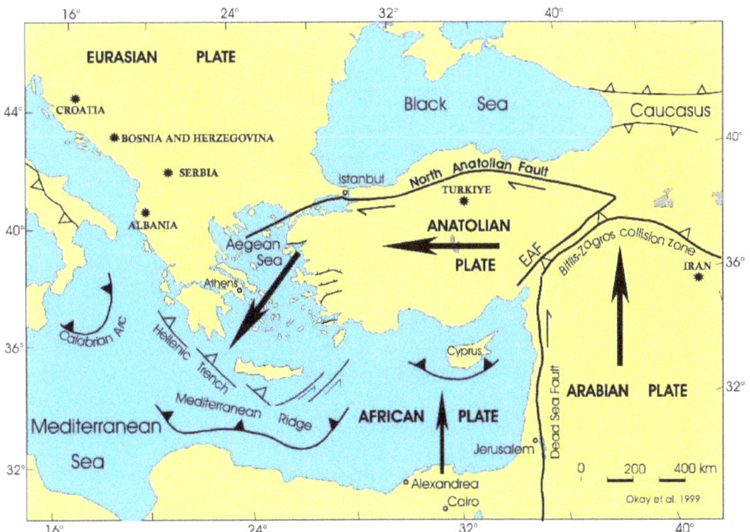

Figure 1. Location of selected countries in the active tectonic map (adopted from [49,50]).

2.1. Albania

Albania is a country with moderate seismicity in the Western Balkans. Located on the Alpine-Mediterranean plate, this region has historically been affected by high-intensity earthquakes. Albanian seismic activity is characterized by intense seismic microactivity (3.0 > M > 1.0) by lots of small earthquakes (5.0 > M > 3.0), few mid-sized earthquakes (7.0 > M > 5.0) and very rarely by large earthquakes (M > 7.0). The most important tremors in the last century are given in Table 1.

Table 1. Major earthquakes in Albania [51].

Date	Area affected	M_w	Depth (km)	Causalities	
				Dead	Injured
26.11.2019	Durres	6.4	20	52	3000+
21.09.2019	Durres	5.6	10	-	108
09.01.1988	Tirana	5.4	24	-	-
16.11.1982	Fier	5.6	22	1	12
15.04.1979	Shkoder	6.9	10	136	1000+
30.11.1967	Diber	6.6	20	12	174
18.03.1962	Fier	6.0	-	5	77
26.05.1960	Korce	6.4	-	7	127
01.09.1959	Fier	6.2	20	2	-
27.08.1942	Diber	6.0	33	43	110
21.11.1930	Vlore	6.0	35	30	100
26.11.1920	Tepelene	6.4	-	36	102
06.01.1905	Shkoder	6.6	-	200	500

Albania and its neighborhood are in a rather complicated seismotectonic region and are prone to earthquakes. A high frequency of earthquakes has been experienced, resulting in loss of life and property destruction in the region (Table 1). According to available records, this region sits in a high rate of seismicity, ranging from moderate to a high seismic risk level. It is characterized by noticeable micro-seismicity (a high number of small earthquakes), sparse mid-sized earthquakes, and very rare large earthquakes. Considering the recorded earthquakes from the accessible data, the earthquakes given in Table 1 were selected by the authors [51].

The first seismic zone intensity map of Albania dates back to 1952. Since then, it has been updated many times until 1979, which is at the moment that the map for seismic evaluation is enforced by the law. The KTP-1963 and KTP-1978 seismic guides were based on the pre-1979s map, which had lower seismic load requirements than the updated values due to a lack of information at the time. Few authors have studied this issue [52]. The largest earthquake in Albania occurred on June 1, 1905, in the North-Western part of Albania with a magnitude of Ms = 6.6. The duration of the tremor was 10–12 s and caused extensive damage to the built environment. In Shkodra alone, around 1500 residential buildings were completely destroyed and all other buildings were severely damaged. In addition, the walls of the historical Shkodra fortress were damaged and partially destroyed. The 15.04.1979 earthquake is one of the strongest earthquakes to occur in the Balkan Peninsula with a moment magnitude of 6.9. The epicenter of this tremor was the coastal area near Petrovac/Montenegro. Several tremors occurred about two weeks before the main shock, and aftershocks lasted for more than nine months. A strong aftershock of Ms = 6.3 occurred on May 24 [53]. This earthquake was one of the main reasons that led to amendments to the earthquake code and seismic zoning maps. Today's seismic zonation map is still based on regions of maximum intensity, not peak ground acceleration. Another strong earthquake occurred in Durrës on November 26, 2019, with a magnitude of M_s = 6.4 [14]. The fact that the epicenter of the earthquake was so close to Albania's most populated and urban area increased the loss of life and injuries. In particular, the old masonry structures in the region were severely damaged and some of them were completely demolished. In this study, this earthquake and its losses will be examined and the results of all analyzes will be compared with the actual damage to the buildings.

The seismic source zones of Albania, characterized by active faults and tectonic regimes, are the essential primary inputs for the estimation of seismic hazards [53]. The following nine earthquake zones have been defined in and around Albania:

1. Zone of Lezha-Ulqin
2. Zone of Peri-Adriatic Lowland
3. Zone of Ionian Coast
4. Zone of Korca-Ohrid
5. Zone of Elbasan-Diber/Tetova
6. Zone of Kukes-Peshkopi
7. Zone of Shkodra-Tropoja
8. Zone of Peja-Prizren
9. Zone of Skopje

The compiled Albania earthquake catalog comprises earthquakes of magnitude Ms > 4.5 that struck the territory between 39.0° N and 43.0° N and between 18.5° E and 21.5° E spanning a timeline 1958–2005 [53]. The best assessments of maximum magnitude are done by taking into account the biggest seismic activity identified and observed in similar tectonic locations. All this data input is processed by utilizing a probabilistic methodology and appropriate attenuation relationships to develop the Probabilistic Hazard Map of Albania.

The seismic zonation map of Albania is based on the intensity values [54], whereas new modern seismic guidelines like Eurocode 8 use probabilistic seismic hazard maps utilizing the peak ground acceleration values derived by probabilistic approaches with different return periods. In many modern codes, Damage Limitation (DL) is expected to be satisfied for an earthquake with peak ground acceleration for a return period of 95 years. Meanwhile, for an earthquake with PGA within the return period of 475 years, buildings should perform as per the limit state of Significant Damage (SD). Seismic hazard maps for maximum horizontal ground acceleration with recurrence periods of 95 and 475 years, respectively, are given for hard rock conditions (Figure 2).

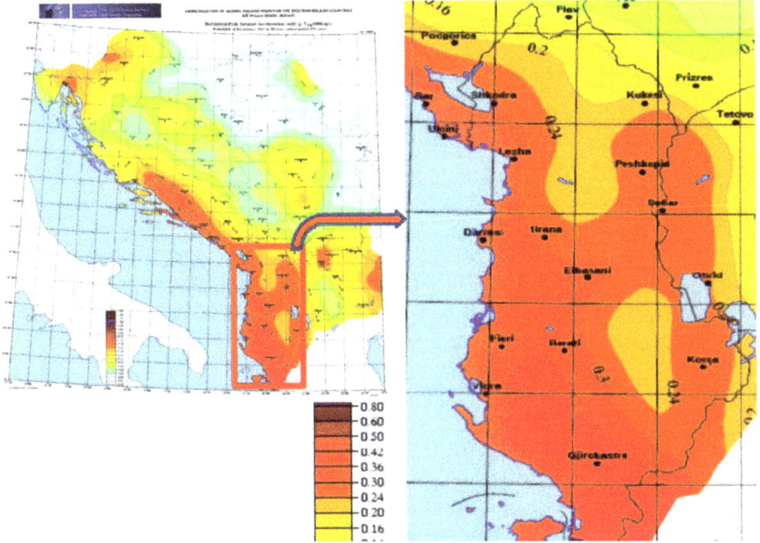

Figure 2. Horizontal peak ground acceleration values for a return period of 475 years (probabilistic seismic hazard map of Albania [55].

As shown in Figures 2 and 4, in many cities with dense masonry structures, such as Durrës, Shkodra, Elbasan, Tirane, and Vlora, the expected PGA for an earthquake with a recurrence period of 95 years is around 0.20 g, whereas this value is around 0.30–0.40 g with a recurrence period of 475 years. If these values are compared to the recordings of the 26 November 2019 shakings, in most of the regions these values are near the values of a 95 year return period. The data for the selected earthquakes in Albania is shown in Table 2.

Table 2. Data of selected earthquakes in Albania.

No	Date	Lat.	Lon.	Magnitude			Loss of Life	Damaged Buildings	Loss of Life/Damaged Buildings	Location
				Mb	Ms	Mw				
1	26/11/2019	41.51°	19.52°			6.4	52	~90,000	0.0006	Durrës/Albania
2	26/11/2019	41.51°	19.52°			6.4	52	~90,000	0.0006	Durrës/Albania
3	21/09/2019	41.43°	19.71°			5.6	-	120	-	Durrës/Albania
4	09/01/1988	41.20°	19.80°			5.9	-	188	-	Tirana/Albania
5	15/04/1979	42.096°	19.209°			6.9	136	~1000	0.14	Shkoder/Albania

A comparison of the measured and suggested PGA values of the selected earthquakes for Albania are given in Table 3.

Table 3. Comparison of the measured and suggested PGA values of the selected earthquakes for Albania.

No	Earthquake Location	Station Name	Year	Earthquake Magnitude (Mw)	PGA(g)	Seismic Risk Zone	A(g)
						Seismic Risk Zone (As Per KTP-N.2-89)	(A(g) (Expected Design Base Acceleration) (from the Probabilistic Map of Albania)
1	Albania	Tirana	2019	6.4	0.11	High	0.30
2	Albania	Durrës	2019	6.4	0.12	High	0.28–030
3	Albania	Tirana	2019	5.6	0.18	High	0.32–0.36
4	Albania	Tirana	1988	5.4	0.40	High	0.28–0.30
5	Albania	Shkoder	1979	6.9	0.46	High	0.30

While the number of damaged buildings in the first two earthquakes considered for Albania was quite high, the loss of life was quite low. In addition, the highest loss of life/damaged buildings ratio for this country was obtained for the fifth earthquake, and this ratio was 0.14. The measured PGA values in these earthquakes that have occurred in these regions with high earthquake risk were considerably lower than the suggested PGA values for the first three earthquakes. However, the measured PGA values for the third and fourth earthquakes are considerably higher than the recommended PGA values. For this country, the seismic risk can be expressed adequately by considering the earthquake ground motion levels for different probabilities of exceedance.

2.2. Bosnia and Herzegovina

Bosnia and Herzegovina is located in the central part of the Dinaridic Mountain System [56]. The location of Mediterranean is characterized by various types of faults that have been identified in this region. The Adriatic coast and the Dinarides are specific for reverse faults, while normal faults are mainly identified in the Apennine Peninsula. The fault plane solution for major earthquakes in Adria has been presented by Slejko et al. [57], while obtaining data from various sources; Gasparini et al. [58], Herak et al. [59], Louvari et al. [60], Sulstarova et al. [61], and Harvard [62].

As stated in the article in [63], quote: "It is evident that with the increase of population in seismically prone areas, urban areas are becoming more vulnerable to seismic risk. Record losses were registered in 2011 [64] after earthquakes that hit Japan and New Zealand, for developed countries with a high degree of earthquake disaster awareness and preparedness. In absolute terms, the costliest disasters happen in the most developed countries, however, with respect to their GDP, it was limited to a few percentage points [65]. The analysis showed that countries of middle income in the last two decades were at a higher risk in comparison to the countries with low and high GDP. From the available data [65], Bosnia and Herzegovina falls into lower-middle-income."

Taking into account the high density of the population, high level of vulnerability of buildings, and moderate to high in some locations PGA results in a high risk of earthquakes in Bosnia and Herzegovina. After the Zagreb 2020 earthquakes, the engineering community awakened regarding the potential risk and level of devastation to the existing building stock in Bosnia and Herzegovina. It should be mentioned that the last devastating earthquake that hit Zagreb was in 1880. Then, 140 years later, the Zagreb 2020 earthquake and Petrinja earthquake occurred and had a major effect on the building and clearly showed the high vulnerability of the existing stock. It is important to state that the Pokupsko- Petrinja Fault is oriented in the NW-SE direction within the Eurasian plate. This is the strongest earthquake that occurred since the 1880 Great Zagreb earthquake (magnitude of 6.3). The seismicity of this region (Croatia and the upper part of Bosnia and Herzegovina-Banja Luka region) is given in Figure 4. Looking at the map, it is believed that the Petrinja fault is the same as the Banja Luka fault, as indicated in Figure 3 and indicated as PKBL = Pokuplje-Kostajnica-Banja Luka right-lateral fault.

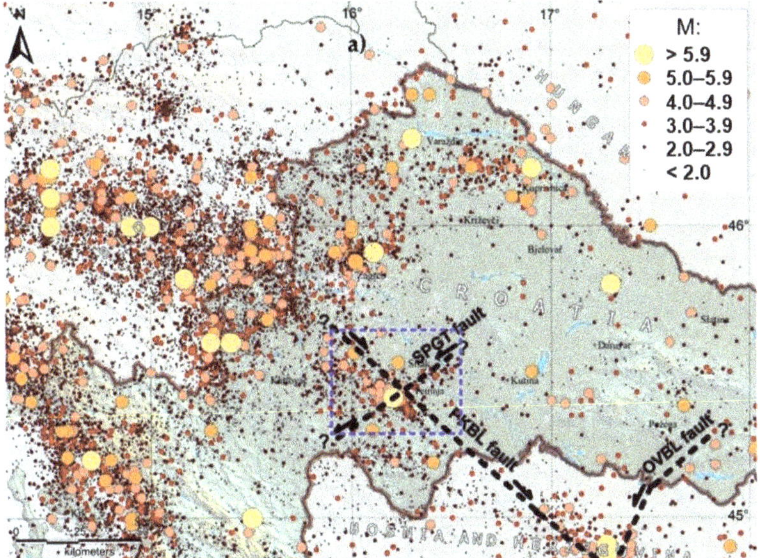

Figure 3. The spatial distribution of earthquakes in Croatia (373 BC–2019, according to the Croatian Earthquake Catalogue (CEC), of which an updated version was first described in [67], with the Pokupsko-Petrinja epicentral area indicated (blue rectangle). Thick, black-dashed lines mark regional active faults: PKBL = Pokuplje-Kostajnica-Banja Luka right-lateral fault, SPGT = Sisak-Petrinja-Glina-Topusko left-lateral fault, and OVBL = Orljava-Vrbas-Banja Luka left-lateral fault.

After the Petrinja earthquake, a quick field inspection revealed that fresh fault planes in the outcrops on the Hrastovička gora appeared mostly along the longitudinal NW–SE-striking Pokupsko–Kostajnica–Banja Luka Fault and showed clear dextral coseismic strike-slip displacements and a 20 km long section of the Pokupsko Fault was (re)activated. It is assumed by Markušić et al. [68] that the creeping sinistral Sisak–Petrinja–Glina–Topusko Fault is locking the dextral Pokupsko–Kostajnica–Banja Luka Fault and a similar complex fault mechanism is also proposed for the Banja Luka area. According to Markušić et al. [68], the dextral Pokupsko–Banja Luka Fault could be one of the main inherited active faults between the crustal segments of Adria.

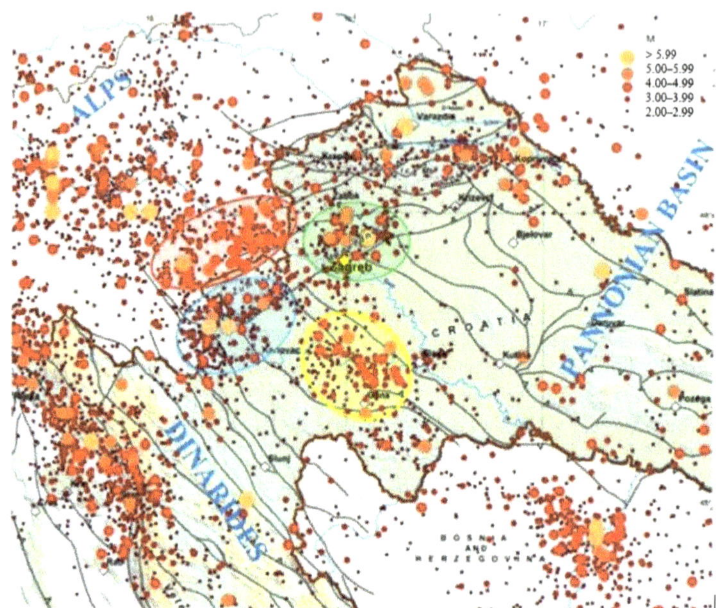

Figure 4. Map of earthquake epicenters in Croatia and part of Bosnia and Herzegovina in the period from BC to 2015 according to the Catalog of Earthquakes in Croatia and the neighboring areas [66–71].

Taking this all into account, it is of the utmost importance to take Bosnia and Herzegovina into account regarding the effects of earthquakes on target displacement in RC structures and other structures as well. Other than the Peak Ground Acceleration, it is necessary to take into account the vulnerability of structures and the exposure of the population during the assessment of the seismic risk. After the Petrinja earthquake, the seismic community in Bosnia and Herzegovina discussions started, and at the moment, there are initiatives for a revision of the interactive seismic map of Bosnia and Herzegovina.

Papeš [72] gave the most comprehensive picture of the tectonic structure in Bosnia and Herzegovina. The longest fault is the Sarajevo Fault, which spreads in the direction of NW-SW, followed by the Banja Luka fault and Konjic Fault. Sarajevo fault with a low to moderate seismic activity level is under-passed by all the transversal deep faults, where the highest seismic motions are noted. Ademović et al. [73] presented that 64% of all earthquakes have a focal depth of up to 10 km and that this is one of the causes of the damaging impact on the structures. Bosnia and Herzegovina in the last 50 years was hit by more than a few medium-sized earthquakes of magnitude M_w up to 6.1 [74]. The earthquake, which had the most devastating impact on the structures, was the 1969 Banja Luka earthquake. According to the MSK-64, the Banja Luka earthquake was marked as a VIII intensity scale [75]. The aftermath of this earthquake was 15 fatalities, 1117 injured people, and over $300 million in damage [74,76].

The second-largest earthquake that should be mentioned is the 1962 Treskavica earthquake with a magnitude Mw= 5.9, and a focal depth of 15 km. As the epicenter of the earthquake was in an abandoned area of Mount Treskavica, there were no major casualties, nor significant damage to the buildings due to the low level of population and construction in this region at that time [77]. Several structures have been damaged in Sarajevo by this earthquake activity (Building of the Executive Council, the Main Post Office, Faculty of Medicine) [73]. The damage caused by this earthquake in the financial means was equal to 396 million dinars [78]. Looking at the period from 306 to 2015, 66.9% of all earthquakes had a magnitude between 3.6–4.5, while 20.5% of the earthquakes had a magnitude in the range of 4.6 to 6. This region was not often hit (4.2% of all earthquakes) by an earthquake of larger magnitudes, while only 8.5% of all earthquakes that hit this region had a magnitude between 3.1 to 3.5 [73].

Figure 5 shows epicentres of regional north-western Balkan earthquakes observed between 1900 and April 2021 with Mw $\geq$ 3.0 [79], as well as the boundaries of Croatia, Bosnia and Herzegovina, and Serbia. It also shows epicentres of the earthquakes from which PGA values have been recorded on rock, as well as the recording sites. In 2018, new seismic hazard maps were compiled for Bosnia and Herzegovina and incorporated into the National Annex to Eurocode 8 [80]. It should be noted that the reference PGA values in these maps are given for ground type A, i.e., for the rock sites. Recently, in all three countries (Croatia, Bosnia and Herzegovina, and Serbia), current official seismic hazard maps are part of the respective National Annexes to Eurocode 8 and the PGA values for rock sites (ground type A) are used to express the hazard. Hence, in Table 5, we have presented only the PGA values recorded on rock (i.e., sites with shear wave velocity in the top 30 m of the soil larger than or equal to 800 m/s). This has unfortunately posed a challenge, since for some devastating historical earthquakes there were very few accelerograph stations on rock sites, while for others we could not find any available data.

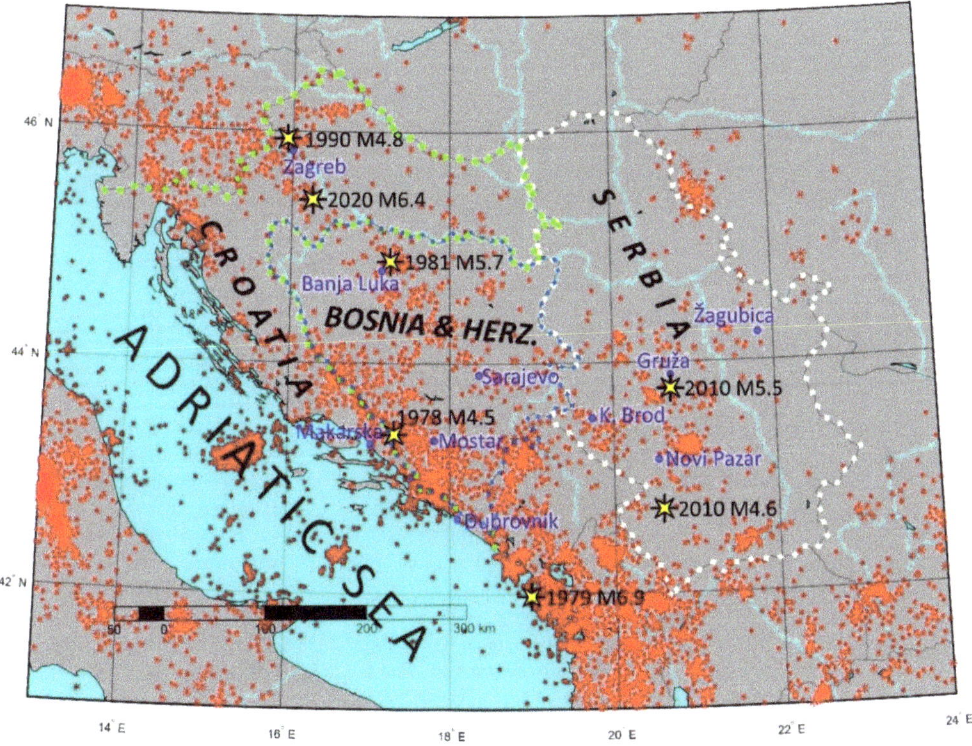

Figure 5. Epicentres of the north-western Balkan earthquakes observed between 1900 and 2021 [79], including the epicentres of the earthquakes that were recorded in Croatia, Bosnia and Herzegovina, and Serbia on rock (blue circles show locations of the corresponding recording stations).

During the analysis, we have chosen the countries of the Balkan as two years ago several earthquakes hit Croatia, which, even though not of "extreme" magnitude, had a major impact on the building stock and community as a whole. The data on the 1981 Banja Luka earthquake are shown in Table 4.

Table 4. Data of the selected earthquake in Bosnia and Herzegovina [81,82].

No	Date	Lat.	Lon.	Magnitude			Loss of Life	Damaged Buildings	Loss of Life/ Damaged Buildings	Location
				Mb	Ms	Mw				
1	13.08.1981	44.82	17.26	5.3	5.5	5.7	-	-	-	Banja Luka (Bosnia and Herzegovina)

The comparison of the PGA values for selected earthquakes in Bosnia and Herzegovina is shown in Table 5. In Table 5, all PGA values were taken from the EQINFOS database [83]. All given PGA values were recorded at rock sites (corresponding to ground type A according to Eurocode 8).

Table 5. Comparison of the measured and suggested PGA values of selected earthquakes for Bosnia and Herzegovina [83].

No	Earthquake Location	Station Name	Year	Earthquake Magnitude (Mw)	Distance to Epicentre to Station (km)	PGA(g)	Seismic Risk Zone	A(g)
							Seismic Risk Zone	(A(g) (Expected Design Base Acceleration) (from the Probabilistic Map of B&H)
1	Banja Luka, B&H	Banja Luka	1981	5.7	7.1	0.29	High	0.17
2	Banja Luka, B&H	Banja Luka	1981	5.7	7.4	0.36	High	0.17
3	Banja Luka, B&H	Banja Luka	1981	5.7	6.5	0.43	High	0.17
4	Montenegro	Sarajevo	1979	6.9	215	0.01	High	0.18
5	Montenegro	Mostar	1979	6.9	177	0.04	High	0.26

For Bosnia and Herzegovina, the loss of life/damaged buildings ratio for the first earthquake, whose data can be accessed, was 0.09. The measured PGA values for the three earthquakes were considerably higher than the predicted PGA values, however, it should be noted that these values were recorded at very short epicentral distances—7.1, 7.4, and 6.5 km, respectively—while the hypocentral depth was only 10 km. Smaller measured PGA values are recorded at large distances of 215 and 177 km, respectively.

2.3. Croatia

As part of the Mediterranean–Trans-Asian belt, the territory of the Republic of Croatia is located in a seismically active area. The territory of Croatia consists of several tectonic units: The Pannonian basin in the north, the eastern part of the Alps in the northwest, the Dinarides, the transition zone between the Dinarides and the Adriatic plate, and the Adriatic plate [84,85]. Structural-geological data on recently active faults, combined with data on seismic activity, form the basis for the interpretation of seismotectonic activity, seismic hazard, and risk in seismically active areas.

The majority of earthquakes in Croatia occur around the Adriatic coast due to the interaction (collision) of the Adriatic Platform and the Dinarides (see Figure 6). However, the north-east parts of Croatia are located in an intraplate low to moderate seismicity region of the Pannonian Basin [85]. Moho depths in Croatia range from 25 km beneath the Pannonian Basin to 45 km beneath the Dinarides [86,87]. Since 2011, current official seismic hazard maps (for a return period of 95 and 475 years) for Croatia are part of the Croatian National Annex to Eurocode 8 [88]. Hazard maps for the return period of 475 years for Croatia are presented in Figure 6. In Table 6, data on selected earthquakes in Croatia are given. All given PGA values were recorded at rock sites (corresponding to ground type A according to Eurocode 8). The ratio of loss of life/damaged buildings was 0.01 for the first earthquake and 0.09 for the third earthquake. Here, the first earthquake is the 6.4 Mw earthquake that devastated the village of Petrinja on 29 December 2020, with the epicentre 40 km south of the capital of Croatia, Zagreb [89]. The focal depth of the earthquake was around 10 km. Another earthquake that should be mentioned here, and which caused damages in Bosnia and Herzegovina as well as in Croatia and even Albania, is the 1979 Montenegro earthquake, which was the strongest earthquake recorded in the area of the former Yugoslavia, with the epicentre offshore in the Adriatic Sea (see Figure 5). While this earthquake was felt up to 900 km from the epicentre, it had destructive consequences only in a 100 km coastal zone and a 25 km stretch from the shore to the mountains [90]. Montenegro suffered 101 and Albania 35 fatalities as a result of the earthquake [90]. This earthquake contributed to the last two PGA values in Table 5, and the third and fourth PGA values in Table 6.

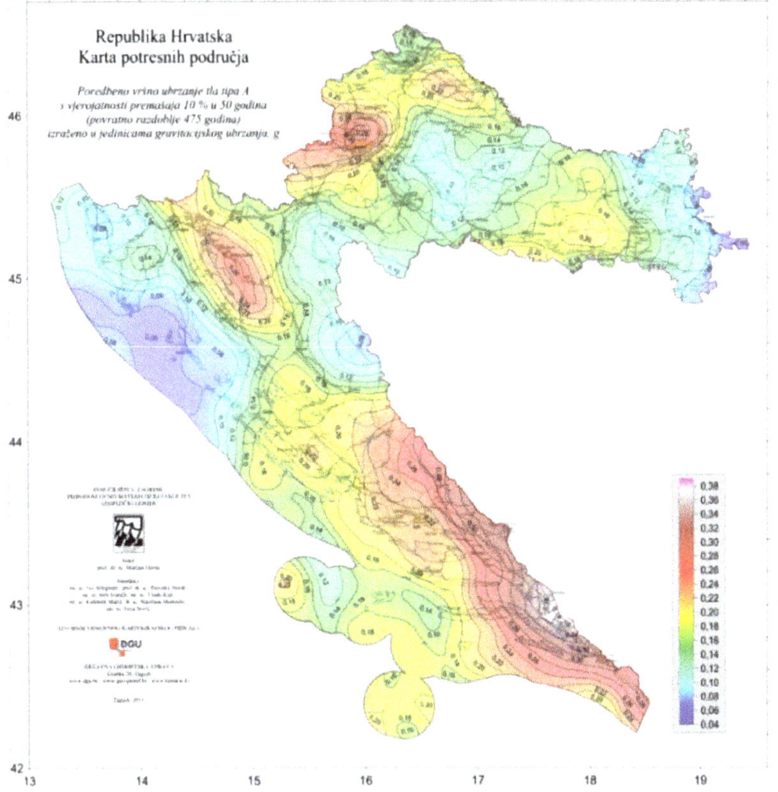

Figure 6. Seismic hazard maps for Croatia for a return period of 475 years (http://seizkarta.gfz.hr/karta.php, accessed 30 August 2022) [91].

Table 6. Data of selected earthquakes in Croatia [79,82,83].

No	Date	Lat.	Lon.	Magnitude			Loss of Life	Damaged Buildings	Loss of Life/ Damaged Buildings	Location
				Mb	Ms	Mw				
1	29.12.2020	45.40	16.22	6.0		6.4	7	8300[*]	0.01	Petrinja, Croatia
2	03.09.1990	45.92	15.92	4.8	4.7	4.9	-	-	-	Kraljev Vrh, Croatia
3	17.12.1978	43.38	17.29	4.5	3.7	4.7	-	-	-	Imotski, Croatia

[*] UNICEF Country Office for Croatia, Earthquake Situation Report #5, 3 February 2021 [92].

The comparison of measured and suggested PGA of the selected earthquakes for Croatia is given in Table 7.

Table 7. Comparison of the measured and suggested PGA values of the selected earthquakes for Croatia [82,83,93,94].

No	Earthquake Location	Station Name	Year	Earthquake Magnitude (Mw)	Distance to Epicentre to Station (km)	PGA(g)	Seismic Risk Zone	A(g) (A(g) (Expected Design Base Acceleration) (from the Probabilistic Map of Croatia)
1	Petrinja, Croatia	Zagreb-Puntijarka	2020	6.4	60	0.04	High	0.279
2	Kraljev Vrh, Croatia	Zagreb	1990	4.9	12	0.06	High	0.259
3	Montenegro	Dubrovnik	1979	6.9	105	0.08	High	0.305
4	Montenegro	Makarska	1979	6.9	208	0.04	High	0.276
5	Imotski, Croatia	Makarska	1978	4.7	24	0.03	High	0.276

Table 7 shows all the PGA values recorded on rock sites in Croatia that could be found at the moment. The first PGA value corresponds to the 2020 Petrinja earthquake [94]. The second PGA value was taken from the ISESD database [81,83]. The last three PGA values were taken from the EQINFOS database [82]. It is interesting to see from Table 5 that, although there were no casualties in Croatia, the PGA values recorded from this earthquake on rock sites at distances of 105 and 218 km are very similar to those recorded in Croatia at much smaller distances, but during moderate size events. From what can be seen from Table 7, the presented PGA values are very, very low compared to the corresponding PGA values given in the Croatian hazard map. However, it should be noted that some of these values were recorded at relatively large epicentral distances. For example, the first value was recorded at a distance of 60 km, while the third and fourth PGA values were recorded at distances of 105 and 218 km, respectively (the hypocentral depth was 12 km). The second value was recorded at the epicentral distance of 12 km while the epicentral depth was 13 km. The fifth value was recorded at the epicentral distance of 24 km, while the epicentral depth was 10 km.

2.4. Serbia

The major part of Serbia is located in intraplate low to moderate seismicity regions. To the north, Serbia comprises the Pannonian Basin's southern part, with a rare occurrence of larger earthquakes [95]. To the southwest, Serbia is surrounded by Dinaric Alps and borders the Mediterranean-Trans-Asian belt, known for its frequent occurrence of stronger earthquakes. To the northeast, Serbia is surrounded by the Carpathian Mountains, and to the southeast by the Balkan Mountains and Rhodopes. The range of the Moho depths is similar to that in Croatia (shallowest beneath the Pannonian Basin and deepest beneath the Dinarides) [86,87]. Normal faults are, however, more common in Serbia than thrusts and strike-slip faults, which do account for practically all occurrences in the External Dinarides.

A series of earthquakes struck central Serbia in the twentieth century, causing largely rural devastation, such as the 1922 M6.0 Lazarevac, 1927 M = 5.9 Rudnik, 1980 M = 5.8 Kopaonik, and 1998 M = 5.7 Mionica earthquakes. The most recent devastating earthquake in Serbia was the M = 5.5 Kraljevo Earthquake, which occurred on 3 November 2010, with an epicentral intensity of VII-VIII °MCS. Two individuals died, 180 people were injured, and numerous buildings were damaged [96].

Data of selected earthquakes that are available for Serbia [97] is given in Table 8 and the comparison of PGA's is given in Table 10. All given PGA values were recorded at rock sites (corresponding to ground type A according to Eurocode 8).

Table 8. Data of selected earthquakes in Serbia [97].

No	Date	Lat.	Lon.	Magnitude			Loss of Life	Damaged Buildings	Loss of Life/ Damaged Buildings	Location
				Mb	Ms	Mw				
1	03.11.2010	43.76	20.73	5.3	5.0	5.5	2	1689	0.01	Kraljevo, Serbia
2	10.03.2010	42.77	20.56	5.0	4.0	4.6	-	-	-	Peć

In 2018, new seismic hazard maps were compiled for Serbia and incorporated into the National Annex to Eurocode 8. Similar to Croatia and Bosnia and Herzegovina, for Serbia, it was also a challenge to find PGA records on rock sites, especially because Serbia did not experience an event with Mw larger than 5.9 in the past 100 years. The values presented in Table 9 are the only ones we could find for the rock sites, and which were recorded by the Seismological Survey of Serbia's (2021) [98] accelerograph network in Serbia.

Table 9. Comparison of the measured and suggested PGA values of the selected earthquakes for Serbia [98].

No	Earthquake Location	Station Name	Year	Earthquake Magnitude (Mw)	Distance to Epicentre to Station (km)	PGA(g)	Seismic Risk Zone	A(g) (A(g) (Expected Design Base Acceleration) (from the Probabilistic Map of Serbia)
1	Kraljevo, Serbia	Gruža	2010	5.5	13	0.06	High	0.20
2	Kraljevo, Serbia	Novi Pazar	2010	5.5	69	0.01	High	0.20
3	Kraljevo, Serbia	Radoinja, Kokin Brod	2010	5.5	83	0.01	Medium	0.15
4	Kraljevo, Serbia	Žagubica	2010	5.5	102	0.01	Medium	0.15
5	Peć	Novi Pazar	2010	4.6	46	0.01	High	0.20

Data of selected earthquakes that are available for Serbia is given in Table 8 and the comparison of PGA's is given in Table 9. All given PGA values were recorded at rock sites (corresponding to ground type A according to Eurocode 8).

For Serbia, the ratio of loss of life to damaged buildings was 0.01 for the first earthquake. The recorded PGA values considered for Serbia are very, very low compared to the corresponding PGA values given in the Serbian official seismic hazard map. However, most of these values shown here were also recorded at relatively large epicentral distances. The epicentral distances for the last four PGA values were 69, 83, 102, and 46 km, respectively, while for the first value the distance was 13 km (the hypocentral depth was 13 km for the first four records and 12 km for the last record).

2.5. Türkiye

Türkiye is situated within the Alpine-Himalayan orogenic belt and is among the most seismically active areas in the world [99,100]. The distribution of seismicity is focused on high-strain regions, many of which are major strike-slip faults, such as the North Anatolian Fault Zone (NAFZ), the East Anatolian Fault Zone (EAFZ), and the Western Anatolian Graben Zones (WAGZ). The NAFZ is a 1200 km long strike-slip fault zone that connects the East Anatolian convergent zone to the Hellenic subduction zone [101–103]. The distribution of earthquakes that dominate the seismic pattern of the northern part of Türkiye is mostly parallel to the NAFZ [104–106]. The NAFZ is a continuous and narrow fault system that cuts the Anatolian Peninsula in an E-W direction from Karlıova in the east to the northern Aegean in the west. The NAFZ, which is the northern plate boundary of the Anatolian Plate with the N-S extensional regime of the Aegean region, spreads as a complex fault system in the eastern part of the Marmara region, in contrast to the simple structure of the NAFZ (Figure 7).

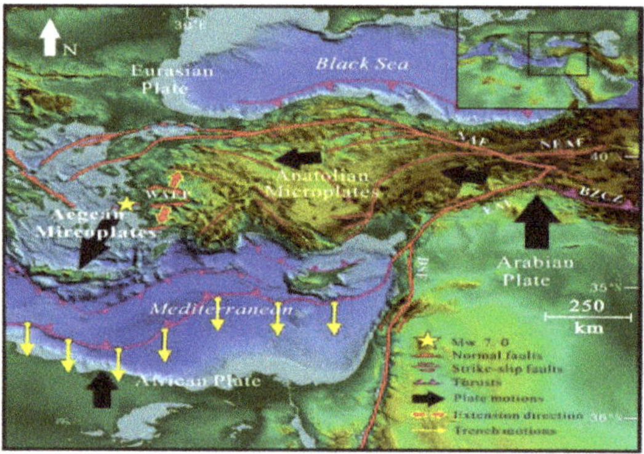

Figure 7. Main tectonics elements for Türkiye [107].

Distribution of the epicenters (M ≥ 3.0) and main fault zones in Türkiye was given in Figure 8.

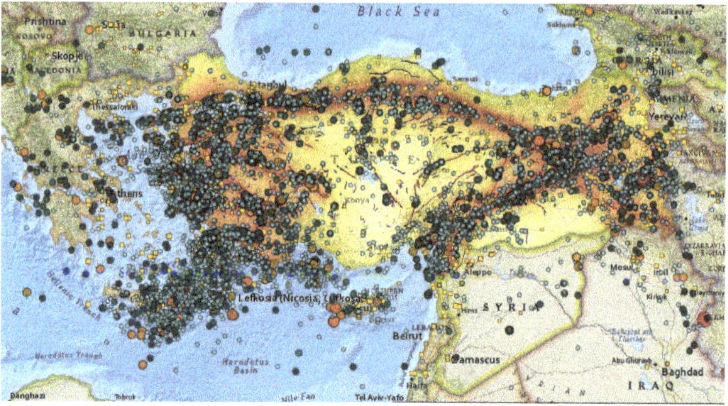

Figure 8. Distribution of epicenters (M ≥ 3.0) and main fault zones in Türkiye.

The earthquakes taken into account for Türkiye are the 1992 Erzincan, 1999 Kocaeli, 1999 Düzce, 2003 Bingöl, and 2011 Van. The 1992 Erzincan earthquake occurred in the eastern half of the Erzincan basin, and two days after this earthquake, the largest aftershock occurred in Pülümür [108]. After this 6.8 magnitude earthquake, many engineering structures were damaged [109]. The 1999 Gölcük (Kocaeli) earthquake, which was felt in and around the Marmara region, caused different levels of structural damage in these settlements. This earthquake, which occurred on the northern branch of the North Anatolian Fault Zone (NAFZ), is associated with a 145 km long surface rupture extending from the southwest of Düzce in the east to the west of the Hersek delta in the west [110]. The 1999 Düzce earthquake, which took place three months after the 1999 Gölcük earthquake, was felt in many different settlements and caused huge structural damage [111]. The surface rupture of this earthquake, which occurred on the Düzce Fault, which is an extension of the North Anatolian Fault Zone in the Bolu Basin, was 40 km long and the maximum right lateral deviation was measured as 500 ± 5 cm [112]. The 2003 earthquake that occurred in Bingöl, one of Türkiye's provinces with high seismicity, occurred approximately 60 km southwest of the triple junction near Karlıova, where the North Anatolian Fault Zone (NAFZ) and the East Anatolian Fault Zone (EAFZ) intersect [113]. The earthquake-causing fault is a right-lateral strike-slip fault and it is stated that the earthquake depth is in the range of 5–15 km [114]. The last earthquake considered in the study is the 2011 Van earthquake that happened in the Lake Van Basin. The epicentral depth of this earthquake, which was centered in Tabanlı village between Van and Erçiş, was measured as 5 km [115]. The large aftershock of 9.11. 2011 ($M_W = 5.7$) was caused by additional damage, especially in the city center of Van, and more than 40 fatalities [116,117]. The settlements where the epicentres of these five different earthquakes, which are considered for Türkiye, have high seismic risk.

The loss of life and property of a total of selected earthquakes and their locations are shown in Table 10. Data on these earthquakes were obtained from the databases of two main institutions that record instrumental earthquakes in Türkiye such as the Republic of Türkiye Prime Ministry Disaster and Emergency Management Presidency (DEMP) and the Kandilli Observatory Earthquake Research Institute of Bogaziçi University (KOERI) and [118,119].

Table 10. Data of selected earthquakes in Türkiye [118,119].

No	Date	Lat.	Lon.	Magnitude			Loss of Life	Damaged Buildings	Loss of Life/ Damaged Buildings	Location
				Mb	Ms	Mw				
1	13.03.1992	39.72	39.63	6.1	6.8		653	8057	0.08	Erzincan
2	17.08.1999	40.76	29.96	6.1	7.4		17,480	73,342	0.24	Gölcük (Kocaeli)
3	12.11.1999	40.81	31.19	6.2	7.2		763	35,519	0.02	Düzce
4	01.05.2003	39.00	40.46	5.7	6.3		176	6000	0.03	Bingöl
5	23.10.2011	38.76	43.36			7.2	644	17,005	0.04	Van

Among the selected earthquakes in this study, the greatest damage occurred in the 1999 Kocaeli (Gölcük) earthquake. The loss of life per building was obtained as 0.24 for this earthquake. The lowest loss of life per building occurred in the 1999 Düzce earthquake. These five different earthquakes caused a total of 19,716 deaths in a total of 139,923 damaged buildings. This data is sufficient to clearly demonstrate Türkiye's earthquake hazard. The loss of life per building for five earthquakes was calculated as 0.14. The measured and recommended PGA values for these earthquakes are given in Table 11. The standard design earthquake ground motion level was selected to determine the suggested PGA values. This level is opposed to probabilities of exceedance of 10% in 50 years, which has a 475-year repetition period.

Table 11. Comparison of the measured and suggested PGA values of the selected earthquakes for Türkiye.

No	Earthquake Location	Station Name	Year	Earthquake Magnitude	Magnitude Type	PGA(g)	Seismic Risk Zone	PGA(g)
							As Per TSDC-2007	As Per TBEC-2018
1	Türkiye	Van	2011	7.2	Mw	0.182	High	0.399
2	Türkiye	Bingöl	2003	6.3	Ms	0.511	Very High	0.633
3	Türkiye	Düzce	1999	7.2	Ms	0.823	Very High	0.588
4	Türkiye	Erzincan	1992	6.8	Ms	0.485	Very High	0.432
5	Türkiye	Kocaeli	1999	7.4	Ms	0.399	Very High	0.690

Except for the third and fourth Düzce earthquakes, the recommended PGA values for the design earthquake were not exceeded for the other three earthquakes. For Türkiye, the recommended PGA values for the first, second and fifth earthquake locations are lower than the predicted PGA values, and the seismic risk for these locations has been adequately taken into account. All earthquake hazard maps used in Türkiye until 2018 were prepared on a regional basis. However, the earthquake hazard is specified specifically for the geographical location with the map currently used after this date. In addition, while there was only one earthquake ground motion level in the previous seismic design code, four different exceedance probabilities are taken into account with the updated code. With the earthquake hazard specific to the geographical location, the expected target displacements from the structures under the effect of the earthquake could be obtained more realistically. Considering the earthquake ground motion levels for different probabilities of exceedance, it is possible for the structures to provide the desired performance levels under the influence of larger earthquakes.

2.6. Iran

The Iranian plateau is located on the Alpine–Himalayan seismic belt, which is considered to be one of the most seismic zones of the world [120,121] and the source of major and destructive earthquakes that occurred in this country throughout history. Some of the most catastrophic earthquakes recorded in the seismic history of Iran include 1960 Lar (Ms = 6.5), 1962 Buin-Zahra (Ms = 7.2), 1978 Tabas (Mw = 7.35), 1990 Manjil (Mw = 7.37), and 2003 Bam (Mw = 6.6) [122]. One of the first elaborate attempts at research on the tectonics and seismicity of Iran was conducted by Ambraseys and Melville (1982) [123]. Berberian (1994) [124] published the first earthquake catalogue of Iran. Updated earthquake catalogues and seismic zoning maps of Iran are regularly published by the seismic zoning sub-committee of the Iranian Seismic Code's permanent committee and are provided by the Iranian Strong Motion Network (ISMN) as the major source of seismology and earthquake engineering in Iran [125]. Figure 9 shows the epicenter of earthquakes that occurred in Iran in 2017 recorded by ISMN (ISMN, 2017) [126], while Figure 10 represents records of large earthquakes that occurred in Iran and adjacent countries from 1900 up to recent years [127]. This figure shows 17 earthquakes with $M_w > 7$, 103 earthquakes with $6 < M_w < 7$, and more than 1700 earthquakes with $M_w > 5$ that have occurred in the recorded seismic history of Iran [125]. It is also demonstrated from Figure 10 that the Zagros zone in the western and southwestern part of Iran is the most seismically active zone which also confirms the major seismic zone categorization proposed by Shoja-Taheri and Niazi (1981) [128]. Based on Figures 11 and 12, as well as the earthquake zonation map of Iran, most of the provinces with large populations are located within high or very high seismic zone areas. As mentioned in the study by Izadkhah and Amini [129], more than 70 percent of cities in Iran are in the vicinity or within the route of active faults, which poses a great risk of seismic hazards to such cities.

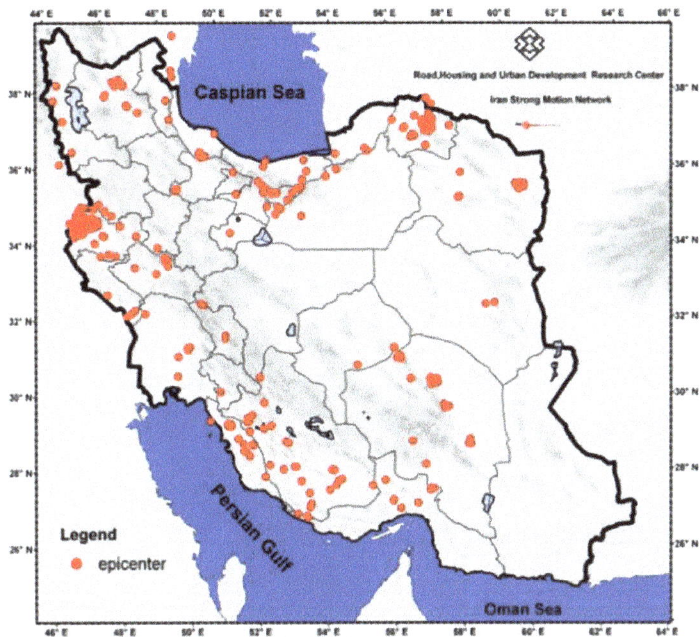

Figure 9. Epicentre of earthquakes occurred in Iran in 2017, recorded by ISMN [126].

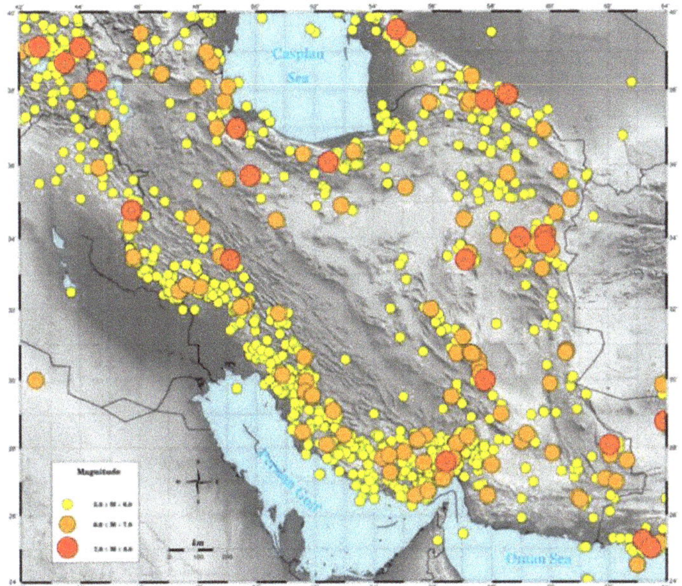

Figure 10. Large earthquakes in Iran and adjacent countries (1900–2019) [126].

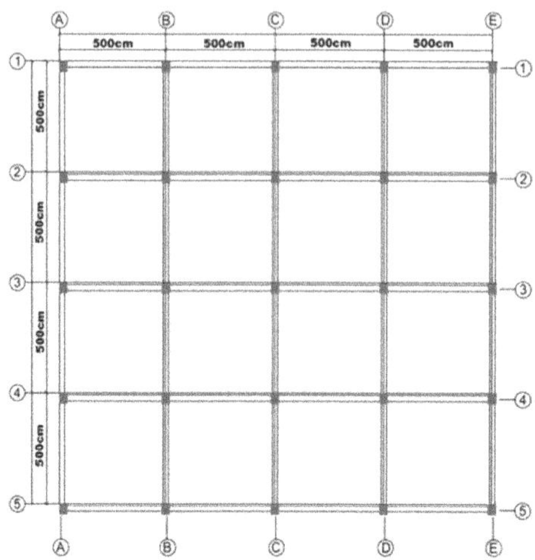

Figure 11. The blueprint of the sample RC building.

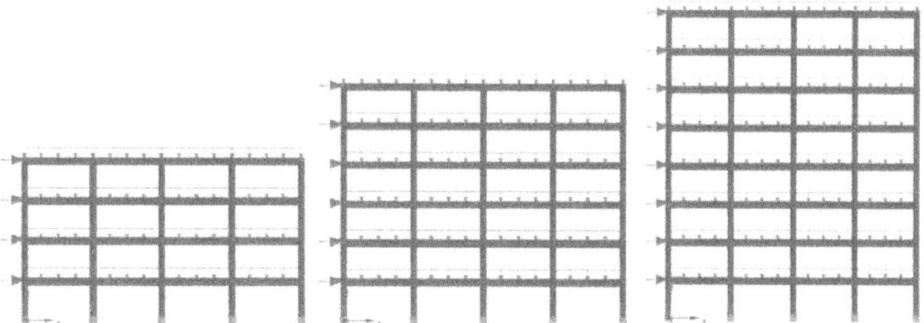

Figure 12. 2D models of the sample RC building for different numbers of stories.

Epicentre locations, loss of life and properties, and magnitudes for some of the most destructive earthquakes that occurred in the seismic history of Iran have been presented in Table 12. Major sources of these data include USGS, ISMN, IIEES, and IRIS.

Table 12. Data of the selected earthquakes in Iran [130–135].

No	Date	Lat.	Lon.	Magnitude			Loss of Life	Damaged Buildings	Loss of Life/ Damaged Buildings	Location
				Mb	Ms	Mw				
1	2003/12/26	29.04	58.33	-	-	6.6	35,000	85%	-	Bam, Iran
2	1990/06/20	36.96	49.41	6.4	7.7	-	40,000–50,000	Nearly all buildings	-	Manjil-Rudbar, Iran
3	1978/09/16	33.37	57.44	6.4	7.4	-	11,000–13,000	>15,000	-	Tabas, Iran
4	1968/08/31	34.02	58.96	-	-	7.2	15,000	>12,000	-	Dasht-e Bayaz, South Khorasan, Iran

Table 13 shows magnitudes, PGA values, and Design Base Accelerations (A(g)) for the calculation of base shear for building structures recommended by the Iranian Code of Practice for Earthquake Resistant Design of Buildings, Standard 2800 [136] for some of the most destructive earthquakes and corresponding seismic zones of Iran.

Table 13. Measured magnitude, PGA values, seismic risk zones, and recommended design base acceleration of selected earthquakes for Iran [126,130–135].

No	Earthquake Location	Station Name	Year	Earthquake Magnitude	Magnitude Type	PGA(g)	Seismic Risk Zone (As Per IS-2800)	A(g) (Design Base Acceleration) (As Per IS-2800)
1	Manjil, Iran	Qazvin	1990	7.37	Mw	0.130	Very High	0.35
2	Manjil, Iran	Rudsar	1990	7.37	Mw	0.086	Very High	0.35
3	Manjil, Iran	Rudsar	1990	7.37	Mw	0.538	Very High	0.35
4	Tabas, Iran	Tabas	1978	7.35	Mw	0.641	Very High	0.35
5	Bam, Iran	Bam	2003	6.60	Mw	0.970	High	0.30

The measured PGA values for the first two recorded earthquakes in the considered stations for Iran are considerably lower than the recommended PGA values, and it can be said that the seismic risk for these locations represents a sufficient level. However, the measured PGA values for the last three earthquakes exceeded the recommended PGA values considerably. This clearly reveals Iran's high potential for seismic risk, taking into account the high population of the selected cities.

3. RC Building Models for Numerical Analysis

Earthquake-resistant rules aim to construct buildings that do not experience damage under an expected ground motion level. Structural analyses for a total of five earthquake locations from each country, whose PGA values can be reached. The Seismostruct software was used for numerical analysis [137]. Pushover analyses were used in these analyses for the sample RC building models with four-storey, six-storey, and eight-storey using obtained data. The story plan was taken in the same way in all analyzed buildings and is shown in Figure 11.

The infrmFBPH (force-based plastic hinge frame elements) were used for structural elements such as beams and columns while creating all building models. These elements model force-based extensional flexibility and limit plasticity to only a finite length. The ideal number of fibers in the section should be sufficient to model the stress-strain distribution in the section [138]. A total of 100 fiber elements are defined for the selected sections. This value is sufficient for such partitions. Plastic-hinge length (Lp/L) was selected as 16.67%. The boundary conditions of the column were set in accordance with the cantilever boundary conditions, which resulted in a fully fixed column footing and a free top end. The boundary condition of the footings was fixed on the ground.

The storey height in all building models is considered as 3 m. The sample RC building was chosen symmetrically in the X and Y directions, and each of this span is 5 m in each direction was considered. The applied loads and 2D and 3D building models are shown for four-storey, six-storey, and eight-storey in Figures 12 and 13, respectively.

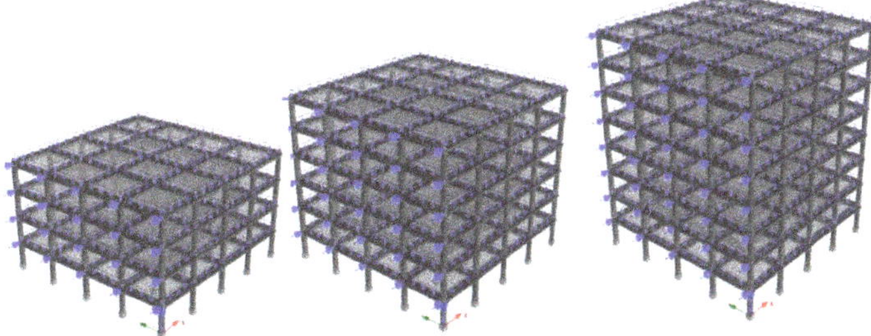

Figure 13. 3D models of the RC building for different numbers of stories.

The structural properties of the RC building model are shown in Table 14.

Table 14. Analysis of input data for the structural models.

Parameter		Value
Concrete Grade		C20
Reinforcement Grade		S420
Beams		250 × 600 mm
Floor height		120 mm
Cover thickness		25 mm
Columns		400 × 500 mm
Longitudinal reinforcement	Corners	4Φ20
	Top bottom side	4Φ16
	Left right side	4Φ16
Transverse reinforcement		Φ10/100
Material model (steel)		Menegotto-Pinto [139]
Material model (Concrete)		Mander et al. nonlinear [140]
Constraint type		Rigid diaphragm
Local soil class		ZD
Incremental loads		5.0 kN
Permanent loads		5.0 kN/m
Target-displacement (4-storey)		0.24 m
Target-displacement (6-storey)		0.36 m
Target-displacement (8-storey)		0.48 m
Importance class		IV
Damping ratio		5%

In performance-based earthquake engineering, it is critical to estimate target displacements for damage estimation when certain performance limits of structural members are reached. The limit states envisaged in Eurocode 8 (Part 3) [141,142] were taken into account for damage estimation in this study. The target displacements are presented in Figure 14 and the description of these states are shown in Table 15.

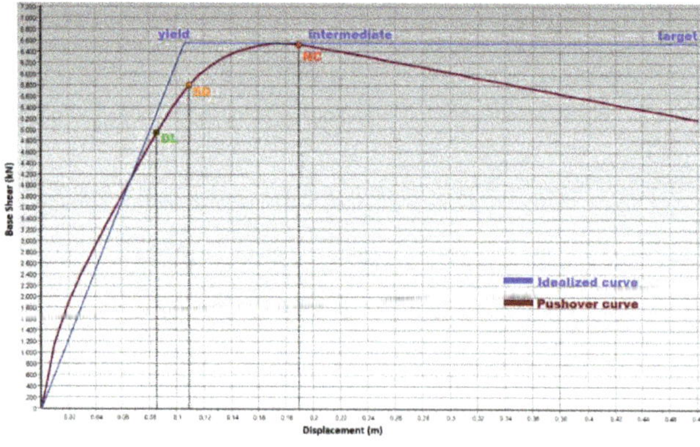

Figure 14. Target displacements on idealized curves/typical pushover.

Table 15. Suggested limit states in Eurocode 8 (Part 3) [141,142].

Limit State	Description	Return Period (Year)	Probability of Exceedance (in 50 Years)
Damage Limitation (DL)	Only lightly damaged, damage to non-structural components is economically repairable	225	0.20
Significant Damage (SD)	Uneconomic to repair, significantly damaged, some residual strength and stiffness, non-structural components damaged,	475	0.10
Near collapse (NC)	Very low residual strength and stiffness, large permanent drift but still standing, heavily damaged	2475	0.02

4. Structural Analyses Results

Within the scope of this study, firstly, the natural fundamental periods for the sample building models were obtained from the eigenvalue analysis. The target displacements and base shear forces for all the structural models were obtained for each country, respectively.

4.1. Comparison of Natural Fundamental Periods

In this part, the period values obtained according to the eigenvalue analyses were compared with the empirical ones predicted for each country. The time required for the undamped system to complete one vibration cycle is called the natural vibration period of the system. The more rigid one of the same mass with a single degree of freedom system will have a shorter natural period and a higher natural frequency. Similarly, of two structures of the same stiffness, the heavier (greater mass) has a lower natural frequency and a longer natural period. This value can be obtained both with approximate formulas and as a result of numerical analysis [143–146]. The empirical relations and explanations stipulated in the corresponding design code for each country are given in Table 16. The comparison of these periods with the ones obtained from structural analyses is shown in Table 16. The empirical formulas used in Table 16 are directly taken from the seismic design codes currently used by countries.

Table 16. Comparison of the natural fundamental periods for selected countries.

Country	Number of Storey	Empirical Formula	Empirical Period (s)	Structural Analyses Period (s)	Description
Albania	4 6 8	$T_1 = (0.09.h)/b^{1/2}$	0.242 0.362 0.483	0.402 0.597 0.796	h—the height of the structure (in meters). b—dimension of the building in parallel to the applied forces (in meters).
Bosnia and Herzegovina	4 6 8	$T = C_t \cdot H^{3/4}$	0.484 0.655 0.813	0.402 0.597 0.796	C_t is 0.075 for RC frame structures and H is the total height of the building
Croatia	4 6 8	$T = C_t \cdot H^{3/4}$	0.484 0.655 0.813	0.402 0.597 0.796	C_t is 0.075 for RC frame structures and H is the total building height
Iran	4 6 8	$T = 0.05H^{0.9}$	0.468 0.674 0.873	0.402 0.597 0.796	H is the total building height
Serbia	4 6 8	$T = C_t \cdot H^{3/4}$	0.484 0.655 0.813	0.402 0.597 0.796	C_t is 0.075 for RC frame structures and H is the total building height
Türkiye	4 6 8	$T_{PA} = C_t \cdot H_N^{3/4}$	0.645 0.874 1.084	0.402 0.597 0.796	H_N is the building's total height; C_t is the correction coefficient. $C_t = 0.1$ for RC building frames that built only beams and columns

The fundamental periods obtained from the structural analyses for all countries were constant since the structural characteristics of the sample RC buildings models did not change. Empirically, the smallest period values were obtained for Albania, while the highest periods were obtained for Türkiye. The empirical periods suggested for Albania were lower than the periods obtained from the structural analysis. For the other five countries, the empirically suggested period values were higher than the period values obtained from the structural analyses.

4.2. Comparisons of Limit States

In this study, the target displacement values for the different number of storeys were obtained from the structural analyses for each country, considering the limit states in Eurocode 8 for six different countries. The comparison of target displacements of sample RC models for Albania is given in Table 17.

Table 17. The obtained target displacements of sample RC models for Albania.

No	Date	Location	Number of Storeys	Code Suggested			Measured		
				DL (m)	SD (m)	NC (m)	DL (m)	SD (m)	NC (m)
1	26 November 2019	Tirana	4	0.115	0.157	0.295	0.035	0.045	0.088
			6	0.203	0.262	0.459	0.071	0.092	0.164
			8	0.272	0.348	0.604	0.099	0.128	0.221
2	26 November 2019	Durres	4	0.115	0.157	0.295	0.039	0.049	0.099
			6	0.203	0.262	0.459	0.078	0.101	0.179
			8	0.272	0.348	0.604	0.109	0.139	0.242
3	21 September 2019	Tirana	4	0.135	0.182	0.338	0.058	0.082	0.164
			6	0.231	0.298	0.521	0.119	0.155	0.272
			8	0.308	0.395	0.685	0.163	0.209	0.362
4	9 January 1998	Tirana	4	0.115	0.157	0.295	0.164	0.219	0.403
			6	0.203	0.262	0.459	0.273	0.351	0.614
			8	0.272	0.348	0.604	0.362	0.465	0.805
5	15 April 1979	Shkoder	4	0.115	0.157	0.295	0.193	0.257	0.468
			6	0.203	0.262	0.459	0.314	0.405	0.707
			8	0.272	0.348	0.604	0.416	0.534	0.926

The target displacements suggested by the seismic design code for the first three earthquakes in Albania provide target displacements in which the acceleration values measured in earthquakes are taken into account. However, the target displacements predicted for the structure for the last two earthquakes and the target displacements obtained under the effect of the earthquake were exceeded. This suggests that the target displacements are adequately represented for some earthquakes, while it is not sufficient for others.

The comparison of target displacements of sample RC models for Bosnia and Herzegovina is given in Table 18.

Table 18. The obtained target displacements of sample RC models for Bosnia and Herzegovina.

No	Date	Location	Number of Storeys	Suggested			Measured		
				DL (m)	SD (m)	NC (m)	DL (m)	SD (m)	NC (m)
1	1969	Bosnia and Herzegovina	4	0.055	0.076	0.154	0.111	0.151	0.284
			6	0.112	0.146	0.257	0.196	0.253	0.443
			8	0.154	0.197	0.342	0.263	0.337	0.584
2	1962	Bosnia and Herzegovina	4	0.055	0.076	0.154	0.145	0.194	0.360
			6	0.112	0.146	0.257	0.245	0.316	0.552
			8	0.154	0.197	0.342	0.326	0.418	0.725
3	1981	Bosnia and Herzegovina	4	0.055	0.076	0.154	0.179	0.237	0.436
			6	0.112	0.146	0.257	0.293	0.378	0.661
			8	0.154	0.197	0.342	0.389	0.499	0.866
4	1969	Bosnia and Herzegovina	4	0.058	0.082	0.164	0.003	0.004	0.007
			6	0.119	0.155	0.272	0.006	0.008	0.014
			8	0.163	0.209	0.362	0.009	0.012	0.020
5	2019	Bosnia and Herzegovina	4	0.096	0.132	0.251	0.013	0.016	0.029
			6	0.175	0.226	0.397	0.026	0.033	0.058
			8	0.235	0.302	0.524	0.036	0.046	0.081

The target displacements suggested by the seismic design code for the first three earthquakes in Bosnia and Herzegovina do not provide target displacements in which the acceleration values measured in earthquakes are taken into account. However, the target displacements predicted for the structure for the last two earthquakes and the target displacements obtained under the effect of the earthquake were not exceeded. This suggests that the target displacements are adequately represented for some earthquakes, while it is not sufficient for others.

The comparison of target displacements of sample RC models for Croatia is given in Table 19.

Table 19. The comparison of target displacements of sample RC models for Croatia.

No	Date	Location	Number of Storeys	Suggested			Measured		
				DL (m)	SD (m)	NC (m)	DL (m)	SD (m)	NC (m)
1	2020	Petrinja	4	0.105	0.144	0.272	0.013	0.016	0.029
			6	0.188	0.243	0.426	0.026	0.033	0.058
			8	0.253	0.324	0.562	0.036	0.046	0.081
2	1990	Kraljev Vrh	4	0.095	0.131	0.250	0.019	0.025	0.043
			6	0.174	0.225	0.395	0.039	0.050	0.087
			8	0.234	0.301	0.522	0.054	0.070	0.121
3	1979	Montenegro	4	0.112	0.160	0.300	0.026	0.033	0.057
			6	0.206	0.266	0.467	0.052	0.067	0.118
			8	0.276	0.352	0.614	0.072	0.093	0.161
4	1979	Montenegro	4	0.104	0.142	0.269	0.013	0.016	0.029
			6	0.186	0.241	0.422	0.026	0.033	0.058
			8	0.250	0.321	0.556	0.036	0.046	0.081
5	1978	Imotsk	4	0.104	0.142	0.269	0.010	0.012	0.021
			6	0.186	0.241	0.422	0.019	0.025	0.043
			8	0.250	0.321	0.556	0.027	0.035	0.060

The target displacements suggested by the seismic design code for all earthquakes in Croatia provide target displacements in which the acceleration values measured in earthquakes are taken into account. It shows that the seismic hazard is adequately taken into account in the structural analysis for all the selected earthquakes.

The comparison of target displacements of sample RC models for Serbia is given in Table 20.

The target displacements suggested by the seismic design code for all earthquakes in Serbia provide target displacements in which the acceleration values measured in earthquakes are taken into account. It shows that the seismic hazard is adequately taken into account in the structural analysis for all selected earthquakes.

The comparison of target displacements of sample RC models for Türkiye is given in Table 21. As seen in Table 21, the target displacements suggested by the seismic design code for the first, second, and fifth earthquakes for Türkiye provide target displacements in which the acceleration values measured in earthquakes are taken into account. However, the target displacements predicted for the other earthquakes for these structures were exceeded. This suggests that the target displacements are adequately represented for some earthquakes, while it is not sufficient for others.

Table 20. The obtained target displacements of sample RC models for Serbia.

No	Date	Location	Number of Storeys	Suggested			Measured		
				DL (m)	SD (m)	NC (m)	DL (m)	SD (m)	NC (m)
1	2010	Gruža	4	0.067	0.094	0.186	0.019	0.025	0.043
			6	0.133	0.173	0.304	0.039	0.050	0.087
			8	0.181	0.232	0.403	0.054	0.070	0.121
2	2010	Novi Pazar	4	0.067	0.094	0.186	0.003	0.004	0.007
			6	0.133	0.173	0.304	0.006	0.008	0.014
			8	0.181	0.232	0.403	0.009	0.012	0.020
3	2010	Kokin Brod	4	0.048	0.063	0.132	0.003	0.004	0.007
			6	0.098	0.128	0.226	0.006	0.008	0.014
			8	0.136	0.174	0.302	0.009	0.012	0.020
4	2010	Žagubica	4	0.048	0.063	0.132	0.003	0.004	0.007
			6	0.098	0.128	0.226	0.006	0.008	0.014
			8	0.136	0.174	0.302	0.009	0.012	0.020
5	1990	Novi Pazar	4	0.067	0.094	0.186	0.003	0.004	0.007
			6	0.133	0.173	0.304	0.006	0.008	0.014
			8	0.181	0.232	0.403	0.009	0.012	0.020

Table 21. The obtained target displacements of sample RC models for Türkiye.

No	Date	Location	Number of Storeys	TBEC-2018, Suggested			Measured		
				DL (m)	SD (m)	NC (m)	DL (m)	SD (m)	NC (m)
1	23 Octorber 2011	Van	4	0.164	0.219	0.402	0.058	0.083	0.167
			6	0.272	0.351	0.612	0.121	0.156	0.276
			8	0.361	0.463	0.803	0.165	0.211	0.366
2	1 May 2003	Bingöl	4	0.278	0.365	0.656	0.218	0.289	0.524
			6	0.435	0.560	0.975	0.350	0.451	0.786
			8	0.573	0.735	1.275	0.463	0.594	1.029
3	12 November 1999	Düzce	4	0.256	0.337	0.607	0.371	0.484	0.862
			6	0.404	0.520	0.906	0.568	0.730	1.270
			8	0.532	0.683	1.184	0.745	0.956	1.657
4	13 March 1992	Erzincan	4	0.180	0.239	0.439	0.206	0.273	0.495
			6	0.295	0.380	0.664	0.332	0.428	0.746
			8	0.391	0.502	0.870	0.439	0.563	0.977
5	17 Augst 1999	Kocaeli	4	0.306	0.401	0.718	0.164	0.219	0.402
			6	0.475	0.611	1.064	0.272	0.351	0.612
			8	0.625	0.801	1.389	0.361	0.463	0.803

The comparison of target displacements of sample RC models for Iran is shown in Table 22.

The target displacements suggested by the seismic design code for the first two earthquakes for Iran provide target displacements in which the acceleration values measured in earthquakes are taken into account. However, the target displacements predicted for the structure for the last three earthquakes and the target displacements obtained under the effect of the earthquake were exceeded. This suggests that the target displacements are adequately represented for some earthquakes, while it is not sufficient for others.

In addition to the structural analysis according to the number of stories, the local soil class change was taken into account. The structural analyzes were carried out only for the four-storey RC building model since it is aimed to reveal the soil class effects. In the previous structural analyses, the ZD soil class envisaged in Eurocode-8 was taken into account. In this section, structural analyzes were made separately for each earthquake by choosing the ZA class in the same code. The recommended properties in the code for these two soil types are given in Table 23. The target displacements for selected earthquakes for each country for the ZA soil class type were given in Table 24.

Table 22. The obtained target displacements of sample RC models for Iran.

No	Date	Location	Number of Storeys	IS-2800 (Suggested)			Measured		
				DL (m)	SD (m)	NC (m)	DL (m)	SD (m)	NC (m)
1	1990	Qazvin	4	0.140	0.188	0.349	0.031	0.039	0.072
			6	0.238	0.307	0.536	0.062	0.079	0.141
			8	0.317	0.407	0.705	0.086	0.110	0.191
2	1990	Rudsar	4	0.140	0.188	0.349	0.028	0.035	0.062
			6	0.238	0.307	0.536	0.056	0.072	0.127
			8	0.317	0.407	0.705	0.078	0.100	0.173
3	1990	Rudsar	4	0.140	0.188	0.349	0.231	0.306	0.553
			6	0.238	0.307	0.536	0.369	0.475	0.828
			8	0.317	0.407	0.705	0.487	0.625	1.083
4	1978	Tabas	4	0.140	0.188	0.349	0.282	0.370	0.665
			6	0.238	0.307	0.536	0.441	0.567	0.988
			8	0.317	0.407	0.705	0.580	0.745	1.291
5	2003	Bam	4	0.115	0.157	0.294	0.442	0.576	1.022
			6	0.203	0.262	0.459	0.670	0.861	1.498
			8	0.272	0.348	0.604	0.878	1.127	1.953

Table 23. The characteristics of local soil types considered in this study [147].

Ground-Type	Description of Stratigraphic Profile	Parameters		
		$V_{s,30}$ (m/s)	N_{SPT} (Blows/30 cm)	C_u (kPa)
A	Rock or other rock-like geological formation, including at most 5 m of weaker material at the surface	>800	—	—
D	Deposits of loose-to-medium cohesionless soil (with or without some soft cohesive layers), or of predominantly soft-to-firm cohesive soil.	<180	<15	<70

Table 24. Comparison of target displacements for the ZA soil class type.

Earthquake No	Country	Location	Number of Storeys	Suggested			Measured		
				DL (m)	SD (m)	NC (m)	DL (m)	SD (m)	NC (m)
1		Tirana	4	0.054	0.070	0.121	0.020	0.025	0.044
2		Durres	4	0.054	0.070	0.121	0.022	0.028	0.048
3	Albania	Tirana	4	0.061	0.079	0.137	0.033	0.042	0.072
4		Tirana	4	0.054	0.070	0.121	0.072	0.093	0.161
5		Shkoder	4	0.054	0.070	0.121	0.083	0.107	0.185
1		Banja Luka,	4	0.030	0.040	0.068	0.052	0.067	0.117
2		Banja Luka	4	0.030	0.040	0.068	0.065	0.083	0.145
3	Bosnia and Herzegovina	Banja Luka,	4	0.030	0.040	0.068	0.078	0.100	0.173
4		Montenegro	4	0.033	0.042	0.072	0.002	0.002	0.004
5		Montenegro	4	0.047	0.060	0.104	0.007	0.009	0.016
1		Petrinja	4	0.050	0.065	0.112	0.007	0.009	0.016
2		Kraljev Vrh	4	0.047	0.060	0.104	0.011	0.014	0.024
3	Croatia	Montenegro	4	0.055	0.071	0.123	0.014	0.019	0.032
4		Montenegro	4	0.050	0.064	0.111	0.007	0.009	0.016
5		Imotsk	4	0.050	0.064	0.111	0.005	0.007	0.012
1		Gruža	4	0.036	0.046	0.080	0.011	0.014	0.024
2		Novi Pazar	4	0.036	0.046	0.080	0.002	0.002	0.004
3	Serbia	Kokin Brod	4	0.027	0.035	0.060	0.002	0.002	0.004
4		Žagubica	4	0.027	0.035	0.060	0.002	0.002	0.004
5		Novi Pazar	4	0.036	0.046	0.080	0.002	0.002	0.004
1		Van	4	0.072	0.092	0.160	0.033	0.042	0.073
2		Bingöl	4	0.114	0.147	0.254	0.092	0.118	0.205
3	Türkiye	Düzce	4	0.106	0.136	0.236	0.149	0.191	0.331
4		Erzincan	4	0.078	0.100	0.174	0.087	0.112	0.195
5		Kocaeli	4	0.125	0.160	0.277	0.072	0.092	0.160
1		Qazvin	4	0.063	0.081	0.141	0.023	0.030	0.052
2		Rudsar	4	0.063	0.081	0.141	0.016	0.020	0.035
3	Iran	Rudsar	4	0.063	0.081	0.141	0.097	0.125	0.216
4		Tabas	4	0.063	0.081	0.141	0.116	0.149	0.258
5		Bam	4	0.054	0.070	0.121	0.175	0.225	0.390

Another parameter chosen in order to put the effect of different structural conditions in common was the importance class of the structure. While the IV class was selected in the previous analysis, it was considered as the II class in the new analysis. The only difference in the initial analysis is the building importance class, all other features remained the same. Selected building importance class characteristics are given in Table 25. The target displacements for selected earthquakes for each country for the II class were given in Table 26.

Table 25. Selected importance classes for buildings [147].

Importance Class	Buildings
II	Ordinary buildings, not belonging to the other categories.
IV	Buildings whose integrity during earthquakes is of vital importancefor civil protection, e.g., hospitals, fire stations, power plants, etc

Table 26. Comparison of target displacements for building important class II.

Earthquake No	Country	Location	Number of Storeys	Suggested			Measured		
				DL (m)	SD (m)	NC (m)	DL (m)	SD (m)	NC (m)
1		Tirana	4	0.074	0.103	0.202	0.025	0.032	0.056
2		Durres	4	0.074	0.103	0.202	0.027	0.035	0.062
3	Albania	Tirana	4	0.088	0.121	0.233	0.041	0.053	0.109
4		Tirana	4	0.074	0.103	0.202	0.109	0.147	0.279
5		Shkoder	4	0.074	0.103	0.202	0.129	0.175	0.326
1		Banja Luka,	4	0.039	0.05	0.101	0.07	0.099	0.194
2		Banja Luka	4	0.039	0.05	0.101	0.094	0.130	0.248
3	Bosnia and Herzegovina	Banja Luka,	4	0.039	0.05	0.101	0.119	0.161	0.302
4		Montenegro	4	0.041	0.053	0.109	0.002	0.003	0.005
5		Montenegro	4	0.06	0.085	0.171	0.009	0.011	0.02
1		Petrinja	4	0.066	0.094	0.185	0.009	0.011	0.02
2		Kraljev Vrh	4	0.059	0.085	0.170	0.014	0.018	0.031
3	Croatia	Montenegro	4	0.075	0.105	0.205	0.018	0.024	0.041
4		Montenegro	4	0.065	0.093	0.183	0.009	0.011	0.02
5		Imotsk	4	0.065	0.093	0.183	0.007	0.009	0.015
1		Gruža	4	0.046	0.059	0.124	0.014	0.018	0.031
2		Novi Pazar	4	0.046	0.059	0.124	0.002	0.003	0.005
3	Serbia	Kokin Brod	4	0.034	0.044	0.085	0.002	0.003	0.005
4		Žagubica	4	0.034	0.044	0.085	0.002	0.003	0.005
5		Novi Pazar	4	0.046	0.059	0.124	0.002	0.003	0.005
1		Van	4	0.108	0.147	0.278	0.042	0.054	0.11
2		Bingöl	4	0.190	0.252	0.46	0.147	0.198	0.365
3	Türkiye	Düzce	4	0.174	0.232	0.425	0.256	0.337	0.607
4		Erzincan	4	0.12	0.162	0.304	0.138	0.186	0.345
5		Kocaeli	4	0.21	0.278	0.504	0.108	0.147	0.278
1		Qazvin	4	0.091	0.126	0.240	0.030	0.038	0.700
2		Rudsar	4	0.091	0.126	0.240	0.020	0.025	0.044
3	Iran	Rudsar	4	0.091	0.126	0.24	0.157	0.21	0.386
4		Tabas	4	0.091	0.126	0.240	0.192	0.256	0.466
5		Bam	4	0.074	0.103	0.202	0.307	0.403	0.721

In addition to all these different structural conditions, the concrete class is also considered as a variable. While previous analyzes were performed for the C20 concrete class, new structural analyzes considered the C12 concrete class with lower properties for all load-bearing elements. The target displacements for selected earthquakes for each country for the C12 concrete class were given in Table 27.

Table 27. Comparison of target displacements for the C12 concrete class.

Earthquake No	Country	Location	Number of Storeys	Suggested			Measured		
				DL (m)	SD (m)	NC (m)	DL (m)	SD (m)	NC (m)
1		Tirana	4	0.124	0.167	0.311	0.038	0.049	0.096
2		Durres	4	0.124	0.167	0.311	0.042	0.054	0.107
3	Albania	Tirana	4	0.144	0.193	0.356	0.063	0.089	0.175
4		Tirana	4	0.124	0.167	0.311	0.175	0.232	0.424
5		Shkoder	4	0.124	0.167	0.311	0.205	0.272	0.492
1		Banja Luka,	4	0.059	0.082	0.164	0.119	0.161	0.3
2		Banja Luka	4	0.059	0.082	0.164	0.154	0.206	0.379
3	Bosnia and Herzegovina	Banja Luka,	4	0.059	0.082	0.164	0.19	0.252	0.458
4		Montenegro	4	0.063	0.088	0.175	0.004	0.005	0.008
5		Montenegro	4	0.104	0.141	0.266	0.014	0.018	0.031
1		Petrinja	4	0.113	0.153	0.287	0.014	0.018	0.031
2		Kraljev Vrh	4	0.103	0.14	0.264	0.021	0.027	0.047
3	Croatia	Montenegro	4	0.126	0.17	0.316	0.028	0.036	0.062
4		Montenegro	4	0.112	0.151	0.284	0.014	0.018	0.031
5		Imotsk	4	0.112	0.151	0.284	0.01	0.013	0.023
1		Gruža	4	0.073	0.102	0.198	0.021	0.027	0.047
2		Novi Pazar	4	0.073	0.102	0.198	0.004	0.005	0.008
3	Serbia	Kokin Brod	4	0.052	0.069	0.141	0.004	0.005	0.008
4		Žagubica	4	0.052	0.069	0.141	0.004	0.005	0.008
5		Novi Pazar	4	0.073	0.102	0.198	0.004	0.005	0.008
1		Van	4	0.174	0.232	0.423	0.064	0.09	0.177
2		Bingöl	4	0.293	0.385	0.688	0.231	0.305	0.55
3	Turkey	Düzce	4	0.271	0.355	0.637	0.39	0.509	0.903
4		Erzincan	4	0.191	0.253	0.46	0.218	0.288	0.52
5		Kocaeli	4	0.322	0.422	0.752	0.174	0.232	0.423
1		Qazvin	4	0.149	0.2	0.367	0.045	0.058	0.118
2		Rudsar	4	0.149	0.2	0.367	0.03	0.039	0.069
3	Iran	Rudsar	4	0.149	0.2	0.367	0.245	0.323	0.58
4		Tabas	4	0.149	0.2	0.367	0.297	0.389	0.697
5		Bam	4	0.124	0.167	0.311	0.465	0.605	1.069

4.3. Evaluation of Existing Building Stocks

4.3.1. Albania

According to the Albanian Institute of Statistics (INSTAT) 2001, the Albanian building stock primarily consists of four typologies, namely brick and stone, prefabricated, wood, and other building materials. However, when referring to the recent census of 2011 (IN-STAT), information on building materials is not included and houses are classified based on their heights and construction period. Table 28 presents a summary of the Albanian building stock based on existing information (INSTAT 2001) [148]. Accordingly, the 'RC and masonry' type represents the biggest part of the current building stock.

Table 28. Albanian building stock [147].

Material Type	<1945	1945–1960	1961–1980	1981–1990	1991–1995
RC and masonry structures	37416	63870	141170	102198	43324
Prefabricated concrete	-	-	4601	5993	4575
Wooden	462	-	1821	1273	743
Other types	2560	3393	7105	6263	4238

According to the Albanian Institute of Statistics (INSTAT) 2011, one-storey buildings account for 85% of the total building stock corresponding to the accommodation of the half population of the country. They were mostly built with unreinforced masonry and reinforced concrete frames with infill walls. However, the total number of multi-storey houses in Albania is significantly lower compared to one-storey houses, but they shelter the remaining half of the population. During the recent earthquake sequences in 2019, multi-storey buildings were significantly affected, resulting in higher damage in the stricken areas. While the available data given in Table 24 is outdated, they highlight an important indicator (design code) on the construction year of the housings. An important portion of the current building stock was built before 1990 showing a lack of adequacy to the modern code requirements [17]. Therefore, it is likely that there were deficiencies affecting the seismic performance of buildings constructed in this time period.

4.3.2. Bosnia and Herzegovina

According to the available data for Bosnia and Herzegovina (CBS 2013) [149], it is noted that, from the total of 1,078,156 buildings, 60.72% are structures made of brick, stone, and concrete, 35.08% of reinforced concrete and steel frames and only 4.20% of wood and light material [73]. The majority of the structures are either confined masonry buildings or RC buildings constructed according to the regulations from 1981 (35%), considering age distribution from 1981–1990. Since 1991 Prestandards (ENV) have been applied and this accounts for 18.9% of all buildings being either RC buildings or confined masonry buildings. The application of Eurocode 8 started after 2006 accounting for 2.0% of all buildings (as well as RC buildings or confined masonry buildings). Masonry structures with rigid floors were mainly constructed in the period from 1971 to 1980, amounting to 33.6% of all structures built in Bosnia and Herzegovina. Brick masonry structures with rigid RC slabs were built in the period from 1946 to 1970, amounting to 6.6% of all the structures built in Bosnia and Herzegovina. The remaining 4% is devoted to stone masonry buildings with wooden floors constructed before 1945. Seismic vulnerability assessment of structures in Bosnia and Herzegovina is mainly done by individual researchers [150–153]. At the moment, Bosnia and Herzegovina does not have a well-organized and efficient database of structures and building's typologies. Several studies were conducted to determine the vulnerability of buildings in several cities of Bosnia and Herzegovina, like Banja Luka and Sarajevo [74]), Visoko [154], and Tuzla [155]. For the first time, the specific site and its influence on the vulnerability were taken into consideration in the Tuzla region [155]. Currently, 700 structures in the city of Sarajevo are being examined and a database is being created [156]. Based on all this preliminary analysis, it is clear that most of the existing building stock in Bosnia and Herzegovina does not possess sufficient resistance to ground motions that may be expected in this region. It is necessary to construct a detailed database taking into account all the data required to conduct adequate seismic assessments and perform a seismic risk assessment. Without this database and conducted calculations, it is not possible to construct an effective disaster management plan.

4.3.3. Croatia

According to the 2011 Census, the total number of dwellings in Croatia by year of construction was 1,496,558. Of that, 13.2% were built before 1945, which means that they did not follow any building codes. Building design and construction in Croatia did not follow earthquake-resistant building rules until 1948 [157].

Masonry houses used timber floor constructions until 1920. Most of these structures were constructed between 1860 and 1920 and are now part of Croatia's historic town centers, most of which are categorized as historical heritage. These structures were not intended to withstand significant horizontal ground motions (e.g., earthquakes). After 1930, the first semi-prefabricated RC floors were installed, followed by monolithic RC floors in 1964. After the Skopje earthquake in 1963, the first seismic building codes were developed and later modified. In addition, following the earthquake in Skopje in 1963, masonry structures throughout the former Yugoslavia were erected systematically using horizontal tie-beams and vertical tie-columns to achieve confined masonry. The load-bearing system in reinforced concrete structures (RC frames and RC shear walls) was built in accordance with the seismic regulations enacted in 1964 (following the 1963 Skopje earthquake) and 1981 (following the 1979 Montenegro (coast) earthquake). Eurocodes were gradually adopted as voluntary structural design norms between 1992 and 1998. Due to the challenges associated with the harmonization of new standards with old national legislation at the time, they kept a pre-standards status (ENV label). The final version was introduced in 1998, with the European standard (EN label), but the ultimate implementation began in 2005 with the adoption of the technical standards for concrete buildings (NN 101/05). Eurocodes were ultimately made a requirement in official usage in 2011, however, pre-standards were still used until the end of 2012 [147].

Predominant structural systems for the buildings in one of the Croatian cities (Osijek) can be summarized as follows [147]: Unreinforced masonry buildings made of old bricks with flexible floors, unreinforced masonry structures with rigid floors, confined masonry structures, RC frame structures, RC shear walls, and RC dual structures. For RC structures, the level of earthquake resistance design should be taken into account.

On December 29, 2020, an earthquake of magnitude 6.4 MW hit Sisak-Moslavina county with an epicentre 3 km southwest of the Croatian city of Petrinja. In the preliminary report on the consequences of the earthquake, a detailed description of the damage to residential low-rise and multi-family residential buildings is presented [158]. The following are the primary sources of damage and failure in low-rise residential buildings: Excessive lateral displacements of flexible timber flooring caused out-of-plane damage or failure of exterior masonry walls at the upper/top levels of older URM structures (built before World War II). Recent masonry structures have rigid floors, but they also suffered damage owing to the lack of vertical reinforcement at the bottom floor level. Due to the extremely high seismic demand, the in-plane damage pattern took the form of diagonal tension cracks in the walls. The major tensile stresses in the walls created by the earthquake surpassed the masonry tensile strength, resulting in the formation of inclined cracks (diagonal tension cracks). In certain situations, the quality of masonry materials and construction appeared to be poor, which was also a source of damage. The primary sources of damage and failure in multi-family residential buildings can be summarized as follows: Excessive lateral displacements of flexible timber flooring caused out-of-plane damage or failure of exterior masonry walls at the upper/top levels of older URM structures (built before World War II). In general, the failure process in low-rise and mid-rise structures is relatively similar.

Many older URM buildings were not properly maintained, and as a result, their condition was poor before the earthquake [159]. The level of damage is thought to have been impacted by the degradation of building materials and components (such as wooden floors and roofing) as well as the use of weak mortar.

Due to extremely high seismic demand, masonry structures with rigid floors built in the 1960s developed in-plane shear cracking at the building's base. The major tensile stresses in the walls created by the earthquake surpassed the masonry tensile strength, resulting in the formation of inclined cracks (diagonal tension cracks). The earthquake caused no structural damage to RC structures; nevertheless, minor damage to non-structural components such as chimneys occurred in several buildings.

4.3.4. Serbia

According to the 2011 Serbian Census of population, household, and dwellings, 85% of all dwellings in Serbia were constructed after 1945, i.e., in the period when there were at least some seismic design codes. Before 2019, the seismic design codes created and implemented in the former Yugoslavia were used. The first seismic design code was published in 1948, but it lacked detailed detailing guidelines for RC and masonry construction. The disastrous earthquakes that struck Skopje in 1963 and Montenegro in 1979 served as turning points in the creation of Yugoslavian seismic design codes. Following the 1963 Skopje Earthquake, the first complete seismic design code was published in 1964. Two years after the 1979 Montenegro earthquake, a new, much more advanced code was published. The seismic design of new structures in Serbia must comply with Eurocode 8—Part 1 [147] as of 2019 [160]. However, although most buildings were constructed after 1945, a recent moderate-size Mw = 5.5 2010 Kraljevo earthquake revealed the vulnerability of the Serbian building stock. It should be noted that reinforced-concrete structures accounted for only 10% of the building stock in the affected area. As expected, buildings with unreinforced brick masonry walls and flexible diaphragms sustained notable damages while properly constructed modern confined masonry buildings remained undamaged. However, numerous one- or two-storey masonry buildings with rigid RC floors and horizontal ring beams suffered severe damage and/or partial collapse due to the inadequate design and construction and/or low-quality building materials and many multi-storey masonry buildings were damaged due to poorly planned and executed renovations and extensions [161]. These findings call for increased efforts toward a more realistic estimation of the vulnerability of the existing building stock in Serbia if one would like to obtain a realistic estimate of the seismic risk.

4.3.5. Türkiye

Most of the existing building stock in Türkiye does not have sufficient resistance to earthquakes. This is clearly seen from the observed damage caused by the recent earthquakes in Türkiye. Earthquake regulations are renewed over time and put into effect, especially after the large-scale loss of life and property [162–164]. Insufficient structural features of the existing building stock play an active role in losses in earthquakes. For this reason, these uninspected buildings, which constitute the majority of the building stock, should be examined, some of them should be strengthened and others should be evaluated within an urban transformation project. The need for low-cost housing as a result of unplanned urbanization due to population growth and migration to big cities in Türkiye has caused both the shift from residential areas to areas with high earthquake hazards and the growth of building stock with weak earthquake safety. Knowing the characteristics of both new buildings and relatively old buildings with weak earthquake safety is of great importance in order to make accurate earthquake risk and loss calculations of settlements. To reduce the damage caused by earthquakes and for effective disaster management, the earthquake risks of existing structures should be calculated realistically. In this context, it is of great importance to know the properties of the building stock that affect the earthquake behavior well, to make the risk and loss calculations correctly. In this respect, examining the Turkish building stock in terms of time and space is of great importance in earthquake risk calculations.

4.3.6. Iran

After the 1990 Manjil mega-earthquake, one of the biggest and most fatal incidents in the seismic history of Iran that claimed the lives of more than 40,000 people, several investigations were initiated by many researchers on the analysis of the damages and vulnerability of the building stocks in the similar earthquake-stricken areas. The catastrophe caused a turning point in the analysis and design approaches of buildings and several modifications to the Iranian Code of Practice for Earthquake Resistant Design of Buildings, Standard 2800, and the definition of many research programs [165]. A study was conducted on three earthquakes in Iran, including the 2003 Bam earthquake, the 2005 Zarand earthquake, and the 2006 Silakhor earthquake by Mahdi and Mahdi [166]. The Bam earthquake has been the biggest earthquake with the highest rate of fatalities in the country after the 1990 Manjil earthquake, in which more than 53,000 buildings were destroyed while the remaining structures were severely damaged [167]. Damage analysis of buildings after the 2003 Bam earthquake by Mostafaei and Kabeyasawa [168] showed that building stock in this city at the time of the earthquake was comprised of adobe, masonry (reinforced and unreinforced), steel, and concrete buildings. As shown in this study, the major building system type has been unreinforced masonry (for around 68% of the buildings), and only 24% of the buildings in the city had been designed seismic-resistant, having a structural system as per the Iranian seismic design code of practice. This is while even the remaining steel or reinforced concrete (RC) damaged buildings suffered from inappropriate structural design, low-quality construction practices, and insufficient implementation controls. Buildings that have been destroyed or heavily damaged in the other two earthquakes have been mostly adobe or unreinforced masonry buildings, while inadequate seismic-resistant structural systems or unsuitable construction practices were recognized as the main reason for damages. According to the 2016 national census conducted by the Statistical Centre of Iran (SCI) on the residential building stock and different building types, it is understood that the five main types of building systems, i.e., concrete, steel, masonry, adobe, and wooden, could be recognized in Iran. The census shows that masonry, steel, and concrete structures contain 39%, 30%, and 27% of the building stock, respectively. The remaining building types were either wooden or adobe, with 0.14% and 4%, respectively, while the remaining 0.26% had no recognizable system [169]. The study by Bastami et al. [169] which proposes new seismic vulnerability models for building stocks in Iran, shows a considerable change in newer versions of the Iranian Code of Practice for Earthquake Resistant Design of Buildings, Standard 2800, in terms of response factor for calculating base shear in the equivalent static method. These revisions as well as other stricter regulations and modifications for the design and construction of seismic-resistant buildings are in line with the above-mentioned started programs and initiatives for more protection of structures in Iran.

5. Results and Conclusions

Both the seismic parameters and the expected target displacements from the structures have been obtained by considering five earthquakes that occurred in six different countries with different seismic risks within the scope of the study. The highest PGA value for all considered earthquakes was obtained in the 2003 Bam (Iran) earthquake and is 0.970 g. The lowest measured PGA was obtained as 0.01 g for the Serbian earthquakes. While the measured PGA's for Croatia and Serbia provided the recommended PGA's, the recommended PGA values for Türkiye, Bosnia and Herzegovina, Iran and Albania were exceeded. PGA values were compared, considering the standard design earthquake with a 10% exceedance of probability in 50 years (repetition period of 475 years). Therefore, it is possible that these values can be met with the consideration of earthquakes with a larger repetition period. Considering the largest earthquake data as the ground motion level in regions with high seismicity risk means that the seismicity risk can be adequately represented.

The selected earthquake range in Albania is between 5.4–6.9. Medium-sized earthquakes are mostly in the range of 0.1–0.2 g. However, the acceleration of the 5.4 magnitude earthquake that occurred in Tirana in 1988 was recorded as very high (0.4 g), exceeding the expected acceleration value (0.28–0.3 g). Two earthquakes, one moderate (5.7) and the other large (6.9) were selected in Bosnia and Herzegovina. The first selected earthquake exceeded the expected acceleration value (0.17 g) at nearby stations and took values in the range of 0.29–0.43 g. However, the 6.9 magnitude earthquake created an acceleration value (0.01–0.4 g) far below the expected acceleration value (0.18–0.26 g) at stations approximately 200 km away. In three different earthquakes selected in Croatia, a small (4.9) earthquake, however, close to the recording station (6.4 km) created an acceleration of 0.0 g. The acceleration value created by the 6.4 magnitude earthquake 60 km away from the station was 0.04 g. The 6.9 magnitude earthquake that occurred in Montenegro in 1979 had an acceleration of 0.08 g in Dubrovnik, 105 km away, and 0.04 g in Makarska, 208 km away. All of the recorded accelerations are considerably lower than the expected acceleration values. The earthquakes considered in Serbia are medium-sized, and the distances of the earthquakes to the acceleration stations are also high. Moreover, the places where the stations are located are rocky. For this reason, the acceleration values formed were quite low. The lowest of the five earthquakes selected that occurred in Türkiye is 6.3 and the highest is 7.4. It is quite interesting that three earthquakes greater than 7 produce very different accelerations from each other. The lowest acceleration was recorded in Van (Mw= 7.2) with 0.182 g and Düzce (Mw = 7.2) with 0.823 g. However, very high accelerations were observed in Erzincan (Mw = 6.8) earthquakes in 1992 and Bingöl (Mw = 6.3) earthquakes in 2003 (Bingöl 0.511 g and Erzincan 0.485 g). One of the main factors in recording very high acceleration values is the proximity of the recording station to the earthquake focus and the other is ground conditions. It is seen that very high acceleration values were recorded in three different earthquakes in Iran. The 7.37 magnitude earthquake that occurred in Manjil in 1990 was recorded differently at different stations. Accordingly, the acceleration value of 0.13 g in Qazvin, 0.086 g in Rudsar, and 0.538 g at the other station in Rudsar show how different geological conditions affect the acceleration value. The acceleration of the 7.35 magnitude earthquake that occurred in Tabas in 1978 was recorded as 0.64 g, and the acceleration of the 6.6 magnitude earthquake that occurred in Bam in 2003 was recorded as an extraordinarily large 0.97 g. It is clear that the very loose soil structure of the city of Bam played the most important role in such magnification of the acceleration value.

The highest loss of life/damaged buildings ratio was found as 0.24 for 17.08.1999 Türkiye (İzmit) and the lowest value was determined as 0.0006 for 26.11.2019 Albania (Durrës). While the highest loss of life among all earthquakes was 17,480 in the 17.08.1999 Türkiye (İzmit) earthquake, the most building damage occurred in the Albania (Durrës) earthquake of 26.11.2019 with ~90000 buildings.

In order to reveal the effect of different structural conditions within the scope of this study, the number of storeys, local soil class, building importance class, and concrete class were chosen as variables. There is complete agreement between the target displacements obtained for all variables. The target displacements increased for three different limit conditions as the number of storeys in the building increased. In the case of weak local soil properties, the target displacements were obtained larger. The values obtained for ZA are lower than the values obtained for ZD. At the same time, target displacements were found to be larger in buildings that were required to be used after the earthquake. The displacements obtained for the building importance class IV are larger than those obtained for the II. class. As the strength of the concrete decreased, the target displacements expected from the structure increased. This once again reveals that buildings with weak earthquake vulnerability require larger displacements.

However, it was examined whether the seismic risks taken into account for different countries are adequately represented. In this context, since the seismicity elements of each country differ, the losses resulting from the earthquakes vary. Therefore, it is obvious that the realistic determination of the seismic risk will result in a more realistic result with the performance levels expected from the structures. In this respect, the building stock characteristics and local ground conditions also directly affect the losses. The vulnerability of the existing building stock increases the structural damage.

Earthquakes occur in fragile parts of the earth's crust due to their formation mechanism. Loose layers near the surface cannot be a source of earthquakes in this sense. However, since such areas are areas of weakness, they allow the incoming tremor to reach the surface easily and stand out because they are geologically highly impacted areas. In this study, it is observed that the earthquakes selected in countries other than Bosnia and Herzegovina and Croatia occur under the influence of geological conditions. Selected earthquakes in Albania are mostly in Durres, Shkoder, and Tirana. The depression areas formed with the neotectonic uplift that started in the Pliocene period in Albania led to the formation of Quaternary lakes and plains. In these grabens, the thickness of which reaches 200 m, unstable soil formations at the swamp level cause earthquakes to be more effective [170]. In Serbia, mainly earthquakes occurred in Kraljevo. This area is in the current alluvial and Tertiary flysch structure. The second important earthquake zone is the Pec zone, which is also the flysch zone. Therefore, it can be said that earthquakes occurring in these regions are based on weak geological conditions. Almost all the earthquakes selected in Türkiye have occurred in the current alluvial areas (Erzincan, Kocaeli, Düzce, Bingöl, Van). Plain regime areas created by very thick alluvial structures and active tectonism continue to be sources of earthquakes. The Bam and Manjil earthquakes, which were selected from the earthquakes that occurred in Iran, were effective in the current alluvial basin-type areas.

The reason for the occurrence of earthquakes in these areas can be considered as specific geological conditions.

Buildings constructed in loose and unstable ground conditions are the most vulnerable to earthquakes. For this purpose, it is necessary to choose the soil-building interaction correctly. Almost all the earthquakes selected in this article, which are considered important in the country where they occurred due to the damage caused, have resulted in severe damage due to incompatibility. One of the main purposes of this study is to show that the damages caused by earthquakes without borders are based on similar faults.

It is important to construct buildings in accordance with earthquake-resistant building design guidelines in earthquake-prone regions against the possibility of the recurrence of earthquakes that cause significant damage. This depends on the correct application of earthquake-resistant building design principles during the design and construction stages. The application of earthquake-resistant building design principles together with adequate supervision can be seen as the first step in minimizing the problems, both during the project and construction phases. In addition, the existing building stock should be determined quickly and reliably, and then strengthening and demolition procedures should be decided in buildings that do not have sufficient earthquake performance. At this point, the number of weak buildings under the effect of earthquakes should be minimized by utilizing urban transformation.

Author Contributions: Conceptualization, E.I., M.H.-N., H.B., N.A., A.B., E.H., B.B., H.B.Ö. and S.E.A.H.; methodology, E.H., S.E.A.H., E.I., A.B. and E.H.; software, E.I., N.A., H.B. and M.H.-N.; validation, E.H., A.B., H.B.Ö. and S.E.A.H.; formal analysis, B.B.; investigation, E.I., M.H.-N., H.B., N.A., A.B., E.H., B.B., H.B.Ö. and S.E.A.H.; resources, E.I., M.H.-N., H.B., N.A., A.B., E.H., B.B., H.B.Ö. and S.E.A.H.; data curation, E.I., M.H.-N., H.B., N.A., A.B., E.H., B.B., H.B.Ö. and S.E.A.H.; writing—original draft preparation, N.A., A.B., E.I., B.B. and M.H.-N.; writing—review and editing, S.E.A.H., N.A., M.H.-N., E.I. and E.H.; visualization, H.B.; supervision, E.I., N.A., M.H.-N. and B.B.; project administration, E.I.; funding acquisition, M.H.-N. All authors have read and agreed to the published version of the manuscript.

Funding: This research received no external funding.

Institutional Review Board Statement: Not applicable.

Informed Consent Statement: Not applicable.

Data Availability Statement: Data sharing is not applicable.

Conflicts of Interest: The authors declare no conflict of interest.

References

1. Kumar, S.; Gupta, V.; Kumar, P.; Sundriyal, Y.P. Coseismic landslide hazard assessment for the future scenario earthquakes in the Kumaun Himalaya, India. *Bull. Eng. Geol. Environ.* **2021**, *80*, 5219–5235. [CrossRef]
2. Işik, M.F.; Işik, E.; Harirchian, E. Application of IOS/Android rapid evaluation of post-earthquake damages in masonry buildings. *Gazi Mühendislik Bilimleri Derg.* **2021**, *7*, 36–50.
3. Sandhu, M.; Sharma, B.; Mittal, H.; Chingtham, P. Analysis of the site effects in the North East region of India using the recorded strong ground motions from moderate earthquakes. *J. Earthq. Eng.* **2022**, *26*, 1480–1499. [CrossRef]
4. Halder, L.; Dutta, S.C.; Sharma, R.P.; Bhattacharya, S. Lessons learnt from post-earthquake damage study of Northeast India and Nepal during last ten years: 2021 Assam earthquake, 2020 Mizoram earthquake, 2017 Ambasa earthquake, 2016 Manipur earthquake, 2015 Nepal earthquake, and 2011 Sikkim earthquake. *Soil Dyn. Earthq. Eng.* **2021**, *151*, 106990. [CrossRef]
5. Işik, E.; Sağır, Ç.; Tozlu, Z.; Ustaoğlu, Ü.S. Determination of Urban Earthquake Risk for Kırşehir, Turkey. *Earth Sci. Res. J.* **2019**, *23*, 237–247. [CrossRef]
6. Hoveidae, N.; Fathi, A.; Karimzadeh, S. Seismic damage assessment of a historic masonry building under simulated scenario earthquakes: A case study for Arge-Tabriz. *Soil Dyn. Earthq. Eng.* **2021**, *147*, 106732. [CrossRef]
7. Tabrizikahou, A.; Hadzima-Nyarko, M.; Kuczma, M.; Lozančić, S. Application of shape memory alloys in retrofitting of masonry and heritage structures based on their vulnerability revealed in the Bam 2003 earthquake. *Materials* **2021**, *14*, 4480. [CrossRef] [PubMed]
8. Ertuncay, D.; Malisan, P.; Costa, G.; Grimaz, S. Impulsive signals produced by earthquakes in Italy and their potential relation with site effects and structural damage. *Geosciences* **2021**, *11*, 261. [CrossRef]

9. Ditommaso, R.; Iacovino, C.; Auletta, G.; Parolai, S.; Ponzo, F.C. Damage detection and localization on real structures subjected to strong motion earthquakes using the curvature evolution method: The Navelli (Italy) case Study. *Appl. Sci.* **2021**, *11*, 6496. [CrossRef]

10. Ozmen, H.B. A view on how to mitigate earthquake damages in turkey from a civil engineering perspective. *Res. Eng. Struct. Mater.* **2021**, *7*, 1–11. [CrossRef]

11. Zuo, H.; Bi, K.; Hao, H.; Ma, R. Influences of ground motion parameters and structural damping on the optimum design of inerter-based tuned mass dampers. *Eng. Struct.* **2021**, *227*, 111422. [CrossRef]

12. Felicetta, C.; Mascandola, C.; Spallarossa, D.; Pacor, F.; Hailemikael, S.; Di Giulio, G. Quantification of site effects in the Amatrice area (Central Italy): Insights from ground-motion recordings of the 2016–2017 seismic sequence. *Soil Dyn. Earthq. Eng.* **2021**, *142*, 106565. [CrossRef]

13. Mase, L.Z.; Likitlersuang, S.; Tobita, T. Ground motion parameters and resonance effect during strong earthquake in northern Thailand. *Geotech. Geol. Eng.* **2021**, *39*, 2207–2219. [CrossRef]

14. Bilgin, H.; Shkodrani, N.; Hysenlliu, M.; Ozmen, H.B.; Isik, E.; Harirchian, E. Damage and performance evaluation of masonry buildings constructed in 1970s during the 2019 Albania earthquakes. *Eng. Fail. Anal.* **2022**, *131*, 105824. [CrossRef]

15. Kamal, M.; İnel, M. Correlation between Ground motion parameters and displacement demands of mid-rise rc buildings on soft soils. *Buildings* **2021**, *12*, 125. [CrossRef]

16. Gijini, A.; Cullufi, H.; Deneko, E.; Xhika, P. Behavior of structure type 82/2 (RC frame), during the earthquake of 26 November 2019 in Durrës, Albania. *Res. Eng. Struct. Mater.* **2021**, *7*, 595–615. [CrossRef]

17. Bulajić, B.Đ.; Pavić, G.; Hadzima-Nyarko, M. PGA vertical estimates for deep soils and deep geological sediments–A case study of Osijek (Croatia). *Comput. Geosci.* **2022**, *158*, 104985. [CrossRef]

18. Işık, E. Comparative investigation of seismic and structural parameters of earthquakes (M $\geq$ 6) after 1900 in Turkey. *Arab. J. Geosci.* **2022**, *15*, 971. [CrossRef]

19. Nayak, C.B. A state-of-the-art review of vertical ground motion (VGM) characteristics, effects and provisions. *Innov. Infrast. Solut.* **2021**, *6*, 124. [CrossRef]

20. Bhanu, V.; Chandramohan, R.; Sullivan, T.J. Influence of ground motion duration on the dynamic deformation capacity of reinforced concrete frame structures. *Earthq. Spectra* **2021**, *37*, 2622–2637. [CrossRef]

21. Todorov, B.; Billah, A.M. Seismic fragility and damage assessment of reinforced concrete bridge pier under long-duration, near-fault, and far-field ground motions. *Structures* **2021**, *31*, 671–685. [CrossRef]

22. Silva, V.; Crowley, H.; Pagani, M.; Monelli, D.; Pinho, R. Development of the OpenQuake engine, the Global Earthquake Model's open-source software for seismic risk assessment. *Nat. Hazards* **2014**, *72*, 1409–1427. [CrossRef]

23. Hancilar, U.; Tuzun, C.; Yenidogan, C.; Erdik, M. ELER software–a new tool for urban earthquake loss assessment. *Nat. Hazards Earth Syst. Sci.* **2010**, *10*, 2677–2696. [CrossRef]

24. Crowley, H.; Pinho, R.; Bommer, J.J. A probabilistic displacement-based vulnerability assessment procedure for earthquake loss estimation. *Bull. Earthq. Eng.* **2004**, *2*, 173–219. [CrossRef]

25. EZ-FRISK. Available online: https://www.ez-frisk.com/ (accessed on 13 November 2022).

26. Nahar, T.T.; Rahman, M.M.; Kim, D. Effective safety assessment of aged concrete gravity dam based on the reliability index in a seismically induced site. *Appl. Sci.* **2021**, *11*, 1987.

27. McGuire, R.K. FRISK: Computer program for seismic risk analysis using faults as earthquake sources. *US Geol. Surv.* **1978**, *78*, 1007.

28. Ordaz, M.; Martinelli, F.; D'Amico, V.; Meletti, C. CRISIS2008: A flexible tool to perform probabilistic seismic hazard assessment. *Seismol. Res. Lett.* **2013**, *84*, 495–504. [CrossRef]

29. Bender, B.; Perkins, D.M. *SEISRISK III: A Computer Program for Seismic Hazard Estimation*; (No. 1772); US Government Printing Office: Washington, DC, USA, 1987.

30. Field, E.H.; Jordan, T.H.; Cornell, C.A. A developing community-modeling environment for seismic hazard analysis. *Seismol. Res. Lett.* **2003**, *74*, 406–419. [CrossRef]

31. Özener, P. Dinamik Yükler, Yer Hareketi Parametreleri ve İvme Spektrumları, Yıldız Tenk Üniversitesi, Ders Notları. Available online: https://studylibtr.com/doc/1406240/ders-2.1-dinamik-y%C3%BCkler--yer-hareketi-parametreleri-ve-i%CC%87vme (accessed on 11 November 2022).

32. Büyüksaraç, A.; Över, S.; Geneş, M.C.; Bikçe, M.; Kacin, S.; Bektaş, Ö. Estimating shear wave velocity using acceleration data in Antakya (Turkey). *Earth Sci. Res. J.* **2004**, *18*, 87–98.

33. Pejovic, J.R.; Serdar, N.N.; Pejovic, R.R. Optimal intensity measures for probabilistic seismic demand models of RC high rise buildings. *Earthq. Struct.* **2017**, *13*, 221–230.

34. Tao, D.; Ma, Q.; Li, S.; Xie, Z.; Lin, D.; Li, S. Support vector regression for the relationships between ground motion parameters and macroseismic intensity in the Sichuan–Yunnan Region. *Appl. Sci.* **2020**, *10*, 3086. [CrossRef]

35. Işık, E.; Peker, F.; Büyüksaraç, A. The effect of vertical earthquake motion on steel structures behaviour in different seismic zones. *J. Adv. Res. Nat. Appl. Sci.* **2022**, *8*, 527–542. [CrossRef]

36. Liu, J.; Wang, W.; Dasgupta, G. Pushover analysis of underground structures: Method and application. *Sci. China Technol. Sci.* **2014**, *57*, 423–437. [CrossRef]

37. Elnashai, A.S. Advanced inelastic static (pushover) analysis for earthquake applications. *Struct. Eng. Mech.* **2001**, *12*, 51–69. [CrossRef]
38. Shendkar, M.R.; Kontoni, D.P.N.; Işık, E.; Mandal, S.; Maiti, P.R.; Harirchian, E. Influence of masonry infill on seismic design factors of reinforced-concrete buildings. *Shock Vib.* **2022**, *2022*, 5521162. [CrossRef]
39. Chopra, A.K.; Goel, R.K. A modal pushover analysis procedure for estimating seismic demands for buildings. *Earthq. Eng. Struct. Dyn.* **2002**, *31*, 561–582. [CrossRef]
40. Rofooeil, F.; Attari, N.K.; Shodja, A.; Rasekh, A. Comparison of static and dynamic pushover analysis in assessment of the target displacement. *Int. J. Civ. Eng.* **2006**, *4*, 212–225.
41. Krawinkler, H.; Seneviratna, G.D.P.K. Pros and cons of a pushover analysis of seismic performance evaluation. *Eng. Struct.* **1998**, *20*, 452–464. [CrossRef]
42. Işık, E.; Karaşin, İ.B.; Karaşin, A. The effect of different earthquake ground motion levels on the performance of steel structures in settlements with different seismic hazards. *Struct. Eng. Mech.* **2022**, *84*, 85–100.
43. Tso, W.K.; Moghadam, A.S. Pushover procedure for seismic analysis of buildings. *Prog. Struct. Eng. Mater.* **1998**, *1*, 337–344. [CrossRef]
44. Pinho, R.; Casarotti, C.; Antoniou, S. A comparison of single-run pushover analysis techniques for seismic assessment of bridges. *Earthq. Eng. Struct. Dyn.* **2007**, *36*, 1347–1362. [CrossRef]
45. Kim, S.; D'Amore, E. Push-over analysis procedure in earthquake engineering. *Earthq. Spectra.* **1999**, *15*, 417–434. [CrossRef]
46. Papanikolaou, V.K.; Elnashai, A.S. Evaluation of conventional and adaptive pushover analysis I: Methodology. *J. Earthq. Eng.* **2005**, *9*, 923–941. [CrossRef]
47. Işık, E.; Kutanis, M. Determination of local site-specific spectra using probabilistic seismic hazard analysis for Bitlis Province, Turkey. *Earth Sci. Res. J.* **2015**, *19*, 129–134. [CrossRef]
48. Kutanis, M.; Ulutaş, H.; Işik, E. PSHA of Van province for performance assessment using spectrally matched strong ground motion records. *J. Earth Syst. Sci.* **2018**, *127*, 99. [CrossRef]
49. Available online: https://seismo.berkeley.edu/gifs/blog_20200126_Figure1.jpg (accessed on 16 November 2022).
50. Okay, A.I.; Kaşlılar-Özcan, A.; İmren, C.; Boztepe-Güney, A.; Demirbağ, E.; Kuşçu, İ. Active faults and evolving strike-slip basins in the Marmara Sea, northwest Turkey: A multichannel seismic reflection study. *Tectonophysics* **2000**, *321*, 189–218. [CrossRef]
51. Bilgin, H.; Hysenlliu, M. Comparison of near and far-fault ground motion effects on low and mid-rise masonry buildings. *J. Build. Eng.* **2020**, *30*, 101248. [CrossRef]
52. Fundo, A.; Ll, D.; Kuka, S.; Begu, E.; Kuka, N. Probabilistic seismic hazard assessment of Albania. *Acta Geod. Geophys. Hung.* **2012**, *47*, 465–479. [CrossRef]
53. Sulstarova, E.; Peçi, V.; Shuteriqi, P. Vlora-Elbasan-Diber transversal fault and its seismic activity. In *29th General Assembly of IASPEI*; Abstract; P. Ziti & Co. Thessaloniki: Thessaloniki, Greece, 1997; p. 115.
54. *KTP-1978 Kusht Teknike te Projektimit. KTP-1978,Technical Design Code KTP*; Akademia e Shkencave: Tirana, Albania, 1978.
55. *NATO SfP Project 983054, Harmonization of Seismic Hazard Maps for the Western Balkan Countries (BSHAP), Final Report*; NATO: Ankara, Türkiye, 2011.
56. Miošić, N.; Samardžić, N.; Hrvatović, H. The current status of geothermal energy use and development in Bosnia and Herzegovina. In Proceedings of the Proceedings World Geothermal Congress, Bali, Indonesia, 25–29 April 2010; pp. 25–29.
57. Slejko, D.; Camassi, R.; Cecic, I.; Herak, D.; Herak, M.; Kociu, S.; Kouskouna, V.; Lapajine, J.; Makropoulos, K.; Meletti, C.; et al. Seismic hazard assessment for Adria. *Ann. Geofis.* **1999**, *42*, 1085–1107. [CrossRef]
58. Gasparini, C.; Iannaccone, G.; Scarpa, R. Fault-plane solution and seismicity of the Italian peninsula. *Tectonophysics* **1985**, *117*, 59–78. [CrossRef]
59. Herak, M.; Herak, D.; Markušić, S. Fault-plane solutions for earthquakes (1956-1995) in Croatia and neighbouring regions. *Geofizika* **1995**, *12*, 43–56.
60. Louvari, H.K.; Kiratzi, A.A.; Papazachos, B.C. Further evidence for strike-slip faulting in the Ionian Islands: The Lefkada fault. In Proceedings of the IASPEI 29th General Assembly, Thessaloniki, Greece, 18–28 August 1997; pp. 18–29.
61. Sulstarova, E.; Peçi, V.; Shuteriqi, P. Vlora-Elbasani-Dibra (Albania) transversal fault zone and its seismic activity. *J. Seismol.* **2000**, *4*, 117–131. [CrossRef]
62. Harvard. CMT Focal Mechanisms. 1998. Available online: http://www.seismology.harvard.edu (accessed on 20 October 2022).
63. Šipoš, T.K.; Hadzima-Nyarko, M. Rapid seismic risk assessment. *Int. J. Dis. Risk Reduct.* **2017**, *24*, 348–360. [CrossRef]
64. MunichRe. Topics Geo: Natural Catastrophes 2011 Analyses Assessments Positions, Munich. Germany. 2012. Available online: http://www.munichre.com/natcatservice/ (accessed on 20 September 2022).
65. Cummins, J.D.; Mahul, O. *Catastrophe Risk Financing in Developing Countries: Principles for Public Intervention*; World Bank Publications: Washington DC, USA, 2009.
66. Archives of the Department of Geophysics, Faculty of Science, University of Zagreb. Available online: https://www.pmf.unizg.hr/geof/en (accessed on 12 November 2022).
67. Herak, M.; Herak, D.; Markušić, S. Revision of the earthquake catalogue and seismicity of Croatia, 1908–1992. *Terra Nova* **1996**, *8*, 86–94. [CrossRef]
68. Markušić, S.; Herak, D.; Ivančić, I.; Sović, I.; Herak, M.; Prelogović, E. Seismicity of Croatia in the period 1993–1996 and the Ston-Slano earthquake of 1996. *G Eofizika* **1998**, *15*, 83–102.

69. Ivančić, I.; Herak, D.; Markušić, S.; Sović, I.; Herak, M. Seismicity of Croatia in the period 1997–2001. *Geofizika* **2002**, *18*, 17–29.
70. Ivančić, I.; Herak, D.; Markušić, S.; Sović, I.; Herak, M. Seismicity of Croatia in the period 2002–2005. *Geofizika* **2006**, *23*, 87–103.
71. Markušić, S.; Stanko, D.; Penava, D.; Ivančić, I.; Bjelotomić Oršulić, O.; Korbar, T.; Sarhosis, V. Destructive M6. 2 petrinja earthquake (Croatia) in 2020—Preliminary multidisciplinary research. *Remote Sens.* **2021**, *13*, 1095. [CrossRef]
72. *Papeš, Tektonska Građa Teritorije SR BiH, Report*; Geoinstitut Ilidža: Sarajevo, Bosnia and Herzegovina, 1998.
73. Ademović, N.; Kalman Šipoš, T.; Hadzima-Nyarko, M. Rapid assessment of earthquake risk for Bosnia and Herzegovina. *Bull. Earthq. Eng.* **2020**, *18*, 1835–1863. [CrossRef]
74. Ademović, N.; Hadzima-Nyarko, M.; Zagora, N. Seismic vulnerability assessment of masonry buildings in Banja Luka and Sarajevo (Bosnia and Herzegovina) using the macroseismic model. *Bull. Earthq. Eng.* **2020**, *18*, 3897–3933. [CrossRef]
75. Lee, V.W.; Manić, M.I.; Bulajić, B.Đ.; Herak, D.; Herak, M.; Trifunac, M.D. Microzonation of Banja Luka for performance-based earthquake-resistant design. *Soil Dyn. Earthq. Eng.* **2015**, *78*, 71–88. [CrossRef]
76. Trukulja, D. Seizmogenetska obilježja oblasti zahvaćene zemljotresima u Banjoj Luci. Međunarodni simpozijum povodom 30 godina zeljotresa u Banjoj Luci. International Symposium on the 30 years 1969 Banja Luka Earthquake. *Univ. Banja Luci* **1999**, 28–41.
77. Janković, M. *Quelques Observations sur les Consé- Quences du Tremblement de terre du 11 juin 1962. Rapport Multigraphié, une Carte*; Sarajevo, Bosnia and Herzegovina, 1963.
78. Petković, K.V. Neue Erkenntnisse über den Bau der Dinariden. Jahrbuch der Geol. Bundesanstalt.(Wien). *Bundesanstalt* **1963**, *101*, 1–24.
79. USGS. Earthquake Catalogue for all Earthquakes with Mw $\geq$ 3 in the Period between 1900 and April 2021 for the Geographic Region between 41.0° N and 47.0° N, and 12.5° E and 23.0° E, as Reported by the United States Geological Survey. 2021. Available online: https://earthquake.usgs.gov/earthquakes/search/ (accessed on 23 May 2021).
80. *BAS EN 1998-1/NA: 2018 Eurocode 8: Design of Structures for Earthquake Resistance—Part 1: General Rules, Seismic Actions and Rules for Buildings—National Annex*; Institut za standardizaciju Bosne i Hercegovine: Istočno Sarajevo, Bosnia and Hercegovina.
81. Jordanovski, L.R.; Lee, V.W.; Manić, M.I.; Olumčeva, T.; Sinadnovski, C.; Todorovska, M.I.; Trifunac, M.D. *Strong Earthquake Ground Motion Data in EQINFOS: Yugoslavia. Part I*; Report No. 87-05; Department of Civil Engineering, University of Southern California: Los Angeles, CA, USA, 1987.
82. Ambraseys, N.; Douglas, J.; Margaris, B.; Sigbjörnsson, R.; Smit, P.; Suhadolc, P. Internet site for European strong motion data. In Proceedings of the 12th European Conference on Earthquake Engineering, London, UK, 9–13 September 2002; p. 837.
83. Ambraseys, N.; Douglas, J.; Margaris, B.; Sigbjörnsson, R.; Berge-Thierry, C.; Suhadolc, P.; Costa, G.; Smit, P. Dissemination of European strong-motion data. In Proceedings of the 13th World Conference on Earthquake Engineering, Vancouver, BC, Canada, 1–6 August 2004; Volume 2, p. 32.
84. Markušić, S.; Herak, M. Seismic zoning of Croatia. *Nat. Hazards* **1998**, *18*, 269–285. [CrossRef]
85. Prevolnik, S. Analiza Akcelerograma Petrinjskih Potresa (Accelerogram Analysis of Petrinja Earthquakes). Available online: https://www.pmf.unizg.hr/geof/seizmoloska_sluzba/potresi_kod_petrinje_2020 (accessed on 11 June 2022).
86. Medak, D.; Pribičević, B.; Prelogović, E. Recent geodynamical GPS-project in Croatia, raziskave s područja geodezije in geofizike 2006. In Proceedings of the 12th Strokovno Srečanje Slovenskega Združenja za Geodezijo in Geofiziko, Ljubljana, Slovenija, 18 January 2007.
87. Morales-Esteban, A.; Martinez-Alvarez, F.; Scitovski, S.; Scitovski, R. A fast partitioning algorithm using adaptive Mahalanobis clustering with application to seismic zoning. *Compute. Geosci.* **2014**, *73*, 132–141. [CrossRef]
88. Skoko, D.; Prelogović, E.; Aljinović, B. Geological structure of the Earth's crust above the Moho discontinuity in Yugoslavia. *Geophys. J. Int.* **1987**, *89*, 379–382. [CrossRef]
89. Bielik, M.; Makarenko, I.; Csicsay, K.; Legostaeva, O.; Starostenko, V.; Savchenko, A.; Šimonová, B.; Dérerová, J.; Fojtíková, L.; Pašteka, R.; et al. The refined Moho depth map in the Carpathian-Pannonian region. *Contrib. Geophys. Geod.* **2018**, *48*, 179–190.
90. *Hrvatski zavod za norme: HRN EN 1998-1:2011/NA:2011. Eurocode 8: Design of Structures for Earthquake Resistance—Part 1: General Rules, Seismic Actions and Rules for Buildings—National Annex*; Hrvatski Zavod za Norme: Zagreb, Croatia, 2011.
91. Ganas, A.; Elias, P.; Valkaniotis, S.; Tsironi, V.; Karasante, I.; Briole, P. Petrinja earthquake moved crust 10 feet. *Temblor* **2021**. [CrossRef]
92. ZHMS. 35 Godina od Katastrofalnog Zemljotresa u Crnoj Gori (35 Years since the Catastrophic Earthquake in Montenegro). Seismological Survey of Montenegro, Podgorica. 2014. Available online: http://www.seismo.co.me/documents/35%20GODINA%20OD%20KATASTROFALNOG%20ZEMLJOTRESA%20U%20CRNOJ%20GORI.pdf (accessed on 15 August 2022).
93. Available online: http://seizkarta.gfz.hr/karta.php (accessed on 30 August 2022).
94. UNICEF Country Office for Croatia. *Earthquake Situation Report #5*. Croatia, 2021. Available online: https://www.unicef.org/media/92246/file/UNICEF%20Croatia%20Situation%20Report%20No.%205%20(Earthquake) (accessed on 31 October 2022).
95. Morales-Esteban, A.; Martinez-Alvarez, F.; Scitovski, S.; Scitovski, R. Mahalanobis clustering for the determination of incidence-magnitude seismic parameters for the Iberian Peninsula and the Republic of Croatia. *Comput. Geosci.* **2021**, *156*, 104873. [CrossRef]
96. RTS. 2012. Available online: https://www.rts.rs/page/stories/sr/story/125/drustvo/1204516/dve-godine-od-zemljotresa-u-kraljevu.html (accessed on 20 April 2022).

97. SSS. Catalog of M≥3 Earthquakes of the Republic of Serbia. Seismological Survey of Serbia, Belgrade, Serbia, 2013. Available online: https://www.seismo.gov.rs/Seizmicnost/Katalog-zemljotresa.pdf (accessed on 10 November 2022).

98. Seismological Survey of Serbia. Accelerograms Recorded during 10 March 2010 Peć and 3 November 2010 Kraljevo Earthquakes. Seismological Survey of Serbia, Republic of Serbia; 2021. Available online: http://www.seismo.gov.rs/O%20zavodu/Infol.htm (accessed on 21 May 2021).

99. Bozkurt, E. Neotectonics of Turkey—A synthesis. *Geodin. Acta* **2001**, *14*, 3–30. [CrossRef]

100. Işık, E.; Büyüksaraç, A.; Ekinci, Y.L.; Aydın, M.C.; Harirchian, E. The effect of site-specific design spectrum on earthquake-building parameters: A case study from the Marmara region (NW Turkey). *Appl. Sci.* **2020**, *10*, 7247. [CrossRef]

101. Tatar, O.; Poyraz, F.; Gürsoy, H.; Cakir, Z.; Ergintav, S.; Akpinar, Z.; Koçbulut, F.; Sezen, F.; Türk, T.; Hastaoğlu, K.Ö.; et al. Crustal deformation and kinematics of the Eastern Part of the North Anatolian Fault Zone (Turkey) from GPS measurements. *Tectonophysics* **2012**, *518–521*, 55–62. [CrossRef]

102. Şengör, A.M.C.; Grall, C.; Imren, C.; Le Pichon, X.; Görür, N.; Henry, P.; Karabulut, H.; Siyako, M. The geometry of the North Anatolian transform fault in the Sea of Marmara and its temporal evolution: Implications for the development of intracontinental transform faults. *Can. J. Earth Sci.* **2014**, *51*, 222–242. [CrossRef]

103. Alkan, H.; Büyüksaraç, A.; Bektaş, Ö.; Işık, E. Coulomb stress change before and after 24.01. 2020 Sivrice (Elazığ) Earthquake (Mw = 6.8) on the East Anatolian Fault Zone. *Arab. J. Geosci.* **2021**, *14*, 2648. [CrossRef]

104. Örgülü, G. Seismicity and source parameters for small-scale earthquakes along the splays of the North Anatolian Fault (NAF) in the Marmara Sea. *Geophys. J. Int.* **2011**, *184*, 385–404. [CrossRef]

105. Bohnhoff, M.; Martínez-Garzón, P.; Bulut, F.; Stierle, E.; Ben-Zion, Y. Maximum earthquake magnitudes along different sections of the North Anatolian Fault Zone. *Tectonophysics* **2016**, *674*, 147–165. [CrossRef]

106. Poyraz, F. Determining the strain upon the eastern section of the North Anatolian fault zone (NAFZ). *Arab. J. Geosci.* **2015**, *8*, 1787–1799. [CrossRef]

107. Meng, J.; Sinoplu, O.; Zhou, Z.; Tokay, B.; Kusky, T.; Bozkurt, E.; Wang, L. Greece and Turkey Shaken by African tectonic retreat. *Sci. Rep.* **2021**, *11*, 6486. [CrossRef]

108. Barka, A.; Eyidoğan, H. The Erzincan earthquake of 13 March 1992 in eastern Turkey. *Terra Nova* **1993**, *5*, 190–194. [CrossRef]

109. Saatcioglu, M.; Bruneau, M. Performance of structures during the 1992 Erzincan earthquake. *Can. J. Civ. Eng.* **1993**, *20*, 305–325. [CrossRef]

110. Barka, A.; Akyuz, H.S.; Altunel, E.; Sunal, G.; Cakir, Z.; Dikbas, A.; Page, W. The surface rupture and slip distribution of the 17 August 1999 Izmit earthquake (M 7.4), North Anatolian fault. *Bull. Seismol. Soc. Am.* **2002**, *92*, 43–60. [CrossRef]

111. Ghasemi, H.; Cooper, J.D.; Imbsen, R.A.; Piskin, H.; Inal, F.; Tiras, A. *The November 1999 Duzce Earthquake: Post-Earthquake Investigation of the Structures on the TEM (No. FHWA-RD-00-146)*; Federal Highway Administration: Washington, DC, USA, 2000.

112. Akyuz, H.S.; Hartleb, R.; Barka, A.; Altunel, E.; Sunal, G.; Meyer, B.; Armijo, V.R. Surface rupture and slip distribution of the 12 November 1999 Duzce earthquake (M 7.1), North Anatolian fault, Bolu, Turkey. *Bull. Seismol. Soc. Am.* **2002**, *92*, 61–66. [CrossRef]

113. Öztürk, S.; Çinar, H.; Bayrak, Y.; Karsli, H.; Daniel, G. Properties of the aftershock sequences of the 2003 Bingöl, MD= 6.4, (Turkey) earthquake. *Pure Appl. Geophys.* **2008**, *165*, 349–371. [CrossRef]

114. Aydan, Ö.; Ulusay, R.; Miyajima, M. *The Bingöl Earthquake of May 1, 2003*; Japan Society of Civil: Tokyo, Japan, 2003.

115. Erdik, M.; Kamer, Y.; Demircioğlu, M.; Şeşetyan, K. 23 October 2011 Van (Turkey) earthquake. *Nat. Hazards* **2012**, *64*, 651–665. [CrossRef]

116. Utkucu, M.; Durmus, H.; Yalçin, H.; Budakoglu, E.; Isik, E. Coulomb static stress changes before and after the 23 October 2011 Van, eastern Turkey, earthquake (MW = 7.1): Implications for the earthquake hazard mitigation. *Nat. Hazards Earth Syst. Sci.* **2013**, *13*, 1889. [CrossRef]

117. Toker, M.; Sengor, A.C.; Schluter, F.D.; Demirbag, E.; Cukur, D.; Imren, C. The structural elements and tectonics of the Lake Van basin (Eastern Anatolia) from multi-channel seismic reflection profiles. *J. Afr. Earth Sci.* **2017**, *129*, 165–178. [CrossRef]

118. Anonymous. Historical Earthquakes. 2021. Available online: http://www.koeri.boun.edu.tr (accessed on 15 May 2022).

119. Anonymous. Historical Earthquakes. 2021. Available online: http://www.deprem.afad.gov.tr (accessed on 15 May 2022).

120. Gupta, H.K. Seismic hazard assessment in the Alpide belt from Iran to Burma. *Ann. Geofis.* **1993**, *36*, 61–82.

121. Hamzehloo, H.; Alikhanzadeh, A.; Rahmani, M.; Ansari, A. Seismic hazard maps of Iran. In Proceedings of the 15th World Conference on Earthquake Engineering, Lisbon, Portugal, 24–28 September 2012; pp. 24–28.

122. Moinfar, A.A.; Naderzadeh, A.; Nabavi, M.H. New Iranian seismic hazard zoning map for new edition of seismic code and its comparison with neighbor countries. In Proceedings of the 15th World Conference on Earthquake Engineering, Lisbon, Portugal, 24–28 September 2012.

123. Ambraseys, N.; Melville, C. *A History of Persian Earthquakes*; Cambridge University Press: New York, NY, USA, 1982.

124. Berberian, M. *Natural Hazards and the First Earthquake Catalogue of Iran. Volume 1: Historical Hazards in Iran Prior to 1900*; Internatiaonal Instituteof Earthquake Engineering and Seismology: Tehran, Iran, 1994; 603p.

125. Shahvar, M.P.; Farzanegan, E.; Eshaghi, A.; Mirzaei, H. i1-net: The Iran Strong Motion Network. *Seismol. Res. Lett.* **2021**, *92*, 2100–2108. [CrossRef]

126. ISMN. Catalogue of Earthquake Strong Motion Records, Iran Strong Motion Network. Road, Housing, and Urban Development Research Center. 2017. Available online: https://ismn.bhrc.ac.ir/en (accessed on 4 July 2022).

127. IRSC. Iranian Seismological Center, Institute of Geophysics, University of Tehran. 2022. Available online: https://irsc.ut.ac.ir/seismicity.php (accessed on 4 July 2022).
128. Shoja-Taheri, J.; Niazi, M. Seismicity of the Iranian plateau and bordering regions. *Bull. Seismol. Soc. Am.* **1981**, *71*, 477–489.
129. Izadkhah, Y.O.; Hosseini, K.A. An evaluation of disaster preparedness in four major earthquakes in Iran. *J. Seismol. Earthq. Eng.* **2010**, *12*, 61.
130. Ambraseys, N.N.; Tchalenko, J.S. The Dasht-e Bayāz (Iran) earthquake of August 31, 1968: A field report. *Bull. Seismol. Soc. Am.* **1969**, *59*, 1751–1792. [CrossRef]
131. Berberian, M. Tabas-e-Golshan (Iran) catastrophic earthquake of 16 September 1978; a preliminary field report. *Disasters* **1979**, *2*, 207–219. [CrossRef] [PubMed]
132. USGS. United States Geological Survey (USGS), Bam Earthquake, Impact. 2022. Available online: https://earthquake.usgs.gov/earthquakes/eventpage/usp0004arq/impact (accessed on 6 July 2022).
133. ISMN. *Iran Strong Motion Network*; Road, Housing, and Urban Development Research Center: Tehran, Iran, 2022.
134. IIEES. International Institute of Earthquake Engineering and Seismology (IIEES): Tabas Earthquake of 16 September 1978 Mw=7.4. 2022. Available online: https://www.iiees.ac.ir/en/tabas-earthquake-of-16-september-1978-mw7-4-2/ (accessed on 6 July 2022).
135. IRIS. 1968 Dasht-e-Bayaz (Iran) Earthquake Archive, Incorporated Research Institutions for Seismology (IRIS). 2022. Available online: https://ds.iris.edu/seismo-archives/quakes/1968dasht-e-bayaz/ (accessed on 6 July 2022).
136. ICSRDB. *Iranian Code of Practice for Earthquake Resistant Design of Buildings (Standard 2800)*, 4th ed.; PN S 253; Building and Housing Research Center of Iran: Tehran, Iran, 2014.
137. Seismosoft. SeismoStruct 2021—A Computer Program for Static and Dynamic Nonlinear Analysis of Framed Structures. 2021. Available online: http://www.seismosoft.com (accessed on 10 June 2022).
138. Antoniou, S.; Pinho, R. *Seismostruct–Seismic Analysis Program by Seismosoft*; Technical Manual and User Manual; Seismosoft: Pavia, Italy, 2022.
139. Menegotto, M.; Pinto, P.E. Method of analysis for cyclically loaded RC plane frames including changes in geometry and non-elastic behavior of elements under combined normal force and bending. In *Symposium on the Resistance and Ultimate Deformability of Structures Acted on by Well-defined Repeated Loads*; International Association for Bridge and Structural Engineering: Zurich, Switzerland, 1973; pp. 15–22.
140. Mander, J.B.; Priestley, M.J.N.; Park, R. Theoretical stress-strain model for confined concrete. *J. Struct. Eng.* **1998**, *114*, 1804–1825. [CrossRef]
141. *EN 1998-3*; Eurocode-8: Design of Structures for Earthquake Resistance-Part 3: Assessment and Retrofitting of Buildings. European Committee for Standardization: Bruxelles, Belgium, 2005.
142. Pinto, P.E.; Franchin, P. Eurocode 8-Part 3: Assessment and retrofitting of buildings. In Proceedings of the Eurocode 8 Background and Applications, Dissemination of Information for Training, Lisbon, Portugal, 10–11 February 2011.
143. Kutanis, M.; Boru, E.O.; Işık, E. Alternative instrumentation schemes for the structural identification of the reinforced concrete field test structure by ambient vibration measurements. *KSCE J. Civ. Eng.* **2017**, *21*, 1793–1801. [CrossRef]
144. Aksoylu, C.; Mobark, A.; Arslan, M.H.; Hakkı Erkan, İ. A comparative study on ASCE 7-16, TBEC-2018 and TEC-2007 for reinforced concrete buildings. *Rev. Construcción* **2020**, *19*, 282–305. [CrossRef]
145. Hadzima-Nyarko, M.; Morić, D.; Draganić, H.; Štefić, T. Comparison of fundamental periods of reinforced shear wall dominant building models with empirical expressions. *Teh. Vjesn.* **2015**, *22*, 685–694. [CrossRef]
146. Aksoylu, C.; Arslan, M.H. Çerçeve+ perde türü betonarme binalarin periyod hesaplarinin TBDY-2019 yönetmeliğine göre ampirik olarak değerlendirilmesi. *Uludağ Univ. J. Fac. Eng.* **2019**, *24*, 365–382. [CrossRef]
147. CEN. *Eurocode 8-Design of Structures for Earthquake Resistance-Part 1: General Rules, Seismic Actions and Rules for Buildings, EN 1998-1:2005*; European Committee for Standardization: Bruxelles, Belgium, 2005.
148. INSTAT. *Population and Housing Census of Albania, 2011—Regjistrimi në Harta—Albania 2001 Census Atlas*; Seria e Studimeve: Tirana, Albania, 2011; p. 11.
149. Bureau of Statistics [CBS]. *National Population and Housing Census 2013 (National Report)*; Agency for Statistics of Bosnia and Herzegovina: Sarajevo, Bosnia and Herzegovina, 2013.
150. Ademović, N. Structural and Seismic Behavior of Typical Masonry Buildings from Bosnia and Herzegovina. Master Thesis, The University of Minho, Guimaraes, Portugal, 2011.
151. Ademovic, N.; Hrasnica, M.; Oliveira, D.V. Pushover analysis and failure pattern of a typical masonry residential building in Bosnia and Herzegovina. *Eng. Struct.* **2013**, *50*, 13–29. [CrossRef]
152. Ademović, N.; Oliveira, D.V. (2017)—Damage Indicators for Unreinforced Masonry Building Walls Subjected to Seismic Actions, UDK: 692.2.042.7. *Građevinski Mater. I Konstr.* **2017**, *60*, 17–32. [CrossRef]
153. Ademović, N.; Oliveira, D.V.; Lourenco, P.B. Seismic evaluation and strengthening of an existing masonry building in Sarajevo, BiH. *Buildings* **2019**, *9*, 30. [CrossRef]
154. Ademovic, N.; Zagora, M.; Hadzima-Nyarko, M. Seismic Vulnerability Analysis in Urban and Rural Regions of Visoko, BIH, The International Symposium on Civil Engineering—ISCE 2018 TLTH (12th) days of Bhaaas in Bosnia and Herzegovina mostar 2021, Advanced Technologies, Systems, and Applications VI. In Proceedings of the International Symposium on Innovative and Interdisciplinary Applications of Advanced Technologies (IAT) 2021, Lecture Notes in Networks and Systems, Sarajevo, Bosnia and Herzegovina, 24–27 June 2021; pp. 421–4299.

155. Ademovic, N.; Hadzima-Nyarko, M.; Zagora, N. Influence of site effects on the seismic vulnerability of masonry and reinforced concrete buildings in Tuzla (Bosnia and Herzegovina). *Bull. Earthq. Eng.* **2022**, *20*, 2643–2681. [CrossRef]
156. Piljug, A.; Medanović, Ć.; Ademović, N.; Hadzima-Nyarko, M.; Zagora, N. Quick visual seismic assessment of existing buildings in Sarajevo (BiH). In Proceedings of 3rd European Conference on Earthquake Engineering & Seismology, pp-1300-1306, Bucharest, Romania, 4–9 September 2022.
157. Pavić, G.; Hadzima-Nyarko, M.; Bulajić, B.; Jurković, Ž. Development of seismic vulnerability and exposure models—A case study of Croatia. *Sustainability* **2020**, *12*, 973. [CrossRef]
158. Miranda, E.; Brzev, S.; Bijelic, N.; Arbanas, Ž.; Bartolac, M.; Jagodnik, V.; Robertson, I. *Petrinja, Croatia December 29, 2020, Mw 6.4 Earthquake Joint Reconnaissance Report (JRR)*; ETH Zurich: Zurich, Switzerland, 2021.
159. Pavić, G.; Hadzima-Nyarko, M.; Bulajić, B. A contribution to a uhs-based seismic risk assessment in Croatia—A Case Study for the City of Osijek. *Sustainability* **2020**, *12*, 1796. [CrossRef]
160. *SRPS EN 1998-1/NA:2018; Evrokod 8-Projektovanje Seizmički Otpornih Konstrukcija Deo 1: Opsta Pravila, Seizmicka Dejstva i Pravila za Zgrade (Eurocode 8-Design of Structures for Earthquake Resistance-Part 1: General Rules, Seismic Actions and Rules for Buildings)*; Institute for Standardization of Serbia: Belgrade, Serbia, 2018.
161. Manić, M.I.; Bulajić, B.Đ. Examples of typical damages to masonry buildings for individual housing in Kraljevo region during the November 03, 2010 earthquake (in Serbian with English abstract). In Proceedings of the Fourth International Conference Earthquake Engineering and Engineering Seismology, Tehran, Iran, 19–21 May 2014; Association of Civil Engineers of Serbia: Belgrad, Serbia, 2014; pp. 315–324.
162. Işık, E.; Karaşin, İ.B.; Demirci, A.; Büyüksaraç, A. Seismic risk priorities of site and mid-rise RC buildings in Turkey. *Chall. J. Struct. Mech.* **2020**, *6*, 191–203. [CrossRef]
163. Dogan, G.; Ecemis, A.S.; Korkmaz, S.Z.; Arslan, M.H.; Korkmaz, H.H. Buildings damages after Elazığ, Turkey earthquake on 24 January 2020. *Nat. Hazards* **2021**, *109*, 161–200. [CrossRef]
164. Yel, N.S.; Arslan, M.H.; Aksoylu, C.; Erkan, İ.H.; Arslan, H.D.; Işık, E. Investigation of the Earthquake Performance Adequacy of Low-Rise RC Structures Designed According to the Simplified Design Rules in TBEC-2019. *Buildings* **2022**, *12*, 1722. [CrossRef]
165. Ghafory-Ashtlany, M.; Jafari, M.K.; Tehranizadeh, M. Earthquake hazard mitigation achievement in Iran. In Proceedings of the 12th World Conference on Earthquake Engineering, Auckland, New Zealand, 30 January–4 February 2000; Volume 30.
166. Mahdi, T.; Mahdi, A. Reconstruction and retrofitting of buildings after recent earthquakes in Iran. *Procedia Eng.* **2013**, *54*, 127–139. [CrossRef]
167. Astaneh-Asl, A.; Saeedikia, M.; Havaii, M.H.; Fathi, M.; Fatemi-Aghda, S.M.; Mir Ghaderi, S.R.; Heidarinejad, G. Reconstruction of housing destroyed in the 2003 Bam-Iran Earthquake. In Proceedings of the 100th Anniversary Earthquake Conference: Commemorating the 1906 San Francisco Earthquake, California, CA, USA, 18–22 April 2006.
168. Mostafaei, H.; Kabeyasawa, T. Investigation and analysis of damage to buildings during the 2003 Bam earthquake. *Bull. Earthq. Res. Inst.* **2004**, *79*, 107–132.
169. Bastami, M.; Abbasnejadfard, M.; Motamed, H.; Ansari, A.; Garakaninezhad, A. Development of hybrid earthquake vulnerability functions for typical residential buildings in Iran. *Int. J. Disaster Risk Reduct.* **2022**, *77*, 103087. [CrossRef]
170. Aliaj, S.H.; Baldassarre, G.; Shkupi, D. Quaternary subsidence zones in Albania: Some case studies. *Bull. Eng. Geol. Environ.* **2001**, *59*, 313–318. [CrossRef]

Article

Flood Risk Assessment Using GIS-Based Analytical Hierarchy Process in the Municipality of Odiongan, Romblon, Philippines

Jerome G. Gacu [1,2,3], Cris Edward F. Monjardin [1,2,4,*], Delia B. Senoro [1,2,4] and Fibor J. Tan [1,2,4]

1 Masters Program in Civil Engineering, School of Graduate Studies, Mapua University, Manila 1002, Philippines
2 School of Civil, Environmental and Geological Engineering, Mapua University, Manila 1002, Philippines
3 Civil Engineering Department, College of Engineering and Technology, Romblon State University, Liwanag, Odiongan, Romblon 5505, Philippines
4 Resiliency and Sustainable Development Center, Yuchengco Innovation Center, Mapua University, Manila 1002, Philippines
* Correspondence: cefmonjardin@mapua.edu.ph

Abstract: The archipelagic Romblon province frequently experiences typhoons and heavy rains that causes extreme flooding, this produces particular concern about the severity of damage in the Municipality of Odiongan. Hence, this study aimed to assess the spatial flood risk of Odiongan using the analytical hierarchy process (AHP), considering disaster risk factors with data collected from various government agencies. The study employed the geographic information system (GIS) to illustrate the spatial distribution of flooding in the municipality. Sendai Framework was the basis of risk analysis in this study. The hazard parameters considered were average annual rainfall, elevation, slope, soil type, and flood depth. Population density, land use, and household number were considered parameters for the exposure assessment. Vulnerability assessments considered gender ratio, mean age, average income, number of persons with disabilities, educational attainment, water usage, emergency preparedness, type of structures, and distance to evacuation area as physical, social, and economic factors. Each parameter was compared to one another by pairwise comparison to identify the weights based on experts' judgment. These weights were then integrated into the flood risk assessment computation. The results led to a flood risk map which recorded nine barangays (small local government units) at high risk of flooding, notably the Poblacion Area. The results of this study will guide local government units in developing prompt flood management programs, appropriate mitigation measures, preparedness, and response and recovery strategies to reduce flood risk and vulnerability to the population of Odiongan.

Keywords: AHP; digital elevation model; flood; GIS; risk assessment

Citation: Gacu, J.G.; Monjardin, C.E.F.; Senoro, D.B.; Tan, F.J. Flood Risk Assessment Using GIS-Based Analytical Hierarchy Process in the Municipality of Odiongan, Romblon, Philippines. *Appl. Sci.* **2022**, *12*, 9456. https://doi.org/10.3390/app12199456

Academic Editors: Andrea Chiozzi, Željana Nikolić and Elena Benvenuti

Received: 17 August 2022
Accepted: 19 September 2022
Published: 21 September 2022

Publisher's Note: MDPI stays neutral with regard to jurisdictional claims in published maps and institutional affiliations.

1. Introduction

Floods are caused by the failure of natural paths and drainage systems to hold excess water during and immediately following excessive rainfall [1]. This condition is among the disastrous natural hazards that can cause tremendous economic loss, damage to infrastructures and natural ecosystems, as well as death. The Organization for Economic Cooperation and Development (OECD) reported that floods trigger more than USD 40 billion in destruction worldwide [2]. The United States loses about USD 8 billion a year due to flooding. Recently, casualties have risen to roughly 100 deaths annually [3] and about 6.8 million were adversely affected by excessive flooding in the northeastern part of India. Nepal, Indonesia, and Japan [4].

According to a study by Monjardin et al. [5], flooding is a dangerous natural phenomenon that has taken numerous lives and caused enormous economic damage in the Philippines. Flood is considered the second most frequent calamity in the Philippines, representing 31.9% of annual natural disasters [6]. The National Council for Disaster Risk

Reduction and Management (NDRRMC) of the Philippines reported on 19 April 2021 that 68,490 individuals were evacuated in Bicol and Eastern Visayas regions due to risk from Tropical Cyclone Surigae (Bising) [7]. Additionally, the Mindanao Island that was formerly considered as a region free from typhoons was devastated by consecutive typhoons, e.g., Sendong (international name, Washi) and Pablo (international name, Bopha). These typhoons altered the usual typhoon pathway and made a new typhoon route. These two typhoons landed in 2011 and 2012, respectively, and caused devastation that killed more than 1000 individuals and 100 people went missing [8]. Further, as mentioned in the study of Siddayao et al. [9], Typhoon Haiyan distressed the Philippines telecommunication signals, power, and water lines on 8 November 2013. In the province of Romblon, floods frequently occur, resulting in losses to the affected municipalities. All rivers and tributaries in the Romblon province overflowed [10] during typhoons. Flood occurrences are frequent in the Municipality of Odiongan, being a low-lying area of Tablas Island, Romblon province.

In the Philippines, flood risk maps are essential for the safety of communities and ecosystems [11]. Decision-makers are looking for longer-term mitigation of the adverse effects of floods and some natural tragedies; hence, confidence criteria in engineered solutions such as flood protection systems are important [8]. Furthermore, the assessment and evaluation of flood hazards must be constructed on accurate flood hazard guides to show the real impact of urban development [12]. Risk assessment is vital in formulating decisions guidelines, policies, and mitigations based on meteorological, hydrological, and socioeconomic factors [13]. Comprehensive flood risk assessment and the enhancement of efficient flood mitigation actions need systematic information regarding flood occurrences at points in a catchment basin [14]. However, specific factors of population, society, economy, environment, transportation, and other disaster-bearing elements in different parts of mountain cities are remarkably varied, which increases the doubt of risk assessment index weight and risk assessment reliability [15]. Hence, accurate hazard maps and least-error indices are important tools in risk assessment.

The GIS tool plays a vital component of flood risk assessment due to the evaluation process that needs spatial information. The practice of a standard approach for evaluation and merging distinctive data affect the precision and comparability of assessment outcomes. Some nations have established national guidelines to assess flood risk potential [16]. In addition, GIS can be utilized to study international, regional, and local flood risks and guide the implementation of a risk mitigation plan [17]. GIS in the Philippines is a primary distinctive tool used in countrywide flood risk modeling. However, existing high-resolution flood risk models have come to be very important. These tools can be used for flood readiness by improving these maps' data levels [18]. ArcGIS, developed by ESRI, is a GIS-based tool that can produce standard Web Services and make numerous network GIS uses [19].

The Digital Elevation Model (DEM) is widely used in GIS modeling, and the enhancement, development, and processing of DEMs are vital in many environmental aspects. It is in the form of a grid as a digital illustration of land with a corresponding pixel value equal to an elevation from the datum [20,21]. According to Suguruman et al. [22], DEMs are used more often in flood risk management, including flood plain models, visualization, flood hazard assessment, and identification of floodplain altitudes. There are numerous sources of DEM information, including Advanced Space Borne Thermal Emission and Reflection Radiometer (ASTER), Synthetic Aperture Radar (SAR), Global Positioning System (GPS), and Light Detection and Ranging (LiDAR) [23]. In the Philippines, hydraulic and hydrologic tools for flood risk analysis are very limited In line with topographic, geometric, and hydrologic river information [24]. The Philippines assimilated geospatial data LiDAR and IfSAR (Interferometric Synthetic Aperture Radar) with excellent resolution Digital Terrain Models (DTMs) covering 300,000 square kilometers of the terrestrial area [25]. This is to deal the insufficient high-resolution topographic maps.

Flooding needs considerable attention, studies have evaluated the connection between urban/rural services, flood history, and disaster readiness in local communities living in safety [26]. The Sendai Framework acknowledged the critical role played by the community in disaster risk reduction [27]. This framework is used in disaster risk management delivers quantifiable parameters for a national and local scale to calculate the reduction in disaster damages. The compilation and evaluation of disaster damages under the Sendai Framework enhance our knowledge of the efficiency of disaster risk reduction approaches [28].

There is a need to understand the spatial extent of flood zones by utilizing multiple data to show a possible baseline for consistent flood risk management and mitigation measures [29]. The methodology using multicriteria analysis (MCA), also known as multicriteria decision-making (MCDM), supports a basis that can hold distinctive assessment on determining the factors of a composite decision, arrange the aspects into a hierarchical configuration, and analyze the relations amid elements of the identified hazard [30]. All MCA methods make the options and their influence on the different criteria clear. They vary, however, in how they associate all the data needed. The method's primary role is to solve the difficulties that decision-makers have encountered when handling a large quantity of complex information. MCA can be used to recognize a single most preferred option, rank options, shortlist a limited number of options for subsequent detailed evaluation, or differentiate conventional from unconventional possibilities [31].

Several approaches have been suggested for MCA, but the Analytical Hierarchy Process (AHP) is being used most frequently to resolve different flood risk assessments [32]. The AHP provides the same advantage as MCA models in focusing decision-maker consideration on developing a structure to gather all the significant factors expected to differentiate the best option [31,33]. AHP represents the problem in three parts where the first part is the matter that needs to be fixed, and the second part is the alternative solutions available to resolve the problem. The third and most important process is the criteria expended to assess the alternative solutions [34]. Studies on flood risk assessment in Thailand [26,35], Bangladesh [36], and Indonesia [37] used GIS and AHP. Additionally, in the Philippines, identified relevant flood factors and judgments of decision-makers were analyzed using AHP judgments to weigh each parameter in estimating flood hazards in the study [38] at the central business district of Tuguegarao City, Philippines. Another study was conducted in Infanta, Quezon Province, Philippines, aiming to give the municipality options and models for flood mitigation. The drainage system in said municipality is at risk of causing flood-related problems deliberating identified relative factors via AHP [6]. The evaluation of flood zones and flood problems for Davao Oriental, Philippines, were analyzed by the AHP and Maxent tool which reduce the subjectivity and uncertainty in selecting and weighting criteria [8]. The rareness of using AHP-based research made it easier to make a model of indecision without compromising the subjective and objective aspects of the assessment process [29]. Hence, the number of flood events in the Municipality of Odiongan that caused property damage to the community explicitly need the output of this research study. The results of this assessment will be used as the basis for the municipality's flood mitigation and risk management. Additionally, the information will useful in areas with similar topography and weather conditions.

2. Materials and Methods

2.1. Study Area

The study area is the Municipality of Odiongan located in the middle west portion of Tablas Island, Romblon province with coordinated of 22°04′ East Longitude and 12°19′ North Latitude. Odiongan has a land area of 185.67 square kilometers representing 12.11% of Romblon province. The town proper lies in the low-lying plains, and the interior part of the municipality is composed of hills and mountainous forests. Odiongan consists of 25 barangays and 1 anchorage, which is linked to other neighboring islands. Figure 1 shows the imagery map of Odiongan with barangay boundaries.

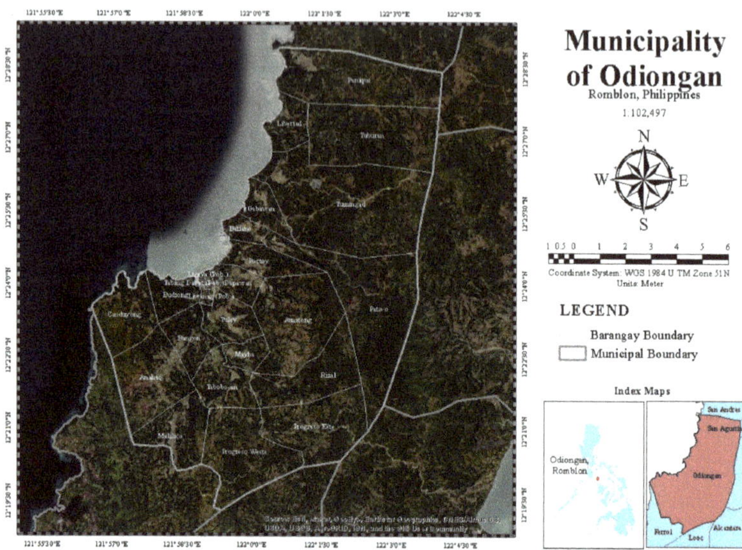

Figure 1. Imagery map with barangay and municipal boundaries of the study area.

2.2. Data Collection and Identification of Factors

The Sendai Framework was followed to identify flood indicators in assessing flood disaster risk, hence identifying the parameters and the data that need to be collected. There were three identified categories of flood risks parameters, such as (a) hazard, (b) vulnerability, and (c) exposure. The hazard parameters considered were: Average annual rainfall, slope, elevation, soil type, and flood depth. The parameters for vulnerability that were considered were: Gender ratio, age, average income, physical health of the individual, educational attainment, water usage, emergency preparedness, types structures, and proximity to the evacuation center. The parameters for flood exposure were: Population density, number of households, and land use/details are shown in Table 1.

Table 1. Parameters with the type of data used, duration/year, and source used for hazard, vulnerability, and exposure assessment.

References	Parameter	Data Type	Duration/Year	Source
Flood Hazard Parameters				
[8,17,35,39–41]	Average Annual Rainfall	Interpolated Climatological Normal using Isohyetal Method	2020	PAGASA and web search for weather station coordinates
[8,17,26,32,35,41]	Slope	Derived from IfSAR Data using Slope Tool in ArcMap	2013	(NAMRIA-DENR)
[6,8,26,32,35,41]	Elevation	Derived from IfSAR Data using Field Contour Tool in ArcMap	2013	(NAMRIA-DENR)
[8,9,17,42]	Soil Type	Shapefile from the archive of CLUP	2011	Municipality of Odiongan, Romblon—(CLUP)
[37,43–45]	Flood Depth	100-year period of flood model simulated in HEC-HMS and HEC-RAS and MGB Flood Susceptibility Map	2018	MGB
Flood Vulnerability Parameters				
[39,43]	Gender Ratio	Men to women gender ratio	2020	Barangay Profile
[43]	Average Age	Mean age of the individual	2020	Barangay Profile
[31,43]	Average Income	Annual average income per household	2020	Barangay Profile
[39,46]	Number of Persons with Disabilities	Number of PWD in barangay	2022	Barangay Management System (BMS)
[43,46]	Highest Educational Attainment	Average educational attainment of individuals in barangay	2020	Barangay Profile

Table 1. *Cont.*

References	Parameter	Data Type	Duration/Year	Source
[46–48]	Water Usage	Primary source of water	2022	Survey Questionnaire
	Emergency Preparedness	Emergency preparedness during unexpected situations like natural disasters	2022	Survey Questionnaire
[29]	Types of Built-up Structures	Classification of structures of every household	2020	Barangay Profile
	Distance from the nearest Evacuation Area	Distance of identified evacuation area using buffer tool in Arcmap	2022	Site Investigation and Survey Questionnaire
Flood Exposure Parameters				
[9,29,40,42,49]	Population Density	Computed from the population over the covered area of the barangay	2020	PSA
[43]	Household Number	Number of households of every barangay	2020	PSA
[6,17,26,35,40,41]	Land Use/Land Cover	Land cover map from CLUP	2011	Municipality of Odiongan, Romblon—(CLUP)

2.2.1. Flood Hazard Parameters

Flood management cannot be adequately completed without assessing flood hazards [48]; therefore, details of indicators are elaborated below.

1. Average Annual Rainfall

Precipitation values were plotted on a suitable base map at their respective stations using isohyetal method, and isohyets were drawn to create an isohyetal map. The study used the climatological normal records [50] from long-term averages over 30 years of PAGASA weather stations (Figure 2) with corresponding coordinates. Spatial interpolation employing the isohyetal method was applied to obtain dimensional rainfall patterns for projections of Romblon.

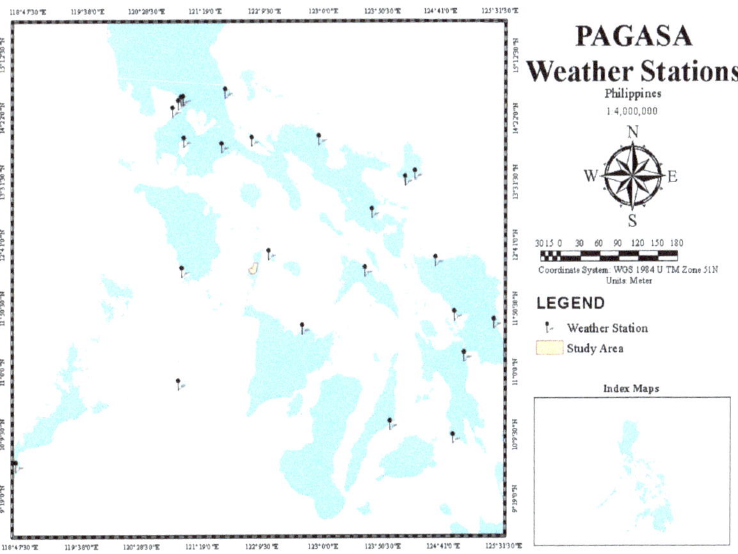

Figure 2. Weather stations considered in the interpolation for average annual rainfall using the Isohyetal Method.

2. Slope

The slope is a critical factor contributing to the intensity of destructive forces of floods in a particular area. The study prepared the slope map using the IfSAR DTM from National Mapping and Resource Information Authority (NAMRIA) and the spatial tool in the GIS application platform.

3. Elevation

Ground elevation is one of the main factors that should be considered in assessing flood hazards. IfSAR data were utilized and processed in the GIS tool.

4. Soil Type

The study used the soil map based on the map of NAMRIA stipulated in the Comprehensive Land Use Plan (CLUP) [51] of the Municipality of Odiongan. These data were correlated to the soil's water holding capacity and infiltration rate.

5. Flood Depth

ArcGIS, HEC-HMS (Hydrologic Engineering Center's—Hydrologic Modeling System) [52] and HEC-RAS (Hydrologic Engineering Center's—River Analysis System) [53] were the tools used for flood hazard simulation. The most important data used in the simulation were the DEM, which were provided by NAMRIA with a resolution of five-by-five (5×5) meters. A combination of simulated maps and Flood Susceptibility Maps [54] from the Mines and Geoscience Bureau (MGB) were used in the study.

2.2.2. Flood Vulnerability Parameters

The vulnerability factor includes social, economic, and personal safety [44]. Demographics and disaster risk reduction data of the Municipality of Odiongan were gathered through actual surveys (questionnaire) and existing records of the local government of Odiongan.

1. Demographics

The demographic data were gathered from the database of Philippine Statistics Authority (PSA). The period considered was 2015 to 2020. The on-site survey was conducted in every barangay of the Municipality of Odiongan.

2. Disaster Risk Reduction Data

All data were extracted from the survey conducted in every barangay and CLUP of the Municipality of Odiongan. The barangay identified evacuation facilities where coordinates and floor areas were recorded using GPS and area measuring tools.

2.2.3. Flood Exposure Parameters

The exposure analysis was aimed at identifying the life and property elements exposed in flooding events [41]. The identified exposure elements were population density, number of households, and land use/cover. The data were taken from the PSA record and municipal zoning maps archived from CLUP of the Municipality of Odiongan.

2.3. Modeling, GIS Mapping, and Validation

Generated models and maps from ArcGIS were the primary basis in the computation and analysis of final flood risk indices.

2.3.1. Basin Model Pre-Processing

In creating a basin model of Odiongan River channels (Bangon River), IfSAR-DEM with a 5 m $\times$ 5 m resolution was used. Data were processed using the GeoHMS10.7 tool plugin in ArcGIS 10.7. This is to generate a basin model and incorporated with the available soil and land cover data of 2004 from NAMRIA to assign curve numbers (CN) for each sub-basin. Soil type and land cover classification were represented as CN for each sub-basin. Initial abstraction (IA), time of concentration (TC), Storage Coefficient (SC), River Length, and sub-watershed area were derived during the pre-processing of the basin model.

2.3.2. Basin Model Calibration and RIDF Simulations

The pre-processing output of the HMS Basin Model was calibrated under the HEC-HMS 4.9 software to model the hydrologic response of the watershed to a specified hydrometeorological input. The parameters' values were tuned to attain an at least acceptable result in all the statistical measures recommended for model evaluation. As the model was calibrated, the simulations of rainfall scenarios of 25-, 50-, and 100-year followed. The rainfall intensity duration frequency (RIDF) data were acquired from PAGASA Romblon, Romblon province rain gauge station with 48-year rainfall records. These data were entered as the meteorological model file using the frequency storm precipitation method in HEC-HMS performed with calibrated basin model. The outputs of the simulations were then calibrated basin model with precipitation and outflow data of the three (3) return periods.

2.3.3. Two-Dimensional (2D) RAS Model Simulations

The processed DEM was used to create the river analysis model (RAS) model using the HEC-RAS 6.2, a practical river hydraulic simulation and analysis software. The RAS model was processed through unsteady flow analysis, and the boundary conditions used were flow hydrograph in the upstream and normal depth in the downstream which considers both the frictional resistance and slope of the channel. The calibrated outflow in HMS and precipitation were incorporated into the model. Flood depth considering a 100-year return period was regarded as one of the parameters in hazard mapping; this was exported as raster files and translated into spatial data in the GIS.

2.4. Evaluation and Assessment of Parameters Using AHP

Contributing factors were identified and assessed in which the weights of each parameter were determined using AHP based on the knowledge of experts composed of end-users, hydrologists, meteorologists, water resource engineers, and persons with comprehensive expertise in disaster risk reduction. Experts from government agencies such as PAGASA, Bureau of Soils and Water Management (BSWM), and MGB participated in the survey. A specialist from academic institutions (University of the Philippines, Mapúa University, Central Luzon State University, and Asian Institute of Technology) and an end-user (LGU-Odiongan) also responded to the survey. The survey for pairwise comparison was delivered and requested thru an online and printed-out questionnaire. A risk assessment was proceeded using the weights of each factor derived in AHP through a pairwise comparison questionnaire.

2.4.1. Determination of the Priorities among the Decision Elements of the Hierarchy

The feature weights were assigned parameters, where levels were reclassified and normalized into 1 for the least priority and 5 for the most focused. This step gathered the weight for each criterion and option using a pairwise comparison technique. Ten (10) experts on-field and end-users participated in determining the relevance of one alternative over the other with a pairwise comparison method presented in a matrix.

Each comparison was graded by experts and end-users using the pairwise comparison technique scale. The procedure usually contains a questionnaire for comparing all the elements and a geometric mean to arrive at a final solution [32] specifying the nine points intensity matrix, as shown in Table A1 of Appendix A.

2.4.2. Derivation of the Overall Relative Weights

The relative significance or weight of the factor after a pairwise comparison matrix was computed based on systematic AHP assessment and expert's inputs. This step was conducted by calculating the normalized values for each criterion and alternative, and choosing the normalized main priority vectors. Normalized values for each criterion and alternative in their respective matrices were derived by dividing each cell into its column and producing a total column of 1 for each criterion and alternative. Weights were calculated by averaging the rows of the matrix. The resulting value will give relative weight to every criterion concerning the best goal, and provide relative weight for the alternatives with respect to the criteria. The final relative weights of the alternatives were defined by computing the product's linear combination (LC) between the relative weight of each criterion and the alternative for the specific criterion. The decision-makers choose the best according to the alternatives' overall weights if the experts' judgments are proven consistent. This is mathematically expressed using Equation (1).

$$C = \{C_j | j = 1, 2, \ldots, n\} \tag{1}$$

The pairwise comparison on the criteria can be generalized using an evaluation matrix A, as shown as Equation (2), in which every element is the quotient of weights of the criteria given in Equation (3) [32].

$$A = \begin{bmatrix} a_{11} & a_{12} & \cdot & a_{1n} \\ a_{21} & a_{22} & \cdot & a_{2n} \\ \cdot & \cdot & \cdot & \cdot \\ a_{n1} & a_{n2} & \cdot & a_{nn} \end{bmatrix}, \ a_{ii} = 1, a_{ji} = \frac{1}{a_{ji}}, \ a_{ij} \neq 0 \tag{2}$$

2.4.3. Verification of the Consistency of Judgments and Conclusions according to Results

AHP's quality output was related to the consistency of the pairwise comparison judgments. This step was essential to identify the consistency of the assessment by computing the consistency ratio (CR) before a decision was completed. However, if the problem was expected during deliberation for choosing the best alternative, the CRs for matrices were computed initially before the alternatives' overall relative weights were calculated. After which, calculations were performed to obtain the largest eigenvalue, consistency index (CI), CR, and normalized values for each criterion and alternative.

The last mathematical process normalized and identified the relative weights per matrix. The right eigenvector gave the relative weights (w) conforming to the highest eigenvalue (λ_{max}), as shown in Equation (3).

$$A_w = \lambda_{max} \tag{3}$$

If the pairwise comparisons were consistent, the matrix A was ranked one and $\lambda_{max} = n$, so the weights can be taken by normalizing any of the rows or columns of A [32]. The relativeness between the entries determines the consistency, and the CI was calculated using the equation below:

$$CI = (\lambda_{max} - n)/(n - 1) \tag{4}$$

The final CR, which enables the decision-maker to accomplish whether the assessments were adequately coherent, was computed as the CI's and the random index (RI) quotient using Equation (5).

$$CR = CI/RI \tag{5}$$

One recommendation for this step was: if the proportion exceeds 0.1, the judgment was considered inconsistent. Therefore, a consistency ratio must be below 0.1 or 10%. The process was reiterated if the evaluation was unpredictable until the CR was within the wanted scale. The user formulated a conclusion according to the assessment results [32].

2.5. Development of Flood Risk Map

In this study, Sendai Framework was the basis to evaluate the flood risk by integrating the three (3) criteria, e.g., hazard, vulnerability, and exposure. The result of the flood risk assessment was laid into a map for a better comprehension of it. The last phase of the methodology was to overlay the analysis technique using ArcGIS. The GIS tool generated two or more different thematic maps of a similar area. It overlapped them on top of one another resulting in a new map using the weighted overlay tool. This technique results in a calculation matrix that defined the primary change forms in a study location [26]. The weighted overlay analysis results were developed employing equal intervals with four (4) levels (very low, low, moderate, and high). Flood risk map results were also validated by doing actual ground assessment of the localities in Odiongan, Romblon, and by reviewing identified flood zones in the area using records of historical flood events.

3. Results

3.1. Data Analysis for Identified Parameter

The result of data collection gathered through related literature, and past research studies were put into maps and analyzed. Each map has a scale of 1:100,000 and mainly focuses on identified parameters enumerated in Table 1 and detailed in Section 2.2. Subsequent sections elaborated the results.

3.1.1. Flood Hazard Parameters

Figure 3 shows the maps for every parameter of the Municipality of Odiongan, Romblon province, based on hazard criteria. It was noted that 2203.9 mm was the recorded average annual rainfall by the Romblon Weather Station. The average annual rainfall of the municipality was classified according to interpolation. Figure 3a shows the generated map from ArcMap using the Isohyetal Method with an average annual rainfall ranging from 2200 to 2250 mm. The amount of rain intensifies from the eastern part to the western part of the municipality.

The result of the reclassified slope layer was presented in Figure 3b and categorized in degrees where the green color means the lowest elevation. At the same time, the red part indicates the highest slope. Most maps show a higher slope ranging from 18 degrees to 50 and above. Residential areas were located in the plain areas (green part) where water accumulated during excessive rainfall.

Figure 3c shows the elevation map of the study location extracted using the IfSAR DTM. The elevation was classified into five (5) levels ranging from 0 to 600 m. Most of the map shows a high elevation of 21 to 600 m. The eastern part of the municipality, where the residential and commercial area was located, has the lowest elevation value, varying from 0 to 20 m. These elevations affect how rapidly stormwater could be drained into the catchment based on its slopes.

The slope map shown in Figure 3d was prepared as a shapefile from the Odiongan CLUP 2015. The map was sorted according to its Hydrologic Soil Group (HSG). Soils were classified by the Natural Resource Conservation Service based on the soil's runoff potential. Group A-class (sand, loamy sand, or sandy loam types of soils) has low runoff potential and high infiltration levels, even when fully saturated. They contain chiefly deep, well-drained to excessively drained sands or gravels, and have a high rate of water transmission. Group B is silt loam or loam. It has a moderate penetration rate when fully saturated and consists of moderately deep to deep, well-drained soils with relatively fine to coarse textures. Group C soils are sandy clay loam. They have low infiltration levels when thoroughly wetted. They consist chiefly of soils with a layer that impedes the downward movement of water and soils with moderately fine to fine structure. It is observed that most of the map falls under Group C. The eastern part of the municipality is mainly silt loam or loam. The map shown in Figure 3e is the overlayed flood depth map combined with the MGB Flood Susceptibility Map (see Figure A1) and further discussed below.

In the simulation or modeling process, as a result of delineation, there were 32 watersheds, 16 junctions, and 16 reaches extracted, as shown in Figure 4.

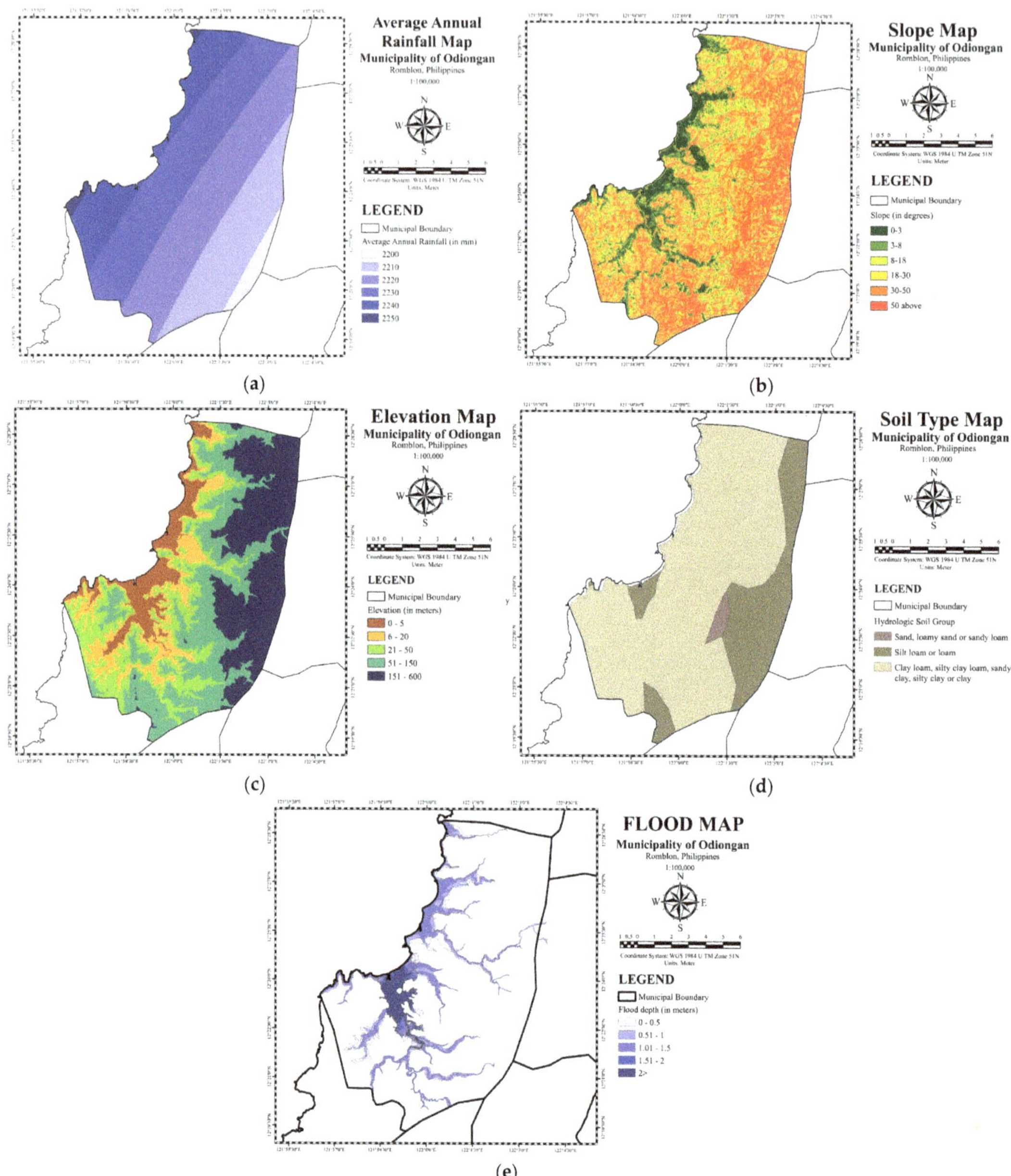

Figure 3. Maps generated using ArcGIS in flood hazard parameters: (**a**) Average Annual Rainfall Map; (**b**) Slope Map; (**c**) Elevation Map; (**d**) Soil Type Map; and (**e**) Flood Depth map of Odiongan, Romblon.

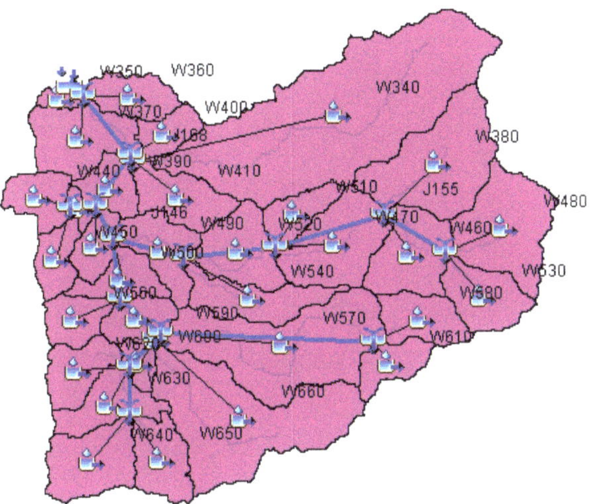

Figure 4. Extracted Basin Model of Odiongan, Romblon using GeoHMS10.7 tool in ArcpMAp.

The results from the basin model for 25-, 50-, and 100-year (Figure 5) simulation were exported to excel for the data preparation for hydraulic modeling in HEC-RAS and recorded 2.9 m^3/s as its total highest inflow.

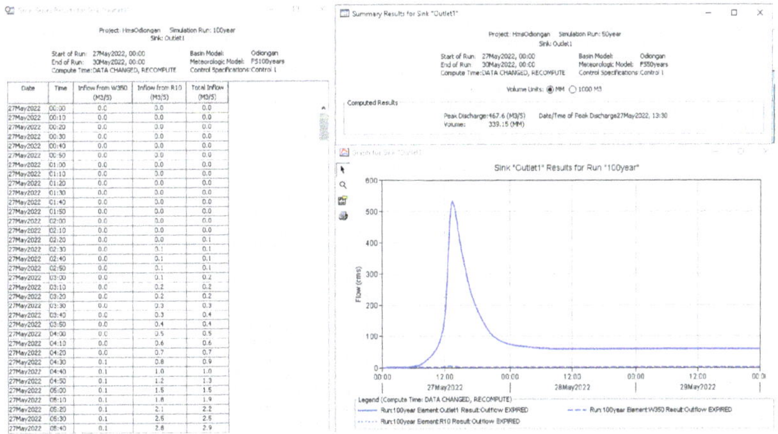

Figure 5. Simulation of total inflow results in 100-year using HEC-HMS.

Figure 6 shows that flood depth for all the periods led to a drastic result in a possible flood for the Poblacion area and nearby barangays from the watershed. Based on the simulation, about 5.87 m of flood height were recorded in the worst-case scenario of a 100-year return period. Flood depth with the 100-year model, considered one of the parameters, was exported as raster files from HEC-RAS and mapped to the layers in GIS.

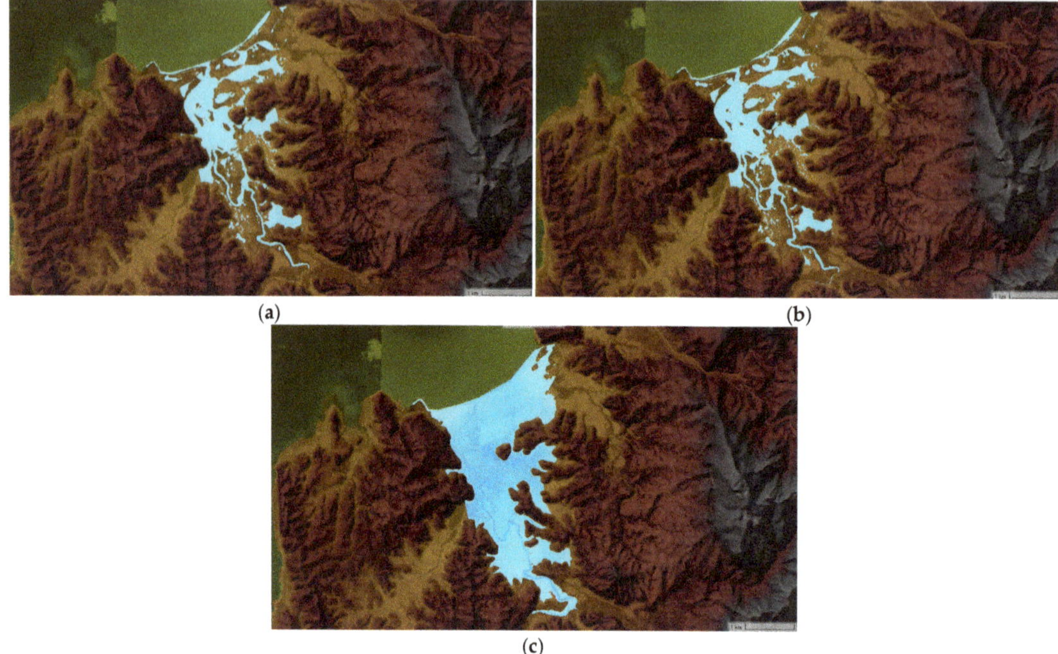

Figure 6. Simulation of maximum flood depth in the main river (Bangon River) of Odiongan, Romblon using HEC-RAS for (**a**) 25 years, (**b**) 50 years, and (**c**) 100-year period.

Modeled flood depths were categorized into five (5), e.g., (1) 0–0.5 m (low), (2) 0.5–1 m (moderate), (3) 1.01–1.5 m (high), (4) 1.51–2 m very high, and (5) 2 m and above is considered extremely high. These ranges were based on the Mines and Geosciences Bureau's Flood Susceptibility maps wherein low susceptibility can experience flood heights of less than 0.5 m and a flood duration of less than one (1) day. These included low hills and gentle slopes. It has also spared moderate drainage density. Moderate susceptibility areas were expected to experience flood depths of 0.5 m to 1 m. These spaces are prone to widespread inundation (flooding) throughout long and extensive heavy rainfall and extreme weather conditions. In areas of high susceptibility, where flood height is 1 meter or more with a time of recession of 3 days, are immediately flooded during heavy rains of several hours. The map indicated that most of the area has an adequate slope where only exposure to flood happens in the low-lying zone. Based on the 100-year flood model, the flood surge was concentrated in the town proper of Odiongan with a depth of 3 m for rainfall that occurred in two (2) days based on simulations.

3.1.2. Flood Vulnerability Parameters

This study considered the demographics and disaster risk reduction data of the Municipality of Odiongan. Figure 7 shows the maps from the available archival and survey data showing the population age, gender ratio, average income, physical health of the individual, educational attainment, emergency preparedness, and variety of built-up structures.

The men to women gender ratio in the Municipality of Odiongan, as shown in Figure 7a, has recorded more women than men. Barangay Amatong, Rizal and Progresso Este showed only a majority number of men to women. For the total men-to-women gender ratio of Odiongan, it was recorded that the population of women and men is almost the same with a ratio of 0.99991.

The mean age of an individual was considered a parameter. The numbers were taken from each barangay database. Mean age was calculated for all the group data as per age category. As per the results, as shown in Figure 7b, the majority of participants were between 30 and 39. Barangay Pato-o, Amatong, Dapawan, and Bangon have the lowest age range of 29 to 30, while Anahao obtained the highest mean age with 37 to 38 age level.

The average income of individuals was mapped per barangay stipulated in each barangay profile. As shown in Figure 7c, the average annual income per barangay was classified under six (6) levels. Most of the average income ranges from 100,000 to 500,000. However, Barangay Amatong, Bangon, Anahao, Malilico, and Progresso Este have the lowest income, having 40,000 and below.

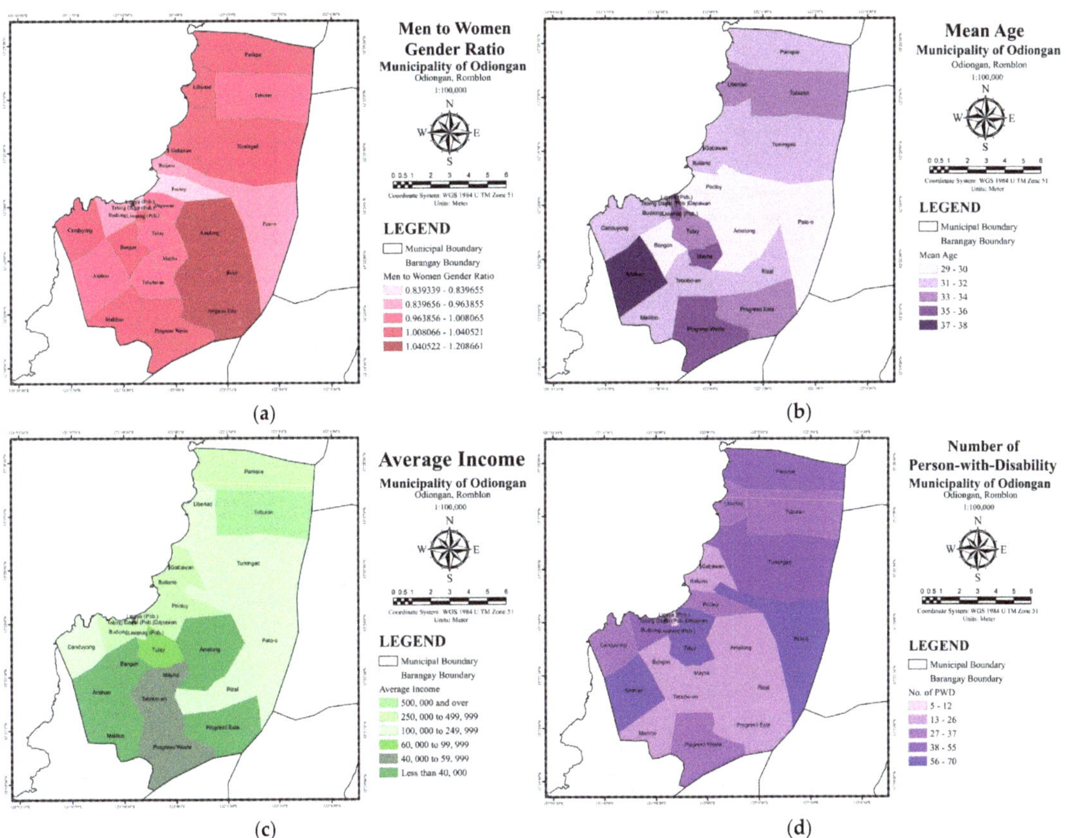

Figure 7. *Cont.*

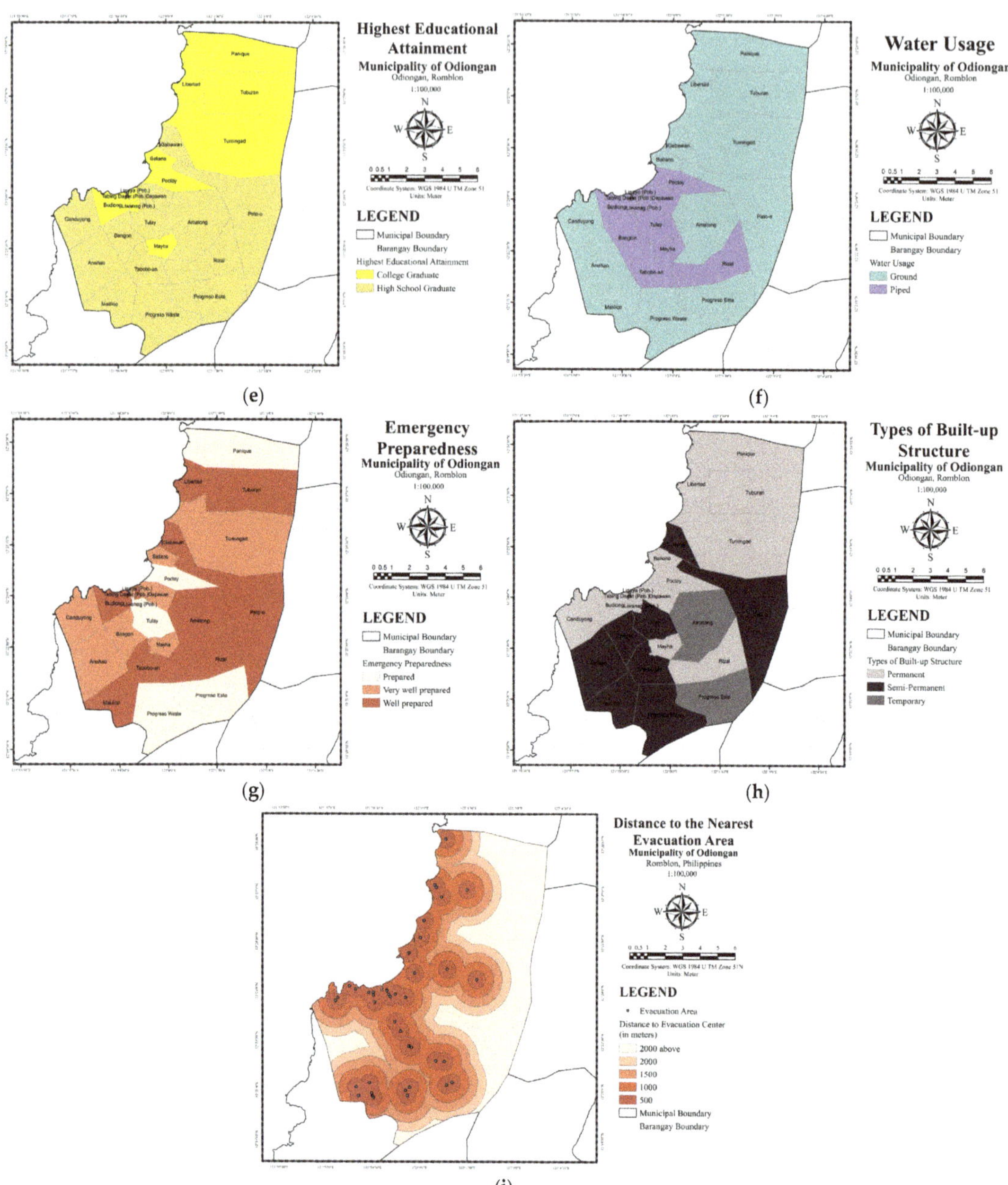

Figure 7. Result maps for flood vulnerability parameters in Odiongan, Romblon using ArcMap: (**a**) Gender Ratio; (**b**) Mean Age; (**c**) Average Annual Income; (**d**) Number of PWD; (**e**) Highest Educational Attainment; (**f**) Water Usage; (**g**) Emergency Preparedness; (**h**) Types of Built-up Structure; and (**i**) Distance to the Nearest Evacuation Center.

1. The floods that hit the areas have disrupted public health and are a subject that has become increasingly important daily due to society's reactions to hazards [55,56].

With a lack of health information, the study considered the number of PWD in each barangay. Under RA 10524, it refers to individuals who agonize long-term physical, mental, intellectual, or sensory impairments that may obstruct their full and practical involvement in society on an equal basis upon interaction with various barriers. The seven types of disabilities mentioned in RA No. 7277 are psychosocial disability, disability due to chronic illness, learning disability, mental disability, visual disability, orthopedic disability, and communication disability [56]. The number of PWD was based on the Barangay Management System (BMS). Records of the number of PWD in Odiongan, Romblon, as shown in Figure 7d, signify a risk in flood events or other natural disasters. The highest number of PWD are in Barangay Anahao, Pato-o, Tulay, and Dapawan, where Tulay and Dapawan experience flood events every year. However, many PWDs were observed in the southern and northern parts of the municipality, excluding barangay Anahao.

2. The higher the level of education a respondent from a household has, the more likely the individual evacuates [57]. The data used in the study were based on a questionnaire survey conducted in each barangay. Only two categories recorded the highest educational attainment in Odiongan, Romblon, as shown in Figure 7e. More than half of the barangays indicate some high school graduate, and almost half were categorized as college graduates.

The research also incorporated water usage (Figure 7f) as a vulnerability parameter as an additional parameter in disaster risk reduction information. The source of information was a questionnaire survey where the head of barangays was asked for the primary source of water supply. Information was also verified in the records and data of Odiongan Water District. From Barangay Rizal down to the Poblacion (Dawapan, Liwanag, Liwayway, Ligay, Tabin-Dagat) area have access to pipe water. According to the data validated from the Odiongan Water District, 11 out of 25 barangays were supplied by piped water, where the main tank and reservoir are from Barangay Rizal. However, as Romblon is given such a water source, water supply from electric pumps and wells was installed for some barangays.

The emergency prepared data were based on a survey questionnaire's knowledge and input from the head of barangays. The emergency preparedness map is shown in Figure 7g and was classified into three (3) categories, e.g., prepared, well prepared, and very well prepared. It was noticed that only 5 out of 25 barangays, namely, Progresso Weste, Progresso Este, Tulay, Poctoy, and Panique, have been categorized as "prepared" barangays during calamities.

3. Type of Built-up Structures

The combination of information was taken from the barangay profile as of 2018, and a questionnaire survey was conducted. Figure 7h shows the types of built-up structures and are classified into three (3) categories. These are the (a) permanent, (b) semi-permanent, and (c) temporary shelters. Permanent buildings are structures with concrete foundations and walling, GI sheets as roofing, and other solid materials. Semi-permanent structures are a combination of lumber and concrete elements. Temporary shelters use sawali, bamboo, nipa, and cogon as construction materials. The map shows that most of the barangay have permanent and semi-permanent structures. This indicated that most of the homes in Odiongan are more resilient in terms of flood events. However, eight (8) barangays in the elevated area have residential structures categorized as temporary shelter. Consistent findings were proven in the previous research regarding households' housing types as a significant factor in flood vulnerability [43].

Coordinates were noted and listed during site investigations and measured the floor area to estimate the capacity of every room area during calamity (see Figure A2). Figure 7i shows the ideal coverage of every evacuation area in the municipality on which 500 to 2000 m circles around each evacuation center were drawn and categorized into five (5) levels. The map shows the number of people who can reach the facility within an acceptable walking distance. In this map, the ideal number of people have been identified for each evacuation center during a disaster.

The Sphere standards imply that in the instant aftershock of a disaster, particularly in dangerous climatic conditions where quarter materials are not readily available, an area of no less than 3.5 square meters per person is suitable to save lives and provide adequate short-term shelter. As an evacuation center is utilized preferably only for a short duration, a center's maximum 'event sheltering' capacity should permit no less than 1.5 square meters per person [55].

3.1.3. Flood Exposure Parameters

There were three (3) identified parameters for flood exposure assessment. Figure 8 shows the population density data, land use map, and household numbers from the CLUP Odiongan and PSA Database.

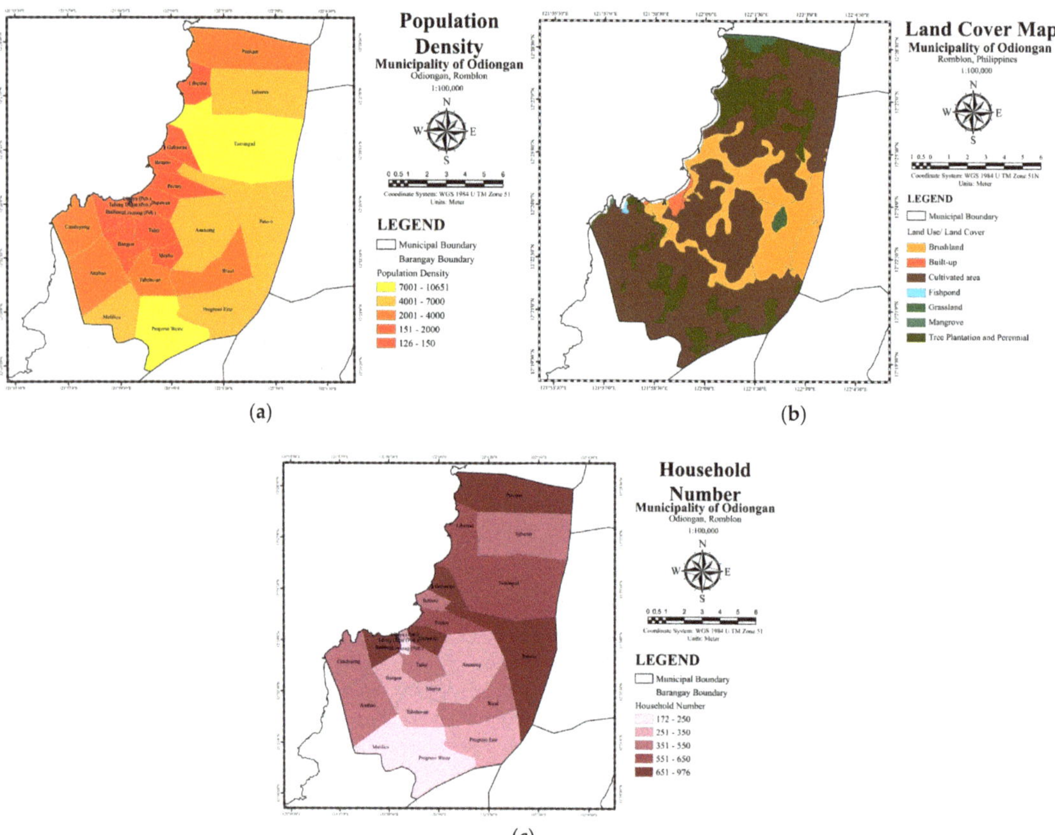

Figure 8. Result maps for flood exposure parameters in using ArcMap: (a) Population Density; (b) Land Use Map; and (c) Household Number.

Figure 8a shows the Population Density in the Municipality of Odiongan. The map demonstrates that population density is low in the nearby Poblacion area or where the center of the municipality is located. Barangay Ligaya and Liwayway as the lowest population density followed by Tabin-Dagat, Dapawan, Budiong, Liwanag, Bangon, Tulay,

Mayha, Batiano, Gabawan, and Libertad. However, Barangay Tumingad and Progresso Weste displayed a high population density.

Land use and land cover type are vital factors responsible for flood incidence. The land cover represents the physical (water, bare ground, and artificial structures) and biological (grass and trees) land cover. In contrast, land use describes how men utilized the land to improve their state of living [58,59]. The occurrence of flooding was inversely related to vegetation density. The study area's land cover classes were prepared from the municipal's CLUP. The land cover map was reassigned by categorizing the land-use types into seven (7) general categories. The map shown in Figure 8b indicates that most of the municipality's land area was cultivated. Brushland is observed in the center part of the map, while the scattered site of tree plantation and perennial were seen on the map. The built-up zone is located in the town proper of the municipality.

Evacuation decision before (preemptive) or during (forced) a disaster indicates the choice of households to evacuate or stay in the area at risk of impending hazard [43]. The household number was considered one of the parameters in assessing exposure. The study obtained the data from PSA's last 2020 census. Presented in Figure 8c is the range of the number of households for every barangay in the municipality. It was recorded that a high number of households in Baranagay, Panique, Pato-o, Gabawan, Dapawan, and Tabin-Dagat were exposed to flood events. Between 172 and 250 households, which was the least number, were observed in Liwanag, Liwayway, Barangay Malilico, and Progresso Weste. However, Liwanag and Liwayway have small land areas that cater only to some houses.

3.2. Evaluation and Assessment of Parameters

Contributing factors were evaluated and assessed.

The decision was segregated into its independent components. It was presented in a hierarchy diagram of at least three levels: goal, criteria, and indicators. The study structure using AHP was shown in Figure 9, wherein the uppermost place of the hierarchy is the primary goal of having a flood risk map. The lower level of the order contains the criteria contributing to attaining the goal: flood hazard map, flood vulnerability map, and flood exposure map. Finally, the lowest level included the indicators: average annual rainfall, elevation, slope, flood depth, soil type, gender ratio, individual age, average income, number of PWD, highest educational attainment, water usage, emergency preparedness, types of built-up structures, distance to evacuation area, population density, land cover, and number of households. The featured weight was assigned for each parameter, where were reclassified and normalized. Assigned values depend on the type of level or category. Table 2 indicates the feature weight of every indicator. The results of the weights computed using the AHP based on experts' inputs are shown in Table 3. These are the final weights of each parameters identified through AHP and was ensured to pass the consistency index requirement for it to be considered as valid.

Table 2. Standard matrix for hazard parameters.

	AAR	E	S	ST	FD	Weights	Percentage Weights
AAR	0.224979	0.221826	0.259802	0.194982	0.228034	0.225925	23%
E	0.208256	0.205338	0.197392	0.202930	0.208920	0.204567	20%
S	0.154522	0.185622	0.178439	0.193329	0.188299	0.180042	18%
ST	0.196749	0.172539	0.157384	0.170516	0.156328	0.170703	17%
FD	0.215493	0.214675	0.206983	0.238243	0.21842	0.218763	22%
							100%

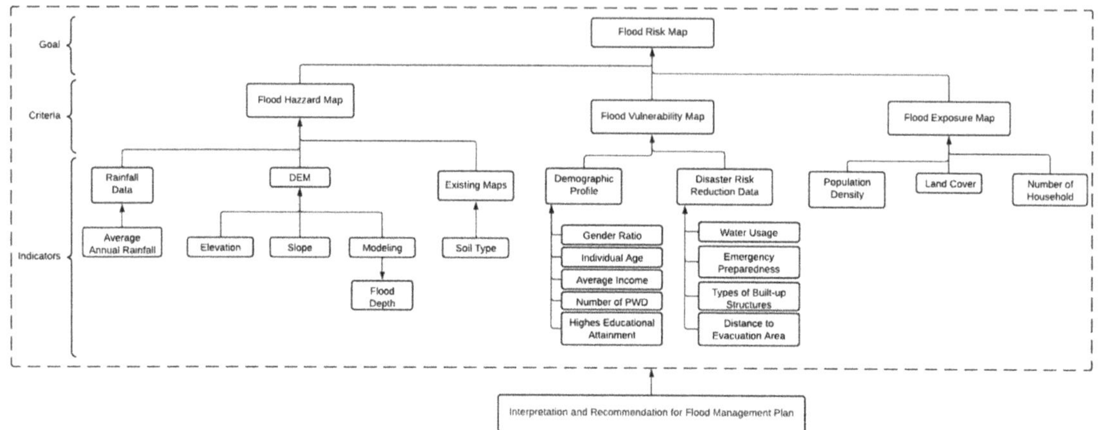

Figure 9. Multicriteria data analysis for flood risk assessment in AHP framework.

Table 3. Standard matrix vulnerability parameters.

	GR	MA	AI	NPWD	HEA	WU	EP	TBS	DEA	Weights	Percentage Weights
GR	0.091812	0.072630	0.083258	0.095499	0.093304	0.088862	0.116943	0.095699	0.080989	0.091000	9%
MA	0.136601	0.108062	0.134171	0.116910	0.076277	0.083485	0.110578	0.090066	0.125462	0.109068	11%
AI	0.124171	0.090691	0.112602	0.090232	0.105573	0.129437	0.126444	0.135126	0.096882	0.112351	11%
NPWD	0.101895	0.097966	0.132265	0.105988	0.102919	0.107756	0.086492	0.107974	0.106007	0.105474	11%
HEA	0.078704	0.113313	0.085309	0.082368	0.079983	0.095524	0.073935	0.096588	0.092532	0.088695	9%
WU	0.109121	0.136708	0.091879	0.103882	0.088433	0.105616	0.085829	0.111776	0.117633	0.105653	11%
EP	0.108148	0.134616	0.122671	0.168801	0.149019	0.169508	0.137751	0.134197	0.130969	0.139520	14%
TBS	0.114932	0.143735	0.099830	0.117594	0.099203	0.113196	0.122971	0.119798	0.130779	0.118004	12%
DEA	0.134616	0.102279	0.138016	0.118726	0.205288	0.106616	0.139056	0.108777	0.118748	0.130236	13%
											100%

The final relative weights of the alternatives which were defined by computing the product's linear combination (LC) between the relative weight of each criterion and the alternative for that specific criterion are shown in Tables 2–4, whereas Tables 5–7 are the computation of CI and CR for hazard, vulnerability, and exposure, respectively. A consistency ratio of 1.32%, 3.31%, and 6.85% were noticed in the hazard, vulnerability, and exposure below 10%. The experts made repeated responses to obtain the acceptable CR for all judgments. Further, final weights for hazard, vulnerability, and exposure are shown in Table 8. These weights are integrated into ArcGIS to generate hazard, vulnerability, and exposure maps with the corresponding index value.

Table 4. Standard matrix for exposure parameters.

	PD	LC	NH	Weights	Percentage Weights
PD	0.326245	1/3	1/3	0.327369	33%
LC	0.37159	0.34364	0.31783	0.344357	34%
NH	0.302161	0.354651	0.328012	0.328274	33%
					100%

Table 5. Computation of CR and CI of hazard parameters for consistency of AHP.

	AAR	E	S	ST	FD	Sum	Crit. Weigths
AAR	0.225925	0.244066	0.328938	0.258341	0.235870	1.293139	5.72376382
E	0.189362	0.204567	0.226295	0.243454	0.195670	1.059347	5.17847996
S	0.123658	0.162756	0.180042	0.204129	0.155214	0.825800	4.58669759
ST	0.149284	0.143437	0.150561	0.170703	0.122176	0.736161	4.31251769
FD	0.209539	0.228710	0.253757	0.305653	0.218763	1.216422	5.56046026
						y_{max}	5.07238386
						CI	0.018095966
						CR	0.016157112

Table 6. Computation of CR and CI of vulnerability parameters for consistency of AHP.

	GR	MA	AI	NPWD	HEA	WU	EP	TBS	DEA	Sum	Crit. Weights
GR	0.091000	0.061162	0.067285	0.081994	0.106155	0.076564	0.077254	0.072694	0.062064	0.696171	7.650275
MA	0.162276	0.109068	0.129960	0.120308	0.104014	0.086214	0.087554	0.081999	0.115235	0.996628	9.137668
AI	0.151949	0.094290	0.112351	0.095649	0.148296	0.137692	0.103129	0.126726	0.091663	1.061744	9.450256
NPWD	0.117057	0.095620	0.123891	0.105474	0.135719	0.107612	0.066225	0.095063	0.094157	0.940818	8.919941
HEA	0.076032	0.093005	0.067196	0.068929	0.088695	0.080220	0.047605	0.071511	0.069114	0.662308	7.467245
WU	0.125572	0.133660	0.086208	0.103554	0.116815	0.105653	0.065829	0.098578	0.104661	0.940530	8.902074
EP	0.164345	0.173804	0.151996	0.222206	0.259944	0.223923	0.139520	0.156289	0.153880	1.645906	11.79692
TBS	0.147721	0.156959	0.104619	0.130927	0.146361	0.126474	0.105343	0.118004	0.129960	1.166368	9.884114
DEA	0.190954	0.123266	0.159630	0.145889	0.334267	0.131470	0.131470	0.118254	0.130236	1.465436	11.25217
										y_{max}	9.384519
										CI	0.048065
										CR	0.033148

Table 7. Computation of CR and CI of exposure parameters for consistency of AHP.

	PD	LC	NH	Sum	Crit. Weights
PD	0.327369	0.287416	0.353462	0.968247	2.957665
LC	0.392224	0.344357	0.33367	1.070251	3.10797
NH	0.30404	0.338789	0.328274	0.971104	2.958207
			y_{max}		3.007947
			CI		0.003974
			CR		0.006851

Table 8. Final weights and percentage weights of every parameter for hazard, vulnerability, and exposure.

Parameters	Weights	Percentage Weights
Hazard Parameters		
Average Annual Rainfall (AAR)	*0.225925*	22.59%
Elevation (E)	*0.204567*	20.46%
Slope (S)	*0.180042*	18.0%
Soil Type (ST)	*0.170703*	17.07%
Flood Depth (FD)	*0.218763*	21.88%
Vulnerability Parameters		
Gender Ratio (GR)	*0.091000*	9.1%
Mean Age (MA)	*0.109068*	10.91%
Average Income (AI)	*0.112351*	11.24%
Number of PWD (NPWD)	*0.105474*	10.55%
Highest Educational Attainment (HEA)	*0.088695*	8.87%

Table 8. *Cont.*

Parameters	Weights	Percentage Weights
Water Usage (WU)	*0.105653*	10.57%
Emergency Preparedness (EP)	*0.139520*	13.95%
Types of Build-up Structures (TBS)	*0.118004*	11.8%
Distance to the nearest Evacuation Area (DEA)	*0.130236*	13.02%
Exposure Parameters		
Population Density (PD)	*0.327369*	32.74%
Land Use/Land Cover (LULC)	*0.344357*	34.44%
Household Number (HN)	*0.328274*	32.83%

3.3. Development of Flood Risk Map

The visualization outputs for hazard, vulnerability, and exposure are shown in Figure 10. These maps were generated after computing the criteria weights using AHP and incorporating these weights with a GIS-based process consisting of overlays, raster conversion, and layer clipping.

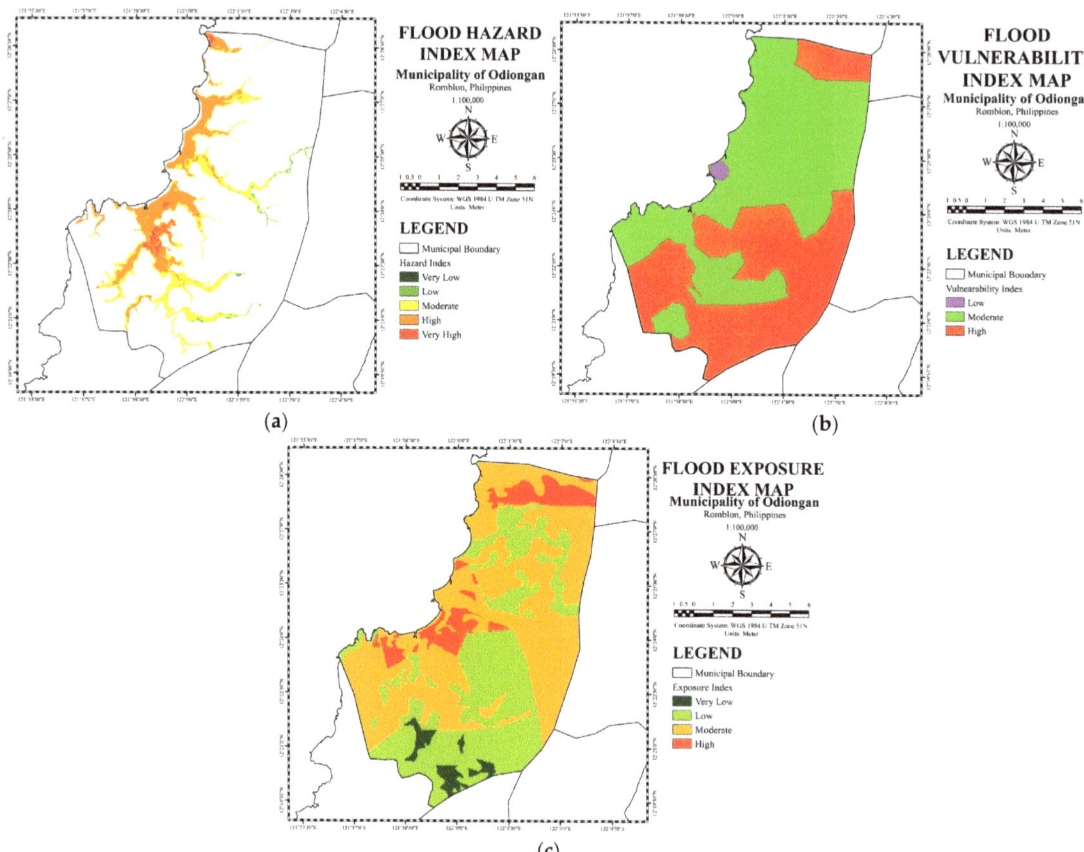

Figure 10. (**a**) Flood Hazard, (**b**) Flood Vulnerability, (**c**) Flood Exposure Index Maps of the Odiongan, Romblon, integrating all parameters in ArcGIS.

As shown in Figure 10a, the flood hazard index map, which combined all five (5) factors, was developed using the overlay tool in ArcGIS. The hazard map was classified into five (5) levels: very low (green), low (yellow-green), moderate (yellow), high (orange), and very high (red) and covers 0.009, 1.07, 8.74, 11.71, and 0.76 square kilometers, respectively. It was observed that areas with hazard index values were mostly affected according to the flood depth map, where river bodies are also located. Moderate to very high hazard is seen in areas of Poblacion, including Poctoy, Bangon, and Anahao. However, a moderate to high index was presented in some parts of Batiano, Gabawan, Libertad, and Paniques.

The vulnerability index map obtained by combining all nine (9) parameters highlights five areas (low, moderate, and high), as shown in Figure 10b, using the overlay tool. The flood vulnerability map generated from ArcGIS determines the degree of susceptibility of the flood-prone zone [29]. Purple connotes low vulnerability, yellow-green for moderate, and red for high vulnerability. Low, moderate, and high classes cover 0.56%, 54.52%, and 44.92% of Odiongan. As can be seen, only barangay Batiano is in low vulnerability and barangays Mayha, Tabobo-an, Canduyong, Malilico, Poctoy, Gabawan, Tumingad, Tuburan, Libertad, and a portion of Poctoy and Budiong is in moderate vulnerability. Tulay, Amatong, Pato-o, Rizal, Progresso Este, Progresso Weste, Anahao, Bangon, Tulay, and Panique is observed as highly vulnerable.

Exposure parameters (population density, number of households, and land cover) were overlaid in ArcGIS to develop a Flood Exposure Index Map. The resulting map derived four categories (very low, low, moderate, and high) in flood exposure using experts' weights, as shown in Figure 10c. The green symbolizes very low exposure, yellow-green for low, orange for moderate, and red for high exposure, which covers 2.61%, 36.39, 53%, and 8.01%, respectively, of the total land area of Odiongan. The Poblacion area and parts of Budiong, Gabawan, Batiano, and Panique were at high exposure to flood. However, more than half of the map is scattered orange illustrating moderate exposure.

The result of the flood risk assessment is laid into a map for a better comprehension. As shown in Figure 11, the analysis result was a map combining flood hazard, flood vulnerability, and flood exposure index maps utilizing ArcGIS. Equal weights were employed in three (3) maps. The flood risk map is categorized using equal intervals with four (4) levels (very low, low, moderate, and high). In total, 93.92 square kilometers (green) are classified as very low risk, comprising 83.78% of the total land area. The yellow-green color as low risk covers approximately 0.198 square kilometers (0.15% of land area) and is seen in a small part of Barangay Rizal and Tumingad. Overall, 12.86% of the total area was categorized as moderate risk (yellow) and noted on the map as 17.56 square kilometers. A portion of barangays Rizal, Progresso Este, Progresso Weste, Malilico, Amatong, Anahao, Canduyong, Pato-o, Tumingad, Mayha, Tuburan, and Panique were observe with moderate risk. Gabawan, Batiano, Tabobo-an, and Libertad were also at moderate risk, with more than half of their respective barangay boundaries. The Poblacion area, Tulay, Bangon, Anahao, Dapawan, and Poctoy were at high risk to flood occurrence, covering 3.26% of land area (4.46 square kilometers). Parcels with high risk were also sighted in Canduyong, Gabawan, and Panique. Through this flood risk map, the municipal councils, planning agencies, and other stakeholders can prepare Flood Management Plan to reduce the threat to lives due to flooding and anticipate future infrastructure development in the municipality.

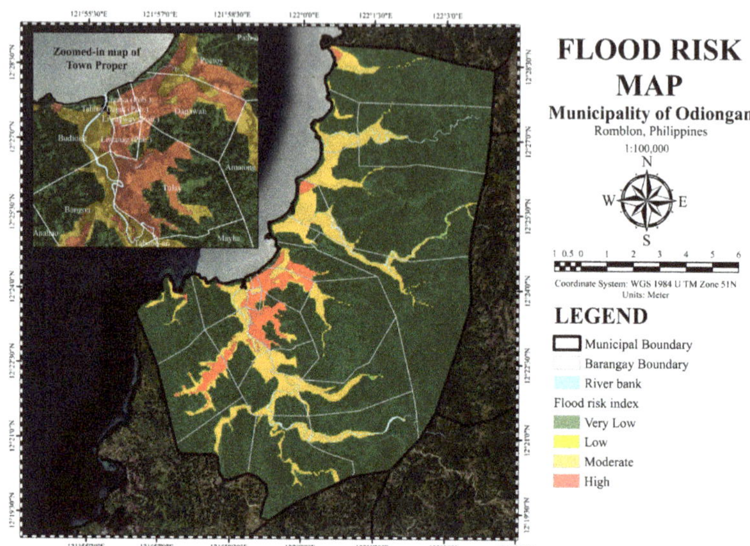

Figure 11. Imagery and flood risk map (50% transparency) with 1:50,000 scale zoomed-in map of town proper of the Municipality of Odiongan.

4. Discussion

The harmful effects of disasters in flood-prone areas have amplified in severity over the years. The damage ensuing from these events are also exponentially growing, where serious land use and climate change impacts are alarming. The development of a flood risk map utilizing elevation models, demographics, awareness, and disaster-risk-related data integrated into GIS and analyzed using AHP is an effective tool for evaluating risk. Integrating available data through the archival, automated process, and survey data were possible in deriving each criterion. It is noted in the flood risk map that approximately 83.78% of the total area map is at very low risk, 0.15% is at low, 12.86% at moderate, and 3.26% for high flood risk. 93.92, 0.198, 17.56, and 4.46 square kilometers are very low, low, moderate, and high risk to flood. A small portion of Barangay Rizal and Tumingad are at low risk (yellow-green). Moderate risk was noted on the map covering a part of barangays Rizal, Progresso Este, Progresso Weste, Malilico, Amatong, Anahao, Canduyong, Pato-o, Tumingad, Mayha, Tuburan, and Panique. Then, Gabawan, Batiano, Tabobo-an, and Libertad were at moderate risk covering half of their respective barangay boundaries. Moreover, based on the zoomed-in map of the town proper, the Poblacion area, Tulay, Bangon, Anahao, Dapawan, and Poctoy are at high risk to flood occurrence. The possible reason for this is the rapid urbanization and infrastructure development in the town proper of Odiongan, as shown in Figure 11. In addition, moderate to high-risk indexes were observed along the riverbanks of Odiongan. In validation, areas at high risk are known to have flood events. The results of the risk maps urge the municipality to plan for flood mitigation or develop a comprehensive flood management plan as a countermeasure during for future flood events. Flood risk assessment is required for the flood management and mitigation of cities and municipalities. However, the specific parameters available in such areas are significantly different, which differs the risk assessment index weight. The study of Cai et al. [15] adopted the frequency of rainstorms and average annual precipitation of counties as hazard parameters, similar to what were used in this study. Average annual rainfall represented 22.59% of the weight of the total hazard parameters which showed its importance or role in identifying hazard level in a certain area. For vulnerability factors, the study considered the following: population density, average area GDP, per capita disposable income, road network density, and land use type. Average income was basically identified to have the highest weight among all other factors considered in the vulnerability index with 11.24%, this just shows that the capability to spend affects how people could prepare and protect themselves during a disaster. A risk assessment could also be performed, even in mega-cities, as based on the study of Lyu et al. [60]. Parameters such as rainy season, average rainfall, average rainy day for hazard and elevation, slope, river proximity, river density as exposure parameters; and land use, metro line proximity, metro line density, road network proximity and road network density as vulnerability indicators were considered, similarly to what was considered in this study. This shows that methodology used in this study could also be applied in mega-cities here in the Philippines with some revision of factors to be considered which will be based on the city's characteristics. A local study [46] used Confirmatory Factor Analysis (CFA) to simultaneously investigate the interrelationship between vulnerability to natural hazards composed of exposure, sensitivity, and resilience. Additionally, a study [61] developed a comprehensive framework for vulnerability assessment to determine vulnerability that would deliver a transparent understanding and improve community competency leading to the development of methodologies to assess factors and indicators of vulnerability. For this assessment, the combination of hazard, vulnerability, and exposure assessment established a clear definition of risk levels [36,62]. Compared to other related studies, the evaluation used the Sendai Framework to clear out the true meaning of disaster risk based on all dimensions of vulnerability, exposure, and hazard characteristics of the environment. Regarding hazard assessment, many studies used factors such as topographical [63], natural, and anthropogenic [62] factors. In this study, factors used were average annual rainfall, slope, elevation, soil type, and simulated flood depth, which were decided according to the data availability. In terms of vulnerability assessment, flood vulnerability is affected by factors such as settlement conditions, infrastructure, policy and capacities of the authorities, social inequities, and economic patterns. The study effectively generated a vulnerability map integrating age, gender ratio, average income, individual physical health, educational attainment, emergency preparedness, and types of built-up structures as factors. Exposure assessment of the study is based on the analysis of [64], which used land use and population density as parameters. The study added the additional factor, household number, to justify the flood exposure.

The accuracy of generating flood hazard maps is highly dependent on the quality of topographical data [62]. Topographical information such as DEMs is an excellent source to derive topographic factors responsible for flood activity [65]. Although local studies [63,64] utilized LiDAR-derived DEM due to its inherent high vertical accuracy and resolution, If-SAR DTM from NAMRIA, as one of the highest resolution available DEM in the location, is used in the study, which showed reasonably consistent with the generated maps.

The flood risk assessment map demonstrates flood risk areas that must be managed on a priority basis. In some studies, different methodologies were established for assessing flood risk. One uses a hybrid intelligence model [14], probability [66,67], polygon approach [68], Quantitative risk assessment methods (e.g., Fine Kiney [55] and Maxent model [8]), and MCDA techniques, such as fuzzy majority approach [30], fuzzy variable set theory [44], multi-attribute value, frequency ratio, artificial neural network [65], fuzzy analytical hierarchy process (FAHP) [41], and decision tree [69] to assess the flood risk. Most government units, municipal planners, and other concerned agencies use AHP as an MCA in terms of disaster risk reduction, land use plans, and decisions requiring a comprehensive judgment and recommendation from experts to benefit the community. Using the AHP decision-making method for the multiple flood-related factors is extensively adopted. From the result of the study, it is noticed that AHP proposes a flexible, stepwise, and precise process of analyzing complicated problems in an MCDA environment. In addition, the resolution of complications in multicriteria methods is realized as the primary use of AHP. In this research, where three (3) criteria (hazard, vulnerability, and exposure) with multiple parameters (5, 9, 3, respectively) have different dimensions, it makes a simple MCDA dilemma more complicated. In this study, the AHP-based flood risk assessment method is considered relatively practical, convenient, and promotes interactive usage by flood managers for continuing improvement. AHP is a useful method for selecting contending options in light of a range of objectives to be convened. The computations were not complex, and the MCA did not need to understand the calculation to use the procedure. Nonetheless, the AHP has established considerations because it highlights the knowledge of decision-makers' preferences [32].

Compared to other flood maps available online and offline, the study highlighted how comprehensive the methodology is. Government agencies' flood maps consider only one to two criteria (hazard, susceptibility, or only topographical related data) that limit the assessment's accuracy. For this research, several social, economic, environmental, and technical factors were considered to develop a comprehensive flood risk map. Future studies could be performed to improve this risk map where environmental quality could be incorporated [70], such as flooding, was found to be correlated with the increasing concentration of manganese in Marinduque. Considering environmental quality in risk assessment would be useful.

5. Conclusions

The number of flooding events in the Municipality of Odiongan caused property damage to the community and has put lives at risk based on historical documents. This showed how vital flooding risk assessments using GIS-based, and multi-criteria decisions are. The results of this study are useful for improving the municipality's flood mitigation and risk management strategies.

This study assessed the flood risk in the Municipality of Odiongan, Romblon, considering relevant factors in floods using AHP and following the Sendai Framework. The study used ArcGIS, HEC-RAS, and HEC-HMS to map and model the primary and secondary data as parameters that mainly contribute to flooding. The study considered the following parameters: average annual rainfall, elevation, slope, soil type, and flood depth for hazard criteria; gender ratio, mean age, average income, PWD, educational attainment, water usage, emergency preparedness, type of built-up structures, and distance to evacuation area in vulnerability and population density, land cover and household number for exposure, respectively. Each parameter was compared to one another by pairwise comparison to identify its weights based on experts' judgment and integrate these weights of factors into AHP. Weights were computed as follows: average annual rainfall with 23%, elevation—20%, slope—18%, soil type—17%, and flood depth—22% for hazard criteria; gender ratio—9%, mean age—11%, average income—11%, PWD—11%, educational attainment—9%, water usage—11%, emergency preparedness—14%, type of built-up structures—12%, and distance to evacuation area—13% in vulnerability and population density with 33%, land use—34% and household number—33% for exposure. It was noted that approximately 83.78% of the total area map was at very low risk, 0.15% is at low, 12.86% at moderate, and 3.26% for high flood risk. Then, 93.92, 0.198, 17.56, and 4.46 square kilometers were very low, low, moderate, and high risk to flood. The risk assessment results derived a flood risk map which found out the nine (9) barangays were at high risk of flooding, notably the Poblacion Area, Tulay, Bangon, Tabobo-an, Dapawan, and Anahao. The flood risk map developed in this study considered the social, economic, environmental, and technical factors that represent those factors in actual scenarios. Highlighting its difference compared to the flood depth map or available flood maps online and released by the different concerned agencies incorporating the topographic aspect of the area.

In conclusion, the result of this flood risk assessment is essential for the municipality to improve their flood management strategies considering the risk factors: hazard, vulnerability, and exposure. It can help the planning agencies and other stakeholders anticipate flood risk, especially the LGU, by integrating the output into their CLUP. This technique can also be employed by other local government units to come up with more practical and effective strategies. Moreover, future studies must be conducted to enhance and update the flood risk assessment methods and management.

Author Contributions: Conceptualization, J.G.G. and C.E.F.M.; methodology, J.G.G. and C.E.F.M.; software, J.G.G.; validation, J.G.G.; formal analysis, J.G.G., C.E.F.M. and D.B.S.; investigation, J.G.G.; resources, J.G.G. and C.E.F.M.; data curation, J.G.G. and C.E.F.M.; writing—original draft preparation, J.G.G.; writing—review and editing, J.G.G., C.E.F.M., D.B.S. and F.J.T.; visualization, J.G.G. and C.E.F.M.; supervision, C.E.F.M., D.B.S. and F.J.T.; project administration, J.G.G. and C.E.F.M. All authors have read and agreed to the published version of the manuscript.

Funding: The study was funded by the Department of Science and Technology—Engineering Research and Development for Technology and Mapúa University.

Informed Consent Statement: Informed consent was obtained from all subjects involved in the study.

Data Availability Statement: All data are contained in the manuscript.

Acknowledgments: The author would like to acknowledge the in-kind support of Romblon State University, Local Government of Odiongan, National Mapping and Resource Information Authority (NAMRIA), Philippine Atmospheric, Geophysical, and Astronomical Services Administration (PA-GASA), Mines and Geosciences Bureau (MGB), Bureau of Soils and Water Management (BSWM), and the Philippine Statistics Authority (PSA) for the conduct of the study.

Conflicts of Interest: The authors declare no conflict of interest.

Appendix A

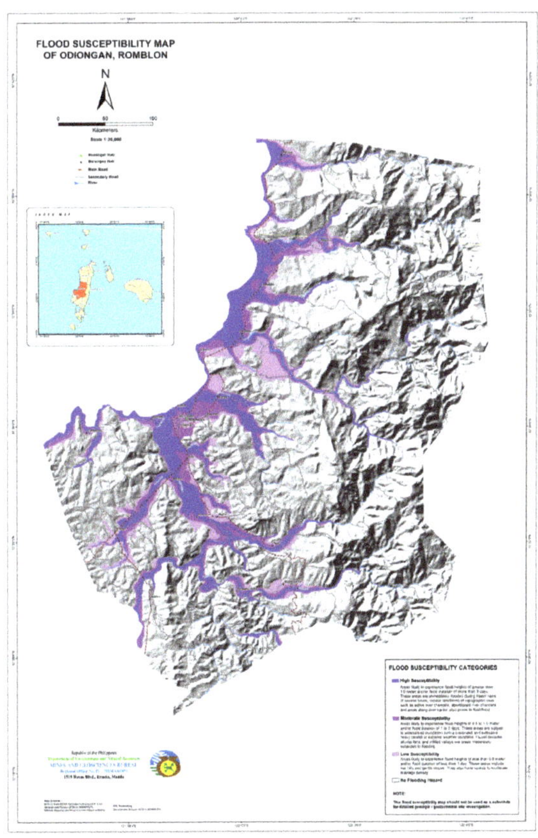

Figure A1. Flood Susceptibility Map from MGB-DENR.

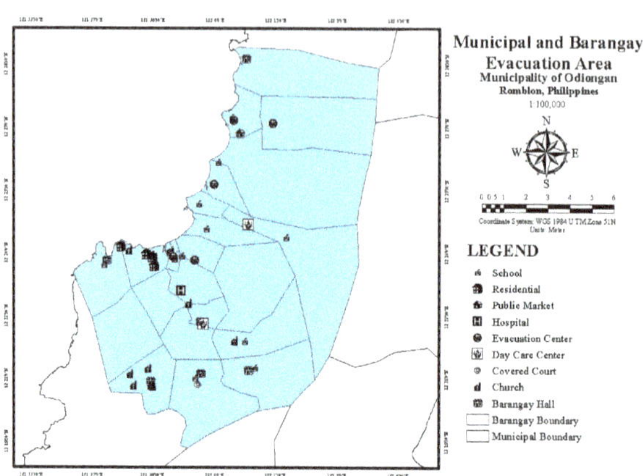

Figure A2. Identified evacuation area of barangay in Odiongan, Romblon, during emergency events.

In identifying the weights of factors, the pairwise comparison procedure usually contains a questionnaire for comparing all the elements and a geometric mean to arrive at a final solution. Psychologists conclude that the nine points shown in Table A1 are the most used comparison matrix individuals can compare simultaneously and consistently rank.

Table A1. The nine-point intensity of importance scale was modified from Schoenherr. Copyright 2008 Elsevier.

Intensity of Importance	Definition	Description
1	Equally important	Two factors contribute equally to the objective
3	Moderately more important	Experience and judgment slightly favor one over the other
5	Strongly more important	Experience and judgment strongly favor one over the other
7	Very strong, more important	Experience and judgment very strongly favor one over the other. Its importance is demonstrated in practice.
9	Extremely more important	The evidence favoring one over the other is of the highest possible validity.
2, 4, 6, 8	Intermediate values	When compromise is needed.
Reciprocals of above	If an element i has one of the above numbers assigned to it when compared with element j, then j has the reciprocal value when compared with i	
Ratios (1.1–1.9)	If the activities (elements) are very close.	It may be challenging to assign the best value, but when compared with other contrasting activities (elements), the size of the small numbers would not be too noticeable, yet they can still indicate the relative importance of the activities (elements)

The feature weight was assigned to each parameter, where levels were reclassified and normalized into 1, 2, 3, 4, and 5 (1 for the least priority and 5 for the most prior). Assigned values depend on how primary the level or category is. Table A2 indicates the feature weight of every indicator.

Table A2. Parameters with their designated feature weight.

Indicators	Feature Class	Feature Weight
Flood Hazard Parameters		
Average Annual Rainfall (in mm)	2200	1
	2210	1
	2220	2
	2230	3
	2240	4
	2250	5
Elevation (in meters)	0–5	5
	6–20	4
	21–50	3
	51–150	1
	151–600	0
Slope (in degrees)	0–3	5
	3–8	4
	8–18	3
	18–30	2
	30–50	1
	50 above	0
Soil Type	Sandy, loamy sand, or sandy loam	1
	Silt loam or loam	3
	Clay loam, silty clay loam, sandy clay, or clay	5
Flood Depth (in meters)	0–0.5	1
	0.51–1	2
	1.01–1.5	3
	1.51–2	4
	2>	5
Flood Vulnerability Parameters		
Gender Ratio (men to women ratio)	0.839339–0.839655	1
	0.839656–0.963855	2
	0.963856–1.008065	3
	1.008066–1.040521	4
	1.040522–1.208661	5
Mean Age	29–30	1
	31–32	2
	33–34	3
	35–36	4
	37–38	5
Average Income	500,000 and over	1
	250,000 to 499,999	1
	100,000 to 249,999	2
	60,000 to 99,999	3
	40,000 to 59,999	4
	Less than 40,000	5
Number of PWD	5–12	1
	13–26	2
	27–37	3
	38–55	4
	56–70	5
Highest Educational Attainment	College Graduate	3
	High School Graduate	5

Table A2. *Cont.*

Indicators	Feature Class	Feature Weight
Water Usage	Ground	4
	Piped	5
Emergency Preparedness	Prepared	5
	Well prepared	4
	Very well prepared	3
Types of Build-up Structures	Permanent	3
	Semi-permanent	4
	Temporary	5
Distance to the nearest Evacuation Area (in meters)	2000 above	5
	2000	4
	1500	3
	1000	2
	500	1
Flood Exposure Parameters		
Population Density	7001–10,651	1
	4001–7000	2
	2001–4000	3
	151–2000	4
	126–150	5
Land Use and Land Cover	Brushland	1
	Built-up	4
	Cultivated Area	3
	Fishpond	5
	Grassland	2
	Mangrove	0
	Tree Plantation and Perennial	0
Household Number	172–250	1
	251–350	2
	351–550	3
	551–650	4
	651–976	5

Pairwise comparison was based on adequate information, expert knowledge, and experience using a questionnaire. Ten (10) experts on-field and end-users determined the relevance of one alternative over the other with a pairwise comparison method presented in a matrix. Gathered weight for each criterion and option used a pairwise comparison technique. Then, each comparison is graded by expert respondents and end-user using the nine-point scale of importance. Eligible respondents and their credentials are shown in Table A3.

Table A3. Respondent's credentials for pairwise comparison technique.

Respondent	Field of Expertise/Project Involvement	Agency/Institution/Project	Years in Service
1	Water Resource Engineering/Disaster Risk	Mapua University	10
2	Meteorology/Hydrology	PAGASA-DOST	30
3	Project Staff	FRAMER—Mapua University	4
4	Researcher	FRA Project—Asian Institute of Technology	3
5	Disaster Risk/Municipal Engineer	LGU—Odiongan	30

Table A3. *Cont.*

Respondent	Field of Expertise/Project Involvement	Agency/Institution/Project	Years in Service
6	Meteorology/Hydrology	Visayas State University—Department of Meteorology	8
7	Meteorology/Hydrology	Central Luzon State University	3
8	Senior Research Specialist	UP Training Center for Applied Geodesy and Photogrammetry	5
9	Agriculturist II/Regional Head	Department of Agriculture—Bureau of Soil and Water Management	3
10	Supervising Geologist/Expert in Landslide and flood susceptibility mapping	Department of Environment and Natural Resources—Mines and Geosciences Bureau MIMAROPA	15

References

1. Osei, B.K.; Ahenkorah, I.; Ewusi, A.; Fiadonu, E.B. Assessment of flood prone zones in the Tarkwa mining area of Ghana using a GIS-based approach. *Environ. Chall.* **2021**, *3*, 100028. [CrossRef]
2. Flood—UN-SPIDER Knowledge Portal. Available online: https://www.un-spider.org/category/disaster-type/flood (accessed on 7 September 2022).
3. Nunez, C. Floods—Facts and Information. Available online: https://www.nationalgeographic.com/environment/article/floods (accessed on 27 June 2021).
4. Malhotra, S. Flooded Cities and Millions Displaced in Pictures—Greenpeace International. Available online: https://www.greenpeace.org/international/story/44296/flooded-cities-and-millions-displaced-in-pictures/ (accessed on 27 June 2021).
5. Monjardin, C.E.F.; Tan, F.J.; Uy, F.A.A.; Bale, F.J.P.; Voluntad, E.O.; Batac, R.M.N. Assessment of the existing drainage system in Infanta, Quezon province for flood hazard management using analytical hierarchy process. In Proceedings of the 2020 IEEE Conference on Technologies for Sustainability (SusTech), Santa Ana, CA, USA, 23–25 April 2020. [CrossRef]
6. ESCAP IDD. *Disasters in Asia and the Pacific: 2015 Year in Review*; ESCAP IDD: Bangkok, Thailand, 2015.
7. Davies, R. Philippines—Thousands Hit by More Floods in Central Regions—FloodList. Available online: http://floodlist.com/asia/philippines-negros-occidental-floods-january-2021 (accessed on 27 June 2021).
8. Cabrera, J.S.; Lee, H.S. Flood risk assessment for Davao Oriental in the Philippines using geographic information system-based multi-criteria analysis and the maximum entropy model. *J. Flood Risk Manag.* **2020**, *13*, e12607. [CrossRef]
9. Siddayao, G.P.; Valdez, S.E.; Fernandez, P.L. Analytic Hierarchy Process (AHP) in Spatial Modeling for Floodplain Risk Assessment. *Int. J. Mach. Learn. Comput.* **2014**, *4*, 450–457. [CrossRef]
10. Teves, C. Romblon Waterways at Risk of Overflow due to 'Quinta' Rains—Philippine News Agency. Available online: https://www.pna.gov.ph/articles/1119823 (accessed on 27 June 2021).
11. Alfonso, C.D.Q.; Sundo, M.B.; Zafra, R.G.; Velasco, P.P.; Aguirre, J.J.C.; Madlangbayan, M.S. Flood risk assessment of major river basins in the philippines. *Int. J. GEOMATE* **2019**, *17*, 201–208. [CrossRef]
12. Rahman, M.Z.A.; Alkema, D. Digital surface model (DSM) construction and flood hazard simulation for Development Plans in Naga City, Philippines. *GIS Dev. Malaysia* **2006**, 1–15.
13. Ali, K.; Bajracharya, R.M.; Koirala, H.L. A Review of Flood Risk Assessment. *Int. J. Environ. Agric. Biotechnol.* **2016**, *1*, 1065–1077. [CrossRef]
14. Pham, B.T.; Luu, C.; Phong, T.V.; Nguyen, H.D.; Le, H.V.; Tran, T.Q.; Ta, H.T.; Prakash, I. Flood risk assessment using hybrid artificial intelligence models integrated with multi-criteria decision analysis in Quang Nam Province, Vietnam. *J. Hydrol.* **2021**, *592*, 125815. [CrossRef]
15. Cai, S.; Fan, J.; Yang, W. Flooding Risk Assessment and Analysis Based on GIS and the TFN-AHP Method: A Case Study of Chongqing, China. *Atmosphere* **2021**, *12*, 623. [CrossRef]
16. Eleutério, J.; Martinez, D.; Rozan, A. Developing a GIS tool to assess potential damage of future floods. *WIT Trans. Inf. Commun. Technol.* **2010**, *43*, 381–392. [CrossRef]
17. Noamen, B.; Taoufik, H.; Arfa, S.B. Flood risk assessment and mapping using multi-criteria analysis (AHP) model and GIS: Case of the Jendouba Governorate—Northwestern Tunisia. *Int. J. Water Sci. Environ. Technol.* **2020**, *2*, 139–149.

18. Santillan, J.R.; Makinano-Santillan, M. Vertical accuracy assessment of 30-M resolution ALOS, ASTER, and SRTM global DEMS over Northeastern Mindanao, Philippines. *Int. Arch. Photogramm. Remote Sens. Spat. Inf. Sci.—ISPRS Arch.* **2016**, *41*, 149–156. [CrossRef]

19. Chen, B.; Ge, Y. The building of network geographic information system based on ArcGIS. In Proceedings of the International Conference on Computer Application and System Modeling, Taiyuan, China, 22–24 October 2010; Volume 14, pp. 90–93. [CrossRef]

20. Hawker, L.; Bates, P.; Neal, J.; Rougier, J. Perspectives on Digital Elevation Model (DEM) Simulation for Flood Modeling in the Absence of a High-Accuracy Open Access Global DEM. *Front. Earth Sci.* **2018**, *6*, 3389. [CrossRef]

21. Jeon, Y.W.; Bae, Y.; Ra, J.B. Error detection in digital elevation model using a camera image. In Proceedings of the International Geoscience and Remote Sensing Symposium, Melbourne, VIC, Australia, 21–26 July 2013; pp. 2517–2519. [CrossRef]

22. Sugumaran, R.; Davis, C.H.; Meyer, J.; Prato, T. High resolution digital elevation model and a web-based client-server application for improved flood plain management. In Proceedings of the International Geoscience and Remote Sensing Symposium, Honolulu, HI, USA, 24–28 July 2000; Volume 1, pp. 334–335. [CrossRef]

23. Ballado, A.H.; Bentir, S.A.P.; Lazaro, J.B.; Macawile, M.J.P. Depth perception analysis of LiDAR digital elevation model for low lying areas using delaunay triangulation algorithm. In Proceedings of the HNICEM 2017—9th International Conference on Humanoid, Nanotechnology, Information Technology, Communication and Control, Environment and Management, Manila, Philippines, 1–3 December 2017; pp. 1–5. [CrossRef]

24. Ternate, J.R.; Celeste, M.I.; Pineda, E.F.; Tan, F.J.; Uy, F.A.A. Floodplain Modelling of Malaking-Ilog River in Southern Luzon, Philippines Using LiDAR Digital Elevation Model for the Design of Water-Related Structures. *IOP Conf. Ser. Mater. Sci. Eng.* **2017**, *216*, 012044. [CrossRef]

25. Lagmay, A.M.F.A.; Racoma, B.A.; Aracan, K.A.; Alconis-Ayco, J.; Saddi, I.L. Disseminating near-real-time hazards information and flood maps in the Philippines through Web-GIS. *J. Environ. Sci.* **2017**, *59*, 13–23. [CrossRef] [PubMed]

26. Kittipongvises, S.; Phetrak, A.; Rattanapun, P.; Brundiers, K.; Buizer, J.L.; Melnick, R. AHP-GIS analysis for flood hazard assessment of the communities nearby the world heritage site on Ayutthaya Island, Thailand. *Int. J. Disaster Risk Reduct.* **2020**, *48*, 101612. [CrossRef]

27. Abe, Y.; Zodrow, I.; Johnson, D.A.K.; Silerio, L. Risk informed and resilient development: Engaging the private sector in the era of the Sendai Framework. *Prog. Disaster Sci.* **2019**, *2*, 100020. [CrossRef]

28. Wilkins, A.; Pennaz, A.; Dix, M.; Smith, A.; Vawter, J.; Karlson, D.; Tokar, S.; Brooks, E. Challenges and opportunities for Sendai framework disaster loss reporting in the United States. *Prog. Disaster Sci.* **2021**, *10*, 100167. [CrossRef]

29. Danumah, J.H.; Odai, S.N.; Saley, B.M.; Szarzynski, J.; Thiel, M.; Kwaku, A.; Kouame, F.K.; Akpa, L.Y. Flood risk assessment and mapping in Abidjan district using multi-criteria analysis (AHP) model and geoinformation techniques, (cote d'ivoire). *Geoenviron. Disasters* **2016**, *3*, 10. [CrossRef]

30. Boroushaki, S.; Malczewski, J. Using the fuzzy majority approach for GIS-based multicriteria group decision-making. *Comput. Geosci.* **2010**, *36*, 302–312. [CrossRef]

31. Dodgson, J.S.; Spackman, M.; Pearman, A.; Phillips, L.D. *Multi-Criteria Analysis: A Manual*; Department for Communities and Local Government: London, UK, 2009; Volume 11, pp. 1–16. ISBN 978-1-4098-1023-0.

32. Ouma, Y.O.; Tateishi, R. Urban flood vulnerability and risk mapping using integrated multi-parametric AHP and GIS: Methodological overview and case study assessment. *Water* **2014**, *6*, 1515–1545. [CrossRef]

33. Juneja, P. What Is Analytical Hierarchy Process (AHP) and How to Use It? Available online: https://www.managementstudyguide.com/analytical-hierarchy-process.htm (accessed on 30 June 2021).

34. Siddayao, G.P.; Valdez, S.E.; Fernandez, P.L. Modeling Flood Risk for an Urban CBD Using AHP and GIS. *Int. J. Inf. Educ. Technol.* **2015**, *5*, 748–753. [CrossRef]

35. Rahadianto, H.; Fariza, A.; Hasim, J.A.N. Risk-level assessment system on Bengawan Solo River basin flood prone areas using analytic hierarchy process and natural breaks: Study case: East Java. In Proceedings of the 2015 International Conference on Data and Software Engineering, ICODSE 2015, Yogyakarta, Indonesia, 25–26 November 2015; pp. 195–200. [CrossRef]

36. Lyu, H.M.; Zhou, W.H.; Shen, S.L.; Zhou, A.N. Inundation risk assessment of metro system using AHP and TFN-AHP in Shenzhen. *Sustain. Cities Soc.* **2020**, *56*, 102103. [CrossRef]

37. Weerasinghe, K.M.; Gehrels, H.; Arambepola, N.M.S.I.; Vajja, H.P.; Herath, J.M.K.; Atapattu, K.B. Qualitative Flood Risk assessment for the Western Province of Sri Lanka. *Procedia Eng.* **2018**, *212*, 503–510. [CrossRef]

38. Seejata, K.; Yodying, A.; Wongthadam, T.; Mahavik, N.; Tantanee, S. Assessment of flood hazard areas using Analytical Hierarchy Process over the Lower Yom Basin, Sukhothai Province. *Procedia Eng.* **2018**, *212*, 340–347. [CrossRef]

39. Cabrera, J.S.; Lee, H.S. Impacts of climate change on flood prone areas in Davao Oriental, Philippines. *Water* **2018**, *10*, 893. [CrossRef]

40. Lim, M.B.B.; Lim, H.R.; Piantanakulchai, M. Flood evacuation decision modeling for high risk urban area in the Philippines. *Asia Pac. Manag. Rev.* **2019**, *24*, 106–113. [CrossRef]

41. Cai, T.; Li, X.; Ding, X.; Wang, J.; Zhan, J. Flood risk assessment based on hydrodynamic model and fuzzy comprehensive evaluation with GIS technique. *Int. J. Disaster Risk Reduct.* **2019**, *35*, 101077. [CrossRef]

42. Robielos, R.A.C.; Lin, C.J.; Senoro, D.B.; Ney, F.P. Development of vulnerability assessment framework for disaster risk reduction at three levels of geopolitical units in the Philippines. *Sustainability* **2020**, *12*, 8815. [CrossRef]

43. Prasetyo, Y.T.; Senoro, D.B.; German, J.D.; Robielos, R.A.C.; Ney, F.P. Confirmatory factor analysis of vulnerability to natural hazards: A household Vulnerability Assessment in Marinduque Island, Philippines. *Int. J. Disaster Risk Reduct.* **2020**, *50*, 101831. [CrossRef]

44. Chen, Y.R.; Yeh, C.H.; Yu, B. Integrated application of the analytic hierarchy process and the geographic information system for flood risk assessment and flood plain management in Taiwan. *Nat. Hazards* **2011**, *59*, 1261–1276. [CrossRef]

45. Gigović, L.; Pamučar, D.; Bajić, Z.; Drobnjak, S. Application of GIS-interval rough AHP methodology for flood hazard mapping in Urban areas. *Water* **2017**, *9*, 360. [CrossRef]

46. Cabrera, J.S.; Lee, H.S. Flood-prone area assessment using GIS-based multi-criteria analysis: A case study in Davao Oriental, Philippines. *Water* **2019**, *11*, 2203. [CrossRef]

47. PAGASA. Climatological Normals. Available online: https://www.pagasa.dost.gov.ph/climate/climatological-normals (accessed on 10 September 2022).

48. LGU Odiongan. *Comprehensive Land Use Plan (CLUP)*; LGU Odiongan: Romblon, Philippines, 2017.

49. HEC-HMS. Available online: https://www.hec.usace.army.mil/software/hec-hms/ (accessed on 10 September 2022).

50. HEC-RAS. Available online: https://www.hec.usace.army.mil/software/hec-ras/ (accessed on 10 September 2022).

51. GEOHAZARD MAPS. Available online: https://region4b.mgb.gov.ph/28-geohazard-maps/98-geohazard-maps#romblon-2 (accessed on 10 September 2022).

52. Kokangül, A.; Polat, U.; Dağsuyu, C. A new approximation for risk assessment using the AHP and Fine Kinney methodologies. *Saf. Sci.* **2017**, *91*, 24–32. [CrossRef]

53. Employment of PWDs. Available online: https://www.pwc.com/ph/en/taxwise-or-otherwise/2017/employment-of-pwds.html (accessed on 2 June 2022).

54. Febrianto, H.; Fariza, A.; Hasim, J.A.N. Urban flood risk mapping using analytic hierarchy process and natural break classification (Case study: Surabaya, East Java, Indonesia). In Proceedings of the 2016 International Conference on Knowledge Creation and Intelligent Computing, KCIC 2016, Manado, Indonesia, 15–17 November 2016; pp. 148–154. [CrossRef]

55. National Disaster Management Office (NDMO). *Republic of Vanuatu National Guidelines for the Selection and Assessment of Evacuation Centres*; National Disaster Management Office: Port Vila, Vanatu, 2016.

56. Engay-Gutierrez, K.G. Land cover change in the silang-santa rosa river subwatershed, Laguna, Philippines. *J. Environ. Sci. Manag.* **2015**, *18*, 34–46. [CrossRef]

57. Stefanidis, S.; Stathis, D. Assessment of flood hazard based on natural and anthropogenic factors using analytic hierarchy process (AHP). *Nat. Hazards* **2013**, *68*, 569–585. [CrossRef]

58. Shrestha, B.B.; Okazumi, T.; Miyamoto, M.; Sawano, H. Development of flood risk assessment method for data-poor river basins: A case study in the Pampanga River Basin, Philippines. In Proceedings of the 6th International Conference on Flood Management, Sao Paolo, Brazil, 16–18 September 2014; Volume ii, pp. 1–12.

59. Shrestha, B.B.; Sawano, H.; Ohara, M.; Nagumo, N. Improvement in flood disaster damage assessment using highly accurate IfSAR DEM. *J. Disaster Res.* **2016**, *11*, 1137–1149. [CrossRef]

60. Lyu, H.M.; Sun, W.J.; Shen, S.L.; Arulrajah, A. Flood risk assessment in metro systems of mega-cities using a GIS-based modeling approach. *Sci. Total Environ.* **2018**, *626*, 1012–1025. [CrossRef]

61. Bera, R.; Maiti, R. Multi hazards risk assessment of Indian Sundarbans using GIS based Analytic Hierarchy Process (AHP). *Reg. Stud. Mar. Sci.* **2021**, *44*, 101766. [CrossRef]

62. Puno, G.R.; Amper, R.A.L.; Talisay, B.A.M. Flood simulation using geospatial models in Manupali Watershed, Bukidnon, Philippines. *J. Biodivers. Environ. Sci.* **2018**, *12*, 294–303.

63. Kia, M.B.; Pirasteh, S.; Pradhan, B.; Mahmud, A.R.; Sulaiman, W.N.A.; Moradi, A. An artificial neural network model for flood simulation using GIS: Johor River Basin, Malaysia. *Environ. Earth Sci.* **2012**, *67*, 251–264. [CrossRef]

64. Talisay, B.A.M.; Puno, G.R.; Amper, R.A.L. Flood hazard mapping in an urban area using combined hydrologic-hydraulic models and geospatial technologies. *Glob. J. Environ. Sci. Manag.* **2019**, *5*, 139–154. [CrossRef]

65. Pornasdoro, K.P.; Silva, L.C.; Munárriz, M.L.T.; Estepa, B.A.; Capaque, C.A. Flood Risk of Metro Manila Barangays: A GIS Based Risk Assessment Using Multi-Criteria Techniques. *J. Urban Reg. Plan.* **2014**, *1*, 51–72.

66. Clark, C. Flood risk assessment. *Int. Water Power Dam Constr.* **2007**, *59*, 1–25. [CrossRef]

67. Åström, H.L.A. *An Urban Flood Risk Assessment Method Using the Bayesian Network Approach*; DTU Environment, Technical University of Denmark: Lyngby, Denmark, 2015.

68. Feloni, E.; Mousadis, I.; Baltas, E. Flood vulnerability assessment using a GIS-based multi-criteria approach—The case of Attica region. *J. Flood Risk Manag.* **2020**, *13*, e12563. [CrossRef]

69. Tehrany, M.S.; Pradhan, B.; Jebur, M.N. Spatial prediction of flood susceptible areas using rule based decision tree (DT) and a novel ensemble bivariate and multivariate statistical models in GIS. *J. Hydrol.* **2013**, *504*, 69–79. [CrossRef]

70. Monjardin, C.E.F.; Senoro, D.B.; Magbanlac, J.J.M.; de Jesus, K.L.M.; Tabelin, C.B.; Natal, P.M. Geo-Accumulation Index of Manganese in Soils Due to Flooding in Boac and Mogpog Rivers, Marinduque, Philippines with Mining Disaster Exposure. *Appl. Sci.* **2022**, *12*, 3527. [CrossRef]

applied
sciences

MDPI

Article

A Database for Tsunamis and Meteotsunamis in the Adriatic Sea

Alessandra Maramai *, Beatriz Brizuela and Laura Graziani

Istituto Nazionale di Geofisica e Vulcanologia, 00143 Rome, Italy; beatriz.brizuela@ingv.it (B.B.);
laura.graziani@ingv.it (L.G.)
* Correspondence: alessandra.maramai@ingv.it; Tel.: +39-06-51860210

Abstract: In the frame of the Interreg Italy-Croatia program, the EU has funded the PMO-GATE project, focusing on the prevention and mitigation of the socioeconomic impact of natural hazards in the Adriatic region. The Database of Adriatic Tsunamis and Meteotsunamis (DAMT) is one of the deliverables of this project. DAMT is a collection of data documenting both meteotsunami and tsunami effects along the Eastern and Western Adriatic coasts, and it was realized by starting from the available database and catalogues, with the inclusion of new data gained from recent studies, newspapers and websites. For each tsunami and meteotsunami, the database provides an overview of the event and a detailed description of the effects observed at each affected location and gives a picture of the geographical distribution of the effects. The database can be accessed through a GIS WebApp, which allows the user to visualize the georeferenced information on a map. The DAMT WebApp includes three layers: (1) Adriatic Tsunami Sources, (2) Adriatic Tsunami Observation Points and (3) Adriatic Meteotsunamis Observation Points. The database contains 57 observations of tsunami effects related to 27 tsunamis along the Italian, Croatian, Montenegrin and Albanian coasts and 102 observations of meteotsunami effects related to 33 meteotsunamis.

Keywords: Adriatic Sea; database; tsunami; meteotsunami; ArcGis; WebApp

Citation: Maramai, A.; Brizuela, B.; Graziani, L. A Database for Tsunamis and Meteotsunamis in the Adriatic Sea. *Appl. Sci.* **2022**, *12*, 5577. https://doi.org/10.3390/app12115577

Academic Editors: Andrea Chiozzi, Željana Nikolić and Elena Benvenuti

Received: 31 March 2022
Accepted: 25 May 2022
Published: 31 May 2022

Publisher's Note: MDPI stays neutral with regard to jurisdictional claims in published maps and institutional affiliations.

1. Introduction

Due to the increasing number of extreme events that are being experienced around the world, the interest of the scientific community in natural hazards has grown significantly in recent years, and one of the main targets is the prevention and reduction of risks related to natural events. The Sendai Framework for Disaster Risk Reduction (2015–2030) outlines the overall objectives to substantially reduce disaster risk and losses of lives, livelihoods, and health. It clearly states that in order to diminish the frequency and impact of disasters, it is required to better understand disaster risk (exposure to hazards, vulnerability and capacity and the hazard's characteristics) and furthermore to improve risk governance and increase resilience.

In this regard, many international projects related to the reduction of risk have been funded in the last decade, and the European Union has supported and financed some projects as well. In particular, a wide-ranging cross-border cooperation program between Italy and Croatia called Interreg has been established, focusing on the sea basin, coastal landscapes, green areas and urban areas as well.

The Adriatic Sea is the core center of the Italy-Croatia cooperation area, and it is a joint economic and environmental asset and a natural platform for combined efforts. The coastal area, both in Italy and in Croatia, is exposed to a range of natural hazards, particularly floods, strong winds, drought, earthquakes, tsunamis and meteotsunamis. Taking into account the coastal vulnerability, disaster risk reduction is a critical factor for the social and economic development of the involved countries.

In the framework of the Interreg Italy-Croatia program, Preventing, Managing and Overcoming Natural-Hazards Risks to mitiGATE economic and social impact (PMO-GATE)

has been funded, and it aims at increasing safety from natural and man-made disasters along both the Italian and Croatian coasts of the Adriatic Sea. Making the most of the capitalization on the expertise gained by all partners in previous projects, and using the data and the results available from previous studies, PMO-GATE's purpose is to enhance the level of protection and resilience against natural disasters specific of the region, such as river and sea floods, earthquakes, meteotsunamis and tsunamis. In particular, the project addresses the vulnerability of the Adriatic Sea system with its coasts and islands, implementing cross-border actions in the field of prevention and management regarding the exposure to floods and meteotsunamis in a seismically vulnerable context.

The final outcome of the project is an innovative methodology for preventing, managing and overcoming multi-hazard natural disasters, and climate-induced hazards (such as floods and meteotsunamis) will be combined with non-climate-induced hazards (earthquakes and tsunamis).

PMO-GATE addresses the general public, local and national public authorities, emergency services, education and training centers, universities and research institutes. An effective communication strategy to increase awareness and perception of risk in the population and public authorities is one of the pillars of the project, and the dissemination of results is one of the main goals of the project.

Among the different deliverables of PMO-GATE is the realization of a database for tsunamis and meteotsunamis occurring in the Adriatic region. The Database of Adriatic Tsunamis and Meteotsunamis (DAMT) is the result of the analysis of tsunami and meteotsunami databases existing in the literature for the Adriatic area. Although the two phenomena have different origins, they affect the coasts with similar characteristics and effects, and therefore, in terms of coastal hazard, similar prevention measures must be taken. The two phenomena have always been treated separately, and the creation of a single database is, therefore, important to optimize the hazard and risk studies in the area.

DAMT contains the Adriatic tsunamis present in the Italian Tsunami Effect Database (ITED) [1] and in the Euro-Mediterranean Tsunami Catalogue (EMTC) [2], while as far as meteotsunamis are concerned, the data are essentially those of the Catalogue of Meteorological Tsunamis in Croatian coastal waters (CMTC) [3], integrated with the results of new studies conducted during the PMO-GATE project. Therefore, DAMT is a tool that can be a starting point for a better understanding of the characteristics of such phenomena in the Adriatic area, and it can contribute to hazard and risk assessment for natural events.

PMO-GATE is also a contribution to delivering new data to educational initiatives in the future. In this frame, DAMT is a tool that has also been used for educational purposes during special events addressed to both the general public and scholars, such as World Earth Day [4] and the National Conference in Science Communication 2021 [5], in order to enhance risk perception, which is one of the main aims of the project.

2. Tsunamis and Meteotsunamis in the Adriatic Region

Among the various natural hazards to which the Adriatic coasts are exposed, the Istituto Nazionale di Geofisica e Vulcanologia (INGV), as a partner of PMO-GATE, has focussed its efforts on tsunamis and meteotsunamis with the aim of developing and implementing a database of these events: DAMT.

The geographical position of the Adriatic region is peculiar. It lies between the Apennine and Dinaric mountain chains, and it is mostly surrounded by active fold-and-thrust belts and strike-slip faults [6–8]. The presence of microplates and the complex system of faults in the area explain the seismicity which affects the Adriatic region.

Both Italy and Croatia experienced several earthquakes in the past, and a large number of studies were carried out with the aim of better assessing the seismicity of this region. It has been underlined that historical seismicity is poorly documented, and the Adriatic Sea is characterized by quite low seismic activity mainly consisting of earthquakes of moderate magnitudes, but important seismic sequences with main shocks of relevant magnitudes have also been observed [9–16]. Frequent earthquakes occur along the well-known fault zones, most of which run close to the coastlines or in the open sea and are thus potential sources for tsunamis [17].

In the Adriatic region, the strongest earthquakes (M > 7) have occurred near the eastern margin of the central Adriatic Sea and at the southern end of the basin near the Ionian Islands. The majority of the remaining structures are potentially capable of generating earthquakes of magnitudes $6 \leq M \leq 7$, therefore having significant potential for causing tsunamis [6].

Tsunami is a well-known term mainly used in reference to earthquakes generated ocean waves, and they have a typical period range from a few to tens of minutes to hours. As they approach the coast, these waves can produce severe damage to coastal structures and can cause loss of human life.

Tsunamis are usually generated by submarine earthquakes. However, as the literature demonstrates, in the Adriatic region, as well as along the other Italian coasts, tsunamigenic sources are located both offshore and very close to the coast [1,18]. In this region, a number of tsunamis related to earthquake activity were observed [1,2,18–26]. Most of them had low intensities, but a few events, such as the 1627 Gargano and 1930 Ancona tsunamis, were classified as "very strong".

Italian tsunamis are quite well-documented, and tsunami effects have been observed in many places from the northern to the southern Adriatic region, while in Croatia, tsunami effects have been reported along the northern and southern coasts, with no evidence along the central coast of the country. In Montenegro, tsunamis have been located along the whole coast, and along the Albanian coast, they were mainly located in the area of Valona. Twelve Adriatic tsunamis originated along the Italian coasts, being observed from 1348 to 1930 [18].

Tsunamis generated in the eastern Adriatic region are less frequent; four tsunamis, mainly of low intensities, were ascertained to have occurred since 1667 along the Croatian coast, while the Montenegro coast has been affected by three events since 1667. Six tsunamis have occurred along the Albanian coasts since 1833, all of which were concentrated in the coastal area of Valona. The Croatian coast is much more prone to meteotsunamis, which are meteorologically generated long ocean waves with characteristics similar to those of tsunamis, and they represent a significant hazard for the eastern Adriatic coast [23–25].

Meteotsunamis are formed by storm systems moving rapidly across the water, such as a squall line, and their development depends on several factors such as the intensity, direction and speed of the disturbance as it travels over a water body [23–25]. They can affect localized areas when they reach land. Theb wave heights and spatial extent of meteotsunamis are smaller compared with tsunamis, but they can cause sea level oscillations of several meters and human losses and injuries. Although meteotsunamis are not catastrophic to the extent of major seismically induced events, their occurrence in time and space are higher than those of seismic tsunamis, as the atmospheric disturbances responsible for the generation of meteotsunamis are much more common.

According to recent studies [26], extreme weather has been a common result of the planet's rising temperature, and climate change, like global warming and the sea level rising, may have an impact on the future occurrence and likelihood of meteotsunamis.

The meteotsunami phenomenon is a relatively recent scientific discovery, considering that the first description of tsunami-like effects produced by atmospheric disturbances appeared only in 1931 [27] and that the term "meteotsunamis" was introduced only in 1961 [27].

Thus far, it has been shown that Mediterranean meteotsunamis tend to be stronger in summer. Despite calm conditions at ground level, fast winds of dry air from Africa in the atmosphere 1500 m up seem to trigger atmospheric waves. Mostly during summertime, small-scale strong atmospheric disturbances take place in the Adriatic area, and the coast has a quite complex topography, with a large number of funnel-shaped bays and harbors which have high amplification factors. In fact, the strength of a meteotsunami is also largely dependent on both the topography and the bathymetry of the affected area, and in the

Adriatic Sea, recurrent meteotsunami events are known to strongly impact the lifestyles of the coastal communities, particularly in the Dalmatian islands, where meteotsunamis can generate serious flooding. Several destructive events affected the eastern Adriatic shore in the last few decades, mainly involving Mali Lošinj, Ist, Stari Grad, Vela Luka and Mali Ston. The strongest event hit Vela Luka in Croatia in June 1978, with a wave height of 6 m (crest to trough) and a period of 18 min. This meteotsunami, the most powerful recorded in the Mediterranean, lasted several hours and caused USD 7 million in damage [28,29]. Siroka Bay in the island of Ist and Mali Losinj Bay in the island of Losinj are the two bays in the northern Adriatic where destructive meteotsunamis have occurred recently. Four-meter waves struck Siroka Bay on 22 August 2007, injuring one person and damaging local infrastructure [26]. This locality also experienced meteotsunami effects on 4 October 1984. Mali Losinj Bay was hit on 15 August 2008 by 2-m waves, flooding a forefront of the most populated town of the Adriatic Sea islands and causing a panic during tourist season [30,31].

3. The Database of the Adriatic Meteotsunamis and Tsunamis (DAMT)

The understanding of natural phenomena, including tsunamis and meteotsunamis, is essentially based on the deep knowledge of past events. In fact, knowing how many events occurred in the past in a specific region and studying their characteristics help to encourage the study of the more prone coastal areas to assess the hazard and to calibrate models of propagation and inundation. A proper and systematic dissemination of the acquired knowledge is also fundamental to increase the awareness of people living in vulnerable areas.

From this perspective, the availability of a database of events is very important because it is a useful tool for increasing public awareness and, when robust information is available, to validate hazard and risk assessments. In the frame of the PMO-GATE project, one of the deliverables is the creation of the database of Adriatic Tsunamis and Meteotsunamis (DAMT), which was carried out by starting from the catalogs available in the literature, particularlt the Italian Tsunami Effects Database (ITED) [1], the Euro-Mediterranean Tsunami Catalogue [2] and the Catalogue of Meteorological Tsunamis in Croatian coastal waters [3].

DAMT was created by selecting and analyzing the information already available in both of the above-mentioned catalogs, as well as including new data acquired in the frame of the PMO-GATE project from the analysis of recent studies, newspapers and websites. As far as tsunamis are concerned, the data available in the catalogues cover a period of time ranging from 1348 to present day [1,2], while for meteotsunamis, the first information is only available from 1931 onward because for previous events, usually only the year of occurrence was known, and therefore, they were not verifiable or reliable [3]. For each tsunami and meteotsunami observation, the database provides a general description of the event, together with a detailed georeferenced description of the effects observed in each affected location, and offers a complete picture of the geographical distribution of the effects on the coast. In Table 1, the list of tsunamis and meteotsunamis included in DAMT is presented, reporting the main parameters for each event.

The data included in DAMT can be retrieved by the public through a web application similar to that used for ITED [1] which allows the user to visualize the georeferenced information on a map. The WebApp, developed in a freely accessible ESRI ArGis online environment, is accessible through this link (https://ingv.maps.arcgis.com/apps/webappviewer/index.html?id=0f465d51001146d79a6c89884a8e5d8c) (accessed on 20 May 2022) without the need for an Esri user account.

The DAMT WebApp includes three layers: (1) Adriatic Tsunami Sources (ATS), containing the information on the tsunamis that occurred in the Adriatic Basin, (2) Adriatic Tsunami Observation Points (ATOPs), namely the localities where tsunami effects were observed, and (3) Adriatic Meteotsunami Observation Points (AMOPs), which are the localities where meteotsunami effects were observed. In Figure 1, the main screen of the

DAMT WebApp is visible, showing the geographical distribution of the whole dataset contained in the database: tsunamis, where the location coincides with the epicenter of the earthquake that triggered the tsunami (blue squares), tsunami observation points (colored dots in shades of beige) and meteotsunami observation points (red points). Using the layers widget at the top right of the screen, the user can choose the layers to be displayed (*Layer List*).

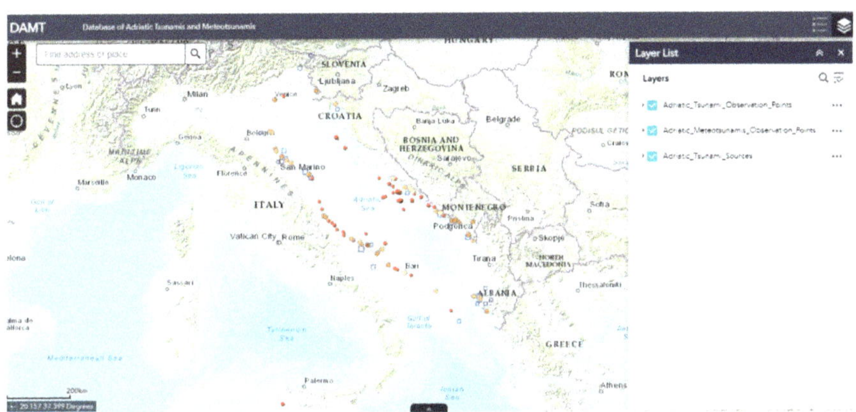

Figure 1. Main screen of the DAMT WebApp. On the right, the layer list is shown. Check marks indicate that all levels are highlighted in the map. Blue squares are tsunamigenic sources. Beige-shaded points are tsunami observation points. Red points are meteotsunami observation points.

The data contained in DAMT also populates an attribute table, which can be retrieved through the arrow located at the bottom of the main screen (Figure 2). The tabular information contained in each layer of the WebApp can be exported in csv format, and it can also be filtered by using user-customized expressions in the tab options of the database table (i.e., selecting data from the extent viewed, by date, reliability, cause, region, etc. or by a combination of several of these parameters).

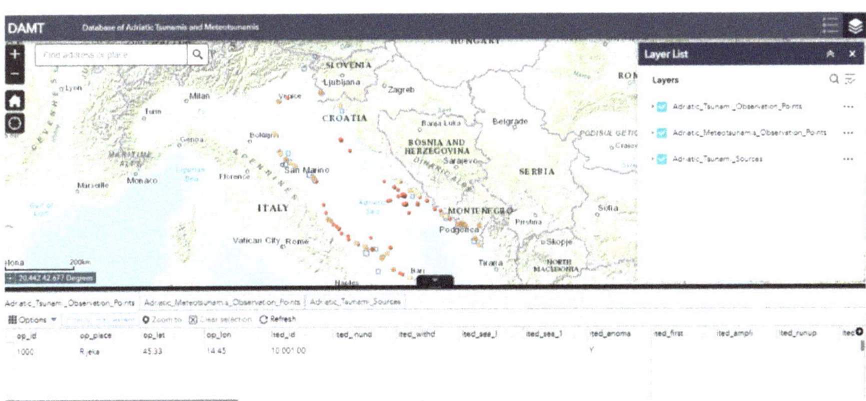

Figure 2. The attribute table allows the user to filter the data contained in the layers of the WebApp according to different selections.

As far as tsunamis are concerned, the ATS layer contains the 24 events with the general description of the effects, mainly taken from the EMTC [2].

Twelve tsunamis were located in the Italian coasts, four in the Croatian ones, three in Montenegro and six along the Albanian coasts. By clicking on one of the blue squares on the screen, a pop-up allows the user to obtain general information on the main parameters of the selected tsunami (earthquake parameters and tsunami intensity), and it can link to the general description of the event. In Figure 3a,b, the example of the 30 October 1930 Ancona tsunami is reported.

(a)   (b)

Figure 3. (**a**) Example of pop-up of the ATS layer, showing the main parameters of the 30 October 1930 Ancona tsunami. (**b**) Example of the 30 October 1930 Ancona tsunami, where the green arrow indicates the link to click on for the general description of the event.

With regard to the Adriatic Tsunami Observation Points (ATOPs) layer, it hosts 57 observation points where tsunami effects were observed: 28 on the Italian coast, 10 in Croatia, 9 in Montenegro and 9 in Albania, respectively (Figure 4a). The Italian ATOPs came from ITED [1], while ATOPs located on the eastern coasts of the Adriatic Sea came from the analysis of the descriptions of the tsunamis contained in EMTC [2]. Additional information has enriched the knowledge on the tsunamis reported in the catalogue. For eastern coast events, additional information was included, and for each observation point, a "local" tsunami intensity value was assessed as "ex novo" on both the Ambraseys–Sieberg [32] and Papadopoulos–Imamura [33] scales. The beige-shaded points are the locations where tsunami effects were observed, while the darker points correspond to more severe effects. For each point, descriptive and, when available, quantitative information (inundation, run-up and wave height values) is provided, along with the corresponding bibliographical references. By clicking on each point, through a pop-up window, the user can obtain information about the effects observed, as well as the main info on the generating tsunami (Figure 4b). Among the Adriatic tsunamis, the maximum run-up observed was 2.5 m at Manfredonia (Apulia region in Italy) during the event on 30 July 1627, while the largest inundation was 50 m at Boka Kotorska (Montenegro) for the event on 15 April 1979, during which one person died [1].

The Adriatic Meteotsunami (AM) layer contains 33 meteotsunami events observed or recorded in 54 places along both the eastern and western Adriatic coasts. Most of the data came from [3], with new information, new events and images being added. Among these 54 places, 8 experienced meteotsunami effects more than once. In particular, from 1931 to the present, Vela Luka has been affected by 18 events, and Stari Grad has been affected by 11 events. In Figure 5, the geographical distribution of the places where meteotsunami effects have been observed is visible. As already mentioned above, meteotsunamis are mainly localized along the coasts of Croatia. As for the northern Adriatic region, although it is a region prone to Proudman resonances, due to the lack of bays or harbors with high amplification factors, no major meteotsunamis occur in that area [23]. Coversely, the Croatian coast of the central Adriatic, characterized by many islands, channels and narrow bays, is the area where the most meteotsunamis occur (Figure 4), most of which are very powerful. On the Italian coasts, most of the effects were due to the 21 June 1978 event, the strongest in the Mediterranean, but some locations were also affected by the 25 June 2014 event.

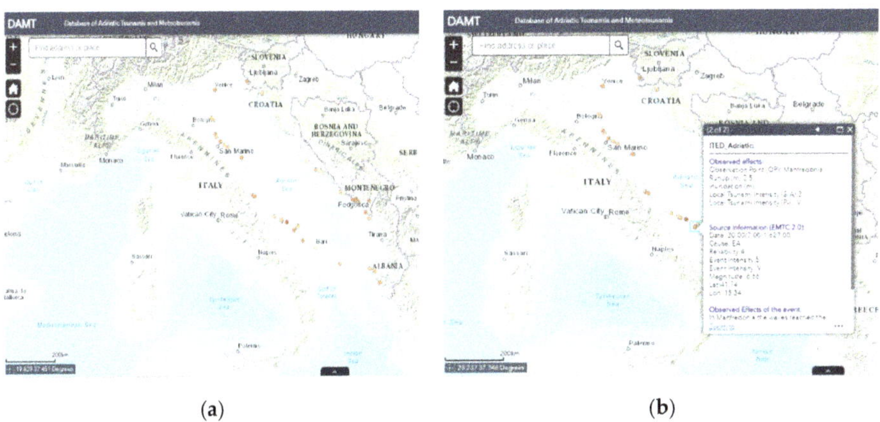

(**a**)   (**b**)

Figure 4. (**a**) Geographical distribution of the Tsunami Observation Points (ATOP layer). (**b**) An example of the pop-up for the Manfredonia (Apulia) observation point.

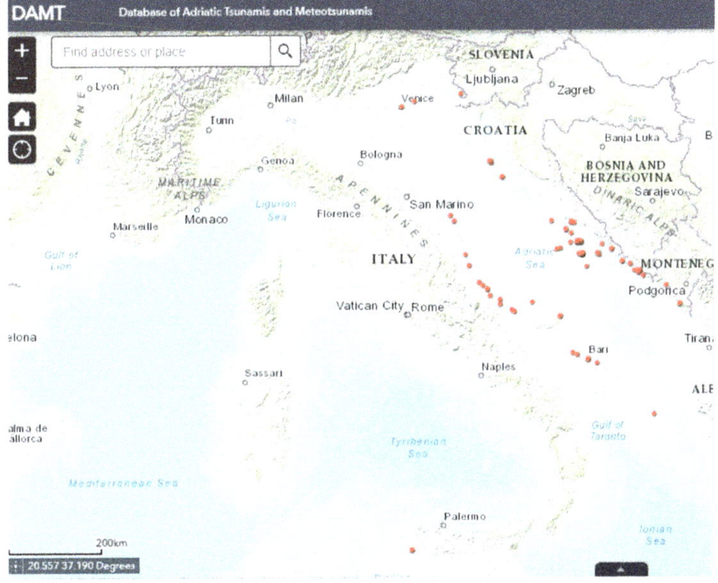

Figure 5. Geographical distribution of the meteotsunami effects.

As for tsunamis, and also for meteotsunamis reported in the WebApp, the user can click on each point where the effects were observed. The pop-up window shows the date of the event (or several dates if there is more than one event at the same point), a short description of the observed effects, a link to the general description of the event and the intensity. Since meteotsunamis are a very complex phenomenon involving meteorological, hydrological and bathymetric factors, for their classification, Orlić and Šepić [3] introduced a value, the QIndex, that indicates how detailed the information about the event is. The QIndex ranges from 1 (elementary description of sea-level variability) to 5 (analysis of oceanographic and meteorological data combined with both oceanographic and meteorological modeling), depending on what kind of bibliographic sources support the available data. In DAMT, the same type of classification has been used, maintaining the QIndex values assigned by [3] and assigning the QIndex to the new events inserted. Figure 6 shows an example of the pop-up for the 11 May 2020 event.

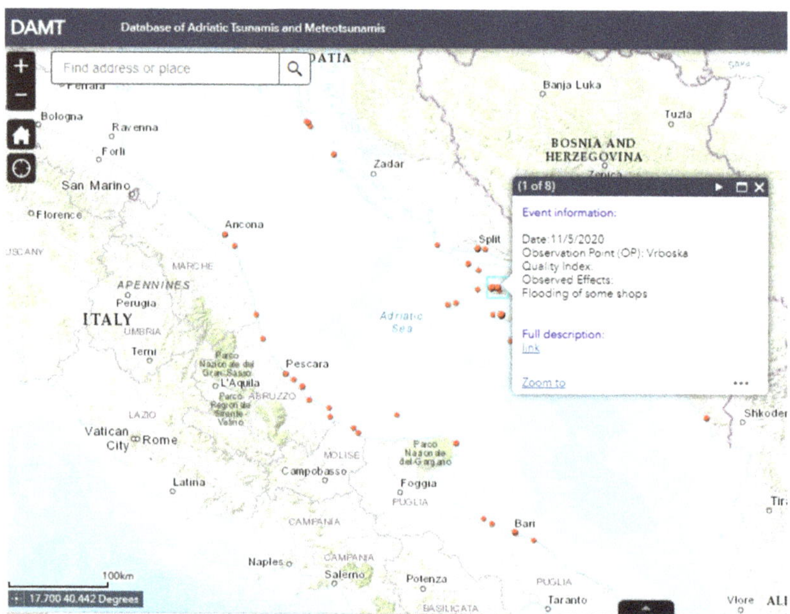

Figure 6. Example of a pop-up of the AM layer, showing the 15 May 2020 at Vrboska observation point. By clicking on the full description link, the user can retrieve detailed information on all effects observed at that location over time.

4. Discussion

The analysis of the characteristics of past events provides clues to what might happen in the future. In this perspective, a database can be an important and useful starting point to better characterize a region in terms of these phenomena. In the frame of the PMO-GATE project, the realization of a database which includes tsunamis and meteotsunamis in the Adriatic region aims at providing a comprehensive picture of the events in order to highlight that the Adriatic region is prone to these events.

DAMT, for the first time, brings together two phenomena, tsunamis and meteotsunamis, which although originating from different causes produce very similar effects when they hit the coast. Therefore, the measures to be implemented for the prevention and reduction of coastal hazards could be similar.

DAMT was built by starting from the data included in the ITED database [1], in the Euro-Mediterranean Tsunami Catalogue [2] and in the Catalogue of Meteorological Tsunamis in Croatian coastal waters by Orlić and Šepić [3], with the insertion of new events and new info and parameters for the events which had already been catalogued. In Table 1, short descriptions of the events included in the database are reported. DAMT aims to enhance the usability of data and is displayed by means of an ESRI WebApp that allows the user to query the database in order to retrieve general and detailed information about events, select the events of interest and know the tsunami and meteotsunami history for each observation point. As far as tsunamis, DAMT contains observations of 24 events that have been reported from 1348 to present day, all of which were caused by earthquakes. The analysis of the database puts forth the evidence that in the Adriatic region, the most affected coasts are the Italian ones, which have been affected by 12 events since 1348. Among them, the central northern coast (Emilia Romagna) and the southern one (Apulia) experienced the most relevant events. Four tusnamis have occurred on the coasts of Croatia from 1667 to 1962, three events have affected the coasts of Montenegro in 1667, 1780 and 1979, and finally, the coasts of Albania experienced six tsunamis from 1833 to 1920.

The tsunamis included in DAMT are classified by a *reliability index,* a value indicating the quality and reliability of the data [21]. Fourteen events (about 60%) have the maximum reliability (four = definite tsunami), and only three events, all of which occurred in Albania in the 1800s, have a reliability of one (very improbable tsunami). Inserting in the database events with low reliability as well allows for keeping a record of those for which little information is available and of which one could otherwise lose track. At the same time, it allows the user to know that the information available for that event is not sufficiently robust to attribute a high reliability value to the event, and therefore, it must be considered with due attention.

Regarding tsunami intensity, to each observation point was assigned a local intensity value based on both the Ambraseys–Sieberg and Papadopoulos–Imamura scales. Most of the events (80%) had intensities that were medium-low, light (2) or rather strong (3), but there were four events with strong (4) and very strong (5) intensities. Two of these tsunamis occurred in Italy in 1627 and 1930, one which took place in Albania in 1920 while the other occurred on the Croatian coast in 1979.

As highlighted by a detailed study of Adriatic tsunamis [34], the analysis of DAMT confirmed that there is no evidence of tsunami effects on the eastern Adriatic side related to the western Adriatc events. However, during two events in the eastern Adriatic region, light effects were observed on the Italian coasts. In particular, after the devastating 1667 Dubrovnik earthquake, anomalous movements were observed in the waters of the canals of Venice, while the tsunami generated by the 1979 Montenegro earthquake was recorded by the tide gauge of Bari in Apulia.

Modeling studies of tsunami propagation from different source areas and tsunami hazards in the Adriatic Sea have shown that earthquakes in the northern Adiatic Sea are not very efficient in generating tsunamis due to the shallow water depth, while the central Adriatic Sea has low seismicity and shallow water, and tsunami waves arriving from the southern Adriatic Sea are partly reflected by the Palagruža rocks [6,34]. Furthermore, the Croatian island chain protects both coasts from tsunami waves propagating from the opposite side of the Adriatic. According to [20], earthquakes located in front of the Montenegrin coast are more efficient at generating tsunamis, and the seismically active region in front of the Albanian coast can generate severe tsunamis, but the modeling results are contradictory as to whether or not tsunamis generated in this area can propagate toward the Italian coasts [6,34]. At the moment, the evidence of the recording of sea level variation during the 1979 Montenegro earthquake in the tide gauge of Bari suggests a possible propagation from the eastern to the western coasts [34]. However, further studies are recommended.

Concerning meteotsunamis, DAMT reports 33 events that occurred from 1931 to the present date, with 103 observations related to 58 places. Compared with tsunamis, for meteotsunamis, the time interval covered by DAMT is much shorter, since the first meteotsunami included occurred in 1931. This is essentilly due to two factors: (1) as highlighted in [3], for previous events, usually only the year of occurrence was known, and therefore, they were not verifiable or reliable, and (2) meteotsunamis as a phenomenon are a recent discovery. Just think that the first description of tsunami-like effects produced by atmospheric disturbances appeared only in 1931 [27], and the term "meteotsunami" was introduced only in 1961 [27].

As also highlighted in [31], for the characterization and proper definition of a meteotsunami event, long-term, high-frequency measurements are necessary. Unfortunately, as mentioned above, historical observations of meteotsunamis in the Adriatic region were extremely scarce and poorly detailed until the early 1930s. In addition, only in recent decades have instruments capable of analyzing such phenomena been installed.

Gathering the information and cataloguing these events is crucial to identify "hotspots", which intensify observational networks in order to obtain, in the future, a robust dataset number to characterize these events, both in terms of effects produced and generated characteristics.

Although the number of events present in DAMT does not allow for performing statistical analysis, it enables us to make general observations.

The exam of the meteotsunami events included in DAMT shows that all of them originated from the Croatian coasts, mainly in the central part of the region. Unlike the tsunamis, the effects of three of them—the 19 September 1977, 21 June 1978 and 25 June 2014 events—were also clearly observed along the Italian coasts. The latter, the strongest meteotsunami in the Mediterranean that was destructive in Vela Luka, affected 14 locations along the Italian coast from Ancona to Otranto, causing serious damage in some places and some injuries in Vieste.

For the classification of meteotsunamis, in the DAMT, we used the criterion proposed by the authors of [3], who assigned a QIndex to each event based on how detailed the bibliographic sources were according to different parameters. In addition, considering the description of meteotsunami effects as if they were tsunami descriptions, in the DAMT, a local intensity value was also assigned for each observation point (OP) according to both Ambraseys–Sieberg [32] and Papadopoulos–Imamura [33] scales. During this process, it was pointed out that the two scales, created for a very similar but nonetheless different phenomenon, are not properly suited for meteotsunami intensity assessment. This is particularly evident for the Papadopoulos–Imamura scale, which has a higher number of degrees (12) with respect to the Ambraseys–Sieberg scale (6 degrees), and the more devastating effects that are described in the higher degrees cannot be produced by meteotsunamis. In fact, even in the case of a very strong meteotsunami, such as the one on 21 June 1978, the observed effects reached a maximum of five on the Ambraseys–Sieberg scale, while they reached a maximum of seven on the Papadopoulos–Imamura scale.

The DAMT contains several reliable data, and it is a tool that can contribute to improving the knowledge of tsunami and meteotsunami activity in the Adriatic area and to increase public awareness along these coasts which, especially in the peak season, are some of most densely populated areas in the Mediterranean. The database has already been used several times for educational purposes to increase the awareness of public authorities, citizens and students toward coastal hazards, and it will be one of the main tools for dissemination of the PMO-GATE project results.

As mentioned earlier, the availability of long-term observations is essential to characterizing the events in a region. Therefore, particularly with regard to meteotsunamis, in the Adriatic area, two factors could be beneficial in the near future: an increase in tide gauge and barometric instrumentation, especially in "hot-spots" that are particularly prone to meteotsunamis on one hand, and on the other hand, in-depth research of bibliographic sources on events that occurred in the past, mainly before 1931. Regarding the latter, a collaboration with Croatian and Albanian researchers is going to start in order to consult local archives and libraries.

Table 1. List of the events included in DAMT with the main parameters. Y = year, M = month, D = day of the event, Ev = type of event (T = tsunami; M = meteotsunami), Lat.-Lon. = coordinates of the tsunamigenic earthquake, Description = short description of the main effects, OP = number of observation points where effects were observed; Int. = maximum intensity of the event, where the first number is the value according to the Ambraseys–Sieberg scale and the roman number is the value according to the Papadopoulos–Imamura scale, QI/Rel. = quality index (meteotsunami) and relibility (tsunami), and Source = catalogue from which the data were mainly derived.

Y	M	D	Ev	Lat.	Lon.	Description	OP	Int.	QI/Rel.	Source
1348	1	25	T	46.500	13.580	Agitation in canals in Venice	1	2/III	2	EMTC and ITED
1511	3	26	T	46.210	13.220	Large sea level rise at Trieste	2	3/V	2	EMTC and ITED
1624	3	19	T	44.640	11.850	Strong agitation in Po River and coastal lagoons	3	2/III	4	EMTC and ITED
1627	7	30	T	41.740	15.340	Large sea withdrawal and flooding in Lesina	7	5/V	4	EMTC and ITED
1667	4	6	T	42.600	18.100	Sea withdrawal at Dubrovnik	3	3/V	4	EMTC and ITED
1672	4	14	T	43.940	12.580	Sea withdrawal and flooding at Rimini	1	3/IV	4	EMTC and ITED
1690	12	23	T	43.550	13.590	Boats stranded in Ancona	1	3/IV	2	EMTC and ITED
1731	3	20	T	41.270	15.760	Sea rise at Siponto and Barletta	2	3/IV	4	EMTC and ITED
1743	2	20	T	39.850	18.770	Sea withdrawal at Brindisi	1	2/III	2	EMTC and ITED
1780	9	21	T	42.400	18.500	Sea withdrawal in Kotor	2	3/IV	4	EMTC and ITED
1833	1	19	T	40.400	19.900	Tsunami at Saseno Island	1	2/V	1	EMTC and ITED
1838	8	10	T	45.200	14.500	Sea oscillations at Fiume	1	2/III	2	EMTC and ITED
1845	8	16	T	42.640	18.110	Sea level rise at Gruž	2	3/IV	4	EMTC and ITED
1851	10	12	T	40.700	19.700	Sea level rise at Valona	1	2/IV	4	EMTC and ITED
1866	1	2	T	40.300	19.400	Tsunami at Valona and Kanina	2	2/VII	3	EMTC and ITED
1866	3	3	T	40.400	19.500	Tsunami at Valona, Kanina and Himara	3	3/IV	1	EMTC and ITED
1875	3	17	T	44.210	12.660	Sea flooding at Rimini and Cervia	5	3/IV	4	EMTC and ITED
1889	12	8	T	41.830	15.690	Sea agitation in Termoli and Mattinata	2	2/III	2	EMTC and ITED
1893	6	14	T	40.300	19.700	Tsunami at Valona	1	2/V	1	EMTC and ITED
1916	8	16	T	44.020	12.740	At Tavollo, tsunami waves observed	1	2/IV	4	EMTC and ITED
1920	12	18	T	40.500	19.500	Tsunami at Saseno Island	1	5/VIII	4	EMTC and ITED
1930	10	30	T	43.690	13.380	Sudden high tide at Ancona	1	4/VI	4	EMTC and ITED
1931	7	21	M			Flooding and ebb at Vela Luka	1	3/IV	1	CMCT
1935	5	28	M			Strong seiches at Vela Luka	1	3/IV	1	CMCT
1937	9	12	M			Large seiches at Vela Luka	1	3/IV	1	CMCT
1951	11	11	M			Strong flood of houses and shops	4	4/VII	1	CMTC
1956	7	21	M			Large flood at Vela Luka	1	2/IV	1	CMTC
1962	1	11	T	43.150	16.940	Sea level oscillation at Split	4	2/II	4	EMTC and ITED
1966	8	27	M			Flooding, boats damaged at Korčula	1	4/V	2	CMTC
1972	2	10	M			Flooding of roads at Vela Luka	1	3/IV	2	CMTC and DAMT
1977	8	21	M			Severe flooding, boats damaged at Vela Luka	1	4/V	2	CMTC
1977	9	19	M			Severe flooding, damage, 5 injured at Jesolo	4	4/VI	2	CMTC and DAMT
1978	6	21	M			Inundation, heavy damage, injured people	30	5/VII	5	CMTC and DAMT
1978	7	6	M			Damage on the waterfront at Vela Luka	1	4/V	2	CMTC
1979	2	12	M			Flooding of roads in Vela Luka	1	3/IV	2	CMTC
1979	4	15	T	42.020	19.070	Damaging wave at Kotor Bay	9	4/VII	4	EMTC and ITED

Table 1. *Cont.*

Y	M	D	Ev	Lat.	Lon.	Description	OP	Int.	QI/Rel.	Source
1980	7	10	M			Strong inundation and heavy damage	2	4/VI	2	CMTC and DAMT
1984	10	5	M			Big waves in Siroka Bay	1	3/IV	2	CMTC
1987	7	26	M			No description	3	1/I	1	CMTC
2003	6	27	M			Large inundation and damage to shops at Stari Grad	6	4/VI	2	CMTC and DAMT
2006	5	24	M			Strong waves at Vrboska	1	2/III	1	CMTC
2007	8	22	M			Damaging wave in Siroka Bay	2	4/V	2	CMTC and DAMT
2008	8	15	M			Flood and damage at Mali Lošinj	1	4/VI	2	CMTC and DAMT
2010	2	19	M			Severe damage to houses and shops at Stari Grad	1	4/VI	2	CMTC and DAMT
2014	6	25	M			Inundation and many boats destroyed	20	4/V	5	CMTC and DAMT
2017	6	28	M			Inundation of the waterfront at Stari Grad	1	3/IV	2	CMTC and DAMT
2017	6	30	M			Inundation of the promenade, slight damage	2	4/V	2	CMTC and DAMT
2017	7	11	M			Flooding of streets and shops	3	4/VI	2	CMTC and DAMT
2018	3	31	M			Flooding of buildings and shops at Stari Grad	1	4/V	2	CMTC and DAMT
2018	10	29	M			Tide gauge record	1	1/II	2	CMTC
2019	7	9	M			Inundation of piers	2	2/II	2	DAMT
2019	11	12	M			Strong flooding of the quays at Vrboska	1	3/III	2	DAMT
2020	2	16	M			Some boats stranded in the Bay of Stobreč	1	3/IV	2	DAMT
2020	2	23	M			No description	3	1/I	1	DAMT
2020	5	11	M			Flooding of shops in Vrboska	3	3/V	5	DAMT
2020	5	14	M			Flooding of the waterfront at Vela Luka	2	3/V	5	DAMT
2020	5	16	M			Flooding of shops and cafes at Vela Luka	1	3/V	3	DAMT

Author Contributions: Conceptualization, A.M. and L.G.; methodology, A.M.; software, B.B.; validation, A.M.; formal analysis, A.M.; investigation, A.M.; resources, A.M.; data curation, A.M.; writing—original draft preparation, A.M.; writing—review and editing, A.M., B.B. and L.G.; visualization, A.M.; supervision, A.M.; project administration, A.M.; funding acquisition, A.M. All authors have read and agreed to the published version of the manuscript.

Funding: This research was funded by the EUROPEAN UNION, Programme Interreg Italy-Croatia, Project "Preventing, managing and overcoming natural-hazards risks to mitigate economic and social impact" PMO-GATE ID 10046122.

Institutional Review Board Statement: Not applicable.

Informed Consent Statement: Not applicable.

Data Availability Statement: The datasets presented in this study can be found in online repositories. The name of the repository and accession number can be found below: https://ingv.maps.arcgis.com/apps/webappviewer/index.html?id=0f465d51001146d79a6c89884a8e5d8c (accessed on 20 May 2022).

Acknowledgments: The authors thank Mario Locati, (INGV, Milan) for the technical support during the realization of the DAMT database.

Conflicts of Interest: The authors declare no conflict of interest.

References

1. Maramai, A.; Graziani, L.; Brizuela, B. *Italian Tsunami Effects Database (ITED): The First Database of Tsunami Effects Observed along the Italian Coasts*; Istituto Nazionale di Geofisica e Vulcanologia: Milan, Italy, 2019. [CrossRef]
2. Maramai, A.; Brizuela, B.; Graziani, L. The Euro-Mediterranean Tsunami Catalogue. *Ann. Geophys.* **2014**, *57*, S0435. [CrossRef]
3. Orlić, M.; Šepić, J. Meteorological Tsunamis in the Adriatic Sea—Catalogue of Meteorological Tsunamis in Croatian Coastal Waters. Available online: http://jadran.izor.hr/~sepic/meteotsunami_catalogue/ (accessed on 15 March 2022).
4. The Meteotsunami Risk in the Adriatic Sea. Available online: https://www.youtube.com/watch?v=xJaxMxTsu_s (accessed on 26 April 2022).
5. Italian Conference on Science Communication, SISSA Trieste. 17–20 November 2021. Available online: https://www.facebook.com/pmogate (accessed on 26 April 2021).
6. Tiberti, M.M.; Lorito, S.; Basili, R.; Kastelic, V.; Piatanesi, A.; Valensise, G. Scenarios of Earthquake-Generated Tsunamis for the Italian Coast of the Adriatic Sea. *Pure Appl. Geophys.* **2008**, *165*, 2117–2142. [CrossRef]
7. Ollier, C.D.; Pain, C.F. The Apennines, the Dinarides, and the Adriatic Sea: Is the Adriatic microplate a reality? *Geogr. Fis. Dinam. Quat.* **2009**, *32*, 167–175.
8. Piccardi, L.; Sani, F.; Moratti, G.; Cunningham, D.; Vittori, E. Present-day geodynamics of the circum-Adriatic region: An overview. *J. Geodynam.* **2011**, *51*, 81–89. [CrossRef]
9. Bada, G.; Horvath, F.; Gemer, P.; Fejes, I. Review of the present-day geodynamics of the Pannonian basin: Progress and problems. *J. Geodynam.* **1999**, *27*, 501–527. [CrossRef]
10. Babbucci, D.; Tamburelli, C.; Viti, M.; Mantovani, E.; Albarello, D.; D'Onza, F.; Cenni, N.; Mugnaioli, E. Relative motion of the Adriatic with respect to the confining plates: Seismological and geodetic constraints. *Geophys. J. Int.* **2004**, *159*, 765–775. [CrossRef]
11. Battaglia, M.; Murray, M.H.; Serpelloni, E.; Bürgmann, R. The Adriatic region: An independent microplate within the Africa Eurasia collision zone. *Geophys. Res. Lett.* **2004**, *31*, L09605. [CrossRef]
12. Venisti, N.; Calcagnile, G.; Pontevivo, A.; Panza, G.F. Tomographic Study of the Adriatic Plate. *Pure Appl. Geophys.* **2005**, *162*, 311–329. [CrossRef]
13. Pinter, N.; Grenerczy, G. Recent advances in Peri-Adriatic geodynamics and future research directions. In *The Adria Microplate: GPS Geodesy, Tectonics and Hazards*, 1st ed.; Pinter, N., Grenerczy, G., Weber, J., Stein, S., Medak, D., Eds.; Springer: Dordrecht, The Netherlands, 2006; pp. 1–20. [CrossRef]
14. Altiner, Y.; Bačić, Z.; Bačić, T.; Coticchia, A.; Medved, M.; Mulić, M.; Nurçe, B. Present-day tectonics in and around the Adria plate inferred from GPS measurements. In *Postcollisional Tectonic and Magmatism in the Mediterranean Region and Asia*; Dilek, Y., Pavlides, S., Eds.; The Geological Society of America: Boulder, CO, USA, 2006; pp. 43–55. [CrossRef]
15. Thouvenot, F.; Fréchet, J. Seismicity along the northwestern edge of the Asia Microplate. In *The Asia Microplate: GPS Geodesy, Tectonics and Hazards*, 1st ed.; Pinter, N., Grenerczy, G., Weber, J., Stein, S., Medak, D., Eds.; Springer: Dordrecht, The Netherlands, 2006; pp. 335–349. [CrossRef]
16. Muço, B. Seismicity of the Adriatic Microplate and a possible triggering: Geodynamic implication. In *The Asia Microplate: GPS Geodesy, Tectonics and Hazards*, 1st ed.; Pinter, N., Grenerczy, G., Weber, J., Stein, S., Medak, D., Eds.; Springer: Dordrecht, The Netherlands, 2006; pp. 351–367. [CrossRef]
17. Lorito, S.; Tiberti, M.M.; Basili, R.; Piatanesi, A.; Valensise, G. Earthquake-generated tsunamis in the Mediterranean Sea: Scenarios of potential threats to Southern Italy. *J. Geophys. Res.* **2008**, *113*, B01301. [CrossRef]

18. Maramai, A.; Graziani, L.; Tinti, S. Investigation on tsunami effects in the central Adriatic Sea during the last century—A contribution. *Nat. Hazards Earth Syst. Sci.* **2007**, *7*, 15–19. [CrossRef]
19. Caputo, M.; Faita, G. Primo catalogo dei maremoti delle coste italiane. *Atti Accad. Naz. Lincei Mem. Cl. Sci. Fis. Mat. Nat. Rend.* **1986**, *80*, 213–356.
20. Bedosti, B.; Caputo, M. Primo aggiornamento del catalogo dei maremoti delle coste italiane. *Atti Accad. Naz. Lincei Rend. Cl. Sci. Fis. Mat. Nat.* **1987**, *80*, 570–584.
21. Tinti, S.; Maramai, A.; Graziani, L. The New Catalogue of Italian Tsunamis. *Nat. Hazards* **2004**, *33*, 439–465. [CrossRef]
22. Paulatto, M.; Pinat, T.; Romanelli, F. Tsunami hazard scenarios in the Adriatic Sea domain. *Nat. Hazards Earth Syst. Sci.* **2007**, *7*, 309–325. [CrossRef]
23. Monserrat, S.; Vilibić, I.; Rabinovich, A.B. Meteotsunamis: Atmospherically induced destructive ocean waves in the tsunami frequency band. *Nat. Hazards Earth Syst. Sci.* **2006**, *6*, 1035–1051. [CrossRef]
24. Rabinovich, A.B.; Vilibić, I.; Tinti, S. Meteorological tsunamis: Atmospherically induced destructive ocean waves in the tsunami frequency band. *Phys. Chem. Earth* **2009**, *34*, 891–893. [CrossRef]
25. Vilibić, I.; Denamiel, C.; Zemunik, P.; Monserrat, S. The Mediterranean and Black Sea meteotsunamis: An overview. *Nat. Hazards* **2021**, *106*, 1223–1267. [CrossRef]
26. Bechle, A.J.; Wu, C.H.; Kristovich, D.A.R.; Anderson, E.J.; Schwab, D.J.; Rabinovich, A.B. Meteotsunamis in the Laurentian Great Lakes. *Sci. Rep.* **2016**, *6*, 37832. [CrossRef]
27. Pattiaratchi, C.B.; Wijeratne, E.M.S. Are meteotsunamis an underrated hazard? *Philos. Trans. A Math. Phys. Eng. Sci.* **2015**, *373*, 20140377. [CrossRef]
28. Vučetić, T.; Vilibić, I.; Tinti, S.; Maramai, A. The Great Adriatic flood of 21 June 1978 revisited: An overview of the reports. *Phys. Chem. Earth* **2009**, *34*, 894–903. [CrossRef]
29. Orlić, M.; Belusić, D.; Janekivić, I.; Pasarić, M. Fresh evidence relating the great Adriatic surge of 21 June 1978 to mesoscale atmospheric forcing. *J. Geophys. Res.* **2010**, *115*, C06011. [CrossRef]
30. Sepić, J.; Vilibić, I.; Monserrat, S. Teleconnections between the Adriatic and the Balearic meteotsunamis. *Phys. Chem. Earth* **2009**, *34*, 928–937. [CrossRef]
31. Belušić, D.; Strelec Mahović, N. Detecting and following atmospheric disturbances with a potential to generate meteotsunamis in the Adriatic. *Phys. Chem. Earth* **2009**, *34*, 918–927. [CrossRef]
32. Ambraseys, N.N. Data for the investigation of seismic sea waves in the Eastern Mediterranean. *Bull. Seism. Soc. Am.* **1962**, *52*, 895–913.
33. Papadopoulos, G.A.; Imamura, F. A proposal for a new tsunami intensity scale. *Environ. Sci.* **2001**, *5*, 569–577.
34. Pasarić, M.; Brizuela, B.; Graziani, L.; Maramai, A.; Orlić, M. Historical tsunamis in the Adriatic Sea. *Nat. Hazards* **2012**, *61*, 281–316. [CrossRef]

Article

Shifting of Meteorological to Hydrological Drought Risk at Regional Scale

Awais Naeem Sarwar [1], Muhammad Waseem [1,*], Muhammad Azam [2], Adnan Abbas [3], Ijaz Ahmad [1], Jae Eun Lee [4,*] and Faraz ul Haq [1]

[1] Centre of Excellence in Water Resources Engineering, University of Engineering & Technology, GT-Road, Lahore 54890, Pakistan; ranaawais094@gmail.com (A.N.S.); ijaz.ahmad@cewre.edu.pk (I.A.); engraraz@uet.edu.pk (F.u.H.)

[2] Faculty of Agricultural Engineering and Technology, PMAS Arid Agriculture University, Rawalpindi 46000, Pakistan; mazammakram@gmail.com

[3] Land Science Research Center, Nanjing University of Information Science & Technology, Nanjing 210044, China; adnanabbas@nuist.edu.cn

[4] National Crisisonomy Institute, Chungbuk National University, Cheongju 28644, Korea

* Correspondence: waseem.jatoi@cewre.edu.pk (M.W.); jeunlee@chungbuk.ac.kr (J.E.L.)

Citation: Sarwar, A.N.; Waseem, M.; Azam, M.; Abbas, A.; Ahmad, I.; Lee, J.E.; Haq, F.u. Shifting of Meteorological to Hydrological Drought Risk at Regional Scale. *Appl. Sci.* **2022**, *12*, 5560. https://doi.org/10.3390/app12115560

Academic Editors: Andrea Chiozzi, Željana Nikolić and Elena Benvenuti

Received: 24 March 2022
Accepted: 27 May 2022
Published: 30 May 2022

Publisher's Note: MDPI stays neutral with regard to jurisdictional claims in published maps and institutional affiliations.

Abstract: The drought along with climate variation has become a serious issue for human society and the ecosystem in the arid region like the Soan basin (the main source of water resources for the capital of Pakistan and the Pothohar arid region). The increasing concerns about drought in the study area have brought about the necessity of spatiotemporal analysis and assessment of the linkage between different drought types for an early warning system. Hence, the streamflow drought index (SDI) and standard precipitation index (SPI) were used for the analysis of the spatiotemporal variations in hydrological and meteorological drought, respectively. Furthermore, statistical approaches, including regression analysis, trend analysis using Mann Kendall, and moving average, have been used for investigation of the linkage between these drought types, the significance of the variations, and lag time identification, respectively. The overall analysis indicated an increase in the frequency of both hydrological and meteorological droughts during the last three decades. Moreover, a strong linkage between hydrological and meteorological droughts was found; and this relationship varied on the spatiotemporal scale. Significant variations between hydrological and meteorological droughts also resulted during the past three (3) decades. These discrepancies would be because of different onset and termination times and specific anthropogenic activities in the selected basin for the minimization of hydrological drought. Conclusively, the present study contributes to comprehending the linkage between hydrological and meteorological droughts and, thus, could have a practical use for local water resource management practices at the basin scale.

Keywords: drought SDI; SPI; linkage; propagation

1. Introduction

Drought is a condition considered as the deficit of water, including surface, ground, or atmospheric water, for a long period [1]. Irrespective of climatic sites, the drought could occur worldwide even in humid and wet environments [2]. Drought is generally classified into meteorological, hydrological, socioeconomic, and agricultural drought [3]. Among these four types of droughts, hydrological is the most important form as sustainable water resources management is heavily dependent on hydro-meteorological information. Various numbers of factors are associated with the onset of hydrological drought, and meteorological drought is one of the main influencing factors. The root cause of drought onset is the deficiency of rainfall across a large area for a very long period and is notated as a meteorological drought [4]. A meteorological drought can develop quickly because it is primarily caused by a lack of precipitation [5], and if the lack of precipitation spreads to specific regions, the meteorological drought can be transformed into a hydrological drought, and then into an agricultural drought. The drought propagation process refers to the transition of different types of droughts.

The relationships between different types of droughts are studied by several authors, for example, [6] resulted that hydrological drought events occurred approximately seven months after meteorological drought events. The streamflow drought can be investigated using meteorological drought information at the annual time scale [7]. The interrelationship between runoff and meteorological drought has also been investigated and it was concluded that the significant dependency of hydrological drought on meteorological drought [8]. The impact of meteorological forcing on hydrological droughts has been computed [9]. A copula-based joint meteorological-hydrological drought index has been used to model the relationship between meteorological and hydrological droughts upstream and downstream of the Kasilian basin [10]. The transmission of meteorological droughts to hydrological droughts and the influence factors have been explored and investigated [11]. The entropy theory was used to create a hybrid drought index that combines hydrological, meteorological, and agricultural data and was used to investigate the drought condition in Northwest China [12]. In literature, the relationships between hydro-meteorological droughts, climatic variables, and human activities were also explored [13,14]. For example, [15] examined the relationship between various hydrologic and meteorological drought indices considering natural and human factors for different basins in the contiguous United States. Similarly, [16–19] studied the dependence structure between meteorological drought and hydrological drought using different approaches and resulted that the occurring time lag between meteorological and hydrological droughts is critical for sustainable drought management.

Most studies in the literature focused on determining hydro-meteorological drought and drought propagation; however, it remains unexplored. Besides that, the lack of understanding considering hydrological drought response to meteorological drought in different regions raises an unanswered question for basin-scale drought risk management [20]. To find answers to these questions, a comprehensive study of droughts that span multiple geographic areas and last for extended periods is required. Strategic water resources can be implemented more efficiently with accurate water-based information based on regional drought characteristics. Furthermore, given Pakistan's erratic, scarce, unstable climate and current drought situation [20–29], drought propagation information is critical at the regional and national levels, as it can provide the appropriate and consistent information required for efficient water management and drought early warning systems. Drought propagation information can aid in the prevention of significant economic losses as well as decision-making. Early warning systems are estimated to save hundreds of lives, save 2.7 billion Euros in natural disaster losses, and generate billions of supplementary benefits per year in Europe based on the optimization of economic production in the energy and agriculture sectors. In the near past, Pakistan has also experienced recurrent droughts and more severe droughts are anticipated in near future due to climate change [22].

The Soan River is a tributary of Pakistan's mighty Indus River Basin. It flows from the Murree Mountains into the Indus via the Dhoke Pathan hydrological station. It is the main hydrological unit for the Pothohar arid region of Pakistan. The importance of the Pothohar region cannot be repudiated as the region of Pothohar is rich in agriculture and agriculture is the backbone of the country's economy [30]. However, there are canal irrigation resources in this area and the crop production depends entirely on rainfall at suitable timings. Pothohar region gets rainfall during both the winter and summer seasons, which is the main source of crop germination, flowering, and maturity. If any of the seasons fail to bring rain, then soil moisture depletes resulting in drought and ultimately significant crop damage. More specifically, the crops in these areas depend on monsoon rains in summer and on western rains in winter. If the summer season fails to bring rains in the area, then it will cause huge crop yield loss in that specific year. Similarly, if the westerly system also fails to bring rains to the area, crops are affected badly, resulting in severe droughts. Historically, this Pothohar region was severely affected by substantial variation in the rainfall and prolonged rain shortage especially from 1999 to 2002 [31]. Hence, the objectives of the present study include the exploration of spatiotemporal evolutions of meteorological and hydrological droughts in the study basin, evaluation of the link between meteorological and hydrological droughts, identifying the lag time, and the investigation of the differences between hydrological and meteorological droughts.

2. Materials and Methods

2.1. Study Area

The Soan River, initiating from the Murree highlands, is a major tributary of the Indus River and a key hydrological entity of the Potohar region of Pakistan. The Soan River flows through Chirah and Dhoke Pathan hydrological gauging stations (Figure 1) before being a part of the Indus River. The total drainage area of the Soan River basin is 6842 km^2 with an elevation ranging from 265 to 2274 m. The Soan basin is characterized by gentle to the steep slope and monsoon fed streams which generate almost all of the basin flow with mean annual precipitation of 1465 mm, and the mean annual temperature ranges from 8 to 22 °C. The Simly Dam, which spans the Soan River, provides drinking water to Islamabad, Pakistan's capital, as well as water for irrigation activities in the Pothohar region [32]. Furthermore, for a better understanding and evaluation of the spatial variation of drought, the study area was divided into two sub-basins: Chirah (hereafter sub-basin 1) and Dhoke Pathan (hereafter sub-basin 2). In the current study, monthly precipitation data of five meteorological stations within the basin (Murree, Zero Point, airport, Dhamial, and SAWCRI) from 1983 to 2015 was collected from Pakistan Meteorological Department (PMD). The average value of four stations lying in sub-basin 2 was used to estimate the meteorological drought in sub-basin 2, whereas the data of the remaining one station, i.e., Murree was used for meteorological drought analysis at sub-basin 1. Similarly, the monthly streamflow data of two hydrological gauging stations, i.e., Chirah and Dhoke Pathan was collected from Water and Power Development Authority (WAPDA) for the same period.

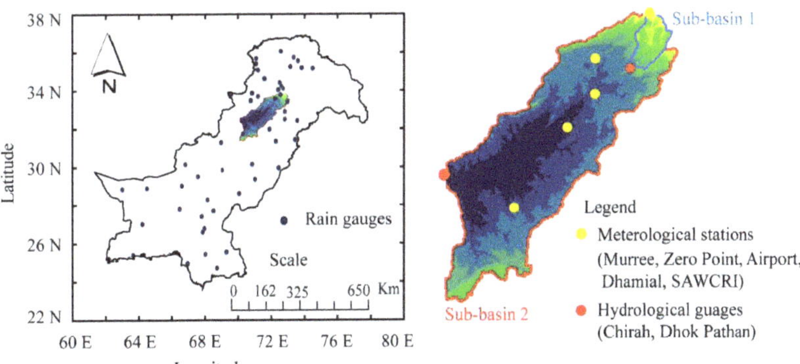

Figure 1. Location map of the study area.

2.2. Spatiotemporal Analysis of Drought

2.2.1. Meteorological Drought Assessment

For the assessment of meteorological drought, the Standardized Precipitation Index (SPI) was used due to its effectiveness, applicability, and suitability at different time scales for many case studies, e.g., [28,33,34]. Considering the hydrological year (October to September), SPI was calculated at four different time scales, i.e., SPI-3 (October–December), SPI-6 (October–March), SPI-9 (October–June), and SPI-12 (October–September). Furthermore, for the calculation of SPI at different time scales, the following expression was used for cumulative precipitation $(R_{i,K})$.

$$R_{i,K} = \sum_{j=1}^{3k} P_{i,j} \; i = 1,2,3,\ldots,n \; j = 1,2,\ldots,12 \; and \; k = 1,2,3,4 \tag{1}$$

where $(P_{i,j})$ corresponds to monthly precipitation, j denotes the particular month of the hydrological year, I indicates the hydrological year, and k reveals the timescale (e.g., SPI-3, $k = 1$ indicating 3-month time scale October-December, and similarly $k = 4$ is for October to September).

SPI computation includes fitting distributions to precipitation data $(R_{i,K})$ and estimating a probability density function (PDF) and cumulative distribution function (CDF). These functions are further transformed into standardized distributions with unit standard deviations and zero means yielding the SPI value using the expression below (Equation (2)). The SPI value could range from <-2 to $>+2$ [35], whereas the positive value of SPI indicates a wet condition while the negative value represents a drought condition. Moreover, in this study, the drought thresholds were as follows: weak drought ($-1.0 \leq$ index < 0), moderate drought ($-1.49 \leq$ index < -1.0), severe drought ($-1.99 \leq$ index < -1.50), and extreme drought (index ≤ -2).

$$\text{SPI}_{i,k} = \frac{R_{i,k} - \overline{R}_{i,k}}{S_k} \tag{2}$$

where S_k and $\overline{R}_k$ are the standard deviation and average value of the precipitation, respectively.

2.2.2. Hydrological Drought Assessment

The Streamflow Drought Index (SDI) was developed by [36] and is widely used to characterize hydrological drought events. SDI computation also includes fitting distributions to runoff data and estimation of PDF and CDF and then transforming into a standardized distribution, yielding the SDI value [13]. Similarly, the positive SDI value represents wet conditions, while the negative value represents drought conditions. Furthermore, SPI thresholds were adopted for hydrological drought analysis.

$$V_{i,k} = \sum_{j=1}^{3k} Q_{i,j} \; i = 1,2,3,\ldots,n \; j = 1,2,3,\ldots,12 \; and \; k = 1,2,3,4 \tag{3}$$

$$\text{SDI}_{i,k} = \frac{V_{i,k} - \overline{V}_{i,k}}{S_k} \tag{4}$$

where $V_{i,k}$ the cumulative runoff for reference period k and the i is the hydrological year; S_k is the standard deviation and $\overline{V}_k$ is the mean value of runoff.

The SDI can also be calculated at various time scales, such as one, three, six, and twelve months. For the estimation of SPI-3, SDI-3, SPI-6, SDI-6, SPI-9, SDI-9, SPI-12, and SDI-12, the cumulative sums of precipitation and runoff for 3, 6, 9, and 12-month timescales were used. When calculating SPI-3 and SDI-3, for example, the cumulative sum for October was obtained by adding the following two-month data (November and December) to the October data [20]. SPI-6 and SDI-6 cumulative sums were calculated by adding six months from October to March.

Furthermore, the Mann-Kendall method [37] was used to estimate the trend in SPI-3, SPI-6, SPI-9, and SPI-12 at a significance level of 5% and similarly for SDI values at different time scales.

2.3. Identification of Linkage between Meteorological and Hydrological Drought and Lag Time

A regression analysis is a statistical technique for estimating the magnitude of the effect of any independent time series (hereafter the meteorological drought index) on any dependent time series (hereafter the hydrological drought index) [38]. In particular, regression analysis is used to better understand the relationships between independent and dependent time series. As a result, a linear regression was used to evaluate the relationship and temporal changes in drought. Coefficient of determination (R^2), slope, and *p*-value were calculated for each relationship using the two-tailed t-test, and R2 was used to indicate the best relationship between SDI and SPI [38]. Equation (5) depicts the mathematical expression of regression between any independent and any dependent time. In the current study, several simple linear regression models were developed between SPI and SDI of the same time scale, e.g., SPI-3 versus SDI-3; SPI-6 versus SDI-6; and similarly for other time scales.

$$Y = \beta_0 + aX + \varepsilon \tag{5}$$

where Y is the dependent time series, a is the regression coefficient, X is the independent time series, β_0 is the intercept, and ε is an error term.

Furthermore, in this study, moving average (running average) analysis was performed at multiple time scales to better assess the correlation between SPI and SDI. It is a quite simple statistical analysis tool used as the fluctuation, trend, or lagging indicator. For a given time series and a fixed size subset, the first value of the moving average is calculated by averaging the fixed subset of the given time series. After this, the subset is shifting forward by including the next number and excluding the first number in the subset. For example, Let (X_1, t_1), (X_2, t_2), ... , (X_n, t_n) represents SPI or SDI time series, X_1, X_2, . . . , X_n are the values of SPI or SDI; against time periods t_1, t_2, . . . , t_n, respectively. The first two moving averages of order k (here 2) can be calculated as given in Equations (6) and (7). A similar procedure was adopted up to the nth value.

$$\text{For 1st MA}: \\ MA_K = \frac{X_1 + X_2 + \ldots + X_k}{K} \tag{6}$$

$$\text{For 2nd MA}: \\ MA_K = \frac{X_2 + X_3 + \ldots + X_{k+1}}{K} \tag{7}$$

3. Results

3.1. Spatiotemporal Analysis of Droughts

3.1.1. Analysis of Meteorological Drought

Figure 2 illustrates the meteorological drought index calculated at selected time scales for both sub-basins of the study area. Based on the analysis, it was observed that there was an increase in the frequency of drought after 1998 specifically in the case of sub-basin-1 when SPIs mostly remained negative. These results are in good agreement with the results previously published on droughts in Pakistan [19,20,23,29,39]. The variations in drought between both sub-basins are due to the more relative decrease in precipitation in subbasin-1, compared to sub-basin-2 [30,32]. Moreover, the numbers of drought events were different at different computed time scales. Significant inconsistencies were found when the 3-month time scale was compared with the 6-month time scale, and a comparison between the 6-month and 12-month time scales gave slight variations in results. Results reveal the frequent occurrence of severe drought events at each sub-basin of the study area, and these were more sensitive to short-term drought compared to long-term droughts. According to the results, the recurrence of drought is more frequent and consecutive at 1 to 6 months' timescale in sub-basin 1 and during the 12-month timescale, there exists a sharp difference between wet and dry episodes. while the sub-basin 2 is found to be more sensitive to drought (mild to severe) at a 1-month timescale as compared to 3, 6, and 12-month timescales. The results also specified that the drought events state varied with the increase in SPI time scales, mostly a declining trend in sub-basin 2. This is due to the reason that computation of 12-month time scale SPI involves the aggregation of total precipitation from October to September and includes both wet and dry seasons. However, the 3-month SPI time scale considers only the sum of three-month precipitation, and the 3-month time scale (March-May) of the study basin was normally a dry season.

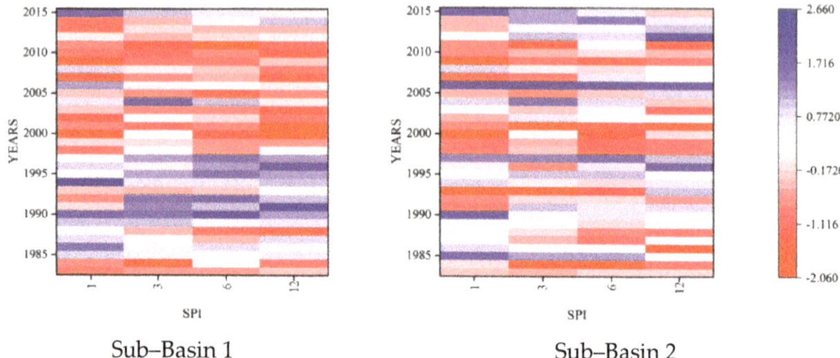

Figure 2. Temporal variations of meteorological drought periods during 1983–2015 in sub–basins 1 and 2.

Figure 3 illustrates the severity of meteorological drought based on the 12-month time scale for both hydrological sub-basins and the difference in their SPI-12 values during the study period (i.e., 1983–2015). Based on the analysis, it resulted that the years from 1998 to 2004 were the driest years of the selected time series. Moreover, the difference in meteorological drought (SPI-12) severity at both sub-basins represents the notable spatial variations across the Soan river basins. The Mann-Kendall test was also used on SPI-12 to figure out the trend in meteorological drought at sub-basin 1 and sub-basin 2. The analysis showed an increase in meteorological drought severity during the past 33 years, and a significant decreasing trend in SPI–12 values was observed downstream, i.e., sub-basin 2. The Z-value of the Mann-Kendall test was 1.78 at upstream sub-basin 1, and 1.71 at downstream sub-basin 2.

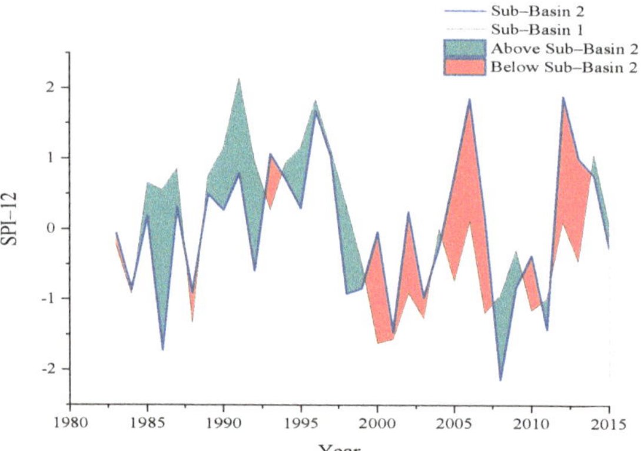

Figure 3. Difference in meteorological drought (SPI–12) severity at both sub-basins representing the spatial variations across the Soan river basins.

3.1.2. Analysis of Hydrological Droughts

Figure 4 depicts the temporal variations of hydrological drought periods in sub-basins 1 and 2 at different time scales, indicating the occurrence of hydrological drought over the last three decades. Hydrological drought, like meteorological drought, was more sensitive to short-term droughts than long-term droughts. Drought events occurred continuously from 1983 to 2015, with the driest period for the study area being 1998–2004. Drought patches were also randomly observed in 1993, 1998–2004, and 2009. Sub-basin 1 was stressed by hydrological drought, and the number of drought events was high from 1985 to 2004, with a slight wet condition from 2005-to 2015. However, in the case of sub-basin 2, the results showed that the frequency of hydrological drought events was increasing in the basin from 1983 to 2015. Figure 5 shows the illustration of the difference in hydrological drought severity (SDI-12) for two sub-basins and indicates notable variations across the two sub-basin. Furthermore, based on the MK test, a statistically significant trend was observed for downstream sub-basins 2 with a Z-value of 1.83, whereas, upstream sub-basin 1 showed opposite behavior with an insignificant statistical trend. However, there still exists an increasing trend of drought at upstream sub-basin 1.

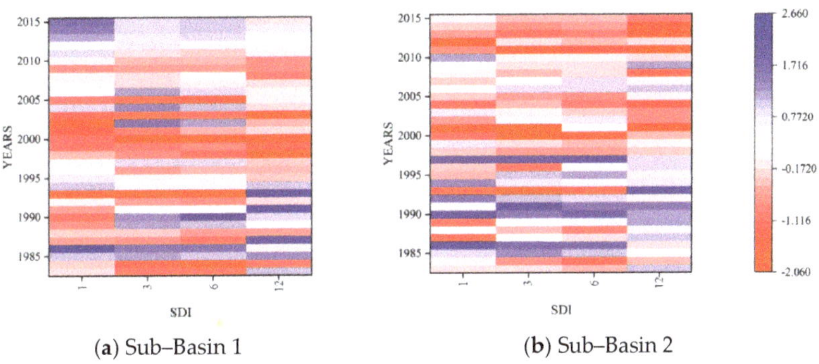

(**a**) Sub–Basin 1   (**b**) Sub–Basin 2

Figure 4. Temporal variation of hydrological drought periods in sub-basins 1 and 2 at different time scales.

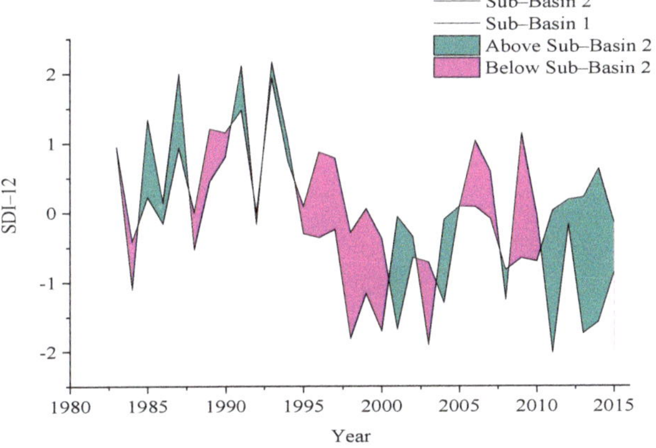

Figure 5. Difference in Hydrological drought (SDI–12) severity at both sub-basins representing the spatial variations across the Soan river basin.

3.2. Link between Meteorological and Hydrological Droughts

3.2.1. Establishing Regression Function

The coefficient of determination (R2) and regression coefficient (a) extracted from linear regression equations (Table 1) showed that a significant relationship existed between SDI and SPI, and it increased up to the 9-month time scale and then decrease in the case of the 12-month time scale. The coefficient of determination (R^2) of the 6-months and the 9-month time scale was relatively higher as compared to the other two time scales. Moreover, the correlation of SPI and SDI was higher at upstream sub-basin 1 of the Soan River Basin as compared to downstream. For instance, sub-basin 1 and sub-basin 2 had 0.66 and 0.63 values of R^2 based on a 9-month time scale respectively. The regression lines with the related data points are shown in Figure 6 which may be utilized for the prediction of hydrological droughts using the meteorological droughts information.

Table 1. Results of Regression modeling developed between SDI on SPI.

K		Sub-Basin 1	Sub-Basin 2
3-Months	a	0.35	0.37
	R^2	0.125	0.14
6-months	a	0.67	0.64
	R^2	0.45	0.4
9-Months	a	0.66	0.63
	R^2	0.44	0.4
12-Months	a	0.54	0.45
	R^2	0.29	0.21

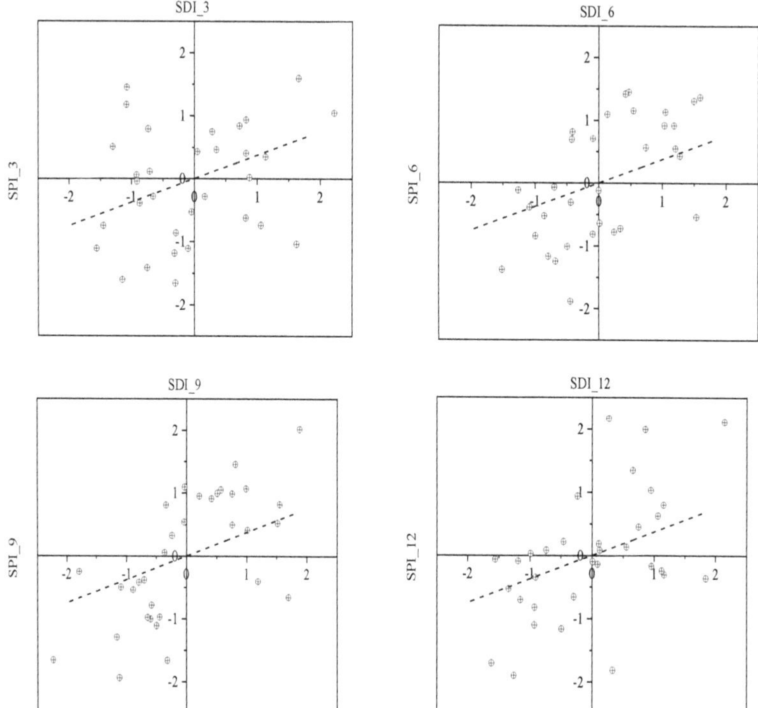

(**a**) For Sub-Basin 1

Figure 6. *Cont.*

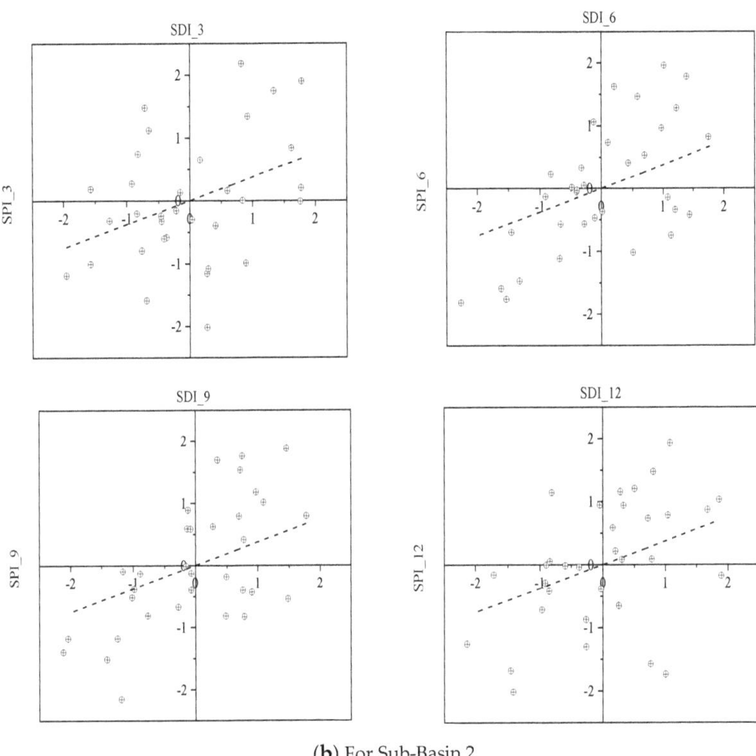

(**b**) For Sub-Basin 2

Figure 6. The linear relationship between SDI and SPI at the four selected time scales.

3.2.2. Moving Average Analysis and Lag Time Identification

A moving average analysis was performed at multiple time scales to better assess the correlation between SPI and SDI, see Figure 7. Results show that the variations of SPI and SDI were quite comparable; however, the differences between them do exist. In general, a strong correlation was observed between SPI and SDI at the upstream sub-basin 1, i.e., CC = 0.66. Overall based on the correlation analysis with different lag times, it was observed that at both sub-basins, the SDI was 1 month lagging behind SPI at 3-month and 12-month time scales and 2 months lagging for 6–month and 9-month time scales. This period of lagging time was in accordance with the prior study [7].

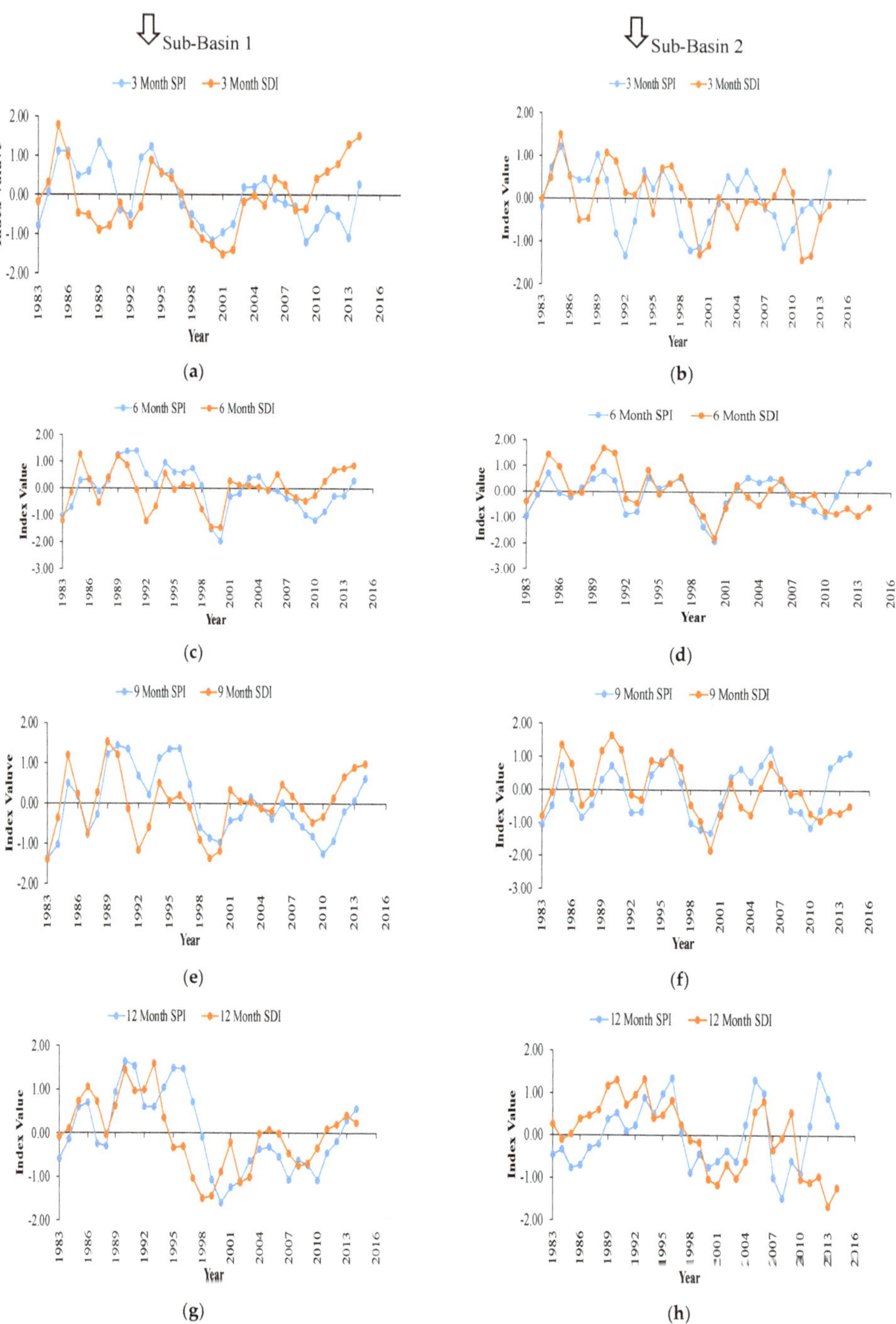

Figure 7. Moving average analysis for lag time identification between SPI and SDI at sub-basins 1 and 2 for different time scales. (**a,b**) 3 months, (**c,d**) 6 months, (**e,f**) 9 months, (**g,h**) 12 months.

4. Discussion

Climate change includes the variations in the behavior of climatic parameters, e.g., precipitation, evapotranspiration, etc. [40]. Whereas anthropogenic activities include land-use/land cover changes, irrigation area expansion, increased water diversion, water conservation practices, etc. During the last decades, the Soan basin experienced precipitation and runoff variation (with a change point in 1998 [30]) due to both climate change and anthropogenic activities [30,32]. Based on the drought analysis performed in the current study, notable variations between SPI and SDI values were found over the past three decades. It could be mainly because of major changes in agricultural land, forest, settlement, water bodies, and bare land in addition to natural climate variability [41].

To further justify, a trend analysis of precipitation, runoff, and runoff coefficient had been performed. The analysis resulted, in the case of sub-basin 1 a declining trend in precipitation was observed during the years 1983–2015, as shown in Figure 8a. For instance, the average annual precipitation decreased from 1983–1997 (1950 mm) to 1998–2015 (1570.50 mm). This decline in precipitation was significant in sub-basin 1, while little variation (no significant trend) was found in the case of sub-basin 2 with an average value of 1115 mm and 1065 mm for 1983–1997 and 1998–2015 respectively. For sub-basin 1, the annual runoff depth had a decreasing trend (i.e., an average of 174.44 mm for 1983–1997, and 84.00 mm for 1998–2015) during the study period (left panel Figure 8b) and, the runoff coefficient had no significant trend (left panel Figure 8c). Similarly, in the case of sub-basin 2, runoff depth has a decreasing trend, while runoff coefficient has no significant trend as the p-value is greater than alpha. The variations in the trend direction of runoff, and runoff coefficient in the case of sub-basins 1 and 2 could be mainly due to land-use changes, e.g., 71 km^2 and 1611 km^2 increase in agriculture land was observed in sub-basins 1 and 2 respectively from 1983–1997 to 1998–2015 [30].

Moreover, the higher decreasing rate in runoff compared to precipitation and change point showed the possible effect of anthropogenic activities [32]. It was anticipated because of a significant increase in the intensity of crops and the construction of large numbers of small dams from 1998- onward [27]. These significant variations were also anticipated due to the significant number of development projects under the Government of Punjab, Pakistan, which comprises the development of ponds and mini-dams for rainwater harvesting and to increase the intensity of agriculture. The initiative of these developments was to increase the storage for crop and drinking purposes, which might have caused the increase in evaporation, and water use for the growing population, ultimately decrease in surface runoff and increasing in probability of drought occurrence. In addition to that, a reservoir could decrease the frequency, duration, and severity of drought events downstream of the dam, and irrigation practices can primarily influence hydrological droughts by consuming streamflow and groundwater, which typically results in a decrease in streamflow and groundwater levels [16,42]. Hence, the Simly reservoir operation, as well as seasonal irrigation practices, could have an impact on the statistical relationship among drought indices.

The difference in SPI and SDI trends could be due to evapotranspiration or lag time between rainfall and runoff, which could predict the propagation of meteorological to hydrological droughts. Drought propagation may also be influenced by basin characteristics such as soil moisture, land use, and the relationships between streamflow and groundwater. A time lag was used to perform cross-correlation between SPI and SDI, which acknowledged the sequence between SPI and SDI and demonstrated that meteorological drought events could be used to predict hydrological droughts in relatively small watersheds with less anthropogenic activities.

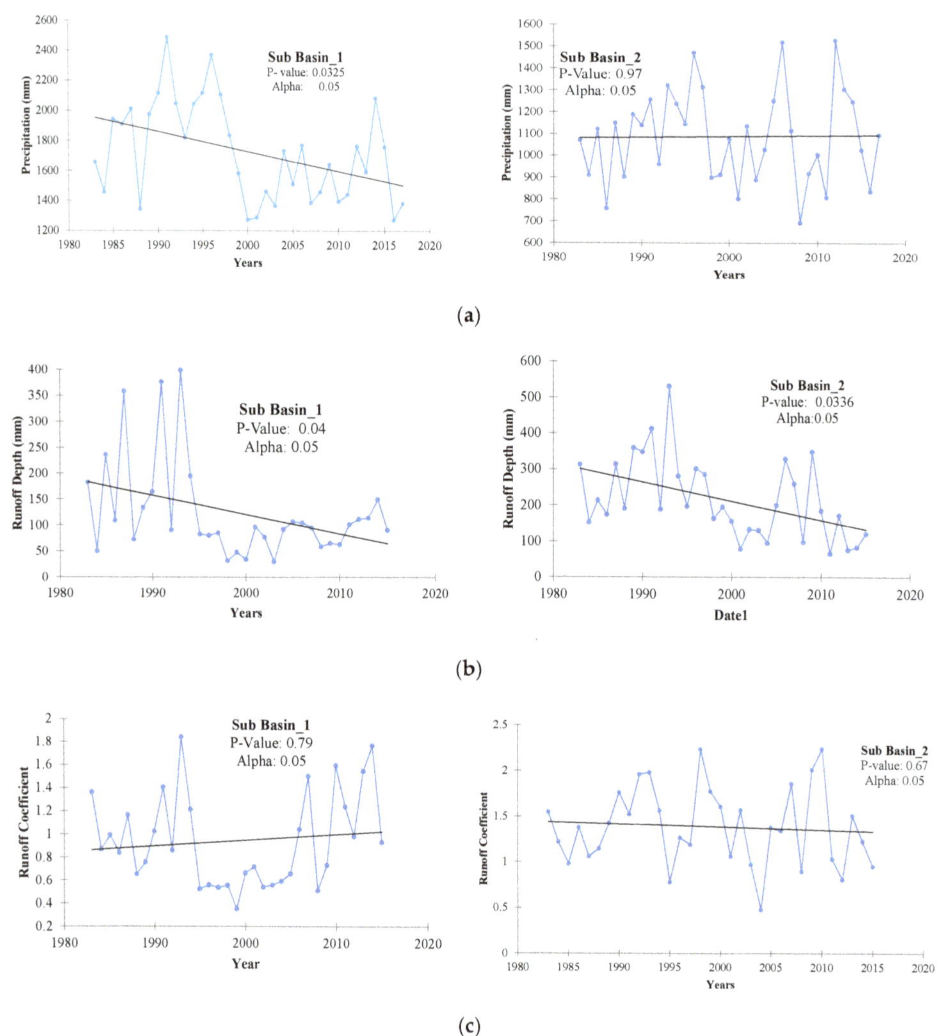

Figure 8. Temporal Variation of (**a**) Precipitation, (**b**) Runoff depth, and (**c**) Runoff Coefficient at sub-basin 1 and 2.

5. Conclusions

This study concludes that the frequency of both hydrological and meteorological droughts increased in the Soan River Basin during the study period. At various time scales, Sub-basin 1 was subjected to more frequent meteorological moderate drought and hydrological drought events. The current study was also designed to investigate the relationship between meteorological and hydrological drought events and SPI and SDI by developing a simple linear function between them. The results of a linear regression between SPI and SDI show an increase in regression coefficients with increasing time scale and became stronger until the ninth month. Climate change and anthropogenic activities (i.e., land use/land cover changes) are the main reasons that cause the variations between these two types of droughts. Moreover, the hydrological drought events commonly lagged 1–3 months (subject to the time scale and sub-basin) from the meteorological drought events. The dissimilarities between these two types of droughts became larger due to climatic variation and might be due to human activities as well. Conclusively, this study provided drought propagation and the basis for long-term drought forecasting and, thus, can be employed for early warning water resources management and as an extension of this current study can be to assess the climate change impacts on hydrological drought at the basin scale.

Author Contributions: Conceptualization, A.N.S. and M.W.; methodology, M.A. and A.A.; software, M.A.; validation, M.W., J.E.L. and M.A.; formal analysis, I.A. and A.A.; investigation, F.u.H.; data curation, A.N.S. and M.W.; writing—original draft preparation, J.E.L.; funding acquisition. All authors have read and agreed to the published version of the manuscript.

Funding: This work was supported by the Ministry of Education of the Republic of Korea and the National Research Foundation of Korea (NRF-2020S1A5B8103910).

Institutional Review Board Statement: Not applicable.

Informed Consent Statement: Not applicable.

Data Availability Statement: All data is provided in the form of tables and figures.

Acknowledgments: The Authors appreciate the NRPU-HEC projects.

Conflicts of Interest: The authors declare no conflict of interest.

References

1. Dai, A. Drought under Global Warming: A Review. *Wiley Interdiscip. Rev. Clim. Chang.* **2011**, *2*, 45–65. [CrossRef]
2. Van Loon, A.F.; Stahl, K.; Di Baldassarre, G.; Clark, J.; Rangecroft, S.; Wanders, N.; Gleeson, T.; Van Dijk, A.I.J.M.; Tallaksen, L.M.; Hannaford, J. Drought in a Human-Modified World: Reframing Drought Definitions, Understanding, and Analysis Approaches. *Hydrol. Earth Syst. Sci.* **2016**, *20*, 3631–3650. [CrossRef]
3. Wilhite, D.A. Drought as a Natural Hazard. In *Droughts*; Routledge: London, UK, 2021; p. 33.
4. Vidal, J.-P.; Martin, E.; Franchistéguy, L.; Habets, F.; Soubeyroux, J.-M.; Blanchard, M.; Baillon, M. Multilevel and Multiscale Drought Reanalysis over France with the Safran-Isba-Modcou Hydrometeorological Suite. *Hydrol. Earth Syst. Sci.* **2010**, *14*, 459–478. [CrossRef]
5. Peters, E.; Bier, G.; Van Lanen, H.A.J.; Torfs, P. Propagation and Spatial Distribution of Drought in a Groundwater Catchment. *J. Hydrol.* **2006**, *321*, 257–275. [CrossRef]
6. Edossa, D.C.; Babel, M.S.; Das Gupta, A. Drought Analysis in the Awash River Basin, Ethiopia. *Water Resour. Manag.* **2010**, *24*, 1441–1460. [CrossRef]
7. Tabrizi, A.A.; Khalili, D.; Kamgar-Haghighi, A.A.; Zand-Parsa, S. Utilization of Time-Based Meteorological Droughts to Investigate Occurrence of Streamflow Droughts. *Water Resour. Manag.* **2010**, *24*, 4287–4306. [CrossRef]
8. Haslinger, K.; Koffler, D.; Schöner, W.; Laaha, G. Exploring the Link between Meteorological Drought and Streamflow: Effects of Climate-catchment Interaction. *Water Resour. Res.* **2014**, *50*, 2468–2487. [CrossRef]
9. Jörg-Hess, S.; Griessinger, N.; Zappa, M. Probabilistic Forecasts of Snow Water Equivalent and Runoff in Mountainous Areas. *J. Hydrometeorol.* **2015**, *16*, 2169–2186. [CrossRef]
10. Cheraghalizadeh, M.; Ghameshlou, A.N.; Bazrafshan, J.; Bazrafshan, O. A Copula-Based Joint Meteorological–Hydrological Drought Index in a Humid Region (Kasilian Basin, North Iran). *Arab. J. Geosci.* **2018**, *11*, 300. [CrossRef]
11. Huang, S.; Li, P.; Huang, Q.; Leng, G.; Hou, B.; Ma, L. The Propagation from Meteorological to Hydrological Drought and Its Potential Influence Factors. *J. Hydrol.* **2017**, *547*, 184–195. [CrossRef]
12. Zhu, J.; Zhou, L.; Huang, S. A Hybrid Drought Index Combining Meteorological, Hydrological, and Agricultural Information Based on the Entropy Weight Theory. *Arab. J. Geosci.* **2018**, *11*, 91. [CrossRef]
13. Gumus, V.; Algin, H.M. Meteorological and Hydrological Drought Analysis of the Seyhan– Ceyhan River Basins, Turkey. *Meteorol. Appl.* **2017**, *24*, 62–73. [CrossRef]
14. Huang, S.; Huang, Q.; Leng, G.; Liu, S. A Nonparametric Multivariate Standardized Drought Index for Characterizing Socioeconomic Drought: A Case Study in the Heihe River Basin. *J. Hydrol.* **2016**, *542*, 875–883. [CrossRef]
15. Tijdeman, E.; Barker, L.J.; Svoboda, M.D.; Stahl, K. Natural and Human Influences on the Link between Meteorological and Hydrological Drought Indices for a Large Set of Catchments in the Contiguous United States. *Water Resour. Res.* **2018**, *54*, 6005–6023. [CrossRef]
16. Wu, J.; Chen, X.; Gao, L.; Yao, H.; Chen, Y.; Liu, M. Response of Hydrological Drought to Meteorological Drought under the Influence of Large Reservoir. *Adv. Meteorol.* **2016**, *2016*, 2197142. [CrossRef]
17. Wong, G.; Van Lanen, H.A.J.; Torfs, P. Probabilistic Analysis of Hydrological Drought Characteristics Using Meteorological Drought. *Hydrol. Sci. J.* **2013**, *58*, 253–270. [CrossRef]

18. Sattar, M.N.; Lee, J.-Y.; Shin, J.-Y.; Kim, T.-W. Probabilistic Characteristics of Drought Propagation from Meteorological to Hydrological Drought in South Korea. *Water Resour. Manag.* **2019**, *33*, 2439–2452. [CrossRef]
19. Abbas, A.; Waseem, M.; Ullah, W.; Zhao, C.; Zhu, J. Spatiotemporal Analysis of Meteorological and Hydrological Droughts and Their Propagations. *Water* **2021**, *13*, 2237. [CrossRef]
20. Waseem, M.; Khurshid, T.; Abbas, A.; Ahmad, I.; Javed, Z. Impact of Meteorological Drought on Agriculture Production at Different Scales in Punjab, Pakistan. *J. Water Clim. Chang.* **2022**, *13*, 113–124. [CrossRef]
21. Ahmed, K.; Shahid, S.; Harun, S.B.; Wang, X. Characterization of Seasonal Droughts in Balochistan Province, Pakistan. *Stoch. Environ. Res. Risk Assess.* **2016**, *30*, 747–762. [CrossRef]
22. Sheikh, M.M. Drought Management and Prevention in Pakistan. In Proceedings of the COMSATS 1st Meeting on Water Resources in the South: Present Scenario and Future Prospects, Islamabad, Pakistan, 1–2 November 2001; Volume 1.
23. Ahmed, K.; Shahid, S.; Nawaz, N. Impacts of Climate Variability and Change on Seasonal Drought Characteristics of Pakistan. *Atmos. Res.* **2018**, *214*, 364–374. [CrossRef]
24. Ahmed, K.; Shahid, S.; Wang, X.; Nawaz, N.; Khan, N. Spatiotemporal Changes in Aridity of Pakistan during 1901–2016. *Hydrol. Earth Syst. Sci.* **2019**, *23*, 3081–3096. [CrossRef]
25. Zahid, M.; Rasul, G. Frequency of Extreme Temperature and Precipitation Events in Pakistan 1965–2009. *Sci. Int.* **2011**, *23*, 313–319.
26. Usman, M.; Nichol, J.E. A Spatio-Temporal Analysis of Rainfall and Drought Monitoring in the Tharparkar Region of Pakistan. *Remote Sens.* **2020**, *12*, 580. [CrossRef]
27. Adnan, S.; Ullah, K.; Gao, S.; Khosa, A.H.; Wang, Z. Shifting of Agro-climatic Zones, Their Drought Vulnerability, and Precipitation and Temperature Trends in Pakistan. *Int. J. Climatol.* **2017**, *37*, 529–543. [CrossRef]
28. Xie, H.; Ringler, C.; Zhu, T.; Waqas, A. Droughts in Pakistan: A Spatiotemporal Variability Analysis Using the Standardized Precipitation Index. *Water Int.* **2013**, *38*, 620–631. [CrossRef]
29. Lee, J.E.; Azam, M.; Rehman, S.U.; Waseem, M.; Anjum, M.N.; Afzal, A.; Cheema, M.J.M.; Mehtab, M.; Latif, M.; Ahmed, R. Spatio-Temporal Variability of Drought Characteristics across Pakistan. *Paddy Water Environ.* **2022**, *20*, 117–135. [CrossRef]
30. Shahid, M.; Cong, Z.; Zhang, D. Understanding the Impacts of Climate Change and Human Activities on Streamflow: A Case Study of the Soan River Basin, Pakistan. *Theor. Appl. Climatol.* **2018**, *134*, 205–219. [CrossRef]
31. Shahid, M.; Dumat, C.; Khalid, S.; Schreck, E.; Xiong, T.; Niazi, N.K. Foliar Heavy Metal Uptake, Toxicity and Detoxification in Plants: A Comparison of Foliar and Root Metal Uptake. *J. Hazard. Mater.* **2017**, *325*, 36–58. [CrossRef]
32. Muhammad, W.; Muhammad, S.; Khan, N.M.; Si, C. Hydrological Drought Indexing Approach in Response to Climate and Anthropogenic Activities. *Theor. Appl. Climatol.* **2020**, *141*, 1401–1413. [CrossRef]
33. Liu, C.; Yang, C.; Yang, Q.; Wang, J. Spatiotemporal Drought Analysis by the Standardized Precipitation Index (SPI) and Standardized Precipitation Evapotranspiration Index (SPEI) in Sichuan Province, China. *Sci. Rep.* **2021**, *11*, 1280. [CrossRef] [PubMed]
34. Qin, Y.; Yang, D.; Lei, H.; Xu, K.; Xu, X. Comparative Analysis of Drought Based on Precipitation and Soil Moisture Indices in Haihe Basin of North China during the Period of 1960–2010. *J. Hydrol.* **2015**, *526*, 55–67. [CrossRef]
35. Zhao, P.; Lü, H.; Wang, W.; Fu, G. From Meteorological Droughts to Hydrological Droughts: A Case Study of the Weihe River Basin, China. *Arab. J. Geosci.* **2019**, *12*, 364. [CrossRef]
36. Nalbantis, I.; Tsakiris, G. Assessment of Hydrological Drought Revisited. *Water Resour. Manag.* **2009**, *23*, 881–897. [CrossRef]
37. Waseem, M.; Ahmad, I.; Mujtaba, A.; Tayyab, M.; Si, C.; Lü, H.; Dong, X. Spatiotemporal Dynamics of Precipitation in Southwest Arid-Agriculture Zones of Pakistan. *Sustainability* **2020**, *12*, 2305. [CrossRef]
38. Waseem, M.; Ajmal, M.; Kim, T.-W. Improving the Flow Duration Curve Predictability at Ungauged Sites Using a Constrained Hydrologic Regression Technique. *KSCE J. Civ. Eng.* **2016**, *20*, 3012–3021. [CrossRef]
39. Waseem, M.; Ajmal, M.; Ahmad, I.; Khan, N.M.; Azam, M.; Sarwar, M.K. Projected Drought Pattern under Climate Change Scenario Using Multivariate Analysis. *Arab. J. Geosci.* **2021**, *14*, 544. [CrossRef]
40. Fu, Q.; Johanson, C.M.; Warren, S.G.; Seidel, D.J. Contribution of Stratospheric Cooling to Satellite-Inferred Tropospheric Temperature Trends. *Nature* **2004**, *429*, 55–58. [CrossRef]
41. He, Y.; Wang, F.; Mu, X.; Yan, H.; Zhao, G. An Assessment of Human versus Climatic Impacts on Jing River Basin, Loess Plateau, China. *Adv. Meteorol.* **2015**, *2015*, 478739. [CrossRef]
42. Wu, J.; Chen, X.; Yao, H.; Gao, L.; Chen, Y.; Liu, M. Non-Linear Relationship of Hydrological Drought Responding to Meteorological Drought and Impact of a Large Reservoir. *J. Hydrol.* **2017**, *551*, 495–507. [CrossRef]

Article

Analysis of the Seismic Properties for Engineering Purposes of the Shallow Subsurface: Two Case Studies from Italy and Croatia

Federico Da Col *, Flavio Accaino, Gualtiero Böhm and Fabio Meneghini

Istituto Nazionale di Oceanografia e di Geofisica Sperimentale-OGS, 34010 Sgonico, Italy; faccaino@ogs.it (F.A.); gbohm@ogs.it (G.B.); fmeneghini@ogs.it (F.M.)
* Correspondence: fdacol@ogs.it

Featured Application: Civil Engineering, Seismic Characterization.

Abstract: We present two case studies of the application of seismic surveys to estimate the elastic properties of soil and rock in the shallow subsurface. The two sites present very different geological characteristics. The first test site is a town on the Croatian coast, not far from the city of Split, built on hard rock, where we acquired three seismic lines. The second site is located in the outskirts of the city of Ferrara, in Italy, in an alluvial plain, where two lines were acquired. In both sites, for detailed characterization, we acquired surface-, compressional- and shear-waves, further distinguishing the latter between horizontally (SH) and vertically (SV) polarized wavefields. We processed the data by performing a Multichannel Analysis of Surface Waves to compute a preliminary one-dimensional shear wave velocity profile. Then, we performed first-break tomography to compute P-, SH- and SV-velocity profiles. Such unusual acquisition allowed us to compute not only basic engineering parameters such as the equivalent shear-wave velocity of the first 30 m of subsurface (V_{S30}) from the SH profiles but also other useful parameters such as the V_P/V_S and estimate the anisotropy of the medium thanks to the V_{SV}/V_{SH}. Given the level of detail of the results and their engineering value, we conclude that the method of investigation we applied in the two test sites is a valuable tool for characterizing the shallow subsurface.

Keywords: geophysical surveying; seismic tomography; geotechnical characterization

---

Citation: Da Col, F.; Accaino, F.; Böhm, G.; Meneghini, F. Analysis of the Seismic Properties for Engineering Purposes of the Shallow Subsurface: Two Case Studies from Italy and Croatia. *Appl. Sci.* **2022**, *12*, 4535. https://doi.org/10.3390/app12094535

Academic Editors: Andrea Chiozzi, Elena Benvenuti and Željana Nikolić

Received: 30 March 2022
Accepted: 21 April 2022
Published: 29 April 2022

Publisher's Note: MDPI stays neutral with regard to jurisdictional claims in published maps and institutional affiliations.

1. Introduction

Geotechnical characterization of the shallow geological structures of a site is an essential tool to evaluate the response of the terrain to a macroseismic event. Typically, the geotechnical properties can be estimated by analyzing samples in a laboratory [1] or with in situ measurements such as the core penetration test and the load-bearing test.

However, these have the main disadvantage that they only provide a punctual estimation of the parameters. For this reason, geophysical surveys are becoming increasingly popular in seismic engineering thanks to their ability to accurately compute the main geotechnical parameters of the terrain and therefore its ability to amplify the seismic waves over a relatively wide area [2]. Specifically, the seismic method is able to evaluate the seismic velocities of the medium, which are a proxy for its elastic moduli [3]. More precisely, the seismic velocities are related to the bulk and shear moduli of the terrain, which describe the stiffness of the medium and therefore its ability to amplify seismic waves [4]. Based on this, in the past several decades, seismologists have established the vital importance of near-surface shear-wave velocity characterization both from a theoretical [5] and observational point of view [6]. Legislation followed these studies, and currently, the most commonly used geotechnical parameter is the equivalent shear-wave velocity of the first 30 m of subsoil (V_{S30}) [7–9]. The most widely used geophysical method to compute V_{S30} is the Multichannel Analysis of Surface Waves (MASW) [10]. Technically, MASW relies on the dispersive behavior of the surface waves and consists in extracting a dispersion curve from a recorded seismogram and inverting it, obtaining a 1D shear-wave velocity profile. Yust et al. (2018) [11] show that the method is able to estimate accurately V_s profiles and therefore compute the V_{S30} with a relatively small error. Another method that relies on the surface waves is the Horizontal to Vertical Spectral Ratio (HVSR, [12]). The method consists in processing the data from 3C single-station data by computing the ratio between the horizontal and vertical spectra and thereby extracting a 1D shear-wave velocity profile together with the resonant frequency of the terrain.

A method that is gaining popularity is the travel-time tomography of first breaks on an SH active seismic survey [13,14]. The main advantage of this method is that it provides a 2D (potentially even 3D) S-wave velocity profile and therefore a computation of the V_{S30} along the entire seismic line.

A more accurate characterization of the shallow subsurface, beyond the computation of the V_{S30}, can be of high interest for engineers. For instance, from the computation of Poisson's ratio or V_P/V_S, information regarding the fracturing of the rocks [1] and its fluid saturation [15] can be inferred.

In this work, we present the application of an innovative survey technique in two case studies from two very different sites. The first was recorded along the roads of a small town lying on hard rock in the area of Split in Croatia. The second case study consists of a dataset recorded in an alluvial plain in an intensively cultivated countryside close to the city of Ferrara. The novelty of the method proposed lies in the fact that we recorded separately compressional- and shear-wavefields, distinguishing in the latter between the components orthogonal (SH) and parallel (SV) to the seismic line. In fact, while integrating P- and SH-waves is increasingly popular in the near-surface geophysics community [16,17], the distinction between SH- and SV-wavefields is much rarer. Seismologists performed several numerical simulations [18] and analyzed field data [19] aiming at the study of shear-wave splitting and anisotropy, but hardly any field example using controlled sources can be found in the literature.

In the same surveys, data specifically for surface-wave analysis was also recorded. The final aim of our analyses is to provide detailed engineering information of the shallow subsurface without invasive tests at a relatively large scale. Successful testing of the method in two such different locations should allow us to confirm whether the method is replicable in most situations.

2. Geological Context

The Croatian site lies in the village of Kaštela Kambelovac, which is a town lying on the shores of the homonymous bay, just northwest of the city of Split. Geologically, the Bay lies in a compressional environment [20], where the rocks are subject to strong thrusting and folding. This caused the thrusting of the Senonian limestone on top of the Eocene flysch [21,22]. The town therefore is built on top of such flysch, the thickness of which has been estimated to be several hundreds of meters and the composition of which is known to be mainly of alternating layers of marls and calcarenites/calcirudites [23]. Previous geological and geotechnical investigations in the area showed that the bedding of the flysch is subvertical because of the tectonic compressional context. More detailed geological information about the Croatian site can be found in [24]

As for the Italian site, the acquisition took place in the southeastern suburbs of the city of Ferrara, in the central Po Valley, an alluvial, subsiding plain. Specifically, the area where the studies were carried out lies at the heart of such a plain, filled with olocenic sediments of marine, lagunal and riverine origin [25]. These are horizontally layered and composed of sand mixed with clay and silt, the proportion of which is variable both depending on the location and on the depth. These sediments are often not consolidated, giving place to phenomena of liquefaction during macroseismic events [26]. At the base of the olocenic sediments, located at approximately 1000 m depth, lie the so-called "Ferrara folds", pre-pleistocene rocks folded by the Appennine orogeny ("Buried Appennine" [27]). In fact, tectonically, the area lies at the margin of the compressional environment that created the Appennine mountain range.

3. Data Acquisition

In both sites, we acquired three wavefields separately (P, SH and SV). To do this, we used as source a wheelbarrow-mounted vibrator, which is capable of generating both P- and S-waves, reorienting it orthogonal and parallel to the line to generate SH- and SV-waves, respectively. As for the receivers, we used 10 Hz vertical geophones to record the P-waves and 14 Hz horizontal geophones to record the S-waves. The latter were re-oriented to record SH and SV, respectively, orthogonal and parallel to the survey line. We vibrated twice at each shot point, to perform stacking of the seismograms and therefore increase the signal–noise ratio.

3.1. Ferrara Site

We acquired two high-resolution seismic lines, the location of which can be seen in Figure 1a, while the UTM coordinates of the two lines are reported in Table 1. For both lines, we deployed 150 active channels, spaced every 2 m, and we shot every 4 m. The length of both lines is therefore approximately 300 m.

Figure 1. (**a**) Location of the lines acquired in the area of Ferrara, Italy. (**b**) Location of the lines acquired in Kaštela, Croatia. Images created on Google Earth Pro on 30 March 2022.

Table 1. UTM coordinates of the first and last receiver of each acquired seismic line.

Line Name	First Rec. N	First Rec. E	Last Rec. N	Last Rec. E	UTM Zone
Ferrara Line 1	4964800	713828	4964503	713850	32
Ferrara Line 2	4964552	712300	4964258	712347	32
Kastela Line 1	4822696	611466	4822770	611167	33
Kastela Line 2	4822643	612232	4822660	611943	33
Kastela Line 3	4822683	611984	4822703	611821	33

Along Line 1 (highlighted in red in Figure 1a), we acquired surface-wave data by deploying 48 4.5 Hz receivers, which were spaced every 4 m. As a source, we used a 100 kg weight-drop. We shot four times: twice at the beginning and twice at the end of the line.

The data of Line 1 were acquired along a gravel road. The main sources of noise were the agricultural activities ongoing in the nearby fruit farms (tractors, walking, irrigation pumps) and the car traffic.

The data of Line 2 were acquired in a fruit farm along a line of pear trees. Therefore, the geophones were planted in the soft soil. The main sources of noise are, similarly to Line 1, the activities going on in the fruit farm, including tractors and irrigation pumps.

3.2. Kaštela Site

We acquired three seismic lines along the roads of the village, planting the sensors in holes drilled in the tarmac. The details of the survey are outlined in [14], and the location of the lines can be seen in Figure 1b, while the UTM coordinates are reported in Table 1. Line 1 and 2 are 300 m long, while Line 3 is 150 m long. The receiver spacing in all lines is 2 m, and the shot spacing is 4 m. It is to be noted that during the acquisition of Line 1 and Line 2, traffic was quite intense in the village, and the weather was windy and even rainy at times. On the other hand, Line 3 was less affected by traffic noise, and the weather was fine.

4. Processing Methods

Multichannel Analysis of Surface Waves [10] provided information only on the Ferrara site, as no dispersive event could be identified in the Kaštela dataset, which was probably due the presence of tubes or other voids just below the road.

In Figure 2a, we show the frequency-wavenumber (f–k) spectrum of the Ferrara dataset, from which we picked the most energetic linear (i.e., dispersive) event to compute the dispersion curve shown in Figure 2b. The frequency range is from 3.5 to 15 Hz, giving a maximum wavelength of 58 m, i.e., a penetration depth of 29 m ($\lambda/2$, where λ is the wavelength). Finally, we inverted such a dispersion curve using the neighborhood algorithm in the GeoPsy software [27]. The initial model consists of three layers + Halfspace, the properties of which are described in Table 2. One million profiles were computed; in Figure 2c, we show the 500 best—that is, those generating a synthetic dispersion curve having the lowest misfit with respect to the picked one.

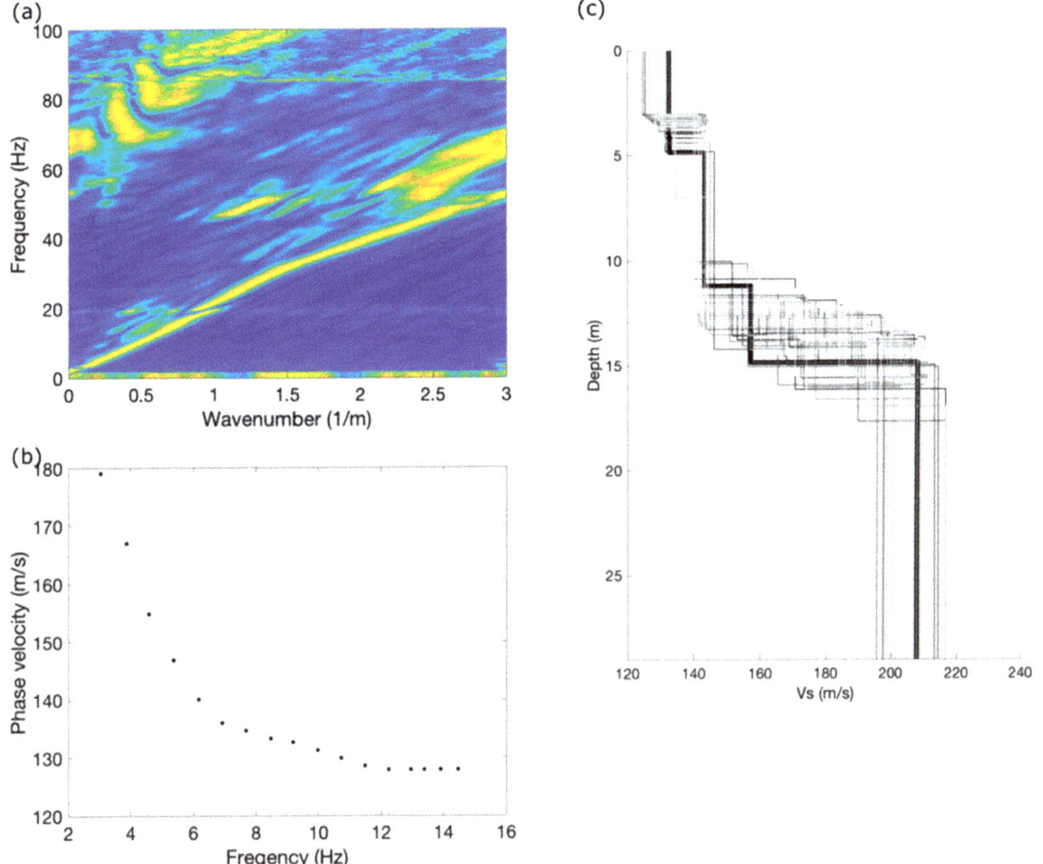

Figure 2. (**a**) f-k spectrum. (**b**) Dispersion curve. (**c**) The 1000 best fitting S-wave velocity profiles. The color of the lines represents the misfit, as indicated in the color bar.

Table 2. Properties of the initial model used for the inversion of the surface waves.

Layer Number	Thickness	Vs	Density
1	3–15 m	100–200 m/s	
2	3–15 m	120–220 m/s	2000 kg/m^3
3	3–15 m	140–240 m/s	
Halfspace		150–250 m/s	

We performed first break tomography on all the lines for all of the wavefields. Before this, some pre-processing was performed. Specifically, to increase the signal-to-noise ratio, for each shot, we applied a predictive deconvolution to the uncorrelated signals as in [28]. Furthermore, to remove random noise, we stacked in the time domain the seismograms relative to the same shot point.

The first breaks were manually picked on the seismograms. Examples of first-break picking can be seen in Figure 3.

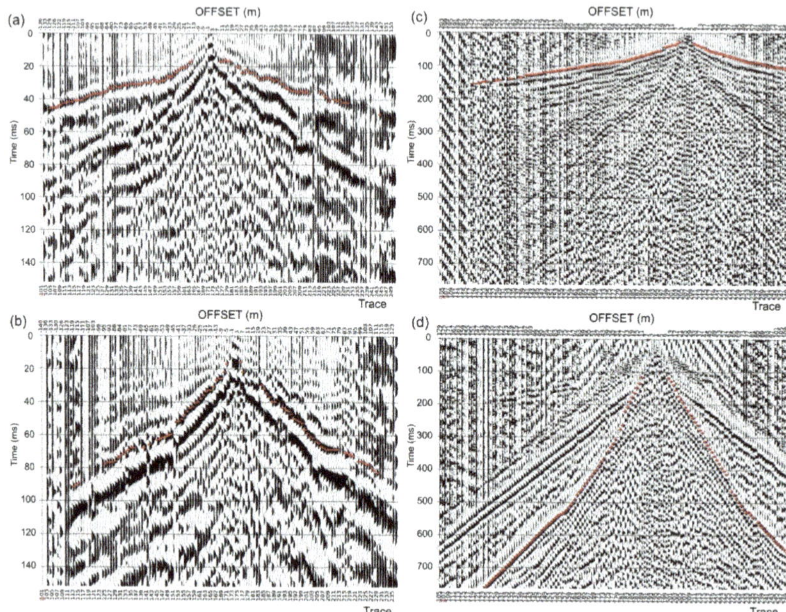

Figure 3. Examples of first-break picking on common shot gathers. Highlighted in red are the picked travel times. (**a**) P-waves from the Croatian site. (**b**) SH-waves from the Croatian site. (**c**) P-waves from the Italian site. (**d**) SH-waves from the Italian site.

Then, the travel times were inverted with Cat3D software [29], which is a tomographic package developed at the National Institute of Oceanography and Applied Geophysics-OGS that uses a ray-tracing algorithm based on a minimum time principle and SIRT (Simultaneous Iterative Reconstruction Technique) as an inversion method [30].

5. Results and Discussion

In Figure 4, we show the velocity profiles from the tomographic inversion of all wavefields relative to the three lines acquired in Kaštela, while in Figure 5, we show those relative to the two lines acquired in Ferrara. The two velocity profiles show very different features. In Kaštela, below a slower weathered layer, the velocity rapidly increases to 4.5 km/s and 3.0 km/s for P- and S-waves, respectively. Furthermore, we observe an overall increase in the velocities in the eastern lines (Lines 2 and 3) compared to the western line 1. In Figure 4d–f, we show the SH-velocity profiles and the corresponding V_{S30}. Since the velocity is >800 m/s along all three profiles, we conclude that the soil can be classified as A-class following the Eurocode-8 provisions [9]. In the bottom-right corner of each velocity profile, the mean RMS error is reported. All profiles present a mean RMS lower than 5%, except for the P-wave profile of line 2. This is probably due to the low signal-to-noise ratio of the data caused by the rainfall during the acquisition. Having said this, all other profiles show low mean RMS and can therefore be considered reliable

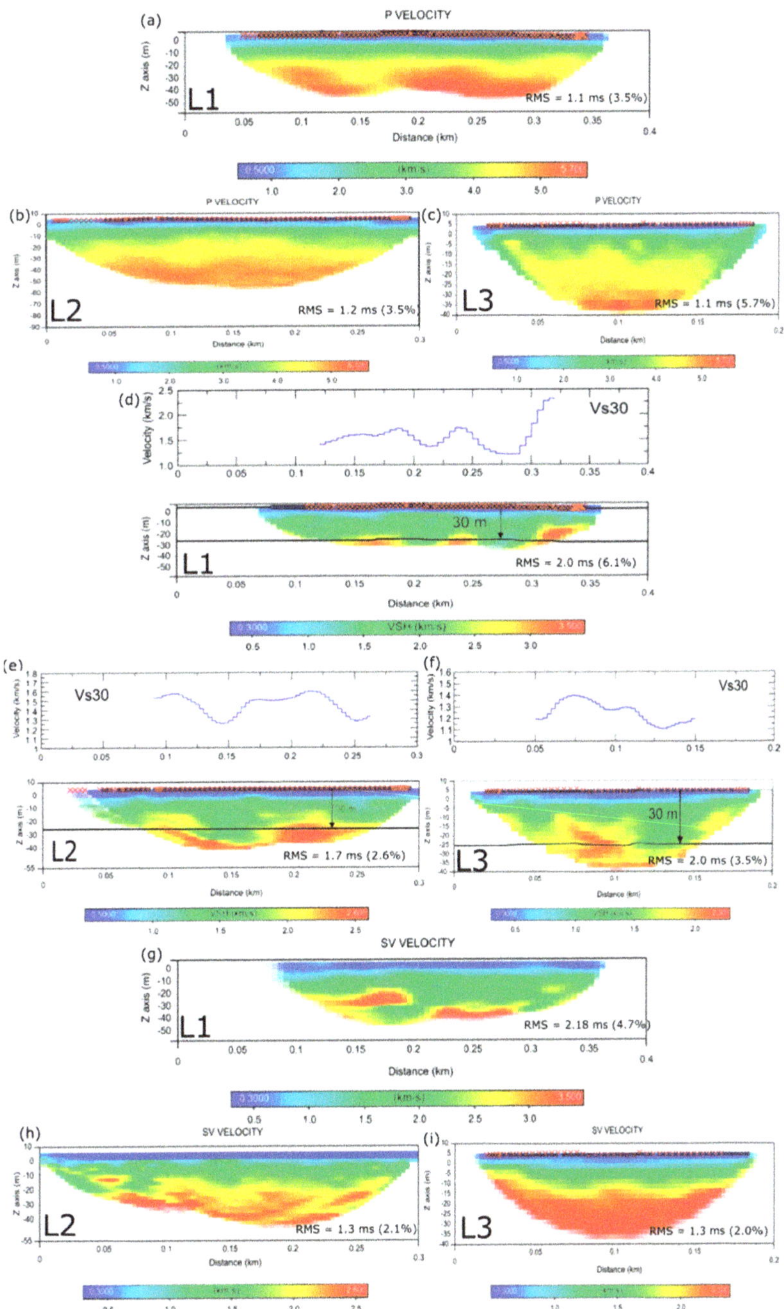

Figure 4. Results of the tomographies in Croatia; in the bottom-right corner, the value of the mean RMS error is shown. (**a**) Line 1, P-waves, (**b**) Line 2, P-waves, (**c**) Line 3, P-waves, (**d**) Line 1, SH-waves and V_{S30}, (**e**) Line 2 SH-waves and V_{S30}, (**f**) Line 3 SH-waves and VS30, (**g**) Line 1 SV-waves, (**h**) Line 2 SV-waves, (**i**) Line 3 SV-waves.

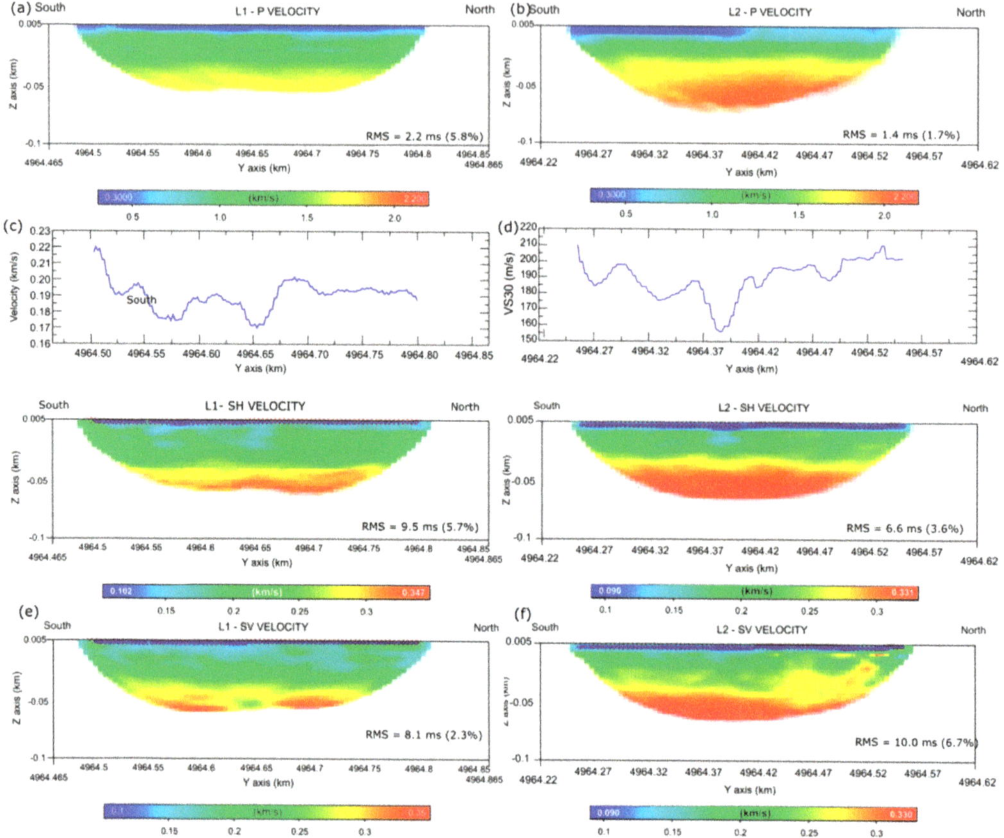

Figure 5. Velocity profiles from the Italian site; in the bottom-right corner, the value of the mean RMS error is shown. (**a**) Line 1 P-waves, (**b**) Line 2 P-waves, (**c**) Line 1 SH-waves with V_{S30}, (**d**) Line 2 SH-waves with V_{S30}, (**e**) Line 1 SV-waves, (**f**) Line 2 SV-waves.

In Figure 4, we show the velocity profiles relative to all wavefields for the two lines acquired in Ferrara. We observe a horizontal layering of the sediments, which is typical of alluvial plains. The highest velocities are reached in Line 2, where they exceed 2 km/s and 0.3 km/s for P- and S-waves, respectively. The V_{S30} is shown in Figure 4c,d, and it shows values for which the soil can be identified as C-class ($V_{S30} > 180$ m/s) following the Eurocode-8 provisions [9], matching the estimates performed in the area by previous large-scale studies [31]. This value and the velocity profile at the center of the line (where the 1D S-wave velocity profile from the MASW is ideally located) is compatible with the results of the inversion of the surface waves, confirming the reliability of our results. The sharp transition at approximately 3 m depth from blue to green in both P-wave velocity profiles most likely indicates the water table. This is consistent with previous geotechnical investigations in the area [32]. Similarly to Figure 4, the mean RMS error is shown in the bottom-right corner of each velocity profile. The values are quite variable, depending on the signal-to-noise ratio of the datasets, which depended on the amount of activities ongoing in the fields. Having said this, they usually show quite a low value, in the range of 5%, and therefore, the velocity profiles can be considered reliable.

Further information about the sites can be obtained by computing the V_P/V_S ratio (proxy for Poisson's ratio). In Figure 6a–c, we show the V_P/V_S profiles relative to the three lines acquired in Croatia, while in Figure 6d,e, we show those relative to the lines acquired in Italy. The profiles acquired in Croatia show an overall increase in the V_P/V_S ratio in the eastern lines (especially Line 3) compared to the western line. This indicates a more fractured medium that is possibly saturated with water. As for the Italian site, the V_P/V_S ratio is very high in both profiles, reaching a maximum in Line 2. The sharp increase in both lines at a depth of approximately 3 m confirms the position of the water table. Furthermore, such high V_P/V_S probably indicate the presence of water-saturated unconsolidated clay sediments.

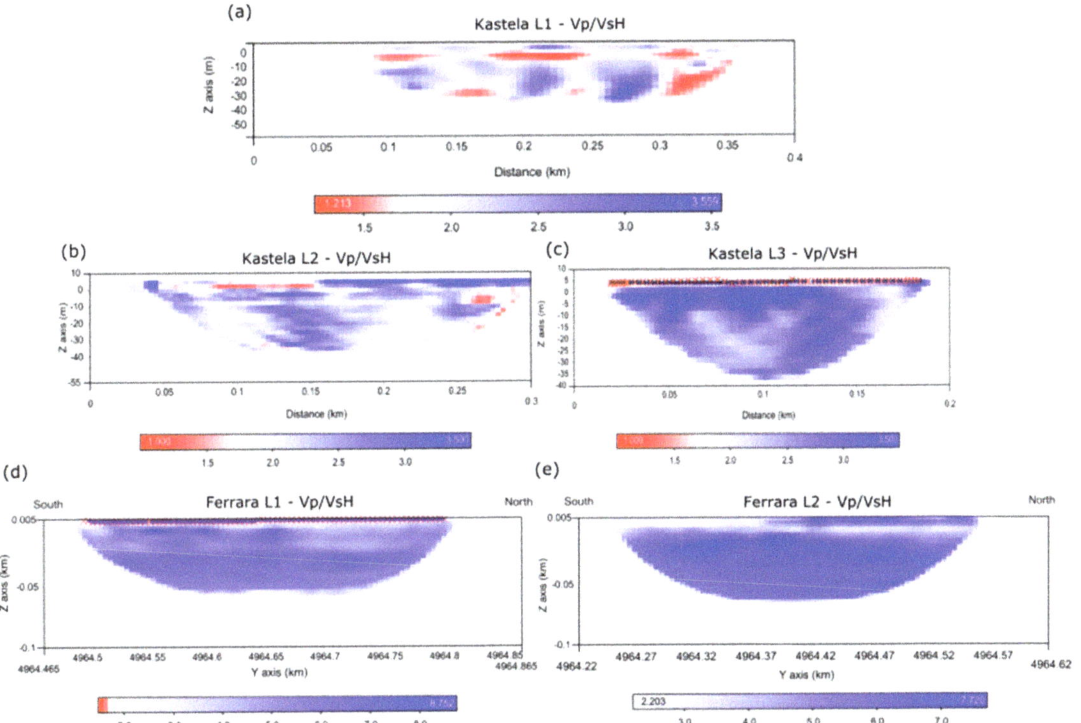

Figure 6. V_P/V_{SH} (**a**) Kaštela Line 1. (**b**) Kaštela Line 2. (**c**) Kaštela Line 3. (**d**) Ferrara Line 1. (**e**) Ferrara Line 2.

In Figure 7, we show profiles of the V_{SV}/V_{SH}. In the Croatian site, shown in Figure 7a–c, the presence of anisotropy is due to the layering of the Eocene flysch [14]. From the geometrical considerations applied to the raypaths, it is possible to estimate the bedding angles (dip and strike), as shown in [14]. As for the Italian site, shown in Figure 7d,e the presence of anisotropy confirms that the sediments are composed of horizontal layers of clay and silt, which are known for producing anisotropy [33,34]

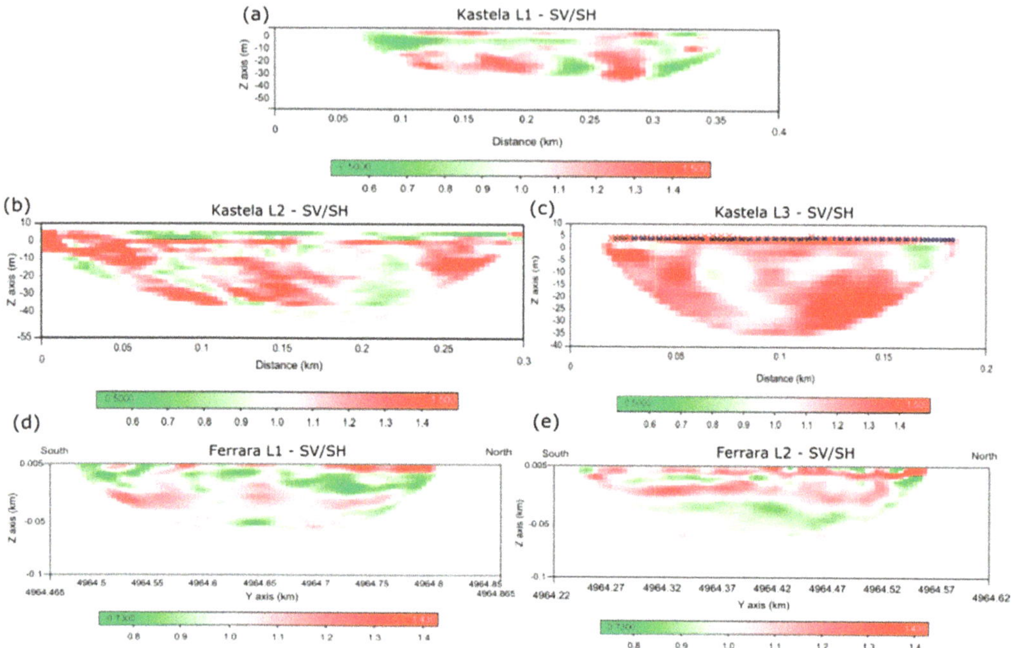

Figure 7. V_{SV}/V_{SH} (**a**) Kaštela Line 1. (**b**) Kaštela Line 2. (**c**) Kaštela Line 3. (**d**) Ferrara Line 1. (**e**) Ferrara Line 2.

6. Conclusions

We presented two case studies from two very different sites in order to validate the method in such different circumstances. The first consisted of three high-resolution seismic lines acquired along the roads of a village built on hard rock in Croatia. The second consisted of two high-resolution seismic lines acquired in a cultivated area in northern Italy. In both cases, we acquired P-waves as well as S-waves, the latter in two different polarizations: parallel (SV) and orthogonal (SH) to the seismic line. Furthermore, surface-waves were acquired in the two sites. However, in Croatia, no dispersive event could be identified and therefore no results could be obtained. This highlights the importance of acquiring SH-waves. We then performed first break tomography on all lines, for all wavefields.

The results output by the presented survey method are of high value for engineering purposes, as well as for the overall geological and geotechnical characterization of the investigated areas. Specifically, from an engineering point a view, the sites were classified based on their V_{S30} as required by the current legislation. The method allows to investigate also other important properties of the subsoil, such as the V_P/V_S and the anisotropy of the medium. From the former, it is possible to infer the fracturing of the medium and its water saturation (i.e., the position of the water table). From the latter in the Croatian site, it was possible to compute the dip and strike angles of the bedding of the flysch, providing therefore a geologically valuable information, while in the Italian site, it confirmed the silty/clay composition of the sediments and their horizontal layering.

The method proposed is non-invasive, replicable and relatively cheap, as the acquisition required only 2–3 days for each seismic line. Furthermore, it provides an in-depth characterization of the composition of the subsurface up to a depth of several tens of meters without the need for expensive procedures such as drilling and coring. Finally, these parameters are given along entire profiles, a few hundred meters long, making it attractive for the extensive characterization of geologically heterogenous areas.

Given the value of the results and all the above considerations, we conclude that the method proposed is an optimal tool to perform an in-depth characterization of the shallow layers of the subsurface.

Author Contributions: Conceptualization, F.A., F.D.C. and G.B.; methodology, F.A. and G.B.; software, F.D.C. and G.B.; validation, F.D.C., G.B. and F.A.; formal analysis, G.B. and F.D.C.; investigation, F.A., F.D.C. and G.B.; resources, F.A. and F.M.; data curation, F.M.; writing—original draft preparation, F.D.C.; writing—review and editing, F.D.C.; visualization, F.D.C., G.B. and F.M.; supervision, F.A.; project administration, F.A.; funding acquisition, F.A. All authors have read and agreed to the published version of the manuscript.

Funding: The research presented in this paper has been carried out within the PMO-GATE (Preventing, Managing and Overcoming Natural-Hazards Risks to mitiGATE economic and social impact) project. Such project falls in the Interreg Italy-Croatia programme of the European Commission under project id 10046122.

Institutional Review Board Statement: Not applicable.

Informed Consent Statement: Not applicable.

Data Availability Statement: The data presented in this study are available on request from the corresponding author.

Acknowledgments: We thank the crew involved in the two data acquisitions, in the names of Stefano Maffione, Gino Cristofano, Massimo Lovo and Andrea Schleifer. Special thanks also to Zeljana Nikolic of the University of Split, who gave us important feedback on the results we obtained. A final thank to the Municipalities both of Ferrara and Kaštela, who gave important help with the logistics.

Conflicts of Interest: The authors declare no conflict of interest.

References

1. Lancellotta, R. *Geotechnical Engineering*; CRC Press: London, UK, 2008.
2. Cardarelli, E.; Cercato, M.; De Donno, G. Characterization of an earth-filled dam through the combined use of electrical resistivity tomography, P-and SH-wave seismic tomography and surface wave data. *J. Appl. Geophys.* **2014**, *106*, 87–95. [CrossRef]
3. Ansal, A.; Kurtulus, A.; Tönük, G. Seismic microzonation and earthquake damage scenarios for urban areas. *Soil Dyn. Earthq. Eng.* **2010**, *30*, 1319–1328. [CrossRef]
4. Aki, K.; Richards, P.G. *Quantitative Seismology*; W.H. Freeman: San Francisco, CA, USA, 1980; Volume 1.
5. Stein, S.; Wysession, M. *An Introduction to Seismology, Earthquakes and Earth Structure*; Blackwell Publishing Ltd.: Malden, MA, USA, 2009.
6. Borcherdt, R.D. Estimates of site-dependend response spectra for design (methodology and justification). *Earth Spectra* **1994**, *10*, 617–654. [CrossRef]
7. Borcherdt, R.D. Simplified site classes and empirical amplification factors for site-dependent code provisions: NCEER, SEADC, BSSC. In Proceedings of the Workshop on Site Response during Earthquakes and Seismic Code Provisions, Los Angeles, CA, USA, 18–20 November 1992.
8. Dobry, R.; Borcherdt, R.D.; Crouse, C.B.; Idriss, I.M.; Joyner, W.B.; Martin, G.R.; Seed, R.B. New site coefficients and site classification system used in recent building seismic code provisions. *Earthquale Spectra* **2000**, *16*, 41–67. [CrossRef]
9. Sabetta, F.; Bommer, J. Modification of the spectral shapes and subsoil conditions in Eurocode 8. In Proceedings of the 12th European Conference on Earthquake Engineering, London, UK, 9–13 September 2002.
10. Park, C.B.; Miller, R.D.; Xia, J. Multichannel analysis of surface waves. *Geophysics* **1999**, *64*, 800–808. [CrossRef]
11. Yust, M.B.S.; Cox, R.B.; Cheng, T. Epistemic Uncertainty in Vs Profiles and Vs30 Values Derived from Joint Consideration of Surface Wave and H/V Data at the FW07 TexNet Station. In *Geotechnical Earthquake Engineering and Soil Dynamics V: Seismic Hazard Analysis, Earthquake Ground Motions, and Regional-Scale Assessment*; American Society of Civil Engineers: Reston, VA, USA, 2018; pp. 387–399.
12. Nakamura, Y. A method for dynamic characteristics estimation of subsurface using microtremor on the ground surface. *Railway Tech. Res. Inst. (Q. Rep.)* **1989**, *30*.
13. Chan, J.H.; Catchings, R.D.; Goldman, M.R.; Criley, C.J. *VS30 at Three Strong-motion Recording Stations in Napa and Napa County, California—Main Street in Downtown Napa, Napa Fire Station Number 3, and Kreuzer Lane—Calculations Determined From s-Wave Refraction Tomography and Multichannel Analysis of Surface Waves (Rayleigh and Love)*; US Geological Survey: Reston, VA, USA, 2018.
14. Da Col, F.; Accaino, F.; Böhm, G.; Meneghini, F. Characterisation of shallow sediments by processing of P, SH and SV wavefields in Kaštela (HR). *Eng. Geol.* **2021**, *293*, 106336. [CrossRef]

15. Uhlemann, S.; Hagedorn, S.; Dashwood, B.; Maurer, H.; Gunn, D.; Dijkstra, T.; Chambers, J. Landslide characterization using P- and S-wave seismic refraction tomography—The importance of elastic moduli. *J. Appl. Geophys.* **2016**, *134*, 64–76. [CrossRef]
16. Chen, J.; Zelt, C.A.; Jaiswal, P. Detecting a known near-surface target through application of frequency-dependent traveltime tomography and full-waveform inversion to P- and SH-wave seismic refraction data. *Geophysics* **2017**, *82*, R1–R17. [CrossRef]
17. Wang, C.; Shi, Z.; Yang, W.; Wei, Y.; Huang, M. High-resoultion shallow anomaly characterization using cross-hole P-and S-wave tomography. *J. Appl. Geophys.* **2022**, *201*, 104649. [CrossRef]
18. Fishman, K.L.; Ahmad, S. Seismic response for alluvial valleys subjected to SH, P and Sv waves. *Soil Dyn. Earthq. Eng.* **1995**, *14*, 249–258. [CrossRef]
19. Yong-Gang, L. Seismic wave propagation in anisotropic rocks with applications to defining fractures in earth crust. In *Rock Anisotropy, Fracture and Earthquake Assessment*; Walter de Gruyter: Berlin, Germany, 2016.
20. Herak, M. *A New Concept of Geotectonics of the Dinarides*; Jugoslavenska Akademija Znanosti i Umjetnosti: Zagreb, Croatia, 1986.
21. Marincic, S.; Magas, N.; Borovic, I. *Osnovna Geoloska Karta SFRJ 1: 100.000, List Split i Pripadaju i Tumac Karte, K33-21*; Institute of Geology: Zagreb, Croatia, 1973.
22. Buljan, R.; Pollack, D.; Pest, D. Engineering geological properties of the rock mass along the Kastela Bay sewage system. *Geol. Soc. Lond.* **2006**, *740*, 467–477.
23. Babic, L.; Zupancic, J. Evolution of a river-fed foreland basin fill: The north Dalmatia flysch revisited (Eocene, outer dinarides). *Nat. Croat.* **2008**, *17*, 357–374.
24. Gargini, A.; Bondesan, M.; Pasini, M.; Messina, A.; Piccinini, L.; Zanella, A.; Oddone, E. *Supporto Tecnico Geologico-Idrologico Alla Procedura di Valutazione e Sostenibilità Ambientale per Il Nuovo Piano Regolatore del Comune di Ferrar—Zona via Bologna—Direttrice per Cona. Relazione n. 1/03.01*; Comune di Ferrara: Ferrara, Italy, 2003.
25. Fontana, D.; Lugli, S.; Marchetti Dori, S.; Caputo, R.; Stefani, M. Sedimentology and composition of sands injected during the seismic crisis of May 2012 (Emilia, Italy): Clues for source layer identification and liquefaction regime. *Sediment. Geol.* **2015**, *325*, 158–167. [CrossRef]
26. Pieri, M.; Groppi, G. *Subsurface Geological Structure of the Po Plain, Italy. Pubbl.414. P.F. Geodinamica*; C.N.R.: Milano, Italy, 1981; pp. 1–23.
27. Wathelet, M. An improved neighborhood algorithm: Parameter conditions and dynamic scaling. *Geophys. Res. Lett.* **2008**, *35*, L09301. [CrossRef]
28. Baradello, L.; Accaino, F. Vibroseis deconvolution: A comparison of pre and post correlation vibroseis deconvolution data in real noisy data. *J. Appl. Geophys.* **2013**, *92*, 50–56. [CrossRef]
29. Böhm, G.; OGS Research Group. *Cat3D. Computer Aided Tomography for 3-D Models. User Manual*; OGS: Trieste, Italy, 2014.
30. Stewart, R. Exploration Seismic Tomography: Fundamentals. In *Course Note Series*; Domenico, S.N., Ed.; SEG—Society of Exploration Geophysicists: Tulsa, OK, USA, 1993; Volume 3.
31. Condotta, M. *Progetto di Adeguamento Funzionale del Sistema Irriguo Della Valli Giralda, Gaffaro e Falce in Comune di Codigoro (FE). Relazione Geologica e Geotecnica. Elaborato 1.2*; Consorzio di Bonifica Pianura di Ferrara: Ferrara, Italy, 2011.
32. Rossi, F.; Tumiati, D.; Bassi, A.; Perelli, P. *Piano Particolareggiato di Iniziativa Pubblica—Sottozona F2—Polo Ospedaliero di Cona. Rapporto di Valutazione Ambientale del Piano Particolareggiato in Località Cona di Ferrara*; Comune di Ferrara: Ferrara, Italy, 2011.
33. Wang, Z. Seismic anisotropy in sedimentary rocks. Part 2, Laboratory tests. *Geophysics* **2002**, *67*, 1348–1672. [CrossRef]
34. Sayers, C.M.; Den Boer, L.D. The elastic anisotropy of clay minerals. *Geophysics* **2016**, *81*, C193–C203. [CrossRef]

 *applied sciences*

Article

Seismic Risk Assessment of Urban Areas by a Hybrid Empirical-Analytical Procedure Based on Peak Ground Acceleration

Željana Nikolić [1],*, Elena Benvenuti [2] and Luka Runjić [3]

1   Faculty of Civil Engineering, Architecture and Geodesy, University of Split, 21000 Split, Croatia
2   Engineering Department, University of Ferrara, 44121 Ferrara, Italy; bnvlne@unife.it
3   Projektni Biro Runjić, 21000 Split, Croatia; lrunjic.ured@gmail.com
*   Correspondence: zeljana.nikolic@gradst.hr

Abstract: The seismic risk assessment of existing urban areas provides important information for the process of seismic risk reduction in different phases of planning and emergency management. Between different large-scale assessment approaches, a vulnerability index method is often used for the first screening of the buildings and vulnerability classification. However, this method cannot fully predict the effects of a specific seismic action on buildings. This paper fully extends the scale of the settlement and properly upgrades a methodology previously proposed by authors to predict seismic damage and the risk to a restricted number of masonry buildings in the Croatian settlement Kaštel Kambelovac located along the Adriatic coast. The proposed approach is based on a hybrid empirical-analytical procedure that combines seismic vulnerability indices with critical peak ground accelerations for different limit states computed through a non-linear pushover analysis. The procedure's outcomes are the computation of a relationship linking vulnerability indices to peak ground acceleration for a series of states, corresponding to damage limitation, significant damage, and near collapse. The described methodology is used to estimate seismic risk in terms of damage and the index of seismic risk for selected return periods. The general methodology has allowed a full seismic vulnerability assessment of the whole Croatian settlement of Kaštel Kambelovac.

Keywords: seismic risk assessment; pushover analysis; vulnerability index; damage index; index of seismic risk; masonry buildings

Citation: Nikolić, Ž.; Benvenuti, E.; Runjić, L. Seismic Risk Assessment of Urban Areas by a Hybrid Empirical-Analytical Procedure Based on Peak Ground Acceleration. *Appl. Sci.* **2022**, 12, 3585. https://doi.org/10.3390/app12073585

Academic Editor: Amadeo Benavent-Climent

Received: 12 March 2022
Accepted: 30 March 2022
Published: 1 April 2022

Publisher's Note: MDPI stays neutral with regard to jurisdictional claims in published maps and institutional affiliations.

1. Introduction

The main reason for excessive human losses and material damage during a seismic event is the insufficient seismic resistance of buildings. The assessment of seismic performance of buildings in an existing urban area is a demanding task for civil engineers, especially in old cities that have been gradually growing and expanding over the course of centuries. The heterogeneous distribution of buildings with different architectural, material and structural characteristics, accompanied by different ages of buildings, material degradation over time, various structural and non-structural interventions and, generally, the lack of knowledge about the performance of the structure, lead to numerous uncertainties in the analysis of such structures. Given the complexity of the problem, the assessment of seismic vulnerability and the risk to large areas is usually performed by simplified methods.

The approaches for the evaluation of structural vulnerability can be generally classified as empirical, analytical, or hybrid. Among them, empirical methods are often used for the first screening of buildings and vulnerability classification. The vulnerability index method [1,2] and the damage probability index method [3] are the most common approaches to assess a building's vulnerability at the urban scale. Different versions of the vulnerability index method have been derived from the approach developed by the Italian Defense National Group against Earthquakes (GNDT) for the seismic vulnerability assessment of masonry and RC buildings located in historical centers [2] by calibrating the weights of vulnerability parameters using information about the damage induced from past earthquakes [4–6]. The damage probability index method predicts the damage pattern caused by the given intensity of an earthquake using different macroseismic scales [7–9]. The advantage of empirical methods is primarily to reduce the computational efforts in comparison with more complex detailed approaches. Empirical methods are based on qualitative evaluations and can be used for setting priorities in reconstruction or undertaking measures of prevention, mitigation, preparedness, and response as part of a seismic risk management.

Analytical methods aim to represent seismic vulnerability through the analysis of the mechanical behavior of the structure. These methods result in the quantitative evaluation of the seismic performance of buildings based on models of different complexity. Among them, the most detailed are the models based on non-linear methods such as non-linear static (pushover) [10] or incremental dynamic analysis [11], which are very demanding even for a single building and cannot be exploited to assess seismic vulnerability on a wider, urban scale. However, these methods can be used to derive fragility curves for certain typologies of buildings, whichcan, in turn, be used as starting points to subsequently proceed with the evaluation of seismic vulnerability and damage scenarios at the urban scale [12–20].

Finally, seismic vulnerability at the urban scale can be assessed by hybrid methods, namely methods combining empirical approaches with detailed analytical ones, and leading to the quantitative representation of the behavior of buildings under certain seismic actions.

In this regard, most state-of-the-art contributions containing assessment methodologies, and their related case studies, rely upon the concept that the levels of vulnerability and damage shall depend on the earthquake intensity [21–23]. In particular, a reference can be performed by the study [24] for a thorough discussion of the most relevant vulnerability assessment methods applicable at different scales. Essentially, the choice of the approach should depend on a series of aspects. In addition to the area of study, as settlements will require methods that cannot be used to analyse cities or entire regions, a discriminant comes from data availability about the building stock. A further important aspect is the purpose of the study, while information of the utmost importance will afford a better understanding of both the actual seismic hazard and the structural damage caused by previous earthquakes.

The study of seismic vulnerability and risk proposed in the present paper is based on the vulnerability index method derived from the original Italian GNDT approach [2]. The method provides a vulnerability index as a sum of vulnerability scores, representing a main material and structural and non-structural characteristics important for the seismic behavior of building. Thus, the vulnerability of an urban area can be represented by the vulnerability index map, which informs civil protection bodies of territorial vulnerability and contributes to the planning and managing of emergency actions.

More information about seismic capacity and risk of the buildings can be obtained by linking the vulnerability indices with the intensity of a seismic event. The knowledge about seismic capacity expressed by intensity, peak ground acceleration, or damage is especially important in the prevention of activities aimed to determine priorities in structural interventions and reconstructions. There have been several studies that established vulnerability–damage–peak ground acceleration relationships on an observational basis starting from information about the pre-existing damage levels triggered by past earthquakes [1,25]. The main problem with the application of this relationships is the scarcity of data on previous earthquakes, which are needed to calibrate the model in another area. Non-linear computational approaches, such as the static non-linear (pushover) method and the incremental dynamic analysis, can compute the critical states of the structure, both in terms of capacity represented by peak ground acceleration and associated damage. Therefore, using these non-linear analytical approaches, post earthquake damage data can be replaced with those obtained from the numerical tests. The consequence is that empirical information from the form of the vulnerability index can be linked with quantitative numerical results obtained by means of a non-linear approach. This is an important step in the calibration of the vulnerability model and in establishing relationships between the vulnerability index based on qualitative empirical estimation and the quantitative indicators of structural capacities. A few recent investigations have been performed to assess seismic capacity and/or damage based on vulnerability indices and critical peak ground acceleration obtained by pushover analyses [26,27]. One such hybrid approach has been developed for establishing seismic vulnerability in the Mediterranean urban center of Lampedusa Island in Italy [27]. The procedure combined experimental data and numerical results obtained for a class of buildings representative of the most widespread typology with the purpose of calibrating the vulnerability curves previously obtained by Guagenti and Petrini [4]. Peak ground accelerations corresponding to the life safety limit state have been also analyzed by pushover analysis and used for the seismic fragility assessment of masonry buildings [17].

The hybrid seismic risk assessment procedure adopted in the present paper combines the vulnerability index method with the non-linear pushover analysis of buildings. The methodology has been applied to the entire settlement of Kaštel Kambelovac, a small Mediterranean urban settlement along the Croatian side of the Adriatic Sea and consisting of a historical core constituted by stonemasonry. The historical core was erected between the 15th and the 19th century, while the periphery outside of the historical core includes more modern buildings. In particular, five main categories of construction data of the modern buildings have been recognized: before 1948, 1949–1964, 1964–1982, 1982–2005, and modern buildings erected from 2005 onwards. All these buildings exhibit different seismic performance depending on the period of construction and applied technical regulation.

It should be noted that there are only a few published studies focusing on the seismic behavior of buildings typical of the Adriatic coastal area. They concern experimental research about the behavior of protected buildings inside Diocletian's Palace in Split [28,29] or propose their numerical modeling using finite-discrete element models [30]. Lattice models are also being developed for the precise modeling of energy dissipation under dynamic actions [31,32] which, due to the possibility of modelling heterogeneity in materials, can be highly suitable for the numerical simulation of masonry buildings, whether of regular or irregular blocks. Although complex, all these models are not yet suitable for the analysis of the complex geometries featuring the buildings placed in the area chosen as the test site.

While the aforementioned studies analyzed single buildings in the Adriatic coastal area, a first version of the present hybrid procedure [33] was recently applied by the authors for the assessment of seismic vulnerability and damage of a limited number of stonemasonry buildings in the historical core of Kaštel Kambelovac. With respect to the previous contribution [33], where only two limit states were considered, the hybrid methodology in the present paper is fully generalized and extended to define vulnerabilitypeak ground acceleration relations for three limit states of the buildings of the entire settlement, which comprises 400 buildings. In particular, the investigation of vulnerability indices and capacities of the buildings obtained by nonlinear pushover analyses has been extended from the historical part of the test site, with stone masonry buildings, to the whole test site including more recent buildings constructed in the 20th and 21st centuries. Furthermore, a general procedure to define seismic risk in terms of damage index and seismic risk index is hereby presented. In fact, the vulnerability curves in terms of the relationships damage index, vulnerability index, and peak ground accelerations, have been generalized in order to account for the influence of the peripherical modern buildings. The damage index has been computed for three return periods for the whole test site and a damage map for convenient visualization is provided. Finally, in the present paper, an entirely new index of seismic risk has been defined and computed for three return periods for the considerable number of 111 buldings.

A massive campaign of field investigations has been purposely performed to gain a full understanding of the material and structural characteristics, including analyzing the available technical documentation and examining the influence of building codes on the design and construction of the buildings. A sample of eighteen masonry buildings representative of the typical buildings located in the test site has been analyzed by a pushover analysis. The critical peak ground accelerations for three limit states (damage limitation, significant damage, and near collapse) have been determined and used to calibrate vulnerability curves at the test site. The described methodology has been used to estimate the damage index and the index of seismic risk for the selected return periods.

2. Methodology for Seismic Risk Evaluation

The seismic risk of buildings is usually defined as a function of seismic hazard of the area, vulnerability of buildings, and exposure. The seismic hazard of the area expresses the probability of the occurrence of a certain intensity earthquake in a given area and at a certain period of time. The vulnerability of buildings represents the susceptibility of the structure to suffer damage due to a seismic event of a given intensity. The exposure measures the quality and quantity of elements exposed to the risk.

In order to completely understand vulnerability and risk of the urban area exposed to seismic action, the comprehensive procedure for seismic risk evaluation and visualization in this paper is structured as follows:

- Documentation of architectural, structural and material features by examining building codes, historical and archival sources, on-site visual inspection, and thermographic imaging;
- Creation of a database of buildings and visualization of input data by the web map in the GIS environment;
- Geophysical survey of the soil type;
- Definition of seismic hazard for the test site using available seismic hazard maps of Croatia and the results of geophysical survey;
- Seismic vulnerability assessment by vulnerability index method for the sample of the buildings;
- Extrapolation of the results for the seismic vulnerability index for the entire test site;
- Non-linear static analysis of the relevant buildings located in the test site and determination of the peak ground accelerations for damage limitation, significant damage and near collapse states;
- Development of vulnerability–peak ground acceleration curves for three limit states (damage limitation, significant damage, and near collapse) for the test site;
- Development of vulnerability curves that establish relations between damage, vulnerability and peak ground acceleration for the test site, and serve to estimate the structural damage for a given seismic action;
- Risk evaluation in terms of seismic damage for three return periods;
- Risk evaluation in terms of the index of seismic risk for three return periods;
- Visualization of hazard, vulnerability indices, damage indices and indices of seismic risk of the buildings in the web map.

3. Investigation of the Test Site

3.1. Architectural, Material, and Structural Characteristics of Buildings

The proposed method has been applied to Kaštel Kambelovac, one of the seven settlements forming the City of Kaštela (Figure 1a). The structure of each settlement from the aspect of architectural, urban and construction feature is similar. Each settlement was formed around an old historical center built between the 15th and the 19th century. The settlements gradually expanded over the years to the surrounding area. In the course of their development, the settlements merged and the entire area forms today's agglomeration of the City of Kaštela. Nowadays, the city has seven separated historical centers, each composed of stone masonry buildings, which are represented by the combination of smaller family houses, old mansions and public facilities.

(a)   (b)

Figure 1. City of Kaštela: (**a**) geographical position of the city in the Kaštela Bay; (**b**) historical center of Kaštel Kambelovac [34].

The test site of Kaštel Kambelovac (Figure 1b) consists of an old historical center dating from the 15th and the 19th century, while the peripherical buildings were built from the beginning of the 20th century to the present day (Figure 2). The relevant area includes more than 400 buildings.

Figure 2. Characteristic parts of the test site.

The masonry walls of the more ancient buildings, located in the historical center (Figure 3), are made of stone blocks and mortar joints, and are from 45 cm to 75 cm thick. The wall textures are variable: roughly shaped stone blocks of various size arranged in a haphazard way alternate to masonry consist of blocks of homogeneous size, well-shaped or cut [33]. The quality of mortar is overall poor. Floors are made of timber beams and wooden floor coverings. Confining elements are lacking, and connections between the walls and floors are generally weak. Some of these buildings were reconstructed and, astypically happens, monolithic reinforced concrete plates replaced the wooden floors.

Figure 3. Buildings in the historical center.

Outside of the historical center (the northern, eastern, and western parts shown in Figure 2), the buildings (Figure 4) were mostly made as masonry structures consisting of stone, concrete, or brick blocks, unreinforced or reinforced with RC confining elements (only with ties or with ties and columns) depending on the construction period and technical regulations.

Figure 4. Buildings outside of the historical center.

In fact, the masonry buildings constructed before 1964 are not earthquake-resistant because they have been built as unreinforced masonry structures [35]. Since 1964, seismic regulations required that all buildings have horizontal confining elements and rigid horizontal diaphragms or horizontal and vertical confining elements and rigid horizontal diaphragms depending on the seismic zone and the number of floors. After 1980, stricter regulations for the construction in earthquake areas were applied and the use of unreinforced masonry was not allowed in the areas of medium and high seismicity. In the Kaštela area, it is allowed to build a two-story masonry structure without vertical confining elements and a three-story structure with vertical and horizontal confining elements. Buildings erected from 2005 onwards are seismically resistant structures due to the application of modern design standards based on the European regulations (Eurocode 8), firstly implemented through the pre-standards (HRN ENV 1998-1:2005 [36]) and finally by introducing the full European standard (Eurocode 8) in 2011 in the Croatian national legislation (HRN EN 1998-1:2011 [37]). They are made as confined masonry.

In addition to the analysis of technical regulations, material and structural characteristics have been investigated using historical documentation and literature [38], archival documentation of the City of Kaštela, field survey by a visual inspection, and thermographic examination in a conspicuous number of cases where, due to non-documented reconstructions and external plaster covering the walls, it was impossible to identify material and structural characteristics of the building. In fact, the texture of the walls, the presence of horizontal and vertical confinement, floor covering material and roof structures, and heterogeneities of materials can be successfully detected by thermographic examination. An example of such an investigation is shown in Figure 5.

Figure 5. Thermographic examination of unreinforced stone masonry building partially covered with plaster, and significant cracks in a gable wall [39].

The mechanical properties of materials for stone masonry buildings (stone blocks, walls, and mortar) were deduced from the literature [40]. Past design rules were exploited to estimate the mechanical properties of buildings erected from 1900 to nowadays.

Additional assistance was obtained from precise geodetic map of the test site that allowed identification of planimetric dimensions, from Google Maps, Street View, as well as from a map dating back to 1968 that made it possible to identify the subsequently reconstructed sites.

3.2. Geophysical Survey of the Test Site

The characterization of soil type has been determined by a geophysical survey. A detailed description of investigation, performed in May 2019, is presented in [41]. The investigation aims to determine shear wave velocity $V_{S,30}$ of the shallow subsurface along three seismic lines in Kaštel Kambelovac. The velocity $V_{S,30}$ was calculated as the average of the V_{sH} velocities, measured from the surface to a depth of 30 m. The $V_{S,30}$ velocity, higher than 800 m/s, allowes us to classify the soil as A class according EN 1998-1:2011 [37] in all three lines. Considering that the investigated test site is relatively small, the soil type A was considered for all buildings in the test area.

4. Seismic Hazard of the Area

The seismic hazard for Croatia is presented in terms of the horizontal peak ground acceleration with two maps for the return periods of 475 and 95 years. The maps have been accepted as a part of the Croatian National Annex of HRN EN 1998-1:2011 [37]. Recently, a new hazard map for T = 225 years has been developed. According these maps, the peak ground acceleration a_g in the Kaštela area, is equal to 0.22 g, 0.17 g, and 0.11 g for the return periods of 475, 225, and 95 years, respectively, and ground type A.

The seismic hazard for soil types different from A increases. A simple engineering way to calculate the hazard for local ground conditions is by multiplying the peak ground acceleration for ground type A with the soil factor S [37] for observed location.

Considering the results of the geophysical survey which indicated the ground type A at the investigated area, the seismic hazard for all buildings at the test site has been assumed to be constant.

5. Seismic Vulnerability Assessment of the Area

5.1. Vulnerability Index Method

We perform a vulnerability assessment analysis by the vulnerability index method developed from GNDT in collaboration with the Italian National Research Council from 1984 onwards [1,2]. The present study further includes the modifications of the GNDT method proposed for the Tuscany region in [42] and considers the replacement of light timber floors with heavier RC floors, which are often used in the reconstruction of old masonry buildings. In fact, such replacement induces a significant increase in the mass at the top of the floors, consequently enhancing the overall in-plane stiffness and causing a different dynamic behavior of the modified structures [33].

The vulnerability index method is here used to calculate the vulnerability index for the building based on the calculation of 11 geometrical, structural and non-structural vulnerability parameters of the building. They consider the influence of the type and quality of the structural system, the shear resistance in two horizontal directions, the position and the foundations, the properties of floors, the configuration in plan and elavation, the maximum wall spacing, the roof's typology and weight, the existence of non-structural elements, and the state of preservation. Four possibilities for each parameter were decided: from "A", indicating an optimal state, to "D", indicating a poor state. Furthermore, the method numerically scores each option. The relative importance of each parameter in the overall vulnerability is computed by using weight coefficients relating to each parameter. Ultimately, a vulnerability index I_V is obtained as follows:

$$I_V = \sum_i s_{vi} w_i \tag{1}$$

where s_{vi} is the numerical score for each class, and w_i is the weight of each parameter. The vulnerability index is normalized in a 0–100% range; the low index indicates high seismic resistance and low vulnerability, while a high vulnerability index is characteristic of the buildings with low seismic resistance and high vulnerability.

Table 1 displays the vulnerability parameters and their weight coefficients used in this paper. The upper value of the vulnerability index I_V is 438.75.

Table 1. Vulnerability parameters and their weights.

Parameter	Score (s_{vi})				Weight (w_i)
	A	**B**	**C**	**D**	
Type and organization of the resistant system (P1)	0	5	20	45	1.50
Quality of the resistant system (P2)	0	5	25	45	0.25
Conventional resistance (P3)	0	5	25	45	1.50
Position of the building and foundation (P4)	0	5	25	45	0.75
Typology of floors (P5)	0	5	15	45	0.50–1.25
Planimetric configuration (P6)	0	5	25	45	0.50
Elevation configuration (P7)	0	5	25	45	0.50–1.00
Maximum distance among the walls (P8)	0	5	25	45	0.25
Roof (P9)	0	15	25	45	0.5–1.5
Non-structural elements (P10)	0	0	25	45	0.25
State of conservation (P11)	0	5	25	45	1.00

5.2. Application of Vulnerability Index Method at the Test Site

The vulnerability indices for 111 buildings with known architectural, structural, and material features (75 in the old city center and 35 outside of the center) were calculated by the vulnerability index method. The distribution of the vulnerability index is shown in Figure 6. The vulnerability indices were included into a web map based on the geographical information system (GIS).

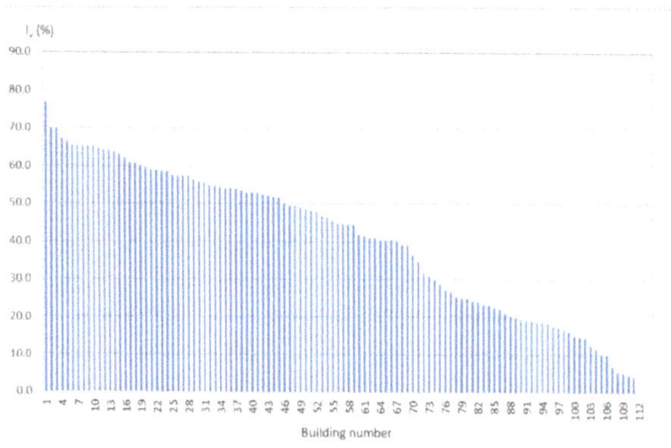

Figure 6. Distribution of I_V for a sample of 111 buildings.

The vulnerability index map is shown in Figure 7. Figure 7a shows the buildings divided into four vulnerability classes: low vulnerability for $I_V < 30$, from medium to low vulnerability for $30 < I_v < 45$, from medium to high vulnerability for $45 < I_v < 60$, and high vulnerability for $I_v > 60$. The range of the values of vulnerability indices are taken according to the GNDT vulnerability classification used in the study of reconstruction of the Municipality of Arsita following the seismic event of 6 April 2009 [2,43]. More insight into the vulnerability of buildings erected in the period 1950 to 2000 can be obtained from the division of the vulnerability index map into 10% intervals (Figure 7b). The lowest vulnerability index, equal to 2.6, belongs to a two-story house with regular layout and elevation, made as a confined masonry structure with horizontal and vertical confining elements (RC ties and RC columns) and rigid horizontal diaphragms, all designed according to EC-8. The highest vulnerability index, equal to 76.9, was obtained for the Cambi Tower, a stone masonry building dating from the 15th century. The Cambi Tower is characterized by having poorly connected walls, flexible floors, an irregular layout and elevation. A vulnerability index of 45 and beyond is ascribed to medium-high and high vulnerability buildings, respectively. Typically, high vulnerability buildings turn out being mainly located in the old city center and made of stonemasonry.

In the northern part of the settlement, which does not belong to the historical center, there is a number of stone masonry buildings. Therefore, most of the buildings belong to high, medium-high and medium-low vulnerability classes. Only two buildings belong to the low vulnerability class.

In the eastern and western parts of the test site, the buildings are built with concrete or clay blocks, with different confining methods: (1) without confinement; (2) with horizontal tie beams, (3) with horizontal and vertical confinement. They mostly belong to the low vulnerability class ($I_v < 30\%$). Within this class, however, there are visible differences in vulnerability. Newer buildings with brick blocks and horizontal and vertical confinement generally have the lowest vulnerability (less than 10%). Older buildings with concrete blocks and horizontal confinement approximately have an index between 10% and 20%. Buildings without confinement or the aforementioned ones, but irregular in elevation and/or layout and with several annexes and additions, have an index mostly between 20% and 30%.

The vulnerability indicess have been computed in detail for 111 buildings with known geometry, structural system, and types of material. The vulnerability of other buildings at the test site without the available technical documentations was determined based on the estimated geometric and structural characteristics of the building using a geodetic survey of the area, a street view map, and a visual inspection of the area. Therefore, a lower precision of vulnerability index results can be expected for these buildings.

5.3. Vulnerability Index Method as a Basis for Seismic Risk Evaluation

Although the vulnerability index is an important indicator of seismic risk, it should be noted that it is useful for a relative comparison of the seismic performance of buildings in the case of equal seismic action. According to modern seismic regulations, such as Eurocode 8 [37], seismic action varies even for constant seismic hazard of the area because it depends on the soil type and the importance factor of the building. More precise information on the behavior of the building subjected to a certain earthquake action can be obtained either by evaluating the damage or by assessing the risk induced by an earthquake of a specific intensity. The former approach establishes a relation between the vulnerability index, the intensity of seismic action and structural damage using the post-earthquake damage observations [1,25] or, alternatively, performing non-linear pushover analyses [27,33]. Seismic risk can also be expressed in terms of the index of seismic risk (or index of seismic safety), expressed as the ratio of the peak ground acceleration achieved to the structural collapse and the demand ground acceleration.

(a)

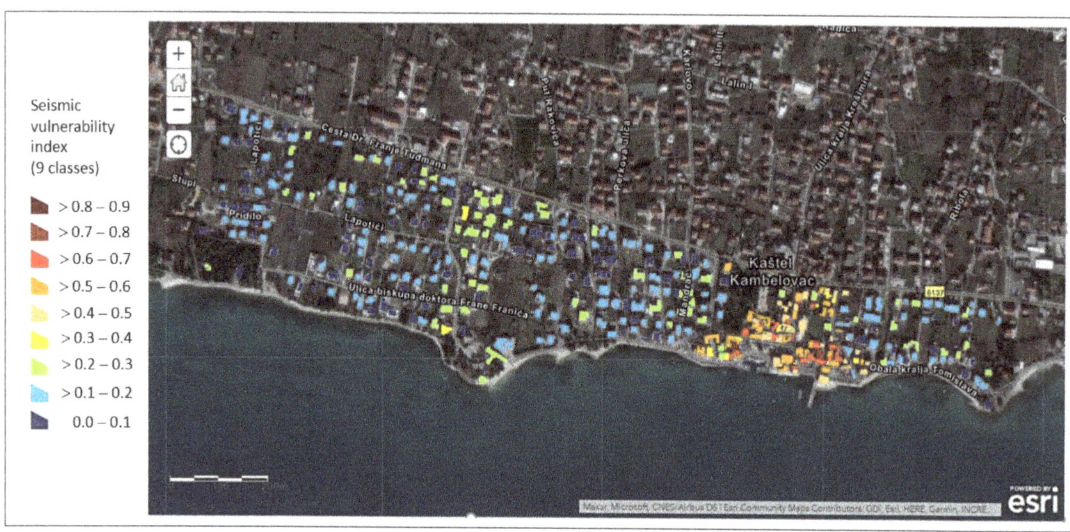

(b)

Figure 7. Vulnerability index map of the test site. (**a**) Division into four vulnerability classes. (**b**) 10% division intervals.

The present study aims to evaluate seismic risk both in terms of the damage index and the index of seismic risk using information about vulnerability of the buildings obtained by vulnerability index method. In order to calculate the damage index, the relation between the vulnerability index and the seismic capacity represented by peak ground acceleration for the sample of 18 buildings has been established. Non-linear static (pushover) analysis has been applied for the calculation of peak ground acceleration for different limit states of buildings (early damage, significant damage, and near collapse). Vulnerability index—peak ground relations enable the calculation of the damage index of the building, but also the calculation of the index of seismic risk for each limit states.

6. Evaluation of PGA Values for Specific Limit States by Pushover Analysis

6.1. Detection of Specific Limit States

A static non-linear pushover method [10,37] is used to evaluate seismic behavior and the capacity of the building for three limit state (LS) conditions that have been taken into account according to Eurocode 8, part 3 [44], as follows:

- Near collapse NC—global capacity of the building is taken to be equal to the ultimate displacement capacity;
- Significant damage SD—global capacity of the building is taken to be equal to $\frac{3}{4}$ of the ultimate displacement capacity;
- Damage limitation DL—the capacity for global assessment is defined as a yield point of the idealized elasto-perfectly plastic force-displacement relationship of the equivalent SDOF.

The pushover analysis has been performed by applying the gradually increasing load up to the structural collapse, using a uniform and modal pattern, according to the N2 method of Eurocode 8 [37]. The analysis allows the determination of capacity curves and collapse load as well as damage monitoring, which continuously increases because of the non-linear deformation.

Pushover analysis results in the MDOF capacity curve, which is transformed into the SDOF capacity curve, and a bilinear force–displacement curve is obtained. Peak ground accelerations for three mentioned limit states have been determined as follows.

We specify that henceforth, the forthcoming equations have been modified with respect to the homologous ones reported in [33] to consider three damage levels.

Given that the displacement in the yield point d_y^* of SDOF is linked with the DL state, the ultimate displacement d_u^* with the NC state and SD state with the displacement equal to $\frac{3}{4} d_u^*$, the corresponding ductilities are expressed as:

$$\mu_{DL} = \mu_y = 1 \; ; \; \mu_{SD} = \frac{\frac{3}{4} d_u^*}{d_y^*} \; ; \; \mu_{NC} = \mu_u = \frac{d_u^*}{d_y^*} \tag{2}$$

where $\mu_{DL} = \mu_y$, μ_{SD} and $\mu_{NC} = \mu_u$ represents damage limitation, significant damage, and near collapse ductility coefficients.

The associated elastic spectral displacements can be calculated as follows:

$$S_{de,i}(T^*) = \frac{d_y^* \overline{R}_\mu(\mu_i)}{[\overline{R}_\mu(\mu_i) - 1] \frac{T_c}{T^*} + 1} \; , \; i = DL, SD, NC \tag{3}$$

where $\overline{R}_\mu$ is a reduction factor depending on the ductility coefficient μ of SDOF system [35].

The spectral accelerations are given as:

$$S_{ae,i}(T^*) = \frac{4\pi}{T^{*2}} S_{de,i}(T^*) \; , \; i = DL, SD, NC \tag{4}$$

The periods T_B, T_C, and T_D divide the elastic response spectrum [37] into four spectral acceleration branches represented with the functions f_i ($i = 1, \ldots, 4$). Therefore, it can be expressed as follows:

$$S_{ae}(T) = PGA \cdot f_i(T) \tag{5}$$

where $PGA = a_g$ is peak ground acceleration and T is the period of the structure. Each limit state (DL, SD, and NC) is characterized with the following peak ground accelerations:

$$PGA_{DL} = \frac{S_{ae,DL}(T^*)}{f_i(T)} \; ; \; PGA_{SD} = \frac{S_{ae,SD}(T^*)}{f_i(T)} \; ; \; PGA_c = \frac{S_{ae,NC}(T^*)}{f_i(T)} \tag{6}$$

Eighteen buildings in the settlement were modelled using 3MURI software [45] following the equivalent frame model approach. Thus, the structural response is checked along two horizontal axes, in the positive and the negative direction. Accidental eccentricity equal to ±5% of the maximum floor dimension is considered to model the non-regular mass distribution.

Pushover analysis provides reliable results for the structures that oscillate predominantly in the first mode. In the presence of both horizontal and vertical irregularities, multi-modal non-linear static analysis can be applied [46]. Due to significant irregularities of buildings, a threefold lateral load distribution, namely uniform, linear, and modal, has been applied in this study. Considering the eccentricities in positive and negative directions, this resulted in a total of 36 analyses.

The evaluation procedure of peak ground accelerations for three limit states is shown for the Cambi Tower (Figure 8).

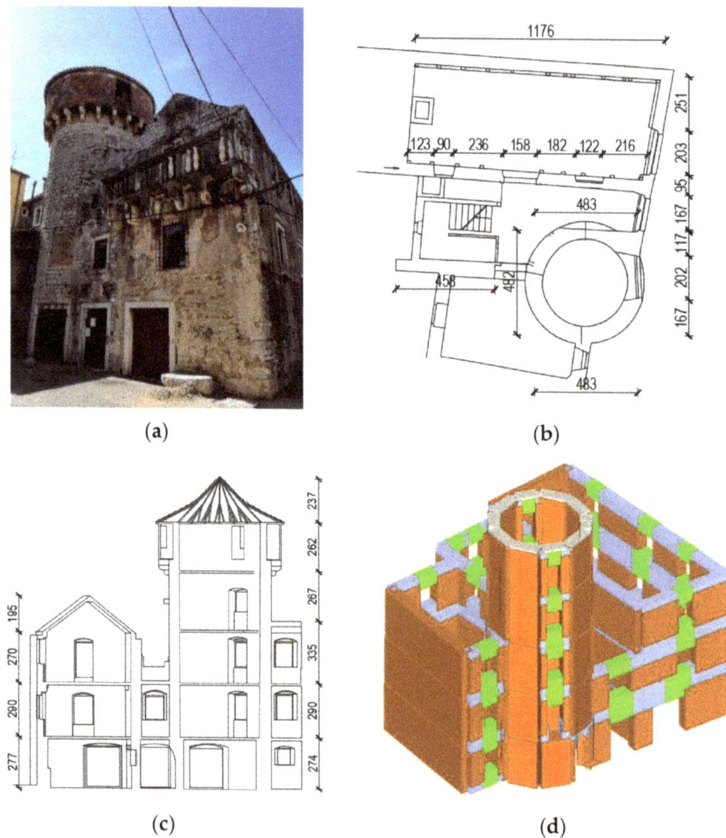

Figure 8. Cambi Tower: (**a**) photo of the building; (**b**) ground floor plan; (**c**) section view; (**d**) structural model.

The floors were modeled as flexible. The walls' mechanical properties were taken according to [10] as follows: compressive strength of 3.20 MPa, tensile strength of 0.10 MPa, modulus of elasticity of 1700 MPa, shear modulus of 580 MPa, and specific weight of 21 kN/m^3.

Seismic demand was deduced from the elastic response acceleration spectrum. A soil class A and type 1 response spectrum [37] have been adopted. Other assumptions include the importance factor $\gamma_1 = 1.2$ and the design ground acceleration $a_g = 0.22$ g. Figure 9 illustrates the results of the pushover analyses for x and y direction.

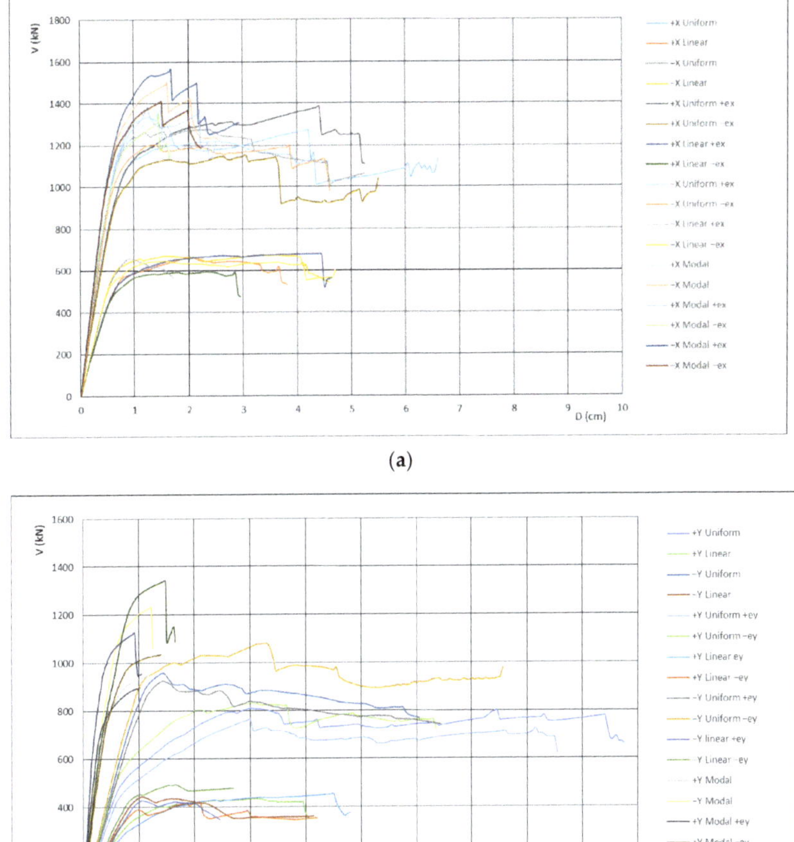

(a)

(b)

Figure 9. Pushover curves for the Cambi Tower: (**a**) x-direction; (**b**) y-direction.

Seismic capacity for two orthogonal directions was evaluated by a pushover analysis, comparing the displacement capacity and the displacement demand for the same control point. The calculation was repeated for all of the 36 loading cases.

The critical peak ground accelerations associated with the DL, SD, and NC limit states were computed as follows: (a) x direction—$PGA_{DL} = 0.093$ g $= 0.422a_g$, $PGA_{SD} = 0.116$ g $= 0.527a_g$, $PGA_{NC} = 0.147$ g $= 0.668a_g$; (a) y direction—$PGA_{DL} = 0.030$ g $= 0.136a_g$, $PGA_{SD} = 0.059$ g $= 0.268a_g$, $PGA_{NC} = 0.078$ g $= 0.355a_g$. For completeness, the design ground acceleration $a_g = 0.22$ g has been obtained based on the seismic hazard map for the return period of 475 years.

6.2. Results of Pushover Analysis of the Buildings

A static non-linear (pushover) method is used for a detailed analysis of 18 buildings at the test site: 10 stone masonry buildings in the historical center (Figure 10) and 8 masonry buildings outside of the historical center (Figure 4).

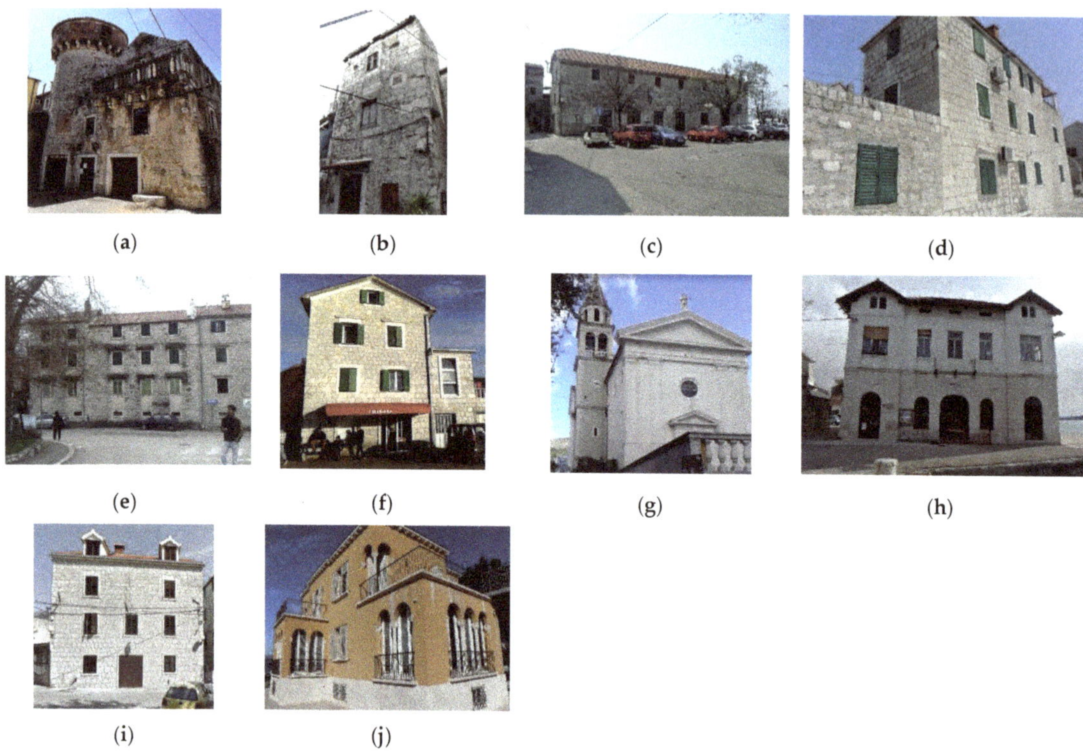

Figure 10. Analyzed buildings in the historical centre: (**a**) Cambi Tower; (**b**) Cumbat Towers; (**c**) Public Library; (**d**) Folk Castle; (**e**) Dudan Palace; (**f**) Perišin house; (**g**) St. Mihovil Church; (**h**) rowing club; (**i**) residential building; (**j**) ballet school.

The buildings in the historical center have different floor plans and height configurations, dimensions, and number of floors. Many have been upgraded over time. Their common feature is that they are all built of stone blocks. Pushover analysis was conducted on ten buildings that have historical or cultural value. Today they are intended for public use or housing. Some of the buildings have very poor mechanical properties, while the others have been reconstructed and show higher seismic resistance. The basic idea in the selection was to include as many different types of buildings as possible.

The masonry buildings outside of historical center are typical for the constructions built of concrete or brick hollow blocks after 1948 and can be classified according to the construction period. They belong to the following categories: (1) Type 1—unreinforced concrete masonry built before the first seismic regulation in 1964; (2) Type 2—concrete masonry with horizontal RC confining elements typical for the period between 1964 and 1980; (3) Type 3—confined concrete masonry with horizontal RC ties and RC columns built between 1980 and 2005, and (4) Type 4—confined brick masonry with horizontal RC ties and RC columns, which are seismically resistant structures due to the applications of modern design standards based on Eurocode 8. The buildings have rigid RC slabs, while the roof is mainly wooden with roof tiles. A large number of buildings, especially the older ones, have a similar floor plan and the ratio of the shear surface area of the walls to the floor area. Since two configurations of buildings prevail in terms of height, with two floors and a roof and three floors and a roof, two buildings of different storeys were selected for each period of construction for detailed pushover analysis. Therfore, two different elevation configurations have been analyzed: (a) P + 1 which consists of ground, one floor, and a roof; and (b) P + 2 which consists of ground, two floors, and a roof (Figure 11).

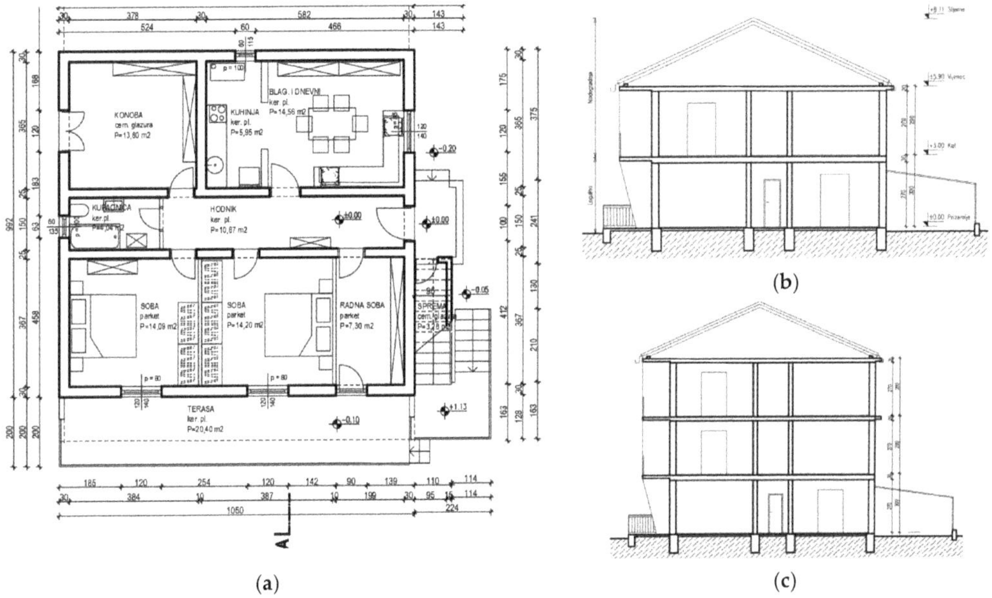

(a)

(b)

(c)

Figure 11. Typical configurations outside of the historical center: (**a**) plan; (**b**) section view P + 1; (**c**) section view P + 2.

Peak ground accelerations for the DL, SD, and NC limit states were computed in the x and y directions. The lowest PGA values were identified for each building and limit state. The critical PGA results and vulnerability indices for the considered buildings are presented in Table 2.

Table 2. Vulnerability index and critical peak ground accelerations of the buildings.

No.	Building	I_V [%]	PGA_{DL} [g]	PGA_{SD} [g]	PGA_{NC} [g]
1	Cambi Tower	76.9	0.030	0.059	0.078
2	Kumbat Towers	65.2	0.057	0.087	0.103
3	Public Library	59.0	0.028	0.061	0.079
4	Folk Castle	58.7	0.081	0.061	0.080
5	Dudan Palace	50.1	0.051	0.068	0.083
6	Perišin house	48.7	0.058	0.061	0.121
7	St. Mihovil Church	40.5	0.057	0.086	0.102
8	Rowing club	40.2	0.064	0.110	0.141
9	Residential building	34.8	0.081	0.095	0.152
10	Ballet school	23.9	0.103	0.142	0.183
11	Type 1 building P + 2	29.1	0.083	0.114	0.142
12	Type 1 building P + 1	29.1	0.061	0.144	0.173
13	Type 2 building P + 2	13.4	0.098	0.145	0.175
14	Type 2 building P + 1	13.4	0.115	0.187	0.220
15	Type 3 building P + 2	6.0	0.065	0.158	0.189
16	Type 3 building P + 1	6.0	0.075	0.175	0.206
17	Type 4 building P + 2	4.3	0.103	0.188	0.243
18	Type 4 building P + 1	2.6	0.130	0.218	0.270

The distribution of vulnerability index I_v is shown in Figure 12, while peak ground accelerations for the NC, SD, and DL limit states calculated by the pushover analysis for 18 buildings are presented in Figure 13.

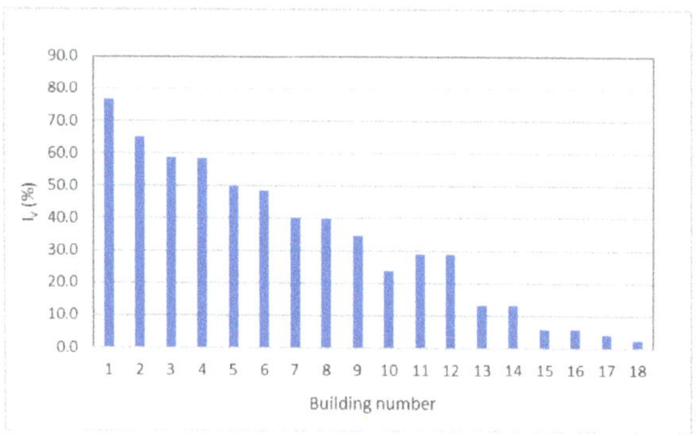

Figure 12. Distribution of I_v for 18 buildings analyzed by pushover analysis.

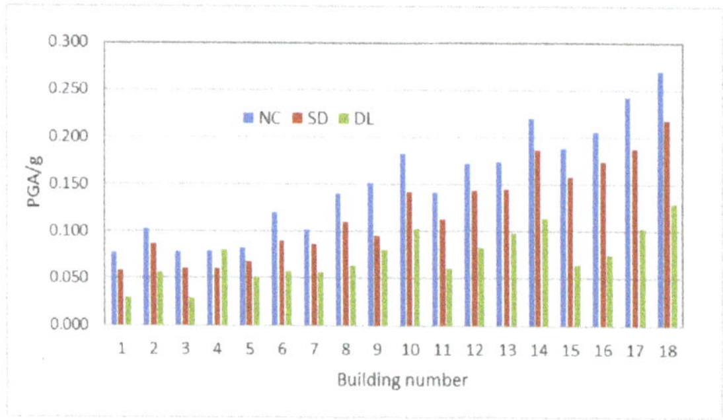

Figure 13. Peak ground accelerations for NC, SD, and DL limit states calculated by pushover analysis.

7. Vulnerability Index—PGA Relations

The emphasis of this part of the paper is to investigate whether the seismic behavior of the building can be estimated from the vulnerability index parameters. Therefore, 18 buildings, which represent 16% of total 111 buildings with calculated vulnerability indices, were analyzed by the non-linear static (pushover) method. The buildings have been chosen considering different material and structural characteristics as well as the period of construction. Additionally, there are buildings among them with different number of the floors and regular and non-regular layout and elevation.

The non-linear pushover analysis carried out on the stone masonry buildings in the old city center indicated a low global capacity in terms of the collapse peak ground acceleration, as well as low global accelerations of significant damage and damage limitation states. Numerical predictions of the acceleration achieved for structural collapse have indicated that no building meets the seismic demand of $a_g = 0.22$ g for T = 475 yearsin both the x and y directions. Moreover, in the simulations, a conspicuous number of buildings reached the collapse at accelerations that are lower than the demand acceleration of $a_g = 0.11$ g for a return period T = 95 years. The local mechanism failure induced by a lack of connection among perpendicular walls, and poor connections between floors/roofs and walls, was also analyzed for few stone masonry buildings in the old city center, where out-of-plane effect can be expected. The lowest acceleration was achieved for the global response of the buildings [33]. Buildings outside of the center are made of concrete or brick masonry with horizontal RC confining elements or both with horizontal and vertical confining elements and rigid horizontal diaphragms. Therefore, the failure of the structure caused by local mechanisms is not expected and the behavior of the buildings represented by capacity accelerations will be analyzed assuming the global failure of the structure.

The results for 18 analyzed buildings (10 in the historical center and 8 outside of the center) presented in Table 2 and Figures 12 and 13, are used to establish the vulnerability index—peak ground acceleration relation for the DL, SD, and NC limit states at the entire test site. Figure 14 shows a cloud of points representing the relationship between the vulnerability index calculated on the basis of 11 parameters I_v and the critical peak ground accelerations associated with the DL, SD, and NC limits. The trend lines I_v–PGA_{DL}, I_v–PGA_{SD} and I_v–PGA_{NC} for three limit states were obtained and are shown in Figure 14. The exponential functions were chosen as the most representative. They are used to approximate the yield, significant damage, and collapse peak ground accelerations for the entire test site. The values of yield and collapse accelerations are the basis for deriving vulnerability curves.

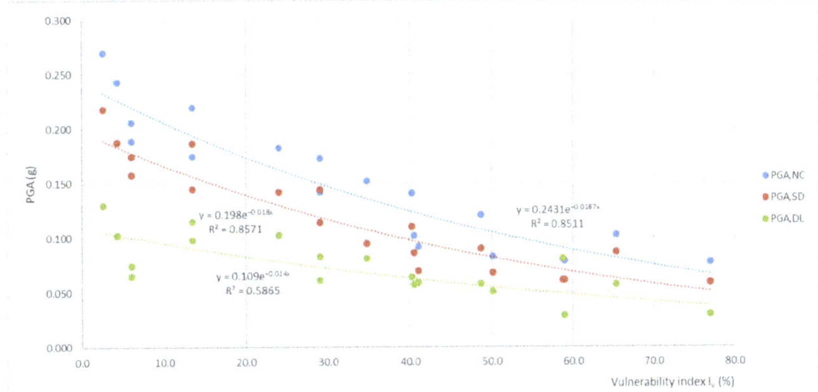

Figure 14. Trend lines I_v–PGA_{DL}, I_v–PGA_{SD}, and I_v–PGA_{NC}.

The quality of the approximation of the results obtained by the pushover analysis and those represented by trend lines for the three limit states are validated by standard deviation. The comparison shows a significantly better quality of peak ground acceleration aproximations for the NC and SD limit states than for the DL state. It is obvious that the most vulnerable buildings have a certain seismic load-bearing capacity expressed with peak acceleration. The derived trend lines are used to estimate peak ground accelerations for three limit states of the buildings using their vulnerability index. It should be noted that the results for near collapse and significant damage states approximate critical accelerations much better than the state of damage limitation.

8. Vulnerability Curves and Damage Index Distribution

The vulnerability curve allows to correlate the vulnerability index, damage index, and peak ground acceleration. Two limit-levels of acceleration are key to the analysis of damage: the acceleration associated to the beginning of the damage and the acceleration associated to the structural collapse. We recall that the damage value varies in the (0, 1) interval. The present investigation relies upon the study devised by Guagenti and Petrini [4], who obtained a relation between vulnerability index, acceleration, and damage, by observing the damage levels of masonry buildings subjected to real earthquakes. Corresponding acceleration/damage relation can be modelled with a smooth vulnerability curve (Figure 15). For simplicity purposes, Guagenti and Petrini, instead of using a vulnerability curve, proposed to exploit a tri-linear law parametrized in terms of the values that the peak ground acceleration takes at early damage, PGA_i, and at the collapse, PGA_c. In this study, instead of a post-earthquake damage observation, the acceleration for the yield and collapse states were calculated by the pushover analysis [27,33]. Then, using the vulnerability indices and yield and collapse accelerations, a new damage–vulnerability–peak ground acceleration relationship was derived. The damage index is expressed in the (0–1) interval via a tri-linear law, analogously to [4] though defined through two parameters: yield acceleration PGA_y, which corresponds the beginning of the damage (d = 0), and collapse acceleration PGA_c (d = 1).

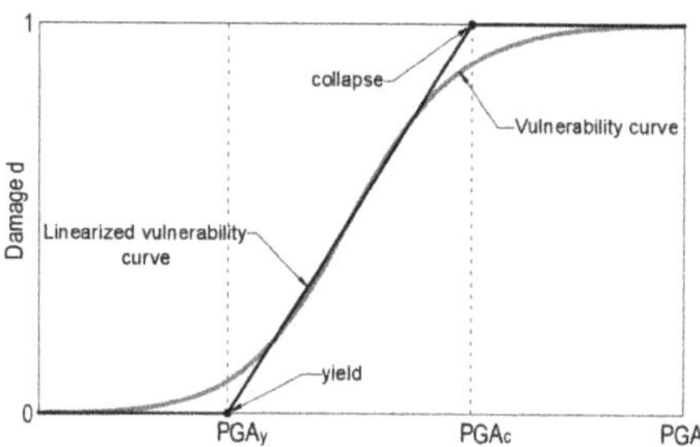

Figure 15. Vulnerability curve and its idealization.

The developed procedure has been applied to the observed test site. The basis for defining vulnerability curves are vulnerability indices, yield peak ground acceleration PGA_y, and collapse peak ground acceleration PGA_c, obtained by the pushover analysis for 18 analyzed buildings, as shown in Table 2. Yield acceleration PGA_y is assigned to PGA_{DL} and collapse acceleration PGA_c to PGA_{NC} limit states, respectively. As PGA_y and PGA_c depend on the vulnerability index I_v, the values of PGA_y, associated with damage d = 0, and PGA_c, associated with damage d = 1, can be calculated for each value of I_v.

Figure 16 shows the vulnerability curves used for the estimation of the damage index of the buildings at the investigated area. These vulnerability curves are partially changed in comparison to those derived from the authors for the sole historical center [31], as here the study was extended to the entire test site. In fact, the influence of the buildings outside of the historical center with vulnerability indices mainly up to 30% changed the relationships between vulnerability index and peak ground accelerations, as well as the vulnerability curves for the low vulnerability buildings.

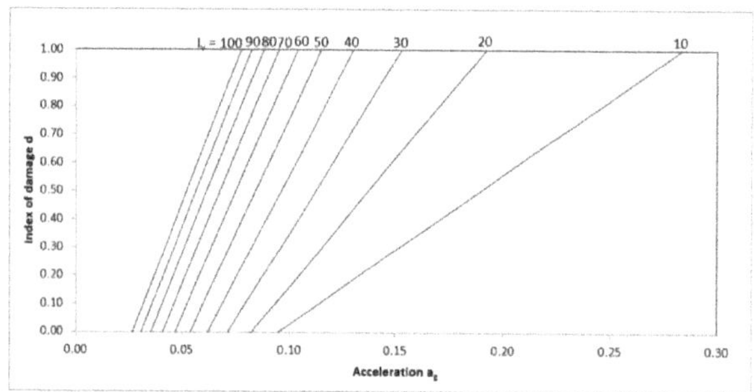

Figure 16. Vulnerability curves for the test site.

The spatial distribution of the damage is represented by the damage index maps of the investigated area for the given intensity of the earthquake. Three seismic scenarios corresponding to return periods of 95, 225, and 475 years and demand peak ground accelerations of 0.11 g, 0.17 g, and 0.22 g, respectively, have been chosen. The damage to the buildings for different scenarios is presented in Figure 17.

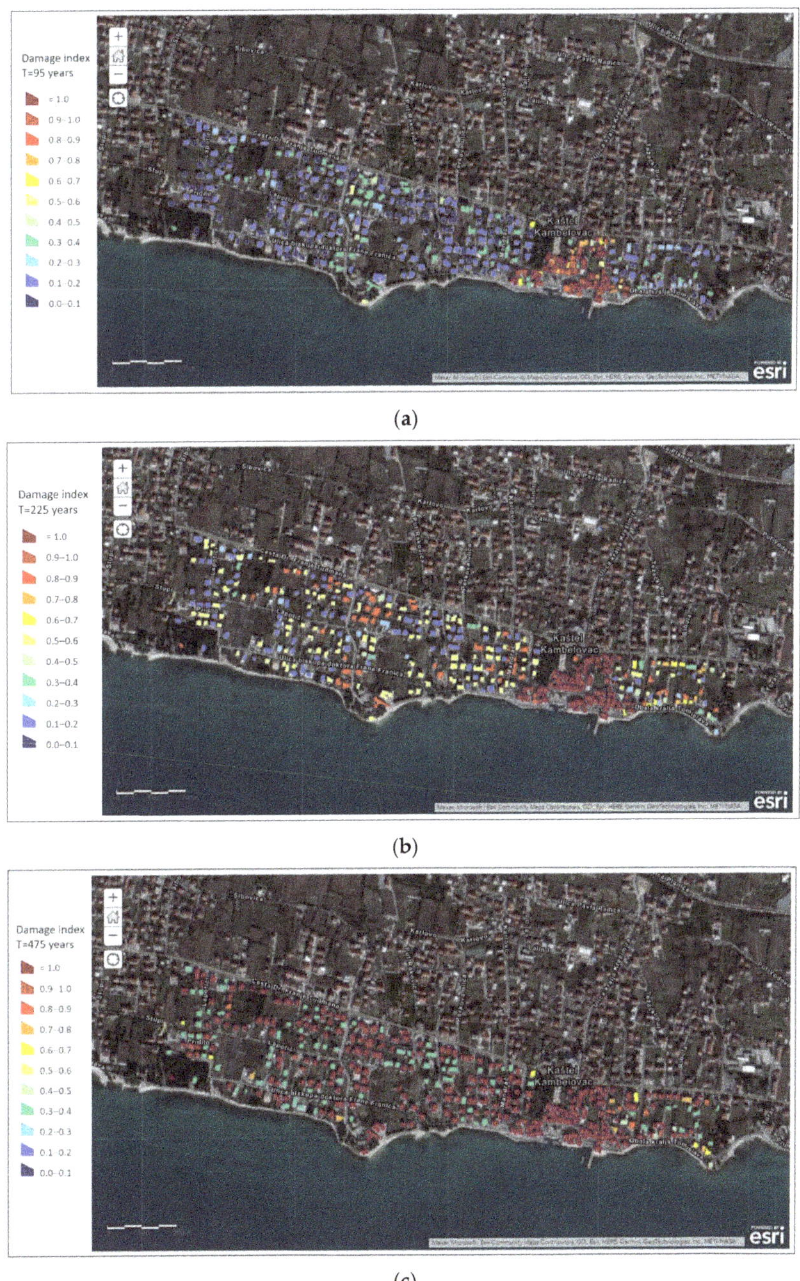

(**a**)

(**b**)

(**c**)

Figure 17. Damage index distribution at the test site: (**a**) T = 95 years; (**b**) T = 225 years; (**c**) T = 475 years.

9. Seismic Risk Distribution in Terms of the Index of Seismic Risk

In this section, the methodology for the assessment of seismic risk in terms of the index of seismic risk is presented. The methodology uses vulnerability index–peak ground acceleration relations for the DL, SD, and NC limit states, as presented in Figure 14, to estimate critical peak ground accelerations from the vulnerability index.

The index of seismic risk is calculated as the ratio of the peak ground acceleration PGA_C associated to the structural capacity and the demand ground acceleration PGA_D. It is expressed in the following form:

$$\alpha_{PGA,C} = \frac{PGA_C}{PGA_D} \tag{7}$$

The capacity of the structure represents the minimum value of PGA for which a certain limit state is achieved. In the further analysis, capacity of the structure leading to the structural collapse (i.e., NC limit state) is analyzed. Therefore, PGA_C is equal to PGA_{NC}.

Seismic hazard is defined according to EC-8 with the following parameters:

- a_g—peak ground horizontal acceleration on type A soil, $a_g = \gamma_I a_{gR}$, where γ_I depends on the importance of the building;
- S—soil parameter.

Demand ground acceleration PGA_D obtained from the seismic hazard map for the selected return period is given as $PGA_D = a_g S$.

Indices of seismic risk are used to validate the safety of the structure. The values $\alpha_{PGA} > 1$ refer to safe structures, while the values $\alpha_{PGA} < 1$ refer to non-safe structures. The indices of seismic risk that evaluate safety for the NC limit state of the buildings for three return periods are presented in Figure 18. Indices of seismic risk for other limit states can be calculated in a similar manner.

(**a**)

Figure 18. *Cont.*

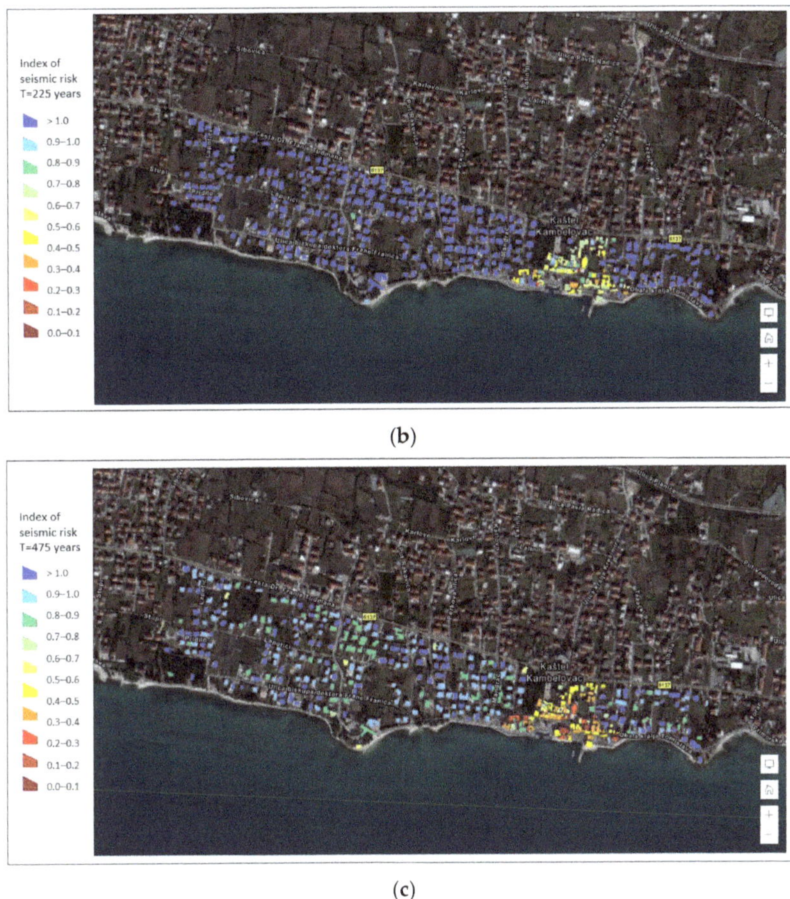

(b)

(c)

Figure 18. Risk maps in terms of index of seismic risk: (**a**) T = 95 years; (**b**) T = 225 years; (**c**) T = 475 years.

10. Discussion

The methodology for large-scale seismic vulnerability and risk assessment has been developed and applied to Kaštel Kambelovac, one of seven settlements of the City of Kaštela. The hybrid procedure combining the vulnerability index method with the pushover analysis of the selected buildings has been used to calculate vulnerability indices, critical peak ground accelerations for three limit states (early damage, significant damage, and near collapse), seismic risk in terms of damage, and indices of seismic risk. Vulnerability and risk of the test site have been demonstrated through the vulnerability index map, damage index map and map of indices of seismic risk. All results have been integrated into a web-map based on the GIS tool.

The main findings of the study are summarized below.

Vulnerability indices have been calculated for 111 buildings with known geometrical, structural, and material characteristics (75 in the old city center and 35 outside of the center). The vulnerability of other buildings at the test site without the available technical documentations was estimated using geometric and structural characteristics of the buildings obtained from a geodetic survey of the area, a street view map, and a visual inspection of the area. The distribution of vulnerability indices shows the medium-high and high vulnerability of the historical center and the part north of the center. Such high vulnerability is associated with stone masonry buildings built between the 16th and the beginning of the 20th century, which are mostly made with walls poorly connected, flexible floors, and irregular layout and elevation. The medium-low and low vulnerability has been achieved for the buildings outside of the historical center, which were built starting from the second half of the 20th century to the present day. They are made of concrete or brick blocks. Some of them are unreinforced, while the others are confined with horizontal RC ring ties or both with horizontal RC ring ties and RC columns which, together with rigid RC floors, contributed to high seismic resistance and low vulnerability.

The pushover analysis has been conducted on 18 typical buildings (10 in the historical center and 8 outside of the center). The buildings in the historical center have reached the collapse for a low peak ground acceleration (between 0.078 g and 0.183 g). The acceleration for significant damage ranges from 0.059 g to 0.142 g, while the lowest acceleration has been achieved for the early damage state (between 0.028 g and 0.103 g). No buildings in the center meet the seismic requirement of a_g = 0.22 g for T = 475 years in either directions. Moreover, several buildings reached the structural collapse at peak ground accelerations that are lower than the demand acceleration of a_g = 0.11 g for a return period T = 95 years. It should be noted that capacity accelerations have been obtained for the global structural response. Although local failure mechanisms can be critical for the buildings with flexible floors and weak connection between the walls, the analysis of local mechanisms of many buildings showed that the lowest accelerations have been reached for the global analysis of the buildings. Considering that the aim of the study is a vulnerability assessment of the settlement, critical acceleration has been determined on all buildings for the global response of the structure. The capacity of the buildings typical for the structures outside of the historical center varied depending on the applied materials and construction rules. Unreinforced concrete masonry buildings built before the first seismic regulation in 1964, concrete masonry with horizontal RC ring ties typical for the period between 1964 and 1980, and confined concrete masonry both with horizontal and vertical RC confining elements built between 1980 and 2005 do not meet seismic demands of a_g = 0.22 g for T = 475 years. Only the buildings made of confined brick masonry, designed according to Eurocode 8, meet the seismic demand.

Vulnerability index–PGA accelerations for the DL, SD, and NC limit states according to EC8-3 have been derived and validated by standard deviation. Satisfactory approximations have been obtained for the NC and SD limit states, while the deviations for the DL state had low accuracy.

The seismic risk has been formulated in terms of the damage index and the index of seismic risk. Damage indices have been calculated from derived vulnerability curves for the test site. Spatial distribution of the damage, represented by the damage index maps, shows a high level damage of the historical center already for the return period of 95 years. High-level damage spreads well beyond the center by increasing peak acceleration. Therefore, almost the entire area is damaged in the return period of 475 years.

Indices of seismic risk have been defined as a ratio of the peak ground acceleration associated to the NC, SD, or DL states, respectively, and the demand ground acceleration. Maps of seismic risk index indicate a number of non-safe buildings in the historical center already for the return period of 95 years. With the increase in peak ground accelerations, the number of non-safe buildings also increases. For T = 475 years, all buildings in the historical center and many buildings outside of the center are considered non-safe.

The developed methodology for seismic risk assessment is a demanding process that carries a number of uncertainties, due to the diversity of materials and performance, a series of construction interventions and upgrades over time, and, generally, a lack of knowledge about the buildings. Additional problems arise from the limitations of vulnerability index method and the non-linear static pushover analysis in predicting the behavior of the buildings under seismic action. Despite these shortcomings, the approach presented in this study gives valuable information on the vulnerability and risk of the area, which can be applied to preventive activities for improving the seismic resistance of buildings and emergency situations in the case of seismic action.

11. Conclusions

This paper presents a comprehensive hybrid approach to large-scale seismic vulnerability and risk assessment of existing urban areas. The methodology combines the advantages of the vulnerability index method in assessing the vulnerability of a large number of buildings with a detailed analytical approach based on the nonlinear static pushover method, which allows determination of the capacity curves and detection of the specific limit states of the buildings.

The proposed methodology includes all phases leading to the final risk estimation, starting from data collection and detecting characteristics of buildings and ground type, vulnerability index evaluation, non-linear analysis of the selected buildings, defining vulnerability–peak ground acceleration curves, deriving vulnerability curves for the selected urban area and, finally, risk representation in terms of the damage index and the index of seismic risk.

The methodology has been applied to a whole urban settlement placed along the Croatian Adriatic coast, which has expanded over past centuries, resulting in a heterogeneous distribution of the buildings with different architectural, material, and structural characteristics. The present study covered various types of masonry buildings, ranging from stone masonry buildings several centuries old to the buildings made of concrete and brick blocks built from the mid-20th century to the present day in compliance with the technical regulations or common building rules during the construction period.

The findings of the present research are crucial for several reasons.

Firstly, the paper presents a methodological approach for a full consideration of seismic risk that can be applied to any settlement. It is not necessary to dispose with the data on damage from the previous earthquakes to conduct the research, although it can, if any, be used to calibrate the results. The methodology provides the relationship between the vulnerability index and peak ground accelerations of early damage, significant damage, and near collapse states. It also allows the derivation of vulnerability curves that serve to determine the damage index of the buildings for a specific seismic action.

Secondly, the pushover analysis conducted on 18 buildings with different characteristics provided valuable results of their behavior up to the failure, as well as peak ground accelerations for specific limit states, contributing to scientific knowledge of the behavior of such buildings. This is of particular importance given the seismic vulnerability of Croatia, and especially the southern part of the Croatian coast affected by devastating earthquakes (Dubrovnik, Ston, Makarska) in the past, and the similarity of the presented test area with many settlements and cities along the Adriatic coast.

Finally, the outcomes of the developed methodology are important for the observed test site and can be immediately applied to improve the seismic risk-coping capacity of the community of the relevant municipality. Indeed, the present procedure has resulted in seismic vulnerability indices, damage indices, and critical accelerations for different limit states, and indices of seismic risk for three return periods (95, 225, and 475 years). The results have been presented in a web map made in the GIS environment, which enables the visualization of vulnerability and risk of the area. Therefore, the developed vulnerability, damage, and risk maps have important operational outcomes in the seismic risk management of the investigated area.

Author Contributions: Conceptualization, Ž.N.; Data curation, L.R.; Funding acquisition, Ž.N. and E.B.; Investigation, Ž.N., E.B. and L.R.; Methodology, Ž.N. and E.B.; Software, L.R.; Supervision, Ž.N. and E.B.; Validation, Ž.N., E.B. and L.R.; Visualization, L.R.; Writing—original draft, Ž.N., E.B. and L.R. All authors have read and agreed to the published version of the manuscript.

Funding: This research was funded by the EUROPEAN UNION, Programme Interreg Italy–Croatia, Project "Preventing, managing and overcoming natural-hazards risks to mitigate economic and social impact"—PMO-GATE ID 10046122. The research is also partially supported through project KK.01.1.1.02.0027, co-financed by the CROATIAN GOVERNMENT and the EUROPEAN UNION through the European Regional Development Fund—the Competitiveness and Cohesion Operational Programme.

Institutional Review Board Statement: Not applicable.

Informed Consent Statement: Not applicable.

Data Availability Statement: Not applicable.

Conflicts of Interest: The authors declare no conflict of interest.

References

1. Benedetti, D.; Petrini, V. Vulnerability of masonry buildings: Proposal of a method of assessment (in Italian). *L'industria Delle Costr.* **1984**, *149*, 66–74.
2. GNDT-SSN. *Scheda di Esposizione e Vulnerabilità e di Rilevamento Danni di Primo e Secondo Livello (Muratura e Cemento Armato)*; GNDT-SSN: Rome, Italy, 1994. Available online: https://protezionecivile.regione.abruzzo.it/files/rischio%20sismico/verificheSism/Manuale_e_scheda_GNDT_II_livello.pdf (accessed on 26 April 2021).
3. Whitman, R.V.; Reed, J.W.; Hong, S.T. Earthquake Damage Probability Matrices. In Proceedings of the 5th World Conference on Earthquake Engineering, Rome, Italy, 25–29 June 1973; Volume II, pp. 2531–2540.
4. Guagenti, E.; Petrini, V. The case of old buildings: Towards a new law damage-intensity (in Italian). In Proceedings of the IV ANIDIS Convention, Milan, Italy, 1989; Volume I, pp. 145–153.
5. Bernardini, A. *The Vulnerability of Buildings-Evaluation on the National Scale of the Seismic Vulnerability of Ordinary Buildings*; CNR-GNDT: Rome, Italy, 2000.
6. Giovinazzi, S. The Vulnerability Assessment and the Damage Scenario in Seismic Risk Analysis. Ph.D. Thesis, Faculty of Engineering Department of Civil Engineering of University of the Florence, Florence, Italy, 2005.
7. Di Pascuale, G.; Orsini, G.; Romeo, R.W. New developments in seismic risk assessment in Italy. *Bull. Earthq. Eng.* **2005**, *3*, 101–128. [CrossRef]
8. Lagomarsino, S.; Giovinazzi, S. Macroseismic and mechanical models for the vulnerability and damage assessment of current buildings. *Bull. Earthq. Eng.* **2006**, *4*, 415–443. [CrossRef]
9. Grunthall, G. *European Macroseismic Scale 1998 (EMS-98)*; Cahiers du Centre Européen de Géodynamique et Séismologie: Luxembourg, 1998; Volume 15.
10. Fajfar, P.; Eeri, M. A nonlinear analysis method for performance based seismic design. *Earthq. Spectra* **2000**, *16*, 573–592. [CrossRef]
11. Vamvatsikos, D.; Cornell, C.A. Incremental Dynamic Analysis. *Earthq. Eng. Struct. Dyn.* **2002**, *31*, 491–514. [CrossRef]
12. Rossetto, T.; Elnashai, A. Derivation of vulnerability functions for European-type RC structures based on observational data. *Eng. Struct.* **2003**, *25*, 1241–1263. [CrossRef]
13. Maniyar, M.M.; Khare, R.; Dhakal, R.P. Probabilistic seismic performance evaluation of non-seismic RC frame buildings. *Struct. Eng. Mech.* **2009**, *33*, 725–745. [CrossRef]
14. Ripepe, M.; Lacanna, G.; Deguy, P.; De Stefano, M.; Mariani, V.; Tanganelli, M. Large-Scale Seismic Vulnerability Assessment Method for Urban Centres. An Application to the City of Florence. *Key Eng. Mater.* **2014**, *628*, 49–54. [CrossRef]
15. Salgado-Galvez, M.A.; Zuloaga, R.D.; Velasquez, C.A.; Carreno, M.L.; Cardona, O.D.; Barbat, A.H. Urban seismic risk index for Medellin, Colombia, based on probabilistic loss and casualties estimations. *Nat. Hazards* **2016**, *80*, 1995–2021. [CrossRef]
16. Salazar, L.G.F.; Ferreira, T.M. Seismic Vulnerability Assessment of Historic Constructions in the Downtown of Mexico City. *Sustainability* **2020**, *12*, 1276. [CrossRef]
17. Battaglia, L.; Ferreira, T.M.; Lourenço, P.B. Seismic fragility assessment of masonry building aggregates: A case study in the old city Centre of Seixal, Portugal. *Earthq. Eng. Struct. Dyn.* **2021**, *50*, 1358–1377. [CrossRef]
18. Lagomarsino, S.; Cattari, S.; Ottonelli, D. The heuristic vulnerability model: Fragility curves for masonry buildings. *Bull. Earthq. Eng.* **2021**, *19*, 3129–3163. [CrossRef]
19. Giordano, N.; De Luca, F.; Sextos, A. Analytical fragility curves for masonry school building portfolios in Nepal. *Bull. Earthq. Eng.* **2021**, *19*, 1121–1150. [CrossRef]
20. Capanna, I.; Aloisio, A.; Di Fabio, F.; Fragiacomo, M. Sensitivity Assessment of the Seismic Response of a Masonry Palace via Non-Linear Static Analysis: A Case Study in L'Aquila (Italy). *Infrastructures* **2021**, *6*, 8. [CrossRef]
21. Vicente, R.; Parodi, S.; Lagomarsino, S.; Varum, H.; Da Silva, M. Seismic vulnerability and risk assessment: Case study of the historic city centre of Coimbra, Portugal. *Bull. Earthq. Eng.* **2011**, *9*, 1067–1096. [CrossRef]
22. Achs, G.; Adam, C. Rapid seismic evaluation of historic brick-masonry buildings in Vienna (Austria) based on visual screening. *Bull. Earthq. Eng.* **2012**, *10*, 1833–1856. [CrossRef]
23. Hadzima-Nyarko, M.; Mišetić, V.; Morić, D. Seismic vulnerability assessment of an old historical masonry building in Osijek, Croatia, using Damage Index. *J. Cult. Herit.* **2017**, *28*, 140–150. [CrossRef]
24. Ferreira, T.M.; Mendes, N.; Silva, R. Multiscale Seismic Vulnerability Assessment and Retrofit of Existing Masonry Buildings. *Buildings* **2019**, *9*, 91. [CrossRef]

25. Angeletti, P.; Bellina, A.; Guagenti, E.; Moretti, A.; Petrini, V. Comparison between vulnerability assessment and damage index, some results. In Proceedings of the 9th World Conference on Earthquake Engineering, Tokyo, Japan, 2–9 August 1988.
26. Ciavattone, A.; Vignoli, A.; Matthies, H.G. Seismic vulnerability analysis for masonry hospital structures: Expeditious and detailed methods. In *Brick and Block Masonry—Trends, Innovations and Challenges*; da Porto, M., Valluzzi, Eds.; Taylor & Francis Group: London, UK, 2016; ISBN 978-1-138-02999-6.
27. Cavaleri, L.; Di Trapani, F.; Ferroto, M.F. A new hybrid procedure for the definition of seismic vulnerability in Mediterranean cross-border urban areas. *Nat. Hazards* **2017**, *86*, 517–541. [CrossRef]
28. Nikolić, Ž.; Krstevska, L.; Marović, P.; Smoljanović, H. Experimental investigation of seismic behaviour of the ancient Protiron monument model. *Earthq. Eng. Struct. Dyn.* **2019**, *48*, 573–593. [CrossRef]
29. Krstevska, L.; Nikolić, Ž.; Kustura, M. Shake table testing of two historical masonry structures for estimation of their seismic stability. *Int. J. Archit. Herit.* **2021**, *15*, 45–63. [CrossRef]
30. Smoljanović, H.; Živaljić, N.; Nikolić, Ž.; Munjiza, A. Numerical Simulation of the Ancient Protiron Structure Model Exposed to Seismic Loading. *Int. J. Archit. Herit.* **2021**, *15*, 779–789. [CrossRef]
31. Nikolić, M.; Do, X.N.; Ibrahimbegovic, A.; Nikolić, Ž. Crack propagation in dynamics by embedded strong discontinuity approach: Enhanced solid versus discrete lattice model. *Comput. Methods Appl. Mech. Eng.* **2018**, *340*, 480–499. [CrossRef]
32. Čarija, J.; Nikolić, M.; Ibrahimbegovic, A.; Nikolić, Ž. Discrete softening-damage model for fracture process representation with embedded strong discontinuities. *Eng. Fract. Mech.* **2020**, *236*, 107211. [CrossRef]
33. Nikolić, Ž.; Runjić, L.; Ostojić Škomrlj, N.; Benvenuti, E. Seismic Vulnerability Assessment of Historical Masonry Buildings in Croatian Coastal Area. *Appl. Sci.* **2021**, *11*, 5997. [CrossRef]
34. Available online: https://marinas.com/view/marina/eyc39lv_Kastel_Kambelovac_Harbour_Kastel_Gomilica_Croatia (accessed on 21 November 2020).
35. Stepinac, M.; Lourenço, P.B.; Atalić, J.; Kišiček, T.; Uroš, M.; Baniček, M.; Šavor Novak, M. Damage classification of residential buildings in historical downtown after the ML5.5 earthquake in Zagreb, Croatia in 2020. *Int. J. Disaster Risk Reduct.* **2021**, *56*, 102140. [CrossRef]
36. *HRN ENV 1998-1 Eurocode 8: Design Provisions for Earthquake Resistance of Structures—Part 1: General Rules, Seismic Actions and General Requirements for Structures*; Croatian Standards Institute: Zagreb, Croatia, 2005.
37. *HRN EN 1998-1:2011. Eurocode 8: Design of Structures for Earthquake Resistance. Part 1: General Rules, Seismic Actions and Rules for Buildings*; Croatian Standards Institute: Zagreb, Croatia, 2011.
38. Marasović, K. Kaštel Kambelovac. *Kaštela J.* **2003**, *7*, 35–61. (In Croatian)
39. Opara, L.K. *Termographic Report—Buildings in Kaštel Kambelovac*; University of Split, Faculty of Electrical Engineering, Mechanical Engineering and Naval Architecture: Split, Croatia, March 2020.
40. Uranjek, M.; Žarnić, R.; Bokan-Bosiljkov, V.; Bosiljkov, V. Seismic resistance of stone masonry building and effect of grouting. *Građevinar* **2014**, *66*, 715–726.
41. Da Col, F.; Accaino, F.; Bohm, G.; Meneghini, F. Characterization of shallow sediments by processing of P, SH and SV wave-fields in Kaštela (HR). *Eng. Geol.* **2021**, *293*, 106336. [CrossRef]
42. Ferrini, M.; Melozzi, A.; Pagliazzi, A.; Scarparolo, S. Rilevamento della vulnerabilita sismica degli edifici in muratura. In *Manuale per la Compilazione Della Scheda GNDT/CNR di II Livello*; Versione modificata dalla Regione Toscana. S.l.; Direzione Generale delle Politiche Territoriale e Ambientali, Settore—Servizio Sismico Regionale: Regione Toscana, Italy, 2003.
43. Piano di Ricostruzione del Comune di Arsita (TE), a Seguito Dell'evento Sismico del 6/04/2009, Document: 3A_01_f_Aspetti Strutturali e di Vulnerabilità_Rev.2 . Available online: http://www.pdr-arsita.bologna.enea.it/wordpress/wp-content/uploads/ arsita/pdr_finale/terza_fase/3A_01_Elaborati%20tecnici/3A_01_f_Aspetti%20strutturali%20e%20di%20vulnerabilit%C3%A0 /3A_01_f_Aspetti%20strutturali%20e%20di%20vulnerabilita%CC%80_Rev.%202.pdf (accessed on 26 March 2019).
44. *HRN EN 1998-3 Eurocode 8: Design of Structures for Earthquake Resistance—Part 3: Assessment and Retrofiting of Buildings*; Croatian Standards Institute: Zagreb, Croatia, 2011.
45. *3MURI Software*, Professional version; S.T.A. DATA: Torino, Germany, 2019.
46. Kreslin, M.; Fajfar, P. The extended N2 method considering higher mode effects in both plan and elevation. *Bull. Earthq. Eng.* **2012**, *10*, 695–715. [CrossRef]

*applied
sciences*

Article

A Systematic Revision of the NFIP Claims Hazard Data in Florida for Flood Risk Assessment

D. W. Shin [1], Steven Cocke [1] and Baek-Min Kim [2,*]

[1] Center for Ocean-Atmospheric Prediction Studies, Florida State University, Tallahassee, FL 32306-2840, USA; shin@coaps.fsu.edu (D.W.S.); scocke@fsu.edu (S.C.)

[2] Department of Environmental Atmospheric Sciences, Pukyung National University, Busan 48513, Korea

* Correspondence: baekmin@pknu.ac.kr; Tel.: +82-10-4194-9404

Abstract: The hazard components of National Flood Insurance Program (NFIP) claims data in Florida were systematically analyzed and revised in the current study. The provided fields in NFIP claims data are not always complete or accurate and are often missing. The authors associated each claim to a proper flood hazard event by updating the provided catastrophe number using the National Hurricane Center HURDAT2 (best track data). The claims with presumably incorrect *cause of loss* fields were identified and revised by adding a variety of other available information to claims data. These datasets included tropical cyclone events, rainfall maxima, and distances to the nearest coast. The enhanced information assisted in identifying the cause or likelihood of a hazard event or attributing a particular hazard event to a loss claim. The revised NFIP claims data will be intensively used to validate the outcomes from flood hazard (i.e., surge, wave and inland flooding) models and to develop a flood vulnerability model in the forthcoming Florida Public Flood Loss Model (FPFLM).

Keywords: NFIP claims; flood risk; flood insurance; flood cause of loss; HURDAT; catastrophe model

Citation: Shin, D.W.; Cocke, S.; Kim, B.-M. A Systematic Revision of the NFIP Claims Hazard Data in Florida for Flood Risk Assessment. *Appl. Sci.* **2022**, *12*, 3537. https://doi.org/10.3390/app12073537

Academic Editor: Andrea Chiozzi

Received: 18 February 2022
Accepted: 29 March 2022
Published: 30 March 2022

Publisher's Note: MDPI stays neutral with regard to jurisdictional claims in published maps and institutional affiliations.

1. Introduction

The Florida Commission on Hurricane Loss Projection Methodology (FCHLPM), under legislative direction, published Flood Standards Reports of Activities [1] similar in concept to the current hurricane wind hazard standards [2]. Accordingly, the Florida Office of Insurance Regulation (FLOIR) requested the Florida Public Hurricane Loss Model (FPHLM) [3] group to undertake a project to develop a flood risk model officially called the Florida Public Flood Loss Model (FPFLM). The flood catastrophe model is to estimate and/or predict aggregated insured losses for properties in the form of annual expected losses and probable maximum losses, which can be employed by Florida state regulators and insurance companies to help evaluate premium rate filings. A developed model will be delivered to the FCHLPM, which would be acceptable under the current flood standards. The FPFLM (a catastrophe model) is to estimate possible losses caused by coastal and inland flood events. It includes three main components: (1) hazard, (2) vulnerability, and (3) actuarial components. During the development process of a risk model, flood claim (and/or exposure) data are necessary for calibration and validation of flood model outputs.

Because the Department of Homeland Security's (DHS's)/Federal Emergency Management Agency's (FEMA's) National Flood Insurance Program (NFIP) claim data have not been publicly available, there have not been many in-depth analysis studies on flood insurance claims until recently. Kousky and Michel-Kerjan [4] examined the NFIP flood insurance claims in the continental United States and found six claim characteristics (e.g., claims are lower for elevated properties and higher from storm surges). A series of studies have also concentrated on flood losses, all of which involved a translation from inundation depth to its economic impacts, e.g., [5]. Wing et al. [6] used over two million NFIP claims to assess flood depth and damage functions (or curves) for economic evaluation. They showed that the NFIP damage data could partially remedy the uncertainties in the currently employed flood depth and damage functions, which involve disparate relationships that match poorly with observations. However, these studies used the original NFIP claim data without any modification attempts in their analysis and/or development of vulnerability curves.

The widely used standard flood depth and damage curves were compiled by the United States Army Corps of Engineers (USACE) [7]. The USACE developed these curves beginning in 1970 using early NFIP loss data, local Corps studies, and the collective judgment of experts [8]. They were assumed to be well calibrated and universally applicable when used in the majority of engineering communities, even though their simple relationships do not match well with observations. Therefore, Pinelli et al. [9,10] developed a better set of flood vulnerability curves using some modified NFIP claim data for the FPFLM project. However, they did not present the detailed analysis and revision process for the NFIP loss data in their paper.

The FLOIR provided the FPFLM group with the NFIP claim data in Florida. Benefiting from this special access, the authors analyzed the NFIP claims filed between 19 July 1975 and 10 January 2014 in the current study. The claim data contained flood hazard information (e.g., date and type of event, cause of loss) and building information (e.g., house location, values, elevation, age). However, the provided NFIP claim data were not always complete or accurate. Thus, it was necessary to make significant modifications to the NFIP claim data to properly assess the risk of all significant hazard events leading to flooding. The authors undertook a major task to identify the hazard event associated with each claim in the NFIP database. While the NFIP claim data included a *catastrophe number* that purports to identify the event, it was found that this information was often incorrect or identified as unknown. Furthermore, the NFIP database listed a *cause of loss*, which the authors also found to be inaccurate or unknown.

The paper is organized as follows: the provided NFIP claim data and additional data used to revise the claims hazard information are presented in Section 2; Section 3 shows the revision process of the NFIP claim hazard data, the results thereof, and further analyses for flood risk assessments; and the conclusions follow in Section 4.

2. Materials and Methods

2.1. NFIP Claim Data in Florida

Florida accounts for roughly 37% of all NFIP policies according to FEMA. The number of total claims in the provided NFIP loss data was 240,469 for the period of 19 July 1975 to 10 January 2014. Among them, only 153,751 (63.9%) were paid claims and were hence used in the current study. The paid claim locations are shown in Figure 1.

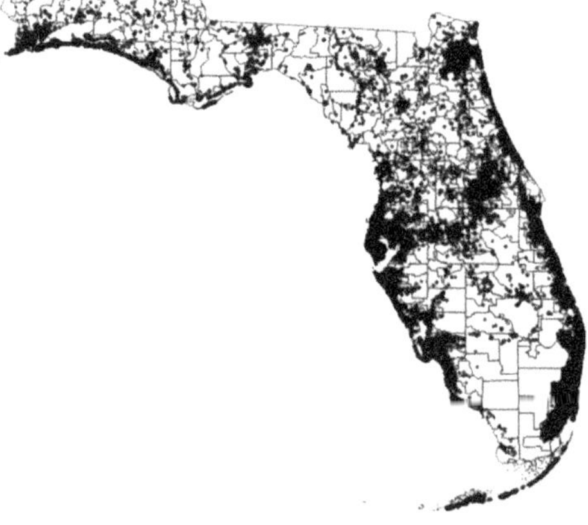

Figure 1. The NFIP paid claim data in Florida (claims period: 19 July 1975–10 January 2014).

The majority of flood damages were claimed along coastal and urban areas (e.g., Miami, Orlando, Jacksonville, and Tampa) as expected, partly because of the high frequency of storm surge hazards and partly because of the high density of the population. Although tropical cyclones are the major natural hazards, other frequently occurring intensive nontropical storm events can cause inundation damages in both coastal and inland areas of Florida.

The provided NFIP flood damage claim data contained 100 fields for hazard-, building, and actuarial-related information. These fields included *date of loss*, policy number, address (*latitude and longitude*), flood risk zone, occupancy type, *catastrophe number*, *cause of loss*, lowest floor elevation, base flood elevation, building claim payment, contents claim payment, etc. The hazard-related fields (italic font) were the main target fields to be assessed in the current study. To respect privacy, sensitive information (e.g., homeowner's address) was neither exposed nor explicitly used.

Brief descriptions of several claim fields are as follows:

- occupancy type: single-family homes or other type;
- catastrophe number: the flood event with which it is associated;
- base flood elevation (BFE): 100 year flood elevation (i.e., 1% chance of flooding);
- building and contents claim payment: how much was paid by the NFIP;
- flood risk zone: to indicate the risks in different parts of the United States, FEMA has assigned a character from the alphabet to each zone. The most hazardous flood zones are V (usually the first row of beachfront properties) and A (usually, but not always, properties near a lake, river, stream, or other body of water). Any building located in an A or V zone is considered to be in a Special Flood Hazard Area and lower than the BFE;
- cause of loss: the natural hazard cause identification for the claimed loss, as follows—0: other causes, 1: tidal water overflow, 2: stream, river, or lake overflow, 3: alluvial fan overflow, 4: accumulation of rainfall, 7: erosion—demolition, 8: erosion—removal, 9: earth movement, landslide, land subsidence, sinkholes, etc.

2.2. Additional Data to Claims

To revise inaccurate flood hazard information in the provided NFIP claim data, additional hazard-related information and land surface characteristics were added to the claim data for each claim location from a variety of resources. This information could help in identifying the cause or likelihood of a hazard event or attributing a particular hazard event to a loss claim.

2.2.1. HURDAT2

Atlantic HURDAT2 comprises the best track data available from the National Oceanic and Atmospheric Administration National Weather Service National Hurricane Center (NOAA NWS NHC; available at http://www.nhc.noaa.gov/data/#hurdat, 11 February 2022). This dataset has a comma-delimited text format with six-hourly information on the location, maximum winds, central pressure, and size of all known tropical cyclones and subtropical cyclones. It is updated at least once per year to include the previous year's best tracks and/or historical reanalysis tracks [11].

Since many of the NFIP claims are due to tropical storms or hurricanes, HURDAT2 might be the best resource for matching historical storms to loss claims properly in Florida. The date and time of the storm track positions can be compared to the claim location and time of loss using spatial and temporal metrics to determine possible storm influence. Additional information concerning the storm intensity level (depression, tropical storm, or hurricane) can be added to the provided claims as well.

2.2.2. PRISM Rainfall

The PRISM (Parameter-Elevation Regression on Independent Slopes Model) analysis, from the PRISM climate group at Oregon State University, is particularly interesting because of its high spatial resolution (800 m to 4 km). It provides daily rainfall from 1 January 1981 to recent [12]. As a demonstration of the quality of the PRISM, rainfall total and frequency statistics for the period of 1981–2014 are shown in Figure 2. The area averaged rainfall total amount was approximately 46,000 mm for these 34 years. The right panel shows a frequency count map of rainfalls of more than 100 mm/day (i.e., ~4 inches; extremely high chance of flooding events). The Western Panhandle region of Florida appeared to be most susceptible to heavy rainfall events, along with coastal areas. Indeed, one notable recent event was heavy rain around Pensacola in 2014. In addition, Hurricane Ivan (2004) made landfall around this area, producing a strong inundation event by surge and inland flooding. Figure 3 shows the total rainfall for the Hurricane Ivan event.

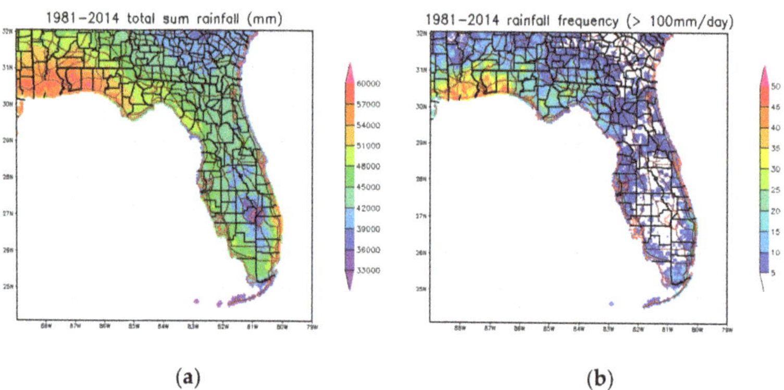

(**a**)   (**b**)

Figure 2. Rainfall total (**a**, unit is mm) and frequency (**b**) statistics for the period of 1981 to 2014 derived from the PRISM dataset.

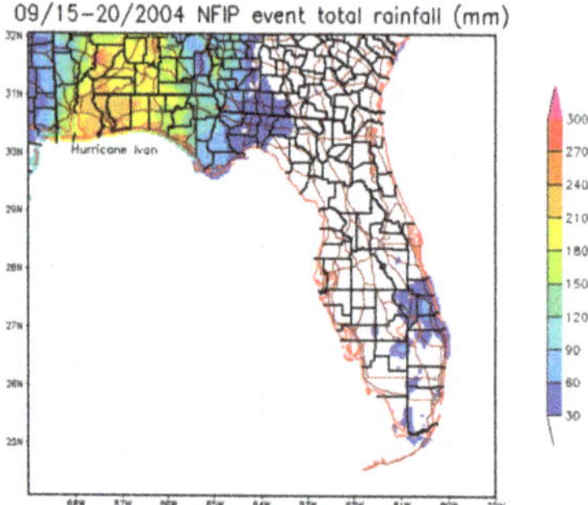

Figure 3. Cumulative rainfall for the Hurricane Ivan (2004) heavy rain event near Pensacola, Florida based on the PRISM dataset. Unit is mm. The provided event period of NFIP claims was from 15 to 20 September 2004.

Precipitation maximum data were derived from the 4 km PRISM daily rainfall database. The largest precipitation amount that occurred with ±1 day and within 10 km of the claim property location was recorded. This information helped determine whether the cause of loss could have been accumulation of rainfall.

2.2.3. NLCD Distances

The distance to coast was the distance of each claim location to the nearest coast. This information was helpful to decide whether a coastal surge flood was likely to occur. The distance was calculated using the 2011 Multiresolution Land Characteristics Consortium (MRLC) National Land Cover Database (NLCD) [13] to identify water and land locations. The NLCD has approximately 30 m resolution, allowing for a reasonably accurate calculation. Figure 4 demonstrates the quality of computed distance to coast for the Florida coastal lines, the area around Tampa, and the area around Miami.

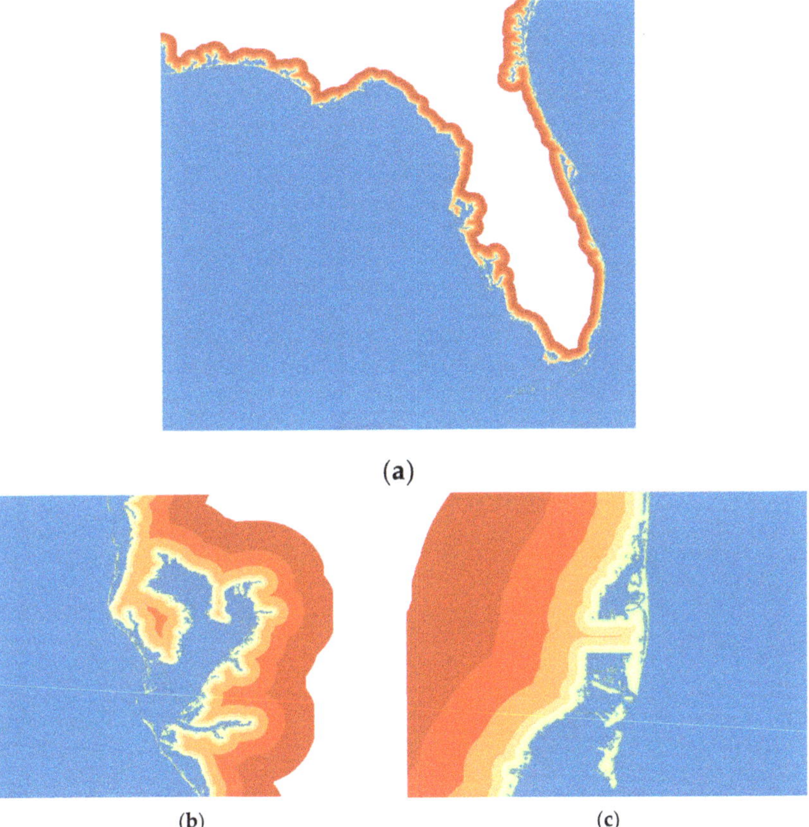

Figure 4. Distance to coast maps for the Florida coastline (**a**), the area around Tampa (**b**), and the area around Miami (**c**). Intervals: 20 m, 100 m, 500 m, 1 km, and 2 km.

The distance to the nearest body of water was computed in the same manner as the distance to the coast, but the distance was to the nearest body of water, regardless of water body type. This computed distance data could help to determine whether flooding could be due to river, stream, or lake overflow. However, these data were not explicitly used in the current analysis.

3. Results and Discussion

3.1. NFIP Claims Hazard Revision

The NFIP claim data had several fields that contained information that could be used to associate each claim to a flood hazard event. Among these fields were the *date of loss*, *property location, catastrophe number*, and *cause of loss*. The information provided by these fields was not always complete or accurate and often did not provide sufficient detail concerning the flood hazard event. For example, the catastrophe number was often missing, or multiple hazard events may have been assigned the same number if they occurred around the same time period. The cause of loss often appeared to be incorrect based on other information. There were cases, for example, in which the loss was listed as *tidal water overflow* even though the property was too far from the coast to be affected by tidal water.

The claim data often did not provide important details of the hazard, such as the flood elevation or wave conditions.

3.1.1. Catastrophe Event Number Matching

Our first task was to attribute meteorological events to claim data that were grouped by a *catastrophe number* or *date of loss*. Using a summary table provided by NFIP, the authors were able to determine the causative event for most of the groups of claim data. The results of this work were used to determine which historical events should be focused on in our modeling studies. These results, however, did not provide details of the cause of loss on the individual claims level. Our initial focus was on determining the types of events that we may need to consider in order to update the FPFLM hazard models.

Among the paid claims, 34,257 (22.3%; money-wise USD 498,827,000 (13.4%)) did not include catastrophe event numbers. The HURDAT2 tropical cyclone information was used to match tropical events to the NFIP claim data. If loss claims were located within 700 km of a storm center during the event date, it was considered that the claims belonged to that storm event. Of the 153,751 claims studied, 108,244 were detected by the HURDAT2 storm name check within claim event dates to keep and/or modify the original catastrophe event number. In addition to the provided claim count and paid amount, the revised statistics are shown in Table 1 for selected catastrophe events. The event number zero was assigned for the unidentified events. While the distinguishable number of claims were now attributed to several of the last century's hurricanes (Major Hurricane (MH) Frederic (1979), MH Georges (1998), MH Mitch (1998), etc.), not many claims were modified for recent hurricanes, perhaps because of a better collection system for claim data in the NFIP. An example of catastrophe event number updates is shown in Figure 5.

Table 1. NFIP catastrophe event number revision (selected events among total 158 events).

Catastrophe Event Number	Provided		Revised		Event
	Claim Count	Paid Amount (Thousands of USD)	Claim Count	Paid Amount (Thousands of USD)	
0	34,257	498,828	27,080	374,330	
22	32	814	897	8686	MH Frederic (1979)
131	3248	40,163	4258	51,425	MH Georges (1998)
133	1466	20,518	854	14,323	MH Georges (1998)
134	8	51	206	2471	MH Mitch (1998)
182	2492	49,429	2515	49,656	MH Charley (2004)
564	3247	139,821	3278	140,967	MH Andrew (1992)
615	241	3386	779	11,787	H Earl (1998)
641	3278	106,816	4789	139,913	MH Frances (2004)
643	10,353	945,500	10,619	959,392	MH Ivan (2004)
646	3800	90,497	3918	93,128	MH Jeanne (2004)
649	3375	109,045	3698	116,507	MH Dennis (2005)
659	9411	355,267	9616	363,625	MH Wilma (2005)

H: hurricane, MH: major hurricane.

Many claims around Miami and Keys were attributed to the event of MH Mitch (1998). The majority of claims were assigned to more reliable and accurate meteorological events in the NFIP flood loss data.

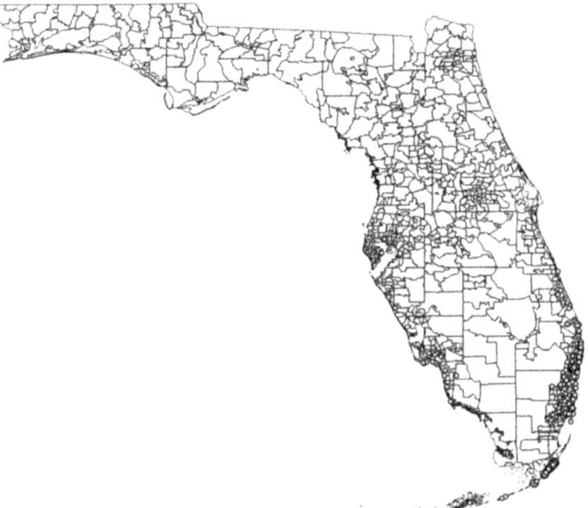

Figure 5. NFIP catastrophe event number update: Major Hurricane Mitch (period: 4 November 1998–6 November 1998); event number = 134 (see Table 1). Provided claim count = 8 (**red dots**) vs. revised claim count = 206 (**green dots**).

3.1.2. Cause of Loss Revision

The following fields were added to the original NFIP claim data: HURDAT2 storm identification, storm flag (tropical depression (TD), tropical storm (TS), hurricane (HU) or nontropical storm (NA)), revised catastrophe number, rainfall maximum, and distances to the nearest coast and body of water. Using these collected information permitted a revision of the cause of loss when the original loss cause was in clear error or ambiguous.

The algorithms for the revision were as follows:

- if the provided cause of loss was tidal water overflow (surge; cause number 1), but the distance to coast was greater than 20 km (based on personal communication with surge modeling experts), the rainfall maximum for the location was checked. If it was greater than 25 mm/day, the cause was revised to be accumulation of rainfall (cause number 4). Otherwise, it was designated as questionable, likely wrong address (see Figure 6);
- for hurricane events, if the original cause of loss was unknown for a property that was less than 20 km to the coast, and the precipitation maximum was greater than 25 mm/day, the cause of loss was marked as undetermined (cause number 5) but was either tidal water overflow or accumulation of rainfall, and further investigation was warranted. Otherwise, the cause of loss was marked as cause number 1;
- for nonhurricane events, if the original cause of loss was unknown for a property that was greater than 20 km to the coast and the precipitation maximum was greater than 25 mm/day, the cause of loss was marked as cause number 4;
- if the original cause of loss was 4 in inland areas, but the precipitation maximum was less than 5 mm/day, then the cause of loss was marked as questionable, possible error in address or date;
- for hurricane-only events, if the distance to the coast was less than 1 km, claims with cause numbers 2, 3, and 4 were revised to tidal water overflow (cause number 1). Figure 7 shows why this might be a practical revision algorithm. Hurricane Ivan (2014) hit the northwest coast of Florida (see Figure 3) and caused widespread surge damages according to FEMA. However, the cause of loss in many NFIP paid claims was marked as accumulation of rainfall (cause number 4), even in the apparent isolated island areas. All these areas were damaged by the storm surge according to the FEMA observed flood estimates.

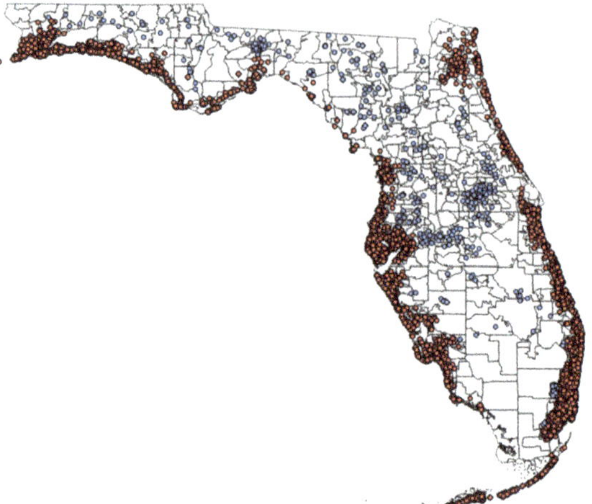

Figure 6. NFIP cause of loss revision—incorrect surge claims. All circles denote the provided NFIP surge claims. Red circles denote revised surge loss claims, while blue circles denote incorrect surge claims.

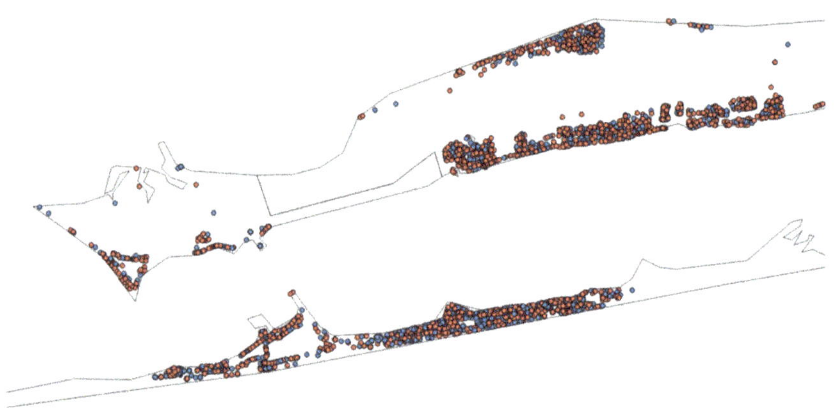

Figure 7. NFIP cause of loss—Hurricane Ivan (2004). Red dots denote surge claims, and blue dots denote accumulation of rainfall claims. Domain: −87.21 W to −87.04 W and 30.3 N to 30.4 N.

3.1.3. Summary of Revised Claims

The provided and revised NFIP claim counts and paid loss amounts with percentages are shown in Tables 2 and 3, respectively. Before our revision, the majority of flood loss causes in Florida were considered due to the accumulation of rainfall (53.1% in claim count and 46.6% in paid loss), followed by surges (29.4% in claim count and 36.6% in paid loss). Hence, it can be misinterpreted that the main flood cause of loss in Florida was accumulation of rainfall. After our systematic revision, this misinformation was dramatically changed to the following: the main cause of flood loss in Florida was storm surge events. Tidal water overflow now explained 46.4% of claims by count and 65.0% of paid losses. While there was just a ca. 17% increase in claim count, there was an increase of more than USD 1 billion (USD 2415 million out of 3714 million) in the paid loss amount.

Table 2. The NFIP claims count summary in accordance with the cause of loss.

Cause	Provided		Revised	
	Count	%	Count	%
0	11,069	7.2	215	0.1
1	**45,222**	**29.4**	**71,386**	**46.4**
2	11,303	7.4	7401	4.8
3	1919	1.2	1522	1.0
4	**81,614**	**53.1**	**59,364**	**38.6**
5	0	0	9596	6.2
6	2624	1.7	4267	2.8
all	153,751	100	153,751	100

0: other causes; 1: tidal water overflow; 2: stream, river, or lake overflow; 3: alluvial fan overflow; 4: accumulation of rainfall; 5: indeterminate, cause 1 or 4; 6: all other causes.

Table 3. The NFIP claims paid loss amount summary in accordance with the cause of loss (see Table 2 for the cause number reference).

Cause	Provided		Revised	
	Paid Loss (Thousands of USD)	%	Paid Loss (Thousands of USD)	%
0	318,190	8.6	4046	0.1
1	**1,358,612**	**36.6**	**2,414,944**	**65.0**
2	260,414	7.0	132,077	3.6
3	42,905	1.2	32,738	0.9
4	**1,729,295**	**46.6**	**810,876**	**21.8**
5	0	0	286,971	7.7
6	4770	0.1	32,534	0.9
all	3,714,186	100	3,714,186	100

A summary of the general types of meteorological events and their respective portions of the claim data is provided in Table 4 for the revised NFIP events. In this table, it can be seen that hurricanes were by far the primary cause of flooding damage in Florida, accounting for 72.1% of the revised claim losses and 54.5% of claims by count. The losses labeled tropical storm and depression did not reach hurricane strength while damage occurred. These events accounted for approximately 13% of the losses in the revised claims. While the losses for tropical storms were lower than for hurricanes, the number of potential events that may need to be considered could be larger. Tropical storms may produce lower surges than hurricanes, but the rainfall from such storms can often be quite significant. The third class of general flood hazard events in Table 4 was nontropical systems. These events are usually heavy rainfall events that are not caused by tropical systems. These events, however, do include nontropical cyclones, such as the *Storm of the Century* (March, 1993). These cyclones can cause surges as well as heavy rain. Currently, the FPFLM does not have any model to take these types of events into account.

Table 5 shows the detailed breakdown of NFIP paid losses due to each type of meteorological events and cause of loss. For hurricane events, approximately 80% of paid losses were due to storm surges. Meanwhile only 21.6% were due to surges in tropical storm events. The accumulation of rainfall explained 9.5% for hurricane, 58.5% for tropical storms, 81.9% for tropical depressions, and 49.1% for nontropical events. The importance of nontropical cyclone events to storm surges and associated coastal inundation was examined through the NFIP claim data. Based on the revised NFIP, the number of nontropical surge claims was 10,564 (~7% of the paid claims and ~15% of storm surge claims), accounting for paid losses of USD 179,719,000 (~5% of the paid claims and ~7.6% of storm surge claims). The Storm of the Century accounted for 5875 surge claims out of total 8846 claims, amounting to paid losses of USD 132,764,000 (79%) for coastal surge claims.

Table 4. The revised NFIP paid loss amounts and corresponding claim counts in accordance with the type of meteorological event.

Cause	Revised NFIP Events			
	Paid (Thousands of USD)	%	Claim Count	%
Nontropical	574,612	15.5	38,733	25.2
Tropical Depression	160,431	4.3	9467	6.2
Tropical Storm	302,267	8.1	21,733	14.1
Hurricane	**2,676,876**	**72.1**	**83,818**	**54.5**
All	3,714,186	100	153,751	100

Table 5. Breakdown of the NFIP claim paid loss amounts due to each type of event as a function of cause of loss (see Table 2 for the cause number reference).

Cause	Hurricane Only		Tropical Storm Only	
	Paid Loss (Thousands of USD)	%	Paid Loss (Thousands of USD)	%
0	182	0	61	0
1	**2,128,099**	**79.5**	**65,284**	**21.6**
2	42,602	1.6	38,657	12.8
3	1106	0	920	0.3
4	**253,644**	**9.5**	**176,714**	**58.5**
5	247,293	9.2	19,970	6.6
6	3950	0.1	661	0.2
all	2,676,876	100	302,267	100

Cause	Tropical Depression Only		Nontropical Only	
	Paid Loss (Thousands of USD)	%	Paid Loss (Thousands of USD)	%
0	9	0	3794	0.7
1	**1207**	**0.8**	**179,719**	**31.3**
2	7633	4.8	50,562	8.8
3	184	0.1	30,933	5.4
4	**131,427**	**81.9**	**282,374**	**49.1**
5	19,709	12.3	0	0
6	262	0.2	27,230	4.7
all	160,431	100	574,612	100

3.2. Further Analyses for Flood Risk Assessments

3.2.1. Nontropical Flood Claims

In order to determine an appropriate method or model for estimate losses due to nontropical inland rainfall flood events, the analysis of the revised NFIP claim data was expanded. It was previously reported that approximately USD 574 million in losses were due to nontropical events (Table 4), specifically events that could not be directly attributed to an event in the HURDAT2 database. The authors further broke down the losses by excluding non-rainfall-related events, such as surges, and determining whether the claims were within 1 km (0.6 miles) of the coast. The results are summarized in Table 6. As shown in Table 6, the losses due to nontropical rainfall for inland locations totaled about USD 228 million. These were losses due to events that could be potentially modeled by a nontropical rainfall model (or set of models).

Table 6. Breakdown of the revised NFIP paid loss amount due to nontropical events.

	Loss (Millions of USD)	% of NFIP	Remarks
Nontropical flood	574	15	Includes surge due to nontropical cyclones
Nontropical nonsurge	394	11	Excludes all surge
Nontropical rain-related	372	10	Excludes surge and other/unknown
Nontropical rain—inland	**228**	**6**	Greater than 1 km to coast
Total NFIP all causes	3714	100	

To gain further insight into these types of only-rainfall-related losses, the authors show the top five loss events in Table 7. The largest loss, USD 29.7 million, was due to a heavy frontal rainfall event in December, 2009. The second largest event was due to the record El Nino winter of 1997–1998. That event lasted several months, and there were claims with dates of loss for nearly every day from 24 December 1997 to 10 April 1998. The fourth largest loss, about USD 12 million, was due to the *Storm of the Century* in March, 1993, which was a very unusually strong nontropical cyclone that entered Florida from the Gulf. Note that these losses did not include surge losses, which were larger. The third and fifth largest losses were possibly due to remnants of tropical disturbances or storms that were not fully recorded in the HURDAT2 database.

Table 7. The top five NFIP nontropical loss events in order of paid loss amount.

Claim Count	Loss (Thousands of USD)	Event Duration		Remarks
		Begin	End	
702	29,671	9 December 2009	20 December 2009	17 December 2009 heavy rain
1154	14,302	24 December 1997	10 April 1998	El Nino record event
558	13,618	28 October 2011	2 November 2011	Possible related to tropical disturbance
529	11,999	12 March 1993	24 March 1993	Storm of the Century
467	11,244	12 May 2009	29 May 2009	Possible remnant of tropical depression 1

Examination of these losses indicated that there were a number of unique and varied meteorological conditions that lead to these heavy rainfall events. This presents a challenge in terms of developing a robust rainfall model. Since the total rainfall-related losses that can be modeled accounted for less than 6% of the NFIP losses historically, a decision needs to be made as to whether these types of losses should be explicitly modeled as opposed to making an actuarial adjustment of the losses to include them implicitly.

3.2.2. Loss Convergence in 30 Zones

The Flood Standards [1] require that modelers create a minimum of 30 geographic zones and verify that any Monte Carlo simulations required by the flood model to estimate losses converge to within a standard error of 5% in each zone of Florida (unless otherwise justified). The authors developed a preliminary set of zones based on the nearest coastal target location using a carefully selected set of 30 target coastal city locations. All zones were developed to have coastal exposure, so that storm surge risk affected every zone. The zones were made somewhat smaller in more vulnerable regions and larger in less vulnerable regions, to ensure a sufficient loss in all zones, while maintaining a reasonable size and structure, so that risk convergence was suitably ensured over all of Florida. These zones are subject to revision once stochastic losses are available and we have better understanding of the distribution of the vulnerability. As stochastic losses had not become available, the NFIP claims data was used for guidance. Figure 8 shows the preliminary set of 30 zones, along with the target coastal locations that define the zones. In addition, the NFIP losses are shown for each zone in millions of USD (paid losses). The Pensacola zone had the largest loss, at USD 592 million, mainly due to Hurricane Ivan (2004). Next were Miami (USD 423 million) and Navarre (USD 415 million). The least vulnerable zone was St. Augustine Beach, at USD 11 million. Note that these losses were heavily influenced by singular events and thus should not be interpreted as an overall measure of risk. If a zone were to have no or negligible losses, then convergence may not be possible.

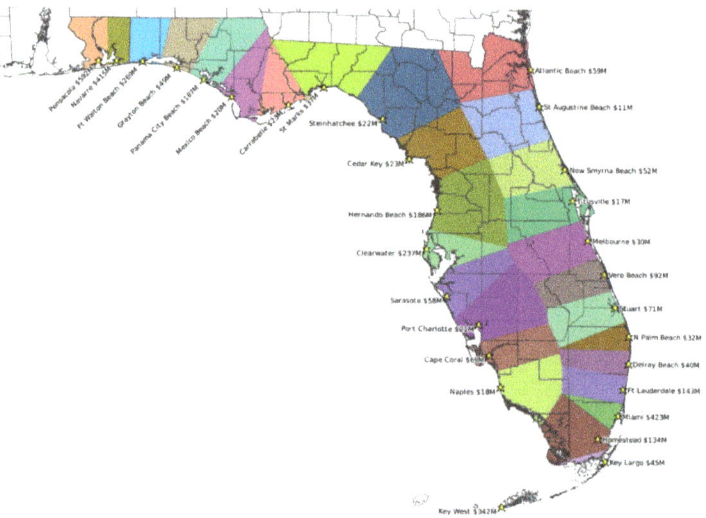

Figure 8. Preliminary creation of 30 zones to be used for Monte Carlo convergence tests. Yellow stars indicate target locations that define the zones. Losses for each zone were based on the analysis of NFIP claim data.

3.2.3. Historical Reconstruction

In general, observations of flood elevations and wave conditions are not available or very limited. To obtain estimates of these data, the authors used the FPFLM hazard models and other data to reconstruct select historical flood events. For storm surges, the Coastal Estuarine Storm Tide (CEST) model [14] was used. The CEST is forced by estimated observed winds from the H*Wind hurricane wind analyses [15] where available; otherwise, modeled winds are used. Wave conditions were determined by a wave model from the FPFLM that was based on the STWAVE (steady-state spectral wave) model [16]. For inland flooding, a simple two-dimensional modeling framework [17] was used. The modeled hazard data was interpolated to the claim locations, taking into account very-high-resolution 5 m DEM (Digital Elevation Model) elevation data. Based on the modeled results, each claim in the revised NFIP portfolio could be assigned to one of four hydrological states: inland flood with no waves; coastal flood with minor waves; coastal flood with moderate waves; and coastal flood with severe waves. A better set of flood vulnerability curves can be developed based on these hydrological states and can be validated using the modified NFIP claim data for the FPFLM project [9,10]

4. Conclusions

The NFIP claim data constitute an essential resource for developing, calibrating, and validating a flood risk model. They can be intensively used, for example, in the assessment of the flood outcomes from hazard (i.e., surge, wave and inland flooding) models and in the development process of a set of vulnerability curves. The provided original (especially hazard) fields in NFIP claim data are not always complete or accurate and often missing. In order to determine significant types of meteorological events that can cause flooding, the authors tackled the revision of the NFIP claim data in Florida filed between 19 July 1975 and 10 January 2014. The hazard fields in the NFIP claim data were systematically revised using existing meteorological and geographical data in Florida. Tropical cyclone events, rainfall maxima, and distance to coast information permitted revisions of the cause of loss when the provided claim loss causes were in clear error or ambiguous. After our revision, the number and paid loss amount of surge claims in Florida increased dramatically, while those of accumulation of rainfall claims decreased. Sample usages of the revised NFIP claim data are presented in this study as well in the analyses of nontropical flood claims, loss convergence, and historical reconstruction of hydrological states.

The types of meteorological events that can lead to flooding are varied. For hurricane events, the currently developing FPFLM has both a surge model and an inland flood model that have been implemented and are now being evaluated using the revised NFIP claim data. However, for nonhurricane events, we do not have such models. There are unusual cases, such as the tropical disturbance in southeast Florida in 2000 (2 October 2020–6 October 2000). At that time, there was no tropical cyclone present, but heavy rain from a disturbance caused significant flooding. A tropical storm later developed from this disturbance after it left the region. Based on the NFIP claim data, the damage due to these rainfall events appeared to exceed the flood damage due to Hurricane Andrew (1992), which impacted the same region. Since the event was not a cyclone at that time, we could not model this event.

It should be noted that our criteria used for the revision of cause of loss in the current paper were somewhat arbitrary, such as the radius of hurricane influence (700 km), rainfall criteria (25 or 5 mm/day), distance to coast for surge influence (20 km), and surge assignment for hurricane event (<1 km). If these criteria were selected more scientifically, the reliability of revised claim counts and loss amounts might be improved. For example, in lieu of a <1 km rule for surge assignment, we could use Florida DEM surge zones based on NHC SLOSH model runs with FEMA observed flood estimates. If there is still doubt about the revised cause of loss, individual claims should be checked manually with added meteorological and georeferencing information to attain a proper selection of cause of loss.

Author Contributions: Conceptualization, D.W.S. and S.C.; methodology, D.W.S. and S.C.; software, D.W.S. and B.-M.K.; validation, D.W.S., S.C. and B.-M.K.; formal analysis, D.W.S. and S.C.; investigation, B.-M.K.; resources, S.C. and B.-M.K.; data curation, D.W.S.; writing—original draft preparation, D.W.S.; writing—review and editing, S.C. and B.-M.K.; visualization, D.W.S. and S.C.; supervision, S.C.; project administration, S.C. and B.-M.K.; funding acquisition, S.C. and B.-M.K. All authors have read and agreed to the published version of the manuscript.

Funding: This research was funded by the Florida Office of Insurance Regulation (FLOIR). B.K. was partially supported by "Investigation and prediction system development of marine heatwave around the Korean peninsula originated from the sub-Arctic and western Pacific" project supported by the Ministry of Ocean (No. 2019044).

Institutional Review Board Statement: Not applicable.

Informed Consent Statement: Not applicable.

Data Availability Statement: Data used in this study are not available for public use. However, redacted NFIP claim data can be obtained from the FEMA website (https://www.fema.gov/openfema-data-page/fima-nfip-redacted-claims, accessed on 11 February 2022).

Acknowledgments: FLOIR provided the authors with the NFIP claim data. The views expressed in this paper are those of the authors and do not necessarily reflect the views of FLOIR or any of its associated agencies.

Conflicts of Interest: The authors declare no conflict of interest.

References

1. Florida Commission on Hurricane Loss Projection Methodology. Flood Standards Report of Activities. Available online: www.sbafla.com/methodology (accessed on 15 January 2022).
2. Florida Commission on Hurricane Loss Projection Methodology. Hurricane Standards Report of Activities. Available online: www.sbafla.com/methodology (accessed on 15 January 2022).
3. Hamid, S.; Pinelli, J.P.; Cheng, S.C.; Gurley, K. Catastrophe model based assessment of Hurricane risk and estimates of potential insured losses for the State of Florida. *Nat. Hazard Rev.* **2011**, *12*, 171–183. [CrossRef]
4. Kousky, C.; Michel-Kerjan, E. Examining flood insurance claims in the United States: Six key findings. *J. Risk Insur.* **2015**, *84*, 819–850. [CrossRef]
5. McGrath, H.; El Ezz, A.A.; Nastev, M. Probabilistic depth-damage curves for assessment of flood-induced building losses. *Nat. Hazards* **2019**, *87*, 1–14. [CrossRef]
6. Wing, O.E.J.; Pinter, N.; Bates, P.D.; Kousky, C. New insights into US flood vulnerability revealed from flood insurance big data. *Nat. Commun.* **2020**, *11*, 1444. [CrossRef] [PubMed]
7. Davis, S.A.; Skaggs, L.L. *Catalog of Residential Depth-Damage Functions Used by the Army Corps of Engineers in Flood Damage Estimation*; IWR-92-R-3; USACE Institute for Water Resources: Fort Belvoir, VA, USA, 1992.
8. Skaggs, L.L.; Davis, S.A. *Variations in Residential Depth-Damage Functions Used by the US Army Corps of Engineers in Flood Damage Estimation*; Association of State Floodplain Managers: Atlanta, GA, USA, 2014.
9. Pinelli, J.-P.; Rodriguez, D.; Roueche, D.; Gurley, K.; Baradaranshoraka, M.; Cocke, S.; Shin, D.W.; Lapaiche, L.; Gay, R. Data management for the development of a flood vulnerability model. In *Safety and Reliability–Safe Societies in a Changing World*; Taylor & Francis: Boca Raton, FL, USA, 2018; pp. 2781–2788. [CrossRef]
10. Pinelli, J.-P.; Cruz, J.D.; Gurley, K.; Paleo-Torres, A.S.; Baradaranshoraka, M.; Cocke, S.; Shin, D.W. Uncertainty reduction through data management in the development, validation, calibration, and operation of a hurricane vulnerability model. *Int. J. Disaster Risk Sci.* **2020**, *11*. [CrossRef]
11. Landsea, C.W.; Franklin, J.L. Atlantic hurricane database uncertainty and presentation of a new database format. *Mon. Weather Rev.* **2003**, *141*, 3576–3592. [CrossRef]
12. Daly, C.; Halbleib, M.; Smith, J.I.; Gibson, W.P.; Doggett, M.K.; Taylor, G.H.; Curtis, J.; Pasteris, P.P. Physiographically sensitive mapping of climatological temperature and precipitation across the conterminous United States. *Int. J. Climatol.* **2008**, *28*, 2031–2064. [CrossRef]
13. Homer, C.G.; Dewitz, J.; Yang, L.; Jin, S.; Danielson, P.; Xian, G.Z.; Coulston, J.; Herold, N.; Wickham, J.; Megown, K. Completion of the 2011 national land cover database for the conterminous United States-Representing a decade of land cover change information. *Photogramm. Eng. Remote Sens.* **2015**, *81*, 345–354.
14. Zhang, K.; Xiao, C.; Shen, J. Comparison of the CEST and SLOSH models for storm surge flooding. *J. Coast. Res.* **2008**, *24*, 489–499. [CrossRef]
15. Powell, M.D.; Houston, S.H.; Amat, L.R.; Morisseau-Leroy, N. The HRD real-time hurricane wind analysis system. *J. Wind. Eng. Ind. Aerodyn.* **1998**, *77–78*, 53–64. [CrossRef]
16. Massey, T.C.; Anderson, M.E.; Smith, J.M.; Gomez, J.; Jones, R. *STWAVE: Steady-State Spectral Wave Model. User's Manual for STWAVE, Version 6.0*; Report ERDC/CHL SR-11-1; US Army Corps of Engineers: Washington, DC, USA, 2011.
17. Jamali, B.; Bach, P.M.; Cunningham, L.; Deletic, A. A cellular automata fast flood evaluation (CA-ffe) model. *Water Resour. Res.* **2019**, *55*, 4936–4953. [CrossRef]

applied sciences

Article

Prediction of Liquefaction-Induced Lateral Displacements Using Gaussian Process Regression

Mahmood Ahmad [1,2], Maaz Amjad [3], Ramez A. Al-Mansob [1,*], Paweł Kamiński [4], Piotr Olczak [5], Beenish Jehan Khan [6] and Arnold C. Alguno [7]

[1] Department of Civil Engineering, Faculty of Engineering, International Islamic University Malaysia, Jalan Gombak, Selangor 50728, Malaysia; ahmadm@iium.edu.my
[2] Department of Civil Engineering, University of Engineering and Technology Peshawar (Bannu Campus), Bannu 28100, Pakistan
[3] Department of Civil Engineering, University of Engineering and Technology, Peshawar 25120, Pakistan; maazamjad.civ@uetpeshawar.edu.pk
[4] Faculty of Civil Engineering and Resource Management, AGH University of Science and Technology, 30-059 Krakow, Poland; pkamin@agh.edu.pl
[5] Mineral and Energy Economy Research Institute, Polish Academy of Sciences, Wybickiego St. 7A, 31-261 Krakow, Poland; olczak@min-pan.krakow.pl
[6] Department of Civil Engineering, CECOS University of IT and Emerging Sciences, Peshawar 25000, Pakistan; beenish@cecos.edu.pk
[7] Department of Physics, Mindanao State University-Iligan Institute of Technology, Iligan City 9200, Philippines; arnold.alguno@g.msuiit.edu.ph
* Correspondence: ramez@iium.edu.my

Citation: Ahmad, M.; Amjad, M.; Al-Mansob, R.A.; Kamiński, P.; Olczak, P.; Khan, B.J.; Alguno, A.C. Prediction of Liquefaction-Induced Lateral Displacements Using Gaussian Process Regression. *Appl. Sci.* **2022**, *12*, 1977. https://doi.org/10.3390/app12041977

Academic Editors: Andrea Chiozzi, Elena Benvenuti and Željana Nikolić

Received: 31 December 2021
Accepted: 8 February 2022
Published: 14 February 2022

Publisher's Note: MDPI stays neutral with regard to jurisdictional claims in published maps and institutional affiliations.

Abstract: During severe earthquakes, liquefaction-induced lateral displacement causes significant damage to designed structures. As a result, geotechnical specialists must accurately estimate lateral displacement in liquefaction-prone areas in order to ensure long-term development. This research proposes a Gaussian Process Regression (GPR) model based on 247 post liquefaction in-situ free face ground conditions case studies for analyzing liquefaction-induced lateral displacement. The performance of the GPR model is assessed using statistical parameters, including the coefficient of determination, coefficient of correlation, Nash–Sutcliffe efficiency coefficient, root mean square error ($RMSE$), and ratio of the $RMSE$ to the standard deviation of measured data. The developed GPR model predictive ability is compared to that of three other known models—evolutionary polynomial regression, artificial neural network, and multi-layer regression available in the literature. The results show that the GPR model can accurately learn complicated nonlinear relationships between lateral displacement and its influencing factors. A sensitivity analysis is also presented in this study to assess the effects of input parameters on lateral displacement.

Keywords: lateral displacement; liquefaction; Gaussian process regression; sensitivity analysis; machine learning

1. Introduction

Loss of life and property remains an unavoidable consequence of major earthquakes. Studies of the consequences of major earthquakes have attempted to analyze the damage and make recommendations for reducing loss in the event of future earthquakes throughout history [1–3]. Liquefaction-induced lateral displacement is one of the most prevalent and damaging of these effects. It can cause enormous blocks of soil to move by a few millimeters to 10 m or more, inflicting substantial damage to lifeline networks, buried utilities, and a variety of other subsurface and civil engineering projects. Liquefaction-induced lateral displacement is most common on gentle slopes built on loose sand with a groundwater table close to the surface of the ground; however, open faces such as stream channels can also be susceptible [4].

Various approaches have been presented to estimate the magnitude of lateral displacement to date, and from the technical perspective, they can be classified as: (1) numerical analysis based on finite element or finite difference approaches (e.g., Finn et al. [5], Liao et al. [6] and Arulanandan et al. [7], (2) simplified analytical methods, e.g., Newmark [8], Towhata et al. [9], and Kokusho and Fujita [10], (3) empirical methods based on either laboratory testing set or analytical methods of lateral spreading case history records (e.g., Hamada et al. [11] and Youd et al. [12]) and (4) machine learning approaches (e.g., Wang and Rahman [13]). These different approaches are reviewed herein, with particular emphasis on empirical models and soft computing techniques.

1.1. Finite Element Analysis

To simulate different aspects of liquefaction and lateral spreading, including seismic loads, rapid loss of shear strength, redistribution of pore water pressure, and soil softening, Liao et al. [6] reported that very complex finite element and finite difference approaches are required. Very intricate numerical techniques, large computer skills, and extensive resources are necessary to create a realistic three-dimensional simulation inside the real-time domains. A number of well-known finite element method (FEM) and finite difference (FD) software programmes are used for liquefaction-induced lateral displacement assessments and earthquake soil dynamic analysis. Other finite element models provided by Hamada et al. [14] and Orense and Towhata [15] to determine the lateral ground deformations generated by earthquakes. Gu et al. [16,17] estimated liquefaction deformation using a planar strain model. It successfully anticipated the pattern of displacements at a wildlife site in California [17], but overestimated the magnitude of displacements by around 30%.

1.2. Simplified Analytical Models

1.2.1. Sliding Block Model

Newmark [8] proposed a model based on strategy of sliding block on a frictional sloping surface that predicted seismically induced ground deformations by integrating accelerations above the sliding block's yield acceleration to obtain its velocities. The angle of inclination and the factor of safety over sliding are associated to yield acceleration. When the driving force (seismic acceleration) equal to or greater than resisting force (yield acceleration), block will begin to slide. The total cumulative resulting deformation is then determined by integrating the sliding block velocity. Yegian et al. [18] used Newmark's concept to introduce their model for predicting the permanent ground displacement expressed as

$$D = N_{eq}T^2a_pf\left(\frac{a_y}{a_p}\right)$$ (1)

where D is the lateral ground deformation, N_{eq} denote cycles number equivalent to uniform base motion, T denotes time interval (s), a_y denotes yield acceleration (g), a_p denotes peak acceleration (g), and f denotes dimensionless function that depends on base motion. Baziar et al. [19] also used Newmark's concept, assuming an equivalent sinusoidal base acceleration record, to propose their model for predicting the permanent ground displacement expressed as:

$$\log D = 1.46\log I_a - 6.642a_y + 1.546$$ (2)

D denotes lateral ground deformation (cm), I_a presents arias intensity (m/s), and a_y presents yield acceleration (g).

1.2.2. Minimum Potential Energy Model

This model was proposed by Towhata et al. [9] depending on the results of shaking table testing. The final position of soil layers was found by the principle of minimal potential energy, using the Lagrangian equations of motion, and assuming the variation of lateral ground deformation with depth as a sine function and with neglecting inertial effects during dynamic loading. Tokida et al. [20] used the same principle to establish equations for predicting the maximum lateral displacement at the center of a slide as:

$$D = 1.73 \times 10^{-5} L^{1.99} H^{0.298} T^{-0.275} \theta^{0.963} \text{ for } (10\ m \leq L \leq 100\ m) \tag{3}$$

$$D = 1.29 \times 10^{-5} L^{1.99} H^{0.28} T^{-0.243} \theta^{0.995} \text{ for } (100\ m < L \leq 1000\ m) \tag{4}$$

where D is horizontal displacement (m), L = length of slide (m), H represents average thickness of liquefied layer (m), T represents average thickness of liquefied surface layer (m), and θ is the slope of ground surface express in percentage.

1.2.3. Shear Strength Loss and Strain Re-Hardening Model

Bardet et al. [21] reported that in 1997, Byrne proposed a method to find the final position of a liquefying slope using the finite element software tool Fast Lagrangian Analysis of Continua. In liquefaction region, it is assumed that the liquefied material is initially free of shear, and is subjected to isotropic pressure. After such immediate melting of liquefied soil, the shear stress (τ) was supposed to rise with shear strain unless reached a certain residual shear strength (τ_{ST}). Although the liquefied soil regains shear strength, the shear modulus was supposed to take a constant value G_{LIQ}. The final position of the slope is determined using the dynamic equation of motion.

1.2.4. Viscous Models

Hadush et al. [22] reported that Aydan [23] considered the liquefied subsoil to act as a visco-elastic object and used an upgraded Lagrangian numerical approach to find the deformation velocities for the liquefied soil sub-layers. They also proposed a numerical method based on cubic interpolated pseudoparticles for liquefaction-induced lateral displacement analysis in the context of fluid dynamics. Liao et al. [6] reported that Hamada et al. [24] recommended to use viscous models to estimate the liquefaction-induced lateral displacement. Kokusho and Fujita [10] studied the role of water film in lateral flow failure during earthquakes, on the basis of field survey results collected from Niigata (1964) earthquake. It was reported that the water films produced under the fine soil sub layers did not actually have shear resistance, and a significant factor for the large lateral flow displacement.

1.3. Empirical Models and Soft Computing Techniques

Hamada et al. [11] provided a preliminary relation of measuring horizontal ground displacement in meter's relying on 60 case histories, majority of them are obtained in Niigata and Noshiro, Japan. It can be seen from Table 1 that the equations are very common and easy to apply that contains only two parameters of site geometry and not considered seismic and geotechnical parameters but it has been suggested for limited dataset making it insufficiently broad to be extended to additional lateral displacement sites.

Youd and Perkins [25] suggested "liquefaction severity index" (*LSI*) to estimate maximum horizontal ground displacement generated by an earthquake. The *LSI* (inches) was calculated using distance to seismic energy source (R) (km); and moment magnitude (M_w) with maximum range of horizontal ground displacement as 2.5 m. This model assumes that the value of *LSI* depends on only seismic parameters (R, M_w). At the time, the proposed equation drew the attention of engineers. Although this method may have been useful for assessing lateral spreads inside the western United States, but lacks applicability and hence didn't receive widespread use.

Bardet et al. [21] used multiple linear regression (MLR) for developing relation to estimate lateral ground deformation for free face and sloping ground situations, respectively, utilizing data gathered by Bartlett and Youd [26,27], including three kinds of input variables:

1. Seismic parameters—seismic source distance (R, km) and earthquake magnitude, (M).
2. Topographic characteristics (in percent)—gradient of ground surface (S) and free face ratio.

3. Geotechnical parameters (in percent)—average mean particle size within T_{15} ($D50_{15}$, mm) and averaged fines contents in T_{15} (F_{15})

The MLR approach was used to create the Youd et al. [12] model, presented in 1992, was built on upgraded results of Bartlett and Youd [26,27] for estimating lateral ground displacement, D_H (m). As indicated in Table 1 the models include free face and sloping ground conditions equations. This model gained attraction amongst geotechnical engineers due to its utilization of a huge dataset from various earthquakes, as well as geometry of site, geotechnical data, and seismic characteristics. Although, it does have some limits in terms of applications. For example, the free face equation was used when $5 \leq W \leq 20\%$, Jafarian and Nasri [28] gathered the latest dataset of liquefaction-induced lateral ground deformation based on uncertainties of different boreholes, which outperformed Hamada et al. [11] Kanibir [29], Al Bawwab [30], Javadi et al. [31], Youd et al. [12], and Baziar and Azizkandi [32] models.

Table 1. Empirical and machine learning approaches for liquefaction-induced lateral displacement.

Method and Technique		Model	Reference
Empirical model	Regression Analysis	$D_H = 0.75H^{1/2}\theta^{1/3}$	Hamada et al. [11]
		$\log LSI = -3.49 - 1.86 \log R + 0.98 M_w$	Youd and Perkins [25]
		$\begin{aligned}\log(D_H + 0.01) = {}& -17.372 + 1.248 M_w - 0.923 \log R - 0.014R \\ & +0.685 \log W + 0.3 \log T_{15} + 4.826 \log(100 - F_{15}) \\ & -1.091 D50_{15} \log(D_H + 0.01) \\ = {}& -14.152 + 0.988 M_w - 1.049 \log R - 0.011R \\ & +0.318 \log S + 0.619 \log T_{15} \\ & +4.287 \log(100 - F_{15}) - 0.705 D50_{15}\end{aligned}$	Bardet et al. [21]
		$\begin{aligned}\log(D_H) = {}& -16.713 + 1.532M - 1.406 \log R^* - 0.012R + 0.592 \log W \\ & +0.540 \log T_{15} + 3.413 \log(100 - F_{15}) - 0.795 \log(D50_{15} + 0.1\text{mm}) \\ \log(D_H) = {}& -16.213 + 1.532M - 1.406 \log R^* - 0.012R + 0.338 \log S \\ & +0.540 \log T_{15} + 3.413 \log(100 - F_{15}) - 0.795 \log(D50_{15} + 0.1\text{mm}) \\ & R^* = R_0 + 10^{0.98M - 5.64}\end{aligned}$	Youd et al. [12]
		$\begin{aligned}\log(D_H) = {}& -17.95 + 1.605 M_w - 1.8673 R^* - (\log(R + 20))^{-3.3836} \\ & +0.547 \log W + 0.4431 \log T_{15} + 4.1873 \log(100 - F_{15}) \\ & -0.7666 \log(D50_{15} + 0.1\text{mm}) \log(D_H) \\ = {}& -19.63 + 2.0137 M_w - 2.6124 \log R^* \\ & -(\log(R + 20))^{-2.7004} + 0.3147 \log S \\ & +0.6985 \log T_{15} + 4.1954 \log(100 - F_{15}) \\ & -0.6772 \log(D50_{15} + 0.1\text{mm})\end{aligned}$	Jafarian and Nasri [28]
Soft computing methods	ANN	$D_H = f(M, R, D50_{15}, T_{15}, F_{15}, W, S, N1_{60s})$	Wang and Rahman [13]
		$D_H = f(M, R, D50_{15}, T_{15}, F_{15}, W, S)$	Baziar and Ghorbani [33]
	GP	$\begin{aligned}D_H = {}& -163.1\frac{1}{M^2} + 57\frac{1}{R \cdot F_{15}} - 0.0035\frac{T_{15}^2}{W \cdot D50_{15}^2} + 0.02\frac{T_{15}^2}{F_{15} \cdot D50_{15}^2} \\ & -0.26\frac{T_{15}^2}{F_{15}^2} + 0.006 T_{15}^2 - 0.0013 W^2 + 0.0002 M^2 \cdot W \cdot T_{15} + 3.7 \\ D_H = {}& -0.8\frac{F_{15}}{M} + 0.0014 F_{15}^2 + 0.16 T_{15} + 0.112 S + 0.04\frac{S \cdot T_{15}}{D50_{15}} \\ & -0.026 R \cdot D50_{15} + 1.14\end{aligned}$	Javadi et al. [31]
	ANFIS	$D_H = f(M, R, D50_{15}, T_{15}, F_{15}, W, S)$	Javdanian [34]

Note: $N1_{60s}$: $(N1)_{60}$ value corresponds to Js, Js is the lowest factor of safety below water table using simplified approach; θ: larger slope of either ground surface or the base of liquefied soil (%); H: thickness of liquefied zone (m); R^*: modified source distance factor that is a function of earthquake magnitude.

Soft computing is made up of a variety of techniques that function together, such as: artificial neural network, genetic algorithm, neuro-computing etc. Wang and Rahman [13] reported that new area of machine learning has arisen for handling decisions, modeling, and control issues. Baziar and Ghorbani [33] and Wang and Rahman [13] both used artificial neural networks (ANN) to estimate horizontal ground displacement. Javadi et al. [31] computed lateral displacement for free face and sloping ground using genetic programming (GP) using upgraded case data from Youd et al. [12]. In comparison to the MLR approach, the proposed GP approach has some advantages. Table 1 shows the proposed equations for free face and sloping ground. Adaptive neuro-fuzzy inference system (ANFIS) based approach was suggested by Jadanian [34] using 426 case histories data and shows an improvement to the Youd et al. [12], Kanibir [29], Bardet et al. [21], and Rezania et al. [4] models.

Soft computing methodologies are more accurate than analytical formulas, according to all of these studies. The findings revealed that the ML models mentioned above are capable of obtaining the experimental observations with acceptable accuracy. However, this field continues to be further explored.

The Gaussian process regression (GPR) approach has been successfully applied in many domains, but its application in geotechnical engineering is limited based on literature surveys. Considering the improved performance of GPR, it is, however, used for the first time in this study to predict the liquefaction-induced lateral spread displacement for free face condition. To demonstrate the efficacy of the proposed GPR-based model, the results are compared with various well-known models for calculating the D_H.

2. Gaussian Process Regression

Gaussian process regression (GPR) is one of the appropriate and newly-proposed methods that have been employed for various machine learning examples. GPR is a stochastic, non-parametric technique for addressing complicated and non-linear challenges. GPR assumes that the target variable m is determined as follows:

$$m = f(n(k)) + \varepsilon \tag{5}$$

where f represents unidentified functional dependency, n represents the number input parameters, and ε represents Gaussian noise with variance $\sigma_a{}^2$. It's a method of indicating precedence straight over function space. The mean and covariance of a Gaussian distribution are matrices and vectors, respectively. The GPR model can determine the prediction distributions, which is similar to ensuring input knowledge [35]. The GPR approach is based on the idea that surrounding data informs neighbours.

GPR makes use of a number of kernel functions. A restriction of GPR regression is the selection of a suitable kernel function. Pearson VII kernel function (*PUK*) is utilized for GPR proposed model in this work.

$$PUK = \left(1 / \left[1 + \left(2\sqrt{\left\| x_i - x_j \right\|}^2 \sqrt{2^{(1/\omega)} - 1} / \sigma \right)^2 \right]^\omega \right) \tag{6}$$

where ω and σ are the Person's width, and peak tailing factor, respectively.

3. Case-History Database

The case-study dataset used for this work was collected using three sources (Chu et al. [36], Youd et al. [12], and Cetin et al. [37]) which contains a total of 247 records of lateral displacement related to free face ground conditions.

The input parameters chosen by Youd et al. [12] have been largely acknowledged amongst researchers as a full and acceptable set for controlling lateral displacement. As a result, several other scholars have chosen the same characteristics as important indicators (e.g., Javadi et al. [31]; Jafarian and Nasri [28]; Baziar and Saeedi Azizkandi [32]). In addition, with inclusion of a ground's intensity measure, peak ground acceleration (PGA, a_{max}) is employed in the present study to increase data set, making it more competent and effective in accounting for earthquake causes. By considering the causative fault types of all earthquakes, Sadigh et al. [38] employed attenuation equation to predict the PGA.

In this research, the following seven key parameters have been used to evaluate lateral displacement: earthquake magnitude (M), peak ground acceleration (a_{max}, g), horizontal distance to seismic energy source (R, km), average particle size in T_{15} ($D50_{15}$, mm), average fines material (particles < 0.075 mm) in T_{15} (F_{15}, %), accumulative thickness of saturated layers with adjusted SPT number $(N_1)_{60} < 15$ (T_{15}, m), free-face ratio (W, %), while the output is liquefaction-induced lateral displacement (D_H, m).

In this work, training datasets are based on 80% of the data available (198 sets of data in free face characteristics). The testing dataset has been used to evaluate the proposed models' prediction abilities. The 49 historical records data are used as testing datasets in this study. The training and testing datasets were partitioned depending on statistical features of the datasets, such as mean and standard deviation. The model efficiency is enhanced by the statistical consistency of the training and testing datasets, which makes it easier to evaluate them. Table 2 shows the evaluation metrics of input and output variables in training and testing datasets for free face. Summary of liquefaction-induced lateral ground deformation database is presented in Appendix A, Table A1.

Table 2. Statistical parameters for free-face condition.

Dataset	Statistical Parameters	Seismic Parameter			Geotechnical Parameter			Topographic Parameter	Output
		M	a_{max}	R	$D50_{15}$	F_{15}	T_{15}	W	D_H
		-	g	km	mm	%	m	%	m
Training	Minimum	6.4	0.15	0.5	0.04	1	0.2	1.64	0
	Average	7.26	0.41	15.10	0.36	18.83	7.80	11.69	2.45
	Maximum	9.2	0.68	100	7.7	70	16.7	57.7	10.16
	Standard deviation	0.51	0.15	11.61	0.65	13.71	5.16	9.96	2.26
Testing	Minimum	6.4	0.15	0.5	0.07	2	0.5	2.11	0
	Average	7.3	0.38	16.10	0.39	13.96	7.98	10.04	2.17
	Maximum	9.2	0.68	60	1.98	66	16	48.98	8.39
	Standard deviation	0.49	0.13	10.61	0.42	12.14	5.20	9.78	2.21

4. Correlation Analysis

Correlation coefficients (ρ) have been used to test the significance of the relation between different factors (see Table 3). The equation for ρ is as:

$$\rho(u,\, v) = \frac{cov(u,\, v)}{\sigma_u \sigma_v} \tag{7}$$

where cov indicates covariance, σ_u represents standard deviation of u, and σ_v defines standard deviation of v. $|\rho| > 0.8$ signifies a strong relation among u and v, values from 0.3–0.8 represents a moderate relationship, while $|\rho| < 0.30$ signifies a weak relation [39]. According to Song et al. [40], a relation is considered as "strong" if $|\rho| > 0.8$. M, a_{max}, R, $D50_{15}$, F_{15}, T_{15} and W have moderate to weak relations, as seen in Table 3. As a result, no variables from the lateral displacement estimation model were eliminated. Table 3 reveals that the correlation coefficient has a maximum absolute value of 0.761 and there is no "strong" link between different pairs of components.

Table 3. Correlation between parameters.

Parameters	M	a_{max}	R	$D50_{15}$	F_{15}	T_{15}	W	D_H
M	1.000							
a_{max}	−0.341	1.000						
R	0.761	−0.722	1.000					
$D50_{15}$	0.033	−0.112	0.013	1.000				
F_{15}	−0.370	0.560	−0.371	−0.230	1.000			
T_{15}	0.208	−0.573	0.360	0.237	−0.591	1.000		
W	0.003	0.178	−0.046	0.025	0.245	−0.145	1.000	
D_H	0.179	−0.250	0.230	−0.078	−0.354	0.518	0.146	1.000

5. Construction and Evaluation of Prediction Model

Figure 1 illustrates the prediction model's creation process. In this case, 80% and 20% of the dataset were chosen as training and test sets, respectively, based on statistical integrity. Second, the predictive model was constructed using the trial-and-error approach based on training set utilizing the optimum hyperparameters configurations. Iterative method was utilized to find optimum values for the hyperparameters after setting them to random values (within a reasonable range). The values of the key kernel parameters, omega (ω) and sigma (σ) are 0.4 while noise is 0. 3in the GPR model after multiple trials. Finally, the testing data was used to evaluate the proposed GPR model's performance using four common evaluation metrics: coefficient of determination (R^2), coefficient of correlation (r), mean absolute error (MAE), root mean square error ($RMSE$), ratio of the root mean square error (RSR) to the standard deviation of measured values, and Nash–Sutcliffe coefficient (NSE). The R^2 and NSE values that are higher, and RSR values that are lower, imply that proposed model's prediction accuracy is better. Waikato Environment for Knowledge Analysis software was used throughout the whole calculation process The Pearson VII function-based kernel [41] was employed in this study for the GPR model.

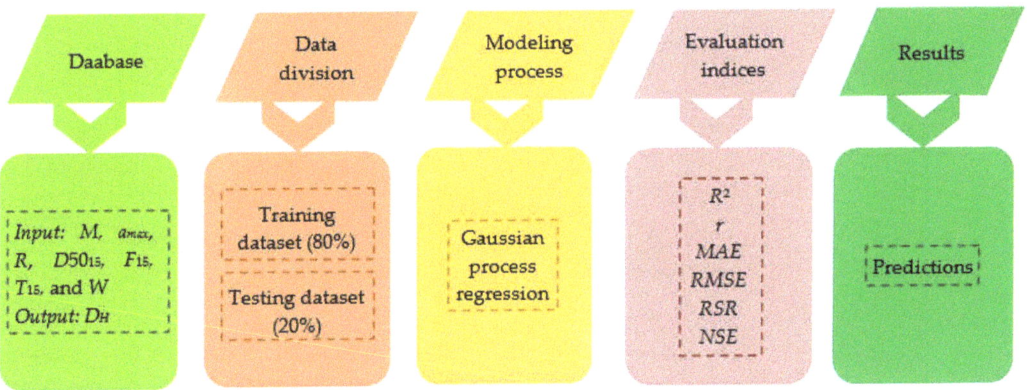

Figure 1. The flowchart for GPR based model to predict liquefaction induced lateral displacement.

The generated model's performance was assessed using R^2, r, MAE, $RMSE$, RSR, and NSE.

$$R^2 = 1 - \frac{\sum_{i=1}^{n}(x_i - y_i)^2}{\sum_{i=1}^{n}(x_i - \overline{x})^2} \tag{8}$$

$$r = \frac{\sum_{i=1}^{n}(x_i - \overline{x})(y_i - \overline{y})}{\sqrt{\sum_{i=1}^{n}(x_i - \overline{x})^2}\sqrt{\sum_{i=1}^{n}(y_i - \overline{y})^2}} \tag{9}$$

$$MAE = \frac{1}{n}\sum_{i=1}^{n}(x_i - y_i) \tag{10}$$

$$RMSE = \sqrt{\frac{1}{n}\sum_{i=1}^{n}(x_i - y_i)^2} \tag{11}$$

$$RSR = \frac{\sqrt{\sum_{i=1}^{n}(x_i - y_i)^2}}{\sqrt{\sum_{i=1}^{n}(x_i - \overline{x})^2}} \tag{12}$$

$$NSE = 1 - \frac{\sum_{i=1}^{n}(x_i - y_i)^2}{\sum_{i=1}^{n}(x_i - \overline{x})^2} \tag{13}$$

where n denotes the set of data points, x_i and y_i denotes the actual and estimated output of data's ith sample, respectively; $\overline{x}$ and $\overline{y}$ represents the mean actual and estimated output of the dataset, respectively. The r value varies from -1 to 1. A perfect distribution between actual and estimated values is represented by value of r equal to 1, whereas a value of 0 shows no relation [42]. For $MAE = 0$, the model's value is perfectly aligned with the real value, and the model is deemed "ideal." The MAE value is between 0 and $+\infty$. The mean squared difference between outputs and targets is termed as $RMSE$, and its value ranges from 0 to $+\infty$. The NSE scale ranges from $-\infty$ to 1, with 1 representing the ideal match. A strong relation is indicated by an NSE score of more than 0.65 [43,44]. The RSR ranges from a perfect 0 to a significant positive number. A smaller RSR indicates low RMSE, indicates that the model is more predictive. The RSR and NSE categorization ranges are shown in Table 4 as very good, good, adequate, and inadequate [44].

Table 4. Statistical indicators for model performance evaluation.

Performance	RSR	NSE
Very Good	$0 \leq RSR \leq 0.5$	$0.75 < NSE \leq 1$
Good	$0.5 < RSR \leq 0.6$	$0.65 < NSE \leq 0.75$
Adequate	$0.6 < RSR \leq 0.7$	$0.5 < NSE \leq 0.65$
Inadequate	$RSR > 0.7$	$NSE \leq 0.5$

6. Result and Discussion

6.1. Performance of GPR Model

The GPR model's efficiency were assessed using coefficient of correlation (r), mean absolute error (MAE), root mean square error ($RMSE$), ratio of root mean square error (RSR) and Nash-Sutcliffe coefficient (NSE). The trend line for GPR in training and testing phases has been drawn by comparing the observed regression in Figure 2 scatter plot, and the GPR findings have the maximum inclination to the line $y = x$ (i.e., $R^2 = 0.9402$ in training and $R^2 = 0.894$ in testing phases). Table 5 shows clearly that for the training model, $r = 0.9697$, $MAE = 0.3403$, $RMSE = 0.5597$, $RSR = 0.248$ and $NSE = 0.938$. Whereas for the testing model $r = 0.9455$, $MAE = 0.5443$, $RMSE = 0.8438$, $RSR = 0.387$ and $NSE = 0.851$. The trend line for GPR in training and testing phases has been drawn by comparing the observed regression in Figure 2 scatter plot, and the GPR findings have the maximum inclination to the line $y = x$ (i.e., $R^2 = 0.9402$ in training and $R^2 = 0.894$ in testing phases).

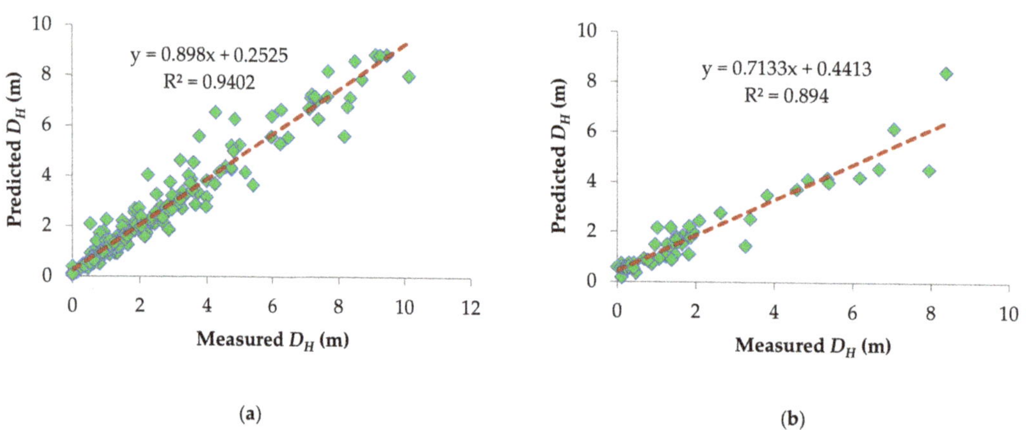

Figure 2. Scatter plot presenting the measured D_H values versus the predicted D_H (**a**) training and (**b**) testing model.

Table 5. Performance statistics of GPR model in comparison with extra available models in literature.

Model	Indicators	R^2	r	*MAE* (m)	*RMSE* (m)	*RSR*	*NSE*	Reference
GPR	Training model	0.9402	0.9697	0.3403	0.5597	0.248	0.938	Present study
	Testing model	0.894	0.9455	0.544	0.8438	0.387	0.851	
EPR	Training model	0.913	-	0.537	1.003	-	-	
	Testing model	0.883	-	0.291	1.158	-	-	
ANN	Training model	0.875	-	0.702	1.074	-	-	[4]
	Testing model	0.872	-	0.82	1.21	-	-	
MLR	Training model	0.868	-	0.81	1.24	-	-	
	Testing model	0.875	-	0.43	1.196	-	-	

Note: - represents that this performance statistic is not included in the reference.

The performance of the developed GPR model was compared to the evolutionary polynomial regression (EPR), MLR and ANN models in literature based on the R^2, *MAE*, and *RMSE* criteria and the results are summarized in Table 5. Furthermore, in terms of the *MAE* and *RMSE* statistical measures in training, the lowest value was found for GPR (*MAE* = 0.3403 m, *RMSE* = 0.5597 m) compared to EPR (*MAE* = 0.537 m, *RMSE* = 1.003 m), ANN (*MAE* = 0.702 m, *RMSE* = 1.074 m), and MLR (*MAE* = 0.81 m, *RMSE* = 1.24 m). Whereas the prediction results in the testing, the MAE and RMSE values was found less for GPR (*MAE* = 0.544 m, *RMSE* = 0.8438 m) compared to EPR, ANN, and MLR except the MAEs values of EPR (=0.291 m) and MLR (=0.43 m) models. The superiority may be owing to the fact that the GPR model excellently captures the nonlinear relationships between lateral displacement and its influencing factors. It can therefore be concluded that, based on statistical indices, the GPR model had the best results. However, due to the use of different number of datasets, a comparison between these results is unwarranted. A project that uses different datasets is needed to gives generalized model to geotechnical earthquake engineering.

6.2. Sensitivity Analysis

The sensitivity results of the GPR model were examined using Yang and Zang's approach for determining the impact of input variables on D_H. This strategy, which has been used in a number of research [44–47], is as follows:

$$r_{ij} = \frac{\sum_{k=1}^{n}(x_{ik} \times x_{ok})}{\sqrt{\sum_{k=1}^{n} x_{ik}^2 \sum_{k=1}^{n} x_{ok}^2}} \tag{14}$$

x_{ik} and x_{ok} are the actual and estimated variables, respectively, and n represents number of datasets (i.e., 198 data samples). For each input variable, the r_{ij} value varies from zero to one, with the greatest r_{ij} values indicating the most efficient output factor (i.e., D_H). Figure 3 displays the r_{ij} values for all input factors. The accumulative thickness of saturated layers with adjusted SPT number, T_{15} (r_{ij} = 0.843) has the greatest effect on the D_H. Furthermore, Table 3 shows that the accumulative thickness of saturated layers with adjusted SPT number, T_{15} has the highest ρ of 0.518 in all other parameters validating the sensitivity analysis results.

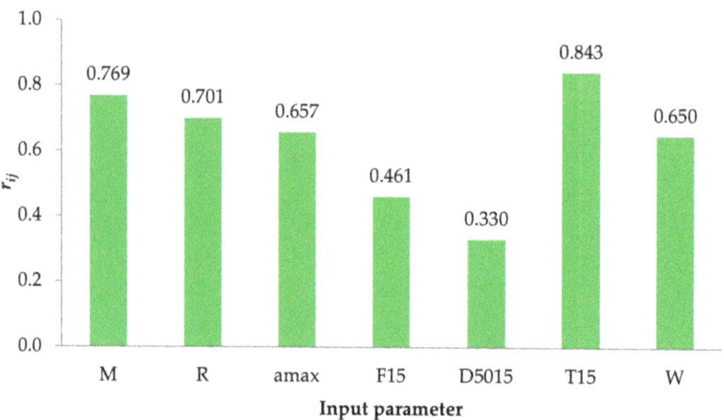

Figure 3. Sensitivity analysis of input variables.

7. Conclusions

The GPR model was used to estimate liquefaction-induced lateral displacement in this work. The predictive model was built using seven input parameters and one output parameter. Performance measures such as R^2, r, *RMSE*, *MAE*, *RSR*, *NSE*, and visual inspection such as scatter plots were used to assess the effectiveness of the developed model. This study's findings can be summarized as follows:

1. With respect to the values of GPR with $R^2 = 0.9402$, $r = 0.9697$, $MAE = 0.3403$, $RMSE = 0.5597$, $RSR = 0.248$ and $NSE = 0.938$ in training phase whereas for testing phase it performed equally well with $R^2 = 0.894$, $r = 0.9455$, $MAE = 0.5443$, $RMSE = 0.8438$, $RSR = 0.387$ and $NSE = 0.851$, In comparison to the EPR, ANN, and MLR models in literature, the GPR model was found to be more accurate and stable than the other models.
2. The results of sensitivity analysis show that the degree of importance of different input parameters on lateral displacement is as $T_{15} > M > R > a_{max} > W > F_{15} > D50_{15}$.
3. The developed Pearson VII kernel function-based GPR model makes predictions accurate and outperforms the others for this dataset and may be applied to a range of geotechnical engineering situations involving uncertainties.

The GPR approach has the advantage of becoming easily modified as new data becomes available, reducing need for expertise and time to modify an existing design aid or equation and/or suggest a new equation.

Author Contributions: Conceptualization, M.A. (Mahmood Ahmad); methodology, M.A. (Mahmood Ahmad), M.A. (Maaz Amjad); software, M.A. (Maaz Amjad) and M.A. (Mahmood Ahmad); validation, R.A.A.-M., P.K., P.O., M.A. (Maaz Amjad) and B.J.K.; formal analysis, M.A. (Mahmood Ahmad) and M.A. (Maaz Amjad); investigation, P.K., P.O., A.C.A. and B.J.K.; resources, P.K.; data curation, M.A. (Maaz Amjad) and M.A. (Mahmood Ahmad); writing—original draft preparation, M.A. (Maaz Amjad); writing—review and editing, M.A. (Maaz Amjad) and M.A. (Mahmood Ahmad); supervision, M.A. (Mahmood Ahmad), R.A.A.-M. and P.K.; project administration, M.A. (Mahmood Ahmad), A.C.A.; funding acquisition, R.A.A.-M., P.K., P.O. and A.C.A. All authors have read and agreed to the published version of the manuscript.

Funding: This research received no external funding.

Institutional Review Board Statement: Not applicable.

Informed Consent Statement: Not applicable.

Data Availability Statement: The data used to support the findings of this study are included within the article.

Conflicts of Interest: The authors declare no conflict of interest.

Appendix A

Table A1. Summary of Liquefaction-Induced Lateral Ground Deformation Database.

Earthquake	M	R (km)	a_{max} (g)	F_{15} (%)	$D50_{15}$ (mm)	T_{15} (m)	W (%)	D_H (m)
1906, San Francisco	7.9	27	0.24	23	0.25	7.2	22.02	1.84
1906, San Francisco	7.9	24	0.26	30	0.16	1.5	17.76	0.92
1964, Alaska	9.2	60	0.3	21	1.35	3.4	24.59	1.86
1964, Alaska	9.2	100	0.2	13	1	10.4	7.03	1.38
1964, Alaska	9.2	60	0.3	23	1.47	3.8	16.07	1.58
1964, Alaska	9.2	60	0.3	66	0.07	3.1	48.98	1.92
1964, Niigata	7.5	21	0.32	5	0.35	12.7	3.06	1.01
1964, Niigata	7.5	21	0.32	4	0.34	13.6	3.15	5.2
1964, Niigata	7.5	21	0.32	24	0.19	8.6	5.36	0.82
1964, Niigata	7.5	21	0.32	6	0.35	0.5	3.43	1.1
1964, Niigata	7.5	21	0.32	32	0.1	2.4	2.03	0.54
1964, Niigata	7.5	21	0.32	26	0.16	2.5	20.61	0.91
1964, Niigata	7.5	21	0.32	6	0.35	0.5	22.37	0.88
1964, Niigata	7.5	21	0.32	10	0.25	11.3	29.7	5.03
1964, Niigata	7.5	21	0.32	6	0.29	7.5	7.32	3.75
1964, Niigata	7.5	21	0.32	24	0.19	8.6	8.78	0.93
1964, Niigata	7.5	21	0.32	12	0.27	12.2	5.01	2.36
1964, Niigata	7.5	21	0.32	24	0.19	8.6	24.02	3.07
1964, Niigata	7.5	21	0.32	9	0.26	11.3	19.62	10.16
1964, Niigata	7.5	21	0.32	12	0.26	12.3	5.76	1.49
1964, Niigata	7.5	21	0.32	31	0.12	2.4	3.26	1.25
1964, Niigata	7.5	21	0.32	10	0.39	9	3.27	2.48
1964, Niigata	7.5	21	0.32	32	0.11	2.4	2.09	1.32
1964, Niigata	7.5	21	0.32	14	0.36	7.1	19.62	3.34
1964, Niigata	7.5	21	0.32	4	0.57	8.6	2.82	1.23
1964, Niigata	7.5	21	0.32	11	0.26	11.9	5.93	2.97
1964, Niigata	7.5	21	0.32	5	0.32	15.6	4.94	7.36
1964, Niigata	7.5	21	0.32	16	0.22	9.6	3.06	2.41
1964, Niigata	7.5	21	0.32	24	0.19	8.6	18.49	1.78
1964, Niigata	7.5	21	0.32	11	0.27	12	4.83	1.84
1964, Niigata	7.5	21	0.32	6	0.32	12.4	4.82	3.66
1964, Niigata	7.5	21	0.32	12	0.26	12.4	5.01	1.75
1964, Niigata	7.5	21	0.32	31	0.12	2.4	3.35	0.69
1964, Niigata	7.5	21	0.32	7	0.35	9.8	4.5	0.53
1964, Niigata	7.5	21	0.32	5	0.32	15.6	7.86	8.37
1964, Niigata	7.5	21	0.32	5	0.32	13.9	5.77	4.58
1964, Niigata	7.5	21	0.32	12	0.26	12	9.18	4.4
1964, Niigata	7.5	21	0.32	12	0.24	11.8	5.54	4
1964, Niigata	7.5	21	0.32	12	0.26	12.2	5.36	2.38
1964, Niigata	7.5	21	0.32	5	0.44	10.1	2.42	1.25
1964, Niigata	7.5	21	0.32	11	0.28	12.1	3.68	2.09
1964, Niigata	7.5	21	0.32	6	0.35	0.5	3.39	0.86
1964, Niigata	7.5	21	0.32	14	0.25	12.6	13.73	6.27
1964, Niigata	7.5	21	0.32	5	0.4	7.9	3.59	1.46
1964, Niigata	7.5	21	0.32	5	0.32	15.6	17.75	9.15
1964, Niigata	7.5	21	0.32	6	0.35	0.5	4.26	0.72
1964, Niigata	7.5	21	0.32	6	0.29	14.3	6.51	3.61
1964, Niigata	7.5	21	0.32	8	0.23	6.8	1.85	0.91
1964, Niigata	7.5	21	0.32	24	0.19	8.6	5.29	1.64
1964, Niigata	7.5	21	0.32	5	0.36	13.6	8.52	4.77
1964, Niigata	7.5	21	0.32	5	0.5	10.9	4.77	0.81
1964, Niigata	7.5	21	0.32	5	0.35	12.7	9.12	6
1964, Niigata	7.5	21	0.32	15	0.25	9.6	2.68	1.89
1964, Niigata	7.5	21	0.32	24	0.19	8.6	8.19	2.2
1964, Niigata	7.5	21	0.32	3	0.35	13.3	4.05	4.76
1964, Niigata	7.5	21	0.32	11	0.26	12	6.53	2.51

Table A1. *Cont.*

Earthquake	M	R (km)	a_{max} (g)	F_{15} (%)	$D50_{15}$ (mm)	T_{15} (m)	W (%)	D_H (m)
1964, Niigata	7.5	21	0.32	5	0.32	15.6	17.75	9.49
1964, Niigata	7.5	21	0.32	11	0.24	11.6	11.06	8.19
1964, Niigata	7.5	21	0.32	13	0.29	13.6	2.76	1.01
1964, Niigata	7.5	21	0.32	7	0.34	10.5	6.03	5.43
1964, Niigata	7.5	21	0.32	6	0.45	10.5	5.84	1.86
1964, Niigata	7.5	21	0.32	5	0.32	15.5	9.98	6.02
1964, Niigata	7.5	21	0.32	6	0.39	9.2	4.87	1.86
1964, Niigata	7.5	21	0.32	12	0.24	11.9	5.06	3.98
1964, Niigata	7.5	21	0.32	5	0.32	15.6	17.05	9.29
1964, Niigata	7.5	21	0.32	15	0.32	11.3	2.86	1.41
1964, Niigata	7.5	21	0.32	16	0.31	11	3.06	1.3
1964, Niigata	7.5	21	0.32	18	0.21	6.7	4.45	0.9
1964, Niigata	7.5	21	0.32	15	0.32	7	7.72	1.92
1964, Niigata	7.5	21	0.32	11	0.28	12.1	2.88	1.56
1964, Niigata	7.5	21	0.32	6	0.38	11.6	3.22	2.71
1964, Niigata	7.5	21	0.32	5	0.32	15.6	5.25	7.19
1964, Niigata	7.5	21	0.32	16	0.3	10.8	3.68	0.71
1964, Niigata	7.5	21	0.32	13	0.38	7.2	20.55	3.28
1964, Niigata	7.5	21	0.32	9	0.4	13	2.05	1.11
1964, Niigata	7.5	21	0.32	13	0.25	12.5	16.07	7.4
1964, Niigata	7.5	21	0.32	6	0.37	12.7	7.05	3.54
1964, Niigata	7.5	21	0.32	6	0.35	0.5	2.68	0.82
1964, Niigata	7.5	21	0.32	9	0.39	9.3	3.72	1.96
1964, Niigata	7.5	21	0.32	5	0.39	7.3	2.76	1.23
1964, Niigata	7.5	21	0.32	24	0.19	8.6	12.86	2.74
1964, Niigata	7.5	21	0.32	12	0.25	12.1	16.72	4.88
1964, Niigata	7.5	21	0.32	11	0.27	12.1	3.38	1.83
1964, Niigata	7.5	21	0.32	5	0.32	15.6	5.77	7.21
1964, Niigata	7.5	21	0.32	2	0.33	10.4	8.89	4.76
1964, Niigata	7.5	21	0.32	13	0.25	12.4	35	7.67
1964, Niigata	7.5	21	0.32	7	0.32	9.4	2.99	1.31
1964, Niigata	7.5	21	0.32	4	0.34	13.5	3.36	3.46
1964, Niigata	7.5	21	0.32	11	0.26	11.6	11.32	3.78
1964, Niigata	7.5	21	0.32	3	0.44	11.3	3.82	1.52
1964, Niigata	7.5	21	0.32	13	0.29	13	3.1	0.56
1964, Niigata	7.5	21	0.32	28	0.14	2.5	4.79	0.88
1964, Niigata	7.5	21	0.32	5	0.32	15.6	19.62	7.7
1964, Niigata	7.5	21	0.32	3	0.44	11.3	4.87	1.9
1964, Niigata	7.5	21	0.32	9	0.37	10	16.4	6.5
1964, Niigata	7.5	21	0.32	7	0.35	9.8	3.9	2.87
1964, Niigata	7.5	21	0.32	12	0.25	12.2	12.47	4.83
1964, Niigata	7.5	21	0.32	5	0.34	13.8	12.01	8.73
1964, Niigata	7.5	21	0.32	24	0.19	8.6	25.93	3.57
1964, Niigata	7.5	21	0.32	14	0.25	12.6	11.32	3.51
1964, Niigata	7.5	21	0.32	17	0.24	6.8	4.26	1.37
1964, Niigata	7.5	21	0.32	11	0.25	11.6	17.05	8.29
1964, Niigata	7.5	21	0.32	5	0.31	14.1	16.4	8.52
1964, Niigata	7.5	21	0.32	24	0.19	8.6	5.76	1.27
1964, Niigata	7.5	21	0.32	15	0.25	9.5	3.04	2.68
1964, Niigata	7.5	21	0.32	13	0.27	11.8	2.27	1.56
1964, Niigata	7.5	21	0.32	13	0.25	12.4	12.47	3.21
1964, Niigata	7.5	21	0.32	24	0.19	8.6	7.14	2.15
1964, Niigata	7.5	21	0.32	24	0.19	8.6	5.15	1.06
1964, Niigata	7.5	21	0.32	7	0.43	10.4	15.25	2.25
1964, Niigata	7.5	21	0.32	5	0.35	16.7	11.06	2.91
1964, Niigata	7.5	21	0.32	8	0.15	3.7	1.64	0.62
1964, Niigata	7.5	21	0.32	5	0.32	15.6	7.72	7.31
1964, Niigata	7.5	21	0.32	13	0.25	12.5	55.68	7.13

Table A1. *Cont.*

Earthquake	M	R (km)	a_{max} (g)	F_{15} (%)	$D50_{15}$ (mm)	T_{15} (m)	W (%)	D_H (m)
1964, Niigata	7.5	21	0.32	6	0.31	15.2	12.44	6.3
1964, Niigata	7.5	21	0.32	13	0.35	11.9	2.86	1.11
1964, Niigata	7.5	21	0.32	11	0.25	11.6	19.37	4.28
1964, Niigata	7.5	21	0.32	10	0.28	12.1	3.09	1.66
1964, Niigata	7.5	21	0.32	5	0.36	9.6	3.72	3.26
1964, Niigata	7.5	21	0.32	13	0.25	12.3	16.07	7.06
1964, Niigata	7.5	21	0.32	2	0.33	10.4	13.65	5.35
1964, Niigata	7.5	21	0.32	13	0.26	12.6	6.23	1.87
1964, Niigata	7.5	21	0.32	3	0.44	11.3	3.27	0.96
1964, Niigata	7.5	21	0.32	7	0.33	10.6	12.01	7.95
1964, Niigata	7.5	21	0.32	3	0.44	11.3	5.12	1.36
1964, Niigata	7.5	21	0.32	30	0.13	2.4	4.18	0.68
1964, Niigata	7.5	21	0.32	4	0.34	13.6	2.99	4.85
1964, Niigata	7.5	21	0.32	28	0.14	2.5	20.61	1.06
1964, Niigata	7.5	21	0.32	11	0.28	12.1	3.86	1.93
1964, Niigata	7.5	21	0.32	2	0.33	10.4	7.58	4.57
1964, Niigata	7.5	21	0.32	11	0.28	12.2	2.9	1.65
1964, Niigata	7.5	21	0.32	8	0.34	11.4	3.1	2.09
1964, Niigata	7.5	21	0.32	5	0.45	10	2.86	1.82
1964, Niigata	7.5	21	0.32	12	0.24	11.9	4.45	3.38
1964, Niigata	7.5	21	0.32	5	0.35	12.7	2.79	1.01
1964, Niigata	7.5	21	0.32	13	0.29	12.9	3.04	0.42
1964, Niigata	7.5	21	0.32	5	0.31	14.1	17.75	8.39
1964, Niigata	7.5	21	0.32	11	0.24	12.1	2.11	1.27
1964, Niigata	7.5	21	0.32	7	0.28	8.1	17.05	6.18
1964, Niigata	7.5	21	0.32	6	0.35	0.5	5.96	0.77
1964, Niigata	7.5	21	0.32	6	0.35	0.5	2.29	1.38
1964, Niigata	7.5	21	0.32	5	0.31	14.1	4.55	6.67
1964, Niigata	7.5	21	0.32	11	0.27	12	3.98	1.83
1964, Niigata	7.5	21	0.32	6	0.29	7.9	17.05	5.39
1964, Niigata	7.5	21	0.32	11	0.27	12.1	2.97	1.49
1971, San Fernando	6.4	0.5	0.68	47	0.08	5.3	19.96	2.93
1971, San Fernando	6.4	0.5	0.68	47	0.08	5.6	4.7	0.47
1971, San Fernando	6.4	0.5	0.68	47	0.08	6.5	5.08	0.52
1971, San Fernando	6.4	0.5	0.68	47	0.08	4.6	20.3	3.16
1971, San Fernando	6.4	0.5	0.68	47	0.08	3.6	20.34	3.18
1971, San Fernando	6.4	0.5	0.68	47	0.08	3	17.07	1.81
1971, San Fernando	6.4	0.5	0.68	47	0.08	2.3	13.59	2.14
1971, San Fernando	6.4	0.5	0.68	47	0.08	1.6	20.41	2.45
1971, San Fernando	6.4	0.5	0.68	47	0.08	4.8	19.61	2.78
1971, San Fernando	6.4	0.5	0.68	47	0.08	2.7	15.43	2.02
1971, San Fernando	6.4	0.5	0.68	47	0.08	2	13.59	1.46
1971, San Fernando	6.4	0.5	0.68	47	0.08	4	18.87	3.26
1971, San Fernando	6.4	0.5	0.68	47	0.08	2.7	20.47	3.16
1971, San Fernando	6.4	0.5	0.68	47	0.08	5.9	4.89	0.54
1971, San Fernando	6.4	0.5	0.68	47	0.08	4.5	19.26	1.99
1971, San Fernando	6.4	0.5	0.68	47	0.08	1	20.27	1
1971, San Fernando	6.4	0.5	0.68	47	0.08	3.1	18.26	2.04
1971, San Fernando	6.4	0.5	0.68	47	0.08	5.2	19.96	2.63
1979, Imperial Valley	6.5	2	0.49	20	0.12	3	8.57	2.63
1979, Imperial Valley	6.5	2	0.49	32	0.09	1.5	6.25	0.37
1979, Imperial Valley	6.5	2	0.49	23	0.11	2	7.89	2.04
1979, Imperial Valley	6.5	2	0.49	17	0.12	3.6	3.08	0.92
1979, Imperial Valley	6.5	2	0.49	15	0.12	3.8	6.56	2.02
1979, Imperial Valley	6.6	6	0.36	70	0.04	0.2	4.26	0.01
1979, Imperial Valley	6.6	6	0.36	54	0.12	1.8	10.66	0.01
1979, Imperial Valley	6.5	2	0.49	17	0.12	3.7	9.6	4
1979, Imperial Valley	6.5	2	0.49	22	0.11	2.6	3.68	0.31

Table A1. *Cont.*

Earthquake	M	R (km)	a_{max} (g)	F_{15} (%)	$D50_{15}$ (mm)	T_{15} (m)	W (%)	D_H (m)
1979, Imperial Valley	6.5	2	0.49	23	0.11	2.4	6.35	1.41
1979, Imperial Valley	6.5	2	0.49	23	0.11	2	6.15	1.1
1979, Imperial Valley	6.5	2	0.49	22	0.11	2.7	6.45	1.53
1979, Imperial Valley	6.5	2	0.49	23	0.11	2.9	7.02	1.43
1979, Imperial Valley	6.5	2	0.49	25	0.1	2.5	6.78	0.72
1979, Imperial Valley	6.5	2	0.49	15	0.12	4	6.56	1.48
1979, Imperial Valley	6.5	2	0.49	17	0.12	3.7	6.78	2.3
1979, Imperial Valley	6.5	2	0.49	21	0.11	1.6	4.8	0.67
1979, Imperial Valley	6.5	2	0.49	25	0.1	2.5	9.84	2.63
1979, Imperial Valley	6.5	2	0.49	22	0.11	1.8	6.67	1.13
1979, Imperial Valley	6.5	2	0.49	30	0.09	1.8	8.05	1.03
1979, Imperial Valley	6.5	2	0.49	22	0.11	2.7	8.05	2.12
1979, Imperial Valley	6.5	2	0.49	25	0.11	2.2	3.68	0.47
1979, Imperial Valley	6.5	2	0.49	16	0.12	3.8	9.37	4.25
1979, Imperial Valley	6.5	2	0.49	22	0.11	3	10.08	3.21
1979, Imperial Valley	6.5	2	0.49	16	0.12	3.7	3.72	1.23
1979, Imperial Valley	6.5	2	0.49	18	0.12	3.4	9.16	3.82
1979, Imperial Valley	6.5	2	0.49	19	0.12	3.3	6.15	1.51
1979, Imperial Valley	6.5	2	0.49	21	0.11	1.4	4.69	0.87
1979, Imperial Valley	6.5	2	0.49	25	0.11	2.2	3.52	0.47
1987, Superstition Hills	6.6	23	0.15	27	0.09	3.5	17.91	0.19
1987, Superstition Hills	6.6	23	0.15	22	0.09	3.3	41.38	0.21
1987, Superstition Hills	6.6	23	0.15	43	0.07	1.7	17.52	0.11
1987, Superstition Hills	6.6	23	0.15	44	0.07	3.6	7.5	0.01
1987, Superstition Hills	6.6	23	0.15	38	0.08	2.7	13.11	0.11
1987, Superstition Hills	6.6	23	0.15	25	0.09	3.4	41.38	0.24
1989, Loma Prieta	7	27.2	0.2	1	0.6	3.4	29.73	0.26
1989, Loma Prieta	7	27.2	0.2	2	0.8	2.7	33.54	0.29
1995, Hyogo-Ken Nanbu	6.8	7.5	0.35	12.6	0.47	14.2	13.95	1.18
1995, Hyogo-Ken Nanbu	6.8	6	0.38	13.4	0.94	12.5	9.25	1.01
1995, Hyogo-Ken Nanbu	6.8	8	0.34	14.6	1.98	16	6.67	0.45
1995, Hyogo-Ken Nanbu	6.8	8	0.34	14.6	1.98	16	16.82	0.93
1995, Hyogo-Ken Nanbu	6.8	7.5	0.35	12.6	0.47	14.2	10.4	0.89
1995, Hyogo-Ken Nanbu	6.8	5.5	0.39	10	1.36	15	14.56	1.34
1995, Hyogo-Ken Nanbu	6.8	5.5	0.39	10	1.36	15	30.21	2.83
1995, Hyogo-Ken Nanbu	6.8	6.5	0.37	10	1.88	12.5	5.16	0.34
1995, Hyogo-Ken Nanbu	6.8	5.5	0.39	10	1.36	15	56.8	2.48
1995, Hyogo-Ken Nanbu	6.8	8	0.34	14.6	1.98	16	18	0.97
1995, Hyogo-Ken Nanbu	6.8	8	0.34	14.6	1.98	16	20.69	0.9
1995, Hyogo-Ken Nanbu	6.8	7.5	0.35	12.6	0.47	14.2	18.56	1.33
1995, Hyogo-Ken Nanbu	6.8	8	0.34	14.6	1.98	16	14.63	0.66
1995, Hyogo-Ken Nanbu	6.8	6.5	0.37	10	1.88	12.5	9.84	1.03
1995, Hyogo-Ken Nanbu	6.8	5.5	0.39	10	1.36	15	14.34	1.31
1995, Hyogo-Ken Nanbu	6.8	6.5	0.37	10	1.88	12.5	14.63	1.47
1995, Hyogo-Ken Nanbu	6.8	6	0.38	13.4	0.94	12.5	15	1.48
1995, Hyogo-Ken Nanbu	6.8	5.5	0.39	10	1.36	15	9.79	1.47
1995, Hyogo-Ken Nanbu	6.8	8	0.34	14.6	1.98	16	8.45	0.41
1999, Chi-Chi	7.6	5	0.67	20.8	0.11	0.5	7.4	0
1999, Chi-Chi	7.6	5	0.67	20.8	0.11	0.8	13.7	0.15
1999, Chi-Chi	7.6	5	0.67	20.8	0.11	0.8	18.4	0.55
1999, Chi-Chi	7.6	5	0.67	20.8	0.11	0.8	25.2	0.8
1999, Chi-Chi	7.6	5	0.67	20.8	0.11	0.8	37.3	1.05
1999, Chi-Chi	7.6	5	0.67	20.8	0.11	0.8	49.9	2.05
1999, Chi-Chi	7.6	5	0.67	13	0.18	0.75	21.2	0.49
1999, Chi-Chi	7.6	5	0.67	20.8	0.11	1.1	11.9	0
1999, Chi-Chi	7.6	5	0.67	20.8	0.11	1.1	26.3	0
1999, Chi-Chi	7.6	5	0.67	30	0.13	0.45	12.2	0.4
1999, Chi-Chi	7.6	5	0.67	30	0.13	0.45	14.3	0.65

Table A1. *Cont.*

Earthquake	M	R (km)	a_{max} (g)	F_{15} (%)	$D50_{15}$ (mm)	T_{15} (m)	W (%)	D_H (m)
1999, Chi-Chi	7.6	5	0.67	30	0.13	0.45	24.6	1
1999, Chi-Chi	7.6	5	0.67	30	0.13	0.45	57.7	1.24
1999, Chi-Chi	7.6	5	0.67	31.4	0.1	1	8	0.35
1999, Chi-Chi	7.6	5	0.67	31.4	0.1	1	10.5	0.61
1999, Chi-Chi	7.6	5	0.67	31.4	0.1	1	19	0.96
1999, Chi-Chi	7.6	5	0.67	31.4	0.1	1	31.3	2.96
1999, Chi-Chi	7.6	5	0.67	48.5	0.1	1.8	9.6	0.35
1999, Chi-Chi	7.6	5	0.67	48.5	0.1	1.8	11.7	0.52
1999, Chi-Chi	7.6	5	0.67	48.5	0.1	1.8	13.3	0.62
1999, Chi-Chi	7.6	5	0.67	48.5	0.1	1.8	23.7	1.62
1999, Chi-Chi	7.6	5	0.67	13	0.18	0.5	5.7	0
1999, Chi-Chi	7.6	5	0.67	13	0.18	0.75	6.6	0.1
1999, Chi-Chi	7.6	5	0.67	13	0.18	0.75	7.9	0.17
1999, Chi-Chi	7.6	5	0.67	13	0.18	0.75	9	0.23
1999, Chi-Chi	7.6	5	0.67	13	0.18	0.75	15	0.29
1999, Kocaeli	7.4	0.5	0.57	11	7.7	1.2	8	0.9
1999, Kocaeli	7.4	0.5	0.57	31	0.55	1.7	6	0.1

References

1. Huang, W.; Zou, M.; Qian, J.; Zhou, Z. Consistent damage model and performance-based assessment of structural members of different materials. *Soil Dyn. Earthq. Eng.* **2018**, *109*, 266–272. [CrossRef]
2. Ma, Y.; Gong, J.X. Probability Identification of Seismic Failure Modes of Reinforced Concrete Columns based on Experimental Observations. *J. Earthq. Eng.* **2017**, *22*, 1881–1899. [CrossRef]
3. Liu, C.; Fang, D.; Zhao, L. Reflection on earthquake damage of buildings in 2015 Nepal earthquake and seismic measures for post-earthquake reconstruction. *Structures* **2021**, *30*, 647–658. [CrossRef]
4. Rezania, M.; Faramarzi, A.; Javadi, A.A. An evolutionary based approach for assessment of earthquake-induced soil liquefaction and lateral displacement. *Eng. Appl. Artif. Intell.* **2011**, *24*, 142–153. [CrossRef]
5. Finn, W.; Ledbetter, R.; Wu, G. Liquefaction in silty soils: Design and analysis. 1994; pp. 51–76. In *Ground Failures under Seismic Conditions*; ASCE: Reston, VA, USA; p. 51.
6. Liao, T.; McGillivray, A.; Mayne, P.; Zavala, G.; Elhakim, A. *Seismic Ground Deformation Modeling Final Report for MAE HD-7a (Year 1)*; Geosystems Engineering/School of Civil & Environmental Engineering, Georgia Institute of Technology: Atlanta, GA, USA, 2002.
7. Arulanandan, K.; Li, X.S.; Sivathasan, K. (Siva) Numerical Simulation of Liquefaction-Induced Deformations. *J. Geotech. Geoenviron. Eng.* **2000**, *126*, 657–666. [CrossRef]
8. Newmark, N.M. Effects of earthquakes on dams and embankments. *Geotechnique* **1965**, *15*, 139–160. [CrossRef]
9. Towhata, I.; Sasaki, Y.; Tokida, K.I.; Matsumoto, H.; Tamari, Y.; Yamada, K. Prediction of Permanent Displacement of Liquefied Ground by Means of Minimum Energy Principle. *Soils Found.* **1992**, *32*, 97–116. [CrossRef]
10. Kokusho, T.; Fujita, K. Site Investigations for Involvement of Water Films in Lateral Flow in Liquefied Ground. *J. Geotech. Geoenviron. Eng.* **2002**, *128*, 917–925. [CrossRef]
11. Hamada, M. Study on liquefaction induced permanent ground displacements. *Report of Association for the Development of Earthquake Prediction* **1986**. [CrossRef]
12. Youd, T.L.; Hansen, C.M.; Bartlett, S.F. Revised Multilinear Regression Equations for Prediction of Lateral Spread Displacement. *J. Geotech. Geoenviron. Eng.* **2002**, *128*, 1007–1017. [CrossRef]
13. Wang, J.; Rahman, M.S. A neural network model for liquefaction-induced horizontal ground displacement. *Soil Dyn. Earthq. Eng.* **1999**, *18*, 555–568. [CrossRef]
14. Hamada, M.; Towhata, I.; Yasuda, S.; Isoyama, R. Study on permanent ground displacement induced by seismic liquefaction. *Comput. Geotech.* **1987**, *4*, 197–220. [CrossRef]
15. Orense, R.; Towhata, I. Prediction of liquefaction—induced permanent ground displacements: A three—dimensional approach. *Tech. Rep. NCEER* **1992**, *1*, 335–349.
16. Gu, W.H.; Morgenstern, N.R.; Robertson, P.K. Progressive failure of lower San Fernando dam. *J. Geotech. Eng.* **1993**, *119*, 333–349. [CrossRef]
17. Gu, W.H.; Morgenstern, N.R.; Robertson, P.K. Postearthquake Deformation Analysis of Wildlife Site. *J. Geotech. Eng.* **1994**, *120*, 274–289. [CrossRef]
18. Yegian, M.K.; Marciano, E.A.; Ghahraman, V.G. Earthquake-induced permanent deformations: Probabilistic approach. *J. Geotech. Eng.* **1991**, *117*, 35–50. [CrossRef]

19. Baziar, M.H.; Dobry, R.; Elgamal, A.-W.M. Engineering Evaluation of Permanent Ground Deformations Due to Seismically Induced Liquefaction. Tech. Rep. NCEER-92-0007. 1992; 306.
20. Tokida, K.; Matsumoto, H.; Azuma, T.; Towhata, I. Simplified Procedure to Estimate Lateral Ground Flow by Soil Liquefaction. *WIT Trans. Built Environ.* **1993**, *3*, 1–16.
21. Bardet, J.; Mace, N.; Tobita, T. *Liquefaction-Induced Ground Deformation and Failure, a Report to PEER/PG&E. Task 4A-Phase 1*; Civil Engineering Department, University of Southern California: Los Angeles, CA, USA, 1999.
22. Hadush, S.; Yashima, A.; Uzuoka, R.; Moriguchi, S.; Sawada, K. Liquefaction induced lateral spread analysis using the CIP method. *Comput. Geotech.* **2001**, *28*, 549–574. [CrossRef]
23. Aydan, Ö. The stress state of the earth and the earth's crust due to the gravitational pull. In Proceedings of the 35th US Rock Mechanics Symposium, Lake Tahoe, CA, USA, 4–7 June 1995; pp. 237–243.
24. Hamada, M.; Sato, H.; Kawakami, T. A consideration of the mechanism for liquefaction-related large ground displacement. In Proceedings of the Fifth US-Japan Workshop on Earthquake Resistant Design of Lifeline Facilities and Countermeasures Against Soil Liquefaction, Technical Report NCEER-94-0026, Snowbird, UT, USA, 29 September–1 October 1994; pp. 217–232.
25. Youd, T.L.; Perkins, D.M. Mapping of liquefaction severity index. *J. Geotech. Eng.* **1987**, *113*, 1374–1392. [CrossRef]
26. Bartlett, S.F.; Youd, T.L. Empirical prediction of lateral spread displacement. In Proceedings of the Fourth Japan-U.S. Workshop on Earthquake Resistant Design of Lifeline Facilities and Countermeasures for Soil Liquefaction, Honolulu, HI, USA, 27–29 May 1992; pp. 351–365.
27. Bartlett, S.F.; Leslie Youd, T. Empirical prediction of liquefaction-induced lateral spread. *J. Geotech. Eng.* **1995**, *121*, 316–329. [CrossRef]
28. Jounrnal, A.; Jafarian, Y.; Nasri, E. Evaluation of uncertainties in the existing empirical models and probabilistic prediction of liquefaction-induced lateral. *AJSR-Civil Environ. Eng.* **2016**, *48*, 107–110.
29. Kanibir, A. *Investigation of the Lateral Spreading at Sapanca and Suggestion of Empirical Relationships for Predicting Lateral Spreading*; Department of Geological Engineering, Hacettepe University: Ankara, Turkey, 2003.
30. Bawwab, W. Al Probabilistic assessment of liquefaction-induced lateral ground deformations. P.h.D. Thesis, Middle East Technical University, Ankara, Turkey, November 2005.
31. Javadi, A.A.; Rezania, M.; Nezhad, M.M. Evaluation of liquefaction induced lateral displacements using genetic programming. *Comput. Geotech.* **2006**, *33*, 222–233. [CrossRef]
32. Baziar, M.; Saeedi Azizkandi, A. Evaluation of lateral spreading utilizing artificial neural network and genetic programming. *Int. J. Civ. Eng. Trans. B Geotech. Eng.* **2013**, *11*, 100–111.
33. Baziar, M.H.; Ghorbani, A. Evaluation of lateral spreading using artificial neural networks. *Soil Dyn. Earthq. Eng.* **2005**, *25*, 1–9. [CrossRef]
34. Javdanian, H. Field data-based modeling of lateral ground surface deformations due to earthquake-induced liquefaction. *Eur. Phys. J. Plus* **2019**, *134*, 297. [CrossRef]
35. Williams, C.K.; Rasmussen, C.E. *Gaussian Processes for Machine Learning*; MIT Press: Cambridge, MA, USA, 2006; Volume 2, p. 4.
36. Chu, D.B.; Stewart, J.P.; Youd, T.L.; Chu, B.L. Liquefaction-Induced Lateral Spreading in Near-Fault Regions during the 1999 Chi-Chi, Taiwan Earthquake. *J. Geotech. Geoenviron. Eng.* **2006**, *132*, 1549–1565. [CrossRef]
37. Cetin, K.O.; Youd, T.L.; Seed, R.B.; Bray, J.D.; Stewart, J.P.; Durgunoglu, H.T.; Lettis, W.; Yilmaz, M.T. Liquefaction-Induced Lateral Spreading at Izmit Bay During the Kocaeli (Izmit)-Turkey Earthquake. *J. Geotech. Geoenviron. Eng.* **2004**, *130*, 1300–1313. [CrossRef]
38. Sadigh, K.; Chang, C.Y.; Egan, J.A.; Makdisi, F.; Youngs, R.R. Attenuation Relationships for Shallow Crustal Earthquakes Based on California Strong Motion Data. *Seismol. Res. Lett.* **1997**, *68*, 180–189. [CrossRef]
39. van Vuren, T. Modeling of transport demand—Analyzing, calculating, and forecasting transport demand. *Transp. Rev.* **2020**, *40*, 115–117. [CrossRef]
40. Song, Y.; Gong, J.; Gao, S.; Wang, D.; Cui, T.; Li, Y.; Wei, B. Susceptibility assessment of earthquake-induced landslides using Bayesian network: A case study in Beichuan, China. *Comput. Geosci.* **2012**, *42*, 189–199. [CrossRef]
41. Üstün, B.; Melssen, W.J.; Buydens, L.M.C. Facilitating the application of Support Vector Regression by using a universal Pearson VII function based kernel. *Chemom. Intell. Lab. Syst.* **2006**, *81*, 29–40. [CrossRef]
42. Ly, H.B.; Nguyen, T.A.; Pham, B.T. Estimation of Soil Cohesion Using Machine Learning Method: A Random Forest Approach. *Adv. Civ. Eng.* **2021**, *2021*, 8873993. [CrossRef]
43. Nash, J.E.; Sutcliffe, J.V. River flow forecasting through conceptual models part I—A discussion of principles. *J. Hydrol.* **1970**, *10*, 282–290. [CrossRef]
44. Ahmad, M.; Ahmad, F.; Wróblewski, P.; Al-Mansob, R.A.; Olczak, P.; Kamiński, P.; Safdar, M.; Rai, P.; Ahmad, M.; Ahmad, F.; et al. Prediction of Ultimate Bearing Capacity of Shallow Foundations on Cohesionless Soils: A Gaussian Process Regression Approach. *Appl. Sci.* **2021**, *11*, 10317. [CrossRef]
45. Ahmad, M.; Kamiński, P.; Olczak, P.; Alam, M.; Iqbal, M.; Ahmad, F.; Sasui, S.; Khan, B.; Ahmad, M.; Kamiński, P.; et al. Development of Prediction Models for Shear Strength of Rockfill Material Using Machine Learning Techniques. *Appl. Sci.* **2021**, *11*, 6167. [CrossRef]

46. Ahmad, M.; Hu, J.-L.; Ahmad, F.; Tang, X.-W.; Amjad, M.; Iqbal, M.; Asim, M.; Farooq, A.; Ahmad, M.; Hu, J.-L.; et al. Supervised Learning Methods for Modeling Concrete Compressive Strength Prediction at High Temperature. *Materials* **2021**, *14*, 1983. [CrossRef]
47. Chen, W.; Hasanipanah, M.; Nikafshan Rad, H.; Jahed Armaghani, D.; Tahir, M.M. A new design of evolutionary hybrid optimization of SVR model in predicting the blast-induced ground vibration. *Eng. Comput.* **2021**, *37*, 1455–1471. [CrossRef]

Article

A PROMETHEE Multiple-Criteria Approach to Combined Seismic and Flood Risk Assessment at the Regional Scale

Arianna Soldati [1], Andrea Chiozzi [2], Željana Nikolić [3], Carmela Vaccaro [2,4] and Elena Benvenuti [1,*]

[1]   Department of Engineering, University of Ferrara, 44122 Ferrara, Italy; arianna.soldati@edu.unife.it
[2]   Department of Environmental and Prevention Sciences, University of Ferrara, 44122 Ferrara, Italy; andrea.chiozzi@unife.it (A.C.); carmela.vaccaro@unife.it (C.V.)
[3]   Faculty of Civil Engineering, Architecture and Geodesy, University of Split, 21000 Split, Croatia; zeljana.nikolic@gradst.hr
[4]   OGS, Istituto Nazionale di Oceanografia e di Geofisica Sperimentale, Borgo Grotta Gigante 42/C, 34010 Sgonico, Italy
*   Correspondence: elena.benvenuti@unife.it

Abstract: Social vulnerability is deeply affected by the increase in hazardous events such as earthquakes and floods. Such hazards have the potential to greatly affect communities, including in developed countries. Governments and stakeholders must adopt suitable risk reduction strategies. This study is aimed at proposing a qualitative multi-hazard risk analysis methodology in the case of combined seismic and flood risk using PROMETHEE, a Multiple-Criteria Decision Analysis technique. The present case study is a multi-hazard risk assessment of the Ferrara province (Italy). The proposed approach is an original and flexible methodology to qualitatively prioritize urban centers affected by multi-hazard risks at the regional scale. It delivers a useful tool to stakeholders involved in the processes of hazard management and disaster mitigation.

Keywords: risk assessment; multi hazard; seismic risk; flood risk; multiple-criteria decision analysis; PROMETHEE algorithm

Citation: Soldati, A.; Chiozzi, A.; Nikolić, Ž.; Vaccaro, C.; Benvenuti, E. A PROMETHEE Multiple-Criteria Approach to Combined Seismic and Flood Risk Assessment at the Regional Scale. *Appl. Sci.* **2022**, *12*, 1527. https://doi.org/10.3390/app12031527

Academic Editor: Igal M. Shohet

Received: 30 November 2021
Accepted: 27 January 2022
Published: 31 January 2022

Publisher's Note: MDPI stays neutral with regard to jurisdictional claims in published maps and institutional affiliations.

1. Introduction

Many areas in Europe and worldwide are increasingly subjected to catastrophic events. These events intensify the exposure of these territories to multi-risk events and make societies more vulnerable to entangled risks [1–7]. Globalization and climate changes are the main culprits of these multi-risk dynamics. Globalization, indeed, makes countries closely linked and interdependent, so communities are not only vulnerable to local extreme events but also to those occurring outside their national territories. Climate change increases, among others, the frequency and intensity of extreme meteorological phenomena, hydrological and flood risk, as well as the risk of fires. The awareness of this worrying trend has determined the need for adequate tools to address and mitigate these risks, as well as information campaigns to foster resilience and coping capacity of communities [5–7].

Understanding risks involving vast inhabited areas is therefore paramount, particularly when assessing potential losses produced by a combination of multiple hazards. Hereafter, a hazard refers to the probability of occurrence in a specified period of a potentially damaging event of a given magnitude in a given area [8]. Total risk is a measure of the expected human (casualties, injuries) and economic (damage to property, activity disruption) losses due to adverse natural phenomena. Such a measure is assumed to be the product of hazard, vulnerability, and exposure instances [9]. Many areas on Earth are subjected to the effects of coexisting multiple hazards, among which floods [3,8] and earthquakes are some of the most widespread [5–7]. Though inhabited environments are affected by multiple hazardous processes, most studies focus on a single hazard [8].

The choice to adopt a multi-risk analysis approach has the potential to play a fundamental role in increasing urban resilience, an essential factor for sustainable development, enabling cities to prepare, respond, and recover when hit by catastrophic events, and therefore prevent or contain economic, environmental, and social losses [1]. However, performing a multi-risk analysis with the tools and methodologies available today raises numerous challenges and difficulties [10–20]. For instance, an updated analysis of multi-hazard aggregated risk for infrastructures considering multiple potential threats has recently been proposed in reference [5].

Risk assessment is indeed carried out through independent procedures that adopt different estimation metrics. This makes comparisons difficult and precludes considering correlations or cascading effects [11]. On the contrary, the Multiple-Criteria Decision Analysis (MCDA) technique is a promising approach in multiple-hazard risk analysis, even if this route has been scarcely explored to date [21–24].

To pave the way for sustainable land-use plans and risk-mitigation strategies, we must analyze, quantify, and, especially, compare all concurrent risks [25]. To date, single-risk assessment is generally performed by means of independent procedures, whose results cannot be compared. The purpose of this paper is to devise an approach for the qualitative assessment of combined risks at the regional scale. In particular, the objective is to jointly analyze the flood and seismic risk for the Ferrara province area. The proposed approach is based on the suitable use of the Preference Ranking Organization Method for Enrichment Evaluations (PROMETHEE), a Multiple-Criteria Decision Analysis technique [26–29]. The province of Ferrara is in a flatland area in the northern part of Italy. Historically, it has been mainly hit by floods and seismic events. Though floods are exogeneous processes, whereas earthquakes are exogenic, we assume flood and seismic hazards to be the two relevant hazards for determining a priority list. This priority list is meant to be useful to stakeholders and public agencies called to rapidly implement investment plans aimed to prevent economic and life losses and foster the coping capacity of communities to manage the adverse conditions induced by natural disasters. Particularly, the present objective is to prioritize this among the different municipalities. Therefore, the adopted level of observation is at the scale of the area included within each municipality.

Assuming the municipalities of the province of Ferrara as the alternatives of the multiple-criteria analysis, the proposed approach defines a priority ranking among all the alternatives. The outcome is represented by qualitative risk maps. These maps are useful tools for stakeholders involved in community management and risk prevention.

Among the Multi-Risk Methodologies applied in Italian territories, we recall here the works by Gallina et al. [23,24] for the assessment of the impact of sea-level rise, coastal erosion, and storm surge induced by climate changes in coastal zones in North Italy. Flood and seismic risks have been multi-assessed through a Machine Learning framework recently devised by the authors for the Emilia Romagna region [30]. Up to now, the present contribution is the very first to use an MCDA approach for multi-risk analysis of combined flood and earthquake risks, while no other relevant contributions exist dealing with multi-hazard analyses of the Province of Ferrara.

2. Materials and Methods

2.1. Geographical Context and Single Risk Description

To introduce the concept of multi-risk assessment, it is first necessary to discuss the concept of single risk. Risk is basically defined as the product of three parameters: Hazard, vulnerability, and exposure [9]. A hazard represents the probability that an adverse event will occur in a specific area and in a specific time interval. Vulnerability, on the other hand, is an intrinsic characteristic of a system; it represents its propensity to suffer a certain level of damage following the occurrence of a hazard event. Finally, exposure indicates the presence of people, critical infrastructures, natural and cultural heritage, and much more still in hazard zones that are thereby subject to potential losses [4].

The concept of multi-risk follows as the overall risk from a multi-hazard and multi-vulnerability perspective. The term multi-hazard indicates several hazards affecting the same exposed elements (with or without space–time coincidence) or the occurrence of a hazard event that triggers another one giving rise to a domino or cascade effect. Furthermore, the term multi-vulnerability indicates those circumstances where several elements are sensitive to different possible vulnerabilities towards the various hazards affecting them or vulnerabilities that vary over time [10,11].

The territory of the province of Ferrara is located at the north-eastern extremity of the Padana Plain, a flat land area in the north part of Italy crossed by the Po River and bathed by the Adriatic Sea on the east side. It is characterized by minimum land slopes and its altimetry is mainly under the mean sea level, as almost half of its area is below the mean sea level, as shown in Figure 1. Moreover, the eastern part of the territory is affected by subsidence phenomena as well. These ground-level modifications, caused mainly by anthropogenic actions as well as by geological and neotectonic factors [31,32], produced a subsidence rate of up to −2.5 mm/year [31]. The main watercourses that flow through the Ferrara province are the Po River, which marks the northern border of the Reno River, and the Idice and Sillaro streams, which are not tributaries of the Po River, and cross the province in their last stretch. Furthermore, numerous artificial canals flow through the Ferrara Province, including the Cavo Napoleonico, which connects the Po and Reno rivers, and the Idrovia Ferrarese.

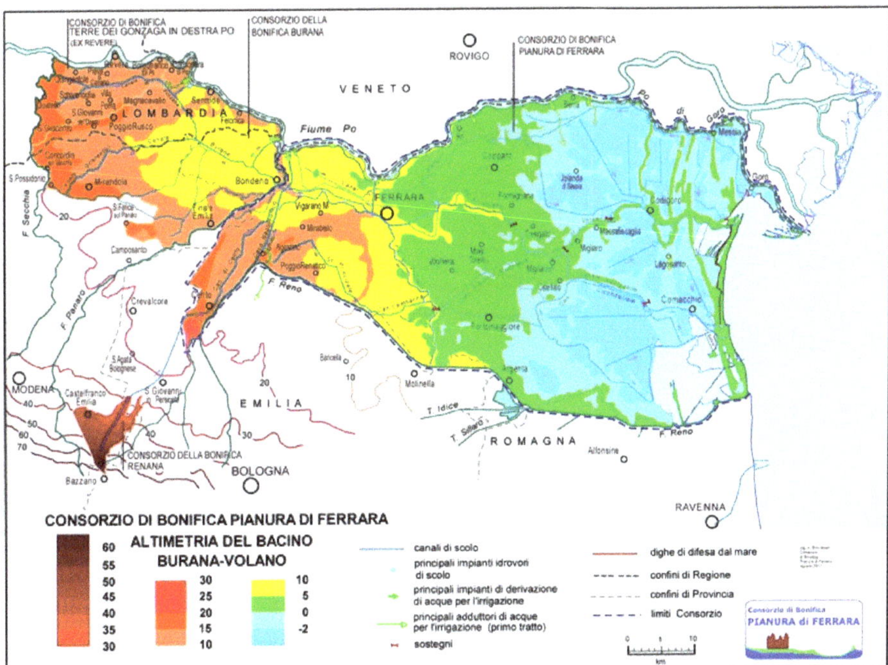

Figure 1. Altimetric map of Ferrara province (free source https://www.bonificaferrara.it/images/Allegati/SITL/4d-3-altimetria(100).pdf, accessed on 3 January 2022, made available by Consorzio di Bonifica Pianura di Ferrara). The minimum and maximum extremal values of the ground level over the sea in the legend are −2 m (dark blue) and 60 m (dark red), respectively.

The province of Ferrara includes 23 municipalities. Attention is hereafter restricted to the two main risks of the area under study, namely flood and seismic risks. Site effects associated with inherent geological morphology and instability issues such as liquefaction were not considered, for simplicity. Desertification is another risk that has been emerging in recent years in the Po delta plain [10]. However, it has not been considered in the present contribution. Hereafter, flood risk refers to the risk that depends on the probability of occurrence of a flood, evaluated concerning the different typologies of watercourses that flow through the territory. The flood risk for the selected region was quantified by the Land Reclamation Authority of the province of Ferrara (Consorzio di Bonifica Pianura di Ferrara), and accounts for flood hazard, exposure, and vulnerability parameters.

Seismic risk depends on the peak ground acceleration (PGA) as well as on the vulnerability of the built environment and the exposure of people and economic activities. We exploited the map of seismic hazard provided by the Italian Institute of Volcanology and Geophysics (INGV), and the seismic classification of municipalities in Emilia (free source https://ambiente.regione.emilia-romagna.it/en/geologia/seismic-risk/seismic-classification, accessed on 26 January 2022), shown in Figure 2a. In Figure 2b, Italy is divided into different areas according to peak ground acceleration values [33] (free source http://zonesismiche.mi.ingv.it/, accessed on 26 January 2022).

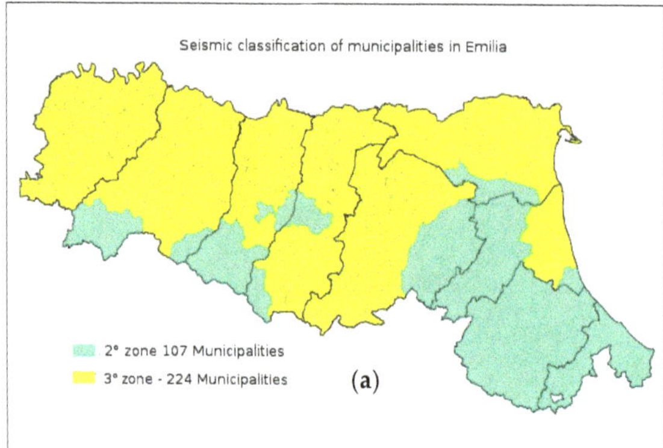

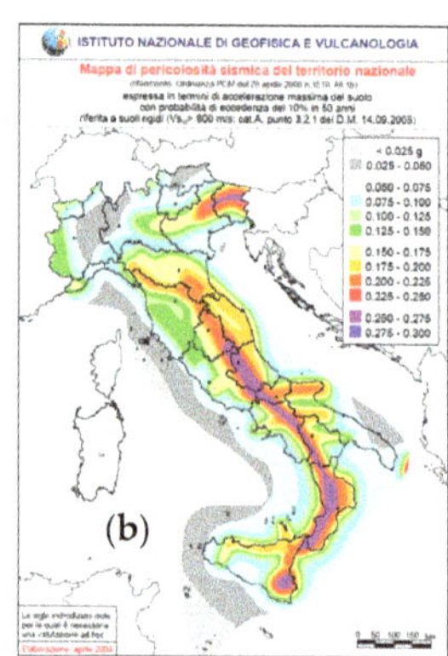

Figure 2. (a) Seismic classification of municipalities in Emilia (https://ambiente.regione.emilia-romagna.it/en/geologia/seismic-risk/seismic-classification, accessed on 26 January 2022). (b) Seismic Hazard Map of Italy (free source from INGV webpage http://zonesismiche.mi.ingv.it/, accessed on 26 January 2022).

Finally, we used the database made available by the Italian National Institute of Statistics (Istat). This database was used in 2018 by the Italian Superior Institute for Environmental Protection and Research (ISPRA) to produce seismic, hydrogeological, volcanic, and social vulnerability hazard maps for the entire Italian peninsula. The reader is referred to the pertinent report by Trigila et al. [34] to obtain a detailed description of ISPRA's methodology for the processing of the data.

2.2. The PROMETHEE Method

The proposed multi-hazard risk analysis procedure for the region under study is based on PROMETHEE [26–29], a Multiple-Criteria Decision Analysis method. It belongs to the class of aggregation methods based on outranking relationships. It is known for its simplicity and the ability to analyze information from multiple sources. PROMETHEE allows one to jointly compare data originally expressed in different units and scales. A flux diagram explaining the various steps of the PROMETHEE-based analysis can be found in reference [29].

PROMETHEE deals with maximization or minimization problems with k different criteria of the kind

$$\max(or\min)\{g_1(a), g_2(a), \ldots, g_k(a) | a \in A\}, \tag{1}$$

where A is a finite set of possible alternatives and function $g_j(a)$ represents the performance of the j-th criterion. Let us consider two alternatives, $(a,b) \in A$. We have the following cases:

$$\begin{cases} \forall j : g_j(a) \geq g_j(b) \\ \exists k : g_k(a) > g_k(b) \end{cases} \Leftrightarrow aPb,$$

$$\forall j : g_j(a) = g_j(b) \Leftrightarrow aIb, \tag{2}$$

$$\begin{cases} \exists s : g_s(a) > g_s(b) \\ \exists r : g_r(a) < g_r(b) \end{cases} \Leftrightarrow aRb,$$

where P, I, and R denote preference (P), indifference (I), or incompatibility relations (R) of one alternative over the other, respectively.

By comparing all the alternatives for each criterion, a hierarchy of alternatives belonging to the starting space A will be obtained. When comparing two actions, $(a,b) \in A$. the result of this comparison is expressed in terms of the preference function $\wp : A \times A \to (0,1)$ that represents the intensity of the preference of alternative a towards alternative b. Therefore, $\wp(a,b) = 0$ indicates no preference of a over b (or indifference), $\wp(a,b) \simeq 0$ indicates a weak preference of a over b, $\wp(a,b) \simeq 1$ indicates a strong preference of a over b, and $\wp(a,b) = 1$ indicates a strict preference of a over b. In practice, the preference function will often be a function of the difference between the evaluations of the two alternatives considered:

$$\wp(a,b) = P(g(a) - g(b)) = P(d), \tag{3}$$

where P is a non-decreasing function, equal to zero for negative values of d. PROMETHEE offers six types of preference functions (see Table 1).

Table 1. Types of preference function.

Generalized Criterion	Definition	Parameters to Fix
Type 1: usual criterion	$P(d) = \begin{cases} 0 & d \leq 0 \\ 1 & d > 0 \end{cases}$	-
Type 2: U-shape criterion	$P(d) = \begin{cases} 0 & d \leq q \\ 1 & d > q \end{cases}$	q
Type 3: V-shape criterion	$P(d) = \begin{cases} 0 & d \leq p \\ \frac{d}{p} & 0 \leq d \leq p \\ 1 & d > p \end{cases}$	p
Type 4: Level criterion	$P(d) = \begin{cases} 0 & d \leq q \\ \frac{1}{2} & q \leq d \leq p \\ 1 & d > p \end{cases}$	p, q
Type 5: V-shape with indifference criterion	$P(d) = \begin{cases} 0 & d \leq q \\ \frac{d-q}{p-q} & q \leq d \leq p \\ 1 & d > p \end{cases}$	p, q
Type 6: Gaussian criterion	$P(d) = \begin{cases} 0 & d \leq 0 \\ 1 - e^{-\frac{d^2}{2s^2}} & d > 0 \end{cases}$	s

Therefore, a preference index is defined as follows:

$$\begin{cases} \pi(a,b) = \sum_{j=1}^{k} P_j(a,b) w_j \\ \pi(b,a) = \sum_{j=1}^{k} P_j(b,a) w_j \end{cases}, \tag{4}$$

where $\pi(a,b)$ expresses the degree to which a is preferred to b over all criteria and vice versa, and w_j is the weight of each criterion and expresses a measure of the importance of the relative criterion.

For all the criteria, a classification is available for the various alternatives necessary to define the so-called outranking flows, which are the fundamental units for the PROMETHEE methodology. Each alternative a faces $(n-1)$ other alternatives that belong to the generic space A. The two following outranking flows are defined:

$$\begin{cases} \Theta^+(a) = \frac{1}{n-1} \sum\limits_{x \in A} \pi(a, x) \\ \Theta^-(a) = \frac{1}{n-1} \sum\limits_{x \in A} \pi(x, a) \end{cases} , \tag{5}$$

where x represents the deviation of the specific preference function with respect to the same function of preference for the other alternatives. $\Theta^+(a)$ expresses how alternative a outranks all the others, otherwise $\Theta^-(a)$ expresses how alternative a is outranked by all the others. The higher $\Theta^+(a)$ (lower $\Theta^-(a)$) is, the more likely alternative a is strongest; otherwise, alternative a, compared to the others, is weakest when $\Theta^+(a)$ assumes small values. Once these two flows have been defined, it becomes very simple to make comparisons between alternatives and subsequently establish their order.

PROMETHEE offers several ways to view the results; the main ones are illustrated below:

- PROMETHEE I Partial Ranking: This is a partial ranking of the alternatives, based on positive and negative flows, and includes preferences, indifference, and incomparability. This scheme allows, therefore, to compare, where possible, the alternatives and establish their partial order of preference through the indices and the related outranking flows.
- PROMETHEE II Complete Ranking: This is useful when the decision maker needs a complete hierarchy among the alternatives of the problem. In this case, the alternatives will be compared in relation to their net flow $\Theta(a) = \Theta(a)^+ - \Theta^-(a)$. PROMETHEE II allows a complete classification of the alternatives; however, it is less realistic and poor in information as it eliminates any possible factor of incomparability between the different alternatives.
- PROMETHEE Table: This displays the Θ, Θ^+, and Θ^- scores. The actions are ranked according to the PROMETHEE II complete ranking.
- PROMETHEE Rainbow: This is a diagram that allows one to highlight, for each alternative, the criteria that positively or negatively affect the final result.
- Profile of alternatives: This is a diagram that shows, for each alternative, the net flow Θ of each criterion.

2.3. Data Collection and Processing

Both flood and seismic risks have been included in PROMETHEE as criteria according to their components (hazard, exposure, and vulnerability), while the municipalities, i.e., the object on which to evaluate the criteria, are the alternatives. Risk parameters for each municipality are made available by the National Institute of Vulcanology and Geophysics (INGV), the Italian National Institute of Statistics (Istat), and the Land Reclamation Authorities of the Province of Ferrara. Accordingly, we have drawn from the aforementioned databases a simplified map of the flood risk. In particular, Figure 3 displays the flood hazard for the Province of Ferrara in terms of the probability of floods. In this map, the classification is based on Italian Government Decree n. 49/2010 [35]. Accordingly, frequent floods are defined as those having a high probability of occurrence, with a return period of $20 \leq T < 50$ years (P3); infrequent floods have an average probability of occurrence with a return period of $100 \leq T \leq 200$ years (P2); finally, low-probability floods have a return period of $200 < T \leq 500$ years (P1).

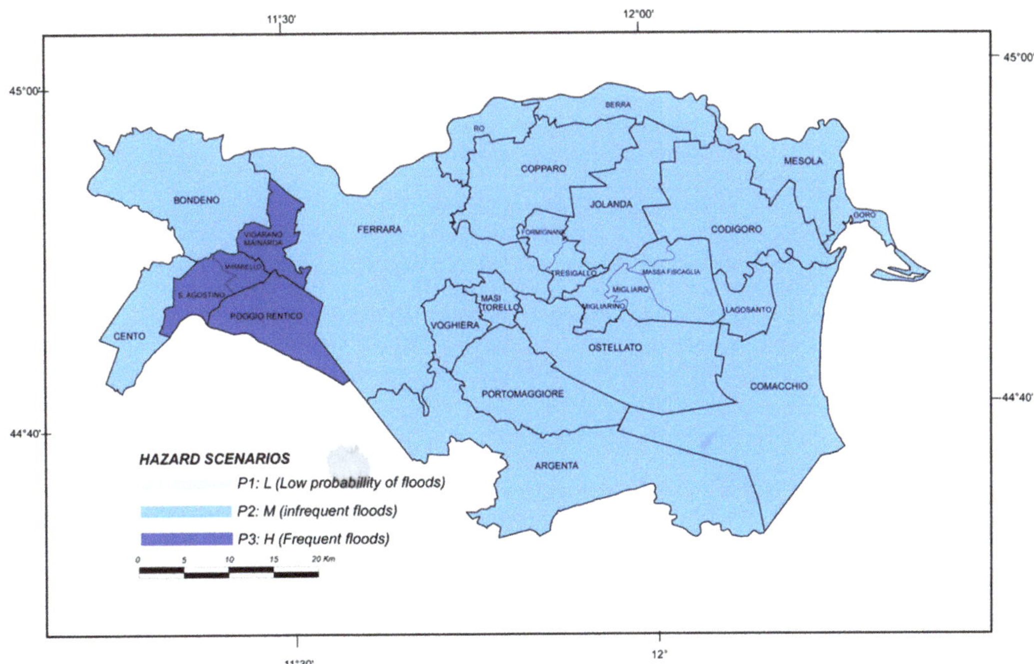

Figure 3. Map of the flood hazard for the province of Ferrara in terms of probability of flood.

Figure 4 provides a map of the seismic hazard for the province of Ferrara in terms of peak ground acceleration (PGA). The PGA-intervals are indicated in the legend.

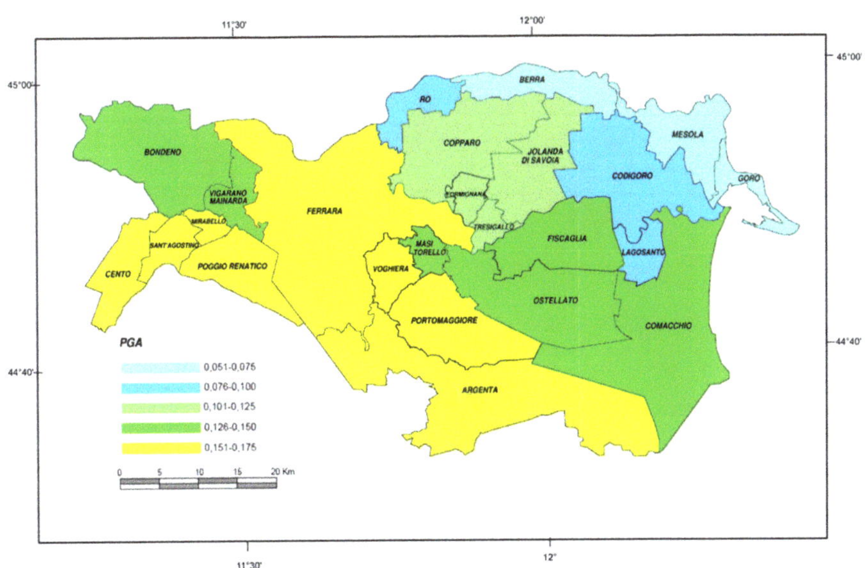

Figure 4. Map of the seismic hazard for the province of Ferrara in terms of peak ground acceleration.

As for exposure-related criteria, for each municipality, we adopted three parameters: Land use percentage, the number of strategic buildings, and population density. All of them were drawn from the Istat database. The strategic buildings were defined based on the presence and number of halls, police stations, fire brigade buildings, schools, universities, water lifting plants, hospitals, and civil protection centers. This information was obtained from the website of the province of Ferrara (http://www.provincia.fe.it/, 1 October 2021), as per educational and public institutions and centers, and from the website of the Consorzio di Bonifica Pianura di Ferrara as per water lifting plants (https://www.bonificaferrara.it/, 1 October 2021).

Specifically, four classes of land use percentages were obtained based on the ratio between the urbanized area divided by the total area. In synthesis, we collected the municipalities into four land use classes (Figure 5), four classes in terms of the number of strategic buildings (Figure 6), and four classes of population density (Figure 7).

As for the vulnerability criteria, we adopted a single non-dimensionalized parameter, which accounts for the average age of buildings. Knowing the age of construction and the corresponding number of buildings, we computed the following vulnerability index:

$$I_v = \frac{A\,\alpha_1 + B\,\alpha_2 + C\,\alpha_3 + D\,\alpha_4}{A + B + C + D},$$

where A, B, C e D represent the number of buildings built between the end of 1800 and 1945; the number of buildings built between 1946 and 1980; the number of buildings built between 1981 and 2000, and finally, the number of buildings built from 2001 up to now. α_1, α_2, α_3 e α_4 are coefficients equal to 1, 0.75, 0.5, and 0.25, respectively. The vulnerability index I_v results in being mainly related to the age of buildings, and its map is shown in Figure 8.

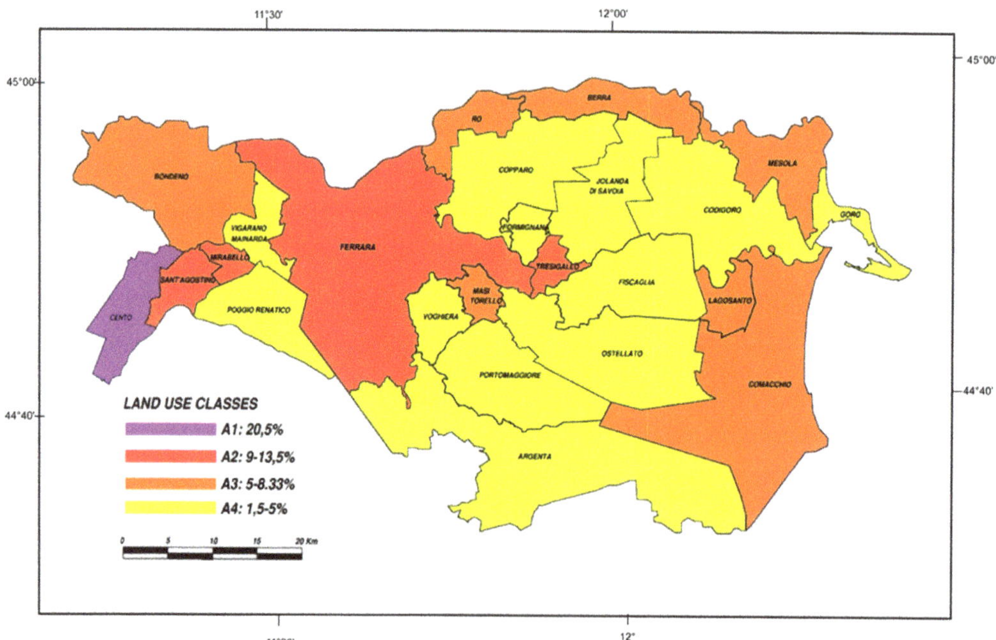

Figure 5. Land use map for the province of Ferrara.

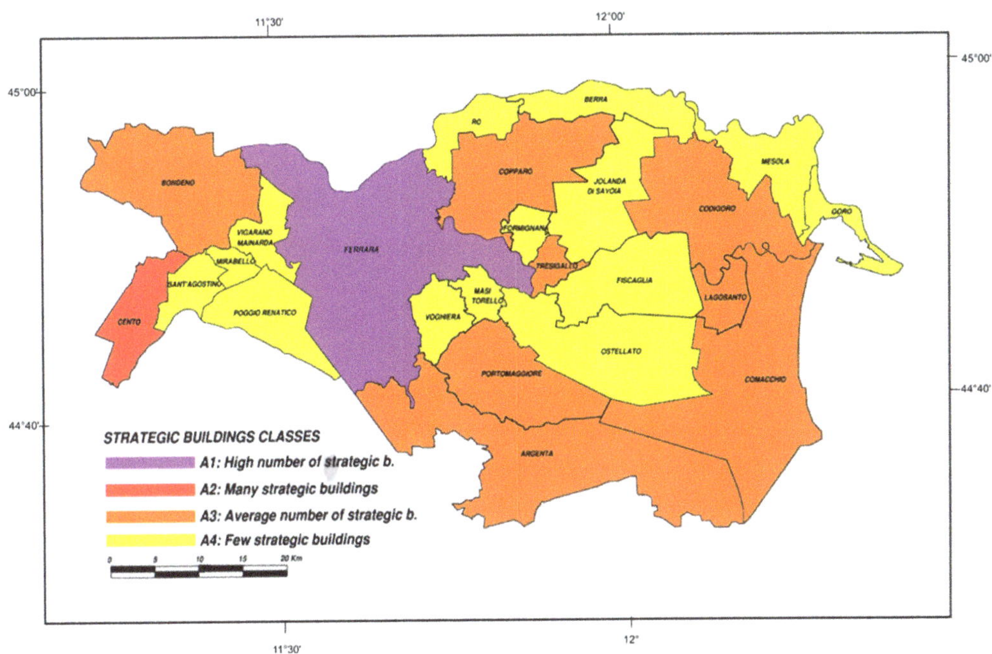

Figure 6. Map of strategic buildings incidence for the province of Ferrara.

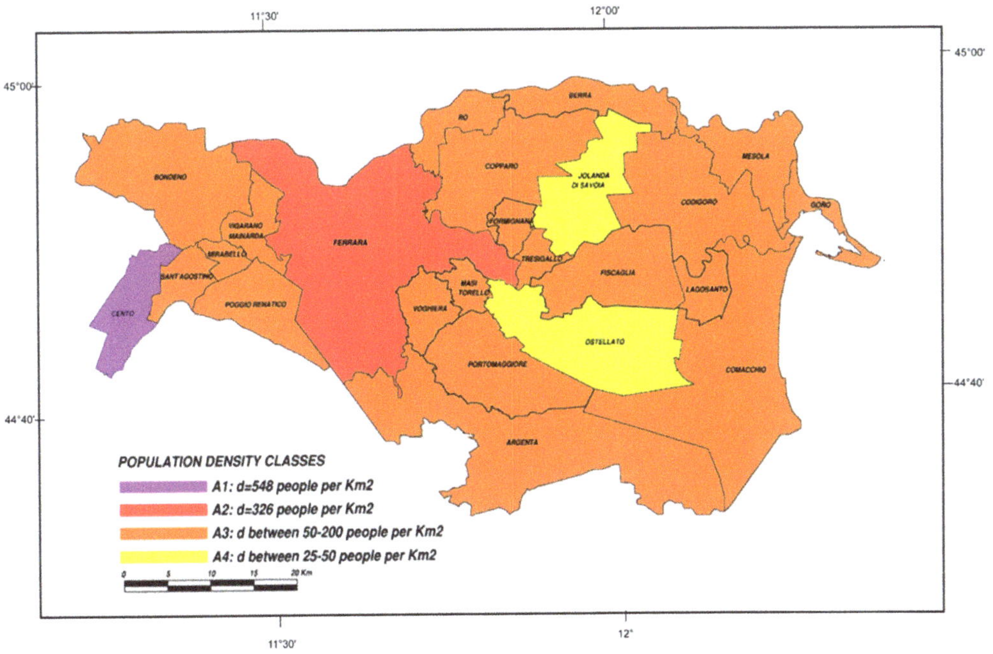

Figure 7. Map of the population density for the province of Ferrara.

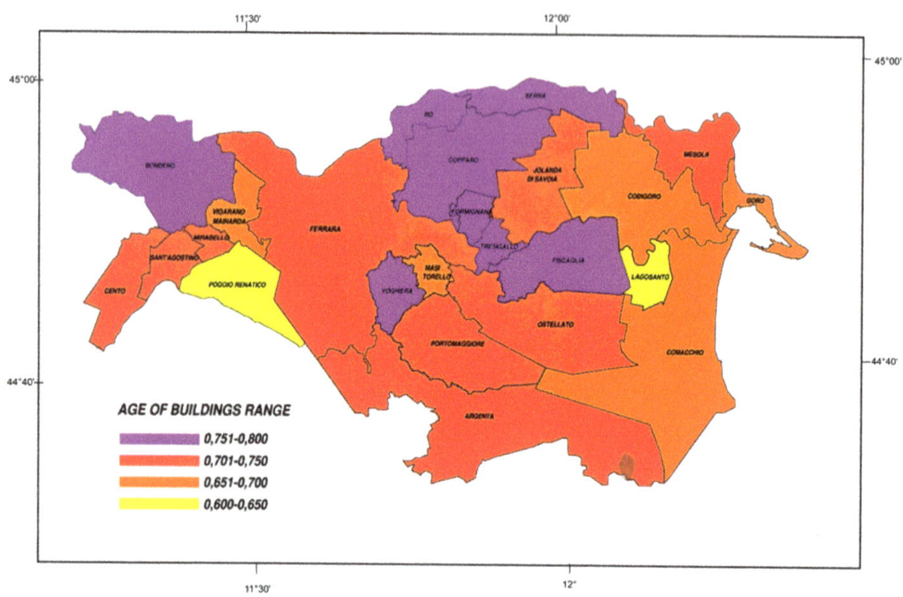

Figure 8. Map of the vulnerability parameter for the province of Ferrara computed as a function of the building age.

2.4. Normalization and Weight Assignment

All the data have been collected in an evaluation matrix, whose rows correspond to each alternative (i.e., each municipality), while each column corresponds to each selected criterion. In other words, the i,j-th element of the evaluation matrix expresses the value of the i-th alternative relating to the attribute of the j-th criterion and describes the performance of each alternative regarding each criterion.

It should be noted that criteria are represented through different scales and units. This precludes mutual comparisons. Thus, it is necessary to further homogenize the data contained in the evaluation matrix and proceed with comparisons through normalization. Through the preference function, the performance of the alternatives is transformed into a dimensionless value, ranging from o to 1. As a first attempt, we adopted the Type 1 preference function described in Table 1, which does not require the definition of any threshold. Subsequently, the linear preference function was also used.

Finally, we attributed weights to each criterion. Through this step, decision makers can make their preferences explicit, since it is not ensured that all the criteria take on the same importance. We first decided to attribute the same weight to each criterion. Then, a sensitivity analysis was performed with varying weights.

The risk maps shown in the following sections indicate three classes of risk levels, namely low, medium, high. It is emphasized that this classification must be intended as a pure ranking in terms of the relative urgency of investments. It does not at all intend to indicate the level of safety in absolute terms of the various municipalities. This classification answers the question as to whether the method can provide the priority level associated with a certain municipality and help to decide how to distribute investments over various municipalities.

2.5. Sensitivity Analysis

To verify the reliability of the results obtained, a sensitivity analysis was carried out. During this sensitivity analysis, we retraced the procedure by which the results were obtained and identified the steps most affected by uncertainties and subjectivity, considering their influence on the final ranking. Specifically, the choice of the preference function and the choice of weights appeared to be the most subjective. As for the choice of the preference function, a previous study [26] recommends assuming a linear preference function endowed with the definition of p, q thresholds. Two approaches are adopted for the determination of p and q: The so-called zero-max method, which imposes that the indifference threshold q is assigned the value of zero while the preference threshold p is set to be equal to the maximum difference between the evaluations of the criteria.

The mean-std method requires the calculation of the average value and the standard deviation of a set of differences between the evaluations of the criteria. In the mean-std method, the indifference threshold is assigned the value of the difference between the average value and standard deviation, while the sum between the average value and standard deviation is assigned to the preference threshold. Following [26–28], we adopted the preference function of the linear type for the quantitative criteria, that is flood hazard, land use, the age of buildings, and population density. However, the algorithm was also run by choosing the usual preference function, which is the simplest possible one. The thresholds were computed as shown in Table 2. As for the sensitivity on the weights, the four scenarios described in Table 3 have been considered.

Table 2. Preference functions and the associated thresholds p, q, and s.

	Criteria					
	Flood Hazard	PGA	Land Use	Strategic Buildings	Age of Buildings	Population Density
Min/Max	max	Max	max	max	max	max
Weight	1	1	1	1	1	1
Preference function	Usual	Linear	Linear	Usual	Linear	Linear
Thresholds	absolute	Absolute	absolute	absolute	absolute	absolute
q: Indifference, zero-max	n/a	0.000	0.0000	n/a	0.000	0.000
p: Preference (zero-max)	n/a	0.098	0.1896	n/a	0.158	523.00
s: Gaussian (zero-max)	n/a	n/a	n/a	n/a	n/a	n/a
q: Indifference (mean-std)	n/a	0.093	0.0261	n/a	0.0676	16.10
p: Preference (mean-std)	n/a	0.155	0.1081	n/a	0.766	238.60
s: Gaussian (mean-std)	n/a	n/a	n/a	n/a	n/a	n/a

Table 3. Sensitivity analysis on the weights of the criteria.

Sensitivity Analysis: Increase of Single Criteria Weights	
Scenario 0	All criteria have the same weight. $p = 17\%$
Scenario 1	Increase the weight of the i-th criterion by 50% compared to its initial value. $p_i = 25.5\%$; $p_{\text{other criteria}} = 14.9\%$
Scenario 2	Increase the weight of the i-th criterion by 50% compared to its previous value. $p_i = 38.2\%$; $p_{\text{other criteria}} = 12.3\%$
Scenario 3	Increase the weight of the i-th criterion by 50% compared to its previous value. $p_i = 57.4\%$; $p_{\text{other criteria}} = 8.5\%$

3. Results

In the following Section, we describe the outcomes of the multiple-criteria analysis for the usual and linear preference function as well as the results of the sensitivity analysis performed for varying weight changes.

3.1. Usual Preference Function

When the usual preference function is used, the algorithm assumes equal weights. We recall that, here, thresholds p and q are not required. Basically, what is provided to the analyst is an order of priority where the municipalities in the province of Ferrara are ordered from the most sensitive to combined flood and seismic risk to the one that is least affected. Table 4 shows the final ranking of the alternatives.

Table 4. Ranking of alternatives for the usual preference function.

Rank	Alternatives	Θ	Θ^+	Θ^-
1	Ferrara	0.6111	0.7302	0.119
2	Cento	0.5873	0.7222	0.1349
3	Tresigallo	0.4127	0.6111	0.1984
4	Vigarano Mainarda	0.2857	0.5873	0.3016
5	Mirabello + Sant'Agostino	0.2698	0.5794	0.3095
6	Argenta + Portomaggiore	0.2381	0.5238	0.2857
7	Bondeno	0.1825	0.4921	0.3095
8	Copparo	0.0238	0.4127	0.3889
9	Poggio Renatico	0.0238	0.4524	0.4286
10	Comacchio	0.0000	0.4048	0.4048
10	Formignana	0.0000	0.381	0.381
12	Voghiera	−0.0238	0.3651	0.3889
13	Lagosanto	−0.0317	0.3889	0.4206
14	Berra	−0.1587	0.3016	0.4603
15	Masi Torello	−0.1746	0.2937	0.4683
16	Ro	−0.1905	0.2857	0.4762
17	Fiscaglia	−0.2063	0.2778	0.4841
18	Mesola	−0.2857	0.2381	0.5238
19	Ostellato	−0.3571	0.1984	0.5556
20	Goro	−0.3651	0.1984	0.5635
21	Codigoro	−0.3651	0.2222	0.5873
22	Jolanda di Savoia	−0.4762	0.1429	0.619

This is not the only way to visualize the results: The PROMETHEE rainbow plot, shown in Figure 9, allows one to highlight, for each alternative, the criteria that positively or negatively affect the results. In Figure 9, the colors are representative of the criterion: Yellow indicates the criteria relating to exposure, red is used for seismic hazard, green for vulnerability, and blue for flood hazard. For example, for the municipality of Ferrara (first in the ranking), it can be observed that the criterion that has a negative effect is the one relating to the flood hazard, whereas the other criteria have a positive effect on the Ferrara municipality. On the contrary, in the municipality of Jolanda di Savoia (last in the ranking), the only criterion that has a positive influence is the one relating to vulnerability, while all the others have a negative influence.

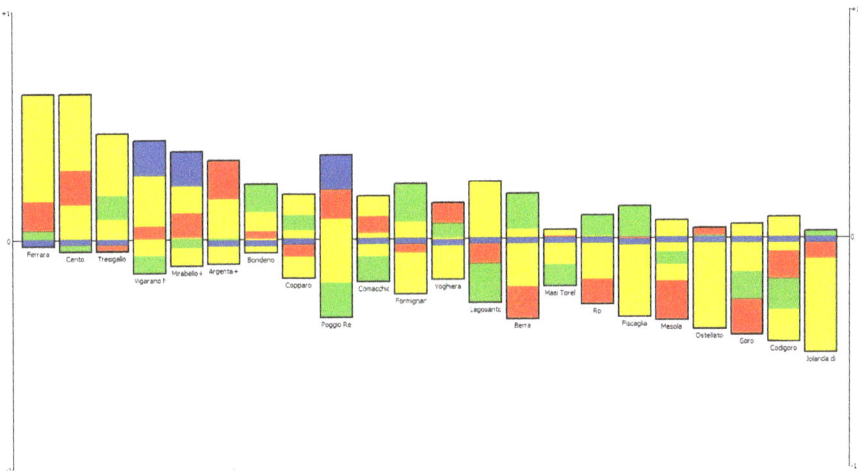

Figure 9. PROMETHEE rainbow plot for the usual preference function. On the vertical axis, the preference function Θ is reported. The yellow bar indicates the criteria relating to exposure, red is used for seismic hazard, green for vulnerability, and blue for hydraulic hazard.

Based on the ranking provided by PROMETHEE, it is possible to create a risk map of the municipalities of the province of Ferrara that highlights high-priority areas as those with a high level of combined flood and seismic risk, medium priority areas as the areas characterized by a medium combined-risk level, and, finally, low combined-risk areas.

This map is shown in Figure 10. It can be seen that the three risk levels are identified by three different colors: Red is used for high risk, orange for medium risk, and yellow for low risk.

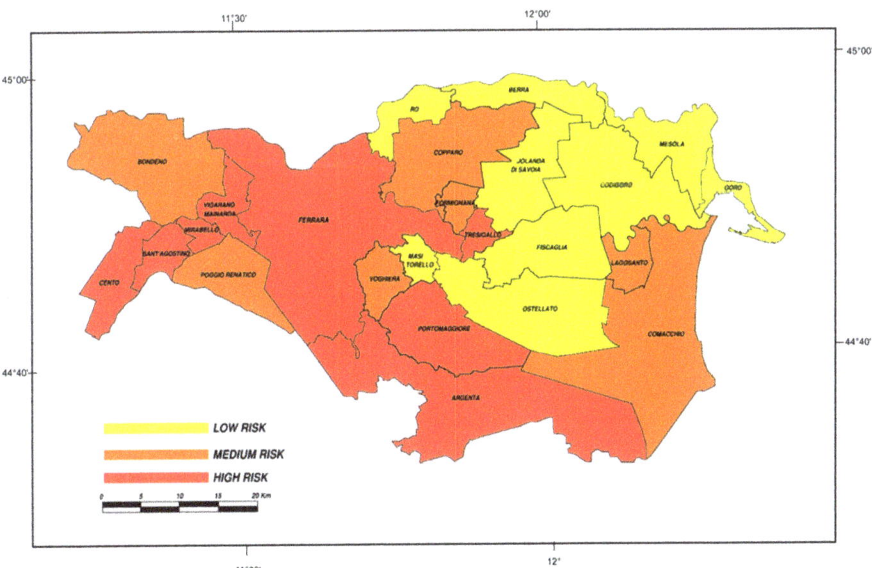

Figure 10. Multiple-risk map for the Ferrara province obtained for the usual preference function (Type 1 in Table 1). The risk levels strictly indicate the relative priority ranking for decision-makers and do not indicate the effective safety level of the various municipalities.

3.2. Linear Preference Function

The ranking of alternatives for the linear preference function and zero-max method is shown in Table 5, while the corresponding multi-risk map is shown in Figure 11. These maps were obtained by associating the quantitative criteria, i.e., flood hazard, land use, age of buildings, and population density, with a linear preference function, while thresholds q and p were determined with the zero-max method.

Table 5. Ranking of alternatives for the linear preference function and zero-max method.

Rank	Alternatives	Θ	Θ^+	Θ^-
1	Cento	0.459	0.5086	0.0496
2	Ferrara	0.3545	0.393	0.0385
3	Tresigallo	0.1821	0.2642	0.0821
4	Mirabello + Sant'Agostino	0.1444	0.2622	0.1179
5	Argenta + Portomaggiore	0.1352	0.2182	0.0829
6	Bondeno	0.1257	0.2069	0.0812
7	Vigarano Mainarda	0.112	0.255	0.143
8	Copparo	0.0505	0.1684	0.1179
9	Poggio Renatico	0.031	0.2216	0.1906
10	Comacchio	0.0224	0.1631	0.1406
10	Voghiera	−0.0422	0.0984	0.1406
12	Formignana	−0.0721	0.0937	0.1657
13	Fiscaglia	−0.0761	0.0868	0.1628
14	Lagosanto	−0.0898	0.1385	0.2283
15	Codigoro	−0.101	0.1155	0.2164
16	Ostellato	−0.1092	0.0736	0.1828
17	Ro	−0.1362	0.0589	0.195
18	Masi Torello	−0.1371	0.0557	0.1928
19	Berra	−0.1551	0.0688	0.2239
20	Jolanda di Savoia	−0.1947	0.0408	0.2355
21	Mesola	−0.2203	0.0353	0.2556
22	Goro	−0.283	0.0137	0.2968

For a linear preference function of the aforementioned quantitative criteria, and thresholds q and p determined by the mean-std method, we obtained the results shown in Table 6 and Figure 12.

By comparing the results obtained from the usual and the linear preference functions, it can be understood that changes of the preference function do not reflect large changes of the final risk maps. The only difference is that the risk levels of the municipality of Vigarano Mainarda swap with Bondeno, and Fiscaglia swaps with Lagosanto.

By comparing the maps in Figures 11 and 12, obtained with the thresholds chosen with the zero-max and mean-std methods, respectively, we observe that the risk levels of Lagosanto, Vigarano Mainarda, and Codigoro increase. Particularly, we observe that the choice of the preference function affects the final ranking of the alternatives especially when the thresholds are chosen according to the mean-std method.

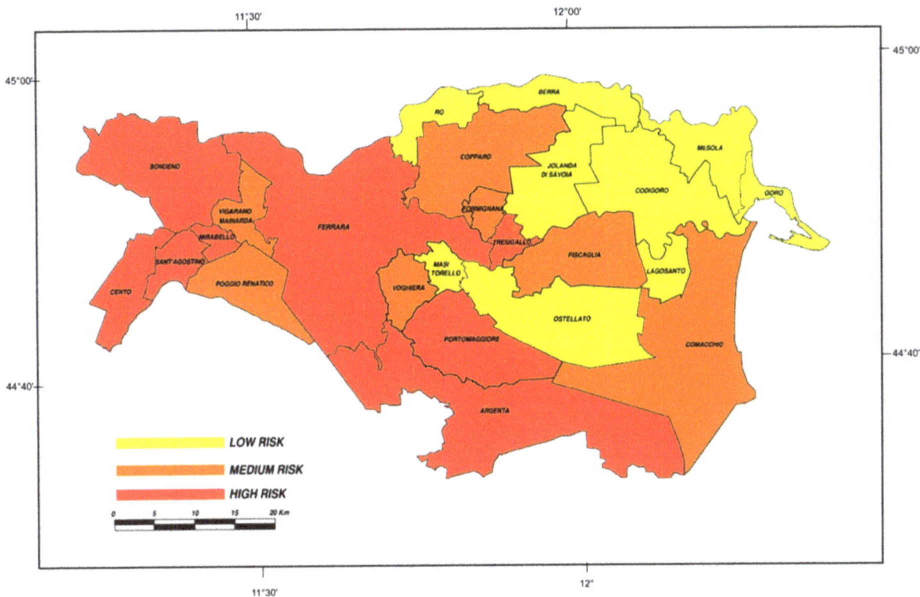

Figure 11. Multiple-risk map for the Ferrara province for the linear preference function with the thresholds chosen with the zero-max method. The risk levels strictly indicate the relative priority ranking for decision-makers and do not indicate the effective safety level of the various municipalities.

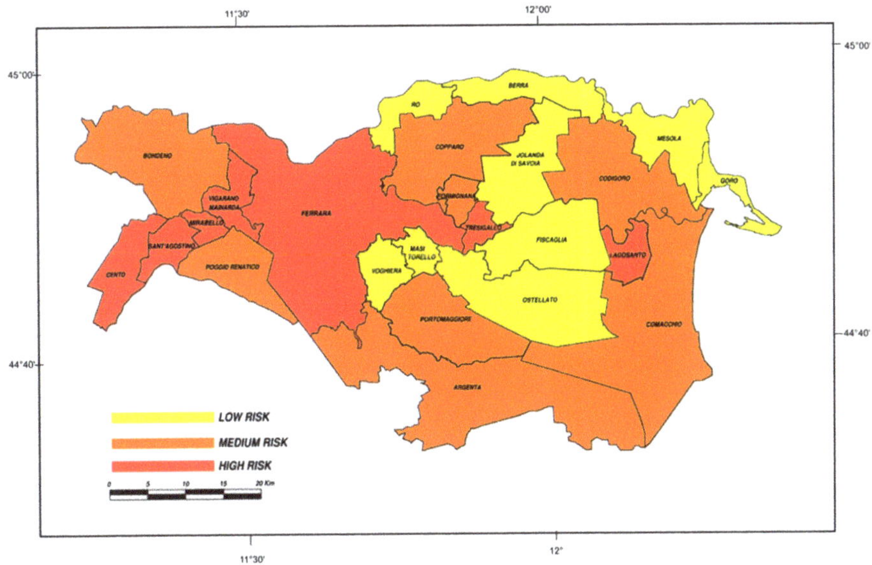

Figure 12. Multiple-risk map for the Ferrara province for the linear preference function with the thresholds chosen with the mean-std method. The risk levels strictly indicate the relative priority ranking for decision-makers and do not indicate the effective safety level of the various municipalities.

Table 6. Ranking of alternatives for the linear preference function and std-mean method.

Rank	Alternatives	Θ	Θ^+	Θ^-
1	Cento	0.4532	0.4849	0.0317
2	Ferrara	0.3769	0.4123	0.0354
3	Tresigallo	0.2051	0.2613	0.0562
4	Vigarano Mainarda	0.1219	0.2185	0.0966
5	Mirabello+ Sant'Agostino	0.078	0.1835	0.1056
6	Lagosanto	0.0597	0.1319	0.0722
7	Poggio Renatico	0.0523	0.167	0.1147
8	Argenta + Portomaggiore	0.0307	0.1133	0.0826
9	Copparo	0.0261	0.1107	0.0846
10	Bondeno	0.0222	0.1075	0.0853
11	Comacchio	0.0217	0.1075	0.0857
12	Codigoro	0.0125	0.1053	0.0928
13	Formignana	−0.1232	0.0146	0.1378
14	Masi Torello	−0.1275	0.0086	0.1361
15	Goro	−0.1278	0.0106	0.1384
16	Mesola	−0.1317	0.0069	0.1386
17	Berra	−0.1383	0.0046	0.1429
18	Voghiera	−0.1383	0.0041	0.1424
19	Ro	−0.1385	0.004	0.1425
20	Fiscaglia	−0.1497	0.002	0.1517
21	Ostellato	−0.1839	0	0.1839
22	Jolanda di Savoia	−0.2014	0	0.2014

Regardless of the preference function chosen, the maps obtained present a similar risk trend, i.e., the territory is divided into two parts: The municipalities of the western part of the territory of the province of Ferrara, plus Ferrara and Tresigallo, are characterized by a medium–high risk level; the upper-eastern part of the province is characterized by a medium–low risk level.

3.3. Sensitivity Analysis on the Choice of Weights

As introduced in Section 2, the sensitivity analysis on the weights was performed by first increasing the weight of each individual criterion at a time, and then assuming the simultaneous increase in the weights of the three "exposure"-related criteria, namely land use, population density, and strategic buildings. Specifically, the weights were changed according to Scenarios 0, 1, 2, and 3 described in Table 3.

For the sake of brevity, we present hereafter the results obtained by assuming the usual preference function. For the reader's convenience, the results are reported as maps, as in the previous sections.

In the first part of the analysis, the criteria are changed according to the following order: Flood hazard, PGA, land use, strategic buildings, age of buildings, and population density. Hereafter, we omit the maps obtained for the changes of the weights relating to the criteria of strategic buildings and age of buildings, for brevity. Figures 13 and 14 illustrate the various risk maps obtained by increasing the weights.

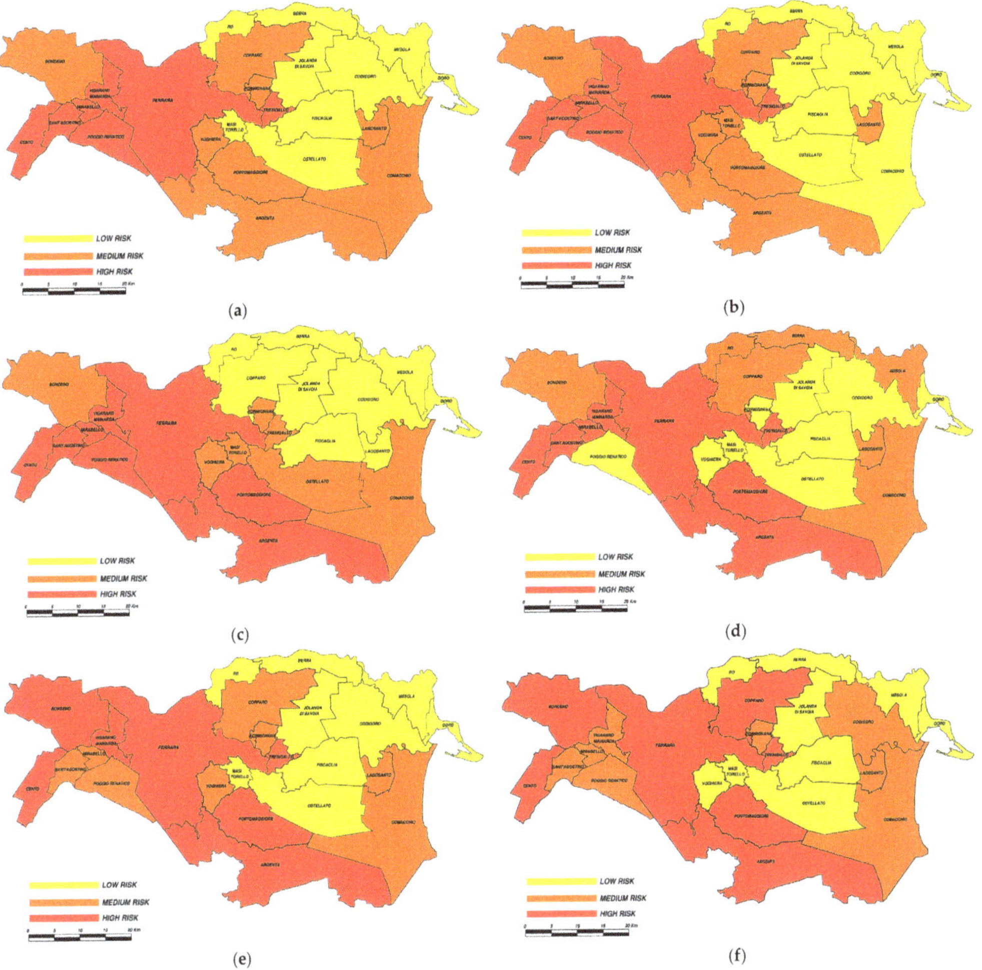

Figure 13. Weights sensitivity analysis, multi-risk maps; (**a**) flood hazard weight increase assuming Scenarios 2 or 3; (**b**) PGA weight increase assuming Scenario 1; (**c**) PGA weight increase assuming Scenario 2; (**d**) land use weight increase assuming Scenarios 2 or 3; (**e**) strategic buildings weight increase assuming Scenario 1; (**f**) strategic buildings weight increase assuming Scenarios 2 or 3. The risk levels strictly indicate the relative priority ranking for decision-makers and do not indicate the effective safety level of the various municipalities.

With Scenario 1 based on the variation of the flood hazard weight, the ranking of the municipalities remains almost unchanged, as the multi-risk map is identical to that of Scenario 0. On the other hand, the maps change when Scenarios 2 and 3 are adopted, as shown in Figure 13a. More marked differences can be observed when the weight of the PGA (Figure 13b,c) is changed. By increasing the weight of the land-use criterion, an increase in Scenario 1 does not reflect evident changes in the multi-risk map (Figure 13d). On the other hand, changes in Scenarios 2 and 3 affect the multi-risk map. Looking back at the strategic building criterion, we observe differences in the risk map when the first change of Scenario 1 is applied, compared to Scenario 0 (Figure 13e), and more so with the last two increases of Scenario 2 (Figure 13f). For the criterion of population density, the first increase in the weight according to Scenario 1 does not affect the map (see Figure 14a), while the subsequent increases in the weight bring about noticeable modifications. In particular, the multi-risk map reported in Figure 14b assigns a comparatively low level of attention to the Argenta municipality, which, however, is associated with a medium seismicity level according to the territorial classification of Figure 2a. Depending on the stakeholders' expectations, this might suggest that weights should not be varied to the extent of downgrading the seismic risk level of certain municipalities classified at medium to high seismic risk.

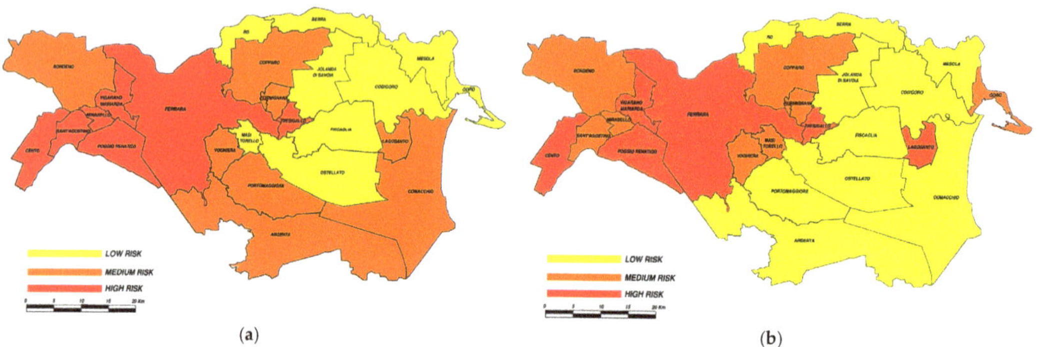

Figure 14. Weights sensitivity analysis, multi-risk maps; (**a**) Scenario 2, population density weight increase; (**b**) Scenario 3, population density weight increase. The risk levels strictly indicate the relative priority ranking for decision-makers and do not indicate the effective safety level of the various municipalities.

Lastly, results of the sensitivity on the weight choice for the criteria related to exposure are presented in Table 7. Proceeding as illustrated in Section 2, the first increase in the weight does not alter the risk map, which remains the same as in Scenario 0, whereas with the changes of Scenario 2, greater variations can be observed.

Therefore, concluding the sensitivity analysis of weights, it can be inferred that, in general, the results are sensitive to the increase in the weights of the criteria, determining a risk map that varies from case to case, causing the risk of some municipalities to decrease while that of others increased. However, these variations do not upset the overall trend, which highlights a territory divided into two parts, that of the municipalities of the western part of the territory of the province of Ferrara characterized by a medium–high risk level, and the municipalities of the north-eastern area characterized by a medium–low risk level.

3.4. Remarks on the Limitations of the Analysis

The proposed methodology requires the definition of several parameters, criteria, and weights, whose choice resulted in being strongly dependent on the expectations of stakeholders and end-users. Thus, the obtained results should be seen as a first attempt towards the proposal of an MCDA methodology that does not require great mathematical expertise, is flexible, and can be easily adapted to many situations. Nevertheless, further efforts are necessary in order for the tool to be readily exploited by public authorities and decision makers. Furthermore, it is worth highlighting other limitations inherent in the present analysis.

The first type of limitation is mainly related to the availability of data. Indeed, the choice of the criteria was based on the availability of the relevant information, which led, for some criteria, to a purely qualitative evaluation. Greater availability, accuracy, and ease of retrieval of the data would lead to the creation of a more complete and more precise analysis, and it could also contribute to the development of operational tools and software.

Table 7. Sensitivity analysis on the exposure factor, ranking of alternatives (EXP: EXPOSURE).

Rank	Scenario 1: WEIGHT = 0.22; OTHERS = 011				Scenario 2: WEIGHT = 0.32; OTHERS = 0.01			
	Alternativa	Θ	Θ^+	Θ^-	Alternativa	Θ	Θ^+	Θ^-
1	Ferrara	0.7153	0.8064	0.0911	Cento	0.9452	0.9683	0.0231
2	Cento	0.7093	0.8061	0.0968	Ferrara	0.9168	0.9538	0.037
3	Tresigallo	0.5092	0.6746	0.1654	Tresigallo	0.696	0.7975	0.1015
4	Vigarano Mainarda	0.2908	0.5771	0.2863	Lagosanto	0.4603	0.6797	0.2193
5	Argenta + Portomaggiore	0.2279	0.534	0.306	Vigarano Mainarda	0.3006	0.5575	0.2569
6	Mirabello + Sant'Agostino	0.219	0.5413	0.3222	Argenta + Portomaggiore	0.2083	0.5536	0.3454
7	Bondeno	0.1597	0.4971	0.3375	Mirabello + Sant'Agostino	0.1207	0.4675	0.3468
8	Lagosanto	0.1359	0.488	0.352	Bondeno	0.1154	0.507	0.3915
9	Copparo	0.0416	0.4381	0.3965	Comacchio	0.1044	0.5017	0.3973
10	Comacchio	0.0356	0.4378	0.4022	Copparo	0.076	0.4873	0.4113
11	Poggio Renatico	−0.0499	0.4041	0.454	Mesola	−0.0919	0.3574	0.4493
12	Formignana	−0.0661	0.3555	0.4216	Masi Torello	−0.1448	0.3309	0.4757
13	Voghiera	−0.1127	0.3295	0.4422	Goro	−0.1563	0.3252	0.4815
14	Masi Torello	−0.1644	0.3064	0.4708	Codigoro	−0.1861	0.3564	0.5426
15	Berra	−0.2045	0.2863	0.4908	Poggio Renatico	−0.1924	0.3107	0.5031
16	Mesola	−0.2197	0.2787	0.4984	Formignana	−0.1938	0.3064	0.5002
17	Ro	−0.226	0.2756	0.5016	Voghiera	−0.2848	0.2607	0.5455
18	Goro	−0.2939	0.2416	0.5355	Berra	−0.2929	0.2569	0.5498
19	Codigoro	−0.3041	0.268	0.5721	Ro	−0.2949	0.2559	0.5507
20	Fiscaglia	−0.3385	0.2193	0.5578	Fiscaglia	−0.594	0.1063	0.7003
21	Ostellato	−0.4816	0.1451	0.6267	Ostellato	−0.7225	0.0418	0.7643
22	Jolanda di Savoia	−0.5829	0.0971	0.68	Jolanda di Savoia	−0.7893	0.0087	0.798

Secondly, this analysis neglected cascade effects, an aspect that deserves further investigation in the future [11].

Thirdly, the present contribution does not consider the impact of modeling assumptions on the seismic risk assessment. At the relevant scale of observations of the present analysis, specific structural aspects connected to the vulnerability levels of the buildings cannot be easily considered. In this regard, we recall that specific structural aspects and modeling assumptions play, among others, a key role for seismic risk evaluation at both the building and the urban scale [36]. A recent study focusing on South America has shown the uncertainties and biases that the use of simplified models or heterogenous data may produce in the determination of seismic vulnerability [36]. For completeness, seismic risk evaluation is extensively discussed, for instance, in the aforementioned contributions [36–39] and the references cited therein.

Finally, we recall that the seismic classification shown in Figure 2 has been merely used as a technical-administrative reference for establishing the priority of actions and measures aimed at preventing and mitigating seismic risk. It must not be used to determine the local seismic action or for the structural design of buildings, which, instead, rely upon more detailed maps highlighting, for instance, the presence of site effects due to the inherent geological structure of the ground or instability effects such as liquefaction. Therefore, the present analysis should be purposefully extended in order to consider the aforementioned local effects [40].

4. Conclusions

For the present case study, the application of the Multiple-Criteria Decision Analysis (MCDA) methodology through the PROMETHEE algorithm has proved an innovative and promising operational tool. Its potential derives from the ability to both analyze information from various sources and jointly systematize data expressed in different units and scales. The application of this methodology has made it possible to rank the various municipalities in terms of the relative proneness to joint flood and seismic hazards. We recall that the objective of the methodology is not to quantify the safety level in absolute terms of the various municipalities. Its scope is, indeed, to provide useful information for decision makers and public authorities to define future intervention priorities. We further emphasize that, in the authors'opinion, the present study is original as it applies the PROMETHEE algorithm for the first time to a multi-risk assessment of seismic and flood hazards.

Depending on the territory to be studied, the relevant risks could be different, and therefore, different criteria must be used to express them. Nevertheless, the generalization to other multi-risk analyses and different case studies deserves further considerable efforts and thoughtful insights. Full validation of the present methodology is also of utmost importance and calls for new developments. However, the proposed methodology is flexible. This suggests that, with due precautions and adaptations, it is possible to apply it to different risk scenarios, such as scenarios including coastal floods and landslides, while keeping the same applicative scheme.

Finally, the obtained results have shown that the proposed methodology is an operational tool that, once further validated, can be used by end users, whether modelers or decision makers, to urgently allocate resources and increase the coping capacity of communities in the case of catastrophic events.

Author Contributions: Conceptualization, Ž.N., A.C., C.V. and E.B.; methodology, Ž.N., A.S. and A.C.; software, A.S.; validation, A.S., A.C. and C.V.; formal analysis A.S., A.C., Ž.N. and C.V.; resources E.B.; data curation, A.S.; writing—original draft preparation, A.S. and A.C.; writing—review and editing, A.C., C.V., Ž.N. and E.B.; visualization, A.S. and A.C.; supervision, E.B.; project administration, E.B.; funding acquisition, E.B. All authors have read and agreed to the published version of the manuscript.

Funding: This research was funded by the EUROPEAN UNION, Programme Interreg Italy-Croatia, Project "Preventing, managing and overcoming natural-hazards risks to mitigate economic and social impact"—PMO-GATE ID 10046122.

Institutional Review Board Statement: Not applicable.

Informed Consent Statement: Not applicable.

Data Availability Statement: Publicly available datasets and figures were analyzed in this study. This data can be found here: https://www.bonificaferrara.it/images, https://www4.istat.it/it/mappa-rischi/documentazione, https://ambiente.regione.emilia-romagna.it/en/geologia/seismic-risk/seismic-classification, http://zonesismiche.mi.ingv.it, http://www.provincia.fe.it. (accessed on 26 January 2022).

Acknowledgments: The authors are very grateful to Consorzio di Bonifica Pianura di Ferrara for making the flood maps freely available, and, particularly, to Eng. Alessandro Bondesan for his kind support in georeferencing the maps.

Conflicts of Interest: The authors declare no conflict of interest.

References

1. UN-ISDR. Sendai Framework for Disaster Risk Reduction 2015–2030. In Proceedings of the UN world Conference on Disaster Risk Reduction, Sendai, Japan, 14–18 March 2015; United Nations Office for Disaster Risk Reduction: Geneva, Switzerland. Available online: http://www.unisdr.org/files/43291_sendaiframeworkfordrren.pdf (accessed on 1 October 2021).
2. Poljanšek, K.; Ferrer, M.M.; De Groeve, T.; Clark, I. Preface. In *Science for Disaster Risk Management 2017: Knowing Better and Losing Less*; Publications Office of the European Union: Luxembourg, 2017; ISBN 978-92-79-60678-6. [CrossRef]
3. Topics Geo: Natural Catastrophes 2013: Analyses, Assessments, Positions, Münchener Rückversicherungs-Gesellschaft, Munich. 2014. Available online: https://www.munichre.com/content/dam/munichre/contentlounge/website-pieces/documents/30 2-08121_en.pdf/_jcr_content/renditions/original./302-08121_en.pdf (accessed on 26 January 2022).

4. UN-ISDR. Terminology: Basic Terms of Disaster Risk Reduction. 2009. Available online: http://www.unisdr.org./we/inform/terminology (accessed on 1 October 2021).
5. Urlainis, A.; Ornai, D.; Levy, R.; Vilnay, O.; Shohet, I.M. Loss and damage assessment in critical infrastructures due to extreme events. *Saf. Sci.* **2022**, *147*, 105587. [CrossRef]
6. Kanamori, H.; Hauksson, E.; Heaton, T. Real-time seismology and earthquake hazard mitigation. *Nature* **1997**, *390*, 461–464. [CrossRef]
7. Quesada-Román, A.; Villalobos-Chacón, A. Flash flood impacts of Hurricane Otto and hydrometeorological risk mapping in Costa Rica. *Geogr. Tidsskr.-Dan. J. Geogr.* **2020**, *120*, 142–155. [CrossRef]
8. Quesada-Román, A.; Ballesteros-Cánovas, J.A.; Granados-Bolaños, S.; Birkel, C.; Stoffel, M. Improving regional flood risk assessment using flood frequency and dendrogeomorphic analyses in mountain catchments impacted by tropical cyclones. *Geomorphology* **2022**, *396*, 108000. [CrossRef]
9. Kron, W. Reasons for the increase in natural catastrophes: The development of exposed areas. In *Topics 2000: Natural Catastrophes, the Current Position*; Munich Reinsurance Company: Munich, Germany, 1999; pp. 82–94.
10. Barredo, J.I. Major flood disasters in Europe: 1950–2005. *Nat. Hazards* **2007**, *42*, 125–148. [CrossRef]
11. Zuccaro, G.; De Gregorio, D.; Leone, M. Theoretical model for cascading effects analyses. *Int. J. Disaster Risk Reduct.* **2018**, *30*, 199–215. [CrossRef]
12. Zschau, J. Where are we with multihazards, multirisks assessment capacities? In *Science for Disaster Risk Management 2017: Knowing Better and Losing Less*; Poljansek, K., Marin Ferrer, M., De Groeve, T., Eds.; Publications Office of the European Union: Luxembourg, 2017; ISBN 978-92-79-60678-6. [CrossRef]
13. Fuchs, S.; Keiler, M.; Zischg, A. A spatiotemporal multi-hazard exposure assessment based on property data. *Nat. Hazard. Earth Syst. Sci.* **2015**, *15*, 2127–2142. [CrossRef]
14. Komentova, N.; Scolobig, A.; Garcia-Aristizabal, A.; Monfort, D.; Fleming, K. Multi-risk approach and urban resilience. *Int. J. Disast. Res. Built Environ.* **2016**, *7*, 114–132. [CrossRef]
15. Marzocchi, W.; Garcia-Aristizabal, A.; Gasparini, P.; Mastellone, M.L.; Di Ruocco, A. Basic principles of multi-risk assessment: A case study in Italy. *Nat. Hazards* **2012**, *62*, 551–573. [CrossRef]
16. Kappes, M.S.; Keiler, M.; von Elverfeldt, K.; Glade, T. Challenges of analyzing multi-hazard risk: A review. *Nat. Hazards* **2012**, *64*, 1925–1958. [CrossRef]
17. Bell, R.; Glade, T. Multi-hazard analysis in natural risk assessments. *WIT Trans. Ecol. Environ.* **2004**, *77*, 1–10.
18. Schmidt, J.; Matcham, I.; Reese, S.; King, A.; Bell, R.; Henderson, R.; Smart, G.; Cousins, J.; Smith, W.; Heron, D. Quantitative multi-risk analysis for natural hazards: A framework for multi-risk modelling. *Nat. Hazards* **2011**, *58*, 1169–1192. [CrossRef]
19. Neri, A.; Aspinall, W.P.; Cioni, R.; Bertagnini, A.; Baxter, P.J.; Zuccaro, G.; Andronico, D.; Barsotti, S.; Cole, P.D.; Esposti Ongaro, T.; et al. Developing an Event Tree for probabilistic hazard and risk assessment at Vesuvius. *J. Volcanol. Geotherm. Res.* **2008**, *178*, 397–415. [CrossRef]
20. Barthel, F.; Neumayer, E. A trend analysis of normalized insured damage from natural disasters. *Clim. Chang.* **2012**, *113*, 215–237. [CrossRef]
21. Skilodimou, H.D.; Bathrellos, G.D.; Chousianitis, K.; Youssef, A.M.; Pradhan, B. Multi-hazard assessment modeling via multi-criteria analysis and GIS: A case study. *Environ. Earth Sci.* **2019**, *78*, 47. [CrossRef]
22. Brans, J.P.; Mareschal, B. Promethee Methods. In *Multiple Criteria Decision Analysis: State of the Art Surveys*; Figueira, J., Greco, S., Ehrogott, M., Eds.; Springer: Berlin/Heidelberg, Germany, 2005.
23. Gallina, V.; Torresan, S.; Critto, A.; Sperotto, A.; Glade, T.; Marcomini, A. A review of multi-risk methodologies for natural hazards: Consequences and challenges for a climate change impact assessment. *J. Environ. Manag.* **2016**, *168*, 123–132. [CrossRef]
24. Gallina, V.; Torresan, S.; Zabeo, A.; Critto, A.; Glade, T.; Marcomini, A. A Multi-Risk Methodology for the Assessment of Climate Change Impacts in Coastal Zones. *Sustainability* **2020**, *12*, 3697. [CrossRef]
25. Peduzzi, P.; Dao, H.; Herold, C.; Mouton, F. Assessing global exposure and vulnerability towards natural hazards: The Disaster Risk Index. *Nat. Hazards Earth Syst. Sci.* **2009**, *9*, 1149–1159. [CrossRef]
26. Brans, J.P.; Vincke, P.; Mareschal, B. How to select and how to rank projects: The PROMETHEE method. *Eur. J. Oper. Res.* **1986**, *24*, 228–238. [CrossRef]
27. Mladineo, M.; Jajac, N.; Rogulj, K. A simplified approach to the PROMETHEE method for priority setting in management of mine action projects. *Croat. Oper. Res. Rev.* **2016**, *7*, 249–268. [CrossRef]
28. Crnjac, M.; Aljinovic, A.; Gjeldum, N.; Mladineo, M. Two-stage product design selection by using PROMETHEE and Taguchi method: A case study. *Adv. Prod. Eng. Manag.* **2019**, *14*, 39–50. [CrossRef]
29. Savic, M.; Nikolic, D.; Mihajlovic, I.; Zivkovic, Z.; Bojanov, B.; Djordjevic, P. Multi-Criteria Decision Support System for Optimal Blending Process in Zinc Production. *Miner. Process. Extr. Metall. Rev.* **2015**, *36*, 267–280. [CrossRef]
30. Rocchi, A.; Chiozzi, A.; Nale, M.; Nikolic, Z.; Riguzzi, F.; Mantovan, L.; Gilli, A.; Benvenuti, E. A Machine Learning Framework for Multi-Hazard Risk Assessment at the Regional Scale in Earthquake and Flood-Prone Areas. *Appl. Sci.* **2022**, *12*, 583. [CrossRef]
31. Carminati, E.; Martinelli, G. Subsidence rates in the Po Plain, northern Italy: The relative impact of natural and anthropogenic causation. *Eng. Geol.* **2002**, *66*, 241–255. [CrossRef]

32. Salvati, L.; Mavrakis, A.; Colantoni, A.; Mancino, G.; Ferrara, A. Complex Adaptive Systems, soil degradation and land sensitivity to desertification: A multivariate assessment of Italian agro-forest landscape. *Sci. Total Environ.* **2015**, *521–522*, 235–245. [CrossRef] [PubMed]
33. Stucchi, M.; Meletti, C.; Montaldo, V.; Akinci, A.; Faccioli, E.; Gasperini, P.; Malagnini, L.; Valensise, G. Pericolosità Sismica di Riferimento Per il Territorio Nazionale MPS04 [Data Set]. Istituto Nazionale di Geofisica e Vulcanologia (INGV). 2004. Available online: https://data.ingv.it/en/dataset/70#additional-metadata (accessed on 1 October 2021). [CrossRef]
34. Trigila, A.; Iadanza, C.; Bussettini, M.; Lastoria, B. *Dissesto Idrogeologico in Italia: Pericolosità e Indicatori di Rischio—Edizione 2018*; Rapporti 287/2018; ISPRA. Roma, Italy, 2018.
35. Decreto Legislativo n. 49/2010. Available online: https://www.mite.gov.it/sites/default/files/archivio/allegati/vari/documento_definitivo_indirizzi_operativi_direttiva_alluvioni_gen_13.pdf (accessed on 3 January 2022).
36. Dolce, M.; Prota, A.; Borzi, B.; da Porto, F.; Lagomarsino, S.; Magenes, G.; Moroni, C.; Penna, A.; Polese, M.; Speranza, E.; et al. Seismic risk assessment of residential buildings in Italy. *Bull. Earthquake Eng.* **2021**, *19*, 2999–3032. [CrossRef]
37. Hoyos, M.C.; Hernández, A.F. Impact of vulnerability assumptions and input parameters in urban seismic risk assessment. *Bull. Earthq. Eng.* **2021**, *19*, 4407–4434. [CrossRef]
38. Asadi, E.; Salman, A.M.; Li, Y.; Yu, X. Localized health monitoring for seismic resilience quantification and safety evaluation of smart structures. *Struct. Saf.* **2021**, *93*, 102127. [CrossRef]
39. Joyner, M.D.; Gardner, C.; Puentes, B.; Sasani, M. Resilience-Based seismic design of buildings through multiobjective optimization. *Eng. Struct.* **2021**, *246*, 113024. [CrossRef]
40. CTMS. Linee Guida per la Gestione del Territorio in Aree Interessate da Faglie Attive e Capaci (FAC). Commissione Tecnica Per la Microzonazione Sismica, Gruppo di Lavoro FAC. Dipartimento Della Protezione Civile e Conferenza Delle Regioni e Delle Province Autonome. 2015. Available online: http://www.protezionecivile.gov.it/resources/cms/documents/LineeGuidaFAC_v1_0.pdf (accessed on 3 January 2022).

Article

Risk Assessment of Riverine Terraces: The Case of the Chenyulan River Watershed in Nantou County, Taiwan

Ji-Yuan Lin [1], Jen-Chih Chao [1,*] and Yung-Ming Hsu [2]

[1] Department of Landscape and Urban Design, Chaoyang University of Technology, Taichung 413310, Taiwan; jylin@cyut.edu.tw

[2] Department of Civil & Construction Engineering, Chaoyang University of Technology, Taichung 413310, Taiwan; s9911625@cyut.edu.tw

* Correspondence: jenchihchao@gmail.com; Tel.: +866-423323000 (ext. 7668)

Abstract: The purpose of this study is to establish a method of hazard assessment for the river terraces along the Chenyulan River and use 40 of them as protected objects. Using a geographic information system, the researchers extracted nine parameters for such terraces. These are length to attack shore, distance away from fault, distance from river channel, number of creeks and streams with possibility of debris flows, height above stream level, average slope degree, geology, number of erosion ditches, and distance from landslide area behind. Next, the weightings identified by analytic hierarchy process analysis were used as the basis for grading the various factors affecting river terraces. Hazard assessment for the river terraces then proceeded via totaling of the potential trends of the various factors and the protected objects, as well as comparison of historical disaster conditions and satellite images. The results showed that there were 8 high-risk river terraces, 14 medium–high-risk river terraces, 14 medium–low-risk river terraces and 4 low-risk river terraces. The evaluation of the current conditions of the settlement environment through parameter weighting has a certain accuracy and reference value in reducing the disaster impact of the riverine terrace settlement.

Keywords: geographic information system; hazard assessment; river terraces; risk assessment

Citation: Lin, J.-Y.; Chao, J.-C.; Hsu, Y.-M. Risk Assessment of Riverine Terraces: The Case of the Chenyulan River Watershed in Nantou County, Taiwan. *Appl. Sci.* **2022**, *12*, 1375. https://doi.org/10.3390/app12031375

Academic Editors: Andrea Chiozzi, Elena Benvenuti and Željana Nikolić

Received: 16 November 2021
Accepted: 19 January 2022
Published: 27 January 2022

Publisher's Note: MDPI stays neutral with regard to jurisdictional claims in published maps and institutional affiliations.

1. Introduction

Taiwan is located at the junction of the Eurasian continental plate and the Philippine Sea plate. Formed through the Penglai orogeny, the Nanao orogeny and crustal changes, it is a mountainous terrain with flat land accounting for only 25% of its total area. Due to population growth and rapid industrial and commercial development in Taiwan in recent years, the use of flat land has become saturated, and development of hillside areas, especially river terraces, has become common. The existence of river terraces indicates frequent geological changes, high erosion rates, abundant sources of silt, and strong river scour [1], but their formation is also affected to some extent by climate change and human activities [2]. As Taiwan is surrounded by the sea on all sides, it receives abundant rainfall throughout the year, about 2500 mm, more than two and a half times the world annual average of 970 mm. During the rainy season (from 1 May to 30 November each year), the region is prone to typhoons, each of which tends to increase the intensity of rainfall within a short period of time, and this can result in landslides and mudslides in mountainous areas, as well as rapid rises in the water levels in rivers, which often results in flooding and the erosion and collapse of riverbanks. To prevent loss of life, residents of river terraces have to evacuate when typhoons occur. On 7 August 2009, when Taiwan was struck by a moderate-strength typhoon, Morakot, heavy rainfall led to a series of disasters in southern Taiwan. Due to the collapse of Xiandu (Xianto) Mountain, Xiaolin Village was destroyed, and a short-term barrier lake that was formed endangered the lives and property of residents downstream.

Due to undercutting and erosion by rivers, river terraces remain above the water surface during normal floods and are distributed in steps on the slopes of the river valleys [3]. Chang and Shi defined river terraces as land along rivers, consisting of terraces and cliffs [4]. The terraces' surfaces were the riverbed or floodplain surfaces in a former period, while the cliffs below them, facing the valley axis, were formed by both down erosion and lateral erosion [2]. The terraces of Xiaolin Village were formed by ancient and recent landslides. Their geology is extremely unstable, and the village was destroyed largely due to its location on dangerous low-lying ones. According to our survey, among the 144 mountain settlements and aboriginal tribal villages in southern Taiwan, 92 are located on river terraces, similar to the situation of Xiaolin. Therefore, the hazard and risk assessment of river terraces demand special attention to avoid similar disasters.

Various explorations of the topographic evolution of river terraces and the reasons they are formed in different regions have been conducted [5–8]. Scholars have also investigated the risks to river-terrace settlements posed by rainfall-induced landslides, based on historical data regarding the potential risk range of debris flows and the areas where landslides occur [9–11]. The Chenyulan River, in particular, has been the subject of multiple studies focused on collapses' potential indicators and locations [12–14], due to the multiple disasters that have struck the terraces of its lower reaches. However, due in part to strong variation in the reasons for the formation of river terraces and the hazards they face, residents of river-terrace settlements tend to have low awareness of disaster risks and are thus unable to effectively mitigate them. The people who live in settlements on the river terraces in Chenyulan today could face disaster at any time. Therefore, levels of danger to such terraces are estimated by risk assessment, so that when a typhoon is about to strike Taiwan, local residents can be quickly moved to safe places and disaster-relief facilities.

This study focuses on factors that may harm river terraces, derived from special questionnaires to establish index weights, and uses a geographic information system (GIS) overlap to allocate these factors to particular river terraces. Then, the scores of these potential factors and preservation factors are summed to estimate the terraces' risk, and establish a risk map of the area, with the wider aim of disaster prevention and reduction in disaster losses. This study uses the analytical features of AHP multilevel evaluation to decompose the elements of the river terraces' environment and construct a model of potential factors of the river terraces. Using the GIS data and AHP model, a matrix of judgement is established based on the corresponding criteria to derive the corresponding element weights, and a spatial analysis of the river terraces' hazard trend map is used to provide a solution to reduce the impact caused by the disaster.

Study Area

Li's survey of the Chenyulan River noted that its inland river terraces, alluvial fans and landslides were highly developed, and that there were 46 fan-shaped terraces and alluvial fans. The large number of these features implies rapid geological change [1].

According to the Bureau of Soil and Water Conservation, part of Taiwan's Council of Agriculture, there are more than 1700 potential soil and rock flows [15]. Within our study area, as shown in Figure 1, there are 49 such potential flows.

2. Literature Review

The potential hazard factors affecting river terraces can be summarized into three latent-sensing categories. The first, in front of the terrace, comprises four factors: attack shore, distance from fault, distance from river, and potential stream-impact quantity. The second, of the river terrace itself, consists of three factors: minimum ratio, average slope and geology. Additionally, the third, behind the river terrace, includes two factors: number of erosion ditches and number of collapses from the rear. Each category is analyzed in the following sections.

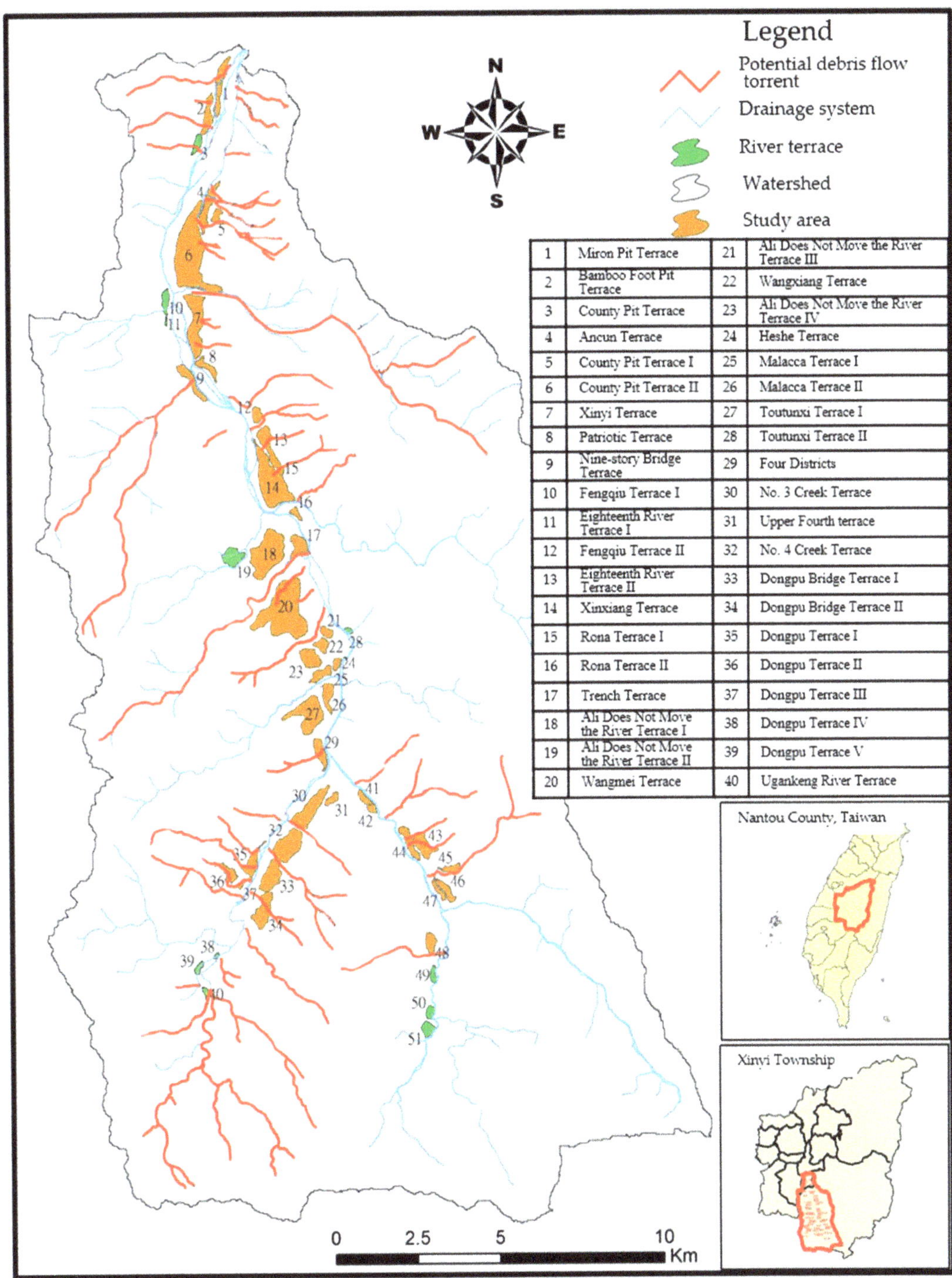

1	Miron Pit Terrace	21	Ali Does Not Move the River Terrace III
2	Bamboo Foot Pit Terrace	22	Wangxiang Terrace
3	County Pit Terrace	23	Ali Does Not Move the River Terrace IV
4	Ancun Terrace	24	Heshe Terrace
5	County Pit Terrace I	25	Malacca Terrace I
6	County Pit Terrace II	26	Malacca Terrace II
7	Xinyi Terrace	27	Toutunxi Terrace I
8	Patriotic Terrace	28	Toutunxi Terrace II
9	Nine-story Bridge Terrace	29	Four Districts
10	Fengqiu Terrace I	30	No. 3 Creek Terrace
11	Eighteenth River Terrace I	31	Upper Fourth terrace
12	Fengqiu Terrace II	32	No. 4 Creek Terrace
13	Eighteenth River Terrace II	33	Dongpu Bridge Terrace I
14	Xinxiang Terrace	34	Dongpu Bridge Terrace II
15	Rona Terrace I	35	Dongpu Terrace I
16	Rona Terrace II	36	Dongpu Terrace II
17	Trench Terrace	37	Dongpu Terrace III
18	Ali Does Not Move the River Terrace I	38	Dongpu Terrace IV
19	Ali Does Not Move the River Terrace II	39	Dongpu Terrace V
20	Wangmei Terrace	40	Ugankeng River Terrace

Figure 1. Study area.

2.1. The Front of the Terrace

2.1.1. The Attack Shore

Lin et al. showed that riverbank erosion mainly occurs at the bends in the river courses and tends to be most severe at the outer edges of such bends, called the attack shore (or cutting slope) [16]. The riverbank located on the attack shore has been eroded by the river for a long time, and the soil and rocks detached by such erosion are constantly being carried away by the river water, causing the toe to be gradually emptied; over time, this makes the riverbank steeper until it collapses. Roads and building foundations on the top are then damaged, or in some cases completely destroyed, due to loss of support. Lin et al. also noted that the movement of sandbars is mainly affected by five factors, i.e., the presence or absence of bends in rivers, the degree of such bends' curvature (known as meander), the presence or absence of confluent lateral structures, the way the river is scrubbed, and the supply of soil and sand [17]. Generally, convex-bank sandbars in curved sections of a river are more developed, and water flows mostly to the concave bank—i.e., the attacking shore—to erode it. Su concluded that, when the river flows through the attacking shore, flow velocity increases, and the centrifugal vortex water is formed there, washing it and transporting soil and sand to the convex bank [18].

2.1.2. Distance from Fault

Being located at the junction of the Eurasian continental plate and the Philippine Sea plate, Taiwan experiences frequent seismic activity. According to data collected by its Central Meteorological Bureau from 1991 to 2006, there are about 18,500 earthquakes in Taiwan each year, of which around 1000 are felt earthquakes. Major earthquakes often result in surface ruptures, rock folds and new faults. At present, scientists are unable to determine whether such faults cause earthquakes or vice versa, but their locations are identified in areas where seismic energy is strong [19]. Geological conditions in such areas are very fragmented due not only to the presence of fault gouges but also to broken zones near them and changes in the Earth's crust. The physical properties of the filler between the discontinuous surfaces of fault gouge or broken zone are usually poor, as is the degree of cementation, and this often causes engineering problems [16]. Lin observed that the rock mass on both sides of the Chenyulan fault is relatively broken, and weathered slate and metamorphic sandstone there, respectively provide fine-grained and coarse-grained material for earth-rock flows. The effects of faulting and river erosion also contribute to such flows [20]. The slopes of the terrain along both sides of the river are relatively steep, so the original weathered-soil layer collapsed due to by heavy rain and formed debris flows. During these flows, the rock plate was broken, and the broken pieces in the rock mass were drawn into them.

2.1.3. Distance from River

Wang's study of the Shaolai River concluded that the collapse percentages of susceptibility increased with proximity to the river's course. Specifically, the percentage of collapses within 200 m from the river channel was 42.7%; between 200 m and 400 m, 26.6%; between 400 m and 600 m, accounted for 19.4% [21]. Beyond 1600 m from the river channel, there were no collapses at all. Therefore, it can be inferred that large increases in the collapsed areas of adjacent rivers may be related to heavy typhoon rain causing water levels to surge, which in turn eroded the slope foot of the Shaolai River and accelerated the collapse of the riverbank [22]. In short, a closer distance to the river entails a higher level of risk. It is also a principle of hydrology that when the slope is closer to the river, it is nearer to groundwater; thus, water seeping into the ground will cause seepage pressure in the slope. If the soil structure is highly permeable, this process will greatly increase the probability that the stability of the slope will be negatively affected [23]. As Chang et al. observed, distance from the river channel determines flood impact [24]. The smoothness of a river channel determines its flood-discharge capacity, and the reclamation-area ratio of a reclaimed lake determines the flood-regulation capacity of the lake in the protection zone.

2.1.4. Potential Stream-Impact Quantity

According to a comprehensive assessment by Taiwan's Bureau of Soil and Water Conservation [15], natural streams or pits are likely to cause debris-flow disasters, though this likelihood is affected by local conditions including the presence or absence of protected objects. The Bureau uses two main criteria for judging the probable impact of potential debris flows. The first is that the slope of the stream bed is greater than 10 degrees, and that the catchment area above this point is greater than three hectares. The second is that, at the downstream exit or overflow point of the stream, there are more than three households or important bridges or roads that need to be protected. Assessment should be divided into four levels—"high", "medium", "low", and "continuous observation"—based on the characteristics of the site.

2.2. The River Terrace Itself

2.2.1. Minimum Ratio

Specific height is defined as the height of the riverbank relative to that of the riverbed surface. Yoshiro divided Taiwan's terrain into eight types; from high to low, these were Highest Peneplain (HP), Old Piedmont (OP), Elevated Highland (EH), Young Piedmont (YP), Lateritic Highland (LH), Lateritic Terrace (LT), Fluvial Terrace (FT), and Fluvial Plain (FP) [25]. Through this perspective, the topographic evolution of the Taiwan River Valley can be explored. Lin subsequently provided a general description of the topographical features of Taiwan's important river systems, along with more detailed descriptions of the topographical categories defined by Tomita [25,26].

2.2.2. Average Slope

According to statistics from the Bureau of Soil and Water Conservation, the most frequent damage from debris flows occurs on Taiwan's mountain slopes above 30 degrees, and especially at 40 degrees or more, while the least damage is suffered where slopes are 15 degrees or less [15]. This is because steep slopes provide greater driving force and also reduce slope resistance, making them conducive not only to the development of shallow landslides but also to the fluidization of landslides and the formation of sloping debris flows [27]. Cheng used GIS and a conditional-probability method to analyze four factors—bare land, eroded gullies, slope, and lithology—and established that, among them, slope had the greatest influence [28]. Kao showed that unstable Index method achieved good accuracy in predicting slope collapses, with 92% of actual collapsed land falling within the areas it identified as being at medium or high risk [29,30]. Liulater used a neural network-like sensitivity analysis to establish that the most important factors in this type of damage were rainfall, slope, slope type, elevation, lithology, fault, slope direction, roads, folds, and erosion gullies [30]. After controlling for rainfall, however, the greatest influence was slope, irrespective of whether Kao's or Liu's analysis method was applied.

2.2.3. Geology

The right bank of the main channel of the Chenyulan River consists of Paleogene submetamorphic rock strata, with interbedded argillite, slate, meta sandstone and quartzite, among other types of rock; on the left bank are Miocene sedimentary rock strata, with interbedded sandstone, shale and sand shale. Other strata include platform accumulation, four-sided sandstone layers, and hsichun, shihti, alluvial, nanchung, kueichoulin and kankou formations [31]. The wider area is dominated by thick-bedded sandstone, shale (argillite), and sandstone and shale formed together. When thick-bedded sandstone is subjected to tectonic stress, the rock mass is often cut into large blocks because it is thick and strong, but the density of the fractured surface is low. This type of rock is also relatively easy to weather. When the degree of weathering is slight, shale often forms smaller cuttings; when the degree of weathering is severe, a weathered soil layer forms. Due to the sharp difference in water permeability and resistance between sand and shale interbeds, the interface between them is often a stratum-slip surface, and the exposed area of interbeds often forms a single-sided mountain topography [32]. Chang and Lin investigated the Chenyulan River after Typhoon Huber and found that the contact between the upper mountain belt and the submetamorphic belt of Taiwan's geological structure was a fault. That is to say, near the Chenyulan River, the rock mass is abnormally broken, and a considerable amount of broken rock and soil accumulates on the surfaces of slopes and in the river itself, which may cause disasters [33].

2.3. Rear of the Terrace

2.3.1. Number of Erosion Ditches

Chang showed that erosion ditches are mainly caused by rain, surface runoff and wind, which causes the original soil to loosen or move; this process removes fine particles, and the resultant slope appears to be grooved [34]. Taiwan's Water Resources Agency, MOEA, on the other hand, defined an erosion ditch as a slender, linear drainage route from the top of a slope to its foot, usually caused by incision and erosion by concentrated runoff on the slope's surface. At the same time, the ditch wall is emptied and collapses, forming an obvious drainage pipe [35]. Chang suggested that the degree of slope erosion is a dynamic topographic effect on slope and is judged by the degree of contour curvature on a topographic contour map, supplemented by field surveys. Such curvature can also be used to determine the grade of a slope-erosion gully, i.e., a trough-shaped depression formed by the removal of vegetation by runoff on a hillside, excluding stream beds [36]. Hung noted that the debris-flow disasters caused by Typhoon Huber in the Chenyulan River Basin mainly occurred in the large erosion ditches (some of which are large enough to be named "Stream") and the flat reclaimed land of the community at the intersection and Provincial Highway 21 [37].

2.3.2. Number of Collapses from the Rear

When the combination of hydrological and geological conditions exceeds its damage threshold, a hillside will collapse. Hydrological conditions include rainfall intensity, rainfall delay, the soil's water content, pore water pressure, etc. Geological conditions include soil cohesion, anti-friction angle, soil slope, surface vegetation and whether there has been a recent earthquake or not. Tang conducted simulations of the Xiaolin Village disaster using PFC 3D. Their preliminary results show that just 60 s after the landslide was triggered, some of the houses in the village may have been covered by falling rocks or pushed to the opposite bank of the Qishan River [38]. Certainly, at its maximum sliding speed of 50 m per second, the kinetic energy of soil and rock is sufficient to cross the river entirely at this point, and a barrier lake was formed by this process in this vicinity. Ji investigated landslides in Caoling over a period from 1862 to 1999 and identified five large-scale ones linked to earthquakes or heavy rain. The landslides directly or indirectly caused disaster to the Caolingtan dyke breach, and a total of 170 people were killed and injured. Additionally, during the "921" earthquake of 1999, Caoling Mountain collapsed rapidly, its soil and rock moving up to 4 km, and the impact area of the collapse was nearly 500 hectares. Such cases of large-scale rock mass sliding are extremely rare, in Taiwan or anywhere else [39].

2.4. Preservation-Factor Assessment

Preservation factors include households, schools, hostels, public buildings (if residential), roads, bridges, farmland, orchards and other such sites. The Bureau of Soil and Water Conservation noted that the streams' debris-flow potential should be evaluated and prioritized according to the formula (natural potential factor affecting the risk level of debris flow × 50%) + (preservation hazard factor × 50%) [15]. The individual scores for the following three factors were added together to obtain the hazard degree score for each preservation object. (1) Building factor: The more buildings there are, the more people live in them, so the damage score is higher. (2) Traffic factor: Damage to the bridge is more harmful to the traffic, so a higher score is given. (3) Effective factors of on-site remediation: After many disasters, there have been many remediation facilities for potential debris flows.

If the remediation facilities are effective, damage to preservation objects by such flows can be reduced.

3. Methods

This study used Li's Chenyulan River terrace map data, modified to reflect the current shape of river terraces there, and purposively selected 40 potential river terraces for further analysis [1]. An analytic hierarchy process was used to analyze the strength of the mutual influences of the various elements, as well as of the high-level elements on the low-level elements; the levels of risk to each focal river terrace were derived through weighting the latent perception factors [40].

3.1. Questionnaire Design

In this study, following the methodology laid out by the Bureau of Soil and Water Conservation, the priority-order score of the potential unearthed rock flows was calculated according to the formula set forth in Section 2.4 above. Therefore, the risk-scoring method for the river terraces in this study equals (the potential factor of river terraces × 50%) + (the preservation hazard factor of the river terraces × 50%) [15]. The questionnaire design can be divided into the two hierarchical-structure diagrams—one for latent factors and the other for preservation factors, shown in Figures 2 and 3.

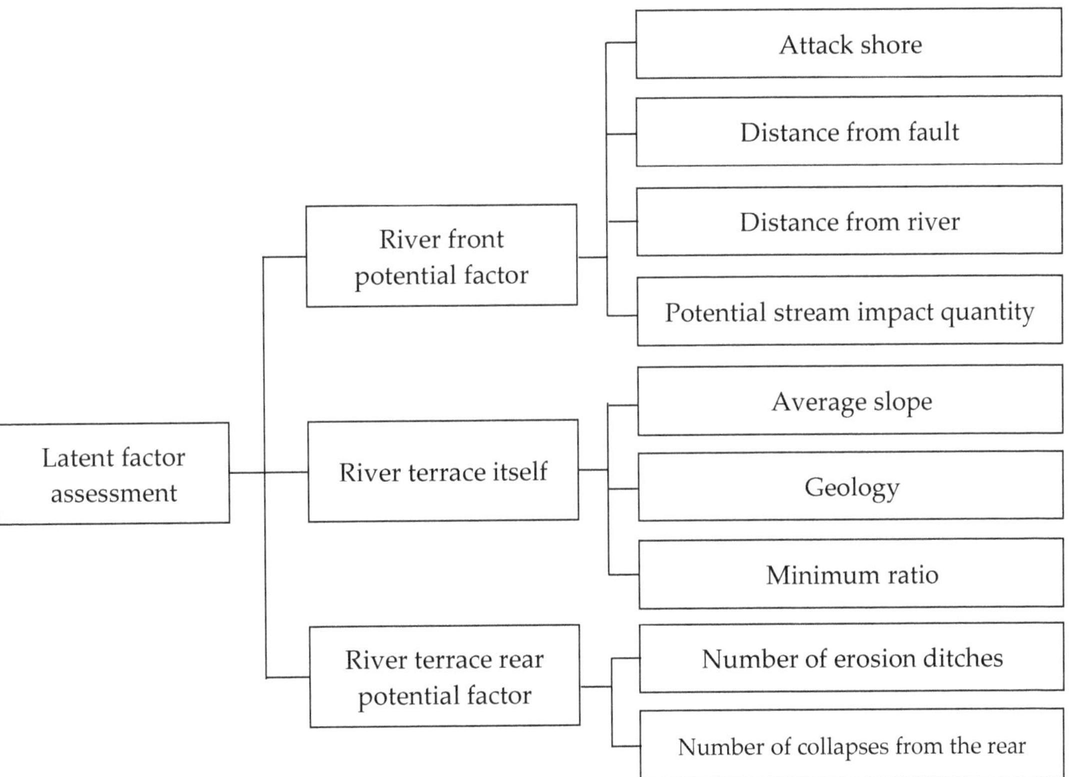

Figure 2. Evaluation conditions of latent factors and hierarchy of related factors.

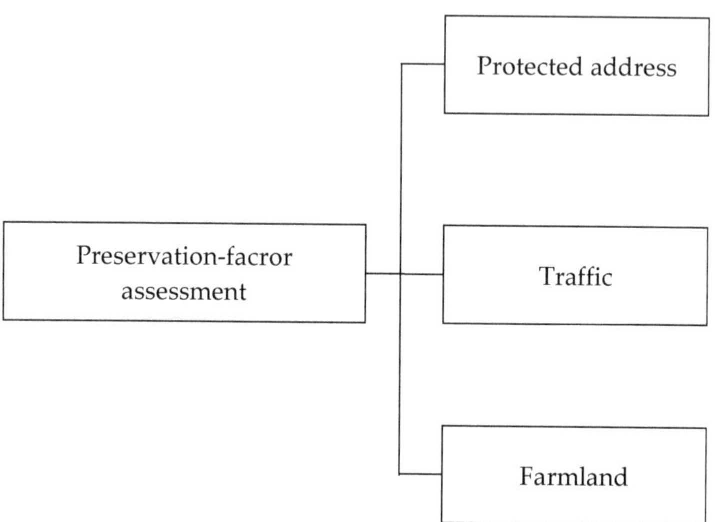

Figure 3. Evaluation conditions of preservation factors and hierarchy of related factors.

3.2. Questionnaire Survey Subjects

The participants in the questionnaire survey were mainly professors from the fields of land, water conservancy, soil and water conservation, geology, and environment and disaster prevention, employed by National Taiwan University, Chung Hsing University, Kaohsiung University, Tamkang University, Hua Fan University, Pingtung University of Science and Technology, and Chaoyang University of Science and Technology.

3.3. Statistical Results of Questionnaire Recovery

A total of 23 questionnaires were sent out and 17 were returned, resulting in a response of 79.9%. After eliminating five questionnaires due to repeated answers or omitted, unanswered items, which led them to have a consistency index (C.I.) and a consistency ratio (C.R.) greater than 0.1, 12 valid questionnaires remained for analysis. The C.R. value achieved a margin <0.1, indicating a strong degree of consistency among the pairwise comparisons, and proved it did not require a statistically significant sample size [40]. Shrestha et al. pointed out that AHP is usually used to survey people who have knowledge about the topic under investigation and a large sample size is not needed [41].

4. Results

Based on AHP principles, the scale indicates the level of relative importance from equal, moderate, strong, very strong to extreme level by 1, 3, 5, 7, and 9, respectively. The intermediate values between two adjacent comparisons are denoted by 2, 4, 6, and 8. The diagonal of the matrix of the comparison is equal to 1, since each criterion is compared to itself. When the number of alternatives is n, a total of $n(n-1)/2$ comparisons are made [40]. As shown in Table 1, our expert respondents' ranking of our three categories of latent-sensing factors was in front of river terrace (0.352) > river terrace itself (0.342) > behind river terrace (0.306).

In the experts' evaluation of the preservation factors of river terraces, as shown in Table 2, the ranking was households (0.599) > traffic (0.292) > farmland (0.109).

Table 1. Evaluation results, river terraces' latent-sensing factors.

Level	Evaluation Project		Inter-Level Weight	Overall Weight	Rank
Level one	Latent factor assessment		1.000	1.000	
Level two	Latent factor assessment	In front of river terrace	0.352	0.352	
		River terrace itself	0.342	0.342	
		Behind river terrace	0.306	0.306	
Level three	In front of river terrace	Attack shore	0.379	0.133	3
		Distance from fault	0.096	0.034	9
		Distance from river	0.288	0.101	7
		Potential impact of debris-flow quantity	0.237	0.083	8
Level three	River terrace itself	Average slope	0.321	0.110	6
		Geology	0.354	0.121	4
		Minimum ratio	0.325	0.111	5
Level three	Behind river terrace	Number of erosion ditches	0.563	0.172	1
		Distance from the collapse	0.438	0.134	2

Table 2. Evaluation results of river terraces' preservation factors.

Level	Evaluation Project		Inter-Level Weight	Overall Weight
Level one	Preservation factor assessment		1.000	1.000
Level two	Preservation factor assessment	Protected address	0.599	0.599
		Traffic	0.292	0.292
		Farmland	0.109	0.109

4.1. Distribution Method

Due to the large gaps between the various factors, to avoid extreme values, we first used statistical methods to find the average value and standard deviation of each factor and set reasonable parameter ranges X_{max} and X_{min}. If a parameter was greater than X_{max}, X_{max} was used, and if one was less than X_{max}, X_{min} was used. Then, an interval mapping method was conducted with AHP weightings to determine the score for each factor. In this study, the maximum and minimum range (0.1–1) of the interval mapping method was multiplied by the overall weight of each factor in the AHP to obtain the maximum and minimum range values.

4.2. Factor-by-Factor Allocation

First, we assumed that each factor was normally distributed, and its average value and standard deviation were ascertained, using one standard deviation to determine its reduction range. In this way, the extreme value of each parameter had less influence when calculating the weight, as shown in Table 3.

Table 3. Average value and standard deviation of each factor.

Factors	Average Value	Standard Deviation	X_{max}–X_{min}
Minimum height ratio (m)	58.6	64.4	0–123
Attack shore length (m)	213.9	375.1	0–589
Distance from river (m)	182.2	202.3	0–384.4
Average slope (degrees)	10.4	4.5	5.9–14.8
Distance from fault (m)	388.6	455.2	0–843.7
Number of potential streams affected	2.2	2.1	1.1–5.3
Number of erosion ditches	3.2	2.5	0.7–5.6
Distance from the collapse (m)	654.2	388.7	285.5–1022.9
Protected address	91.1	130.1	0–221.1

According to Juang et al., to improve the learning rate and accuracy of a similar neural network, the inconsistency of the difference between the numerical ranges of the parameters should be calculated, as before analysis, the input parameters must be normalized to avoid the problem of temporary instability of the network and difficulty in convergence [42]. Accordingly, this study utilized a modified version of the interval-mapping method in Juang et al.'s normalization formula, and the maximum and minimum values obtained were between 0.1 and 1.

$$\text{"X"}\_\text{"norm"} \text{"="} (\text{"X"} + a\text{")"}/b \quad X_{norm} = (X + a)/b \tag{1}$$

Among them:

$$\text{"a"} = \text{"(("X"}\_\text{"max"} \text{"-"} \llbracket\text{"10X"}\rrbracket\_\text{"min"})\text{"}/\text{"9"} \quad a = (X_{max} - 10X_{min})/9 \tag{2}$$

$$b = (X_{max} - X_{min})/0.9 \tag{3}$$

In the above three formulae, X_{norm} is the normalized value; X is the actual input parameter value; X_{max} is the maximum actual input parameter; X_{min} is the minimum actual input parameter.

For ease of calculation, a full score was held to be 100 points. In this study, the maximum and minimum range of the interval mapping method (0.1–1) is multiplied by the overall AHP weight of each factor and then multiplied by 100 to obtain the factors' respective maximum and minimum ratio ranges, as shown in Table 4. Substituting the data of each factor into Equation (1), the calculation formula of each factor can be calculated, and the factor weight of each river terrace calculated.

Table 4. Weight distribution of each factor.

Factors		Maximum Weighting	Minimum Weighting
	Minimum height ratio (m)	11.1	1.11
	Attack shore length (m)	13.3	1.33
	Distance from river (m)	10.1	1.01
	Average slope (degrees)	11.0	1.1
Latent susceptibility factors	Distance from fault (m)	3.4	0.34
	Number of potential streams affected	8.3	0.83
	Geology	12.1	1.21
	Number of erosion ditches	17.2	1.72
	Distance from the collapse (m)	13.4	1.34
	Total	100	10
	Protected address	59.9	5.99
Preservation factors	Traffic	29.3	2.93
	Farmland	10.9	1.09
	Total	100	10

In its geological aspects, this research is based on the results of a survey by the Civil Engineering Research Institute of the Ministry of Construction of Japan regarding where earth-rock flows occur, along with the characteristics of Taiwan geology. Adopting a predetermined risk standard of geological lithology, this study divides such geology into three broad categories, based on the maximum, minimum and intermediate values. Among the preservation factors, traffic and farmland were deemed to be either "present" or "not present" and also allocated based on the minimum values.

5. Discussion

5.1. Risk Assessment of River Terraces

The evaluation results for each river terrace are shown in Table 5. The average value (57.70) and standard deviation (13.346) were calculated by statistical methods, with the average value as the center plus or minus one standard deviation. After adjusting with the concept of rounding to integers, the boundaries were 70, 55 and 40. Thus, risk was divided into four categories: high risk (70–100), medium–high risk (55–69), medium risk (41–54) and low risk (0–40). These categories are also presented in map form in Figure 4.

After the risk assessment, 8 of the 40 focal river terraces were deemed to be high risk, 14 at medium-high risk, another 14 at medium risk, and the remaining 4 at low risk.

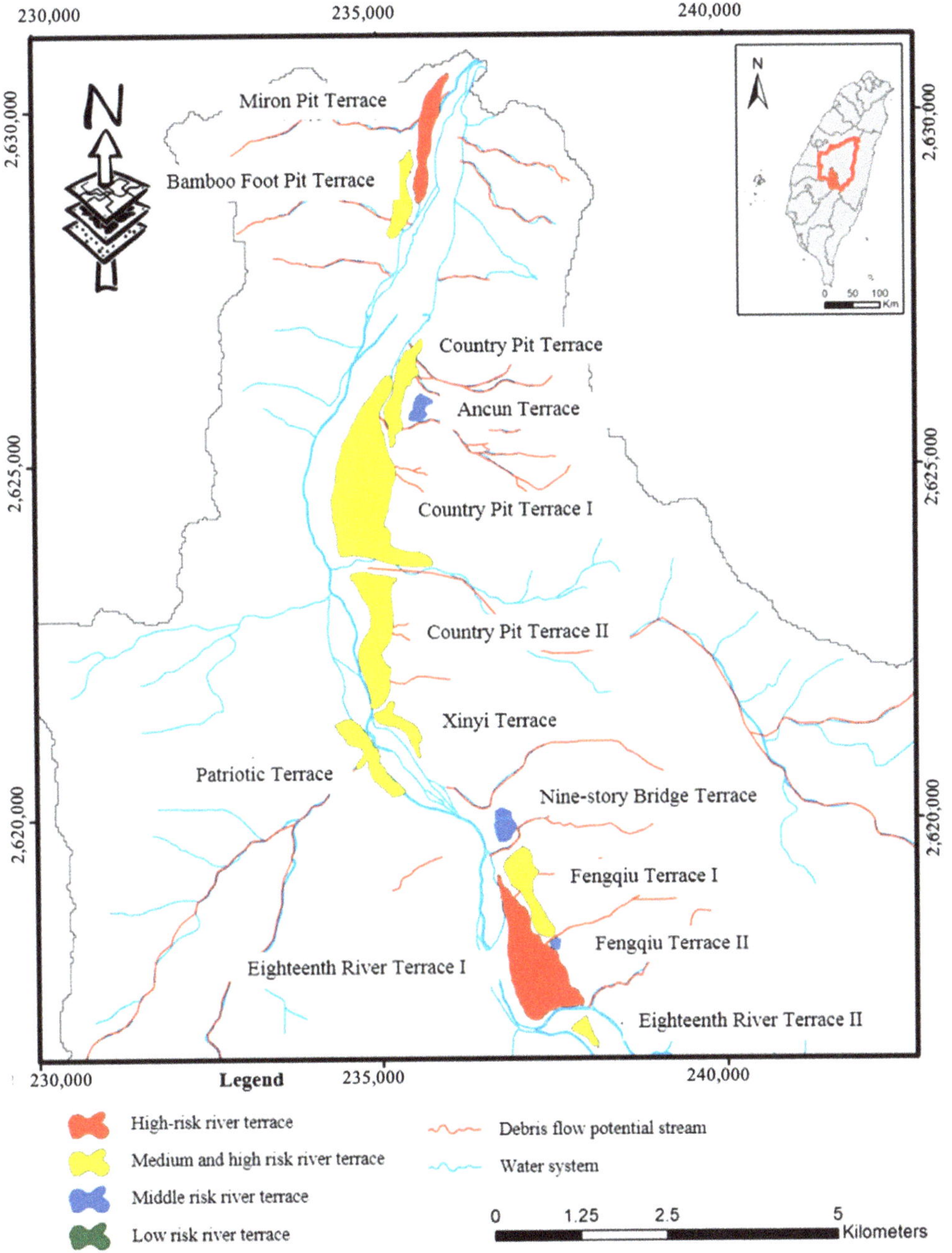

Figure 4. The distribution map of the danger degree of the river terrace.

Table 5. Risk assessment of river terraces.

No.	River Terrace Name	Risk	Hazard Classification
1	Miron Pit Terrace	67	Medium high
2	Bamboo Foot Pit Terrace	59	Medium high
3	County Pit Terrace	72	High
4	Ancun Terrace	49	Medium
5	County Pit Terrace I	79	High
6	County Pit Terrace II	85	High
7	Xinyi Terrace	81	High
8	Patriotic Terrace	78	High
9	Nine-story Bridge Terrace	49	Medium
10	Fengqiu Terrace I	66	Medium high
11	Eighteenth River Terrace I	76	High
12	Fengqiu Terrace II	45	Medium high
13	Eighteenth River Terrace II	53	Medium high
14	Xinxiang Terrace	53	Medium high
15	Rona Terrace I	55	Medium
16	Rona Terrace II	70	High
17	Trench Terrace	38	Low
18	Ali Does Not Move the River Terrace I	35	Low
19	Ali Does Not Move the River Terrace II	65	Medium high
20	Wangmei Terrace	50	Medium
21	Ali Does Not Move the River Terrace III	39	Low
22	Wangxiang Terrace	49	Medium
23	Ali Does Not Move the River Terrace IV	68	Medium high
24	Heshe Terrace	78	High
25	Malacca Terrace I	59	Medium high
26	Malacca Terrace II	42	Medium
27	Toutunxi Terrace I	67	Medium high
28	Toutunxi Terrace II	59	Medium high
29	Four Districts	44	Medium
30	No. 3 Creek Terrace	65	Medium high
31	Upper Fourth terrace	37	Low
32	No. 4 Creek Terrace	41	Medium
33	Dongpu Bridge Terrace I	55	Medium high
34	Dongpu Bridge Terrace II	52	Medium
35	Dongpu Terrace I	50	Medium
36	Dongpu Terrace II	62	Medium high
37	Dongpu Terrace III	47	Medium
38	Dongpu Terrace IV	63	Medium high
39	Dongpu Terrace V	57	Medium high
40	Ugankeng River Terrace	49	Medium

5.2. Verification

Three comparison methods were used in this research to verify our approach. These were: (1) unsupervised classification using the SPOT-3 satellite multi-spectral state (XS) image map of Chen Youlanxi in each period to determine changes in river-terrace area; (2) comparison of river terraces in various periods with satellite images and aerial photos; (3) comparison of historical disaster data from the Chenyulan River, covering a total of 87 floods linked to 42 discrete weather events from August 1959 through October 2009 [43].

5.3. Historical Disaster Comparison

5.3.1. Dangerous River Terraces

As noted above, eight river terraces were deemed high-risk by our approach because they scored above 70 points. This indicated that the frequency of disasters there is high, damage to buildings and crops is noteworthy, and the area affected is relatively large. From historical disaster data, it can be seen that debris flows struck these eight terraces at 1.36 times the average rate; dike destruction (by number of occurrences) and land loss (in hectares) were both 5 times the average; houses totally destroyed, 4.44 times the average; damaged houses, 4.09 times the average; number of deaths, 2.74 times the average.

5.3.2. County Pit I

County Pit I is a high-risk river terrace located on the right bank of the lower reaches of Chenyulan River. Figure 5 was obtained by extracting and overlapping images from five periods and shows little change in this area before and after Typhoon Hebo, whereas after Typhoon Tochigi, a shrinking trend in its land area can be observed. After the 72nd flood, the terrace's area was obviously reduced, but after Typhoon Morakot five years later, it had increased.

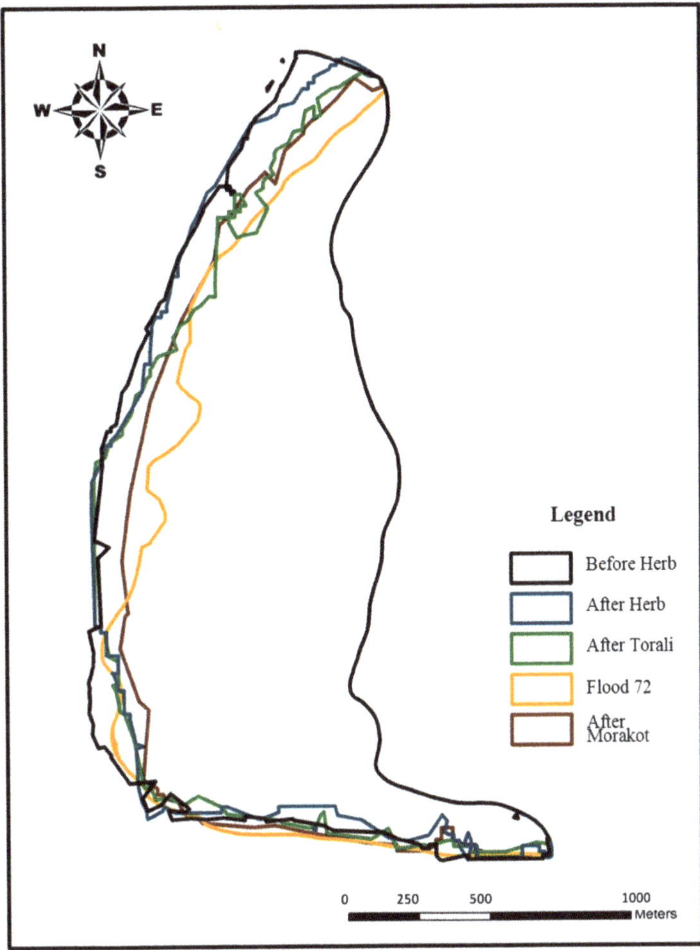

Figure 5. Area changed to County Pit I across five flooding events.

County Pit Terrace I was destroyed by Typhoon Mintouli in 2004, which in turn caused the embankment of Junkengxi Terrace I to be washed away. The disaster area was very large, as shown in aerial photographs obtained from the Fourth River Bureau, Water Resources Department, Ministry of Economic Affairs (Figures 6–9).

Figure 6. The County Pit embankment in 2003 (the year before it was destroyed).

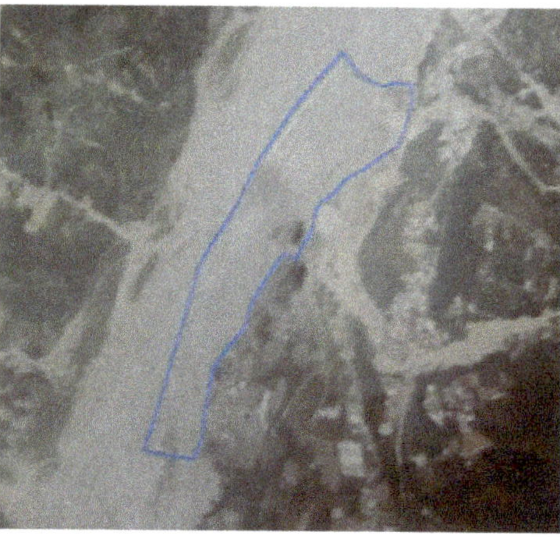

Figure 7. The former area of the County Pit embankment, marked in blue, in 2004.

Figure 8. The Shang'an embankment in 2003 (before its destruction).

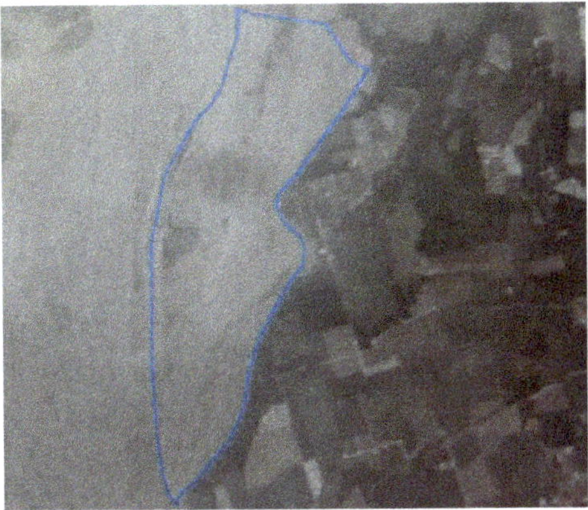

Figure 9. The former area of the Shang'an embankment, marked in blue, in 2004.

5.3.3. Toutunxi Terrace I

Toutunxi I is a medium–high-risk river terrace located on the right bank of the middle reaches of Heshe River. A schematic diagram of its changes, based on image extraction and overlap from five flooding events, is presented in Figure 10. It is obvious from Figure 10 that the area of the terrace was broadly unchanged after Typhoon Toraji and the 72nd flood, but after Typhoon Morakot, it was significantly reduced.

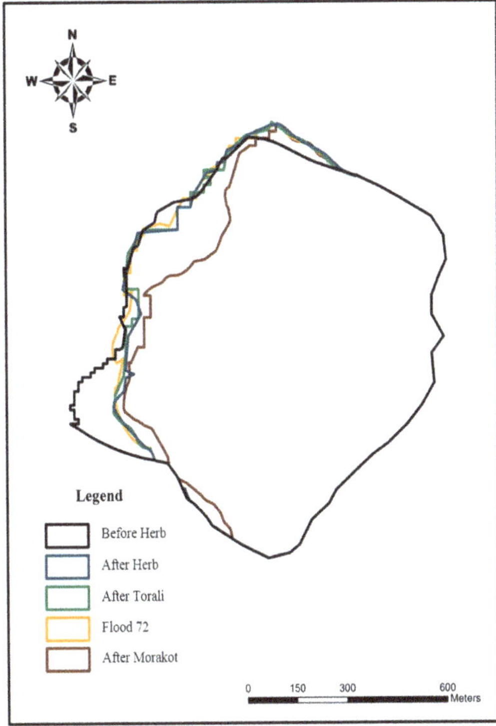

Figure 10. Area changed to Toutunxi Terrace I across five flooding events.

During Typhoon Morakot in 1998, three large landslides in the upper reaches of Toutunxi Creek indirectly formed soil and rock flows, which caused both banks of the downstream terraces to be washed away. In a satellite image from before this disaster (Figure 11, left), the channel of the Toukeng River is not obvious, being then only 20–30 m wide, but one taken after it (Figure 11, right) shows the channel clearly, as it had widened to 120 m.

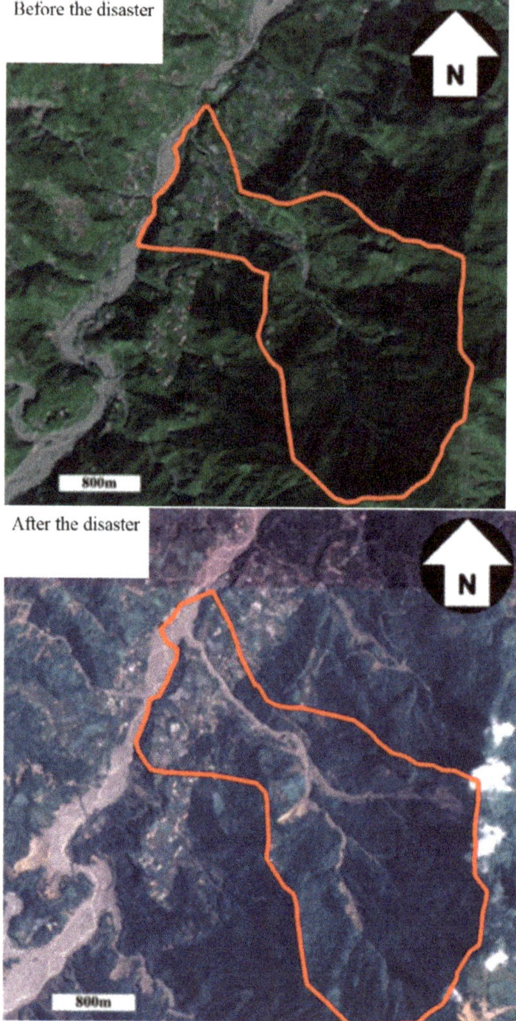

Figure 11. The confluence of Toutunxi Creek and the Toukeng River before and after Typhoon Morakot in 1998, with study area marked in red.

6. Conclusions

The latent-susceptibility factors selected in this study through the literature were the length of the attacking bank, distance from fault, distance from river course, number of potential debris flows, minimum specific height, average slope, geology, number of erosion ditches, distance to collapsed land, etc. The three preservation factors selected were the preservation of households, transportation and farmland; the risk to particular river terraces was established from the latent-susceptibility factors and preservation factors. The results of the AHP analysis indicated that, among the latent-susceptibility factor conditions, those affecting the front of the river terrace were considered more important than those affecting the river terrace itself or the area behind it. Additionally, the results indicated that the preservation of households was deemed more important than traffic or farmland factors. The research also established that, of the 40 focal river terraces, the highest-risk ones were County Pit Terrace, County Kengxi Terraces I and II, Xinyi Terrace, Patriotic Terrace, Eighteenth Terrace I, Rona Terrace II and Heshe Terrace. In this study, a potential factor assessment framework was established to examine whether the research results were contradictory by comparing SPOT satellite images and historical hazards. Unlike other risk assessments of riverine terraces [10,11], the current environmental conditions of riverine terraces can be used to assess the risk of disasters. This study's risk-level designation has important implications for both disaster prevention and the evacuation of local residents when disasters occur, and, if generally adopted, it should mitigate loss of life and property. However, this study was not without its limitations. Chief among these was that it relied on the 5m × 5m DTM of Taiwan surveyed and mapped by the Chengda Satellite Center in 2004 to conduct its river-terrace risk assessments and stability analyses. The use of a more recent DTM would undoubtedly increase the accuracy of the analysis results. Additionally, when conducting stability analysis, due to limited drilling data from the Chenyulan River Basin, such data from Toutunxi River Terrace I were used as a proxy for it. Thus, more geological data would, therefore, improve our approach's ability to identify potential river-terrace collapses.

Author Contributions: Conceptualization, J.-Y.L.; data curation, Y.-M.H.; writing—original draft, Y.-M.H.; writing—review and editing, J.-C.C.; supervision, J.-Y.L. All authors have read and agreed to the published version of the manuscript.

Funding: This study is part of the research results of the Ministry of Science and Technology number NSC100-2218-E-005-001. I would like to thank the staff for their hard work during the planning period and to express their greatest gratitude to the National Science Council for their support research. Funding has enabled this study to be successfully completed.

Institutional Review Board Statement: Not applicable.

Informed Consent Statement: The authors declare no conflict of interest.

Data Availability Statement: The data presented in this study are available in the article.

Acknowledgments: We would like to thank the anonymous reviewers for their helpful suggestions and feedback.

Conflicts of Interest: The authors declare no conflict of interest.

References

1. Li, X.D. From the point of view of topography, Chen Youlan's He Bo Feng disaster. *Geotech. Technol.* **1996**, 17–24. Available online: http://www.geotech.org.tw/purchase-inner.php?id=131 (accessed on 15 November 2021).
2. Liu, M.X. Topographical Research on the River Steps in the Northwest of Taiwan. Ph.D. Thesis, Department of Geography, National Taiwan Normal University, Taipei, Taiwan, 2004.
3. Du, H.Y.; Chen, H.H.; Cao, B.X. *Geomorphology and Quaternary Geology*; one edition of Three Brushes; Geological Publishing House: Beijing, China, 1991.
4. Chang, R.J.; Shi, Z.T. An overview of the study of the rivers. *J. Chin. Geogr. Soc.* **1990**, *18*, 1–8.
5. Ger, M.L. A Study of the Relations between the Terraces and Subsurface Structure in the Hsinshe Area. Ph.D. Thesis, Department of Geography, National Taiwan Normal University, Taipei, Taiwan, 2005.
6. Wang, F.T. Comparison of River Terraces in Hua-Tung Coast Area: A Case Study in Streams of Shuilian, Shuimuding and Sanxian. Master's Thesis, Department of History and Geography, University of Taipei, Taipei, Taiwan, 2016.
7. Chang, T.Y. Fluvial Terrace Landscape in Tai-Yuan Basin and Its Implications. Master's Thesis, Department of Natural Resources and Environmental Studies, National Dong Hwa University, Hualien County, Taiwan, 2016.
8. Wu, T.Y. Holocene Landscape Evolution of the Jinlun, Dazhu and Dawu River Basins in the Southeastern Taiwan. Master's Thesis, Department of Geography, National Kaohsiung Normal University, Kaohsiung, Taiwan, 2019.

9. Chen, Y.C. Application of Geographic Information System in Risk Asssessment of River Terrace in Chenyulan River Watershed, Nantou County. Master's Thesis, Department of Civil and Construction Engineering, Chaoyang University of Technology, Taichung, Taiwan, 2010.
10. Zhuang, Z.Z. Apply Artificial Intelligence and Weight of Evidence to Assess Vulnerability for River Terrace Communities. Master's Thesis, Department of Land Management and Development, Chang Jung Christian University, Tainan, Taiwan, 2020.
11. Lin, J.R. Risk Assessment of Rainfall-induced Landslides for River Terrace Communities. Master's Thesis, Department of Land Management and Development, Chang Jung Christian University, Tainan, Taiwan, 2020.
12. Chang, C.F. Effects of Watershed Runoff Factors on Debris Flow Occurrence–Using the Watershed of Chenyoulan Stream as An Example. Master's Thesis, Department of Bioenvironmental Systems Engineering, National Taiwan University, Taipei, Taiwan, 2015.
13. Lin, S.Y. Study on the Factors of Landslide Susceptibility Analysis in Chen- You-Lan Watershed. Master's Thesis, Department of Soil and Water Conservation, National Chung Hsing University, Taichung, Taiwan, 2018.
14. Thi-To, N.N. Early Warning of Landslide Hazard by Integrating Spatial Statistical Analysis Methods to Calculate the Landslide Susceptibility Index in the Chen-Yu-Lan River Watershed. Taiwan. Ph.D. Thesis, Department of Earth Sciences, National Cheng Kung University, Tainan, Taiwan, 2019.
15. Debris Flow Disaster Prevention Information, Soil and Water Conservation Bureau. Available online: https://246.swcb.gov.tw/?lang=en (accessed on 15 November 2021).
16. Lin, M.L.; Wu, W.L.; Zhou, K.X.; Yang, Z.S.; Wang, J.P. Discussion on Geographical Location Types and Remediation Cases of Taiwan Slope Disasters. *CECI Eng. Technol.* **2008**, *77*. Available online: https://www.ceci.org.tw/Upload/Download/F8950026-8 5C8-4386-9075-13CCB737881D.pdf (accessed on 15 November 2021).
17. Lin, D.G.; Lai, Y.C.; Liu, W.T. Design and analysis of natural ecological method for river and stream improvement. In Proceedings of the Natural Ecological Practices Conference, Taitpi, Taiwan, November 2002. Available online: https://www.m.ishare.iask.sina.com.cn/f/34860363.html (accessed on 15 November 2021).
18. Su, K.H. Investigating the Slope Failure of Interbedded Sandstone and Mudstone Slope: Case of Highway R174 on 50 k + 650. Master's Thesis, Institute of Resources and Environment, University of Kang Ning, Taipei, Taiwan, 2007.
19. Central Weather Bureau. Available online: https://www.cwb.gov.tw/eng/ (accessed on 15 November 2021).
20. Lin, C.W. Geological Factors to Influence the Landslides in The Hoshe Area, Nantou Hsien. *Sino-Geotech.* **1996**, *57*, 5–16.
21. Wang, M.H. Application of SPOT Satellite Images and GIS for Landslide Potential Analysis-A Case Study of Shao-Lai River. Master's Thesis, Department of Civil and Construction Engineering, Chaoyang University of Technology, Taichung, Taiwan, 2008.
22. Chu, Y.T.; Hsu, M.L. The Relationship between Discharge and Suspended Sediment Concentration at Typhoon Events in Yu-Feng Catchment. *J. Geogr. Sci.* **2007**, *49*, 1–22.
23. Huang, T.H. The Study of Slope Landslide Disaster Management-Take Tainan County for Example. Master's Thesis, Department of Land Management and Development, Chang Jung Christian University, Tainan, Taiwan, 2004.
24. Zhang, L.; Weng, Y.; Chen, X.H. Research on evaluation model for vulnerability of flood control based on SVM. *Yangtze River* **2009**, 40.
25. Lin, C.K. Topography of Taiwan, *Taiwan Provincial General Records 1957*, 1. (Land Records, Geography, Volume 1), published by the Taiwan Provincial Literature Committee, 424 pp. Available online: https://tm.ncl.edu.tw/article?u=006_001_0000395034&lang=chn (accessed on 15 November 2021).
26. Lin, C.C. Topography of Taiwan, Draft of the General Gazetteer of Taiwan Province, Taiwan Provincial Literature Committee. 1, 1957.
27. Wang, L.C. The Characteristics of Topography and Occurrence of Hillslope Debris Flows in Shitou area. Master's Thesis, Department of Earth Sciences, National Cheng Kung University, Tainan, Taiwan, 2005.
28. Cheng, Y.C. Application of Geographic Information System to Regional Slope Stability Analysis—Tianxiang-Tailuege Section of Zhongheng Highway. Master's Thesis, Department of Materials Science and Engineering, National Cheng Kung University, Tainan, Taiwan, 1992.
29. Kao, C.L. A Study on the Failure Potential of Slopes along 31~75 K Section of the 14th Provential Highway. Master's Thesis, Department of Civil and Construction Engineering, Chaoyang University of Technology, Taichung, Taiwan, 2003.
30. Liu, H.F. Evaluation of Slope Failures along the 14th Provincial Highway Using Artificial Neural Networks. Master's Thesis, Department of Civil and Construction Engineering, Chaoyang University of Technology, Taichung City, Taiwan, 2005.
31. Hou, C.F. Application of SPOT Remote Sensing Images and GIS on Landslide potential Analysis Using Chen Yu Lan Creek Basin. Master's Thesis, Department of Civil and Construction Engineering, Chaoyang University of Technology, Taichung, Taiwan, 2006.
32. Luo, S.Y.; Huang, H.Y. Chenyulan. Available online: http://www1.geo.ntnu.edu.tw/~{}shensm/Course/CourseWork/TaiGeom_Stu90/24.25%E9%99%B3%E6%9C%89%E8%98%AD%E6%BA%AA/ (accessed on 15 November 2021).
33. Jeng, F.S.; Lin, M.L. Engineering Deficiencies Exposed by Typhoon Herb. *Sino-Geotech.* **1996**, *57*, 65–74.
34. Chang, W.C. Research on the Sustainability of Debris Flow Influenced Areas in Yilan County with the Application of the Fuzzy Backpropagation Neural Network Deduction Pattern. Master's Thesis, Department of Civil Engineering, National Ilan University, Ilan, Taiwan, 2006.
35. Taiwan's Water Resources Agency, MOEA. Available online: https://eng.wra.gov.tw/ (accessed on 15 November 2021).
36. Chang, S.J. Hillside Survey Planning, Evaluation and Slump Prediction and Management. Master's Thesis, Department of Geography, National Taiwan Normal University, Taibei, Taiwan, 1993.

37. Hung, J.J. Typhoon Herb, The New-Central-Cross-Island-Highway and Slopeland Failures in Central Taiwan. *Sino-Geotechnics* **1996**, *57*, 25–30.
38. Tarng, H.R. The Transportation and Deposition of Catastrophic Landslides in Taiwan: Insight from Granular Discrete Element Simulation. Ph.D. Thesis, Department of Geosciences, National Taiwan University, Taibei, Taiwan, 2010.
39. Chi, C.C. Investigation and Interpretation of Rock Slide Potential. *Nation Sci. Technol. Cent. Disaster Reduct.* **2010**, 62. Available online: https://www.ncdr.nat.gov.tw/Epaper/MessageView?itemid=5061&mid=39 (accessed on 15 November 2021).
40. Saaty, T.L.; Vargas, T.L. Comparision of igenvalue, logarithmic least squares and least squares medthods in estimationg ratios Mathematic. *Modelling* **1984**, *5*, 309–324. [CrossRef]
41. Shrestha, R.K.; Alavalapati, J.R.; Kalmbacher, R.S. Exploring the potential for silvopasture adoption in south-central Florida: An application of SWOT–AHP method. *Agric. Syst.* **2004**, *81*, 185–199. [CrossRef]
42. Juang, C.H.; Chen, C.J. A rational method for development of limit state for liquefaction evaluation based on shear wave velocity measurements. *Int. J. Numer. Anal. Methods Geomech.* **2000**, *24*, 1–27. [CrossRef]
43. Lai, M.F. Application of GIS to Spatial Characteristic of Rainfall—Chenyulan River Watershed as an Example. Master's Thesis, Department of Civil and Construction Engineering, Chaoyang University of Technology, Taichung, Taiwan, 2010.

 *applied sciences*

Article

Quantifying the Occurrence of Multi-Hazards Due to Climate Change

Diamando Vlachogiannis *, Athanasios Sfetsos, Iason Markantonis, Nadia Politi, Stelios Karozis and Nikolaos Gounaris

Environmental Research Laboratory (EREL), INRASTES, NCSR "Demokritos", Agia Paraskevi, 15341 Attiki, Greece; ts@ipta.demokritos.gr (A.S.); jasonm@ipta.demokritos.gr (I.M.); nadiapol@ipta.demokritos.gr (N.P.); skarozis@ipta.demokritos.gr (S.K.); gounaris@ipta.demokritos.gr (N.G.)
* Correspondence: mandy@ipta.demokritos.gr; Tel.: +30-210-650-3417

Abstract: This paper introduces a climatic multi-hazard risk assessment for Greece, as the first-ever attempt to enhance scientific knowledge for the identification and definition of hazards, a critical element of risk-informed decision making. Building on an extensively validated climate database with a very high spatial resolution (5×5 km^2), a detailed assessment of key climatic hazards is performed that allows for: (a) the analysis of hazard dynamics and their evolution due to climate change and (b) direct comparisons and spatial prioritization across Greece. The high geographical complexity of Greece requires that a large number of diverse hazards (heatwaves—TX, cold spells—TN, torrential rainfall—RR, snowstorms, and windstorms), need to be considered in order to correctly capture the country's susceptibility to climate extremes. The current key findings include the dominance of cold-temperature extremes in mountainous regions and warm extremes over the coasts and plains. Extreme rainfall has been observed in the eastern mainland coasts and windstorms over Crete and the Aegean and Ionian Seas. Projections of the near future reveal more warm extremes in northern areas becoming more dominant all over the country by the end of the century.

Keywords: climate change; multi-hazards; WRF-ARW; EC-Earth GCM; Greece

Citation: Vlachogiannis, D.; Sfetsos, A.; Markantonis, I.; Politi, N.; Karozis, S.; Gounaris, N. Quantifying the Occurrence of Multi-Hazards Due to Climate Change. *Appl. Sci.* **2022**, *12*, 1218. https://doi.org/10.3390/app12031218

Academic Editor: Jason K. Levy

Received: 15 December 2021
Accepted: 21 January 2022
Published: 24 January 2022

1. Introduction

A new era has unequivocally emerged that has brought climate change and its impacts to the foreground of scientific research. There is growing evidence that weather and climate extremes (i.e., hazards) are increasing in frequency, intensity, spatial coverage and duration, indicating the need for a more meticulous investigation and a better physical understanding of the processes governing the state of the climate and its future evolution [1–3]. The adverse impacts of extreme events, evidenced in the reported data of disaster implications, e.g., [4–6], are also an active subject of climate research of paramount importance [7–11]. Noteworthily, anthropogenic effects are emerging as the underlying cause of the weather and climate extremes [1,12–18].

Over recent years, research works have consistently reported that climate change aggravates climate hazards, amplifying the risks of various impacts (river and coastal floods, wildfires, droughts, landslides, etc.) [19–23]. Several studies on temperature and precipitation extremes have provided important findings on the regional variability of the impacts of climate change across Europe, e.g., [24–29]. Forzieri et al. [30] reported that the risks of wildfires, windstorms and inland flooding would increase in Europe, with varying degrees of change across regions, while the most dramatic rise is predicted to be in damages in southern Europe caused by heatwaves, droughts and coastal floods. The report of PESETA IV [31] consolidated those findings and indicated "a clear north–south divide, with the southern regions in Europe being much more impacted by the effects of extreme heat, water scarcity, drought, forest fires and agriculture losses". The estimated patterns of climate-change developments urge for more efficient risk management of climate-related extremes and disasters in order to significantly advance climate-change adaptation, particularly in the most vulnerable regions. This consequently implies the need for a reliable quantification of the probability of extremes in the current and future climate. Identifying the climate vulnerabilities of key societal systems should be based on a detailed knowledge of projected climate-change hazards and the factors affecting the likelihood of each one for the selected assessment of the region of interest.

According to the report of the United Nations Office for Disaster Risk Reduction [32], the term "multi-hazard" is used to promote risk reduction and disaster management, and denotes hazardous events that may occur simultaneously or cumulatively over time. One of the most challenging research questions is the harmonization of risk metrics to allow the comparison of risks across hazards, regions, time, assets, or sectors [33]. Establishing a harmonized risk understanding would pave the way to a multi-hazard risk assessment, introducing interactions and cascading effects as well as providing some analytical interpretations of the compound and systemic risks. This would lead to more credible scenarios for describing future disaster events in terms of their magnitude and probability based on the validated scientific knowledge that can benefit from high-resolution climate projections.

The single-hazard risk assessment is a proven methodology, but shifting to multi-hazards is not a linear or easily understood process, as a multi-hazard risk analysis is not just the sum of single hazard risk examinations and thus, comparability of the single-hazard results is strongly needed [34]. Due to the diversity of the hazard characteristics' complex relationships, triggering effects, climate-changing mechanisms, compounds and interactions could be potentially established [35–37].

In this work, the occurrence of hazards due to climate change was determined for Greece using data that were dynamically downscaled to a very high resolution by the Advanced Weather Research and Forecasting (WRF-ARW) model [38], initially produced by the EC–Earth Global Climate Model (GCM) [39]. The hindcast period covered the years from 1980 to 2004, while for the future projections, two different periods, i.e., 2025–2049 (near future) and 2075–2099 (far future), were studied using the Intergovernmental Panel on Climate Change (IPCC) Representative Concentration Pathways (RCPs), RCP4.5 and RCP8.5, following the recommendations of the EU National Risk Assessment [33] and similar studies in the US [40]. The RCP scenarios demonstrated a significant convergence in their emission pathways in the near future and considerable deviations towards the end of the century [41].

It has been established that high-resolution, dynamic-downscaling models applied to regional climate assessments can be implemented to assess the climate-change impacts on extremes, especially in areas with complex topography and local scale effects [42–45]. Here, we sought to provide the first step towards a comprehensive multi-hazard risk assessment for the country based on high-resolution model data to support training and preparatory activities for disaster risk reduction (DRR). The analysis focused on four critical climate hazards for Europe: heat and cold extremes, flash floods, and windstorms, each one described by a climate indicator (Section 2.2).

The scope of this work was to carry out a very detailed assessment of the most significant hazards that have occurred in Greece in the past and to predict their evolution in the future considering the impact of climate change. This is a highly valuable process as disaster management should also take into account the (non-)stationary characteristics of climate change. The current study examined these parameters for Greece using very-high-resolution climate simulations at 5 km. Furthermore, one of the goals was to identify a common categorization framework across different climate hazards, which allowed a direct and coherent prioritization of the hazards and their evolution due to climate change considering complex patterns due to local geographic conditions. The produced hazard data could readily be applied to the generation of multiple scenarios with various likelihoods of occurrence in order to obtain a more complete picture of risk [32], accounting for climate-change projections (IPCC). In addition, this work was based on the recommendations set by the UNDRR/ISC Sendai Hazard Definition and Classification Review Technical Report [46] and intended to introduce the climate dimensions and dynamic evolution of risk harmonization that was missing from such assessments [33].

Section 2 focuses on the details of the data used and the methodology that was developed to estimate the probability of the occurrence of extreme values of the variables. Section 3 presents the results and discussion of the analysis applied to the quantification of risks and the likelihood of hazard evolution due to climate change. Finally, the final section concludes the paper.

2. Materials and Methods

2.1. Area of the Study and Model Datasets

The study area included the country of Greece. The country presents several climatic variations, always in the Mediterranean climate frame, due to the influence of its vivid geomorphologic complexity (interplay of mountainous regions and plains, extended coastline, and numerous islands) on the different atmospheric-pressure dependencies from the Atlantic, central Mediterranean area, Eurasia and North Africa. This enhances the need for higher-resolution climate modeling to resolve the topography features more effectively.

In the present work, climate data of EC–Earth (1.125° horizontal resolution originally) downscaled by the WRF-ARW version 3.6.1 model to 5×5 km^2 were employed at a temporal resolution of 6-h. The hindcast climate simulations have been extensively evaluated in our previous works, whereby exhaustive quantitative validation of the highly resolved fields of temperature, precipitation, wind speed and solar radiation were performed for our observations [45,47–50]. The WRF-ARW modeling domain covering Greece comprises a grid of 185×185 cells in the horizontal and 40 levels in the vertical that are arranged according to terrain, following the hydro-static-pressure vertical coordinates (up to ~50 mbars). A more detailed description of the WRF model setup and physical parameterization schemes can be found in [47].

For the simulations of future years under the influence of climate change, the two IPCC greenhouse-gas-emission scenarios, RCP4.5 and RCP8.5, were selected as they constitute the most commonly used scenarios by impact-assessment modelers. In particular, RCP4.5 represents an increase in the radiative forcing of the atmosphere of 4.5 W/m^2 relative to the pre-industrial era with a profile of greenhouse-gas emissions increasing until the mid-century (~2050) and stabilizing thereafter until the end of the century (2100). On the other hand, RCP8.5 is considered to be the most extreme scenario with greenhouse-gas emissions increasing sharply until the end of the century, implying at its end a radiative forcing of 8.5 W/m^2 relative to the pre-industrial era. The future time periods in the RCP4.5 and RCP8.5 simulations were selected in order to study the projected climate-change effects on hazard dynamics, both in the middle and near the end of the 21st century. For the present analysis, the model-downscaled data that were used and the corresponding 25-year-period slots are presented in Table 1.

Table 1. The global (EC–Earth) model datasets downscaled by WRF model used in the study and corresponding periods.

Dataset	Period
EC–EARTH–WRF	1980–2004 (historical)
EC–EARTH–WRF RCP4.5	2025–2049 (near future)
EC–EARTH–WRF RCP8.5	2025–2049 (near future)
EC–EARTH–WRF RCP4.5	2075–2099 (far future)
EC–EARTH–WRF RCP8.5	2075–2099 (far future)

2.2. Statistical Tools and Data Processing

All model-data processing and figure drawing were executed with R software and the ARC.GIS (MAP) 10.0 environment. Figure 1 depicts the basic steps of the process followed for the assessment of the extreme values of the variables, namely those of temperature (maximum and minimum), precipitation rate, snowfall and wind speed. Firstly, the maximum and minimum temperature values were retrieved for the summer and winter seasons, respectively, while data were extracted throughout the year from the gridded datasets to determine the extremes of the precipitation rate, snowfall and wind speed. In this manner, at each grid cell and for each period (see Table 1), the 25 maximum values of each variable were obtained. The process was applied to all three time periods of interest and both RCP scenarios (see Table 1).

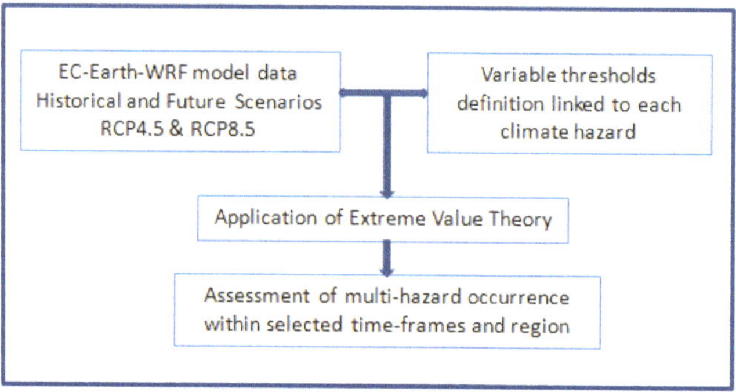

Figure 1. Schematic representation of the main methodological steps for determining a quantified climatic multi-hazard assessment.

Then, the modeled values of the variables of interest to this study were categorized in terms of likelihood, i.e., probability of occurrence, according to Table 2. The likelihood categories, six in total, and the threshold values of Table 2 were retrieved from the EU-CIRCLE project report [51] for the characterization of the hazard (maximum (summer) temperature, minimum (winter) temperature, precipitation rate, snow rate and wind speed). The probability of occurrence of each variable was calculated for all six categories. As the study was aimed at determining the likelihood of extremes, the focus was placed only on the specific categories characterizing the highest threshold values and hence, the probabilities of occurrence were calculated as totals of the three classes "High", "Very High" and "Exceptional".

In this work, the method for calculating the probability of exceeding a value was based on the Extreme Value Theory (EVT). The determination of the extreme values of the studied variables required the selection and definition of the underlying distribution functions. The estimations were made using the R package "Extremes" [52], fitting a Generalized Extreme Value distribution (GEV) to block maxima data (annual maxima) under the assumption of non-stationarity [53]. The GEV distribution has three parameters: the shape factor ξ, the scale or dispersion parameter σ and the location or mode parameter μ. The GEV-distribution function, $G(y)$, is given by:

$$\text{For } \xi \neq 0, \ G(y) = \exp\left(-\left[1 + \xi\left(\frac{y-\mu}{\sigma}\right)\right]^{-1/\xi}\right) \tag{1}$$

$$\text{For } \xi = 0, \ G(y) = \exp\left(-\exp\left(-\frac{y-\mu}{\sigma}\right)\right) \tag{2}$$

The GEV has three types depending on shape parameter ξ, as follows:

1. When $\xi = 0$, GEV is known also as Type I Extreme Value Distribution (or Gumbel Distribution, light tail)
2. When $\xi > 0$, GEV is known also as Type II Extreme Value Distribution (or Frechet Distribution, heavy tail)

3.   When $\xi < 0$, GEV is known also as Type III Extreme Value Distribution (or Weibull Distribution, upper finite end point).

Table 2. Likelihood categories and threshold values for maximum and minimum temperature, maximum precipitation rate, maximum snow rate and maximum wind speed. In bold, the threshold values to which the probability of occurrence was applied in the current study.

Variables	Likelihood Categories					
	Very Low	Low	Medium	High	Very High	Exceptional
Daily Minimum Temperature [°C]	0<	0–(−2)	(−2)–(−5)	**(−5)–(−10)**	**(−10)–(−15)**	**<(−15)**
Daily Maximum Temperature [°C]	<30	30–33	33–35	**35–39**	**39–42**	**>42**
Daily Maximum Precipitation rate [mm/h]	<2.5	2.5–7.6	7.6–10.0	**10–50**	**50–100**	**>100**
Daily Maximum Snowfall [mm/h]	<2.5	2.5–12.7	12.7–25.4	**25.4–76.2**	**76.2–127**	**>127**
Daily Maximum wind speed value [m/s]	0–3	3–12	12–15	**15–20**	**20–30**	**>30**

The final step of the applied methodology included the calculation of the most frequently appearing hazard at each grid cell, time period and RCP scenario. In this manner, we performed the spatial assessment of the occurrence of multi-hazards in the historical period and in the two studied future periods according to the two RCP projections in order to illustrate the areas susceptible to hazards over long time scales.

3. Results and Discussion

In this part, we present the results of the previously described applied approach. For reasons of clarity, each subsection provides the results pertinent to each studied hazard. Figure 2a presents the highly resolved topography of the EC-Earth global model downscaled by WRF to the high horizontal resolution of 5×5 km². Additionally, highlighted in Figure 2b are the regions of the country, where some important findings are more extensively discussed.

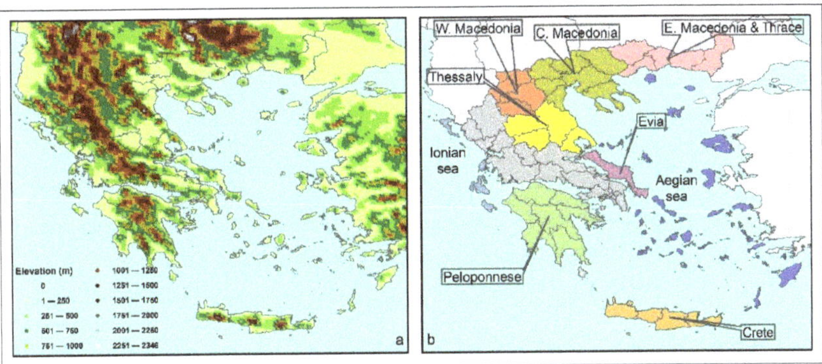

Figure 2. (**a**) EC–Earth–WRF topography of the simulated domain of Greece with horizontal resolution (5×5 km²). (**b**) Highlighted regions of the country of particular interest for discussion.

3.1. Maximum Temperature

Figure 3a shows the probability of the occurrence of TX exceeding the threshold value (i.e., the probability of exceedance) of 35 °C during the summer season of the historical period of 1980–2004. Figure 3b–e depict the differences in the probability of exceedance between the future projections and the historical values (i.e., future–historical) for both RCPs and the studied periods. During the historical period, we may deduce that the majority of areas show low probabilities of exceedance below 2.5% (Figure 3a). On the other hand, the highest occurrence of maximum temperature values with the probability of exceedance above 10% is seen in the plains of the regions of Thessaly, central Macedonia, Peloponnese, the western mainland, the eastern Aegean islands and southern Crete, which are well known as summer hot-spot areas [54].

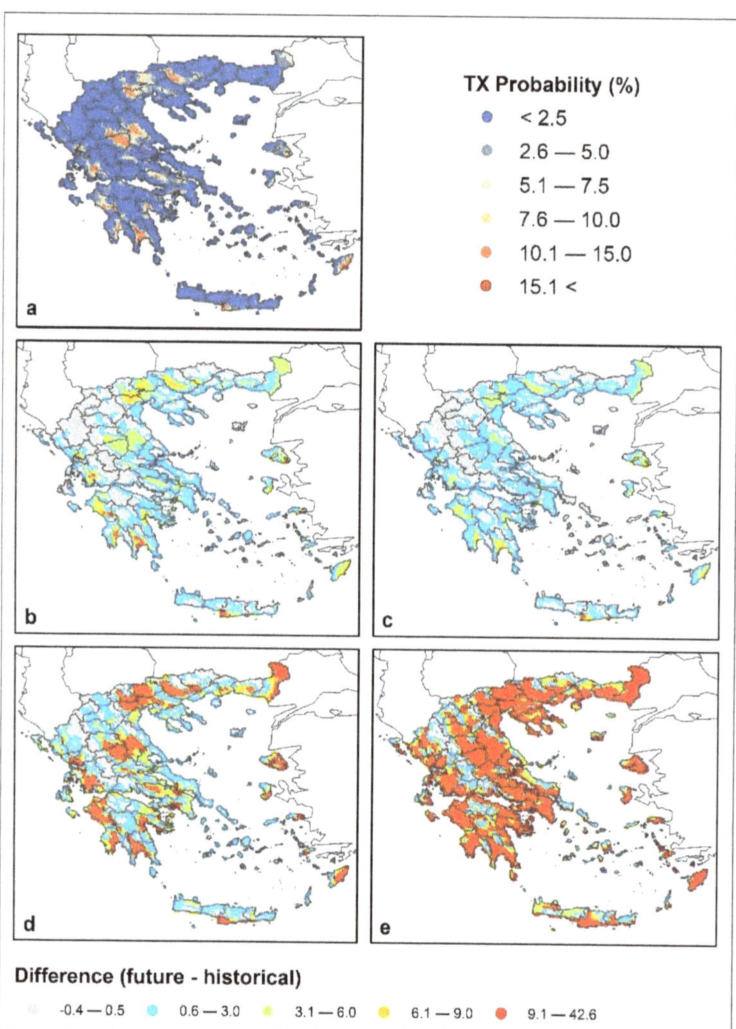

Figure 3. (**a**) Spatial distribution of probability of TX exceedance above 35 °C calculated using EC–Earth–WRF downscaled data for the historical summer period 1980–2004. Differences (future–historical) in the probability of TX exceedance for: (**b**) RCP4.5 in near future (2025–2049), (**c**) RCP8.5 in near future (2025–2049), (**d**) RCP4.5 in far future (2075–2099) and (**e**) RCP8.5 in far future (2075–2099).

Overall, in the near future, more areas compared to the historical period were found to be exposed to extreme TX values (Figure 3b,c). In addition, it appears that the probability increase in hot-spot areas is higher in RCP4.5, with these values exceeding 3%. The differences in the referenced areas become much larger, exceeding 10%, and even extend spatially into the far future and more profoundly in RCP8.5 (Figure 3d,e). In fact, the probability of exceedance according to RCP8.5 in the far future increases not only in the plains areas with low topographic heights but also in areas with heights in the range of 500 to 1000 m (Figure 3e).

This could be considered an important change predicted by the worst-case emissions scenario that would cause adverse conditions near the end of the century. Nevertheless, in the near future, insignificant changes are expected over the high mountainous regions according to both RCPs concerning the historical period (Figure 3b,c). In the far future, the insignificant changes are still found over the highest mountains with RCP4.5 (Figure 3d) but the RCP8.5 scenario diminishes these (around zero values), showcasing detectable differences in the probability of exceedances of 3–6% with respect to the historical period, even over the highest summits of the mainland (Figure 3e).

3.2. Minimum Temperature

Figure 4a presents the probability of TN exceeding $-5\,^{\circ}\mathrm{C}$ towards lower values during the winters of the historical period. Overall, the probability values remain below 5% in most areas of the mainland and the islands. Over the higher topographic heights of the central and northeastern mainland (see Figure 2a), the probability increases to noticeable levels in the range of 20 to 40% that are consistent with the occurrence of very low temperatures and the extreme winter climatology of the country [54]. The differences in the probabilities of exceedance of TN between the two future and historical periods (future–historical) are shown in Figure 4b–e for both RCPs. We may observe that under all scenarios and periods, areas that have historically had very low probability values of extremely cold temperatures preserve these characteristics in the future.

In general, we observe a strong decrease in the probabilities of exceedance of TN over the mountainous areas of the central and northeastern mainland in both future periods with respect to the historical period, which denotes a reduction in the future occurrence of extreme values of TN and thus, fewer winter extremes. In the near-future period, the highest decreases in the central mountainous areas are more intense in RCP4.5 (Figure 4b) than in RCP8.5 (Figure 4c) while in eastern Macedonia, Thrace and Peloponnese there are no noticeable differences between the two scenarios. Stronger decreases in the occurrence of winter extremes are estimated for both scenarios in the far future (Figure 4d,e). The impacted areas according to RCP4.5 remain the same during both periods (Figure 4b,d) but they extend more spatially in the far future with RCP8.5 (Figure 4c,e).

3.3. Precipitation Rate

Figure 5a presents, for the historical period, the probability of precipitation extremes calculated as the probability of precipitation rates exceeding 10 mm/h. The pattern with persisting probabilities over the majority of the domain of less than 0.025% does not yield noticeable spatial variability. Yet, increased probabilities of extreme precipitation with values greater than 0.075% are observable in the very high mountainous areas of the central and eastern mainland and over the summits of Crete and Peloponnese. In addition, some areas known for high-precipitation rates such as the Ionian islands and parts of central and eastern Macedonia and Rhodes reasonably exhibit distinguishable contours of the probability of exceeding 10 mm/h up to 0.075%.

The differences between the future projections and historical simulations indicate a reduction in future extreme precipitation mostly in the eastern parts of the mainland, central Aegean islands and mountainous areas of Crete (Figure 5b–e). The maximum decrease is obtained in the near future (Figure 5b,c). Additionally, with reference to the historical period and both RCPs, we may observe minute and insignificant changes in extreme precipitation rates, more extensively in the regions of Thessaly and central Macedonia and predominantly in the near future (Figure 5b,c). However, interesting patterns of increased probabilities of extreme precipitation can be seen in highly mountainous areas primarily in the western and northeastern mainland (in the range of 0.051–0.285%) and more vividly for RCP8.5 (Figure 5d,e). The RCP4.5 projections do not show a noticeable change between the two future periods. On the contrary, according to RCP8.5, the far-future period shows on average an increase in the probability of rainfall extremes compared with the near future, both in magnitude and spatial extent. Overall, the patterns of differences showcase increased probabilities of extreme precipitation rates in the studied future periods even at low topographic heights, which may highlight a bothersome climate-change effect for the agricultural economy.

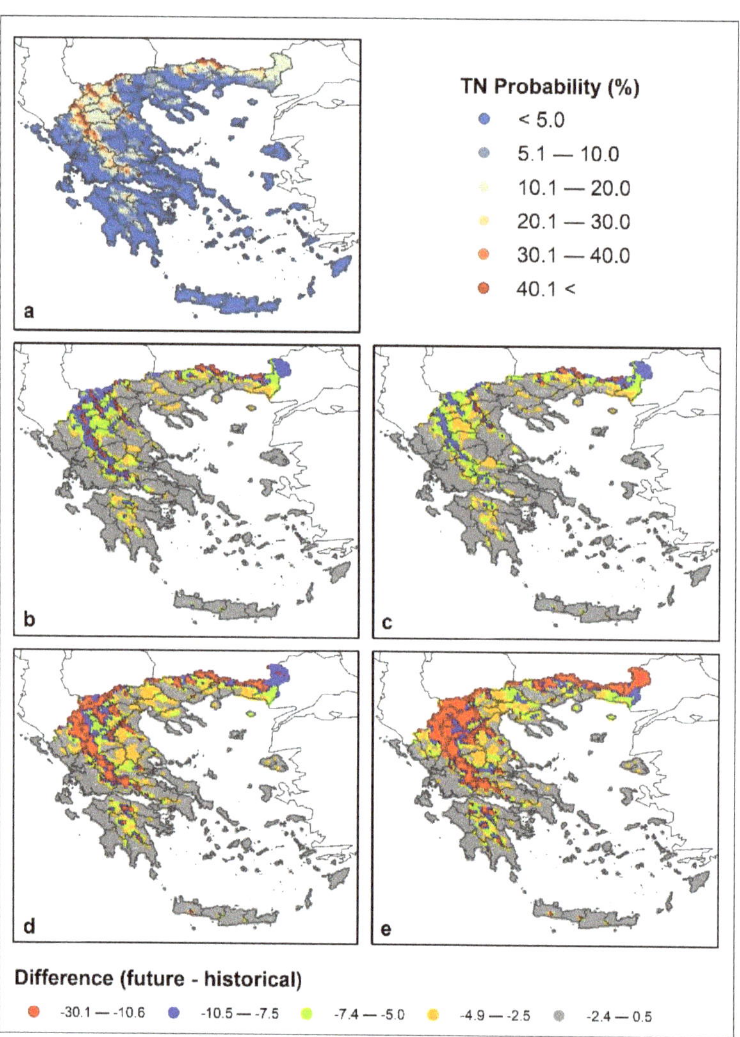

Figure 4. (**a**) Spatial distribution of probability of TN exceedance below −5 °C calculated using EC–Earth–WRF downscaled data for the historical winter period 1980–2004. Differences (future–historical) in the probability of TN exceedance for: (**b**) RCP4.5 in near future (2025–2049), (**c**) RCP8.5 in near future (2025–2049), (**d**) RCP4.5 in far future (2075–2099) and (**e**) RCP8.5 in far future (2075–2099).

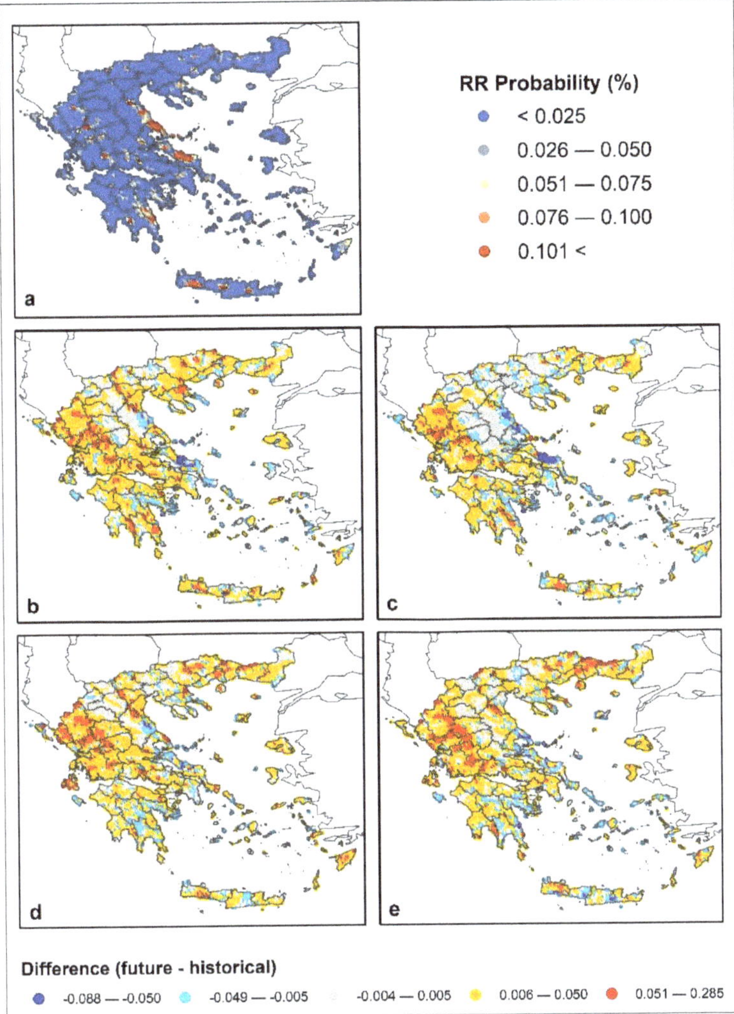

Figure 5. (**a**) Spatial distribution of probability of RR exceedance above 10 mm/h calculated using EC–Earth–WRF downscaled data for the historical period 1980–2004. Differences (future–historical) in the probability of RR exceedance for: (**b**) RCP4.5 in near future (2025–2049), (**c**) RCP8.5 in near future (2025–2049), (**d**) RCP4.5 in far future (2075–2099) and (**e**) RCP8.5 in far future (2075–2099).

3.4. Wind

During the historical period, the areas of high wind-speed (exceeding 15 m/s) probability are found to be consistent with other studies in the literature [45,55,56] (Figure 6a) and in accordance with the known synoptic atmospheric systems associated with the prevailing wind patterns. More particularly, the areas of the northeastern Aegean Sea are impacted by north-easterlies and the central Aegean by the Etesians, whereas the topography of Crete significantly amplifies the patterns of the extreme winds.

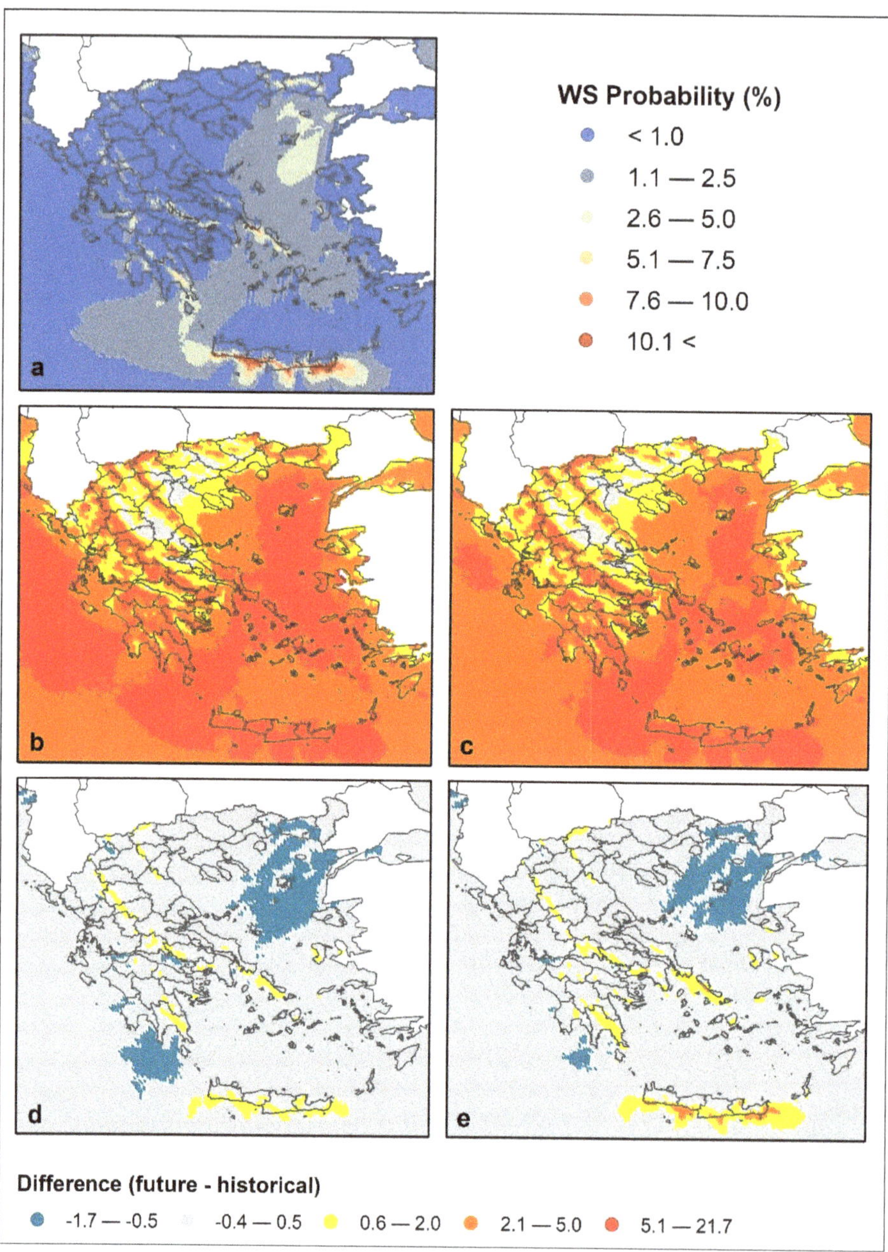

Figure 6. (**a**) Spatial distribution of probability of wind speed exceedance over the threshold (15 m/s) calculated using EC–Earth–WRF downscaled data for the historical period 1980–2004. Differences (future–historical) in the probability of wind speed exceedance for: (**b**) RCP4.5 in near future (2025–2049), (**c**) RCP8.5 in near future (2025–2049), (**d**) RCP4.5 in far future (2075–2099) and (**e**) RCP8.5 in far future (2075–2099).

In the near future and under both scenarios, the probability of the occurrence of windstorms is amplified almost all over the domain and mostly over the sea areas and mountain ridges (Figure 6b,c). The Ionian and central Aegean Seas are projected to exhibit the highest changes in the probabilities in the RCP4.5 scenario.

In the far future, both scenarios show a decrease in the probability of extreme winds over the northeastern Aegean and south Ionian Seas (Figure 6d,e). On the other hand, the mountainous areas, the central Aegean Sea and Crete present an increase in the probability of extreme winds. The increases in extreme winds associated with the Etesians over the Aegean Sea projected by both RCPs are in agreement with the findings of [57].

3.5. Multi-Hazard Probability Maps

Figure 7 depicts the spatial distribution of the probability of occurrence of the most dominant hazard calculated for the historical period and both future periods and scenarios. The dominant-hazard map for the historic period indicates the dominance of extreme TX over plains and coastal areas and the dominance of extreme TN at topographic heights higher than ~500 m a.m.s.l (Figure 7a). Over the seas, the islands of the central Aegean and the southeastern corners of Evia and Peloponnese, the extreme winds become dominant. However, extreme temperatures preside in the islands of the Ionian and eastern Aegean Seas, while in the northern parts of some of them, the extremely low temperatures or extreme winds dominate. In the case of Crete in particular, we observe the dominance of the three hazards with extreme TX and TN over the plains and high mountainous regions, respectively, and windstorms over the remaining areas of the island. Moreover, the dominance of extreme rainfall is observed in parts of the eastern coasts of the central mainland and Peloponnese.

The projected changes according to RCP4.5 in the near future yield a more extended dominance of extreme TX, particularly in the areas of Peloponnese, the central-eastern mainland and in the north of the country over the plains of central and eastern Macedonia and Thrace (Figure 7b). The same effect is observed in the central Aegean islands and northern parts of Crete. In addition, the extreme rainfall is seen to prevail more extensively in eastern coastal areas of the mainland. On the other hand, according to RCP8.5, the dominance of extreme TX becomes more profound in Thessaly and the plains areas of northern Greece (Figure 7c). Moreover, an interesting pattern of prevailing extreme winds is revealed in parts of the mainland and mostly in the Peloponnese where in the historical period the dominance of extreme temperatures was evident. Furthermore, RCP8.5 presents extremely windier conditions for the Ionian islands and Crete. This result agrees with the findings of Karozis et al. [56] that highlighted the reduction in the persisting anticyclonic activities over Greece and the Balkans in the near future, a change that denotes less frequent stagnant atmospheric conditions. The same study indicated a possible reduction in the passage of cyclones over Greece originating from the cyclogenesis region of the Central Mediterranean and the Adriatic Sea that could explain the reduced extreme-rainfall findings.

In the far future, the projected changes due to RCP4.5 show the dominance of extreme TX in areas of low altitude and more vividly in western Greece, Peloponnese, Thessaly and the northern mainland (Figure 7d). Additionally, in Crete and the central Aegean islands the extreme winds become less dominant and extreme TX predominates. In addition, the extreme rainfall is noted to persist in the far future over coastal parts of the eastern mainland. All the same, the extreme TN dominates in the mountainous areas. The high occurrence of extreme-winter-cold events under high-emission scenarios agrees with other climate-model projections that estimated increases in mid-latitude westerlies and northerly cold-air flow due to the influence of the upper-tropospheric equator-to-pole temperature difference in the storm-track response to climate change [58].

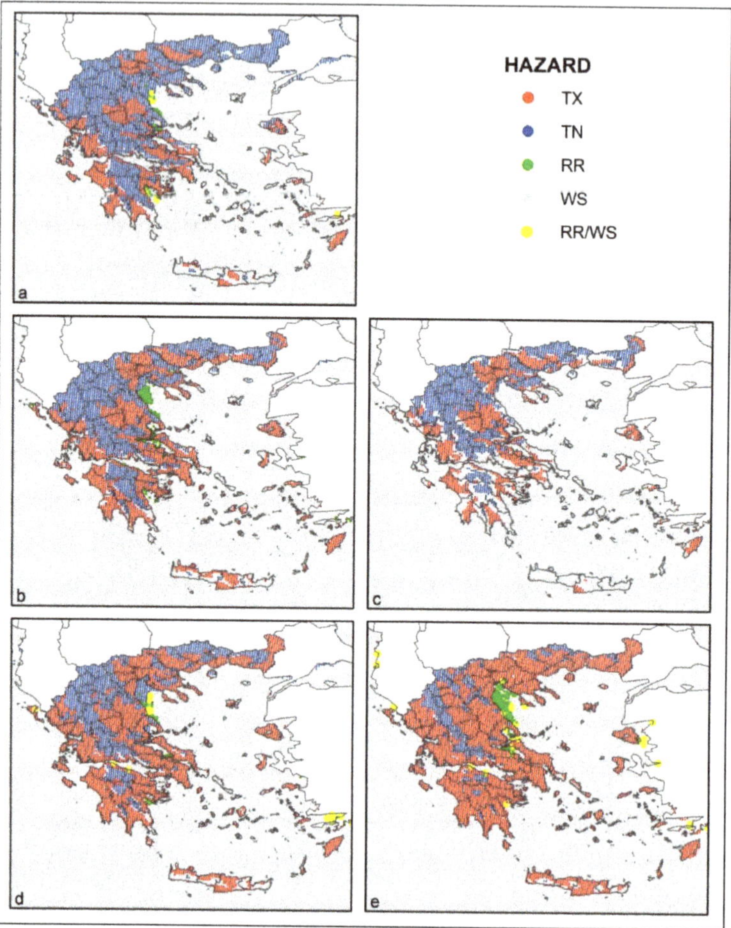

Figure 7. Dominant-hazard maps using EC–Earth–WRF downscaled data for: (**a**) the historical period 1980–2004, (**b**) RCP4.5 in near future (2025–2049), (**c**) RCP8.5 in near future (2025–2049), (**d**) RCP4.5 in far future (2075–2099) and (**e**) RCP8.5 in far future (2075–2099).

Furthermore, the RCP8.5 projections of the far future showcase the extended dominance of extreme TX in the islands and mainland except for the highest altitudes (circa 1500 m a.m.s.l.) (Figure 7e). The future intensification of extreme-temperature events is in agreement with other studies due to non-linear interactions, which are presently not quantified, between Arctic teleconnections and other remote and regional feedback processes [59,60]. Furthermore, the extreme winds remain the dominant hazard over the seas while the eastern coasts of the mainland would mostly experience extreme rainfall events that only persist locally in the far future. This outcome is associated with the Arctic amplification and possible connection to the weakening of mid-latitude storm tracks [61].

Figure 8 presents the multi-hazard occurrence in Greece for the examined periods. The value indicates the cumulative annual probability that at least one of the studied extreme hazards will occur. It demonstrates that patterns of highly exposed areas in Greece over the historic period will be spatially shifted in the far future as the increase in hotter climate regimes will be dominating the risk landscape.

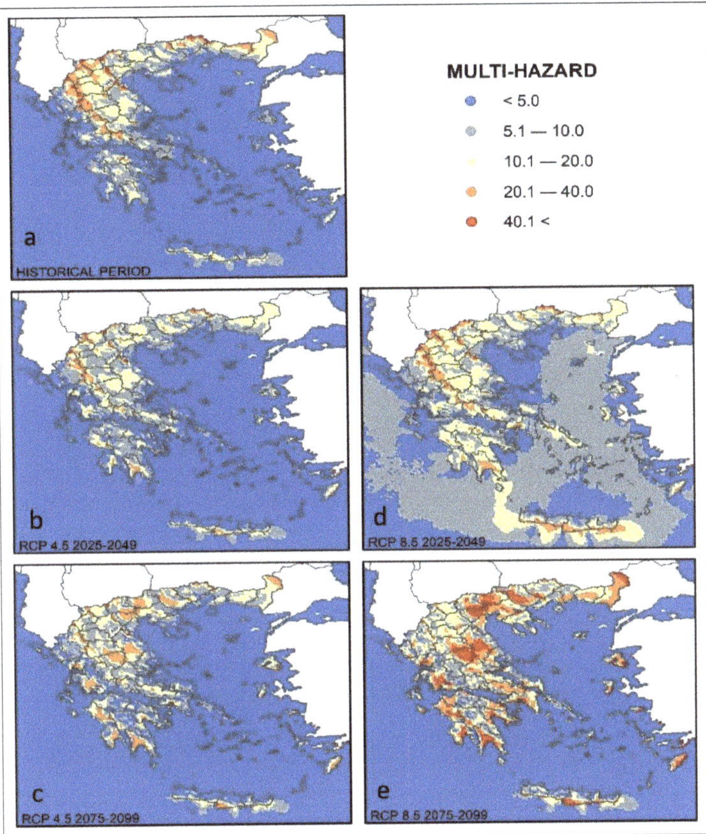

Figure 8. Annual percentage of multi-hazard using EC–Earth–WRF downscaled data for: (**a**) the historical period 1980–2004, (**b**) RCP4.5 in near future (2025–2049), (**c**) RCP8.5 in near future (2025–2049), (**d**) RCP4.5 in far future (2075–2099) and (**e**) RCP8.5 in far future (2075–2099).

In the historic period (Figure 8a), mountainous areas in the northern parts appear to be the most exposed regions in Greece, with TN and RR being the most significant risks (Figure 7a). The plains areas in the mainland and Crete exhibit lower risk levels through a combination of TX and extreme winds. The coastal zones and the Aegean islands appear to lie at the lower risk level.

In the near future, the RCP4.5 scenario (Figure 8b) appears to demonstrate the smallest variability in the risk levels when compared to RCP8.5 (Figure 8c). The mountainous areas will be exposed to lower risk levels by a factor between 5 and 14% compared to present times, contrasting a similar increase due to TX in the lowlands and western Greece. These relative changes were determined to be statistically significant at the 95% confidence level using the student t-test. The southern parts of the Eastern Aegean islands and Crete will also experience increased level of hazards. For the RCP8.5 scenario, in the near future, a greater number of regions are projected to be exposed to higher levels of risk due to both higher TX values and stronger winds (Figure 8c). These appear to be located in the southern and the western parts of the mainland and the Aegean Sea. For the far future (Figure 8d,e), both scenarios will exhibit similar patterns of risk changes compared to the present period, although RCP8.5 will be associated with higher hazard risk, often exceeding a 5% increase. Plains, agricultural lands and islands will be especially exposed to increased risk.

4. Conclusions

The study presented here aimed to elucidate the highly dynamic changing patterns of climate risk in Greece, a European climate hot spot [62]. The findings highlighted the areas that are exposed to multiple climate hazards in the country, considering the influence of the highly complex topography. In addition, the generated multi-hazard risk maps could be used to support disaster-risk-prevention activities such as avoiding future human, natural and material losses and generating economic benefits by reducing climate-related risks. The introduced method could be easily transferred to other geographical regions provided that the climate simulations are available. In addition, based on the perceived risk values or values identified in national risk assessments, the likelihood categories of the hazards (Table 2) could be adjusted accordingly. It should be mentioned that a limitation in the approach emanates from the complex topography of a domain or parts of it. In such a case, it may be required to downscale the climate data to even higher than a 5 km resolution over the complex topography areas, where there is the need to study the occurrence of (multi-)hazard(s) in more detail and accuracy.

Overall, the analysis demonstrated that climate change is a highly non-stationary process and the exposed areas, risk level and dominant risk will be significantly changed in the future under both RCP4.5 and RCP8.5 scenarios. More particularly, the impact of global warming on the country will become more evident in the far future (end of the century) when the extreme maximum temperature will dominate all other hazards.

According to the RCP4.5 scenario, a gradual expansion of the extreme maximum temperature can be anticipated from the coastal regions to the higher altitude areas, and this trend will become more persistent towards the end of the century. Under the RCP8.5 scenario, the extreme wind speed was found to be the dominant hazard in the near future, while afterwards, near the end of the century, the extreme maximum temperature becomes the most significant hazard.

Author Contributions: Conceptualization, D.V. and A.S.; methodology, D.V. and A.S.; software, I.M., S.K. and N.P.; validation, A.S. and N.P.; formal analysis, D.V. and N.G.; investigation, D.V.; resources, D.V.; writing—original draft preparation, D.V.; writing—review and editing, A.S.; supervision, D.V. All authors have read and agreed to the published version of the manuscript.

Funding: This research received no external funding.

Institutional Review Board Statement: Not applicable.

Informed Consent Statement: Not applicable.

Data Availability Statement: Data used in this study can be obtained by request to the authors.

Acknowledgments: This work was supported by computational time granted from the Greek Research & Technology Network (GRNET) in the National HPC facility—ARIS—under project ID HRCOG (pr004020 and pr006028).

Conflicts of Interest: The authors declare no conflict of interest.

References

1. IPCC; Masson-Delmotte, V.; Zhai, P.; Pirani, A.; Connors, S.L.; Péan, C.; Berger, S.; Caud, N.; Chen, Y.; Goldfarb, L.; et al. *Climate Change 2021: The Physical Science Basis. Contribution of Working Group I to the Sixth Assessment Report of the Intergovernmental Panel on Climate Change*; Cambridge University Press: Geneva, Switzerland, 2021. Available online: https://www.ipcc.ch/report/ar6/wg1/ (accessed on 1 December 2021).
2. Fischer, E.M.; Sippel, S.; Knutti, R. Increasing probability of record-shattering climate extremes. *Nat. Clim. Chang.* **2021**, *11*, 689–695. [CrossRef]
3. Guerreiro, S.B.; Fowler, H.J.; Barbero, R.; Westra, S.; Lenderink, G.; Blenkinsop, S.; Lewis, E.; Li, X.F. Detection of continental-scale intensification of hourly rainfall extremes. *Nat. Clim. Chang.* **2018**, *8*, 803–807. [CrossRef]
4. WMO Climate and Weather Related Disasters Surge Five-Fold over 50 Years, but Early Warnings Save Lives. Available online: https://news.un.org/en/story/2021/09/1098662 (accessed on 30 September 2021).

5. NOAA National Centers for Environmental Information (NCEI) Billion-Dollar Weather and Climate Disasters: FAQ | National Centers for Environmental Information (NCEI). Available online: https://www.ncdc.noaa.gov/billions/ (accessed on 30 September 2021).
6. European Environment Agency Economic Losses from Climate-Related Extremes in Europe. Available online: https://www.eea.europa.eu/data-and-maps/indicators/direct-losses-from-weather-disasters-4/assessment (accessed on 30 September 2021).
7. Gampe, D.; Zscheischler, J.; Reichstein, M.; O'Sullivan, M.; Smith, W.K.; Sitch, S.; Buermann, W. Increasing impact of warm droughts on northern ecosystem productivity over recent decades. *Nat. Clim. Chang.* **2021**, *11*, 772–779. [CrossRef]
8. van Vliet, M.T.H.; van Beek, L.P.H.; Eisner, S.; Flörke, M.; Wada, Y.; Bierkens, M.F.P. Multi-model assessment of global hydropower and cooling water discharge potential under climate change. *Glob. Environ. Chang.* **2016**, *40*, 156–170. [CrossRef]
9. Schewe, J.; Gosling, S.N.; Reyer, C.; Zhao, F.; Ciais, P.; Elliott, J.; Francois, L.; Huber, V.; Lotze, H.K.; Seneviratne, S.I.; et al. State-of-the-art global models underestimate impacts from climate extremes. *Nat. Commun.* **2019**, *10*, 1005. [CrossRef]
10. Monier, E.; Paltsev, S.; Sokolov, A.; Chen, Y.-H.H.; Gao, X.; Ejaz, Q.; Couzo, E.; Schlosser, C.A.; Dutkiewicz, S.; Fant, C.; et al. Toward a consistent modeling framework to assess multi-sectoral climate impacts. *Nat. Commun.* **2018**, *9*, 660. [CrossRef]
11. Sillmann, J.; Sippel, S.; Russo, S. *Climate Extremes and Their Implications for Impact and Risk Assessment*; Elsevier: Amsterdam, The Netherlands, 2019; pp. 1–9. [CrossRef]
12. Otto, F.E.L.; Van Oldenborgh, G.J.; Eden, J.; Stott, P.A.; Karoly, D.J.; Allen, M.R. The attribution question. *Nat. Clim. Chang.* **2016**, *6*, 813–816. [CrossRef]
13. Hoegh-Guldberg, O.; Jacob, D.; Taylor, M.; Bindi, M.; Abdul Halim, S.; Achlatis Australia, M.; Alexander, L.V.; Allen, M.R.; Berry, P.; Boyer, C.; et al. Impacts of 1.5 °C global warming on natural and human systems. In *Global Warming of 1.5 °C. An IPCC Special Report on the Impacts of Global Warming of 1.5 °C above Pre-Industrial Levels and Related Global Greenhouse Gas Emission Pathways, in the Context of Strengthening the Global Response to the Threat of Climate Change, Sustainable Development, and Efforts to Eradicate*; IPCC Secretariat: Geneva, Switzerland, 2018; pp. 175–311.
14. Lehmann, J.; Coumou, D.; Frieler, K. Increased record-breaking precipitation events under global warming. *Clim. Change* **2015**, *132*, 501–515. [CrossRef]
15. Ribes, A.; Zwiers, F.W.; Azaïs, J.M.; Naveau, P. A new statistical approach to climate change detection and attribution. *Clim. Dyn.* **2017**, *48*, 367–386. [CrossRef]
16. King, A.D. Attributing Changing Rates of Temperature Record Breaking to Anthropogenic Influences. *Earth's Futur.* **2017**, *5*, 1156–1168. [CrossRef]
17. Van Der Wiel, K.; Kapnick, S.B.; Jan Van Oldenborgh, G.; Whan, K.; Philip, S.; Vecchi, G.A.; Singh, R.K.; Arrighi, J.; Cullen, H. Rapid attribution of the August 2016 flood-inducing extreme precipitation in south Louisiana to climate change. *Hydrol. Earth Syst. Sci.* **2017**, *21*, 897–921. [CrossRef]
18. Diffenbaugh, N.S.; Singh, D.; Mankin, J.S.; Horton, D.E.; Swain, D.L.; Touma, D.; Charland, A.; Liu, Y.; Haugen, M.; Tsiang, M.; et al. Quantifying the influence of global warming on unprecedented extreme climate events. *Proc. Natl. Acad. Sci. USA* **2017**, *114*, 4881–4886. [CrossRef] [PubMed]
19. Marsooli, R.; Lin, N.; Emanuel, K.; Feng, K. Climate change exacerbates hurricane flood hazards along US Atlantic and Gulf Coasts in spatially varying patterns. *Nat. Commun.* **2019**, *10*, 3785. [CrossRef] [PubMed]
20. Ali, H.; Modi, P.; Mishra, V. Increased flood risk in Indian sub-continent under the warming climate. *Weather Clim. Extrem.* **2019**, *25*, 100212. [CrossRef]
21. Coogan, S.C.P.; Robinne, F.N.; Jain, P.; Flannigan, M.D. Scientists' warning on wildfire—A canadian perspective. *Can. J. For. Res.* **2019**, *49*, 1015–1023. [CrossRef]
22. Mezősi, G.; Bata, T.; Meyer, B.C.; Blanka, V.; Ladányi, Z. Climate Change Impacts on Environmental Hazards on the Great Hungarian Plain, Carpathian Basin. *Int. J. Disaster Risk Sci.* **2014**, *5*, 136–146. [CrossRef]
23. Cook, B.I.; Mankin, J.S.; Anchukaitis, K.J. Climate Change and Drought: From Past to Future. *Curr. Clim. Chang. Rep.* **2018**, *4*, 164–179. [CrossRef]
24. Pereira, S.C.; Carvalho, D.; Rocha, A. Temperature and Precipitation Extremes over the Iberian Peninsula under Climate Change Scenarios: A Review. *Climate* **2021**, *9*, 139. [CrossRef]
25. Savi, S.; Comiti, F.; Strecker, M.R. Pronounced increase in slope instability linked to global warming: A case study from the eastern European Alps. *Earth Surf. Processes Landf.* **2021**, *46*. [CrossRef]
26. Tijdeman, E.; Hannaford, J.; Stahl, K. Human influences on streamflow drought characteristics in England and Wales. *Hydrol. Earth Syst. Sci.* **2018**, *22*, 1051–1064. [CrossRef]
27. Cammalleri, C.; Naumann, G.; Mentaschi, L.; Bisselink, B.; Gelati, E.; De Roo, A.; Feyen, L. Diverging hydrological drought traits over Europe with global warming. *Hydrol. Earth Syst. Sci.* **2020**, *24*, 5919–5935. [CrossRef]
28. Sieck, K.; Nam, C.; Bouwer, L.M.; Rechid, D.; Jacob, D. Weather extremes over Europe under 1.5 and 2.0 °C global warming from HAPPI regional climate ensemble simulations. *Earth Syst. Dyn.* **2021**, *12*, 457–468. [CrossRef]
29. Suarez-Gutierrez, L.; Li, C.; Müller, W.A.; Marotzke, J. Internal variability in European summer temperatures at 1.5 °C and 2 °C of global warming. *Environ. Res. Lett.* **2018**, *13*, 064026. [CrossRef]
30. Forzieri, G.; Bianchi, A.; Silva, F.B.; Marin Herrera, M.A.; Leblois, A.; Lavalle, C.; Aerts, J.C.J.H.; Feyen, L. Escalating impacts of climate extremes on critical infrastructures in Europe. *Glob. Environ. Chang.* **2018**, *48*, 97–107. [CrossRef]

31. Feyen, L.; Ciascar, J.; Gosling, S.; Ibarreta, D.; Soria, A.; Dosio, A.; Naumann, G.; Russo, S.; Formetta, G.; Forzieri, G.; et al. *JRC Science for Policy Report*; Joint Research Centre: Petten, The Netherlands, 2020.

32. The United Nations Office for Disaster Risk Reduction Proposed Updated Terminology on Disaster Risk Reduction: A Technical Review. Available online: http://www.unisdr.org/we/inform/terminology (accessed on 8 October 2021).

33. EUR 30596 EN. *JRC Recommendations for National Risk Assessment for Disaster Risk Management in EU: Where Science and Policy Meet*; Publications Office of the European Union: Luxembourg, 2021; Version 1; p. 273. Available online: https://data.europa.eu/doi/10.2760/43449 (accessed on 1 December 2021).

34. Kappes, M.S.; Keiler, M.; von Elverfeldt, K.; Glade, T. Challenges of analyzing multi-hazard risk: A review. *Nat. Hazards* **2012**, *64*, 1925–1958. [CrossRef]

35. Tilloy, A.; Malamud, B.D.; Winter, H.; Joly-Laugel, A. A review of quantification methodologies for multi-hazard interrelationships. *Earth-Science Rev.* **2019**, *196*, 102881. [CrossRef]

36. Gallina, V.; Torresan, S.; Critto, A.; Sperotto, A.; Glade, T.; Marcomini, A. A review of multi-risk methodologies for natural hazards: Consequences and challenges for a climate change impact assessment. *J. Environ. Manag.* **2016**, *168*, 123–132. [CrossRef]

37. Sperotto, A.; Molina, J.L.; Torresan, S.; Critto, A.; Marcomini, A. Reviewing Bayesian Networks potentials for climate change impacts assessment and management: A multi-risk perspective. *J. Environ. Manag.* **2017**, *202*, 320–331. [CrossRef]

38. Skamarock, W.C.; Skamarock, W.C.; Klemp, J.B.; Dudhia, J.; Gill, D.O.; Barker, D.M.; Wang, W.; Powers, J.G. *A Description of the Advanced Research WRF Version 3. NCAR Technical Note -475+STR*; University Corporation for Atmospheric Research: Boulder, CO, USA, 2008.

39. Doblas Reyes, F.; Acosta Navarro, J.C.; Acosta Cobos, M.C.; Bellprat, O.; Bilbao, R.; Castrillo Melguizo, M.; Fuckar, N.; Guemas, V.; Lledó Ponsati, L.; Menegoz, M. *Using EC-Earth for Climate Prediction Research*; European Centre for Medium-Range Weather Forecasts (ECMWF): Reading, UK, 2018; Volume 154, pp. 35–40.

40. Kc, B.; Shepherd, J.M.; King, A.W.; Johnson Gaither, C. Multi-hazard climate risk projections for the United States. *Nat. Hazards* **2021**, *105*, 1963–1976. [CrossRef]

41. Jay, A.D.R.; Reidmiller, C.W.; Avery, D.; Barrie, B.J.; DeAngelo, A.; Dave, M.; Dzaugis, M.; Kolian, K.L.M.; Lewis, K.; Reeves, D. Winner, Overview. In *Impacts, Risks, and Adaptation in the United States: Fourth National Climate Assessment, Volume II*; Reidmiller, D.R., Avery, C.W., Easterling, D.R., Kunkel, K.E., Lewis, K.L.M., Maycock, T.K., Stewart, B.C., Eds.; U.S. Global Change Research Program: Washington, DC, USA, 2018; pp. 33–71. [CrossRef]

42. Lhotka, O.; Kyselý, J.; Plavcová, E. Evaluation of major heat waves' mechanisms in EURO-CORDEX RCMs over Central Europe. *Clim. Dyn.* **2018**, *50*, 4149–4262. [CrossRef]

43. Cardoso, R.M.; Soares, P.M.M.; Lima, D.C.A.; Miranda, P.M.A. Mean and extreme temperatures in a warming climate: EURO CORDEX and WRF regional climate high-resolution projections for Portugal. *Clim. Dyn.* **2019**, *52*, 129–157. [CrossRef]

44. Tian, L.; Jin, J.; Wu, P.; Niu, G.Y.; Zhao, C. High-resolution simulations of mean and extreme precipitation with WRF for the soil-erosive Loess Plateau. *Clim. Dyn.* **2020**, *54*, 3489–3506. [CrossRef]

45. Katopodis, T.; Markantonis, I.; Vlachogiannis, D.; Politi, N.; Sfetsos, A. Assessing climate change impacts on wind characteristics in Greece through high-resolution regional climate modelling. *Renew. Energy* **2021**, *179*, 427–444. [CrossRef]

46. UNDRR International Science Council Sendai. *Hazard Definition & Classification Review: Technical Report*; UNDRR: Panama City, Panama, 2020.

47. Politi, N.; Vlachogiannis, D.; Sfetsos, A.; Nastos, P.T. High-resolution dynamical downscaling of ERA-Interim temperature and precipitation using WRF model for Greece. *Clim. Dyn.* **2021**, *57*, 1–27. [CrossRef]

48. Politi, N.; Sfetsos, A.; Vlachogiannis, D.; Nastos, P.T.; Karozis, S. A sensitivity study of high-resolution climate simulations for Greece. *Climate* **2020**, *8*, 44. [CrossRef]

49. Politi, N.; Nastos, P.T.; Sfetsos, A.; Vlachogiannis, D.; Dalezios, N.R. Evaluation of the AWR-WRF model configuration at high-resolution over the domain of Greece. *Atmos. Res.* **2018**, *208*, 229–245. [CrossRef]

50. Katopodis, T.; Markantonis, I.; Politi, N.; Vlachogiannis, D.; Sfetsos, A. High-resolution solar climate atlas for greece under climate change using the weather research and forecasting (WRF) model. *Atmosphere* **2020**, *11*, 761. [CrossRef]

51. Habermann Nadine; Thanasis Sfetsos; Ralh Hedel; Albert Chen EU-CIRCLE, Report on Climate Related Critical Event Parameters: Deliverable 3.2. Available online: https://www.eu-circle.eu/wp-content/uploads/2018/10/D3.2.pdf (accessed on 8 November 2021).

52. Gilleland, E.; Katz, R.W. New software to analyze how extremes change over time. *Eos Trans. Am. Geophys. Union* **2011**, *92*, 13–14. [CrossRef]

53. Cooley, D. Return periods and return levels under climate change. In *Extremes in a Changing Climate*; Springer: Berlin/Heidelberg, Germany, 2013; pp. 97–114.

54. Hellenic Meteorological Service Climate Atlas of Greece, 1971–2000. Available online: http://climatlas.hnms.gr/sdi/ (accessed on 18 November 2021).

55. Katopodis, T.; Vlachogiannis, D.; Politi, N.; Gounaris, N.; Karozis, S.; Sfetsos, A. Assessment of climate change impacts on wind resource characteristics and the wind energy potential in Greece. *J. Renew. Sustain. Energy* **2019**, *11*, 066502. [CrossRef]

56. Karozis, S.; Sfetsos, A.; Gounaris, N.; Vlachogiannis, D. An assessment of climate change impact on air masses arriving in Athens, Greece. *Theor. Appl. Climatol.* **2021**, *145*, 501–517. [CrossRef]

57. Dafka, S.; Toreti, A.; Zanis, P.; Xoplaki, E.; Luterbacher, J. Twenty-First-Century Changes in the Eastern Mediterranean Etesians and Associated Midlatitude Atmospheric Circulation. *J. Geophys. Res. Atmos.* **2019**, *124*, 12741–12754. [CrossRef]
58. Harvey, B.J.; Shaffrey, L.C.; Woollings, T.J. Deconstructing the climate change response of the Northern Hemisphere wintertime storm tracks. *Clim. Dyn.* **2015**, *45*, 2847–2860. [CrossRef]
59. Russo, S.; Dosio, A.; Graversen, R.G.; Sillmann, J.; Carrao, H.; Dunbar, M.B.; Singleton, A.; Montagna, P.; Barbola, P.; Vogt, J.V.; et al. Magnitude of extreme heat waves in present climate and their projection in a warming world. *J. Geophys. Res. Atmos.* **2014**, *119*, 12500–12512. [CrossRef]
60. Coumou, D.; Di Capua, G.; Vavrus, S.; Wang, L.; Wang, S. The influence of Arctic amplification on mid-latitude summer circulation. *Nat. Commun.* **2018**, *9*, 2959. [CrossRef]
61. Chang, E.K.M.; Ma, C.G.; Zheng, C.; Yau, A.M.W. Observed and projected decrease in Northern Hemisphere extratropical cyclone activity in summer and its impacts on maximum temperature. *Geophys. Res. Lett.* **2016**, *43*, 2200–2208. [CrossRef]
62. Cos, J.; Doblas-Reyes, F.; Jury, M.; Marcos, R.; Bretonnière, P.-A.; Samsó, M. The Mediterranean climate change hotspot in the CMIP5 and CMIP6 projections. *Earth Syst. Dyn. Discuss.* **2021**, 1–26. [CrossRef]

applied
sciences

Article

A Machine Learning Framework for Multi-Hazard Risk Assessment at the Regional Scale in Earthquake and Flood-Prone Areas

Alessandro Rocchi [1], Andrea Chiozzi [2], Marco Nale [1], Zeljana Nikolic [3], Fabrizio Riguzzi [4], Luana Mantovan [1], Alessandro Gilli [1] and Elena Benvenuti [1,*]

[1] Department of Engineering, University of Ferrara, 44122 Ferrara, Italy; alessandro.rocchi@edu.unife.it (A.R.); marco.nale@unife.it (M.N.); luana.mantovan@edu.unife.it (L.M.); alessandro.gilli@edu.unife.it (A.G.)

[2] Department of Environmental and Prevention Sciences, University of Ferrara, 44122 Ferrara, Italy; andrea.chiozzi@unife.it

[3] Faculty of Civil Engineering, Architecture and Geodesy, University of Split, 21000 Split, Croatia; zeljana.nikolic@gradst.hr

[4] Department of Mathematics and Computer Science, University of Ferrara, 44122 Ferrara, Italy; fabrizio.riguzzi@unife.it

* Correspondence: elena.benvenuti@unife.it

Abstract: Communities are confronted with the rapidly growing impact of disasters, due to many factors that cause an increase in the vulnerability of society combined with an increase in hazardous events such as earthquakes and floods. The possible impacts of such events are large, also in developed countries, and governments and stakeholders must adopt risk reduction strategies at different levels of management stages of the communities. This study is aimed at proposing a sound qualitative multi-hazard risk analysis methodology for the assessment of combined seismic and hydraulic risk at the regional scale, which can assist governments and stakeholders in decision making and prioritization of interventions. The method is based on the use of machine learning techniques to aggregate large datasets made of many variables different in nature each of which carries information related to specific risk components and clusterize observations. The framework is applied to the case study of the Emilia Romagna region, for which the different municipalities are grouped into four homogeneous clusters ranked in terms of relative levels of combined risk. The proposed approach proves to be robust and delivers a very useful tool for hazard management and disaster mitigation, particularly for multi-hazard modeling at the regional scale.

Keywords: risk assessment; multi hazard; seismic risk; hydraulic risk; machine learning; principal component analysis

Citation: Rocchi, A.; Chiozzi, A.; Nale, M.; Nikolic, Z.; Riguzzi, F.; Mantovan, L.; Gilli, A.; Benvenuti, E. A Machine Learning Framework for Multi-Hazard Risk Assessment at the Regional Scale in Earthquake and Flood-Prone Areas. *Appl. Sci.* **2022**, *12*, 583. https://doi.org/10.3390/app12020583

Academic Editor: Salvador García-Ayllón Veintimilla

Received: 21 November 2021
Accepted: 1 January 2022
Published: 7 January 2022

Publisher's Note: MDPI stays neutral with regard to jurisdictional claims in published maps and institutional affiliations.

1. Introduction

The frequency of natural extreme events is increasing worldwide [1–9], and human activities often interact with devastating effects, affecting people and natural environments, and producing great economic losses, especially in developing countries. On the other hand, in some developed countries, disasters have been decreasing since the beginning of the 20th century [3,4]. Understanding risk involving vast inhabited areas is, therefore, paramount, particularly when assessing potential losses produced by a combination of multiple hazards, which are defined as the probability of occurrence in a specified period of a potentially damaging event of a given magnitude on a given area [5]. In fact, total risk is a measure of the expected human (casualties and injuries) and economic (damage to property and activity disruption) losses due to a particular adverse natural phenomenon. Such a measure is conceptually assumed as the product of hazard, vulnerability, and exposure instances [6]. Exposure of people to the consequences of extreme natural phenomena could be reduced if predictive models based on new approaches and deeper knowledge of effective factors were employed [7].

Many areas on Earth are subjected to the effects of coexisting multiple hazards, among which floods and earthquakes are some of the most widespread [8,9] and even if it is well established that inhabited environments are affected by multiple hazardous processes, most studies focus on a single hazard [10]. However, hazards usually interact with each other and contribute to the overall risk in a complex way. For this reason, the development of multi-hazard risk assessment approaches is of first importance [11] and multi-hazard mapping is receiving increasing attention [12,13]. In particular, Schmidt et al. proposed a multi-hazard risk assessment methodology in New Zealand, devising an adaptable computational tool allowing its users to input the natural phenomena of interest [11]. Still, relatively scarce are the studies exploiting machine learning techniques to assess multi-hazard risks [14–16], albeit machine learning is especially useful when dealing with the huge amount of data encountered in risk analysis, particularly at the regional scale.

In this study, machine learning is used to construct a risk assessment framework in which the combined effects of two major natural events (flood and earthquakes) are analyzed for the Emilia Romagna test region (Italy). A large input dataset containing, for each municipality of the test region, a wide number of quantitative variables related to hazard, exposure, and vulnerability instances for both flood and earthquake hazards is adopted. Then, the number of variables is suitably reduced by means of Principal Component Analysis (PCA) [17–19], and the municipalities are subsequently grouped into four approximately risk-wise homogeneous clusters using a K-means clustering algorithm [20,21]. Finally, a qualitative overall risk level is assigned to each cluster. The proposed methodology represents a robust tool for the qualitative multi-hazard risk assessment at the regional scale, which enables suitable extraction of risk-related information from a large input dataset and provides a useful instrument that assists stakeholders in decision-making processes, especially with respect to intervention prioritization.

2. Materials and Methods

The proposed multi-hazard risk assessment approach is based on the analysis of available data using logical, mathematical, and statistical tools. It was applied to the Emilia Romagna region, which is located in the Northern part of Italy. Our analysis focused on seismic and hydraulic risks associated with this territory. A map of the seismic classification of municipalities in Emilia is shown in Figure 1. A hot-spot of hydraulic risk in Emilia Romagna, Ferrara possesses an altimetry below the sea level over a large part of its territory, as illustrated in Figure 2.

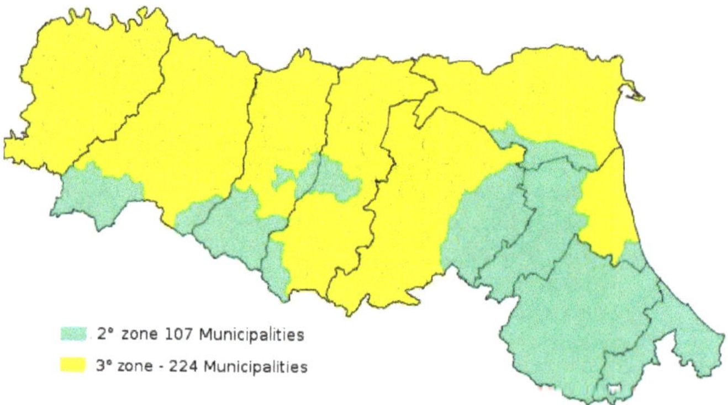

Figure 1. Seismic classification of municipalities in Emilia (https://ambiente.regione.emilia-romagna.it/en/geologia/seismic-risk/seismic-classification, accessed on 15 October 2021).

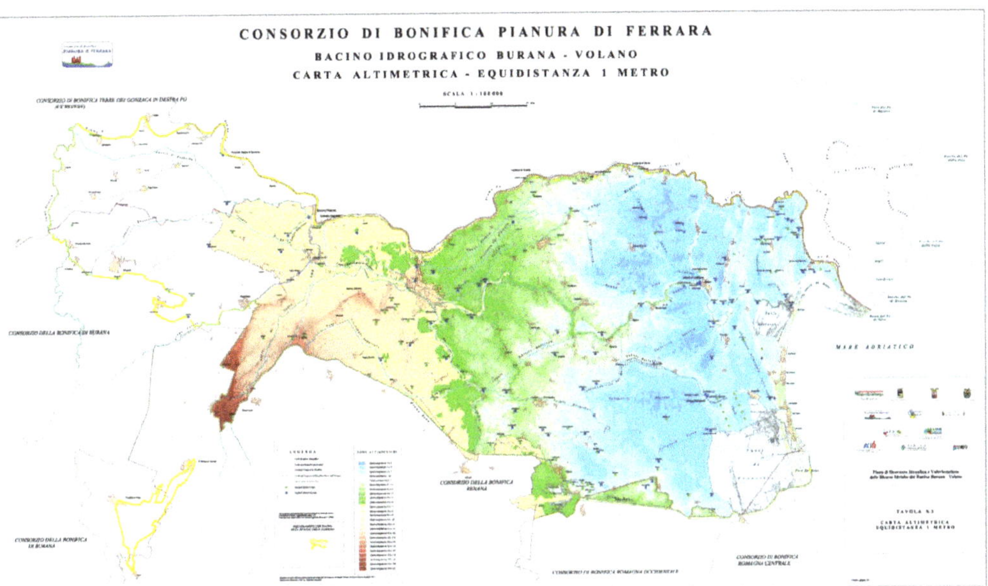

Figure 2. Ferrara territory altimetry (The map can be downloaded from https://www.bonificaferrara. it and has been released from "Consorzio di Bonifica Pianura di Ferrara"; accessed on 15 October 2021).

To evaluate the overall combined risk for the different municipalities in the test region, several intermediate steps were necessary. At first, the reliability of the method was tested on a smaller data sample given by the municipalities in the Province of Ferrara (Italy), then on a slightly larger one, considering municipalities from other provinces in the test region, and then, finally, expanding the data sample to each municipality of the Emilia Romagna region. This type of approach improved control on both the algorithm and its calibration, as well as the initial dataset, leading to a significant reduction in terms of computational time. In what follows, we omit the description of the intermediate steps and directly present the analysis for the whole test region.

2.1. Dataset

Choosing the correct amount of data is paramount. The data employed for our analysis have been obtained from the Italian National Institute of Statistics (ISTAT) database, which was used in 2018 by the Italian Superior Institute for Environmental Protection and Research (ISPRA) to produce seismic, hydrogeological, volcanic, and social vulnerability hazard maps for the entire Italian peninsula as shown in the report by Trigila et al. [22]. These maps constitute a fundamental tool of support to national risk mitigation policies, allowing the identification of intervention priorities, the allocation of funds, and the planning of soil protection interventions.

The input dataset was organized as a matrix in which the rows corresponded to each of the 331 municipalities of the Emilia-Romagna region and the columns corresponded to quantitative variables associated with different aspects of seismic and flood risk. Hence, we had 331 rows or observations and hundreds of columns or variables. For instance, we adopted as variables the number of buildings sharing certain features (such as building material, the period of construction, or the state of conservation), superficial extension, number of inhabitants, population density, seismic peak ground acceleration, etc. Overall, all the variables can be grouped into three macro-categories: variables related to vulnerability instances, variables related to exposure instances, and variables related to hazard instances for both seismic and hydraulic risks.

Since hydraulic risk, as a combination of hydraulic vulnerability, exposure, and hazard, has previously been evaluated for each observation by the Italian National Institute of Geophysics and Vulcanology (INGV), it was represented in the proposed analysis as a unique variable, which condensed all the variables related to hydraulic risk.

The relative importance between some variables and the relation among them is quantified by means of the PCA method, which will be described in the next subsections.

For instance, some of the crucial variables were identified as follows:

- agMAX_50: maximum value of the peak ground acceleration about the grid data point;
- DENSPOP: Population density (n. of inhabitants/kmq);
- E1-E31: Type of Buildings (e.g., residential, masonry, and state of conservation);
- IDR_AreaP1/P2/P3: Hydraulic risk surface, respectively, low/medium/high;
- IDR_PopP1/P2/P3: Population living in, respectively, low/medium/high hydraulic risk surface.

An extensive table reporting the explanations of all acronyms associated with the relevant variables is reported in Appendix A.

2.2. Initial Exploratory Analysis

Exploratory analysis is a typical analytical approach in statistics that is suitable for defining and synthesizing the main characteristics of a group of data. This type of approach enables preliminarily evaluating, searching, and finally, analyzing possible notable patterns within the data, in a phase where possible interactions among variables are not known yet. Again, graphics techniques for data visualization are quite useful in this step, producing diagrams such as box plots, scatter plots, histograms, etc. More analytical techniques, such as PCA, are very useful. The whole proposed analysis has been implemented and performed in a MATLAB computing environment [23].

2.2.1. Standardization

The first step of the exploratory analysis is data standardization. As usual [15,16], the metric of standard deviation was adopted to test the machine learning model's accuracy and to measure confidence in the obtained statistical conclusions. This allows us to compare variable data with different units of measure, scaling all the variables such that each scaled variable will have mean value equal to 0 and standard deviation equal to 1, referred to the data distribution for each variable. To attain this outcome, for each variable x of the dataset, mean μ and the standard deviation σ have been calculated. Then the z-score formula has been applied:

$$z = \frac{x - \mu}{\sigma}. \tag{1}$$

2.2.2. PCA

Once the entire dataset was standardized, PCA was applied. One of the main targets of PCA is to reduce the dimensionality of the initial dataset without losing the amount of information belonging to it. A dimensionality reduction technique is a process that takes advantage of linear algebraic operations to convert an n-dimensional dataset to an n-k dimensional one. Clearly, this transformation comes at the cost of a certain loss of information, but it also gives the benefit of being able to graphically visualize the data, while keeping good accuracy.

The idea behind PCA is to find the best subspace, which explicates the highest possible variance in the dataset. Using linear transformations, starting from an initial standardized matrix in the n-dimensional space, changes in variables are carried out that makes possible to identify observations in the space generated from the principal components, which have the particularity to catch the maximum possible variance of the initial dataset, thus reducing the loss of information.

Given p random standardized variables $\widetilde{X}_1, \widetilde{X}_2, \ldots, \widetilde{X}_p$, collected into the matrix $\widetilde{X}$, the analysis allows determining $k < p$ variables $Y_1, Y_2, \ldots, Y_k$, each of them a linear combination of the p starting variables, having maximum variance. To find Y_i, also known as the i-th principal component, we need to find the vector V_i such that

$$Y_i = \widetilde{X} V_i \tag{2}$$

by maximizing the variance relative to the first principal component. In other words, vectors V_i are the eigenvectors of the covariance matrix C of $\widetilde{X}$, i.e., the $n \times p$ matrix whose generic element C_{hk} is equal to $COV(\widetilde{X}_h, \widetilde{X}_k)$.

The j-th element of Y_i represents the *score* of the i-th principal component for j-th statistical unit. The j-th element of V_i represents the *weight* that the j-th variable $\widetilde{X}_j$ has in the definition of the i-th principal component. Vectors V_i can be collected as columns in the matrix of weights V.

Lastly, axis rotations are applied, which mean a change of position of the dimensions obtained during the factor's extraction phase, keeping the initial variance fixed as much as possible. The axis can be rigidly rotated (orthogonal rotation) or interrelated (oblique rotation). The result is a new matrix of rotated factors.

Once the dimension of the dataset has been reduced, it is possible to plot the observations in the new space generated by the principal components, space where the coordinates of the observations have undergone linear transformation, in accordance with the variables as mentioned before.

The scatter plot represented in Figure 3, depicts the observations after variable reduction. One can notice the presence of elements defined as outliers, i.e., abnormal values, far from the average observations. These disturbing elements could generate unbalanced compensations inside the analytical model, and that is why they will be handled with care, modifying the algorithm's settings whenever possible or, in extreme cases, removed from the dataset. In this case, the outliers were almost all the administrative centers of Emilia-Romagna region, far away, in terms of the quantitative variables, from the rest of the observations.

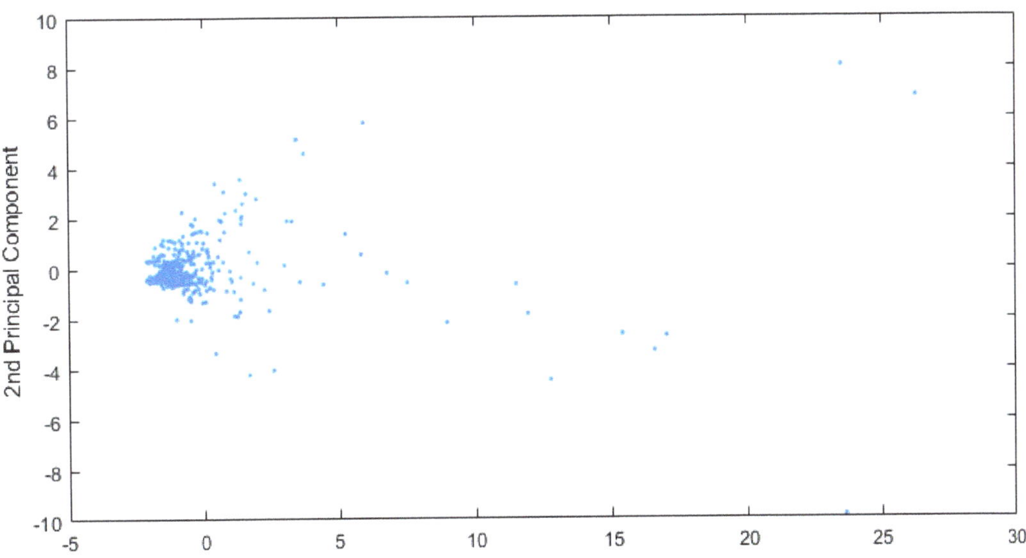

Figure 3. Observations scatter plot which depicts the observations after variable reduction.

It is a good rule to consider the principal components that catch at least 80% of the variance of the starting dataset. The more the considered variables, the higher the number of principal components necessary to reach that quote. Whenever the amount of variance reached is not sufficient, an additional reduction in variables is performed by iterating the process.

One of PCA's main purposes is to delete the noise due to non-useful data, which is evaluated in terms of how much information and how much variance they carry inside the dataset. Figure 4 represents variance for each principal component before variable reduction. Loading plots have been generated as histograms representing the weight of the variables transformed after the PCA and are reported in Figures 5 and 6. The variables reported along the abscissa have been selected among all the available data for being the most meaningful as per the multi-risk evaluation. For instance, AGMAX_50 denotes the maximum ground acceleration (fiftieth percentile) calculated on a grid with a 0.02° step, with the maximum and minimum of the values of the grid points falling within the municipal area. IDR_POPP3 indicates the resident population at risk in areas with high hydraulic hazard (P3). From Figures 5 and 6, the variables with the highest coefficients have been extrapolated, the higher the coefficient of the variable, the higher the weight of the variable on the principal component. Along the first principal component, the difference between observations will be led by the different values referred to the variables with highest coefficient in the histogram depicted in Figure 5.

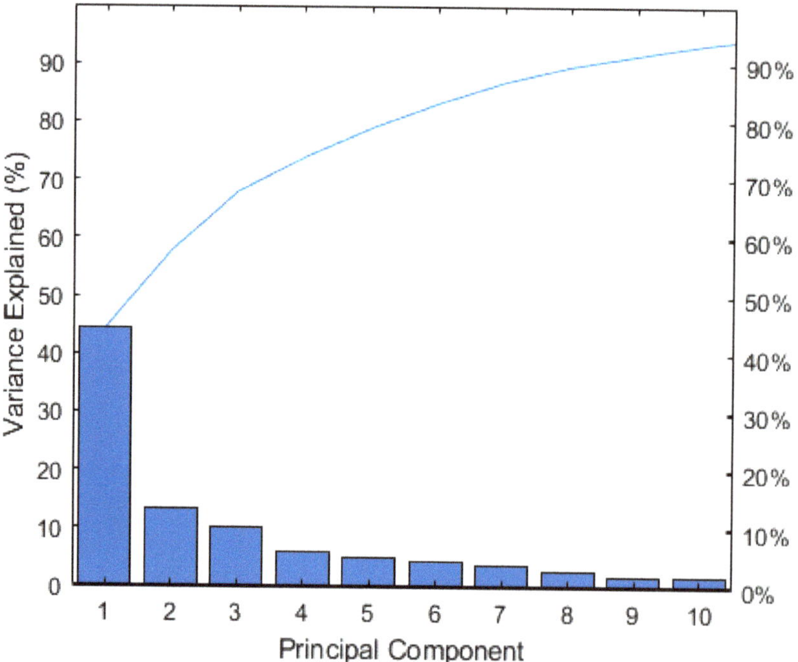

Figure 4. Variance for each principal component before variable reduction.

We chose to assess the weight of the coefficient of the variables referring to the first two principal components only, because they explicated more than 70% of the variance and are the most significant of the combined risk assessment. Figure 7 depicts the variance explicated by the first 10 principal components after the PCA.

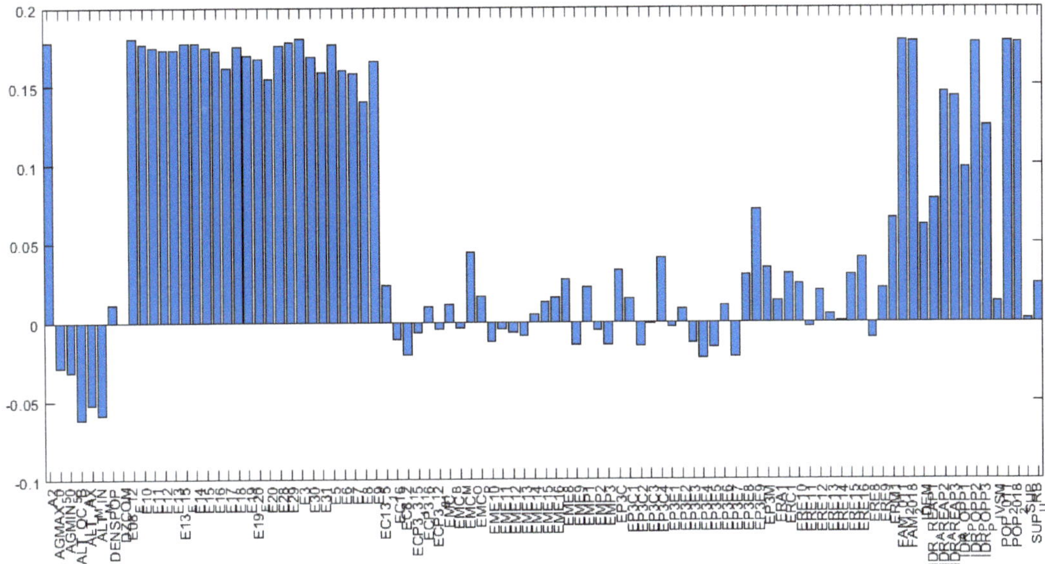

Figure 5. Loading plot of the variable coefficients along the first principal component (see Appendix A for an explanation of the acronyms).

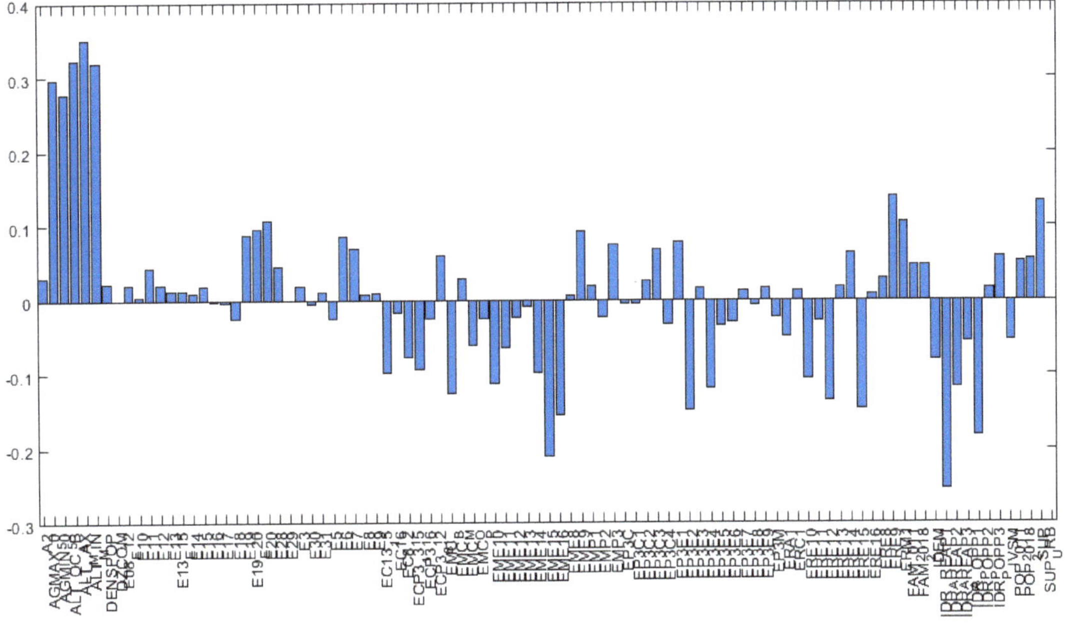

Figure 6. Loading plot of the variables coefficients along the second principal component (see Appendix A for an explanation of the acronyms).

Fundamental to the visualization of both observations and the relation between the variables is the biplot in Figure 8.

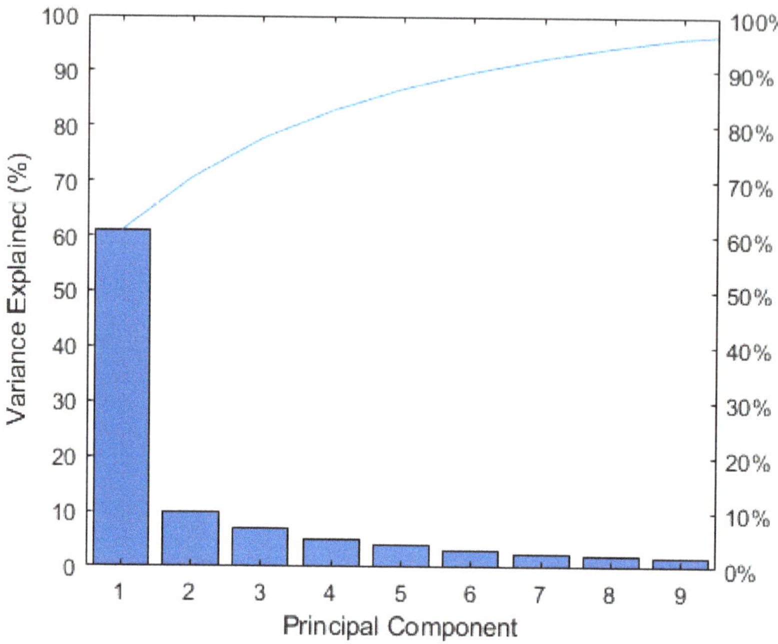

Figure 7. Variance of the first 10 principal components after variable reduction.

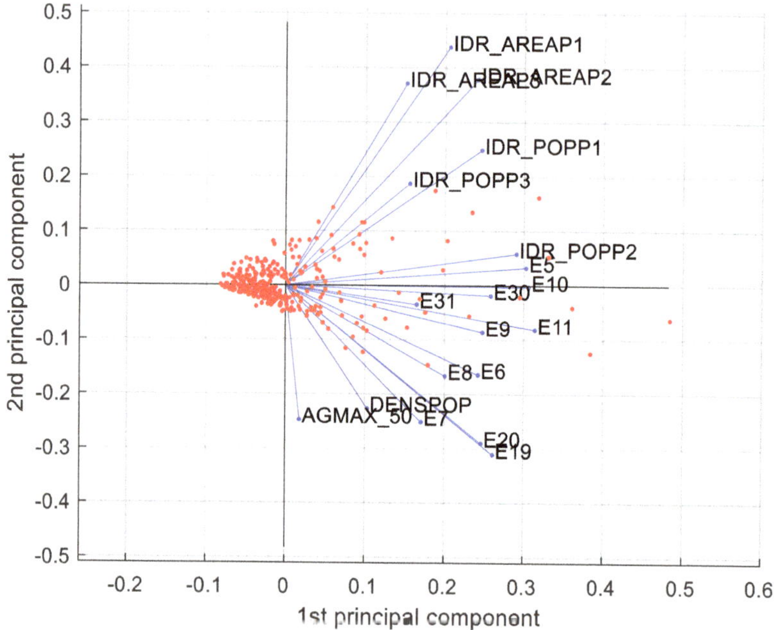

Figure 8. Biplot along the first two principal components.

This plot allows catching at an early stage any pattern within the dataset, such as the separation between observations and deep relation among variables. In general:

- the projection of the values on each principal component shows how much weight those values have on that principal component;
- when two vectors are close, in terms of angle, the two represented variables have a positive correlation;
- if two vectors create a 90 angle, the respective variables are not correlated;
- when they diverge and create an angle of almost 180, they are negatively correlated.

Outliers differ from the other observations in terms of vulnerability and the population at hydraulic risk. It is reasonable because, remembering the outliers are the provincial administrative centers, they present higher values in terms of population and built environment. Moreover, along the vertical axis the observations differ in terms of seismic hazard and exposition.

Moreover, vulnerability and exposure to hydraulic risk variables are quite correlated and differentiate the observations along the horizontal axis, whereas seismic hazard and exposition variables are not correlated with the variables representing surfaces at hydraulic risk. These remarks will come in handy later, at a post-clustering stage, a level of multi-risk will be attributed to each cluster.

2.3. K-Means Clustering Algorithm

The PCA allowed us to reduce the dimensionality of the dataset and plot the observations, i.e., the municipalities of the Emilia Romagna region, in the new sub-space identified by the principal components, while retaining the majority of information, which identified the observations in the initial n-dimensional space before the linear transformations.

To suitably group the observations according to homogeneous levels of overall risk, we used an unsupervised machine learning algorithm, known as *k-means clustering*.

In general, cluster analysis is a technique to group data where the main purpose is to gather observations according to the features selected by the user. The analysis allows splitting a set of observations into clusters according to similar or non-similar features. Cluster analysis does not require knowing the classes in advance, as in the case of supervised algorithms.

In the k-means clustering algorithm, we assumed N observations $x_1, x_2, \ldots, x_n$ and partitioned them into k clusters, each defined by a centroid $c_1, c_2, \ldots, c_k$. We assigned the x_i observation to the cluster, such that the distance among the observation and the cluster center was minimum.

The algorithm began by randomly choosing k centroids. After measuring the distance of each observation to each centroid, the observation was assigned to the closest cluster. Then, centroids were updated, as the average of the observations in each centroid. The procedure was repeated iteratively, each time minimizing the distance between observation and centroid.

Different choices for such distance function are possible and readily available in many scientific computing software packages such as MATLAB: the squared Euclidean distance, one minus the cosine of the included angle between points (treated as vectors), or one minus the sample correlation between points (treated as sequences of values).

In particular, the squared Euclidean metric does not allow keeping the outlier in the dataset because of the square of the distance. By doing so, the algorithm will place a specific cluster just for the outlier, influenced by its distance from the other observations. Later, we will propose a comparison among the distances in terms of the quality of clustering.

To legitimate the clusterization carried out with the k-means algorithm, the *silhouette method* was employed. The technique provided a succinct graphical representation of how well each observation has been classified. The silhouette value is a measure of how similar an object is to its own cluster (cohesion) compared to other clusters (separation). The silhouette ranges from -1 to $+1$, where a high value indicates that the object is well matched to its own cluster and poorly matched to neighboring clusters. If most objects have a high value, then the clustering configuration is appropriate. If many points have a low or negative value, then the clustering configuration may have too many or too few clusters. The silhouette can be calculated with any distance metric, such as the Euclidean distance or the so-called Manhattan distance.

To decide which metrics to adopt, a comparison based on the silhouette of each method was performed (see Figure 9). Correlation metrics appear to be the most reliable, whereas the squared Euclidean would be as good if it were not for the outliers.

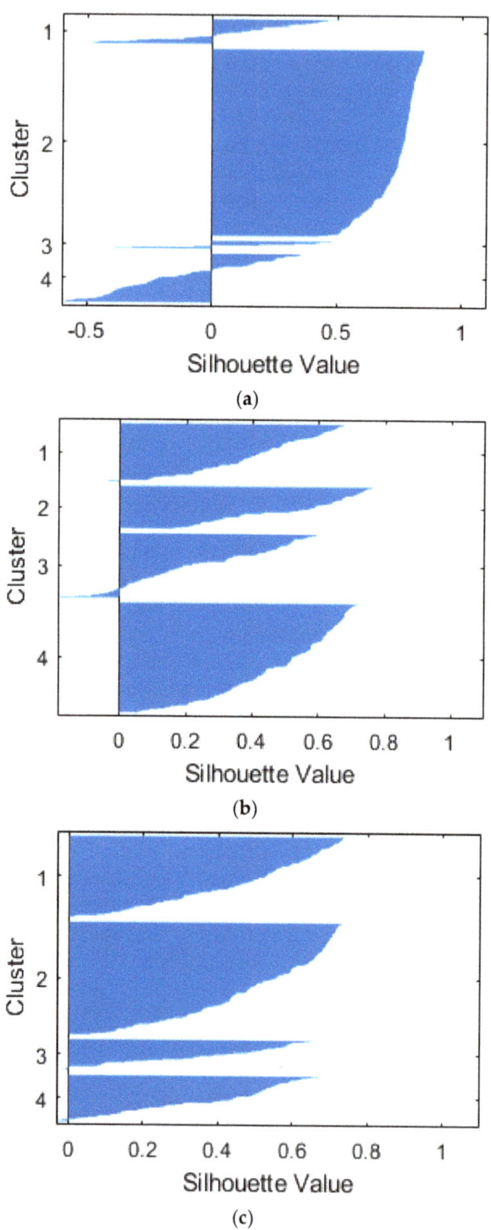

Figure 9. Silhouette for different metrics: squared Eculidean (**a**), cosine (**b**), and correlation (**c**).

A four clusters grouping was chosen for the proposed analysis. Figure 10 represents the final clusterization of Emilia-Romagna municipalities. Cluster evaluation was conducted considering the weight and the distribution of the variables. All the outliers belonged to cluster 4, which was developed both on the horizontal axis, led by seismic vulnerability and hydraulic risk variables, and slightly on the vertical one, led by seismic hazard and by hydraulic risk variables. The great majority of the municipalities presented similar quantitative values of variables, in particular, those belonging to clusters 2 and 3. Silhouette values relative to this clusterization were good, reinforcing the reliability of the method proposed.

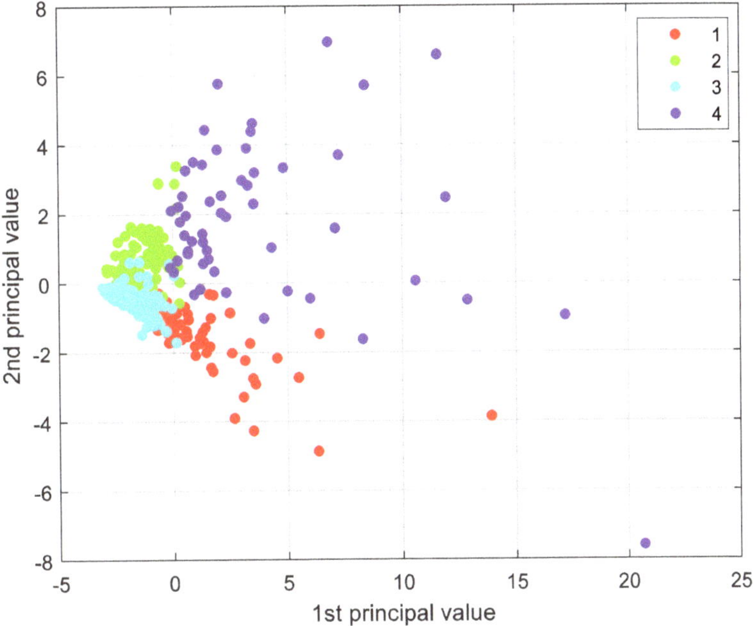

Figure 10. Grouping of the Emilia Romagna municipalities into four clusters.

3. Results

In this section, we show how to assign to each observation and, more generally, to each cluster, a label which identifies the associated level of overall risk.

3.1. Variables Label Assignment

First, we set intervals in an objective way, in order to suitably define labels for the variables. To this aim, we set interval extremals in correspondence of quartile percentages Q1, Q2, and Q3 as indicated in Table 1.

Table 1. Labels and intervals for cluster definition.

Intervals	Label
first element: Q1	Low
Q1: Q2	Medium-to-low
Q2: Q3	Medium-to-high
Q3: last element	High

The chosen labels referred, respectively, to the presence of low, medium-to-low, medium-to-high and high amounts regarding that specific variable. Such subdivision was allowed because the variables were quantitative types and sorted by normal distribution. Furthermore, sorting out variables, the information within them was unaffected.

We analyzed the variable with the greater value from the previous analysis, as a component that defined a risk, with the risk as the combined result of three factors, hazard, exposure, and vulnerability. We illustrate how to assign a label to each cluster for each variable considered, among the most relevant ones.

We first considered the variable $a_{g,max}$, i.e., the peak ground acceleration for the site, with a return period of 475 years. The first step was the extrapolation of observables in the initial dataset. Subsequently, we associated each observation with the respective cluster indexes and the respective values of $a_{g,max}$. Then, we rearranged the observables in ascending order of $a_{g,max}$, and defined the quartile as the extreme point of the interval. The cluster composition in terms of $a_{g,max}$ is reported in Table 2, together with the resulting assigned labels.

Table 2. Quartile distribution of the $a_{g,max}$ variable in the four clusters.

	%Q1	%Q2	%Q3	%Q4	Label
CL1	74	20	6	0	Low
CL2	0	25	25	50	Medium-to-high
CL3	3	18	68	11	Medium-to-low
CL4	7	16	13	64	High

The labels were assigned based on the percentage prevalence of the cluster for each quartile. A prevalence allocated in the fourth quartile for one of the clusters indicated that the selected cluster gathered the most dangerous municipalities in terms of $a_{g,max}$ On the other hand, a prevalence in the first quartile indicated that the cluster gathered the less dangerous municipalities in terms of seismic hazard.

The same operation was carried out for the hydraulic risk component IDR_POPP2, the prevailing seismic vulnerability variable, i.e., the percentage of buildings under poor maintenance conditions E_30, and the main exposure variable, i.e., density population DENS_POP (see Tables 3–5).

Table 3. Quartile distribution of the IDR_POPP2 variable in the four clusters.

	%Q1	%Q2	%Q3	%Q4	Label
CL1	17	20	54	9	Medium-to-low
CL2	46	41	12	1	Low
CL3	3	5	18	74	Medium-to-high
CL4	4	5	9	82	High

Table 4. Quartile distribution of the E_30 variable in the four clusters.

	%Q1	%Q2	%Q3	%Q4	Label
CL1	41	27	22	9	Low
CL2	25	30	29	15	Medium-to-low
CL3	18	29	37	16	Medium to high
CL4	0	4	11	86	High

Table 5. Quartile distribution of the $a_{g,max}$ variable in the four clusters.

	%Q1	%Q2	%Q3	%Q4	Label
CL1	16	29	40	14	Medium-to-low
CL2	46	28	19	7	Low
CL3	0	0	3	97	High
CL4	5	25	27	43	Medium-to-high

3.2. Overall Risk Definition

Once the variables were rearranged, the incidence of clusters for each variable were calculated and a label for each variable and cluster was assigned (based on the distribution of the cluster indexes within the variable); each cluster was assigned an overall risk label based on their score for each rearranged variable (Table 6).

Table 6. Overall risk quantification for each cluster of municipalities.

	Hydraulic Risk	Seismic Exposition	Seismic Vulnerability	Seismic Hazard	Label
CL1	Medium-to-low	Low	Medium-to-low	Low	Low
CL2	Low	Medium-to-low	Low	Medium-to-high	Low-to-medium
CL3	Medium-to-high	Medium-to-high	High	Medium-to-low	Medium-to-high
CL4	High	High	Medium-to high	High	High

The significance of the assigned risk labels was strictly dependent on the starting population, i.e., from the region under study and do not have absolute value.

This means that the obtained labels cannot be extrapolated to a larger scale without losing their significance. As shown in Figure 11, it is also possible to represent the population of each risk cluster by the main administrative province in the Emilia Romagna region. Obviously, frequency values for each province depend on the number of municipalities, which constitute each province. Therefore, this plot allows analyzing risk clusters from the same province, but comparing clusters from different provinces may be inappropriate. It is worth noting that the proposed methodology has recognized Piacenza as the province with most low-risk municipalities, while the main cluster featuring Parma, Modena, Bologna, Forlì-Cesena, and Rimini is the low-to-medium risk cluster. Most municipalities of the Reggio-Emilia province are associated with low and low-to-medium clusters. Finally, each of the provinces of Ferrara and Ravenna result being equally split in two main clusters, namely the low and the high-risk clusters in the former case, and the low-to-medium and high-risk clusters in the latter case.

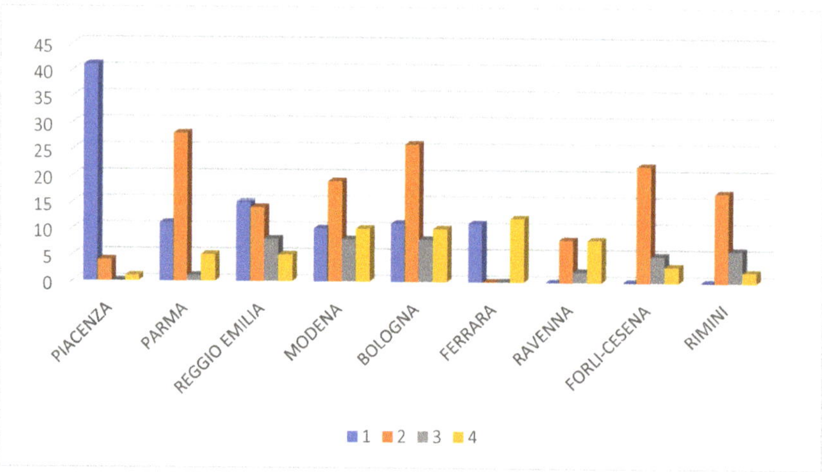

Figure 11. Population of each risk cluster by province in the Emilia Romagna region. In the *y*-axis, the number of municipalities has been reported.

4. Discussion

Ensuring ethical, inclusive, and unbiased machine learning tools is one of the new epistemic frontiers in the application of artificial intelligence technologies to disaster risk management. We recall that this paper discusses an individual application of machine learning tools to a multi-risk assessment of a Northern Italy case study. For this purpose, we had at our disposal a massive amount of data from the ISTAT database containing indicators and data on seismic, hydrogeological, and volcanic risk as well as demographic, housing, territorial and geographical information, obtained through the integration of various institutional sources such as Istat, INGV, ISPRA, Italian Ministry for Cultural Heritage. Like all big data technologies, the adopted machine learning model proved effective in reducing CPU time and model-development costs, owing to its ability to process quantities and sources of data that could not have been otherwise simply elaborated [24,25]. We expect that the model can be used to devise mitigation measures, prepare emergency response, and plan flood recovery measures. The proposed tool has, indeed, the potential for being an operational instrument for land use managers and planners. However, misuse should be avoided, and, for this purpose, crucial issues such as applicability, bias, and ethics should be carefully considered [24–26]. The ethical issues pertaining to a possible misuse of Artificial Intelligence technologies are several [25], including the loss of human decision making, the potential for criminal and malicious use, the emergence of problems of control and use of data and systems, the dependence of the outcomes on users' bias, and the possible prioritization of the "wrong" problems with respect to stakeholder expectations.

Prioritization in disaster multi-risk management, additionally, is markedly affected by needs and expectations of private users, public agencies, and final stakeholders. For instance, a water level management company will be expectedly more inclined to consider flood risk as the most important risk to cope with, while any public agency that is called to reduce the seismic vulnerability of a certain region will tend to consider seismic risk as a priority. Thus, the labeling of the clusterization will be intrinsically permeated with the end-user's intentions. A further aspect is that one should understand that publicizing the results of a multi-risk algorithm might inadvertently touch sensitive aspects from a privacy point of view [27].

In many cases, criticalities rely upon an inherent disconnect between the algorithm's designers and the communities where the research is conducted [26], while users may complain about a lack of transparency and accountability. Furthermore, immature machine learning tools might be used in safety-critical situations for which they are not yet ready.

As suggested by Gevaert et al. [26], disaster-risk-management specialists constantly seek expertise on how to clearly communicate the results and uncertainties of machine learning algorithms to reduce inflated expectations. Furthermore, sensitive groups should be identified and audited for overcoming bias. Therefore, we suggest that, before being systematically applied, the present machine learning methodology is validated against established computational modeling tools. We also believe that the obtained results are very promising, but further efforts are necessary to assess the proneness of the proposed machine learning tool to the aforementioned ethical and bias issues.

5. Conclusions

The purpose of this work is to illustrate a sound methodology for the qualitative multi-risk analysis at the regional scale by means of machine learning techniques that allow dealing with large and heterogeneous amounts of data. The initial dataset, made of variables carrying information about hazard, exposure, and vulnerability for both seismic and hydraulic risk for each municipality of the Emilia Romagna region, has been suitably normalized and reduced through the PCA, whereas observations have been clustered through a machine-learning algorithm.

Then, risk labels were individually assigned to clusters for each variable. Finally, based on the score of each variable an overall risk label was assigned to each cluster. Results confirmed previous risk classifications for the case study analyzed. Both provinces with a moderate risk level and high-risk level have been correctly detected by the proposed approach. The reliability of the obtained results is dependent on the existence of valid quantitative initial data for the region under study. In fact, the proposed methodology does not allow qualitative data, whether they are fundamental or not.

In conclusion, the proposed analysis delivers useful information: municipalities with major priority of intervention are identified so that stakeholders can take advantage of this tool to prioritize any preventive measures. Moreover, the procedure also allows identifying the most important variables to consider in a combined seismic and hydraulic multi-risk analysis. In other words, this tool allows evaluating the variables most suited to categorize the observations in terms of combined risk. Indeed, from the analysis, variables have emerged relative to different types of risks, which better communicate with each other and carry most information. By contrast, the methodology also allows identifying variables, which do not collaborate with variables of different nature and, therefore, cannot be usefully employed.

Author Contributions: Conceptualization, A.R., M.N., A.C., and E.B.; methodology, A.R., M.N., A.C., F.R., and E.B.; software, A.R., M.N., L.M., A.G., and F.R.; validation, A.R., A.C., M.N., F.R., L.M., A.G., and Z.N.; formal analysis A.R., A.C., M.N.; resources, E.B., and Z.N.; data curation, A.R., M.N., L.M., and A.G.; writing—original draft preparation, A.R., A.C., E.B., and F.R.; writing—review and editing, A.C., F.R., Z.N., and E.B.; visualization, A.R., M.N., and F.R.; supervision, A.C., M.N., Z.N., and E.B.; project administration, E.B., and Z.N.; funding acquisition, E.B., and Z.N. All authors have read and agreed to the published version of the manuscript.

Funding: This research was funded by: (1) the EUROPEAN UNION, Programme Interreg Italy-Croatia, Project "Preventing, managing and overcoming natural-hazards risks to mitigate economic and social impact"—PMO-GATE ID 10046122.

Institutional Review Board Statement: Not applicable.

Informed Consent Statement: Not applicable.

Data Availability Statement: The Istat dataset used for the simulations is public and download-able from the web page https://www4.istat.it/it/mappa-rischi/documentazione (accessed on 15 October 2021).

Acknowledgments: The support of Eng. Alessandro Bondesan from Consorzio di Bonifica Pianura di Ferrara is gratefully acknowledged.

Conflicts of Interest: The authors declare no conflict of interest.

Appendix A

We provide hereafter a table with the acronyms of the variables used for Figures 5–7:

Table A1. Description of the variables used in Figures 5–7.

DENSPOP	Population Density
AGMAX_50	Maximum ground acceleration (50th percentile) calculated on a grid with a 0.02° step, maximum (MAX) and minimum (MIN) of the values of the grid points falling within the municipal area.
IDR_POPP3	Resident population at risk in areas with high hydraulic hazard-P3
IDR_POPP2	Resident population at risk in areas with medium hydraulic hazard-P2
IDR_POPP	Resident population at risk in areas with low hydraulic hazard-P1
IDR_AREAP1	Areas with low hydraulic hazard P1 (low probability of floods or extreme event scenarios))–D.Lgs. 49/2010 (km^2)
IDR_AREAP2	Areas with average hydraulic hazard P2 (return time between 100 and 200 years)–D.Lgs. 49/2010 (km^2)
IDR_AREAP3	Areas with high hydraulic hazard P3 (return time between 20 and 50 years)-D.Lgs. 49/2010 (km^2)
E5	Residential buildings in load-bearing masonry
E6	Residential buildings in load-bearing reinforced concrete
E7	Residential buildings in other load-bearing materials (steel, wood, …)
E8	Residential buildings made before 1919
E9	Residential buildings made between 1919 and 1945
E10	Residential buildings made between 1946 and 1960
E11	Residential buildings made between 1961 and 1970
'E19	Residential buildings with three floors
E20	Residential buildings with more than three floors
E30	Residential buildings with a poor state of conservation
'E31	Residential buildings with a very poor state of conservation

References

1. Fuchs, S.; Keiler, M.; Zischg, A. A spatiotemporal multi-hazard exposure assessment based on property data. *Nat. Hazards Earth Syst. Sci.* **2015**, *15*, 2127–2142. [CrossRef]
2. Kron, W. Reasons for the increase in natural catastrophes: The development of exposed areas. In *Topics 2000: Natural Catastrophes, the Current Position*; Munich Reinsurance Company: Munich, Germany, 1999; pp. 82–94.
3. Zschau, J. Where are we with multihazards, multirisks assessment capacities? In *Science for Disaster Risk Management: Knowing Better and Losing Less*; Poljansek, K., Marin Ferrer, M., De Groeve, T., Clark, I., Eds.; European Union: Luxembourg, 2017; pp. 98–115.
4. Barthel, F.; Neumayer, E. A trend analysis of normalized insured damage from natural disasters. *Clim. Chang.* **2012**, *113*, 215–237. [CrossRef]
5. Munich, R.E.; Kron, W.; Schuck, A. Analyses, assessments, positions. In *Topics Geo: Natural Catastrophes*; Münchener Rückversicherungs-Gesellschaft: Munich, Germany, 2014.
6. Peduzzi, P.; Dao, H.; Herold, C.; Mouton, F. Assessing global exposure and vulnerability towards natural hazards: The Disaster Risk Index. *Nat. Hazards Earth Syst. Sci.* **2009**, *9*, 1149–1159. [CrossRef]
7. Bell, R.; Glade, T. Multi-hazard analysis in natural risk assessments. *WIT Trans. Ecol. Environ.* **2004**, *77*, 1–10.
8. Barredo, J.I. Major flood disasters in Europe: 1950–2005. *Nat. Hazards* **2007**, *42*, 125–148. [CrossRef]
9. Kanamori, H.; Hauksson, E.; Heaton, T. Real-time seismology and earthquake hazard mitigation. *Nature* **1997**, *390*, 461–464. [CrossRef]
10. Kappes, M.S.; Keiler, M.; von Elverfeldt, K.; Glade, T. Challenges of analyzing multi-hazard risk: A review. *Nat. Hazards* **2012**, *64*, 1925–1958. [CrossRef]
11. Schmidt, J.; Matcham, I.; Reese, S.; King, A.; Bell, R.; Henderson, R.; Smart, G.; Cousins, J.; Smith, W.; Heron, D. Quantitative multi-risk analysis for natural hazards: A framework for multi-risk modelling. *Nat. Hazards* **2011**, *58*, 1169–1192. [CrossRef]

12. Gruber, F.E.; Mergili, M. Regional-scale analysis of high-mountain multi-hazard and risk indicators in the Pamir (Tajikistan) with GRASS GIS. *Nat. Hazards Earth Syst. Sci.* **2013**, *13*, 2779–2796. [CrossRef]

13. Tyagunov, S.; Vorogushyn, S.; Jimenez, C.M.; Parolai, S.; Fleming, K. Multi-hazard fragility analysis for fluvial dikes in earthquake- and flood-prone areas. *Nat. Hazard. Earth Syst. Sci.* **2018**, *18*, 2345–2354. [CrossRef]

14. Yousefi, S.; Pourghasemi, H.R.; Emami, S.N.; Pouyan, S.; Eskandari, S.; Tiefenbacher, J.P. A machine learning framework for multi-hazards modeling and mapping in a mountainous area. *Sci. Rep.* **2020**, *10*, 12144. [CrossRef] [PubMed]

15. Boniolo, F.; Dorigatti, E.; Ohnmacht, A.J.; Saur, D.; Schubert, B.; Menden, M.P. Artificial intelligence in early drug discovery enabling precision medicine. *Expert Opin. Drug Discov.* **2021**, *16*, 991–1007. [CrossRef]

16. Pouyan, S.; Pourghasemi, H.R.; Bordbar, M.; Rahmanian, S.; Clague, J.J. A multi-hazard map-based flooding, gully erosion, forest fires, and earthquakes in Iran. *Sci. Rep.* **2021**, *11*, 14889. [CrossRef] [PubMed]

17. Pearson, K. On Lines and Planes of Closest Fit to Systems of Points in Space. *Phil. Mag.* **1901**, *2*, 559–572. [CrossRef]

18. Hotelling, H. Analysis of a complex of statistical variables into principal components. *J. Edu. Psychol.* **1933**, *24*, 417–441. [CrossRef]

19. Jolliffe, I.T. *Principal Component Analysis*; Springer: New York, NY, USA, 2002.

20. MacQueen, J.B. Some methods for classification and analysis of multivariate observations. In *Proceedings of the 5th Berkeley Symposium on Mathematical Statistics and Probability, Berkeley, CA, USA, 1967*; University of California Press: Berkeley, CA, USA, 1967; pp. 281–297.

21. Ding, C.; Xieofeng, H. K-means clustering via principal components analysis. In *Proceedings of the 21st International Conference on Machine Learning, Banff, AB, Canada, 4–8 July 2004*; ACM Press: New York, NY, USA, 2004.

22. Trigila, A.; Iadanza, C.; Bussettini, M.; Lastoria, B. *Dissesto Idrogeologico in Italia: Pericolosità e Indicatori di Rischio—Edizione 2018*; Rapporti 287/2018; ISPRA: Roma, Italy, 2018.

23. Mantovan, L.; Gilli, A. Supplementary Material, Open Access. Available online: https://github.com/alessandrogilli/analisi-multirischio (accessed on 15 October 2021).

24. Wagenaar, D.; Curran, A.; Balbi, M.; Bhardwaj, A.; Soden, R.; Hartato, E.; Sarica, G.M.; Ruangpan, L.; Molinario, G.; Lallemant, D. Invited perspectives: How machine learning will change flood risk and impact assessment. *Nat. Hazards Earth Syst. Sci.* **2020**, *20*, 1149–1161. [CrossRef]

25. Stahl, B.C. Ethical Issues of AI. Artificial Intelligence for a Better Future: An Ecosystem Perspective on the Ethics of AI and Emerging Digital Technologies. In *SpringerBriefs in Research and Innovation Governance*; Springer: Berlin/Heidelberg, Germany, 2021; pp. 35–53.

26. Gevaert, C.M.; Carman, M.; Rosman, B.; Georgiadou, Y.; Soden, R. Fairness and accountability of AI in disaster risk management: Opportunities and challenges. *Patterns* **2021**, *2*, 100363. [CrossRef] [PubMed]

27. GFDRR. *Machine Learning for Disaster Risk Management*; GFDRR: Washington, WA, USA, 2018.

Article

Population Bias on Tornado Reports in Europe

Răzvan Pîrloagă [1,2,*], Dragoş Ene [1] and Bogdan Antonescu [1]

[1] Remote Sensing Department, National Institute of Research and Development for Optoelectronics INOE 2000, Str. Atomiştilor 409, 077125 Măgurele, Romania; dragos.ene@inoe.ro (D.E.); bogdan.antonescu@inoe.ro (B.A.)

[2] Faculty of Physics, University of Bucharest, 077125 Măgurele, Romania

* Correspondence: razvan.pirloaga@inoe.ro

Abstract: Tornadoes are associated with damages, injuries, and even fatalities in Europe. Knowing the spatial distribution of tornadoes is essential for developing disaster risk reduction strategies. Unfortunately, there is a population bias on tornado reporting in Europe. To account for this bias, a Bayesian modeling approach was used based on tornado observations and population density for relatively small regions of Europe. The results indicated that the number of tornadoes could be 53% higher that are currently reported. The largest adjustments produced by the model are for Northern Europe and parts of the Mediterranean regions.

Keywords: tornadoes; climatology; Bayesian

Citation: Pîrloagă, R.; Ene, D.; Antonescu, B. Population Bias on Tornado Reports in Europe. *Appl. Sci.* **2021**, *11*, 11485. https://doi.org/10.3390/app112311485

Academic Editor: Ben J. Anthony

Received: 2 November 2021
Accepted: 2 December 2021
Published: 3 December 2021

Publisher's Note: MDPI stays neutral with regard to jurisdictional claims in published maps and institutional affiliations.

1. Introduction

Tornadoes in Europe can be associated with severe damage and can result in injuries and fatalities. Despite this, until recently their threat has been underestimated. Antonescu et al. [1] showed that European tornadoes reported between 1995–2015 resulted in 4462 injuries and 316 fatalities and damages estimated at more than €1 billion. They also indicated that the density of tornado reports was the highest over Belgium, Germany, the Netherlands, and the southeastern United Kingdom . For these countries, there is a reporting bias, as central and western Europe have a high population density compared with other regions in Europe (e.g., Eastern Europe). Coastal areas were also hot-spots for tornado reports (e.g., western Italy, eastern Spain), as these areas tend to have higher population density compared with inland areas and also because of the waterspouts that move from the Mediterranean Sea inland [2].

This difference in the population density (i.e., different regions of Europe, coastal area versus inland area) introduces a bias in the reporting of tornadoes. This is because tornadoes (and also other types of severe weather events like hail or extreme winds) are "targets of opportunity" [3]. Thus, an observer needs to witness the event and then to report it and systems need to exist for collecting and verifying the reports (i.e., tornado database). Very few countries in Europe have developed and maintained such databases, which resulted in a lack of information about tornadoes [2]. One reason for not developing tornado databases is that they do not seem justified. Compared with the United States, the impact of tornadoes in European countries (given their relative small area) is relatively low, and thus there is no need to develop tornado databases for individual countries [4]. Only when considered from a pan-European perspective does the impact of tornadoes in Europe start to emerge. The collection of tornado reports and other types of severe weather reports at the pan-European level started in 2006 with the development of the European Severe Weather Database (ESWD) by the European Severe Storms Laboratory [5]. Currently, ESWD contains more than 16,500 tornado reports collected between 1800–2020.

Given the differences in population density across Europe, the real number of tornadoes is an unknown quantity. Even in relatively high populated areas, tornadoes might not be reported because of their small spatial extent and short life time, obstruction of the observer view point (e.g., forest, hills, buildings) or if the tornado occurred during the nighttime. As indicate in previous studies for the United States [6] and Canada [7], the population density is the key factor, besides the meteorological factors, in determining the bias in tornado reports.

Several studies have addressed the issue of population bias on tornado reports. Anderson et al. [6] used a hierarchical Bayesian model to account for the population bias on tornado occurrence using historical tornado reports for the United Stated from the Storm Prediction Center between 1953–2001. Their results for the central and eastern United States indicated that F0–F1 tornado reports vary less with population density compared with F2–F5 tornadoes. Starting from the hypothesis that the number of tornado reports in Canada is significantly lower than the actual number of tornadoes, Cheng et al. [7] also used a Bayesian modeling approach that considered the population bias on tornado reports. Their model also included the occurrence of cloud-to-ground lightning, as cloud-to-ground lightning can be used to quantify convective storms activity and thus can be used as precursor for tornadoes. Their results showed that in areas with low population density, the probability of tornado occurrence is significant higher compared with the observed tornado climatology for Canada. More recently, Potvin et al. [8] developed a Bayesian hierarchical modeling framework for correcting the reporting bias in the United States tornado database. Compared with other covariates (e.g., distance from the nearest city, terrain ruggedness index, road density) population density explained more of the variance in the number of reported tornadoes. Their model indicated that approximately 45% of the tornadoes that occurred in the study domain were reported.

The aim of this article is to analyze the effects of population bias on tornado reports in Europe using a Bayesian modeling approach. The expected tornado counts over Europe can be used to better understand the societal and economic impact of tornadoes and to be included in national disaster risk reduction strategies. This article is structured as follows. Section 2 details the tornado and the population density datasets. Section 3 describes the Bayesian modeling approach. The results and discussions are presented in Sections 4 and 5, respectively. Finally, Section 6 summarizes the results.

2. Tornado and Population Datasets

Tornado reports were obtained from the European Severe Weather Database. The ESWD collects information on severe storms over Europe (i.e., tornadoes, severe wind, large hail, heavy rain, heavy snowfall, damaging lightning) using a citizen-science approach [9] and through collaborations with national weather services and volunteer severe weather spotter networks. Before the inclusion in the ESWD, each report is verified and receives a quality control level. In this article, tornado reports with a quality control level Q0+ (i.e., validated with meteorological data such as radar and/or satellite imagery) have been used. Unlike the United States tornado database that contains only reports for tornadoes, the ESWD also contains reports for waterspouts [2]. The ESWD include information about the surface type (e.g., land, forest, sea, lake) over which tornadoes have been observed and the surface types crossed during the event. Thus, all the waterspouts that moved inland were included in the analyses presented in this article.

Data on population density in Europe were obtained from Eurostat [10]. The data were extracted for NUTS3 regions. NUTS (Nomenclature of territorial units for statistics) classification is a system for dividing the economic territory of European Union and the United Kingdom. The NUTS 2021 [11] classification valid from 1 January 2021 contains 1166 regions at NUTS3 level. NUTS3 level represent small regions for specific diagnoses. For example, for Romania there are 42 NUTS3 regions (41 counties and Bucharest, the capital city). In previous studies, for the United States, was argued that the rural population density at the county level is a more appropriate measure for tornado reporting compare with total population density [12]. Here, we follow [6] and use the total population density as population tends to be distributed over much of the counties and not concentrated in isolated towns. Based on the data availability and overlap with the tornado dataset, for each NUTS3 region the population density was average for the period 2006–2019 and the area for the period 2006–2015. These data together with the number of tornadoes reported between 2006–2020 at NUTS3 level were included in the Bayesian model.

3. Hierarchical Bayesian Model

3.1. Model

The Bayesian model used in this article starts with the hypothesis that the population density is the main influence on tornado reporting and thus, that tornadoes are underreported in Europe. As indicated by [6,7], the occurrence of tornadoes can be described as a series of conditional models linked using the Bayes' rule. Considering the t_n as number of observed tornadoes and T_n as the true number of tornadoes ($T_n \geq t_n$), a binomial model can specified in which

$$t_n | T_n, p_n(\beta) \sim Binomial[T_n, p_n(\beta)] \tag{1}$$

where $p_n(\beta)$ represents the probability to observe a tornado and n indicates the NUTS3 region. The probability to observe a tornado is a function of β, which is related to population density (x_n). Anderson et al. (2007) used an exponential model for p_n assuming that the probability of detection increases with population density

$$p_n(\beta) = exp(-\beta/x_n) \tag{2}$$

The true number of tornadoes (T_n) in the n NUTS3 region is modeled as Poisson process, which is conditioned on the climatological frequency λ

$$T_n | \lambda \sim Poisson(\lambda a_n) \tag{3}$$

where a_n is the area of the NUTS3 region and λ (i.e., Poisson intensity) is a measure the tornado frequency per unit area [6].

3.2. Estimation

For the Bayesian approach, the prior distribution for β and λ need to be specified. These prior distributions are non-informative (i.e., large variance) because there is no prior knowledge that can inform the distributions. Thus, the population parameter β is specified as

$$exp(\beta) \sim N\left(\mu_\beta, \sigma_\beta^2\right) \tag{4}$$

The distribution of $exp(\beta)$ is a normal distribution characterized by the mean μ_β and variance σ_β^2. In the model developed in [6], μ_β was set to 0.5 and σ_β^2 to 10,000. For the climatological frequency parameter λ, a prior gamma distribution was used

$$\lambda \sim gamma(q, r) \tag{5}$$

Following [6], the shape parameter q was set to 0.001 and the scale parameter r to 0.001 corresponding to a prior mean of 1 and a prior variance of 1000 (non-informative). Using the Bayes' rule

$$\underbrace{p(\beta, \lambda, T_1, ..., T_N | t_1, ..., t_N)}_{\text{Posterior model}} \propto \prod_{n=1}^{N} \underbrace{p(t_n | T_n, \beta)}_{\text{Data model (Equation (1))}} \times$$
$$\underbrace{p(T_n | \lambda)}_{\text{Explanatory model (Equation (3))}} \times \underbrace{p(\beta)[(\lambda)}_{\text{Parameter model (Equations (4) and (5))}} \tag{6}$$

A Markov Chain Monte Carlo analysis of the Bayesian model was applied to obtain a sequence of realizations from the posterior model ([6,7]) using the WinBUGS software [13]

(Available online at https://www.mrc-bsu.cam.ac.uk/software/bugs/the-bugs-project-winbugs/, accessed on 23 October 2021).

4. Results

Based of the data from ESWD a total number of 2319 tornadoes were observed in Europe between 2006–2020 over the NUTS3 regions considered in this article. The predicted number of tornadoes by the model considering the population bias is 3563 tornadoes (Table 1). Thus, 65% of tornadoes predicted by the model were reported.

Table 1. Posterior parameters and model predictions.

	Mean	Standard Deviation
β	14.91	1.050
λ	0.0006008	0.00001669
	Total (15 yr)	**Mean (yr^{-1})**
$\sum T_n$	3563	237.5
$\sum t_n$	2319	154.6

The difference between predicted and observed number of tornadoes is low (<0.027 tornadoes 10,000 km^{-2} yr^{-1}) over parts of the United Kingdom, the Netherlands, Belgium, Germany, and Italy (Figure 1). As indicated previously [1,14], these are regions characterized by both a high number of tornado reports and high population density. For some NUTS3 regions there were no tornadoes reported during the study period, for example, parts of north-western and southeastern France, central Romania, central Italy, Albania, Northern Macedonia, southern Bulgaria, Finland. For these regions the predicting values of tornado counts are less then 2 tornadoes over the 15 years study period. The highest difference between predicted and observed tornadoes (between 0.33–0.41 tornadoes 10,000 km^{-2} yr^{-1}) is over Iceland and northern parts of the United Kingdom, Norway, Sweden, and Finland (Figure 1). This is not surprising given that the population density of these regions is low compared with other regions of Europe and also given that the number of observed tornadoes by NUTS3 regions in this area is less than 10 tornadoes over the entire study period.

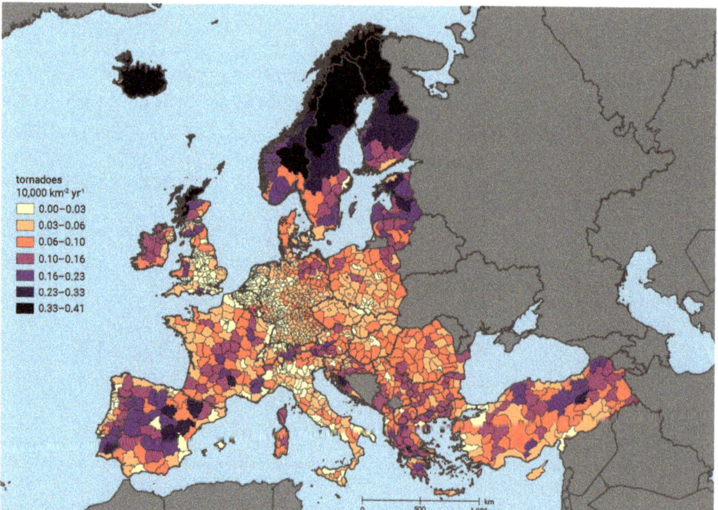

Figure 1. Difference between posterior adjusted number of tornadoes and the number of observed tornadoes normalized by the area of NUTS3 regions and year (shaded according to the scale, 10,000 km^{-2} yr^{-1}).

Low values for standard deviation (<0.13 tornadoes 10,000 km^{-2} yr^{-1}) of the tornado occurrence from the posterior distribution are found over most of Europe (Figure 2), with the exception of an area stretching from eastern Germany over Austria, Croatia, Bulgaria, and Greece characterized by values between 0.13–0.30 tornadoes 10,000 km^{-2} yr^{-1}. In these regions the NUTS3 area are characterized by relatively small area and low population density and also a low number of observed tornadoes during the study period.

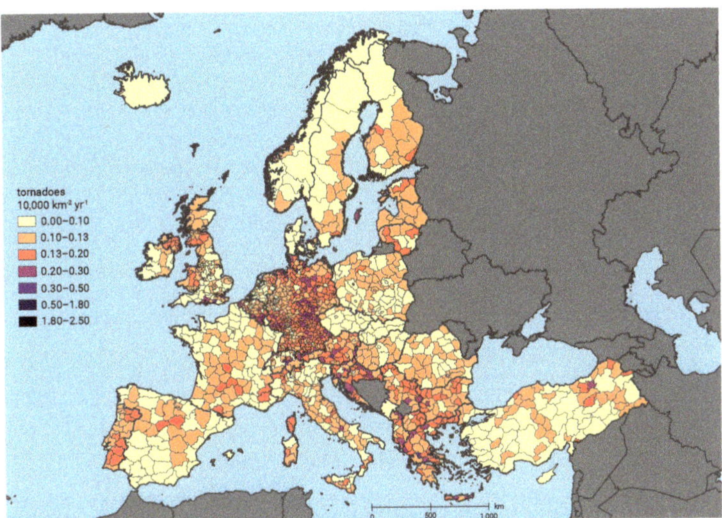

Figure 2. As in Figure 1, but for the standard deviation of tornado occurrence from posterior distribution.

5. Discussion

The Bayesian model used in this article only included the population effects, but previous studies using a similar approach have also included meteorological data [7]. These are large areas were the data collection is unreliable and thus can reduce the predictive capacity of the model. In their study of the probability of tornado occurrence across Canada, Cheng et al. [7] included the cloud-to-ground lightning climatology to account of for the spatial variability of tornadoes. For Europe, an indication regarding the predictive capacity of the model can be obtained by comparing the results from Figure 1 with lightning density over Europe between 2008–2012 developed by Anderson and Klugmann [15] using data from the Arrival Time Differing NETwork. Over northern Norway, Sweden, and Finland, where NUTS3 have a large area and the collection of data is not as reliable as for other regions of Europe, the Bayesian model is introducing large adjustments. For these regions, the lightning density (Figure 4 from [15]) is lower (<0.4 flashes km^{-2} yr^{-1}) compared with almost any other regions in Europe. Thus, the adjustments in the number of tornadoes are less realistic for the regions from a meteorological point of view. For the Baltic states (i.e., Estonia, Latvia, Lithuania) the corrections from the model are realistic due to the relatively large lightning density (i.e., 0.4–2.5 flashes km^{-2} yr^{-1}) in this region. The predictive capacity of the model in these areas can be improved by improving data collection (e.g., ref. [16] used satellite data to obtain information on unreported tornadoes that occurred in forested regions) or by considering as covariate meteorological factors related to tornado occurrence [17].

Future research will develop the model by considering meteorological covariates such as lightning density (e.g., from Arrival Time Difference long-range lightning detection network [18]) and tornadic environments (e.g., ERA5 reanalysis data [19]).

6. Conclusions

In this article, a Bayesian model was applied to adjust the number of observed tornadoes over Europe. The hypothesis was that the number of observed tornadoes in Europe is lower that the real number due to the differences in population density. The model indicated that the average annual number of tornadoes during the study period (2006–2020) was 237.5 compared with an annual average of 154.6 observed tornadoes. The largest adjustments occur over northern Europe (e.g., Iceland, Norway, Sweden, Finland), but also parts of the Mediterranean region (e.g., Spain, Greece).

The corrected distribution of tornadoes in Europe can be used to better understand the risk posed by European tornadoes. Compared with the previous distribution of tornadoes in Europe, the distribution obtained in this article is more relevant for risk reduction strategies, as it is including a correction for the population bias on tornado reporting. Furthermore, the current results can be used by decision-makers and emergency managers to develop disaster risk reduction strategies for tornadoes. Very few countries in Europe have developed tornado preparedness and response programs.

Author Contributions: Conceptualization, B.A.; methodology, R.P.; formal analysis, R.P., D.E. and B.A.; original draft preparation, R.P., D.E. and B.A.; visualization, R.P.; funding acquisition, B.A. All authors have read and agreed to the published version of the manuscript.

Funding: This research was funded by the Romanian Ministry of Education and Research, CNCS-UEFISCDI (Project No. PN-III-P1-1.1-TE-2019-0649) within PNCDI III.

Institutional Review Board Statement: Not applicable.

Informed Consent Statement: Not applicable.

Data Availability Statement: Data used in this study can be obtained by request to the authors.

Acknowledgments: This work was supported by the Romanian National Core Program (Contract No. 18N/2019) and by a grant of the Romanian Ministry of Education and Research, CNCS-UEFISCDI (Project No. PN-III-P1-1.1-TE-2019-0649) within PNCDI III. Further support was provided by the European Regional Development Fund through the Competitiveness Operational Programme 2014–2020, Action 1.1.3 Creating synergies with H2020 Programme, project H2020 Support Centre for European project management and European promotion, MYSMIS code 107874, project Strengthen the participation of the ACTRIS-RO consortium in the pan-European research infrastructure ACTRIS, ACTRIS-ROC, MYSMIS code 107596 (contract No.337/2021). The authors thanks Cristina Marin for her comments on an earlier version of the manuscript.

Conflicts of Interest: The authors declare no conflict of interest.

References

1. Antonescu, B.; Schultz, D.M.; Holzer, A.; Groenemeijer, P. Tornadoes in Europe: An underestimated threat. *Bull. Am. Meteorol. Soc.* **2017**, *98*, 713–728. [CrossRef]
2. Antonescu, B.; Schultz, D.M.; Lomas, F.; Kühne, T. Tornadoes in Europe: Synthesis of the Observational Datasets. *Mon. Weather Rev.* **2016**, *144*, 2445–2480. [CrossRef]
3. Brooks, H.E. Severe thunderstorm and climate change. *Atmos. Res.* **2013**, *123*, 129–138. [CrossRef]
4. Doswell, C.A. Societal impacts of severe thunderstorms and tornadoes: Lessons learned and implications for Europe. *Atmos. Res.* **2003**, *67–68*, 135–152. [CrossRef]
5. Groenemeijer, P.; Púčik, T.; Holzer, A.M.; Antonescu, B.; Riemann-Campe, K.; Schultz, D.M.; Kühne, T.; Feuerstein, B.; Brooks, H.E.; Doswell, C.A.; et al. Severe convective storms in Europe: Ten years of research and education at the European Severe Storms Laboratory. *Bull. Am. Meteorol. Soc.* **2017**, *98*, 2641–2651. [CrossRef]
6. Anderson, C.J.; Wikle, C.K.; Zhou, Q.; Royle, J.A. Population influences on Tornado Reports in the United States. *Weather Forecast.* **2007**, *22*, 571–579. [CrossRef]
7. Cheng, V.Y.S.; Arhonditsis, G.B.; Sills, D.M.L.; Auld, H.; Shephard, M.W.; Gough, W.A.; Klaassen, J. Probability of tornado occurrence across Canada. *J. Clim.* **2013**, *26*, 9415–9428. [CrossRef]
8. Potvin, C.K.; Broyles, C.; Skinner, P.S.; Brooks, H.E.; Rasmussen, E. A Bayesian hierarchical modeling framework for correcting reporting bias in the U.S. tornado database. *Weather Forecast.* **2019**, *34*, 15–30. [CrossRef]
9. Muller, C.L.; Chapman, L.; Johnston, S.; Kidd, C.; Illingworth, S.; Foody, G.; Overeem, A.; Leigh, R.R. Crowdsourcing for climate and atmospheric sciences: Current status and future potential. *Int. J. Climatol.* **2015**, *35*, 3185–3203. [CrossRef]

10. Eurostat Database. Available online: https://ec.europa.eu/eurostat/data/database (accessed on 1 November 2021).
11. Nomenclature of Territorial Units for Statistics (NUTS). Available online: https://ec.europa.eu/eurostat/documents/345175/6 29341/NUTS2021.xlsx (accessed on 1 November 2021).
12. Changnon, S.A. Trends in tornado frequencies: Fact or fallacy. In Proceedings of the Preprints 12th Conference on Severe Local Storms, San Antonio, TX, USA, 12–15 January 1982; pp. 42–44.
13. Lunn, D.J.; Thomas, A.; Best, N.; Spiegelhalter, D. WinBUGS—A Bayesian modelling framework: Concepts, structure, and extensibility. *Stat. Comput.* **2000**, *10*, 325–337. [CrossRef]
14. Groenemeijer, P.; Kühne, T. A climatology of tornadoes in Europe: Results from the European Severe Weather Database. *Mon. Weather Rev.* **2014**, *142*, 4775–4790. [CrossRef]
15. Anderson, G.; Klugmann, D. A European lightning density analysis using 5 years of ATDnet data. *Nat. Hazards Earth Syst. Sci.* **2014**, *14*, 815–829. [CrossRef]
16. Shikhov, A.; Chernokulsky, A. A satellite-derived climatology of unreported tornadoes in forested regions of northeast Europe. *Remote Sens. Environ.* **2018**, *204*, 553–567. [CrossRef]
17. Taszarek, M.; Allen, J.T.; Púčik, T.; Hoogewind, K.A.; Brooks, H.E. Severe convective storms across Europe and the United States. Part II: ERA5 environments associated with lightning, large hail, severe wind, and tornadoes. *J. Clim.* **2020**, *33*, 10263–10286. [CrossRef]
18. Enno, S.E.; Sugier, J.; Alber, R.; Seltzer, M. Lightning flash density in Europe based on 10 years of ATDnet data. *Atmos. Res.* **2020**, *235*, 104769. [CrossRef]
19. Hersbach, H.; Bell, B.; Berrisford, P.; Hirahara, S.; Horányi, A.; Muñoz–Sabater, J.; Nicolas, J.; Peubey, C.; Radu, R.; Schepers, D.; et al. The ERA5 global reanalysis. *Q. J. R. Meteorol. Soc.* **2020**, *146*, 1999–2049. [CrossRef]

MDPI

St. Alban-Anlage 66

4052 Basel

Switzerland

Tel. +41 61 683 77 34

Fax +41 61 302 89 18

www.mdpi.com

Applied Sciences Editorial Office

E-mail: applsci@mdpi.com

www.mdpi.com/journal/applsci